Diagnostic Ultrasound Imaging: Inside Out

Diagnostic Ultrasound Imaging: Inside Out

Second Edition

Thomas L. Szabo

Boston University, Boston, MA, USA

AMSTERDAM • BOSTON • HEIDELBERG • LONDON
NEW YORK • OXFORD • PARIS • SAN DIEGO
SAN FRANCISCO • SINGAPORE • SYDNEY • TOKYO
Academic Press is an imprint of Elsevier

Academic Press is an imprint of Elsevier
The Boulevard, Langford Lane, Kidlington, Oxford OX5 1GB, UK
Radarweg 29, PO Box 211, 1000 AE Amsterdam, The Netherlands
225 Wyman Street, Waltham, MA 02451, USA
525 B Street, Suite 1800, San Diego, CA 92101-4495, USA

First edition 2004
Second edition 2014

British Library Cataloguing-in-Publication Data
A catalogue record for this book is available from the British Library

Library of Congress Cataloging-in-Publication Data
A catalog record for this book is available from the Library of Congress

ISBN: 978-0-12-396487-8

For information on all Academic Press publications
visit our website at http://store.elsevier.com/

Typeset by MPS Limited, Chennai, India
www.adi-mps.com

Printed and bound in the USA

14 13 12 11 10 11   10 9 8 7 6 5 4 3 2 1

Contents

Companion Website for this Book:
http://booksite.elsevier.com/9780123964878

Preface

In planning to update the first edition, I was in for a surprise because the medical ultrasound landscape had shifted dramatically. In the last decade, new developments and concurrent disruptive technologies overwhelmed nearly fifty years of slow, steady growth of what seemed to be an already mature technology on a stable trajectory. Even though the fundamental science underlying medical ultrasound has remained the same, innovations have expanded the capabilities of diagnostic ultrasound in the following directions; tissue contrast (elastography), 3-D real-time volume imaging (matrix arrays), detailed blood flow characteristics (vector Doppler and ultrafast Doppler and Color Flow), high speed and throughput (plane wave compounding, Fourier transform imaging and synthetic aperture), and widespread, low cost ultrasound (pocket ultrasound) . These new capabilities come at a time in which the improvements in other medical imaging modalities are static or at best, incremental. Furthermore, diagnostic ultrasound is gaining ground by providing real-time interventional surgery and image fusion and therefore, it is offsetting its former limitations of restricted volume coverage. Pocket ultrasound is widening rapidly the number of ultrasound imaging system users in smaller clinics and its affordability is expanding its global reach.

Medical ultrasound is sowing the seeds of its own evolution. New imaging system architectures, the wide availability of research systems and imaging system chips are enabling explosive growth in new applications. In addition, the fastest growing new segment of medical ultrasound is therapeutic ultrasound including high intensity focused ultrasound, cosmetic ultrasound and neuro-ultrasound.

This edition builds on the foundation of the first edition and presents the essentials of new state-of-the-art developments. The new material serves as a gateway to deeper exploration. Over 250 pages of new text and hundreds of seminal references have been added, including 138 new figures.

Some of the new topics are summarized here. Chapter 1 has a more up to date medical ultrasound including the latest developments and an in-depth comparison of diagnostic ultrasound to other medical imaging modalities. Chapter 4 examines and assesses different viscoelastic models. New transducer materials have been added to Chapter 5. Additional simulations and holey transducers for high intensity focused ultrasound applications have been added to Chapter 6. Information on diffraction field simulation methods and conformable arrays is now included in Chapter 7. Speckle tracking is included in Chapter 8. Chapter 9 has been expanded to cover three dimensional multi-parameterized tissue characterization, time-reversal techniques and aberration correction through the skull. Chapter 10 is full of new developments: ultrasound system chips, more on 3D imaging, interventional imaging and image fusion, more on harmonic imaging, micro-beamforming, new matrix arrays, plane wave compounding, Fourier transform imaging, synthetic aperture and research system architectures. Included in Chapter 11 are plane wave and ultrafast Doppler, transverse oscillation and vector Doppler and functional ultrasound. Chapter 12 has been updated by sections on nonlinear wave simulators and acoustic radiation force calculations. Chapter 13 now includes new test system methodology, linear and nonlinear forward and backward field simulation and new force balance approaches. More on opacification, perfusion, therapeutic ultrasound contrast agents, ultrasound-Induced bioeffects related to contrast agents, targeted contrast agent applications, monodispersed microbubbles and enhanced contrast agent visualization can be found in Chapter 14. Chapter 15 on bioeffects has been thoroughly rethought and offers a new comprehensive synthesis of ultrasound-induced bioeffects as well as new perspectives on a thermal bioeffects continuum which spans both diagnostic and therapeutic ultrasound. A new chapter 16 provides an introduction to major forms of elastography including 1D elastography, quasi-static elastography, sonoelastography, shear wave elasticity imaging, acoustic radiation force impulse imaging, vibro-acoustography imaging, harmonic motion imaging, supersonic shear wave imaging and natural or physiological imaging and presents the key physical principles including viscoelasticity, strain imaging, nonlinearity, acoustic radiation forces and model-based inversion. In Chapter 17 can be found a comprehensive sweeping view of this large and fast growing field including the following : HIFU simulation, histotripsy, boiling and hemostasis, cavitation-enhanced HIFU, monitoring, transcranial ultrasound, sonothrombolysis, cosmetic ultrasound, lithotripsy, ultrasound-mediated drug delivery and gene therapy, ultrasound-induced neurostimulation and neuromodulation and bone and wound healing.

Many of the things included in this edition which were thought to be impossible a decade ago are now technological triumphs and ingenious solutions to challenging problems. As implied in Chapter 1, medical ultrasound continues to grow rapidly by expanding its diagnostic and therapeutic capabilities and offering new data-rich views of anatomy and physiological functioning. New methods such as transcranial ultrasound functional ultrasound show promise of fulfilling the potential of unlocking and healing the brain, goals that date back to the earliest known diagnostic ultrasound images of the brain by Dussik brothers and the early therapeutic ultrasound work of the Fry brothers.

Thomas L. Szabo
Newburyport, Massachusetts
October 2013

Acknowledgments

This revision builds on the work of thousands of people who contributed to advancing the science of diagnostic imaging. Those of you reading this may become, if you are not already, part of those who are expanding the circle of medical ultrasound which is rapidly eclipsing all other imaging modalities. Many articles, books, conversations, presentations, visits and amazing feats of ingenuity inspired the writing of this revision.

Special thanks are due to colleagues at Boston University, Paul Barbone, Gynn Holt, Kenneth Lutchen, Ron Roy, Bela Suki, Adam LaPrad, Brian Harvey and Tyler Wellman, Rathan Subramanian and Gustavo Mercier for introducing me to new vistas of imaging without ultrasound. I am indebted to Peder Pedersen for his far-sighted vision of the growing global importance and challenges of portable and now pocket ultrasound and our years of collaboration on these topics. Thanks to Jonathan Newell, David Isaacson and Gary J. Saulnier for initiating me into multimodal imaging design opportunities. I appreciate alert readers and students who helped clarify and correct the presentation of the material of the first edition. I am thankful to Jacques S. Abramowicz, Gerald Harris, Thomas Nelson, the late Wesley Nyborg and Marvin Ziskin for plunging me deeper into the world of bioeffects. Over the years, my associations with the working groups of Technical Committee 87 of the International Electrotechnical Commission, in particular Working Group 6, have enriched my knowledge of diagnostic and therapeutic ultrasound.

There is a special group of people to whom I am particularly grateful for their patience with my questions and/or their willingness to review and recommend improvements for the second edition: Javier de Ana Arbeloa, Jean- Francois Aubry, Paul Barbone, Jeremy Bercoff, Ronald Daigle, Diane Dalecki, Francis Duck, Caleb Farny, Leonid Gavrilov, Sverre Holm, Robert McGough, Timothy Hall, Sam Howard, Jørgen A. Jensen, Jian-yu Lu, Oleg Sapozhnikov and Joshua Soneson.

Notable conversations and talks and sometimes chance meetings with the following led me to fascinating areas of medical ultrasound: Javier de Ana Arbeloa, Jean-Francois Aubry, Jeffrey Bamber, Peter Barthe, Stefan Catheline, Kris Dickie, Mathias Fink, Stuart Foster, Kullervo Hynynen , Vera Khokhlov, Elisa Konofagou, Peter A. Lewin, Kathryn Nightingale, Mickael Tanter, Gail ter Haar, Robert Muratore, Oleg Sapozhnikov, Mike Sekins and Dr. Feng Wu. There were many others that contributed to my understanding as well as meetings of the Acoustical Society of America, American Institute of Ultrasound in Medicine (AIUM), the AIUM Bioeffects Committee, the IEEE Ultrasonics Symposia, and the International Society of Therapeutic Ultrasound.

Thanks go to those who gave advance or less known materials and figures: Olivier le Baron, Klaus Beissner, Muyinatu Bell, Mingzhu Lu, Jeffery Ketterling, Ernest Madsen, Pierre Maréchal, Ali Sadeghi-Naini, and T. Nelson. In addition, I appreciate those who gave me permission to reproduce their work: J.-F. Aubry, J. Bamber, P. Barthe, P. Burns, S. Catheline, C. Church, R. Cleveland, C. Coussios, N. de Jong, C. Desilets, M. Doyley, K. Ferrara, D. Miller, E. Feleppa, M. Fink, F. Forsberg, Freeman, T. Gallot, General Electric Healthcare, J. Greenleaf, T. Hasegawa, J. Hassock, G. Hesley, S. Holm, V. Humphrey, K. Hynynen, R. L. King, E. Konofagou, H. S. Lee, E. Madsen, Mobisante, K. Nightingale, Y. Okamura, R. McGough, G. Norton, M. Oelze, J. Ophir, K. Parker, M. Roubidoux, A. Sarvazyan, A. Shaw, K. Shung, Siemens Healthcare, R. Skyba, R. Subramaniam, T. Sugimoto, M. Tanter, Toshiba America Medical Systems, G. Trahey, G. Treece, W. Walker, V. Wilkens, F. Wu, J. Wu, and Q. Zhu.

I am indebted to Amy Lex of Philips Healthcare whose patience, generosity and support of my many requests for images and information; these superb figures have enhanced the book greatly and led to a book cover. I am thankful to several people at Philips for their discussions, explanations, figures and am particularly grateful for the contributions of Daniel Cote, Loriann Davidsen, Mike Peszynski, Rick Snyder, and Karl Thiele of Philips Healthcare.

Thanks are also due to N. Carter, F. Geraghty, and C. Owen whose careful reading, persistence, patience, and suggestions resulted in many improvements to the quality of the book. Readers also helped by finding errors in the first edition. Those errors that remain despite all their best efforts are my own.

Finally, I thank my children, Sam and Vivien for their understanding of my disappearances and for their fun and grounding engagement. More than ever, I am deeply grateful to my wife Deborah whose patience, wisdom, sacrifice, understanding, good cheer and support made this work possible and enjoyable.

CHAPTER 1

Introduction

Chapter Outline

1.1 Introduction

Ultrasound, a type of sound we cannot hear, has enabled us to see a world otherwise invisible to us. The purpose of this chapter is to explore medical ultrasound from its antecedents and beginnings, relate it to sonar, describe the struggles and discoveries necessary for its development, and provide the basic principles and reasons for its success.

The development of medical ultrasound was a great international effort involving thousands of people over sixty years, and the field is still expanding rapidly, so it is not possible to include many of the outstanding contributors in the short space that follows. Only the fundamentals of medical ultrasound and representative snapshots of key turning points are given here, but additional references are provided. The critical relationship between the growth of the science of medical ultrasound and key enabling technologies is also examined. Why these allied technologies will continue to shape the future of ultrasound is described. Finally, the unique role of ultrasound imaging is compared to other diagnostic imaging modalities.

1.1.1 Early Beginnings

Robert Hooke (1635–1703), the eminent English scientist responsible for the theory of elasticity, pocket watches, compound microscopy, and the discovery of cells and fossils, foresaw the use of sound for diagnosis when he wrote (Tyndall, 1875):

> *It may be possible to discover the motions of the internal parts of bodies, whether animal, vegetable, or mineral, by the sound they make; that one may discover the works performed in the several offices and shops of a man's body, and therby [sic] discover what instrument or engine is out of order, what works are going on at several times, and lie still at others, and the like. I could proceed further, but methinks I can hardly forbear to blush when I consider how the most part of men will look upon this: but, yet again, I have this encouragement, not to think all these things utterly impossible.*

Many animals in the natural world, such as bats and dolphins, use echolocation, which is the key principle of diagnostic ultrasound imaging. The connection between echolocation and the medical application of sound, however, was not made until the science of underwater exploration matured. Echolocation is the use of reflections of sound to locate objects.

Humans have been fascinated with what lies below the murky depths of water for thousands of years. The naval term to "sound" means to measure the depth of water at sea. The ancient Greeks probed the depths of the seas with a "sounding machine," which was a long rope knotted at regular intervals with a lead weight on the end. American naturalist and philosopher Henry David Thoreau measured the depth profiles of Walden Pond near Concord, Massachusetts, with this kind of device. Recalling his boat experiences as a young man, American author and humorist Samuel Clemens chose his pseudonym, Mark Twain, from the second mark or knot on a sounding lead line. While sound may or may not have been involved in a sounding machine, except for the thud of a weight hitting the sea bottom, the words "to sound" set the stage for the later use of actual sound for the same purpose.

The sounding-machine method was in continuous use for thousands of years until it was replaced by ultrasound echo-ranging equipment in the twentieth century. Harold Edgerton

(1986), famous for his invention of stroboscopic photography, related how his friend, Jacques-Yves Cousteau, and his crew found an ancient Greek lead sounder (250 BC) on the floor of the Mediterranean sea by using sound waves from a side scan sonar. After his many contributions to the field, Edgerton used sonar and stroboscopic imaging to search for the Loch Ness monster (Rines, Wycofff, Edgerton, & Klein, 1976).

1.1.2 Sonar

The beginnings of sonar and ultrasound for medical imaging can be traced to the sinking of the Titanic. Shortly after the Titanic tragedy, British scientist L. F. Richardson (1913) filed patents for detecting icebergs with underwater echo ranging. In 1913, there were no practical ways of implementing his ideas. However, the discovery of piezoelectricity (the property by which electrical charge is created by the mechanical deformation of a crystal) by the Curie brothers in 1880 and the invention of the triode amplifier tube by Lee de Forest in 1907 set the stage for further advances in pulse–echo range measurement. The Curie brothers also showed that the reverse piezoelectric effect (voltages applied to certain crystals cause them to deform) could be used to transform piezoelectric materials into resonating transducers. By the end of World War I, C. Chilowsky and P. Langevin (Biquard, 1972), a student of Pierre Curie, took advantage of the enabling technologies of piezoelectricity for transducers and vacuum tube amplifiers to realize practical echo ranging in water. Their high-power echo-ranging systems were used to detect submarines. During transmissions, they observed schools of dead fish that floated to the water surface. This shows that scientists were aware of the potential for ultrasound-induced bioeffects from the early days of ultrasound research (O'Brien, 1998). More details can be found in Duck (2008).

The recognition that ultrasound could cause bioeffects began an intense period of experimentation and hopefulness. After World War I, researchers began to determine the conditions under which ultrasound was safe. They then applied ultrasound to therapy, surgery, and cancer treatment. The field of therapeutic ultrasound began and grew erratically until its present revival in the forms of lithotripsy (ultrasound applied to the breaking of kidney and gallstones) and high-intensity focused ultrasound (HIFU) for surgery. However, this branch of medical ultrasound, which is concerned mainly with ultrasound transmission, is distinct from the development of diagnostic applications, which is the focus of this chapter.

During World War II, pulse–echo ranging applied to electromagnetic waves became radar (radio detection and ranging). Important radar contributions included a sweeping of the pulse–echo direction in a 360-degree pattern and the circular display of target echoes on a plan position indicator (PPI) cathode-ray-tube screen. Radar developments hastened the evolution of single-direction underwater ultrasound ranging devices into sonar with similar PPI-style displays.

1.2 Echo Ranging of the Body

After World War II, with sonar and radar as models, a few medical practitioners saw the possibilities of using pulse–echo techniques to probe the human body for medical purposes. In terms of ultrasound in those days, the body was vast and uncharted. In the same way that practical underwater echo ranging had to wait until the key enabling technologies were available, the application of echo ranging to the body had to wait for the right equipment. A lack of suitable devices for these applications inspired workers to do amazing things with surplus war equipment and to adapt other echo-ranging instruments.

Fortunately, the timing was right in this case because F. Firestone's invention of the supersonic reflectoscope in 1940 applied the pulse–echo ranging principle to the location of defects in metals in the form of a reasonably compact instrument (Firestone, 1945). A diagram of a basic echo-ranging system of this type is shown in Figure 1.1. A transmitter excites a transducer, which sends a sequence of repetitive ultrasonic pulses into a material or a body. Echoes from different target objects and boundaries are received and amplified so they can be displayed as an amplitude versus time record on an oscilloscope. This type of display became known as the "A-line," (or "A-mode" or "A-scope"), with "A" representing amplitude.

When commercialized versions of the reflectoscope were applied to the human body in Japan, the United States, and Sweden in the late 1940s and early 1950s (Goldberg & Kimmelman, 1988), a new world of possibility for medical diagnosis was born. Rokoru Uchida in Japan was one of the first to use flaw detectors for medical A-line pulse–echo ranging. In Sweden in 1953, Dr I. Edler and Professor C. H. Hertz detected heart motions with a flaw detector and began what later was called "echocardiography," the application of ultrasound to the characterization and imaging of the heart (Edler, 1991).

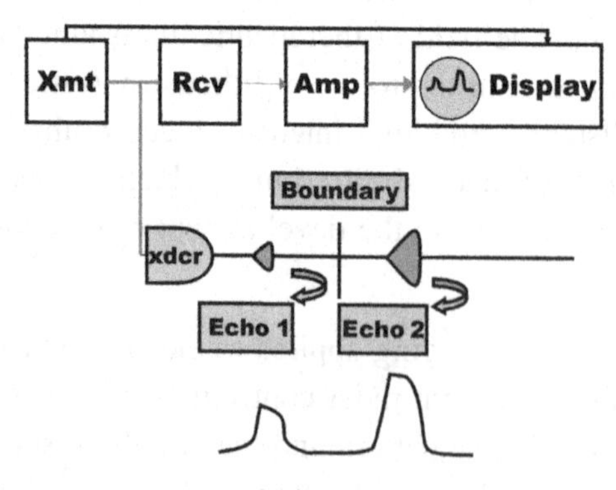

Figure 1.1
Basic Echo-ranging system showing multiple reflections and an a-line trace at the bottom. Xmt = transmitter, Rcv = receiver, Amp = amplifier and xdcr = transducer

Medical ultrasound in the human body is quite different from many sonar applications that detect hard targets, such as metal ships in water. At the Naval Medical Research Institute, Dr George Ludwig, who had underwater-ranging experience during World War II, and F. W. Struthers embedded hard gallstones in canine muscles to determine the feasibility of detecting them ultrasonically. Later, Ludwig (1950) made a number of time-of-flight measurements of sound speed through arm, leg, and thigh muscles. He found the average to be $c_{av} = 1540$ m/s, which is the standard value still used today. The sound speed, c, can be determined from the time, t, taken by sound to pass through a tissue of known thickness, d, from the equation, $c = d/t$. Ludwig found the sound speeds to be remarkably similar, varying in most soft tissues by only a few percent. Normalized speed-of-sound measurements taken more recently are displayed in Figure 1.2.

The remarkable consistency among sound speeds for the soft tissues of the human body enables a first-order estimation of tissue target depths from their round trip (pulse–echo) time delays, t_{rt}, and an average speed of sound, c_{av}, from $z = c_{av}t_{rt}/2$. This fact makes it possible for ultrasound images to be faithful representations of tissue geometry.

In the same study, Ludwig also measured the characteristic acoustic impedances of tissues. He found that the soft tissues and organs of the body have similar impedances because of their high water content. The characteristic acoustic impedance, Z, is defined as the product of density, ρ, and the speed of sound, c, or $Z = \rho c$. The amplitude reflection factor of acoustic plane waves normally incident at an interface of two tissues with impedances Z_1 and Z_2 can be determined from the relation, $RF = (Z_2 - Z_1)/(Z_2 + Z_1)$.

Fortunately, amplitude reflection coefficients for tissue are sensitive to slight differences in impedance values so that the reflection coefficients relative to blood (Figure 1.3) are quite different from each other as compared to small variations in the speed of sound

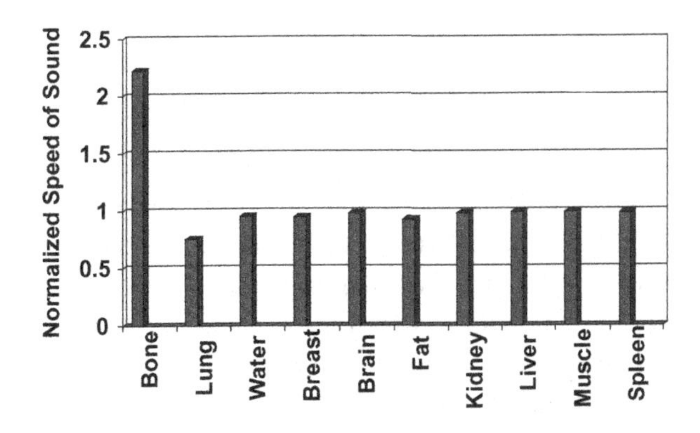

Figure 1.2
Acoustic speed of sound of various tissues normalized to the speed of sound in blood.

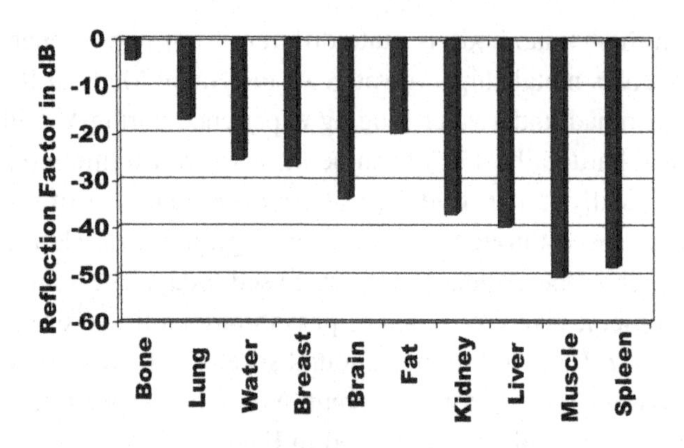

Figure 1.3
Amplitude reflection factor coefficients in dB for various tissues relative to blood.

values for the same tissues (see Figure 1.2). This fortuitous range of reflection coefficient values is why it is possible to distinguish between different tissue types for both echo ranging and imaging. Note that the reflection coefficients are plotted on a dB, or logarithmic, scale (explained in Section 4.1.1). For example, each change of -10 dB means that the reflection coefficient value is a factor 3.2 less in amplitude or a factor 10 less in intensity.

Around the same time in 1949, Dr D. Howry of Denver, Colorado, who was unaware of Ludwig's work, built a low-megahertz pulse–echo scanner in his basement from surplus radar equipment and an oscilloscope. Howry and other workers using A-line equipment found that the soft tissues and organs of the body, because of their small reflection coefficients and low absorption, allowed the penetration of elastic waves through multiple tissue interfaces (Erikson, Fry, & Jones, 1974). This is illustrated in Figure 1.1. In Minnesota, Dr John J. Wild, an English surgeon who also worked for some time in his basement, applied A-mode pulse echoes for medical diagnosis in 1949, and shortly thereafter he developed imaging equipment with John M. Reid, an electrical engineer.

When identifying internal organs with ultrasound was still a novelty, Wild used a 15-MHz navy radar trainer to investigate A-lines for medical diagnosis. He reported the results for cancer in the stomach wall in 1949. In 1952, Wild and Reid analyzed data from 15-MHz breast A-scans. They used the area underneath the echoes to differentiate malignant from benign tissue, as well as to provide the first identification of cysts. These early findings triggered enormous interest in diagnosis, which became the most important reason for the application of ultrasound to medicine. Later this topic split into two camps: diagnosis, findings directly observable from ultrasound images, and tissue characterization, findings about the health of tissue and organ function determined by parameterized inferences and calculations made from ultrasound data.

1.3 Ultrasound Portrait Photographers

The A-mode work described in the previous section was a precursor to diagnostic ultrasound imaging just as echo ranging preceded sonar images. The imaginative leap to imaging came in 1942 in Austria, when Dr Karl Dussik and his brother published their through-transmission ultrasound attenuation image of the brain, which they called a "hyperphonogram." In their method, a light bulb connected to the receiving transducer glowed in proportion to the strength of the received signal, and the result was optically recorded (Figure 1.4). This transcranial method was not adopted widely because of difficult refraction and attenuation artifacts in the skull, but it inspired many others to work on imaging with ultrasound. Their work is even more remarkable because it preceded the widespread use of radar and sonar imaging.

Despite the problems caused by refraction through varying thicknesses of the skull, others continued to do ultrasound research on the brain. This work became known as "echoencephalography." Dr Karl Dussik met with Dr Richard Bolt, who was then inspired to attempt to image through the skull tomographically. Bolt tried this in 1950 with his group and Dr George Ludwig at the MIT Acoustic Laboratory, but he later abandoned the project. In 1953, Dr Lars Leksell, of Lund University in Sweden, used flaw detectors to detect midline shifts in the brain caused by disease or trauma. Leksell found an acoustic window through the temples. Equipment for detecting midline shifts and cardiac echoes became available in the 1960s.

Dussik's work, as well as war developments in pulse–echo imaging, motivated others to make acoustic images of the body. For example, Dr D. Howry and his group were able to show that highly detailed pulse–echo tomographic images of cross-sections of the body

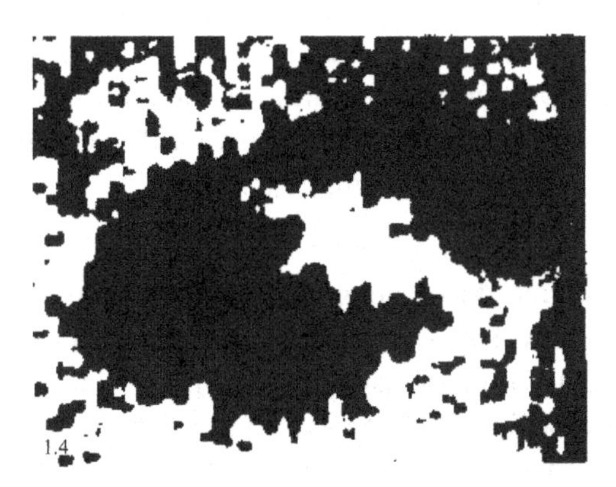

Figure 1.4
The Dussik transcranial image, which is one of the first ultrasonic images of the body ever made. Here, white represents areas of signal strength and black represents complete attenuation. *From Goldberg and Kimmelman, 1988; reprinted with permission of the AIUM.*

correlated well with known anatomical features (Holmes, 1980). Their intent was to demonstrate that ultrasound could show accurate pictures of soft tissues that could not be obtained with X-rays. Howry and his group transformed the parts of a World War II B-29 bomber gun turret into a water tank. A subject was immersed in this tank, and a transducer revolved around the subject on the turret ring gear. See Figure 1.5 for pictures of their apparatus and results.

The 1950s were a period of active experimentation with both imaging methods and ways of making contact with the body. Many versions of water-bath scanners were in use. Dr John J. Wild and John M. Reid, both affiliated with the University of Minnesota, made one of the earliest handheld contact scanners. It consisted of a transducer enclosed in a water column and sealed by a condom. Oils and eventually gels were applied to the ends of transducers to achieve adequate coupling to the body (Wells, 1969a).

The key element that differentiated pulse–echo-imaging systems (Figure 1.6) from echo-ranging systems is a means of either scanning the transducer in a freehand form with the detection of the transducer position in space or by controlling the motion of the transducer. As shown, the position controller or position sensor is triggered by the periodic timing of the transmit pulses. The display consists of time traces running vertically (top to bottom) to indicate depth. Because the brightness along each trace is proportional to the echo amplitude, this display presentation came to be known as "B-mode," with "B" meaning brightness. However, it was first used by Wild and Reid, who called it a "B-scan." In an alternative (A) in Figure 1.6, a single transducer is scanned mechanically at intervals across an elliptically shaped object. At each controlled mechanical stopping point, sound (shown as a line) is sent across an object and echoes are received. For the object being scanned linearly upward in the figure, the bright dots in each trace on the display indicate the front and back wall echoes of the object. By scanning across the object, multiple lines produce an "image" of the object on the display. This method, which was used in early scanners, corresponds to the detection of transducer position and attitude. The modern equivalent of controlled motion is to switch sequentially from one transducer element to another in an array of transducers, as shown in (A) in Figure 1.6.

Various scanning methods are shown in Figure 1.7. A straightforward method is linear scanning, or translation of a transducer along a flat surface or straight line. Angular rotation, or sector scanning, involves moving the transducer in an angular arc without translation. Two combinations of the translation and angular motions are compound (both motions are combined in a rocking, sliding motion) and contiguous (angular motion switches to translation and back to angular). An added twist is that the scanning surface may not be flat but may be curved or circular instead. Dr Howry's team, along with Dr Ian Donald and his group at the University of Glasgow, developed methods to display each scan line in its correct geometric attitude. For example, the first line in an angular scan

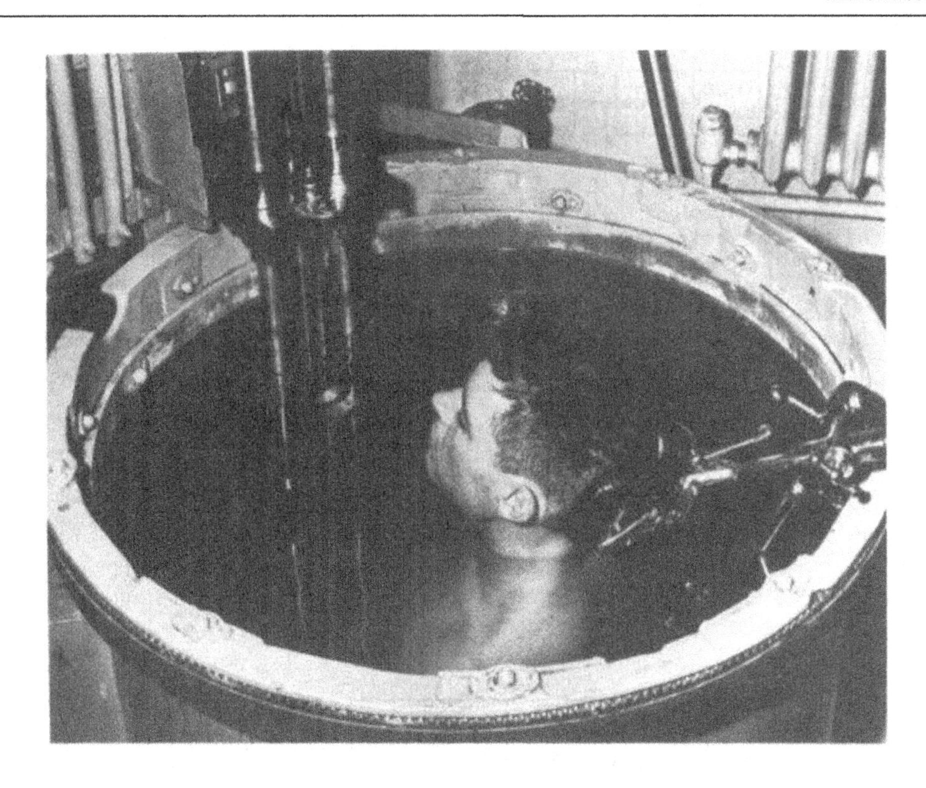

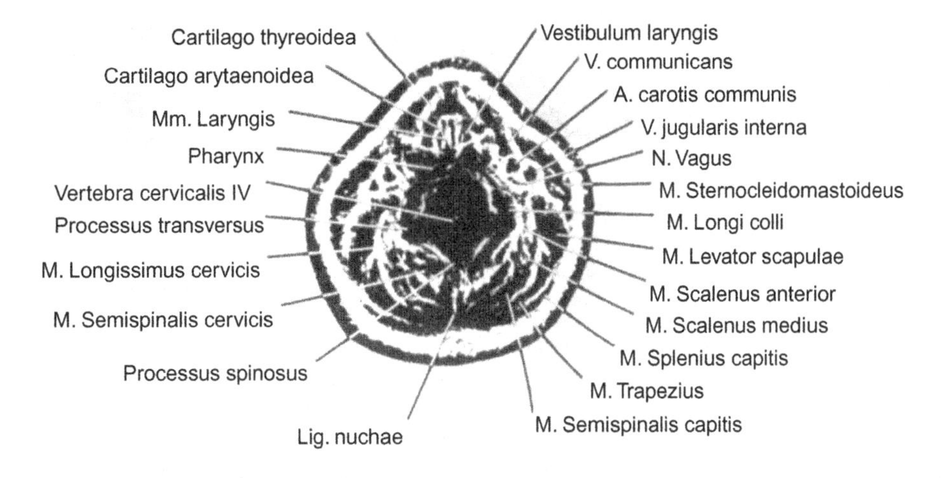

Figure 1.5

Howry's B-29 gun-turret ultrasonic tomographic system and resulting annotated image of the neck. *From Goldberg and Kimmelman, 1988; reprinted with permission of the AIUM.*

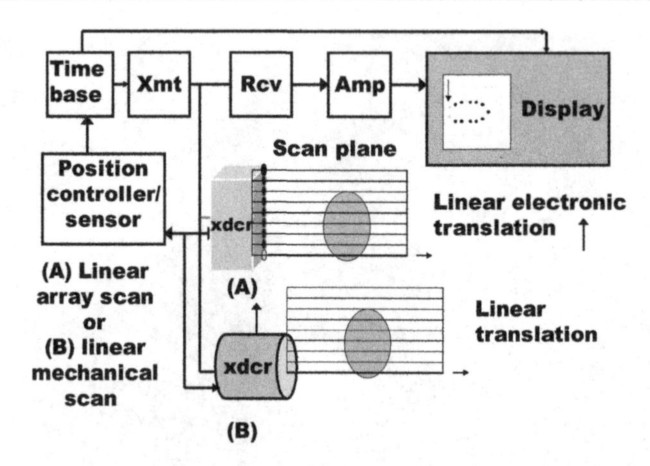

Figure 1.6
Basic elements of a pulse–echo-imaging system. Shown with linear scanning of two types: (A) electronic linear-array scanning, which involved switching from one element to another, and (B) mechanical scanning, which involved controlled translation of a single transducer.

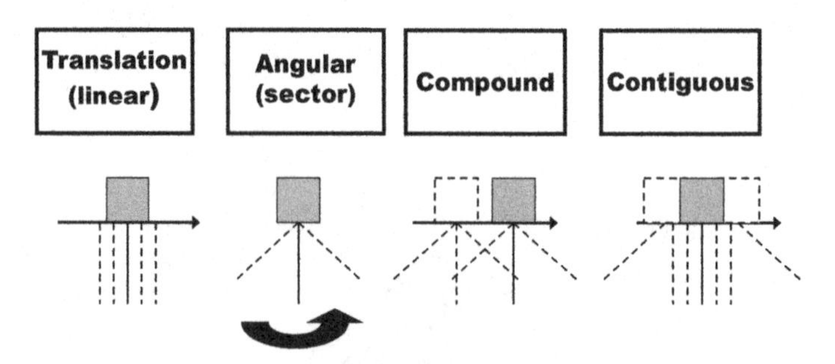

Figure 1.7
Scanning methods: translation, angular rotation, compound (Translation and Rotation), and contiguous (Rotation, Translation, and Rotation).

at −45 degrees would appear on the scope display as a brightness-modulated line at that angle with the depth increasing from top to bottom.

The most popular imaging method from the 1950s to the 1970s became freehand compound scanning, which involved both translation and rocking. Usually, transducers were attached to large articulated arms that both sensed the position and attitude of the transducer in space and communicated this information to the display. In this way, different views (scan lines) contributed to a more richly detailed image because small, curved interfaces were better defined by several transducer positions rather than one. This approach is a precursor to spatial compounding, described in Section 10.11.4.

Sonography in this time period was like portrait photography. Different patterns of freehand scanning were developed to achieve the "best picture." For each position of the transducer, a corresponding time line was traced on a cathode-ray-tube (CRT) screen. The image was not seen until scanning was completed or later because the picture was usually in the form of either a storage scope image or a long-exposure photograph (Devey & Wells, 1978; Goldberg & Kimmelman, 1988). Of course, the "subject" being imaged was not to move during scanning. In 1959, the situation was improved by the introduction of the Polaroid scope camera, which provided prints in minutes.

Around the same time, seemingly unrelated technologies were being developed that would revolutionize ultrasound imaging (*Electronic Engineering Times*, October 30,1997). The invention of the transistor and the digital computer in the late 1940s set profound changes in motion. In 1958, Jack Kilby's invention of an integrated circuit accelerated the pace by combining several transistors and circuit elements into one unit.

In 1965, Gordon Moore predicted that the density of integrated circuits would grow exponentially (now revised to double every 24 months), as illustrated in Figure 1.8 (Wikipedia, 2012). By 1971, 2300 transistors on a single chip had as much computational power as the ENIAC (Electronic Numerical Integrator and Computer) computer that, 25 years before, was as big as a boxcar and weighed 60,000 lb. Hand calculators, such as the Hewlett Packard scientific calculator HP-35, speeded up chip miniaturization. Digital memories and programmable chips were also produced.

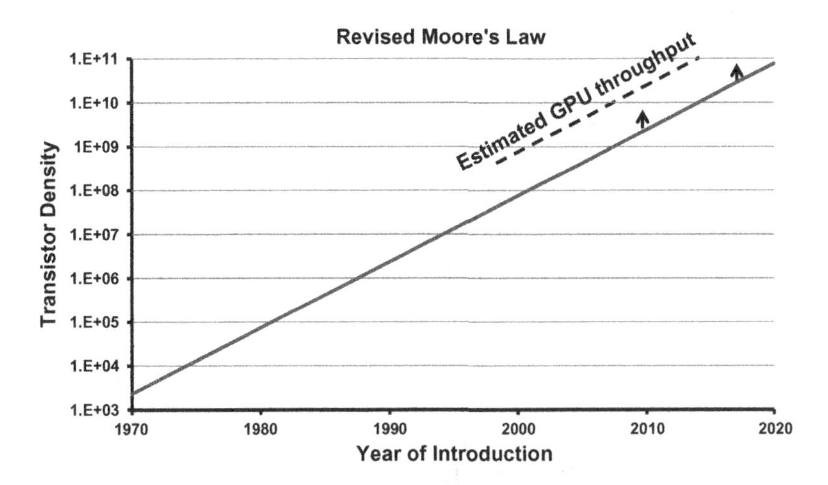

Figure 1.8

Revised Moore's law predicts exponential growth of microprocessor density that doubles every two years. *Solid curve based on data from product introductions compiled by W. G. Simon (Wikipedia, 2012). Upper curve is estimated equivalent GPU (graphical processing unit) improved throughput. Arrows indicate estimated 3D architecture improvements.*

By the early 1960s, the first commercialized contact B-mode static scanners became available. These consisted of a transducer mounted on a long, moveable articulated arm with spatial position encoders, a display, and electronics (Goldberg & Kimmelman, 1988). An early scanner of this type, called the "Diasonograph" and designed by Dr Ian Donald and engineer Tom G. Brown (1968) in Scotland, achieved commercial success. For stable imaging, the overall instrument weighed one ton and was sometimes called the "Dinosaurograph." Soon other instruments, such as the Picker unit, became available, and widespread use of ultrasound followed.

These instruments, which began to incorporate transistors (Wells, 1969b), employed the freehand compound scanning method and produced still (static) pictures. The biphasic images were black and white. Whereas A-mode displays had a dynamic range of 40 dB, B-mode storage scopes had only a 10-dB range (a capability to display 1–10 × in intensity), and regular scopes had a 20-dB (1–100 ×) range (Wells, 1977). Storage scopes and film had blooming and exposure variations, which made consistent results difficult to obtain.

At the time of biphasic imaging, interest was focused on tissue interfaces and boundaries. During an extended stay at W. J. Fry's focused ultrasound surgery laboratory at the University of Illinois, George Kossoff observed that the pulse echoes from boundaries were strongly dependent on the angle of insonification. Because the transducer had a large focal gain and power, Kossoff also noticed that the lower amplitude scattering from tissue was much less sensitive to angular variation. These insights led to his methods to image soft tissues more directly. By emphasizing the region of dynamic range for soft-tissue scattering and implementing logarithmic amplifiers to better display the range of information, Kossoff (1974) and his coworkers at the Commonwealth Acoustic Laboratory in Australia published work on implementing gray-scale imaging though analog methods. Gray-scale became widespread because of the availability of digital electronic programmable read-only memories (EPROMs), random-access memories (RAMs), microprocessors, and analog/digital (A/D) converters. These allowed the ultrasound image to be stored and scan-converted to the rectangular format of CRTs at video rates. By 1976, commercial gray-scale scan converters revolutionized ultrasound imaging by introducing subtle features and an increased dynamic range for better differentiation and resolution of tissue structures.

One of the most important applications of ultrasound diagnosis is obstetrics. A study by Alice Stewart in England in 1956 linked deaths from cancer in children to their prenatal exposure to X-rays (Kevles, 1997). Dr Ian Donald foresaw the benefit of applying ultrasound to obstetrics and gynecology, and his Diasonograph became successful in this area. Eventually, ultrasound imaging completely replaced X-rays in this application and provided much more diagnostic information. By 1997, a estimated 70% of pregnant women in the United States had prenatal ultrasounds (Kevles, 1997).

Ultrasound was shown to be a safe noninvasive methodology for the diagnosis of diseased tissue, the location of cysts, and identifying fetal abnormalities and heart irregularities. By the late 1970s, millions of clinical exams had been performed by diagnostic ultrasound imaging (Devey & Wells, 1978).

1.4 Ultrasound Cinematographers

Gray-scale was not enough to save the static B-scanner (the still portrait camera of ultrasound) because the stage was set for movies, or ultrasound cinematography. In the early 1950s, Dr John J. Wild and John Reid worked on an alternative method: a real-time handheld array-like scanner, in which they used mechanically scanned (controlled position) transducers (Figure 1.9). In this figure, the rectangular B-scan image format is a departure from the plan position indication (PPI) format of sonar B-mode images and earlier tomographic (circular) images. Wild and Reid's vision of real-time scanning was a few years ahead of its time.

The year 1965 marked the appearance of Vidoson from Siemens, the first real-time mechanical commercial scanner. Designed by Richard Soldner, the Vidoson consisted of a revolving transducer and a parabolic mirror. By the early 1970s, real-time contact mechanical scanners with good resolution were beginning to replace the static B-scanners.

Radar and sonar images, and eventually ultrasonic images, benefited from the maturing of electronically scanned and focused phased-array technology for electromagnetic applications in the late 1950s and 1960s. In 1971, Professor N. Bom's group in Rotterdam, the Netherlands, built linear arrays for real-time imaging (Bom, Lancee, van Zwieten, Kloster, & Roelandt, 1973). An example of an early linear-array imaging system is illustrated in Figure 1.6. The position controller takes the form of a multiplexer, which is an electronic switch that routes the input/output channel to different transducer array elements sequentially. As each transducer element is fired in turn, a pulse–echo image line is created. These efforts produced the Minivisor (Ligtvoet, Rijsterborgh, Kappen, & Bom, 1978; Roelandt, Wladimiroff, & Bars, 1978), which was the first portable ultrasound system including a built-in linear array, electronics, display, and a 1.5-hr battery, with a total weight of 1.5 kg (shown in Figure 1.13).

J. C. Somer (1968) of the Netherlands reported his results for a sector (angular) scanning phased array for medical ultrasound imaging. Shown in Figure 1.10 are two views of different steering angles from the same array. On the left are Schlieren measurements (an acousto-optic means of visualizing sound beams) of beams steered at different angles. They are depicted as acoustic lines on the oscilloscope images on the right. By 1974, Professor Thurstone and Dr von Ramm and Thurstone (1975) of Duke University obtained live images of the heart with their 16-channel phased-array imaging system called the "Thaumascan."

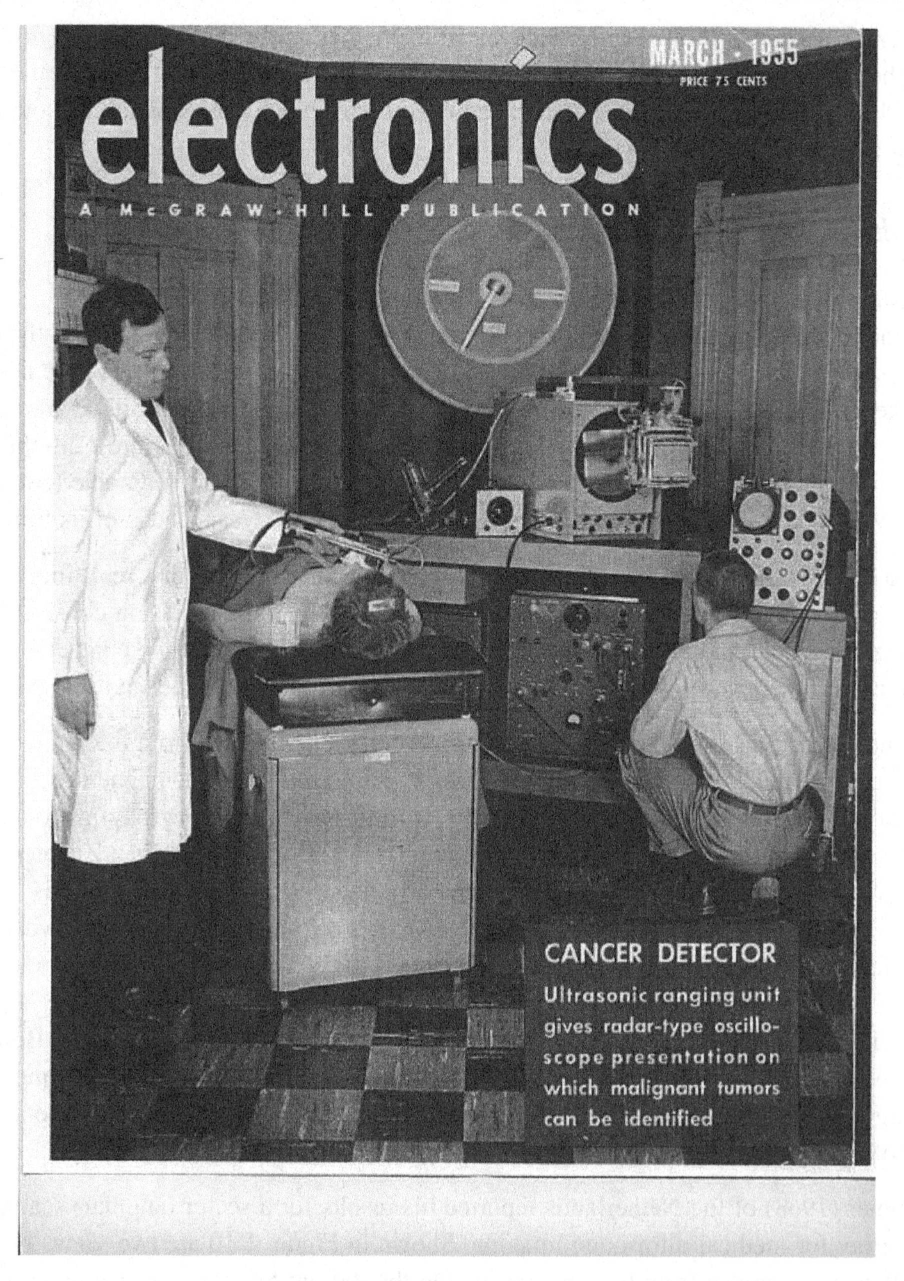

Figure 1.9
Dr John J. Wild Scans a patient with a handheld, linearly scanned 15-MHz contact transducer. John Reid (later Professor Reid) adjusts modified radar equipment to produce a B-scan image on a large-diameter scope display with a recording camera (The reader is referred to the Web version of the book to see the figure in color.). *Courtesy of J. Reid; reprinted with permission of VNU Business Publications.*

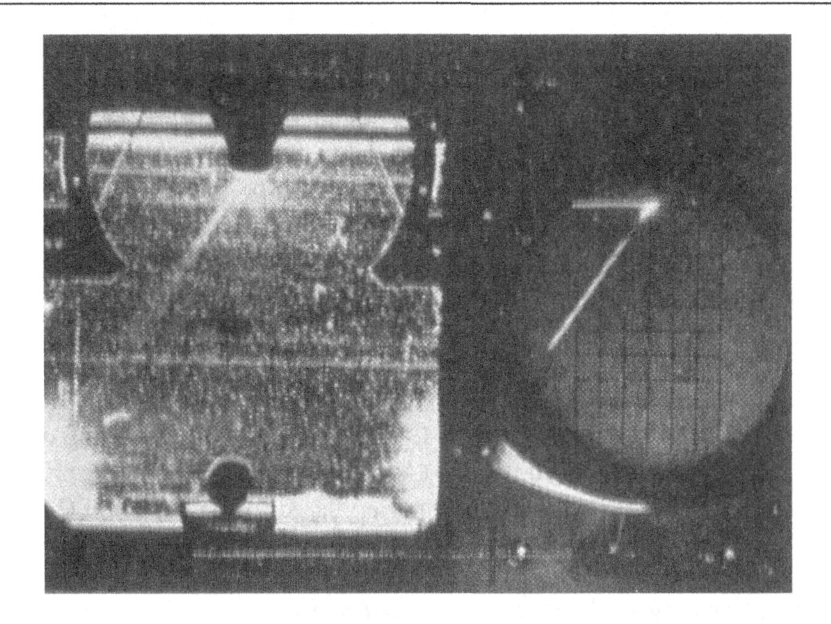

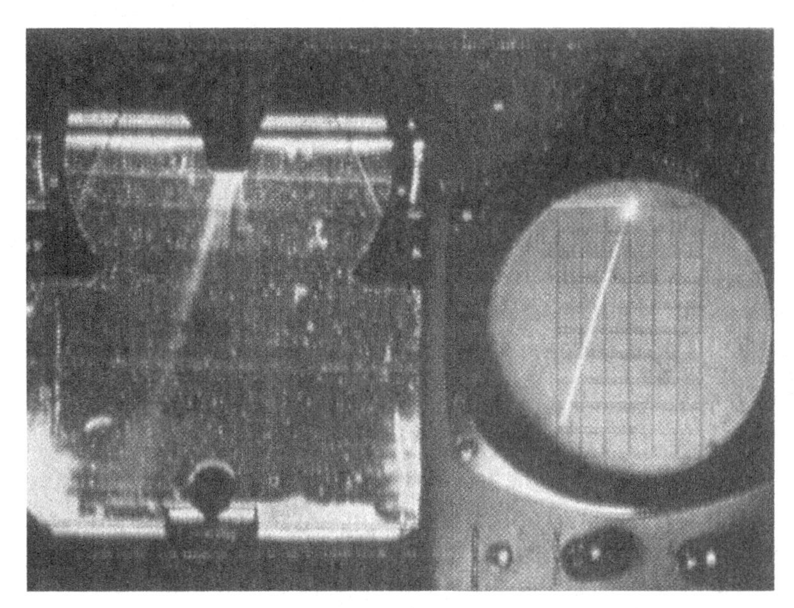

Figure 1.10
Acoustic scan lines from one of the first phased arrays for medical ultrasound imaging. (Right) Pulse–echo acoustic lines at two different angles on an oscilloscope from a phased array designed by J. C. Somer (1968); (left) Schlieren pictures of the corresponding acoustic beams as measured in a water tank. *Courtesy of N. Bom.*

The appearance of real-time systems with good image quality marked the end of the static B-scanners (Klein, 1981). Parallel work on mechanically scanned transducers resulted in real-time commercial systems by 1978. By 1980, commercial real-time phased-array imaging systems were made possible by recent developments in video, microprocessors, digital memory, small delay lines, and the miniaturization offered by programmable integrated circuits. In 1981, the Hewlett Packard 70020A phased-array system (see Figure 1.11) became a forerunner of future systems, which had wheels, modular architectures, microprocessors, programmable capabilities, and upgradeability (Hewlett Packard, 1983).

During the 1980s, array systems became the dominant imaging modality. Several electronic advancements (*Electronic Engineering Times*, 1997) rapidly improved imaging during this decade: application-specific integrated circuits (ASICs), digital signal processing chips (DSPs), and the computer-aided design (CAD) of very large scale integration (VLSI) circuits.

1.5 Modern Ultrasound Imaging Developments

The concept of deriving real-time parameters other than direct pulse–echo data by signal processing or by displaying data in different ways was not obvious at the very beginning of medical ultrasound. M-mode, or a time–motion display, presented new time-varying information about heart motion at a fixed location when I. Elder and C. H. Hertz introduced it in 1954. In 1955, S. Satomura, Y. Nimura, and T. Yoshida reported experiments with Doppler-shifted ultrasound signals produced by heart motion. Doppler signals shifted by blood movement fall in the audio range and can be heard as well as seen on a display. By 1966, D. Baker and V. Simmons had shown that pulsed spectral Doppler was possible (Goldberg & Kimmelman, 1988). P N. T. Wells (1969b) invented a range-gated Doppler to isolate different targets.

In the early 1980s, Eyer, Brandestini, Philips, and Baker (1981) and Namekawa, Kasai, Tsukamoto, and Koyano (1982) described color flow-imaging techniques for visualizing the flow of blood in real time. During the late 1980s, many other signal-processing methods for imaging and calculations began to appear on imaging systems. Concurrently, sonar systems evolved to such a point that Dr Robert Ballard was able to discover the Titanic at the bottom of the sea with sonar and video equipment in 1986 (Murphy, 1986).

Also during the 1980s, transducer technology underwent tremendous growth. Based on the Mason equivalent circuit model and waveguide, as well as the matching-layer design technology and high coupling piezoelectric materials developed during and after World War II, ultrasonic phased-array design evolved rapidly. Specialized phased and linear arrays were developed for specific clinical applications: cardiology; radiology (noncardiac internal organs); obstetrics/gynecology and transvaginal; endoscopic (transducer manipulated on the tip of an endoscope); transesophageal (transducer down the esophagus) and transrectal;

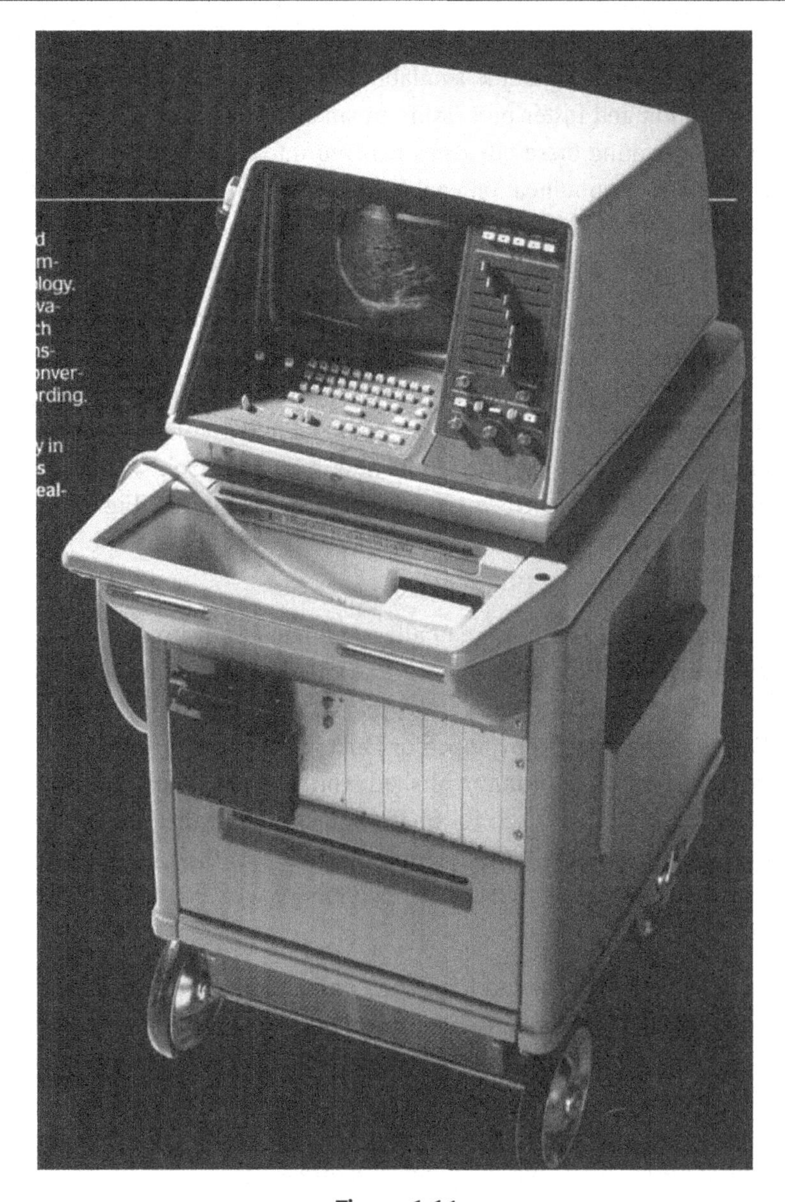

Figure 1.11
The first Hewlett Packard phased-array system, the 70020A. *Courtesy of Philips Healthcare.*

surgical, intraoperative (transducer placed in body during surgery), laparoscopic, and neurosurgical; vascular, intravascular, and small parts (discussed in detail in Section 10.7.2). With improved materials and piezoelectric composites, arrays with several hundred elements and higher frequencies became available. Wider transducer bandwidths allowed the imaging and operation of other modes within the same transducer at several frequencies selectable by the user.

By the 1990s, developments in more powerful microprocessors, high-density gate arrays, and surface-mount technology, as well as the availability of low-cost analog/digital (A/D) chips, made greater computation and faster processing in smaller volumes available at lower costs. Imaging systems incorporating these advances evolved into digital architectures and beamformers. Broadband communication enabled the live transfer of images for telemedicine. Transducers appeared with even wider bandwidths and in 1.5D (segmented arrays with limited elevation electronic focusing capabilities) and matrix-array configurations.

By the late 1990s, near-real-time three-dimensional (3D) imaging became possible. Commercial systems mechanically scanned entire electronically scanned arrays in ways similar to those used for single-element mechanical scanners. Translating, angular fanning, or spinning an array about an axis created a spatially sampled volume. Special image-processing techniques developed for movies such as John Cameron's *Titanic* enabled nearly real-time 3D imaging, including surface-rendered images of fetuses. Figure 1.12 shows a survey of fetal images that begins with a black-and-white image from the 1960s and ends with a surface-rendered fetal face from 2002.

True real-time 3D imaging is much more challenging because it involves two-dimensional (2D) arrays with thousands of elements, as well as an adequate number of channels to process and beamform the data. An early 2D array, 3D real-time imaging system with 289 elements and 4992 scan lines was developed at Duke University in 1987 (Smith et al., 1991; von Ramm et al., 1991). Kojima published work on matrix arrays (1986); a non-real-time, 3600-2D-element array was used for aberration studies at the University of Rochester (Lacefield & Waag, 2001). In 2003, Philips introduced a real-time 3D imaging system that utilized fully sampled 2D 2900-element-array technology with beamforming electronics in the transducer handle.

To extend the capabilities of ultrasound imaging, contrast agents were designed to enhance the visibility of blood flow. In 1968, Gramiak and Shah (Shah, Gramiak, & Kramer, 1976) discovered that microbubbles from indocyanine green dye injected in blood could act as an ultrasound contrast agent. By the late 1980s, several manufacturers were developing contrast agents to enhance the visualization of and ultrasound sensitivity to blood flow. To emphasize the detection of blood flow, investigators imaged contrast agents at harmonic frequencies generated by the microbubbles. As imaging system manufacturers became involved in imaging contrast agents at second harmonic frequencies, they discovered that tissues could also be seen. Signals sent into the body at a fundamental frequency returned from tissue at harmonic frequencies. Tissues talked back. P. N. T. Wells (1969a) mentioned indications that tissues had nonlinear properties. Some work on imaging the nonlinear coefficient of tissues directly (called their "B/A" value) was done in the 1980s but did not result in manufactured devices. By the late 1990s, the clinical value of tissue harmonic imaging was recognized and commercialized. Tissue harmonic images have proved to be very useful in imaging otherwise

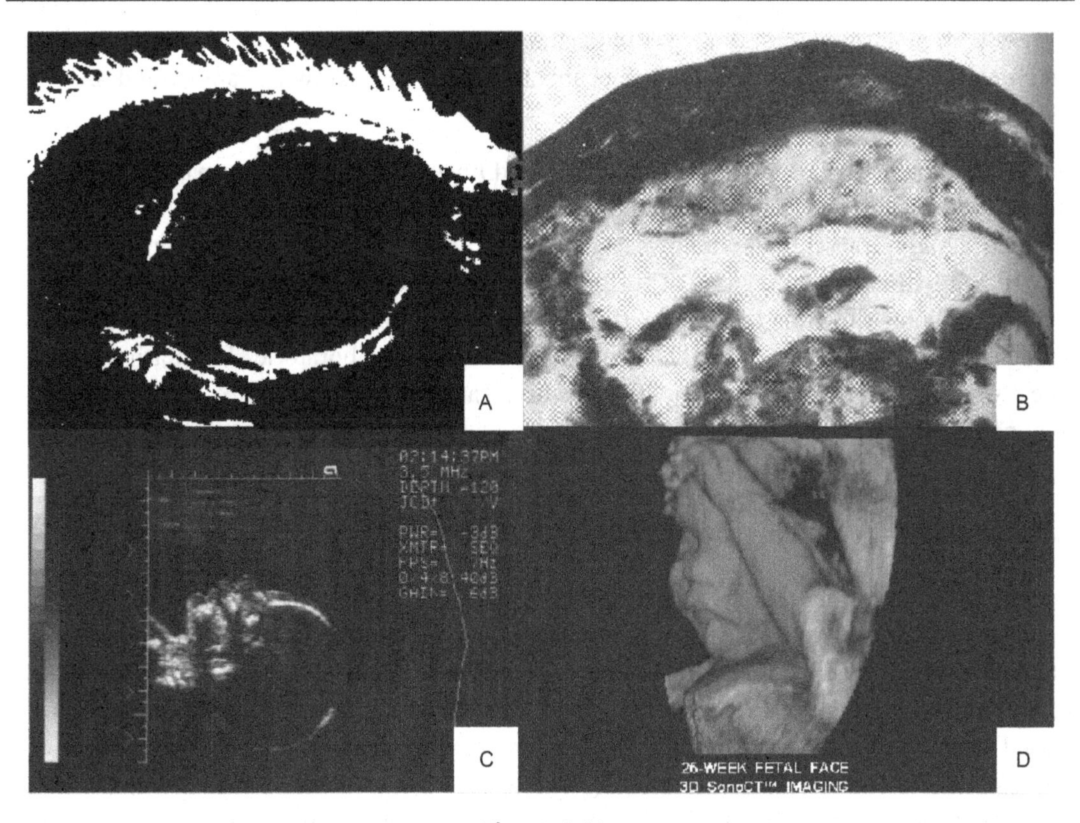

Figure 1.12

The evolution of diagnostic imaging as shown in fetal images. (A) Fetal head black-and-white image (I. Donald, 1960); (B) early gray-scale negative image of fetus from the 1970s; (C) high-resolution fetal profile from the 1980s; (D) surface-rendered fetal face and hand from 2002. *Sources: A) and B) Goldberg and Kimmelman, 1988, reprinted with permission of AIUM. Courtesy of B. Goldberg, (C) Siemens Healthcare, and (D) Philips Healthcare.*

difficult-to-image people, and in many cases, they provide superior contrast resolution and detail compared with images made at the fundamental frequency.

At the beginning of the twenty-first century, different forms of elastography (Lerner & Parker, 1987; Ophir, Cspedes, Ponnekanti, Yazdi, & Li, 1991) and shear wave acoustic radiation force imaging (Fink and Tanter, 2010; Nightingale, Palmeri, Nightingale, & Trahey, 2001; Sarvazyan, Rudenko, Swanson, Fowlkes, & Emelianov, 1998) made large advances in revealing the elastic and viscoelastic properties of tissue for diagnosis and became commercialized in imaging systems (much more in Chapter 16). The production of molecular or targeted ultrasound contrast agents for specialized applications and therapies made significant progress (Chapter 14). High-intensity focused ultrasound (HIFU) and therapeutic ultrasound proved to be a relatively painless way of performing knifeless surgery with precisely focused intense ultrasound beams in the body. Other surgical, interventional and cosmetic applications of this technology are growing rapidly and are discussed in Chapter 9 and Chapter 17.

In the more than 60 years since the first ultrasound image of the head, comparatively less progress has been made in imaging through the skull. Valuable Doppler data have been obtained through transcranial windows. By the late 1980s, methods for visualizing blood flow to and within certain regions of the brain were commercialized in the form of transcranial color-flow imaging. The difficult problems of producing undistorted images through other parts of the skull have been solved at research laboratories but not in real time (Aarnio, Clement, & Hynynen, 2001; Aubry, Tanter, Thomas, & Fink, 2001; Pajek and Hynynen, 2012); more details are in Chapter 9.

1.6 Enabling Technologies for Ultrasound Imaging

Attention is usually focused on ultrasound developments in isolation. However, continuing improvements in electronics, seemingly unrelated, are shaping the future of medical ultrasound. The accelerated miniaturization of electronics, especially ASICs, made possible truly portable imaging systems for arrays with full high-quality imaging capabilities. When phased-array systems first appeared in 1980, they weighed about 800 lbs. The prediction of the increase in transistor density, according to a revised Moore's law, is a factor of 1000 in area-size reduction from 1980 to 2000. Over the years, Moore's law slowed down a bit, and was revised from a doubling every 18 to every 24 months, The more realistic Moore's law (shown in Figure 1.8) has sustained its rate for five decades. Just as many experts thought that Moore's law could not be sustained because of physical limitations, it received a boost from new technological breakthroughs (Ahmed & Schuegraf, 2012). Moore's law, long based on planar real estate, may have to be modified to accommodate higher densities made possible by the vertical growth of three-dimensional architectures such as the tri-gate transistor by Intel. In addition, graphical processing units (GPUs) with parallel processing architectures and improved data routing (Lindholm et al., 2008) offer significant speed opportunities as estimated in Figure 1.8 (see Chapter 10).

While a straightforward calculation in the change of size of an imaging system cannot be made easily, several imaging systems that were available in 2003 have more features than some of the first phased-array systems and yet weigh only a few pounds (shown in Figure 1.13). Towards the end of the decade, even smaller pocket ultrasound systems became available, as displayed in Figure 1.14. Another modern achievement is a handheld 2D array with built-in beamforming, see Section 10.12.1.

Portable systems, because of their affordability, can be used as screening devices in smaller clinics, as well as in many places in the world where the cost of an ultrasound imaging system

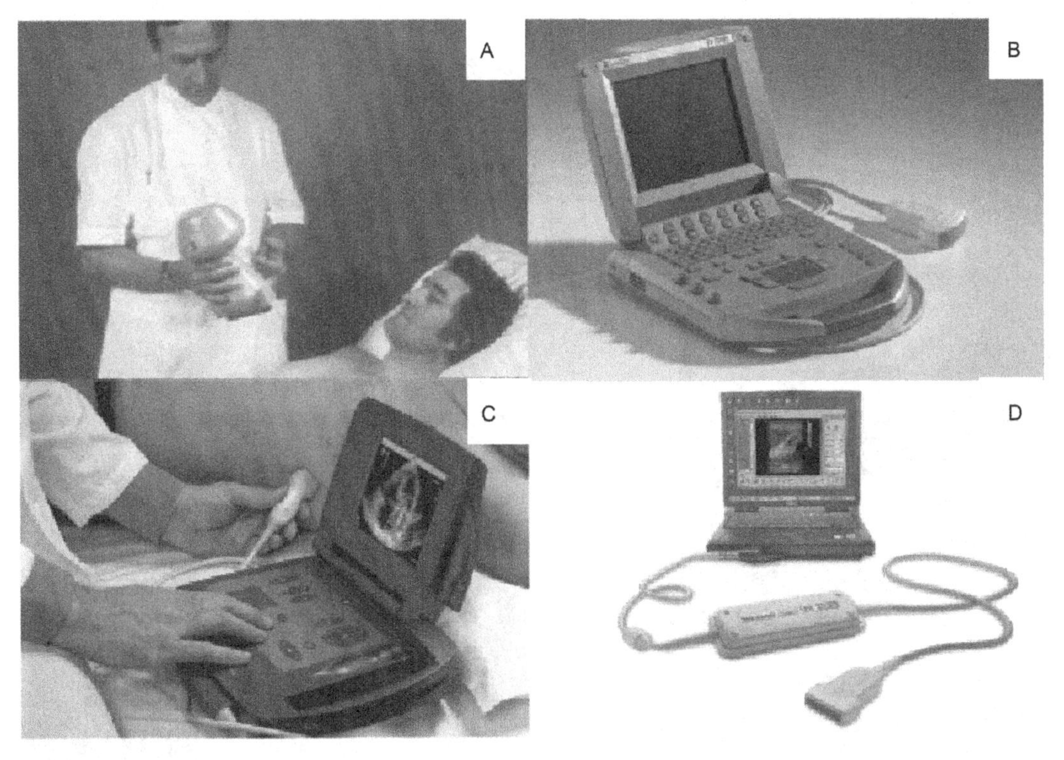

Figure 1.13
Portable ultrasound imaging systems. (A) Minivisor, the self-contained truly portable ultrasound imaging system; (B) a newer version of the Sonosite, the first modern handheld ultrasound portable; (C) OptiGo, a cardiac portable with automated controls; (D) Terason 2000, a laptop-based ultrasound system with a proprietary beamformer box. *Courtesy of (A) N. Bom; (B) Sonosite, Inc.; (C) Philips Healthcare; (D) Terason, Teratech Corp.*

is prohibitive. Figure 1.13 shows four examples of portable systems that appeared on the cover of a special issue of the *Thoraxcentre Journal* on portable cardiac imaging systems (2001). The first portable system, the Minivisor, was self-contained with a battery, but its performance was relatively primitive (this was consistent with the state of the art in 1978). The OptiGo owed its small size to custom-designed ASICs, as well as automated and simplified controls. The Titan, a newer version of the original Sonosite system and one of the first modern portables, had a keyboard and trackball, and it is also miniaturized by several ASICs. The Terason system had a charge-coupled device (CCD)-based proprietary 128-channel beamformer, and much of its functionality is software-based in a powerful laptop. More information on these portables can be found in the December 2001 issue of the *Thoraxcentre Journal.*

Since the first edition of this book was published in 2004, a new generation of even smaller ultrasound imaging systems has emerged. Dubbed "pocket ultrasound" because of the

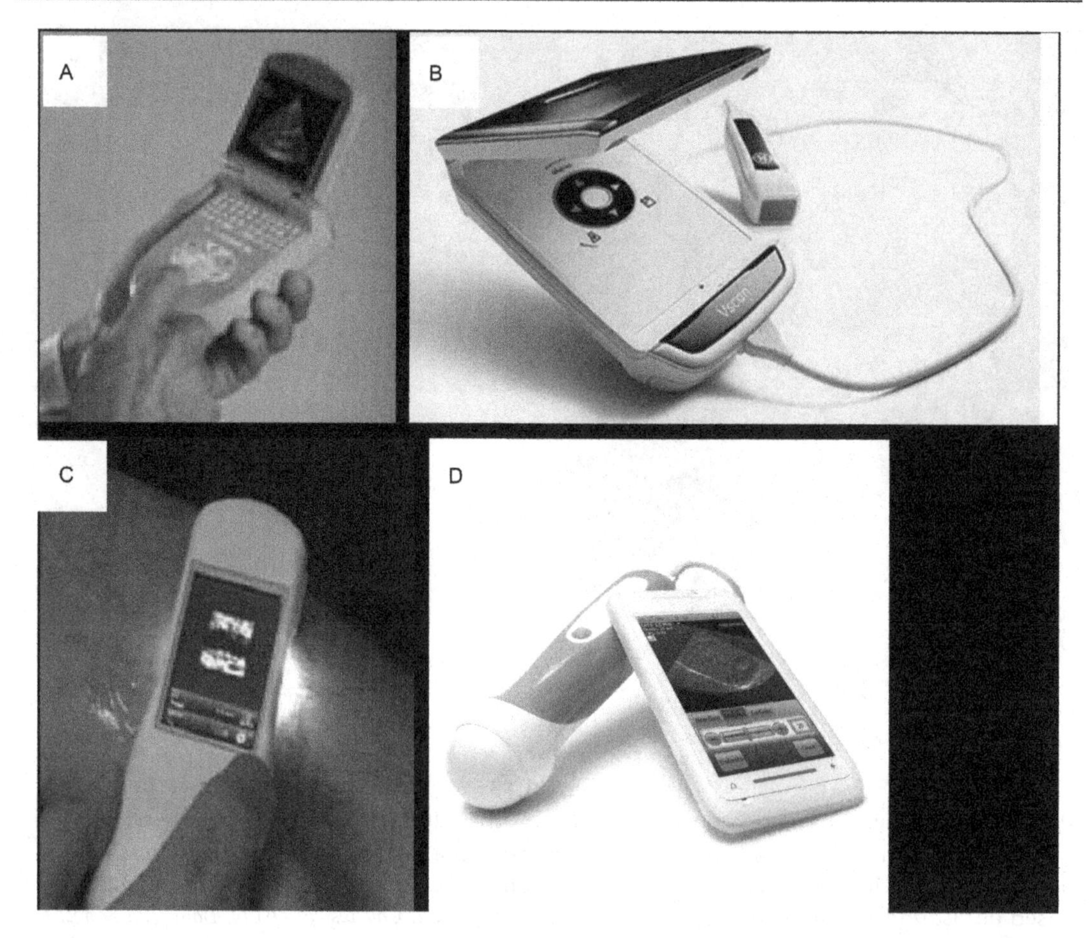

Figure 1.14
Pocket ultrasound imaging systems. (A) P-10, the first pocket personal device accessory (PDA)-style ultrasound device; (B) the V-Scan, with a simplified user interface; (C) Sonic Window, an integrated C-scan imaging prototype device with a 2D array; (D) MobiUS SP1, which operates from a mobile phone. *Courtesy of (A) Siemens Healthcare; (B) General Electric Healthcare; (C) Fuller, Owen, Blalock, Hossack, and Walker, 2009; (D) Mobisante.*

smaller footprint of these devices (they fit in a lab-coat pocket), these devices represent the next phase of miniaturized ultrasound devices. These systems, the beneficiaries of size reduction by Moore's law, clever engineering, automation, and reduced feature sets, are illustrated by the systems shown in Figure 1.14. The Siemens P-10 (Egan & Ionescu, 2008), the first of these pocket devices, had a personal device accessory (PDA)-style interface with a keypad, automatic time gain compensation, and harmonic imaging, and is shown in Figure 1.14A. General Electric's V-Scan, with automated features that facilitated user interaction and training, had a one-handed intuitive control and voice annotation and is depicted in Figure 1.14B (Lafitte et al., 2011). The Sonic Window is a departure from B-mode imaging in that it is primarily a C-mode imaging device (the image plane is parallel to the plane of the array). A prototype of the Sonic Window consists of a 3600-element 2D transducer array integrated with the electronics and battery into one modular unit; designed for vascular and newer applications, is pictured in Figure 1.14C (Fuller, Owen, Blalock, Hossack, & Walker, 2009). MobiSante has developed the first mobile phone-based ultrasound imaging device (Figure 1.14D), employing a self-contained unit including electronics and a mechanically scanned transducer. Pocket ultrasound devices are widening the circle of ultrasound applications because of their low cost and extreme portability. Their novelty has led to a continual re-evaluation of appropriate users and applications, product distribution, training, and diagnostic quality (Sicari et al., 2011).

Change is in the direction of higher complexity at reduced costs. Modern full-sized imaging systems have a much higher density of components and far more computing power than their predecessors. The enabling technologies and key turning points in ultrasound are summarized in Table 1.1.

1.7 Ultrasound Imaging Safety

Diagnostic ultrasound has had an impressive safety record since the 1950s. In fact, no substantiated cases of harm from imaging have been found (O'Brien, 1998). Several factors have contributed to this record. First, a vigilant worldwide community of investigators is looking continuously for possible ultrasound-induced bioeffects. Second, the two main bioeffects (cavitation and thermal heating) are well enough understood so that acoustic output can be controlled to limit these effects. The Output Display Standard provides imaging system users with direct on-screen estimates of relative indices related to these two bioeffects for each imaging mode selected. Third, a factor may be the limits imposed on acoustic output of US systems by the Food and Drug Administration (FDA). All manufacturers of diagnostic ultrasound imaging systems selling in the U.S. measure acoustic output levels of their systems with wide-bandwidth-calibrated hydrophones, force balances, and report their data to the FDA. International standards for ultrasound equipment, safe practices and consistent measurements and characterization, are accepted worldwide. More information on these topics can be found in Chapter 15.

1.8 Ultrasound and Other Diagnostic Imaging Modalities

1.8.1 Imaging Modalities Compared

Ultrasound, because of its efficacy and low cost, is often the preferred imaging modality. Millions of people have been spared painful exploratory surgery by noninvasive imaging.

Table 1.1 Chronology of Ultrasound Imaging Developments and Enabling Technologies

Time	Ultrasound	Enablers
Pre-WWII	Echo ranging	Piezoelectricity, vacuum tube amplifiers
1940s	Dussik image of brain	Radar, sonar
	PPI images	Supersonic reflectoscope
	Therapy and surgery	Colossus and ENIAC computers, transistors
1950s	A-line	Integrated circuits
	Compound scanning	Phased-array antennas
	Doppler ultrasound	
	M-mode	
1960s	Contact static B-scanner	Moore's law
	Real-time mechanical scanner	Microprocessors
	Echoencephalography	VLSI
		Handheld calculators
1970s	Real-time imaging	RAM
	Scan-conversion	EPROM
	Gray-scale	ASIC
	Linear and phased arrays	Scientific calculators
		Altair, first PC
1980s	Commercial array systems	Gate arrays
	Pulsed wave Doppler	Digital signal processing
	Color flow imaging	chips
	Wideband and specialized transducers	Surface mount components, computer-aided design of VLSI circuits
1990s	Digital systems	Low-cost A/D converters
	1.5D and matrix arrays	Powerful PCs
	Harmonic imaging	3D image processing
	Commercialized 3D imaging	0.1 μm fabrication of line widths for electronics
		Signal processing
		Broadband transducers
2000s	Handheld 2D array for real-time 3D imaging	Continued miniaturization
	Pocket Ultrasound Devices	Advanced and faster signal Processing
	Elastography	High speed architectures
	ARFI	
	Shear Wave Imaging	

Their lives have been saved by ultrasound diagnosis and timely intervention, their hearts have been evaluated and repaired, their children have been found in need of medical help by ultrasound imaging, and their surgeries have been guided and checked by ultrasound. Many more people have breathed a sigh of relief after a brief ultrasound exam found no disease or confirmed the health of their future child.

How does ultrasound compare to other imaging modalities? Each major diagnostic imaging method is examined in the following sections, and the overall results are tallied in Table 1.2 and compared in Figure 1.15. The first edition of this book had a graph comparing different imaging modalities for the year 2000; in Figure 1.15, this data is

Table 1.2 Comparison of Imaging Modalities

Modality	Ultrasound	X-ray	CT	MRI
What is imaged	Longitudinal, shear, mechanical properties	Mean X-ray tissue absorption	Local tissue X-ray absorption	Biochemistry (*T*1 and *T*2)
Access	Small windows adequate	2 sides needed	Circumferential around body	Circumferential around body
Spatial resolution	Frequency and axially dependent 0.2 − 3 mm	~1 mm	~1 mm	~1 mm
Penetration	Frequency dependent, 3−25 cm	Excellent	Excellent	Excellent
Safety	Excellent for > 50 years	Ionizing radiation	Ionizing radiation	Very good
Speed	> 100 frames/sec	Minutes	20 minutes	Typical: 45 minutes; fastest-10 frames/sec
Cost	$_ > $-	$	$$	$$$
Portability	Excellent	Good	Poor	Poor
Volume Coverage	Real-time 3D volumes, improving	2D	Large 3D volume	Large 3D volume
Contrast	Increasing (shear)	Limited	Limited	Slightly flexible
Intervention	Real-time 3D increasing	No, Fluoroscopy limited	No	Yes, limited
Functional	Functional ultrasound	No	No	fMRI

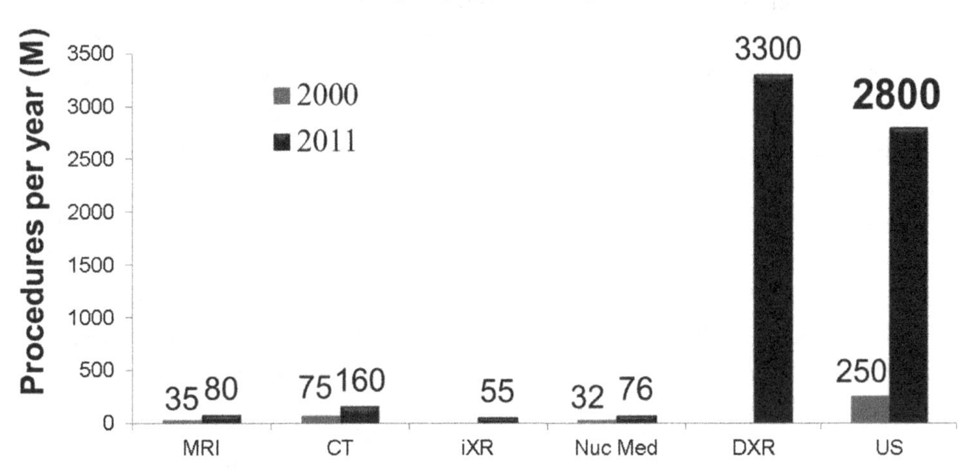

Figure 1.15

Comparison of estimated number of imaging exams given worldwide for the years 2000 and 2011 (data courtesy of Daniel Cote). CT, computed tomography; DXR, digital X-ray; iXR, interventional X-ray; MRI, magnetic resonance imaging; Nuc Med, nuclear medicine; PET, positron emission tomography; US, diagnostic ultrasound.

compared with 2011 numbers. While other imaging modalities made modest gains, diagnostic ultrasound grew by an order of magnitude in the decade since the first edition of this book! Reasons for this change are explored in Section 1.11. Combinations of imaging modalities are discussed in Section 1.10.1. Images produced by multi-wave interactions are covered by Section 1.10.2.

1.8.2 Ultrasound

Ultrasound imaging has a spatially variant resolution that depends on the size of the active aperture of the transducer, the center frequency and bandwidth of the transducer, and the selected transmit focal depth. A commonly used focal-depth-to-aperture ratio is five, so that the half-power beam-width is approximately two wavelengths at the center frequency. Therefore, the transmit lateral spatial resolution in millimeters is 2λ (mm) $= 2c_{av}/f_c \sim 3/f_c$, where f_c is center frequency in Megahertz. For typical frequencies in use, ranging from 1 to 15 MHz, lateral resolution corresponds to 3 mm to 0.2 mm. This resolution is best at the focal length distance and widens away from this distance in a nonuniform way because of diffraction effects caused by apertures on the order of a few to tens of wavelengths. The best axial resolution is approximately two periods, T, of a short pulse, or the reciprocal of the center frequency, which also works out to be two wavelengths in distance, since $z = 2c_{av}T = 2c_{av}/f_c = 2\lambda$.

Another major factor in determining resolution is attenuation, which limits penetration. Attenuation steals energy from the ultrasound field as it propagates and, in the process, effectively lowers the center frequency of the remaining signals, another factor that reduces resolution further. Attenuation also increases with higher center frequencies and depth; therefore, penetration decreases correspondingly so that fine resolution is difficult to achieve at deeper depths. This limitation is offset by specialized probes such as transesophageal (down the throat) and intracardiac (inside the heart) transducers that provide access to regions inside the body. Otherwise, access to the body is made externally through many possible "acoustic windows," where a transducer is coupled to the body with a water-based gel. This "footprint," or window is the area of contact the transducer makes with the body and is typically a few centimeters squared. Except for regions containing bones, air, or gas, which are opaque to imaging transducers, even small windows can be enough to visualize large interior regions.

Ultrasound images are highly detailed and geometrically correct to the first order. These maps of the mechanical structures of the body (according to their "acoustic properties," such as differences in characteristic impedance) depend on density and stiffness or elasticity. The dynamic motion of organs such as the heart can be revealed by ultrasound operating up to hundreds of frames per second.

Diagnostic ultrasound is noninvasive (unless you count the "trans" and "intra" families of transducers which are placed within the body and are very effective in producing better images). Ultrasound is regarded as safe and does not have any cumulative biological side effects (more on this topic can be found in Chapter 15). Two other strengths of diagnostic ultrasound imaging are its relatively low cost and portability. With the widespread availability of miniature portable and pocket ultrasound systems for screening and imaging, these two factors will continue to improve.

A high skill level is needed to obtain good images with ultrasound. This expertise is necessary because of the number of access windows, the differences in anatomy, and the many possible planes of view. Experience is required to find relevant planes and targets of diagnostic significance and to optimize instrumentation. Furthermore, a great deal of experience is required to recognize, interpret, and measure images for diagnosis. The wider use of 3D ultrasound is enabling improved anatomical visualization and new applications in interventions and surgery. Real-time 3D ultrasound imaging is also now providing increased coverage, or larger volume visualization, a notable drawback when 2D or planar ultrasound images were compared to the volume images produced by computed tomography (CT) and magnetic resonance imaging (MRI). This disadvantage is being offset by image fusion, a combination of ultrasound co-registered with other 3D imaging modalities.

1.8.3 Plane X-rays

Conventional X-ray imaging is more straightforward than ultrasound. Digital X-rays (DXR) are replacing the use of film. Because X-rays travel at the speed of light with a wavelength of less than 1 Å (0.0001 mm), they do so in straight ray paths without diffraction effects. As a result of the ray paths, highly accurate images are obtained in a geometric sense. As the X-rays pass through the body, they are absorbed by tissue so that an overall "mean attenuation" image results along the ray path. Three-dimensional structures of the body are superimposed as a two-dimensional projection onto film or a digital sensor array. The depth information of structures is lost as it is compressed into one image plane. Spatial resolution is not determined by wavelength but by focal spot size of the X-ray tube and scatter from tissue. The state of the art is about 1 mm as of 2013. X-rays cannot differentiate among soft tissues but can detect air (as in lungs) and bones (as in fractures). Radioactive contrast agents can be ingested or injected to improve visualization of vessels. Still X-ray images require patients not to move during exposure. Because these are through transmission images, parts of the body that can be imaged are limited to those that are accessible on two sides. Real-time X-ray is available as fluoroscopy.

Most conventional X-ray systems in common use are dedicated systems (fixed in location) even though portable units are commercially available for special applications. Systems tend to be stationary so that safety precautions can be taken more easily. Though exposures are short, X-rays are a form of ionizing radiation, so dosage effects can be cumulative. Extra precautions are needed for sensitive organs such as eyes and for pregnancies.

The taking of X-ray images is relatively straightforward after some training. Interpretation of the images varies with the application, from broken bones to lungs, and in general requires a high level of skill and experience.

1.8.4 Computed Tomography Imaging

Computed tomography (CT), which is also known as computed axial tomography (CAT), scanning also involves X-rays. Actually, the attenuation of X-rays in different tissues varies, so tomographic ways of mathematically reconstructing the interior values of attenuation from those obtained outside the body, have been devised. In order to solve the reconstruction problem uniquely, enough data has to be taken to provide several views of each spatial position in the object.

This task is accomplished by an X-ray fan-beam source on a large ring radiating through the subject's body to an array of detectors working in parallel on the opposite side of the ring. The ring is rotated mechanically in increments until complete coverage is obtained. Rapid reconstruction algorithms create the final image of a cross-section of a body. The latest multislice equipment utilizes a cone beam and a 2D array of sensors. The result has over two orders of magnitude more dynamic range than a conventional X-ray, so subtle shades of the attenuation variations through different tissue structures are seen. The overall dose is much higher than that of a conventional X-ray, but the same safety precautions as those of conventional X-rays apply. CT equipment is large and stationary in order to fit a person inside, and as a result, it is relatively expensive to operate. Consecutive pictures of a moving heart are now achievable through synchronization to electrocardiogram (ECG) signals.

The resolution of CT images is typically 1 mm (though 0.3 mm is possible). CT scanning creates superb images of the brain, bone, lung, and soft tissue, so it is complementary to ultrasound.

While the taking of CT images requires training, it is not difficult. Interpretation of CT cross-sectional images demands considerable experience for definitive diagnosis.

1.8.5 Magnetic Resonance Imaging

Magnetic resonance has been applied successfully to medical imaging of the body because of its high water content. The hydrogen atoms in water (H_2O) and fat make up 63% of the body by weight. Because there is a proton in the nucleus of each hydrogen atom, a small magnetic field or "moment" is created as the nucleus spins. When hydrogen is placed in a large static magnetic field, the magnetic moment of the atom spins around it like a tiny gyroscope at the Larmor frequency, which is a unique property of the material. For imaging, a radiofrequency rotating field in a plane perpendicular to the static field is needed. The frequency of this field is identical to the Larmor frequency. Once the atom is excited, the applied field is shut off and the original magnetic moment decays to equilibrium and emits a signal. This voltage signal is detected by coils, and two relaxation constants are sensed. The longitudinal magnetization constant, $T1$, is more sensitive to the thermal properties of tissue. The transverse magnetization relaxation constant, $T2$, is affected by the local field inhomogeneities. These constants are used to discriminate among different types of tissue and for image formation.

For imaging, the subject is placed in a strong static magnetic field created by a large enclosing electromagnet. The resolution is mainly determined by the gradient or shape of the magnetic field, and it is typically 1 mm. Images are calculated by reconstruction algorithms based on the sensed voltages proportional to the relaxation times. Tomographic images of cross-sectional slices of the body are computed. The imaging process is fast and safe because no ionizing radiation is used. Because the equipment needed to make the images is expensive, exams are costly.

Magnetic resonance imaging (MRI) equipment has several degrees of freedom, such as the timing, orientation, and frequency of auxiliary fields; therefore, a high level of skill is necessary to acquire diagnostically useful images. Diagnostic interpretation of images involves both a thorough knowledge of the settings of the system, as well as considerable experience.

Magnetic Resonance Imaging Applications

Functional MRI (fMRI) has been used extensively to map regions of the brain connected with stimuli, activities, and higher-level cognition. In this method, fMRI detects localized changes in brain activity, usually in the form of changes in cerebral metabolism, blood flow, volume, or oxygenation in response to task activation. These changes are interrelated and may have opposite effects. For example, an increase in blood flow increases blood oxygenation, whereas an increase in metabolism decreases it. The most common means of detection is measuring the changes in the magnetic susceptibility of hemoglobin. Oxygenated blood is dimagnetic and deoxygenated blood is paramagnetic. These differences lead to a detection method called blood-oxygen-level-dependent contrast (BOLD). Changes in blood oxygenation can be seen as differences in the T*2 decay constant, so T2-weighted images are used.

1.8.6 Magnetoencephalography

Magnetoencephalography (MEG) is a form of neuroimaging that maps the tangential components of magnetic fields associated with scalp potentials produced by the brain

(Cohen & Halgren, 2009). These potentials are the same ones that can be recorded as electroencephalograms (EEGs), but the dynamic magnetic components of these potentials contain different information with spatial sampling. Unlike fMRI or positron emission tomography (PET) images that provide indirect or delayed clues to synaptic events through changes in the metabolism or blood flow, MEG locates regions of the brain that respond directly to stimuli and yields corresponding time responses at various spatial locations.

MEG is used to explore the characteristics of mental patterns such as epilepsy and schizophrenia. It is also useful in brain research to understand cognitive functioning and to learn which parts of the brain are involved in different tasks.

1.8.7 Positron Emission Tomography

For positron emission tomography (PET), gamma rays that come from a positron-emitting radionuclide material are detected by pairs of scintillation detectors. These detectors are arranged in an annular ring and 3D images are constructed using tomographic techniques to form a 3D image. Concentrations of tracer materials such as ^{18}F-FDG, a glucose analog found in the body, are taken up by tissues and indicate local metabolic activity. Only concentrations of this contrast agent are displayed, hanging in space, so it is difficult to establish spatial landmarks to determine locations; therefore, PET images are often combined with another imaging modality such as CT to provide spatial localization.

1.9 Contrast Agents

Contrast agents emphasize or magnify physiological features or functions that would otherwise be invisible, weak, or obscured in images. One of the most frequent applications of these agents is in the vasculature. Another is the uptake of an agent by an organ, indicating a degree of functional or metabolic activity. Each agent is designed to work with the particular physics of the intended imaging modality as well as the region of indication in the body.

1.9.1 Computed Tomography Agents

Intravenously administered CT contrast agents containing iodine or gadolinium are employed in the visualization of blood vessels. For example, for a CT pulmonary angiogram, emboli and obstructions can be seen through attenuation of the X-rays or opacification, which brightens the appearance of flow in a vessel. Through the enhancement of blood vessels, other applications include visualization of the internal structures of the brain, spine, liver, kidney, and other organs. Afterward, the kidney and liver normally aid in passing the agents out of the system. In some cases, these agents may involve some risk and possible allergic reactions such as those for the kidney.

For the diagnosis of abdominal diseases and injuries, CT contrast agents are administered either orally or rectally. A dilute suspension of barium sulfate is used most often. Applications include the diagnosis and characterization of cancer and the investigation of sources of severe abdominal pain and trauma.

1.9.2 Magnetic Resonance Imaging Agents

For MRI, agents distort or alter the magnetic fields locally, where they are concentrated so there is a relative change in a region. The effects achieved can either intensify or weaken the signal close by, depending on the image weighting selected and the context of the agent; they work by reducing local T1 and T2 relaxation times nearby. The agents tend to be either paramagnetic or supermagnetic materials that are modified to be ingested or infused into the body safely and passed out. The major type of agent used is chelated gadolinium in various forms. The main applications are intravascular and gastrointestinal.

1.9.3 Ultrasound Agents

Ultrasound contrast agents are small microspheres that are used to highlight the passage of blood in regions otherwise difficult to image, such as blood, which is nearly invisible to ultrasound without special processing. These agents are usually microspheres filled with air or gases such as perfluorocarbon, with thin, flexible shells typically made of human serum albumin or surfactants, and are about 4 μm in diameter, similar to the size of red blood cells. They are injected into the venous system to act as blood-cell tracers. While in the blood, ultrasound agents significantly improve the ability and sensitivity of ultrasound to follow the flow of blood, especially in small vessels, and the perfusion of blood in muscles, especially those of the heart, and to identify walls of the moving chambers of the heart (opacification). After traveling along in blood, the spheres diffuse and are released as they pass through the lungs.

Ultrasound contrast agents are designed for different applications and effects and are matched to operate at certain insonifying frequencies. They act as miniature nonlinear resonators that greatly enhance their reflectivity for ultrasound. Under certain conditions, the unusual properties of these agents are changes in size, cavitation, fragmentation, or directed movement. Much more about these agents can be found in Chapter 14.

1.10 Comparison of Imaging Modalities

How can imaging modalities be combined and compared? Each major diagnostic imaging method is examined in the last sections, and the overall results are tallied in Table 1.2. Examples of three imaging modalities—CT, MRI, and ultrasound—are shown as different images of the right kidney in Figure 1.16. Each imaging modality reveals a different physical characteristic of the kidney. The CT image, based on the X-ray absorption of the

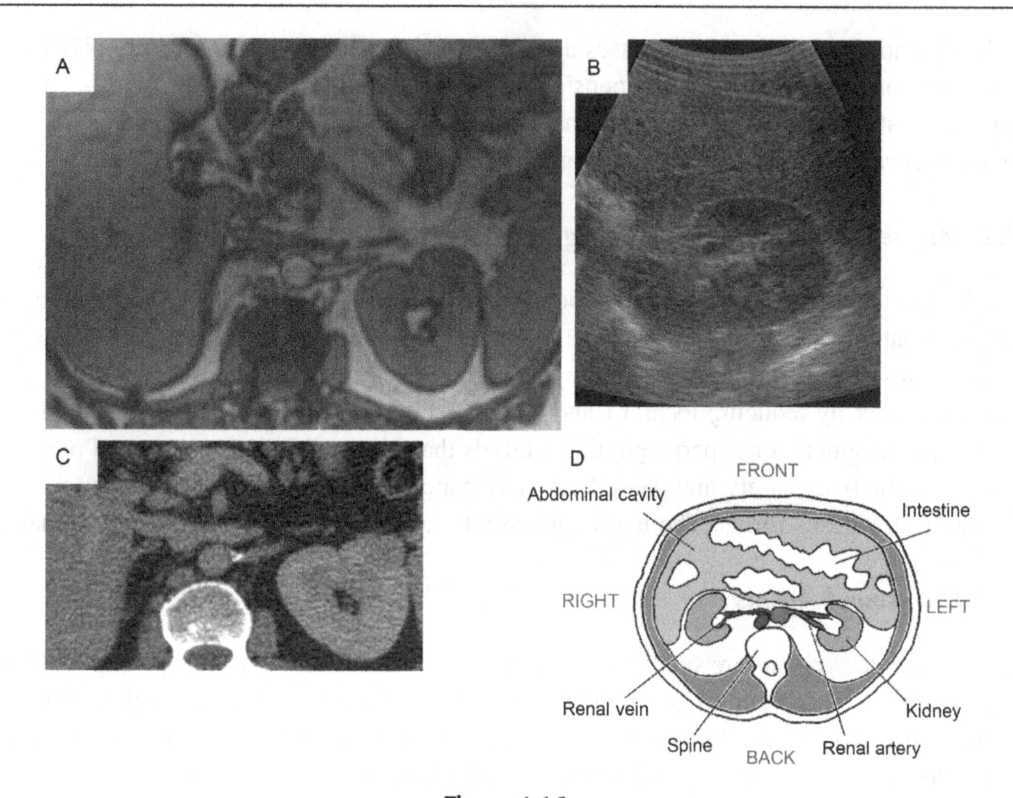

Figure 1.16
Images of the right kidney as viewed by different imaging modalities. (A) MRI; (B) ultrasound; (C) CT; (D) graphical depiction of abdominal cross-section showing location of kidneys.
Courtesy of Dr Marilyn Roubidoux.

kidney, shows a clear outline of the kidney; MRI gives an image related to the chemistry of the kidney, and the ultrasound shows details of the mechanical nature of the kidney.

1.10.1 Image Fusion

As implied by the comparison of images and the discussion of PET imaging in Section 1.8.7, the physics of each imaging modality reveals different characteristics of tissue properties and functions. In order to obtain a more complete picture, two (or more) imaging modalities can be combined. Image fusion is the simultaneous display of two different but usually complementary types of images either side by side or superimposed.

To demonstrate these different points of view, Figure 1.17 compares four types of imaging views of a metastasis. One of the leading types of image fusion is PET/CT, where specialized scanners, spatially coregistered, take two sets of images of the same subject in one convenient instrument (Blodgett, Meltzer, & Townsend, 2007). As an example, images of

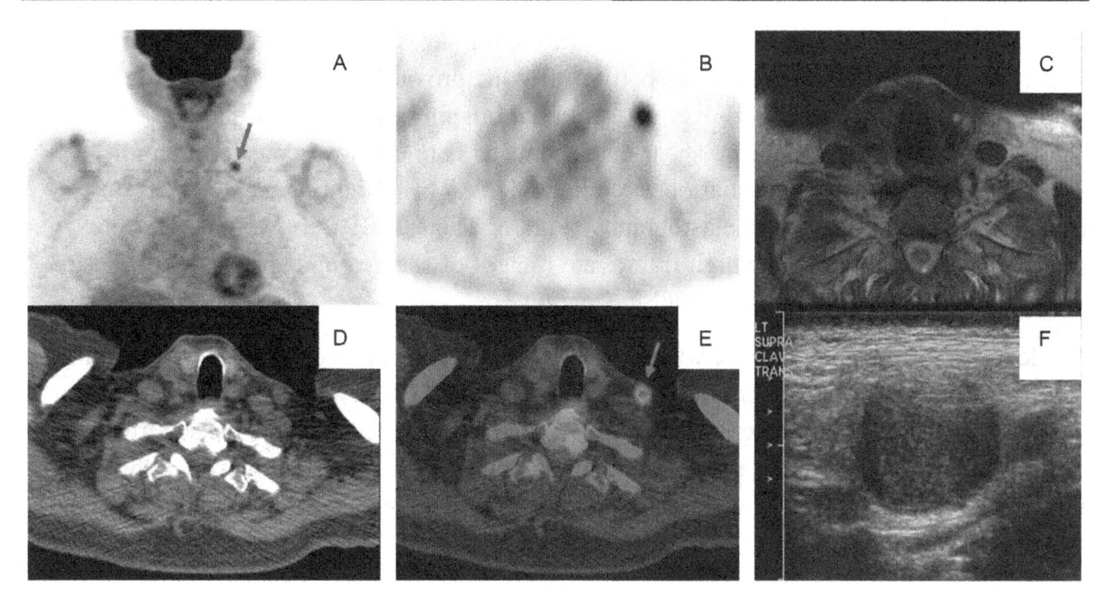

Figure 1.17

Images of different imaging modalities. (A) PET frontal; (B) PET transverse; (C) MRI transverse; (D) CT transverse; (E) PET/CT transverse fused image; and (F) ultrasound. Images of a 74-year-old male patient with nasal cavity esthesioneuroblastoma (a form of nasal cancer) who was restaged on a routine follow-up. PET identified a left supraclavicular hypermetabolic nodal metastasis (A, B, D, and E) that was not identified in the MRI (C). An ultrasound-guided biopsy of the node was positive for esthesioneuroblastoma and then surgically resected (F). The reader is referred to the Web version of the book to see the figure in color. *Courtesy of Dr Rathan Subramaniam, 2010.*

a patient with cancer given FDG are shown in Figure 1.17. The top left PET images in frontal and transverse views (1.17A and B) display the uptake of FDG as a small, dark region (see arrow), signifying hypermetabolic activity and the likelihood of a metastasis; however, the anatomical location of the cancer nodule is ambiguous in these views. The CT view (1.17D), displays good resolution as well as tissue structures, but not a clear indication of the nodule. The fusion image (1.17E), a combination of the PET and CT transverse, clearly emphasizes the metastasis. For comparison, the MRI image (1.17C) of the same view, while providing spatial organization of the organs, is by itself not as informative. Finally, the ultrasound image (1.17F) provides a different view and was useful for a guided biopsy.

1.10.2 Multi-wave and Interactive Imaging

In addition to the combination of individual imaging modalities in complementary ways, different types of waves and/or forces can be combined synergistically to provide new diagnostic information or to improve resolution. An example of a beneficial force–wave interaction is the original version of elastography (Ophir et al., 1991), in which a static force was used to displace tissue to create ultrasound images that revealed the elastic properties of tissue. Imaging shear properties of tissue provides orders of magnitude more contrast (Saravyzan et al., 1998) over conventional imaging with longitudinal waves (see Chapter 16.). One way to achieve shear displacements is to employ acoustic radiation forces. One form is acoustic radiation force imaging (ARFI) (Nightingale et al., 2001); another is transient shear wave elastography (Fink & Tanter, 2010), in which ARFs are used in novel ways to produce high frame rate images of shear sound speed. All of these methods are discussed in greater detail in Chapter 16.

Besides wave–force interactions, wave-to-wave interactions, coined "multi-wave imaging" (Fink and Tanter, 2010), can provide imaging modalities with new capabilities. Here, the synergistic effect of the interaction creates an imaging region that reveals tissue structure otherwise unobtainable. Other interactions include microwaves, thermal waves, light, magnetic fields, and even two different kinds of ultrasound waves (longitudinal and shear).

1.11 Conclusion

The different features and characteristics of ultrasound and three other major imaging modalities are summarized in Table 1.2 and shown in Figures 1.16 and 1.17. The tissue properties being imaged are considerably different among the different modalities and can be complementary. Image fusion offers new opportunities to reconcile limitations of individual imaging modalities. Different fusion combinations of imaging modalities have been attempted, but PET/CT fusion is the rapidly becoming the most popular (Blodgett et al., 2007). Both CT and MRI can image the whole body with consistent resolution and contrast; however, they are expensive methods of imaging and are slow and not portable. Ultrasound, on the other hand, has high but variable resolution and penetration and limited access to certain portions of the body (intestines, lungs, and bones). Ultrasound is used to image soft tissues only, but it has the advantages of low cost, portability, and real-time interactive and interventional imaging; and its ability to image larger 3D volumes is improving.

With the exception of standard X-ray exams, ultrasound is the leading imaging modality worldwide and in the United States. Unlike the relatively stable or incremental improvements of the other imaging modalities, ultrasound is acquiring significantly enhanced imaging capabilities in several areas due to the synergistic effect of disruptive technologies (see Table 1.1). Over the years, ultrasound has adapted to new applications through new arrays suited to specific clinical purposes and to signal processing, measurement, and visualization packages. Key strengths of ultrasound are its abilities to reveal anatomy, the dynamic movement of organs, and details of blood flow in real time. Diagnostic ultrasound continues to evolve by improving in diagnostic capability, image quality, convenience, ease of use, image transfer and management, and portability.

From the tables chronicling ultrasound imaging developments and enabling technologies, it is evident that there is often a time lag between the appearance of a technology and its effect. The most dramatic changes have been through the continual miniaturization of electronics in accordance with a modified Moore's law and new architectures. Smaller-sized components led to the first commercially available phased-array imaging systems as well as to new, pocket imaging systems, which weigh less than 0.5 kg. A revised Moore's law and recent vertical chip integration indicate that imminent manufacturing advances will bring higher chip complexity in smaller packages and at lower costs. New architectures such as those with graphical processing units and new imaging paradigms are rapidly increasing spatial and temporal resolution and providing more diagnostic information at greater speeds (see Chapters 10 and 11); because of the time lag of technology implementation, the latest developments have not had their full impact on ultrasound imaging.

The potential in diagnostic ultrasound imaging seen by early pioneers in the field has been more than fulfilled. The combination of continual improvements in electronics and a better understanding of the interaction of ultrasound with tissues will lead to imaging systems of increased complexity.

Development of new imaging modalities such as elastography and ARFI continue to expand opportunities for improved diagnostic capabilities with effective benefit/cost ratios. Other advantages of ultrasound diagnostic imaging include safety, flexibility, adaptability, portability, mobility, and reduced cost, which increase the range of potential applications and users. The ultrasound growth shown by Fig. 1.15, can be explained by a number of dynamic factors given by Table 1.2. Given these trends, ultrasound technology shows no sign of slowing down in the foreseeable future.

References

Aarnio, J., Clement, G. T., & Hynynen, K. (2001). Investigation of ultrasound phase shifts caused by the skull bone using low-frequency reflection data. *Proceedings of the IEEE ultrasonics symposium*, 2001, Atlanta, GA. 2, 1503–1506.

Ahmed, K., & Schuegraf, K. (2012). Transistor wars. *IEEE Spectrum*, *48*, 50–53.

Aubry, J. F., Tanter, M., Thomas, J. L., & Fink, M. (2001). Pulse echo imaging through a human skull: in vitro experiments. *Proceedings of the IEEE ultrasonics symposium*, 2001, Atlanta, GA. 2, 1499–1502.

Biquard, P. (1972). Paul Langevin. *Ultrasonics*, *10*, 213–214.

Blodgett, T. M., Meltzer, C. C., & Townsend, D. W. (2007). PET/CT: form and function. *Radiology*, *242*, 360–385.

Bom, N., Lancee, C. T., van Zwieten, G., Kloster, F. E., & Roelandt, J. (1973). Multiscan echocardiography. *Circulation*, *XLVIII*, 1066–1074.

Brown, T. G. (1968). Design of medical ultrasonic equipment. *Ultrasonics*, *6*, 107–111.

Cohen, D., & Halgren, E. (2009). Magnetoencephalography. *Encyclopedia of Neuroscience*, *5*, 615–622.

Devey, G. B., & Wells, P. N. T. (1978). Ultrasound in medical diagnosis. *Scientific American*, *238*, 98–106.

Duck, F. A. (2008) Icebergs and submarines: the genesis of ultrasonic detection. Scope September 2008, 46–50.

Edler, I. (1991). Early echocardiography. *Ultrasound in Medicine and Biology*, *17*. 425–131.

Edgerton, H. G. (1986). *Sonar images*. Englewood Cliffs, NJ: Prentice Hall.

Egan, M., & Ionescu, A. (2008). The pocket echocardiograph: a useful new tool? *European Journal of Echocardiography, 9*, 721–725.

Electronic Engineering Times. (October 30, 1997).

Erikson, K. R., Fry, F. J., & Jones, J. P. (1974). Ultrasound in medicine: a review. *IEEE Transactions on Sonics and Ultrasonics SU-21*144–170.

Eyer, M. K., Brandestini, M. A., Philips, D. J., & Baker, D. W (1981). Color digital echo/Doppler presentation. *Ultrasound in Medicine and Biology, 7*, 21.

Fink, M., & Tanter, M. (2010). Multiwave imaging and super-resolution. *Physics Today, 63*, 28–33.

Firestone, F. A. (1945). The supersonic reflectoscope for interior inspection. *Metal Progress, 48*, 505–512.

Fuller, M. I., Owen, K., Blalock, T. N., Hossack, J. A., & Walker W. F. (2009). Real-time imaging with the Sonic Window: a pocket-sized C-scan medical ultrasound device. *Proceedings of the IEEE ultrasonics symposium*, Rome, 2009, 196–199.

Goldberg, B. B., & Kimmelman, B. A. (1988). *Medical diagnostic ultrasound: A retrospective on its 40th anniversary*. New York: Eastman Kodak Company.

Hewlett Packard, Hewlett Packard Journal 34, 3–10. (1983).

Holmes, J. H. (1980). Diagnostic ultrasound during the early years of the AIUM. *Journal of Clinical Ultrasound, 8*, 299–308.

Kevles, B. H. (1997). *Naked to the bone*. New Brunswick, NJ: Rutgers University Press.

Klein, H. G. (1981). Are B-scanners' days numbered in abdominal diagnosis? *Diagnostic Imaging, 3*, 10–11.

Kojima, T. (1986). Matrix array transducer and flexible matrix array transducer. *Proceedings of the IEEE ultrasonics symposium*. Williamsburg, VA, 1986, 649–654.

Kossoff, G. (1974). Display techniques in ultrasound pulse echo investigations: a review. *Journal of Clinical Ultrasound, 2*, 61–72.

Lacefield, J. C., & Waag, R. C. (2001). Time-shift estimation and focusing through distributed aberration using multirow arrays. *IEEE Transactions on Ultrasonics, Ferroelectrics and Frequency Control, 48*, 1606–1624.

Lafitte, S., Alimazighi, N., Reant, P., Dijos, M., Zaroui, A., Mignot, A., et al. (2011). Validation of the smallest pocket echoscopic device's diagnostic capabilities in heart investigation. *Ultrasound in Medicine and Biology, 37*, 798–804.

Lerner, R. M., & Parker, K.J. (1987) Sonoelasticity images derived from ultrasound signals in mechanically vibrated targets. *Proceedings of the 7th european communities workshop* (October 1987,Nijmegen, the Netherlands).

Ligtvoet, C., Rijsterborgh, H., Kappen, L., & Bom, N. (1978). Real-time ultrasonic imaging with a hand-held scanner: Part I, Technical description. *Ultrasound in Medicine and Biology, 4*, 91–92.

Lindholm, E., Nickholls, J., Oberman, S., & Montrym, J. (2008). Nvidia Tesla: A unified graphics and computing architecture. *IEEE Micro*, 39–55.

Ludwig, G. D. (1950). The velocity of sound through tissues and the acoustic impedance of tissues. *Journal of the Acoustical Society of America, 22*, 862–866.

Murphy, J. (1986). Down into the deep. *Time, 128*, 48–54.

Namekawa, K., Kasai, C., Tsukamoto, M., & Koyano, A. (1982). Imaging of blood flow using autocorrelation. *Ultrasound in Medicine and Biology, 8*, 138.

Nightingale, K. R., Palmeri, M. L., Nightingale, R. W., & Trahey, G. E. (2001). On the feasibility of remote palpation using acoustic radiation force. *Journal of the Acoustical Society of America, 110*, 625–634.

O'Brien (1998). Assessing the risks for modern diagnostic ultrasound imaging. *Japanese Journal of Applied Physics, 37*, 2781–2788.

Ophir, J, Cspedes, I. C., Ponnekanti, H., Yazdi, Y., & Li, X. (1991). Elastography: a quantitative method for imaging the elasticity of biological tissues. *Ultrasonic Imaging, 13*, 111–134.

Richardson, L. F. (Filed May 10, 1912, issued March 27, 1913). British Patent No. 12.

Rines, R. H., Wycofff, C. W, Edgerton, H. E., & Klein, M. (1976). Search for the Loch Ness monster. *Technology Review*, *78*, 25–40.

Roelandt, J., Wladimiroff, J. W, & Bars, A. M. (1978). Ultrasonic real-time imaging with a hand-held scanner: Part II, Initial clinical experience. *Ultrasound in Medicine and Biology*, *4*, 93–97.

Sarvazyan, A. P., Rudenko, O. V., Swanson, S. D., Fowlkes, J. B., & Emelianov, S. Y. (1998). Shear wave elasticity imaging: a new ultrasonic technology of medical diagnostics. *Ultrasound in Medicine and Biology*, *24*, 1419–1435.

Shah, P. M., Gramiak, R., & Kramer, D. H. (1976). Ultrasound localization of left ventricular outflow obstruction in hypertrophic obstructive cardiomyopathy. *Circulation*, *40*. 3–11

Sicari, R., Galderisi, M., Voigt, J-U, Habib, G., Zamorano, J. L., Lancellotti, P., et al. (2011). The use of pocket-size imaging devices: a position statement of the European Association of Echocardiography. *European Journal of Echocardiography*, *12*, 85–87.

Simon, W. G., Wikipedia Common, accessed 19 January 2012.

Smith, S. W., Pavy, H. G., Jr., & von Ramm, O. T. (1991). High-speed ultrasound volumetric imaging system I: transducer design and beam steering. *IEEE Transactions on Ultrasonics, Ferroelectrics and Frequency Control*, *37*, 100–108.

Somer, J. C. (1968). Electronic sector scanning for ultrasonic diagnosis. *Ultrasonics*, *6*, 153–159.

The Thoraxcentre Journal 13, No. 4, cover. (2001).

Tyndall, J. (1875). Sound 3, 41. Longmans, Green and Co., London.

von Ramm, O. T., et al. (1991). High-speed ultrasound volumetric imaging system II: parallel processing and image display. *IEEE Transactions on Ultrasonics, Ferroelectrics and Frequency Control*, *38*, 109–115.

von Ramm, O. T., & Thurstone, F. L. (1975). Thaumascan: design considerations and performance characteristics. *Ultrasound in Medicine*, *1*, 373–378.

Wells, P. N. T. (1969a). *Physical principles of ultrasonic diagnosis*. London: Academic Press.

Wells, P. N. T. (1969b). A range-gated ultrasonic Doppler system. *Medical and Biological Engineering*, *7*, 641–652.

Wells, P. N. T. (1977). *Biomedical ultrasonics*. London: Academic Press.

Bibliography

Electronic Engineering Times. (December 30, 1996).

Proceedings of the IEEE on Acoustical Imaging. (April, 1979).

State-of-the-art review of acoustic imagining and holography.

Pajek, D., & Hynynen, K. (2012). Applications of transcranial focused ultrasound surgery. Acoustics Today, 8, 814.

Wells, P. N. T. (1979). Historical Review of Echo-instrumentation, C. T. Lancee (ed.), Martinus Nijhoff, The Hague, Netherlands.

This book, as well as Wells' other books in the References, provide an overview of evolving ultrasound imaging technology.

Woo, J. D. http://www.ob-ultrasound.net.

An excellent Web site for the history of medical ultrasound imaging technology and a description of how it works.

Goldberg, B. B., Wells, P. N. T., Claudon, M., and Kondratas, R. History of Medical Ultrasound, A CD-ROM compiled by WFUMB History/Archives Committee, WFUMB, 2003, 10th Congress, Montreal.

A compilation of seminal papers, historical reviews, and retrospectives.

CHAPTER 2

Overview

Chapter Outline

2.1 Introduction

Ultrasound imaging is a complicated interplay between physical principles and signal-processing methods, so it provides many opportunities to apply acoustic and signal-processing principles to relevant and interesting problems. In order to better explain the workings of the overall imaging process, this book uses a block diagram approach to organize various parts, their functions, and their physical processes. Building blocks reduce a complex structure to understandable pieces. This chapter introduces the overall organization that links upcoming chapters, each of which describes the principles of blocks in more detail. The next sections identify the principles used to relate the building blocks to each other and apply MATLAB programs to illustrate concepts.

2.2 Fourier Transform

2.2.1 Introduction to the Fourier Transform

Signals such as the Gaussian pulse in Figure 2.1 can be represented as either a time waveform or as a complex spectrum that has both magnitude and phase. These forms are alternate but completely equivalent ways of describing the same pulse. Some problems are

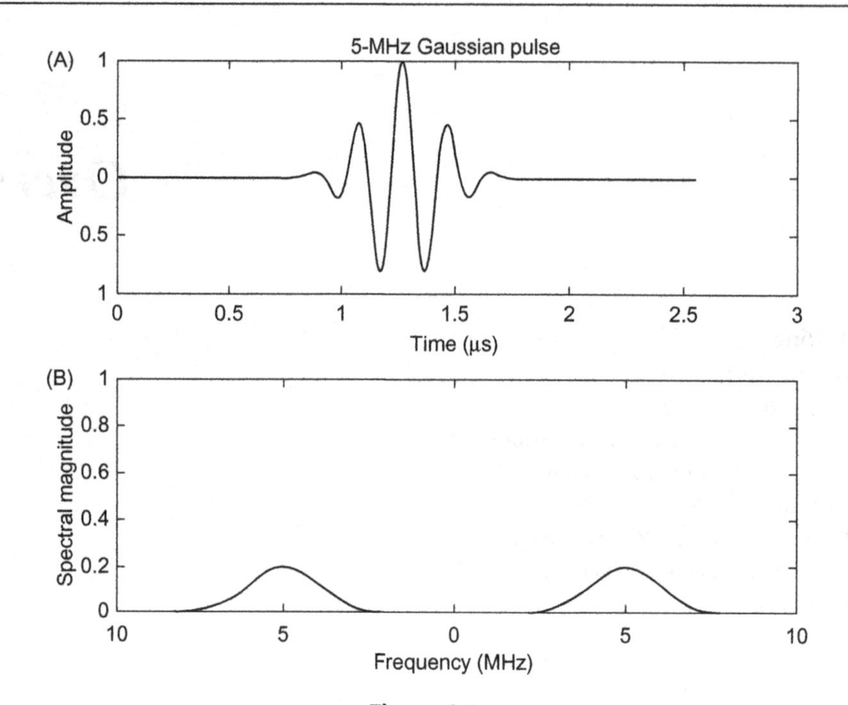

Figure 2.1
Forms of the Gaussian pulse. (A) Short 5-MHz time pulse and its (B) spectrum magnitude and phase.

more easily solved in the frequency domain, while others are better done in the time domain. Consequently, it will be necessary to use a method to switch from one domain to another. Joseph Fourier, a nineteenth century French mathematician, had an important insight that a waveform repeating in time could be synthesized from a sum of simple sines and cosines of different frequencies and phases (Bracewell, 2000). These frequencies are harmonically related by integers: a fundamental frequency (f_0) and its harmonics, which are integral multiples ($2f_0$, $3f_0$, etc.). This sum forms the famous Fourier series.

While the Fourier series is interesting from a historical point of view and its applicability to certain types of problems, there is a much more convenient way of doing Fourier analysis. A continuous spectrum can be obtained from a time waveform through a single mathematical operation called the "Fourier transform." The minus *i* Fourier transform, also known as the Fourier integral, is defined as:

$$H(f) = \Im_{-i}[h(t)] = \int_{-\infty}^{\infty} h(t)e^{-i2\pi ft}dt \tag{2.1}$$

in which $H(f)$ (with an upper-case letter convention for the transform) is the minus *i* Fourier transform of $h(t)$ (lower-case letter for the function), "i" is $\sqrt{-1}$, and $\Im_{-i}$ symbolizes the minus *i* Fourier transform operator. Note that, in general, both $h(t)$ and $H(f)$ may be complex, with both real and imaginary parts. Another operation, the minus i inverse Fourier transform, can be used to recover $h(t)$ from $H(f)$ as follows:

$$h(t) = \Im_{-i}^{-1}[H(f)] = \int_{-\infty}^{\infty} H(f)e^{i2\pi ft}df. \tag{2.2}$$

In this equation, $\Im_{-i}^{-1}$ is the symbol for the inverse minus i Fourier transform. A sufficient but not necessary condition for a Fourier transform is the existence of the absolute value of the function over the same infinite limits; another condition is a finite number of discontinuities in the function to be transformed. If a function is physically realizable, it most likely will have a transform. Certain generalized functions that exist in a limiting sense and that may represent measurement extremes (such as an impulse in time or a pure tone) are convenient and useful abstractions. The Fourier transform also provides an elegant and powerful way of calculating a sequence of operations represented by a series of building blocks, as shown shortly.

For applications involving a sequence of numbers or data, a more appropriate form of the Fourier transform, the discrete Fourier transform (DFT), has been devised. The DFT consists of a discrete sum of N-weighted complex exponents, $\exp(-i2\pi\, mn/N)$, in which m and n are integers. J. W. Cooley and J. W. Tukey (1965) introduced an efficient way of calculating the DFT called the fast Fourier transform (FFT). The DFT and its inverse are now routine mathematical algorithms and have been implemented directly into signal processing chips.

2.2.2 Fourier Transform Relationships

The most important relationships for the Fourier transform, the DFT, and their application are reviewed in Appendix A. This section emphasizes only key features of the Fourier transform, but additional references are provided for more background and details.

A key Fourier transform relationship is that time lengths and frequency lengths are related reciprocally. A short time pulse has a wide extent in frequency, or a broad bandwidth. Similarly, a long pulse, such as a tone burst of n cycles, has a narrow band spectrum. These pulses are illustrated in Figure 2.2 and Figure 2.3. If, for example, a tone burst of 10 cycles in Figure 2.2 is halved to 5 cycles in Figure 2.3, its spectrum is doubled in width. All of these effects can be explained mathematically by the Fourier transform scaling theorem:

$$\Im_{-i}[g(at)] = \frac{1}{|a|}G(f/a). \tag{2.3}$$

For this example, if g(t) is shown in Figure 2.2, then for the shorter length signal in Figure 2.3, if $a = 2.0$, then the spectrum is halved in amplitude and its width is stretched by

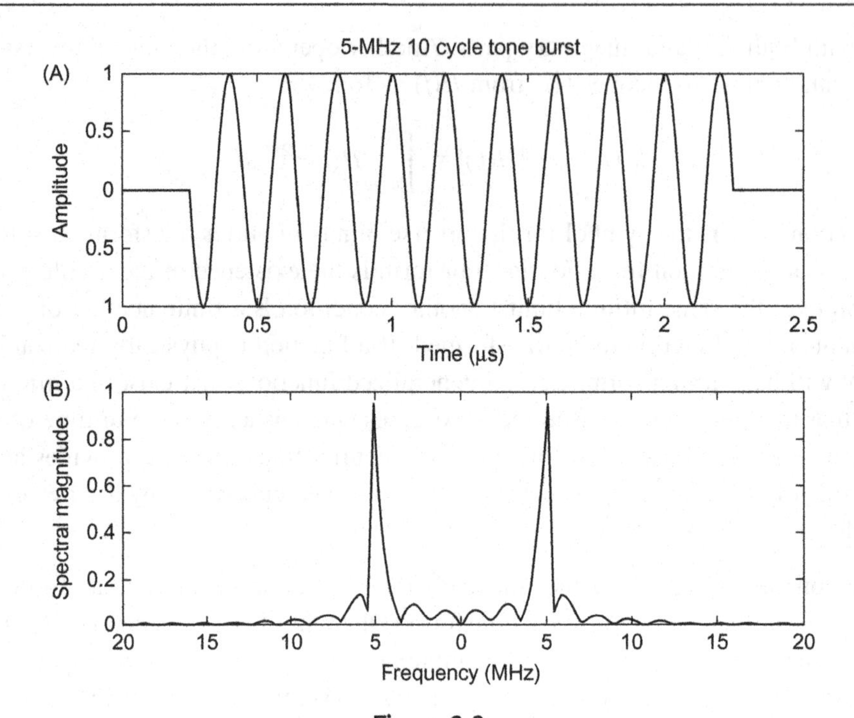

Figure 2.2
A 5-MHz center frequency tone burst of 10 cycles and its spectral magnitude.

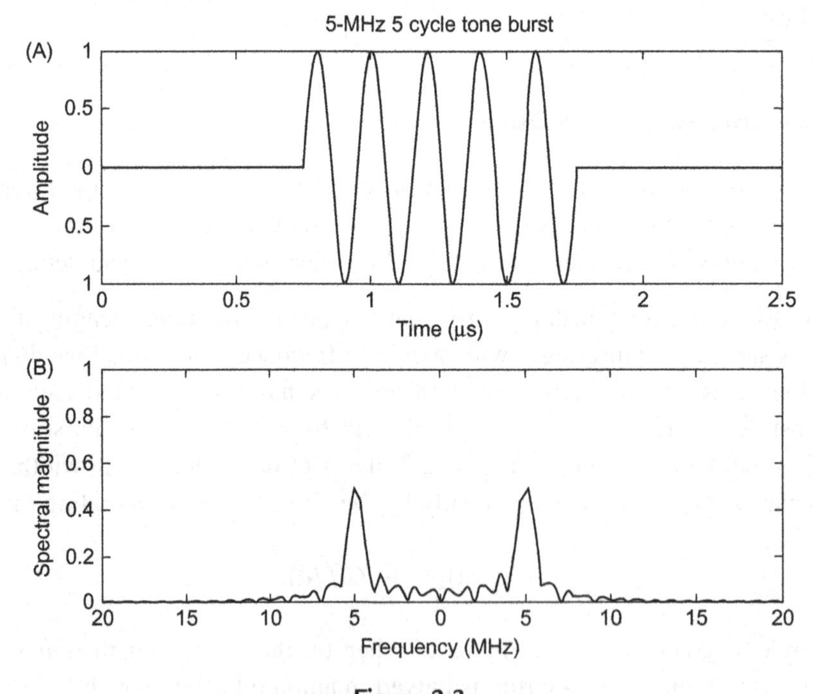

Figure 2.3
A 5-MHz center frequency tone burst of 5 cycles and its spectral magnitude.

a factor of two in its frequency extent. Many other Fourier transform theorems are listed in Table A.1 of Appendix A.

Consider the Fourier transform pair from this table for a Gaussian function:

$$\Im_{-i}[exp(-\pi t^2)] = exp(-?f^2). \tag{2.4}$$

To find the minus i Fourier transform of a following given time domain Gaussian analytically, for example,

$$g(t) = exp(-wt^2), \tag{2.5A}$$

first put it into a form appropriate for the scaling theorem, Eqn 2.3, and the Gaussian, Eqn 2.4,

$$g(t) = exp\left[-\pi\left(t\sqrt{w/\pi}\right)^2\right], \tag{2.5B}$$

so that $a = \sqrt{w/\pi}$. Then by the scaling theorem, the transform is:

$$G(f) = exp\left[-\pi(f/\sqrt{w/\pi})^2\right]/\sqrt{w/\pi} = \sqrt{\pi/w} = \sqrt{\pi/w}\ exp\left[-(\pi^2/w)f^2\right]. \tag{2.6}$$

The Gaussian is well behaved and has smooth time and frequency transitions. Fast time transitions have a wide spectral extent. An extreme example of this characteristic is the impulse in Figure 2.4. This pulse is so short in time that, in practical terms, it appears as a spike or as a signal amplitude occurring only at the smallest measurable time increment. The ideal impulse would have a flat spectrum (or an extremely wide one in realistic terms). The converse of the impulse in time is a tone burst so long that it would mimic a sine wave as in Figure 2.5. The spectrum of this nearly pure tone would appear on a spectrum analyzer (an instrument for measuring the spectra of signals) as either an amplitude at a single frequency in the smallest resolvable frequency resolution cell or as a spectral impulse. Note that instead of a pair of spectral lines representing impulse functions in Figure 2.5, finite width spectra are shown as a consequence of the finite length time waveform used for this calculation by a digital Fourier transform. All of these effects can be demonstrated beautifully by the Fourier transform. The Fourier transform operations for Figures 2.1–2.5 were implemented by MATLAB program chap2figs.m.

2.3 Building Blocks

2.3.1 Time and Frequency Building Blocks

One of the motivations for using the Fourier transform is that it can describe how a signal changes its form as it propagates or when it is sent through a device or filter. Both of these

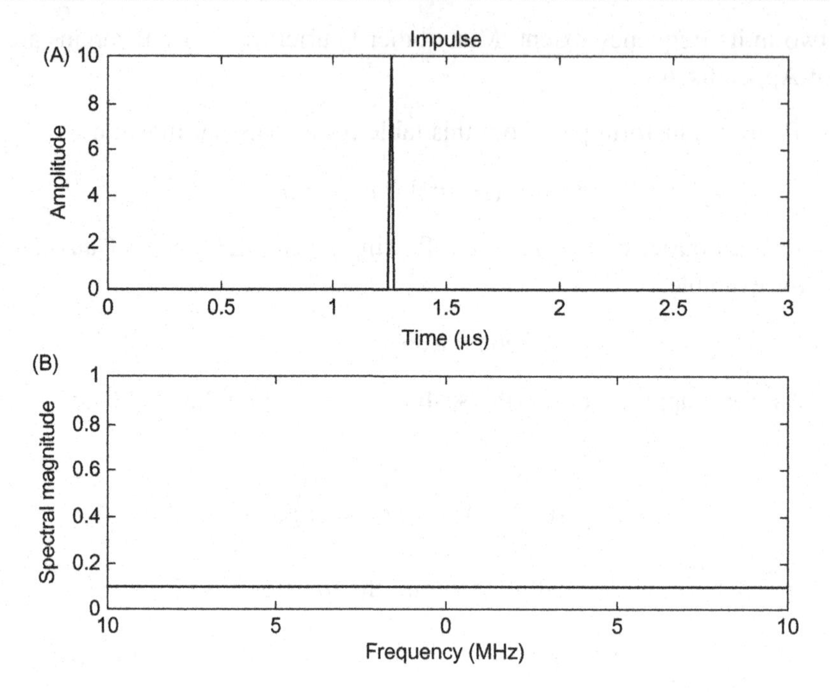

Figure 2.4
A time impulse and its spectral magnitude.

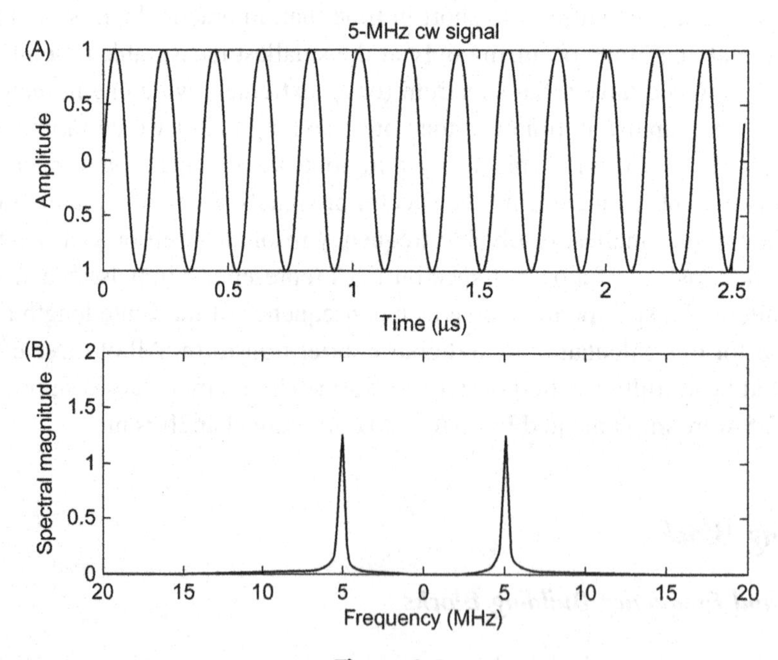

Figure 2.5
A 5-MHz pure tone and its spectral magnitude.

changes can be represented by a building block. Assume there is a filter that has a time response, $q(t)$, and a frequency response, $Q(f)$. Each of these responses can be represented by a building block, as given by Figure 2.6. A signal, $p(t)$, sent into the filter, $q(t)$, with the result, $r(t)$, can be symbolized by building blocks like those of Figure 2.7.

As a general example of a building block, a short Gaussian pulse, $e_1(t)$,is sent into a filter, $g_2(t)$, with a longer Gaussian impulse response (from Figure 2.1). This filtering operation is illustrated in both domains by Figure 2.7. In this case, the output pulse is longer than the original, and its spectrum is similar in shape to the original but slightly narrower. For the same filter in Figure 2.8, the time impulse input of Figure 2.4 results in a replication of the time response of the filter as an output response (also known as "impulse response"). Because the impulse has a flat frequency response, Figure 2.8 also replicates the frequency response of the filter as an output response. In Figure 2.9, a single-frequency input signal of unity amplitude from Figure 2.5 results in a single-frequency output weighted with amplitude and phase of the filter at the same frequency.

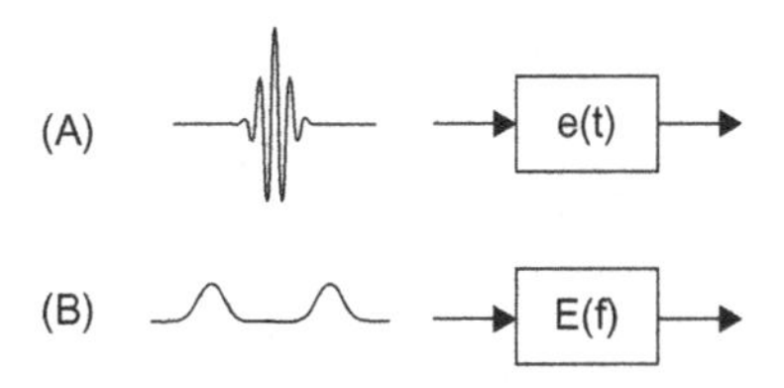

Figure 2.6
(A) A time domain building block and (B) Its frequency domain equivalent.

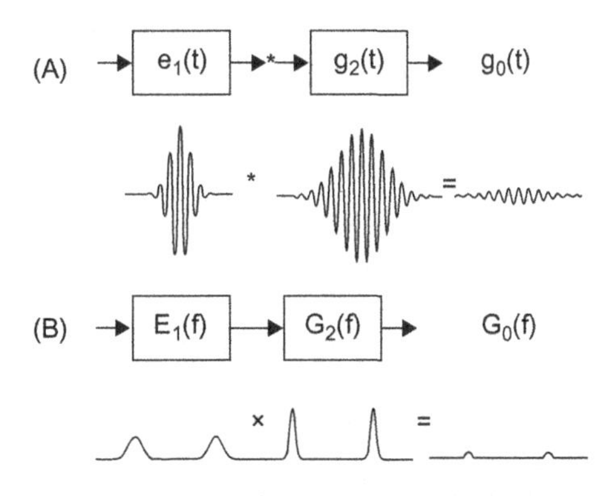

Figure 2.7
(A) Time waveforms for input, filter, and output result represent filter with a time domain convolution operation; (B) Corresponding frequency domain representation includes a multiplication. Both the filter and input have the same 5-MHz center frequency but different bandwidths.

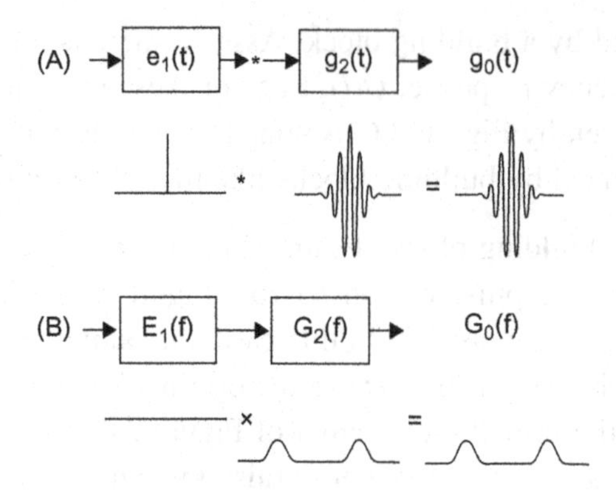

Figure 2.8
(A) Time domain filter output, or impulse response, for a time domain impulse input; (B) Spectrum magnitude of filter output.

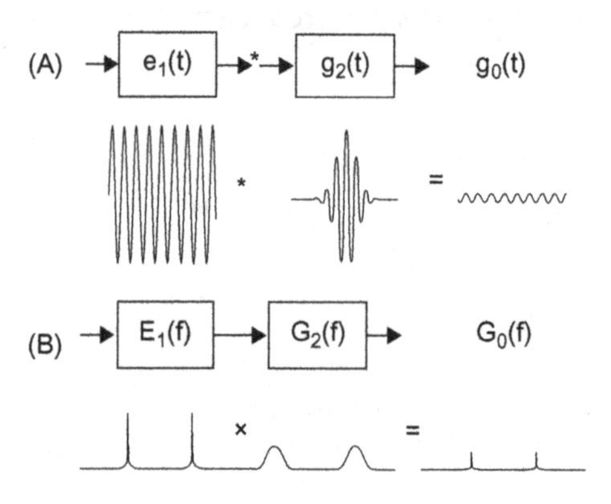

Figure 2.9
(A) Time domain filter output, to a 4.5-MHz tone input; (B) Spectrum magnitude of filter output is also at the input frequency but changed in amplitude and phase. The filter is centered at 5 MHz.

The operations illustrated in Figures 2.7–2.9 can be generalized by two simple equations. In the frequency domain, the operation is just a multiplication:

$$R(f) = P(f)\ Q(f). \tag{2.7A}$$

The three frequency domain examples in these figures show how the products of $P(f)$ and $Q(f)$ result in $R(f)$.

In the time domain, a different mathematical operation called "convolution" is at work. Time domain convolution, briefly stated, is the mathematical operation that consists of flipping one waveform around left to right in time, sliding it past the other waveform, and summing the amplitudes at each time point. The details of how this is done are covered in Appendix A.

Again, this is a commonplace computation that is represented by the symbol $*_t$, meaning time domain convolution. Therefore, the corresponding general relation for the time domain operations of these figures is written mathematically as:

$$r(t) = p(t) *_t q(t). \tag{2.7B}$$

It does not take much imagination to know what would happen if a signal went through a series of filters, $W(f)$, $S(f)$, and $Q(f)$. The end result is:

$$R(f) = P(f)\ Q(f)\ S(f)\ W(f), \tag{2.8A}$$

and the corresponding time domain version is:

$$r(t) = p(t) *_t q(t) *_t s(t) *_t w(t). \tag{2.8B}$$

We are close to being able to construct a series of blocks for an imaging system, but first we have to discuss spatial dimensions.

2.3.2 Space Wave Number Building Block

A Fourier transform approach can also be applied to the problems of describing acoustic fields in three dimensions. Until now, the discussion has been limited to what might be called "one-dimensional" operations. In the one-dimensional sense, a signal was just a variation of amplitude in time. For three dimensions, a source such as a transducer occupies a volume of space and can radiate in many directions simultaneously. Again, a disturbance in time is involved, but now the wave has a three-dimensional spatial extent that propagates through a medium but does not change the structure permanently as it travels.

Spatial Transforms

In the one-dimensional world, there are signals (pulses or sine waves). In the three-dimensional world, waves must have a direction also. For sine waves (the primitive elements used to synthesize complicated time waveforms), the period T is the fundamental unit, and it is associated with a specific frequency by the relation $T = 1/f$. The period is a measure in time of the length of a sine wave from any point to another point where the sine wave repeats itself. For a wave in three dimensions, the primitive element is a plane wave with a wavelength λ, which is also a measure of the distance in which a sinusoidal plane wave repeats itself. A special wavevector ($\boldsymbol{k}$) is used for this purpose; it has a direction and a magnitude equal to the wavenumber, $k = 2\pi f/c = 2\pi/\lambda$, in which "c" is the sound speed of the medium. Just as there is frequency (f) and angular frequency ($\omega = 2\pi f$), an analogous relationship exists between spatial frequency ($\tilde{f}$) and the wavenumber (k) so that $k = 2\pi\tilde{f}$. Spatial frequency can also be thought of as a normalized wavenumber or the reciprocal of wavelength, $\tilde{f} = k/2\pi = 1/\lambda$.

Before starting three dimensions, consider a simple single-frequency plane wave that is traveling along the positive z axis and that can be represented by the exponential, $exp[i(\omega t - kz)] = exp[i2\pi (ft - \tilde{f}z)]$. Note that the phase of the wave has two parts: the first is associated with frequency and time, and the second, opposite in sign, is associated with inverse wavelength and space. In order to account for the difference in sign of the second term, a different Fourier transform operation is needed for $\boldsymbol{k}$, or spatial frequency and space. For this purpose, the plus i Fourier transform is appropriate:

$$G(\tilde{f}) = \Im_{+i}[g(x)] = \int_{-\infty}^{\infty} g(x)e^{i2\pi\tilde{f}x}dx. \tag{2.9A}$$

Of course, there is an inverse plus i Fourier transform to recover $g(x)$:

$$g(x) = \Im_i^{-1}[G(\tilde{f})] = \int_{-\infty}^{\infty} G(\tilde{f})e^{-i2\pi\tilde{f}x}d\tilde{f}. \tag{2.9B}$$

One way to remember the two types of transforms is to associate the conventional minus i Fourier transform with frequency and time. You can also remember to distinguish the plus i transform for wavenumber (spatial frequency) and space as "contrary" to the normal convention because it has an opposite phase or sign in the exponential argument. More information on these transforms and alternate forms are given in Appendix A. To simplify these transform distinctions in general, a Fourier transform will be assumed to be a minus i Fourier transform unless specifically named, in which case it will be called a plus i Fourier transform.

In three dimensions, a point in an acoustic field can be described in rectangular coordinates in terms of a position vector $\boldsymbol{r}$ (Figure 2.10A). In the corresponding three-dimensional world of k-space, projections of the $\boldsymbol{k}$ wavevector corresponding to the $\boldsymbol{x}$, $\boldsymbol{y}$, and $\boldsymbol{z}$ axes are $\boldsymbol{k_1}$, $\boldsymbol{k_2}$, and $\boldsymbol{k_3}$ (depicted in Figure 2.10B). Each projection of $\boldsymbol{k}$ has a corresponding spatial frequency ($\tilde{f}_1 = k_1/2\pi$, etc.). See Table 2.1 for a comparison of the variables for both types

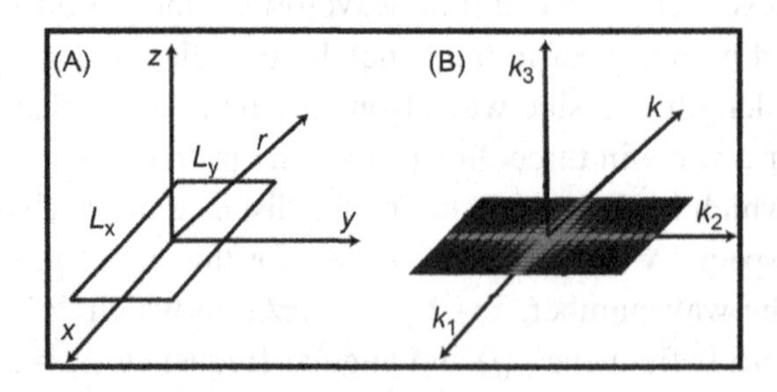

Figure 2.10
(A) Normal space with rectangular coordinates and a position vector $\boldsymbol{r}$ to a field point; (B) Corresponding k-space coordinates and vector $\boldsymbol{k}$.

Table 2.1 Fourier Transform Acoustic Variable Pairs

Variable	Transform Variable	Type
Time t	Frequency f	$-i$
Space	Spatial frequency	
x	$\tilde{f}_1$	$+i$
y	$\tilde{f}_2$	$+i$
z	$\tilde{f}_3$	$+i$

(A)

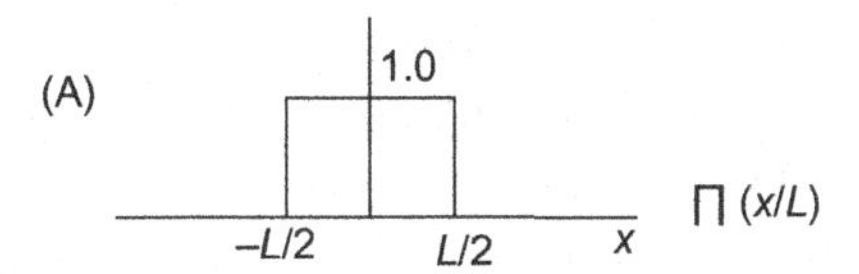

(B)

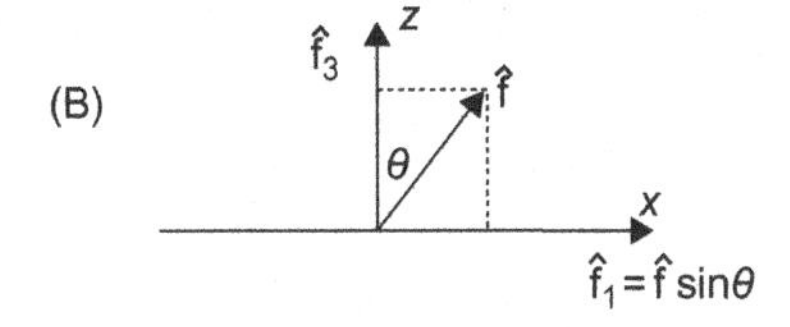

Figure 2.11
(A) Line source of length (L) and amplitude 1.0 lying along the $\boldsymbol{x}$ axis in the xz plane; (B) The plane wave wavevector at an angle θ to the $\boldsymbol{k}_3$ axis and its projections. A plane wavefront would be perpendicular to the wavevector.

of Fourier transforms. To extend calculations to dimensions higher than one, Fourier transforms can be nested within each other as explained in Chapter 6.

Spatial Transform of a Line Source

As an example of how plane waves can be used to synthesize the field of a simple source, consider the two-dimensional case for the xz plane with propagation along z. The xz plane in Figure 2.11A has a one-dimensional line source that lies along the x axis and has a length (L) and an amplitude of one. This shape can be described by the rect function (Bracewell, 2000) shown in Figure 2.11a, and is defined as follows:

$$\prod(x/L) = \left\{ \begin{array}{ccc} 0 & |x| & > L/2 \\ 1/2 & |x| & = L/2 \\ 1 & |x| & < L/2 \end{array} \right\}. \tag{2.10}$$

As the source radiates, plane waves are sprayed in different directions. For each plane wave, there is a corresponding wavevector that has a known magnitude, $\boldsymbol{k} = \omega/c$, and lies at an angle θ to the $\boldsymbol{k}_3$ axis, which corresponds to the z-axis direction. In Figure 2.11B, each $\boldsymbol{k}$

vector has a projection $k_1 = k \sin\theta$ along $\boldsymbol{x}$ and a value $k_3 = \sqrt{k^2 - k_1^2}$ along the z axis. This vector symbolizes the direction and magnitude of a plane wave with a flat wavefront perpendicular to it, as illustrated by the dashed line in Figure 2.11B.

In a manner analogous to a time domain waveform having a spectrum composed of many frequencies, the complicated acoustic field of a transducer can be synthesized from a set of weighted plane waves from all angles (θ), called the "angular spectrum of plane waves" (Goodman, 1968). Correspondingly, there is a Fourier transform relation between the source amplitude and spatial angular spectrum or spatial frequency (proportional to wavenumber) distribution as a function of $\tilde{f}_1 = \tilde{f}\ \sin\ \theta$ in the xz plane. For a rectangular coordinate system, it is easier mathematically to deal with projection $\tilde{f}_1$ rather than θ directly. To find the continuous distribution of plane waves with angle, the $+i$ Fourier transform of the rectangular source function depicted in Figure 2.11A is taken at the distance $z = 0$,

$$G(\hat{f}_1) = \Im_{+i}[g(x)] = \int_{-\infty}^{\infty} \prod(x/L)e^{i2\pi\tilde{f}_1 x}dz = L\mathrm{sinc}(L\tilde{f}_1) = L\mathrm{sinc}(L\tilde{f}\ \sin\ \theta), \tag{2.11}$$

in which the sinc function (Bracewell, 2000), also listed in Appendix A, is defined as:

$$sinc(a\hat{f}) = \frac{sin(\pi a\hat{f})}{(\pi a\hat{f})}, \tag{2.12A}$$

with the properties:

$$\begin{aligned}
&sinc\ \ 0 = 1 \\
&sinc\ \ n = 0\ \ (n = \text{nonzero integer}). \\
&\textstyle\int_{-\infty}^{\infty}\ sinc\ x\ dx = 1
\end{aligned} \tag{2.12B}$$

The simplest situation would be one in which a single plane wave came straight out of the transducer with the $\tilde{f}$ vector oriented along the z axis ($\theta = 0$). Figure 2.12 reveals this situation is not the case for an actual radiating transducer aperture. While the amplitude is a maximum for the plane wave in that direction, and most of the highest spatial frequency amplitudes are concentrated around a small angle near the $\tilde{f}_3$ axis, the rest are plane waves diminishing in amplitude at larger angles. Based on our previous experience with transform pairs with steep transitions, such as the vertical edges of the source function of Figure 2.3A, we would expect a broad angular spectrum weighted in amplitude from all directions or angles. The sinc function, which applies to cases with the steep transitions in the other domain, as well as to the present one, has infinite spectral extent. If the source is halved to $L/2$, for example, the main lobe of the sinc function is broadened by a factor of two, as is predicted by the Fourier transform scaling theorem and as was shown for the one-dimensional cases earlier. More information about the calculation of an acoustic field amplitude in two and three dimensions can be found in Chapter 6.

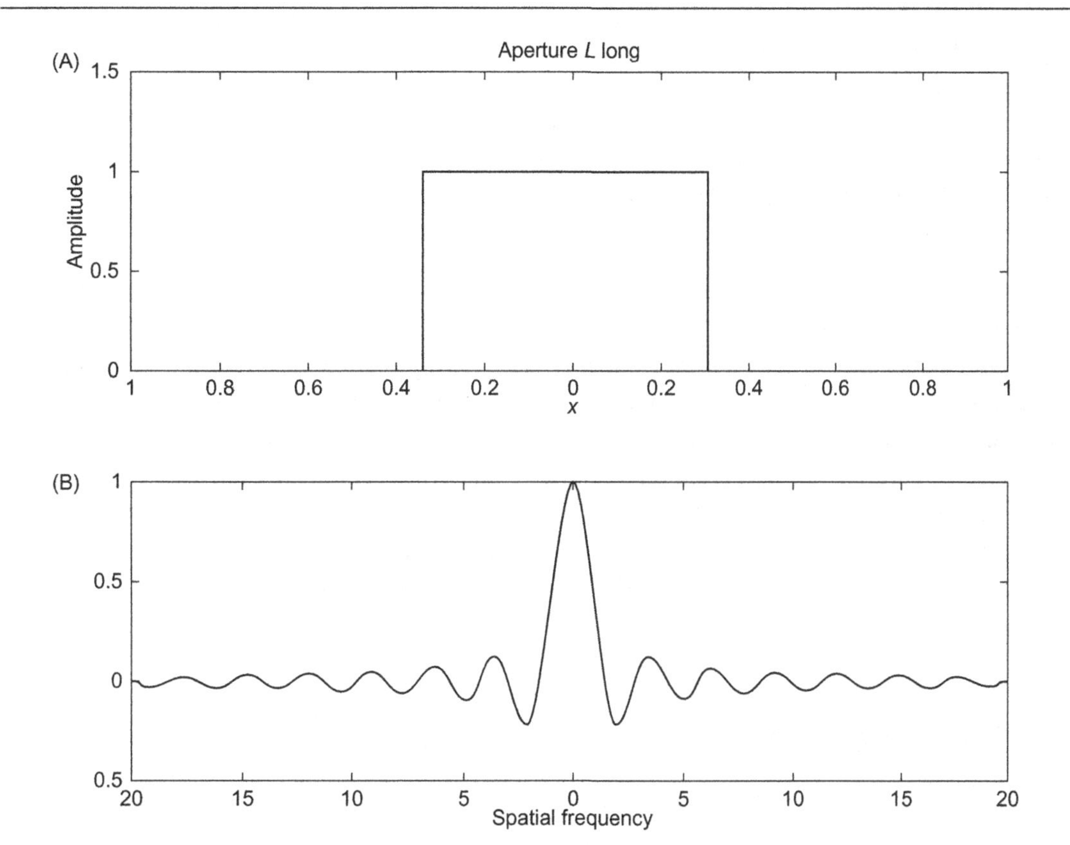

Figure 2.12
(A) A source function of length (L) and amplitude 1 along the x axis; (B) The corresponding spatial frequency distribution from the Fourier transform of the source as a function of $\tilde{f}_1$.

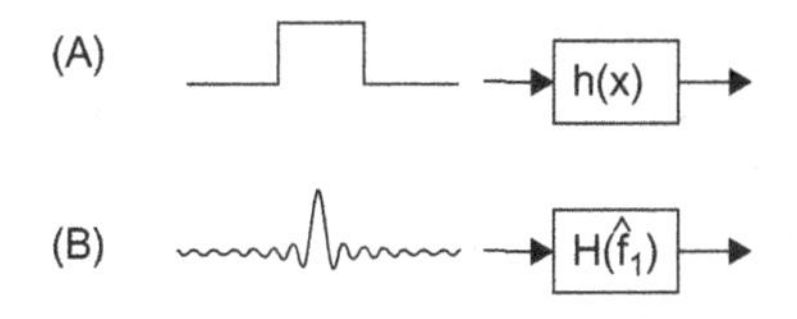

Figure 2.13
(A) An angular spatial frequency domain building block, and (B) Its spatial domain equivalent.

Spatial Frequency Building Blocks

Building blocks can be constructed from spatial frequency transforms (Figure 2.13). Because these represent three-dimensional quantities, it is helpful to visualize a block as representing a specific spatial location. For example, planes for specific values of z, such as a source plane ($z = 0$) and an image plane ($z \neq 0$), are in common use. Functions of spatial frequency can be multiplied in a manner similar to functions of frequency, as is done in the field of Fourier optics (Goodman, 1968). Functions in the space (*xyz*) domain are convolved, and the symbols for convolution have identifying subscripts: $*_x$ for the *xz* plane and $*_y$ for the *yz* plane. A simplifying assumption for most of these calculations is that the medium of propagation is not moving, or is "time invariant."

Recall that the scalar wavenumber k is also a function of frequency (here, $k = 2\pi f/c$). In general, building blocks associated with acoustic fields are functions of both frequency (f) and wavenumber (k), so they can be connected and multiplied. Conceptually, a time domain pulse has a spectrum with many frequencies. Each of these frequencies could interact with an angular frequency block to describe an acoustic field. All frequencies are to be calculated in parallel and involve many parallel blocks (mathematically represented by a sum operation); this process can be messy. Fortunately, a simpler numerical method is to use convolution. Just as there is a time pulse, a time domain equivalent of calculating acoustic fields has been invented, called the spatial impulse response (explained in Chapter 7).

2.4 Central Diagram

Building blocks are assembled into a frequency domain diagram in Figure 2.14. This diagram is not that of an imaging system but of a representation of the major processes that occur when an ultrasound image is made. Shaded blocks such as the first one, $E(f)$, or the transmit waveform generator, are related to electrical signals. The other (unshaded) blocks represent acoustic or electroacoustic events.

This central diagram provides a structure that organizes the different aspects of the imaging process. Future chapters explain each of the frequency domain blocks in more detail. Note that a similar and equivalent time domain block diagram can be constructed with convolution operations rather than the multipliers used here. The list below identifies each block with appropriate chapters, starting with $E(f)$ at the left and proceeding left to right. Finally, there are topics that deal with several blocks together.

$E(f)$ is the transmit waveform generator explained in Chapter 10: Imaging Systems and Applications. Signals from $E(f)$ are sent to $XB(f)$, the transmit beamformer found in Chapter 7: Array Beamforming.

From the beamformer, appropriately timed pulses arrive at the elements of the transducer array. More about how these elements work and are designed can be found in Chapter 5: Transducers. These elements transform electrical signals from the beamformer, $XB(f)$, to pressure or stress waves through their responses, $G_T(f)$.

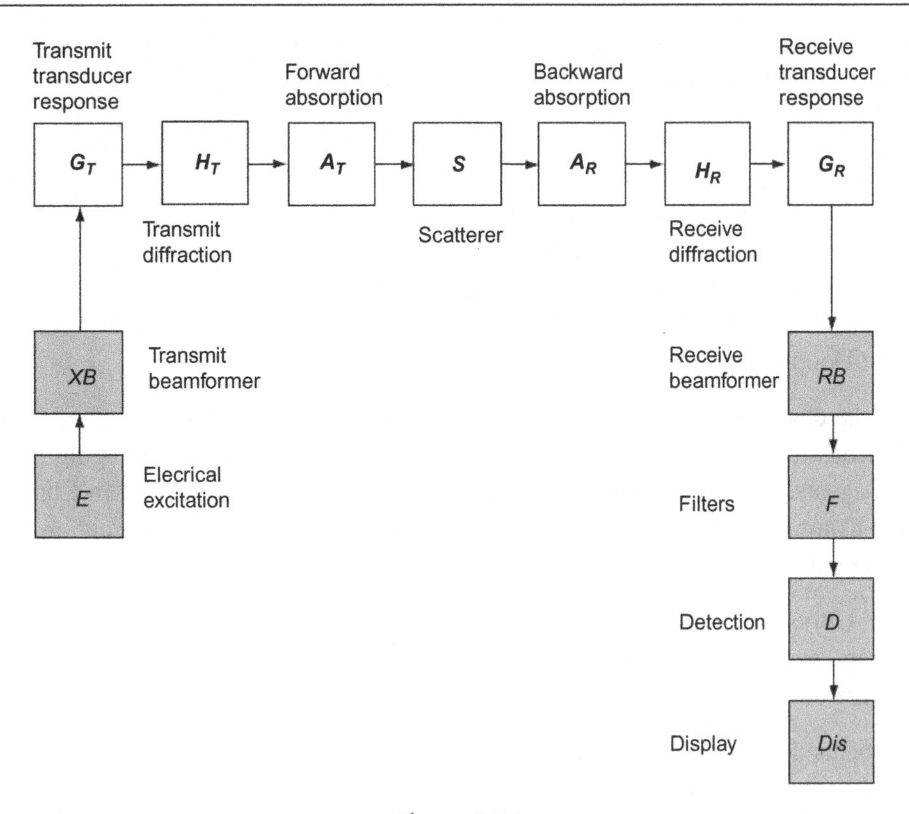

Figure 2.14
The central diagram, including the major signal and acoustic processes as a series of frequency domain blocks.

Acoustic (stress or pressure) waves obey basic rules of behavior that are described in review form in Chapter 3: Acoustic Wave Propagation. Waves radiate from the faces of the transducer elements and form complicated fields, or they diffract as described by three-dimensional (3D) transmit diffraction block H_T (k(f)) and Chapter 6: Beamforming. How the fields of individual array elements combine to focus and steer a beam is taken up in more detail in Chapter 7.

While diffracting and propagating, these 3D waves undergo loss. This is called attenuation or forward absorption and is explained by A_T (k(f)) in Chapter 4: Attenuation. Also, along the way, these waves encounter obstacles large and small that are represented by S(k(f)) and described in Chapter 8 (Wave Scattering and Imaging) and Chapter 9 (Scattering from Tissue and Tissue Characterization).

Portions of the wave fields that are scattered find their way back toward the transducer array. These echoes become more attenuated on their return through factor A_R(k(f)), backward absorption, as is also covered in Chapter 4. The 3D fields are picked by the elements according to principles of diffraction H_R (k(*f*)), as noted in Chapter 6. These acoustic waves pass back through array elements and are converted back to electrical signals through *GR*(*f*), as is explained in Chapter 5. The converted signals are shaped into coherent beams by the receive beamformer, *RB*(*f*), as is described in Chapter 7.

Electrical signals carrying pulse–echo information undergo filtering, *F*(*f*), and detection, *D*(*f*), and display, Dis, processes, which are included in Chapter 10. This chapter also includes the diagram of a generic digital imaging system. In addition, it covers different types of arrays and major clinical applications. Alternate imaging modes are discussed in Chapter 11: Doppler Modes.

In most of the chapters, linear principles apply. Harmonic imaging, based on the science of nonlinear acoustics, is explained in Chapter 12: Nonlinear Acoustics and Imaging. The use of contrast agents, which are also highly nonlinear acoustically, is described in Chapter 14: Ultrasound Contrast Agents. Topics in both these chapters involve beam formation, scattering attenuation, beamforming, and filtering in interrelated ways.

Chapter 13, Ultrasonic Exposimetry and Acoustic Measurements, applies to measurements of transducers, acoustic output and fields, and related effects. Safety issues related to ultrasound are covered in Chapter 15: Ultrasound-Induced Bioeffects.

Two new topics describe newer applications of ultrasound. Elastography has opened up ways of measuring and imaging the elastic properties of tissues more directly and it is described in Chapter 16. Finally, ultrasound has been employed to deliberately induce bioeffects for surgery and diverse applications, as explained in Therapeutic Ultrasound, Chapter 17.

Appendices supplement the main text. Appendix A shows how the Fourier transform and digital Fourier transform (DFT) are related in a review format. It also lists important theorems and functions in tabular form. In addition, it covers the Hilbert transform and quadrature signals. Appendix B lists tissue and transducer material properties. Appendix C derives a transducer model from simple two-by-two matrices and serves as the basis for a MATLAB transducer program. Numerous MATLAB programs, such as program chap2figs.m are used to generate Figures 2.1–2.5, also supplement the text and serve as models for homework problems that are listed by chapter on the main Web site, http://www.books.elsevier.com.

References

Bracewell, R. (2000). *The Fourier Transform and Its Applications*. New York: McGraw-Hill.

Cooley, J. W., & Tukey, J. W (1965). An algorithm for the machine computation of complex Fourier series. *Mathematics of Computation*, *19*, 297–301.

Goodman, J. W. (1968). *Introduction to Fourier Optics*. New York: McGraw-Hill.

CHAPTER 3

Acoustic Wave Propagation

Chapter Outline

3.1 Introduction to Waves

Waves in diagnostic ultrasound carry the information about the body back to the imaging system. Both elastic and electromagnetic waves can be found in imaging systems. How waves propagate through and interact with tissue will be discussed in several chapters, beginning with this one. This chapter also introduces powerful matrix methods for describing the complicated transmission and reflection of plane waves through several layers of homogeneous tissue. It first examines the properties of plane waves of a single frequency along one axis. This type of wave is the basic element that can be applied later to build more complicated wave fields through Fourier synthesis and the angular spectrum of waves method. Second, this chapter compares types of waves in liquids and solids. Third, matrix tools will be created to simplify the understanding and analysis of wave propagation, as well as reflections at boundaries. Fourth, this chapter presents methods of solving two- and three-dimensional wave problems of mode conversion and refraction at the boundaries of different media, such as liquids and solids.

Because tissues have a high water content, the simplifying approximation that waves in the body are like waves propagating in liquids is often made. Many ultrasound measurements are made in water also, so modeling waves in liquids is a useful starting point. In reality, tissues are elastic solids with complicated structures that support many different types of waves. Later in this chapter, elastic waves will be treated with the attention they deserve.

Another convenient simplification is that the waves obey the principles of linearity. Linearity means that waves and signals keep the same shape as they change amplitude and that different scaled versions of waves or signals at the same location can be combined to form or synthesize more complicated waves or signals. This important principle of superposition is at the heart of Fourier analysis and the designs of all ultrasound imaging systems. You may have heard that tissue is actually nonlinear, as is much of the world around you. This fact need not bother us at this time because linearity will allow us to build an excellent foundation for learning not only about how a real imaging system works, but also about how nonlinear acoustics (described in Chapter 12) alters the linear situation.

Finally, in this chapter, materials that support sound waves are assumed to be lossless. Of course, both tissue and water have loss (a topic saved for Chapter 4).

3.2 Plane Waves in Liquids and Solids

3.2.1 Introduction

Three simple but important types of wave shapes are plane, cylindrical, and spherical (Figure 3.1). A plane wave travels in one direction. Stages in the changing pattern of the wave can be marked by a periodic sequence of parallel planes that have infinite lateral extent and are all perpendicular to the direction of propagation. When a stone is thrown into water, a widening circular wave is created. In a similar way, a cylindrical wave has a

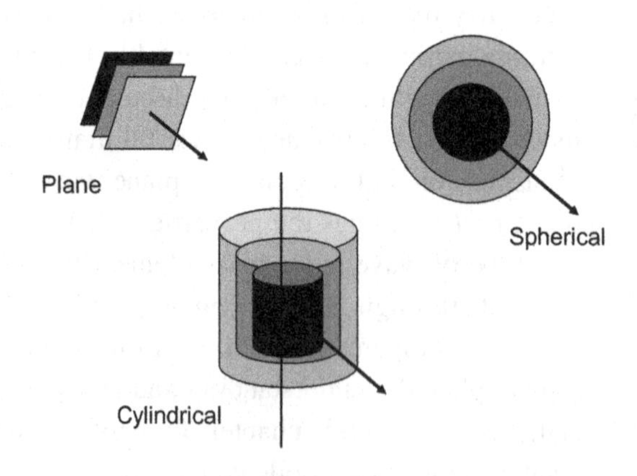

Figure 3.1
Plane, cylindrical, and spherical waves showing surfaces of constant phase.

cross-section that is an expanding circular wave that has an infinite extent along its axial direction. A spherical wave radiates a growing ball-like wave rather than a cylindrical one. In general, however, the shape of a wave will change in a more complicated way than these simple idealized shapes, which is why Fourier synthesis is needed to describe a journey of a wave.

In order to describe these basic wave surfaces, some mathematics is necessary. The next section presents the essential wave equations for basic waves propagating in an unbounded fluid medium. In order to characterize simple echoes, following sections will introduce equations and powerful matrix methods for describing waves hitting and reflecting from boundaries.

3.2.2 Wave Equations for Fluids

In keeping with the common application of a fluid model for the propagation of ultrasound waves, note that fluid waves are of a longitudinal type. A longitudinal wave creates a sinusoidal back-and-forth motion of particles as it travels along in its direction of propagation. The particles are displaced from their original equilibrium state by a distance or displacement amplitude (u) and at a rate or particle velocity (v) as the wave disturbance passes through the medium. This change also corresponds to a local pressure disturbance (p). The positive half cycles are called "compressional," and the negative ones, "rarefactional." If the direction of this disturbance or wave is along the z axis, the time required to travel from one position to another is determined by the longitudinal speed of sound c_L, or $t = z/c_L$. This wave has a wavenumber defined as $k_L = \omega/c_L$, where $\omega = 2\pi f$ is the angular frequency.

In an idealized inviscid (incompressible) fluid, the particle velocity is related to the displacement as:

$$v = \partial u/\partial t, \tag{3.1A}$$

or for a time harmonic or steady-state particle velocity (where capitals represent frequency-dependent variables), as follows:

$$V(\omega) = i\omega U(\omega). \tag{3.1B}$$

For convenience, a velocity potential (ϕ) is defined such that:

$$v = \nabla\phi. \tag{3.2}$$

Pressure is then defined as:

$$p = -\rho\partial\phi/\partial t, \tag{3.3A}$$

or for a harmonic wave:

$$P(\omega) = -i\omega\rho\Phi(\omega), \tag{3.3B}$$

where ρ is the density of the fluid at rest. Overall, wave travel in one dimension is governed by the wave equation in rectangular coordinates,

$$\frac{\partial^2 \phi}{\partial z^2} - \frac{1}{c_L^2}\frac{\partial^2 \phi}{\partial t^2} = 0, \tag{3.4}$$

in which the longitudinal speed of sound is:

$$c_L = \sqrt{\frac{\gamma B_T}{\rho_0}}, \tag{3.5}$$

where γ is the ratio of specific heats, ρ_0 is density, and B_T is the isothermal bulk modulus. The ratio of a forward traveling pressure wave to the particle velocity of the fluid is called the specific acoustic or characteristic impedance, as follows:

$$Z_L = p/v_L = \rho_0 c_L, \tag{3.6}$$

and this has units of Rayls (Rayl = kg/m^{2}s).

Note Z_L is negative for backward-traveling waves. For fresh water at 20 °C, $c_L = 1481$ m/s, $Z_L = 1.48$ MegaRayls (10^6 kg/m^{2}s), $\rho_0 = 998$ kg/m^3, $B_T = 2.18 \times 10^9$ N/m^2, and $\gamma = 1.004$.

The instantaneous intensity is:

$$I_L - pp^*/Z_L = vv^* Z_L. \tag{3.7}$$

The plane wave solution to Eqn 3.4 is:

$$\phi(z,t) = g(t - z/c_L) + h(t + z/c_L), \tag{3.8}$$

in which the first term represents waves traveling along the positive z axis, and the second represents them along the $-z$ axis. One important specific solution is the time harmonic,

$$\phi = \phi_0(exp[i(\omega t - k_L z)] + exp[i(\omega t + k_L z)]). \tag{3.9}$$

In a practical situation, the actual variable would be the real part of the exponential; for example, the instantaneous pressure of a positive-going wave is:

$$p = p_0 RE\{exp[i(\omega t - k_L z)]\} + p_0 \cos(\omega t + k_L z). \tag{3.10}$$

Note that the phase can also be expressed as $i(\omega t \pm k_L z) = i\omega(t \pm z/c_L)$, in which the ratio can be recognized as the travel time due to the speed of sound.

The plane wave (Eqn 3.4) can be generalized to three dimensions as:

$$\nabla^2 \phi - \frac{1}{c^2}\phi_{tt} = 0, \tag{3.11}$$

in which the abbreviated notation $\phi_{tt} = \frac{\partial^2 \phi}{\partial t^2}$ is introduced. Basic wave equations for other geometries include the spherical,

$$\phi_{rr} + \frac{2}{r}\phi_r - \frac{1}{c_2}\phi_{tt} = 0, \tag{3.12}$$

where r is the radial distance, and the cylindrical case, where:

$$\phi_{rr} + \frac{1}{r}\phi_r - \frac{1}{c_2}\phi_{tt} = 0. \tag{3.13}$$

The solution for Eqn 3.11 is:

$$\phi = \phi_0(exp[i(\omega t - \boldsymbol{k}\cdot\boldsymbol{r})] + exp[i(\omega + \boldsymbol{k}\cdot\boldsymbol{r})]), \tag{3.14}$$

where $\boldsymbol{k}$ can be broken down into its projections ($\boldsymbol{k}_1$, $\boldsymbol{k}_2$, and $\boldsymbol{k}_3$) along the $\boldsymbol{x}$, $\boldsymbol{y}$, and $\boldsymbol{z}$ axes, respectively, $\boldsymbol{r}$ is the direction of the plane wave, and:

$$k^2 = k_1^2 + k_2^2 + k_3^2. \tag{3.15}$$

Note Eqn 3.11 can be expressed in the frequency domain as the Helmholtz equation:

$$\nabla^2\Phi + k^2\Phi = 0, \tag{3.16}$$

where Φ is the Fourier transform of ϕ.

The general solution for the spherical wave equation (Blackstock, 2000) is

$$\phi(z,\, t) = \frac{g(t - r/c_L)}{r} + \frac{h(t + r/c_L)}{r}. \tag{3.17}$$

Unfortunately, there is no simple solution for the cylindrical wave equation except for great distances, r:

$$\phi(z,\, t) \approx \frac{g(t - r/c_L)}{\sqrt{r}} + \frac{h(t + r/c_L)}{\sqrt{r}}. \tag{3.18}$$

Finally, it is worth noting that the same wave equations hold if p or v is substituted for ϕ.

Most often the characteristics of ultrasound materials, such as the sound speed (c) and impedance (Z), are given in tabular form in Appendix B, so calculations of these values are often unnecessary. The practice of applying a fluid model to tissues involves using tabular measured values of acoustic longitudinal wave characteristics in the previous equations.

The main difference between waves in fluids and solids is that only longitudinal waves exist inside fluids; many other types of waves are possible in solids, such as shear waves. These waves can be understood through electrical analogies. The main analogs are stress for voltage and particle velocity for current. The relationships between acoustic variables

Table 3.1 Similar Wave Terminology

Sound			Sound					
Liquid			**Solid**			**Electrical**		
Variable	**Symbol**	**Units**	**Variable**	**Symbol**	**Units**	**Variable**	**Symbol**	**Units**
Pressure	P	MPa	Stress	T	MPa	Voltage	V	Volts
Particle velocity	v	m/s	Particle velocity	v	m/s	Current	I	Amps
Particle displacement	u	m	Particle displacement	u	m	Charge	Q	Coulombs
Density	ρ	kg/m^3	Density	ρ	kg/m^3			
Longitudinal speed of sound	C_L	m/s	Longitudinal speed of sound	C_L	m/s	Wave speed	$1\sqrt{LC}$	m/s
Longitudinal impedance	$Z_L(\rho C_L)$	Mega Rayls	Longitudinal impedance	$Z_L(\rho C_L)$	MegaRayls	Impedance	$\sqrt{L/C}$	Ohms
Longitudinal wave number	k_L	m^{-1}	Longitudinal wave number	k_L	m^{-1}			
			Shear vertical speed of sound	c_S	m/s			
			Shear vertical impedance	$Z_S(\rho c_S)$	MegaRayls			
			Shear vertical wave number	k_S	m^{-1}			

and similar electrical terms are summarized in Table 3.1. Correspondence between electrical variables for a transmission line and those for sound waves along one dimension in both fluids and solids enables the borrowing of electrical models for the solution of acoustics problems, as is explained in the rest of this chapter. Note that for solids, stress replaces pressure, but otherwise all the basic relationships of Eqns 3.1A,B–3.4 carry over. Another major difference for elastic waves in solids in Table 3.1 is the inclusion of shear waves. Waves in solids will be covered in more detail in Section 3.3.

3.2.3 One-dimensional Wave Hitting a Boundary

An important solution to the wave equation can be constructed from exponentials like those of Eqn 3.9. Consider the problem of a single-frequency acoustic plane wave propagating in an ideal fluid medium with the characteristics k_1 and Z_1 and bouncing off a boundary of different impedance (Z_2), as shown in Figure 3.2 at z_0. Assume a solution of the form:

$$p = p_0\, exp(i[\omega t - k_L z]) + RFp_0\, exp(i[\omega t - k_L z]), \tag{3.19}$$

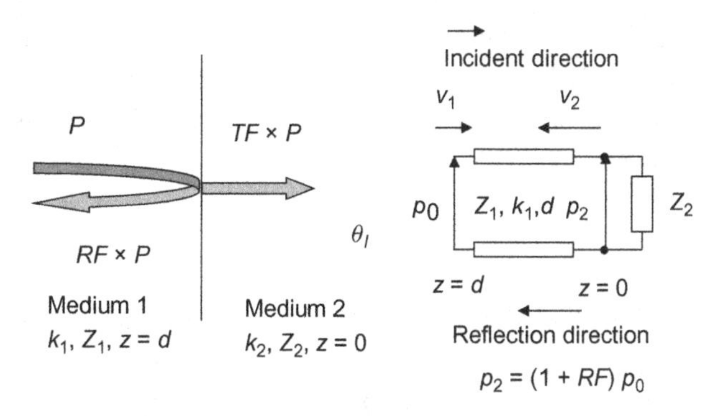

Figure 3.2
One-dimensional model of wave propagation at a boundary.

which satisfies the previous wave equation. *RF* is a reflection factor for the amplitude of the negative-going wave. An electrical transmission line analog for this problem, described in more detail shortly, is symbolized by the right-hand side of Figure 3.2. The transmission line has a characteristic impedance (Z_1), a wavenumber (k_1), and a length (d) equal to $z_0 - z$. The second medium is represented by a real load of impedance (Z_2) located at $z_0 = 0$ and a wavenumber (k_2). By the analogy presented in Table 3.1, the pressure at $z_0 = 0$ is like a voltage drop across Z_2,

$$p_2 = p_0(1 + RF), \tag{3.20A}$$

and the particle velocity there is like the sum of currents flowing in the transmission line in opposite directions, corresponding to the two particle velocity wave components,

$$v_2 = (1 - RF)p_0/Z_1. \tag{3.20B}$$

The impedance (Z_2) can be found from:

$$Z_2 = \frac{p_2}{v_2} = \frac{(1 + RF)Z_1}{1 - RF}. \tag{3.21}$$

Finally, solve the right-hand side of Eqn 3.21 to obtain:

$$RF = \frac{Z_2 - Z_1}{Z_2 + Z_1}. \tag{3.22A}$$

A transmission factor (TF) can be determined from $TF = 1 + RF$,

$$TF = \frac{2Z_2}{Z_1 + Z_2}. \tag{3.22B}$$

Eqn 3.22A tells us that there will be a reflection if $Z_2 \neq Z_1$, but not if $Z_2 = Z_1$. If $Z_2 = 0$, an open circuit or air-type boundary, there will be a 180-degree inversion of the incident wave, or $RF = -1$. Here, the reflected wave cancels the incident, so $TF = 0$. If $Z_2 = \infty$, corresponding to a short circuit condition or a stress-free boundary, the incident wave will be reflected back, or $RF = +1$. In this case, $TF = 2$ because the incident and reflected waves add in phase; however, no power or intensity (see Eqn 3.7) is transferred to medium 2 because $v_2 = p_2/Z_2 = 0$.

3.2.4 ABCD Matrices

Extremely useful tools for describing both acoustic and electromagnetic waves in terms of building blocks are matrices (Matthaei, Young, & Jones, 1980). In particular, the ABCD matrix form (Sittig, 1967) is shown in Figure 3.3 for the electrical case and is given by the following equations:

$$V_1 = AV_2 + BI_2, \quad \text{and} \tag{3.23A}$$

$$I_1 = CV_2 + DI_2, \tag{3.23B}$$

where V is voltage and I is current. The analogous acoustic case is:

$$p_1 = Ap_2 + Bv_2, \quad \text{and} \tag{3.24A}$$

$$v_1 = Cp_2 + Dv_2 \tag{3.24B}$$

The comparisons for these analogies are given in Table 3.1. The variables on the left (subscript 1) are given in terms of those on the right because usually the impedance on the right (Z_M) is known. The input impedance looking in from the left is given by:

$$Z_{IN1} = \frac{AZ_M + B}{CZ_M + D}, \tag{3.25A}$$

and the ratio of output to input voltages or pressures is:

$$\frac{V_2}{V_1} = \frac{Z_M}{AZ_M + B}. \tag{3.25B}$$

$$\begin{bmatrix} V_1 \\ I_1 \end{bmatrix} = \begin{pmatrix} A & B \\ C & D \end{pmatrix} \begin{bmatrix} V_2 \\ I_2 \end{bmatrix}$$

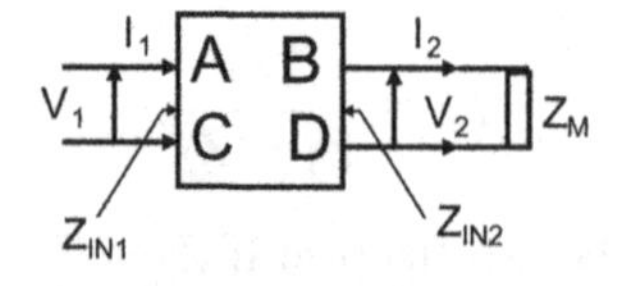

Figure 3.3
General ABCD matrix form.

There are also equations that can be used for looking from right to left:

$$Z_{INR1} = \frac{DZ_G + B}{CZ_G + A}, \tag{3.26A}$$

where a source impedance, Z_G, is assumed and input to output ratios are of the form:

$$\frac{V_1}{V_2} = \frac{Z_G}{DZ_G + B}. \tag{3.26B}$$

What are *A*, *B*, *C*, and *D*? Figure 3.4 shows specific forms of ABCD matrices (Matthaei et al., 1980). With only these four basic matrix types, more complicated configurations can be built up. From these basic matrices, a complete transducer model will be constructed in Chapter 5.

Figure 3.4C is the ABCD matrix for a transmission line (acoustic or electric) with a wavenumber (k_1), impedance (Z_1), and length (d_1) for a medium designated by "1." This important matrix can model continuous wave, one-dimensional wave propagation and scattering. For example, the input impedance of a transmission line loaded by Z_M can be determined from Eqn 3.25A and the ABCD matrix elements from Figure 3.4C:

$$Z_{IN1} = Z_1 \left[\frac{Z_M\cos(k_1 d_1) + iZ_1\sin(k_1 d_1)}{Z_1\cos(k_1 d_1) + iZ_M\sin(k_1 d_1)}\right]. \tag{3.27}$$

This equation can show that a transmission line a quarter of a wavelength long and loaded by Z_M, has an input impedance $Z_{IN1} = Z_1^2/Z_M$, so that the transmission line is an impedance transformer. A half-wavelength line is also curious, $Z_{IN1} = Z_M$, and the effect of the transmission line disappears. Reflection factors similar to Eqn 3.22A can also be

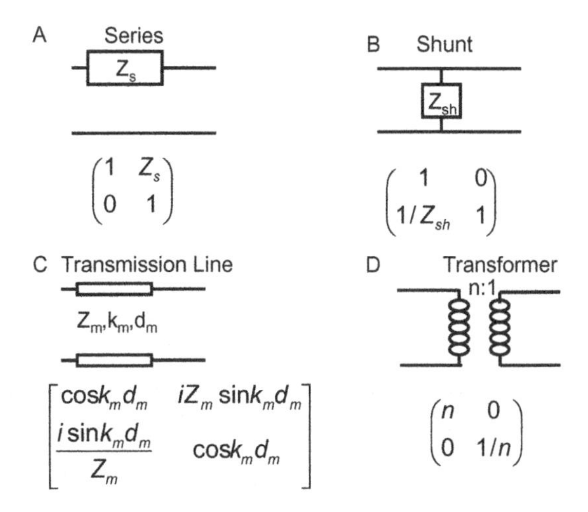

Figure 3.4
Specific forms of ABCD matrices.
(A) Series; (B) shunt; (C) transmission line; (D) transformer.

determined at the load end of the transmission line, designated by "R," for either voltage or pressure (stress):

$$RF_R = \frac{Z_M - Z_{IN2}}{Z_M + Z_{IN2}}. \tag{3.28A}$$

A transmission factor can also be written at the load:

$$TF_R = \frac{2Z_M}{Z_M + Z_{IN2}}. \tag{3.28B}$$

Another set of equations are appropriate for current (electrical model) or particle velocity (acoustical model) reflection and transmission at the left (input) end:

$$RF_i = \frac{1/Z_M - 1/Z_{IN2}}{1/Z_M + 1/Z_{IN2}} \tag{3.29}$$

and

$$TF_i = \frac{2/Z_M}{1/Z_M + 1/Z_{IN2}}. \tag{3.30}$$

A similar set of equations for the other end of the transmission line (looking to the left) mimic those above: Eqns 3.28A,B–3.30 with Z_{IN1} replacing Z_{IN2} and Z_G replacing Z_M.

These transmission lines (shown in Figure 3.4C) can be cascaded and combined with circuit elements shown in the figure. Primitive ABCD circuit elements can be joined to form more complicated circuits and loads. In Figure 3.4A is a series element, Z_S, and as an example, this matrix leads to the equations (Figure 3.3):

$$V_1 = V_2 + Z_S I_2, \text{ and} \tag{3.31}$$

$$I_1 = I_2. \tag{3.32}$$

Figure 3.4B is a shunt element. A transformer with a turns ratio n:1 is depicted in Figure 3.4D. Different types of loads include the short circuit (electrical, $V = 0$) or vacuum load (acoustical, $p = -T_{ZZ} = 0$) and the open circuit (electrical, $I = 0$) or clamped load (acoustical, $v = 0$). In general, $AD - BC = 1$ if the matrix is reciprocal. If the matrix is symmetrical, then $A = D$.

Individual matrices can be cascaded together (illustrated in Figure 3.5). For example, the input impedance to the rightmost matrix loaded by Z_R is given by:

$$Z_{IN1} = \frac{A_1 Z_R + B_1}{C_1 Z_R + D_1}, \tag{3.33}$$

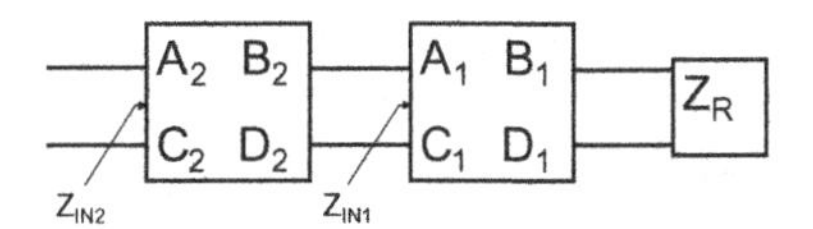

Figure 3.5
ABCD matrices in cascade.

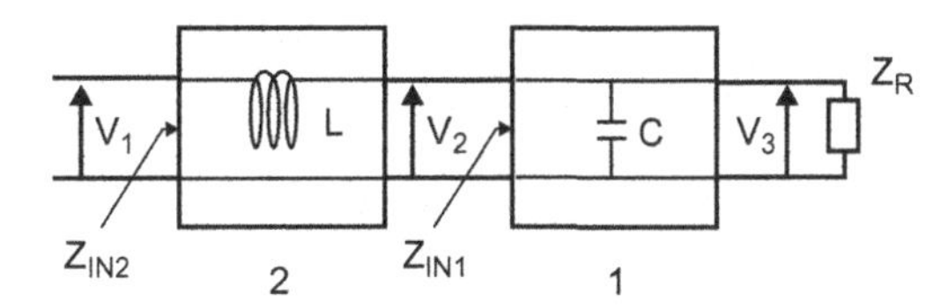

Figure 3.6
An example of two ABCD matrices in cascade terminated by a load (Z_R).

and for the impedance of the leftmost matrix:

$$Z_{IN2} = \frac{A_2 Z_{IN1} + B_2}{C_2 Z_{IN1} + D_2}. \tag{3.34}$$

As an example of cascading, consider the matrices for the case shown in Figure 3.6. Individually, the matrices are

$$\begin{bmatrix} A_1 & B_1 \\ C_1 & D_1 \end{bmatrix} = \begin{bmatrix} 1 & 0 \\ i\omega C & 1 \end{bmatrix}, \text{ and} \tag{3.35A}$$

$$\begin{bmatrix} A_2 & B_2 \\ C_2 & D_2 \end{bmatrix} = \begin{bmatrix} 1 & i\omega L \\ 0 & 1 \end{bmatrix}. \tag{3.35B}$$

The problem could be solved by multiplying the matrices together and by substituting the overall product matrix elements in Eqn 3.33 for those of the first matrix. Instead, the problem can be solved in two steps: Substituting matrix elements from Eqn 3.35A into Eqn 3.33 yields:

$$Z_{IN1} = \frac{Z_R}{i\omega C Z_R + 1}, \tag{3.35C}$$

which, when inserted as the load impedance for Eqn 3.34, provides:

$$Z_{IN2} = \frac{Z_R - \omega^2 L C Z_R + i\omega L}{i\omega C Z_R + 1}. \tag{3.35D}$$

Another important calculation is the overall complex voltage ratio, which, for this case, is:

$$\frac{V_3}{V_1} = \frac{V_2}{V_1}\frac{V_3}{V_2}. \tag{3.35E}$$

From Eqn 3.25B, the individual ratios are:

$$\frac{V_3}{V_2} = \frac{Z_R}{A_2 Z_R + B_2} = \frac{Z_R}{1 \times Z_R + i\omega L} \tag{3.35F}$$

and

$$\frac{V_2}{V_1} = \frac{Z_{IN1}}{A_1 Z_{IN1} + B_1} = \frac{Z_R/(1 + i\omega C Z_R)}{1 \times Z_R/(1 + i\omega C Z_R) + 0} = 1, \tag{3.35G}$$

so that from Eqn 3.35E, $V_3/V_1 = V_3/V_2$ for this example.

3.2.5 Oblique Waves at a Liquid–Liquid Boundary

Because of the common practice of modeling tissues as liquids, next examine what happens to a single-frequency longitudinal wave incident at an angle to a boundary with a different liquid medium 2 in the plane xz (depicted in Figure 3.7). At the boundary, stress (or pressure) and particle velocity are continuous. The tangential components of wavenumbers must also match, so along the boundary:

$$k_{1x} = k_1 \sin\theta_i = k_2 \sin\theta_T = k_1 \sin\theta_R, \tag{3.36A}$$

where k_1 and k_2 are the wavenumbers for mediums 1 and 2, respectively. The reflected angle (θ_R) is equal to the incident angle (θ_i), and an acoustic Snell's law is a result of this equation:

$$\frac{\sin\theta_i}{\sin\theta_T} = \frac{c_1}{c_2}, \tag{3.36B}$$

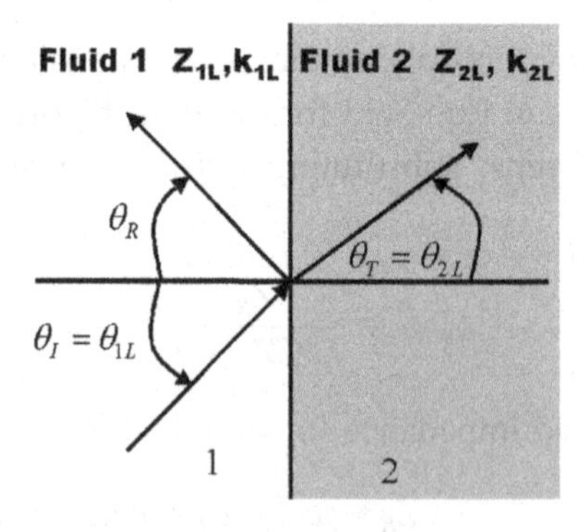

Figure 3.7
Oblique waves at a liquid–liquid interface.

which can be used to find the angle θ_T. Eqn 3.36A can also be used to determine θ_R. The wavenumber components along z are the following:

$$k_{Iz} = k_1 \cos \theta_i, \tag{3.37A}$$

$$\text{Reflected: } k_{Rz} = k_1 \cos \theta_R, \text{ and} \tag{3.37B}$$

$$\text{Transmitted: } k_{Tz} = k_2 \cos \theta_T, \tag{3.37C}$$

which indicate that the effective impedances at different angles are the following:

$$Z_{1\theta} = \frac{\rho_1 c_1}{\cos \theta_i} = \frac{Z_1}{\cos \theta_i} \tag{3.38A}$$

and

$$Z_{2\theta} = \frac{\rho_2 c_2}{\cos \theta_T} = \frac{Z_2}{\cos \theta_T}. \tag{3.38B}$$

Note that impedance is a function of the angle, reduces to familiar values at normal incidence, and otherwise grows with the angle. The incident wave changes direction as it passes into medium 2; this bending of the wave is called refraction. Since we are dealing at the moment with semi-infinite fluid media joined at a boundary, each medium is represented by its characteristic impedance, given by Eqn 3.38A. Then, just before the boundary, the impedance looking towards medium 2 is given by Eqn 3.38B. The reflection coefficient there is given by Eqn 3.22A,

$$RF = \frac{Z_{2\theta} - Z_{1\theta}}{Z_{2\theta} + Z_{1\theta}} = \frac{Z_2 \cos \theta_i - Z_1 \cos \theta_T}{Z_2 \cos \theta_i + Z_1 \cos \theta_T}, \tag{3.39A}$$

where the direction of the reflected wave along θ_R and the transmission factor along θ_T is:

$$TF = \frac{2Z_{2\theta}}{Z_{1\theta} + Z_{2\theta}} = \frac{2Z_2 \cos \theta_i}{Z_2 \cos \theta_i + Z_1 \cos \theta_T}. \tag{3.39B}$$

Note that in order to solve these equations, θ_T is found from Eqn 3.36A,B. Note, for wave components along x, replace the cosines by sines.

3.3 Elastic Waves in Solids

3.3.1 Types of Waves

Stresses (force/unit area) and particle velocities tend to be used for describing elastic waves in solids. If we imagine a force applied to the top of a cube, the dimension in the direction of the force is compressed and the sides are pushed out (exaggerated in Figure 3.8). Not only does the vertical force on the top face get converted to lateral forces, but it is also related to the forces on the bulging sides. This complicated interrelation of stresses in different directions results in a stress field that can be described by naming conventions. For example, the stress on the *xz* face has three orthogonal components: T_{zy} along *z*, T_{xy} along *x*, and T_{yy} along *y*. The first subscript denotes the direction of the component, and the second denotes the normal to the face. Thanks to symmetry, these nine stress components for three orthogonal faces reduce to six unique values in what is called the "reduced form" notation, T_I (Auld, 1990). This notation is given in Table 3.2 and will be explained shortly.

In general, a displacement due to the vibration of an elastic wave is described by a vector (u) having three orthogonal components. Stress along one direction can be described as a vector:

$$\boldsymbol{T}_y = \hat{\boldsymbol{x}}T_{xy} + \hat{\boldsymbol{y}}T_{yy} + \hat{\boldsymbol{z}}T_{zy}. \tag{3.40A}$$

First-order strain is defined as an average change in relative length in two directions, such as:

$$S_{ij} = \frac{1}{2}\left(\frac{\partial u_i}{\partial x_j} + \frac{\partial u_j}{\partial x_i}\right). \tag{3.40B}$$

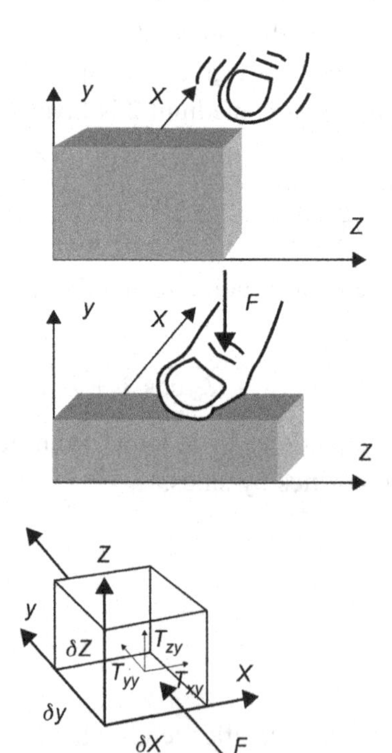

Figure 3.8
Stress conventions.

Table 3.2 Reduced Forms for Stress and Strain

T Reduced	T_{ij} Equivalent	S_{ij} Reduced	S_{ij} Equivalent	Type of Stress or Strain
T_1	T_{xx}	S_1	S_{xx}	Longitudinal along *x* axis
T_2	T_{yy}	S_2	S_{yy}	Longitudinal along *y* axis
T_3	T_{zz}	S_3	S_{zz}	Longitudinal along *z* axis
T_4	T_{yz}	$S_4/2$	S_{yz}	Shear about *x* axis
T_5	T_{zx}	$S_5/2$	S_{zx}	Shear about *y* axis
T_6	T_{xy}	$S_6/2$	S_{xy}	Shear about *z* axis

From Kino (1987).

For example,

$$S_{xy} = \frac{1}{2}\left(\frac{\partial u_x}{\partial y} + \frac{\partial u_y}{\partial x}\right). \tag{3.40C}$$

Sometimes the directions coincide:

$$S_{xx} = \frac{1}{2}\left(\frac{\partial u_x}{\partial x} + \frac{\partial u_x}{\partial x}\right) = \frac{\partial u_x}{\partial x}. \tag{3.40D}$$

Reduced-form notation for strain is given in Table 3.2. For example:

$$S_{xy} = S_{yx} = \frac{S_6}{2}. \tag{3.40E}$$

Overall, the strain relation can be described in reduced notation as follows (Kino, 1987):

$$\begin{bmatrix} S_1 \\ S_2 \\ S_3 \\ S_4 \\ S_5 \\ S_6 \end{bmatrix} = \begin{bmatrix} \frac{\partial}{\partial x} & 0 & 0 \\ 0 & \frac{\partial}{\partial y} & 0 \\ 0 & 0 & \frac{\partial}{\partial z} \\ 0 & \frac{\partial}{\partial z} & \frac{\partial}{\partial y} \\ \frac{\partial}{\partial z} & 0 & \frac{\partial}{\partial x} \\ \frac{\partial}{\partial y} & \frac{\partial}{\partial x} & 0 \end{bmatrix} \begin{bmatrix} u_x \\ u_y \\ u_z \end{bmatrix}. \tag{3.40F}$$

An equivalent way of expressing strain as a six-element column vector, Eqn 3.40F, is in an abbreviated dyadic notation:

$$\mathbf{S} = \nabla_S u, \tag{3.40G}$$

in which each term is given by Eqn 3.40F. Stress and strain are related through Hooke's law, which can be written in matrix form:

$$\begin{bmatrix} T_1 \\ T_2 \\ T_3 \\ T_4 \\ T_5 \\ T_6 \end{bmatrix} = [C] \begin{bmatrix} S_1 \\ S_2 \\ S_3 \\ S_4 \\ S_5 \\ S_6 \end{bmatrix}, \tag{3.41A}$$

where symmetry $C_{IJ} = C_{JI}$ has reduced the number of independent terms. Depending on additional symmetry constraints, the number is significantly less. Eqn 3.41A can be written in a type of symbolic shorthand called dyadic notation for vectors:

$$\boldsymbol{T} = \boldsymbol{C}{:}\boldsymbol{S}. \tag{3.41B}$$

As an example of how these relations might be used, consider the case of a longitudinal wave traveling along the z axis:

$$\boldsymbol{u} = \hat{z}\ \cos(\omega t - kz), \tag{3.42}$$

in which the displacement direction denoted by the unit vector ($\hat{z}$) and the direction of propagation (z) coincide, and $k = \omega/\left(C_{11}/\rho\right)^{1/2}$. Then the strain is:

$$\boldsymbol{S}_3 = \boldsymbol{S}_{zz} = \hat{z}\ k\ \sin(\omega t - kz). \tag{3.43}$$

The corresponding stress for an isotropic medium (one in which k or sound speed, c, is the same in all directions for a given acoustic mode) is given by the isotropic elastic constant matrix:

$$\begin{bmatrix} T_1 \\ T_2 \\ T_3 \\ T_4 \\ T_5 \\ T_6 \end{bmatrix} = \begin{bmatrix} C_{11} & C_{12} & C_{13} & C_{14} & C_{15} & C_{16} \\ C_{21} & C_{22} & C_{23} & C_{24} & C_{25} & C_{26} \\ C_{31} & C_{32} & C_{33} & C_{34} & C_{35} & C_{36} \\ C_{41} & C_{42} & C_{43} & C_{44} & C_{45} & C_{46} \\ C_{51} & C_{52} & C_{53} & C_{54} & C_{55} & C_{56} \\ C_{61} & C_{62} & C_{63} & C_{64} & C_{65} & C_{66} \end{bmatrix} \begin{bmatrix} 0 \\ 0 \\ S_3 \\ 0 \\ 0 \\ 0 \end{bmatrix} = \begin{bmatrix} C_{11} & C_{12} & C_{12} & 0 & 0 & 0 \\ C_{12} & C_{11} & C_{12} & 0 & 0 & 0 \\ C_{12} & C_{12} & C_{11} & 0 & 0 & 0 \\ 0 & 0 & 0 & C_{44} & 0 & 0 \\ 0 & 0 & 0 & 0 & C_{44} & 0 \\ 0 & 0 & 0 & 0 & 0 & C_{44} \end{bmatrix} \begin{bmatrix} 0 \\ 0 \\ S_3 \\ 0 \\ 0 \\ 0 \end{bmatrix}, \tag{3.44}$$

which results in the following nonzero values:

$$\boldsymbol{T}_1 = C_{12}S_3 = \hat{x}C_{12}k\ sin(\omega t - kz), \tag{3.45A}$$

$$\boldsymbol{T}_2 = C_{12}\boldsymbol{S}_3 = \hat{\boldsymbol{y}}C_{12}k\ sin(\omega t - kz), \text{ and} \tag{3.45B}$$

$$\boldsymbol{T}_3 = C_{11}\boldsymbol{S}_3 = \hat{\boldsymbol{z}}C_{12}k\ sin(\omega t - kz). \tag{3.45C}$$

For an isotropic medium, the elastic constants are related:

$$C_{44} = \frac{1}{2}(C_{11} - C_{12}). \tag{3.46}$$

Other often-used constants are Lame's constants, $\overline{\lambda}$ and μ:

$$C_{11} = \overline{\lambda} + 2\mu, \tag{3.47A}$$

$$C_{12} = \overline{\lambda}, \text{ and} \tag{3.47B}$$

$$C_{44} = \mu, \tag{3.47C}$$

where $\overline{\lambda}$ is an elastic constant (not wavelength). Another is Poisson's ratio:

$$\sigma = \frac{C_{12}}{C_{11} + C_{12}}. \tag{3.48}$$

This is the ratio of transverse compression to longitudinal expansion when a static longitudinal axial stress is applied to a thin rod. Poisson's ratio is between 0 and 0.5 for solids, and it is 0.5 for liquids (Kino, 1987). The ratio of axial stress to strain in a thin rod is Young's modulus:

$$E = C_{11} - \frac{2C_{12}^2}{C_{11} + C_{12}}. \tag{3.49}$$

Though there are many types of waves other than longitudinal waves that propagate along the surface between media or in certain geometries, the other two most important wave types are shear. The earlier Eqn 3.42 described a longitudinal wave along z in the xz plane with a sound speed:

$$c_L = \left(\frac{C_{11}}{\rho}\right)^{1/2}. \tag{3.50A}$$

Now consider a shear vertical (SV) wave in an isotropic medium with a sound speed:

$$c_S = \left(\frac{C_{44}}{\rho}\right)^{1/2}, \tag{3.50B}$$

with a transverse displacement along x and a propagation direction along z:

$$\boldsymbol{u}_{SV} = \hat{\boldsymbol{x}}u_{SV0}\cos(\omega t - k_S z), \tag{3.51}$$

as depicted in Figure 3.9. When these *SV* waves travel at an angle θ to the z axis, they can be described more generally by:

$$u_{SV} = (\hat{x}u_{SVX} + \hat{z}u_{SCZ})\cos(\omega t - k_S \cdot r) = (\hat{x}u_{SVX} + \hat{z}u_{SVZ})\cos(\omega t - k_S z\ \cos\theta + k_S x\ \sin\theta) \tag{3.52}$$

A shear horizontal (*SH*) wave, on the other hand, would have a transverse displacement along y perpendicular to the xz plane and a propagation along z:

$$u_{SH} = \hat{y}u_{SH0}\cos(\omega t - k_S z) \tag{3.53}$$

How are these three types of waves interrelated when a longitudinal wave strikes the surface of a solid? Stay tuned to the next section to find out.

3.3.2 Equivalent Networks for Waves

Oliner and colleagues (Oliner, 1969; Oliner, Bertoni, & Li, 1972a, 1972b) developed a powerful methodology for modeling acoustic waves with transmission lines and circuit elements, and it is translated here into ABCD matrix form. This approach can be applied to many different types of elastic waves in solids and fluids, as well as to infinite media and

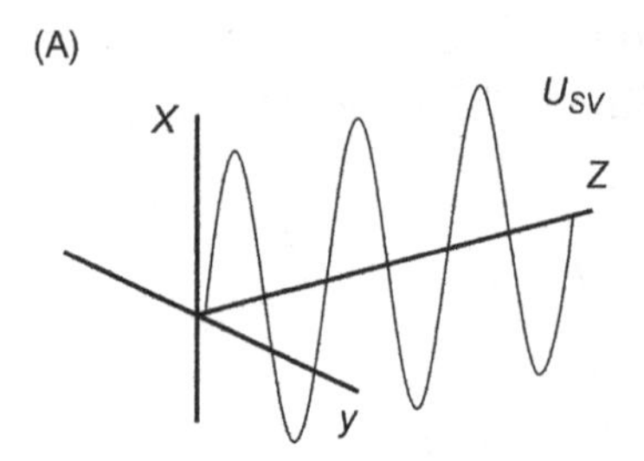

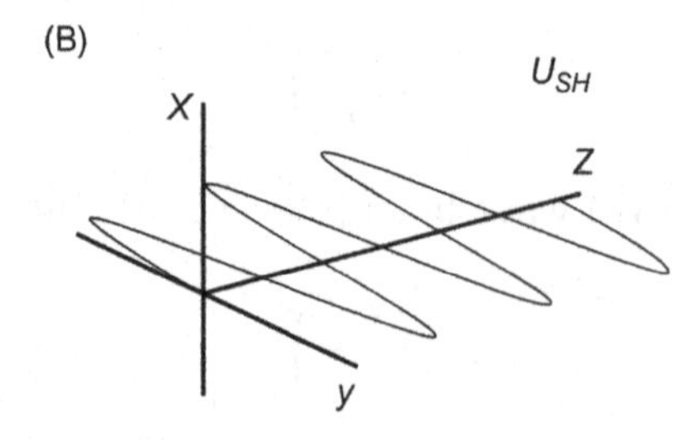

Figure 3.9
Types of basic shear waves.
(A) Shear vertical (*SV*) and (B) shear horizontal (*SH*).

stacks of layers of finite thickness. Rather than re-deriving applicable equations for each case, this method offers a simple solution in terms of the reapplication and combination of already derived equivalent circuits. At the heart of most of these circuits is one or more transmission line, each with a characteristic impedance, wave number, and length.

As an example, we will re-examine an oblique wave at a fluid-to-fluid boundary depicted in Fig. 3.7. From Section 3.2.4, we can construct a transmission line of length d for the first medium by using the appropriate relations for the incident wave from Eqns 3.37A and 3.38A. Figure 3.10 comprises two diagrams: the top diagram shows a general representation of each medium with its own transmission line, and the bottom drawing indicates the second medium as being semi-infinite and as represented by an impedance, $Z_{2\theta}$. Note that different directions are associated with the incident, reflected, and transmitted waves even though the equivalent circuit appears to look one-dimensional; this approach follows that outlined in Section 3.2.5. At normal incidence to the boundary, previous results are obtained. Connecting the load to the transmission line automatically satisfies appropriate boundary conditions.

Applications of different boundary conditions are straightforward, as is illustrated for fluids by transmission lines shown for normal incidence in Figure 3.11. In this figure, the notation is the following: k_f is wavenumber, V_f corresponds to pressure, and I_f corresponds to particle velocity (v). In Figure 3.11A, for an air/vacuum boundary (called a pressure-release boundary), a short circuit for T_{zz} is applied ($T_{zz} = -p$). For a rigid solid or clamped

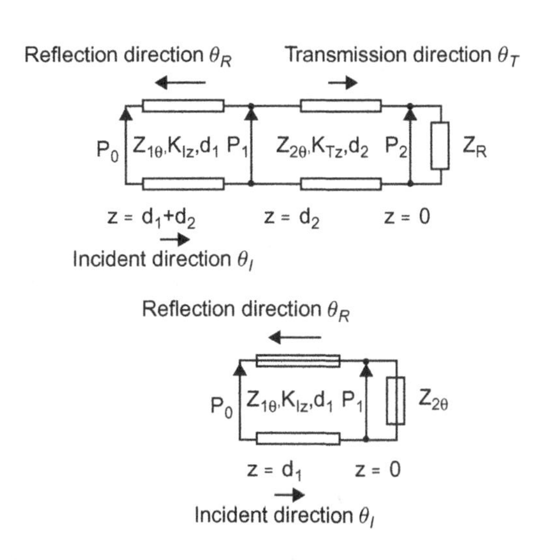

Figure 3.10

Equivalent circuits for acoustic waves in fluids.
(Top) two-transmission-line representation of fluid boundaries; (bottom) transmission line for fluid and semi-infinite fluid boundary.

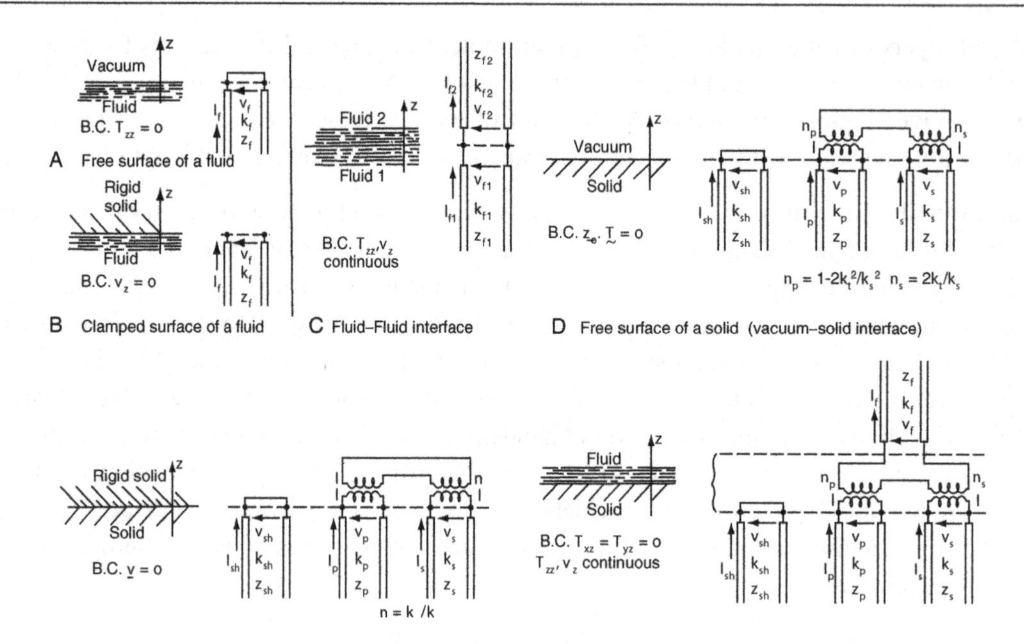

Figure 3.11

Equivalent circuits for acoustic waves at boundaries of solids.
(A) Free surface of a fluid; (B) clamped surface of a fluid; (C) fluid–fluid interface; (D) free surface of a solid; (E) clamped surface of a solid; (F) fluid–solid interface. *Source: Oliner et al., 1972b, 1972a* IEEE.

condition, given by Figure 3.11B, an open circuit load is appropriate. When there is an infinitesimally thin interface between two fluids, the coupling of different transmission lines corresponding to the characteristics of the fluids ensures that the stress and particle velocity are continuous across the boundary (Figure 3.11C). If the waves are at an angle, impedances of the forms given by Eqn 3.38A,B are assumed.

3.3.3 Waves at a Fluid–Solid Boundary

A longitudinal wave incident on the surface of a solid creates, in general, a longitudinal and shear wave as shown in Figure 3.12. A reflected shear wave is not generated because it is not supported in liquids; however, one would be reflected at the interface between two solids. The fluid pressure at the boundary is continuous ($p = -T_3$), as are the particle velocities.

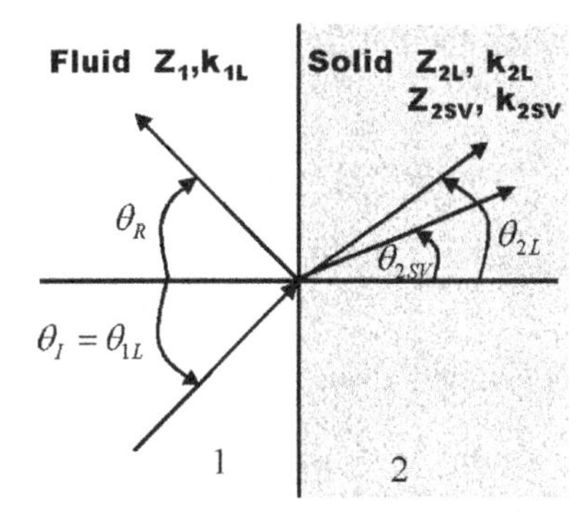

Figure 3.12
Wavevectors in the *xz* plane for fluid–solid interface problem.

Circuits applicable to three types of loading for solids (shown in Figure 3.11) anticipate the discussion of this section. In this figure, the notation is slightly different and corresponds to the following: "*p*" designates a longitudinal wave, "*s*" a shear vertical wave, and "*sh*" a shear horizontal wave. Note that in Figure 3.11F, a wave from a fluid is, in general, related to three types of waves in the solid. In these cases, transformers represent the mode conversion processes. In Figure 3.11D is the circuit for the pressure-release (air) boundary, and in Figure 3.11E is the clamped boundary condition, both for waves traveling upward in the solid. In all three cases, the shear horizontal wave does not couple to other modes. In the more general case of all three types of waves coupling from one solid to another (not shown), a complicated interplay among all the modes exists. This problem, as well as the circuits for many others, are found in Oliner et al. (1972a and 1972b). Derivations and more physical insights for these equivalent circuits are in Oliner (Oliner, 1969; Oliner et al., 1972a, 1972b).

The case of a wave in a fluid incident on a solid (Figure 3.12) is now treated in more detail in terms of an equivalent circuit. This problem is translated into the equivalent circuit representation of Figure 3.13A, which shows mode conversion from the incoming longitudinal wave into a longitudinal wave and a vertical shear wave in the solid. Since the motions of these waves all lie in the *xz* plane, they do not couple into a horizontally polarized shear wave with motion orthogonal to that plane. Also, because an ideal nonviscous fluid does not support transverse motion, none of the shear modes in the solid couple into shear motion in the fluid. Here the solid and fluid are semi-infinite in extent, so characteristic impedances replace the transmission lines. Because the input impedances of the converted waves are transformed via Eqn 3.33 and the ABCD matrix for a transformer, the input impedance at position *a*, looking to the right in Figure 3.13B, is

$$Z_{INA} = n_L^2 Z_{2L\theta} + n_{SV}^2 Z_{2SV\theta}. \tag{3.54}$$

Where the angular impedances used for the fluid–fluid problem are used,

$$Z_{2L\theta} = \frac{\rho_2 c_{2L}}{\cos\theta_{2L}} = \frac{Z_{2L}}{\cos\theta_{2L}}, \text{ and} \tag{3.55A}$$

$$Z_{2SV\theta} = \frac{\rho_2 c_{2S}}{\cos\theta_{2SV}} = \frac{Z_{2SV}}{\cos\theta_{2SV}}, \tag{3.55B}$$

in which these angles can be determined from the Snell's law for this boundary:

$$k_{1x} = k_1 \sin\theta_{1L} = k_{2SV} \sin\theta_{2SV} = k_{2L} \sin\theta_{2L}. \tag{3.56}$$

The stress reflection factor at a is simply:

$$RF_a = \frac{Z_{INA} - Z_{1L\theta}}{Z_{INA} + Z_{1L\theta}}, \tag{3.57A}$$

where

$$Z_{1L\theta} = \frac{\rho_1 c_{1L}}{\cos\theta_i}. \tag{3.57B}$$

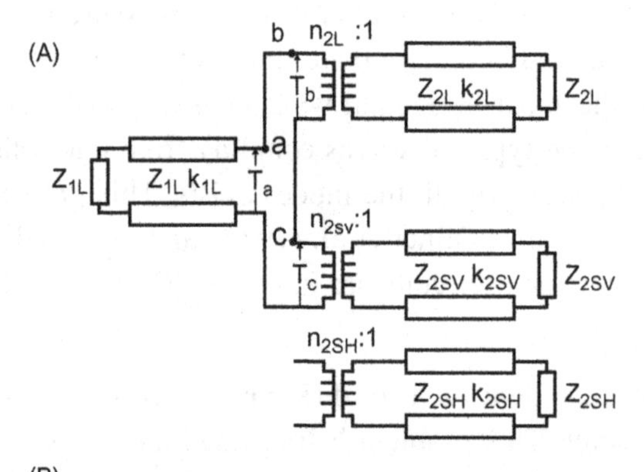

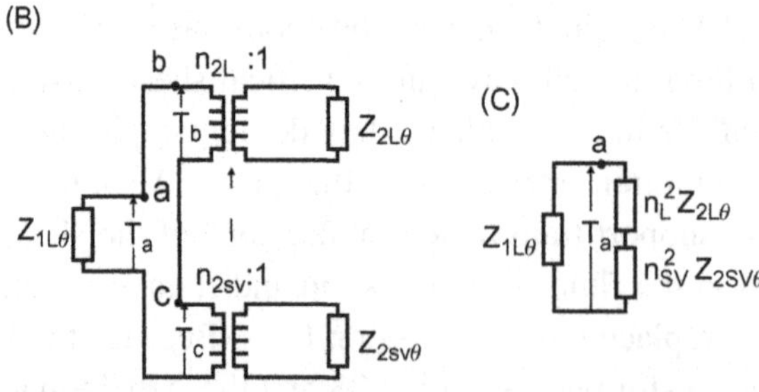

Figure 3.13
Equivalent circuit for fluid–solid interface problem.
(A) Overall equivalent circuit diagram; (B) reduction of circuit to transformed loads; (C) simplified circuit.

The transmission stress factors for each of the two waves in the solids can be found from Eqn 3.28A,B and impedance at each location. First at b in Figure 3.13B:

$$TF_L = \frac{2n_L^2 Z_{2L\theta}}{Z_{1L\theta} + Z_{INA}}; \tag{3.58}$$

second at c:

$$TF_{SV} = \frac{2n_{SV}^2 Z_{2SV\theta}}{Z_{1L\theta} + Z_{INA}}. \tag{3.59}$$

Here these factors represent the ratios of amplitudes arriving at different loads over the amplitude arriving at both loads, position a in Figure 3.13B. Usually, it is most desirable to know the intensity rather than the stress arriving at different locations (e.g. the relative intensities being converted into shear and longitudinal waves). From the early definitions of time average intensity (Eqn 3.7) and the three previous factors, it is possible to arrive at the following intensity ratios relative to the input intensity: first the intensity reflection ratio,

$$r = (RF_a)^2 = \frac{(Z_{INA} - Z_{1L\theta})^2}{(Z_{INA} + Z_{1L\theta})^2}, \tag{3.60}$$

and the intensity ratio for the longitudinal waves,

$$\tau_L = (TF_L)^2 \frac{Z_{1L\theta}}{n_L^2 Z_{2L\theta}} = \frac{4Z_{1L\theta} n_L^2 Z_{2L\theta}}{(Z_{1L\theta} + Z_{INA})^2}, \tag{3.61}$$

and the intensity ratio for the shear waves,

$$\tau_{SV} = (TF_{SV})^2 \frac{Z_{1L\theta}}{n_{SV}^2 Z_{2SV\theta}} = \frac{4Z_{1L\theta} n_{SV}^2 Z_{2SV\theta}}{(Z_{1L\theta} + Z_{INA})^2}. \tag{3.62}$$

An example of an intensity calculation is shown in Figure 3.14.

3.4 Elastic Wave Equations

In order to determine useful relations for waves and derive the wave equation, we employ the useful and elegant shorthand notation from this chapter. Recall basic relations and definitions, pressure **p** in terms of particle velocity **v** and particle velocity and particle displacement **u**:

$$\mathbf{p} = \rho \mathbf{v}, \quad \text{and} \tag{3.63A}$$

$$\mathbf{v} = \frac{\partial \mathbf{u}}{\partial t}. \tag{3.63B}$$

Two useful equations are the following:

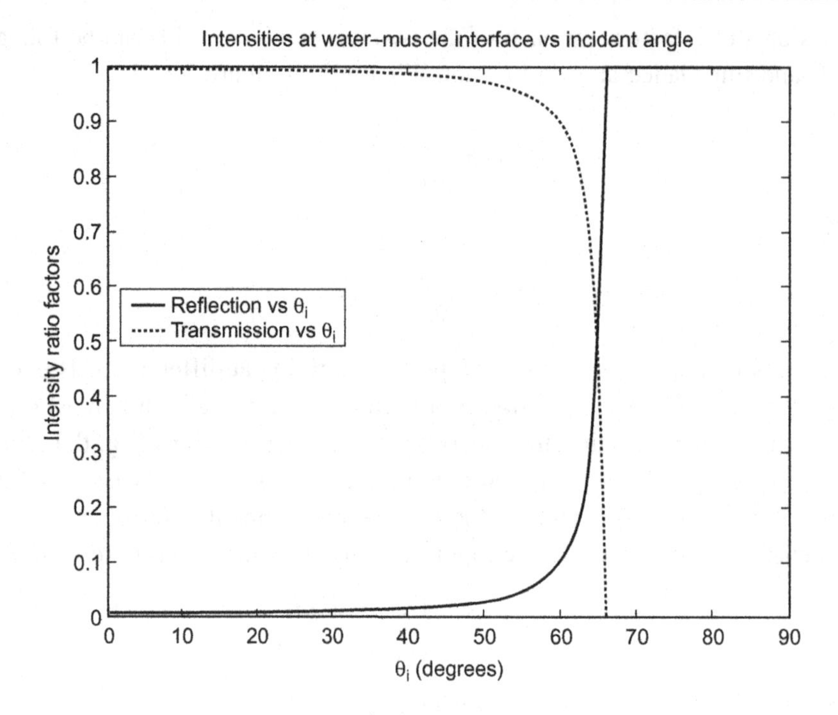

Figure 3.14
Intensity transmission and reflection graphs for water–muscle boundary as an example of a fluid–solid interface.

$$\nabla \cdot \mathbf{T} = \rho \frac{\partial \mathbf{v}}{\partial t} - \mathbf{F}, \text{ and} \tag{3.64}$$

$$\nabla_s \mathbf{v} = \mathbf{s} : \frac{\partial \mathbf{T}}{\partial t}, \tag{3.65A}$$

in which **s**, the compliance matrix, is the inverse of the stiffness matrix **c**, $[\mathbf{s}] = [\mathbf{c}]^{-1}$. A related useful trick is multiplying the last equation by **c**:

$$\mathbf{c} \nabla_s \mathbf{v} = \mathbf{c} : \mathbf{s} : \frac{\partial \mathbf{T}}{\partial t} = \frac{\partial \mathbf{T}}{\partial t}. \tag{3.65B}$$

Equations 3.64 and 3.65B can be written out as matrices and vectors well suited for MATLAB:

$$\begin{bmatrix} 0 & 0 & 0 & 0 & \frac{\partial}{\partial z} & 0 \\ 0 & 0 & 0 & \frac{\partial}{\partial z} & 0 & 0 \\ 0 & 0 & \frac{\partial}{\partial z} & 0 & 0 & 0 \end{bmatrix} \begin{bmatrix} T_1 \\ T_2 \\ T_3 \\ T_4 \\ T_5 \\ T_6 \end{bmatrix} = \rho \frac{\partial}{\partial t} \begin{bmatrix} v_x \\ v_y \\ v_z \end{bmatrix} - \begin{bmatrix} F_x \\ F_y \\ F_z \end{bmatrix}, \text{ and} \tag{3.66}$$

$$\begin{bmatrix} T_1 \\ T_2 \\ T_3 \\ T_4 \\ T_5 \\ T_6 \end{bmatrix} = \begin{bmatrix} C_{11} & C_{12} & C_{12} & 0 & 0 & 0 \\ C_{12} & C_{11} & C_{12} & 0 & 0 & 0 \\ C_{12} & C_{12} & C_{11} & 0 & 0 & 0 \\ 0 & 0 & 0 & C_{44} & 0 & 0 \\ 0 & 0 & 0 & 0 & C_{44} & 0 \\ 0 & 0 & 0 & 0 & 0 & C_{44} \end{bmatrix} \begin{bmatrix} 0 \\ 0 \\ \frac{\partial v_z}{\partial z} \\ \frac{\partial v_y}{\partial z} \\ \frac{\partial v_x}{\partial z} \\ 0 \end{bmatrix} = \frac{\partial}{\partial t} \begin{bmatrix} T_1 \\ T_2 \\ T_3 \\ T_4 \\ T_5 \\ T_6 \end{bmatrix}. \tag{3.67}$$

For waves along z and no force, T can be eliminated by taking the time derivatives of the first set of equations and the spatial derivatives for the second set. Examples for deriving wave equations follow:

$$\frac{\partial^2 T_3}{\partial z \partial t} = \rho \frac{\partial^2 v_z}{\partial t^2} - \frac{\partial F_z}{\partial t}, \tag{3.68A}$$

$$c_{11} \frac{\partial^2 v_z}{\partial z^2} = \frac{\partial^2 T_3}{\partial z \partial t}, \quad \text{and} \tag{3.68B}$$

$$\frac{\partial^2 v_z}{\partial z^2} - \frac{1}{c_L^2} \frac{\partial^2 v_z}{\partial t^2} = 0, \tag{3.68C}$$

where this wave equation is for longitudinal waves along z with a speed of sound $c_L = \sqrt{\frac{c_{11}}{\rho}}$;

$$\frac{\partial^2 T_5}{\partial z \partial t} = \rho \frac{\partial^2 v_x}{\partial t^2}, \tag{3.69A}$$

$$c_{44} \frac{\partial^2 v_x}{\partial z^2} = \frac{\partial^2 T_5}{\partial z \partial t}, \quad \text{and} \tag{3.69B}$$

$$\frac{\partial^2 v_x}{\partial z^2} - \frac{1}{c_{SX}^2} \frac{\partial^2 v_x}{\partial t^2} = 0, \tag{3.69C}$$

where this wave equation is for x-polarized z-propagating shear waves with a speed of sound $c_{SX} = \sqrt{\frac{c_{44}}{\rho}}$.

3.5 Conclusion

In this chapter, wave equations describe three basic wave shapes. When waves strike a boundary, they are transmitted and reflected. For the one-dimensional case, solutions consist of positive- and negative-going waves. Through the application of ABCD matrices, solutions for complicated cases consisting of several layers can be constructed from cascaded matrices rather than by re-deriving the equations needed to satisfy boundary conditions at each interface. This approach will be used extensively in developing a transducer model in Chapter 5 and Appendix C. Matrix methodology has been extended to oblique waves at an interface between different media.

Even though tissues are most often represented as fluid media, they are, in reality, elastic. An important case is the heart, which has muscular fibers running in preferential directions (to be described in Chapter 9). In addition, elastic waves are necessary to describe transducer arrays and piezoelectric materials (to be discussed in Chapter 5 and Chapter 6). An extra level of complexity is introduced by elasticity; namely, the existence of shear and other forms of waves created from both boundary conditions and geometry. Reflection and mode conversions among different elastic modes can be handled in a direct manner with the equivalent approach introduced by A. A. Oliner. His methodology is well suited to the ABCD matrix approach developed here. It also has the capability of handling mode conversions to other elastic modes, such as Lamb waves and Rayleigh waves (as described in his publications). Because tissues are actually viscoelastic not elastic, the effects of viscoelasticity on wave propagation will be the subject of the next chapter.

References

Auld, B. A. (1990). *Acoustic waves and fields in solids* (Vol. 1). Malabar, FL: Krieger Publishing. (Chapter 8).

Blackstock, D. T. (2000). *Fundamentals of physical acoustics*. New York: John Wiley & Sons.

Duck, F. A. (1990). *Physical properties of tissue: A comprehensive reference book*. London: Academic Press.

Kino, G. S. (1987). *Acoustic waves: Devices, imaging, and analog signal processing*. Englewood Cliffs, NJ: Prentice-Hall.

Matthaei, G. L., Young, L., & Jones, E. M. T. (1980). *Microwave filters, impedance-matching networks, and coupling structures*. Dedham, MA: Artech House. pp. 255–354, (Chapter 6).

Oliner, A. A. (1969). Microwave network methods for guided elastic waves. *IEEE Transactions on Microwave Theory and Techniques, 17*, 812–826.

Oliner, A. A., Bertoni, H. L., & Li, R. C. M. (1972a). A microwave network formalism for acoustic waves in isotropic media. *Proceedings of the IEEE, 60*, 1503–1512.

Oliner, A. A., Bertoni, H. L., & Li, R. C. M. (1972b). Catalog of acoustic equivalent networks for planar interfaces. *Proceedings of the IEEE, 60*, 1513–1518.

Sittig, E. K. (1967). Transmission parameters of thickness-driven piezoelectric transducers arranged in multilayer configurations. *IEEE Transactions on Sonics and Ultrasonics, 14*, 167–174.

Bibliography

For more information on elastic waves, see Kino (1987), now available on a CD-ROM archive from the IEEE Ultrasonics, Ferroelectrics, and Frequency Control Group, and Auld (1990).

CHAPTER 4

Attenuation

Chapter Outline

Diagnostic Ultrasound Imaging: Inside Out.

4.1 Losses in Tissues

Waves in actual media encounter losses. Real tissue data indicate that attenuation has a power-law dependence on frequency. As a result of this frequency dependence, acoustic pulses not only become smaller in amplitude as they propagate, but they also change shape. Attenuation in the body is a major effect; it limits the detectable penetration of sound waves in the body or the maximum depth at which tissues can be imaged. In order to compensate for attenuation, all imaging systems have a way of increasing amplification with depth. These methods will be discussed at the end of this chapter.

Attenuation is a general term which needs to be used carefully, for there are several sources of losses: reflection, refraction, 1/r or spreading loss depending on the type of source, scattering from tissue and interfaces, and absorption. This chapter investigates absorption or viscous losses; Chapter 6 and Chapter 7 deal with diffraction loss, and Chapter 8 and Chapter 9 explain sound propagation in heterogeneous tissue. In general, for a clinical imaging situation, the sources of loss may not be identified well enough to quantify them without further testing.

Usually, absorption is treated in the frequency domain. Because imaging is done with pulse echoes, it is important to understand the effect of absorption on waveforms. This chapter introduces models suitable for the kind of losses in tissues that can work equally well in the domains of both time and frequency. When absorption is present, phase velocity usually changes with frequency as well (an effect known as dispersion). The loss model can predict how both absorption and phase-velocity dispersion affect pulse shape during propagation. Absorption and dispersion are related through the principle of causality. Tissues are viscoelastic media, meaning they have both elastic properties and losses. The model can also be extended to cover these characteristics. In addition, appropriate wave equations and stress–strain relations (Hooke's law for lossy media) complete the simulation of acoustic waves propagating in tissue with losses.

This time causal model provides a self-consistent viewpoint which is suitable for most cases in diagnostic imaging. There are other modes, such as those in shear wave elastography and acoustic radiation force imaging, in which other approaches need to be considered. Alternate models are explored at the end of this chapter.

4.1.1 Losses in Exponential Terms and in Decibels

When waves propagate in real media, losses are involved. Just as forces encounter friction, pressure and stress waves lose energy to the medium of propagation and result in weak local heating as well as other loss mechanisms such as diffraction, scattering, reflection, and refraction. These small losses are called "attenuation" and can be described by an exponential law with distance. For a single-frequency (f_c) plane wave, a multiplicative amplitude loss term can be added:

$$A(z,t) = A_0\, exp(i(\omega_c t - kz))\, exp(-\alpha z). \tag{4.1}$$

The attenuation factor (α) is usually expressed in terms of nepers per centimeter in this form. Another frequently used measure of amplitude is the decibel (dB), which is most often given as the ratio of two amplitudes (A and A_0) on a logarithmic scale:

$$Ratio(\text{dB}) = 20\,\log_{10}(A/A_0), \tag{4.2}$$

or in those cases where intensity is simply proportional to amplitude squared ($I_0 \propto A_0^2$),

$$Ratio(\text{dB}) = 10\,\log_{10}(I/I_0) = 10\,\log_{10}(A/A_0)^2. \tag{4.3}$$

Most often, α is given in dB/cm:

$$\alpha_{\text{dB}} = 1/z\{20 \times \log_{10}[exp(-\alpha_{\text{nepers}} z)]\} = 8.6886(\alpha_{\text{nepers}}). \tag{4.4}$$

Graphs for a loss constant α equivalent to 1 dB/cm are given in Figure 4.1 on several scales.

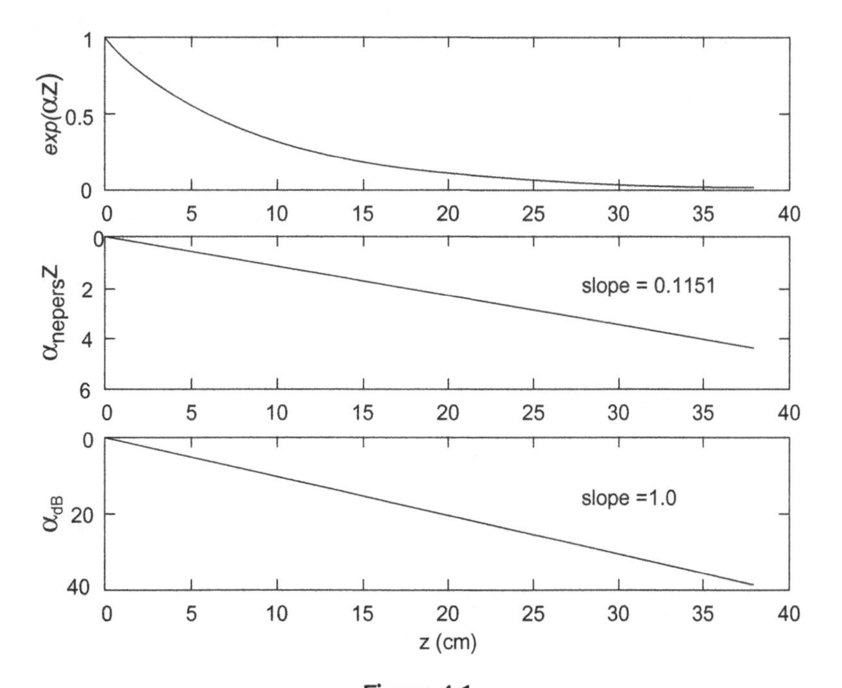

Figure 4.1
Constant Absorption as a Function of Depth on a (Top) Linear Scale, (Middle) dB Scale, and (Bottom) Neper Scale.

A plane wave multiplied by a loss factor that increases with travel distance (z) was shown in Eqn 4.1. This equation for a single-frequency (f_c) plane wave can be rewritten as:

$$A(z,t) = A_0\ exp(-\alpha z)\ exp[i\omega_c(t - z/c_0)], \tag{4.5A}$$

in which $c = c_0$ is a constant speed of sound and $\alpha = \alpha_0$ is a constant. Also, the second exponential argument can be recognized as a time delay. The Fourier transform of this equation is:

$$A(z,f) = A_0\ exp([-\alpha_0 z - i\omega_c z/c])\delta(f - f_c). \tag{4.5B}$$

This result indicates that the exponential term is frequency independent and acts as a complex weighting amplitude for this spectral frequency.

The actual loss per wavenumber is very small, or $\alpha/k \ll 1$, a fact that will be useful later. Even though the loss per wavelength is small, absorption has a strong cumulative effect over many wavelengths. Attenuation for a round-trip echo path usually determines the allowable tissue penetration for imaging.

4.1.2 Tissue Data

These simple loss and delay factors are not observed in real materials and tissues. Data indicate that the absorption is a function of frequency. Many of these losses obey a frequency power law, defined as:

$$\alpha(f) = \alpha_0 + \alpha_1 |f|^y, \tag{4.6A}$$

in which α_0 is often zero and y is a power law exponent. A graph for the measured absorption of common tissues as a function of frequency is given in Figure 4.2A. In addition to absorption loss, the phase velocity of tissue also varies with frequency:

$$c(f) = c_0 + \Delta c(f), \tag{4.6B}$$

where $\Delta c(f)$ is a small change in sound speed with frequency. What is plotted in Figure 4.2B is the change in sound speed with the constant term subtracted out for convenience:

$$\Delta c(f) = c(f) - c_0. \tag{4.6C}$$

This change of phase velocity with frequency is required by causality, as will be shown shortly. Although this velocity dispersion is considered by many to be a small effect, its consequences can be significant, especially for broad bandwidth pulses.

As pointed out in Chapter 1, soft tissues are mostly (60%) composed of water; therefore, values for average velocity are similar, varying only about $\pm$ 10% from a mean value (see

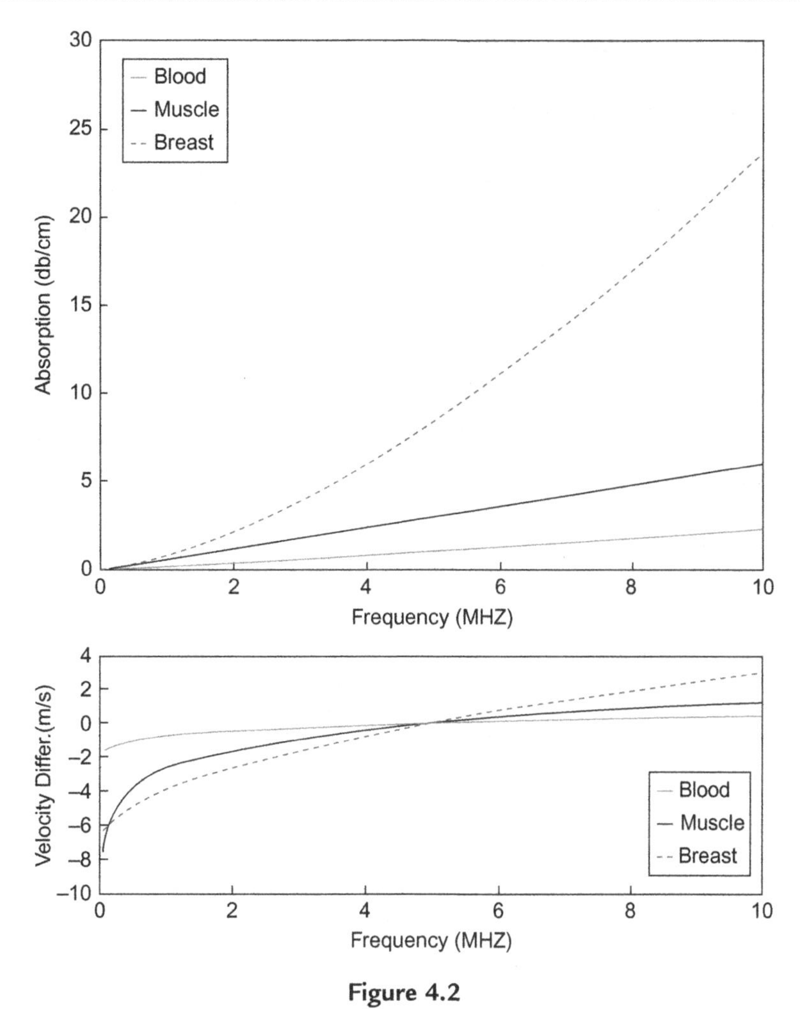

Figure 4.2

(A) Absorption as a Function of Frequency for Muscle, Fat, and Blood; (B) Phase Velocity Dispersion Difference Minus a Midband (5-MHz) Sound Speed Value for the Same Tissues.

Figure 1.2). Lists of common tissue average absorption and average phase velocity values, and characteristic impedances are in Appendix B, and a more complete list is in Duck (1990). Tissue attenuation has two parts: absorption and scattering. At low-MHz frequencies, scattering is typically 10–15% of the total value of attenuation (Bamber, 1986, 1998). The tissue structure causes the scattering to vary with angle. Both absorption and scattering are frequency-dependent because the wavelength changes in relation to the scale of tissue structure. This scaling implies that as the imaging frequency is lowered, greater averaging over the structure occurs. The topic of scattering will be treated in more depth in Chapter 8 and Chapter 9; for now, we neglect scattering and assume that each type of tissue has characteristics that are uniform everywhere (homogeneous).

4.2 Losses in Both Frequency and Time Domains

4.2.1 The Material Transfer Function

The combined effects of absorption and dispersion on pulse propagation can be described by a material transfer function (*MTF*) in the frequency domain:

$$MTF(f,\ z) = exp[\gamma_T(f)z], \tag{4.7A}$$

in which:

$$\gamma_T(f) = -\alpha(f) - i\beta(f) = -\alpha(f) - i[k_0(f) + \beta_E(f)], \tag{4.7B}$$

where $k_0 = \omega/c_0$, a baseline wavenumber (where c_0 is a sound speed value taken in this context to be at the center frequency of the spectrum of a pulse), and $\beta_E(f)$ is an excess dispersion term required by causality (to be presented later). In Eqn 4.7, a frequency-dependent amplitude term is associated with $\alpha(f)$, and an effective frequency-dependent phase velocity is determined by:

$$1/c(\omega) = \beta/\omega = k_0/\omega + \beta_E/\omega = 1/c_0 + \beta_E/\omega, \tag{4.8}$$

which, because the second term is very small relative to the first, can be approximated by:

$$c(\omega) - c_0 = -c_0^2\beta_E/\omega, \tag{4.9}$$

like Eqn 4.6C.

4.2.2 The Material Impulse Response Function

The material transfer function (*MTF*) has its time domain counterpart, called the material impulse response function, *mirf*(t):

$$mirf(t,z) = \Im^{-1}\{exp[\gamma_T(f)z]\}. \tag{4.10}$$

For an initial pulse described by $p_0(t)$ having a spectrum $P_0(f)$, the pulse at a distance z can be described simply by either:

$$p(t,z) = p_0(t) *t\ mirf(t,z), \tag{4.11A}$$

or, equivalently, in the frequency domain:

$$p(f,z) = P_0(f)\ MTF(f,z). \tag{4.11B}$$

Eqn 4.11A was used to calculate pulses at increasing distances (z) in Figure 4.3 for a medium with loss. An initial starting pulse of a Gaussian modulated sinusoid was used. A series of *mirfs* that were used for Figure 4.3 are shown in Figure 4.4 for the case of a

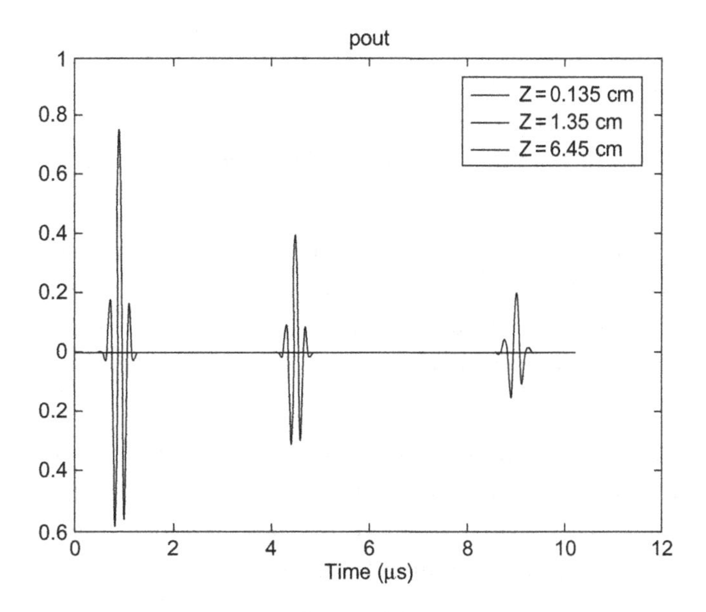

Figure 4.3
Changes in Pressure-pulse Shape of an Initially Gaussian Pulse Propagating in a Medium With a 1 dB/MHz$^{1.5}$-cm Absorption $y = 1.5$ for Three Different Increasing Propagation Distances (z).

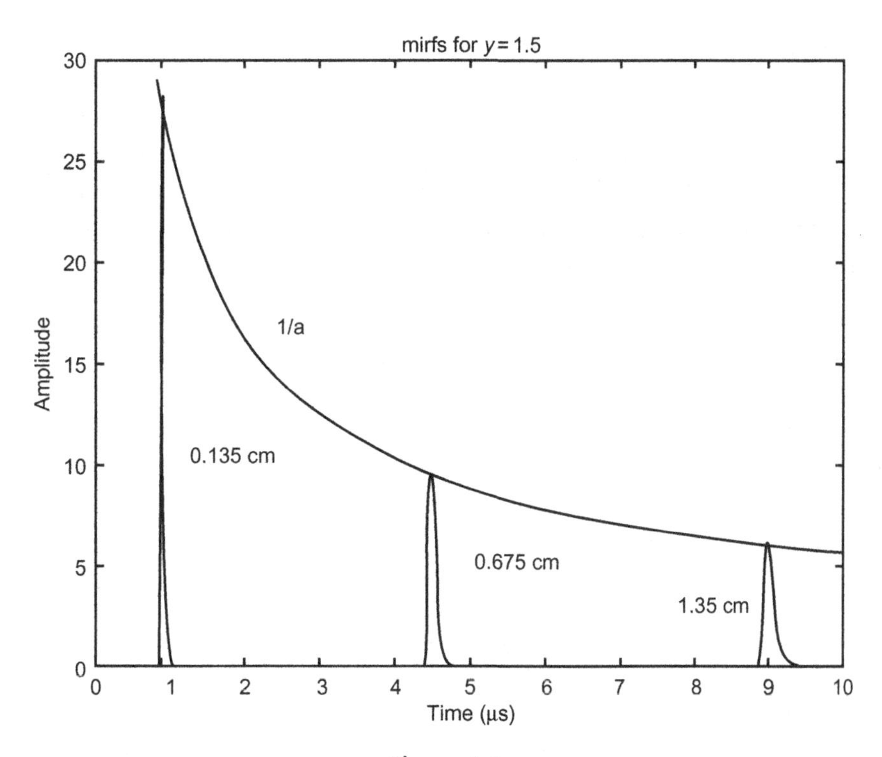

Figure 4.4
Material Impulse Response Functions for a Medium With a 1 dB/MHz$^{1.5}$-cm Absorption for Different Propagation Distances (z).
Note that peaks follow a $1/a$ characteristic. Here, $a = (a_1 z)^{1/y}$ from Section 4.4.1.

frequency dependence for absorption, with $y = 1.5$ (based on the time causal model, which will be introduced shortly). Note that the *mirfs* follow a delay time that is approximately z/c_0, where $c_0 = 1.5$ mm/μs.

From these figures, we can see that the *mirf* function has some very interesting properties. Its amplitude diminishes with increasing distance. As the delay time approaches zero, *mirf* must become impulse-like. For short distances, *mirf*(t) has a steep leading edge, as required by causality, so that its left edge does not extend into negative time.

What is causality? It is natural law that does not allow a response to precede its cause. In order for the pulse response to be zero for $t < 0$, the complex Fourier spectral components must add up to produce an effect in the time domain, which is the equivalent of canceling out the pulse response for times less than zero. This mathematical operation is similar to multiplying the time response by $H(t)$, the step function that is zero for $t < 0$.

4.3 Tissue Models

4.3.1 Introduction

In order to calculate the material transfer function and the material impulse response, a model must be selected. The models available describe both absorption and dispersion as a function of frequency and differ in terms of their convenience and accuracy. These models are not very satisfying in terms of an explanation based on first principles, but they can describe adequately, in an empirical way, the way absorption affects acoustic propagation.

The choices for models are the classic relaxation model, the time causal model, and the Kramers–Kronig relation. The relaxation model (Kinsler, Frey, Coppens, and Sanders, 2000) is the most well known; however, it has its own absorption function that is not exactly the power-law dependence observed for tissues (to be described in Section 4.5.4). An often-used approach is to fit one or more relaxation-absorption functions to match the power law absorption characteristic over a limited frequency range. While this procedure is suitable for valid band or a band-limited starting spectrum (P_0), the responses at very low and high frequencies (which are important in Fourier inversion) must be watched carefully.

The time causal model (Szabo, 1994, 1995) is based on the observed power law absorption characteristic and provides a more direct implementation. A third approach, the application of the Kramers–Kronig relation (Krönig, 1926), is more challenging for an absorption power law type characteristic, which increases exponentially as frequency approaches an infinite limit. This difficult problem has been solved by the method of subtractions and has been found to be equivalent to the time causal method (Waters, Hughes, Brandenburger, and Miller, 2000a; Waters, Hughes, Mobley, Brandenburger, and

Miller, 2000b). These last two methods (really the same result solved in different domains), though more recent, have had extensive experimental validation. All of these models satisfy the laws of causality.

4.3.2 The Time Causal Model

The time casual model is based on a power law that includes macromolecular effects. The relaxation models are causal, which means that an effect cannot precede its cause. Traditionally, the effects of causality have been determined by use of the Kramers–Kronig relation, which relates the excess dispersion characteristic to a convolution integral operator on absorption over all frequencies:

$$\beta_E(f) = \frac{-1}{\pi f} * [-\alpha(f) + \alpha_0]. \tag{4.12}$$

In other words, β_E is the Hilbert transform of $\alpha(f) - \alpha_0$. This approach poses numerical evaluation problems for the Hilbert transform integral of power law absorption because, unlike the characteristics of the thermoviscous model of Section 4.5.4, which decrease with very high frequencies, power law characteristics increase monotonically with frequency. One solution to this problem is to use a time causal relation. In this approach, the power law is assumed to hold for all frequencies; however, this is an approximation which is expected to hold well beyond the frequency range of interest. At extremely high frequencies, other mechanisms are expected to modify the power law behavior. Another way to solve the numerical difficulties is to apply a method of subtractions to the Kramers–Kronig method (Waters et al., 2000a, 2000b). These two approaches give the same results and can be considered to be complementary statements of causality.

For the time causal method, causality can be expressed directly in the time domain (Szabo, 1994, 1995; Szabo and Wu, 2000) through the use of generalized functions, as:

$$L_{\beta_E}(t) = -i\, sgn(t) * L_\alpha(t), \tag{4.13A}$$

where $sgn(t)$ is a signum function (Bracewell, 2000):

$$sgn(t) = 1 \; t > 0, \tag{4.13B}$$

and

$$sgn(t) = 1 \; t < 0, \tag{4.13C}$$

and $L_{\beta E}$ as well as L_α are Fourier transform pairs of $\beta_E(f)$ and $\alpha_0 - \alpha(f)$, respectively. These transforms involve the application of generalized functions (these details need not concern us here and can be found in the references above). The overall propagation operator can be written as:

$$L_y(t) = L_\alpha(t) + iL_{\beta_E}(t) = [1 + sgn(t)]L_\alpha(t) = 2H(t)L_\alpha(t), \tag{4.14}$$

where $H(t)$ is the step function (Bracewell, 2000).

Note that because the Hilbert transform is involved, there is a similarity between this operation and that for creating an analytic signal (described in Section A.2.7 of Appendix A). In this case, the time absorption operator is multiplied by $2H(t)$, ensuring causality, whereas for the analytic signal, the negative frequency components were dropped by a similar operation in the frequency domain (as described in Appendix A). From the time causal relation, an imaginary signal $iL_{\beta E}(t)$ is created from the real one, $L_\alpha(t)$, just as for the analytic signal, an imaginary quadrature signal was derived from the original real signal. An alternate formulation includes a cosine term in the absorption and a sine term in the dispersion term (Ochmann and Makarov, 1993). From the power law, the time causal relation, and the Fourier transform relation between these time operators and α_1 and β_E, the excess dispersion has been found to depend on the kind of power law type of exponent y. The steps in the process are the following:

1. Find $L_\alpha(t)$ from the inverse Fourier transform of $\alpha(f) - \alpha_0$ through the use of generalized functions.
2. Obtain $L_{\beta E}(t)$ from Eqn 4.13A, the time causal relation.
3. Find $\beta_E(f)$ from the Fourier transform of $L_{\beta E}(t)$ through the use of generalized functions.

The results for $\beta_E(f)$ indicate that various functions are necessary for different forms of y. For y as an even integer or noninteger,

$$\beta_E(f) = \alpha_1 \tan(\pi y/2) f |f|^{y-1}, \tag{4.15A}$$

and for y as an odd integer,

$$\beta_E(f) = -(2/\pi)\alpha_1 f^y ln|f|. \tag{4.15B}$$

Versions of these equations that are more appropriate for phase velocity data follow. For y as an even integer or noninteger,

$$1/c(f) = 1/c(f_0) + \alpha_1 \tan(\pi y/2)[|f|^{y-1} - |f_0|^{y-1}], \tag{4.16}$$

in which $c(f_0)$ is the speed of sound at a reference frequency (f_0), usually a midband value. For y as an odd integer,

$$1/c(f) = 1/c(f_0) - (2\alpha_1 f^{y-1}/\pi)\,(ln|f| - ln|f_0|). \tag{4.17}$$

Dispersion is maximum for $y = 1$ or equal to an odd integer, given equivalent absorption loss coefficients. Note that if y is zero or an even constant, $\beta_E = 0$. The validity of these results has been verified by independent analysis and confirmed by experiments at several laboratories (He, 1998a, 1998b, 1999; Szabo, 1995; Szabo & Wu, 2000; Trousil, Waters, & Miller, 2001; Waters et al., 2000b). In Figure 4.5, the attenuation follows a power law with an exponent $y = 1.79$, and the phase velocity increases with frequency. For the

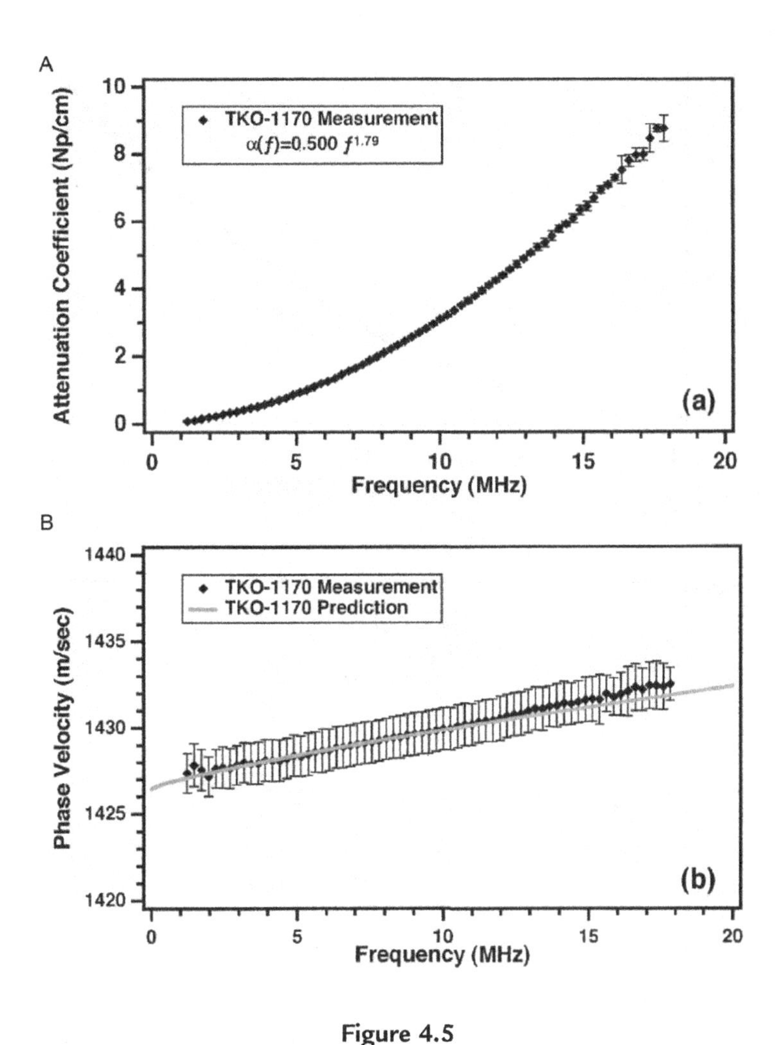

Figure 4.5
(A) Measured Absorption Data as a Function of Frequency for TKO-1170 (High-viscosity Hydrocarbon Oil) With Power Law Fit of $y = 1.79$; (B) Measured Relative Velocity Dispersion with Error Bars Compared to Prediction from Eqn 4.19A Using the Measured Absorption Data Power Law Fit. *From Waters et al. (2000b), Acoustical Society of America.*

second case (Figure 4.6), $y \approx 2$, so very little dispersion is expected (in excellent agreement with data).

The time causal analysis holds for the entire ultrasound imaging range and is valid to very high frequencies, including multiple harmonic frequencies.

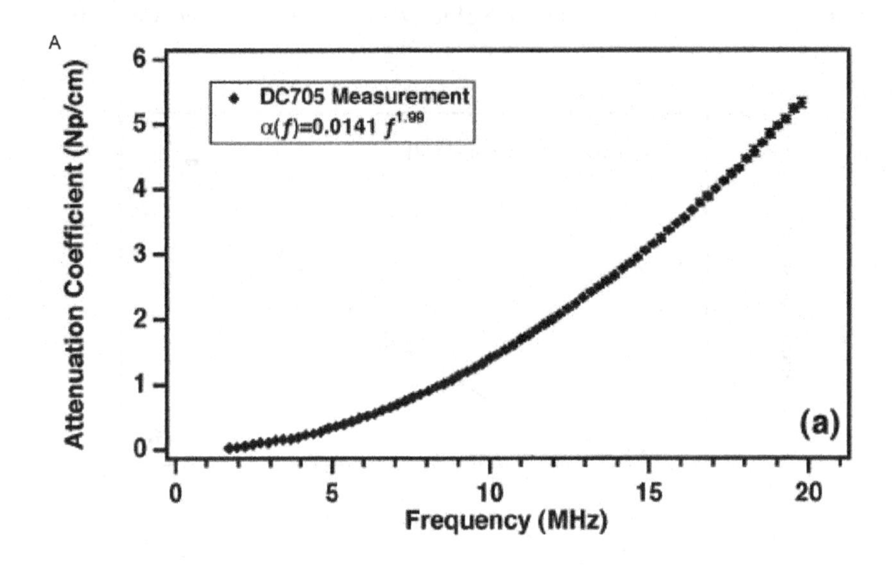

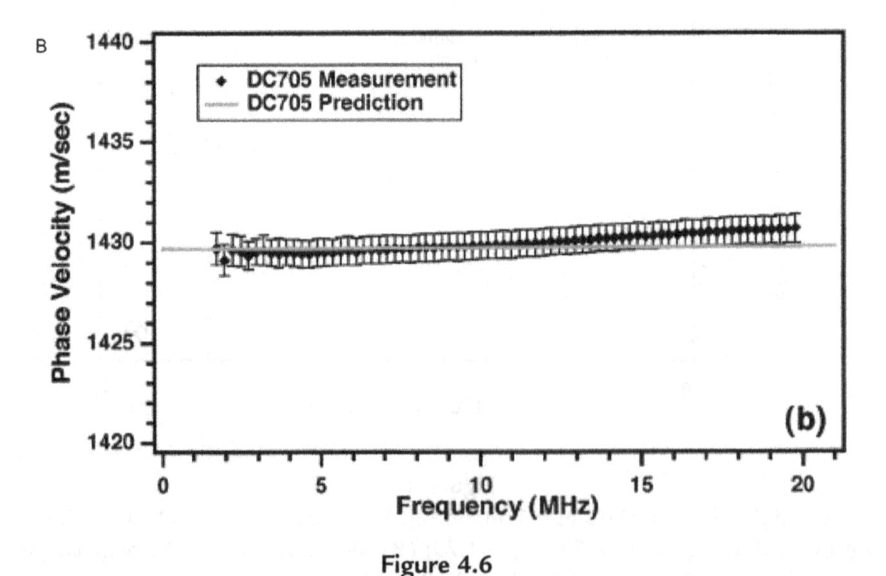

Figure 4.6

(A) Measured Absorption Data as a Function of Frequency for DC 705 (Silicone Fluid) With Power Law Fit of $y \approx 2$; (B) Measured Relative Velocity Dispersion With Error Bars Compared to Prediction from Eqn 4.19A Using the Measured Absorption Data Power Law Fit. Note that dispersion is nearly zero for this quadratic loss case. *From Waters et al. (2000b), Acoustical Society of America.*

For cases where y is equal to one or a noninteger, the Kramers–Kronig approach with the method of subtractions (Waters et al., 2000b) has shown these dispersion results to be valid without a high-frequency limit.

4.4 Pulses in Lossy Media

4.4.1 Scaling of the Material Impulse Response Function

Even though losses are evaluated most often in the frequency domain, it is also possible to examine the combined effects of loss and dispersion in the time domain directly through the material impulse response function. At each depth (z), the *mirf* from Eqn 4.10 encodes all the frequency loss and dispersion into a time waveform. This waveform can be determined from the inverse Fourier transform of the material transfer function, which can be calculated from Eqns 4.6, 4.7, and 4.15. In general, the material impulse response has to be evaluated numerically. When a fast Fourier transform (FFT) of a digital Fourier transform (DFT) is used, enough frequency points need to be taken to ensure that the nearly vertical leading edge of the response is captured accurately.

Because the velocity dispersion is often small, about $\pm 2 \times 10^{-3}$ of c_0 in Figure 4.5, for example, most people assume its effect is negligible. In Figure 4.7, simulations of broadband pulse propagation with and without dispersion are compared to experiment. In this case, the power exponent is the same as that for Figure 4.6 ($y \approx 2$), so no dispersion is expected from the time causal model (in agreement with data). The third comparison is to

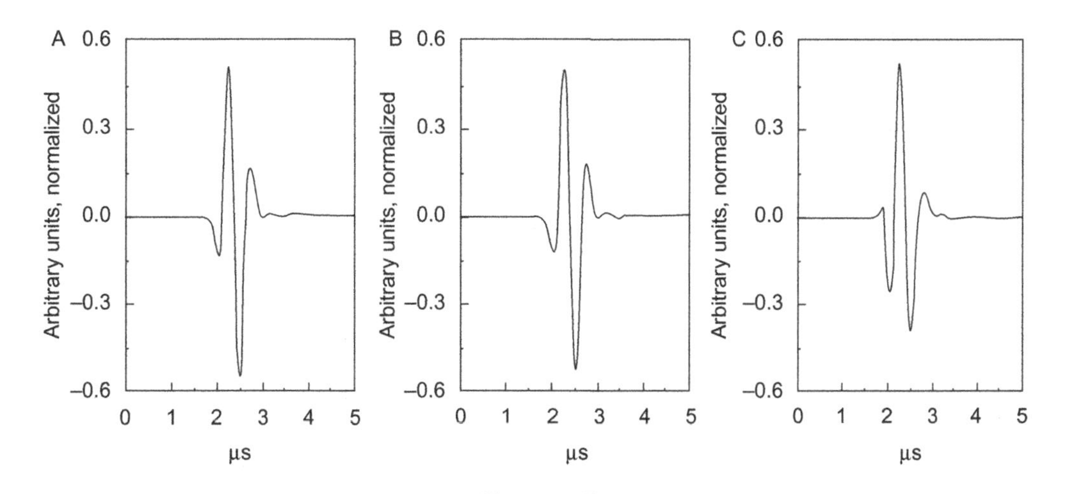

Figure 4.7
Comparison of Simulations to Data for a Broadband Pulse Passing Through an ATS Laboratories, I. Tissue-mimicking Phantom With $\alpha(f_{\mathrm{MHz}}) = 0.21 + 0.20f^{1.98}$ dB/cm.
(A) Pulse data; (B) time causal model; (C) nearly local Kramers–Kronig model. *From He (1998a), IEEE.*

an older, nearly local Kramers–Kronig model that predicts dispersion for this case. This older model is more accurate near $y \approx 1$ and has been replaced by Eqn 4.16 (Szabo, 1995; Waters et al., 2000a). The differences in the pulse predictions are noticeable and show the need for correctly accounting for dispersion, especially for wider bandwidth pulses.

As indicated in the last section, dispersion is maximum for $y = 1$. In Figure 4.8, corresponding time domain calculations for $y = 1$ are shown for a material impulse response for propagation of 1 cm in a tissue with absorption equal to 1 dB/MHz-cm (0.1151 neper/MHz-cm) and a reference phase velocity of $c_0 = 0.15$ cm/μs with and without dispersion calculated from causality. Two striking differences are that the acausal response is symmetrical ($\beta_{EM} = 0$) while the causal one is not, and the causal waveform peaks at $0.972z/c_0$. A simple explanation of the earlier arrival of the causal response is that for most of its frequency range, the phase velocity increases and is greater than its value at zero frequency, whereas in the acausal case, the phase velocity is constant.

A second calculation, for propagation in the same material to a depth of 10 cm, is shown in Figure 4.9. Note that the amplitude has dropped by a factor of 10 and the timescale has been expanded by a factor of 10. Careful observation shows that the time delays for the causal model at the 1-cm distance are slightly earlier (about 1% of the total delay) than the *mirf* for the 10-cm distance because more high frequencies get through the shorter distance.

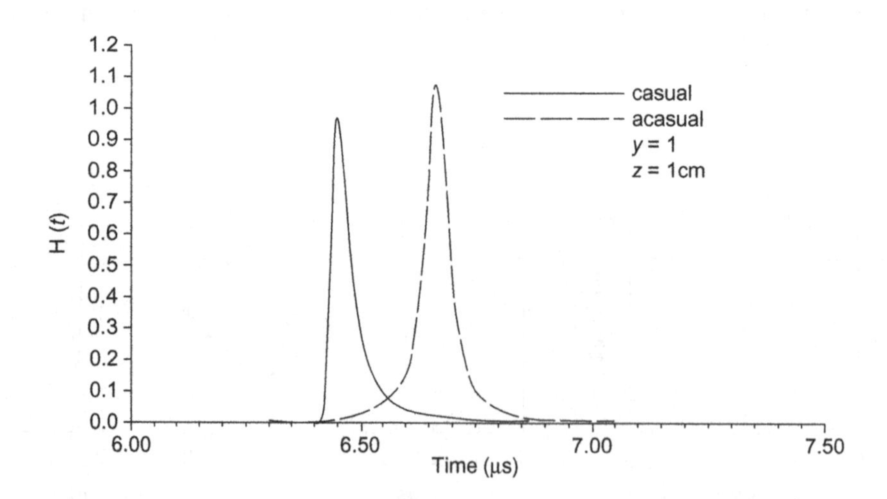

Figure 4.8
Comparison of Causal Time *Mirf* and an Acausal Nondispersive *Mirf* for Linear Loss ($y = 1$), for a Propagation Distance of 1 cm and an Absorption Coefficient of 0.1151 Neper/MHz-cm (1 dB/MHz-cm). *From Szabo (1993).*

It is possible to generalize these results through the scaling theorem. Rewriting the material transfer function (*MTF*) into an equivalent scaled function:

$$MTF(f) = M(af), \tag{4.18A}$$

in which

$$a = (\alpha_1 z)^{1/y} \tag{4.18B}$$

leads to a material impulse response function (*mirf*) through the scaling theorem:

$$mirf(t) = \Im^{-1}[MTF(f)] = \Im^{-1}[M(af)] = (1/a)m(t/a). \tag{4.19}$$

This time relationship, Eqn 4.19, provides a more direct and intuitive understanding of how loss changes with distance than the more often-used *MTF*. For $y = 1$, this relation shows that the amplitude of the *mirf* drops by $1/z$ and expands in length by z. Figures 4.8 and 4.9 show, to first order, how the amplitude drops by a factor of 10 and the response elongates by a factor of 10 in accordance with Eqn 4.19. Interestingly, the *mirf* waveform shape and area are nearly maintained! In another example, for $y = 2$, Eqn 4.19 shows that amplitude falls off more gradually with distance as $1\sqrt{z}$. Figure 4.4 shows how amplitude falls as $1/a$ for $y = 1.5$.

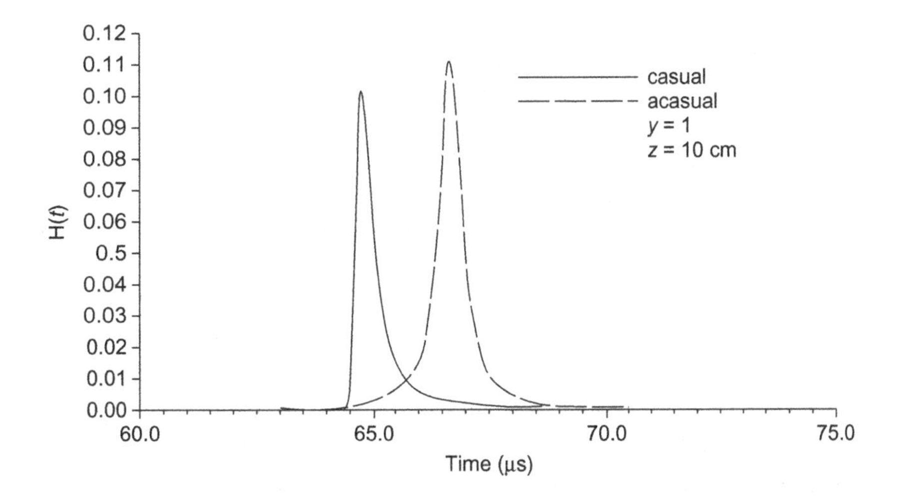

Figure 4.9
Comparison of Causal Time *Mirf* and an Acausal Nondispersive *Mirf* for Linear Loss ($y = 1$) for a Propagation Distance of 10 cm and an Absorption Coefficient of 0.1151 Neper/MHz-cm (1 dB/MHz-cm). *From Szabo (1993).*

More formally, the *MTF* scaling relation, Eqn 4.18A, can be rewritten in the following way:

$$M(af) = exp\left\{\left[-(\alpha_1 z)^{1/y}|f|\right]^y\right\}exp\left\{-i[2\pi(\alpha_1 z)^{1/y}f/[(\alpha_1 z)^{1/y}c_0] + \hat{\beta}_E[(\alpha_1 z)^{1/y}f]]\right\}$$
$$M(af) = exp\left\{-[a|f|]^y\}exp\{-i[2\pi af/(ac_0) + \hat{\beta}_E(af)]\right\} \tag{4.20A}$$

where

$$\hat{\beta}_E[(\alpha_1 z)^{1/y}f] = \beta_E(f), \tag{4.20B}$$

so that the material impulse response is found from the inverse Fourier transform:

$$\Im_{-i}^{-1}\{MTF|f|\} = \Im^{-1}\{M[af]\} = \Im^{-1}\{M[(\alpha_1 z)^{1/y}f]\} = [1/(\alpha_1 z)^{1/y}]m[t/(\alpha_1 z)^{1/y}]. \tag{4.21}$$

Eqns. 4.20 and 4.21 are exact formulations of the scaling relations and include the effects of dispersion. To first order, the shapes of the *mirf* functions at different distances are nearly the same except for scaling, and if we assume the slight differences in delay are negligible (they will be small when applied to finite bandwidth pulses), then losses can be explained through the scaling of the material impulse response functions with distance. In summary, Eqn 4.21 provides an intuitively satisfying picture of loss in the time domain through the *mirf*. The amplitude of the *mirf* falls by $1/a$, and the *mirf* widens by a factor of a with distance.

4.4.2 Pulse Propagation: Interactive Effects in Time and Frequency

In order to apply this scaling principle to the propagation of pulses in lossy media, the interaction of the pulse characteristics, with the constantly changing *mirf* with distance, must be taken into account. For example, the *mirfs* in Figure 4.4 obey the scaling law, but the corresponding pulses in Figure 4.3 undergo changes in shape and a slightly different drop in amplitude with distance.

A perspective that includes both the time and frequency domain viewpoints is helpful in explaining the changes in pulse propagation in a lossy medium. The bandwidth of a pulse can have a considerable effect on pulse shape during propagation. The Fourier theory, Eqn 4.11, can be applied to predict the shape of a pulse at some point in the medium, if its initial shape at $z = 0$ is known in either the frequency or time domain (illustrated by Figure 4.3). Beginning with Eqn 4.11, we can examine the effects of loss on the spectral magnitude for an initial pulse, which is a sine wave amplitude modulated by a Gaussian envelope, that propagates into a medium that has an absorption with linear frequency dependence, as follows:

$$|P(f,z)| = |P_0(f)MTF(f,z)|, \tag{4.22A}$$

and

$$|P(f,z)| = |exp[-b(f-f_c)^2]\, exp(-\alpha_1|f|z - i\beta z)|. \tag{4.22B}$$

The dispersion has a magnitude equal to one, so:

$$|P(f,z)| = exp[-b(f-f_c)^2]\, exp(-\alpha_1|f|z). \tag{4.22C}$$

By differentiating this magnitude with respect to frequency and setting the result to zero, we obtain the position of the spectral peak as a function of propagation distance:

$$\partial|P(f,z)|/\partial z = [-2b(f-f_c) - \alpha_1|f|z]|P(f,z)| = 0, \tag{4.23A}$$

and solving for f gives:

$$f_{peak} = f_c - \frac{\alpha_1 z}{2b}. \tag{4.23B}$$

This result indicates a downshift in peak frequency as depth z is increased. Since a smaller value of b indicates a broader original bandwidth, Eqn 4.23B shows a greater downshift for wider bandwidths, as shown in Figure 4.10.

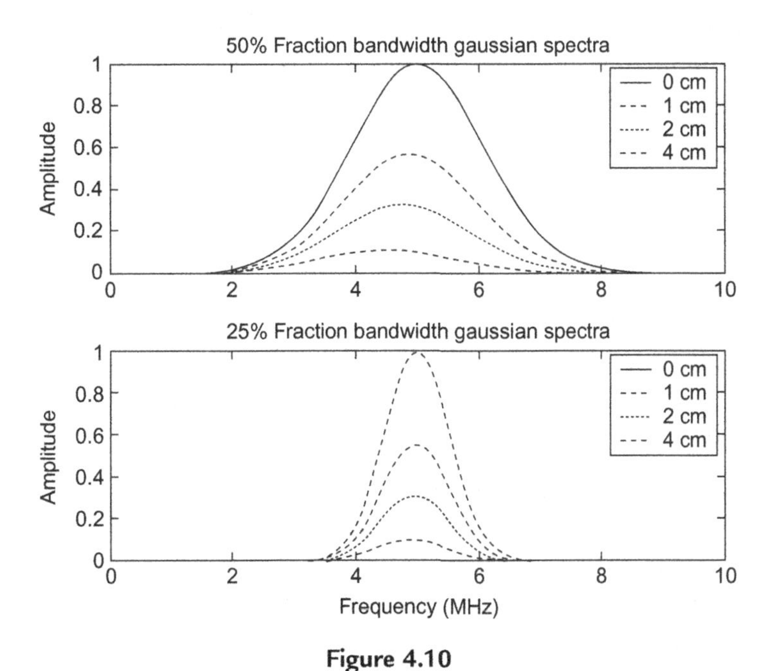

Figure 4.10

(A) Downshift in Peak Frequency of Spectra for a Gaussian-pulse Input Pulse With a 50% fractional bandwidth as a Function of Increasing Depth in a Medium With 1 dB/MHz-cm Linear Frequency Loss Dependence; (B) Spectra for a Gaussian Pulse With 25% fractional bandwidth.

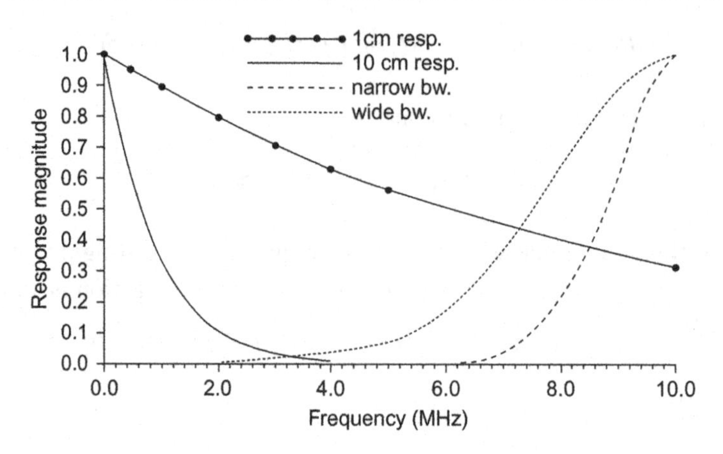

Figure 4.11
(Left) Curves for material transfer functions for $y = 1$, $z = 1$, and $z = 10$ cm, and an absorption coefficient of 0.1151 Neper/MHz-cm (1 dB/MHz-cm); (Right) curves for the spectral magnitudes of a narrowband Gaussian (long dashes, 25% fractional bandwidth) and wideband Gaussian (short dashes, 50% fractional bandwidth).
Both spectra have a center frequency of 10 MHz. *From Szabo (1993).*

Recall from Eqn 4.11A that the overall spectrum at a depth is the product of the *MTF* for that depth multiplied by the spectrum of the initial pulse. To explore these effects in more detail, we plot the magnitude of spectra for 10-MHz-center frequency Gaussian starting pulses with fractional bandwidths of 25% and 50% (on the right side of Figure 4.11). On the left side of this figure are *MTFs* for distances of 1 and 10 cm, for a medium with linear frequency absorption ($y = 1$) and a coefficient of 1 dB/MHz-cm (0.1151 neper/MHz-cm). The underlying causes for spectral downshift are evident in this figure. For the *MTF* for a 1-cm depth, the considerable overlap of spectra shows that the product of spectra will result in a high peak frequency. In contrast, there is little interaction between the *MTF* for 10 cm and the Gaussian spectra. The overlap, and consequently, the resulting peak frequency, is lower for the broader bandwidth.

The effects of bandwidth on pulses can be seen by comparing pulse envelopes with the *mirfs* corresponding to the same propagation distances. This perspective on loss mechanisms is found in Figures 4.12 and 4.13 for the time domain plots of the analytic envelopes of propagated pulses that are compared with the corresponding *mirfs* used for the calculations in Figures 4.8 and 4.9. For the 1-cm distance (Figure 4.12), the *mirf* is narrow, the delays of the peaks remain grouped together, and the original pulse envelopes are relatively unchanged. At a larger distance ($z = 10$ cm), the effects of dispersion alter both the delays of the peaks and the shapes of the envelopes in Figure 4.13.

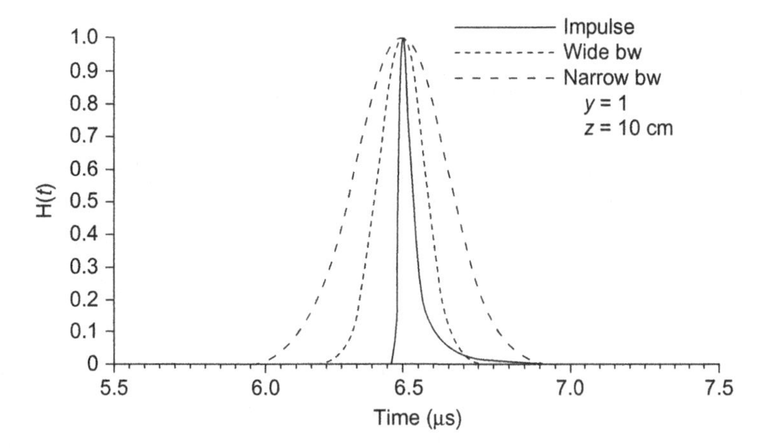

Figure 4.12
(Solid Curve) Material Impulse Response Function (*Mirf*) for $y = 1$, $z = 1$ cm, and an Absorption Coefficient of 0.1151 Neper/MHz-cm (1 dB/MHz-cm); (long dashes) Pulse envelope of a *Mirf* Convolved With Narrowband Gaussian (25% fractional bandwidth) Input Pulse; (short dashes) Pulse Envelope of a *Mirf* Convolved With Wideband Gaussian (50% fractional bandwidth) Input Pulse Curves Normalized to One. *From Szabo (1993).*

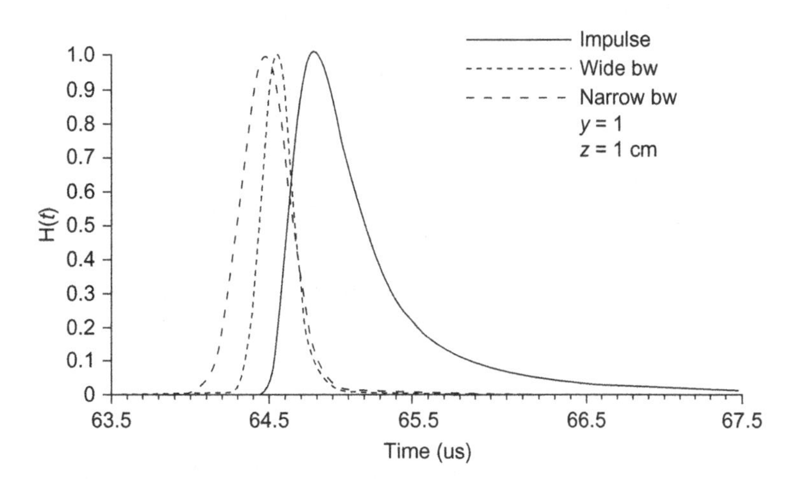

Figure 4.13
(Solid Curve) *Mirf* for $y = 1$, $z = 10$ cm, and an Absorption Coefficient of 0.1151 Neper/MHz-cm (1 dB/MHz-cm); (long dashes) Pulse Envelope of a *Mirf* Convolved With Narrowband Gaussian (25% fractional bandwidth) Input Pulse; (short dashes) Pulse Envelope of a *Mirf* Convolved With Wideband Gaussian (50% fractional bandwidth) Input Pulse Curves Normalized to One. *From Szabo (1993).*

4.4.3 Pulse Echo Propagation

In order to cover the effects of absorption to pulse echoes, both the effect of reflections and the return path must be included. For a round-trip path, the *MTF* for each part of the propagation path is accounted for by multiplying the individual *MTF*s. In general, for paths z_1 and z_2, the exponential nature of the *MTF*s results in an overall *MTF* of $MTF(f,z) = exp[\gamma_T(f)(z_1 + z_2)]$. The corresponding time domain equivalent is the convolution of a *mirf* for z_1 and one for z_2.

To add reflections to a lossy medium, the analysis of the last chapter on reflections and wave propagation can be extended. A simple modification of the ABCD transmission line matrix can be made. For the transmission line matrix for a path length *d*, hyperbolic functions replace the trigonometric ones, and γ replaces *k* in the arguments. Then the matrix elements are:

$$\begin{pmatrix} cosh(\gamma d) & Z_M sinh(\gamma d) \\ \sinh(\gamma d)/Z_M & cosh(\gamma d) \end{pmatrix}, \tag{4.24}$$

where sinh and cosh are the hyperbolic sine and cosine functions of complex argument. With these changes, the matrix approaches for reflections and mode conversion can be combined with losses.

4.5 Modified Hooke's Laws and Tissue Models for Viscoelastic Media

4.5.1 Voigt Model

Another way of describing materials with losses of the type considered so far is that they are viscoelastic. From Chapter 3, elastic constants were sufficient to account for wave behavior in solids and fluids. Hooke's law, which shows that the proportionality of stress to strain holds only for purely elastic media, was described in Chapter 3. Another material property, viscosity, can be added to include the behavior of losses and dispersion. Hooke's law, Eqn 3.41, can be modified by the addition of a second term (Auld, 1990):

$$\boldsymbol{T} = \boldsymbol{C} : \boldsymbol{S} + \eta : \frac{\partial \boldsymbol{S}}{\partial t}, \tag{4.25A}$$

where *C* is an elastic stiffness constant matrix and η is a viscosity stiffness constant matrix of the same size. The Voigt dashpot, or low-frequency version of a thermoviscous model for a single cell, corresponding to this equation is given by Figure 4.14A. The spring on the left of the figure represents the first elastic term of Eqn 4.25A. The dashpot on the right of this figure introduces loss and symbolizes the second term of Eqn 4.25A. The Fourier transform of this modified Hooke's law is:

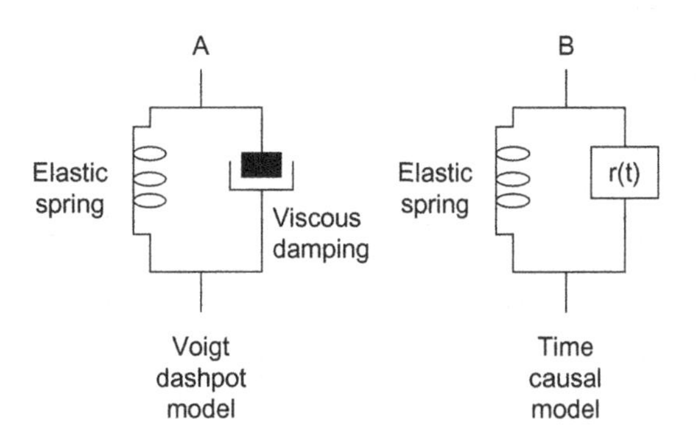

Figure 4.14
Models for Loss in Viscoelastic Materials.
(A) Relaxation or voigt dashpot model; (B) time causal model. *From Szabo and Wu (2000), Acoustical Society of America.*

$$\boldsymbol{T}(f) = \boldsymbol{C} : \boldsymbol{S}(f) + i\omega\boldsymbol{\eta} : \boldsymbol{S}(f). \tag{4.25B}$$

For a single frequency, Eqn 4.28B has led to the concept of Hooke's law with a complex elastic constant, $C + i\omega\eta$ (Auld, 1990).

Equations for absorption and phase velocity (Catheline et al., 2004) are the following:

$$\alpha(\omega) = \sqrt{\frac{\rho\omega^2\sqrt{C^2 + \omega^2\eta^2} - C)}{2(C^2 + \omega^2\eta^2)}}. \tag{4.26A}$$

Even though the absorption dependence on frequency is not a power law, for the low-frequency approximation $(\omega\eta/C)^2 \ll 1$, $\alpha(\omega)$ can be shown be proportional to ω^2,

$$c(\omega) = \sqrt{\frac{2(C^2 + \omega^2\eta^2)}{\rho(C + \sqrt{C^2 + \omega^2\eta^2})}}. \tag{4.26B}$$

Similarly, for the same approximation, the speed of sound approaches a constant value.

4.5.2 Time Causal Model

Another approach that is self-consistent with the time causal model, is depicted in Figure 4.14B, in which the dashpot is replaced by a more general response function, $r(t)$. In this case, the counterpart to Eqn 4.25A is:

$$\boldsymbol{T} = \boldsymbol{C} : \boldsymbol{S} + \boldsymbol{\eta} : r(t) *_t \boldsymbol{S}, \tag{4.27A}$$

and its Fourier transform is:

$$T(f) = C : S(f) + i\eta : R(f)S(f), \tag{4.27B}$$

so that the concept of a complex elastic constant no longer holds except at a single frequency. Eqn 4.27B shows that the imaginary term changes in a more complicated way than that of Eqn 4.25B, as a function of frequency. This second model is the time causal model, which has more general applicability, and Eqn 4.27C reduces to Eqn 4.25B for the case of $y = 2$. Expressions for $r(t)$ can be found in Szabo and Wu (2000). For noninteger and even integer values of y, Eqn 4.27B can be expressed as:

$$T(t) = C : S(t) + \eta : \frac{\partial^{y-1} S(t)}{\partial t^{y-1}}. \tag{4.27C}$$

An explicit form of Eqn 4.27B for the time causal model is:

$$T(f) = C : S(f) + (i\omega)^{y-1}\eta : S(f), \tag{4.27D}$$

for $\omega > 0$.

4.5.3 Maxwell Model

The Maxwell model bears a similarity to the Voigt model in that it is comprised of a dashpot and a spring as shown in Figure 4.15A, except that they are placed in series. The corresponding Hooke's law for this arrangement is the following in the time domain (Catheline et al., 2004; Kumar, Andrews, Jayashankar, Mishra, and Suresh, 2010):

$$C_1 T(t) + \eta_1 \frac{\partial T(t)}{\partial t} = C_1 \eta_1 \frac{\partial S(t)}{\partial t}, \tag{4.28A}$$

and in the frequency domain:

$$(C_1 + \eta_1 i\omega) T(\omega) = i\omega C_1 \eta_1 S(\omega). \tag{4.28B}$$

These results can be related to expressions for absorption and the speed of sound:

$$\alpha(\omega) = \sqrt{\frac{\rho\omega^2\left(\sqrt{1 + \frac{C_1^2}{\omega^2\eta_1^2}} - 1\right)}{2C_1}}, \tag{4.29A}$$

and

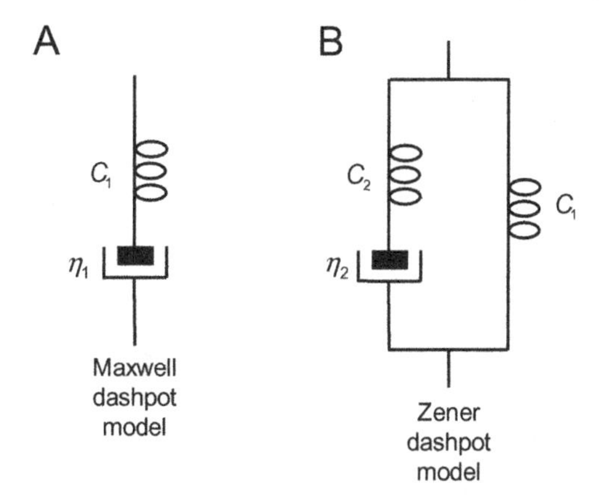

Figure 4.15
Models for Loss in Viscoelastic Materials.
(A) Maxwell model; (B) Zener model.

$$c(\omega) = \sqrt{\frac{2C_1}{\rho\left(\sqrt{1 + \frac{C_1^2}{\omega^2\eta_1^2}} + 1\right)}}, \tag{4.29B}$$

where for $(\omega\eta/C)^2 \ll 1$, these parameters approach constant values.

4.5.4 Thermoviscous Relaxation Model

In the classic thermoviscous model (Blackstock, 2000), Figure 4.14A, the medium is composed of noninteracting molecules that all have an associated sound speed and relaxation constant (τ). For this model, the absorption divided by frequency (α/ω) has the form shown in Figure 4.16A. Also, the excess dispersion divided by frequency (β_E/ω) is plotted in Figure 4.16B, with τ as a relaxation constant. The equations used for calculating the absorption and dispersion in Figure 4.15 are the following:

$$\alpha(\omega) = \alpha_1\omega\left(\frac{1}{\sqrt{1 + (\omega\tau)^2}} - \frac{1}{1 + (\omega\tau)^2}\right), \tag{4.30A}$$

and

$$\beta_E(\omega) = \alpha_1\omega\left(\frac{1}{\sqrt{1 + (\omega\tau)^2}} - \frac{1}{1 + (\omega\tau)^2}\right). \tag{4.30B}$$

In the time domain, the *mirf* can be found from Eqn 4.10 with substitutions from Eqn 4.30.

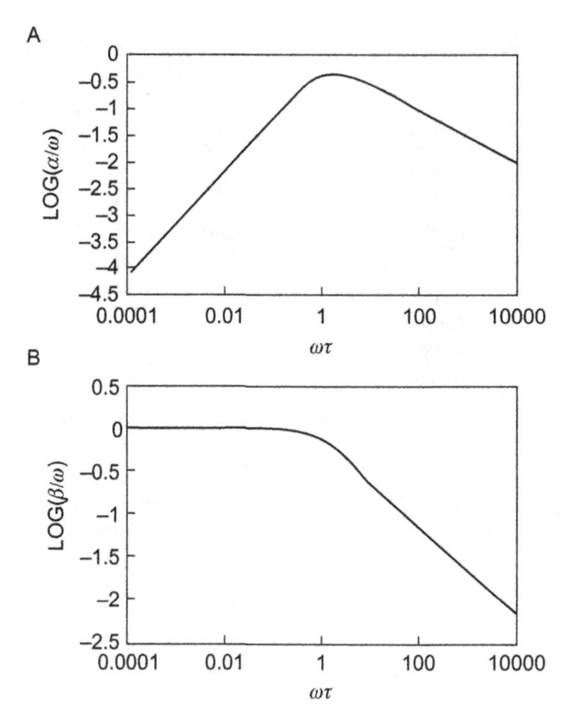

Figure 4.16
(A) Absorption (α) Divided by Angular Frequency (ω) Versus Angular Frequency (ω) Times (τ) for the Thermoviscous Model. Loss Peak at $\omega_p = 1/\tau$, where τ is Relaxation Time. (B) Excess Dispersion (β_E) Divided by Angular Frequency (ω) Versus Angular Frequency (ω) Times τ. *From Szabo and Wu (2000), Acoustical Society of America.*

The graphs for absorption loss for tissues in Figure 4.2 show characteristics that are monotonically increasing because they follow a power law; therefore, they do not have inflection points and negative slopes like those for the relaxation model depicted in Figure 4.16. Because of this fact and the smallness of the loss per wavenumber, $(\alpha/k_0)^2 \ll 1$, a low-frequency approximation of the relaxation model is applied more often to model loss in tissues and other materials. Note that in this region where $\omega\tau < 1$, the absorption depends on the frequency squared and the phase velocity is nearly constant:

$$\alpha \approx \alpha_1 \omega^2, \tag{4.31A}$$

and

$$c(\omega) \approx c_0. \tag{4.31B}$$

4.5.5 Multiple Relaxation Model

One way of overcoming the discrepancy between the thermoviscous model and observed power law absorption characteristics of tissues is to use a fitting procedure to power law data, involving either a superposition or distribution of several relaxation time constants (Bamber, 1986; Nachman, Smith, and Waag, 1990; Wojcik, Mould, Ayter, and Carcione, 1998). This multiple relaxation model corresponds to a tissue model with different, independent, noninteracting molecules, each with its own relaxation constant and associated speed of sound. Typically, two to three relaxation constants are used to fit a measured absorption frequency characteristic for a prescribed frequency range.

As described by Nachman et al. (1990), the multiple relaxation wave equation results in quite complicated expressions for attenuation and phase velocity. At low frequencies, these equations are approximately as follows:

$$\alpha(\omega) \cong \frac{\rho c_0 \omega^2}{2} \sum_{n=1}^{N} \kappa_n \tau_n, \tag{4.32A}$$

and

$$c(\omega) \cong c_0, \tag{4.32B}$$

where constants κ_n and time constants τ_n are associated with each relaxation mode (n). This multiple relaxation model corresponds to the tissue model with different molecules, each with its own relaxation constant and sound speed.

4.5.6 Zener Model

The Zener model, also known as the "standard linear model," can be thought of as a generalization of Maxwell and Voigt models, as illustrated by Figure 4.15B. The stress–strain relations can be written as follows:

$$T(t) + \tau_E \frac{\partial T}{\partial t} = E_0 \left[S(t) + \tau_\sigma \frac{\partial S}{\partial t} \right], \tag{4.33A}$$

where:

$$\tau_E = \frac{\eta}{C_2(C_1 + C_2)}, \tag{4.33B}$$

$$\tau_\sigma = \frac{\eta}{C_2}, \tag{4.33C}$$

and

$$C_0 = \frac{C_1}{C_1 + C_2}. \tag{4.33D}$$

The frequency domain version of this relationship is:

$$T(\omega) + i\omega\tau_E T(\omega) = E_0 S(\omega) + i\omega\tau_\sigma S(\omega). \tag{4.34}$$

4.5.7 Fractional Zener and Kelvin–Voigt Fractional Derivative Models

An extension of the Zener model to include fractional derivatives was proposed by Bagley and Torvik (1983) and Nasholm and Holm (2011):

$$T(t) + \tau_E^b \frac{\partial^b T}{\partial t^b} = E_0\left[S(t) + \tau_\sigma^a \frac{\partial^a S}{\partial t^a}\right]. \tag{4.35}$$

If $\tau_E^b = 0$, the above equation reduces to the Kelvin–Voigt stress–strain relation. The frequency domain version of Eqn 4.35 is:

$$[1 + (i\omega)^b \tau_E^b]T(\omega) = E_0 S(\omega)[1 + (i\omega)^a \tau_\sigma^a]. \tag{4.36}$$

What has become known as the KVFD (Kelvin–Voigt fractional derivative) model is Eqn 4.36 without the $\tau_E^b = 0$ term:

$$T(\omega) = E_0 S(\omega)[1 + (i\omega)^a \tau_\sigma^a]. \tag{4.37A}$$

Note that this equation has the same form as the time causal model for Hooke's law in Section 4.5.2. An alternate form can be obtained from expanding i^a by de Moivre's theorem:

$$T(\omega) = E_0 S(\omega)\{1 + (\omega)^a \tau_\sigma^a[\cos(\pi a/2) + i\,\sin(\pi a/2)]\}. \tag{4.37B}$$

Finally, if the exponents are equal, $b = a$, the absorption and phase velocity can be determined (Nasholm and Holm, 2011):

$$\alpha_Z(\omega) \triangleq -\text{Imag}\left[\frac{\omega}{c_0}\sqrt{\frac{1 + (i\omega\tau_\varepsilon)^b}{1 + (i\omega\tau_\sigma)^b}}\right], \tag{4.38A}$$

and

$$c_Z(\omega) \triangleq -\text{Real}\left\{ c_0\left[\frac{1+(i\omega\tau_\varepsilon)^b}{1+(i\omega\tau_\sigma)^b}\right]^{-1/2}\right\}. \tag{4.38B}$$

At low frequencies, Eqn 4.38A can be approximated by (Nasholm and Holm, 2011):

$$\alpha_Z(\omega) \approx \frac{\omega^{1+b}}{2c_0}(\tau_\sigma^b - \tau_\varepsilon^b)\sin\frac{\pi b}{2}. \tag{4.38C}$$

4.6 Wave Equations for Tissues

4.6.1 Voigt Model Wave Equation

If Eqn 4.25A for the thermoviscous model is applied to the derivation of a wave equation for this type of solid (Auld, 1990), the following equations can be obtained for a plane wave propagating along the z axis of an isotropic solid:

$$\frac{\partial^2 v}{\partial z^2} - \frac{1}{c_0^2}\rho\frac{\partial^2 v}{\partial t^2} + \frac{\eta_{ii}}{C_{ii}}\frac{\partial^3 v}{\partial z^2 \partial t} = 0, \tag{4.39A}$$

where $v = v_m$ is the particle velocity in the direction m such as z, and the speed of sound is:

$$c_0 = \sqrt{C_{ii}/\rho}. \tag{4.39B}$$

Note the addition of a third loss term at the end of the usual wave equation. Here, this type of equation applies equally well to simple cases of elastic longitudinal wave and shear wave propagation in an isotropic medium through the use of the appropriate elastic constants, C_{11} and C_{44}, respectively, or for fluids by the appropriate constants (Szabo, 1994; Szabo and Wu, 2000).

When this equation is converted to the frequency domain and a solution for the particle velocity of the form and direction associated with a specific mode,

$$v(z, \omega) = v_0(0, \omega)\, exp(i\omega t + \gamma z), \tag{4.40}$$

is substituted in Eqn 4.39A, γ can be obtained. Equations 4.30A and 4.30B are the result of this operation. Here $\alpha_1 = \sqrt{\rho/2C_{ii}}$ and $\tau = \eta_{ii}/C_{ii}$.

In order to derive a useful low-frequency approximation to this model, substitute the plane wave approximation,

$$\frac{\partial v}{\partial z} = \frac{1}{c_0}\frac{\partial v}{\partial t}, \tag{4.41}$$

into Eqn 4.30A. The result is:

$$\frac{\partial^2 v}{\partial z^2} - \frac{1}{c_0^2}\frac{\partial^2 v}{\partial t^2} + L_0\frac{\partial^3 v}{\partial t^3} = 0, \tag{4.42A}$$

where:

$$L_0 = \frac{\eta_{ii}}{C_{ii}c_0^2}. \tag{4.42B}$$

This approach results in a quadratic frequency dependence for loss ($\alpha = \alpha_1 f^2$) and a nearly constant sound speed with frequency to first order; therefore, it is quite limited in mimicking the observed behavior of tissues listed in Appendix B and illustrated by Figure 4.2.

4.6.2 Time Causal Model Wave Equations

Are there wave equations with losses that correspond more directly with the frequency power law behavior of absorption for many types of materials? Szabo (1994) and Szabo and Wu (2000) have shown that, in general, the wave equation can be written as:

$$\nabla^2 v - \frac{1}{c_0^2}\frac{\partial^2 v}{\partial t^2} - L_\gamma * v = 0, \tag{4.43}$$

where L_γ is a time convolution propagation operator determined by the constraints of causality in the time domain (explained in Section 4.3.2). Specifically, the form of L_γ depends on whether y is an even or odd integer or a noninteger, so we denote it as $L_\gamma = L_{\alpha,y,t}$. It is helpful to express higher order derivatives in the abbreviated notation $v_{z^n} = \partial^n v/\partial z^n$. For example, if $n = 2$, $v_{zz} = \partial^2 v/\partial.z^2$. For the even-integer case, Eqn 4.43 becomes:

$$\nabla^2 v - \frac{1}{c_0^2} v_{tt} - (-1)^{y/2}\frac{2\alpha_1}{c_0} v_{t^{y+1}} = 0. \tag{4.44}$$

The form of L_γ in this instance of even integers has the specific form:

$$L_{\alpha,y,t} * \upsilon = (-1)^{y/2}\frac{2\alpha_1}{c_0} v_{ty+1} = (-1)^{y/2}\frac{2\alpha_1}{c_0}\delta^{y+1}(t) * \upsilon. \tag{4.45A}$$

For the case of y as an odd integer:

$$L_{\alpha,y,t} * \upsilon = \frac{4h_0}{c_0}\int_{-\infty}^{t}\frac{v(t')}{(t-t')^{y+2}}dt' = \frac{4h_0}{c_0}\frac{H(t)}{t^{y+2}} * \upsilon, \tag{4.45B}$$

in which

$$h_0 = -\alpha_1 (y+1)!(-1)^{(y+1)/2}/\pi, \tag{4.45C}$$

and

$$\alpha_1 = \frac{\eta_{ii}}{2c_0 C_{ii}}. \tag{4.45D}$$

Similarly, for y as a noninteger:

$$L_{\alpha,y,t} * v = \frac{4h_{ni}}{c_0}\int_{-\infty}^{t}\frac{v(t')}{|t-t'|^{y+2}}dt' = \frac{4h_{ni}}{c_0}\frac{H(t)}{|t|^{y+2}} * v, \tag{4.45E}$$

where

$$h_{ni} = \alpha_1 \Gamma(y+2)\sin(\pi y/2)/\pi. \tag{4.45F}$$

More details about the derivation of a wave equation for a viscoelastic solid can be found from equations of the previous forms (Szabo and Wu, 2000). An implementation of the version of these equations for fluids (Szabo, 1994) can be found in Norton and Novarini (2003). They demonstrate that absorption and dispersion can be calculated by these operators directly in the time domain. Kelly, McGough, and Meerschaert (2008) have shown that for the general propagation factor, $\gamma_T(f) = ik(f)$ given a dispersion relation, Eqn 4.15A for the noninteger and even integer case, provides an exact solution to a wave equation like Eqn 4.43 with an extra term:

$$\nabla^2 v - \frac{1}{c_0^2}\frac{\partial^2 v}{\partial t^2} - L_\gamma * v - \frac{\alpha_1^2}{\cos^2(\pi y/2)}\frac{\partial^{2y} v}{\partial t^{2y}} = 0, \tag{4.46}$$

which for small α_1, the usual situation, is an insignificant second-order term compared to the other terms; however, in other circumstances it provides an exact solution. This solution as well as the time causal and related models violate the causality condition for $y > 1$ proposed by Weaver and Pao (1981); however, insofar as practical applications are concerned, the power law seems to hold experimentally over a large frequency range (Szabo and Wu, 2000) for many materials, extending over powers of 10 in frequency. At very high frequencies, the power law is not expected to hold; other physical mechanisms undoubtedly would come into play.

4.6.3 Time Causal Model Wave Equations in Fractional Calculus Form

Fractional calculus has built a considerable body of knowledge and formalism for dealing with problems similar to those encountered here (Mainardi, 2010). The time causal equations can be rewritten in an equivalent fractional derivative form. A fractional derivative operator $\Phi_{-\chi}$ for a positive number $\chi > 0$ and a causal function $f(t)$ that is zero for $t < 0$ can be defined in terms both strange and familiar. A fractional derivative for the case when χ is a positive noninteger can be defined as:

$$\Phi_{-\chi} * f(t) = \left\{\frac{1}{\Gamma(-\chi)}\int_{0-}^{t+}\frac{f(\tau)}{(t-\tau)^{1+\chi}}d\tau\right\}. \tag{4.47A}$$

And when $\chi = m$, a positive integer, it has a derivative form:

$$\Phi_{-m} * f(t) = \left\{ (-1)^m \frac{d^m f}{dt^m} \right\}. \tag{4.47B}$$

By changing the lower limit of integration through incorporating the step function $H(t)$ inherent in the time convolution propagation operators defined in the previous section, we can restate these operators in terms of fractional derivatives. First, for the even-integer case, where $y = 2m$ and $m = 1,2,\ldots,$

$$L_{\alpha,y,t} * v = (-1)^{-y/2-1} \frac{2\alpha_1}{c_0} \Phi_{-y-1} * v = (-1)^{-m-1} \frac{2\alpha_1}{c_0} \Phi_{-2m-1} * v. \tag{4.48}$$

Second, for the noninteger case:

$$L_{\alpha,y,t} * v = \frac{4h_{ni}}{c_0} \Gamma(-y-1) \Phi_{-y-1} * v. \tag{4.49}$$

Third, the odd-integer case poses some mathematical difficulties and does not fit the fractional derivative formalism except in a limiting sense. As y approaches 1 (or an odd integer value), $h_{ni} \rightarrow h_0$ in value. Under this condition, the time convolution propagation operator approximates the noninteger case given by Eqn 4.49. Several groups have recognized that the integral in Eqn 4.47A is divergent because it is not locally absolutely integrable (Mainardi, 2010). Through a process of regularization, alternate forms more suitable for numerical evaluation have been devised (Mainardi, 2010), such as Riemann–Liouville or Caputo fractional time derivatives.

4.7 Discussion

4.7.1 First Principles

The question of why the power law describes the viscoelastic behavior of many different materials and tissues (Szabo and Wu, 2000) naturally arises. Stamenovic (2008) argued that a weak power law is present in cellular movement because of its tensegrity structure. Suki, Barabasi, and Lutchen (1994) discussed the power law behavior of lung tissue based on macromolecular interactions. Chen (2008) traced the power law back to quantum mechanical principles. Buckingham (1997, 2005a, 2005b) has derived wave equations that show absorption is linear with frequency for sediments.

Kelly and McGough (2009) proposed a hierarchical fractal ladder network model based on tissue, cellular, and subcellular structures on different scales for the viscoelastic behavior of soft tissue. Using a low-frequency approximation, they found an absorption power law, dispersion relations, and fractional derivatives with numerical results equivalent to those of the time causal model.

4.7.2 Power Law Wave Equation Implementations

Considerable work has been done either implementing or deriving power law wave equations. Wismer and Ludwig (1995) and Wismer (2006) were among the first to implement power law in a explicit finite element method with fractional time derivatives based on previous time steps. Norton and Novarini (2003, 2004) have had considerable success in numerically implementing the time causal equations in a finite time domain finite difference (FTFD) scheme for two dimensional (2D) spatially varying media and realistic sources and scattering for both linear and nonlinear cases. In their approach, they use a local causal time domain propagation operator based on Eqn 4.17. Chen and Holm (2003) pursued a space fractional time derivative using a modified positive Caputo (1967) integral. Chen and Holm (2004) developed a regularized space-fractional Laplacian operation that predicts power law behavior, and also extended their approach to the nonlinear Burgers equation. Liebler, Ginter, Dreyer, and Riedlinger (2004) introduced a new recursive efficient numerical time domain implementation of the time causal equations. Through the use of a splitting FTFD technique, they obtained excellent agreement with broadband data for both linear and nonlinear cases.

Treeby and Cox (2010) developed a numerically efficient algorithm based on a wave equation with a lossy fractional Laplacian. They demonstrated that their MATLAB implementation predicted power law behavior and the *mirf* time sequence of Figure 4.4. Nasholm and Holm (2011) showed that a continuum of relaxation mechanisms, each with an effective compressibility distribution, leads to a fractional derivative wave equation.

4.7.3 Transient Solutions for Power Law Media

The search for time domain solutions for power law media began with the thermoviscous wave equation (4.39A). Blackstock (1967) derived an analytic solution for a tone-burst excitation that showed the existence of a precursor and a postcursor. This work inspired the author's decades-long interest in solving equations of this type (Szabo, 1969). Nicoletti and Anderson (1997) explored attenuation and power law distributions. Following in this tradition, Cobbold, Sushilov, and Weathermon (2004) examined transient solutions to finite-length sinusoids to examine dispersive effects and obtained exact solutions for some power law cases. Buckingham (2008) also contributed to the understanding of transient solutions to classic wave equations.

For waves in power law media described by fractional derivative wave equations, Section 4.4 provided a time domain perspective of how losses affect pulse propagation through *mirfs* (material impulse response functions). As these functions propagate, their amplitude falls and they lengthen, and yet their area remains relatively constant.

4.7.4 Green Functions for Power Law Media

A next step in understanding more realistic sound sources is to generate a Green's function, which describes how a point source radiates in power law media. The approach taken by Kelly et al. (2008) is to begin with the power law propagation factor of the form in Eqn 4.7B, power law absorption as in Eqn 4.6A in which $\alpha_0 = 0$, and dispersion factor as in Eqn 4.15A. The propagation factor $\gamma_T(f)$ is inserted in a nonhomogeneous Helmholtz wave equation the Fourier transform of which yields:

$$\nabla^2 v - \frac{1}{c_0^2}\frac{\partial^2 v}{\partial t^2} - \frac{2\alpha_1}{c_0 \cos(\pi y/2)}\Phi_{-(y+1)} v - \frac{\alpha_1^2}{c_0 \cos^2(\pi y/2)}\Phi_{-2y} v = -\delta(R)\delta(t), \qquad (4.50A)$$

with an impulse source function on the right-hand side. The part of the equation in brackets below can be recognized as Eqn 4.35 with small extra term and a point source function,

$$\left[\nabla^2 v - \frac{1}{c_0^2}\frac{\partial^2 v}{\partial t^2} - L_\gamma * v\right] - \frac{\alpha_1^2}{c_0 \cos^2(\pi y/2)}\Phi_{-2y} v = -\delta(R)\delta(t). \qquad (4.50B)$$

In this form, it is not necessary to make the assumption that α_1 is small. The extra fractional derivative term of order $2y$ is small and was neglected in the original derivation and approximation for Eqn 4.43, and in this form becomes:

$$\left[\nabla^2 v - \frac{1}{c_0^2}\frac{\partial^2 v}{\partial t^2} - L_\gamma * v\right] = -\delta(R)\delta(t). \qquad (4.50C)$$

Furthermore, the exact solution of Eqn 4.50B is a Green's function:

$$G(R,f) = e^{\gamma_T(f)R}/4\pi R, \qquad (4.51A)$$

which is also an approximate solution to Eqn 4.50C. The numerator is the material transfer function, Eqn 4.7A. The Fourier transform of the Green's function, Eqn 4.51A, is:

$$g(r,t) = \left[\frac{\delta(t - R/c_0)}{4\pi R}\right] * mirf(t + R/c_0, R). \qquad (4.51B)$$

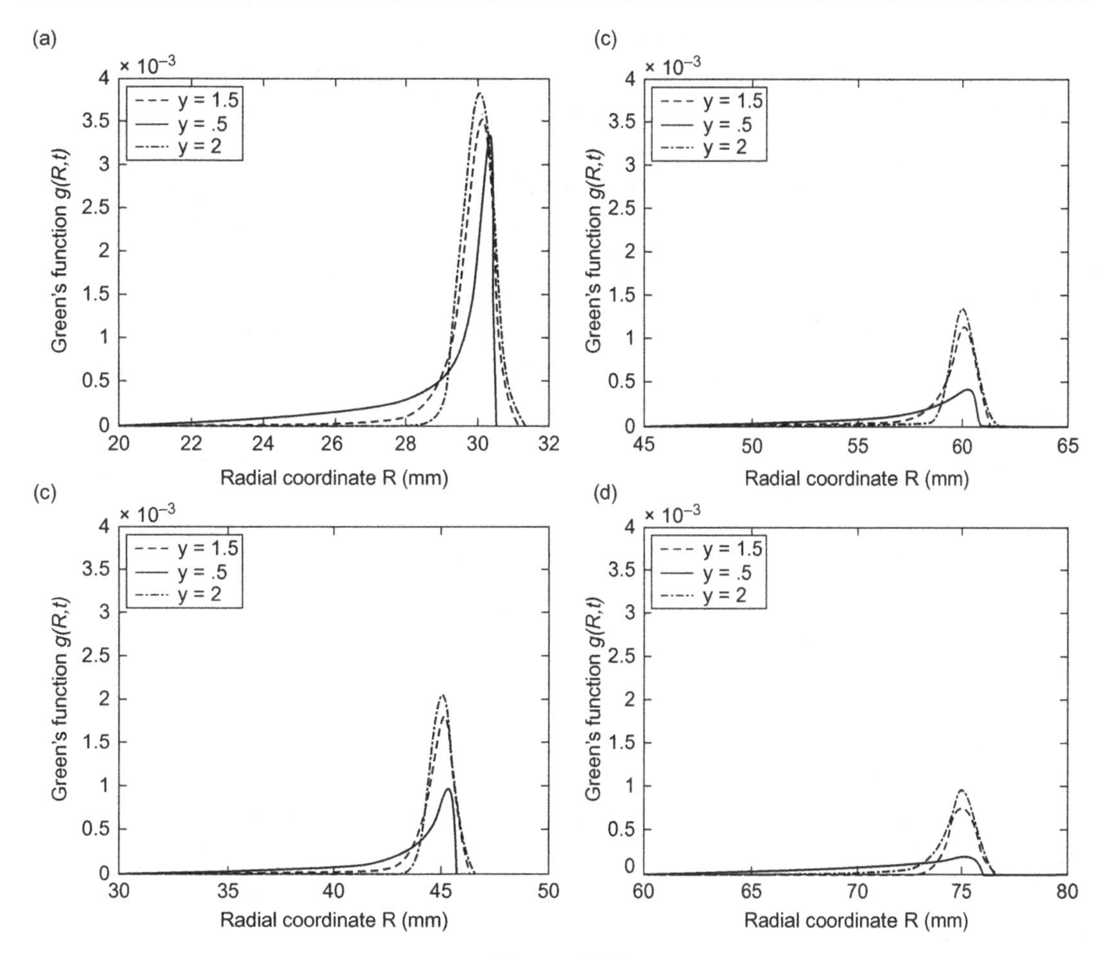

Figure 4.17

Snapshots of the Three-dimensional Power-law Green's Function for $y = 0.5$, 1.5, and 2.0 for $\alpha_1 = 0.05\ \text{mm}^{-1}\ \text{MHz-}y$.

Snapshots of the Green's function are shown for $t = 20$, 30, 40, and 50 s. *From Kelly et al. (2008), Acoustical Society of America.*

in which $mirf(t + R/c_0, R)$is simply the *mirf* function without the delay term, which is now contained in the bracketed term. This *mirf* function can be expressed as:

$$mirf(t + R/c_0, R) = \Im^{-1}\{\exp[-\alpha_1(f)R - i\beta_E(f)R]\}. \tag{4.51C}$$

Kelly et al. (2008) have derived explicit analytic expressions of time domain functions corresponding to the inverse Fourier transform in Eqn 4.51B and different values of the exponent y. A scaling law similar to Eqn 4.19 holds for these functions. By exploring these Green's functions, Kelly et al. (2008) have found *mirf*-like functions when plotted against time (Figure 4.4); however, when they plotted them for different propagation times and values of y as in Figure 4.17, these plots revealed wake-like characteristics. For the Gaussian cases ($y = 2$), the waveforms widen and drop in amplitude as expected (see Section 4.4.1) and remain confined in radial extent. Wavefronts for other values of y have a decaying wake following the primary higher amplitude. This effect can be explained as the slow relaxation of the power law medium behind the main initial wavefront. These phenomena could result in smearing effects on images or the lengthening of axial time responses. Particular effects would depend on the driving pulse shape (Section 4.4.2). Note that the approach of Kelly et al. (2008) shares the same limitation as the time causal model in that it does not hold for extremely high frequencies, but these frequencies are usually well beyond the frequencies of interest as discussed in Section 4.3.2.

4.7.5 Shear Waves in Power Law Media

Despite widespread diagnostic imaging employing longitudinal or compressional waves, recent interest has been focused on the application of shear waves to the characterization and imaging of tissues for diagnosis. One motivation has been the greater dynamic range of shear elastic moduli (Sarvazyan, Rudenko, Swanson, Fowlkes, & Emelianov, 1998), several orders of magnitude compared to the narrow range of less than one order of magnitude for longitudinal waves utilized in conventional diagnostic imaging. In principle, the shear wave equations can be considered to be decoupled from longitudinal waves so that their form is similar, as explained in Section 4.4. Szabo and Wu (2000) have provided experimental confirmation of the power law for shear waves in plastics at megahertz frequencies. Tissues, however, appear to have different shear wave behavior.

In order to measure or image shear properties of tissue, several ingenious methods have been devised which depart considerably from usual longitudinal wave imaging, both in frequency and approach, such as elastography, which is described in more depth in Chapter 16. Shear wave imaging and elastography include methods that involve frequencies below a hundred Hz and shear wave speeds on the order of meters/second, several orders of magnitude from the values in conventional imaging. Waves can be induced by a variety of means including mode conversion, vibrating plates and rods, and acoustic radiation forces.

For these reasons, investigators have employed several tissue models from Section 4.5. Some have found the Maxwell, Voigt, and standard models and their combinations useful in matching data (Brigham, Aquino, Mitri, Greenleaf, & Fatemi, 2007; Catheline et al., 2004; Deffieux et al., 2009). There has been a shift towards viscoelastic power law models, and in many cases the KVFD model has agreed with the functional dependence of data (Kiss, Varghese, & Hall, 2004; Kumar et al., 2010; Zhang et al., 2008). Holm and Sinkus (2010) have extended the range of the KVFD approach through an analysis of both the low- and

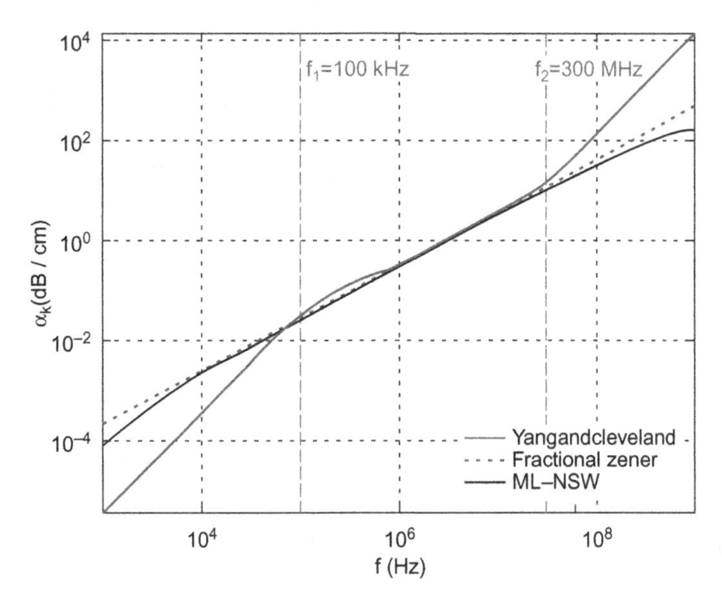

Figure 4.18
Comparison of Resulting Attenuation Modeled by Two Relaxation Constants in Yang and Cleveland (2005), Fractional Zener Model (Dashed Line), and the Approximate Multiple Relaxation Continuum Model (Thin Solid Line). *From Nasholm and Holm (2011), Acoustical Society of America.*

high-frequency approximations based on first principle arguments. Nasholm and Holm (2011) showed that a power law model could be approximated by a continuum of relaxation constants. Examples of model-fitting are shown in Figure 4.18. The fitting of data to the various models of Section 4.5 over a limited range is no guarantee of the validity of the models. For example, what is the physical meaning of relaxation constants in the context of soft tissue? Caution must be taken when transferring fitted results to the time domain. Other nonphysical features of several of the models, such as the speed of sound and derivatives growing at very high frequencies, tend to be ignored if the models satisfactorily describe observed phenomena within the range of interest. Holm and Nasholm (2011) extended their earlier results and attempted to bridge the gap between shear and longitudinal waves and to achieve reasonable high-frequency finiteness for power law media. Their fractional derivative unified Zener model predicts different behavior for low-, intermediate-, and high-frequency regimes.

Another consideration in analyzing wave data is the geometry of the source and measurement region. For shear wave spectroscopy, Deffieux et al. (2009) found that wavefronts could be modeled as expanding cylindrical waves propagating in viscoelastic media. In their study of longitudinal waves in tissue phantom blocks, Baghani, Eskandari, Salcudean, and Rohling (2009) used a viscoelastic model and applied boundary conditions to more accurately predict wave distributions compared to independent rheology measurements.

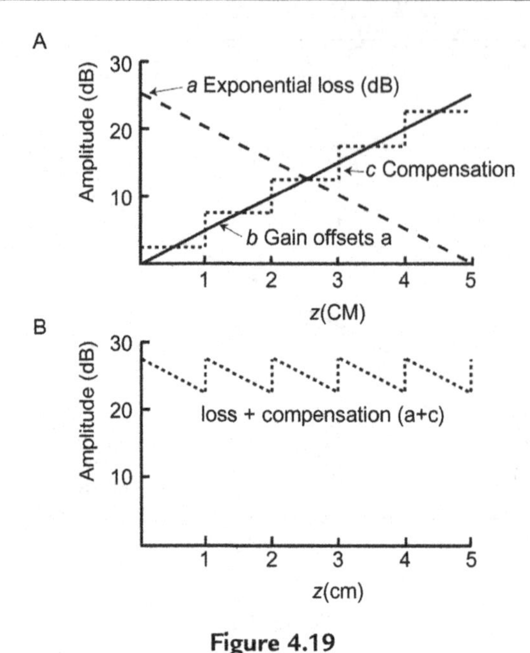

Figure 4.19

(A) Exponential Decay of Round-trip Loss With Depth in dB and (*b*) Gain to Offset Loss in dB. Gain is Approximated by (*c*) Stepwise Five TGC Zones With Depth for Compensation. (B) Net Effect of Compensation (*c*) and Absorption (*a*) on Background Level in Image.

4.8 Penetration and Time Gain Compensation

In order to compensate for the effects of absorption and focusing, imaging systems have a method called time gain compensation built in as a set of controls. The depth dimension of the image is divided into horizontal (linear format) or radial (sector format or curved linear array format) strips, each of which is connected to a separate amplifier stage with a variable gain. These gain controls can be adjusted manually to boost the gain independently in each strip zone. The net effect of these gains is that they provide a means to increase gain with depth in a stepwise manner in order to offset the effects of absorption. Adjustments in time gain compensation (TGC) are made to approximate a uniform background level throughout the field of view (illustrated in Figure 4.19). For this example, the overall absorption as a function of depth (z) is divided into four zones, each of which has an amplification gain adjusted to offset the average loss in the zone.

The penetration depth of an imaging system can be determined from a knowledge of the effective dynamic range of the system and the loss in a phantom or body, as well as from the fact that round-trip absorption decays as $exp(-\alpha 2z)$. For example, if the dynamic range

(DR) of the system was 100 dB, and the one-way loss was 5 dB/cm at 5 MHz, then if $DR = \alpha_{\text{dB}} 2z$, the penetration depth is $z = 10$ cm.

References

Auld, B. A. (1990). *Acoustic fields and waves in solids 2* (Vol. 1). Malabar, FL: Krieger Publishing.

Baghani, A., Eskandari, H., Salcudean, S., & Rohling, R. (2009). Measurement of viscoelastic properties of tissue-mimicking material using longitudinal wave excitation. *IEEE Transactions on Ultrasonics, Ferroelectrics and Frequency Control, 56*, 1405–1418.

Bagley, R. L., & Torvik, P. J. (1983). A theoretical basis for the application of fractional calculus to viscoelasticity. *Journal of Rheology, 27*, 201–210.

Bamber, J. C. (1986). In C. R. Hill (Ed.), *Physical principles of medical ultrasonics* (pp. 118–199). Chichester, UK: John Wiley & Sons.

Bamber, J. C. (1998). Ultrasonic properties of tissue. In F. A. Duck, A. C. Baker, & H. C. Starritt (Eds.), *Ultrasound in medicine*. Bristol, UK: Institute of Physics Publishing.

Blackstock, D. T. (1967). Transient sound radiated into a viscous fluid. *Journal of the Acoustical Society of America, 41*, 1312–1319.

Blackstock, D. T. (2000). *Fundamentals of physical acoustics*. New York: John Wiley & Sons.

Bracewell, R. (2000). *The fourier transform and its applications*. New York: McGraw-Hill.

Brigham, J. C., Aquino, W., Mitri, F. G., Greenleaf, J. F., & Fatemi, M. (2007). Inverse estimation of viscoelastic material properties for solids immersed in fluids using vibroacoustic techniques. *Journal of Applied Physics, 101*, 023509.

Buckingham, M. J. (1997). Theory of acoustic attenuation, dispersion, and pulse propagation in unconsolidated granular materials including marine sediments. *Journal of the Acoustical Society of America, 102*, 2579–2596.

Buckingham, M. J. (2005a). Compressional and shear wave properties of marine sediments: Comparisons between theory and data. *Journal of the Acoustical Society of America, 117*, 137–152.

Buckingham, M. J. (2005b). Causality, Stokes' wave equation, and acoustic pulse propagation in a viscous fluid. *Physical Review E, 72*. 026610 _2005_

Buckingham, M. J. (2008). On the transient solutions of three acoustic wave equations: van Wijngaarden's equation, Stokes' equation and the time-dependent diffusion equation. *Journal of the Acoustical Society of America, 124*, 1909–1920.

Caputo, M. (1967). Linear models of dissipation whose Q is almost frequency-independent, II. *Geophysical Journal of the Royal Astronomical Society, 13*, 529–539.

Catheline, S., Gennissson, J. -L., Delon, G., Fink, M., Sinkus, R., Abouelkaram, S., & Cuiloli, J. (2004). Measurement of viscoelastic properties of homogeneous soft solid using transient elastography: An inverse problem approach. *Journal of the Acoustical Society of America, 116*, 3734–3741.

Chen, W. (2008). An intuitive study of fractional derivative modeling and fractional quantum in soft matter. *Journal of Vibration and Control, 14*, 1651–1657.

Chen, W., & Holm, S. (2003). Modified Szabo's wave equation models for lossy media obeying frequency power law. *Journal of the Acoustical Society of America, 114*, 2570–2574.

Chen, W., & Holm, S. (2004). Fractional Laplacian time–space models for linear and nonlinear lossy media exhibiting arbitrary frequency power-law dependency. *Journal of the Acoustical Society of America, 115*, 1424–1430.

Cobbold, R. S. C., Sushilov, N. V., & Weathermon, A. C. (2004). Transient propagation in media with classical or power-law loss. *Journal of the Acoustical Society of America, 116*, 3294–3303.

Duck, F. A. (1990). *Physical properties of tissue*. New York: Academic Press.

He, P. (1998a). Simulation of ultrasound pulse propagation in lossy media obeying a frequency power law. *IEEE Transactions on Ultrasonics, Ferroelectrics and Frequency Control, 45*, 114–125.

He, P. (1998b). Determination of ultrasonic parameters based on attenuation and dispersion measurements. *Ultrasonic Imaging, 20*, 275–287.

He, P. (1999). Experimental verification of models for determining dispersion from attenuation. *IEEE Transactions on Ultrasonics, Ferroelectrics and Frequency Control, 46*, 706–714.

Holm, S., & Nasholm, S. P. (2011). A causal and fractional all-frequency wave equation for lossy media. *Journal of the Acoustical Society of America, 130*, 2195–2202.

Holm, S., & Sinkus, R. (2010). A unifying fractional wave equation for compressional and shear waves. *Journal of the Acoustical Society of America, 127*, 542–548.

Kelly, J. F., & McGough, R. J. (2009). Fractal ladder models and power law wave equations. *Journal of the Acoustical Society of America, 126*, 2072–2081.

Kelly, J. F., McGough, R. J., & Meerschaert, M. M. (2008). Analytical time domain Green's functions for power-law media. *Journal of the Acoustical Society of America, 124*, 2861–2872.

Kinsler, L. E., Frey, A. R., Coppens, A. B., & Sanders, J. V. (2000). *Fundamentals of acoustics*. New York: John Wiley & Sons.

Kiss, M. Z., Varghese, T., & Hall, T. J. (2004). Viscoelastic characterization of in vitro canine tissue. *Physics in Medicine and Biology, 49*, 4207–4218.

Kronig, R. D. L. (1926). On the theory of dispersion of x-rays. *Journal of the Optical Society of America, 12*, 547–557.

Kumar, K., Andrews, M. E., Jayashankar, V., Mishra, A. K., & Suresh, S. G. (2010). Measurement of viscoelastic properties of polyacrylamide-based tissue-mimicking phantoms for ultrasound elastography applications. *IEEE Transactions on Instrumentation and Measurement, 59*, 1224–1232.

Liebler, M., Ginter, S., Dreyer, T., & Riedlinger, R. E. (2004). Full wave modeling of therapeutic ultrasound: Efficient time-domain implementation of the frequency power-law attenuation. *Journal of the Acoustical Society of America, 116*, 2742–2750.

Nachman, A. I., Smith, J. F., III, & Waag, R. C. (1990). An equation for acoustic propagation in inhomogeneous media with relaxation losses. *Journal of the Acoustical Society of America, 88*, 1584–1595.

Nasholm, S. P., & Holm, S. (2011). Linking multiple relaxation, power-law attenuation, and fractional wave equations. *Journal of the Acoustical Society of America, 130*, 3038–3045.

Näsholm, S. P. and Holm, S.(2013) On a fractional zener elastic wave equation, Fract. Calc. Appl. Anal. Vol. 16, No 1.

Nicoletti, D., & Anderson, A. (1997). Determination of grain-size distribution from ultrasonic attenuation: Transformation and inversion. *Journal of the Acoustical Society of America, 101*, 686–689.

Norton, G. V, & Novarini, J. G. (2003). Including dispersion and attenuation directly in the time domain for wave propagation in isotropic media. *Journal of the Acoustical Society of America, 113*, 3024–3031.

Norton, G. V., & Novarini, J. C. (2004). Including dispersion and attenuation in the time domain modeling of pulse propagation in spatially-varying media. *Journal of Computational Acoustics, 12*, 501–519.

Ochmann, M, & Makarov, S. (1993). Representation of the absorption of nonlinear waves by fractional derivatives. *Journal of the Acoustical Society of America, 94*, 3392–3399.

Sarvazyan, A. P., Rudenko, O. V., Swanson, S. D., Fowlkes, J. B., & Emelianov, S. Y. (1998). Shear wave elasticity imaging: A new ultrasonic technology of medical diagnostics. *Ultrasound in Medicine and Biology, 24*, 1419–1435.

Stamenovic, D. (2008). Rheological behavior of mammalian cells. *Cellular and Molecular Life Sciences, 65*, 3592–3605.

Suki, B., Barabasi, A. L., & Lutchen, K. R. (1994). Lung tissue viscoelasticity: A mathematical framework and its molecular basis. *Journal of Applied Physiology, 76*, 2749–2759.

Szabo, T. L. (1969). Lumped-element transmission-line analog of sound in a viscous medium. *Journal of the Acoustical Society of America, 45*, 124–130.

Szabo, T. L. (1993). Linear and Nonlinear Acoustic Propagation in Lossy Media, Ph.D. thesis. University of Bath, UK.

Szabo, T. L. (1994). Time domain wave equations for lossy media obeying a frequency power law. *Journal of the Acoustical Society of America, 96*, 491–500.

Szabo, T. L. (1995). Causal theories and data for acoustic attenuation obeying a frequency power law. *Journal of the Acoustical Society of America, 97*, 14–24.

Szabo, T. L., & Wu, J. (2000). A model for longitudinal and shear wave propagation in viscoelastic media. *Journal of the Acoustical Society of America, 107*, 2437–2446.

Treeby, B., & Cox, B. (2010). Modeling power law absorption and dispersion for acoustic propagation using the fractional Laplacian. *Journal of the Acoustical Society of America, 127*, 2741–2748.

Trousil, R. L., Waters, K. R., & Miller, J. G. (2001). Experimental validation of the use of Kramers–Kronig relations to eliminate the phase sheet ambiguity in broadband phase spectroscopy. *Journal of the Acoustical Society of America, 109*, 2236–2244.

Waters, K. R., Hughes, M. S., Brandenburger, G. H., & Miller, J. G. (2000a). On a time-domain representation of the Kramers–Kronig dispersion relations. *Journal of the Acoustical Society of America, 108*, 2114–2119.

Waters, K. R., Hughes, M. S., Mobley, J., Brandenburger, G. H., & Miller, J. G. (2000b). On the applicability of Kramers–Kronig relations for ultrasonic attenuation obeying a frequency power law. *Journal of the Acoustical Society of America, 108*, 556–563.

Weaver, R. L., & Pao, Y. -H. (1981). Dispersion relations for linear wave propagation in homogeneous and inhomogeneous media. *Journal of Mathematical Physics, 22*, 1909–1918.

Wismer, M. (2006). Finite element analysis of broadband acoustic pulses through inhomogenous media with power law attenuation. *Journal of the Acoustical Society of America, 120*, 3493–3502.

Wismer, M., & Ludwig, R. (1995). An explicit numerical time domain formulation to simulate pulsed pressure waves in viscous fluids exhibiting arbitrary frequency power law attenuation. *IEEE Transactions on Ultrasonics, Ferroelectrics and Frequency Control, 42*, 1040–1049.

Wojcik, G. L., Mould Jr. J. C., Ayter, S., and Carcione, L. M. (1998). A study of second harmonic generation by focused medical transducer pulses. Proceedings of the IEEE Ultrasonics Symposium, 1998, Sendai, 2, 1583–1588.

Yang, X., & Cleveland, R. O. (2005). Time domain simulation of nonlinear acoustic beams generated by rectangular pistons with application to harmonic imaging. *Journal of the Acoustical Society of America, 117*, 113–123.

Zhang, M., Nigwekar, P., Castaneda, B., Hoyt, K., Joseph, J. V., Di Sant Agnese, A., Messing, E. M., Strang, J. G., Rubens, D. J., & Parker, K. J. (2008). Quantitative characterization of viscoelastic properties of human prostate correlated with histology. *Ultrasound in Medicine and Biology, 34*, 1033–1042.

CHAPTER 5

Transducers

Chapter Outline

Diagnostic Ultrasound Imaging: Inside Out.

5.1 Introduction to Transducers

The one indispensable part of a diagnostic imaging system is the transducer. Transducers come in many shapes, frequencies, and sizes. Specific forms of transducers for different clinical applications and scanning methods will be covered in Chapter 10. This chapter concentrates on the essential questions:

How do transducers work?
What are they made of?
How can piezoelectric materials for transducers be compared?
What are the characteristics of transducers?
How can transducers be modeled and designed?
How are they constructed?
What are promising new developments for transducers of the future?

5.1.1 Transducer Basics

This section presents a basic intuitive model of a piezoelectric transducer to describe its essential acoustic and electrical characteristics. The simplest transducer is a piece of piezoelectric material with electrodes on the top and bottom (depicted in Figure 5.1). Unlike the drawing at the top of this figure, the top has a cross-sectional area (*A*) and sides that are much longer (>10×) than the thickness (*d*). Piezoelectric material is dielectric; therefore, it has a clamped capacitance:

$$C_0 = \varepsilon^S A/d, \tag{5.1}$$

in which ε^S is a clamped dielectric constant under the condition of zero strain. Because of piezoelectricity, the Hooke's law for this capacitor has an extra term:

$$T = C^D S - hD, \tag{5.2}$$

in which h is a piezoelectric constant. The elastic stiffness constant C^D is obtained under a constant dielectric displacement D, and E is the electric field,

$$D = \varepsilon^S E = \frac{\varepsilon^S AV}{dA} = C_0 V/A. \tag{5.3}$$

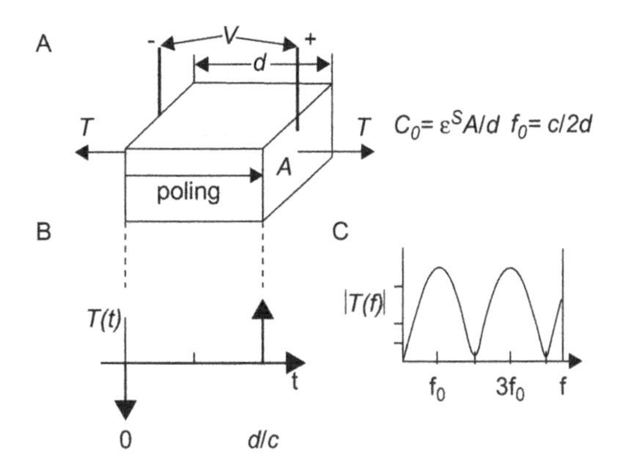

Figure 5.1
A simple transducer model.
(A) Diagram for a thickness expander piezoelectric crystal radiating into a medium matched to its impedance; (B) stress time response; (C) stress frequency response.

If a voltage impulse is applied across the electrodes, the piezoelectric effect creates forces at the top and bottom of the transducer, given by:

$$F(t) = TA = (hC_0V/2)[-\delta(t) + \delta(t - d/c)], \tag{5.4}$$

for which we have assumed that the media outside the electrodes has the same acoustic impedance as the transducer (see Figure 5.1A). To obtain the spectrum of this response, take the Fourier transform of Eqn 5.4:

$$F(f) = -i(hC_0V)e^{-i\pi fd/c}\sin[\pi(2n+1)f/2f_0], \tag{5.5}$$

an expression with maxima at odd harmonics (note that $n = 1, 2, 3$, etc.) of the fundamental resonance $f_0 = c/2d$ (shown in Figure 5.1C). The speed of sound between the electrodes is given by $c = \sqrt{C^D/\rho}$.

5.1.2 Transducer Electrical Impedance

Because of the forces generated by the transducer, the electrical impedance looking through the voltage terminals is affected. A radiation impedance, Z_A, is added to the capacitive reactance so that an equivalent circuit (see Figure 5.2A) for the overall electrical impedance is:

$$Z_T = Z_A - i(1/\omega C_0) = R_A(f) + i[X_A(f) - 1/\omega C_0]. \tag{5.6}$$

Here, Z_A is radiation impedance, of which R_A and X_A are its real and imaginary parts.

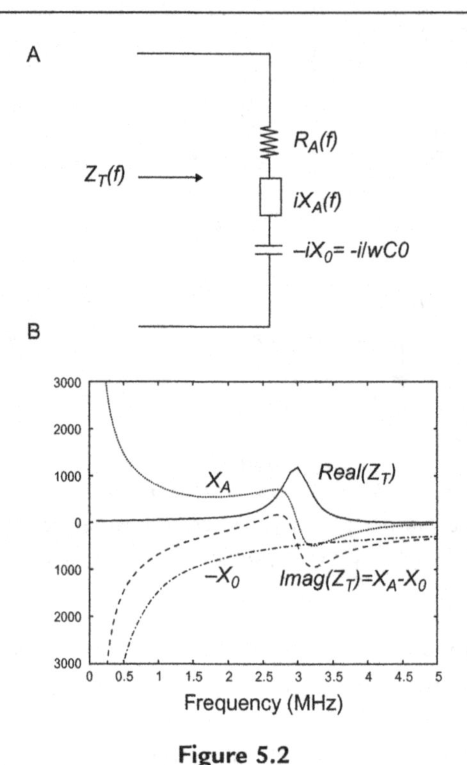

Figure 5.2
(A) Transducer Equivalent Circuit; (B) Transducer Impedance as a Function of Frequency.

What is R_A? To first order, it can be found from the total real electrical power flowing into the transducer for an applied voltage (V) and current (I):

$$W_E = II^* R_A/2 = I^2 R_A/2, \tag{5.7A}$$

where current is $I = i\omega Q = i\omega C_0 V$. The total power radiated from both sides of the transducer into a surrounding medium of specific acoustic impedance, $Z_C = \rho c A$ (equal to that of the crystal), is:

$$W_A = ATT^*/(2Z_C/A) = A^2|F(f)/A|^2/2Z_C = |hC_0V\ \sin(\pi f/2f_0)|^2 2Z_C. \tag{5.7B}$$

Setting the powers of Eqns 5.7A and 5.7B equal, we can solve for R_A:

$$R_A(f) = R_{AC}\ \ \text{sinc}^2(f/2f_0), \tag{5.8A}$$

where $\text{sinc}(x) = \sin(\pi x)/(\pi x)$ and

$$R_{AC} = \frac{k_T^2}{4f_0C_0} = \frac{d^2k_T^2}{2A\varepsilon^S}. \tag{5.8B}$$

The electroacoustic coupling constant is k_T, and $k_T = h/\sqrt{C^D/\varepsilon^S}$. Interesting properties of R_{AC} include an inverse proportionality to the capacitance and area of the transducer and a direct dependence on the square of the thickness (d). Note that at resonance:

$$R_A(f_0) = \frac{k_T^2}{\pi^2 f_0 C_0}. \tag{5.8C}$$

Network theory requires that the imaginary part of an impedance be related to the real part through a Hilbert transform (Nalamwar & Epstein, 1972) (Appendix A), so the radiation reactance can be found as:

$$X_A(f) = \Im_{Hi}[R_A(f)] = R_{AC}\frac{[\sin(\pi f/f_0) - \pi f/f_0]}{2(\pi f/2f_0)^2}. \tag{5.8D}$$

The transducer impedance is plotted as a function of frequency in Figure 5.2B. Here, R_A is maximum at the center frequency, where X_A is zero.

Because both R_A and X_A are complicated functions of frequency, it is useful to understand them in terms of a purely electrical lumped element equivalent circuit (Kino, 1987; Mason, 1964). This circuit mimics the electrical impedance characteristic of a simple transducer with acoustic loads Z_1 and Z_2 on the electroded faces of the piezoelectric material. The lumped element circuit with a series resonance ω_1 and a parallel antiresonance ω_0 is shown in Figure 5.3. The values of these parameters are given as follows:

$$R_S = \frac{\pi(Z_1 + Z_2)}{4k_T^2\omega_1 C_0 Z_C}, \tag{5.9A}$$

$$L_S = \frac{1}{\omega_1^2 C_S}, \quad \text{and} \tag{5.9B}$$

$$C_S = \frac{8C_0 k_T^2/\pi^2}{1 - 8k_T^2/\pi^2}. \tag{5.9C}$$

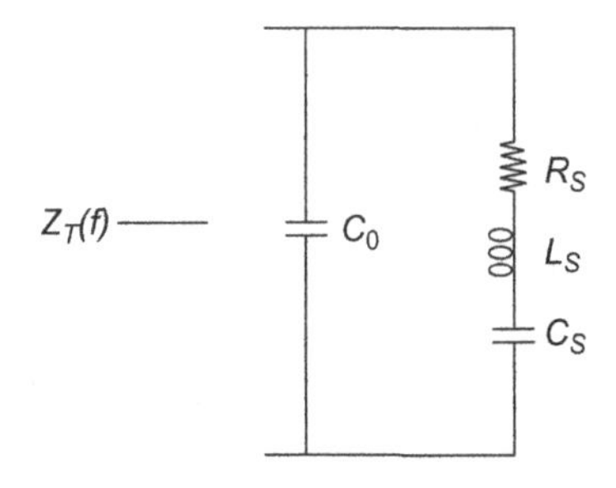

Figure 5.3
Electrical Lumped Element Circuit Mimicking Basic Electrical Transducer Impedance.

This type of representation is limited to the simplest kinds of transducer configurations, but it is useful for explaining and simulating the electrical characteristics of a transducer impedance. In other words, this circuit simulates the real and imaginary parts of the electrical transducer impedance for these simple acoustic loads (shown in Figure 5.2). However, because this circuit is just electrical, it does not describe the acoustic response. Therefore, it is not a replacement for a more complete model.

5.1.3 Summary

In summary, the piezoelectric element sends out two acoustic signals for each electrical excitation in this simple model. When transducers resonate, they have a distinctive electrical impedance signature that can be measured electrically. The spectral response of a transducer peaks at odd multiples of the fundamental frequency. Most piezoelectric transducers are reciprocal, so they act as receivers equally well.

If the acoustic impedances on either side of the crystal differ from that of the crystal, then multiple reflections occur at the interfaces, as described by Redwood (1963). A pair of oppositely signed stress pulses are created at each crystal boundary in this case. In order to develop a more complete transducer equivalent circuit model that accounts for these effects in both the time and frequency domains, it is necessary to discuss transducer geometry, construction, and piezoelectric materials.

5.2 Resonant Modes of Transducers

5.2.1 Resonant Crystal Geometries

Key factors in determining transducer parameters are the geometry or shape of the piezoelectric material, the crystallographic orientation of the piezoelectric crystal with respect to the electrical poling direction, and the placement of the electrodes. Piezoelectric materials are anisotropic, meaning that their properties vary with angle. In other words, depending on which direction the crystal is poled (a process in which a high voltage is applied to opposite sides of a crystal to align domains), the piezoelectric coupling and sound speed can vary. Piezoelectric materials are explained in more detail in Section 5.8.

An important factor in determining transducer performance is the shape or geometry of the piezoelectric material in the transducer. From the simple model of a transducer (explained in Section 5.1.1), the fundamental resonance and odd frequency harmonics, $f_m = (2m + 1)f_0$, correspond to odd multiples of half wavelengths, $(2m + 1)\lambda_0/2$. A practical implementation of this simple resonance idea is the "thickness-expander mode" geometry (see Figure 5.4A), where the lateral dimensions are much greater than the thickness dimension, and *poling* is oriented perpendicular to the electrodes. In other words, the vibrations are dominated by the

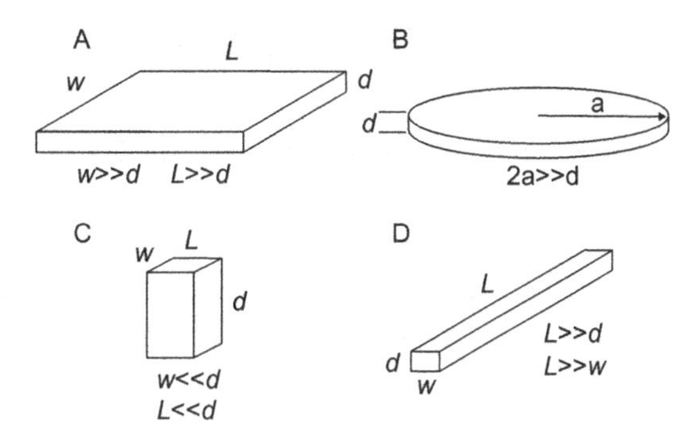

Figure 5.4
Resonator Geometries for Longitudinal Vibration Modes Along the *z* Axis.
(A) Thickness-expander rectangular plate; (B) thickness-expander circular plate disk; (C) length-expander bar; (D) width-extensional bar or beam plate.

thickness direction (z) so that resonances in the lateral directions are so low in frequency that they are negligible (shown in Figures 5.4A and 5.4B for rectangular and circular plates). The appropriate piezoelectric coupling constant for this geometry is the thickness coupling constant (K_T) and the speed of sound (c_T). Electrical polarization is along the z or 001 axis shown as the depth axis (d) for all four geometries in Figure 5.4.

In the early days of ultrasound imaging, transducers were of the thickness-expander type, were usually circular in cross-section, and were used in mechanical scanning; however, most of the transducers in use today are arrays. Among the earliest arrays was the annular type (Melton & Thurstone, 1978; Reid &Wild, 1958), with circular concentric rings on the same disk, phased to focus electronically (Foster et al., 1989).

The two geometries most relevant to one-dimensional (1D) and two-dimensional (2D) arrays are the length-expander bar and the beam or width-extensional mode (shown in Figures 5.4C and 5.4D). In each case, two dimensions are either much smaller or larger than the third so that only one resonance mode is represented by each picture. In reality, these rectangular geometries are limiting cases of a rectangular parallelepiped, in which three orthogonal coupled resonances are possible; each is determined by the appropriate half-wavelength thickness (d, w, or L). In the cases shown in Figure 5.4, the relative disparity in the lateral resonance dimensions compared with the thickness dimension allow them to be neglected relative to a dominant thickness resonance determined by geometry.

The bar geometry (Figure 5.4C) has an antiresonant frequency determined by length, which is the dominant dimension. This shape is the one used as piezoelectric pillars in 1–3 composites (to be described in Section 5.7.7) and is also helpful for 2D arrays. Important constants for design are summarized for different piezoelectric materials and geometries in

Table B.2 in Appendix B. From Figure 5.4C, the appropriate coupling constant for this geometry is k_{33} and the speed of sound is c_{33}.

The geometry most applicable to elements of 1D arrays is the beam mode, in which the length (*L*), corresponding to an elevation direction, is much greater than the lateral dimensions (de Jong, Souquet, & Bom, 1985; Souquet, Defranould, & Desbois, 1979). For this representation to be applicable, the width to thickness ratio (*w*/*d*) must be less than 0.7. Other *w*/*d* ratios will be discussed shortly. One lucky break for transducer designers was that, in general, the coupling constant for this geometry (k_{33}) is significantly greater than k_T (e.g. for PZT-5H, $k_{33} = 0.7$ and $k_T = 0.5$).

The beam mode represents a limiting case. Imagine a steamroller running over a tall piezoelectric element of the beam shape (Figure 5.4D) and changing it into a thickness-expander shape (Figure 5.4A), which is the other extreme. For the cases in between, calculations are necessary to predict characteristics as a function of the ratio *w*/*d* (shown in Figure 5.5), in which two sound speed dispersion curves are indicated for different aspect ratios and vibrational modes. For more precise design for *w*/*d* ratios of less than 0.7, sound speed dispersion and coupling characteristics must be calculated or measured (Selfridge,

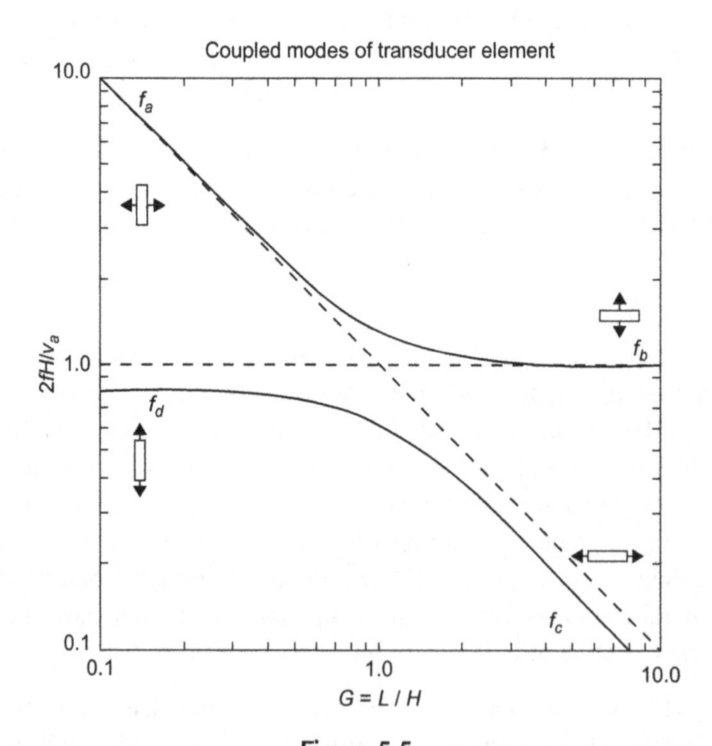

Figure 5.5
Sound Speed Dispersion (v_a) for a Piezoelectric Element as a Function of Aspect Ratio ($G = w/d$). *From Selfridge et al. (1980), IEEE.*

Kino, & Khuri-Yakub, 1980; Szabo, 1982). For *w*/*d* ratios of greater than 0.7, spurious multiple resonant modes can degrade transducer performance (de Jong et al., 1985). In general, the coupling constant and speed of sound vary with this ratio (Onoe & Tiersten, 1963; Sato, Kawabuchi, & Fukumoto, 1979), as shown in Figure 5.5.

5.2.2 Determination of Electroacoustic Coupling Constants

The relevant equations are given in Selfridge et al. (1980). When the input electrical impedance of a crystal of this thickness-expander geometry is measured in air, it has a unique spectral signature. As discussed earlier, the electrical characteristics of a simply loaded crystal are like the circuit of Figure 5.3, which has a resonant and an antiresonant frequency. These frequencies are related to the coupling constant and sound speed through the following equation:

$$f_a = c_T/2d, \tag{5.10A}$$

also known as the antiresonant frequency. The resonant frequency (f_r), can be found from the solution of the transcendental equation:

$$K_T^2 = \frac{\pi f_r}{2f_a}\cot\left(\frac{\pi f_r}{2f_a}\right), \tag{5.10B}$$

where K_T can be calculated from fundamental constants. Alternatively, a resonant and an antiresonant frequency can be measured and used to find the coupling constant experimentally through Eqn 5.10B. Both the electromechanical coupling constant (k_T) and speed of sound (c_T) equations are also given for different geometries in Selfridge et al. (1980) and Kino (1987).

5.2.3 Array Construction

How does a single piezoelectric crystal plate fit into the structure of an array? The array begins as a series of stacked layers with a relatively large area or footprint (e.g. 3 × 1 cm). The crystal and matching layers are bonded together and onto a backing pedestal. This sandwich of materials is cut into rows by a saw or by other means (as Figure 5.6 illustrates). The cut space between the elements is called a "kerf," and the remaining material has a width (w), repeated with a periodicity or pitch (p). Only after the cutting process does an individual crystal element resemble the beam mode shape with a *long* elevation length (L), width (w), and thickness (d). After the elements are cut, they are covered by a cylindrical lens for elevation focusing and are connected electrically to the imaging system through a cable. Figure 5.7 shows the overall look of the array before placement into a handle.

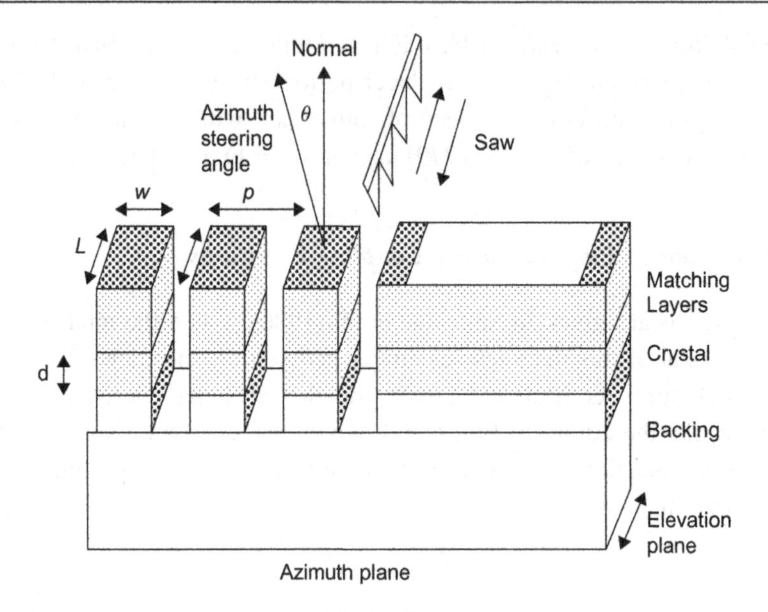

Figure 5.6
A Multilayer Structure Diced by a Saw into One-dimensional Array Elements. *From Szabo (1998), IOP Publishing.*

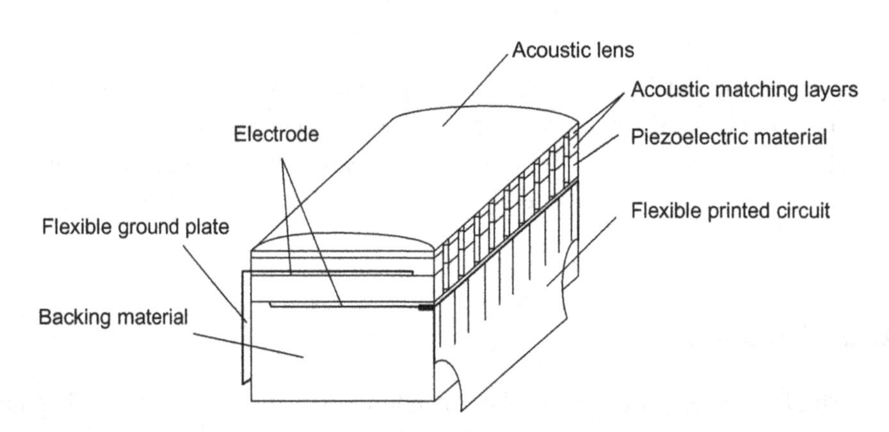

Figure 5.7
Construction of a One-dimensional Array With an Elevation Plane Lens. *From Saitoh et al. (1999), IEEE.*

A typical design constraint for phased arrays is that the pitch (p) between elements be approximately one-half a wavelength in water. The thickness dimension of the element is also close to one-half a wavelength along the depth direction in beam mode in the crystal material, which has a considerably different speed of sound than water. These constraints often determine the allowable w/d ratio. For 2D arrays, elements have small sides. This is a difficult design problem in which strong coupling can exist among all three dimensions.

Models are available for these cases (Hutchens, 1986; Hutchens & Morris, 1985;), and there are materials designed to couple less energy into unwanted modes (Takeuchi, Jyomura, Ishikawa, & Yamamoto, 1982). Finite element modeling (FEM) of these geometries is another alternative that can include other aspects of array construction for accurate simulations (McKeighen, 2001; Mills & Smith, 2002).

5.3 Equivalent Circuit Transducer Model

5.3.1 KLM Equivalent Circuit Model

To first order, the characteristics of a transducer can be well described by a 1D equivalent circuit model when there is one dominant resonant mode. To implement a model for a particular geometry, the same equivalent circuit model can be applied, but with the appropriate constants for the geometry selected. This complete model includes all impedances, both acoustic and electrical, as well as signal amplitudes in both forward and backward directions as a function of frequency. By looping through this single-frequency model a number of times, a complex spectrum can be generated, from which a time waveform can be calculated by an inverse Fourier transform.

To connect acoustic and electrical parameters, use will be made of acoustical–electrical analogs (described in Chapter 3). Warren P. Mason (1964) utilized these analogs to derive several models for different piezoelectric transducer geometries. The most applicable model for medical transducers is the thickness-expander model. Based on exactly the same wave equations, a newer model was introduced by Leedom, Krimholtz, & Matthaei (1978). This "KLM model," named after the initials of the authors, gives exactly the same numerical results as the Mason model but has several advantages for design (shown in Figure 5.8).

One of the main advantages of the KLM model is a separation of the acoustical and electrical parts of the transduction process. Three major sections can be seen in Figure 5.8: an electrical group extending from port 3, and two acoustic groups extending to the left and right from a center junction with the electrical group. This partitioning will allow us to analyze these ports separately to improve the design of the transducer. Port 1 will be used to represent forward transmission into water or the body, whereas port 2 will be for the acoustic backing, a load added to modify the bandwidth, and sensitivity of the transducer. Derivations for the physical basis of the KLM model can be found in Leedom et al. (1978) and Kino (1987). As shown in Figure 5.8B, the entire model can be collapsed into a single ABCD matrix between the electrical port and the forward acoustic load. The derivation of this matrix from the basic 2×2 ABCD forms introduced in Chapter 3 is explained thoroughly in Appendix C.

The piezoelectric element, described by the KLM model in Figure 5.8, is part of the overall representation of a transducer or an array element. The complete model can be represented

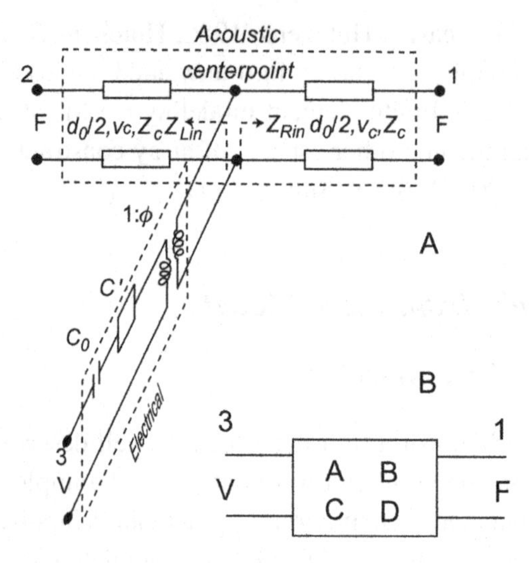

Figure 5.8
(A) Schematic Representation of the KLM Transducer Three-port Equivalent Circuit Model; (B) ABCD Representation of the KLM Model by an ABCD Matrix Between the Electrical Port 3 and Acoustic Port 1.

by a series of simple ABCD matrices cascaded together (Selfridge & Gehlbach, 1985; Sittig, 1967; van Kervel & Thijssen, 1983) as derived in detail in Appendix C. This derivation forms the basis for a numerical ABCD matrix implementation in the form of MATLAB program xdcr.m.

5.3.2 *Organization of Overall Transducer Model*

The organization of the model as a whole is illustrated by Figure 5.9. Physically, this model mimics the layers in an element of an array (see Figure 5.6), in which the layers on top of the piezoelectric element are represented by those on the right side of the piezoelectric element in the model (Figure 5.9). The piece from Figure 5.8 for the piezoelectric element is connected through port 3 to an electrical source. These parameters are needed for the piezoelectric element: a crystal that has a thickness (d_0), a speed of sound (c), an area (A), resonant frequency ($f_0 = c/2d_0$), a clamped capacitance ($C_0 = \varepsilon_R^S \varepsilon_0 A/d_0$), an electromechanical coupling constant (k_T), and a specific acoustic impedance ($Z_C = \rho c A$). In the KLM model, an artificial acoustic center is created by splitting the crystal into halves, each with a thickness of $d_0/2$ (refer to Figure 5.8). Each of these halves, as well as all layers, are represented by an acoustic transmission line. The right-end load, usually to tissue or water, is represented by a real load impedance, Z_R.

Each layer numbered "n," which can be a matching layer, bond layer, electrode, or lens, is represented by the following acoustic transmission line parameters: an area (A), an

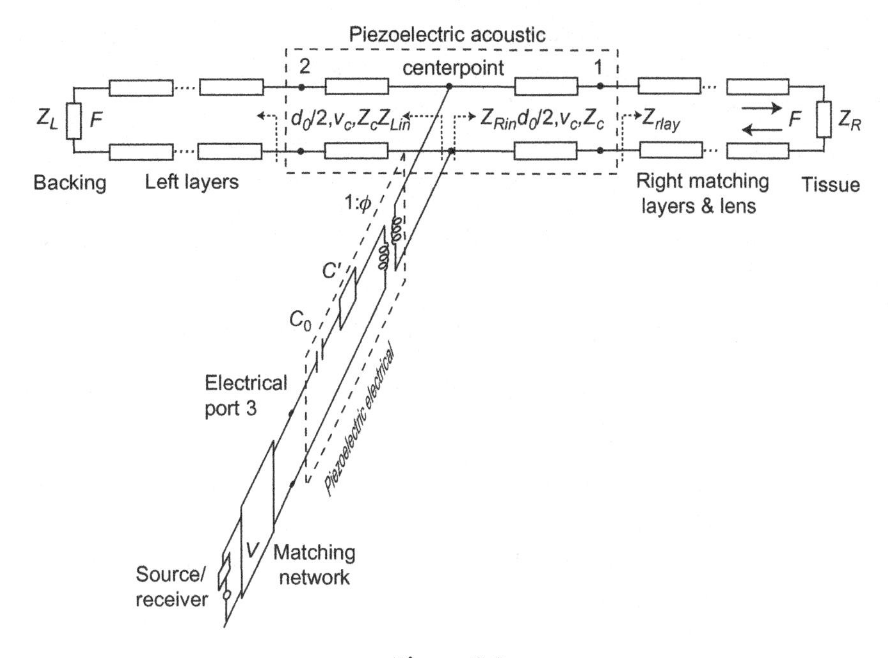

Figure 5.9
Overall Equivalent Circuit Transducer Model.

impedance (Z_{nR}), a sound speed (c_{nR}), a propagation factor (y_{nR}), and physical length (d_{nR}). The acoustic impedance looking from port 1 into the series of layers is called Z_{rlay}. Port 2 is usually connected to a backing, represented by a simple load (Z_B), or the acoustic impedance looking to the left is $Z_{llay} = Z_B$. If layers need to be added to the left side of the crystal, the same layer approach can be followed with indices such as d_{nL}. However, there is usually not a design incentive for doing so. Because force, rather than stress, is a key acoustic variable, all acoustic impedances are multiplied by the area (A), as is done for the definition of Z_C. More details can be found in Appendix C.

5.3.3 Transducer at Resonance

Now that all the pieces are accounted for in the model, they can be used to predict the characteristics of the transducer. This section starts with a more general description of the electrical impedance of the transducer. The key part of the model that connects the electrical and acoustic realms is the electroacoustic transformer. As shown in Figure 5.8A, this transformer has a turns ratio (ϕ), defined as:

$$\phi = k_T \left(\frac{1}{2f_0 C_0 Z_C} \right)^{\frac{1}{2}} \text{sinc}\left(\frac{f}{2f_0} \right), \tag{5.11A}$$

that converts electrical signals to acoustic waves and vice versa. The sinc function is related to the Fourier transform of the dielectric displacement field between the electrodes, which has a rectangular shape. The KLM model also accommodates multiple piezoelectric layers, which can be represented by a single turns ratio related to the transform of the complete field through all the piezoelectric layers together (Leedom et al., 1978).

Other electrical elements of the model includes block C', a strange negative capacitance-like component:

$$C' = -C_0/K_T^2 \,\mathrm{sinc}(f/f_0), \tag{5.11B}$$

that has to do with the acoustoelectric feedback and the Hilbert transform of the dielectric displacement. Finally, there is the ordinary clamped capacitance C_0.

The electrical characteristics of a transducer can be reduced to the simple equivalent circuit (shown earlier in Figure 5.2A). A complex acoustic radiation impedance (Z_A) can be found by looking through the KLM transformer at the combined acoustic impedance found at the center point of the model, $Z_{in}(f)$, as:

$$Z_A(f) = \varphi^2 Z_{in}(f), \tag{5.12}$$

where Z_A is purely electrical. Recall that at the center point, the acoustic impedance to the right is Z_{Rin}, and to the left is Z_{Lin}. By throwing in other components in the electrical leg of the KLM model, we arrive at the overall electrical transducer impedance:

$$Z_T(f) = \phi^2 \mathrm{Real}(Z_{in}) + i\left[\phi^2 \mathrm{Imag}(Z_{in}) - \frac{k_T^2}{\omega_0 C_0}\mathrm{sinc}(f/f_0) - 1/\omega C_0\right]; \tag{5.13A}$$

$$Z_T(f) = R_A(f) + i[X_A(f) - 1/\omega C_0] = Z_A(f) - i/\omega C_0. \tag{5.13B}$$

A typical plot of Z_T was given in Figure 5.2b.

At resonance, the radiation reactance, $X_A(f_0)$, is zero. The radiation resistance, R_A, is:

$$R_A(f) = \left[\frac{k_T^2}{2f_0 C_0 Z_c}\mathrm{sinc}^2(f/2f_0)\right]\left[\frac{Z_{Lin}Z_{Rin}}{Z_{Lin} + Z_{Rin}}\right], \tag{5.14}$$

and at f_0, it becomes:

$$R_A(f_0) = R_{A0} = \left[\frac{k_T^2}{2f_0 C_0 Z_c}\mathrm{sinc}^2(f_0/2f_0)\right]\left[\frac{Z_c^2}{Z_{llay} + Z_{rlay}}\right] = \frac{2k_T^2}{\pi^2 f_0 C_0}\left(\frac{Z_c}{Z_{llay} + Z_{rlay}}\right), \tag{5.15A}$$

where the resonant half crystals have become quarter-wave transformers ($Z_{Rin} = Z_c^2/Z_{rlay}$). The impedance looking from the right face of the crystal to the right is Z_{rlay}, and that

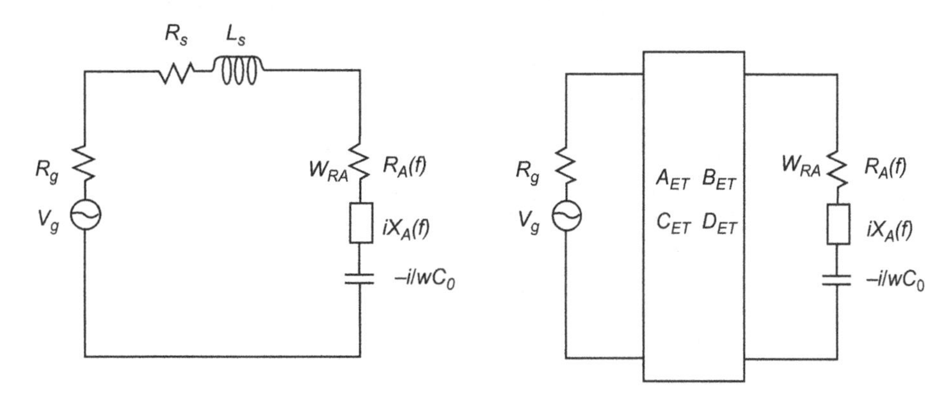

Figure 5.10
Electrical voltage source and electrical matching network.
(Left) simple series inductor and resistor; (right) ABCD representation of a more general network.

looking from the left face of the crystal is Z_{lray}. If there are no other layers, then for a medical transducer (typically $Z_{lray} = Z_B$, the backing impedance, and $Z_{rlay} = Z_w$, the impedance of water or tissue) the radiation resistance at resonance is:

$$R_A(f_0) = R_{A0} = \frac{2k_T^2}{\pi^2 f_0 C_0}\left(\frac{Z_c}{Z_B + Z_w}\right) \tag{5.15B}$$

Note that as a sanity check, if the loads are instead made equal to Z_c, Eqn 5.15B reduces to the simple model result of Eqn 5.8C.

To complete the electrical part of the transducer model, a source and matching network are added, as in Figure 5.10. A convenient way to add electrical matching is a series inductor. A voltage source (V_g) with an internal resistance (R_g) is shown with a series tuning inductance. These components can be represented in a series ABCD matrix (see Chapter 3). A more complicated tuning network can be used instead with the more general matrix elements A_{ET}, B_{ET}, C_{ET}, and D_{ET}, as Figure 5.10 implies.

5.4 Transducer Design Considerations

5.4.1 Introduction

In order to design a transducer, we need criteria to guide us. To make a transducer sensitive, some measure of efficiency is required. For a pulse–echo configuration, two different transducers can be used for transmission and reception (indicated in Figure 5.11), or in the more usual configuration the same transducer is used as a

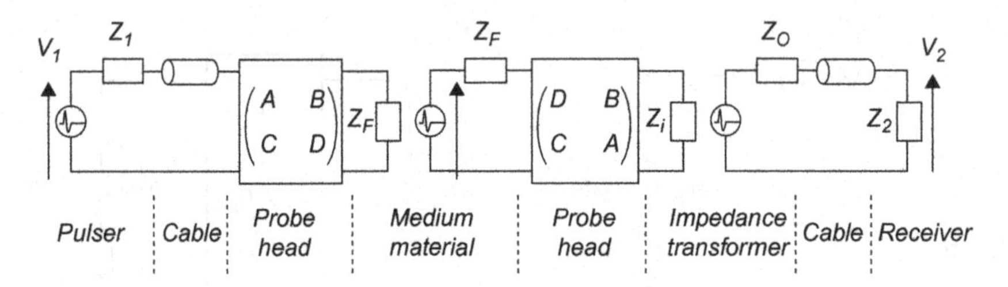

Figure 5.11
Equivalent Circuit for the Round-trip Response of a Transducer With a Cable and Lens. *From Saitoh et al. (1999), IEEE.*

transceiver (Hunt, Arditi and Foster, 1983). In general, there may be two different matching networks: E_T, for transmit, and E_R (each represented by its ABCD matrix). If the transducers, matching networks, and loads R_g and R_f are the same, the transducer efficiencies are identical and reciprocal (Saitoh et al., 1999; Sittig, 1967, 1971).

In this situation, if the transmit transducer has an ABCD matrix relating the electrical and acoustic variables, then the receiver will have a DCBA matrix. From repeated calculations of this model for a range of frequencies, pulses can be calculated using an inverse Fourier transform from the spectrum. If the round-trip pulse length is shorter than the transit time between the transducers, then the models can be decoupled or calculated independently; however, for a longer pulse or a continuous wave transmit situation, the individual transducer models are connected by a transmission line between the transmit and receive sections of the model.

5.4.2 Insertion Loss and Transducer Loss

One measure of overall round-trip efficiency is "insertion loss." As illustrated in Figure 5.12, efficiency is measured by comparing the power in load resistor R_f with the transducer in place to the power there without the transducer. Insertion loss is defined as the ratio of the power in R_f over that available from the source generator, where W_f is the power in R_f, and W_g is that available from the source V_g. The maximum power available is for $R_f = R_g$. In Figure 5.11, $R_f = Z_2$.

$$IL(f) = \frac{W_f}{W_g} = \left[\left| \frac{V_f}{V_g} \right|^2 \left(\frac{R_f + R_g}{R_f} \right) \right], \tag{5.16A}$$

and in dB, it is:

$$IL_{\mathrm{dB}}(f) = 10 \log_{10} IL(f). \tag{5.16B}$$

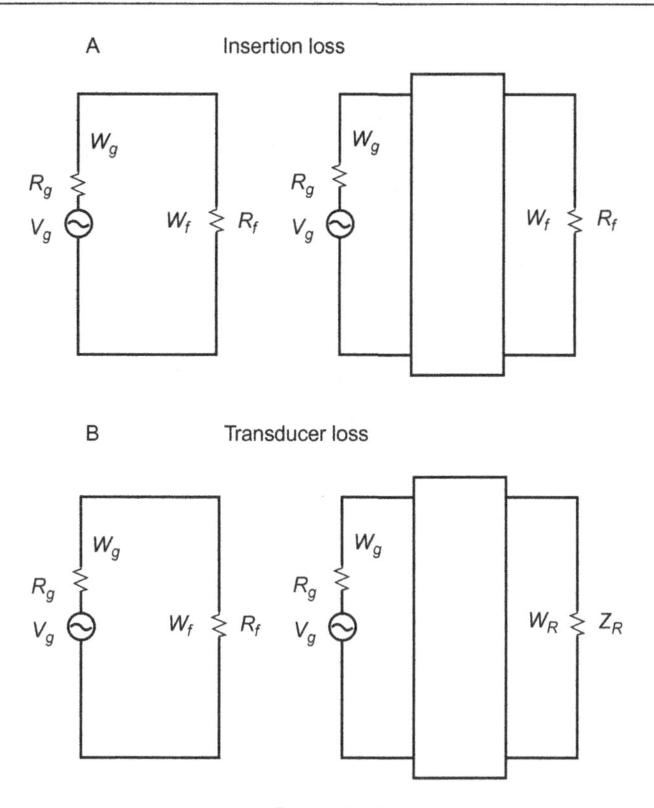

Figure 5.12
(A) Transducer Insertion Loss Shown as a Comparison of the Source and Load With and Without a Device in Between; (B) Similar Transducer Loss Definition for One-way Transducer Loss.

Likewise, it is possible to define a one-way loss, called a "transducer loss" (Sittig, 1971), that is a measure of how much acoustic power arrives in right acoustic load Z_R from a source V_g. Transducer loss (as shown in Figure 5.12B) is:

$$TL(f) = \frac{W_R}{W_g} = \left[\left|\frac{F_{AR}}{V_g}\right|^2 \left(\frac{4R_g}{Z_R}\right)\right], \tag{5.17A}$$

and defined in dB as:

$$TL_{\text{dB}}(f) = 10\log_{10} TL(f), \tag{5.17B}$$

where W_R is the power in Z_R, and F_{AR} is the acoustic force across load Z_R. Note that for identical transducers,

$$TL = \sqrt{IL}(\text{linear}), \tag{5.18A}$$

and

$$TL_{\text{dB}} = IL_{\text{dB}}/2(\text{dB}). \tag{5.18B}$$

5.4.3 Electrical Loss

For highest transducer sensitivity, we would like transducer and insertion losses to be as small as possible. With the KLM model, it is possible to partition the transducer loss into electrical loss (EL) and acoustic loss, (AL):

$$TL(f) = EL(f)AL(f), \tag{5.19}$$

as symbolized by Figure 5.13. By looking at each loss factor individually, we can determine how to minimize the loss of each contribution. From Figure 5.10, the voltage transfer ratio for the specific case in which the matching network (E_T) is a series tuning inductor, $Z_s = R_z + i\omega L_s$, with matrix elements A_{ET}, B_{ET}, C_{ET}, and D_{ET},

$$\frac{V_T}{V_g} = \frac{Z_T}{A_{ET}Z_T + B_{ET}} = \frac{Z_T}{Z_T + Z_S + R_g}. \tag{5.20}$$

Now electrical loss is defined as the power reaching R_A divided by the maximum power available from the source:

$$EL = \frac{W_{RA}}{W_g} = \frac{I^2 R_A/2}{V_g^2/8R_g} = \frac{\left|\frac{V_T}{Z_T}\right|^2 R_A/2}{V_g^2 8/R_g} = \left|\frac{V_T}{V_g}\right|^2 \frac{4R_A R_g}{|Z_T|^2}. \tag{5.21}$$

Combining Eqns 5.20 and 5.21:

$$EL = \frac{4R_A R_g}{|A_{ET}Z_T + B_{ET}|^2}; \tag{5.22A}$$

$$EL = \frac{4R_A R_g}{(R_A + R_g + R_s)^2 + (X_A - 1/\omega C_0 + \omega L_S)^2}. \tag{5.22B}$$

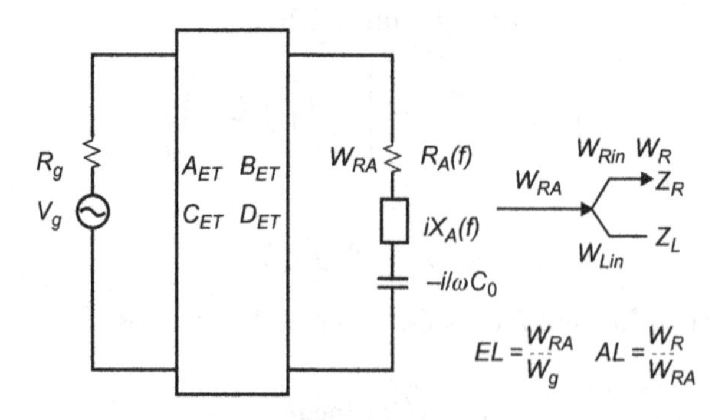

Figure 5.13
Diagram of Electrical Loss as the Power Reaching the Radiation Resistance Divided by Source Power, and Acoustical Loss as the Power Reaching the Right Acoustic Load Divided by the Power Reaching the Radiation Resistance.

If the capacitance is tuned out by a series inductor, $L_s = 1/(\omega_0^2 C_0)$, then at resonance:

$$EL(f_0) = \frac{4R_A R_g}{(R_A + R_g + R_s)^2}. \tag{5.22C}$$

Furthermore, if $R_A = R_g$, and $R_s << R_g$, then $EL(f_0) \sim 1$.

An example of the effect of electrical tuning is given by Figure 5.14A. In this case, a 3-MHz-center frequency transducer is tuned with an inductor at 3 MHz. These curves were generated by the MATLAB program xdcr.m. The effect of tuning is strong and alters both the shape of the transducer loss response and its absolute sensitivity.

5.4.4 Acoustical Loss

Acoustical loss is the ratio of the acoustic power reaching the front load (Z_R), over the total acoustic power converted. In order to determine acoustical loss, we begin with the real electrical power reaching R_A, which, after being converted to acoustical power at the acoustic center of the KLM model, splits into the left and right directions:

$$W_{RA} = W_{Lin} + W_{Rin}. \tag{5.23A}$$

Refer to Figure 5.13. If the equivalent acoustic voltage or force at the center is F_c, then the power (W_{Rin}) to the right side is:

$$W_{Rin} = \frac{1}{2}\left|\frac{F_c}{Z_{Rin}}\right|^2 \mathrm{REAL}(Z_{Rin}), \tag{5.23B}$$

and the power to the left is:

$$W_{Lin} = \frac{1}{2}\left|\frac{F_c}{Z_{Lin}}\right|^2 \mathrm{REAL}(Z_{Lin}). \tag{5.23C}$$

If there is absorption loss along the acoustic path, then the power to the right is instead:

$$W_R = vv^* \mathrm{REAL}(Z_R)/2 = \left|\frac{F_R}{Z_R}\right|^2 \mathrm{REAL}(Z_R)/2, \tag{5.24}$$

where F_R is the force across load Z_R. The acoustical loss is simply the power to the right divided by the total incoming acoustic power:

$$AL(f) = \frac{W_R}{W_{RA}} = \frac{W_R}{W_{Lin} + W_{Rin}}. \tag{5.25}$$

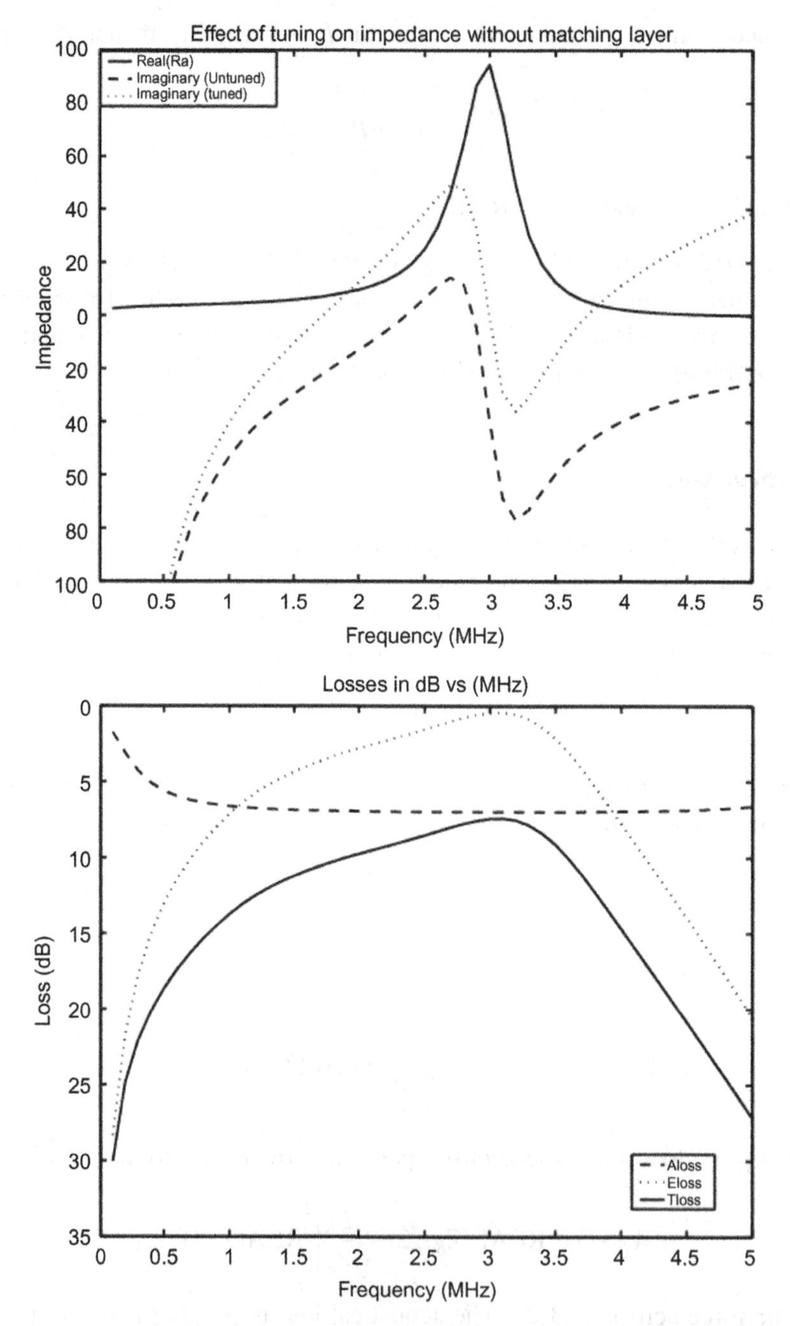

Figure 5.14

Transducer Operating into a Water Load in a Beam Mode With a Crystal of PZT-5H, Having an Area of 5.6 mm^2 and a Backing Impedance of 6 MegaRayls.
(A) Transducer impedance untuned and tuned with a series inductor; (B) Electrical loss, acoustical loss and transducer loss with tuning. aloss, acoustical loss; eloss, electrical loss; tloss, transducer loss.

If there is no absorption loss along the right path, then $W_R = W_{Rin}$. At resonance with no loss, this expression can be shown to be (see Figure 5.9):

$$AL(f_0) = \frac{Z_{rlay}}{Z_{rlay} + Z_{llay}}, \tag{5.26}$$

where these are the acoustic impedances to the right and left of the center. For no layers:

$$AL(f_0) = \frac{Z_R}{Z_R + Z_L} = \frac{Z_W}{Z_B + Z_W}. \tag{5.27}$$

For an air backing, $AL(f_0) = 1$. For a backing matched to the crystal-specific impedance, $Z_B \approx 30A$ (recall A is area), and for a water load, $Z_R = Z_w = 1.5A$, $AL(f_0) = 0.05$. Acoustic loss curves for several back acoustic loads at port 2 are plotted for a 3.5-MHz center frequency in Figure 5.15.

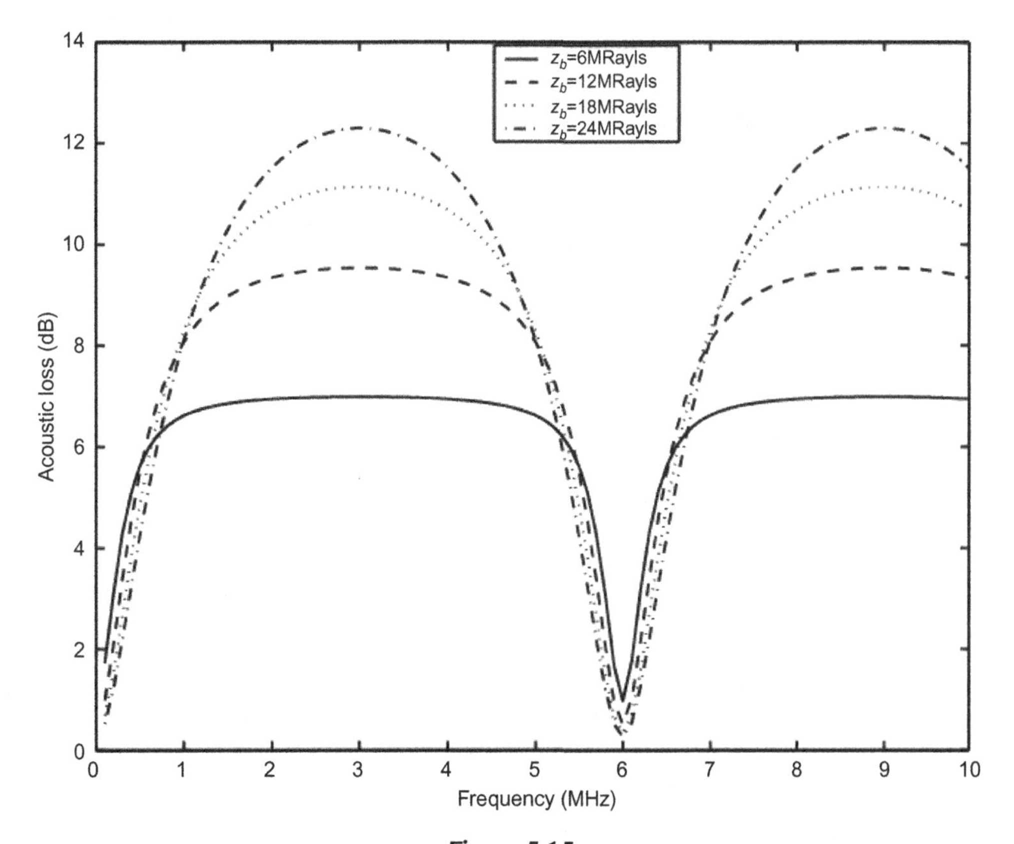

Figure 5.15
Acoustical Loss Versus Frequency for a Water Load and Several Backing (z_b) Values for a Transducer With a 3-MHz Center Frequency.

5.4.5 Matching Layers

To improve the transfer of energy to the forward load, quarter-wave matching layers are used. The simplest matching is the mean of the impedances to be matched:

$$Z_{ml} = \sqrt{Z_1 Z_2}. \tag{5.28}$$

If we interpose this quarter-wave matching layer on the right side for the last case of a matched backing, then since $Z_1 = Z_c = 30A$, $Z_2 = Z_w = 1.5A$, $Z_{ml} = 6.7A$, and $Z_{rlay} = 30A$, so then $AL(f_0) = 0.5$. At the resonant frequency, recall that the value of acoustic loss can be found from the simple formula in Eqn 5.26. The dramatic effect matching layers can have in lowering loss over a wide bandwidth will be demonstrated with examples in Section 5.4.6. The increase in fractional bandwidth as a function of the number of matching layers is shown in Figure 5.16. Note that for a single matching layer, the −3-dB fractional bandwidth is about 60%. More matching layers can be used to increase bandwidth. Philosophies differ as to how the values for matching layer impedances are selected (Desilets, Fraser, & Kino, 1978; Goll & Auld, 1975); however, a good starting point is the maximally flat approach borrowed from microwave design (Matthaei, Young, & Jones, 1980). For two matching layers, for example, the values are $z_1 = z_c^{4/7} z_w^{3/7}$ and $z_2 = z_c^{1/3} z_w^{2/3}$. This approach was used to estimate one-way −3-dB fractional acoustic bandwidths in percent for the right side as a function of the number of matching layers (shown in Figure 5.16).

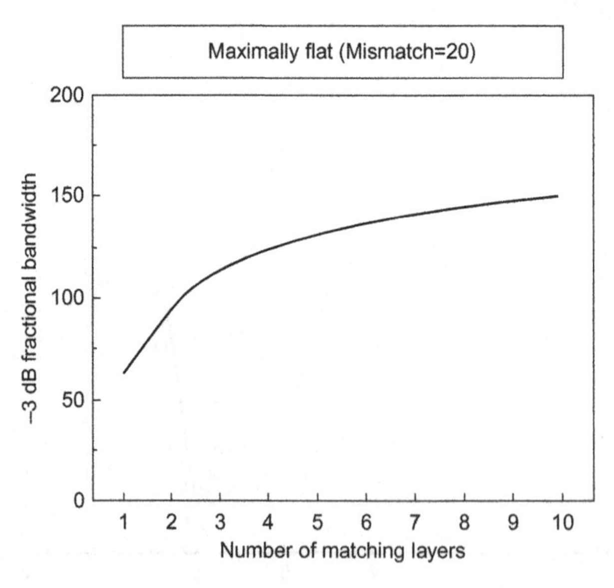

Figure 5.16

The −3-dB fractional bandwidths versus the number of matching layers determined from the maximally flat criteria for an overall mismatch ratio of $20 = Z_c/Z_w$. *From Szabo (1998), IOP Publishing.*

5.4.6 Design Examples

We are now ready to look at two examples. The first case is a transducer element made of lead zirconium titanate (PZT)-5H with a 3-MHz resonant frequency desired. From Table B.2 (in Appendix B), the coupling constant and parameters for the beam mode for this material can be selected. From the crystal sound speed, the crystal thickness is 662 μm ($c/2f_0$). The given area is $A = 7e - 6\ \mathrm{m}^2$, and the backing impedance is $Z_b = 6$ megaRayls. The crystal acoustic impedance is 29.8 megaRayls. This case is the default for the transducer simulation program xdcr.m. The values of these variables can be found by typing the following variable names, one at a time, at the MATLAB prompt: edi, area, zbi, and zoi. Finally, the clamped capacitance can be found from Eqn 5.1 to be $C_0 = 1380\ pf\ (pf = picofarad = e - 12\ farad)$, with the variable name $c0$. The value of reactance at f_0 is tuned out by a series inductor (matching oppositely signed reactances) as $L_s = 1/(\omega_0^2 C_0) = 2.04$ μH (symbol for microHenry), with the variable name ls0. Putting all of these input variables into the program gives a tuned impedance similar in shape to that shown in Figure 5.14A. Transducer, electrical, and acoustical loss curves are given in Figure 5.17.

In all cases, the values of the loss curves at resonance (predictable by simple formulas) provide a sanity check. From Eqn 5.27:

$$AL(f_0) = \frac{1.5}{6 + 1.5} = 0.2, \tag{5.29A}$$

or −7 dB. This checks with the program variable alossdb(30), where index 30 corresponds to 3 MHz. From Eqn 5.15B, $R_{A0} = 94.6$ ohms, variable real ($zt(30)$). Then from the definition of electrical loss at resonance, Eqn 5.22C, since $R_s = 0$, and $R_g = 50$ ohms:

$$EL(f_0) = \frac{4 \times 94.62 \times 50}{(94.62 + 50)^2} = 0.9048, \tag{5.29B}$$

or −0.43 dB, for a total one-way transducer loss of −7.43 dB at the resonant frequency. The points at resonance serve as sanity anchors for the curves in Figure 5.17. Note that the losses in dB can be simply added. Though both the acoustical and electrical losses are interrelated, it is apparent that the acoustical loss has a much wider bandwidth.

Now a matching layer will be used for the forward side. From Eqn 5.28, $Z_{ml} = 6.68$ megaRayls. Assume that a matching layer material with the correct impedance and a sound speed of 3.0 mm/μs can be applied. For a quarter wave at the resonant frequency, the layer thickness is $d = c_{ml}/(4f_0) = 250$ μm. This information can be turned on in the program by setting the parameter $ml = 1$ rather than $ml = 0$ (default). Note that even with a matching layer, the tuning inductor is unchanged. The resulting impedance has a different appearance (shown in Figure 5.18A). The corresponding losses are shown in Figure 5.18B. This time,

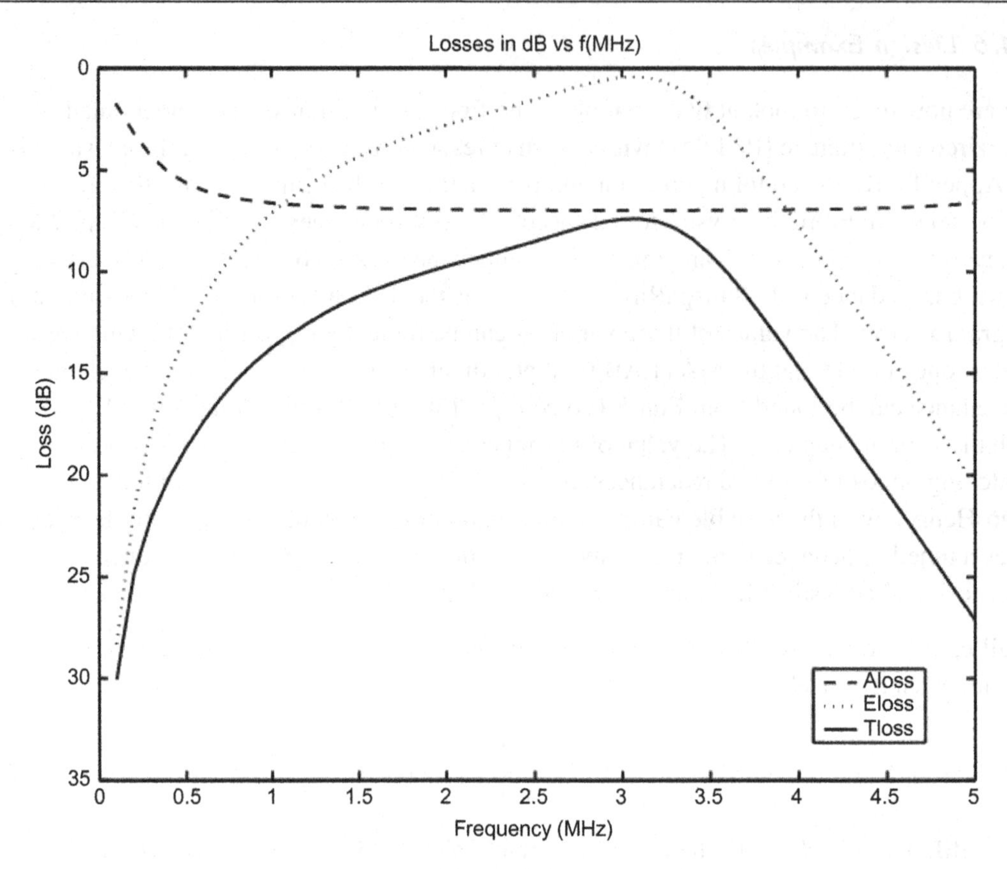

Figure 5.17
Transducer, Acoustical, and Electrical Loss Curves for 3-MHz Tuned Design. aloss, acoustical loss; eloss, electrical loss; tloss, transducer loss.

the acoustical loss is found from Eqn 5.26, in which the acoustic impedance at the right crystal face looking toward the matching layer is $Z_{Rlay} = 29.8A$ megaRayls, so that:

$$AL(f_0) = \frac{29.8}{6 + 29.8} = 0.832, \tag{5.30A}$$

or 0.798 dB. In this case, $R_{A0} = 19.85$ ohms, so from Eqn 5.22C:

$$EL(f_0) = \frac{4 \times 19.85 \times 50}{(19.85 + 50)^2} = 0.8137, \tag{5.30B}$$

or 0.895 dB, for a total one-way transducer loss of -1.69 dB at the resonant frequency.

Comparison of the two cases shows considerable improvement in sensitivity and bandwidth from the inclusion of a matching layer. The overall shape of the transducer loss could be improved because it is related to the pulse shape.

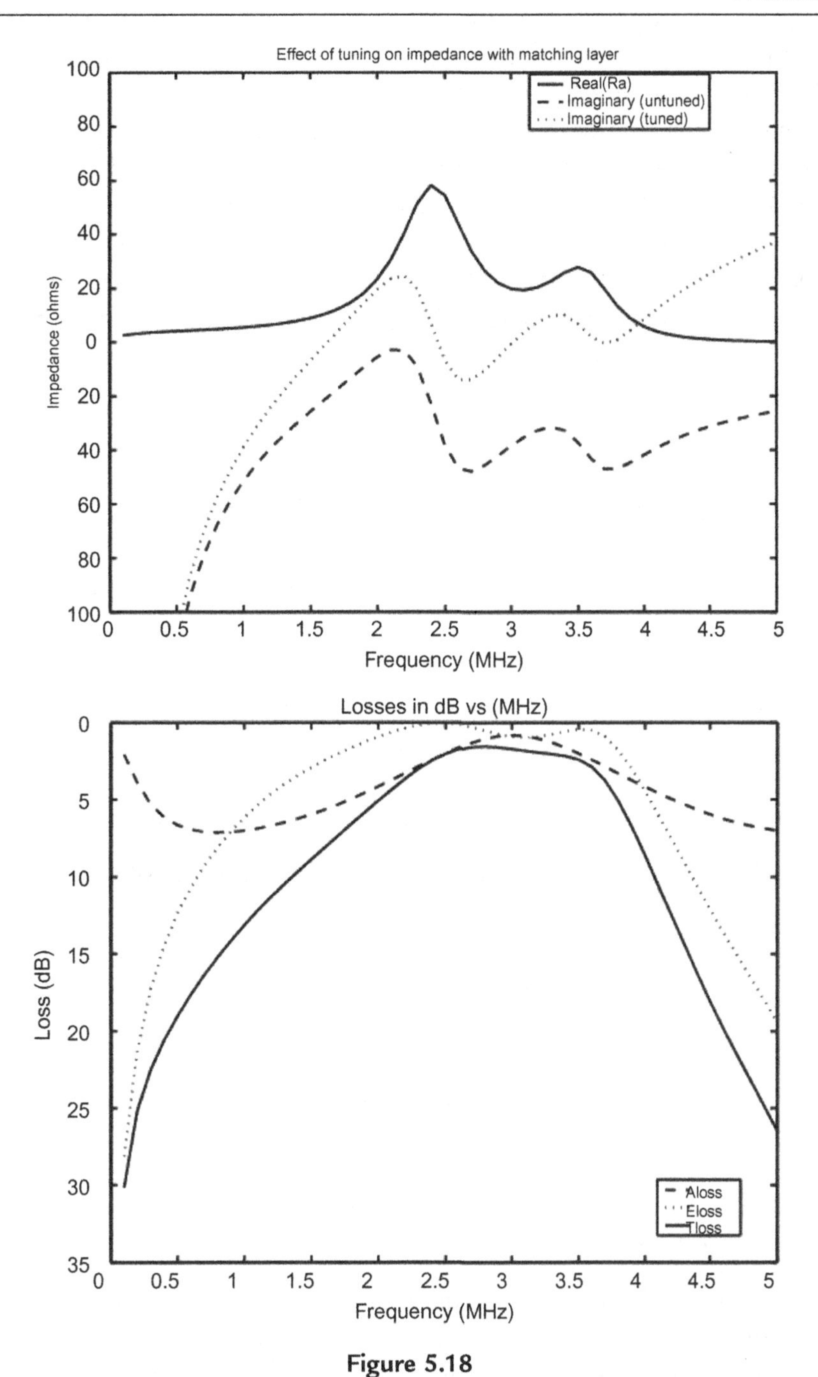

Figure 5.18

(A) Electrical Impedance for Design With Matching Layer, With and Without Tuning; (B) Corresponding Transducer Loss, Acoustical Loss, and Electrical Loss Curves for 3-MHz-tuned Design With a Matching Layer.

aloss, acoustical loss; eloss, electrical loss; tloss, transducer loss.

In order to refine the design, the resonant frequencies of the crystal and matching layer can be adjusted, or more matching layers can be added. Because of constraints beyond the designer's control, transducer design requires adaptability, creativity, and patience. For a typical array element design, nonlinear electronic circuitry and a coaxial cable are added to the electrical port. In addition, a lens with absorption loss as a function of frequency is thrown into the mix to make the design a little more interesting. More information on design can be found in the following references: Desilets et al. (1978); Goll and Auld (1975); Kino (1987); Persson and Hertz (1985); Rhyne (1996); Sittig (1967); Souquet et al. (1979); Szabo (1984); van Kervel and Thijssen (1983).

5.5 Transducer Pulses

5.5.1 Standard Pulse and Spectral Measurements

Because the primary purpose of a medical transducer is to produce excellent images, an ideal pulse shape is the ultimate design goal. Agreement has been reached that the pulse should be as short as possible and with a high-amplitude peak (good sensitivity). Some would argue that the ideal shape is Gaussian because this shape is maintained during propagation in absorbing tissue (only true for attenuation proportional to frequency—Section 4.4.2). Unfortunately, because of causality, a Gaussian shape is not achieved by transducers; instead, the leading edge of a pulse is usually much steeper than its tail.

To get beyond the "looks nice" stage requires quantitative measures of a spectrum and its corresponding pulse. Spectral bandwidths are measured from a certain number of decibels down from the spectral maximum. Typical values are −6-dB, −10-dB, and −20-dB bandwidths. The center frequency of a round-trip spectrum is defined as:

$$f_c = (f_{low} + f_{high})/2, \tag{5.31}$$

where f_{low} and f_{high} are the −6 (or other number)-dB low and high round-trip frequencies, respectively. For the pulse, the pulse widths, as measured in dB levels down from the peak of the analytical envelope (see Appendix A), are usually at the −6-dB, −20-dB, and −40-dB levels. These widths measure pulse "ringdown" and quantify the axial spatial resolution of the transducer.

Another consideration in pulse shaping is the excitation pulse. The overall pulse is the convolution of the excitation pulse and the impulse response of the transducer. Figure 5.19 shows plots for these pulses from a 3.5-MHz linear array design with two matching layers and a PZT-5H crystal operating in the beam mode. They were calculated by a commercially available transducer simulation/design program called PiezoCAD. Here the excitation pulse is 3.5 MHz, 2½ cycle sinusoid.

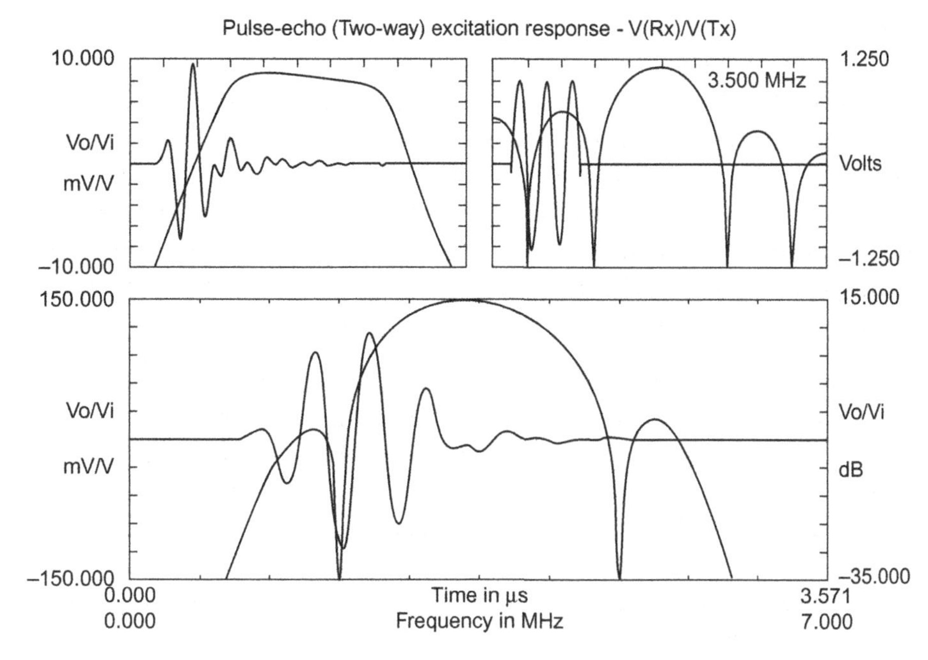

Figure 5.19

(A) Pulse—Echo Impulse Response and Spectrum for a 3.5-MHz Linear Array Design; (B) A 3.5-MHz, 2½-cycle Sinusoid Excitation Pulse and Spectrum; (C) Resultant Output Pulse and Spectrum. All calculations by PiezoCAD transducer design program. *Courtesy of G. Keilman, Sonic Concepts, Inc.*

Table 5.1 Linear Array Design Width Measurements from Figure 5.19.

	−6 dB	**−20 dB**	**−40 dB**
Center frequency (CF) (MHz)	3.404	3.472	3.513
Bandwidth (BW) (MHz)	1.649	2.494	4.704
Fractional BW of CF (%)	48.44	71.83	133.90
Pulse length (µs)	0.511	1.072	1.932

This program calculates the spectral and pulse envelope widths as given by Table 5.1. It has many features that make it convenient for design and has examples and tables of piezoelectric and other materials.

The design problem is to create pulses that are short in the sense that the tail is short and the so-called "time sidelobes" in the tail section after the main lobe are at very low levels. If these time sidelobes are high, a single actual target may appear as a series of targets or an elongated target under image compression, a process that elevates lower image signals for visualization. From the Fourier transform theory of Chapter 2, these restrictions require that the spectrum not contain sharp transitions or corners at the band edges. In other words, a wideband (for short temporal extent) rounded spectrum will do. This requirement presents another design constraint—shaping the spectrum so as to achieve short pulses. Various solutions have been proposed. One of the most widely known solutions is that of Selfridge, Baer, Khuri-Yakub, and Kino (1981). They developed a computer-aided design program that varies acoustic and electric parameters so as to achieve a pleasing pulse shape. Lockwood and Foster (1994) based their computer-aided design algorithm on a generalized ABCD matrix representation of the transducer. Rhyne (1996) devised an optimization program that is based on spectral shaping and the physical limitations of the transducer.

Finally, it is important to remember that pulse design is usually done with the system in mind. The overall shaping of the round-trip pulse after the transducer has been excited by a certain-shaped drive pulse and has passed through receive filters is a primary design goal (McKeighen, 1997). For better inclusion of the effects of electronics, SPICE transducer models (Hutchens and Morris, 1984; Morris and Hutchens, 1986; Puttmer, Hauptmann, Lucklum, Krause, & Henning, 1997) have been developed that marry the transducer more directly to the driver and to receive electronics. Nonlinearities of switching and noise figures can be handled by this approach.

5.6 Equations for Piezoelectric Media

What are the effects of piezoelectricity on material constants? As shown earlier in Section 5.1.1, Hooke's law is different for piezoelectric materials than for purely elastic (see Chapter 3) or viscoelastic (see Chapter 4) materials, and is stated more generally below (Auld, 1990):

$$\boldsymbol{T} = \boldsymbol{C}^{D}{:}\boldsymbol{S} - \boldsymbol{h}{:}\boldsymbol{D}, \tag{5.32A}$$

where $\boldsymbol{C}^D$ is a 6-by-6 tensor matrix of elastic constants taken under conditions of constant $\boldsymbol{D}$, $\boldsymbol{h}$ is a 6-by-3 tensor matrix, and $\boldsymbol{D}$ is a 1-by-3 tensor vector. This type of equation can be calculated by the same type of matrix approach used for elastic media in Chapter 3. The companion constitutive relation is:

$$\boldsymbol{E} = -\boldsymbol{h}{:}\boldsymbol{S} + \boldsymbol{\beta}^{S}{:}\boldsymbol{D}, \tag{5.32B}$$

where $\boldsymbol{\beta}^S$ is dielectric impermeability under constant or zero strain. Pairs of constitutive relations appear in various forms suitable for the problem at hand or the preference of the user, and they are given in Auld (1990).

Alternatively, stress can be put in the following form:

$$\boldsymbol{T} = \boldsymbol{C}^{E}{:}\boldsymbol{S} - \boldsymbol{e}{:}\boldsymbol{E}, \tag{5.33}$$

in which C^E is a set of elastic constants measured under constant or zero electric field and e is another piezoelectric constant. A companion constitutive equation to Eqn 5.33 is:

$$D = \varepsilon^S{:}E + e{:}S, \tag{5.34}$$

where ε^S is permittivity determined under constant or zero strain. If $D = 0$ in this equation, and E is found for the one-dimensional case, then $E = -eS/\varepsilon^S$. With E substituted in Eqn 5.33 (Kino, 1987):

$$T = C^E\left(1 + \frac{e^2}{C^E\varepsilon^S}\right)S = C^D S, \tag{5.35}$$

which is an abbreviated Hooke's law version of Eqn 5.33 with $D = 0$. C^D is called a stiffened elastic constant, with:

$$C^D = C^E(1 + K^2), \tag{5.36}$$

in which K is not wave number but the piezoelectric coupling constant:

$$K = \frac{e^2}{C^E\varepsilon^S}. \tag{5.37}$$

The consequence of a larger stiffened elastic constant is an apparent increase in sound speed caused by the piezoelectric coupling. The net effect of piezoelectric coupling seen from the perspective of Eqn 5.35 is an increased stress over the nonpiezoelectric case for the same strain. Various forms of K exist for specific geometries and crystal orientations 0 (to be covered in the next section). The term K^2 is often interpreted as the ratio of mutual coupling energy to the stored energy.

For the case of a stress-free condition ($T = 0$) in Eqn 5.33, the value of strain S can be substituted in Eqn 5.34 to yield:

$$D = \varepsilon^T E = \varepsilon^S(1 + K^2)E, \tag{5.38}$$

in which the stress-free dielectric constant ε^T is bigger than the often-used strain-free or clamped dielectric constant ε^S.

5.7 Piezoelectric Materials

5.7.1 Introduction

How does piezoelectricity work? What are some of the values for the constants just described, and how can they be compared for different materials?

In 1880, the Curie brothers discovered piezoelectricity, which is the unusual ability of certain materials to develop an electrical charge in response to a mechanical stress on the material. This relation can be expressed for small signal levels as the following:

$$\boldsymbol{D} = \boldsymbol{d}{:}\boldsymbol{T} + \varepsilon^T{:}\boldsymbol{E}. \tag{5.39}$$

There is a converse effect in which strain is created from an applied electric field, given by the companion equation:

$$\boldsymbol{S} = \boldsymbol{s}^E{:}\boldsymbol{T} + \boldsymbol{e}{:}\boldsymbol{S} + \boldsymbol{d}{:}\boldsymbol{E}, \tag{5.40}$$

where $\boldsymbol{s}^E = (\mathbf{C}^{\mathrm{E}})^{-1}$ is determined under a constant electric field condition. All piezoelectric materials are ferroelectric. This kind of material contains random ferroelectric domains with electric dipoles, and the ceramic itself is shown in the photomicrograph of Figure 5.20C. If an electric field is applied, the direction of spontaneous polarization (the alignment of the majority of domains in about the same direction shown in Figure 5.20A) can be switched by the direction of the field. Furthermore, if an appropriately strong field is applied under the right conditions (usually at elevated temperature), the polarization remains even after the polarizing field is removed.

The major types of piezoelectric media are described as follows. Some of these materials can be found in Table B.2 of Appendix B.

5.7.2 Normal Polycrystalline Piezoelectric Ceramics

For polarization to be possible, the material must be anisotropic. A phase diagram for the piezoelectric ceramic lead zirconate titanate (PZT[1]) is given by Figure 5.21. This plot indicates that the type of anisotropic symmetry depends on both composition and temperature. Note that in Figure 5.21, coupling and dielectric permittivity increase rapidly near the phase boundary. These ceramics are poled close to this boundary to get high values. All ferroelectric materials have a Curie temperature (T_C), above which the material no longer exhibits ferroelectric properties. Properties of the ceramic are more stable at temperatures farther below the Curie temperature.

Ceramics such as the polycrystalline PZT family are called normal ferroelectrics and are the most popular materials for medical transducers. Combining high coupling and large permittivity with low cost, physical durability, and stability, they are currently the material of choice for most array applications.

[1] Trademark, Vernitron Piezoelectric Division.

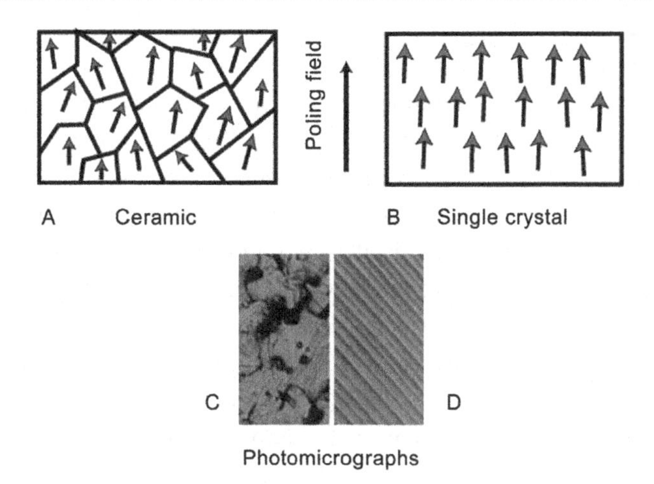

Figure 5.20
Microstructure of piezoelectric materials.
(A) Aligned electric dipoles in domains of a poled polycrystalline ferroelectric; (B) highly aligned dipoles in domain-engineered, poled single-crystal ferroelectric; (C) 800 × photomicrograph of polycrystalline ceramic; (D) 800 × photomicrograph of domain-engineered, poled single crystal.
C and D courtesy of Philips Healthcare.

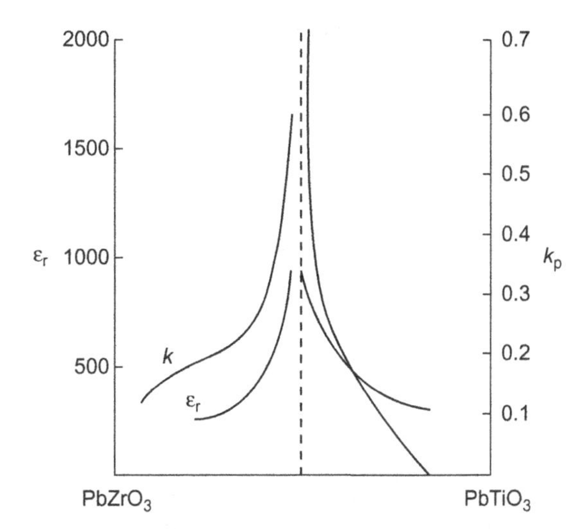

Figure 5.21
PZT Phase Diagram.
On the left scale is the dielectric constant, and on the right scale is electromechanical coupling as a function of chemical composition. The dashed line is the phase boundary. *From Safari, Panda, and Janas (1996).*

5.7.3 Relaxor Piezoelectric Ceramics

Relaxor ferroelectrics have many strange characteristics, as well as more diffuse phase boundaries and lower Curie temperatures (Shrout & Fielding, 1990) than normal ferroelectrics. Their permittivities are usually strongly frequency dependent. While crystals can function as normal piezoelectrics, they can also be electrostrictive under certain conditions. Electrostrictive materials have strains that change with the square of the applied electric fields (a different mechanism from piezoelectricity). This property leads to some unusual possibilities in which the piezoelectric characteristics of a device can be altered or switched on or off via a bias voltage (Chen & Gururaja, 1997; Takeuchi, Masuzawa, Nakaya, & Ito, 1990). All dielectrics can be electrostrictors; however, the relaxor piezoelectrics have large coupling constants because they can be highly polarized.

The Maxwell stress tensor for dielectrics (Stratton, 1941) shows that the stress is proportional to the applied electric field squared:

$$T_{33} = \left(\frac{\varepsilon - a_3}{2}\right)E^2, \tag{5.41}$$

where a_3 is a deformation constant. If the thickness of the dielectric is d and a DC bias voltage (V_{DC}) is applied to electrodes in combination with an AC signal of amplitude A_0, then:

$$T_{33} = \left(\frac{\varepsilon - a_3}{2d^2}\right)(V_{DC} + V_0 \sin \omega_1 t)^2, \tag{5.42A}$$

and

$$T_{33} = \left(\frac{\varepsilon - a_3}{2d^2}\right)\left(V_{DC}^2 + V_0^2/2 + 2V_{DC}V_0 \sin \omega_1 t - \frac{V_0^2 \cos 2\omega_1 t}{2}\right), \tag{5.42B}$$

in which the third term in the second parentheses indicates how the bias can control the amplitude of the original sinusoid at frequency ω_1, and the last term is at the second harmonic of this frequency.

5.7.4 Single-crystal Ferroelectrics

A number of ferroelectrics are termed "single crystal" because of their highly ordered domains, symmetrical structure, very low losses, and moderate coupling. These hard, brittle materials require optical-grade cutting methods, and therefore, they tend not to be used for medical devices, but rather for high-frequency surface and bulk acoustic wave transducers, as well as for optical devices. This group of materials includes lithium niobate, lithium tantalate, and bismuth germanium oxide.

5.7.5 Piezoelectric Organic Polymers

Some polymers with a crystalline phase have been found to be ferroelectric and piezoelectric. Poling is achieved through a combination of stretching, elevating temperature, and applying a high electric field. Two popular piezopolymers are polyvinylidene fluoride, or PVDF (Kawai, 1969), and copolymer PVDF with trifluoroethylene (Ohigashi et al., 1984). Advantages of these materials are their conformability and low acoustic impedance. The low impedance is not as strong an advantage because matching layers can be utilized with higher-impedance crystals. Drawbacks are a relatively low coupling constant (compared to PZT), a small relative dielectric constant (5–10, which is a big drawback for small array-element sizes), a high dielectric loss tangent (0.15–0.25 compared to 0.02 for PZT), and a low Curie temperature (70–100°C). These materials are better as receivers such as hydrophones and are less efficient as transmitters (Callerame, Tancrell, & Wilson, 1978). A special issue of the *IEEE Transactions on Ultrasonics, Ferroelectrics and Frequency Control* has been devoted to many applications of these polymers (Brown & Harris, 2000).

5.7.6 Domain-engineered Ferroelectric Single Crystals

A relatively recent development is the growing of domain-engineered single crystals. Unlike other ferroelectric relaxor-based ceramics, in which domains are randomly oriented with most of them polarized, these materials are grown to have a nearly perfect alignment of domains (shown in Figure 5.20B and photomicrograph, 5.20D). Considerable investments in materials research and special manufacturing techniques were necessary to achieve extremely high coupling constants (Park & Shrout, 1997; Saitoh et al., 1999) and other desirable properties in crystals such as PZN-4.5% PT (lead zinc niobate-lead titanate) and 0.67 PMN-0.33 PT (Yin, Jiang, & Cao, 2000; Zhang, Jiang, & Cao, 2001). Because both sensitivity and bandwidth are proportional to the coupling constant squared, significant improvements are possible (as discussed in Section 5.8).

5.7.7 Composite Materials

Another successful attempt at optimizing transducer materials for applications like medical ultrasound is the work on piezoelectric composites (Gururaja, Schulze, Cross, & Newnham, 1985; Newnham, Skinner, & Cross, 1978). PZT, which has the drawback that its acoustic impedance is about 30 megaRayls, is mismatched to tissue impedances of about 1.5 megaRayls. By imbedding pieces of PZT in a low-impedance polymer material, a composite with both high coupling and lower impedance is achieved. Two of the most common composite structures are illustrated in Figure 5.22. In a 1–3 composite, posts of a piezoelectric material are organized in a grid and backfilled with a polymer such as epoxy. A 2–2 composite consists of alternating sheets of piezoelectric and polymer material. For

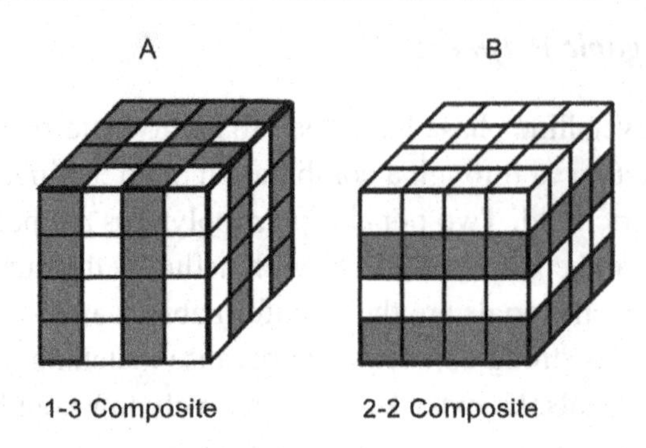

Figure 5.22
(A) 1–3 Composite Structure; (B) 2–2 Composite Structure. *From Safari, Panda, and Janas (1996).*

design purposes, a composite can be described by "effective parameters" as if it were a homogeneous solid structure (Smith, Shaulov, & Singer, 1984). Effective parameters for two 1–3 composites, one with PZT-5H and another with single-crystal PMN (Ritter et al., 2000), are listed in Table B.2 in Appendix B. More on composite high frequency arrays can be found in Bantignies et al., 2009.

5.7.8 Piezoelectric Gels

For high-frequency single transducers and arrays, alternative piezoelectric materials have been developed to replace thin, fragile piezoelectric substrates. Lukacs, Sayer, and Foster (1997, 2000) have made transducers from a piezoelectric PZT composite gel that can be spread thinly on a substrate. Broadband PZT thick film transducers up to 100 MHz have been successfully made by this method (Zhang et al., 2006; Zhou, Shung, Zhang, & Djuth, 2006).

5.7.9 Lead-free Piezoelectrics

The most widely used piezoelectric material is the lead-based ceramics such as PZT. Because of environmental concerns, lead-free piezoelectrics are of interest. Even though there are single-crystal materials such as quartz that are lead-free, their high acoustic impedance and relatively low electromechanical coupling constants make them less suitable for medical imaging applications. Shen, Li, Chen, Zhou, and Shung (2011) describe the use of SPSed (spark plasma sintering) Pb-free NKLNT fine-grain piezoceramics to construct a 1-3 NKLNT/epoxy composite transducer with two matching layers. The composite had a respectable coupling constant of $k_t = 0.655$, a low acoustic impedance of

6.6 MRayls, and a relative dielectric constant of $\varepsilon_r = 302$. The resulting transducer achieved a 90% fractional bandwidth and a center frequency of 29 MHz.

5.8 Comparison of Piezoelectric Materials

Because of the many factors involved in transducer design (Sato, Kawabuchi, & Fukumoto, 1980), it is difficult to select a single back-of-the-envelope criterion for comparing the most important material characteristics. The following are simplifications, but they provide a relative means that agrees with observations. Usually, impedances of transducer elements are high because of their small size; therefore, $R_{A0} >> R_g$. From the electrical side, the -3-dB bandwidth is given approximately by the electrical Q_e:

$$BW = 1/Q_e = \omega_0 C_0 R_{A0} = \frac{4k^2}{\pi}, \tag{5.43}$$

in which matching layers are assumed as well as $Z_B << Z_C$ and $K = K_{33}$ for most materials, except the composites $BaTiO_3$, and PVDF, for which K_T is used. Furthermore, the electrical bandwidth is assumed to be much smaller than the acoustical bandwidth from the acoustical loss factor, and therefore, it dominates. Another important factor in determining acoustic impedance, which is inversely proportional to clamped capacitance, is the relative dielectric constant (ε^S). These two figures of merit are plotted in Figure 5.23 for materials with the constants appropriate for a geometry in common use. Ideally, materials in the upper right of the graph would be best for array applications.

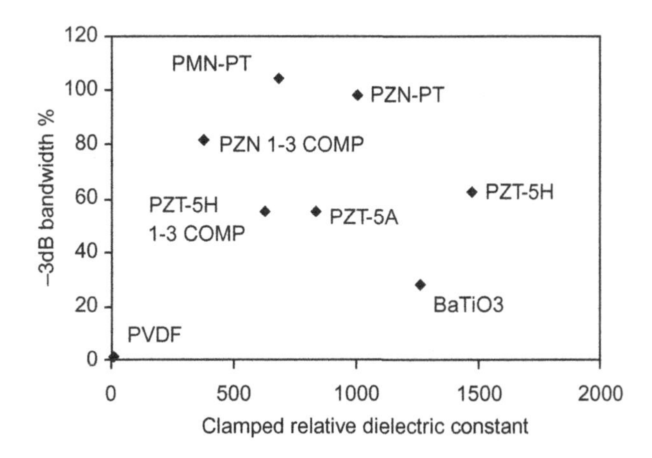

Figure 5.23
Comparison of Piezoelectric Materials.
$-$ 3-dB bandwidth versus relative dielectric constant.

As a specific example, consider the spectrum of a design optimized for a 5-MHz array transducer on PZT-5H compared to that of a design optimized for PZN-M domain-engineered single-crystal material, which is shown in Figure 5.24. The coupling constants and calculated −6-dB round-trip bandwidths for the two cases are 0.66 and 66%, and 0.83 and 90%, which are in good agreement with the bandwidth estimates of 56 and 86%, respectively (note a one-way −3-dB bandwidth is equivalent to a −6-dB round-trip bandwidth). A simple estimate of the relative spectral peak sensitivities is in proportion to their coupling constants to the fourth power (see Eqns 5.15B and 5.22C). In this case, the estimate for the relative −6-dB round-trip spectral peaks is +4 dB compared to the calculated value of 3–5 dB. Additional comparisons of more transducer materials are discussed in Szabo and Lewin (2007).

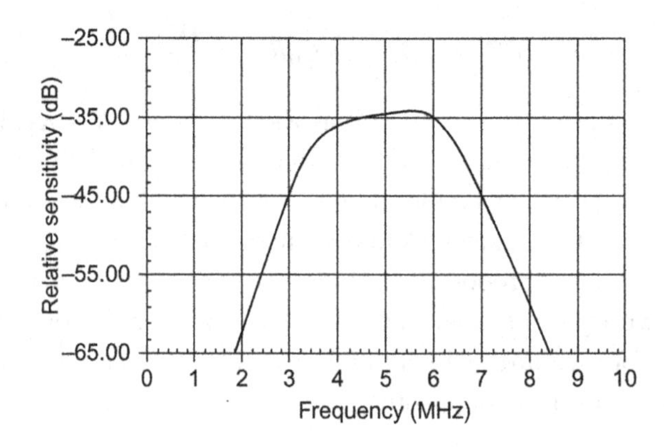

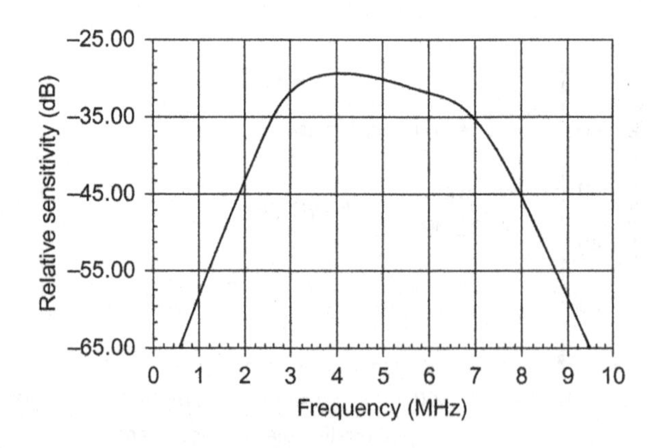

Figure 5.24

Comparison of Round-trip Spectra for 2.5-MHz-center Frequency Designs.
(A) PZT-5H, and (B) single-crystal PZN-M transducers. *From Gururaja, Panda, Chen, and Beck (1999), IEEE.*

5.9 Transducer Advanced Topics

5.9.1 Internal Transducer Losses

Two other effects that often affect transducer performance are losses and connecting cables. Two major types of losses are internal mechanical losses within the crystal element and absorption losses in the materials used. The usually small crystal mechanical loss can be modeled by placing a loss resistance in parallel with the transducer C_0. Piezoelectric material manufacturers provide information about this loss through mechanical Q data. As we found from Chapter 4, all acoustic materials have absorption loss and dispersion. Loss can be easily included in an ABCD matrix notation by substituting the lossless transmission line matrix in Figure 3.4 by its lossy replacement:

$$\begin{pmatrix} A & B \\ C & D \end{pmatrix} = \begin{pmatrix} \cosh(\gamma d) & Z_0 \sinh(\gamma d) \\ \sinh(\gamma d)/Z_0 & \cosh(\gamma d) \end{pmatrix}, \tag{5.44}$$

in which γ is the complex propagation factor from Chapter 4, d is the length of the transmission line, and Z_0 is its characteristic impedance. Finally, array elements are most often connected to a system through a coaxial cable, which can also be modeled by the same lossy transmission line matrix with appropriate electromagnetic parameters. Lockwood and Foster (1994) devised methods for accurately measuring and modeling internal losses especially for high frequency transducers. Signal-to-noise ratios can also be calculated by a modified KLM model (Oakley, 1997). Methods of incorporating the switching directly in the transducer have been accomplished (Busse et al., 1997).

5.9.2 Trends in Transducer Modeling

The 1D transducer model is a surprisingly useful and accurate design tool. Array architectures are not really composed of individual isolated elements because they are close to each other, and as a result, mechanical and electrical cross-coupling effects occur (to be discussed in Chapter 7). In addition to the dispersion of the elements, these effects can be predicted by more realistic finite element modeling (FEM) (Lerch, 1990). A 3D depiction of the complicated vibrational mode of array elements can be predicted by a commercially available FEM program, PZFLEX (Wojcik et al., 1996). To be accurate, a precise knowledge of all the material parameters is required, as discussed by McKeighen (2001).

FEM modeling is especially helpful in predicting the behavior of advanced arrays. These arrays include 1.5D (Wildes et al., 1997) and 2D (Kojima, 1986) arrays.

5.9.3 Matrix or 2D Arrays

Several major problems for 2D arrays are electrically matching and connecting to large numbers of small elements, as well as spurious coupled vibrational modes. Turnbull and Foster (1992) described the modeling and construction of a 2D array and cross-coupling effects. Davidsen and Smith (1998) proposed a multilayer circuit interconnect scheme to ease the problem of a large number of electrical element connections for 2D arrays. Another solution was to integrate the electronics and switch into the transducer structure through the use of multilayer chip fabrication techniques (Erikson, Hairston, Nicoli, Stockwell, & White, 1997). A 16,384-element 2D array has been made by this method for C-scan imaging. Other alternatives are reviewed by von Ramm (2000). Philips Medical Systems introduced a fully populated 2D array with microbeamformers built into the handle for a real-time 3D imaging system (Savord & Solomon, 2003). More information on recent developments in 2D array fabrication and beamforming can be found in Chapter 7 and Chapter 10.

5.9.4 CMUT Arrays

Another approach to the large array fabrication issue is an alternative transduction technology, called capacitive micromachined ultrasonic transducers (CMUTs; see Figure 5.25), which is based on existing silicon fabrication methods (Haller & Khuri-Yakub, 1994; Ladabaum et al., 1996). To first order, the CMUT is a tiny, sealed air-filled capacitor. When a bias voltage is applied to these miniature membrane transducers, a stress is developed proportional to the voltage applied squared, and the top electrode membrane deflects. Like the Maxwell stress tensor equations, Eqns 5.41 and 5.42, if the DC bias includes an AC signal, the pressure or deflection can carry AC signal information. The voltage applied is:

$$V = V_{DC} + V_{AC}, \tag{5.45A}$$

resulting in a vertical deflection:

$$x = x_{DC} + x_{AC}. \tag{5.45B}$$

To first order, the pressure on the membrane is:

$$p_E = \frac{\varepsilon_0 V_{DC}}{d_0^2(r)} V_{AC} + \frac{\varepsilon_0 V_{DC}^2}{d_0^2(r)} x_{AC}, \tag{5.45C}$$

where $d_0(r)$ is a radial displacement. Note the similarity to Eqn 5.42. A model more appropriate for 2D arrays can be found in Caronti, Caliano, Iula, and Pappalardo (1986). This equivalent circuit model is a combination of electrostatics and the acoustics of a

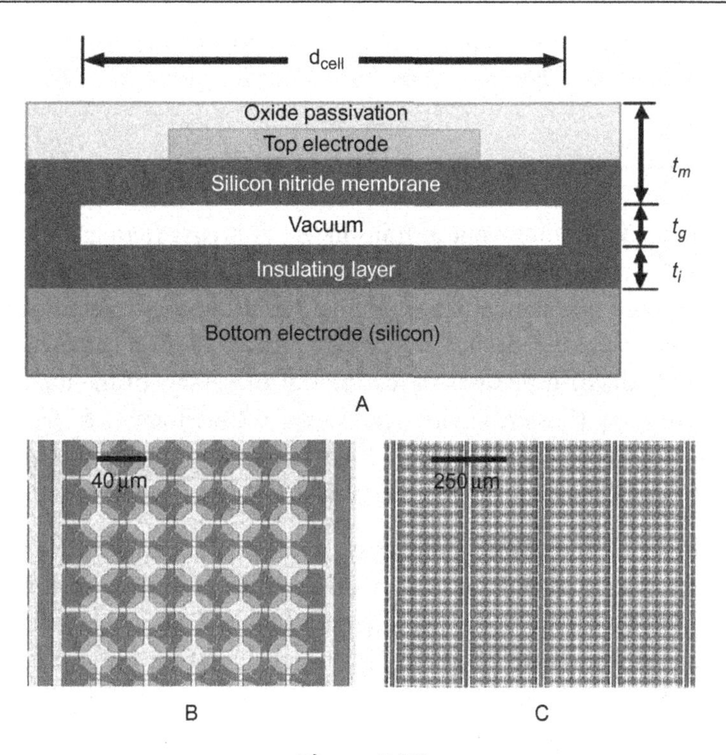

Figure 5.25
One-dimensional CMUT Array.
(A) Schematic cross-section of a CMUT cell; (B) magnified view of a single five-cell-wide 1D array element; (C) a portion of four elements of the 1D CMUT array. *From Oralkan et al. (2002), IEEE.*

miniature resonant drum, and it predicts the radiation impedance and other characteristics of the CMUT. In addition to design improvements, the flexibility of fabrication options (Ergun et al., 2005) enables the replication of individual cells into a variety of geometric active area configurations (Khuri-Yakub & Oralkan, 2011) such as 1D and 2D (Oralkan et al., 1999) (Zhuang et al., 2005) and ring arrays. The scalability of the CMUT approach allows miniature arrays to be made without some of the attendant spurious modes encountered in conventional 2D arrays described in the previous section (methods to minimize unwanted modes unique to CMUTS have been devised). A major advantage is the monolithic integration of the elements with front-end electronics, thereby greatly simplifying interconnect problems and opening the possibility of large-scale integration at lower per-unit cost. CMUTs can have broad bandwidth and high receive sensitivity, and their transmit capabilities have improved to the point of attracting considerable interest in the medical imaging community. Imaging with CMUT arrays has been demonstrated (Caronti et al., 2006; Oralkan et al., 2002; Panda, Daft, & Wagner, 2003; Yeh et al., 2005). A solution to the high DC voltage drive requirement (Machida et al., 2009) led to the first commercialized diagnostic ultrasound array, the "Mappie" (Sako et al., 2009). Other applications include miniature intracardiac imaging arrays and therapeutics. Khuri-Yakub and Oralkan (2011) provide a comprehensive review of recent CMUT developments.

5.9.5 High-Frequency Transducers

An important trend is the development of transducers at higher frequencies. Commercially available intravascular ultrasound (IVUS) imaging systems operate in the 20–40-MHz range. Either a miniature mechanical single-element transducer is rotated or phased or synthetic array elements are electronically scanned on the end of a catheter to obtain circumferential highly detailed pictures of the interior of vessels of the human body. Ultrasound biomicroscopy (Foster, Pavlin, Harasiewicz, Christopher, & Turnbull, 2000; Saijo & Chubachi, 2000) provides extremely high resolution images, as well as new information about the mechanical functioning and structure of living tissue.

Recent advances in the design of miniature arrays at frequencies of 20 MHz and above have enabled the commercialization and availability of these arrays, in conjunction with high-frequency imaging systems. One of the main initiatives of the National Center for Transducers at Pennsylvania State University is the development of high-frequency transducers (Ritter, Shrout, Tutwiler, & Shung, 2002) and arrays and materials. Careful attention in design of these arrays is focused on reducing losses to and within the transducer (Lockwood & Foster, 1994) and developing low loss materials (Wang, Ritter, Cao, & Shung, 1999). Examples of high-frequency arrays can be found in the piezoecomposite 30-MHz array (Michau, Mauchamp, & Dufait, 2004), a 35-MHz array made from a 2–2 composite (Cannata, Williams, Zhou, Ritter, & Shung, 2006), and 40-MHz arrays with a 1–3 composite construction (Bantignies et al., 2011; Brown, Foster, Needles, Cherin, & Lockwood, 2007). An alternative 1–3 composite construction can be found in Yin, Lee, Brown, Chérin, & Foster (2010), and micromachining for high-frequency arrays is in Lukacs et al. (2006) and Zhou et al. (2006). A more detailed and comprehensive review of advances in high-frequency arrays up 2006 is in Cannata et al. (2006). Features of high-frequency arrays and a digital high-frequency system are demonstrated in Foster et al. (2009).

References

Auld, B. A. (1990). *Acoustic waves and fields in solids* (Vol. 1). Malabar, FL: Krieger Publishing.

Bantignies,C., Mauchamp, P., Férin, G., Michau, S., & Dufait,R. (2009) Focused 20-MHz single-crystal piezocomposite ultrasound array. *Proceedings of the IEEE ultrasonics symposium* (pp. 2722–2725). Rome.

Bantignies, C., Mauchamp, P., Dufait, R., Levassort, F., Mateo, T., Grégoire, J.-M., et al. (2011) 40-MHz piezo-composite linear array for medical imaging and integration in a high resolution system. *Proceedings of the IEEE ultrasonics symposium* (pp. 226–229). Orlando, FL.

Brown, J. A., Foster, F. S., Needles, A., Cherin, E., & Lockwood, G. R. (2007). Fabrication and performance of a 40-MHz linear array based on a 1-3 composite with geometric elevation focusing. *IEEE Transactions on Ultrasonics, Ferroelectrics and Frequency Control, 54*, 1888–1894.

Brown, L. F., & Harris, G. R. (Eds.), (2000). Special issue on the 30th anniversary of the discovery of piezoelectric PVDF. *IEEE Transactions on Ultrasonics, Ferroelectrics and Frequency Control*, (6), 47.

Busse, L. J., Oakley, C. G., Fife, M. J., Ranalletta, J. V, Morgan, R. D., & Dietz, D. R. (1997). The acoustic and thermal effects of using multiplexers in small invasive probes. *Proceedings of the IEEE ultrasonics symposium* (Vol. 2) (pp. 1721–1724). Toronto.

Callerame, J. D., Tancrell, R. H., & Wilson, D. T. (1978). Comparison of ceramic and polymer transducers for medical imaging. *Proceedings of the IEEE ultrasonics symposium* (pp. 117–121).

Cannata, J. M., Williams, J. A., Zhou, Q., Ritter, T. A., & Shung, K. K. (2006). Development of a 35-MHz piezo-composite ultrasound array for medical imaging. *IEEE Transactions on Ultrasonics, Ferroelectrics and Frequency Control, 53*, 224–236.

Caronti, A., Caliano, G., Carotenuto, R., Savoia, A., Pappalardo, M., Cianc, E., et al. (2006). Capacitive micromachined ultrasonic transducer (CMUT) arrays for medical imaging. *Microelectronics Journal, 37*, 770–777.

Caronti, A., Caliano, G., Iula, A., & Pappalardo, M. (1986). An accurate model for capacitive micromachined ultrasonic transducers. *IEEE Transactions on Ultrasonics, Ferroelectrics and Frequency Control, 33*, 295–298.

Chen, J., & Gururaja, T. R. (1997). DC-biased electrostrictive materials and transducers for medical imaging. *Proceedings of the IEEE ultrasonics symposium* (Vol. 2, pp. 1651–1658). Toronto.

Davidsen, R. E., & Smith, S. W. (1998). Two-dimensional arrays for medical ultrasound using multilayer flexible circuit interconnection. *IEEE Transactions on Ultrasonics, Ferroelectrics and Frequency Control, 45*, 338–348.

Desilets, C. S., Fraser, J. D., & Kino, G. S. (1978). The design of efficient broad-band piezoelectric transducers. *IEEE Transactions on Sonics and Ultrasonics SU-25*, 115–125.

de Jong, N., Souquet, J., & Bom, N. (1985). Vibration modes, matching layers, and grating lobes. *Ultrasonics*, July 1985, 176–182.

Ergun, A. S., Huang, Y., Zhuang, X., Oralkan, O., Yaralıoglu, G. G., & Khuri-Yakub, B. T. (2005). Capacitive micromachined ultrasonic transducers: Fabrication technology. *IEEE Transactions on Ultrasonics, Ferroelectrics and Frequency Control, 52*, 2242–2258.

Erikson, K., Hairston, A., Nicoli, A., Stockwell, J., & White, T. A. (1997). 128 × 128K (16 k) ultrasonic transducer hybrid array. *Acoustical Imaging, 23*, 485–494.

Foster, F. S. (2000). Transducer materials and probe construction. *Ultrasound in Medicine and Biology, 26* (Suppl. 1), S2–S5.

Foster, F. S., Larso, J. D., Masom, M. K., Shoup, T. S., Nelson, G., & Yoshida, H. (1989). Development of a 12 element annular array transducer for real-time ultrasound imaging. *Ultrasound in Medicine and Biology, 15*, 649–659.

Foster, F. S., Pavlin, C. J., Harasiewicz, K. A., Christopher, D. A., & Turnbull, D. H. (2000). Advances in ultrasound biomicroscopy. *Ultrasound in Medicine and Biology, 26*, 1–27.

Foster, S. F., Mehi, J., Lukacs, M., Hirson, D., White, C., Chaggares, C., et al. (2009). A new 15–50 MHz array-based micro-ultrasound scanner for preclinical imaging. *Ultrasound in Medicine and Biology, 35*, 1700–1708.

Goll, J., & Auld, B. A. (1975). Multilayer impedance matching schemes for broadbanding of water loaded piezoelectric transducers and high Q resonators. *IEEE Transactions on Sonics and Ultrasonics, 22*, 53–55.

Gururaja, T. R., Panda, R. K., Chen, J., & Beck, H. (1999). Single crystal transducers for medical imaging applications. *Proceedings of the IEEE ultrasonics symposium* (Vol. 2, pp. 969–972). Caesars Tahoe, NV.

Gururaja, T. R., Schulze, W. A., Cross, L. E., & Newnham, R. E. (1985). Piezoelectric composite materials for ultrasonic transducer applications, Part 11: Evaluation of ultrasonic medical applications. *IEEE Transactions on Sonics and Ultrasonics, 32*, 499–513.

Haller, M. I., & Khuri-Yakub, B. T. (1994). A surface micromachined electrostatic ultrasonic air transducer. *Proceedings of the IEEE ultrasonics symposium*, 1994, Cannes, France. 2, 1241–1244.

Hunt, J. W, Arditi, M., & Foster, F. S. (1983). Ultrasound transducers for pulse-echo medical imaging. *IEEE Transactions on Biomedical Engineering, 30*, 452–481.

Hutchens, C. G. (1986). A three dimensional equivalent circuit for a tall parallel piped piezoelectric. *Proceedings of the IEEE ultrasonics symposium* (pp. 321–325). Williamsburg, VA.

Hutchens, C. G., & Morris, S. A. (1984). A three port model for thickness mode transducers using SPICE II. *Proceedings of the IEEE ultrasonics symposium* (pp. 897–902). Dallas, TX.

Hutchens, C. G., & Morris, S. A. (1985). A two dimensional equivalent circuit for the tall thin piezoelectric bar. *Proceedings of the IEEE ultrasonics symposium* (pp. 671–676). San Francisco, CA.

Kawai, H. (1969). The piezoelectricity of poly(vinylidene fluoride). *Japanese Journal of Applied Physics, 8*, 975–976.

Kino, G. S. (1987). *Acoustic waves: devices, imaging, and analog signal processing*. Englewood Cliffs, NJ: Prentice-Hall.

Khuri-Yakub, B. T., & Oralkan, O. (2011). Capacitive micromachined ultrasonic transducers for medical imaging and therapy. *Journal of Micromechanics and Microengineering, 21*(5), 054004–054014.

Kojima, T. (1986). Matrix array transducer and flexible matrix array transducer. *Proceedings of the IEEE ultrasonics symposium* (pp. 335–338). Williamsburg, VA.

Ladabaum, I., Jin, X., Soh, H. T., Pierre, F., Atalar, A., & Khuri-Yakub, B. T. (1996). Microfabricated ultrasonic transducers: Towards robust models and immersion devices. *Proceedings of the IEEE ultrasonics symposium* (Vol. 1, pp. 335–338). San Antonio, TX.

Leedom, D. A., Krimholtz, R., & Matthaei, G. L. (1978). Equivalent circuits for transducers having arbitrary even- or odd-symmetry piezoelectric excitation. *IEEE Transactions on Sonics and Ultrasonics, 25*, 115–125.

Lerch, R. (1990). Simulation of piezoelectric devices by two- and three-dimensional finite elements. *IEEE Transactions on Ultrasonics, Ferroelectrics and Frequency Control, 37*, 233–247.

Lockwood, G. R., & Foster, F. S. (1994). Modeling and optimization of high frequency ultrasound transducers. *IEEE Transactions on Ultrasonics, Ferroelectrics and Frequency Control, 41*, 225–230.

Lukacs, M., Sayer, M., & Foster, F.S., (1997) Single element and linear array PZT ultrasound biomicroscopy transducers. *Proceedings of the IEEE ultrasonics symposium* (Vol. 2) (pp. 1709–1712). Toronto.

Lukacs, M., Sayer, M., & Foster, F. S. (2000). Single element high frequency (<50 MHz) PZT sol–gel composite ultrasound transducers. *IEEE Transactions on Ultrasonics, Ferroelectrics and Frequency Control, 47*, 148–159.

Lukacs, M., Yin, J., Pang, G., Garcia, R. C., Cherin, E., Williams, R., et al. (2006). Performance and characterization of new micromachined high-frequency linear arrays. *IEEE Transactions on Ultrasonics, Ferroelectrics and Frequency Control, 53*, 1719–1729.

Machida, S., Kobayashi, T., Degawa, M., Takezaki, T., Tanaka, H., Migitaka, S., et al. (2009). Capacitive micromachined ultrasonic transducer with driving voltage over 100 V and vibration durability over 1011 cycles. *Proceedings of the IEEE international solid-state sensors, actuators and microsystems conference* (pp. 2218–2221), Denver, CO.

Mason, W. P. (Ed.), (1964). *Physical acoustics* (Vol. 1A). New York: Academic Press, Chapter 3.

Matthaei, G. L., Young, L., & Jones, E. M. T. (1980). *Microwave filters, impedance-matching networks, and coupling structures*. Dedham, MA: Artech House. pp. 255–354, Chapter 6.

McKeighen, R. (2001). Finite element simulation and modeling of 2D arrays for 3D ultrasonic imaging. *IEEE Transactions on Ultrasonics, Ferroelectrics and Frequency Control, 48*, 1395–1405.

McKeighen, (1997). Influence of pulse drive shape and tuning on the broadband response of a transducer. *Proceedings of the IEEE ultrasonics symposium* (Vol. 2, pp. 1637–1642). Toronto.

Melton, H. E., & Thurstone, F. L. (1978). Annular array design and logarithmic processing for ultrasonic imaging. *Ultrasound in Medicine and Biology, 4*, 1–12.

Michau, S., Mauchamp, P., & Dufait R. (2004). Piezocomposite 30 MHz linear array for medical imaging: Design challenges and performances evaluation of a 128 element array. *Proceedings of the IEEE ultrasonics symposium* (Vol. 2, pp. 898–901).

Mills, D. M., & Smith, S. W. (2002). Finite element comparison of single crystal vs. multi-layer composite arrays for medical ultrasound. *IEEE Transactions on Ultrasonics, Ferroelectrics and Frequency Control, 49*, 1015–1020.

Morris, S. A., & Hutchens, C. G. (1986). Implementation of Mason's model on circuit analysis programs. *IEEE Transactions on Ultrasonics, Ferroelectrics and Frequency Control, 33*, 295–298.

Nalamwar, A. L., & Epstein, M. (1972). Immitance characterization of acoustic surface-wave transducers. *Proceedings of the IEEE, 60*, 336–337.

Newnham, R. E., Skinner, D. P., & Cross, L. E. (1978). Connectivity and piezoelectric-pyroelectric composites. *Materials Research Bulletin, 13*, 525–536.

Oakley, C. G. (1997). Calculation of ultrasonic transducer signal-to-noise ratios using the KLM model. *IEEE Transactions on Ultrasonics, Ferroelectrics and Frequency Control, 44*, 1018–1026.

Ohigashi, H., Koga, K., Suzuki, M., Nakanishi, T., Kimura, K., & Hashimoto, N. (1984). Piezoelectric and ferroelectric properties of P (VDF-TrFE) copolymers and their application to ultrasonic transducers. *Ferroelectrics, 60*, 264–276.

Onoe, M., & Tiersten, H. F. (1963). Resonant frequencies of finite piezoelectric ceramic vibrators with high electromechanical coupling. *IEEE Transactions on Ultrasonics Engineering, 10*, 32–39.

Oralkan, O., Jin, X. C., Degertekin, F. L., & Khuri-Yakub, B. T. (1999). Simulation and experimental characterization of a 2D, 3-MHz capacitive micromachined ultrasonic transducer (CMUT) array element. *Proceedings of the IEEE ultrasonics symposium* (Vol. 2, pp. 1141–1144). Caesars Tahoe, NV.

Oralkan, O., Ergun, A. S., Johnson, J. A., Karaman, M., Demirci, U., Kaviani, K., et al. (2002). Capacitive micromachined ultrasonic transducers: Next-generation arrays for acoustic imaging? IEEE Transactions on Ultrasonics. *Ferroelectrics and Frequency Control, 49*, 1596–1610.

Panda, S., Daft, C., & Wagner, C. (2003). Microfabricated ultrasound transducer (CMUT) probes: Imaging advantages over piezoelectric probes. *Ultrasound in Medicine and Biology, 29*(5S), S69.

Park, S. E., & Shrout, T. R. (1997). Characteristics of relaxor-based piezoelectric single crystals for ultrasonic transducers. *IEEE Transactions on Ultrasonics, Ferroelectrics and Frequency Control, 44*, 1140–1147.

Persson, H. W., & Hertz, C. H. (1985). Acoustic impedance matching of medical ultrasound transducers. *Ultrasonics, 1985*, 83–89.

Puttmer, A., Hauptmann, P., Lucklum, R., Krause, O., & Henning, B. (1997). SPICE model for lossy piezoceramic transducers. *IEEE Transactions on Ultrasonics, Ferroelectrics and Frequency Control, 44*, 60–67.

Redwood, M. (1963). A study of waveforms in the generation and detection of short ultrasonic pulses. *Applied Materials Research, 2*, 76–84.

Reid, J. M., & Lewin, P. A. (1999). Ultrasonic transducers, imaging. Wiley Encyclopedia of Electrical and Electronics Engineering Online, December 17, 1999. Available at: http://www.mrw.interscience.wiley.com/eeee.

Reid, J. M., & Wild, J. J. (1958). Current developments in ultrasonic equipment for medical diagnosis. *Proceedings of the national electronics Conference XII* (pp. 1002–1015), Chicago IL.

Rhyne, T. L. (1996). Computer optimization of transducer transfer functions using constraints on bandwidth, ripple and loss. *IEEE Transactions on Ultrasonics, Ferroelectrics and Frequency Control, 43*, 136–149.

Ritter, T., Geng, X., Shung, K. K., Lopath, P. D., Park, S. E., & Shrout, T. R. (2000). Single crystal PZN/PT-polymer composites for ultrasound transducer applications. *IEEE Transactions on Ultrasonics, Ferroelectrics and Frequency Control, 47*, 792–800.

Ritter, T. A., Shrout, T. R., Tutwiler, R., & Shung, K. K. (2002). A 30-MHz piezo-composite ultrasound array for medical imaging applications. *IEEE Transactions on Ultrasonics, Ferroelectrics and Frequency Control, 49*, 217–230.

Sachse, W, & Hsu, N. N. (1979). Ultrasonic transducers for materials testing and their characterization. In W. P. Mason, & R. N. Thurston (Eds.), *Physical acoustics* (Vol. XIV). New York: Academic Press, Chapter 4.

Safari, A., Panda, R. K., & Janas, V. F. (1996). Ferroelectricity: Materials, characteristics and applications. *Key Engineering Materials, 35–70*, 122–124.

Saijo, Y., & Chubachi, N. (2000). Microscopy. *Ultrasound in Medicine and Biology, 26*(Suppl. 1), S30–S32.

Saitoh, S., Takeuchi, T., Kobayashi, T., Harada, K., Shimanuki, S., & Yamashita, Y. A. (1999). 3.7 MHz phased array probe using 0.91Pb (Zn1/3Nb2/3)O3 – 0.09PbTiO3 single crystal. *IEEE Transactions on Ultrasonics, Ferroelectrics and Frequency Control, 46*, 414–421.

Sako, A., Ishida, K., Fukada, M., Asafusa, K., Sano, S., & Izumi, M. (2009). Development of ultrasonic transducer "Mappie" with cMUT technology. *Medix, 51*, 31–34 Available online at: http://www.hitachi-medical.co.jp/medix/pdf/vol51/P31-34.pdf

Sato, J-I., Kawabuchi, M., & Fukumoto, A. (1979). Dependence of the electromechanical coupling coefficient on width-to-thickness ratio of plank-shaped piezoelectric transducers used for electronically scanned ultrasound diagnostic systems. *Journal of the Acoustical Society of America, 66*, 1609–1611.

Sato, J-I., Kawabuchi, M., & Fukumoto, A. (1980). Performance of ultrasound transducer and material constants of piezoelectric ceramics. *Acoustical Imaging, 10*, 717–729.

Savord, B., & Solomon, R. (2003). Fully sampled matrix transducer for real time 3D ultrasonic imaging. *Proceedings of the IEEE ultrasonics symposium* (Vol. 1, pp. 945–953).

Selfridge, A. R., Baer, R., Khuri-Yakub, B. T., & Kino, G. S. (1981). Computer-optimized design of quarter-wave acoustic matching and electrical networks for acoustic transducers. *Proceedings of the IEEE ultrasonics symposium* (pp. 644–648). Chicago, IL.

Selfridge, A. R., & Gehlbach, S. (1985). KLM transducer model implementation using transfer matrices. *Proceedings of the IEEE ultrasonics symposium* (pp. 875–877), San Francisco, CA.

Selfridge, A. R., Kino, G. S., & Khuri-Yakub, R. (1980). Fundamental concepts in acoustic transducer array design. *Proceedings of the IEEE ultrasonics symposium* (pp. 989–993).

Shen, Z-Y, Li, J-F, Chen, R., Zhou, Q., & Shung, K. K. (2011). Microscale 1-3-type (Na,K)NbO3-based Pb-free piezocomposites for high-frequency ultrasonic transducer applications. *Journal of the American Ceramic Society, 94*(5), 1346–1349.

Shrout, T. R., & Fielding Jr., J. (1990). Relaxor ferroelectric materials. *Proceedings of the IEEE ultrasonics symposium* (Vol. 2) (pp. 711–720). Honolulu.

Sittig, E. K. (1967). Transmission parameters of thickness-driven piezoelectric transducers arranged in multilayer configurations. *IEEE Transactions on Sonics and Ultrasonics, 14*, 167–174.

Sittig, E. K. (1971). Definitions relating to conversion losses in piezoelectric transducers. *IEEE Transactions on Sonics and Ultrasonics, 18*, 231–234.

Smith, W. A., Shaulov, A. A., & Singer, B. M. (1984). Properties of composite piezoelectric material for ultrasonic transducers. *Proceedings of the IEEE ultrasonics symposium* (pp. 539–544). Dallas, TX.

Souquet, J., Defranould, P., & Desbois, J. (1979). Design of low-loss wide-band ultrasonic transducers for noninvasive medical application. *IEEE Transactions on Sonics and Ultrasonics, 26*, 75–81.

Stratton, J. A. (1941). *Electromagnetic theory*. New York: McGraw-Hill. pp. 97–103.

Szabo, T. L. (1982). Miniature phased-array transducer modeling and design. *Proceedings of the IEEE ultrasonics symposium* (pp. 810–814) , San Diego, CA.

Szabo, T. L. (1984). Principles of nonresonant transducer design. *Proceedings of the IEEE ultrasonics symposium* (pp. 804–808). Dallas, TX.

Szabo, T. L. (1998). Transducer arrays for medical ultrasound imaging. In F. A. Duck, A. C. Baker, & H. C. Starritt (Eds.), *Ultrasound in medicine, medical science series*. Bristol, UK: Institute of Physics Publishing, Chapter 5.

Szabo, T. L. (2010). Medical ultrasound sensors. In D. Jones (Ed.), *Biomedical sensors*. New York: Momentum Press, Chapter X.

Szabo, T., & Lewin, P. (2007). Piezoelectric materials for imaging. *Journal of Ultrasound in Medicine, 26*, 283–288.

Takeuchi, H., Jyomura, S., Ishikawa, Y., and Yamamoto, E. (1982). A 7. 5 MHz linear array ultrasonic probe using modified PbTiO3. *Proceedings of the IEEE ultrasonics symposium* (pp. 849–853). San Diego, CA.

Takeuchi, H., Masuzawa, H., Nakaya, C., & Ito, Y. (1990). Relaxor ferroelectric transducers. *Proceedings of the IEEE ultrasonics symposium* (Vol. 2) (pp. 697–705) Honolulu.

Turnbull, D. H., & Foster, F. S. (1992). Fabrication and characterization of transducer elements in two-dimensional arrays for medical ultrasound imaging. *IEEE Transactions on Ultrasonics, Ferroelectrics and Frequency Control, 39*, 464–475.

van Kervel, S. J. H., & Thijssen, J. M. (1983). A calculation scheme for the optimum design of ultrasonic transducers. *Ultrasonics, 1983*, 134–140.

von Ramm, O. T. (2000). 2D arrays. *Ultrasound in Medicine and Biology, 26*(Suppl. 1), S10–S12.

Wang, H., Ritter, T., Cao, W., & Shung, K. K. (1999). Passive Materials for High Frequency Ultrasound Transducers. *SPIE Proceedings, 3664*, 35–41.

Wildes, D. G., Chiao, R. Y., Daft, C. M. W., Rigby, K. W., Smith, L. S., & Thomenius, K. E. (1997). Elevation performance of 1.25D and 1.5D transducer arrays. *IEEE Transactions on Ultrasonics, Ferroelectrics and Frequency Control, 44*, 1027–1036.

Wojcik, G., DeSilets, C., Nikodym, L., Vaughan, D., Abboud, N., & Mould J., Jr. (1996). Computer modeling of diced matching layers. *Proceedings of the IEEE ultrasonics symposium* (Vol. 2) (pp. 1503–1508). San Antonio, TX.

Yeh, D. T., Oralkan, O., Ergun, A. S., Zhuang, X., Wygant, I. O., & Khuri-Yakub, B. T. (2005). High-frequency CMUT arrays for high-resolution medical imaging. *SPIE Proceedings, 5750*, 87–94.

Yin, J., Jiang, B., & Cao, W (2000). Elastic, piezoelectric, and dielectric properties of 0.995Pb (Zn1/3Nb2/3)O3 –)0.45 PbTiO3 single crystal with designed multidomains. *IEEE Transactions on Ultrasonics, Ferroelectrics and Frequency Control, 47*, 285–291.

Yin, J., Lee, M., Brown, J., Chérin, E., & Foster, F. S. (2010). Effect of triangular pillar geometry on high-frequency piezocomposite transducers, IEEE Transactions on Ultrasonics. *Ferroelectrics and Frequency Control, 57*, 957–968.

Zhang, Q. Q., Djuth, F. T., Zhou, Q. F., Hu, C. H., Cha, J. H., & Shung, K. K. (2006). High frequency broadband PZT thick film ultrasonic transducers for medical imaging applications. *Ultrasonics, 44*, e711–e715.

Zhang, R., Jiang, B., & Cao, W (2001). Elastic, piezoelectric, and dielectric properties of multidomain 0.67 Pb (Mg1/3 Nb2/3)O3 –)0.33 PbTiO3 single crystals. *Journal of Applied Physics, 90*, 3471–3475.

Zhou, Q. F., Shung, K. K., Zhang, Q., & Djuth, F. T. (2006). High frequency piezoelectric micromachined ultrasonic transducers for imaging applications. *SPIE Proceedings, 6223*(62230H), 1–7.

Zhuang, X., Wygant, I. O., Yeh, D. T., Nikoozadeh, A., Oralkan, O., Ergun, A. S., et al. (2005). Two-dimensional capacitive micromachined ultrasonic transducer (CMUT) arrays for a miniature integrated volumetric ultrasonic imaging system. *SPIE Proceedings, 5750*, 37–46.

Bibliography

Overview treatments of transducers can be found in the following: Mason (1964); Sachse and Hsu (1979); Hunt et al. (1983); Kino (1987); Szabo (1998, 2010); Foster (2000); Reid and Lewin (1999).

CHAPTER 6

Beamforming

6.1 What is Diffraction?

Chapter 3 explained that radiation from a line source aperture consists of not just one plane wave but many plane waves being sprayed in different directions. This phenomenon is called diffraction (a wave phenomenon in which radiating sources on the scale of wavelengths create a field from the mutual interference of waves generated along the source boundary). A similar effect occurs when an ultrasound wave is scattered from an object with a size on the order of wavelengths (to be described in Chapter 8).

Acoustic diffraction is similar to what occurs in optics. When the wavelength is comparable to the size of the objects, light does not create a geometric shadow of the object but a more complicated shadow region with fringes around the object. Light from a distant source incident on an opening (aperture) on the scale of wavelengths in an opaque plane will cause a complicated pattern to appear on a screen plane behind it.

The same thing happens with sound waves as is shown in Figure 6.1, which is an intensity plot of an ultrasound field in the xz plane. In the front is the line aperture radiating along the beam axis z. Here, the scale of the z axis is compressed and represents about 1920 wavelengths, whereas the lateral length of the aperture is 40 wavelengths. Figure 6.2 gives a top view of the same field with the aperture on the bottom. Sound spills out beyond the width of the original aperture. Diffraction, in this case, gives the appearance of bending around objects! This phenomenon can be explained by the sound entering an aperture (opening) and reradiating secondary waves along the aperture beyond the region defined by straight geometric projection.

Diffraction is the phenomenon that describes beams from transducers. This chapter emphasizes frequency domain methods of predicting the characteristics of the ultrasound fields radiated by transducer apertures. It examines two major approaches: one involves spatial frequencies (the angular spectrum of plane waves), and the other employs spherical waves. This chapter also covers both circular and rectangular apertures, as well as the important topics of focusing and aperture weighting (apodization). In Chapter 7, complementary time domain methods (spatial impulse response) are applied to simulate focused and steered beams from arrays.

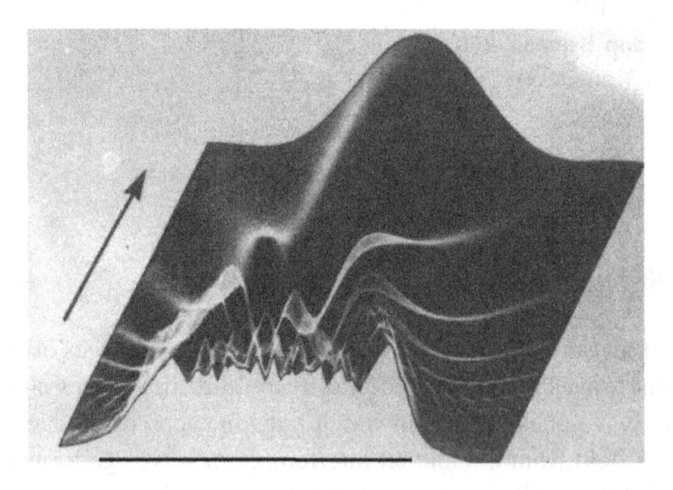

Figure 6.1
Diffracted Field of a 40-wavelength-wide Line Aperture Depicted as a Black Horizontal line. The vertical axis is intensity and shown as a gray scale (maximum equals full white), the beam axis is compressed relative to the lateral dimension, and 1920 wavelengths are shown. *From Szabo and Slobodnik (1973).*

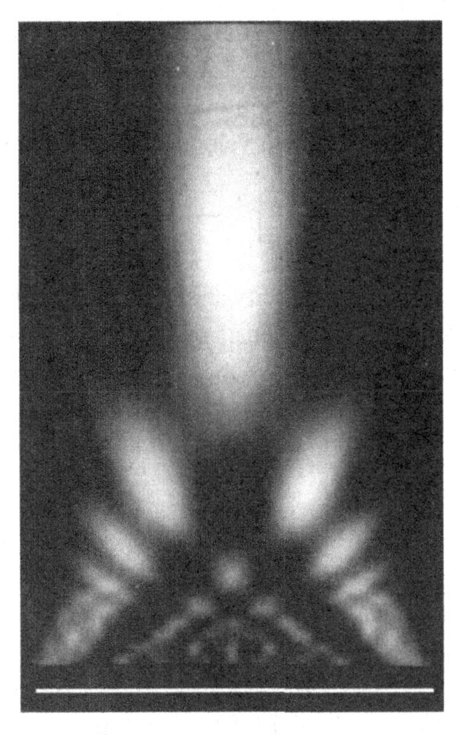

Figure 6.2
Top View of a Diffracted Field from a 40-wavelength-wide Line Aperture on the bottom. The same field from Figure 6.1 is shown in gray scale. The beam axis is compressed relative to the lateral dimension. *From Szabo and Slobodnik (1973).*

6.2 Fresnel Approximation of Spatial Diffraction Integral

Christian Huygens visualized the diffracting process as the interference from many infinitesimal spherical radiators on the surface of the aperture rather than many plane waves, an approach described in Section 2.3.2 (Spatial transform of a line source). His perspective gives an equally valid mathematical description of a diffracted field in terms of spherical radiators, as was shown in Eqn 3.17. Revisiting Figure 6.2, notice the many peaks and valleys near the aperture where the field could be interpreted as full of interference from many tiny sources crowded together. Also, far from the aperture, the spheres of influence have spread out, and the resulting field is smoother and more expansive. The Rayleigh–Sommerfeld integral (Goodman, 1968) is a mathematical way of describing Huygen's diffracting process as a velocity potential produced by an ideal radiating piston set in an (inflexible) hard baffle:

$$\varphi(r,\omega) = \frac{-1}{2\pi}\int_{s}\frac{e^{i[\omega t - k\cdot(r-r_0)]}(\partial v(r_0)/\partial n)dS}{|r-r_0|}, \tag{6.1A}$$

where $v_n = \partial v(r_0)/\partial n$ is the component of particle velocity normal to the element dS. Within the integrand, the frequency domain solution of a spherical radiator can be recognized from Chapter 3. In terms of the field pressure amplitude shown in Eqn 3.3B, this model can be described as a spatial integral of the particle velocity over the source S:

$$p(r,\omega) = \frac{i\rho_0 ckv_0}{2\pi}\int_S \frac{e^{i[\omega t - k\cdot(r-r_0)]}A(r_0)dS}{|r-r_0|}, \tag{6.1B}$$

where $v_n = v_0 A(r_0)$ is the normal particle velocity and $A(r_0)$ is its distribution across the aperture S (shown in Figure 6.3). For a rectangular coordinate system, the Fresnel or paraxial approximation of this integral is an expansion of the vector $|r - r_0|$ as a small-term binomial series:

$$|r-r_0| = \sqrt{z^2 + (x-x_0)^2 + (y-y_0)^2}, \tag{6.2A}$$

$$|r-r_0| \approx z\left[1 + \frac{1}{2}\left(\frac{x-x_0}{z}\right)^2 + \frac{1}{2}\left(\frac{y-y_0}{z}\right)^2\right], \tag{6.2B}$$

where the terms in Eqn 6.2B are small compared to one, and a replacement of $|r - r_0|$ in the denominator by z results in:

$$p(r,\omega) = \frac{i\rho_0 ckv_0}{2\pi z}e^{i(\omega t-kz)}e^{-ik(x^2+y^2)/2z}\iint_S [e^{-ik(x_0^2+y_0^2)/2z}A(x_0,y_0)]e^{ik(xx_0+yy_0)/z}dx_0dy_0. \tag{6.3}$$

If the aperture has a rectangular shape (shown in Figure 6.4), it has sharp transitions along its boundary, and it can be represented by an aperture function:

$$A(x_0,\ y_0,\ 0) = A_x(x_0)A_y(y_0), \tag{6.4A}$$

$$A(x_0, y_0, 0) = \prod(x_0/L_x)\prod(y_0/L_y). \tag{6.4B}$$

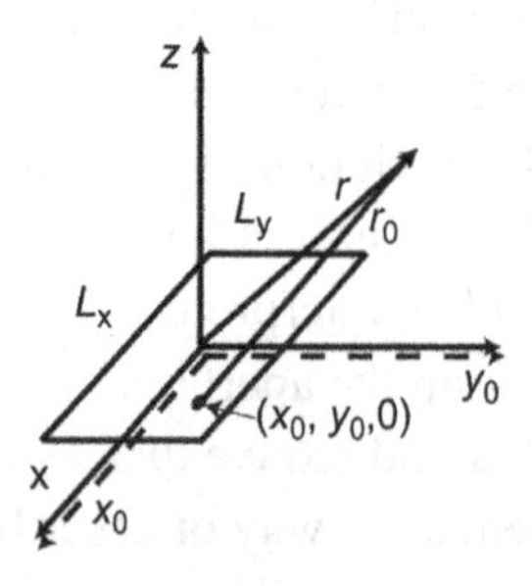

Figure 6.3
Coordinate System of an Aperture in the *xy* Plane and Radiating Along the *z* Axis.
Source coordinates in the aperture plane are denoted by the subscript 0. The rectangular aperture has sides L_x. and L_y. Radial arrows end in a spatial field point.

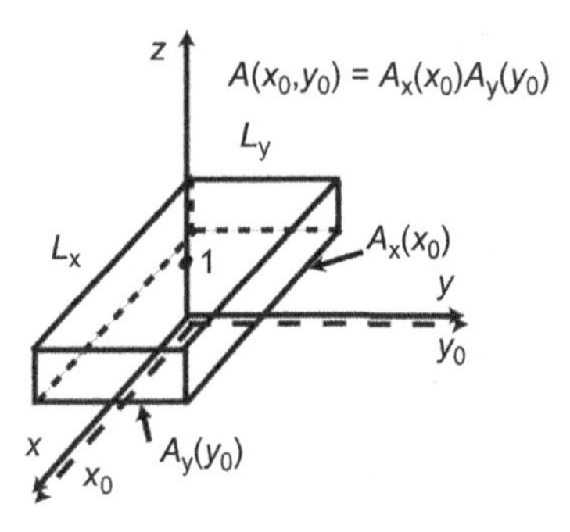

Figure 6.4
A Constant Amplitude Aperture Function for a Rectangular Aperture Consists of Two Orthogonal Rect Functions Multiplied Together.

If the aperture distribution function is separable, as in Eqn 6.4A, then the integration can be performed individually for each plane. As an example, if the plane-wave exponent is neglected, and:

$$A_0 = \frac{\rho_0 ckv_0}{2\pi x} = \frac{p_0 k}{2\pi z} = \frac{p_0}{\lambda z}. \tag{6.5A}$$

The overall integral can be factored as $p(x, y, z, w) = p_x(x, z, w)\, p_y(y, z, w)$ so that each integral is of the form:

$$p_x(x, \omega) = \sqrt{A_0} e^{i\pi/4} e^{-ikx^2/2z} \int_{\infty}^{\infty} [e^{-ikx_0^2/2z} A(x_0)] e^{ik(xx_0)/z} dx_0. \tag{6.5B}$$

If we define $\Gamma = 1/(\lambda z)$, and $\varsigma = \Gamma x_0$, then this integral can be recognized as plus i Fourier transform of the argument in the brackets (Szabo, 1977, 1978):

$$p_x(x, \Gamma, \omega) = \frac{\sqrt{A_0}}{\Gamma} e^{i\pi/4} e^{-i\pi\Gamma x^2} \int_{-\infty}^{\infty} [e^{-i\pi\varsigma/\Gamma} A(\varsigma/\Gamma)] e^{i2\pi\varsigma x} d\varsigma, \tag{6.5C}$$

which can be evaluated by a standard inverse fast Fourier transform (FFT) algorithm.

6.3 Rectangular Aperture

The previous analysis can be applied to the prediction of a field from a solid rectangular aperture, which is the same outer shape as most linear and phased array transducers. These aperture shapes will be helpful in anticipating the fields of arrays. Predictions will be only for a single frequency, yet they will provide insights into the characteristics of beams from any rectangular aperture radiating straight ahead along the beam axis. Here, relations for fields from line sources are derived to clarify the main features of an ultrasound field. For example, a line source can be used to simulate the field in an azimuth or xz imaging plane. Two orthogonal line apertures can be applied to simulate a rectangular aperture, as is given by Eqn 6.4A and Figure 6.4.

For several cases, simple analytic solutions can be found (Szabo, 1978). For example, for the case of a constant normal velocity on the aperture, with A as the rectangular function of Eqn 6.4B, an exact expression for the pressure field under the Fresnel approximation can be found from Eqn 6.5C:

$$p_x(x, z, \omega) = \frac{\sqrt{p_0}}{\sqrt{2}} e^{i\pi/4} \left[F\left(\frac{x + L_x/2}{\sqrt{\lambda z/2}}\right) - F\left(\frac{x - L_x/2}{\sqrt{\lambda z/2}}\right) \right], \tag{6.6}$$

where

$$F(z) = \int_0^z e^{-i\pi t^2/2} dt, \tag{6.7}$$

and $F(z)$ is the Fresnel integral of negative argument (Abramowitz and Stegun, 1968).

Far from the aperture, the quadratic phase terms in Eqn 6.5C are negligible, and the pressure at a field point is simply the plus i Fourier transform of the aperture distribution, as:

$$p_x(x, \Gamma, \omega) = \frac{\sqrt{A_0}}{\Gamma} e^{i\pi/4} \int_{-\infty}^{\infty} [A(\varsigma/\Gamma)] e^{i2\pi\varsigma x} d\varsigma, \tag{6.8}$$

which for a constant amplitude aperture distribution is:

$$p_x(x, \Gamma, \omega) = \frac{L_x}{\lambda z} \left(\frac{\sqrt{A_0}}{\Gamma}\right) e^{i\pi/4} \text{sinc}\left(\frac{L_x x}{\lambda z}\right) = \frac{L_x\sqrt{p_0}}{\sqrt{\lambda z}} e^{i\pi/4} \text{sinc}\left(\frac{L_x x}{\lambda z}\right). \tag{6.9A}$$

The field from a 28-wavelength-wide aperture is presented as a contour plot in Figure 6.5A. The contours represent points in the field that are -3, -6, -10, and -20 dB below the maximum axial value at each depth (z) in the field. This plot was generated by a public beam simulation program developed by Professor S. Holm and his group at the University of Oslo, Norway (see Section 7.8 for more information). The far-field beam profile pattern is given by Eqn 6.9A and shown in Figure 6.11. To determine the -6-dB beam halfwidth far from the aperture, solve for the value of x in the argument of the sinc function of Eqn 6.9A that gives a pressure amplitude value of 0.5 of the maximum value:

$$x_{-6} = 0.603\lambda z/L, \tag{6.9B}$$

and the full width half maximum (FWHM) is twice the -6-dB half beamwidth:

$$\text{FWHM} = 1.206\lambda z/L. \tag{6.9C}$$

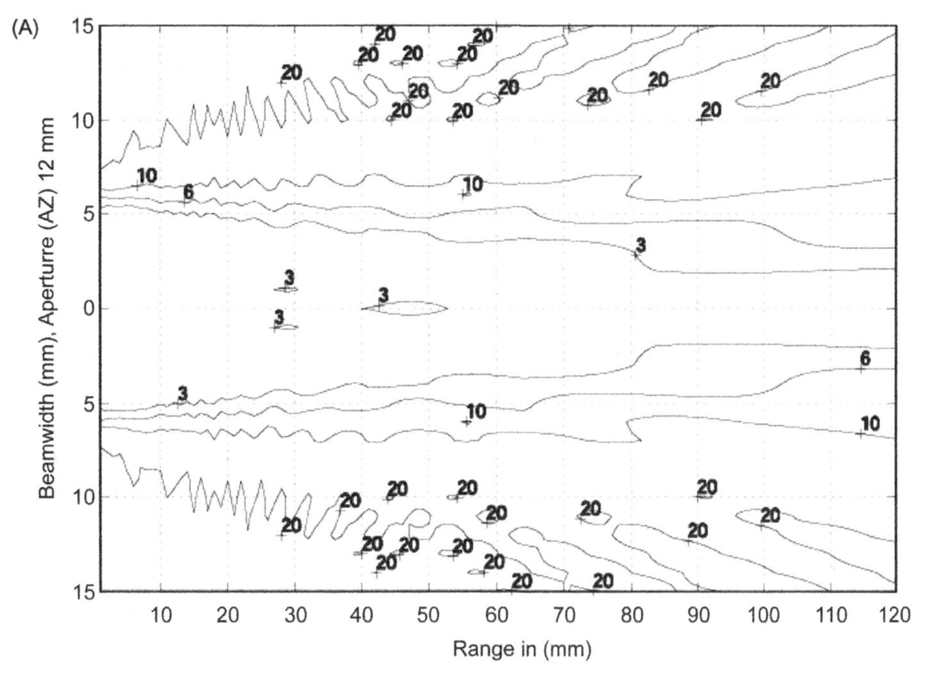

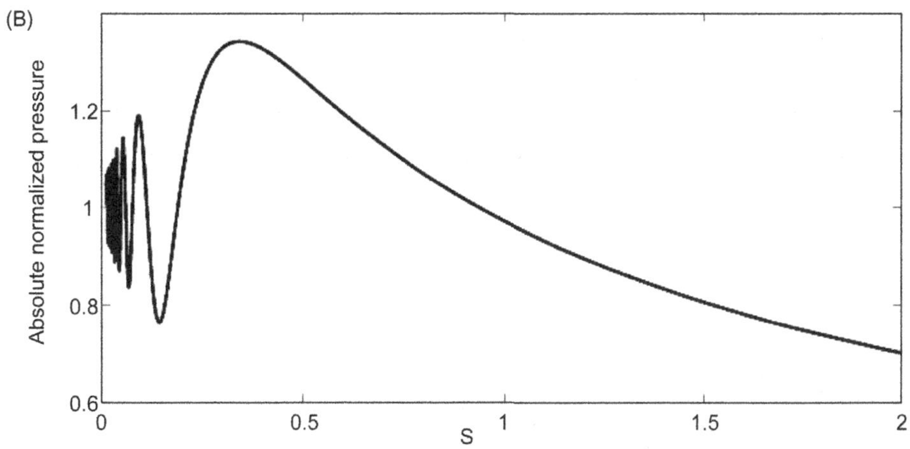

Figure 6.5
Beam contour plot and axial pressure for a nonfocusing line aperture.
(A) Contour beam plot for a 12-mm (28-wavelength), 3.5-MHz line aperture with −3, −6, −10, and −20-dB contours normalized to axial values at each depth. Nonfocusing aperture approximated by setting deep focal depth to 1000 mm ($S = 3$); (B) axial plot of normalized absolute pressure versus S from Eqn 6.10A with $S = 0.36$ corresponding to $Z = 120$ mm. *Plot in (A) generated using ultrasim software developed by professor S. Holm of the University of Oslo.*

The -6-dB beamwidth just calculated can be compared to the actual -6-dB contour in Figure 6.5A to illustrate the good match at longer distances from the aperture. Other beamwidths, such as the -20 dB, can be determined by this approach as well. A decibel plot of the half-beam (symmetry applies) over a larger range is shown in the top right of Figure 6.6. Section 6.4 will explain the effect of changing the amplitude profile of the source on the beam shape.

The beam along the z axis can be found by setting $x = 0$ in Eqn 6.6:

$$p(0, z, \omega) = e^{i\pi/4}\sqrt{2}p_0\left[F\left(\frac{L_x/2}{\sqrt{\lambda z/2}}\right)\right] = e^{i\pi/4}\sqrt{2}p_0\left[F\left(\frac{1}{\sqrt{2S}}\right)\right]. \qquad (6.10A)$$

An axial cross-section of the beam that was calculated from this equation is plotted in Figure 6.5B. Note that for any combination of parameters L_x, z, and λ that have the same argument in F of Eqn 6.10A, identical results will be obtained. This argument leads to a universal parameter (S):

$$S = \lambda z/L^2 = \hat{z}/\hat{L}^2, \qquad (6.10B)$$

where wavelength-scaled parameters are useful, $\hat{z} = z/\lambda$, and $\hat{L} = L/\lambda$. The universal parameter S can be expressed equally well in wavelength-scaled variables, as can previous equations such as Eqns 6.6, 6.9A, and 6.10A. The mathematical substitution of wavelength-scaled variables shows that what matters are the aperture and distance in wavelengths. For example, a 40-wavelength-wide aperture will have the same beam-shape irrespective of frequency.

A second observation is that nearly identical beam-shapes will occur for the same value of S, as shown for beam profiles in Figure 6.7. A definite progression of beam patterns occurs as a function of z, but if these profiles are replotted as a function of the universal parameter S, this same sequence of profile shapes can apply to all apertures and distances except very near the aperture. For the same value of S, the same shapes occur for different combinations of z and L. Look at Figure 6.8, in which different apertures and distances combine to give the same value of $S = 0\ 3$. For example, if $\lambda = 1$ mm and $L_1 = 40$ mm, $z_1 = 480$ mm; if $L_2 = 40$ mm, then $z_2 = 1920$ mm to give the same value of S. This scaling result can be shown by reformulating the arguments of Eqn 6.6 in terms of wavelength-scaled parameters and S:

$$\left(\frac{\hat{x} + \hat{L}_x/2}{\sqrt{\hat{z}/2}}\right) = \left(\frac{\hat{x}/\hat{L}_x \pm 1/2}{\sqrt{S/2}}\right). \qquad (6.10C)$$

From this relation, it is evident that for two combinations of z and L values having the same value of S, the argument will have exactly the same numerical value when $\hat{x}_2 = (\hat{L}_2/\hat{L}_1)\hat{x}_1$. In

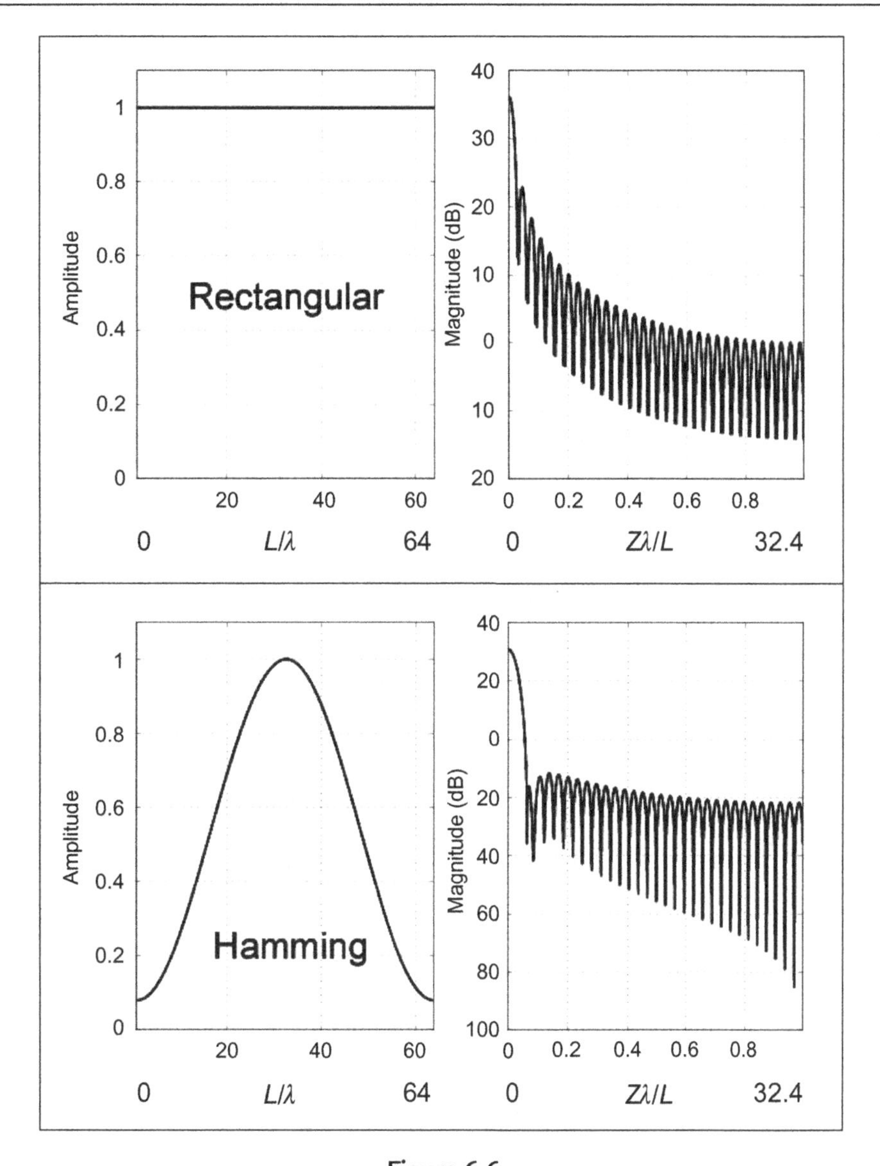

Figure 6.6
Far-field Beam Cross-sections or Beam Profile on a Scale for a (A) Rectangular Constant Amplitude Function Source and (B) Truncated Gaussian Source.

the previous example, this result shows that for the larger aperture, the profile is stretched by a scaling factor of two over the profile for the smaller aperture (shown in Figure 6.8). Remember that a limitation to this approach is that the distance and aperture combinations must satisfy the Fresnel approximation, Eqn 6.2B, on which this result depends.

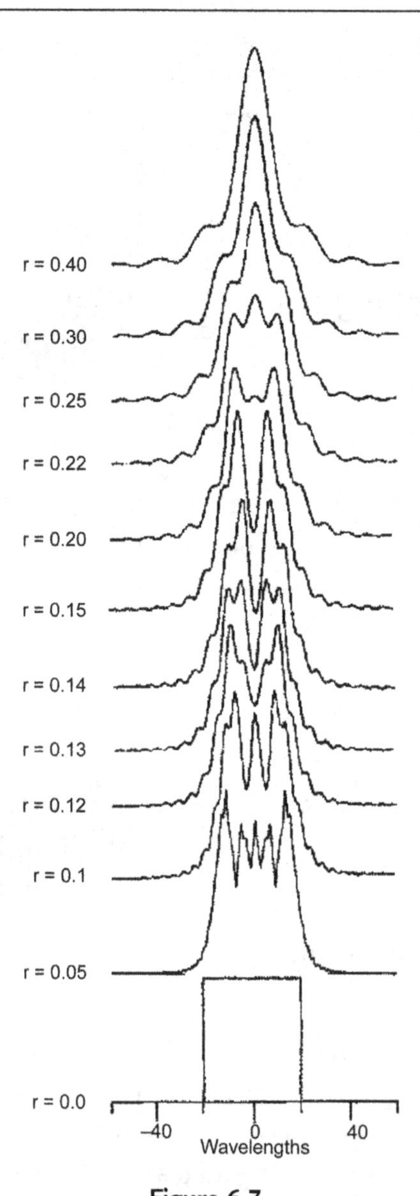

Figure 6.7
Diffraction Beam Profiles for Different Values of $S = r$ With $\hat{\mathbf{L}} = \mathbf{40}(\boldsymbol{\lambda})$.
Profiles for other values of L can be found by scaling the profile at the appropriate value of S.
From Szabo and Slobodnik (1973).

A third realization is that the last axial maximum for a line aperture L_x, shown in Figure 6.5, occurs at a transition distance, $z_t \approx L_x^2/(\pi\lambda)$, or when the argument in Eqn 6.10A is equal to $\sqrt{\pi/2}$. This distance separates the "near" and "far" fields. More exactly, Figure 6.5B shows the beam maximum occurs at $z_{\max} = 0.339L_x^2/\lambda$. The location of

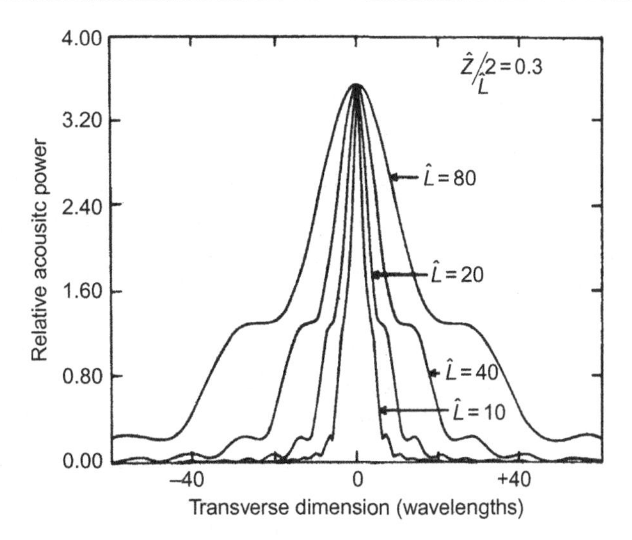

Figure 6.8

Diffraction Beam Profiles Versus Transverse Wavelength-scaled Distance ($\hat{x}$) for Different Values of Wavelength-scaled Apertures $\hat{L}$ and the Same Value of $S = 0.3$. *From Szabo and Slobodnik (1973).*

minimum −6-dB beamwidth is $z_{min} = 0.4L_x^2/\lambda$. Eqn 6.6 describes the whole field along the axis (except perhaps very close to the aperture). A program rectax.m, based on Eqn 6.10A, was used to calculate Figure 6.5B. In the far field, given by Eqn 6.9A, the pressure along the axis falls off as $(\lambda z)^{-1/2}$. The shape of the far field is only approximately given by Eqn 6.9A because, in reality, the transition to a final far-field shape (in this example, a sinc function) occurs gradually with distance from the aperture. For a rectangular aperture with sides L_x and L_y, the contributions from both orthogonal line apertures to the field can be multiplied to determine pressure as:

$$p(x, y, z, \lambda) = \frac{p_0}{2} e^{i\pi/2} \left[F\left(\frac{x + L_x/2}{\sqrt{\lambda z/2}}\right) - F\left(\frac{x - L_x/2}{\sqrt{\lambda z/2}}\right) \right] \left[F\left(\frac{y + L_y/2}{\sqrt{\lambda z/2}}\right) - F\left(\frac{y - L_y/2}{\sqrt{\lambda z/2}}\right) \right] \tag{6.11A}$$

and the on-axis pressure is:

$$p(0, 0, z, \lambda) = e^{i\pi/2} 2p_0 \left[F\left(\frac{L_x/2}{\sqrt{\lambda z/2}}\right) F\left(\frac{L_y/2}{\sqrt{\lambda z/2}}\right) \right], \tag{6.11B}$$

and there can be two on-axis peaks if $L_x \neq L_y$ (one from local beam pressure maximum in the xz plane and another from the pressure maximum in the yz plane). Experimental verification of these equations for rectangular apertures can be found in Sahin and Baker (1994).

6.4 Apodization

Apodization is amplitude weighting of the normal velocity across the aperture. In a single transducer, apodization can be achieved in many ways, such as by tapering the electric field along the aperture, by attenuating the beam on the face of the aperture, by changing the physical structure or geometry, or by altering the phase in different regions of the aperture. In arrays, apodization is accomplished by simply exciting individual elements in the array with different voltage amplitudes.

One of the main reasons for apodization is to lower the "sidelobes" on either side of the main beam. Just as time sidelobes in a pulse can appear to be false echoes, strong reflectors in a beam profile sidelobe region can interfere with the interpretation of on-axis targets. Unfortunately, for a rectangular aperture, the far-field beam pattern is a sinc function with near-in sidelobes only −13 dB down from the maximum on-axis value (shown in Figure 6.6A). A strong reflector positioned in the first sidelobe could be misinterpreted as a weak (−13 dB) reflector on-axis. Shaping is also important because, as we shall discuss later, the beam-shape at the focal length of a transducer has the same shape as that in the far field of a nonfocusing aperture.

A key relationship for apodization for a rectangular aperture is that in each plane (xz or yz), the far-field pattern is the plus i Fourier transform of the aperture function, according to Eqn 6.8. Aperture functions need to have rounded edges that taper toward zero at the ends of the aperture to create low sidelobe levels. Functions commonly used for antennas and transversal filters are cosine, Hamming, Hanning, Gaussian, Blackman, and Dolph–Chebycheff (Harris, 1978; Kino, 1987; Szabo, 1978). There is a trade-off in selecting these functions: The main lobe of the beam broadens as the sidelobes lower (illustrated by Figure 6.6B). A number of window functions can be explored conveniently and interactively through the wintool graphical user interface in the MATLAB signal-processing toolbox; this interface was used to create Figure 6.6, which compares Hamming apodization to no apodization. The effect of apodization on the overall field is given by Figure 6.9, which compares the field from the same truncated Gaussian apodization with that from an unapodized aperture. Here, not only is the apodized beam-shape more consistent, but also the axial variation is less. Note that for any apodization, universal scaling can be still applied even though the beam evolution is different.

6.5 Circular Apertures

6.5.1 Near and Far Fields for Circular Apertures

Many transducers are not rectangular in shape but are circularly symmetric; expressions for their fields can be described by a single integral. The spatial diffraction integral, Eqn 6.3,

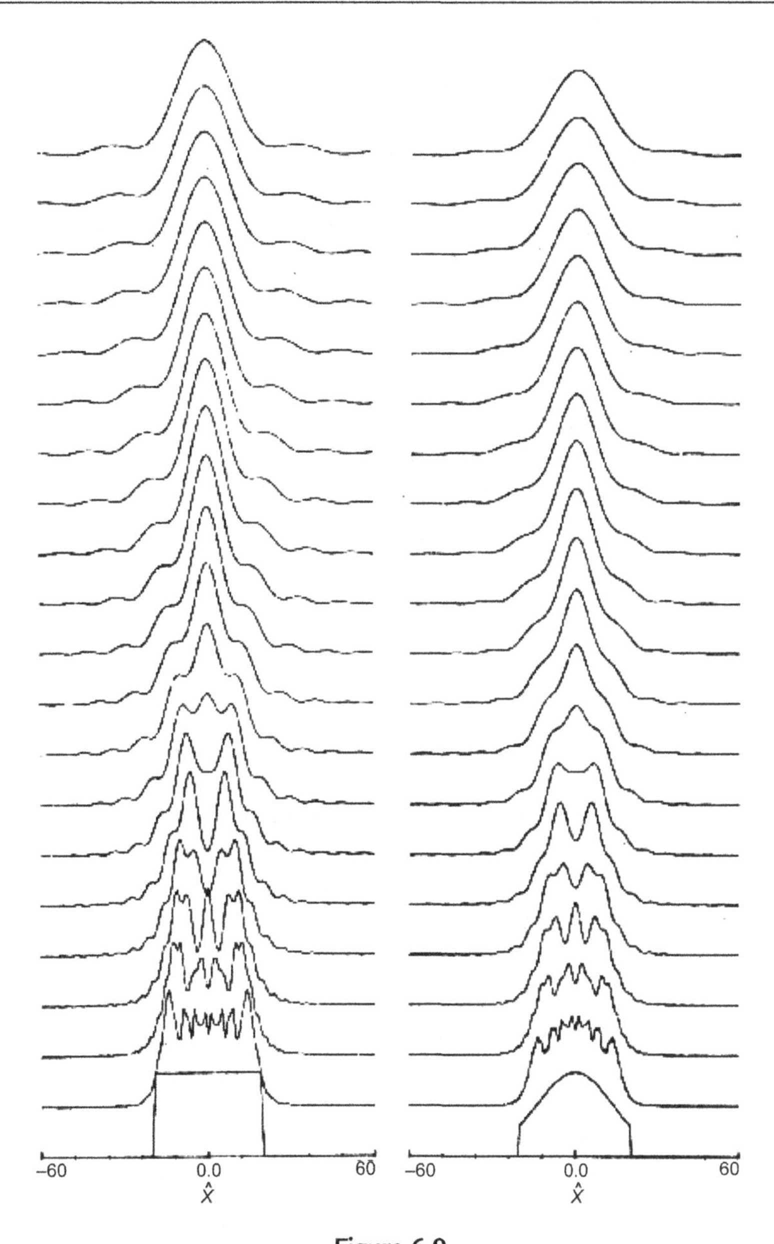

Figure 6.9
Diffraction Beam Profiles for an Unapodized Aperture (on the Left) Compared to a Truncated Gaussian Aperture (on the Right), Both With an Overall Aperture of 40 wavelengths. *From Szabo (1978).*

can be rewritten in polar coordinates for apertures with a circular geometry, neglecting the plane wave factor (Goodman, 1968; Szabo, 1981), as:

$$p(\overline{\rho},\ z) = \frac{i2\pi p_0}{\lambda z} e^{-i\pi\overline{\rho}^2/\lambda z} \int_0^{\infty} A(\overline{\rho}_0) e^{-i\pi\overline{\rho}_0^2/\lambda z} J_0\left(\frac{2\pi\overline{\rho}\overline{\rho}_0}{\lambda z}\right) \overline{\rho}_0 d\overline{\rho}_0 \quad (6.12A)$$

in which $\overline{\rho}$ and $\overline{\rho}_0$ are the cylindrical coordinate radii of the field and source (as distinct from ρ_0 used for density),

$$\overline{\rho}^2 = x^2 + y^2, \quad \text{and} \quad (6.12B)$$

$$\overline{\rho}_0^2 = x_0^2 + y_0^2, \quad (6.12C)$$

for a field point $(\overline{\rho},\ z)$ (given in Figure 6.10), and J_0 is a zero-order Bessel function. By letting $\mathrm{Y} = 2\pi\overline{\rho}/(\lambda z)$, we can transform the integral above through a change in variables:

$$p(\overline{\rho}, z) = \frac{i2\pi p_0}{\lambda z} e^{-iY\overline{\rho}/2} \int_0^{\infty} \left[A\left(\overline{\rho}_0\right) e^{-i\pi\rho_0^2/\lambda z}\right] J_0(Y\overline{\rho}_0)\overline{\rho}_0 d\overline{\rho}_0 \quad (6.13)$$

This equation is a zero-order Hankel transform, defined with its inverse as the following:

$$\mathbf{A}(q) = H[\mathrm{U}(r)] = \int_0^{\infty} U(r) J_0(qr) r dr, \quad \text{and} \quad (6.14A)$$

$$U(q) = H^{-1}[\mathbf{A}(q)] = \int_0^{\infty} \mathbf{A}(q) J_0(qr) q dq. \quad (6.14B)$$

Eqns 6.12A and 6.13 are valid for both the near and far fields. In the far field, as $\overline{\rho}/z$ and $\overline{\rho}_0/z$ become very small, then:

$$p(\overline{\rho}, z, \lambda) \approx \frac{i2\pi p_0}{\lambda z} \int_0^{\infty} A(\overline{\rho}_0) J_0(Y\overline{\rho}_0)\overline{\rho}_0 d\overline{\rho}_0. \quad (6.15)$$

Therefore, the pressure beam pattern in the far field is the Hankel transform of the aperture function. For a constant normal velocity across the aperture of radius a:

$$A(\overline{\rho}_0) = \prod\left(\frac{\overline{\rho}_0}{2a}\right), \quad (6.16)$$

$$p(\overline{\rho}, z, \lambda) \approx \frac{i2\pi p_0}{\lambda z} H\left[\prod\left(\frac{\overline{\rho}_0}{2a}\right)\right], \quad \text{and} \quad (6.17A)$$

$$p(\overline{\rho}, z, \lambda) \approx \frac{ip_0\pi a^2}{\lambda z} \frac{2J_1(2\pi\overline{\rho}a/(\lambda z))}{2\pi\overline{\rho}a/(\lambda z)} = ip_0\left(\frac{\pi a^2}{\lambda z}\right) jinc\left(\frac{\overline{\rho}a}{\lambda z}\right), \quad (6.17B)$$

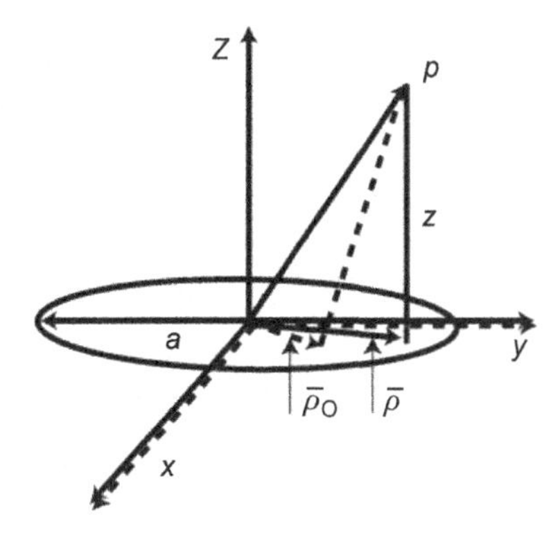

Figure 6.10
Cylindrical Coordinate System for Circularly Symmetric Apertures.

where

$$\text{jinc}(x) = 2J_1(2\pi x)/(2\pi x), \tag{6.18A}$$

and J_1 is a first-order Bessel function. The far-field beam cross-section is shown in Figure 6.11. The FWHM for this aperture is:

$$\text{FWHM} = 0.704\lambda z/a. \tag{6.18B}$$

An exact expression without approximation can be obtained for on-axis pressure (Blackstock, 2000):

$$p(0, z, \lambda) = ie^{-ikz}e^{-i\varsigma}2p_0 \sin\left\{\frac{kz}{2}\left[\sqrt{1 + (a/z)^2} - 1\right]\right\}, \tag{6.19A}$$

where ς is equal to the argument of the sin in the curly brackets and this equation under the Fresnel approximation, $z^2 >> a^2$, and binomial approximation, becomes:

$$p(0, z, \lambda) \approx i2p_0e^{-ikz}e^{-i\pi a^2/2\lambda z} \sin\left(\frac{\pi a^2}{2\lambda z}\right). \tag{6.19B}$$

Note that for large values of z, the phase from the beginning factor of Eqn 6.19B goes to $\pi/2$ as in Eqn 6.11B. A contour plot for a 13.54-mm-diameter aperture is given in Figure 6.12.

6.5.2 Universal Relations for Circular Apertures

The argument of the sine function in Eqn 6.19B has a familiar look. If we define a diffraction parameter for circular apertures as $S_c = z\lambda/a^2$, then 6.19B becomes:

$$|p(0, z, \lambda)| \approx 2p_0 \sin\left(\frac{\pi}{2S_c}\right). \tag{6.19C}$$

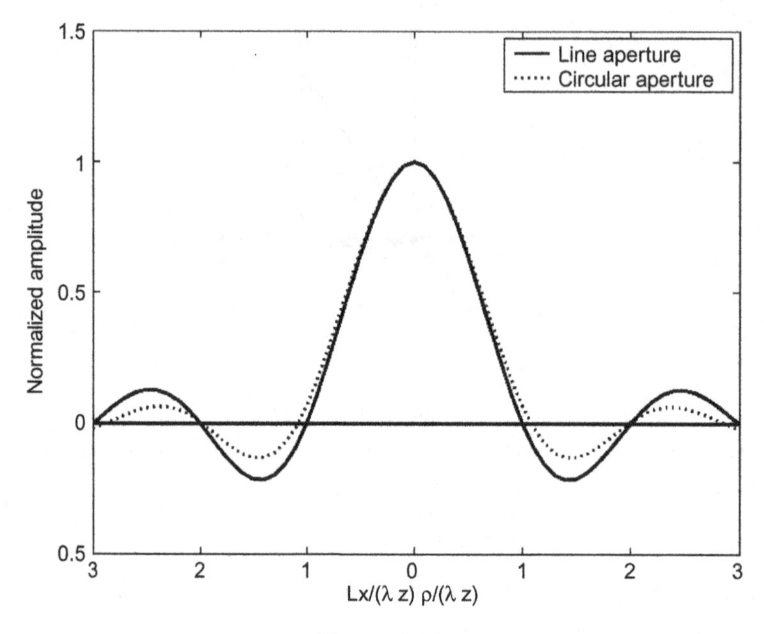

Figure 6.11

Far-field jinc Beam-shape from a Circular Aperture (Dashed Line) Normalized to a Far-field sinc Function From a Line Aperture (Solid Line) With the Same Aperture Area. The scale to the left of center of the horizontal axis goes with the line aperture, and the scale to the right is for the circular aperture.

We can see that the last axial maximum occurs when $S_c = 1$ (the argument $= \pi/2$) or equivalently, when $z_{\max} = z_t = a^2/\lambda$, the transition point is between near and far fields. Note the similarity to the transition distance for a line source, which occurs when the argument of Eqn 6.10A is equal to $\sqrt{\pi/2}$.

As we would expect from linearity and transform scaling, similar beam-shapes occur for the same values of the S_c parameter. Apodization can be applied to circular apertures using Hankel transforms of window functions.

The aperture area also plays a role in determining the axial far-field falloff in amplitude. If the circular aperture area is set equal to a square aperture, then $\pi a^2 = L^2$, or $a = L/\sqrt{\pi}$. Beam profiles of different shapes can be compared on an equivalent area basis as done in Figure 6.11. Figures 6.5A and 6.12 were generated on an equivalent area basis for a square aperture and a circular one. Substitute this equivalent value of a in:

$$z_{\max} = a^2/\lambda \approx L^2/(\pi\lambda), \tag{6.20A}$$

where the distance to the maximum for a line aperture is given approximately. Recall that a more accurate value for the line aperture is $z_{\max} = 0.339L^2/\lambda$. In general, approximately,

$$z_{\max} = AREA/(\pi\lambda). \tag{6.20B}$$

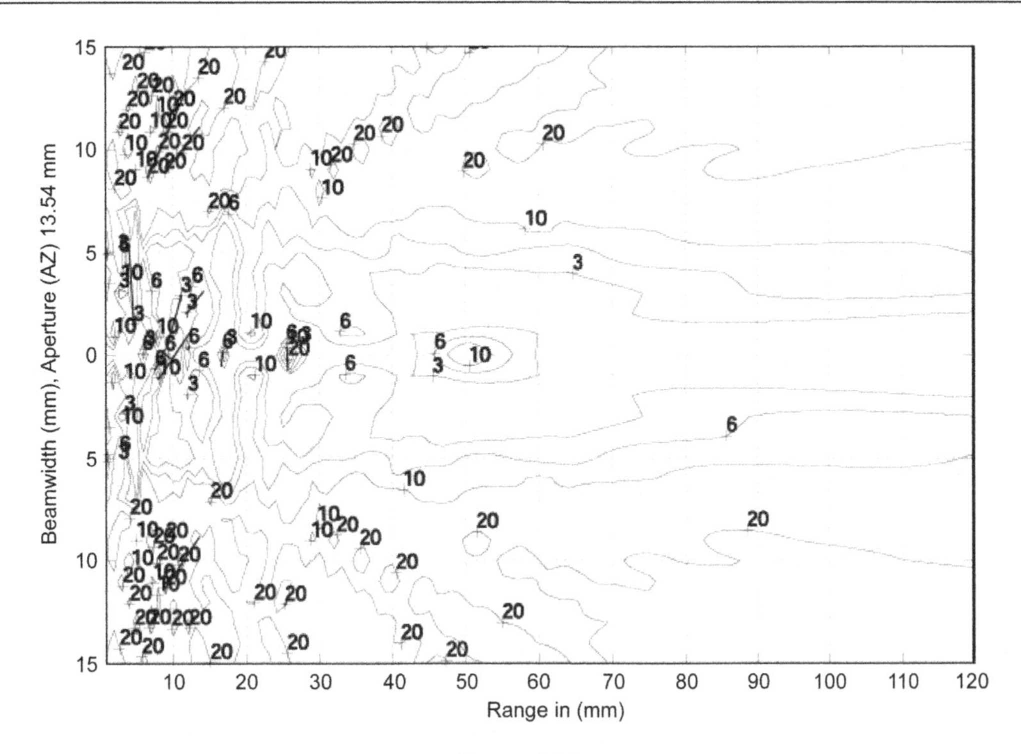

Figure 6.12
Contour Beam Plot for a 13.54-mm-diameter, 3.5-MHz Circular Aperture With −3, −6, −10, and −20-dB Contours Normalized to Axial Values at Each Depth.
Plot generated using ultrasim software developed by Professor S. Holm of the University of Oslo

For large values of z, the replacement of sine by its argument in Eqn 6.19B leads to a far-field falloff in axial pressure amplitude of:

$$|p(0,z,\lambda)| \approx p_0(\pi a^2)/(z\lambda) = p_0 AREA/(z\lambda), \tag{6.21A}$$

and a similar far-field approximation for Eqn 6.11B for a rectangular aperture gives:

$$|p(0,z)| \approx p_0 AREA/(z\lambda) = p_0(L^2)/(z\lambda). \tag{6.21B}$$

This equation would be appropriate for a square aperture with sides of length L. A rectangular aperture with different length sides can be found from the product of the line aperture relations:

$$|p(0,z)| \approx \sqrt{p_0(L_x^2)/(z\lambda)}^* \sqrt{p_0(L_y^2)/(z\lambda)} = p_0 AREA/(z\lambda). \tag{6.21C}$$

This approximation has limited application for rectangular apertures because in general there may be two distinct peaks caused by each in-plane diffraction, xz or yz, and therefore there are possibly two separate peak values at these locations. Under these circumstances, the pressure at each in-plane peak would be approximately $L_{x \text{ or } y}/\sqrt{\lambda z_{\max\, x \text{ or } y}}$.

6.6 Focusing

6.6.1 Introduction to Focusing

Focusing is a term borrowed from optics where it is associated with circular lenses. In photography, image magnification, resolution, depth of field, and focal length are familiar terms associated with focusing. In ultrasound, focusing has somewhat different meanings even though the basic principles from optics apply. One major difference is that, unlike optics, apertures do not have to be circularly symmetric; they can be rectangular, for example. Unlike optics, lenses for acoustics can be made from materials that have the equivalent of an index of refraction of less than one. Another difference covered in the next chapter is focusing through electronic means for arrays.

Focusing is a means of increasing resolution in a selected region. Increased resolution is the narrowing of beamwidth in this region, as illustrated by Figure 6.13 for rectangular aperture focusing in specified plane. Diffraction from a rectangular aperture, as shown in this chapter, implies different beam properties in orthogonal planes including the beam axis. In ultrasound imaging, it is common to have two different means of focusing for these two orthogonal planes, especially for arrays; therefore, it is logical to describe focusing in a

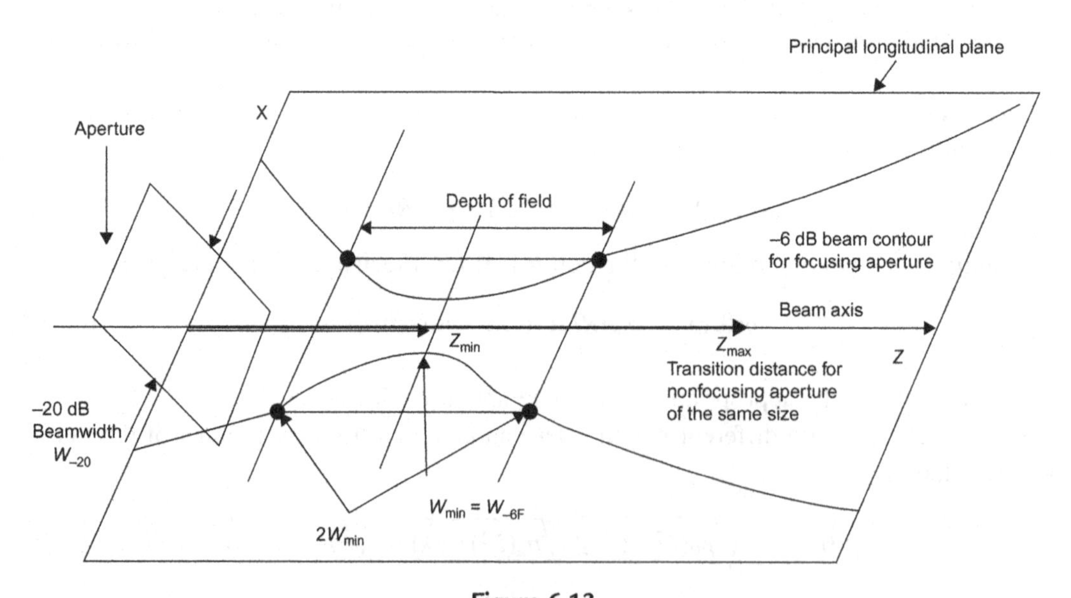

Figure 6.13
Focusing as Defined by the Narrowness of a Beam in a Specified Plane. *Adapted from the International Electrotechnical Commission (2001).*

selected plane as shown in Figure 6.13 where the beam narrows over a depth-of-field region in which the beamwidth is at its minimum −6-dB value. This approach is in contrast to optics, where focusing is described for circularly symmetric apertures and lenses for which all planes rotated about including the beam axis z are the same.

Shown in Figure 6.14 are several plots of axial absolute pressure for spherically focusing transducers of different focal lengths and the same diameter. A trend is evident: the shorter the focal length, the higher the pressure and the narrower the width of the peak. Furthermore, another pattern is that the higher the axial peak the narrower the lateral beamwidth becomes (not shown yet). Typically, for a circular aperture in optics, depth of field is measured as the −3-dB falloff from either side of the peak axial intensity (Kino, 1987). As a result, in either acoustics or optics, there is a trade-off between how high the peak can be (focal gain) and how extended the depth of field can become. In order to predict and understand these effects for design purposes, a few key relationships need to be derived.

6.6.2 Derivation of Focusing Relations

Focusing is usually accomplished by a lens on the outer surface of a transducer, by the curvature of the transducer itself, or by electronic means in which a sequence of delayed pulses sent to array elements to produce the equivalent of a lens. We shall focus our attention on the thin lens. Lenses can be cylindrical (curvature in one plane only) for a geometric line focus, or spherical (curvature the same in all planes around the z axis) for a geometric focus at a point (shown in Figure 6.15).

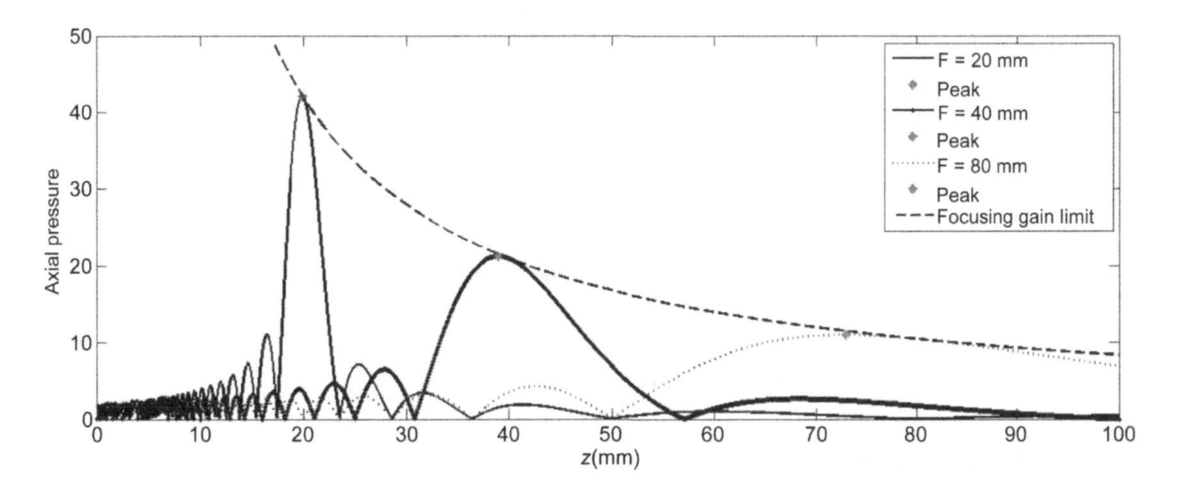

Figure 6.14
Pressure Plots vs Axial Distance for Three Spherically Focusing Transducers.
Shown for the same 20-mm aperture at a frequency of 1 MHz and three focal lengths: 20, 40, and 80 mm. Peak pressures and pressures at the geometric foci are indicated.

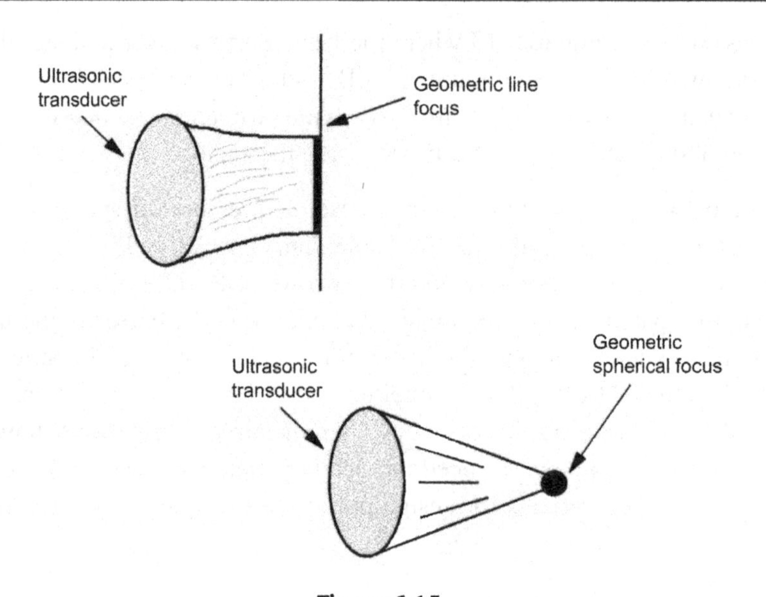

Figure 6.15
(A) Line Focusing for a Cylindrical Lens; (B) Point Focusing by a Spherical Lens. *From the International Electrotechnical Commission (2001).*

By a convention similar to but opposite in sign to that of optics, a thin lens is made of a material with an index of refraction n and a thickness $\Delta(x, y)$, as shown in Figure 6.16. This lens has a phase factor:

$$T_L(x, y) = exp\,(ik\Delta_0)exp\,(ik(n-1)\Delta(x, y)), \tag{6.22}$$

where k is for the medium of propagation (usually water or tissue) and Δ_0 is a constant. For a paraxial approximation (Goodman, 1968):

$$\Delta(x, y) \approx \Delta_0 - \frac{(x^2 + y^2)}{2}\left(\frac{1}{R_2} - \frac{1}{R_1}\right), \tag{6.23}$$

where R_1 and R_2 are the lens radii (shown in Figure 6.15). A geometric focal length is defined (Goodman, 1968) as:

$$\frac{1}{F} \triangleq (n-1)\left(\frac{1}{R_1} - \frac{1}{R_2}\right), \tag{6.24}$$

so that the phase factor for a thin lens is:

$$T_L(x, y) = exp\,(ikn\Delta_0)\,exp\,(ik(x^2 + y^2)/2F), \tag{6.25}$$

and the first factor, $exp(ikn\Delta_0)$, a constant, is dropped or set to equal one.

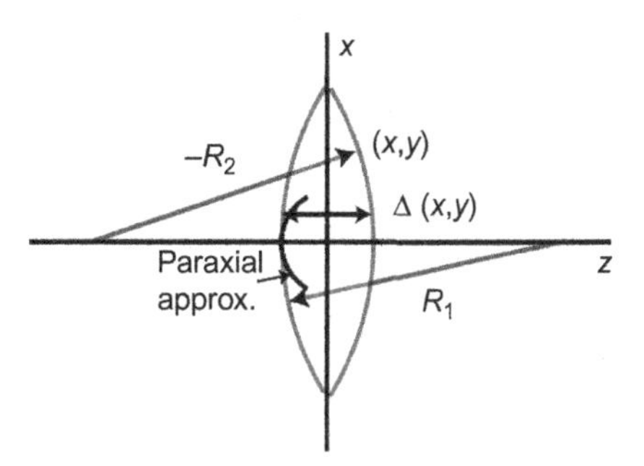

Figure 6.16
Thin Lens Geometry and Definitions.

Unlike optics, ultrasonic lenses can have an index of refraction of less than one (shown in Figure 6.17). Here, common lens shapes are plano-convex or plano-concave, where one side is flat and the corresponding R is infinite. For example, for the plano-concave lens and the convention (opposite of that used in optics) shown in Figure 6.17C, the focal length becomes:

$$\frac{1}{F} \triangleq (n-1)\left(\frac{1}{\infty} - \frac{1}{R_{\text{lens}}}\right), \tag{6.26A}$$

or

$$F = \frac{-R_{\text{lens}}}{n-1} = \left|\frac{R_{\text{lens}}}{n-1}\right|, \tag{6.26B}$$

which numerically is a positive number because n is less than one (the case in Figure 6.17C) and $R_{\text{lens}} = R_1$. By similar reasoning, in the case in Figure 6.17B, which has a positive radius of curvature $R_{\text{lens}} = R_2$ by convention and a positive index of refraction, the focal length also ends up being a positive number:

$$F = \frac{R_{\text{lens}}}{n-1} = \left|\frac{R_{\text{lens}}}{n-1}\right|. \tag{6.26C}$$

If the phase factor, Eqn 6.25, with the understanding of the numerical value of the geometric focal length,

$$T_L(x,y) = exp\,(ik(x^2+y^2)/2F), \tag{6.27A}$$

is put in the diffraction integral, Eqn 6.3, under the Fresnel approximation, the following results:

$$p(r,\omega) = \frac{ip_0k}{2\pi z}e^{i(\omega t - kz)}e^{-ik(x^2+y^2)/2z}\int\int_s [e^{-ik(x_0^2+y_0^2)/2z}e^{ik(x_0^2+y_0^2)/2F}A(x_0y_0)]e^{ik(xx_0+yy_0)/z}dx_0dy_0. \tag{6.27B}$$

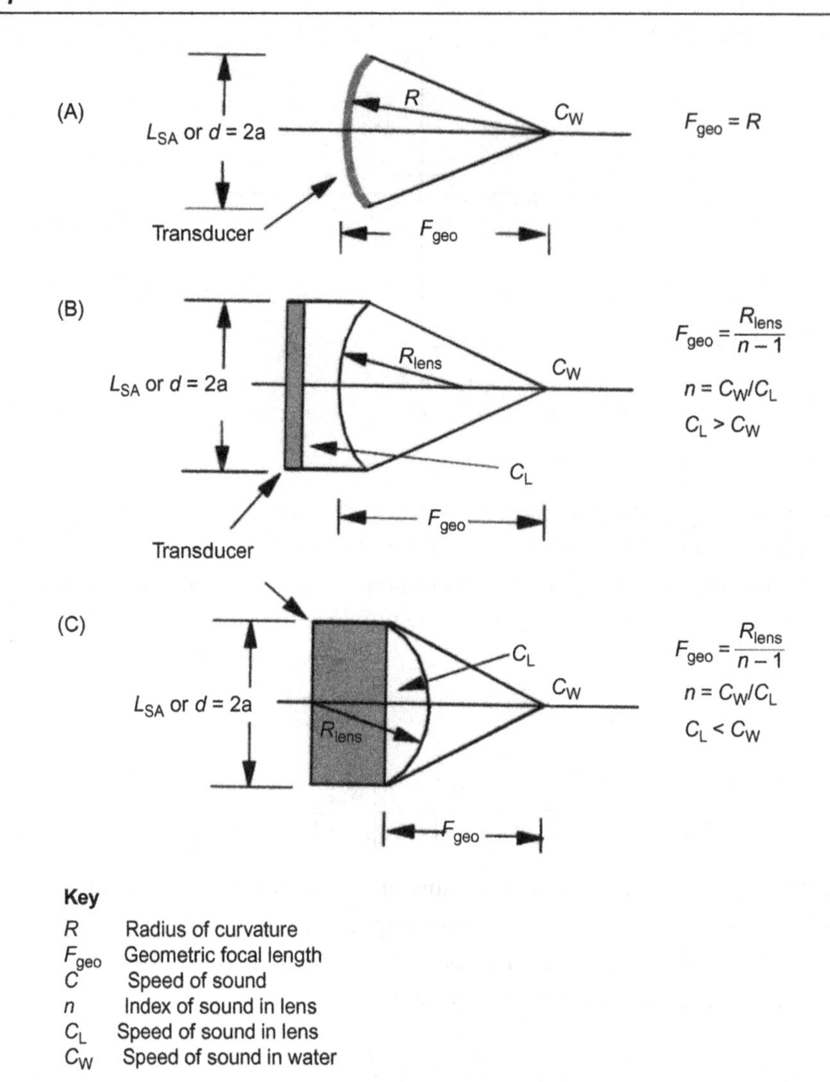

Figure 6.17
Methods of Focusing.
(A) Transducer with a radius of curvature R so that the focal length is equal to R; (B) transducer with a plano-concave lens; (C) transducer with a plano-convex lens. *Adapted from the International Electrotechnical Commission (2001).*

The net effect is replacing the $-1/z$ term in the quadratic term in the integrand by $-(1/z - 1/F)$, or:

$$1/z_e = 1/z - 1/F. \tag{6.28A}$$

This relation can be thought of as replacing the original z in Eqn 6.3 by an equivalent, z_e:

$$z_e = z/(1 - z/F). \tag{6.28B}$$

Recall that, without focusing, a prescribed sequence of beam patterns occurs along the beam axis z (shown in Figure 6.7). With focusing, the same shapes occur but at an accelerated rate at distances given by z_e. Thus, the whole beam evolution that would normally take place for a nonfocusing aperture from near field to extreme far field occurs for a focusing transducer within the geometric focal length F! At the focal distance, $z = F$, the quadratic term in the integrand of Eqn 6.27B is zero, and the beam-shape is the double $+i$ Fourier transform of the aperture function in rectangular coordinates. Note, as before, that the aperture can be factored into two functions, so a single Fourier transform is required for each plane (xz or yz). Similarly, for shallow-bowl spherically focusing transducers, Eqn 6.28 also holds approximately; the Hankel transform of the aperture function occurs at $z = F$.

6.6.3 Zones for Focusing Transducers

To understand the different regions in the field of a focusing transducer, we return to an approximate expression for the on-axis pressure, Eqn 6.19B, from a circularly symmetric transducer, but this time with spherical focusing (Kossoff, 1979) and for $z \neq F$:

$$p(0,z) \approx \frac{i2p_0 e^{-ikz} e^{-i\pi a^2/2\lambda z_e}}{(z/z_e)} \sin\left(\frac{\pi a^2}{2\lambda z_e}\right), \tag{6.29A}$$

and for $z = F$ (note the similarity to Eqn 6.21A):

$$p(0,F) \approx i2p_0 e^{-ikF}\left(\frac{\pi a^2}{2\lambda F}\right). \tag{6.29B}$$

Recall that when the transition distance z_t for the nonfocusing case is substituted in the on-axis pressure equation, an overall phase of $\pi/2$ appears in the argument of the sine function. To obtain this same equivalent phase for the focusing aperture, we set the argument of the sine in Eqn 6.29A to $\pm\pi/2$ and solve for z_e:

$$z_e = \pm a^2/\lambda. \tag{6.30}$$

For a positive value of z_e and the definition of z_e in Eqn 6.28A, as well as the definition $z_t = a^2/\lambda$, Eqn 6.30 can be applied to the determination of the near-transition distance, $z = z_{t1}$, for a focusing transducer, which separates the pre-focal (formerly near-Fresnel) zone from the focal (formerly focal Fraunhofer) zone depicted in Figure 6.18:

$$1/z_{t1} = 1/z_t + 1/F, \tag{6.31A}$$

or:

$$z_{t1} = z_t F/(z_t + F). \tag{6.31B}$$

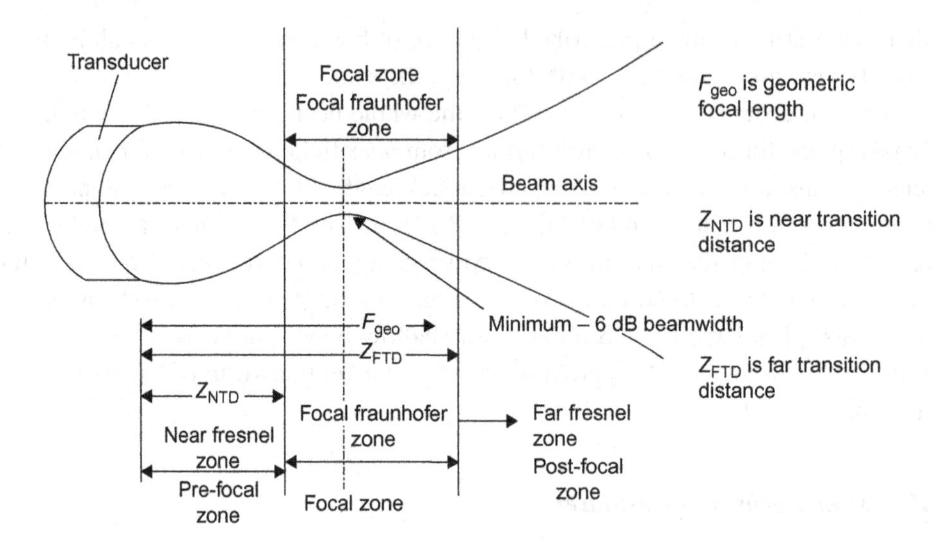

Figure 6.18
Beamwidth Diagram in a Plane Showing the Three Zones of a Focused Field Separated by Transition Distances One and Two. *Adapted from the International Electrotechnical Commission (2001).*

Similarly, through the use of the negative value of z_e in Eqn 6.30, the far-transition distance between the far end of the focal (formerly focal Fraunhofer) zone and the post-focal (formerly far Fresnel) zone:

$$1/z_{t2} = 1/z_t + 1/F, \tag{6.31C}$$

or:

$$z_{t2} = z_t F/(z_t - F). \tag{6.31D}$$

Another way of interpreting Eqn 6.31A is that the location of the overall focal length is reciprocally related to the combined effects of the natural beam maximum without focusing F and geometric focusing through the reciprocal lens law from optics. Note that these comments and Eqns 6.31A–D apply equally well to the focusing of rectangular transducers in a plane with the appropriate value of $z_t \approx L_x^2/(\pi\lambda)$ for the plane considered. More accurate estimates for the location of beam maxima can be found in Section 6.6.4.

From the equivalent distance relation, Eqn 6.28B, it is possible to compare the beam profiles of a focusing aperture to that of a nonfocusing aperture. The beam of a focusing aperture undergoes the equivalent of the complete evolution from near to far field of a nonfocusing aperture within the geometric focal length, because as z approaches F in value, z_e increases to infinity. At the focal length, previous far-field Eqns. 6.8, 6.9A, and 6.17 and can be used with $z = F$. For $z > F$, the phase becomes negative. A curious result is that the focal length loses its effectiveness past the natural transition distance (without focusing). This effect can be seen from a rewriting of the above equations:

$$z_{t1} = F/(1 + S_{cF}) = z_t/(1 + z_t/F), \tag{6.31E}$$

where

$$S_{cF} = F\lambda/a^2 = F/z_t, \tag{6.31F}$$

and, similarly:

$$z_{t2} = F/(1 - S_{cF}), \tag{6.31G}$$

and as F becomes large, these transition distances approach the value of z_t, which is the transition distance for the same aperture without focusing. Another odd consequence of focusing is that for strongly focused apertures, significant axial peaks and valleys may be generated beyond the focal length in the post focal zone (formerly far Fresnel zone). Because these Fresnel interference effects happen much farther from the aperture, they are generally less severe and may not occur at all, depending on the strength of the focusing. These interesting features are shown in the beam contour plot of Figure 6.19 for a strongly focusing spherical aperture.

6.6.4 Focusing Gain and Peak Pressure Values

One measure of the strength of focusing is focusing gain, which is defined as the ratio of the pressure amplitude at the focal length to the pressure amplitude on the face of the aperture. We can rewrite the expression for the axial pressure of a circular focusing transducer, Eqn 6.29, as:

$$p(0, z) \approx \left[\frac{\pi a^2}{\lambda z}\right] i p_0 e^{-ikz} e^{-i\pi a^2/2\lambda z_e} \operatorname{sinc}\left(\frac{a^2}{2\lambda z_e}\right), \tag{6.32}$$

in which the first term in brackets is the focusing gain term. When $z = F$, the sinc function has a value of 1; therefore, for a circularly symmetric unapodized aperture, the focal gain is:

$$G_{focal} = \pi a^2/(\lambda F), \tag{6.33A}$$

as can be seen from the on-axis pressure equation for a circularly symmetric focusing transducer, Eqn 6.29B. For an unapodized line aperture of length L, the gain in a focal plane is:

$$G_{focalx} = \sqrt{L^2/\lambda F}. \tag{6.33B}$$

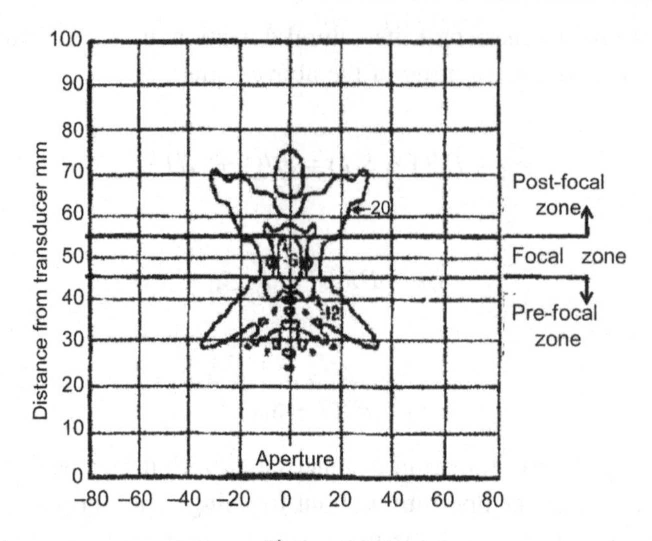

Figure 6.19
Beam Contour Plot for a Strongly Focusing Spherical Aperture. *Adapted from the International Electrotechnical Commission (2001).*

Focal gain for a rectangular aperture is more difficult to define here because noncoincident foci can interfere; nonetheless, the gain can be found from the product of the gains for the line apertures. For the case in which the focal lengths are coincident:

$$G_{focal} = L_x L_y / \lambda F. \tag{6.33C}$$

In general, the gain for coincident foci is:

$$G_{focal} = \text{Aperture area} / \lambda F. \tag{6.33D}$$

Note, for a rectangular aperture with different focusing mechanisms for each plane, a gain can be associated with each focal length and the focal gain is taken as the highest value of the two. This focal gain proportionality to aperture area is a consequence of a Fourier transform principle, which states that the center value of a transform is equal to the area of the corresponding function in the other domain. In other words, the axial (center) value in the focal plane is proportional to the area of the aperture.

Associated with focal gain is the all-important improvement in resolution. The − dB-beamwidth can still be found in the FWHM equations, such as Eqs. 6.9C and 6.18B, but with $z = F$ (the focal length). Since F is much closer to the aperture than a far-field distance for a nonfocusing aperture, an improvement in resolution is obtained.

One of the curious features from Figure 6.14 is that the location of the pressure maximum does not usually coincide with the focal length but precedes it. A simple estimate of the

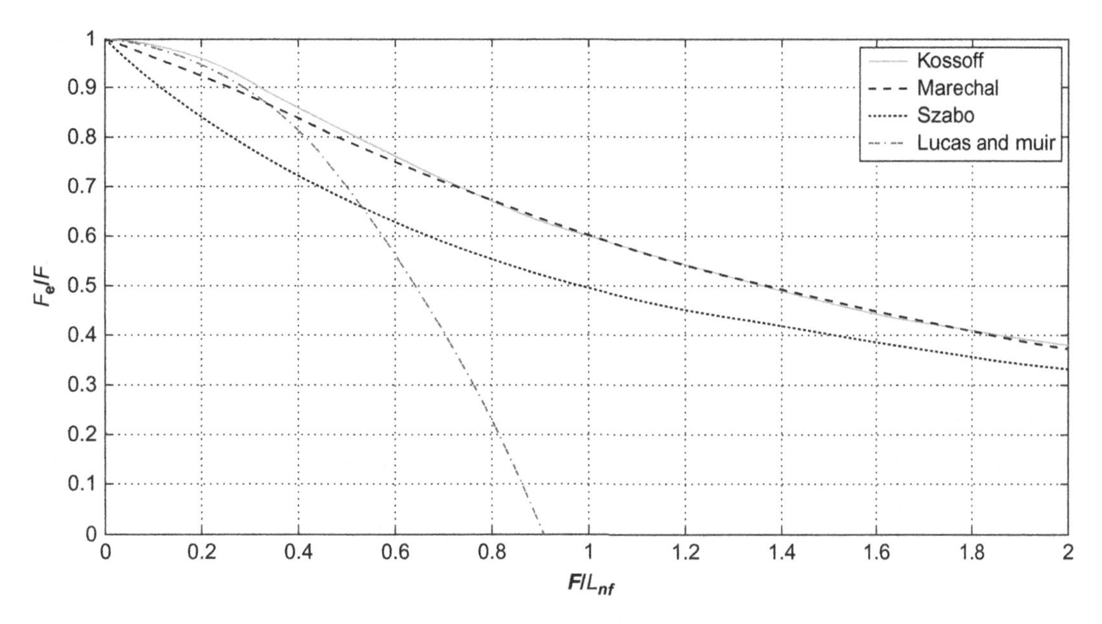

Figure 6.20
Estimates of the Location of the Axial Peak, F_e, Normalized to the Geometric Focus, F, Compared to the Kossoff Equation, 6.29A.
Mare'chal's formula is closest, Eqn 6.34, followed by first-transition distance, Eqn 6.31E, and an estimate from Lucas and Muir, all plotted against $S_{cF} = F/L_{nf}$. *Calculations courtesy of P. Mare'chal*

location of this maximum is Eqn 6.31E, the location of the near-transition distance. A better estimate has been found by Mare'chal, Levassort, Tran-Huu-Hue, and Lethiecq (2007):

$$z_{pk} = F/\left[1 + \frac{2}{3}(S_{cF})^{4/3}\right]. \tag{6.34}$$

Their estimate is more accurate for predicting the location of the pressure maximum of spherically focusing transducers than the transition distance approach or the approximation of Lucas and Muir (1982) as shown in Figure 6.20. An exact solution for the location of the pressure maximum in the form of transcendental equations can be found in Goldstein (2006). The equations produced so far are only approximate for focusing spherically curved transducers. The focal gain area for a deep spherically curved bowl transducer is $A = 2\pi hR$, where h is the depth of the bowl and $F = R$, the radius of curvature. A more exact treatment of these focusing transducers can be found in Section 7.3.

6.6.5 Depth of Field

A measure of the quality of focusing is a quantity called depth of field (DOF). From optics, this term has been taken to mean a falloff in axial intensity around the focal length for a spherically focusing aperture. For example, the difference between locations of the −3-dB points below the axial peak of a spherically focused transducer has been approximated by Mare'chal et al. (2007) as:

$$DOF\ 3dB = 1.69 S_{cF} F. \tag{6.35A}$$

Another similar expression is for DOF in terms of −6-dB points:

$$DOF\ 6dB = 2.35 S_{cF} F. \tag{6.35B}$$

A more general approach to defining DOF is to use the lateral changes in the beam (International Electrotechnical Commission, 2001; Lu, Zou, & Greenleaf, 1994). A definition of DOF more appropriate to rectangular geometries, as well as to circular ones, is the difference between distances where the lateral −6-dB beamwidth has doubled over its minimum value, as illustrated by Figure 6.13. As an example of this approach, an approximate method of calculating beamwidth will be employed.

Kossoff (1979) has shown that for spherically focusing apertures, beamwidths (here we use the example of the −6-dB (FWHM) beamwidth, W_{-6}) can be approximated from the axial intensity, for which a simple equation is available. The premise for this approach is that the intensity of a beam is approximately constant as it flows through each constant z plane. (Note: a more exact method would be to use wavefronts that are shaped like the aperture.) The steps are the following:

1. Find the absolute pressure amplitude A_F and beamwidth, W_{-6F}, in the focal plane. For example, A_F can be found from Eqn 6.29B, and the beamwidth can be found from Eqn 6.17B. Specifically, for the −6-dB beamwidth, use the FWHM value from Eqn 6.18B, or $W_{-6F} = \text{FWHM}$.
2. The intensity beamwidth-squared product is constant in any plane, so the unknown product is set equal to that easily calculated in the focal plane:

$$A^2(z) W^2{}_{-6}(z) = A_F^2 W_{-6F}^2. \tag{6.36A}$$

   Note that this relation is more effective within the main pressure lobe containing the focal length and defined by the region between the nulls on either side of the main peak.
3. The unknown beamwidth at a depth z can be found by solving Eqn 6.36A for W_{-6}, since A and A_F (A in the focal plane) can be found from Eqns 6.29A and 6.29B, and the focal beamwidth can be found from Eqn 6.18B.

For the spherical focusing case, Eqn 6.36A, these steps can be combined with expressions for the focal plane, namely focusing gain (Eqn 6.33A) and focal FWHM beamwidth (Eqn 6.18B) with $z = F$ to give one simple equation for beamwidth:

$$W_{-6}(z) = 0.7047 * \pi * a/p(0, z), \tag{6.36B}$$

in which $p(0,z)$ from Eqn 6.29 or 6.32 can be used. This approximate method is attractive because the calculation of beam profiles at planes other than the focal plane can be computationally involved for spherically focusing apertures. There is less benefit of applying this approach to the rectangular case because calculations involve either straightforward Fresnel integrals or FFTs.

These relations can now be applied to determining depth of field in terms of beamwidth. One of the complications in determining depth of field is that pressure is determined (for a fixed aperture and frequency) by two factors, a z^{-1} falloff and a distorted sinc function, Eqn 6.32. In Figure 6.14, the focal gain or pressure maxima lie on a focal gain curve (Eqn 6.33A with $F = z$). For the middle case, for which $F = 40$ mm, pressure is replotted in Figure 6.21 with actual −6-dB DOF points among others. From Eqn 6.36A, the inverse relation between beamwidth and pressure amplitude is:

$$\frac{W_{-6}(z)}{W_{-6F}} = \frac{A_F}{A(z)}. \tag{6.37}$$

So when this ratio has a value of 2, the expression implies that the beamwidth has doubled its value as needed in the beamwidth definition of DOF, and the axial pressure amplitude has halved as needed in the −6-dB axial pressure falloff for the pressure amplitude definition of DOF. For spherical focusing, the two DOF definitions are approximately equivalent. Another estimate of DOF is the difference in transition

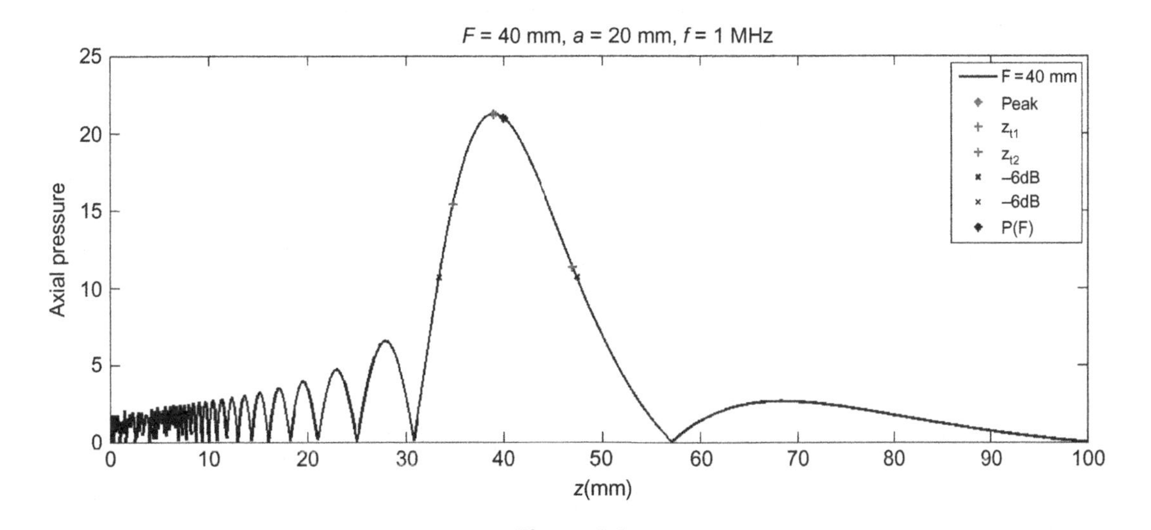

Figure 6.21
Pressure Plot vs Axial Distance for a Spherically Focusing Transducer.
Shown for a spherically focusing transducer with a 20-mm aperture at a frequency of 1 MHz and a focal length of 40 mm with −6-dB DOF points and near- and far-transition distances demarcating the pre-focal, focal, and post-focal zones.

distances, $z_{t2} - z_{t1}$. These points are shown in Figure 6.21. In this example, the numerical −6-dB DOF is 14.1 mm, which agrees well with Eqn 6.35B. The difference in transition distances, also shown in the plot, is 12.3 mm, a less accurate estimate. A similar beamwidth-based −3-dB DOF approach for circularly symmetric Gaussian apodized beams is explained by Lu, Zou, and Greenleaf (1994). Note that a more exact time-based approach for field calculations for spherically curved focusing transducers follows in the next chapter. Field calculations for circularly symmetric transducers can be made by using two-dimensional FFT methods and defining the amplitude and phasing of the source in rectangular coordinates.

6.6.6 Scaling of Beams

This section now examines several examples of these remarkable scaling laws for focusing. Recall that in the far field, the −6-dB half-beamwidth is proportional to the distance divided by the line aperture in wavelengths, as in Eqn 6.9B. This equation can actually be generalized to any distance in terms of wavelength-scaled parameters:

$$\hat{x}_{-6} = b\hat{z}/\hat{L}, \tag{6.38A}$$

where, away from the far-field region, the constant b must be determined numerically. The angle from the origin to this width can be shown to be inversely proportional to the aperture in wavelengths:

$$\tan\theta_{-6} = \hat{x}_{-6}/\hat{z} = b/\hat{L}. \tag{6.38B}$$

These equations can be applied to any beamwidths (such as −20 dB), provided the appropriate constant b is determined.

This series of examples for an aperture of 32 wavelengths will demonstrate how equivalent beam cross-sections can be obtained for a variety of conditions. Beam plots as a function of angle are calculated by the MATLAB focusing program beamplt.m, which uses a numerical FFT calculation of Eqn 6.27B for a one-dimensional unapodized line aperture. The first example is a nonfocusing aperture, and since this is a focusing program, the nonfocusing case can be approximated well by setting the focal distance to a large number (approximating infinity), $\hat{F} = 50,000$. Using the location of the transition distance in wavelengths (see Section 6.3), we obtain $\hat{z}_{t1} = \hat{L}^2/\pi = 326$. The corresponding beam plot and half-beamwidth angles are in Figure 6.22A.

The next example is for a focusing aperture with $\hat{F} = 100$. The first transition distance can be found from Eqn 6.31B to be $\hat{z}_{t1} = 76.5$. The corresponding beam plot is shown in Figure 6.22B, where the beam-shape is that of the nonfocusing case with half-beamwidth angles agreeing within quantization and round-off errors. Similarly, from Eqn 6.31D, the

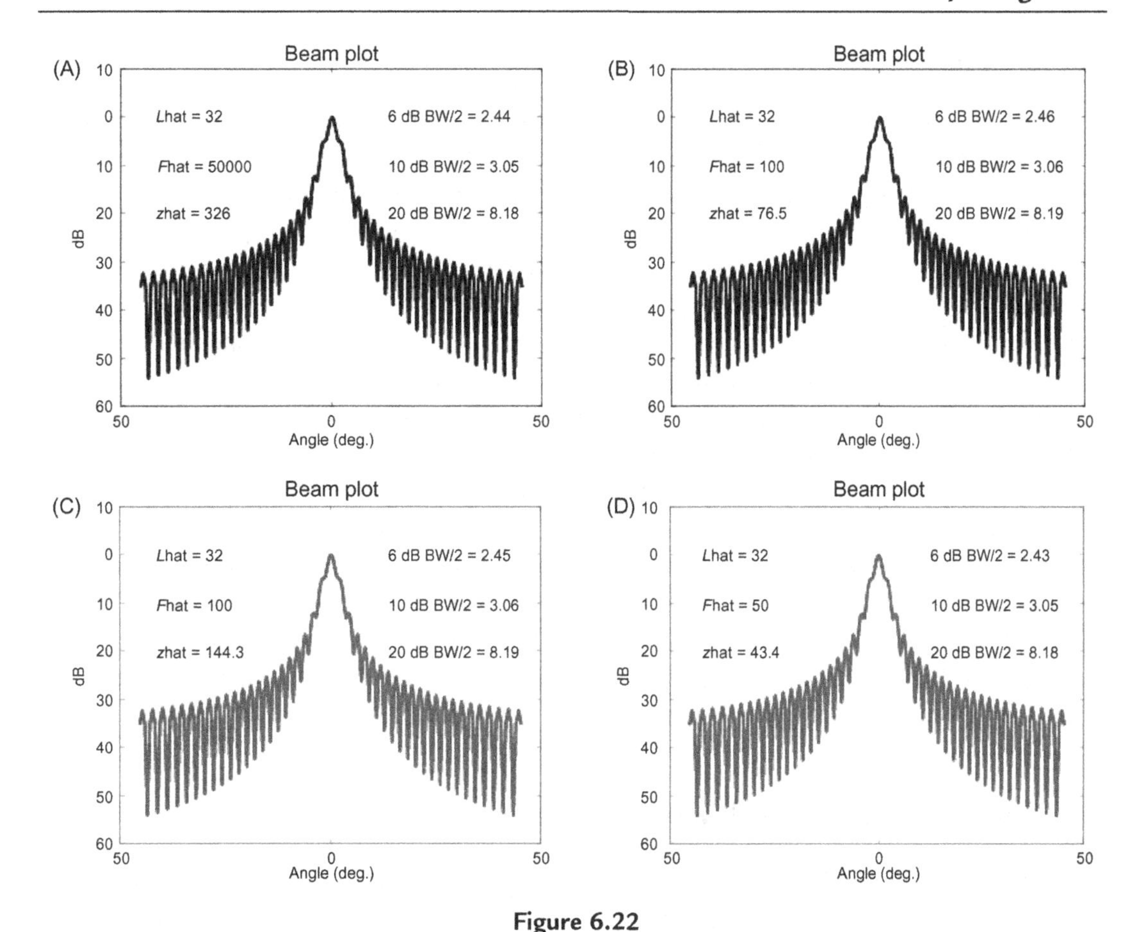

Figure 6.22

Beam Plots and Half-beamwidth Angles for a Nonfocusing Aperture.
(A) Beam plot in dB versus angle from the beam axis for a nonfocusing line aperture of 32 wavelengths ($\hat{L} = 32$) at the transition distance $z_t = 326$ with the half-beamwidth (BW/2) angles shown; (B) beam plot at the first transition distance, $\hat{z}_{t1} = 76.5$, for the same aperture with a focal distance of $\hat{F} = 100$; (C) beam plot for the same case but at the second transition distance, $\hat{z}_{t2} = 144.3$; (D) beam plot at the first transition distance, $\hat{z}_{t1} = 43.4$, for the same aperture with a focal distance of $\hat{F} = 50$.

second transition distance is $\hat{z}_{t2} = 144.3$, and the corresponding plot is Figure 6.22C. Finally, if we keep the aperture the same but switch the focal length to $\hat{F} = 50$, the first transition distance falls to $\hat{z}_{t1} = 43.4$, but the shape is essentially the same. Note that the beam-shapes for all of these cases are the same, but, because of the different axial distances involved for each case, the linear lateral beamwidths along x differ. Another striking illustration of the similarities in scaling can be found in Figures 12.19A and 12.19C, where complete two-dimensional contour plots for focusing beams are compared at one frequency and also at twice the same frequency. Similar relations to Eqn 6.37 hold for circular transducers as indicated by Eqn 6.18B.

6.6.7 Focusing Summary

To summarize, focusing compresses the whole beam evolution, normally expected for a nonfocusing aperture, into the geometric focal length. The universal scaling relationships derived previously for nonfocusing apertures can be combined with the focusing equivalent z relation, Eqn 6.28, to quickly determine beam patterns for a particular case of interest. The same beam-shapes occur as in the nonfocusing cases, but they are compressed laterally and shifted to different axial distances. Focused fields can be divided into three regions: the pre-focal zone, the focal zone, and the post-focal zone. The terms near field and far field are only appropriate for nonfocusing apertures. These zones can be distinguished by transition distances. Focusing has been defined in terms of beamwidth in a plane so that the contributions from different focusing mechanisms can be separated. Focusing creates a beamwidth that is narrower than what would be obtained for the natural narrowing of a beam from a nonfocusing aperture. Depth of field based on the doubling of the narrowest -6-dB beamwidth is approximately equal to a depth of field based on a -6-dB axial pressure drop for a spherically focusing aperture. The greatest pressure (beam maximum) usually occurs before the focal depth distance. Maximum focusing gain is proportional to active aperture area and inversely proportional to the product of wavelength and focal distance.

6.7 Angular Spectrum of Waves

For completeness, we will now review an alternative way of calculating beam patterns, called the "angular spectrum of plane waves," which is based on using a composition of plane waves steered in different directions rather than Huygen's spherical wavelets, as introduced in Section 2.3.2. This approach, which is an exact solution to the wave equation, is a powerful numerical method and can be applied to anisotropic media and mode conversion. A drawback to this method is that it cannot provide as much analytical insight as the spatial diffraction methods can.

By extending the results for the angular spectrum of a single line aperture given in Chapter 3, we take the double $+i$ Fourier transform of Eqn 6.4B, which, in this case, is just two one-dimensional transforms multiplied, since Eqn 6.4B is separable:

$$G(\tilde{f}_1,\tilde{f}_2) = \int_{-\infty}^{\infty} \prod(x/L_x)e^{i2\pi\tilde{f}_1 x}d\tilde{f}_1 \int_{-\infty}^{\infty} \prod(y/L_y)e^{i2\pi\tilde{f}_2 y}d\tilde{f}_2, \tag{6.39A}$$

$$G(\tilde{f}_1,\tilde{f}_2) = L_x L_y \mathrm{sinc}(L_x\tilde{f}_1)\mathrm{sinc}(L_y\tilde{f}_2), \tag{6.39B}$$

in which $\tilde{f}_1$ is a spatial frequency along axis 1 (the x axis, here), $k_1 = 2\pi\tilde{f}_1$, and so forth. Recall that this result from Chapter 3 meant that these apertures radiate plane waves of different amplitudes dependent on their direction. Each of these plane waves can be represented as exp $(i(\mathbf{k}\cdot\mathbf{r} - \omega t)$. Now, if this propagation factor is broken down into Cartesian coordinates and weighted by the directivity of the aperture, all the contributions from the aperture source can be allowed to propagate, so that at any field point the pressure amplitude can be represented by the following integral:

$$p(x,y,z) = p_0 \iiint_{-\infty}^{\infty} G(\tilde{f}_1,\tilde{f}_2,0)e^{i2\pi(\tilde{f}_1 x+\tilde{f}_2 y+\tilde{f}_3 z)}d\tilde{f}_1 d\tilde{f}_2 d\tilde{f}_3, \tag{6.40A}$$

where

$$p_0 = -4\pi\omega\rho_0 L_x L_y v_{30}, \tag{6.40B}$$

in which the normal particle velocity (along axis 3) is v_{30}. Fortunately, the spatial frequencies are related by:

$$\tilde{f}_1^2 + \tilde{f}_2^2 + \tilde{f}_3^2 = \frac{k^2}{4\pi^2} = \frac{f^2}{c^2} = \frac{1}{\lambda^2}, \tag{6.41A}$$

$$\tilde{f}_3^2 = \pm(\tilde{f}^2 - \tilde{f}_1^2 - \tilde{f}_2^2)^{1/2} \quad \text{if} \quad \tilde{f}^2 > \tilde{f}_1^2 + \tilde{f}_2^2, \quad \text{and} \tag{6.41B}$$

$$\tilde{f}_3^2 = i(\tilde{f}_1^2 + \tilde{f}_2^2 - \tilde{f}^2)^{1/2} \quad \text{if} \quad \tilde{f}^2 < \tilde{f}_1^2 + \tilde{f}_2^2, \tag{6.41C}$$

so that Eqn 6.40A can be reduced to two dimensions:

$$p(x,y,z) = p_0 \iint_{-\infty}^{\infty} [G(\tilde{f}_1,\tilde{f}_2)e^{2\pi\tilde{f}_3(\tilde{f}_1,\tilde{f}_2)z}]e^{i2\pi(\tilde{f}_1 x+\tilde{f}_2 y)}d\tilde{f}_1 d\tilde{f}_2. \tag{6.42}$$

Values of $\tilde{f}_3$, which are imaginary in Eqn. 6.41C, represent evanescent waves that die out quickly or attenuate. Note that this integral can be evaluated as a double plus *i* Fourier transform with FFTs. For a one-dimensional calculation in the *xz* plane, only one FFT is needed with $\tilde{f}_3 = +\sqrt{\tilde{f}^2 - \tilde{f}_1^2}$ for propagation in the positive half-plane. Another exactly equivalent form of these spatial transforms uses wavevector components, *k*s (k_1, k_2, etc.), instead of spatial frequencies, $\tilde{f}$s. An alternate version of Eqn 6.42 is applicable to circular apertures and cylindrical coordinates (Christopher & Parker, 1991; Kino, 1987).

6.8 Diffraction Loss

When two transducers act as a transmitter–receiver pair, only a part of the spreading radiated beam is intercepted by the receiver, and this loss of power is called "diffraction loss." A mathematically identical problem is that of a single transducer acting as a transceiver radiating at an infinitely wide, perfect reflector plane. In the first case, the transmitter and receiver are separated by a distance *z*; in the second, they are separated by a distance $2z$, where z is the distance to the reflector. The simplest definition of diffraction loss between a transmitter and a receiver is the ratio of the received acoustic power to that emitted at the face of the transducer (Szabo, 1978):

$$DL(z) = \frac{\int_{\sigma R} p(x, y, z)p^*(x, y, z)dxdy}{\int_{\sigma T} p(x, y, 0)p^*(x, y, 0)dxdy}, \tag{6.43A}$$

and in dB,

$$DL_{\text{dB}}(z) = 10\log_{10}|DL(z)|, \tag{6.43B}$$

and the phase advance is:

$$\varphi_{DL} = \arctan[imag(DL)/real(DL)], \tag{6.43C}$$

where σ_T and σ_R are the areas of the transmitter and receiver, respectively. The pressure is that calculated by diffraction integrals and integrated over the face of the receiving transducer. The transmitted power can be obtained from the known aperture function. For separable functions such as those for rectangular transducers, the integration can be carried out in each plane (xz and yz) separately as line sources, and the results can be multiplied. Calculations for several apodized line sources and unapodized receivers are given in Figure 6.23. Note that the results can be plotted as a function of the universal parameter S, and they are reciprocal (transmitters and receivers can be interchanged to give identical curves). The loss consists of an absolute power loss and a phase advance, which for one plane goes to $\pi/4$ in the extreme far field. The contribution from both planes for a rectangular aperture provides a total phase shift of $\pi/2$ for large distances (z).

For circularly symmetric transducers, the same definition applies in a cylindrical coordinate system and results in a single radial integration (Seki, Granato, and Truell, 1956). Loss is plotted in Figure 6.24 against S_c, and phase advance rises asymptotically to a value $\pi/2$ for large z.

6.9 Limited Diffraction Beams

The curves in the last section show that the variations in the near field of the beam can be smoothed out by apodization. In the far field, even apodized nonfocused beams spread out. Focusing also has a limited effect over a predictable DOF. A way to offset these changes and reduce diffraction loss is by a type of complex apodization that involves both amplitude and phase weighting over the aperture. A class of functions with this type of weighting can produce "limited diffraction beams." These beams have unusual characteristics: They maintain their narrow beamwidths for considerable distances, and they maintain axial amplitudes better than normal beams.

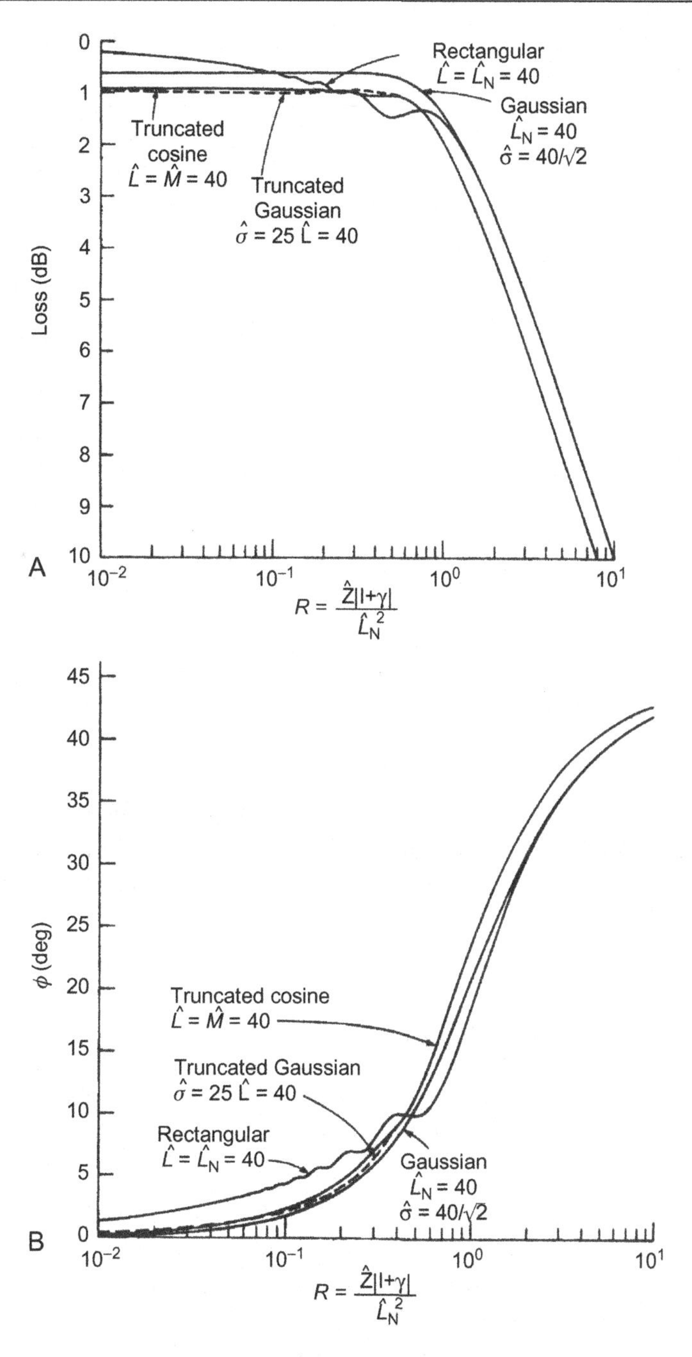

Figure 6.23

(A) Diffraction Loss Curves as a Function of s for Several Different Apodized Transmit Line Source Apertures and Unapodized Receivers; (B) Corresponding Phase Advances. *From Szabo (1978), Acoustical Society of America.*

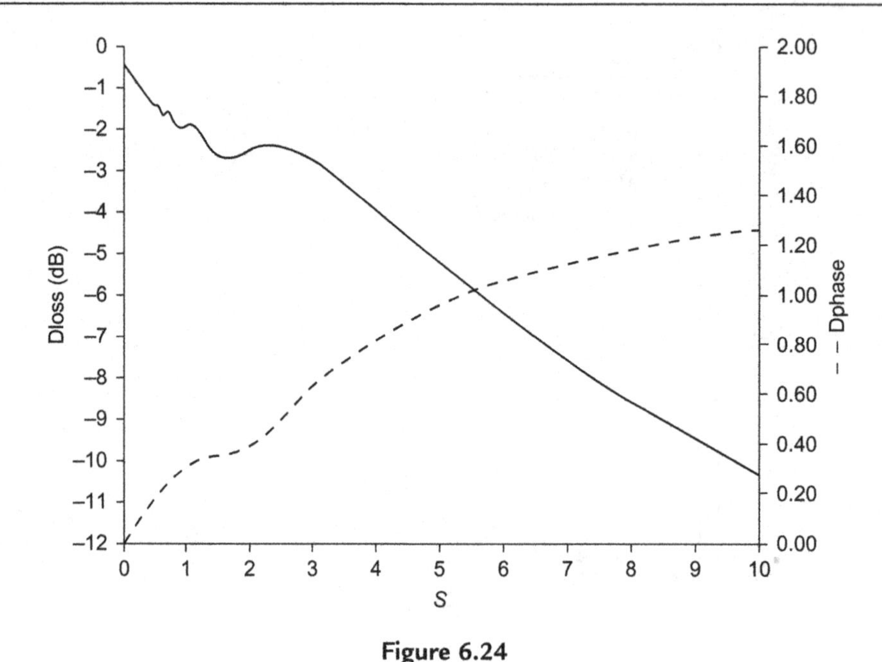

Figure 6.24
Diffraction Loss (dB) and Phase Curves (Radians) as a Function of S_c for an Unapodized Circular Transmitter and Receiver of Radius *a*. *From Szabo (1993).*

Two examples of limited diffraction beams are the zeroth-order Bessel beam and "X beam" shown in Figure 6.25. While the details of these beams are beyond the scope of this chapter, they are reviewed by Lu et al. (1994). An important application of X beams can be found in Section 10.12.3.

6.10 Holey Focusing Transducers

This chapter began with a strange property of diffraction: its apparent ability to bend sound around objects, a topic to which we now return. How can sound appear along the axis of a transducer with a hole in its center? Why would someone put a hole in their transducer? One reason is that those involved with high-intensity focusing ultrasound (HIFU) (covered in Chapter 17) would like to monitor the ultrasound effects of a larger spherically focusing therapeutic transducer with a smaller imaging transducer placed in a hole in its center, as illustrated in Figure 6.26.

To understand the field of this type of therapeutic transducer, consider that complementary diffracting objects have complementary fields, as stated by Babinet's principle. As a consequence of this principle, at the same frequency, the diffracting field of an aperture would be similar to the scattering field of a disk of the same size as the aperture.

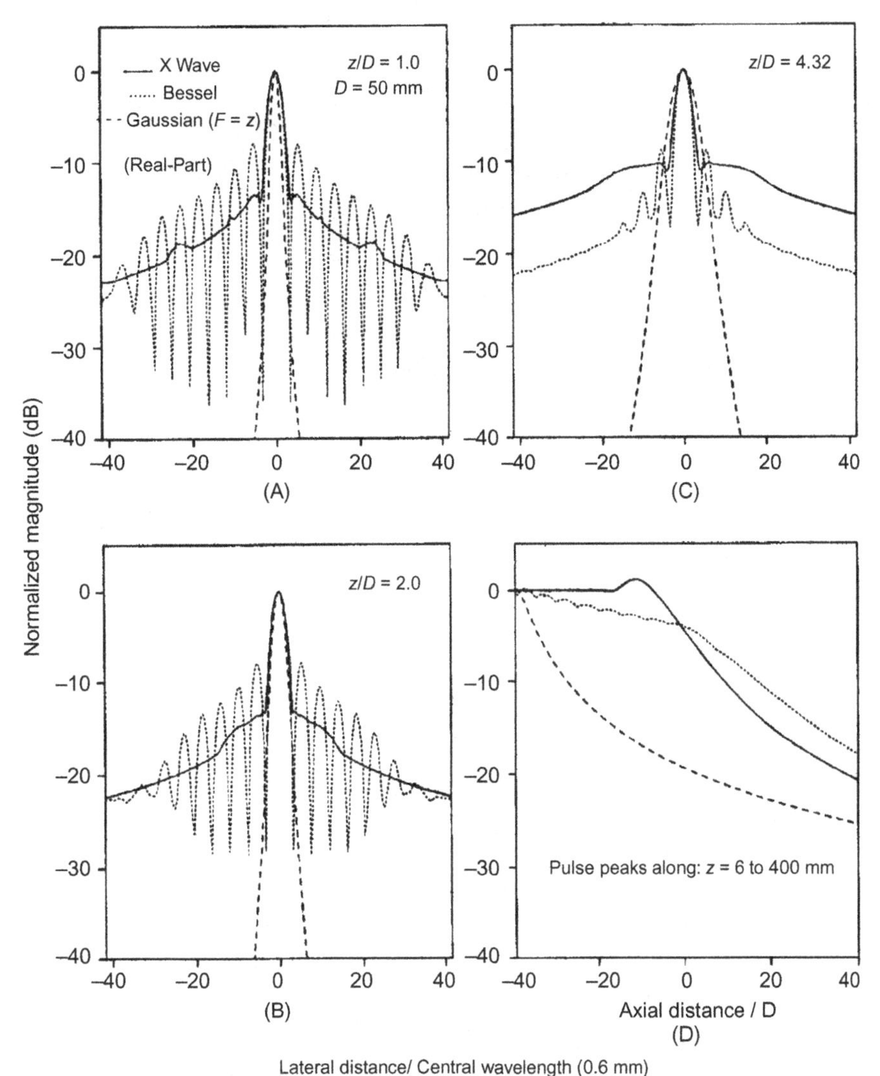

Figure 6.25

Zeroth-order Bessel Beam and X Beam.

Three types of beams compared to a zeroth-order X wave (full lines), J_0 Bessel beam (dotted lines), and dynamically focused ($F = z$) Gaussian beam (dashed lines), all for an aperture diameter of 50 mm and a center frequency of 2.5 MHz. The real parts of complex beams are plotted as lateral beam profiles for three depths: (A) 50, (B) 100, and (C) 216 mm. The peaks of pulses are plotted as pulses propagate from 6 to 400 mm in (D). *From Lu et al. (1994), with permission from the World Federation of Ultrasound in Medicine and Biology.*

Figure 6.26
Spherically Curved Outer Focusing Therapeutic Transducer With a Hole in its Center for an Imaging Transducer. *Courtesy of Professor Lu-Xían, Jiaotong University.*

Consider first a sound source with dimensions much greater than a wavelength, so that it is governed by geometric acoustics as illustrated by Figure 6.27A. The cylindrical extent of the source is interrupted by a disc of radius a on axis. Common sense indicates that a geometric shadow region of radius a will appear behind the disk. In the second case, Figure 6.27B, a circularly shaped plane wave is incident on an opaque flat ring, or hard baffle on which $v = 0$, with a hole of radius a in its center where $p = 0$. This time, the hole or aperture is on the order of wavelengths in diameter so that it is in the diffractive regime. An opaque disk is placed farther down the beam axis of symmetry. This field can be broken down into positive and complementary components according to Babinet's principle. The positive component can consist of an aperture such as the circular one shown and the complementary field can be an opaque disk in a free field (Hitachi & Takata, 2010). When the dimensions of the disk and aperture are in the diffractive regime, Babinet's principle still holds with interesting unexpected results. Sound, in this second case, does appear along the axis behind the disk! Blackstock (2000; pp. 485–489) noted this odd phenomenon: placing an opaque disc centered on the acoustic axis and parallel to a radiating full disc caused what he called the "bright spot behind the penny" effect. Blackstock (2000) wrote an expression for the acoustic pressure for this case as:

$$p_{\text{free field}} = p_{\text{aperture}} + p_{\text{disk}}. \tag{6.44}$$

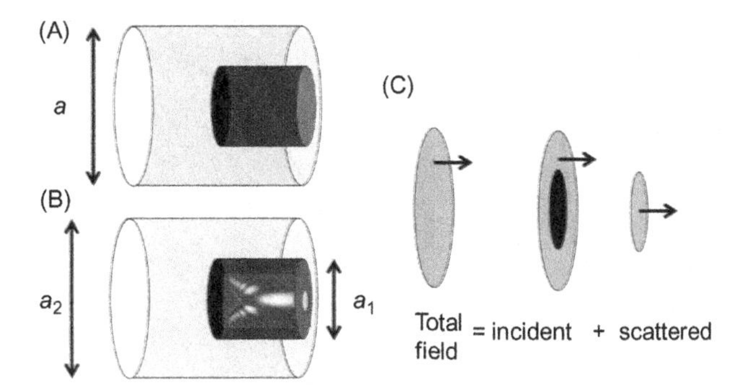

Figure 6.27
(A) Aperture insonifying a disk of radius a under geometric acoustics; (B) an incident acoustic wave insonifying an aperture of radius a_2 and eventually a disk of radius a_1 under diffraction conditions—sound appears behind the disk; (C) the total field of outer radius a_2 consists of the incident wave from an aperture of outer radius a_2 and central hole of radius a_1 and a second complementary aperture of radius a_1.

Another way of expressing these field relations more generally is in terms of the total and incident and scattered fields:

$$p_{\text{totalfield}} = p_{\text{incident}} + p_{\text{scattered}}. \tag{6.45}$$

Back to the case of a flat holey transducer. Here, an aperture of radius a_2 is radiating to the right. A center portion of the aperture, of radius a_1, is missing or equivalently blocked. Even though the central portion of the radiating aperture is missing, acoustic pressure does appear along the z axis, as calculated by Beissner (2012) from first principles and shown in Figure 6.28. These diffraction effects can be explained to first order by an acoustic equivalent of Babinet's principle. Thus, approximately, the field radiated to the right of the holey transducer is:

$$p_{\text{incident}} = p_{\text{totalfield}} - p_{\text{scattered}}, \tag{6.46A}$$

$$p_{\text{incident}} = p_{a_2} - p_{a_1}. \tag{6.46B}$$

Note the phase of the scattered wave can change and the wave can be directed to the right or left. As a consequence of this relationship, the field of this type of transducer can be constructed from known solutions for circular holeless apertures. In this case, the axial pressure can be computed from solutions for solid apertures of radius a_2 and radius a_1.

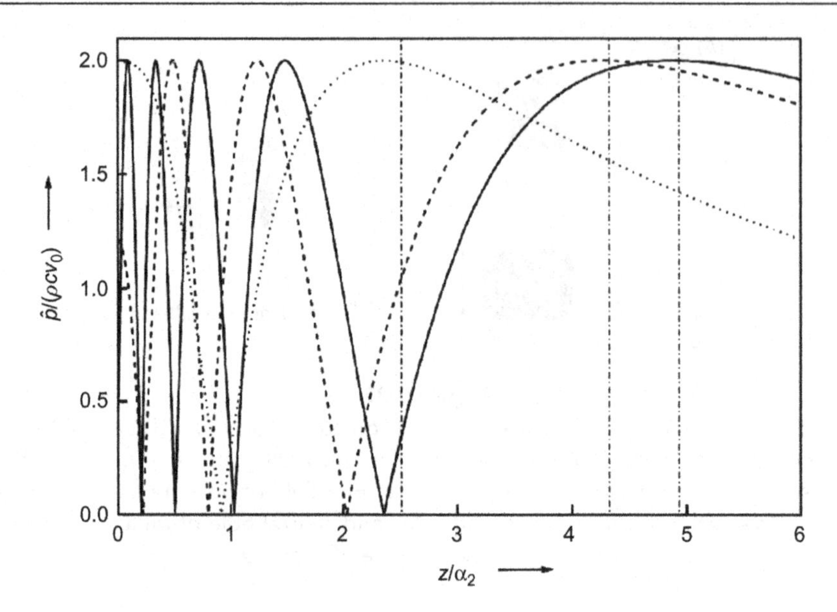

Figure 6.28
Normalized Pressure Amplitude as a Function of the Normalized Distance.
plane piston with $ka_2 = 31$. Solid curve, $a_1 = 0$; dashed curve, $a_1/a_2 = 0.35$; dotted curve, $a_1/a_2 = 0.7$. The three vertical dashed/dotted lines indicate the nominal position of the last axial pressure maximum in each case. *From Beissner (2012), Acoustical Society of America.*

Likewise, the fields can be found from methods in Section 6.5. Because this is a general principle, it holds not only for focusing transducers and transducers of different shapes. Beissner (2012) has derived exact closed-form on-axis solutions for flat pistons and focusing spherically curved transducers with holes in their centers.

References

Abramowitz, M., & Stegun, I. (1968). *Handbook of mathematical functions, Chap. 7, 7th printing*. Washington, D.C: US Government Printing Office.

Beissner, K. (2012). Some basic relations for ultrasonic fields from circular transducers with a central hole. *Journal of the Acoustical Society of America, 131*, 620–627.

Blackstock, D. T. (2000). *Fundamentals of physical acoustics*. New York: John Wiley & Sons. pp. 452–457.

Christopher, P. T., & Parker, K. J. (1991). New approaches to the linear propagation of acoustic fields. *Journal of the Acoustical Society of America, 90*, 507–521.

Goldstein, A. (2006). Steady state spherically focused circular aperture beam patterns. *Ultrasound in Medicine and Biology, 32*, 1441–1458.

Goodman, J. W. (1968). *Introduction to fourier optics*. New York: McGraw-Hill.

Harris, F. J. (1978). On the use of windows for harmonic analysis with the discrete Fourier transform. *Proceedings of the IEEE, 66*, 51–83.

Hitachi, A., & Takata, M. (2010). Babinet's principle in the Fresnel regime studied using ultrasound. *American Journal of Physics, 78*, 678–684.

International Electrotechnical Commission (2001). *IEC 61828. Ed. 1.0 English. Ultrasonics: Focusing transducers definitions and measurement methods for the transmitted fields*. Geneva, Switzerland: Author.

Kino, G. S. (1987). *Acoustic waves: Devices, imaging, and analog signal processing*. Englewood Cliffs, NJ: Prentice-Hall.

Kossoff, G. (1979). Analysis of focusing action of spherically curved transducers. *Ultrasound in Medicine and Biology, 5*, 359–365.

Lu, J. -Y., Zou, H., & Greenleaf, J. F. (1994). Biomedical ultrasound beam forming. *Ultrasound in Medicine and Biology, 20*, 403–428.

Lucas, B. G., & Muir, T. G. (1982). The field of a focusing source. *Journal of the Acoustical Society of America, 72*, 1289–1296.

Mare'chal, P., Levassort, F., Tran-Huu-Hue, L. -P., & Lethiecq, M. (2007). Effect of radial displacement of lens on response of focused ultrasonic transducer. *Japanese Journal of Applied Physics, 46*, 3077–3085.

Sahin, A., & Baker, A. C. (1994). Ultrasonic pressure fields due to rectangular apertures. *Journal of the Acoustical Society of America, 96*, 552–556.

Seki, H., Granato, A., & Truell, R. (1956). Diffraction effects in the ultrasonic field of a piston source and their importance in the accurate measurement of attenuation. *Journal of the Acoustical Society of America, 28*, 230–238.

Szabo, T. L. (1977). Anisotropic surface acoustic wave diffraction. In W. P. Mason, & R. N. Thurston (Eds.), *Physical acoustics* (Vol. XIII, pp. 79–113). New York: Academic Press, Chapter 4.

Szabo, T. L. (1978). A generalized Fourier transform diffraction theory for parabolically anisotropic media. *Journal of the Acoustical Society of America, 63*, 28–34.

Szabo, T. L. (1981). Hankel transform diffraction theory for circularly symmetric sources radiating into parabolically anisotropic (or isotopic) media. *Journal of the Acoustical Society of America, 70*, 892–894.

Szabo, T. L. (1993). *Linear and nonlinear acoustic propagation in lossy media* (Ph.D. Thesis). University of Bath, UK.

Szabo, T. L., & Slobodnik, A. J., Jr. (1973). *Acoustic surface wave diffraction and beam steering*. AFCRL-TR-73-0302, AF Cambridge Research Laboratories, Bedford, MA.

Bibliography

Bracewell, R. (2000). *The Fourier Transform and its Applications*. New York: McGraw-Hill.

Goodman, J. W (1968). *Introduction to Fourier Optics*. New York: McGraw-Hill.

A resource for classic treatments of optical diffraction.

International Electrotechnical Commission (2001). *IEC 61828. Ed. 1.0 English. Ultrasonics: Focusing Transducers Definitions and Measurement Methods for the Transmitted Fields*. Geneva, Switzerland: Author.

An international standard on focusing terms, principles, and related measurements.

Kino, G. S. (1987). *Acoustic Waves: Devices, Imaging, and Analog Signal Processing*. Englewood Cliffs, NJ: Prentice-Hall.

Sections 3.1 to 3.3 introduce diffraction and diffraction loss related to imaging.

Krautkramer, J., & Krautkramer, H. (1975). *Ultrasonic Testing of Materials*. New York: Springer Verlag.

Thorough treatment of diffraction, focusing, and apodization related to scattering.

Lu, J. -Y., Zou, H., & Greenleaf, J. F. (1994). Biomedical beam forming. *Ultrasound in Medicine and Biology, 20*, 403–428.

An excellent review of diffraction and focusing, including limited diffraction beams.

CHAPTER 7

Array Beamforming

Chapter Outline

7.1 Why Arrays?

If, to first order, the beam pattern of an array is similar to that of a solid aperture of the same size, why bother with arrays? Arrays provide flexibility not possible with solid apertures. By the control of the delay and weighting of each element of an array, beams can be focused electronically at different depths and steered or shifted automatically.

Lateral resolution and beam-shaping can also be changed through adjustment of the length and apodization of the active aperture (elements turned on in the array.) Dynamic focusing on receive provides nearly perfect focusing throughout the scan depth instead of the fixed focal length available with solid apertures. Finally, electronically scanned arrays do not have any moving parts, compared with mechanically scanned solid apertures, which require maintenance. Somer (1968) demonstrated that phased array antenna methods could be implemented at low-MHz frequencies for medical ultrasound imaging (illustrated by Figure 1.10). An early phased array imaging system, the Thaumascan, was built at Duke University (Thurstone & von Ramm, 1975; von Ramm & Thurstone, 1975). The technology to make compact delay lines and phase shifters for focusing and steering enabled the first reasonably sized clinical phased array ultrasound imaging systems to be made in the early 1980s.

Because images are formed from pulse echoes, this chapter introduces time domain diffraction approaches that are suited to short pulses. The benefit of the time domain approach is that it involves a single convolution calculation with a pulse instead of the many repeated frequency domain calculations necessary to synthesize a pulse using the frequency domain methods of Chapter 6. Both approaches will be helpful in describing arrays that can be thought of as continuous apertures sampled along spatial coordinates.

As a warm-up, this chapter first applies time domain approaches to the previous results for circular apertures. Next it describes arrays in detail, including how they differ from solid radiating apertures. The chapter also discusses pulse–echo beamforming and focusing, as well as the principles and implementations of two-dimensional (2D) arrays. Finally, it examines factors that prevent arrays from realizing ideal performance.

7.2 Diffraction in the Time Domain

The Rayleigh–Sommerfeld diffraction integral from Eqn 6.1A can be rewritten in a frequency domain form:

$$\Phi(r,f) = \int_A \frac{V_n(r_0,f)X(z,r_0)\,exp\,(-i2\pi f(r-r_0)/c]}{2\pi(r-r_0)}dA_0 \tag{7.1A}$$

An inverse $-i$ Fourier transform leads to its equivalent time domain form:

$$\varphi(r,\ t) = \int_A \frac{\chi(z,t)v_n[r_0,t-(r-r_0)/c]}{2\pi(r-r_0)}dA_0 = v_n(t)*_t h(r_0,t) \tag{7.1B}$$

in which ϕ is the velocity potential, v_n is particle velocity normal to the rigid source plane at $z=0$, dA_0 is an infinitesimal surface area element, A is the surface area of the source, and χ or X is an obliquity factor (see Section 7.6) set equal to one for now. If we factor v_n into a time and aperture distribution function, $v_n(r_0,t) = v_n(t)v_n(r_0)$, and let $v_n\ (r_0)$ be constant over the aperture for the remainder of the chapter, then we can express Eqn 7.1B in a convolution form later.

The geometry for a circularly symmetric radiator is given in Figure 7.1. Here, h is the spatial impulse response function defined as:

$$h(r,t) = \chi(z,t)\int_A \frac{\delta[t-(r-r_0)/c_0]}{2\pi(r-r_0)}dA_0. \tag{7.2}$$

Recall that the instantaneous particle velocity v and pressure p at a position r in a fluid can be found from:

$$v(r,t) = -\nabla\varphi(r,t), \quad \text{and} \tag{7.3}$$

$$p(r,t) = \rho_0\partial\varphi(r,t)/\partial t. \tag{7.4}$$

Just as in the diffraction integrals of the previous chapter, these time domain field expressions are geometry specific. The previous integrals will first be applied to the familiar circular piston radiator and then to array elements with a rectangular shape.

7.3 Circular Radiators in the Time Domain

Fortunately, time domain diffraction integrals have been worked out for simple geometries (Harris, 1981a ; Oberhettinger, 1961; Stephanishen, 1971; Tupholme, 1969). For the geometry given in Figure 7.1 for a circular aperture of radius a, the following delay variables are convenient:

$$\eta_1 = z/c_0, \quad \text{and} \tag{7.5A}$$

$$\eta_2 = \sqrt{(z^2+a^2)}/c_0. \tag{7.5B}$$

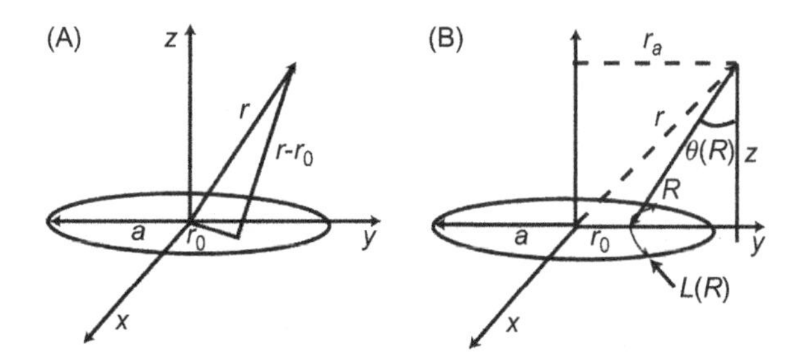

Figure 7.1
Geometries for circularly symmetric radiating elements. (A) Conventional geometry; (B) field-point-centered coordinates.

The local observer approach advocated by Stepanishen (1971) is based on time domain spatial impulse responses that have finite start and stop times defined by the intersections of lines from the field point to the closest and farthest points on the aperture (Figure 7.1). For example, for field points on-axis, the spatial impulse response is a rect function (Kramer, McBride, Mair, & Hutchins, 1988; Stepanishen, 1971):

$$h(r,t) = -c_0 \prod \left[\frac{t-(\eta_1+\eta_2)/2}{\eta_2-\eta_1} \right], \tag{7.6}$$

where η_1 is the delay from the closest point from the center of the aperture, and η_2 is that from the farthest points on the edges. This response, along with Eqns 7.1B and 7.4, lead to an on-axis pressure:

$$p(z,t) = \rho_0[v_n(t) *_t \partial h(z,t)/\partial t] = \rho_0 c_0 v_n(t) *_t [\delta(t-\eta_1) - \delta(t-\eta_2)]. \tag{7.7A}$$

The Fourier transform of Eqn 7.7A can be shown to be:

$$p(z,f) = i2\rho_0 c_0 v_n exp\left[ikz(1+\sqrt{1+(a/z)^2}) \right] \sin\left[\frac{kz}{2} \left(\sqrt{1+(a/z)^2} - 1 \right) \right], \tag{7.7B}$$

in agreement with the earlier exact result of Eqn 6.19A. In Eqn 7.7A, the on-axis pressure has a pulse from the center of the transducer, $\delta(t-\eta_1)$, and an inverted pulse from the edges of the aperture, $\delta(t-\eta_2)$. These contributions, called the "plane wave" and the "edge wave," merge eventually and interfere at half-wavelength intervals on-axis, depending on the pulse shape and length, $v_n(t)$. For broadband excitation, the on-axis pressure can differ remarkably from the continuous wave (CW) case, as illustrated by Figure 7.2.

Off-axis, expressions for the spatial impulse response are:

$$h(r,r_0,z,t) = \begin{cases} 0, & ct < z \text{ for } a > r_a,\ ct < R_1 \text{ for } a \le r, \\ c, & r_a \le r \le R_1 \\ \dfrac{c}{\pi}\cos^{-1} & \left[\dfrac{r_0^2 + c^2t^2 - z^2 - a^2}{2r_a(c^2t^2 - z^2)^{1/2}} \right], R_1 < ct \le R_2, \\ 0, & ct > R_2 \end{cases} \tag{7.8A}$$

in which:

$$R_1 = \sqrt{z^2 + (a - r_a)^2}, \quad \text{and} \tag{7.8B}$$

$$R_2 = \sqrt{z^2 + (a - r_a)^2}, \tag{7.8C}$$

and r_a is the radius from the z axis to the field point so that the field point is at (r_a, z). This expression is far simpler to evaluate numerically than the Hankel transform from Eqn 6.19.

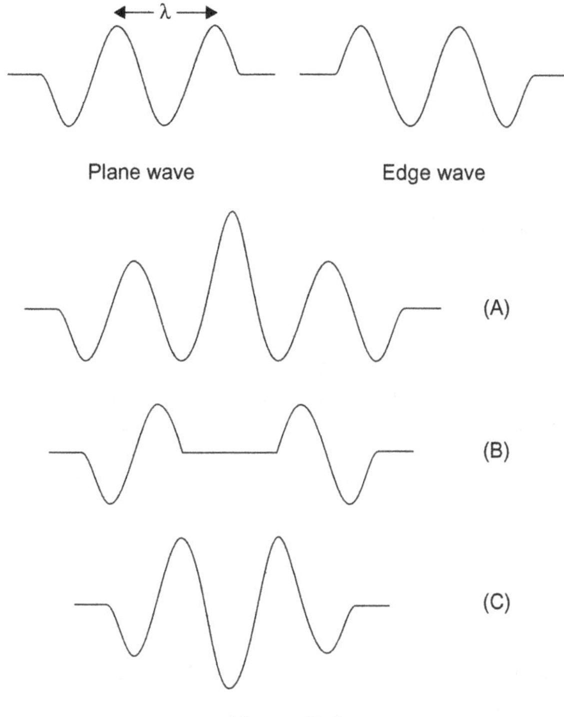

Figure 7.2
Plane and edge wave interference at three axial positions for an excitation function of two sinusoidal cycles. (A) $ct = 3\lambda/2$; (B) $ct = \lambda$; (C) $ct = \lambda/2$. *From Kramer et al. (1988), IEEE.*

Expressions (Arditi, Foster, & Hunt, 1981; Penttinen & Luukkala, 1976) for a concave spherical focusing radiator are similar in form to those above. The geometry for a spherically focusing aperture is given by Figure 7.3. Note the two regions: region I is within the geometric cone of the aperture, and region II lies outside it. Cylindrical symmetry is implied. Key variables are the following:

$$x = r\cos\theta, y = r\sin\theta. \tag{7.9A}$$

The depth (d) of the concave radiator is:

$$d = R\left[1 - \left(1 - \frac{a^2}{R^2}\right)^{1/2}\right], \tag{7.9B}$$

where R is the radius of curvature of the radiator, and a is the radius of the radiator.

For a field point P in region I, r_0 is defined as the shortest (for $z < 0$) or longest (for $z > 0$) distance between P and the source, and it is the line that passes through the origin and P

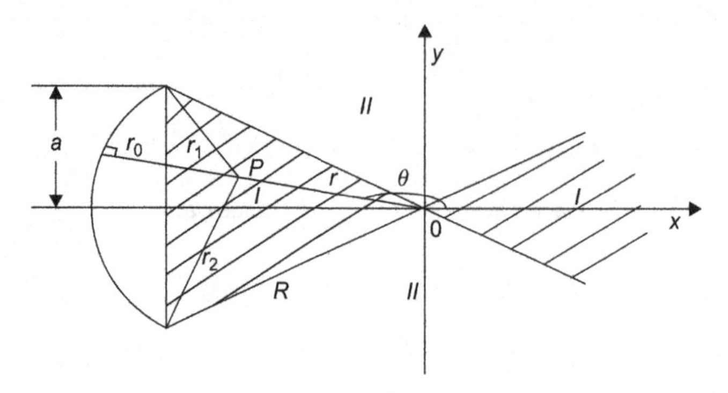

Figure 7.3
Nomenclature for spatial impulse response geometry for spherical focusing transducers. *From Arditi et al. (1981), with permission of Dynamedia, Inc.*

and intersects the surface of the source at normal incidence. Furthermore, r_0 can be expressed as:

$$r_0 = \begin{cases} R - r & for\ z < 0 \\ R + r & for\ z > 0 \end{cases}, \tag{7.9C}$$

where r_1 and r_2 represent the distances from P to the closest and farthest edges of the radiator for both regions I and II:

$$r_1 = [(a-y)^2 + (R-d+z)^2]^{1/2}, \quad \text{and} \tag{7.9D}$$

$$r_2 = [(a-y)^2 + (R-d+z)^2]^{1/2}. \tag{7.9E}$$

The spatial impulse response of a concave radiator is:

$$h(\vec{r}, t) = \begin{cases} 0 \\ \dfrac{c_0 R}{r} \\ \dfrac{c_0 R}{r}\dfrac{1}{\pi}\cos^{-1}\left[\dfrac{\eta(t)}{\sigma(t)}\right] \\ 0 \end{cases} \left| \begin{array}{l} c_0 t < r_0 \\ r_0 < c_0 t < r_1 \\ r_1 < c_0 t < r_2 \\ r_2 < c_0 t \end{array} \right| \begin{array}{l} c_0 t < r_1 \\ r_2 < c_0 t < r_0 \\ r_1 < c_0 t < r_2 \\ r_2 < c_0 t \end{array} \left| \begin{array}{l} c_0 t < r_1 \\ - - - - \\ r_1 < c_0 t < r_2 \\ r_2 < c_0 t \end{array} \right., \tag{7.9F}$$

in which:

$$\eta(t) = R\left[\frac{1 - d/R}{\sin\theta} + \frac{1}{\tan\theta}\left(\frac{R^2 + r^2 - c_0^2 t^2}{2rR}\right)\right], \quad \text{and} \tag{7.9G}$$

$$\sigma(t) = R\left[1 - \left(\frac{R^2 + r^2 - c_c^2 t^2}{2rR}\right)^2\right]^{1/2}. \tag{7.9H}$$

On the beam axis, the spatial impulse response is:

$$h(z,t) = \frac{c_0 R}{|z|} \prod \left(\frac{c_0 t - M}{\Delta(z)} \right), \tag{7.10A}$$

where

$$M = (r_0 + r_1)/2, \Delta(z) = r_1 - r_0. \tag{7.10B}$$

At the geometric focal point, the solution is a δ function multiplied by d:

$$h(0,t) = d\delta(t - R/c_0). \tag{7.10C}$$

Therefore, the pressure waveform at the focal point is a delayed replica of the time derivative of the normal velocity at the face of the aperture from Eqns 7.1B and 7.4. A frequency domain expression corresponding to pressure on-axis can be obtained from a Fourier transform of Eqn 7.10A:

$$H(z,f) = \left(\frac{R\Delta(z)}{|z|c_0} \right) \text{sinc} \left(\frac{2\Delta(z)f}{c_0} \right), \tag{7.10D}$$

and from Eqn 7.10C, the focusing gain is:

$$G_{focal} = \left(\frac{2\pi}{\lambda} \right) \left(\frac{A}{2\pi R} \right) = \frac{A}{\lambda R}, \tag{7.10E}$$

where the surface area is $A = 2\pi R d$.

7.4 Arrays

As opposed to large continuous apertures, arrays consist of many small elements that are excited by signals phased to steer and focus beams electronically (shown in Figure 7.4). The elements scan a beam electronically in the azimuth or xz plane. A molded cylindrical lens provides a fixed focal length in the elevation or yz plane. The nominal beam axis is the z axis (the means of steering the beam in the azimuth plane will be discussed later).

A layout of array element dimensions and steering angle notation are given by Figure 5.6. Here, the pitch or element periodicity is p, the element width is w, and the space between elements, or kerf width, is p-w. Two-dimensional and other array geometries will be discussed later.

7.4.1 The Array Element

This section first examines the directivity of an individual element. These elements are most often rectangular in shape, such as the one depicted in Figure 7.5. For small elements

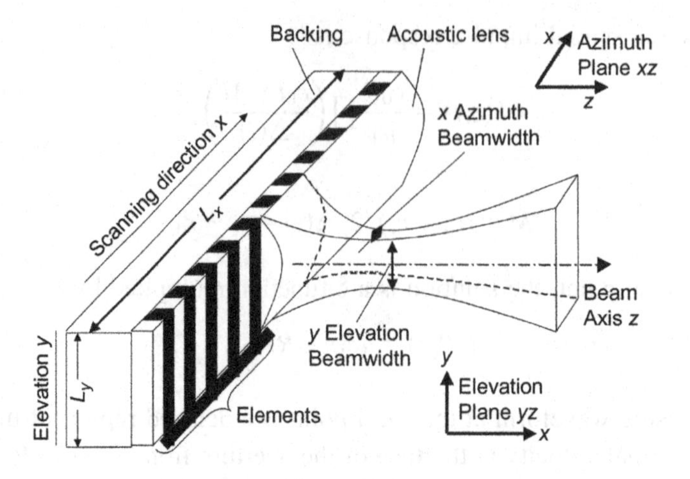

Figure 7.4
Relation of phased array to azimuth (Imaging) and elevation planes. *Adapted from Panda (1998).*

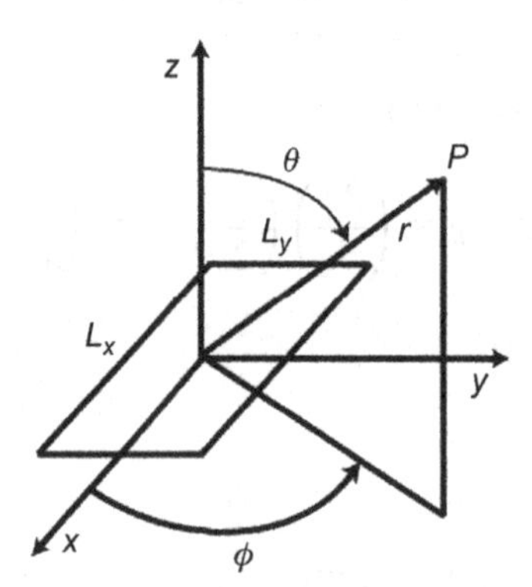

Figure 7.5
Simplified geometry for a rectangular array element in the *xz* plane.

with apertures on the order of a wavelength, the far-field beam pattern can be found from the $+i$ Fourier transforms of the aperture functions:

$$H_e(x, y, z, \lambda) = \frac{c_0}{2\pi z} \int_{-\infty}^{\infty} A_x(x_0) e^{i2\pi x_0 (x/\lambda z)} dx_0 \int_{-\infty}^{\infty} A_y(y_0) e^{i2\pi y_0 (y/\lambda z)} dy_0, \tag{7.11A}$$

which, for line sources of lengths L_x and L_y multiplied together to describe a rectangular aperture with sides L_x and L_y, gives:

$$H_e(x, y, z, \lambda) = H_x H_y = \frac{c_0}{2\pi z} L_x \operatorname{sinc}\left(\frac{L_x x}{\lambda z}\right) L_y \operatorname{sinc}\left(\frac{L_y y}{\lambda z}\right). \tag{7.11B}$$

Recall in the original diffraction integral that the Fresnel approximation was made by a binomial approximation of the difference vector $|r - r_0|$ and the substitution of z for r, so that this approximation was valid only for the xz and yz planes. A more exact result for any field point in the far field can be derived by accounting for the total rectangular shape of the aperture. The direction cosines to the field point are introduced from the spherical coordinate geometry given by Figure 7.5:

$$u = \sin\theta \cos\varphi, \quad \text{and} \tag{7.12A}$$

$$v = \sin\theta \sin\varphi, \tag{7.12B}$$

where θ is the angle between r and the z axis, and ϕ is the angle between r and the x axis.

Stepanishen (1971) has shown that the far-field response for this geometry is:

$$H_e(x, y, z, \lambda) = H_x H_y = \frac{c_0}{2\pi r} L_x \operatorname{sinc}\left(\frac{L_x u}{\lambda}\right) L_y \operatorname{sinc}\left(\frac{L_y v}{\lambda}\right), \tag{7.13}$$

which reduces to the previous expression in the xz plane ($\phi = 0$), and the yz plane ($\theta = 0$), and z is replaced by r. The time domain equivalent of this expression can be found from the inverse Fourier transform of Eqn 7.13 with $\lambda = c/f$:

$$h_e(u, v, r, t) = \frac{c_0}{2\pi r} L_x \left(\frac{c}{L_x u}\right) \prod\left(\frac{t}{L_x u/c}\right) * L_x \left(\frac{c}{L_y v}\right) \prod\left(\frac{t}{L_y v/c}\right). \tag{7.14}$$

This convolution of two rectangles has the trapezoidal shape illustrated by Figure 7.6. For equal aperture sides and $u = v$, a triangle results. For on-axis values, the rect functions reduce to impulse functions, so that:

$$h_e(0, 0, r, t) = \frac{c_0}{2\pi r} L_x L_x \delta(t). \tag{7.15}$$

This equation, in combination with Eqns 7.1B and 7.4, indicates that on-axis pressure in the far field is the derivative of the normal velocity, is proportional to the area of the aperture, and falls off inversely with r.

For two-dimensional beam scanning in the xz plane, a one-dimensional array will extend along the x axis (two-dimensional arrays are covered later in Section 7.6). For this plane, H_x can be expressed as a function of frequency from Eqn 7.13 with $1/\lambda = f/c$ and the convenient substitution $h_{ox} = \sqrt{c_0/2\pi r}$ as:

$$H_x(\theta, r, f) = h_{0x} L_x \operatorname{sinc}\left(\frac{L_x f \sin\theta}{c}\right), \tag{7.16}$$

where, on-axis, as $\theta \to 0$, $H_x \to h_{0x} L_x$ (as shown in Figure 7.7). Note that this function has zeros when u is integral multiples of λ/L_x. This element directivity has been examined

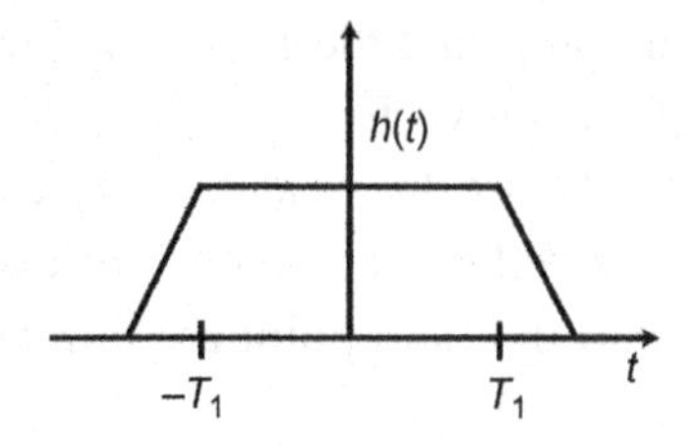

Figure 7.6
Trapezoidal far-field spatial impulse response for a rectangular array element.

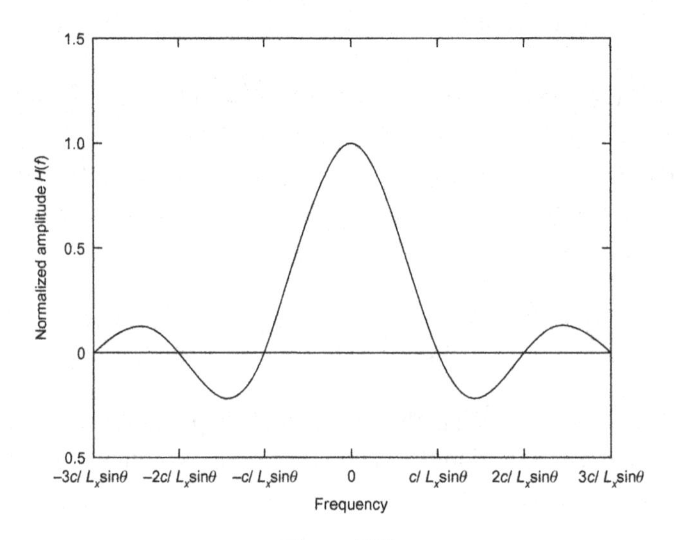

Figure 7.7
Far-field element directivity as a function of frequency for an element length L_x.

(Sato, Fukukita, Kawabuchi, & Fukumoto, 1980; Smith, von Ramm, Haran, & Thurstone, 1979), and it is discussed in more detail in Section 7.5. The far-field time response is the inverse Fourier transform of Eqn 7.7:

$$h_x(\theta, r, t) = h_{0x} L_x \left(\frac{c}{L_x \sin\theta} \right) \prod \left(\frac{t}{L_x \sin\theta / c} \right), \tag{7.17A}$$

which is illustrated by Figure 7.8. The limiting value of this expression on-axis is:

$$h_x(0, r, t) = h_{0x} L_x \delta(t). \tag{7.17B}$$

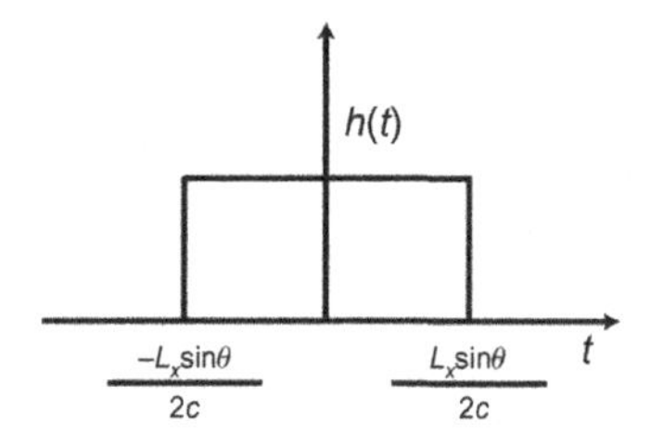

Figure 7.8
Spatial impulse response h_x along the z axis for an element of length L_x oriented along the x axis.

7.4.2 Pulsed Excitation of an Element

To find the pressure pulse in the far field of an element in the scan (xz) plane for a pulse excitation $g(t)$, we convolve the input pulse that we assume is in the form of the normal velocity, $g(t) = \partial v_n/\partial t$, with the time derivative of ψ_x, as given by Eqn 7.2A:

$$p(r,t) = \rho_0 \partial\psi/\partial t = \rho_0 \partial v_n/\partial t *_t h(r,t) = \rho_o g *_t h(r,t). \tag{7.18}$$

As an example (Bardsley & Christensen, 1981), let $g(t)$ have the decaying exponential form shown in Figure 7.9:

$$g(t) = v_{0x} e^{-at} H(t) \cos(\omega_c t), \tag{7.19}$$

in which $H(t)$ is the step function and v_{0x} is the normal particle velocity on the aperture. Then the pressure can be found from:

$$p(r,t) = \rho_0 g(t) *_t h_{0x} L_x \left(\frac{c}{L_x \sin\theta} \right) \prod \left(\frac{t}{L_x \sin\theta / c} \right) \tag{7.20A}$$

off-axis, and from:

$$p(r,t) = \rho_0 g *_t h = \rho_0 g(t) *_t h_{0x} \delta(t) = \rho_0 h_{x0} g(t) \tag{7.20B}$$

for the on-axis value. The pressure response calculated from Eqn 7.20B is plotted in Figure 7.10 over a small angular range.

An equivalent frequency domain expression for pressure at a field point, from Eqn 7.20A, is:

$$P(r,f,\theta) = G(f) H_x(r,f,\theta), \tag{7.21}$$

where $r = \sqrt{x^2 + z^2}$.

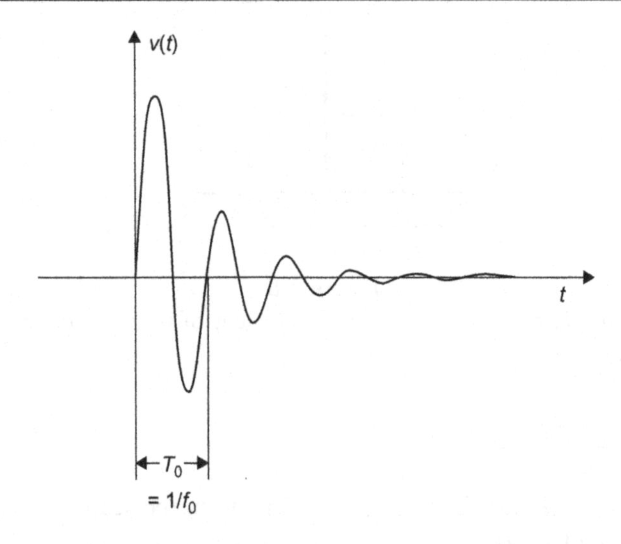

Figure 7.9
Typical short acoustic pulse waveform with $Q = 3.1$ and center frequency of 2.25 MHz used for array calculations for examples. *From Bardsley and Christensen (1981), Acoustical Society of America.*

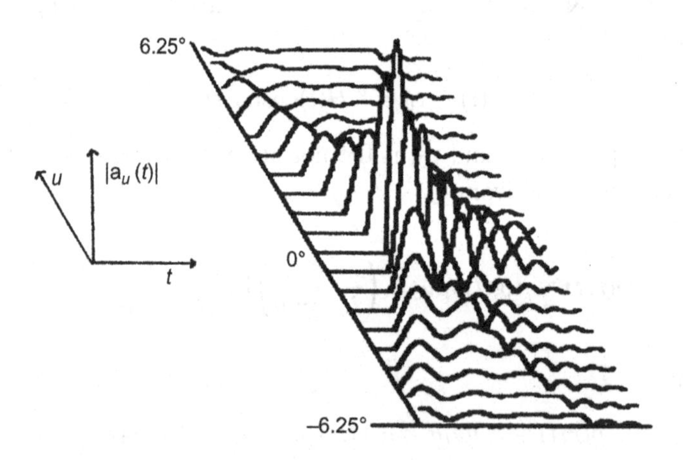

Figure 7.10
Absolute values of pressure waveforms as a function of angular direction (u) and time (t). Plotted in an isometric presentation over a small angular range: −6.25° to 6.25° in 0.625° angular increments for the pulse of Figure 7.9 and a 2.56-cm-long aperture. *From Bardsley and Christensen (1981), Acoustical Society of America.*

7.4.3 Array Sampling and Grating Lobes

In order to find out how an element functions as part of an array, a good starting point is a perfect ideal array made up of spatial point samples. An infinitely long array of these samples (shown in Figure 7.11A) can be represented by a shah function with a periodicity or pitch p (note p does not represent pressure in the following discussions). Since the pressure at a field point is related to a Fourier transform of the aperture or array, the result is another shah function with a periodicity (λ/p), as given by this expression, Figure 7.11B, and (see Section A.2.4 of Appendix A):

$$\Im_i\left[\text{III}\left(\frac{x}{p}\right)\right] = p\text{III}\left(\frac{puf}{c_0}\right) = p\text{III}\left(\frac{u}{\lambda/p}\right) \tag{7.22}$$

For an aperture of finite length L_x, the infinite sum of the shah function is reduced to a finite one in the spatial x domain, as is given in Figure 7.12 and as follows:

$$\Im_i\left[\prod\left(\frac{x}{L_x}\right)\text{III}\left(\frac{x}{p}\right)\right] = pL_x\sum_{-\infty}^{\infty}\text{sinc}\left[\frac{L_x}{\lambda}(u-m\lambda/p)\right] = L_x\text{sinc}\left[\frac{L_xu}{\lambda}\right]*_u p\text{III}\left(\frac{\text{u}}{\lambda/p}\right). \tag{7.23}$$

In this figure, the main lobe is centered at $u=0$, and the other modes for which $m\neq0$ are called grating lobes. Grating lobes are centered on direction cosines u_g at angles:

$$\theta_g = \pm\arcsin(m\lambda/p). \tag{7.24}$$

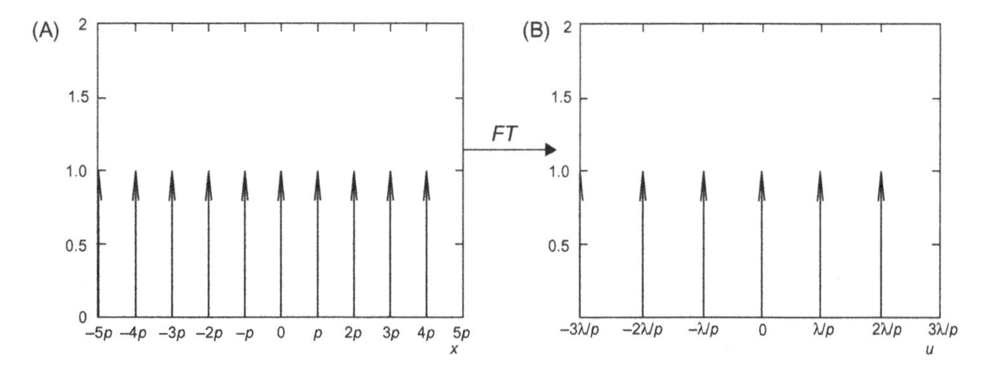

Figure 7.11
Ideal samplers transform into discrete angles. (A) A shah function of ideal samplers spaced along the x axis with a periodicity of p; (B) Normalized Fourier transform of a shah function is another shah function with samplers situated at intervals of u equal to integral multiples of λ/p. The amplitude of the transformed shah function is p.

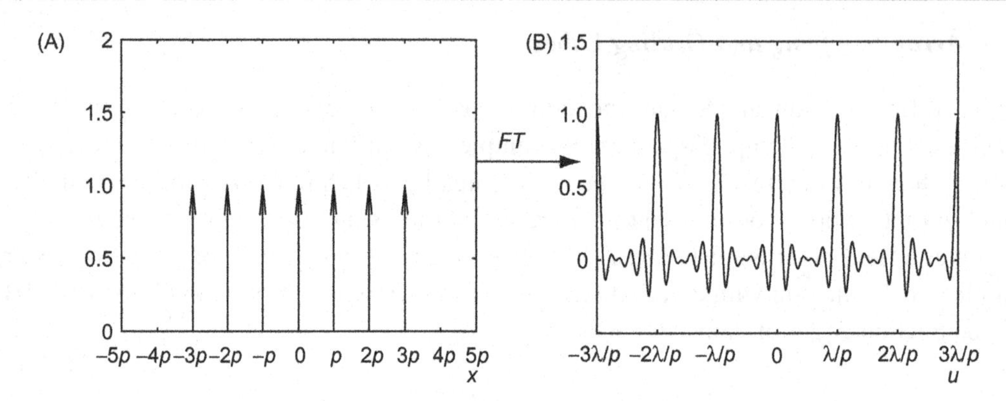

Figure 7.12
(A) An array of $2n_L + 1$ point samples along the x axis with a periodicity of p; (B) Normalized fourier transform of a finite length array of point samplers is an infinitely long array of sinc functions situated at intervals of u equal to integral multiples of λ/p with an actual amplitude of $L_x p$.

The first grating lobe is the most important, or $m = \pm 1$. If the periodicity is set equal to half-wavelength spacing (the usual spacing for phased arrays), which is the Nyquist sampling rate, the grating lobes are out of the picture because they are at $\pm\pi/2$(the endfire directions). If the spacing is larger in terms of wavelengths, then instead of one beam transmitted, three or more are sent. For example, for a two-wavelength spacing, beams appear at 0° and ±30°. For linear arrays, spacing is often one or two wavelengths because steering requirements are minimal, but for phased arrays that create sector scans, grating lobe minimization is important (described in Section 7.4.5).

For the CW case, grating lobes can be as large as the main lobe, but for pulses, grating lobes can be reduced by shortening the pulse. The effect of the transducer bandwidth on the grating lobe can be seen from Eqn 7.21 and Figure 7.13. Shown are the main lobe and grating lobe centered on the center frequency f_0 and with a −3-dB bandwidth, given approximately by $0.88c/N_p u_g$. The fractional bandwidth of $G(f)$ is approximately $0.88f_0/n$, where n is the number of periods (cycles) in the pulse (corresponding to a $Q = \Delta f/f_0 = 1.14n$). Recall the overall response is given by the product of $H(f)$ and $G(f)$ from Eqn 7.21 and that amplitude of the grating lobe will be proportional to the overlap area of these functions from their Fourier transform relation. As a consequence of these factors, the wider the bandwidth of $G(f)$ (the shorter the pulse), the smaller the overlap and the lower the amplitude of the grating lobe in the time domain. An approximate expression for the grating lobe is Q/N, where N is the number of elements (Schwartz & Steinberg, 1998).

Another perspective on grating lobe effects is observed in the time domain for finite length pulses through the convolution operation. The on-axis main lobe pulse contributions add

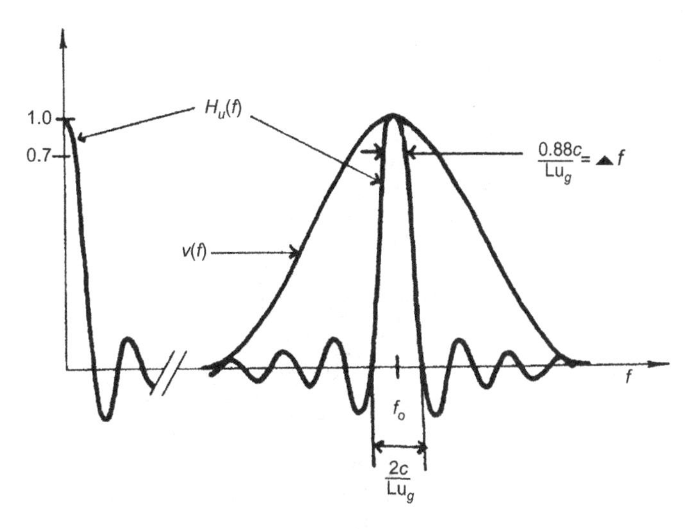

Figure 7.13
The spatial transfer function $H(f,u)$, showing a first-order grating lobe $u_g = \lambda/p = 1/2.4$ at 24.6° with a bandwidth of $0.88c/N_p u_g = 0.124$ MHz as well as the pulse spectrum $G(f) = V(f)$ with a bandwidth of 0.726 MHz. *From Bardsley and Christensen (1981), Acoustical Society of America.*

coherently, and, at grating lobe locations, pulses add sequentially to form a long, lower-level pulse. The overall impact of a grating lobe can be seen over a small angular range in Figure 7.14, in which the long grating lobe pulse builds at larger angles. From this viewpoint, it is evident that the shorter the pulse, the less pulses will overlap and build in amplitude to create a significant grating lobe.

7.4.4 Element Factors

Until now, the array has been treated as having point sources. To include the imperfect sampling effects of rectangular elements described in Section 7.4, we replace the point samples by elements of width w, as shown in Figure 7.15 and by the following:

$$H_0(u,\lambda) = h_{0x}\Im_i\left[\prod\left(\frac{x}{w}\right) * \sum_{-nL}^{nL}\delta(x-np)\right] = \sum_m h_{0x}L_x pw \,\text{sinc}\left(\frac{wu}{\lambda}\right)\text{sinc}\left[\frac{L_x}{\lambda}(u - m\lambda/(p))\right]. \tag{7.25}$$

Here, the first sinc term is called the element factor. In the angle or frequency domain, the small element size translates into a broad directivity modulating the sequence of grating lobes as shown in Figure 7.15. The -3-dB directivity width is approximately $0.88\lambda/w$, as opposed to the width of a main or grating lobe, which is about $0.88\lambda/L$.

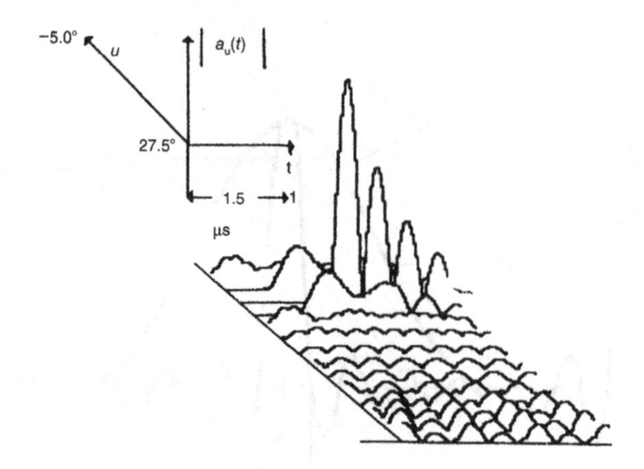

Figure 7.14

An Isometric presentation of pulse on-axis and the long pulse of the grating lobe. The angular range of −5–27.5° in increments of 2.5° for the parameters given in Figures 7.12–7.13. At 0°, the pulses add coherently to give an amplitude N. Near the grating lobe angle of 24.6°, pulses overlap sequentially to create a long pulse with an amplitude approximately equal to Q. *From Bardsley and Christensen (1981), Acoustical Society of America.*

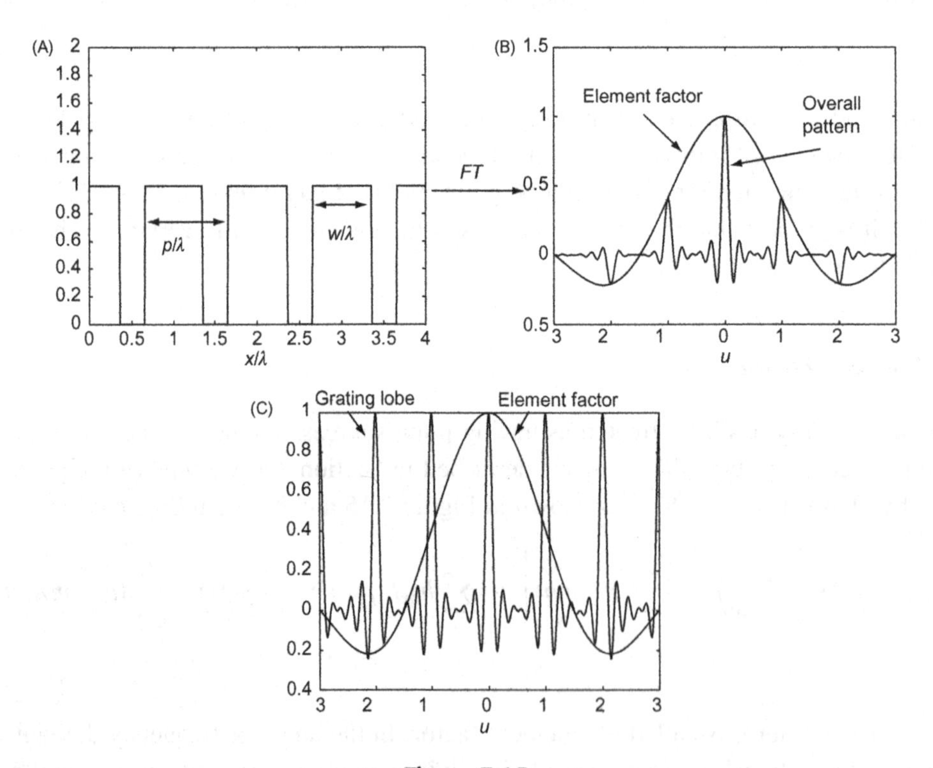

Figure 7.15

Array made up of rectangular elements of width w. (A) A finite length array of elements of width w and periodicity p; (B) Fourier transform of spatial element amplitude results in modulation of grating lobes by broad angular directivity of element factor; (C) factors contributing to overall transform.

7.4.5 Beam Steering

If a linear phase is placed across the array elements, corresponding to a wave front at an angle θ_s from the z axis, the result is a beam steered at an angle θ_s (shown in Figure 7.16). This phase (τ_{sn}) is applied, one element at a time, as a linear phase factor with $u_s = \sin\theta_s$:

$$exp(-i\omega_c\tau_{sn}) = exp(-i2\pi f_c(npu_s)/c) = exp(-i2\pi npu_s/\lambda_c), \tag{7.26}$$

to unsteered array response, Eqn 7.25, then the beams are steered at u_s:

$$\begin{aligned} H_s(u,u_s,\lambda_c) &= \Im_i\left[\prod\left(\frac{x}{w}\right)\sum_{-nL}^{nL} a_n\delta(x-np)\exp(-i2\pi(npu_s)/\lambda_c)\right] \\ &= L_x pw\text{sinc}\left(\frac{wu}{\lambda_c}\right)\text{sinc}\left[\frac{L_x}{\lambda_c}(u-m\lambda_c/p-u_s)\right] \end{aligned}, \tag{7.27}$$

and the amplitude weights (a_n) are equal to one. Figure 7.17 shows the effects of element directivity on the steered beam and grating lobes. In sector or angular scanning, the location of the grating lobe is related to the steering angle:

$$\theta_g = \pm\arcsin(m\lambda/p + u_s), \tag{7.28}$$

where $m = \pm 1$ are the indices of the first grating lobes. As an example, consider a period of one wavelength and a steering angle of $-45°$, then the first grating lobe will be at $\theta_g = \arcsin(1\ -0.707) = 17°$. This result would not be appropriate for a phased array, but it would do for a linear array. What periodicity would be necessary to place the grating lobe at 45° for a steering angle of $-45°$?

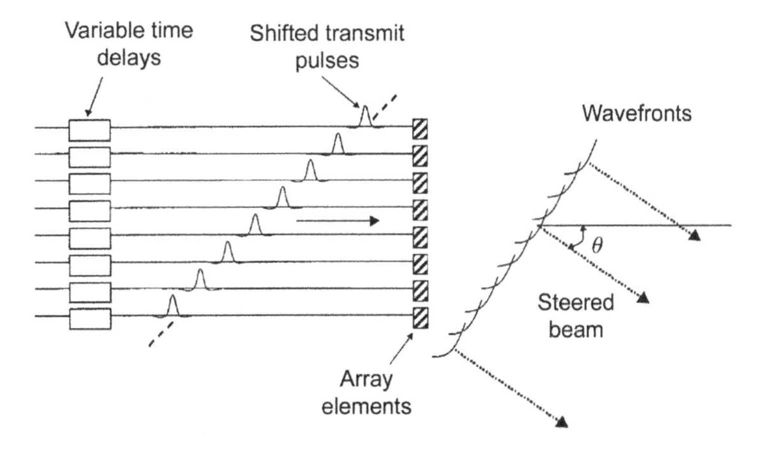

Figure 7.16
Delays for steering an array. *From Panda (1998).*

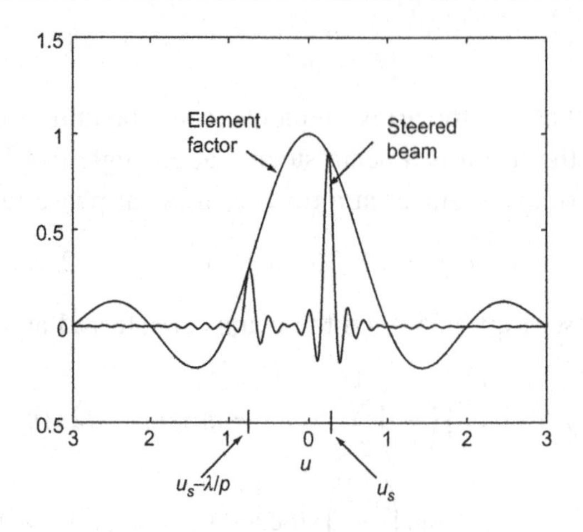

Figure 7.17
Array angular response when steered at θ_s.

7.4.6 Focusing and Steering

Until now, a far-field condition was assumed; however, this is not true in general. For an array aperture of several or many wavelengths in length, a near-field pattern will emerge. Just as lenses were used to focus (as explained in Section 6.6), arrays can be focused by adding time-delayed pulses that simulate the effect of a lens to compensate for the quadratic diffraction phase term. The time delays to focus each element (n) are:

$$\tau_n = \left[r - \sqrt{(x_r - x_n)^2 + z_r^2}\right]/c + t_0, \tag{7.29A}$$

where r is the distance from the origin to the focal point, $r = \sqrt{x_r^2 + z_r^2}$, x_n is the distance from the origin to the center of an element indexed as "n" ($x = np$), and t_0 is a constant delay added to avoid negative (physically unrealizable) delays. The application of a paraxial approximation under the assumption that lateral variations are smaller than the axial distance leads to:

$$\tau_n \approx (x_n u_s - x_n^2/2z_r)/c + t_0 = [npu_s - (np)^2/2z_r] + t_0. \tag{7.29B}$$

From this approximate expression, the first term is recognizable as the steering delay, Eq 7.26, and the second is recognizable as the quadratic phase term needed to cancel the similar term caused by beam diffraction, as shown for a lens in Eqn 6.27B. In practice, usually the exact Eqn 7.29A is used for arrays rather than its approximation. Putting all this together, we start with a modification of Eqn 7.17A for the spatial impulse response of a single element located at position $x_n = np$:

$$h_n \approx (u, r, t) = a_n h_{0x} w\left(\frac{c}{wu}\right) \prod \left(\frac{t}{wu/c}\right), \tag{7.30A}$$

where u is defined in Eqn 7.12A, and then the one-way transmit spatial impulse response for an element with focusing is of the form:

$$h_n\left(t - \frac{1}{c}\sqrt{(x-x_n)^2 + z^2} - \tau_n\right) = h\left(t - \frac{\sqrt{(x-x_n)^2 + z^2}}{c} - r/c + \sqrt{\frac{(x-x_n)^2 + z_r^2}{c}}\right), \tag{7.30B}$$

and when $x = x_r$, and $z = z_r$ at the focus:

$$h_n\left(t - \frac{1}{c}\sqrt{(x-x_n)^2 + z^2} - \tau_n\right) = h(t - r/c). \tag{7.30C}$$

The overall array response (h_a) is simply the sum of the elements:

$$h_a(t) = \sum_{-nL}^{nL} a_n h_n\left(t - \frac{1}{c}\sqrt{(x-x_n)^2 + z^2} - \tau_n\right). \tag{7.30D}$$

The pressure can be found from the convolution of the excitation pulse and array response as in Eqn 7.18. Here, a perfect focus is achieved when the field point at (x, z) is coincident with the focal point (x_r, z_r). However, at all other points, zones corresponding to those described in Section 6.6.3 (a pre-focal zone (formerly near Fresnel zone), focal zone (formerly focal Fraunhofer zone), and post-focal zone (formerly far Fresnel zone)) will be

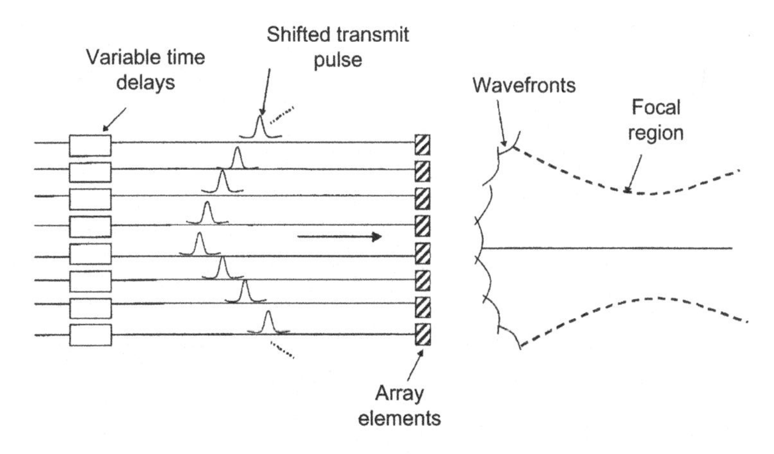

Figure 7.18
Array delays for focusing a beam. *From Panda (1998).*

created. Figure 7.18 illustrates the delays needed for focusing. The same type of delay equations can be used for receive or transmit.

7.5 Pulse–Echo Beamforming

7.5.1 Introduction

Several factors are involved in the ultrasound imaging of the body, as was symbolized by the block diagram in Figure 2.14. In Chapter 5, the response of the transducer to a pulse excitation in a pulse–echo mode was discussed. These are covered by the electrical excitation and are also represented by the electrical excitation block (E), the transmit transducer response (G_T), and the receive response (G_R). A more practical description includes the effects of the transmit pulse, $e_T(t)$. The electroacoustic conversion impulse response of the transducer from voltage to the time derivative of the particle velocity, $g_T(t)$, the derivative operation, and the corresponding receive functions (denoted by R), can all be lumped together as:

$$e_{RT}(t) = e_T(t) *_t g_T(t) *_t g_R(t), \tag{7.31A}$$

or, in the frequency domain, as:

$$E_{RT}(f) = E_T(f) G_T(f) G_R(f). \tag{7.31B}$$

The overall voltage output, including focusing on transmit and receive, can be described by the product of the array transmit and receive spatial responses (shown by Figure 7.16):

$$V_0(r, f, \theta) = H_T(r_T, f, \theta_T) H_R(r_R, f, \theta_R) E_{RT}(f). \tag{7.32A}$$

The equivalent time domain formulation of the pulse–echo signal is:

$$v_o(z, r, t) = h_t *_t h_r *_t e_{RT}. \tag{7.32B}$$

Implicit in the spatial impulse responses are the beamformers, which organize the appropriate sequence of transmit pulses and the necessary sum and delay operations for reception. The beamforming operations, represented by blocks XB (transmit) and RB (receive), reside in the imaging system (to be explained in Chapter 10). Attenuation effects, symbolized by blocks A_T (forward path) and A_R (return path), will be discussed in Section 7.9.4. Chapter 8 and Chapter 9 describe the scattering block (S), as well as the scattering of sound from real tissue and how it affects the imaging process.

The ability of a beamformer to resolve a point target is determined by the spatial impulse response of the transmit and receive beams intercepting the target. A measure of how well an imaging system can resolve a target is called the "point spread function," which is another

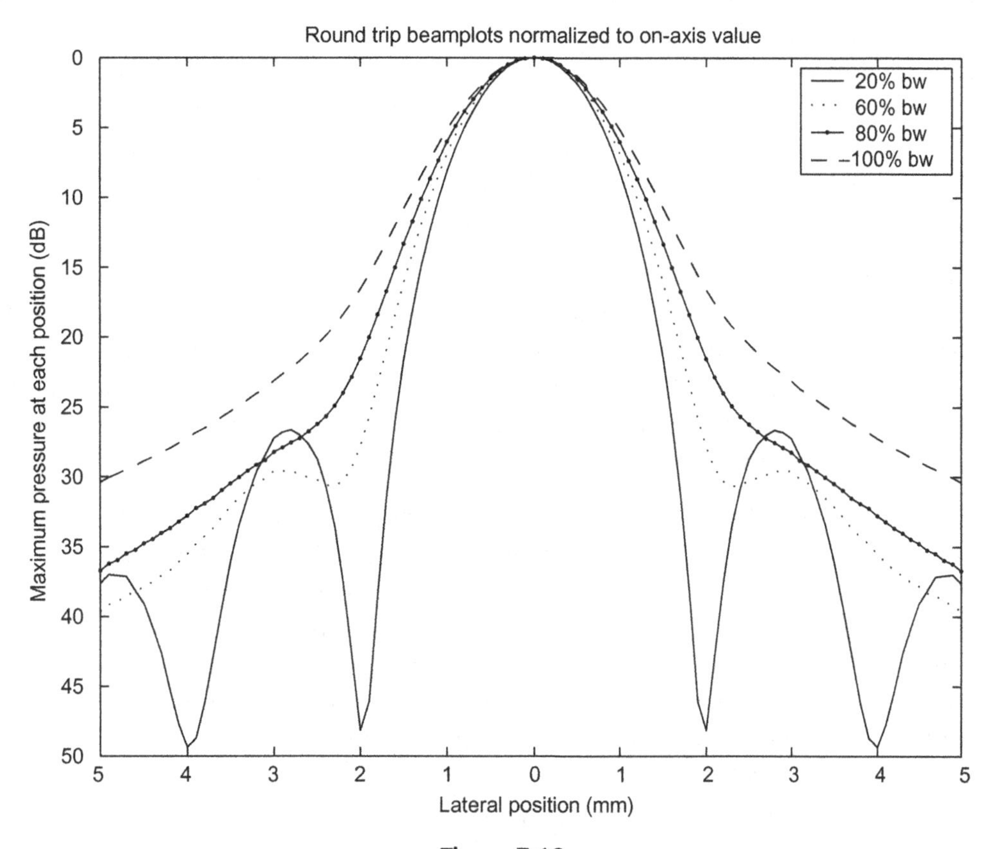

Figure 7.19

Normalized full hamming apodized beams in focal plane for three round-trip gaussian pulses of differing fractional bandwidths: 20, 60, 80, and 100%. *Created with Graphical User interface (GUI) for Field 2 from the Duke University virtual imaging lab.*

name for the function given by Eqn 7.32. This equation shows that the beam-shape is related to the type of pulse applied. For example, the effect of bandwidth on the beam profile can be seen in Figure 7.19. For very short pulses or wider bandwidths, sidelobe levels can rise; this suggests that a moderate fractional bandwidth in the 60–80% is a better compromise between resolution and sidelobe suppression. The shaping of the pulse is also important in achieving a compact point spread function with low time and spatial sidelobes (Wright, 1985).

7.5.2 Beam-shaping

From Eqn 7.32A, the overall beam-shape is the product of the transmit and receive beams, each of which can be altered in shape by apodization ('t Hoen, 1982). So far, each element has had an amplitude weight of one that led to a sinc-shaped directivity in the focal plane. By altering the weight of each element (a_n) (see Eqn 7.27), through a means such as changing the voltage applied to each element, other weighting functions can be obtained, such as those discussed in Section 6.4 to lower sidelobes (Harris, 1978). Individual transmit and receive aperture lengths and apodizations can be combined to complement each other to achieve narrow beams with low sidelobes. The apodization can also increase the depth of field. Two drawbacks of apodization are an increased mainlobe width and a reduction in amplitude proportional to the area of the apodization function.

Two ways of measuring the effectiveness of a beam-shape are its detail resolution and its contrast resolution. Detail resolution, commonly taken as the −6-dB beamwidth, is the ability of the beam to resolve small structures. Point scatterers end up being imaged as blobs. The size of a blob is determined by the point spread function and can be estimated by a −6-dB ellipsoid, which has axes that are the axial resolution (pulse envelope) and lateral resolutions in x and y at −6 dB below the peak value in each dimension (Figure 7.20).

The contrast resolution of a beam (Maslak, 1985; Wright, 1985) is a measure of its ability to resolve objects that have different reflection coefficients and is typically taken to be the −40-dB (or −50-dB) round-trip beamwidth. Pulse–echo imaging is dependent on the backscattering properties of tissue. To first order, the possibility of distinguishing different tissues in an image is related to the reflection coefficients of tissues relative to each other (such as those shown in Figure 1.3). These often subtle differences occur at the −20- to −50-dB level. Consider three scatterers at reflectivity levels of 0, −20, and −40 dB. If the main beam is clear of sidelobes down to the −50 dB level, then these three scatterers can be cleanly distinguished. If, however, the beam has high sidelobes at the −13-dB level, then both weak scatterers would be lost in the sidelobes. The level of the sidelobes sets a range between the strongest scatterers and the weakest ones discernible. In other words, the sidelobe level sets an acoustic clutter floor in the image.

As an example of the effect of apodization, Figure 7.21 compares an image without apodization to one with receive Hanning apodization, both at the same amplitude settings.

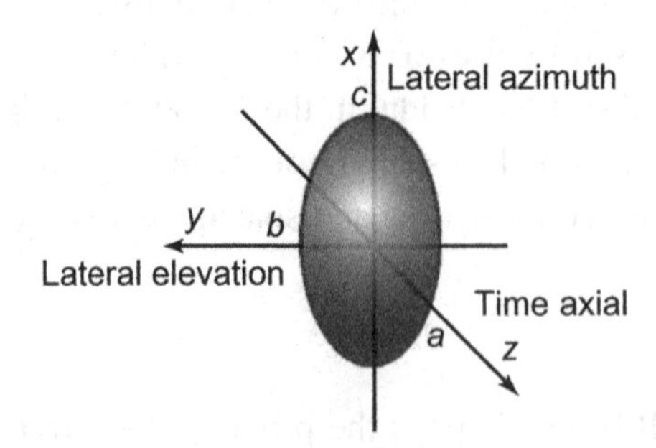

Figure 7.20

A −6-dB-resolution Ellipsoid. The axes represent −6-dB resolution in the lateral directions x and y and the axial pulse resolution along z.

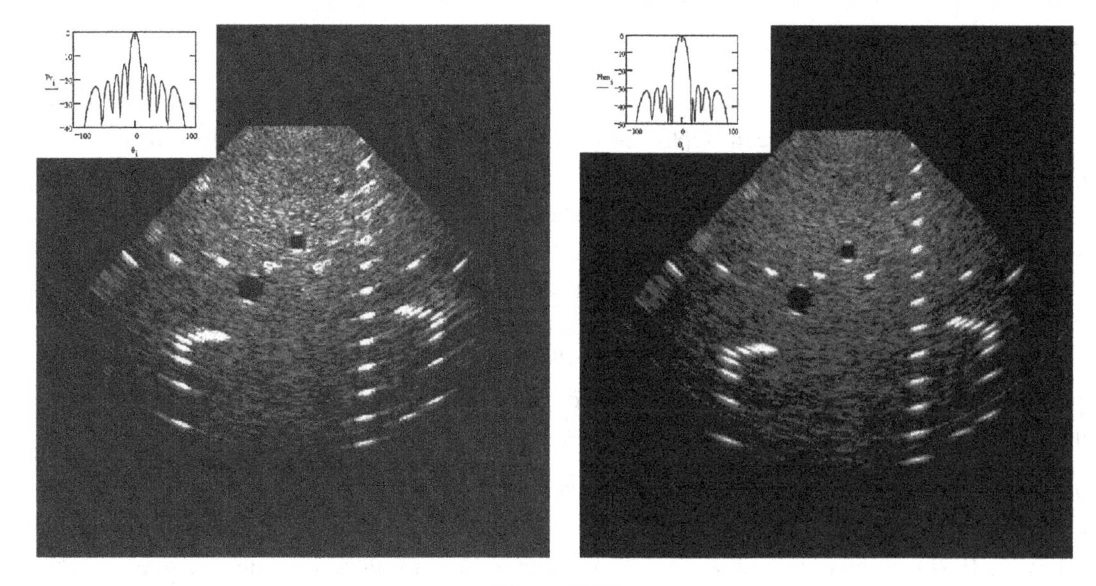

Figure 7.21
Effect of apodization on imaging. (A) Unapodized beam plot insert and corresponding image of phantom with point targets; (B) Hanning apodization on receive beam shown in insert and corresponding image of phantom. *Courtesy of P. Chang, Terason, Teratech Corporation.*

The amplitude apodization functions are graphed above each image (recall that the overall beam pattern is the product of the transmit and receive beam patterns). What is being imaged is a tissue-mimicking phantom with small wire-like objects (slightly smaller than the resolution capability of the imaging system) seen in cross-section against a background of tissue-like material full of tiny unresolvable scatterers. The appearance of the wire objects is blob-like and varies with the detail resolution, as expected, through the field of view. Near the transmit focal length, the blobs are smaller. Careful observation of the wire targets in the image with apodization indicates that they are slightly dimmer and wider, results of less area under the apodization curve and a wider −6-dB beamwidth; therefore, the penetration (the maximum depth at which the background can be observed) is less.

In the image made without apodization, the resolvable objects appear to have more noticeable sidelobe "wings" (a smearing effect caused by high sidelobe levels). Another difference in the image made with apodization is contrast: The wire targets stand out more against a darker background. For extended diffuse targets, such as the tissue-mimicking material, the sidelobes have an integrating effect. For a beam with high sidelobes, the overall level in a background region results in a higher signal level; however, for a beam with low sidelobes, the overall integration produces a lower signal level that gives the appearance of a darker background in the image. The net result is that the difference in gray levels of a bright (wire) target and its background (tissue-mimicking material) is less for the first case than the second, so that the apparent contrast is greater for the second case.

7.5.3 Pulse–Echo Focusing

On transmit, only a single focal length can be selected. However, if the region of interest is not moving too fast, the scan depth can be divided into smaller ranges close to the focal zones of multiple transmit foci. These multiple transmit ranges can then be "spliced" together to form a composite image that has better resolution over the region of interest (see Figure 10.9 for an example). The transmit aperture length can be adjusted to hold a constant F number, ($F\# = F/L$) to keep the resolution constant over an extended depth (Maslak, 1985). For example, from Eqn 6.9C, the one-way, −6-dB full width half maximum (FWHM) beamwidth for an unapodized aperture is $2x_{-6} = 0.384\lambda F\#$. This approach has the disadvantage of slowing the frame rate by a factor equal to the number of transmit foci used.

One way to increase frame rate is to employ "parallel focusing" (Davidsen & Smith, 1993; Shattuck, Weinshenker, Smith, & von Ramm, 1984; Thomenius, 1996; von Ramm, Smith, & Pavey, 1991). In this method, a smaller number of broad transmit beams are sent so that two or more narrower receive beams can fit within each one. On reception, multiple beams are offset in steering angle to fit within the width of each transmit beam (Figure 7.22). In this way, the frame rate, which is normally limited by the round-trip time of the selected scan depth, can be increased by a factor equal to the number of receive beams.

On receive, however, a method called "dynamic focusing" (Vogel, Bom, Ridder, & Lance, 1979) provides nearly perfect focusing throughout the entire scan depth. In this case, the scan depth is divided into many zones, each one of which is assigned a receive focal length. In modern digital scanners, the number of zones can be increased so that the transitions between zones are indistinguishable and focusing tracks the received echo depth. In addition, the receive aperture can be changed and/or apodized with depth to

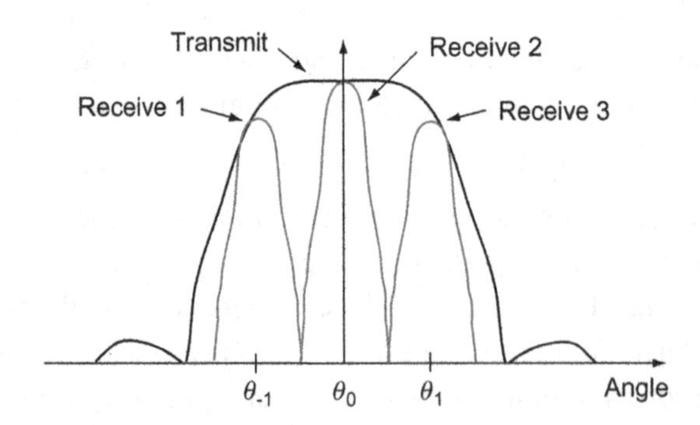

Figure 7.22
Parallel receive beamforming in which the transmit beam is broadened so that two or more receive beams can be extracted. Frame rate is increased by reducing the number of transmit beams.

maintain consistent resolution. Finally, the overall scan depth can be divided into N sections, each one with a separate transmit focus, and the individual sections can be spliced together; however, this approach reduces frame rate by $1/N$. Examples of the resolution improvements attainable are shown in Figure 7.23 for a 12-element, 4.5-MHz annular array with an outer diameter of 30 mm (Foster et al., 1989a, 1989b). Figure 7.23A shows the highly localized short depth of field for a fixed focus on receive, Figure 7.23B demonstrates the benefits of receive dynamic focusing, and Figure 7.23C illustrates the effects of a two-transmit-zone splice with dynamic focusing. Recently, annular arrays have staged a comeback in the high-frequency arena (Cao, Hu, & Shung, 2003; Gottlieb, Cannata, Hu, & Shung, 2006; Ketterling, Aristizábal, Turnbull, & Lizzi, 2005).

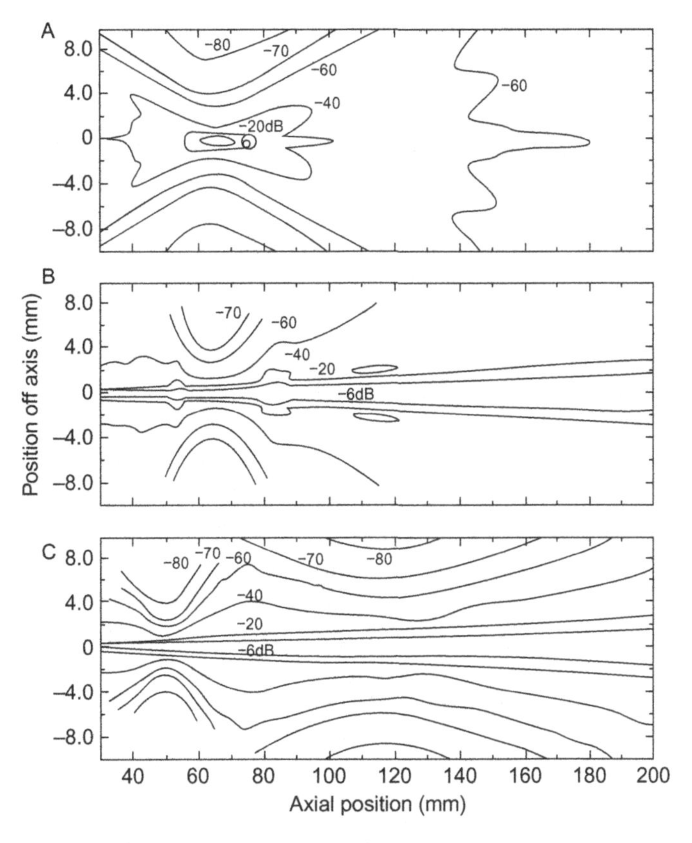

Figure 7.23

Beam contour plots for a 12-element, 4.5-MHz annular array. (A) Fixed transmit focus 65 mm and fixed receive focus 65 mm; (B) fixed transmit focus 65 mm and dynamic receive focusing; (C) first fixed transmit focus 50 mm and second fixed focus 130 mm, both with dynamic focusing and spliced together at 76 mm. *From Foster et al. (1989a), with permission from the World Federation of Ultrasound in Medicine and Biology.*

7.6 Two-dimensional Arrays

One-dimensional (1D) arrays (Figure 7.24) often have 32 to 300 elements and come in many forms (to be described in more detail in Chapter 10). These arrays scan in the azimuth plane, and a mechanical cylindrical lens produces a fixed focal length in the elevation plane. Two-dimensional (2D) arrays (refer to Figure 7.24) are needed to achieve completely arbitrary focusing and steering in any direction. While a typical phased array may have 64 elements, a 2D array might have 64^2 or 4096 elements. The large number of elements of 2D arrays presents challenges for their physical realization (see Section 7.9.2) as well as for efficient simulation of their fields. In Section 7.9.3, 1.5-dimensional (1.5D) arrays, intermediate between 1D and 2D arrays, are described.

The geometry for a 2D array of point sources of period p is shown in Figure 7.25. The diffraction impulse response for this array is:

$$H_s(r,\theta,\varphi,\lambda) = \frac{L_x L_y p^2}{2\pi r} \sum_{n=-\infty}^{\infty} \text{sinc}\left[\frac{L_x}{\lambda}(u - n\lambda/p - u_s)\right] \sum_{m=-\infty}^{\infty} \text{sinc}\left[\frac{L_x}{\lambda}(v - m\lambda/p - v_s)\right], \tag{7.33A}$$

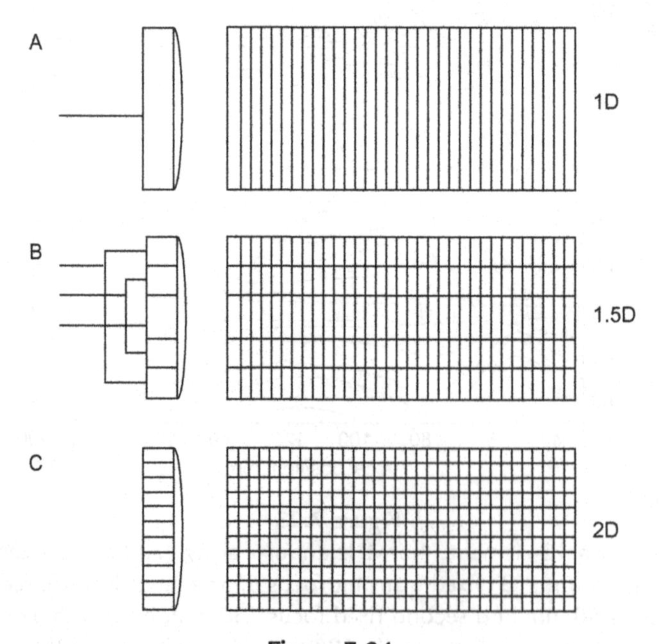

Figure 7.24
Types of arrays in profile and azimuth plane views. (A) 1D array; (B) 1.5D array; (C) 2D array.

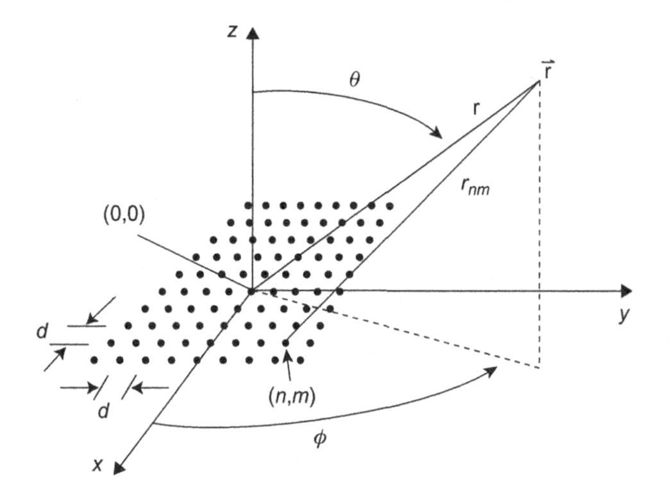

Figure 7.25
Geometry for a square 2D array of point sources with $2N+1$ Elements on a side with d corresponding to p in the text. *From Turnbull (1991).*

in which the directions to the field point are u and v and the steering directions are:

$$u_s = \sin\theta_0 \cos\varphi_0, \quad \text{and} \tag{7.33B}$$

$$v_s = \sin\theta_s \sin\varphi_s, \tag{7.33C}$$

and the overall apertures are the following:

$$L_x = (2N+1)p, \quad \text{and} \tag{7.33D}$$

$$L_y = (2M+1)p. \tag{7.33E}$$

For a 2D array, grating lobes occur at the following locations:

$$u_g = u_s \pm n\lambda/p, \quad \text{and} \tag{7.34A}$$

$$v_g = v_s \pm m\lambda/p. \tag{7.34B}$$

For examples of the effect of spacing, refer to Figure 7.26. For a square array composed of 101 point sources on a side, first-order grating lobes appear when the periodicity is one wavelength according to steering at $(u_0, v_0) = (0.5, 0.5)$ for the main lobe and $(u, v) = (0.5, -0.5)$, $(-0.5, 0.5)$ and $(-0.5, -0.5)$ for the grating lobes for a CW excitation.

The first phased arrays were narrowband, so a CW model was adequate. With the arrival of digital systems, true-time delays for both steering and focusing became practical. For this

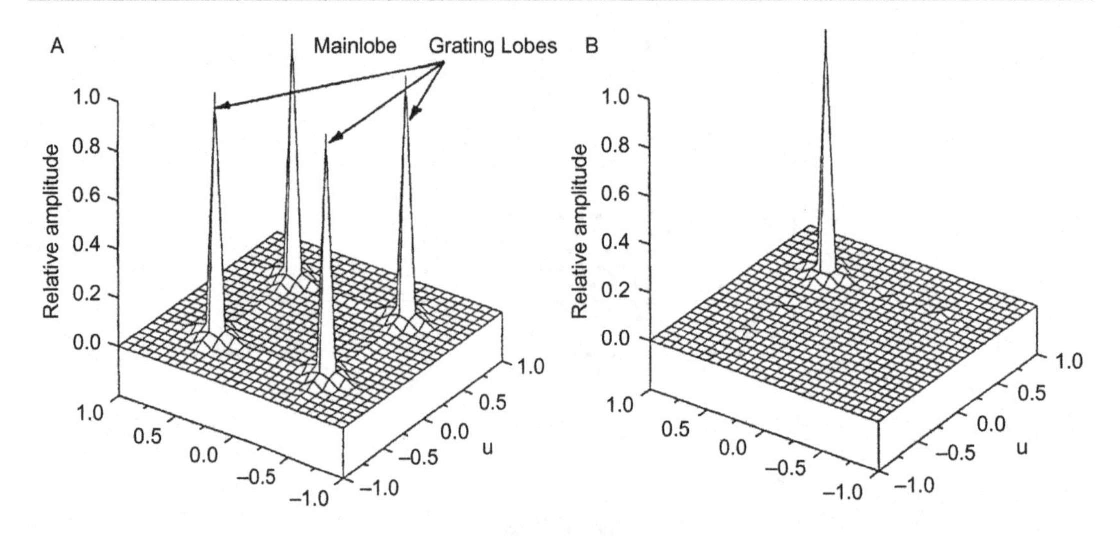

Figure 7.26
Far-field Continuous Wave Pressure Fields for an Array of 101-by-101 Point Sources with Array Steered to $\theta = \phi = 45°$. (A) $p = \lambda$; (B) $p = \lambda/2$. *From Turnbull (1991).*

approach, time domain models are more appropriate for broadband arrays. The method presented here is for 2D and 1D arrays; however, it can be extended to other cases in Section 7.9.3.

A general geometry is given by Figure 7.27, where small square elements with sides w and corresponding period p make up the array. Field positions are assumed to be in the far field of any individual element, or $r >> p^2/(\pi\lambda)$. To determine field pressure, the effects of element directivity can be added to Eqn 7.33 through element factors:

$$P(r, \theta, \phi, f) = E_T(f)H_s(r, \theta, \phi, \lambda)w^2 \operatorname{sinc}\left(\frac{wu}{\lambda}\right)\operatorname{sinc}\left(\frac{wu}{\lambda}\right) \times \text{Obliquity factor.} \quad (7.35)$$

For pulses, Eqn 7.35 must be repeated for many frequencies (a computationally intensive process). An alternative is to develop a spatial impulse response for the array. From the far-field spatial impulse response of a rectangular element in Eqn 7.14, the overall time response of a rectangular element will be the convolution of two rect functions in time, or in general, the trapezoidal time function given by Figure 7.6. Therefore, the spatial impulse response of the central element at the origin to field point position (x, y, z) can be determined by the time delays to the corners given by Figure 7.27. Details can be found in Lockwood and Willette (1973) or Jensen and Svendsen (1992). Focusing and steering for

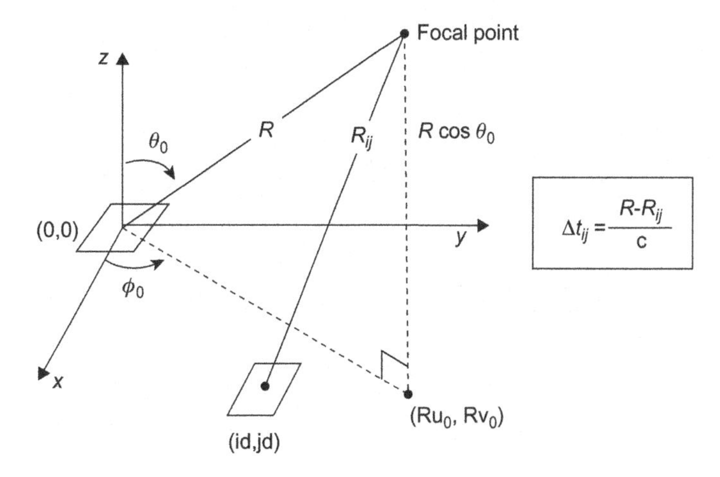

Figure 7.27
Time delay between central element at origin and element *mn* of a 2D array. Indices *i*, *j* correspond to indices *m*, *n* in the text, and $d = p$. *Adapted from Turnbull and Foster (1991), IEEE.*

the beams can be added by introducing the relative delays in Figure 7.27 to the spatial impulse response functions for each element:

$$t_{mn} = (r - r_{mn})/c = \frac{r}{c}\left\{1 - \sqrt{\left[(u_0 - mp_x/r)^2 + (v_0 - np_y/r)^2 + \cos^2\theta_0\right]} - t_0, \right. \quad (7.36\text{A})$$

in which the focal point is defined by r and the direction cosines (u_0 and v_0). The one-way spatial impulse response is therefore:

$$h_a(\bar{r}, t) = \sum_{n=-N}^{N} \sum_{m=-M}^{M} h_{m,n}(\bar{r}, t - t_{mn}). \quad (7.36\text{B})$$

For $\phi = 0$ and $n = 0$, this equation reduces to the 1D array result of Eqn 7.30. For r coincident with the focal point, Eqn 7.36B becomes $h(t - 2r/c + t_0)$. The pulse–echo overall response can be constructed from the transmit and receive array responses, as in Eqn 7.32B,

$$v_0(\bar{r}, t) = h_a^T(\bar{r}, t) *_t h_a^R(\bar{r}, t) *_t e_{RT}(t), \quad (7.37)$$

where superscripts T and R indicate transmit and receive, respectively.

7.7 Baffled

Recall that the element factor has a wide directivity and is an important effect for steered beams; consequently, this topic has received much attention beyond the studies previously mentioned. The directivity of an element is strongly influenced by its surroundings.

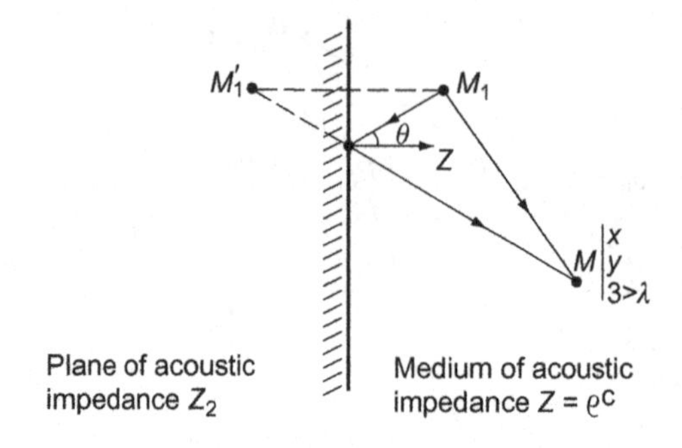

Figure 7.28
Geometry for an aperture, embedded in a medium of impedance Z_2, that radiates in medium Z. A point M_1 and its image M_1 are shown. *From Pesque' and Fink (1984), IEEE.*

In Figure 7.28 is an illustration of a radiating element or active aperture embedded in a material called a baffle that has an impedance Z_2. This baffle determines the boundary conditions for aperture radiation into a medium with a wave number k and an impedance Z and modifies the directivity of the aperture by an obliquity factor that we shall now determine. Even more important is to find out what kind of baffle is most appropriate for medical ultrasound.

The radiation problem has the solution in the form of the Helmholtz–Kirchoff diffraction integral:

$$\psi(r,k) = \int_S \left[G\frac{\partial\psi(r_0)}{\partial n} - \psi\frac{\partial G(r_0)}{\partial n} \right] dS_0, \tag{7.38}$$

in which the Green's function consists of two parts associated with the field point r and its mirror image r':

$$G(k,r,r',k) = \frac{exp(-ik|r-r_0|)}{4\pi|r-r_0|} + R\frac{exp(-ik|r'-r_0|)}{4\pi|r'-r_0|}, \tag{7.39}$$

where R is to be determined and the derivatives above are taken to be normal to the aperture.

Three commonly accepted cases have been studied and experimentally verified by measuring the directivity of a single slotted array element in the appropriate surrounding baffle (Delannoy, Lasota, Bruneel, Torguet, & Bridoux, 1979). All of these can be reduced to the form:

$$\Psi(r,k) = \int_S \frac{X(z,r,r_0)V_n(r_0,k)exp(-i2\pi k(r-r_0))}{2\pi(r-r_0)} dS_0, \tag{7.40}$$

like Eqn 7.1A, where X is an obliquity factor. It is useful to define a direction cosine as:

$$\cos\theta = \frac{z}{|r - r_0|} = \frac{z}{\sqrt{(x-x_0)^2 + (y-y_0)^2 + z^2}}. \tag{7.41A}$$

The first case is when the element is dangling in free space and might be appropriate for an element completely surrounded by water. Here, $Z_2 = Z$, so in Eqn 7.40:

$$X = (1 + \cos\theta)/2. \tag{7.41B}$$

The second case is the hard baffle, for which $Z_2 >> Z$ and:

$$X = 1, \tag{7.41C}$$

so Eqn 7.40 becomes the Rayleigh integral (Strutt, 1945), which we have been using so far in this chapter and is the most common diffraction integral. The third case is the soft baffle, for which $Z_2 << Z$ and:

$$X = \cos\theta, \tag{7.41D}$$

and Eqn 7.40 becomes the Sommerfeld integral used in optics. Delannoy et al. (1979) obtained good experimental agreement with each of these cases and argued that the soft baffle situation might be the most appropriate of the three to represent a transducer held in air against a tissue boundary.

Each of these cases, however, are extreme ones. In general, we would expect the impedances Z_2 and Z to be different and to be within a reasonable range of known materials. Pesque', Coursant, and Me'quio (1983) found a solution for this practical intermediate case. They let the factor in Eqn 7.39 be the reflection factor:

$$R = RF = \frac{Z_2 \cos\theta - Z}{Z_2 \cos\theta + Z}. \tag{7.42}$$

Their approach leads to the following obliquity factor:

$$X = \frac{Z_2 \cos\theta}{Z_2 \cos\theta + Z}. \tag{7.43}$$

Pesque' et al. (1983) and Pesque' and Fink (1984) show that their more general result reduces to the preceding soft and hard baffle cases. Their calculations for the directivity of an element in an array are compared to data in Figure 7.29. Note that this figure demonstrates that it is the impedance in contact with water (tissue) that determines what value of Z_2 to apply. They found that by accounting for the actual impedance at the

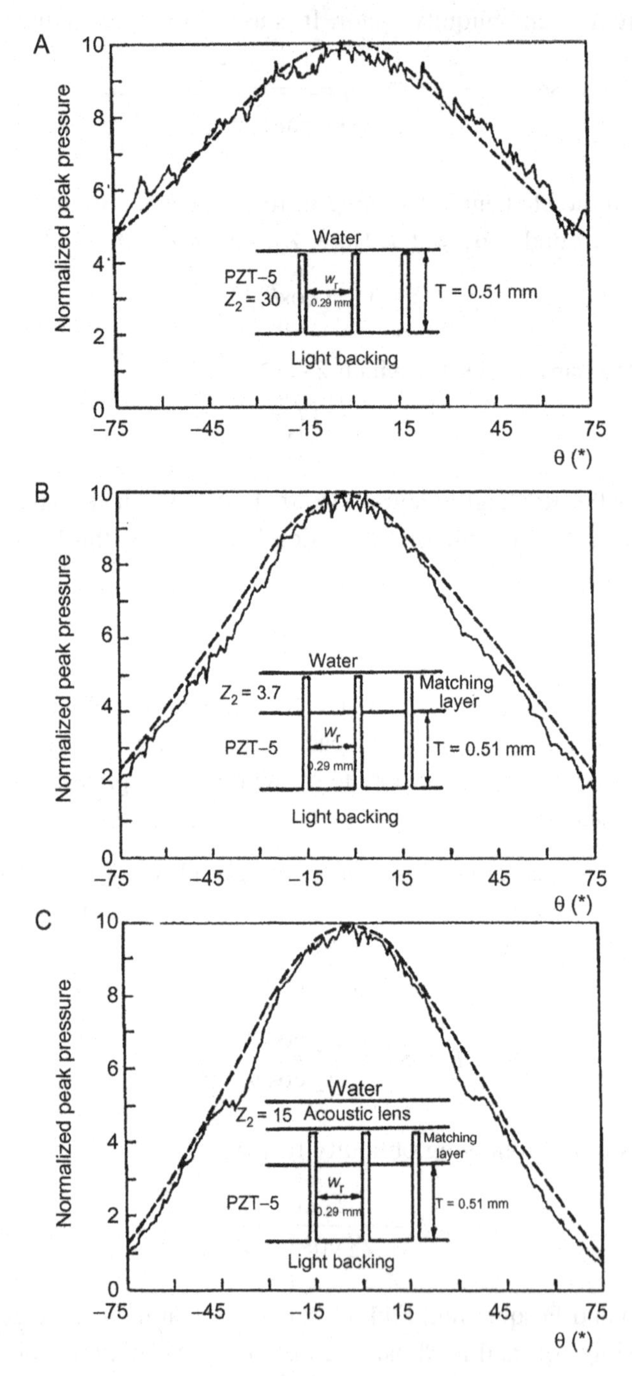

Figure 7.29

Comparison of calculations of element directivity to data. Simulations (solid lines) of the angular radiation pattern of a phased array element $w = 0.29$ mm excited at 3 MHz and radiating into water as a function for three different baffle impedances (Z_2) compared with data (dashed lines). *From Pesque' and Fink (1984), IEEE.*

interface with the body, which normally is a soft mechanical lens, good agreement could be obtained with data. The counterpart of this general result in the time domain is:

$$\psi(r,t) = \int_s \frac{v_n[r_0, t-(r-r_0)/c]Z_2 \cos\theta}{2\pi(r-r_0)(Z_2\cos\theta + Z)} dS_0. \tag{7.44}$$

As explained in Section 5.4, an array element vibrates in a mode dictated by its geometry, so it does not always act like a perfect piston. Smith et al. (1979) realized the nonuniform radiation problem and devised an approximate model. A more exact model was derived by Selfridge, Kino, and Khuri-Yakub (1980), who found that for elements typical in arrays, the element radiated nonuniformly. Delannoy, Bruneel, Haine, and Torguet (1980) examined the problem from the viewpoint of Lamb-like waves generated along elements more than a water wavelength wide. They demonstrated that by subdicing the element, this effect was minimized. Finally, spurious modes and radiation patterns can be created through the architecture of the array, which provides possibilities for different waves to be generated simultaneously with the intended ones. In these cases, finite element modeling (FEM) (Lerch and Friedrich, 1986) is useful.

7.8 Computational Diffraction Methods

Because only a few geometries have been solved for time domain diffraction calculations, more general numerical approaches have been devised. These methods apply to solid transducers of arbitrary shape and apodization, as well as arrays with larger elements. Solutions for solid apertures are discussed first and then applied to arrays.

To some extent, solutions for solid apertures were discussed in Chapter 6 for circular, spherical, and rectangular apertures. Some of these solutions were then applied to the description of element apertures of arrays; other methods include those of: Harris (1981a 1981b); Verhoef, Clostermans, and Thijssen (1984); Piwakowski and Delannoy (1989); and Hossack and Hayward (1993). Exact time domain solutions for the rectangular element in both the near and far field include those of Lockwood and Willette (1973) and San Emeterio and Ullate (1992).Cheng, Lu, Lin, and Qin (2011) have devised a fast algorithm for calculating the spatial impulse response of rectangular planar transducers. In addition, there are 2D Fourier transform numerical calculations as well as those discussed in Section 6.7, the angular spectrum of waves for both Cartesian and polar coordinates.

The most widely used numerical approach generally applicable to arrays, originated by Jensen (Jensen & Svendsen, 1992), is the Field 2 simulation program, which breaks each element aperture down into a mosaic of small squares (or triangles) like those used in a 2D array just described (Jensen, 1996a, 1996b). Each square is assigned an amplitude corresponding to an apodization weighting at that spatial location. Assumptions are that the radius of curvature is large compared to a wavelength and that each rectangular tile is small enough so that the field points are in its far field at the highest frequency in the pulse spectrum used. Even though triangles and a second method using bounding lines and curves have been devised for more general aperture shapes and apodizations (Jensen, 1999), squares are faster and have the flexibility to accommodate a wide range of transducer aperture shapes and arrays.

Other methods include a second approach for arrays created by Holm (1995), the diffraction simulation program Ultrasim, which performs a numerical integration of the Rayleigh–Sommerfeld integral, Eqn 7.1B, by breaking the surface velocity in the integrand into a product of spatial and time functions. An interactive version with hardware computational advantages and portable devices has been implemented (Asen & Holm, 2012). Earlier field calculations also employed a numerical evaluation of this integral with the pulse time response included (Weyns, 1980).

Computational advantages can be obtained from alternate analytical forms of the spatial impulse response. The impulse response has singularities near the edges of the aperture that require high sampling rates to capture faithfully. McGough (2004) has introduced an approach, the "fast nearfield method," that removes the singularities through a combination of analytic and numerical methods for fast calculations of time-harmonic and transient near-field pressures generated by circular (McGough, Samulski, & Kelly, 2004), rectangular (McGough, 2004), triangular (Chen, Kelly, & McGough, 2006), and spherically focusing (McGough, 2013) transducers. Corresponding expressions have also been derived for circular (Kelly & McGough, 2007) and rectangular transducers (Chen & McGough, 2006). An analytic simplification and a time–space decomposition method results in a shorter computation time for transient pressure fields (Kelly & McGough, 2006).

Fortunately, three powerful programs for beamforming simulation are available to the general public. Jensen's program, Field II with MATLAB interfaces, is not only for beam calculations but also can simulate an entire ultrasound imaging process (see Chapter 8), including the creation of artificial phantoms to be imaged. Trahey and coworkers at Duke University have created a useful graphical user interface (GUI) for Field 2 on their virtual imaging lab website. Holm and his team at the University of Oslo have created Ultrasim, an interactive beam simulation program that includes 1D, annular, 1.5D, and 2D arrays. McGough and his group have employed their fast computational methods for transient and time-harmonic pressure fields for arrays and various apertures in the form of algorithms gathered under the name of "FOCUS" (Chen et al., 2006; McGough, Samulski, & Kelly, 2004; Zeng & McGough, 2008). These three programs can be found by doing a web search.

7.9 Nonideal Array Performance

7.9.1 Quantization and Defective Elements

Fields of arrays approach the shape of beams obtained by solid apertures that have the same outer dimensions if Nyquist sampling is achieved. For this case, to first order, array performance can be estimated by a solid aperture with appropriate delay and steering applied. A subtle difference between solid apertures and arrays of the same outer dimensions is that the active area of an array is slightly smaller because of the kerf cuts that isolate each element ($N(p-w)$ smaller for a 1D array). Because of the discrete nature of an array, however, performance is also dependent on the quantization of delay and amplitude that is possible in the imaging system (Thomenius, 1996), as well as individual variations in element-to-element performance and cross-coupling effects.

The first concern about quantization is the spacing of the array itself: Does it meet the Nyquist criteria ($\lambda/2$ spacing) at the highest frequency in the pulses used? In a digital system, phase quantization error is set by the sampling frequency of the system. The effects of phase quantization error (Magnin, von Ramm, & Thurstone, 1981) increase as the number of samples per period near the center frequency decrease, and result in a growth in the width and level of sidelobe structure in the beam (Bates, 1979). Amplitude quantization errors also bring similar effects in beam structure, but they are less severe, in general (Bates, 1979). These effects are caused by round-off errors at the highest number of bits available in the analog-to-digital (A/D) and digital-to-analog (D/A) converters in an imaging system.

While these sorts of errors are straightforward to analyze, second-order effects within the array itself are more troublesome. Unlike an ideal piston source that vibrates in a single longitudinal mode, the vibration of an array element consists of a more complicated combination of longitudinal and transverse modes (described in Section 5.4). Because the element may be physically connected to a backing pedestal, matching layers, and protective foils, other interrelated modes, such as Lamb and Rayleigh waves, can be generated (Larson, 1981). This strange dance of elements causes beam narrowing, ring-down, and other artifacts that can affect the image. Cross-coupling can also be caused by electromagnetic coupling from element to element and through the cable connecting elements to the system. Design solutions to these problems often involve experimental detective work and FEM modeling. An overview of the kinds of problems encountered in the design of a practical digital annular array system can be found in Foster et al. (1989a, 1989b).

Elements can also be defective; they can be completely inoperative or partially so. An element is called "dead" either because of depolarization of the crystal or an electrical disconnect (open) or an unexpected connection (short). Partial element functioning can be due to a number of possible flaws in construction, such as a debonding of a layer. The effect of inoperative elements is straightforward to analyze (Bates, 1979). In Eqn 7.18, for example, the amplitude coefficient of a dead or missing element is set to zero; therefore, the beam pattern is no longer a sinc or the intended function, but a variation of it with higher sidelobes.

7.9.2 Sparse and Thinned Arrays

This topic leads us to the subject of deliberately stolen elements. Can the same beam pattern be achieved with fewer elements? Because channels are expensive, the challenge to do more with less is there for extremely low-cost portable systems, as well as for 2D arrays. What are the issues? Methods for linear arrays will be evaluated and then extended to 2D arrays.

Three main methods are used to decrease the number of elements in an array: periodic, deterministic aperiodic, and random (Schwartz and Steinberg, 1998). The simplest method is to make the elements fewer by increasing the period in terms of wavelengths with the consequence of creating grating lobes. There are also ways of "thinning" an array that usually start with a full half-wavelength-spaced array, from which elements are removed by a prescribed method (deterministic aperiodic) (Skolnik, 1969). A fundamental transform law can be applied to the CW Fourier transform relation between the aperture function and its beam pattern in the focal plane or far field: The gain or on-axis value of the beam is equal to the area of the aperture function. As a result of this law, removing elements decreases the gain of the array and the missing energy reappears as higher sidelobes. If the fraction of elements remaining is P in a normally fully populated array with N elements, then the relative one-way reduced gain to an average far-out sidelobe level is $PN/(1 - P)$. For example, if 70% of 64 elements remain, this relative gain drops from an ideal N squared (4096) to 149 or -22 dB.

The behavior of the main beam and nearby sidelobes is governed by the cumulative area of the thinned array (Skolnik, 1969). An algorithm can be developed to selectively remove elements of unity amplitude to simulate a desired apodization function in a least-squares sense. This method has been automated and extended to arbitrary weighted elements (Laker, Cohen, Szabo, & Pustaver, 1977, 1978). The success of this approach improves as N increases, but the sidelobe level grows away from the main beam. This disadvantage can be compensated for by selecting a complementary (receive or transmit) beam with sidelobes that decrease away from the main beam.

Other approaches also have a similar sidelobe problem. A random method in which the periodicity is deliberately broken up to eliminate sidelobes and to simulate an apodization function statistically results in an average sidelobe level inversely proportional to the number of elements used (Skolnik, 1969; Steinberg, 1976).

One perspective is that the shape of the round-trip beam plot is the primary goal. For fully sampled arrays, the product of the transmit and receive CW beam-shapes provides the desired result. Because of the Fourier transform relation between the aperture function and focal plane beam plot, an equivalent alternative is to tailor the aperture functions so that their convolution yields an effective aperture that gives the desired beam-shape. With this approach, apertures with a few elements can simulate the shape of a fully sampled effective aperture with apodization. A minimum number of elements occurs when each array has the square root of the effective aperture of the final populated array to be simulated. Therefore, for a 64-element array, two differently arranged arrays of eight elements could provide the selected beam-shape.

Images generated by this approach were compared to those made by fully sampled arrays (Lockwood, Li, O'Donnell, & Foster, 1996). While the expected resolution was obtained near the focal zone, grating lobes were seen away from this region. Penetration was also less than a normal array, as would be expected based on arguments described earlier for missing elements. In a follow-up work, Lockwood, Talman, and Brunke (1998) estimated the effect of a decreased signal from a 1D sparse array by a signal-to-noise ratio (SNR) equal to $N_t(N_r)^{1/2}$, where transmit gain (N_t) is proportional to the number of elements, and receive gain (N_r) is related to the square root of elements due to receiver noise. This estimate gives a relative decrease of SNR of -54.9 dB for the 128-element full array, compared to the effective aperture method with only 31 total elements and with different halves (16) used on transmit and receive.

The need for decreasing channel count is even more urgent for 2D arrays for real-time 3D imaging (Thomenius, 1996). At Duke University, early 2D array work was done with a Mills cross and parallel processing to achieve high-speed imaging (Davidsen and Smith, 1993). Later work included a random array employing 192 transmit elements and 64 receive elements with an average sidelobe level of $1/\sqrt{N_t N_r}$ of -41 dB. Other work there (Smith, Davidsen, Emery, Goldberg, & Light, 1995) on a 484-element 2D array, and at the University of Toronto (Turnbull and Foster, 1991) showed that the principal difficulties were the requirement for many hundreds of active channels, severe problems in electrical connection, and the extremely low transducer SNR because of small element size. Alternative methods of 2D array construction also look promising. Kojima (1986) described a 2560-element matrix (2D) array. Greenstein, Lum, Yoshida, and Seyed-Bolorforosh (1996) reported the construction of a 2.5-MHz, 2500-element array. Erikson, Hairston, Nicoli, Stockwell, and White (1997) employed standard integrated circuit packaging to simplify the interconnection of a 30,000-element array for a real-time C-scan imaging system. Smith et al. (1995) discussed the challenges of 2D array construction and presented results for random sparse implementation.

In addition to random 2D arrays, other alternatives have been proposed. Lockwood and Foster (1994, 1996) simulated a radially symmetric array with 517 elements (one sixth the number of a fully populated 65-by-65 array) using the effective aperture approach and found it to be better than a random design. While most array designs are based on CW theory, Schwartz and Steinberg (1998) found that by accounting for pulse shape, ultra-sparse, ultra-wideband arrays can be designed with very low sidelobe levels on the order of N^{-2} one way. As pointed out in Section 7.4.3, grating lobes can be reduced by shortening the exciting pulse. In a similar way, very short pulses in this design do not interfere in certain regions, which leads to very low sidelobes. Inevitably, performance of 2D arrays will be compared to that of conventional 1D arrays in terms of SNR. Schwartz and Steinberg (1998) showed that if the acoustic output of a conventional 1D array of area A is limited in terms of acoustic output by federal regulation, then a 2D array with elements of area a would need to have $N = A/a$ elements to achieve the same output and equivalent SNR. This conclusion returns us to the concept of the gain in a beam on-axis as determined by the area of the active aperture, as given by Eqn 6.33D, G_{focal} = Aperture area/λF.

In 2003, Philips Medical Systems introduced a fully populated 2D array with 2900 elements with an active area comparable to conventional arrays. Highly integrated electronics in the transducer handle accomplished micro-beamforming to provide a true interactive real-time 3D imaging capability. An image from this array is shown in Figure 10.29. More on 2D arrays, 3D and 4D imaging can be found in Sections 10.11.6 and 10.12.1.

7.9.3 1.5-dimensional Arrays

Intermediate between 1D and 2D arrays are 1.5-dimensional (1.5D) arrays (Tournois, Calisti, Doisy, Bureau, & Bernard, 1995; Wildes et al., 1997). This "poor man's" 2D array splits the elevation aperture into a number of horizontal strips, as shown in Figure 7.24B. Elements in each strip can now be assigned at different delays for focusing, and each strip can become an element in a coarsely sampled array along the y axis. Because of symmetry (focusing and no steering), the same delays can be applied to similarly symmetrically positioned strips, so they can be joined together, as shown in side view in the figure, in order to reduce connections. Note that the two central strips merge into a wider combined strip. The individually addressable joined groups are referred to as "Y" groups.

To compare the three types of arrays in Figure 7.24, we start with a 1D array of 64 elements as an example. For the 1.5D array in the figure, there are three Y groups, corresponding to six horizontal strips or an overall element count of $6 \times 64 = 384$ effective elements. However, because of their joined grouping, only $3 \times 64 = 192$ connections are required. These numbers contrast the 64 elements and connections for the 1D array example and the 4096 (N^2) elements and connections for the 2D example. Note that a 1.5D array can combine electronic focusing with the focusing of an attached fixed lens to reduce absolute focusing delay requirements.

Other variants that permit primitive steering are possible (Wildes et al., 1997). Despite their coarse delay quantization in elevation, 1.5D arrays bring improved image quality because the elevation focusing can track the azimuth focusing electronically. Also, 1.5D arrays provide a cost-effective improvement over 1D arrays. A variant of the 1.5D array is an expanding aperture array, which can switch in different numbers of Y groups with or without electronic elevation focusing to alter the *F* number (*F#*) in the elevation plane.

7.9.4 Diffraction in Absorbing Media

A major effect on array performance is attenuation (Foster & Hunt, 1979). Conceptually, the inclusion of attenuation seems straightforward: replace the exponential argument in the diffraction integral (Eqn 7.1A), $-i2\pi k(r-r_0)$, with the complex propagation factor $\gamma_T(r-r_0)$ from Eqn 4.7B. While this change can be done numerically (Berkhoff, Thijssen, & Homan, 1996; Goodsitt & Madsen, 1982; Lerch & Friedrich, 1986), many of the computational advantages of the spatial impulse response approach no longer apply.

Fortunately, Nyborg and Steele (1985) found that by multiplying the Rayleigh integral by an external attenuation factor in the frequency domain for a circular transducer, they were able to obtain good correspondence with a straightforward numerical integration of the Rayleigh integral with attenuation included in the integrand. They improved their agreement when they used a mean distance equal to the maximum and minimum distances from points on the aperture to the field point. Jensen, Ghandi, and O'Brien (1993) explored a time domain alternative, in which a factor containing attenuation and dispersion was convolved with the spatial impulse response and compared to a numerically integrated version of Eqn 7.1B that was modified to include losses. Their findings were similar: very good agreement was obtained overall, and even better results were found using a mean distance for field points close to the transducer.

These findings can be re-evaluated from an analytical vantage point by reconsidering the Rayleigh–Sommerfeld integral (Eqn 6.1B) from Section 6.2 with γ_T replacing $-ik_0$ for attenuating media:

$$p_\alpha(r,\omega) = \frac{i\rho_0 ckv_0}{2\pi}\int_s \frac{e^{i\omega t+\gamma_T\cdot(r-r_0)}A(r_0)dS}{|r-r_0|} \tag{7.45}$$

The spatial exponent can be expressed as the paraxial approximation (Eqn 6.2B):

$$\gamma_T|r-r_0| \approx (-ik_0-\alpha-i\beta_E)z\left[1+\frac{1}{2}\left(\frac{x-x_0}{z}\right)^2+\frac{1}{2}\left(\frac{y-y_0}{z}\right)^2\right], \tag{7.46A}$$

which expands to:

$$(-\alpha-i\beta_E)z-ik_0\left[z+\frac{1}{2}\left(\frac{x-x_0}{z}\right)^2+\frac{1}{2}\left(\frac{y-y_0}{z}\right)^2\right]+(-\alpha-i\beta_E)\left[\frac{1}{2}\left(\frac{x-x_0}{z}\right)^2+\frac{1}{2}\left(\frac{y-y_0}{z}\right)^2\right]. \tag{7.46B}$$

The first z term can be taken outside the integral as the attenuation-dispersion factor, and the second term is the conventional paraxial approximation. The last term, a product of two small factors, can be neglected. These considerations end up as the product of an attenuation-dispersion factor and the conventional paraxial approximation $p(r,\omega)$ from Eqn 6.3,

$$p_\alpha(r,\omega)=e^{[(-\alpha-i\beta_E)z]}p(r,\omega). \tag{7.47}$$

This final product shows that under the limits of the paraxial approximation, the attenuation factor can be multiplied by the diffraction term in the frequency domain, which corresponds to a convolution in the time domain. A convenient means of implementing an attenuation-dispersion factor in either domain is through the analytical power law functions derived by Kelly, McGough, and Meerschaert (2008) and described in Section 4.7.4. Alternatively, if the delay term is included (the first three products containing z in Eqn 7.46B are combined as an exponent outside the integral), the attenuation factor is recognized as the *MTF* (the material transfer function from Section 4.2.1) and its time domain version, the *mirf* (the material impulse response function from Section 4.2.2).

Whichever version of the attenuation factor is employed, with or without delay, they produce equivalent results because the delay must be included in an overall scheme. To represent attenuation and diffraction in the central diagram (from Section 2.4) describing the overall imaging process, the *MTF* for the forward path is represented by block A_T, and that for the return path is represented by $A_R = MTF(r_R, f)$. Then, Eqn 7.32A can be extended to include attenuation:

$$V_o(r,f,\theta)=H_T(r_T,f,\theta_T)A_T(r_T,f)H_R(r_R,f,\theta_R)A_R(r_R,f)E_{RT}(f). \tag{7.48A}$$

Similarly, the corresponding time domain operations are $a_T(r_T,\ t) = mirf(r_T,\ t)$ and $a_R(r_R,\ t) = mirf(r_R,\ t)$, so that Eqn 7.32B becomes:

$$v_o(r,\ t)=h_t*_ta_T*_ta_R*_th_r*_te_{RT}. \tag{7.48B}$$

7.9.5 Body Effects

Finally, the biggest detractor from ideal array performance is the body itself. The gain of an array is based on the coherent summation of identical waveforms. The fact that the paths from elements to the focal point can include different combinations of tissues leads to aberration effects that weaken focusing (to be explained in Section 9.7). An example of the shape and delay differences in pulse propagation for two types of tissue paths is given by Figure 7.30 (Kelly et al., 2008). Simulations of acoustic beams propagating in aberrating tissue are shown in Figures 12.20, 12.21, and 12.22 and are discussed in Section 12.5.5.

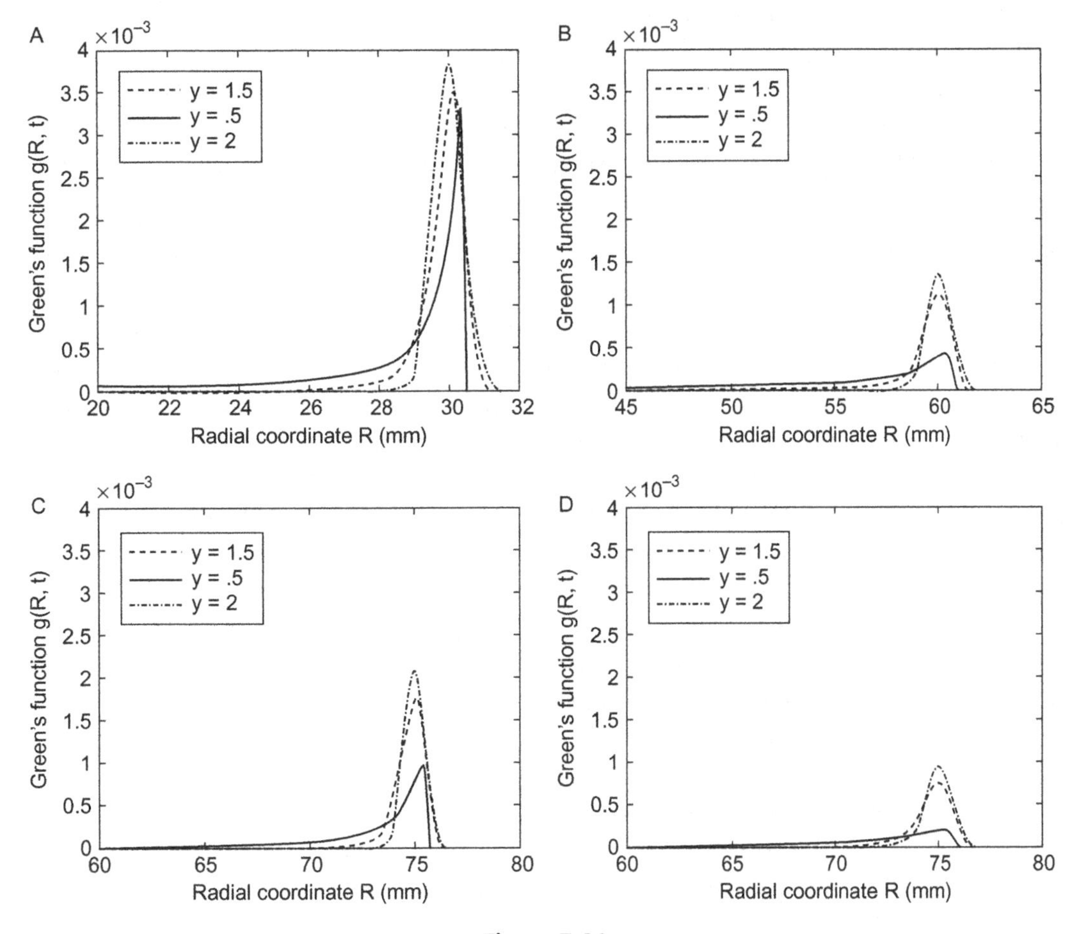

Figure 7.30
Green's function at four depths for water, liver and fat. Comparison of a fat-like medium with $y = 1.5$, $c_0 = 1.432$ mm/μs, and $\alpha_1 = 0.086$ Np/MHz$^{1.5}$/cm labeled as $y = 1.5$ in figures with a liver-like medium with $y = 1.139$, $c_0 = 1.569$ mm/μs, and $\alpha_1 = 0.0459$ Np/MHz$^{1.139}$/cm labeled erroneously as $y = .5$ in figures. The velocity potential is displayed at radial distances (A) 10, (B) 25, (C) 50, and (D) 100 mm. *From Kelly et al. (2008), Acoustical Society of America*

Under real imaging circumstances, unexpected off-axis scatterers do occur. Grating lobes can be sensitive to low-level scatterers (Pesque' & Blanc, 1987). While sparse or thinned arrays appear attractive in simulations or water tank tests, they rely on fewer elements, which means a reduced figure of merit (discussed in Section 7.9.2) and the introduction of grating lobes that can bounce off strong scatterers not included in modeling. Body effects and their influence on imaging will be discussed in detail in Chapter 8, Chapter 9, and Chapter 12.

7.10 Conformable and Deformable Arrays

Besides the capability of flexible transducers to conform to body surfaces, they held the promise of achieving operator-independent operation. Early work with flexible polyvinylidene fluoride(PVDF) transducers showed that they had advantages of conformability and broadband width for high resolution (Howie & Gallantree, 1983); however, their low sensitivity for pulse–echo put them at a disadvantage compared with recent broadband transducer developments.

The attraction of flexible arrays was renewed when built-in intelligence was added. One of the first attempts at operator-independent volumetric imaging involved a 1D mechanically scanned array and a telemedicine system for sending the 3D ultrasound data to a remote location for interpretation (Littlefield, Macedonia and Coleman, 1998). Apparently in this case, the intelligence was also remotely located. Daft (2010) provided an excellent review of progress on this topic. He pointed out the challenges in realizing a practical system which include: elements embedded in a conformable, flexible fabric; low electric power requirement to prevent surface heating; safe limit of acoustic output; a flexible interconnect scheme; a means of determining the spatial (3D) location and attitude of each element; automatic adaptation and compensation for blocked elements; registration and combination of overlapping volumes; the accommodation of extremely high data rates; and, if possible, aberration correction. One of the main issues is that many elements ($m \times n$) are needed for adequate coverage of a large volume. Sekins et al. (2012) reported on conformable arrays in the form of cuffs and patches that automatically adjusted themselves to send high-intensity ultrasound to selected sites to stop bleeding using ultrasound-induced hemostasis. Spatially registered conformable imaging arrays provided 3D compounded images for targeting and treatment monitoring. Electronically steered and focused therapeutic arrays distributed throughout a cuff simultaneously transmitted a therapeutic dose to a target site. Li, Krishnan, and O'Donnell (1994) and Li and O'Donnell (1995) addressed the issues of aberration correction and blocked elements in the design of an adaptive and conformable 2D array. Ries and Smith (1997) designed deformable arrays for aberration correction. In retrospect, an expensive approach that may solve a number of the challenges posed by Daft (2010) would be to employ 2D matrix array technology with microbeamforming; another may be to employ highly integrated capacitive micromachined ultrasonic transducers (CMUT) arrays.

References

Arditi, M., Foster, F. S., & Hunt, J. W. (1981). Transient fields of concave annular arrays. *Ultrasonic Imaging, 3*, 37–61.

Asen, J. P., & Holm, S. (2012). *Huygens on Speed: Interactive Simulation of Ultrasound Pressure Fields* (pp. 1643–1646). *Proceedings of the IEEE Ultrasonics Symposium*. Germany: Dresden.

Bardsley, B. G., & Christensen, D. A. (1981). Beam patterns from pulsed ultrasonic transducers using linear systems theory. *Journal of the Acoustical Society of America, 69*, 25–30.

Bates, K. N. (1979). *Tolerance analysis for phased arrays, Acoustical Imaging* (Vol. 9, pp. 239–262.). New York: Plenum Press.

Berkhoff, A. P., Thijssen, J. M., & Homan, R. J. F. (1996). Simulation of ultrasonic imaging with linear arrays in causal absorptive media. *Ultrasound in Medicine and Biology, 22*, 245–259.

Cao, P. -J., Hu, C. -H., & Shung, K. K. (2003). Development of a real time digital high frequency annular array ultrasound imaging system. *Proceedings of the IEEE Ultrasonics Symposium, 2003*(2), 1867–1870.

Chen, D., Kelly, J. F., & McGough, R. J. (2006). A fast nearfield method for calculations of time-harmonic and transient pressures produced by triangular pistons. *Journal of the Acoustical Society of America, 120*, 2450–2459.

Chen, D., & McGough, R. J. (2006). A 2D fast near-field method for calculating near-field pressures generated by apodized rectangular pistons. *Journal of the Acoustical Society of America, 124*, 1526–1537.

Cheng, J., Lu, J. -Y., Lin, W., & Qin, Y. -X. (2011). A new algorithm for spatial impulse response of rectangular planar transducers. *Ultrasonics, 51*, 229–237.

Daft, C. M. W. (2010) Conformable transducers for large-volume, operator-independent imaging. *Proceedings of the IEEE ultrasonics symposium* (pp. 798–808). San Diego, CA.

Davidsen, R. E. and Smith, S.W. (1993) Sparse geometries for two-dimensional array transducers in volumetric imaging. *Proceedings of the IEEE ultrasonics symposium* (Vol. 2) (pp. 1091–1094). Baltimore, MD.

Delannoy, B., Bruneel, C., Haine, F., & Torguet, R. (1980). Anomalous behavior in the radiation pattern of piezoelectric transducers induced by parasitic Lamb wave generation. *Journal of Applied Physics, 51*, 3942–3948.

Delannoy, B., Lasota, H., Bruneel, C., Torguet, R., & Bridoux, E. (1979). The infinite planar baffles problem in acoustic radiation and its experimental verification. *Journal of Applied Physics, 50*, 5189–5195.

Erikson, K., Hairston, A., Nicoli, A., Stockwell, J., & White, T. A. (1997). 128 × 128 K (16 k) ultrasonic transducer hybrid array. In S. Lees, & L. A. Ferrari (Eds.), *Acoustical imaging* (Vol. 23, pp. 485–494). New York: Plenum Press.

Foster, F. S., & Hunt, J. W. (1979). Transmission of ultrasound beams through human tissue: Focusing and attenuation studies. *Ultrasound in Medicine and Biology, 5*, 257–268.

Foster, F. S., Larson, J. D., Mason, M. K., Shoup, T. S., Nelson, G., & Yoshida, H. (1989a). Development of a 12-element annular array transducer for realtime ultrasound imaging. *Ultrasound in Medicine and Biology, 15*, 649–659.

Foster, F. S., Larson, J. D., Pittaro, R. J., Corl, P. D., Greenstein, A. P., & Lum, P. K. (1989b). A digital annular array prototype scanner for realtime ultrasound imaging. *Ultrasound in Medicine and Biology, 15*, 661–672.

Goodsitt, M. M., & Madsen, E. L. (1982). Field patterns of pulsed, focused, ultrasonic radiators in attenuating and nonattenuating media. *Journal of the Acoustical Society of America, 71*, 318–329.

Gottlieb, E. J., Cannata, J. M., Hu, C. - H., & Shung, K. K. (2006). Development of a high-frequency (> 50 MHz) copolymer annular-array, ultrasound transducer. *IEEE Transactions on Ultrasonics, Ferroelectrics and Frequency Control, 53*, 1037–1045.

Greenstein, M., Lum, P., Yoshida, H., & Seyed-Bolorforosh, M. S. (1996). A 2.5-MHz 2D array with z-axis backing. *Proceedings of the IEEE ultrasonics symposium* (Vol. 2, pp. 1513–1516). San Antonio, TX.

Harris, F. J. (1978). On the use of windows for harmonic analysis with the discrete Fourier transform. *Proceedings of the IEEE, 66*, 51–83.

Harris, G. R. (1981a). Review of transient field theory for a baffled planar piston. *Journal of the Acoustical Society of America, 70*, 10–20.

Harris, G. R. (1981b). Transient field of a baffled planar piston having an arbitrary vibration amplitude distribution. *Journal of the Acoustical Society of America, 70*, 186–204.

Holm, S. (1995). Simulation of acoustic fields from medical ultrasound transducers of arbitrary shape. *Proceedings of the nordic symposium in physical acoustics*, Ustaoset, Norway.

Hossack, J. A., & Hayward, G. (1993). Efficient calculation of the acoustic radiation from transiently excited uniform and apodised rectangular apertures. *Proceedings of the IEEE ultrasonics symposium* (Vol. 2) (pp. 1071–1075), Baltimore, MD.

Howie, P. A., & Gallantree, H. R.(1983) Transducer applications of PVDF. *Proceedings of the IEEE ultrasonics symposium* (pp. 566–569). Atlanta GA.

Jensen, J. A. (1996a). Field: A program for simulating ultrasound systems. *Medical and Biological Engineering and Computing, 34*(Suppl. 1), 351–353.

Jensen, J. A. (1996b). Ultrasound fields from triangular apertures. *Journal of the Acoustical Society of America, 100*, 2049–2056.

Jensen, J. A. (1999). A new calculation procedure for spatial impulse responses in ultrasound. *Journal of the Acoustical Society of America, 105*, 3266–3274.

Jensen, J. A., Ghandi, D., & O'Brien, W. O., Jr. (1993). Ultrasound fields in an attenuating medium. *Proceedings of the IEEE ultrasonics symposium* (Vol. 2) (pp. 943–946). Baltimore, MD.

Jensen, J. A., & Svendsen, N. B. (1992). Calculation of pressure fields from arbitrarily shaped, apodized, and excited ultrasound transducers. *IEEE Transactions on Ultrasonics, Ferroelectrics and Frequency Control, 39*, 262–267.

Kelly, F., & McGough, R. J. (2006). A fast time-domain method for calculating the nearfield pressure generated by a pulsed circular piston. *IEEE Transactions on Ultrasonics, Ferroelectrics and Frequency Control, 53*, 1150–1159.

Kelly, J. F., McGough, R. J., & Meerschaert, M. M. (2008). Analytical time domain Green's functions for power-law media. *Journal of the Acoustical Society of America, 124*, 2861–2872.

Ketterling, J. A., Aristiźabal, O., Turnbull, D. H., & Lizzi, F. L. (2005). Design and fabrication of a 40-MHz annular array transducer, IEEE Transactions on Ultrasonics. *Ferroelectrics and Frequency Control, 52*, 672–681.

Kojima, T. (1986). Matrix array transducer and flexible matrix array transducer. *Proceedings of the IEEE ultrasonics symposium* (pp. 649–654). Williamsburg, VA.

Kramer, S. M., McBride, S. L., Mair, H. D., & Hutchins, D. A. (1988). Characteristics of wide-band planar ultrasonic transducers using plane and edge wave contributions. *IEEE Transactions on Ultrasonics, Ferroelectrics and Frequency Control, 35*, 253–263.

Laker, K. R., Cohen, E., Szabo, T. L., & Pustaver, J. A. (1977). Computer-aided design of withdrawal weighted SAW bandpass transversal filters. *IEEE international symposium on circuits and systems*, Cat. CH1188-2CAS, pp. 126–130.

Laker, K. R., Cohen, E., Szabo, T. L., & Pustaver, J. A. (1978). Computer-aided design of withdrawal weighted SAW bandpass filters. *IEEE Transactions on Circuits and Systems, 25*, 241–251.

Larson, J. D. (1981). Non-ideal radiators in phased array transducers. *Proceedings of the IEEE ultrasonics symposium* (pp. 673–683). Chicago, IL.

Lerch, R., & Friedrich, W (1986). Ultrasound fields in attenuating media. *Journal of the Acoustical Society of America, 80*, 1140–1147.

Li, P.-C., Krishnan, S., & O'Donnell, M.(1994) Adaptive ultrasound imaging systems using large, two-dimensional, conformal arrays. *Proceedings of the IEEE Ultrasonics Symposium* (Vol. 3) (pp. 1625–1628). Cannes, France.

Li, P. -C., & O'Donnell, M. (1995). Phase aberration correction on two-dimensional conformal arrays, IEEE Transactions on Ultrasonics. *Ferroelectrics and Frequency Control, 42*, 73–82.

Littlefield, R. J., Macedonia, C. R., & Coleman, J. D. (1998) MUSTPAC 3-D ultrasound telemedicine/ telepresence system. *Proceedings of the IEEE Ultrasonics Symposium* (Vol. 2) (pp. 1669–1675). Sendai.

Lockwood, G. R., & Foster, F. S. (1994). Optimizing sparse two-dimensional transducer arrays using an effective aperture approach. *Proceedings of the IEEE ultrasonics symposium* (Vol. 3) (pp. 1497–1501). Cannes, France.

Lockwood, G. R., & Foster, F. S. (1996). Optimizing the radiation pattern of sparse periodic two-dimensional arrays. *IEEE Transactions on Ultrasonics, Ferroelectrics and Frequency Control, 43*, 15–19.

Lockwood, G. R., Li, P-C., O'Donnell, M., & Foster, F. S. (1996). Optimizing the radiation pattern of sparse periodic linear arrays. *IEEE Transactions on Ultrasonics, Ferroelectrics and Frequency Control, 43*, 7–13.

Lockwood, G. R., Talman, J. R., & Brunke, S. S. (1998). Real-time 3-D ultrasound imaging using sparse synthetic aperture beamforming. *IEEE Transactions on Ultrasonics, Ferroelectrics and Frequency Control, 45*, 980–988.

Lockwood, J. C., & Willette, J. G. (1973). High-speed method for computing the exact solution for the pressure variations in the nearfield of a baffled piston. *Journal of the Acoustical Society of America, 53*, 735–741.

Magnin, P., von Ramm, O. T., & Thurstone, F. (1981). Delay quantization error in phased array images. *IEEE Transactions on Sonics and Ultrasonics, 28*, 305–310.

Maslak, S. M. (1985). Computed sonography. In R. C. Sanders, & M. C. Hill (Eds.), *Ultrasound annual 1985*. New York: Raven Press.

McGough, R. J. (2004). Rapid calculations of time-harmonic nearfield pressures produced by rectangular pistons. *Journal of the Acoustical Society of America, 115*, 1934–1941.

McGough, R. J. (2013). Numerical modeling and simulation and treatment planning thermal therapy. In Eduardo Moros (Ed.), *Ultrasound in physics of thermal therapy: fundamentals and clinical applications*. Boca Rotan, FL: CRC Press.

McGough, R. J., Samulski, T. V., & Kelly, J. F. (2004). An efficient grid sectoring method for calculations of the nearfield pressure generated by a circular piston. *Journal of the Acoustical Society of America, 115*, 1942–1954.

Nyborg, W. L., & Steele, R. B. (1985). Nearfield of a piston source of ultrasound in an absorbing medium. *Journal of the Acoustical Society of America, 78*, 1882–1891.

Oberhettinger, F. (1961). On transient solutions of the baffled piston problem. *Journal of Research of the National Bureau of Standards, 65B*, 1–6.

Panda, R. K. (1998). *Development of novel piezoelectric composites by solid freeform fabrication techniques* (Dissertation). New Brunswick, NJ: Rutgers University.

Penttinen, A., & Luukkala, M. (1976). The impulse response and pressure nearfield of a curved ultrasound radiator. *Journal of Physics D, 9*, 1547–1557.

Pesque', P., & Blanc, C.. (1987). Increasing of the grating lobe effect in multiscatterers medium. *Proceedings of the IEEE ultrasonics symposium* (pp. 849–852). Denver, CO.

Pesque', P., Coursant, R. H., & Me'quio, C. (1983). Methodology for the characterization and design of linear arrays of ultrasonic transducers. *Acta Electronica, 25*, 325–340.

Pesque', P., & Fink, M.. (1984). Effect of the planar baffle impedance in acoustic radiation of a phased array element theory and experimentation. *Proceedings of the IEEE ultrasonics symposium* (pp. 1034–1038). Dallas, TX.

Piwakowski, B., & Delannoy, B. (1989). Method for computing spatial pulse response: Time domain approach. *Journal of the Acoustical Society of America, 86*, 2422–2432.

San Emeterio, J. L., & Ullate, L. G. (1992). Diffraction impulse response of rectangular transducers. *Journal of the Acoustical Society of America, 92*, 651–662.

Sato, J., Fukukita, H., Kawabuchi, M., & Fukumoto, A. (1980). Farfield angular radiation pattern generated from arrayed piezoelectric transducers. *Journal of the Acoustical Society of America, 67*, 333–335.

Schwartz, J. L., & Steinberg, B. D. (1998). Ultrasparse, ultrawideband arrays. *IEEE Transactions on Ultrasonics, Ferroelectrics and Frequency Control, 45*, 376–393.

Sekins, K. M., Zeng, X., & Barnes, S., et al. (2012) Proceedings of the 11th International Symposium on Therapeutic Ultrasound (ISTU), 2011, AIP Conference Proceedings, 1481, 311–316.

Selfridge, A. R., Kino, G. S., & Khuri-Yakub, B. T. (1980). A theory for the radiation pattern of a narrow-strip transducer. *Applied Physics Letters, 37*, 35–36.

Shattuck, D. P., Weinshenker, M. D., Smith, S. W, & von Ramm, O. T. (1984). Explososcan: A parallel processing technique for high speed ultrasound imaging with linear phased arrays. *Journal of the Acoustical Society of America, 75*, 1273–1282.

Skolnik, M. I. (1969). Nonuniform arrays. In R. E. Collin, & F. J. Zucker (Eds.), *Antenna Theory, Part 1* (pp. 207–234). New York: McGraw-Hill.

Smith, S. W, Davidsen, R. E., Emery, C. D., Goldberg, R. L., & Light, E. D. (1995). Update on 2-D array transducers for medical ultrasound. *Proceedings of the IEEE ultrasonics symposium* (Vol. 2) (pp. 1273–1278). Seattle, WA.

Smith, S. W, von Ramm, O. T., Haran, M. E., & Thurstone, F. I. (1979). Angular response of piezoelectric elements in phased array ultrasound scanners. *IEEE Transactions on Sonics and Ultrasonics, 26*, 186–191.

Somer, J. C. (1968). Electronic sector scanning for ultrasonic diagnosis. *Ultrasonics, 6*, 153–159.

Steinberg, B. D. (1976). *Principles of aperture and array design*. New York: John Wiley & Sons.

Stephanishen, P. R. (1971). Transient radiation from pistons in an infinite planar baffle. *Journal of the Acoustical Society of America, 49*, 1629–1638.

Strutt, J. W. (Lord Rayleigh) (1945) *Theory of Sound*, Vol. 2, Chap. 14. Dover, New York. [Original work published 1896.]

't Hoen, P. J. (1982). Aperture apodization to reduce the off-axis intensity of the pulsed-mode directivity function of linear arrays. *Ultrasonics, 20*, 231–236.

Thomenius, K. E. (1996). Evolution of ultrasound beamformers. *Proceedings of the IEEE Ultrasonics Symposium* (Vol. 2) (pp. 1615–1622). San Antonio, TX.

Thurstone, F. L., & von Ramm, O. T. (1975). A new ultrasound imaging technique employing two-dimensional electronic beam steering. In P. S. Green (Ed.), *Acoustical Holography and Imaging* (Vol. 5, pp. 249–259). New York: Plenum Press.

Tournois, P., Calisti, S., Doisy, Y., Bureau, J. M., and Bernard, F. (1995). A 128*4 channels 1.5D curved linear array for medical imaging. *Proceedings of the IEEE Ultrasonics Symposium* (Vol. 2, pp. 1331–1335). Seattle, WA.

Tupholme, G. E. (1969). Generation of acoustic pulses by baffled plane pistons. *Mathematika, 16*, 209–224.

Turnbull, D. H. (1991). *Two-dimensional transducer arrays for medical ultrasound imaging* (Ph.D. thesis), Department of Medical Biophysics, University of Toronto, Canada.

Turnbull, D. H., & Foster, S. F. (1991). Beam steering with pulsed two-dimensional transducer arrays. *IEEE Transactions on Ultrasonics, Ferroelectrics and Frequency Control, 38*, 320–333.

Verhoef, W. A., Clostermans, M. J. T. M., & Thijssen, J. M. (1984). The impulse response of a focused source with an arbitrary axisymmetric surface velocity distribution. *Journal of the Acoustical Society of America, 75*, 1716–1721.

Vogel, J., Bom, N., Ridder, J., & Lance, C. (1979). Transducer design considerations in dynamic focusing. *Ultrasound in Medicine and Biology, 5*, 187–193.

von Ramm, O. T., Smith, S. W, & Pavey, H. G., Jr. (1991). High speed ultrasound volumetric imaging system II: Parallel processing and image display. *IEEE Transactions on Ultrasonics, Ferroelectrics and Frequency Control, 38*, 109–115.

von Ramm, O. T., & Thurstone, F. L. (1975). Thaumascan: Design considerations and performance characteristics. *Ultrasound in Medicine and Biology, 1*, 373–378.

Weyns, A. (1980). Radiation field calculations of pulsed ultrasonic transducers. *Ultrasonics, 18*, 183–188.

Wildes, D. G., Chiao, R. Y., Daft, C. M. W., Rigby, K. W., Smith, L. S., & Thomenius, K. E. (1997). Elevation performance of 1.25 D and 1.5 D transducer arrays. *IEEE Transactions on Ultrasonics, Ferroelectrics and Frequency Control, 44*, 1027–1036.

Wright, J. N. (1985). Resolution issues in medical ultrasound. *Proceedings of the IEEE ultrasonics symposium* (pp. 793–799). San Francisco, CA.
Zeng, X., & McGough, R. J. (2008). Evaluation of the angular spectrum approach for simulations of nearfield pressures. *Journal of the Acoustical Society of America, 123*, 68–76.

Bibliography

Cobbold, R. S. C. (2007). *Foundations of biomedical ultrasound.* Oxford, UK: Oxford University Press.
Chapter 3, Field Profile Analysis,is a comprehensive self-contained reference on ultrasound fields
Collin, R. E., & Zucker, F. J. (Eds.), (1969). *Antenna theory, part 1* New York: McGraw-Hill.
A general reference for more information on arrays.
Fink, M. A., & Cardoso, J. F. (1984). Diffraction effects in pulse–echo measurement. *IEEE Transactions on Sonics and Ultrasonics, 31*, 313–329.
A helpful review article on arrays and time domain diffraction.
Jensen, J. A. (1996). *Estimation of Blood Velocities Using Ultrasound.* Cambridge, UK: Cambridge University Press.
A book providing a brief introduction to ultrasound imaging, diffraction, and scattering.
Macovski, A. (1983). *Medical imaging systems.* Englewood Cliffs, NJ: Prentice-Hall.
A general reference on arrays and medical imaging.
Sternberg, B. D. (1976). *Principles of aperture and array design.* New York: John Wiley & Sons.
A general engineering reference on arrays.
t' Hoen, P. J. (1983). Design of ultrasonographic linear arrays. *Acta Electronica, 25*, 301–310.
A helpful review article on array design, construction, diffraction, and simulation.
Thomenius, K. E. (1996). Evolution of ultrasound beamformers. *Proceedings of the IEEE ultrasonics symposium* (Vol. 2, pp. 1615–1622). San Antonio, TX.
A review article on beamforming methods.
von Ramm, O., & Smith, S. W. (1983). Beam steering with linear arrays. *IEEE Transactions on Biomedical Engineering, 30*, 438–452.
An informative article on phased arrays.
Wright, J. N. (1985). Resolution issues in medical ultrasound. *Proceedings of the IEEE ultrasonics symposium* (pp. 793–799). San Francisco, CA.
Article on design trade-offs for beamforming.

CHAPTER 8

Wave Scattering and Imaging

Chapter Outline

8.1 Introduction

What is it we see in an ultrasound image? To answer this question, several aspects of the overall imaging process must be understood in a comprehensive way. First, how does sound scatter from an object at typical ultrasound frequencies (Section 8.2)? Second, what is the role of the spatial impulse response of the transducer (Section 8.3)? Third, how does the way the image is organized into multiple acoustic beams affect what is seen (Section 8.4)? The answers to these questions are about how an ultrasound imaging system senses and portrays tissue objects. The actual nature, structure, and acoustic characteristics of tissue are discussed in Chapter 9.

The array acts as an intermediary between the actual tissue and the created image. With ultrasound, the field is spatially variant, so the appearance of the same object depends on its location in the sound beam. In addition, the physical organization of tissue presents

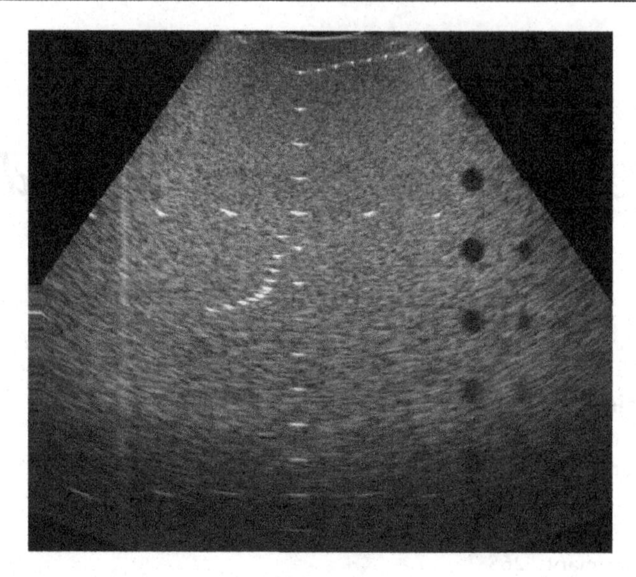

Figure 8.1
Two-dimensional Ultrasound Image of a Tissue-mimicking Phantom with Wire (Point) Targets and Cyst Targets. For this image, the transmit focal length of the 5-MHz convex array is positioned at 6 cm, which is the level of the horizontal wire target group. *Image made with Analogic AN2800 imaging system.*

scatterers on several length scales so that their backscatter changes according to their shape and size relative to the insonifying wavelength.

These effects are apparent in an image of a tissue-mimicking phantom (Figure 8.1), in which three types of scattering objects are seen. Figure 8.2 illustrates the arrangement of scatterers in the phantom. Note the vertical column of nylon filament point targets that appear as dots in the cross-section. To their right are columns of anechoic cylinders of varying diameters that appear as circles in the cross-section. In Figure 8.1, on the left, images of nylon filament targets with a diameter much smaller than the wavelength at a frequency of 5 MHz. This is because the transducer point spread function (see Section 7.4.1), appear larger than their physical size and vary in appearance away from the focal point. On the right are images of columns of cysts (seen as cross-sections of cylinders) of varying diameters on the order of several wavelengths. These cysts have approximately the same impedance as the host matrix material surrounding them, but they have fewer subwavelength scatterers within them and appear black. Note that in the image, the smaller-diameter cysts are more difficult to recognize and resolve. This problem is due in part to the resolving power of the transducer array used, as well as to the interfering effect of the background material, which has its own texture. The targets are suspended in a tissue-mimicking material composed of many subwavelength scatterers per unit volume. The imaging of this matrix material appears as speckle, a grainy texture. Speckle, described

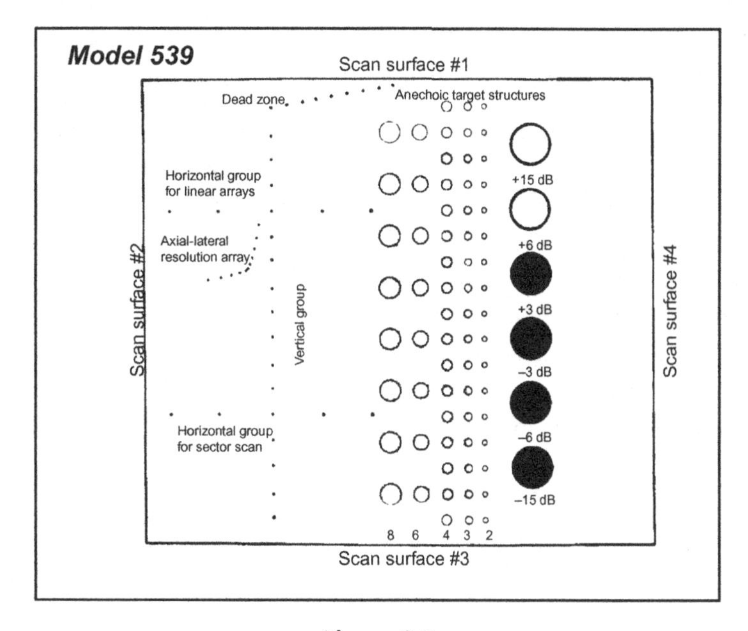

Figure 8.2
Illustration of Arrangement of Scattering Objects in the same Tissue-mimicking Phantom as Shown in Figure 8.1. *Courtesy of ATS Laboratories.*

in more detail later, arises from the constructive and destructive interference of these tiny scatterers, and it appears as a light and dark mottled grainy pattern. This varying background interferes with the delineation of the shapes of the smaller cysts.

In general, there are three categories of scatterers based on length scales: *specular* for reflections from objects whose shapes are much bigger than a wavelength (large-diameter cysts in Figure 8.1), *diffractive* for objects slightly less than a wavelength to hundreds of wavelengths (smaller-diameter cysts), and *diffusive* for scatterers much smaller than a wavelength (background matrix material).

8.2 Scattering of Objects

8.2.1 Specular Scattering

Before examining the complexity of tissue structure, we shall find it easier to deal with the scattering process itself. The type of ultrasound scattering that occurs depends on the relation of the shape or roughness of the object to the insonifying sound wavelength. Objects fall roughly into three groups: those with dimensions either much larger or much smaller than a wavelength, and the rest that fall in between these extremes. Our discussion of backscattering will show how the scattering from a sphere will change its appearance depending on its size relative to the wavelength of the incident wave.

These categories are related to the smoothness of the object relative to a wavelength. If a wavelength is much smaller than any of the object's dimensions, the reflection process can be approximated by rays incident on the object so that the scattered wavefront is approximately a replica of the shape of the object. In the case of a plane wave of radius b illuminating a sphere of radius a much greater than a wavelength, as illustrated in Figure 8.3A, the intercepted sound sees a cross-sectional area of πb^2. It is reflected by a reflection factor (RF) due to the impedance mismatch between the propagating medium and sphere. As the reflected wavefront is backscattered, it grows spherically so that the ratio of overall backscattered intensity (I_r) to the incident intensity (I_i); can be described by (Kino, 1987):

$$\frac{I_r}{I_i} = \frac{\pi b^2}{4\pi r^2}|RF|^2 = |RF|^2\frac{b^2}{4r^2}, \tag{8.1A}$$

in which RF is from Eqn 3.22A (Z_2 is the impedance of the sphere, and Z_1 is the impedance of the surrounding fluid). Note that this result does not depend on the wavelength. In this regime, ray theory holds. The importance of the angle of incidence was apparent for plane waves reflected from and mode converting into a smooth flat boundary in Chapter 3. The consequences of a nearly oblique plane wave striking a boundary are that the returning wave may be reflected away from the source and that the nearly normal components of the wave front are reflected more strongly, according to the impedance cosine variation described in Chapter 3. In the simple case presented here, the sphere is assumed to be rigid so that mode conversion is neglected.

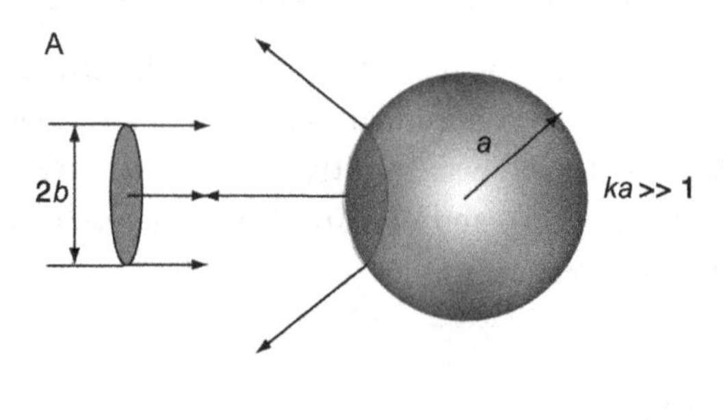

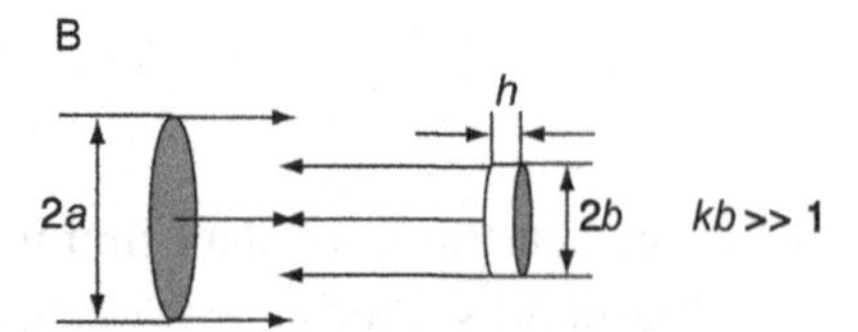

Figure 8.3
(A) Reflections from a Rigid Sphere of Radius a in the $ka >> 1$ regime; (B) Scattering from a Rigid Disk of radius b for $kb >> 1$.

Now consider a disk-shaped object of radius b illuminated by a cylindrical beam of radius a (shown by Figure 8.3B). In this case, the ratio of backscattered intensities is:

$$\frac{I_r}{I_i} = \frac{\pi b^2}{\pi a^2}|RF|^2 = |RF|^2\frac{b^2}{a^2}. \tag{8.1B}$$

Note that for a transducer positioned at one distance from a target, it would be difficult to tell these objects apart or determine their size only from their backscattered reflections.

8.2.2 Diffusive Scattering

At the other extreme, when the wavelength is large compared to a scattering object, individual reflections from roughness features on the surface of the object fail to cause any noticeable interference effects. In other words, the phase differences between reflections from high and low points on the surface are insignificant.

Lord Rayleigh discovered that for this type of scattering, intensity varies as the fourth power of frequency. Amazingly enough, for all the millions of humans who looked up at the sky, he was the first person determined enough to find out why it was blue. In his landmark paper, *On the Light from the Sky, Its Polarization and Colour* (1871), and in a later paper, he showed that the blueness of the sky was due to the predominant scattering of higher-frequency (blue) light by particles much smaller than a wavelength (Strutt, 1871).

Scattering in this regime has important implications in medical imaging. Tissue is often modeled as an aggregate of small subwavelength point scatterers like the one depicted in Figure 8.4. Blood flow, as measured by Doppler methods, is dependent on scattering by many small, spatially unresolved blood cells. Also, most ultrasound contrast agents are tiny gas-filled resonant spheres used as tracers to enhance the scattering of ultrasound from blood pools and vessels. These topics will be covered in more detail in Chapter 11 and Chapter 14.

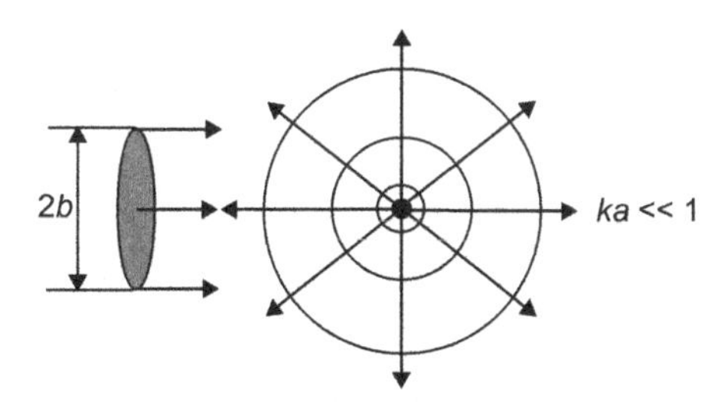

Figure 8.4
Reflections from a Rigid Sphere of Radius a in the $ka << 1$ Regime.

Lord Rayleigh (Strutt, 1871) and Morse and Ingard (1968) derived an expression for the scattering of pressure from a sphere much smaller than a wavelength with different elastic properties in density and compressibility from an exact solution for $ka << 1$:

$$\frac{p_s}{p_i} = \frac{-k^2a^3}{3r}\left[\frac{3(1-\rho_2/\rho_1)\cos\theta}{1+2\rho_2/\rho_1} + \left(1-\frac{\kappa_1}{\kappa_2}\right)\right], \quad \text{and} \tag{8.2A}$$

$$\frac{I_s}{I_i} = \frac{k^4a^6}{9r^2}\left[\frac{3(1-\rho_2/\rho_1)\cos\theta}{1+2\rho_2/\rho_1} + \left(1-\frac{\kappa_1}{\kappa_2}\right)\right]^2, \tag{8.2B}$$

in which subscript 2 indicates the object density ρ and object compressibility κ, and $\theta = 0$ is along the axis of forward propagation (Figure 8.5). For a rigid sphere, $\rho_2/\rho_1 \rightarrow \infty$ and $\kappa_2/\kappa_1 \rightarrow \infty$, Eqn 8.2B becomes:

$$\frac{I_s}{I_i} = \frac{k^4a^6}{9r^2}\left[1-\frac{3\cos\theta}{2}\right]^2. \tag{8.2C}$$

Lord Rayleigh (Strutt, 1871) showed that a small rigid sphere acts like a dipole (two main lobes at 0 and π) in its directivity. Therefore, the intensity backscattered along $\theta = \pi$ is:

$$\frac{I_s}{I_i} = \frac{25k^4a^6}{36r^2}, \tag{8.2D}$$

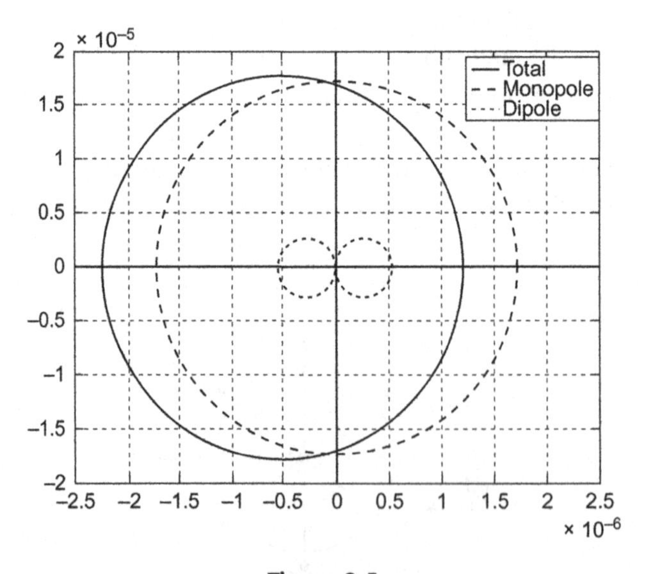

Figure 8.5
Simulated scattering from a red blood cell. Scattered pressure normalized to incident pressure.
From Coussios (2002), Acoustical Society of America.

a result that is frequency dependent (through k), unlike Eqn 8.1. Kino (1987) has pointed out that in this case, the total scattering cross-section is the ratio of the total power scattered divided by the incident intensity:

$$\sigma(total) = \frac{7\pi k^4 a^6}{9}, \tag{8.3}$$

which compares with the specular reflector cross-section of $2\pi a^2$, since the sphere radiates equally in the forward and backward directions. In the specular case, the forward-scattered part of the wave cancels out the incident wave behind the sphere to create a geometric shadow.

Of interest is how the scattered intensity (related to pressure squared) is proportional to the frequency to the fourth power and to the sixth power of the radius in Eqns 8.2B–8.2D. Note that just by changing frequency, the same scatterer will appear to have stronger or weaker reflections. As a result, the intensity reflected from this small sphere radiates outward as an expanding sphere whose intensity is proportional to the difference in compressibility and/or density and the fourth power of frequency. Later, we shall return to the subject of how groups of these small scatterers can have a cumulative effect on imaging.

8.2.3 Diffractive Scattering

The last and largest category of scattering objects are those in between the extremes described earlier. Scattering for these objects is governed by the same Helmholtz–Kirchoff integral, Eqn 7.38, applied to the problem of diffraction of waves from transducer apertures. To first order, the scattered waves can be considered to be originating from the surfaces of the illuminated objects, which act as secondary sources. Actually, the scattering from elastic objects is far more involved, and only solutions for simply shaped objects have been solved analytically. Note that in this regime, scattered waves can be different in shape from the object and they can have maxima and minima that vary with angle and ka number. This conclusion is evident in Figure 8.6, in which the scattering of a rigid sphere for different values of ka and different directions show vastly different results for the same physical object (Jafari, Madsen, Zagzebski, & Goodsitt, 1981). In these polar diagrams, 0° is the incident direction and 180° is the backscattered direction back toward the source. Exact solutions for elastic scattering from solid spheres and cylinders can be found in Faran (1951) and Hickling (1962).

Frequency Domain Born Approximation

An often-used estimate of scattering in this intermediate wavelength to object range is the Born approximation. The starting point is the Helmholtz–Kirchoff integral. The key

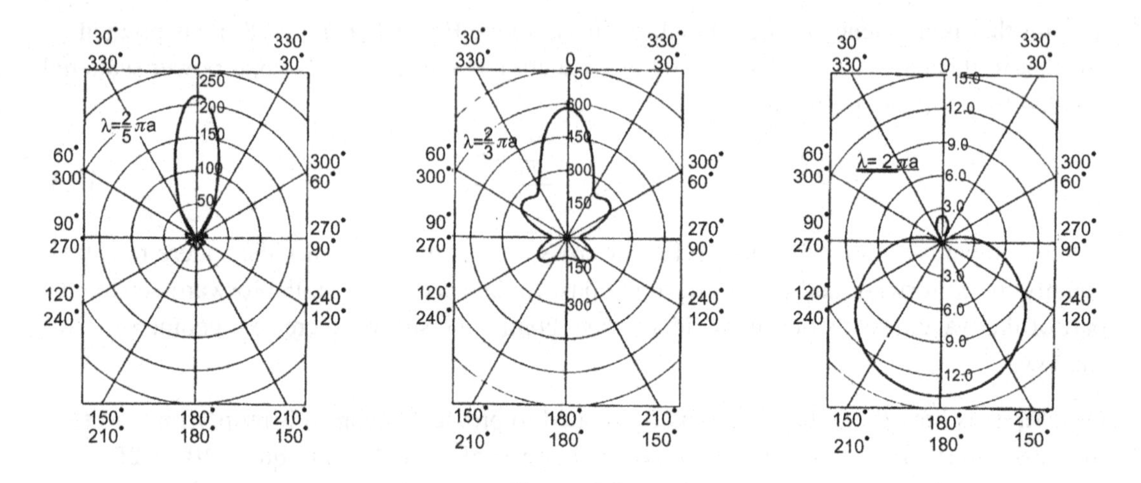

Figure 8.6
Polar Scattering Diagrams for a Rigid Sphere for different *ka* Numbers. *From Jafari, et al. (1981), with permission from the World Federation of Ultrasound in Medicine and Biology.*

assumption in this first-order approach is weak scattering so that the pressure on the surface of the scatterer is approximated by the incident pressure. Elastic mode conversion, multiple scattering, and resonance are neglected. The total pressure is taken to be the sum of the incident (p_i) and scattered fields (p_s):

$$p(r,t) = p_i(r,t) + p_s(r,t). \tag{8.4}$$

If we consider either an object slightly different from its surroundings, such as a piece of tissue suspended in water, or local mild fluctuations from a homogeneous region within tissue, the Born approximation is appropriate. These small fluctuations, denoted by the subscript f, from the local average of the host material a are usually expressed as the following for density and compressibility perturbations:

$$\rho(r) = \rho_a + \rho_f(r), \tag{8.5A}$$

and

$$\kappa(r) = \kappa_a + \kappa_f(r), \tag{8.5B}$$

which are rewritten in the more convenient form,

$$\gamma_\kappa = (\kappa_f - \kappa_a)/\kappa_a, \tag{8.6A}$$

and

$$\gamma_\rho = (\rho_f - \rho_a)/\rho_f. \tag{8.6B}$$

For an incident plane wave and with primed coordinates representing the scattering surface, an expression for the scattered pressure in spherical coordinates (Morse & Ingard, 1968) is:

$$p_s(\boldsymbol{r}) = \int_{V_0} \{k^2\gamma_\kappa(\boldsymbol{r_0})p(\boldsymbol{r_0}) - div[\gamma_\rho(\boldsymbol{r_0})\cdot\nabla_0 p(\boldsymbol{r_0})]\}G(\boldsymbol{r}/\boldsymbol{r_0})dV_0, \quad (8.7A)$$

in which standard $\boldsymbol{r_0}$ coordinates are used for the scatterer, k is the local average wave number, θ is the angle between the incident plane wave direction and the vector $\boldsymbol{r}$ to the receiving point, and G is the usual Green's function:

$$G(\boldsymbol{r}/\boldsymbol{r_0}) = exp\,(ik|\boldsymbol{r}-\boldsymbol{r_0}|)/4\pi|\boldsymbol{r}-\boldsymbol{r_0}|. \quad (8.7B)$$

Here, the Fourier transform of the object function is recognizable in Eqn 8.7A with the exponent of the Green's function helping out. This relationship indicates that objects with sharp edges or corners will be strong scattering centers.

Nassiri and Hill (1986) have derived results from this integral under the Born approximation for a sphere of radius a and a disk of radius a and thickness h respectively:

$$p_{ss}(r) = \frac{e^{ikr}}{r}\frac{k^2}{k_s^3}(\gamma_\kappa + \gamma_\rho \cos\theta)(\sin k_s a - k_s a \cos k_s a), \quad \text{and} \quad (8.8A)$$

$$p_{sd}(r) = \frac{e^{ikr}}{r}\frac{hk^2a^2}{2}(\gamma_\kappa + \gamma_\rho \cos\theta)\frac{J_1(k_s a)}{k_s a}, \quad (8.8B)$$

where $k_s = 2k\sin(\theta/2)$. An often-used term for scattering is the differential scattering cross-section, $\sigma_d(\theta)$, defined as "the fraction of power of a plane progressive wave disturbance incident on the scatterer that is scattered per unit angle." Nassiri and Hill (1986) provide cross-sections for the sphere and disk:

$$\sigma_s(\theta) = \left[\frac{k^2}{k_s^3}(\gamma_\kappa + \gamma_\rho \cos\theta)(\sin k_s a - k_s a \cos k_s a)\right]^2, \quad \text{and} \quad (8.9A)$$

$$\sigma_d(\theta) = \left[\frac{hk^2a^2}{2}(\gamma_\kappa + \gamma_\rho \cos\theta)\frac{J_1(k_s a)}{k_s a}\right]^2. \quad (8.9B)$$

8.2.4 Scattering Summary

In summary, specular reflectors have a reflected pressure that does not vary with frequency. At the other extreme, diffusive scatterers much smaller than a wavelength have a parabolic pressure dependence on frequency, according to Eqn 8.2A in the small ka range ($ka < 0.35$). In between these extremes, the exact solution (Hickling, 1962) for a rigid sphere and the Born approximation are backscattered pressures that rise from a value of zero at $ka = 0$, and undulate with a periodicity related to the interference between the front and back surfaces of the sphere, and asymptotically approach the specular reflection value.

8.3 Role of Transducer Diffraction and Focusing

The previous chapter stressed that the point-spread function (the ability of a transducer to resolve an ideal point scatterer) varies with position and orientation. The discussion in the last section on scattering objects presented results for incident plane waves. In reality, a complicated field pattern from a transducer is incident on a scattering object, not a plane wave. This pattern is then reradiated as a secondary source and scattered in a way dependent on the object's shape and size related to its wave number. In the simplest case, such as an object in the focal plane of a narrowband transducer, where a jinc- or sinc-type beam cross-section function might occur, the amplitude varies across the object and the phase changes sign at each spatial sidelobe. Inclusion of these kinds of effects is necessary in a more accurate model.

A simple approach to include these effects is to build on the model given in Section 7.5.1 by adding a scattering term (s). In the time domain, the overall response can be written as:

$$v_0(z,r,t) = e_{RT}(t) *_t h_t(z,r,t) *_t h_r(z,r,t) *_r s(r,t), \tag{8.10A}$$

where $e_{RT}(t)$ is the round-trip signal of the transducer from Section 7.4.1, and the h's are the diffraction impulse responses for the transmitter and receiver. The frequency counterpart of this equation is:

$$V_0(z,r,f) = E_{RT}(f)H_t(z,r,f)H_r(z,r,f)S(z,r,f). \tag{8.10B}$$

These descriptions, which are reasonably accurate simplifications of the actual process in that it is broken down into identifiable factors, are already becoming rather complicated. For this reason, for simulation purposes, the models of scattering objects are often represented by many ideal rigid point scatterers with different reflection amplitudes organized in the shape of desired scattering objects or tissue. Using the scattered pressure for a rigid sphere, from Eqn 8.2, S can be expressed as an ensemble of ideal point targets each positioned at r_n:

$$S(r,f) = \sum_n s_n \frac{k^2a^3}{3|r-r_n|}\left[1 - \frac{3\cos\theta}{2}\right]e^{-ik(r-r_n)}, \tag{8.11A}$$

where frequency is contained in k, S_n is a constant, and a corresponding time domain version is:

$$s(r,t) = \sum_n S_n \frac{a^3}{3c_0^2|r-r_n|}\left[1 - \frac{3\cos\theta}{2}\right]\frac{\partial^2\delta(t-|r-r_n|/c_0)}{\partial t^2}. \tag{8.11B}$$

Another important pulse–echo case of interest is the plane or mirror scatterer at depth Z_m (Carpenter & Stepanishen, 1984). This pulse–echo configuration (illustrated in

Figure 8.7A) is equivalent to an identical transmitter–receiver pair separated by $2Z_m$ with a reflection factor (*RF*) included for a partially reflecting flat mirror (Chen, Phillips, Schwarz, Mottley, & Parker, 1997; Chen and Schwarz, 1994). From Figure 8.7B, this *RF* becomes a transmission factor (*TF*), $TF = RF(2Z_m)$, for a partially transparent membrane for the equivalent two-transducer configuration. The overall transfer function can be recognized as a modified diffraction loss (see Section 6.8):

$$DL_{eqiv}(2Z_m,f) = exp(-ik2Z_m)RF_m DL(2Z_m,f). \tag{8.12}$$

A few cases have been worked out in the time domain. Rhyne (1977) derived a time domain equivalent for the circular piston and an ideal mirror. He called it the "impulse response of the radiation coupling filter."

8.3.1 Time Domain Born Approximation Including Diffraction

Jensen (1991) has reformulated the Born approximation in the time domain and has included the electromechanical and diffraction field effects of the transmitting and receiving transducers. Jensen points out that the scattered field can be expanded into higher-order terms that represent multiple scattering, but that a first-order approximation is usually sufficient. We can show that with a slight rearrangement of his results below, his formulation is similar in concept to the pulse–echo case of Eqn 8.10A:

$$v_0(t) = e_{RT}(t) * {}_t f_m(r_c) * {}_r h_{pe}(r_c, r, t), \tag{8.13A}$$

where v_0 is the round-trip pulse–echo voltage, r is the vector to the position on the scattering object, and r_c is the vector to the center of the receiving transducer,

$$h_{pe}(r_c, r, t) = \frac{1}{c_0^2}\frac{\partial^2 H_{pe}(r_c, r, t)}{\partial t^2}, \tag{8.13B}$$

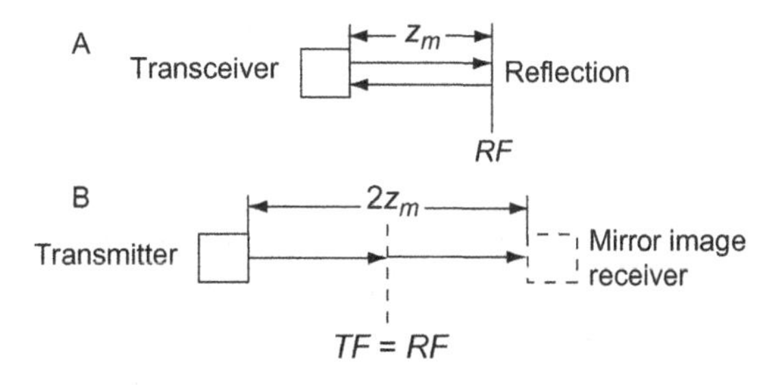

Figure 8.7
(A) Mirror or Plane Reflector Configuration; (B) Equivalent Transducer Pair Configuration.

and $H_{pe}(t) = h_t(t)_t * h_r(t)$, from Section 7.5.1. Also, an inhomogeneity scattering function is:

$$f_m(r) = \Delta\rho(r_c)/\rho_0 - 2\Delta c(r_c)/c_0, \tag{8.13C}$$

that represents the small changes of density and the speed of sound from their nonperturbed average values:

$$\Delta\rho(r) = \rho(r) - \rho_0, \tag{8.14A}$$

and

$$\Delta c(r) = c(r) - c_0. \tag{8.14B}$$

Note that this definition of f_m is slightly different than the standard definitions of Eqns 8.5 and 8.6. However, it can be recognized as the scattering function of Eqn 8.10A:

$$s(r, t) = \frac{f_m}{c_0^2} \frac{\partial^2 \delta(t - r/c_0)}{\partial t^2}, \tag{8.15A}$$

and

$$S(r, f) = \frac{-\omega^2}{c_0^2} f_m, \tag{8.15B}$$

if the double differential operator is shifted from Eqn 8.13B to the scattering function. The function $f_m(r)$ represents the actual physical inhomogeneities or objects that are distorted by a time convolution and a spatial convolution, according to Eqn 8.13A during the scattering process.

The overall scattering process can be rewritten as:

$$V_o(t) = e_{RT}(t) *_t H_{pe}(r_c, r, t) *_r s(r, t), \tag{8.15C}$$

and

$$V_o(f) = E_{RT}(f) H_{pe}(r_c, r, f) S(r, f). \tag{8.15D}$$

As an example of how this approach may be used, consider a mirror placed at the focal plane of a circularly symmetric transducer where $Z_m = F$. Chen et al. (1997) and Chen and Schwartz (1994) have shown that the response is a very weak function of frequency there, so that the round-trip response, $v_0(t)$, would correspond to the function $e_{RT}(t)$, as shown for a measurement in Figure 8.8. A point scatterer placed also at the focus of this transducer would be expected to have the differential scattering characteristic of Eqn 8.15A. In this case, the overall response is proportional to $\partial^2 e_{RT}/\partial t^2$. This result is shown to be the case in Figure 8.8 (solid line) and it is compared to data (crosses) (Szabo, Karbeyaz, Miller, & Cleveland, 2004).

Jensen (1996) has included a point scatterer approach for tissue representation with reflection strengths proportional to values obtained from the Born approximation in his

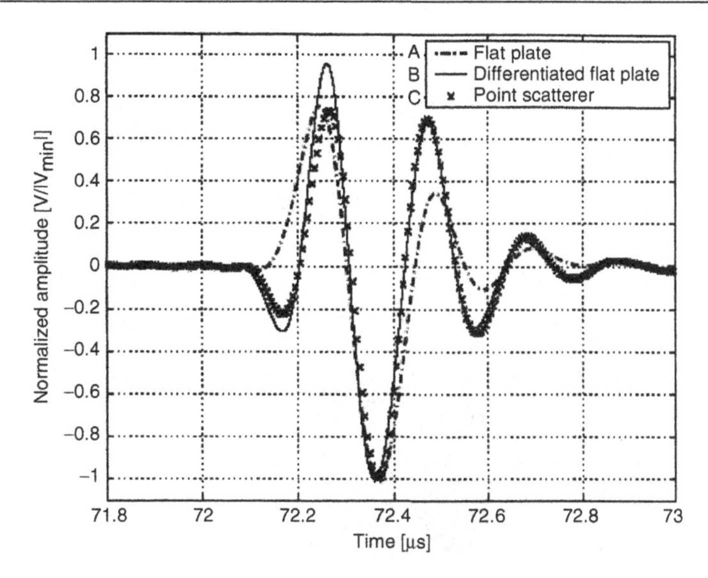

Figure 8.8

Pulse—Echo responses for a Flat Plate and a Point Scatterer at the Focus of a Circularly Symmetric Transducer. Pulse—echo response ($v_0 = e_{RT}$) for a flat plate placed at $Z_m = F$ (dot—dash line), doubly differentiated (solid line) ($\partial^2 e_{RT}/\partial t^2$); normalized pulse—echo to a point scatterer (crosses)—a normalized measurement of a pulse—echo from the tip of an optical fiber also placed at the focal point of the same spherically focused 3.5-MHz transducer. *From Szabo et al. (2004), Acoustical Society of America.*

Field II simulation program. The Field II program is based on Eqn 8.13A and includes methods for implementing different array and transducer geometries with apodization, beamforming (including static and dynamic focusing), and absorption effects, and the ability to create synthetic phantoms made from an organized set of point scatterers (Jensen & Munk, 1997). In Figure 8.9A, an organized set of weighted point scatterers represents an optical image of the right kidney and liver. In Figure 8.9B, an image is simulated as seen by a 128-element, 7-MHz phased array. Details can be found in Jensen and Nikolov (2000). In addition, the user interface for Field II is through a series of MATLAB scripts. This program is available for public use on the World Wide Web and can be found by searching for Field II or on J. A. Jensen's website.

8.4 Role of Imaging

8.4.1 Imaging Process

A comprehensive viewpoint of imaging is necessary to include the major effects involved: beamforming and the spatially varying point-spread function (p.s.f.), the extent of the scattering objects, angles of inclination of beams to these objects, image line-sampling rate,

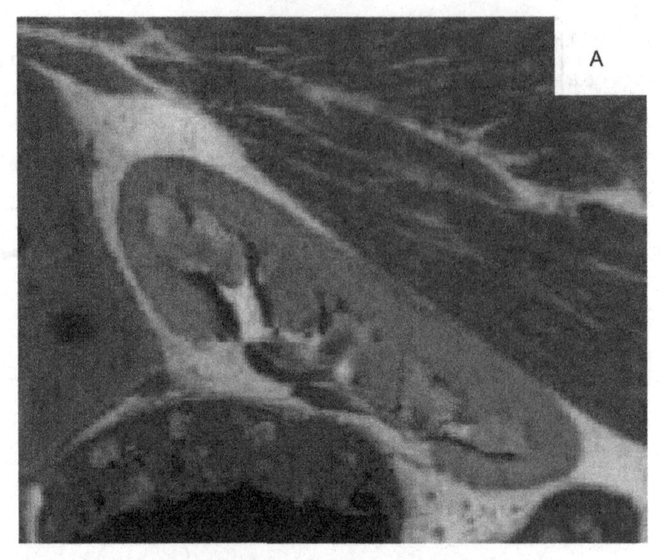

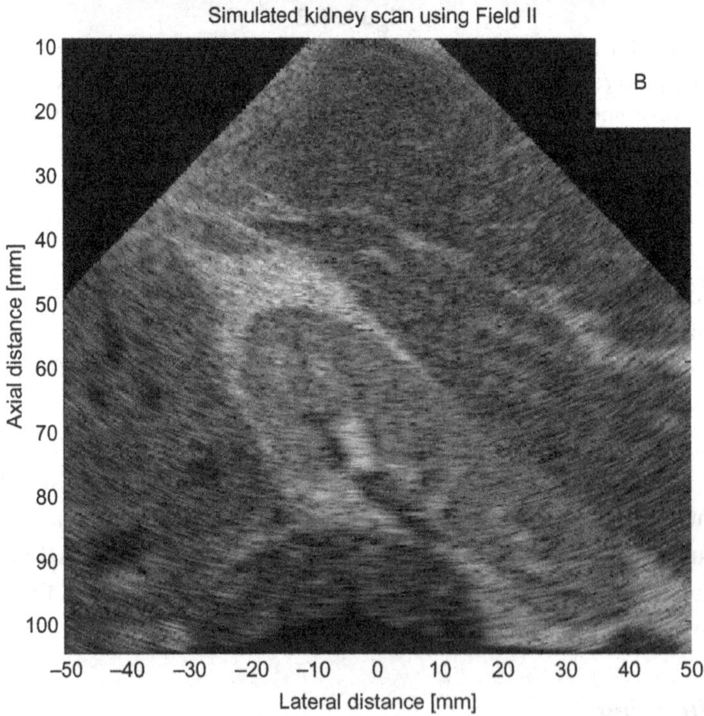

Figure 8.9
(A) Synthetic phantom Scatter map of Weighted point Scatterers Based on Optical image of Right Kidney and Liver; (B) Gray-scale simulated b-mode Image of the Right Kidney and Liver Created by field II program. *(A) based on data from the Visible Human Project; (B) courtesy of J. A. Jensen, Technical University of Denmark.*

interpolation, and presentation. The steps in the imaging process that include all these effects are shown schematically in Figure 8.10 and are listed below.

Images are constructed from a number of acoustic "lines" or vectors usually organized in a sequential pattern. Even though an acoustic vector appears as a thin line in an image after conversion by envelope detection into an image line, each line physically represents a time record of three-dimensional (3D) scattered waves from different depths. The process of image formation is explained by the following sequence of events with reference to letters denoting stages in Figure 8.10:

1. A pulse packet, having 3D spatial extent, travels along the beam vector axis z and changes shape according to its p.s.f. field characteristics (A).
2. After transmission along an acoustic vector direction, the traveling acoustic pulse is scattered over a broad angular range by a series of objects that are each located at a scattering depth r_i and correspond to time delays r_i/c (note r indicates that scattering may include objects off the vector axis z that are captured by an acoustic beam) (B).
3. Angular portions of the series of reflections are intercepted by the receiving pulse–echo transducer. Each echo is at a time delay approximately equal to $2z_i/c$ (C).
4. These intercepted waves are integrated over the surface of the receiving transducer with appropriate weighting and time delays added for focusing and beamforming (D).
5. The integration of step 4 has reduced the 3D scattered signals to a one-dimensional (1D) time record of length $2z_{max}/c$, where z_{max} corresponds to the maximum scan depth selected for the image (E).
6. The time record is envelope detected (F).
7. The amplitude of this envelope-detected time record is logarithmically compressed and processed nonlinearly so that a larger dynamic range of weak to strong echoes can be presented in the same image (G).
8. The next vector in a prescribed sequence of vector directions and spatial increments or directions repeats steps 1–6 (H).
9. The scan lines are placed as vectors in their correct geometric arrangement (I).
10. Once the line sequence is completed, all the lines are interpolated or "scan-converted" to form a filled-in pulse–echo image from a number of image lines arranged in their correct geometrical attitude. The image is converted to gray-scale mapping for final presentation (J).

8.4.2 A Different Attitude

As described in Chapter 1, early ultrasound images were formed by either mechanical translation or freehand with a mechanically sensed movement of a transducer. The vector direction corresponding to each transducer position was controlled or sensed and then

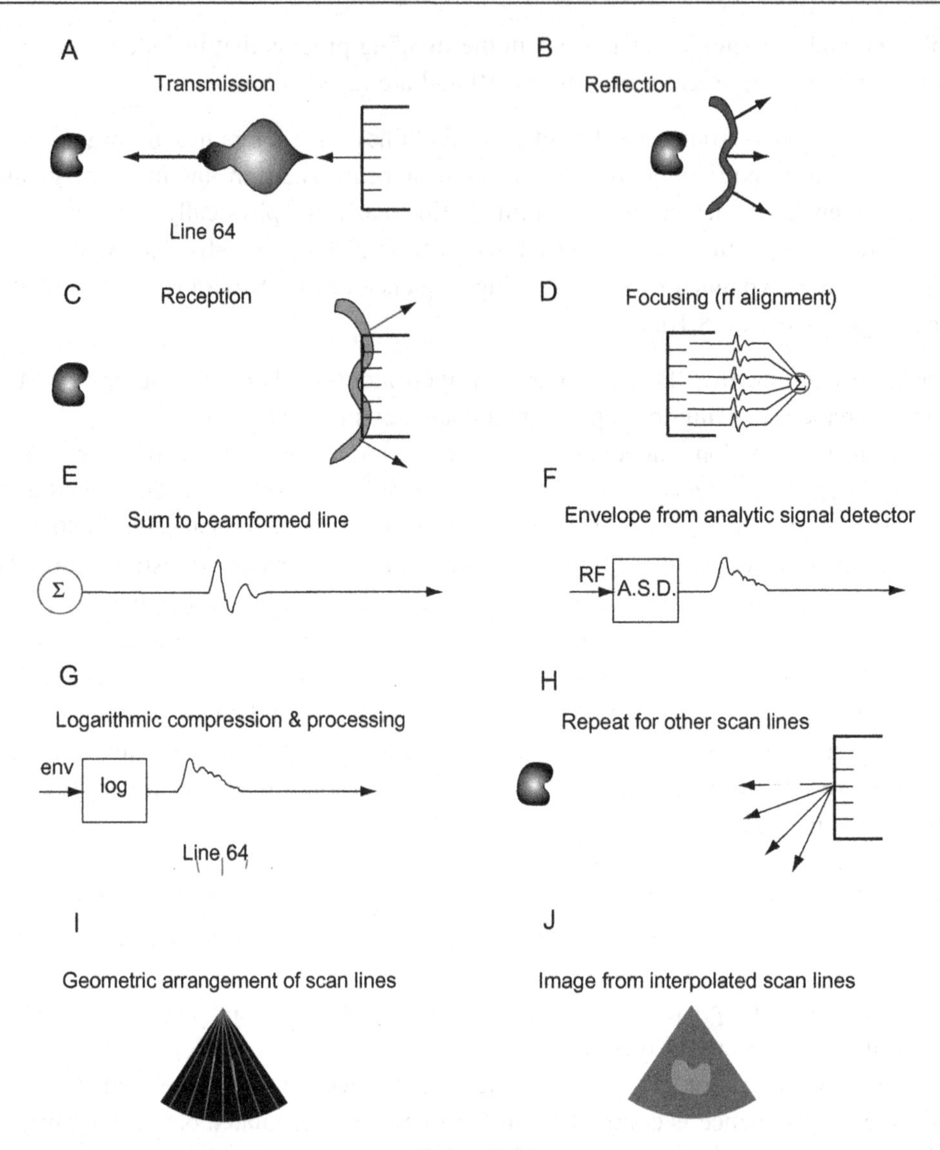

Figure 8.10

Steps in the Imaging Process. (A) Pulse packet (step 1); (B) pulse scattered by objects (step 2); (C) scattered wave fronts intercepted by receiving transducers (step 3); (D) receiving transducers integrate and focus signals (step 4); (E) one-dimensional time record created (step 5); (F) envelope detection (step 6); (G) nonlinear processing of amplitude (step 7); (H) next line repeats previous steps until frame is complete (step 8); (I) scan lines arranged geometrically (step 9); (J) interpolation and gray-scale mapping completes image (step 10).

displayed in its correct attitude (direction) on a cathode-ray tube (CRT) (see Figure 1.10). When the sequence of lines was completed, the image was created either from a long-term persistence of the phosphor on the CRT or through a long-term photographic exposure. As explained in Chapter 1, the two main movements were translational and rotational. The former is now associated with the rectangular format of linear arrays, and the latter is associated with the sector scans of phased arrays. Early on, workers found that the best images were those that combined a rotational (rocking) motion with translation, and this combination became known as compound imaging. Another combination in which rotational movements are added to the ends of translation movements is called contiguous imaging (see Figure 1.3).

As we have seen, in general, scattering from an object occurs over a wide angular extent, but only a portion may be intercepted by a receiving transducer. Furthermore, the direction of insonification is extremely important for determining in which direction sound will be scattered. For example, when sound is nearly parallel (large oblique angle) to the left ventricle wall, very little is reflected back toward the transducer. Figure 8.11 shows another example, in which a cylinder has a radius of curvature that is large relative to a wavelength. The images strongly depend on its curvature and the angle of insonification, so that only the parts of the surface nearly perpendicular to the beam are effective at specularly backscattering sound to the transducer. Each image is a partial view that does not portray the cylinder as a complete circular object.

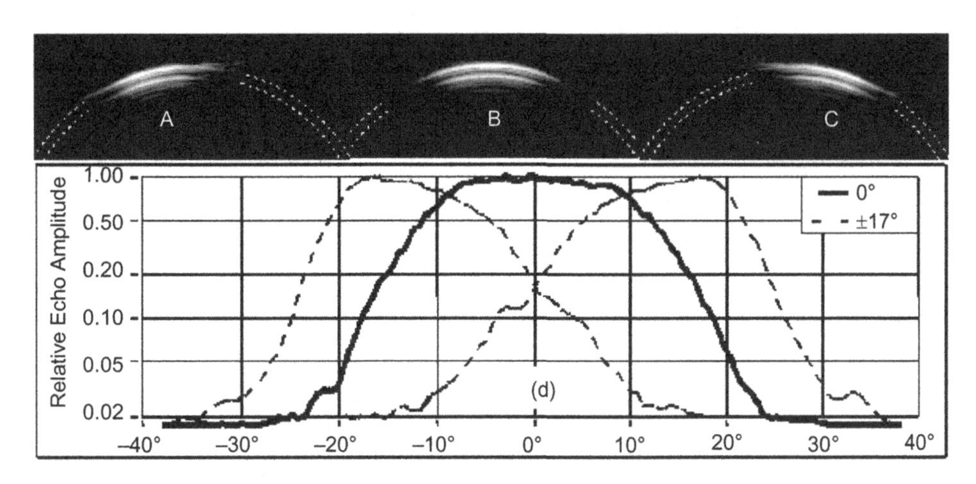

Figure 8.11
Example of a cylinder with a Large Radius of Curvature Relative to a Wavelength. (A–C) Three images of a cylindrical reflector insonified at steering angles of 17°, 0°, and −17°, respectively, by a 5–12-MHz linear array; (D) plots of the corresponding pulse–echo amplitudes for each angle. *From Entrekin, Jago, and Kofoed (2000), with permission of Kluwer Academic/Plenum Publishers.*

Perceived resolution in an image is dependent not only on the extent of the spatial impulse response for a specific beam, but also on the spacing and orientation of the vector lines themselves. To determine adequate sampling, consider a simple sector scan example. For an unapodized round-trip beam in the azimuth plane at the focal length, the beam-shape is approximately a sinc^2 function from Section 7.3. The first nulls of this function occur when the argument is equal to π, or $\theta = \arcsin \lambda/L \approx \lambda/L$, as illustrated in Figure 8.12. To achieve full system resolution, Nyquist sampling at $\theta/2$ should be applied (Steinberg, 1976; von Ramm and Smith, 1983). For example, for a 2.5-MHz transducer with $L = 30$ mm, $\theta = 0.02$ radian or 1.15°, for a scan angular sampling increment of 0.57°, or about 157 lines in a 90° sector.

This approach is an oversimplification because it is based on targets that also have an angular variation; however, targets are better described in linear or rectangular coordinates. For example, an arc length s depends on depth r according to $s = r\Delta\theta$. Because resolution is spatially varying with depth, a linear target can be severely oversampled close to the transducer where many sector lines traverse it. At distances past the focal region, the target may not be resolved, not because of an inadequate sampling rate but because the beamwidth may be too wide. The sampling criterion presented earlier is conservative and provides a reasonable estimate. Apodization produces broader beams that would result in a wider sampling rate requirement. For linear scanning, the same method can be applied with

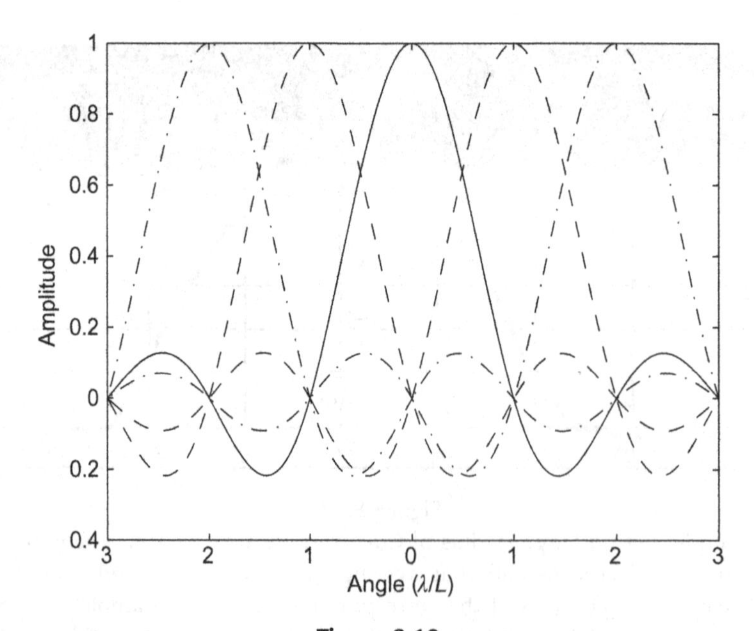

Figure 8.12
Adjacent point Spread Functions for a Focal Plane, centered at Different Angles, Indicate that Angular Separation is Achieved for $\Delta\theta \approx \lambda/L$.

the argument in terms of x and z so that the lateral sampling increment becomes related to the F number (Section 7.5.3), $\Delta x = \lambda F\#/2$.

Unlike the optical cases, acoustic waves have a measurable phase. Sampling approaches based on both magnitude and phase can lead to an improvement in resolution. These methods preserve phase and include it in a different interpolation scheme than the one outlined previously. Phase itself can provide an alternative picture of backscattered signals. Also, phase can provide complimentary information about backscattering. Phase images for medical ultrasound were briefly implemented commercially (Ferrari, Jones, Gonzalez, & Behrens, 1982).

Besides resolution, image contrast is important; it is the ability to identify objects against a background. This thought leads us to the concept of signal-to-noise in an image. The scan depth beyond which the image is lost to electronic noise is called the "penetration distance," as discussed in Chapter 4. This effect is mainly due to tissue absorption and the imaging system (front-end design), and it is somewhat dependent on the efficiency and size (focal gain) of the array. A different type of imaging noise occurs in an image even when there is enough signal; this interfering textural pattern leads us to the curious ultrasound imaging artifact of speckle.

8.4.3 Speckle

Apart from the larger tissue structures in an ultrasound image, there is a textural overlay on different types of tissue, as mentioned in connection with Figure 8.1. This granular texture is called "speckle" after a similar effect in laser optics, even though the physical mechanisms are somewhat different (Abbot and Thurstone, 1979). In optics, intensity plays a dominant role. In ultrasound, however, the phase and amplitude effects are important, as well as the way pulse-envelope data are displayed on a gray scale (no amplitude is assigned to black and maximum amplitude is assigned to white). For many years, users of ultrasound systems assigned a diagnostic value to the appearance of speckle, and they assumed it was tissue microstructure. This discussion will examine the causes of speckle and show that speckle is an illusion more dependent on the measuring system than on the tissue itself (Thijssen & Oosterveld, 1986; Wells & Halliwell, 1981). Also, speckle is detrimental because it reduces both image contrast (the ability to see desired structure against a background) and the distinction of subtle gradations and boundaries in tissue structure. At the end of this section, methods to reduce speckle will be reviewed.

Even though a clinical image contains much more than speckle, in order to understand the effects of speckle in isolation, it is helpful to start with a medium filled with small weak scatterers and no other larger structures. In reality, tissue is filled with small inhomogeneities. Bamber and Dickinson (1980) created a scattering model in which tissue compressibility varied about a mean value. They found that the speckle was determined by the spatial impulse response and not the fluctuations.

A more common alternative is to view the scattering medium as homogeneous but filled with tiny, rigid point scatterers that can be assigned some scattering strength value, as in Eqn 8.11B. The size of each of these ideal scatterers is beyond the resolution capability of the imaging system; nonetheless, these small scatterers can have a profound effect on the image. In a typical pulse–echo situation, a bundle of energy formed by the point-spread function is sent into tissue and is partially scattered along its path. At any instant of time, this energy bundle has a finite extent and weights the scatterers according to the spatial impulse response at that location (as discussed in Section 8.3.1). The extent of the influence of this 3D pulse is called the isochronous volume, which is depicted in two dimensions (2D) in Figure 8.13 (Foster, Arditi, Foster, Patterson, & Hunt, 1983). Note that scatterers in the same isochronous volume produce a backscatter that corresponds to a specific time delay region over a wide angular range. Because individual scatterers in the same volume at different angles have the same time delay, there is an ambiguity in backscattering with angle.

For the simulation of speckle in the second or point scatterers model, the contributions of the spatial impulse responses at each scatterer are summed and added. The spacing between vector or image lines also plays a role. An arrangement of scatterers, along with related

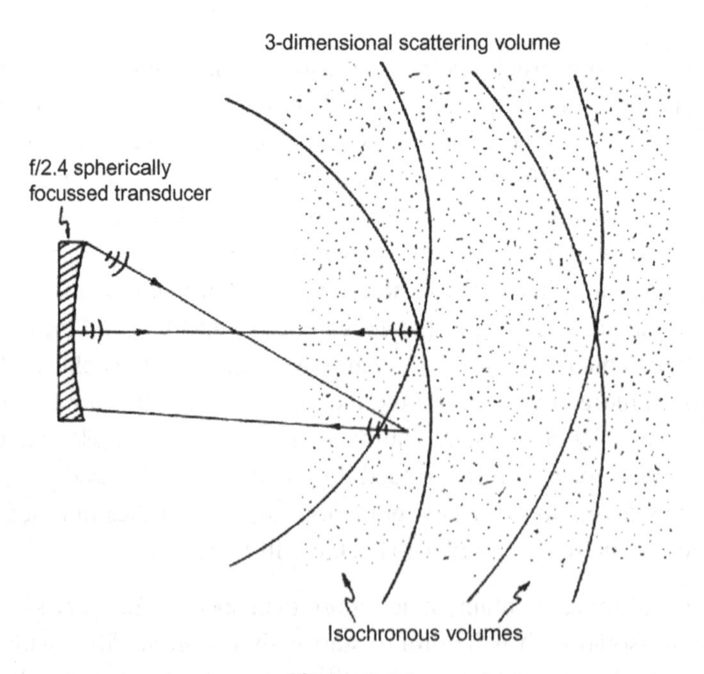

Figure 8.13
Transducers illuminating Point Scatterers in Different 3D is Ochronous Volumes. *From Foster et al. (1983), with permission of Dynamedia, Inc.*

images, is illustrated by Figure 8.14. While an exact reproduction of speckle pattern is impossible without a knowledge of the actual positions of the scatterers, simulations are remarkably effective in recreating the look of speckle (as seen in Figure 8.15). Here, Foster et al. (1983), at the University of Toronto, compared simulated images of exactly the same arrangement of point scatterers with the corresponding images for three different transducer combinations. Speckle is an illusion despite its deceiving appearance as a tissue texture.

Figure 8.14

Stimulated images for Random Scatterers in a Volume, including a Simulated Spherical 2.6-mm-diameter Void. (A) Randomly distributed point scatterers represented with amplitude weights shown as dark and light points against a homogeneous gray background 8 × 8 mm (line shown for scale = 1 mm); (B) gray-scale representation of all the summed and weighted pulse–echo impulse response functions for all points lying in the 3D volume; (C) simulated radio frequency image of the random scattering medium; (D) simulated B-scan image after envelope detection showing the formation of speckle and the presence of a cyst. *From Foster et al. (1983), with permission of Dynamedia, Inc.*

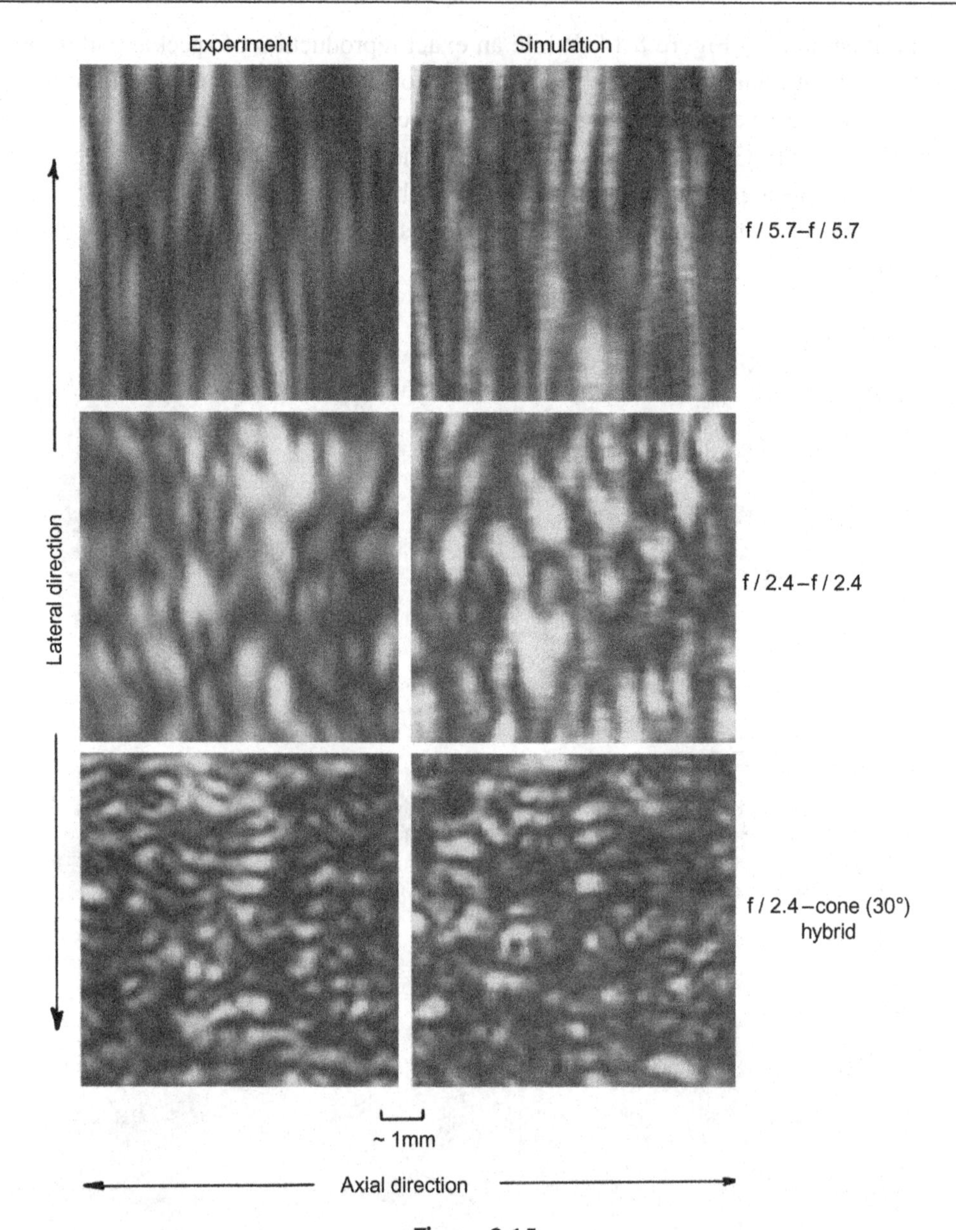

Figure 8.15
Comparison of Simulated Speckle and Data for Three Different Transducer Combinations. *From Foster et al. (1983), with permission of Dynamedia, Inc.*

Speckle in simulation is the constructive and destructive interference of point-spread functions scattered at apparently random specific physical locations. Speckle in a clinical image is generated mainly by constructive and destructive interference of subresolution tissue scatterers at fixed spatial locations. The resulting images of these subresolution scatterers are not random but deterministic, and they can be reproduced exactly if the transducer is returned to the same position, as can be easily demonstrated with a tissue-mimicking phantom. This feature of speckle is used to track tissue movement and displacement, as well as to correct for aberration (to be discussed in Chapter 9).

8.4.4 Contrast

The effect of speckle can be quantified by a "contrast ratio" (CR) and a signal-to-noise ratio (SNR). A classic imaging problem is quantifying the ability to define a cyst object against a speckle background. This contrast ratio is simply the average gray-scale brightness level in the cyst compared to its surround:

$$CR = \frac{\hat{A}_{out} - \hat{A}_{in}}{\hat{A}_{out} + \hat{A}_{in}}, \tag{8.16}$$

where $\hat{A}_{in}$ is the average signal level in the cyst, and $\hat{A}_{out}$ is that in the surrounding material. Note that a value of ± 1 is for good contrast; values close to zero indicate poor contrast. For example, the contrast ratios for the 2.6-mm-diameter cyst of Figure 8.14, is 0.37 ± 0.04. For comparison, contrast ratios for a cyst for the transducer combinations depicted (without a cyst) in the top, middle, and bottom of Figure 8.15 are 0.30 ± 0.04, and 0.37 ± 0.04, respectively. These ratios are governed by the first-order statistics of the speckle (Flax, Glover, & Pelc, 1981).

The probability density function (p.d.f.) for fully developed speckle can be approximated by a Rayleigh distribution based on a random walk assumption that the phase is randomly distributed between 0 and 2π. The Rayleigh probability density distribution, plotted in Figure 8.16, is given by:

$$P(A) = (2A/\overline{A}^2)exp(-A/\overline{A}^2), \tag{8.16A}$$

in which A is brightness or amplitude and $\overline{A}^2$ is the mean of the squared amplitudes.

In general, the probability density function for the envelope amplitude in an image is governed by a K-distribution (Chen, Zagzebski, & Madsen, 1994; Dutt & Greenleaf, 1995; Jakeman & Tough, 1987; Thijssen, 2000; Weng & Reid, 1991) that leads to a signal-to-noise ratio:

$$\text{SNR} = \frac{\sqrt{\pi}\Gamma(\eta + 1/2)}{\sqrt{4\Gamma(\eta + 1) - \pi\Gamma^2(\eta + 1/2)}}, \tag{8.17}$$

where η is the effective number of scatterers within a resolution cell defined by the full width half maximum (FWHM) in axial and lateral dimensions of the point-spread function. This ratio approaches that of a Rayleigh p.d.f. when η is large ($\eta \geq 10$). For this case, the SNR is 1.91, as confirmed by many independent measurements. For tissue with structure, a

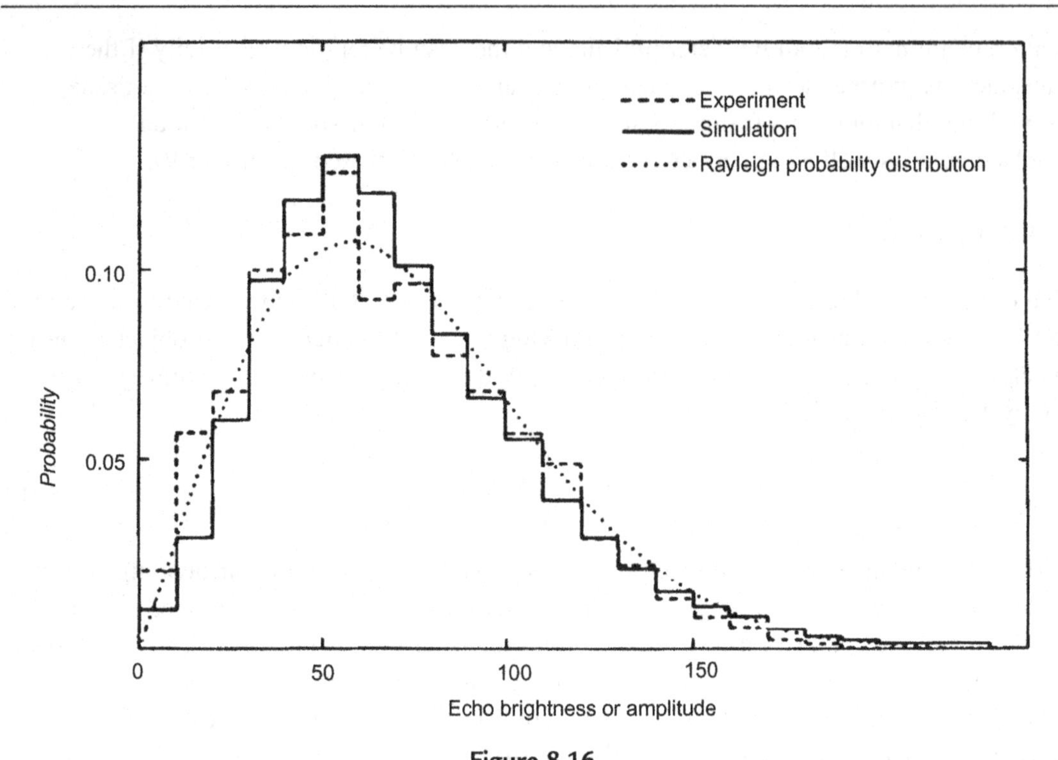

Figure 8.16
Rayleigh distribution Compared to Histograms of Echo Brightness or Amplitude for an Experimental Image and a Simulated One. *From Foster et al. (1983), with permission of Dynamedia, Inc.*

Rician is more appropriate (Insana, Wagner, Garra, Brown, & Shawker, 1986; Thijssen, 1992). It is important to realize that these statistics apply only to video data that have not undergone nonlinear processing. The use of the SNR belongs to first-order analysis of images.

Another measure of contrast is the contrast-to-noise ratio (Lediju, Trahey, Byram, & Dahl, 2011):

$$\mathrm{CNR} = \frac{|S_i - S_o|}{\sqrt{\sigma_i^2 + \sigma_o^2}}, \tag{8.18A}$$

in which S_i and S_o are the signals inside and outside a lesion at the same depth, and σ_i and σ_o are the standard deviations of these signals.

In their study of clutter, Lediju, Pihl, Hsu, Dahl, and Trahey (2008) found that a veil of clutter is distance dependent and fills in expected anechoic regions such as bladders, vessels, and lesions, and therefore reduces contrast in these regions and obscures boundaries. They derived an equation for predicting contrast in clutter-filled regions based on the assumption that clutter is an additive Gaussian random variable:

$$C_{dB} = 20 \log_{10}\sqrt{\frac{S_B^2 + S_C^2}{S_L^2 + S_C^2}}, \tag{8.18B}$$

in which S_B, S_L, and S_C are the background, lesion, and clutter signals.

Line-to-line aspects of images have been analyzed by second-order statistics. Wagner, Smith, Sandrik, and Lopez (1983) and Smith, Sandrik, Wagner, and van Ramm (1987) showed that the average size of speckle is related to an autocovariance function. Their conclusion can be summarized by the following steps. The relation between pressure at two different positions, X_1 and X_2, is conventionally described by an autocorrelation function defined for an incoherent source as (Mallart & Fink, 1991; Wagner et al., 1983):

$$R_p(X_1, X_2, f) = \langle P(X_1, f), P^*(X_2, f)\rangle, \tag{8.19}$$

in which P is pressure at positions X_1 and X_2, f is frequency, the brackets indicate an average over an ensemble of the scattering media, and * denotes complex conjugate. Another useful function is the covariance:

$$C_p(X_1, X_2) = R_p(X_1, X_2) - \langle P(X_1)\rangle\langle P^*(X_2)\rangle. \tag{8.20A}$$

The second term is zero, so that the spatial autocovariance function is:

$$C_P(\Delta X) = R_P(\Delta X), \tag{8.20B}$$

where $\Delta X = X_2 - X_1$ and the value used is often normalized to C_P (0). Pressure must be related to the point-spread function, which for the simplest case, a square piston source, is proportional to a sinc function. In the focal plane of this transducer, $z = F$, if absorption is neglected. Then for the lateral direction:

$$C_{Px}(\Delta X) = B_x \sin c^2(\tilde{f}_{0x}\Delta X)^* \sin c^2(\tilde{f}_{0x}\Delta X), \tag{8.20C}$$

in which the spatial frequency along the lateral x direction is $\tilde{f}_{0x} = L/(\lambda F)$, and B_x is a constant. Similarly, for a Gaussian spectrum varying as $exp(-z^2/2\sigma_z^2)$, axial autocovariance is:

$$C_{Pz}(\Delta X) = B_z exp[-(\Delta z)^2/4\sigma_z^2]. \tag{8.21}$$

These results can provide the average speckle size found from the correlation cell size:

$$S_c = \int_{-\infty}^{\infty} \frac{C_p(\Delta X)}{C_p(0)} d\Delta X. \tag{8.22}$$

For the example of the rectangular transducer, the correlation lateral and axial cell sizes become:

$$S_{cx} = 0.87\,\lambda F/L = 0.87/\tilde{f}_{0x}, \tag{8.23A}$$

and

$$S_{cz} = 0.91 c_0/\Delta f = 1.37/\Delta f, \tag{8.23B}$$

where the last equation is in mm when the −6-dB bandwidth is in MHz and C_0 in mm/μs.

For example, for a 2.5-MHz array that is 25 mm long and focused at 50 mm, the resolution from Chapter 6 is 0.46 mm, and the correlation cell size in the focal plane is 1.04 mm, from Eqn 8.23A. For a 60%, −6-dB fractional bandwidth, the axial correlation cell size is 0.91 mm. Other examples of autocovariance functions for lateral and axial ranges of the transducer combinations from Foster et al. (1983) (discussed in Section 8.4.3), are plotted in Figure 8.17.

Note that the results so far are for focal planes. Speckle size close to the transducer, for example, is much finer than that in the focal plane, yet the resolution is much poorer; this is an interesting counterintuitive result for those who associate speckle size with resolution or tissue microstructure. Later workers (Huisman & Thijssen, 1998; Oostervald, Thijssen, & Verhoef, 1985) found that lateral speckle size is strongly dependent on depth even after correction for diffraction and absorption effects.

8.4.5 Van Cittert–Zernike Theorem

To complete the description of scattering objects, it is necessary to include the properties of random scattering media illuminated by the field of a focusing transducer. Somewhat surprisingly, the backscattered field from a random arrangement of small particles can be correlated. This interesting property can be formalized by the van Cittert–Zernike theorem. From Eqn 8.18, Mallart and Fink (1991) make a number of simplifications that include the Fresnel approximation and $X_1 = (x_1, z)$, $X_2 = (x_2, z)$; the use of a narrow time window (short pulse); and a diffuse random scattering medium that is a separable function of frequency and space:

$$\langle \chi(X_1,f)\chi(X_2,f) \rangle = \chi_0(f)\delta(X_1 - X_2). \tag{8.24}$$

They show that:

$$R_p(x_1,x_2,z,f) = \frac{\chi_0(f)}{z^4}\int\!\!\int \tilde{A}(x_1')\tilde{A}^*[x_1 - (x_1' - x_2')]d^2x_1', \tag{8.25A}$$

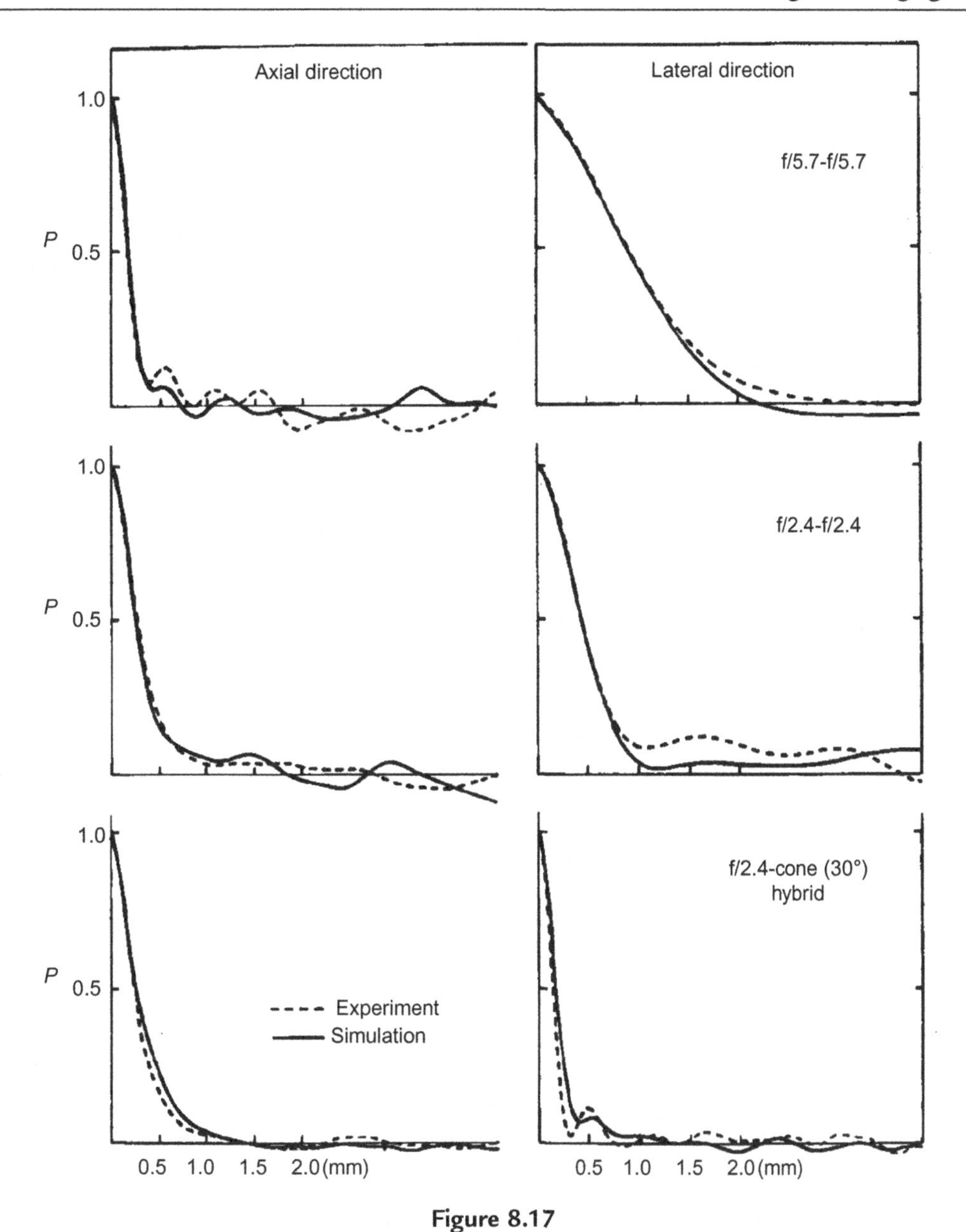

Figure 8.17
Plots of simulated and averaged-experimental lateral and axial covariance coefficients for the three transducer combinations used for Figure 8.15. *From Foster et al. (1983), with permission of Dynamedia, Inc.*

in which $\tilde{A}(x) = A(x)\ exp\{[i2\pi/(z\lambda)]x \cdot x\}$ and A is the aperture weighting function, so:

$$R_p(x_1, x_2, z, f) = \frac{\chi_0(f)}{z^4} R_{\tilde{A}}(x_1 - x_2). \tag{8.25B}$$

The van Cittert–Zernike theorem can be stated as follows for the ultrasound case:

The spatial covariance of the field at points X_1 *and* X_2 *of an observation plane is equal to the Fourier transform of the source aperture function* A(X) *taken at spatial frequency* $\tilde{f} = \Delta X/(\lambda z)$ *where* z *is the distance between the source and the observation plane.*
Mallart and Fink (1991)

Recall that the inverse Fourier transform of a power spectrum is the autocorrelation function:

$$\Im_{-i}[|G(f)|^2] = \int_{-\infty}^{\infty} G(f)G^*(f)e^{i2\pi ft}df = g(t) * g^*(-t) = g \otimes g^*. \tag{8.26}$$

In practical terms, this theorem means that the spatial covariance function in the focal plane is proportional to the autocorrelation of the 2D aperture function. As shown in Figure 8.18 for a 1D aperture with a weighting function $A(x, 0, 0)$, the field is backscattered by random tiny scatterers over a narrow time window as an energy pattern proportional to the autocorrelation of the aperture weighting, $A \otimes A^*$. For a uniformly weighted aperture, $A(x, 0, 0) = \prod(x/L)$, so R_p is proportional to the triangle function, $\Lambda(x/L)$. Here, the physical explanation is as follows: The aperture radiates an energy pattern at its focal plane that is proportional to a sinc function squared. This energy pattern is reflected by random scatterers back to the transducer, where the spatial covariance at the aperture is the Fourier transform of this energy diagram. Because of the way Eqn 8.25B was derived, wave fronts need to be brought into phase (time-aligned) before applying the theorem. Some of the surprising results of this theorem are that the spatial covariance pattern in a focal plane is independent of focal length, F number, and frequency, and depends only on the autocorrelation function of the aperture (Mallart & Fink, 1991; Trahey, Smith, & von Ramm, 1986b). Thus, the longer the aperture, the wider the spatial covariance and the greater the region of spatial coherence of speckle. The signal-to-noise ratio (SNR) at a receiver varies inversely with spatial covariance, so to increase SNR, a wider receiver is needed. Away from the focal zone, decorrelation is more rapid (Trahey et al., 1986b). This theorem is useful for locating regions where the speckle is well correlated for speckle tracking applications (Bamber, 1993). In general, the applications of the van Cittert–Zernike theorem to real tissue needs further work because in tissue characterization, the exceptional deviation from normal tissue structure is of interest (Liu & Waag, 1995).

Two transmit–receive systems can be compared by the correlation between received signals. Walker and Trahey (1995) have shown that correlation $\rho(f)$, as a function of frequency between a system denoted by subscript "0" and another called "1," is:

$$\rho(f) = \frac{\int_{-\infty}^{\infty}[A_{T0}(X,f) \ast A_{R0}(X,f)][A_{T1}(X,f) \ast A_{R1}(X,f)]^{*}dX}{\sqrt{\int_{-\infty}^{\infty}|[A_{T0}(X,f)^{*}A_{R0}(X,f)]|^{2}dX \int_{-\infty}^{\infty}|[A_{T1}(X,f)^{*}A_{R1}(X,f))]|^{2}dX}}, \tag{8.27}$$

where the convolution is in the direction $X = [x\ y]$ in the aperture planes, and T and R denote the transmitter and receiver in the focal plane, respectively.

Extending the work of Walker and Trahey (1995), Lediju et al. (2011) found the following normalized digitized form of spatial correlation useful:

$$\hat{R}(m) = \frac{1}{N-m}\sum_{i=1}^{N-m}\frac{\sum_{n=n1}^{n2} s_i(n)s_{i+m}(n)}{\sqrt{\sum_{n=n1}^{n2} s_i^2(n)\sum_{n=n1}^{n2} s_{i+m}^2(n)}}, \tag{8.28}$$

for a receive aperture of N equally spaced elements, where the time-delayed signal received by the ith element is $s_i(n)$, n is depth or time sample index, and $s_i(n)$ is a zero-mean signal. Note that these signals are already time-delayed so that a signal $s_i(n)$ corresponds to the same location n. For a random diffuse scatterer, a function similar in shape to that depicted in Figure 8.18, this relation would predict a triangular function of length $2N - 1$, of amplitude 1 at the center and amplitude 0 at the two ends, $N - 1$ and $-(N - 1)$. A useful term is spatial "lag" which is the receive element spacing n which extends from 0 to $N - 1$.

8.4.6 Speckle Reduction

Many ways to reduce speckle have been proposed. Some of these have been reviewed by Bamber (1993). Most of these approaches involve a kind of diversity so that speckle effects can be averaged, minimized, or broken up. Several methods of compounding can be explored through Eqn 8.27. Trahey et al. (1986b) show that to reduce speckle effectively, N independent speckle images are needed to reduce speckle by a factor $\sqrt{N}$. Breaking the overall aperture up into subapertures is a kind of spatial compounding that involves a trade-off between speckle reduction and resolution. For example, dividing the aperture in half and averaging images reduces speckle by $\sqrt{2}$ (Trahey et al., 1986b), but this arrangement also degrades resolution at the focus by a factor of two. Frequency compounding (Melton & Magnin, 1984; Trahey, Allison, Smith, & von Ramm, 1986a) involves summing multiple images created from signals filtered at different center frequencies and bandwidths; this approach has enjoyed recent commercial success because of the availability of broad bandwidth transducers. Spatial compound imaging, which creates images from several angular views, was discussed in Chapter 1 and by Trahey et al. (1986b) and Entrekin, Jago,

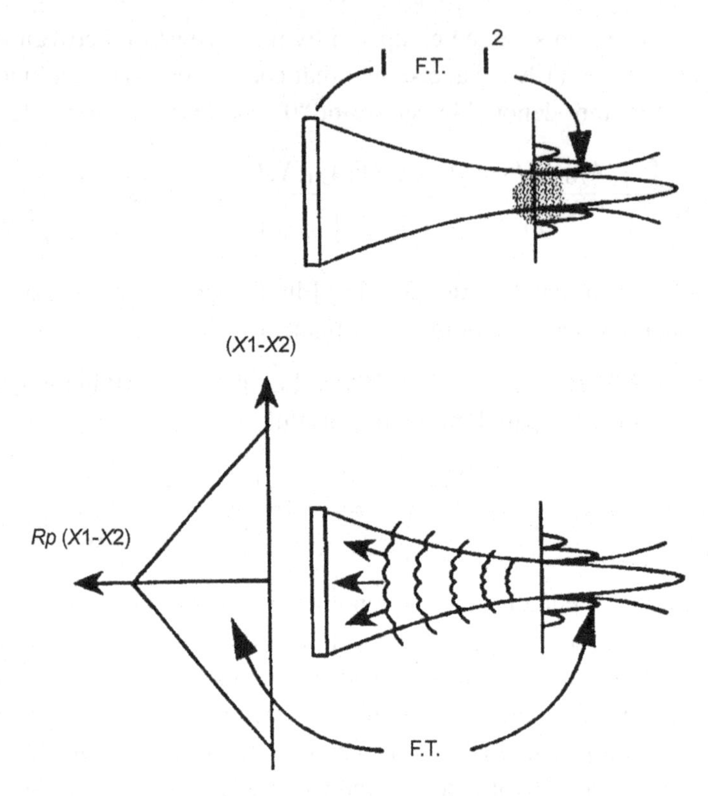

Figure 8.18
Spatial covariance of backscatter from random media (Top) energy pattern in focal plane is proportional to magnitude of Fourier Transform (F.T.) of aperture squared. (Bottom) energy pattern reflected by random scatterers back to the aperture, where the spatial covariance is the Fourier transform of this energy pattern, according to the Van Cittert–Zernike theorem. *From Mallart and Fink (1991), Acoustical Society of America.*

and Kofoed (2000). Examples of the implementation of real-time speckle reduction methods can be found in Chapter 10.

8.4.7 Speckle Tracking

A unique property of speckle is its space invariance. Given the same set of circumstances, i. e. the same object or tissue being imaged with the same transducer in the same position, then the speckle pattern is reproducible. Ironically, even though speckle may not be an accurate depiction of what is being imaged because it is dependent on subresolution scatterers and their locations in a particular transducer pressure field, it can be useful in measuring local spatial changes. In Figure 8.19, a static vertical force is applied to a solid with an inclusion. Pulse echo *RF* lines shown in the figure shift differently in the different regions in response to the force. The relative displacements can be determined by correlating corresponding lines before and after compression. In a second example, through the comparison of one frame of ultrasound imaging sequence to another, changes in the fine-grained texture of speckle can reveal movement. Consider the following example of a spherical transducer focusing into a liver sample, as shown in Figure 8.20. With enough driving voltage and high focal gain, this transducer can generate sufficient intensity to heat the tissue in the focal region through acoustic absorption. This heating will locally increase the speed of sound in the focal region. Through a comparison of individual *RF* lines in each image frame before and after heating, the relative displacement can be determined by cross-correlation, as illustrated by Figure 8.21. With additional processing (Miller, Bamber, & Meaney, 2002), the heating region can be computed and displayed as an image.

This basic process, which is called "speckle tracking," has found many applications. Often, as in the example, displacement with more processing can lead to other parameters that reveal the response of tissue to various types of perturbations or stimuli. In the example, the stimulus was local heating. For elastography, speckle tracking is used widely to examine the local viscoelastic response of different tissues or structures to externally applied forces or vibrations as a means of determining the tissue's hardness or softness. This topic is treated in far more detail in Chapter 16. Speckle tracking is also useful in detecting tissue movement, such as detecting the flow of blood. As an alternative to commonly used Doppler and color Doppler imaging methods of depicting and measuring blood flow (topics found in Chapter 11), these tracking methods have a number of advantages such as angle-independent detection of flow. Other uses of speckle tracking are temperature measurement (Hsu, Miller, Evans, et al., 2005;

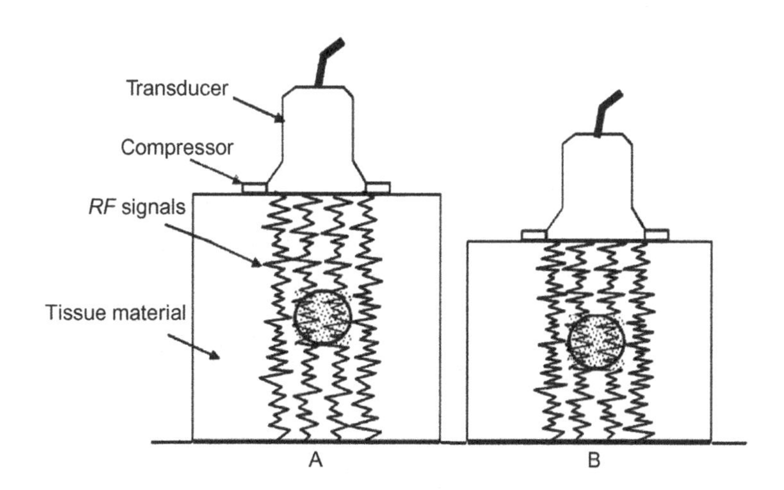

Figure 8.19
The Principle of Elastography.
The tissue is insonified (A) before and (B) after a small uniform compression. Inside the harder tissues (e.g. the circular lesion depicted) the echoes will be less distorted than in the surrounding tissues, thus denoting smaller strain. *From Konofagou (2004), Ultrasonics.*

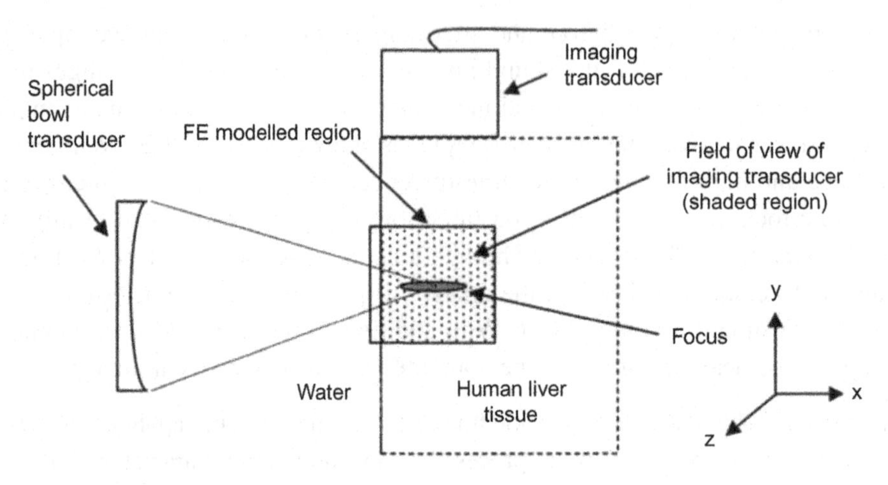

Figure 8.20

Spherical Transducer Focused on a Region in a liver Sample. Schematic diagram of a cross-section through the finite element (FE) modeled region (not to scale). The *xy* plane in this figure occurs at $z = 0$. Heating is expected in the focal region. *From Miller et al. (2002), with permission from the World Federation of Ultrasound in Medicine and Biology.*

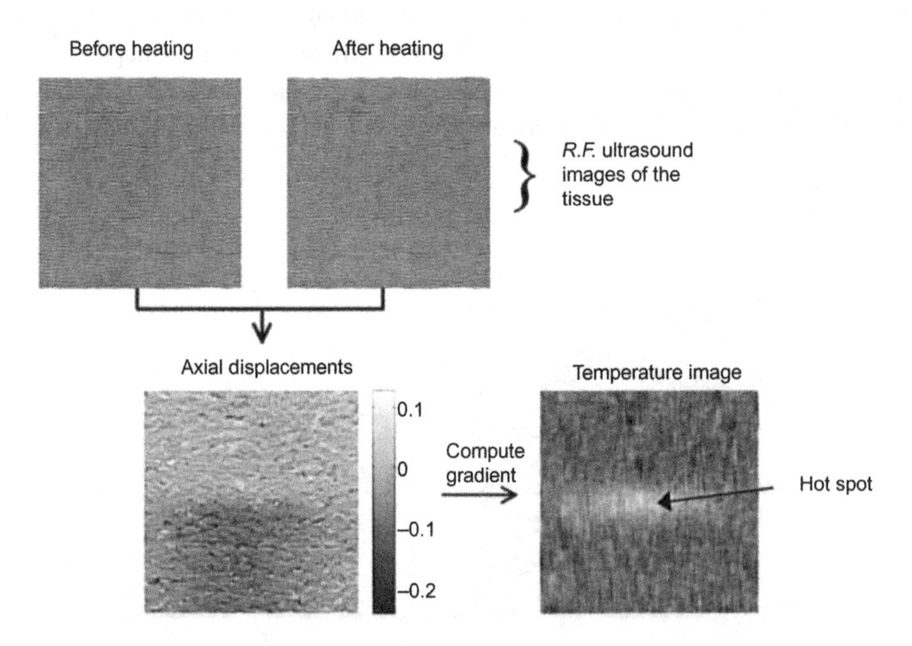

Figure 8.21

Ultrasonic Temperature Imaging Procedure Showing Images Before and After Heating. The images in this figure were generated for liver of normal fat content and for an ultrasonic SNR of 30 dB. Speckle tracking is used to create the displacement image, where displacement units are in pixels; the total size of the radio frequency images was 400 × 400 pixels. Finally, the temperature was calculated from the displacements and a temperature model. *From Miller et al. (2002), with permission from the World Federation of Ultrasound in Medicine and Biology.*

Miller, Bamber, & Meaney, 2002; Shi et al., 2003) for therapy and high-intensity focused ultrasound, monitoring of respiratory and cardiac cycles and arterial and organ motion.

Originally, speckle tracking was employed for angle-independent flow detection comparing frames of data (Bohs, Friemel, McDermott, & Trahey,1993; Trahey, Hubbard, & von Ramm, 1988). Other early applications were the 1D cross-correlation of *RF* pulse–echo lines to calculate displacements for elastography (Ophir, Cespedes, Ponnekanti, Yazdi, & Li, 1991) and the tracking of arterial wall motion (O'Donnell, Skovoroda, & Shapo, 1991). One of the major challenges of applying speckle tracking is the decorrelation of signals (Ramamurthy & Trahey, 1991) because of off-axis and out-of-plane or large displacements, soft tissue heterogeneities, undersampling, poor signal-to-noise ratios, and geometric and other effects. Several solutions to the problem have been proposed, including pattern matching (Bohs, Friemel, & Trahey, 1995) (see Figure 8.22); multilevel blocking to improve matching, as shown in Figure 8.23 (Yeung, Levinson, & Parker, 1998); block matching with error correction (Zhu and Hall, 2002); feature tracking (Bashford & von Ramm, 1996; Yeung, Levinson, Fu, et al., 1998); meshes (Zhu, Chaturvedi and Insana, 1999); complex base-band processing and short-time correlation (Lubinski, Emelianov, & O'Donnell, 1999; O'Donnell, Skovoroda, Shapo, & Emelianov, 1994); block matching in selected regions (Golemati et al., 2003); multidimensions (3D and 4D) (Harris, Miller, Bamber, et al., 2010, 2011); and real-time, compound, and ultrafast imaging (Bohs et al., 1993; Tanter, Bercoff, Sandrin, & Fink, 2002).

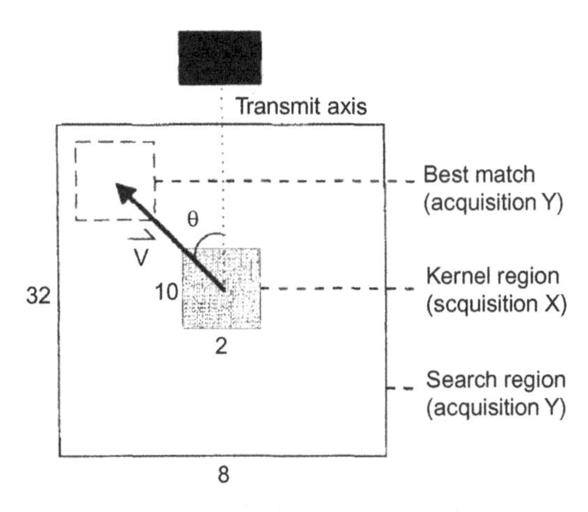

Figure 8.22
2D Speckle Tracking Geometry Utilized in Pattern Matching. A kernel region (shaded) of 2 × 10 pixels was identified in the first acquisition, X, and compared with all possible matching regions in an 8 × 32 search region in a successive acquisition, Y, using the sum-absolute-difference (SAD) algorithm. The minimum SAD value determined the best-match region (dotted) and thereby defined the displacement vector, V. The direction of motion, θ, is defined relative to the transducer axis as shown. *From Bohs et al. (1995), with permission from the World Federation of Ultrasound in Medicine and Biology.*

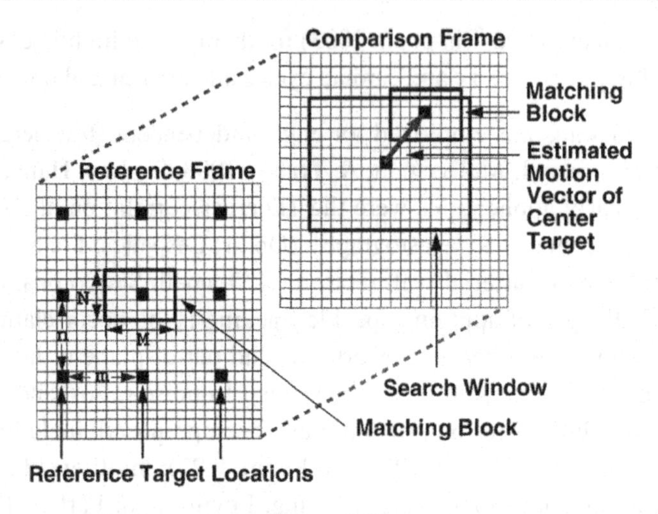

Figure 8.23

Schematic Representation of the Multi-level General Block-Matching Algorithm. *From Yeung, Levinson, and Parker (1998), with permission from the World Federation of Ultrasound in Medicine and Biology.*

The subjects of decorrelation and methodology have been studied by several groups. Ramamurthy and Trahey (1991) examined early *RF* and video correlation methods. Hein and O'Brien (1993) and Viola and Walker (2003) have provided useful analyses and comparisons of different time-correlation algorithms and methods. Chen, Jenkins, & O'Brien (1995) showed that the effectiveness of correlation methods depended on the type of tissues measured and that correlation itself may not be a guarantee of accurate tracking. Part of the decorrelation issues could be attributed to the significant differences in axial and lateral sampling obtained in conventional imaging and out-of-plane effects. From Figure 3.8, depicting a big finger pressing down on an elastic solid, and related text, it is evident that both stress and strain are coupled in three dimensions. In previous sections of this chapter about speckle, there were demonstrations of how dramatically the speckle pattern is affected by spatially variable differences in the point-spread function across the extent of the image, as by the examples in Figures 8.14 and 8.15. These images were for random scatterers in a homogeneous medium; even greater differences can be expected for heterogeneous tissue. One approach to these problems, by adequate sampling, interpolation, and recorrelation, validated by finite-element modeling, can be found in Konafagou and Ophir (1998); Lee et al. (2007); and Luo, Lee, and Konofagou (2009). This whole topic of speckle tracking will be revisited in Chapter 16, on elastography.

References

Abbot, J. G., & Thurstone, F. L. (1979). Acoustic speckle: Theory and experimental analysis. *Ultrasonic Imaging*, *1*, 303–324.

Bamber, J. C. (1993). Speckle reduction. In P. N. T. Wells (Ed.), *Advances on ultrasound techniques and instrumentation* (pp. 55–67). New York: Churchill Livingstone.

Bamber, J. C., & Dickinson, R. J. (1980). Ultrasonic B-scanning: A computer simulation. *Physics in Medicine and Biology*, *25*, 463.

Bashford, G. R., & von Ramm, O. T. (1996). Ultrasound three-dimensional velocity measurements by feature tracking. *IEEE Transactions on Ultrasonics, Ferroelectrics and Frequency Control*, *43*, 376–384.

Bohs, L. N., Friemel, B. H., McDermott, B. A., & Trahey, G. E. (1993). A real-time system for quantifying and displaying two-dimensional velocities using ultrasound. *Ultrasound in Medicine and Biology*, *19*, 751–761.

Bohs, L. N., Friemel, B. H., & Trahey, G. E. (1995). Experimental velocity profiles and volumetric flow via two-dimensional speckle tracking. *Ultrasound in Medicine and Biology*, *21*, 885–898.

Carpenter, R. N., & Stepanishen, P. R. (1984). An improvement in the range resolution of ultrasonic pulse echo systems by deconvolution. *Journal of the Acoustical Society of America*, *75*, 1084–1091.

Chen, E. J., Jenkins, W. K., & O'Brien, W. D., Jr. (1995). Performance of ultrasonic speckle tracking in various tissues. *Journal of the Acoustical Society of America*, *98*, 1273–1278.

Chen, J. -F., Zagzebski, J. A., & Madsen, E. L. (1994). Non-Gaussian versus non-Rayleigh statistical properties of ultrasound echo signals. *IEEE Transactions on Ultrasonics, Ferroelectrics and Frequency Control*, *41*, 435–440.

Chen, X., Phillips, D., Schwarz, K. Q., Mottley, J. G., & Parker, K. J. (1997). The measurement of backscatter coefficient from a broadband pulse–echo system: A new formulation. *IEEE Transactions on Ultrasonics, Ferroelectrics and Frequency Control*, *44*, 515–525.

Chen, X., & Schwarz, K. Q. (1994). Acoustic coupling from a focused transducer to a flat plate and back to the transducer. *Journal of the Acoustical Society of America*, *95*, 3049–3054.

Coussios, C. C. (2002). The significance of shape and orientation in single-particle weak-scatterer models. *Journal of the Acoustical Society of America*, *112*, 906–915.

Dutt, V., & Greenleaf, J. F. (1995). Speckle analysis using signal-to-noise ratios based on fractional order moments. *Ultrasonic Imaging*, *17*, 251–268.

Entrekin, R. R., Jago, J. R., & Kofoed, S. C. (2000). Real-time spatial compound imaging: Technical performance in vascular applications. In Halliwell, & Wells (Eds.), *Acoustical imaging* (Vol. 25, pp. 331–342). New York: Kluwer Academic/Plenum Publishers.

Ferrari, L., Jones, J. P., Gonzalez, V., & Behrens, M. (1982). Acoustical imaging using the phase of echo waveforms. In E. A. Ash, & C. R. Hill (Eds.), *Acoustical imaging* (Vol. 12, pp. 635–641). New York: Plenum Press.

Flax, S. W., Glover, G. H., & Pelc, N. J. (1981). Textural variations in B-mode ultrasonography: A stochastic model. *Ultrasonic Imaging*, *3*, 235–257.

Foster, D. R., Arditi, M., Foster, F. S., Patterson, M. S., & Hunt, J. W. (1983). Computer simulations of speckle in B-scan images. *Ultrasonic Imaging*, *5*, 308–330.

Golemati, S., Sassano, A., Lever, M. J., Bharath, A. A., Dhanjil, S., & Nicolaides, A. N. (2003). Carotid artery wall motion estimated from B-mode ultrasound using region tracking and block matching. *Ultrasound in Medicine and Biology*, *29*, 387–399.

Harris, E. J., Miller, N. R., Bamber, J. C., Evans, P. M., & Symonds-Tayler, J. R. N. (2010). Speckle tracking in a phantom and feature-based tracking in liver in the presence of respiratory motion using 4D ultrasound.. *Physics in Medicine and Biology*, *55:*, 3363–3380.

Harris, E. J., Miller, N. R., Bamber, J. C., Evans, P. M., & Symonds-Tayler, J. R. N. (2011). The effect of object speed and direction on the performance of 3D speckle tracking using a 3D swept-volume ultrasound probe. *Physics in Medicine and Biology*, *56*, 7127–7143.

Hein, I. A., & O'Brien, W. D., Jr. (1993). Current time-domain methods for assessing tissue motion by analysis of reflected ultrasound echoes: A review.. *IEEE Transactions on Ultrasonics, Ferroelectrics and Frequency Control*, *40*, 84–102.

Hickling, R. (1962). Analysis of echoes from solid elastic sphere in water. *Journal of the Acoustical Society of America*, *34*, 1582–1592.

Hsu, A, Miller, N. R, Evans, P. M, Bamber, J. C, & Webb, S (2005). Feasibility of using ultrasound for real-time tracking during radiotherapy. *Medical Physics*, *32*, 1500–1512.

Huisman, H. J., & Thijssen, J. M. (1998). An in vivo ultrasonic model of liver parenchyma. *IEEE Transactions on Ultrasonics, Ferroelectrics and Frequency Control*, *45*, 739–750.

Insana, M. F., Wagner, R. F., Garra, B. S., Brown, D. G., & Shawker, T. H. (1986). Analysis of ultrasound image texture via generalized Rician statistics. *Optical Engineering*, *25*, 743–748.

Jakeman, E., & Tough, R. J. A. (1987). Generalized k-distribution: A statistical model for weak scattering. *Journal of the Optical Society of America*, *4*, 1764–1772.

Jafari, F., Madsen, E. L., Zagzebski, J. A., & Goodsitt, M. M. (1981). Exact evaluation of an ultrasonic scattering formula for a rigid immovable sphere. *Ultrasound in Medicine and Biology*, *7*, 293–296.

Jensen, J. A. (1991). A model for the propagation and scattering of ultrasound in tissue. *Journal of the Acoustical Society of America*, *89*, 182–190.

Jensen, J. A. (1996). Field: A program for simulating ultrasound systems. *Medical and Biological Engineering and Computing*, *34*(Suppl. 1), 351–353.

Jensen, J. A., & Munk, P. (1997). *Computer phantoms for simulating ultrasound B-mode and CFM images, Acoustical imaging* (Vol. 23, pp. 485–494).). New York: Plenum Press.

Jensen, J. A. and Nikolov, S. (2000). Fast simulation of ultrasound images. *Proceedings of the IEEE ultrasonics symposium* (Vol. 2) (pp. 1721–1724). San Juan.

Kino, G. S. (1987). *Acoustic waves: devices, imaging, and analog signal processing*. Englewood Cliffs, NJ: Prentice-Hall. pp. 300–357.

Konofagou, E. E. (2004). Quo vadis elasticity imaging? *Ultrasonics*, *42*, 331–336.

Konafagou, E., & Ophir, J. (1998). A new elastographic method for estimation and imaging of lateral displacement, lateral strains, corrected axial strains and Poisson's ratios in tissues. *Ultrasound in Medicine and Biology*, *24*, 1183–1199.

Lediju, MA, Pihl, MJ, Hsu, SJ, Dahl, JJ, & Trahey, GE. (2008). Quantitative assessment of the magnitude, impact, and spatial extent of ultrasonic clutter. *Ultrasonic Imaging*, *30*, 151–168.

Lediju, MA, Trahey, GE, Byram, BC, & Dahl, JJ. (2011). Short-lag spatial coherence of backscattered echoes: Imaging characteristics, IEEE Transactions on Ultrasonics. *Ferroelectrics and Frequency Control*, *58*, 1377–1388.

Lee, W. -N., Ingrassia, C. M., Fung-Kee-Fung, S. D., Costa, K. D., Holmes, J. W., & Konofagou, E. E. (2007). Theoretical quality assessment of myocardial elastography with in vivo validation. *IEEE Transactions on Ultrasonics, Ferroelectrics and Frequency Control*, *54*, 2233–2245.

Liu, D. -L., & Waag, R. C. (1995). About the application of the van Cittert–Zernike theorem in ultrasonic imaging. *IEEE Transactions on Ultrasonics, Ferroelectrics and Frequency Control*, *42*, 590–601.

Lubinski, M. A., Emelianov, S. Y., & O'Donnell, M. (1999). Speckle tracking methods for ultrasonic elasticity imaging using short-time correlation, IEEE Transactions on Ultrasonics. *Ferroelectrics and Frequency Control*, *46*, 82–96.

Luo, J., Lee, W. -N., & Konofagou, E. E. (2009). Fundamental performance assessment of 2-D myocardial elastography in a phased-array configuration. *IEEE Transactions on Ultrasonics, Ferroelectrics and Frequency Control*, *56*, 2320–2327.

Mallart, R., & Fink, M. (1991). The van Cittert–Zernike theorem in pulse–echo measurements. *Journal of the Acoustical Society of America*, *90*, 2718–2727.

Melton, H. E., & Magnin, P. A. (1984). A-mode speckle reduction with compound frequencies and bandwidths. *Ultrasonic Imaging*, *6*, 159–173.

Miller, N. R., Bamber, J. C., & Meaney, P. M. (2002). Fundamental limitations of noninvasive temperature imaging by means of ultrasound echo strain estimation. *Ultrasound in Medicine and Biology*, *28*, 1319–1333.

Morse, P. M., & Ingard, K. U. (1968). *Theoretical acoustics*. Princeton, NJ: Princeton University Press. pp. 400–411.

Nassiri, D. K., & Hill, C. R. (1986). The use of angular acoustic scattering measurements to estimate structural parameters of human and animal tissues. *Journal of the Acoustical Society of America, 79*, 2048–2054.

O'Donnell M, Skovoroda AR, Shapo BM. (1991)Measurement of arterial wall motion using Fourier-based speckle tracking algorithm. *Proceedings of the IEEE ultrasonics symposium* (Vol. 2) (pp. 1101–1104). Orlando, FL.

O'Donnell, M., Skovoroda, A. R., Shapo, B. M., & Emelianov, S. Y. (1994). Internal displacement and strain imaging using ultrasonic speckle tracking. *IEEE Transactions on Ultrasonics, Ferroelectrics and Frequency Control, 41*, 314–325.

Oosterveld, B. J., Thijssen, J. M., & Verhoef, W. A. (1985). Texture in B-mode echograms: 3-D simulations and experiments of the effects of diffraction and scatter density.. *Ultrasonic Imaging, 7*, 142–160.

Ophir, J., Cespedes, E. I., Ponnekanti, H., Yazdi, Y., & Li, X. (1991). Elastography: A method for imaging the elasticity in biological tissues. *Ultrasonic Imaging, 13*, 111–134.

Ramamurthy, B. S., & Trahey, G. E. (1991). Potential and limitations of angle-independent flow-detection algorithms using radio-frequency and detected echo. *Ultrasonic Imaging, 13*, 252–268.

Rhyne, T. L. (1977). Radiation coupling of a disk to a plane and back or to a disk to disk: An exact solution. *Journal of the Acoustical Society of America, 61*, 318–324.

Shi, Y., Witte, R. S., Milas, S. M., Neiss, J. H., Chen, X. C.; Cain, C. A., & et al. (2003) Ultrasonic thermal imaging of microwave absorption. *Proceedings of the IEEE ultrasonics symposium* (Vol. 1) (pp. 224–227).

Smith, S. W., Sandrik, J. M., Wagner, R. F., & van Ramm, O. T. (1987). Measurements and analysis of speckle in ultrasound B-scans. *Acoustical Imaging, 10*, 195–211.

Steinberg, B. D. (1976). *Principles of aperture and array design.* New York: John Wiley & Sons.

Strutt, J. W. (Lord Rayleigh). (1964). *On the light from the sky, its polarization and colour.* Scientific Papers, Lord Rayleigh, Vol. 1. Dover Publications, New York. [Original work published 1871].

Szabo, T. L., Karbeyaz, B. U., Miller, E., & Cleveland, R. (2004). Comparison of a flat plate and point targets for broadband determination of the pulse–echo electromechanical characteristic of a transducer. *Journal of the Acoustical Society of America, 116*, 90–96.

Tanter, M., Bercoff, J., Sandrin, L., & Fink, M. (2002). Ultrafast compound imaging for 2-D motion vector estimation: Application to transient elastography. *IEEE Transactions on Ultrasonics, Ferroelectrics and Frequency Control, 49*, 1363–1374.

Thijssen, J. M. (1992). Echographic image processing. In P. W. Hawkins (Ed.), *Advances in electronics and electron physics* (Vol. 84, pp. 317–349). Boston: Academic Press.

Thijssen, J. M., & Oosterveld, B. J. (1986). Speckle and texture in echography: Artifact or information? *Proceedings of the IEEE ultrasonics symposium* (pp. 803–809). Williamsburg, VA.

Trahey, G. E., Allison, J. W., Smith, S. W., & von Ramm, O. T. (1986a). Speckle pattern changes with varying acoustic frequency: Experimental measurements and implications for frequency compounding. *Proceedings of the IEEE ultrasonics symposium* (pp. 815–818). Williamsburg, VA.

Trahey, G. E., Hubbard, S. M., & von Ramm, O. T. (1988). Angle-independent ultrasonic blood-flow detection by frame-to-frame correlation of B-mode images. *Ultrasonics, 26*, 271–276.

Trahey, G. E., Smith, S. W., & von Ramm, O. T. (1986b). Speckle pattern correlation with lateral aperture translation: Experimental results and implications for spatial compounding. *IEEE Transactions on Ultrasonics, Ferroelectrics and Frequency Control, 32*, 257–264.

Viola, F., & Walker, W. F. (2003). A comparison of the performance of time-delay estimators in medical ultrasound. *IEEE Transactions on Ultrasonics, Ferroelectrics and Frequency Control, 50*, 392–401.

Wagner, R. F., Smith, S. W., Sandrik, J. M., & Lopez, H. (1983). Statistics of speckle in ultrasound B-scans. *IEEE Transactions on Sonics and Ultrasonics, 30*, 156–163.

Walker, W. F., & Trahey, G. E. (1995). The application of K-space in medical ultrasound. *Proceedings of the IEEE ultrasonics symposium* (Vol. 2) (pp. 1379–1383).

Wells, P. N. T., & Halliwell, M. (1981). Speckle in ultrasonic imaging. *Ultrasonics, 19*, 225–229.

Weng, L., & Reid, J. M. (1991). Ultrasound speckle analysis based on the K distribution. *Journal of the Acoustical Society of America, 89*, 2992–2995.

Yeung, F., Levinson, S. F., Fu, D. S., & Parker, K. J. (1998). Feature-adaptive motion tracking of ultrasound image sequences using a deformable mesh.. *IEEE Transactions on Medical Imaging*, *17*, 945–956.

Yeung, F., Levinson, S. F., & Parker, K. J. (1998). Multilevel and motion model-based ultrasonic speckle tracking algorithms. *Ultrasound in Medicine and Biology*, *24*, 427–441.

Zhu, Y., Chaturvedi, P., & Insana, M. F (1999). Strain imaging with a deformable mesh. *Ultrasonic Imaging*, *21*, 127–146.

Zhu, Y., & Hall, T. J. (2002). A modified block matching method for real-time freehand strain imaging. *Ultrasonic Imaging*, *24*, 100–108.

Bibliography

Angelsen, B. A. J. (2000). *Ultrasound imaging: Waves, signals, and signal processing*. Norway: Emantec.

An excellent and detailed mathematical explanation of scattering and many aspects of ultrasound imaging.

Bamber, J. C., & Dickinson, R. J. (1980). Ultrasonic B-scanning: A computer simulation. *Physics in Medicine and Biology*, *25*, 463.

A review of speckle and speckle reduction methods.

Faran, J. J., Jr. (1951). Sound scattering by solid cylinders and spheres. *Journal of the Acoustical Society of America*, *23*, 405–418.

Provides more information on scattering.

Kino, G. S. (1987). *Acoustic waves: Devices, imaging, and signal processing*. Englewood Cliffs, NJ: Prentice-Hall. pp. 300–357.

A resource for more information on scattering transducers and imaging.

Morse, P. M., & Ingard, K. U. (1968). *Theoretical acoustics*. Princeton, NJ: Princeton University Press. pp. 400–411.

Provides more rigorous derivations for acoustic and elastic scattering.

Thijssen, J. M. (2000). *Ultrasonic tissue characterization, Acoustical Imaging* (Vol. 25, pp. 9–25).). New York: Plenum Press.

A comprehensive review of speckle and texture analysis.

von Ramm, O., & Smith, S. W. (1983). Beam steering with linear arrays. *IEEE Transactions on Biomedical Engineering*, *30*, 438–452.

A useful resource for arrays related to imaging.

CHAPTER 9

Scattering From Tissue and Tissue Characterization

Chapter Outline

9.1 Introduction

Until now, tissues have been treated as homogeneous elastic media with acoustic characteristics such as impedance, speed of sound, absorption and dispersion, and scattering. Tables of values of tissue characteristics are summarized in Appendix B, and many more can be found in Duck (1990). It is fair to say that at this point in time, there is much more that researchers do not know about the acoustic and elastic properties of tissues. This state of knowledge is in part due to simplified views of tissues as either uniform or random in structure and the lack of a consistent, widely based, systematically executed broadband study of elastic tissue properties and structure.

Tissue is much more interesting than that! Living tissue is full of structure, movement, and organization on several length scales. It is nonlinear both elastically and dynamically. Tissue is continually adapting and self-regulating, growing and reproducing, becoming diseased, healing and repairing, altering metabolism, and interacting with other organs.

This chapter addresses some of the problems posed by the complexity of tissue. First, it reviews a method for classifying scattering from tissues on several length scales. Second, it presents actual measurements of heterogeneous tissue structure, as well as their impact on scattering. Third, this chapter examines recent developments in tissue characterization (the science of inferring tissue properties from ultrasound measurements), including dynamic as well as static methods. Fourth, it discusses adaptive means of measuring tissue characteristics and compensating for their undesirable effects, such as aberration. Finally, this chapter gives ways of simulating wave propagation in more realistic tissues.

9.2 Scattering from Tissues

The similarity of the acoustic properties of tissues is primarily due to their high water content. The main constituents of the body are water (60%), protein (17%), and lipids (15%) (Greenleaf & Sehgal, 1992). In addition to blood, cells are bathed in fluid (interstitial) and have fluid within them (intracellular), as well as minerals and ions. Groups of similar cells (the basic building blocks) are organized into tissues. Different types of tissues are combined to perform specific functions as an organ, such as the heart or liver. Greenleaf and Sehgal (1992) provide more details of how tissues function and maintain homeostasis (equilibrium) in response to changing external factors (injury and disease).

They have also proposed a classification scheme for tissue scattering that is quite useful. In their terminology, class 0 scattering is associated with molecular solvent effects on a length scale of 10^4 Å (10^{-10} m). This type of scattering is due to macromolecular effects, which produce absorption and sound speed dispersion (discussed in Chapter 4). Class 1 scattering is caused by the concentrations of living cells being higher than 25 per resolution cell, and it is diffusive according to its length scale, $ka << 1$. Class 2, which is diffractive on a length scale, is scattering from the structure of tissue in concentrations lower than one per resolution cell. While class 1 scatterers would result in speckle or measurable aggregate (combined) effects, class 2 scatterers are independent and distinguishable through their unique space- and frequency-dependent characteristics. Class 3 scattering is specular on a length scale $ka >> 1$, and is associated with organ and vessel boundaries. A fifth category, class 4, applies to tissue in motion, such as blood.

In a typical ultrasound image of the liver, as given by Figure 9.1, are examples of several scattering types. From the need to compensate for absorption, as indicated by the time gain compensation (TGC) profile not shown but discussed (Section 4.5), we conclude that class 0 scatterers are present. The speckle indicates class 1 scatterers ($ka << 1$). Small vessels correspond to class 2 scatterers. Finally, the liver boundaries are class 3 scatterers.

Figure 9.2 introduces frequently used terms for tissue. Tissue regions that can be represented by one value of a parameter at every spatial point are "homogeneous." Global values of parameters are assigned to each region that is homogeneous. Waves crossing region boundaries may experience reflection and transmission effects, possible mode conversion, refraction, and changes in sound speed and absorption, according to the appropriate length scales. For our purposes, the term "inhomogeneous" is used for tissues that are predominantly the same type with small fluctuations about a mean value (as depicted in Figure 9.2B). In Figure 9.2C, a region enclosing a group of contiguous regions with different characteristics is called "heterogeneous." In this case, the tissue

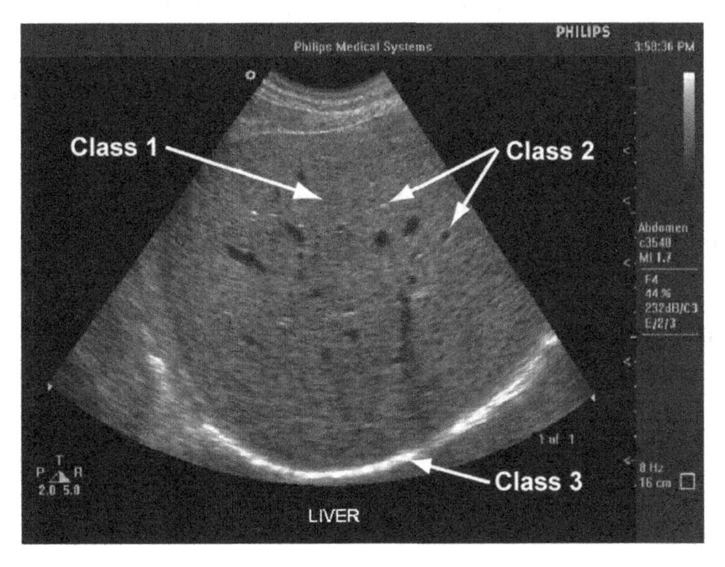

Figure 9.1

Ultrasound image of a liver showing four types of scattering effects. *Image courtesy of Philips Healthcare.*

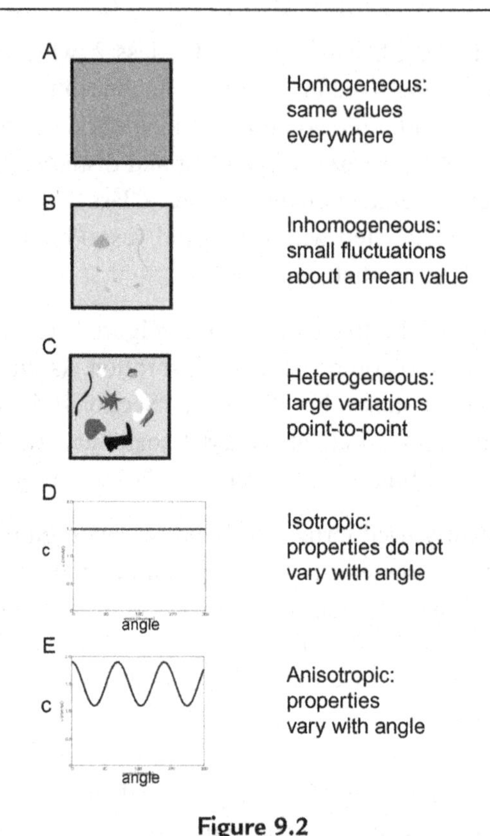

Figure 9.2
General terminology for tissue structures. (A) Homogeneous; (B) inhomogeneous; (C) heterogeneous; (D) isotropic; (E) anisotropic.

properties of the enclosed region vary with spatial position either through smaller subregions or, in the limit, from point to point. The term "isotropic" applies to tissue properties that do not vary with angular orientation (as in Figure 9.2D). If the properties do vary with the angle of insonification, then they are "anisotropic" (as in Figure 9.2E). Anisotropy occurs when tissue structure has a preferential structural orientation, such as muscular fibers.

It is helpful to know the general properties of homogeneous tissue first. The more important ones are listed in Table B.2 in Appendix B. In this table, both the absorption and backscattering loss (at 180°) components of attenuation are listed. Recall that attenuation has both absorption and scattering components:

$$\alpha_T = \alpha + \alpha_s. \tag{9.1}$$

Absorption is considered to be propagating energy that is converted to heat through mechanisms such as viscous and thermal conduction effects (as discussed in Section 4.5).

These characteristics would fall under the class 0 scattering and be assigned to homogeneous tissue regions. The scattering loss is caused by the partial interception by the receiving transducer of the angular distribution of backscattered energy. Another way of interpreting this scattering is that it occurs on several length scales simultaneously owing to tissue structure. In addition to absorption (class 0), this type of scattering may include class 1 or diffusive subwavelength scattering. Typically, scattering loss is less than 20% of the overall attenuation for soft tissue, but there are exceptions (Nassiri & Hill, 1986a).

A way to determine how much energy is scattered away from the receiving transducer is to measure scattering as a function of angle. This approach can also aid in separating the contributions of changes in the density (dipole directivity) from those of the elastic constant (monopole directivity), each of which has a different angular scattering dependence, as was indicated by the Born approximation in Section 8.2.3. Nassiri and Hill (1986a, 1986b) have developed inhomogeneous statistical models for tissue based on the Born approximation and have obtained good experimental agreement. In this case, there are many effective scatterers per resolution cell, so that class 1 scattering, in a statistical way, results in different scattering patterns. In addition, all tissue has loss that increases with frequency; therefore, all tissue includes class 0 scattering.

Another class 1 scatterer is blood. Blood cells have been modeled as subwavelength-sized spheres, cylinders, or disks (explained briefly in Section 8.2.2 and treated in more detail in Section 11.3). Flowing blood also falls in the class 4 category.

An example of a class 2 scatterer is an isolated microcalcification in the breast. Anderson, Soo, and Trahey (1998) used a spherical scatterer as a model for this case. Their calculations for elastic and inelastic spheres are shown in Figure 9.3 for an approximately Gaussian pulse with a 60% fractional bandwidth. By comparing these results with data (Figure 9.4), they concluded that microcalcifications behaved as elastic scatterers. This type of scatterer corresponds to a one-per-resolution cell, or a class 2.

9.3 Properties of and Propagation in Heterogeneous Tissue

9.3.1 Properties of Heterogeneous Tissue

Limited information is available on tissue scattering properties, in part because of the difficulty of making the measurements. Part of the problem in measuring tissue properties is the correction for system effects such as transducer spectral characteristics and directivity. In addition, scattering itself, in the frequency range commonly employed for medical imaging, is frequency dependent. The inferring of tissue properties from transducer measurements is an ongoing worldwide effort called "tissue characterization," which is described more fully in Section 9.5.

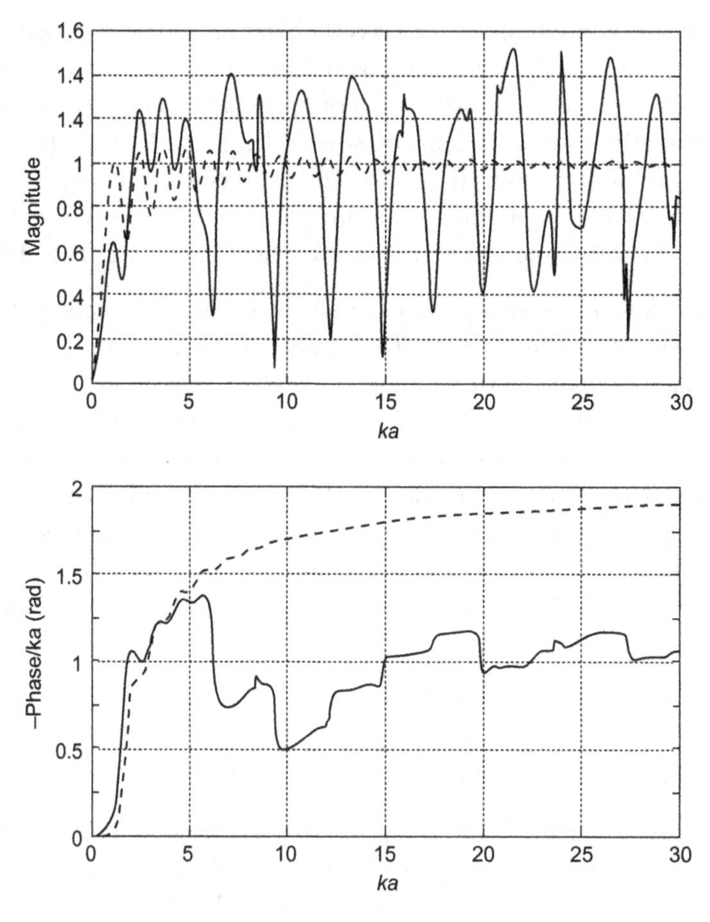

Figure 9.3

Scattering calculations for elastic and inelastic spheres (Top) magnitude of far-field scattering from a hydroxyapatite sphere simulated by a Faran model as an elastic sphere (solid line) and as a rigid sphere (dashed line); (bottom) corresponding phases scaled by $-1/ka$ to allow comparison with Hickling's results. *From Anderson et al. (1998), IEEE.*

Before methods of tissue characterization are described, some of the issues and complexities of acoustic propagation in real tissue need to be examined. A number of early studies are summarized by Li (1997) and Bamber (1998). Because most ultrasound imaging is done through abdominal and chest walls, these have been studied extensively at the University of Rochester. Measurement-corrected data of the acoustic properties of abdominal and chest walls provide more direct information about the local spatial fluctuations of tissue properties.

The measurement of wall properties consisted of placing the wall specimen in a sealed chamber that had acoustically transparent windows and was filled with water. This chamber was aligned in a water tank between a diverging transmitter–transducer with a wide

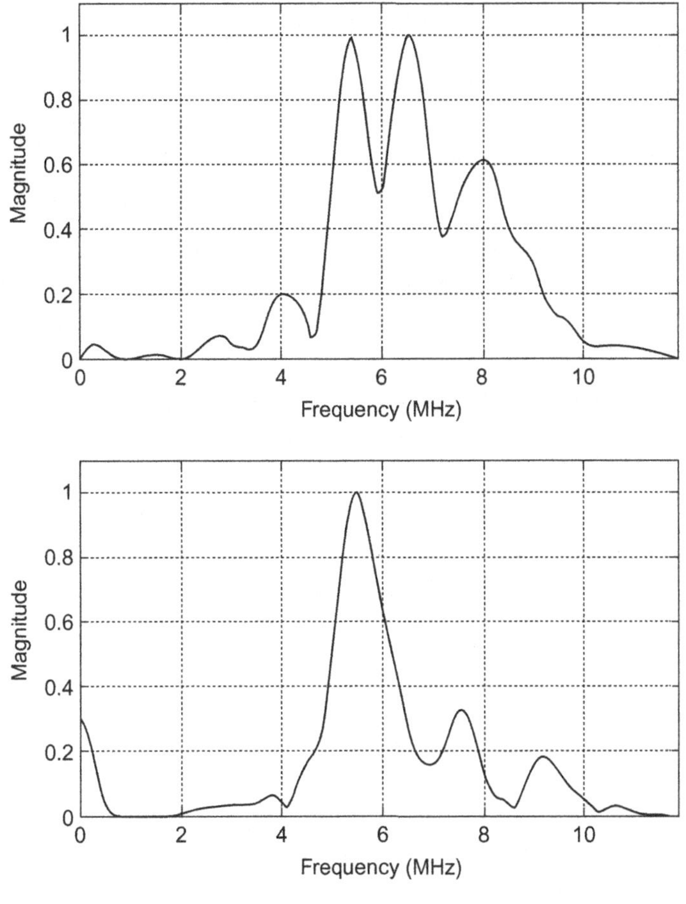

Figure 9.4
The magnitude spectra of radio frequency echoes from suspected in vivo microcalcifications in two different subjects. The similarity of the prominent peaks and nulls in the spectra to those of the elastic simulation of Figure 9.3 indicates elastic behavior. Higher absorption in the second case (bottom) reduces amplitudes at higher frequencies. *From Anderson et al. (1998), IEEE.*

directivity and a mechanically scanned linear array acting as a receiver on the upper side of the specimen. For the chest walls (Hinkelman, Szabo, & Waag, 1997), a 96-element linear array designed to have small elements, 0.21×0.4 mm, was electronically switched, onc element at a time, at each of 50 lateral positions to create a matrix of recorded-through transmission pulses, equal to $96 \times 50 = 4800$ positions. A reference database was also recorded with no specimen in the tank. The effects of measurement geometry were removed by fitting a surface to the arrival-time wavefronts and subtracting the fit from the data. The following features of the data were tabulated: waveform similarity, arrival-time and energy-level fluctuations, and full width half maximum (FWHM) correlation lengths.

An average reference pulse was correlated with each pulse in a set to obtain a measure of waveform similarity, time-delay fluctuations, and correlation lengths. Energy waveform values were calculated by integrating the squared amplitudes of the received signals. The resulting chest wall data are shown in Figure 9.5 as pictures of the two-dimensional spatial fluctuations of 16 samples, presented as pairs depicting time-delay and energy fluctuations. The top row is a water reference. Note the similarity of features in most pairs.

Measurements for abdominal walls are compared to this data in Figure 9.6. The average root mean square (r.m.s.) time-delay fluctuations for 16 chest wall samples is 21 ns, as compared with 56 ns for abdominal walls. Just as speckle has been analyzed statistically, radio frequency (RF) data can be interpreted that way also. Correlation lengths, for example, can reveal both local and statistical information about the spacing between scatterers. Recall that in a convolution operation, the integrand consists of one function flipped from right to left and multiplied by another function (Appendix A); but in correlation there is no flipping involved, so that it is expressed as $r(t) = x(t) * y(-t)$. FWHM correlation lengths for chest walls and abdominal walls, for example, were 2.5 mm and 5.8 mm. The implications of these localized time domain delay differences are discussed in Section 9.3.2.

9.3.2 Propagation in Heterogeneous Tissue

The effect of tissue structure on wave propagation was also studied by the University of Rochester group. Hinkelman, Metlay, Churukian, and Waag (1996) developed a method for staining a cross-section of an abdominal wall to identify tissue types. The stained wall cross-sections were digitized and assigned acoustic values appropriate to each tissue type, as depicted by Figure 9.7. Mast, Hinkelman, Orr, and Waag (1998), Mast, Hinkelman, Metlay, Orr, and Waag (1999), and Hinkelman, Mast, Metlay, and Waag (1998) used these data to model plane wave propagation with a finite-difference time domain program through an abdominal wall, as depicted for sequential times in Figure 9.8.

Multiple scattering, refraction, and aberration are responsible for the complicated exiting wavefront. Smaller scatterers provide a low level of reverberation that trails after the larger, directly transmitted main wavefront; these signals contribute to image clutter. Simulations of focused beams propagating through abdominal walls, based on these data under linear and nonlinear conditions, are examined in Section 12.5.5. A calculated, focused wavefront from an array propagating through the breast is illustrated by Figure 9.9. Note the nearly circular wavefronts scattered off the tips of vertical septa, which appear to be the main disrupters of the plane wavefront. A sequential simulation of a plane wave propagating between ribs in a chest wall (shown in Figure 9.10) reveals multiple scattering from ribs that was observed in data sets (Hinkelman et al., 1997; Mast et al., 1999).

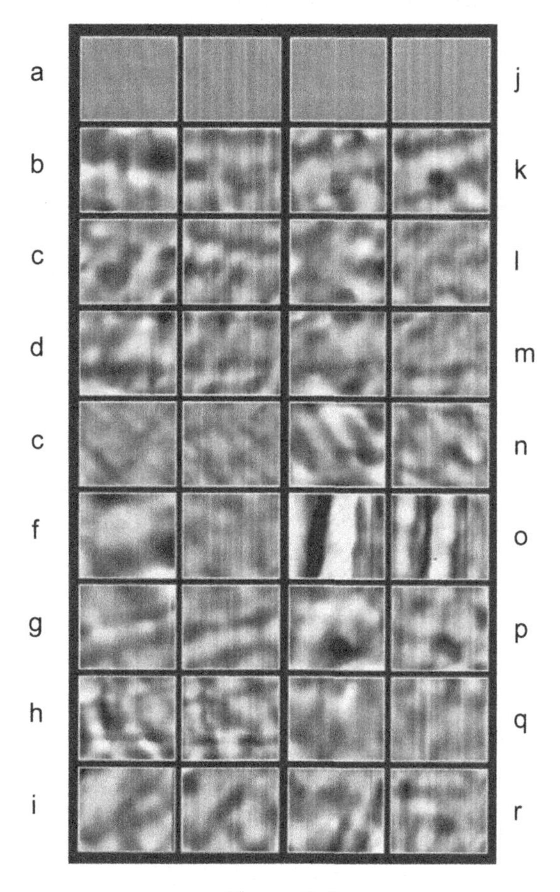

Figure 9.5

Arrival-time and energy-level fluctuations for two water paths and intercostal spaces of 16 chest wall samples. The first and third columns are arrival time panels; the second and fourth columns depict energy-level fluctuations. The top row shows two water measurements for reference. In the left panel of each pair, arrival time difference is shown on a linear scale with a maximum arrival-time fluctuation of 150 ns represented by white, and a minimum arrival-time fluctuation of 250 ns represented by black. In the right panel of each pair, energy-level fluctuations are shown on a logarithmic scale with a maximum positive excursion of 15 dB represented by white, and a maximum negative excursion of 25 dB represented by black. In all panels, the horizontal coordinate is the array direction and spans a distance of 14.28 mm in 0.21-mm increments, while the vertical coordinate corresponds to position of the array in elevation and spans a distance of 11.60 mm with points interpolated from measurements at 0.40-mm intervals to produce data at 0.20-mm increments. *From Hinkelman et al. (1997), Acoustical Society of America.*

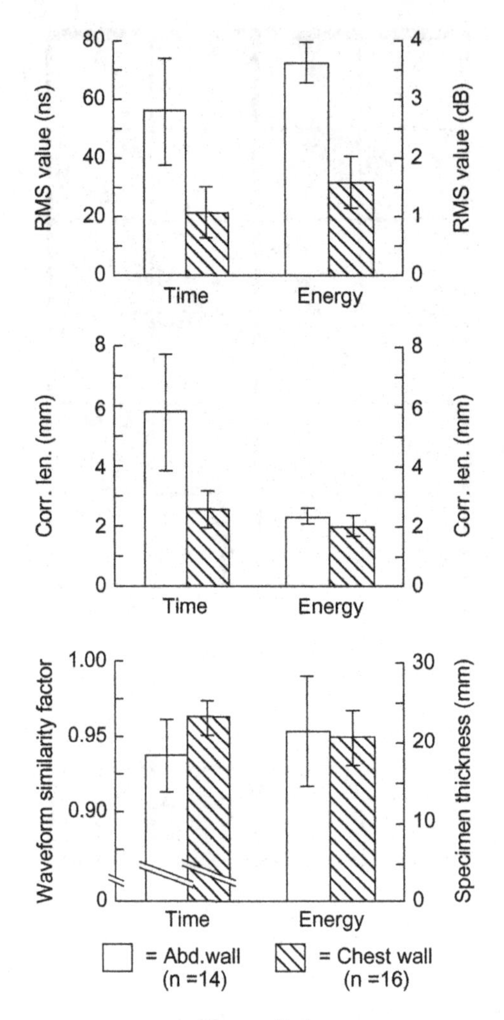

Figure 9.6
Comparison of chest wall and abdominal wall wavefront distortion statistics.
In each chart, the average and standard deviation of the measurements within each group are shown. *From Hinkelman et al. (1997), Acoustical Society of America*

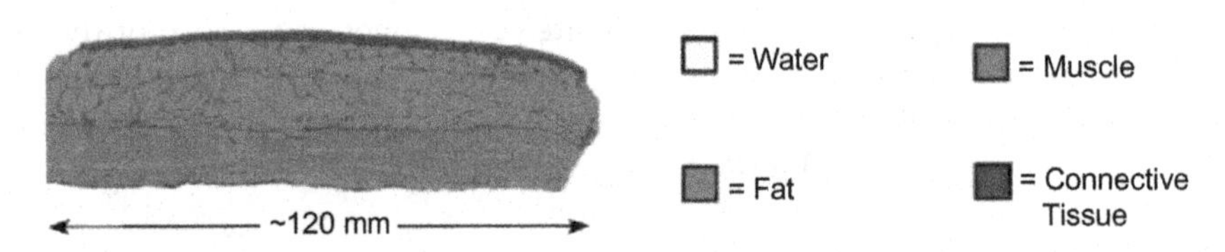

Figure 9.7
Cross-sectional tissue map of an abdominal wall with assigned acoustic properties.
The reader is referred to the Web version of this book to see the figure in color. *From Mast, Hinkelman, Orr, Sparrow, and Waag (1997), Acoustical Society of America.*

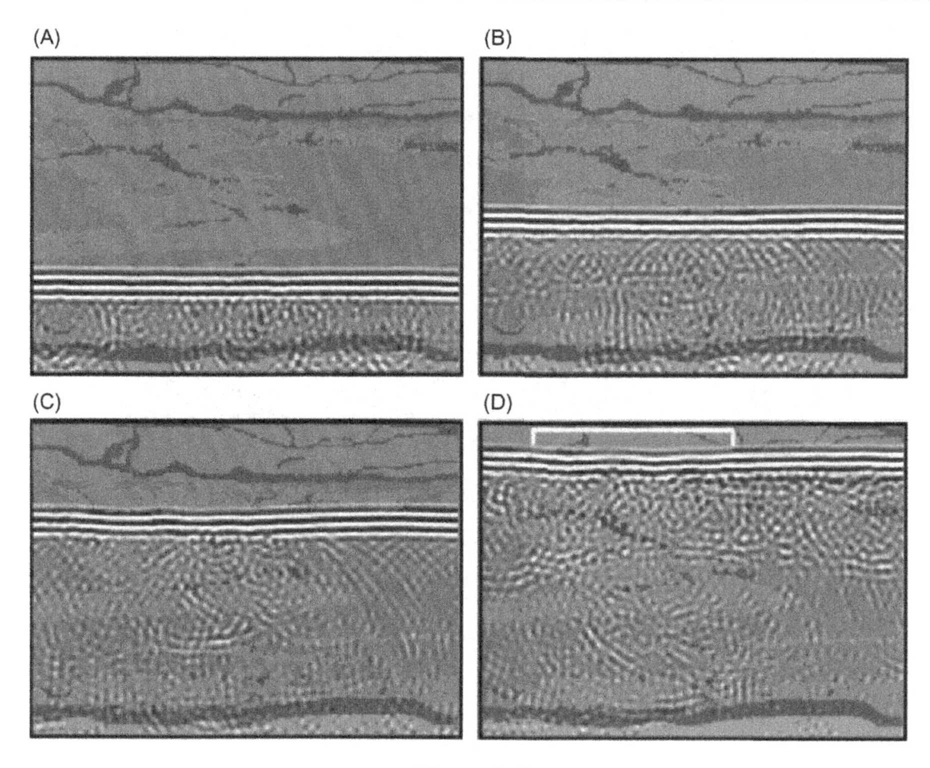

Figure 9.8

Propagation of a plane wave through a section of the abdominal wall sample depicted in Figure 9.7. (A–D) Upward progression of the main wavefront through the muscle layer, including an aponeurosis comprised of fat and connective tissue, resulting in time-shift aberration across the wavefront. The area shown in each frame is 16.0 mm in height and 18.7 mm in width. The temporal interval between frames is 1.7 ms. Tissue is color coded according to the scheme of Figure 9.7, while gray background represents water. Wavefronts are shown on a bipolar logarithmic scale with a 30-dB dynamic range. The wavefront represents a 3.75-MHz tone burst with white representing maximum positive pressure and black representing maximum negative pressure. A cumulative delay of about 0.2 ms, associated with propagation through the aponeurosis, is indicated by the square bracket in panel (D). The reader is referred to the Web version of this book to see the figure in color. *From Mast et al. (1997), Acoustical Society of America.*

Scattering from real tissue includes three contributions. These simulations show only transmitted waves; however, they demonstrate that the main direct wavefronts are followed by a complex pattern of waves diffracted from tissue structures. The calculations provide a more realistic depiction of wave propagation (even though they are planar) than previous treatments of homogeneous tissue layers; however, they do not include the granularity necessary to include the third contribution, which is speckle. In summary, propagation in

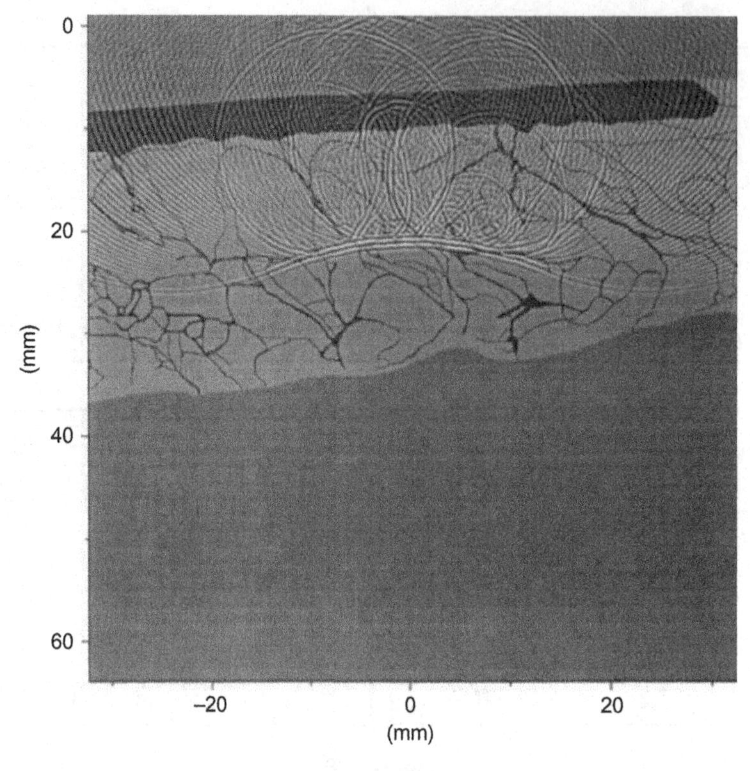

Figure 9.9
A converging focused pulse wavefront from a virtual array and secondary scattered wavefronts at an instant of time during propagation through a representative breast tissue map. The wavefront is superimposed on the map and displayed on a 60-dB bipolar logarithmic gray scale. In the map, dark gray denotes connective tissue or skin and light gray denotes fat. *From Tabei, Mast, and Waag (2003), Acoustical Society of America.*

tissue has three parts: main wavefronts, lower-level waves from diffracting structures, and speckle (diffusive waves from subwavelength tissue structures).

9.4 Array Processing of Scattered Pulse–Echo Signals

If complicated scattering from heterogeneous tissue occurs, what is its effect on imaging? In Section 8.4.1, the steps of the imaging process were identified. First, a three-dimensional pulse packet from a focused beam races through tissue along a designated vector direction, and it changes shape at each depth according to its point-spread function. Second, multiple scattering of the pulse happens at the depth location of each scatterer (z_i/c) over a broad angular range. Third, parts of the scattered wavefronts are intercepted by each element acting as a receiver with wide directivity at times ($2z_i/c$). Each element has a distinct spatial

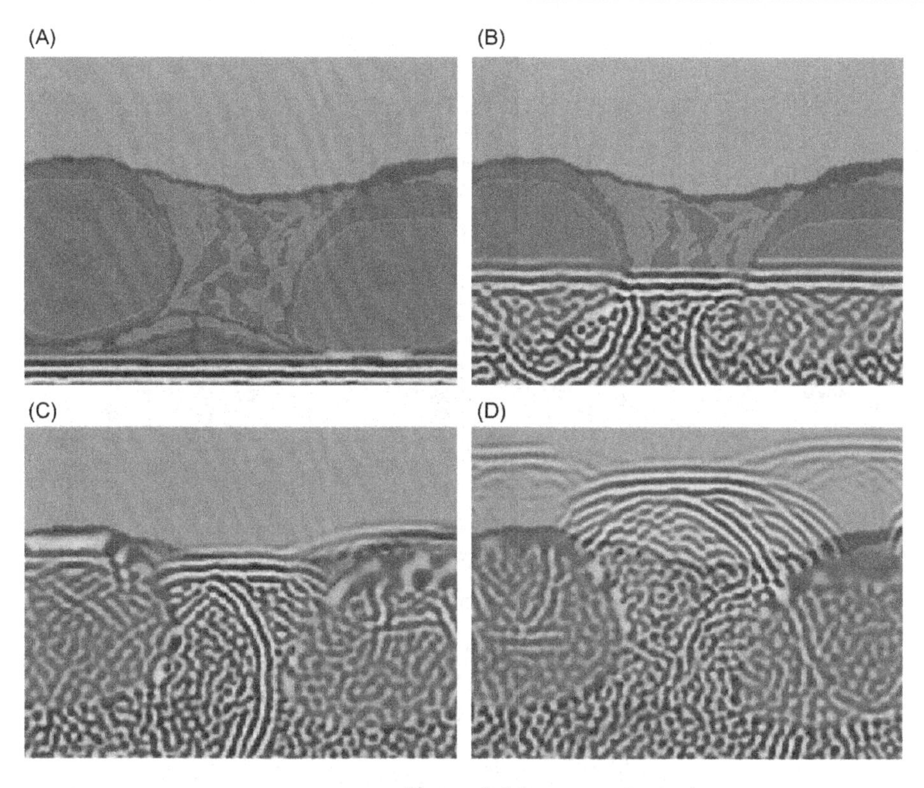

Figure 9.10

Simulation of 2.3-MHz plane wave tone burst wavefront propagating through a chest tissue map. In each map, blue denotes skin and connective tissue, cyan denotes fat, purple denotes muscle, orange denotes bone, and green denotes cartilage. Blood vessels appear as small water-filled (white) regions. Logarithmically compressed wavefronts are shown on a bipolar scale with black representing minimum pressure, white representing maximum pressure, and a dynamic range of 57 dB. Each panel shows an area that spans 28.27 mm horizontally and 21.20 mm vertically. The reader is referred to the Web version of the book to see the figure in color. *From Mast et al. (1999), Acoustical Society of America.*

location and, therefore, intercepts different parts of scattered wavefronts and converts them via the piezoelectric effect into an electrical time record (volts).

We now have a look at a hidden aspect of imaging—what happens inside the beamformer after the three steps in the imaging process described earlier—when real data are received from a heart (Szabo & Burns, 1997). For the formation of an image line, array elements receive a time-stream of data. In this example, 54 time records from active elements of a 2.5-MHz, 64-element phased array are shown in Figure 9.11 for an apical view of a heart. These time traces were captured in real time in the process of forming a cardiac sector image and are shown after focusing time alignment. Events between time samples 3200 and

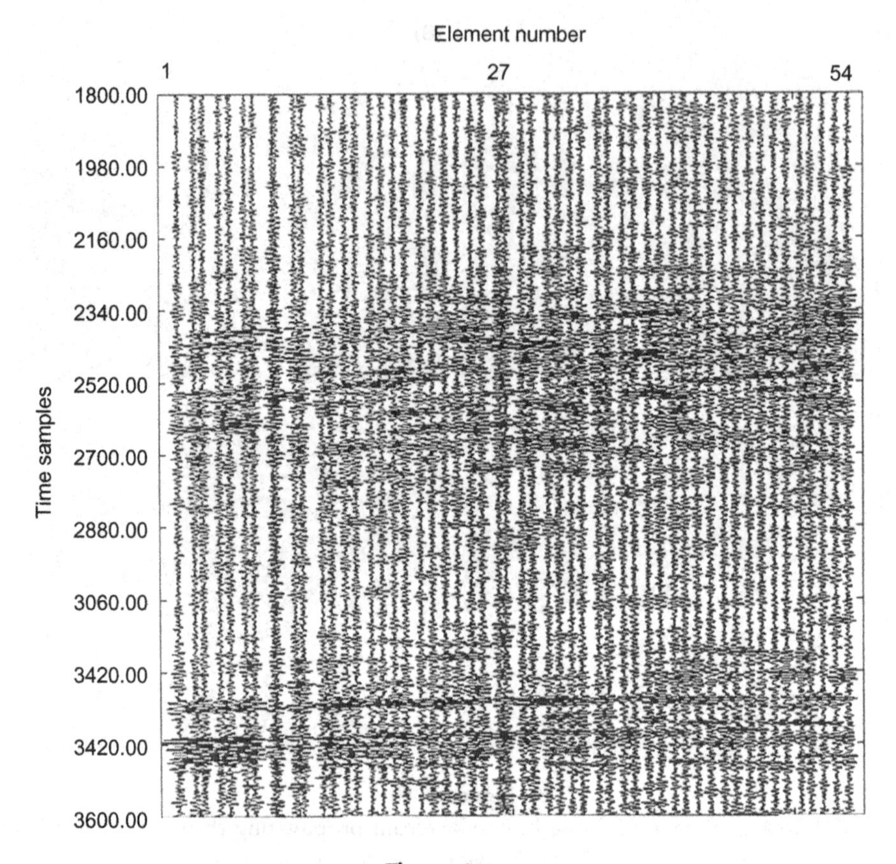

Figure 9.11
Raw RF data from line 72 (+8.25 Degree Steer Direction) of an apical window data set. The data are shown for 54 traces of a 64-element transducer (traces 6–59) and for the time sample range of 1800–3600, or 90 μs. The horizontal events between time samples 3200 and 3500 correspond to specular tissue reflections, while the events between 2200 and 2700 correspond to a lung artifact. The data have been normalized to the maximum value in the plot (maximum = 65). *From Szabo and Burns (1997), reprinted with permission of Kluwer Academic/Plenum Publishers.*

3500 are caused by specular tissue reflections, and those events between 2200 and 2700 belong to a lung artifact. An ideal reflector will be aligned across all traces and will form a horizontal wavefront. The wavefront at time sample 3300 comes close to this ideal, but it neither extends across all elements nor lines up completely across traces. Irregularities in terms of depth (vertical) time alignment along a wavefront are caused by aberration effects. What is surprising are the smaller signals, many of which are aligned across several traces; therefore, they cannot be random. These groups of similar waveforms, extending across a few traces and combining to form small wavefronts not aligned horizontally, indicate multiple scatters illuminated at various angles of incidence and reflecting in other angles according to their scattering shape (similar to those discussed in the last section).

What happens to all this information? The beamformer sums all these traces from elements into a final single time record, which, after envelope detection, becomes a line in the image. This kind of beamformer operates on the principle of coherence (explained in Chapter 7). After focusing delay alignment, each time record is summed. Coherent waveforms, aligned in time, form large signals. Many of the smaller waveforms, which are coherent only across a few traces when summed with adjacent nonaligned signals in the same time slot, are suppressed. This summation processing shows the beauty and simplicity of the linear beamformer; coherent signals add, and noncoherent signals are rejected and appear as random, low-level clutter and speckle. Several research imaging systems have the capability for RF capture from elements as described in Chapter 10.

9.5 Tissue Characterization Methods

9.5.1 Introduction

The science of ultrasonic tissue characterization (UTC) is the untangling of hidden patterns in pulse–echo data to extract more information about tissue function, structure, and pathology than that seen in conventional images (Thijssen, 2000). Tissue characterization, as was originally conceived by Dr J. J. Wild and J. M. Reid in 1952 (see Section 1.2), was a type of remote, painless ultrasound telehistology (a noninvasive way of determining the health of tissue or organ function through calculations and parameterized inferences from ultrasound data). The application of tissue characterization to detection of cirrhosis in the liver by Wells, McCarthy, Ross, and Read (1969) and Mountford and Wells (1972) marked a turning point in the serious application of these methods. Hill and Chivers (1972) contributed to a scientific examination of scattering from tissue with the hope of retrieving quantitative data. Since then, this branch of medical ultrasound has undergone considerable development. Methods now include specialized measurements, signal processing, statistical analysis, and parameterized imaging.

In order to place these diverse UTC approaches in perspective, we can use the diagram in Figure 9.12, which serves as an introduction to the remaining contents of this chapter. The topics included here stretch the conventional use of the term "tissue characterization" but fit the definitions given previously. Most of UTC is based on RF signals either from beamformers prior to detection in imaging systems or from specialized measurements. The methods based exclusively on processing of RF data are the basic spectral method, spectral features, integrated backscatter, and signal processing. Even conventional Doppler (see Chapter 11) can be considered to be a signal processing type of UTC because it calculates parameters based on RF pulse–echo data from flowing blood (Taylor & Wells, 1989). A number of other signal processing methods, such as color flow imaging (which determines blood flow velocity) and others that detect

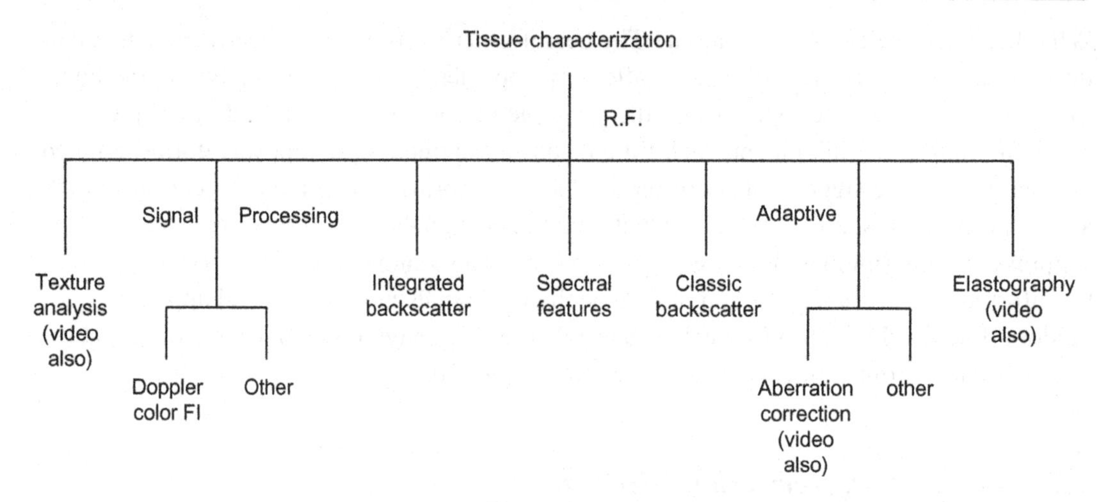

Figure 9.12
Diagram of ultrasound tissue characterization applications. Color FI, color flow imaging.

tissue movement and provide calculations and/or parameterized displays, will be described in Chapter 10 and Chapter 11.

Several approaches can utilize video or RF data, the main one being texture analysis. Elastography, a collection of ways of imaging the stiffness of tissue, can use either video or RF data. On the far right of Figure 9.12 are adaptive signal processing methods, such as aberration correction, that not only sense differences in tissue parameters but also alter the function of the imaging system in response to these sensed changes. Aside from the signal processing methods covered in the next two chapters, the topics in the diagram will be described in the following sections and elastography will be described in Chapter 16.

The breadth of UTC science can only be highlighted briefly in this chapter. Whole books (Greenleaf & Sehgal, 1992; Shung & Thieme, 1993) have been written on these topics, and regular conferences on these subjects, such as the annual Tissue Characterization Symposia organized by Melvin Linzer, have been held since 1976. From the last section, it is obvious that backscatter from real tissue is messy, which is why this field presents interesting challenges. Nonetheless, if we expand the concept of UTC to encompass the areas in Figure 9.12, there are a number of success stories and exciting developments for the future. Section 9.6 presents several UTC applications.

The usual starting point of UTC analysis is acquisition of raw RF signals because, as will be evident in the next chapter, the imaging systems add nonlinear processing on top of envelope detection to make images presentable over a large dynamic range. It is also possible to obtain tissue information from the video information in the image itself after removing this nonlinear processing, as discussed later.

9.5.2 Fundamentals

The foundations for UTC were explained in Chapter 8. A major goal in UTC is inverting or revealing the properties of the tissue through a backscatterer or another measured ultrasound parameter. A second, but even more important, goal is to use this acquired information to distinguish between states of tissue (healthy or diseased) or to detect changes in tissue property in response to a stimulus or, over longer periods of time, in response to natural processes or medication. A key problem in the untangling is undoing or correcting the spatially varying properties of the measurement or imaging process. For example, a homogeneous tissue would appear to be inhomogeneous in an image simply from the variations in pressure from diffraction and absorption effects along the beam axes. These problems will be clarified soon. UTC approaches range from "proof of concept" experiments involving single lines of RF data from tissue samples in vitro (outside the body) to those involving data acquisition and processing from a system imaging a body in real time. A primary motivation is to reach a measure of diagnostic objective truth that is independent of the measuring system employed. This aspect of tissue characterization is now called "quantitative ultrasound" or QUS. The display of a derived quantitative parameter is called "parameterized imaging."

Since many of the physical aspects of these problems overlap, it is easier to start with the simplest approach. A framework for understanding the factors involved is essential (the central block diagram in Figure 2.14 is helpful for identifying factors symbolizing the physical mechanisms involved). These factors can be consolidated into an expanded version of Eqn 8.10B for the received output voltage:

$$V_0(z,r,f) = E_{RT}(f)A_t(z,r,f)H_t(z,r,f)H_r(z,r,f)A_r(z,r,f)F_G(z,f)S(z,r,f), \tag{9.2A}$$

or, in terms of power (within a constant factor):

$$V_0V_0^* = E_{RT}E_{RT}^*A_tA_t^*H_tH_t^*A_rA_t^*F_GF_G^*SS^*, \tag{9.2B}$$

where $F_G(z,f)$, a symbol for a generic filter and/or linear amplifier function, has been added. From a UTC viewpoint, the unknown tissue properties are frequency- and angle-dependent scattering and, sometimes, frequency-dependent absorption, but these characteristics are distorted by the interrogating acoustic beam and the limitations of the electroacoustic transduction process. The beam and transducer act as imperfect samplers, both spatially (including geometric orientation, as well as limited angular range) and spectrally. The scattering properties are the sought-after signatures of the targets, and the effects of the array and transduction are somehow removed by calibration and compensated for, or minimized by, a number of methods. Most UTC approaches depend on comparisons of data to a comprehensive model of backscattering from tissue, such as Eqns 9.2A and 9.2B provide. An alternative is to extract attenuation rather than to backscatter information (Ce'spedes & Ophir, 1990).

Certain objects have unique scattering signatures in the frequency domain and, as a result, most of tissue characterization is done in the frequency domain. The most often used parameter is the "backscattering coefficient." Sigelmann and Reid (1973) proposed a method for extracting a backscattering coefficient from power reflected from a volume containing scatterers such as blood with reference to the power reflected from a known flat plate. One of the first applications of backscatterer measurements was to blood by Shung, Sigelmann, and Reid (1976). A more recent twist on this theme is the use of blood as a reference scatterer to characterize nearby tissue (Pedersen, Chakareski, & Lara-Montalvo, 2003). Measurements of blood are given in more detail in Chapter 11.

9.5.3 Backscattering Definitions

Most of the tissue characterization spectral methods are based on several assumptions to simplify the factors in Eqn 9.2 (Reid, 1993). The incident wave on the scatterer is assumed to be nearly plane insofar as a small volume of the scatterer is concerned (a situation almost satisfied near the focal plane of a focusing transducer or the far field of a nonfocusing transducer). The scatterer or inhomogeneity in tissue is taken to be a small, weak spherical scatterer that can be described well by the Born approximation, Eqn 8.8A. In terms of spectra, it is convenient to work with scattered power (W_s), which is related to the incident intensity (I_i) by:

$$W_s = \sigma_t I_i, \tag{9.3}$$

where σ_t is the scattering cross-section. This reasoning, for a spherical scatterer, leads to the scattered intensity:

$$I_s = \sigma_t I_i / 4\pi r^2. \tag{9.4}$$

For a backscatterer that varies with angle, a differential scattering cross-section is more appropriate, as:

$$\sigma_d I_i = W_s / 4\pi, \tag{9.5}$$

where the power is divided by 4π steradians in a unit sphere. Conversely, the total scattering cross-section is the differential scattering cross-section integrated over the solid angle about a spherical surface. In practice, the total cross-section can be the sum of individual cross-sections or may be integrated over a distribution of scatterers. Finally, there is the scattering cross-section per unit volume, defined by:

$$\sigma = \int_v \eta dv, \quad \text{and} \tag{9.6A}$$

$$\sigma_d = \int_v \eta_d dv, \tag{9.6B}$$

where η has units of inverse length, compared to a cross-section that has units of area.

9.5.4 The Classic Formulation

As discussed in Chapter 8, the scattering process involves the entire volume of tissue intercepted by the acoustic beam; therefore, a volume integration over this volume is involved for transmit. Likewise, the location of the receive beam relative to the scatterer (if it is not coincident with the transmit direction) is involved over the intercepted scattered volume; therefore, in general, two volume integrations are needed. Finally, a calibration method is convenient to obtain a more absolute measurement. In the original approach, a substitution method was applied to the problem. A nearly ideal plane reflector was placed in the position of the scattering volume under identical insonification conditions, and the voltage was recorded. The power ratio of scattered power to reference-reflected power gives an expression for mean value of the backscatter coefficient η_d (Reid, 1993) in an abbreviated form:

$$\overline{\eta}_d(f) = \frac{\iint E_{RT}E_{RT}^*A_tA_t^*H_tH_t^*A_rA_r^*F_GF_G^*SS^*(dr_0')^3(dr_0)^3}{\iint E_{RT}E_{RT}^*A_tA_t^*H_tH_t^*A_rA_r^*F_GF_G^*(dr_0')^3(dr_0)^3} = \frac{\langle V_0(f)V_0^*(f)\rangle \overline{r}^2}{V_{ref}(f)V_{ref}^*(f)4AS_{eff}l}, \tag{9.7}$$

where A_r and A_t are attenuation factors appropriate for the different sample and reference paths, A is an overall attenuation correction factor, S_{eff} is an effective beam cross-sectional area, integration is over pulse length, $l = c(t_2 - t_1)/2$, and the intersection volume of the coincident beams, and $\overline{r}$ is the distance to the center of the sample.

In the original formulation, simplifications were made about the far field and beamwidth of the transducer and gate length of the received pulse, and after correction for the absorption of the sample and the water path, this ratio was shown to be proportional to the squared magnitude of the output voltage from the scattered region to the squared magnitude of the reference voltage. This narrowband approach was subsequently endorsed by the American Institute of Ultrasound in Medicine (1990) and led to further extensions of the original method.

9.5.5 Extensions of the Original Backscatter Methodology

Work at the University of Wisconsin (Madsen, Insana, & Zagzebski, 1984) provided a more general model for broadband applications and focusing transducers. Further work there, based on statistical continuum models (Chen & Zagzebski, 1996; Chen, Zagzebski, & Madsen, 1993), has shown that the backscatter increases with the number density of scatterers per volume fraction as a function of frequency. Waag and Astheimer (1993) developed a generalized backscattering approach for noncoincident transmitters and receivers. A way of extending the methodology to broadband applications (O'Donnell, Bauwens, Mimbs, & Miller, 1979; Thomas et al., 1989), called the integrated backscatter approach, will be explained separately.

A more recent evolution of the original method, illustrated in Figure 9.13, is a general broadband approach to finding the backscatter coefficient (Chen, Phillips, Schwarz,

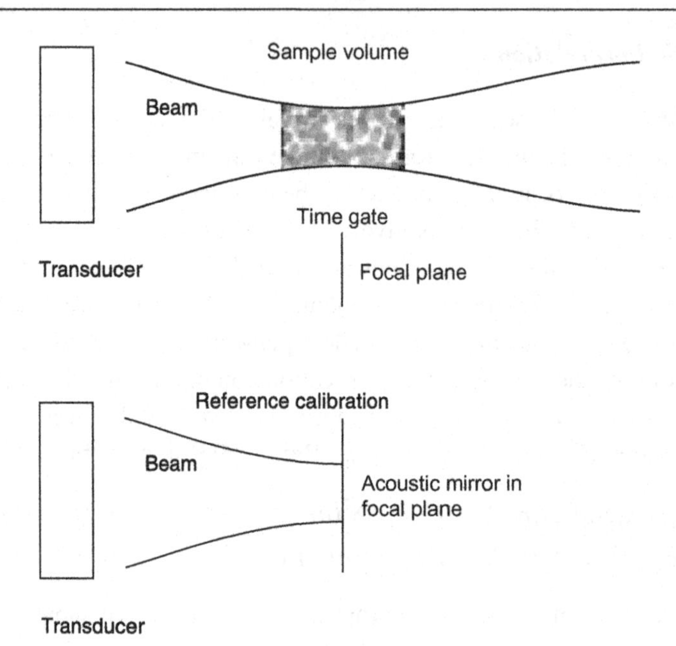

Figure 9.13
Configuration for classic backscatter measurement using the substitution method. (Top) tissue measurement setup shown for a focusing transducer; (bottom) calibration setup with acoustic mirror.

Mottley, & Parker, 1997) that utilizes a calibration waveform from a flat plate to obtain $E_{RT}(f)$ (see Section 8.3, Eqn 8.12) and provides an analytical formulation for the entire pressure distribution in the beam as a function of frequency. Their expression is:

$$\overline{\eta}_d(f) = \frac{\langle V_0(f)V_0^*(f)\rangle}{V_{ref}(2z_{ref},f)V_{ref}^*(2z_{ref},f)}\frac{|D_{ref}(2z_{ref},f)|^2}{\overline{TF}^4 AI\overline{D}_s(\overline{r},f)}, \tag{9.8A}$$

where D_{ref} is the equivalent diffraction loss, DL_{equiv}, from Eqn 8.12; V_{ref} is the Fourier transform of round-trip output voltage $e_{RT}(t)$ at $2z_{ref}$ (explained in Section 8.3.1), A is an attenuation factor for the tissue and water paths taken outside of the diffraction integral (described in Section 7.9.4); and $\overline{TF}^4$ is the overall transmission factor from water to tissue and from tissue to water (on the return path):

$$\overline{TF}^4 = \frac{16(Z_2/Z_1)^2}{(1+Z_2/Z_1)^4}, \tag{9.8B}$$

where Z_1 is the coupling medium or water to the tissue, Z_2 is the acoustic tissue impedance, and $\overline{D}_s$ is a diffraction correction term that is known analytically or can be approximated (Chen et al., 1997). The last correction term is an improvement over other methods because it corrects for diffraction as a function of both frequency and distance in a convenient way; the tissue sample distance does not have to coincide with the reference distance. In addition, it includes the influence of the entire beam rather than a −3-dB beamwidth, as was used in the original method. This method has been found to be in excellent agreement with data, even for strongly focused transducers (Machado & Foster, 1999) after attenuation correction (D'Astous & Foster, 1986) was applied. This general method reduces to the Sigelmann–Reid technique under similar experimental conditions for nonfocusing transducers and to the approach of Madsen et al. (1984) for focusing transducers when the effect of the finite receive gate (a sliding time window, often apodized) is neglected.

9.5.6 Integrated Backscatter

Another approach that has been extremely successful for cardiac and other applications (discussed in more detail in Section 9.6.2) is integrated backscatter, which was developed at Washington University. The relative broadband integrated backscatter can be expressed as the ratio of the power spectrum averaged over the effective bandwidth of the transducer relative to that from a standard plane reflector in the focal zone of the transducer (O'Donnell et al., 1979; Thomas et al., 1989):

$$\overline{S}_r = \int_{f_0-\Delta f}^{f_0+\Delta f} \frac{|V_0(f)|^2}{|V_{ref}(f)|^2} df = \frac{\int_{-\infty}^{\infty} |v_0(t)|^2 dt}{\int_{-\infty}^{\infty} |v_{ref}(t)|^2 dt}, \tag{9.9}$$

where the signal power spectrum is $|V_0(f)|^2$, that of the reference is $|V_{ref}(f)|^2$, and Δf is half the useable bandwidth, and the power or Parseval's theorem from Appendix A has been used to relate the frequency and time domain. Note that $|v_0(t)|$ is the envelope of the output time signal. Integrated backscatter can be corrected for attenuation, depending on the application.

An interesting property of the envelope $|V_0(t)|$ that comes from the Fourier transform property is that the area in one domain is equal to the value of its transform at zero. Consider a simple example such as a Gaussian modulated sinusoidal time waveform, like those considered in Chapter 2, which is centered on delay time $t = t_{\text{delay}} = 2z/c$. If the envelope of the received signal is $e(t) = |v_0(t)|$, then the value of the envelope at $t = t_{\text{delay}}$ or equivalently, at the peak of a delayed signal, is:

$$e(t - t_{delay}) = f^{-1}[E(f)e^{-iz\pi f t_{delay}}], \tag{9.10A}$$

$$e_{peak} = e(0)|_{t=t_{delay}} = \int_{-\infty}^{\infty} E(f)e^{i2\pi f(t-t_{delay})} df|_{t=t_{delay}} = \int_{-\infty}^{\infty} E(f)df, \tag{9.10B}$$

and

$$e_{peak} \approx \int |E(f)|df, \tag{9.10C}$$

where for a useable bandwidth, we assume that $|E(f)| \sim E(f)$ and that, alternatively, a similar result for the integral of the absolute value of the spectrum could be obtained by taking the absolute value of the whole integrand of the inverse Fourier transform and by using Schwartz's inequality.

9.5.7 Spectral Features

Other useful spectral features for tissue characterization have been proposed by Lizzi, Greenbaum, Feleppa, and Elbaum (1983), Lizzi and Feleppa (1993), and Lizzi, Astor, Fellepa, Shao, and Kalisz (1997). By analyzing the frequency response of spherical scatterers of different sizes, they concluded that the slope of the backscattered spectrum, once corrected for attenuation and calibrated to a reflection from a plate (as in the other classical methods), was indicative of the size of the scatterers. Corrected data, in other words, were fit by a least-squares method, and the intercept and slope became useful spectral features. The intercept is related to the acoustic concentration (related to the scatterer volume concentration and the relative acoustic impedance). Finally, the midband slope value is also valuable and related in a statistical sense to integrated backscatter (Lizzi et al., 1997). These features are shown in Figure 9.14. Lizzi's group was also one of the first to appreciate that scattered power over small adjacent distances (Δx, etc.) was a spatial autocorrelation function in three dimensions. These autocorrelation functions are R_D (Δx, Δy), directivity; R_G (Δz), the axial time gate; and R_Q (Δz), the relative impedance. Lizzi and his group modeled soft tissue characteristics as a stochastic spatial distribution of small fluctuations of acoustic impedance Z about a mean value. They developed a Gaussian autocorrelation function along the propagation axis in terms of an effective scatterer radius (Lizzi et al., 1997). These functions could be compared to corrected measurements to determine effective scatter size in different regions. They later went on to determine the statistical distribution of different tissue parameters (Lizzi, Alam, Mikaelian, Lee, & Feleppa, 2006) including the effects of windowing (time gating), the sequence of calculations (such as logarithmic conversion), and the size of the region of interest. They identified two paths, shown in Figure 9.15: one for parameter determination and another for parameterized imaging.

An often-used technique in tissue characterization is to calculate and extract a relevant parameter, such as the midband value of the backscattering coefficient, and create a visual map of this parameter using a color map superimposed on the original B-mode image. Better discrimination is achieved by combining features (Lizzi et al., 1997; Lizzi and Feleppa, 1993). In some cases, even combined features are not enough because of overlapping regions. Feleppa et al. (2001) demonstrated that by using advanced techniques, such as neural network classification, differences between healthy and cancerous tissues in the prostate can still be determined. A combination of prostate-specific antigen (PSA) and spectral features were used to identify the likelihood of cancer occurring in different

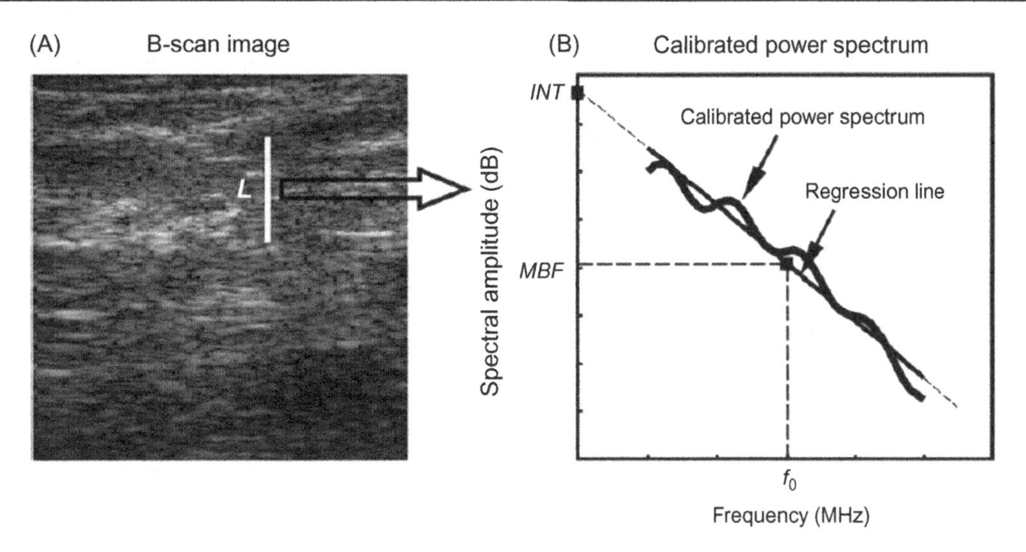

Figure 9.14

Spectrum analysis procedure. A calibrated power spectrum of windowed RF data is evaluated; a window (typically a Hamming function) of length *L* is applied before computing the spectrum. Linear regression defines a line that approximates the spectrum. In this example, *MBF* is the midband-fit value (value of the regression line at center frequency f_0) and *INT* is the spectral intercept (value of the regression line extrapolated to $f = 0$). (A) Section of B-scan image; (B) log of backscatter coefficient magnitude vs log frequency. *From Lizzi et al. (2006), with permission from the World Federation of Ultrasound in Medicine and Biology.*

regions, as shown in the parameterized images of Figure 9.16. Note that the 12-mm tumor was not visible in conventional imaging and not identified by palpitation.

W.O'Brien's group (Oelze and O'Brien, 2004a) showed that the tissue model could be generalized to the form:

$$W_{theor}(f) = B(L, a)C(r_{eff}, \eta_z)f^4 F(f, r_{eff}), \tag{9.11}$$

in which B is a gate function of gate length L and aperture radius a, C is a constant based on the effective scatterer radius r_{eff} and scatterer density in mm^{-3}, and F is a form factor dependent on frequency and effective scatterer radius. In general, F can refer to the ratio of the scattering function of a sphere, shell, Gaussian, or other scattering object relative to that of a point scatterer. The equation can be recognized as the same form as Eqn 8.10B, where the first two terms of Eqn 9.11 correspond to the first three terms of Eqn 8.10B and the last two terms of Eqn 9.11 correspond to the scattering term $S(z,r,f)$ of Eqn 8.10B. This group investigated the effects of attenuation compensation (Oelze and O'Brien, 2002), gate correction factors (Oelze & O'Brien, 2004a,b), and trade-offs in the size of the region of interest (Oelze, O'Brien, Blue, & Zachary, 2004).

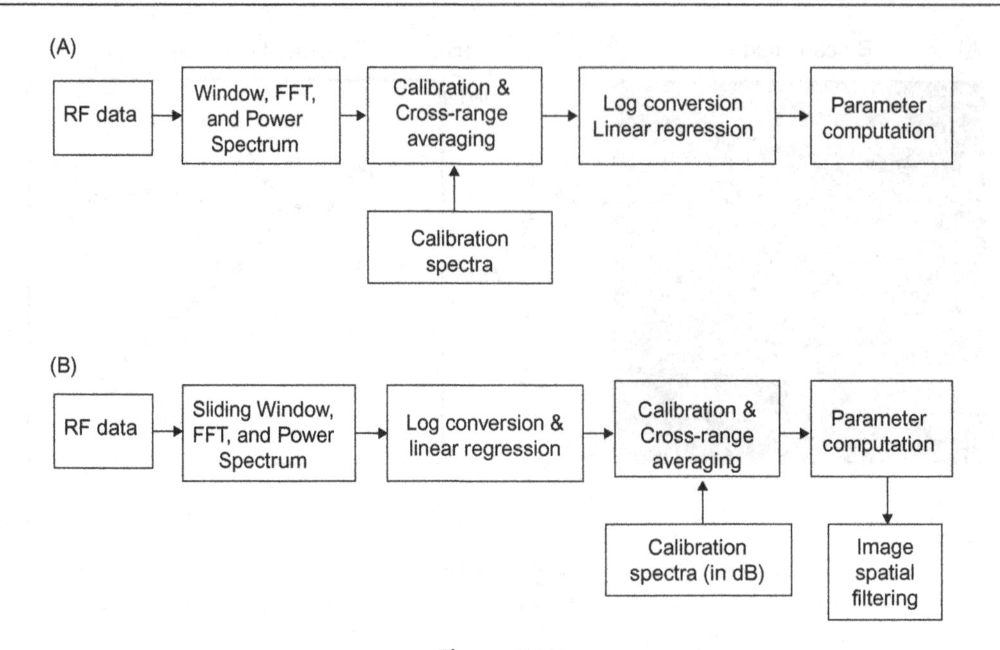

Figure 9.15

Spectral processing. (A) For large-region-of-interest mode, and (B) for spectral-parameter-imaging mode. *From Lizzi et al. (2006), with permission from the World Federation of Ultrasound in Medicine and Biology.*

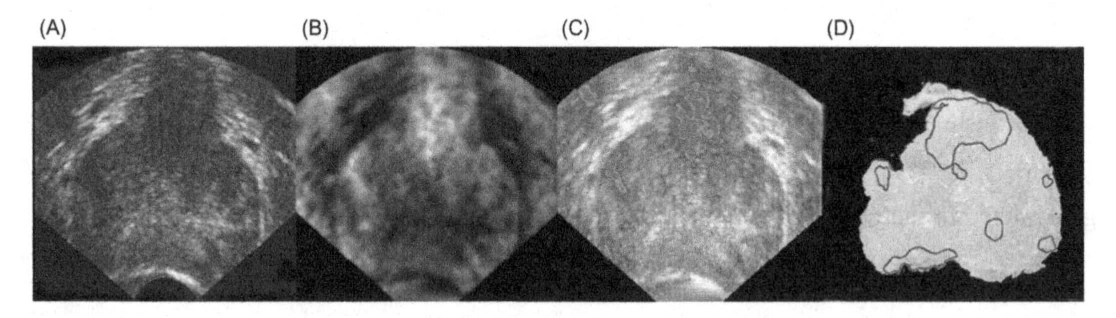

Figure 9.16

Images of the central plane of a prostate gland having an ultrasonically-occult anterior tumor as viewed from the Apex of the prostate. (A) Computer-generated envelope-detected B-mode image; (B) gray-scale cancer-likelihood image (white = maximum likelihood); (C) color-encoded overlay on a midband parameter image depicting the two highest levels of likelihood in red and orange (red areas appear black in the gray-scale image reproduced in the print version of the book, and orange areas are gray (mostly in areas immediately surrounding the red). The reader is referred to the Web version of the book to see the figure in color); (D) corresponding histological section that shows a 12-mm tumor protruding through the anterior surface and several smaller circular intracapsular foci of cancer and neoplasia, as manually demarcated in ink by the pathologist. *From Feleppa et al. (2001), reprinted with permission of Dynamedia, Inc.*

9.5.8 Backscattering Comparisons

A classic UTC experiment is to tell the size of small subwavelength scatterers from the backscatter alone. Well, not quite alone. A reference model for the expected backscatter for a small sphere, such as that derived by J. J. Faran, Jr. (described in Section 8.2.3), is calculated over the frequency range of interest and compared to the backscatter.

Several research groups have studied the backscatter coefficient of many spheres in suspension without multiple scattering (Romijn et al., 1989; Ueda & Ozawa, 1985). Thijssen (2000) has shown that the slope of the backscattering coefficient at low frequencies (small *ka*) is related to the size of the scatterers (20–500 μm), as did Lizzi et al. (1983). In addition, Bridal, Wallace, Trousil, Wickline, and Miller (1996) applied the standard methodology to a −6-dB spatial resolution cell of polystyrene beads in argose over a frequency range of 5–65 MHz with attenuation correction and renormalization. They compared their data to the Faran elastic sphere scattering theory. (It is remarkable that the size of scatterers as small as one sixth of a wavelength can be determined from the slope of their backscatter coefficient (shown in Figure 9.17).) Excellent absolute agreement for the spectral characteristics of spheres was also obtained by Hall et al. (1997). Chaturvedi and Insana (1996) explored the errors and limits of determining the sizes of small scatterers.

Interlaboratory comparisons on identical phantoms were made to see how well a backscatter coefficient (BSC) could be measured independently on four modified different imaging systems and if the size of an object embedded in an attenuating medium could be determined. The tissue-mimicking phantom that was used consisted of a tightly controlled distribution of 41-μm-diameter glass spheres in an agar gel background. Measurements were made on ultrasound imaging research systems (more on these systems can be found in Chapter 10) and results were within 1.5 dB of the Faran sphere model (Nam, Rosado-Mendez, Wirtzfeld, et al., 2012) as indicated by Figure 9.18. This study demonstrated that BSCs could be determined accurately and independently using near-clinical imaging systems.

In a second investigation, a phantom mimicking two different rodent mammary tumors was involved in an interlaboratory comparison (Nam, Rosado-Mendez, Wirtzfeld, et al., 2011). The slopes of the attenuation coefficient and backscatter coefficents were measured using seven transducers and four clinical systems. Overall agreement was good with a few exceptions. For the attenuation coefficient slope, the average and maximum differences between the clinical systems and laboratory determined values were 11 and 29% respectively, and for the backscatter coefficients, the average and maximum differences were 16 and 33%.

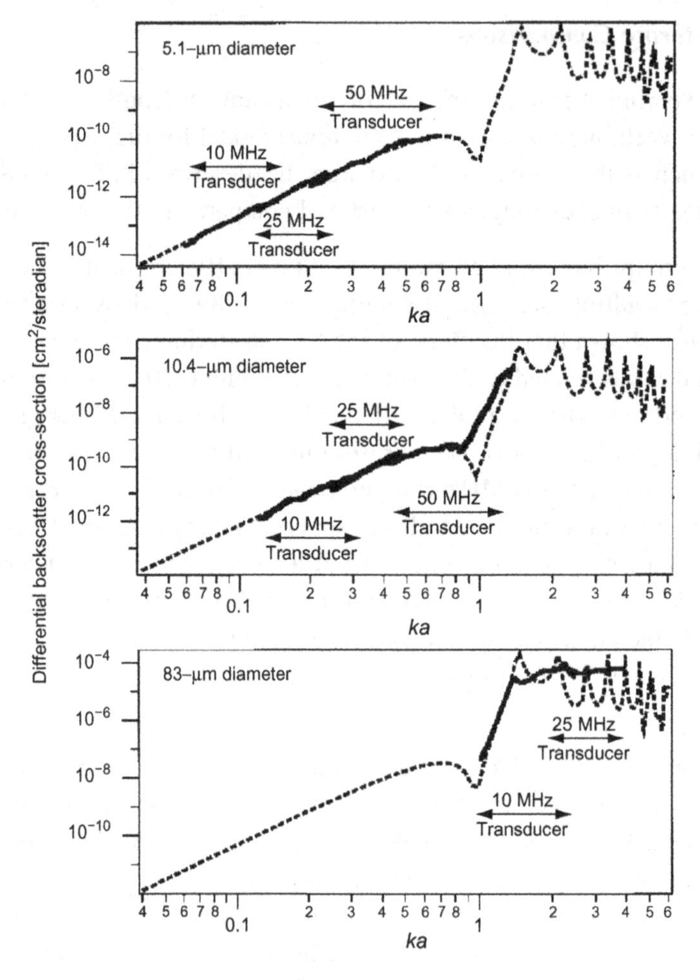

Figure 9.17
Comparison for simulation (Solid Lines) of backscatter from a sphere to backscatter data (Dashed Lines) as a function of *ka*. Theory was used to calculate the differential backscatter cross-section for a single polystyrene sphere. Each data set is renormalized to align it to the single scatterer cross-section curve. *From Bridal et al. (1996), Acoustical Society of America.*

9.6 Applications of Tissue Characterization

9.6.1 Radiology and Ophthalmic Applications

Determining the scatterers per unit volume can be helpful in discriminating between healthy and abnormal tissues. The slope of the backscatter coefficient approach has been successfully applied to eye tumors (Feleppa, Lizzi, Coleman, & Yaremko, 1986; Liu et al., 2004; Romijn, Thijssen, Oosterveld, & Verbeek, 1991; Thijssen, Verbeek, Romijn,

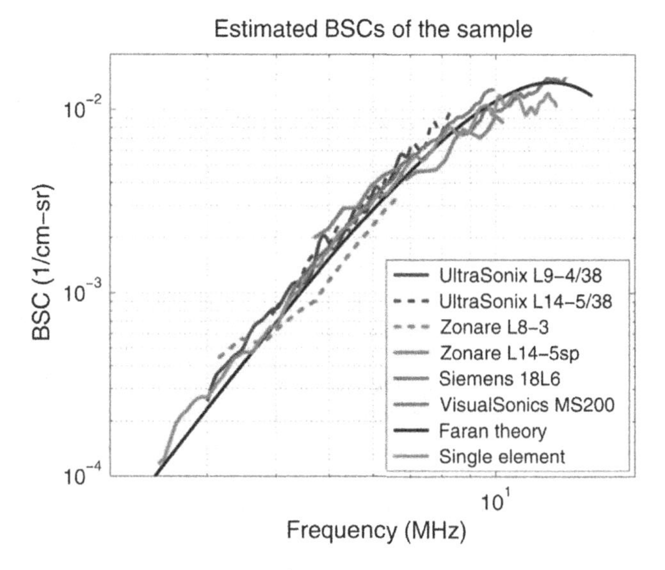

Figure 9.18
Backscatter coefficients vs frequency estimates using each of the clinical ultrasound systems. Each data line represents a different transducer–system combination. Results are presented for two transducers for both the UltraSonix and the Zonare scanners. Also shown are lab measurements employing single-element transducers. The solid black curve is computed using the theory of Faran. The reader is referred to the Web version of the book to see the figure in color. *From Nam et al. (2012), Acoustical Society of America.*

De Wolff-Rouzendaal, & Oosterhuis, 1991); lymph nodes (Mamou et al., 2009); animal model melanomas (Romijn et al., 1989); and glomeruli size estimation, both renal (Hall, Insana, Harrison, & Cox, 1996; Insana, Hall, Wood, & Yan, 1993) and liver (Sommer, Stern, Howes, & Young, 1987); and many other cases. In addition, it is sometimes possible to decipher a scattering component that is related to structural patterns in tissue (Thijssen, 2000).

Fellingham and Sommer (1984) predicted that the regularity of tissue structure would result in periodic peaks in the autocorrelation of backscattered pressure and that disease would disrupt this structure. By examining the RF data from in vitro samples of the liver and spleen, they were able to show some differences in "mean scatterer spacing" on the scale of millimeters for some diseased states.

By comparing backscattered data to an empirical scattering model, Eqn 9.11, Oelze et al. (2004) applied QUS methods to distinguish between rat mammary fibroadenomas and 4T1 mouse carcinomas. They found that by using a Gaussian form factor, the estimated average scatterer diameter and acoustic concentration were both good discriminants. A slight dependence of the parameters on probe center frequency was found. Parameterized QUS images were an aid in visualizing the extent of the features, as indicated in Figure 9.19.

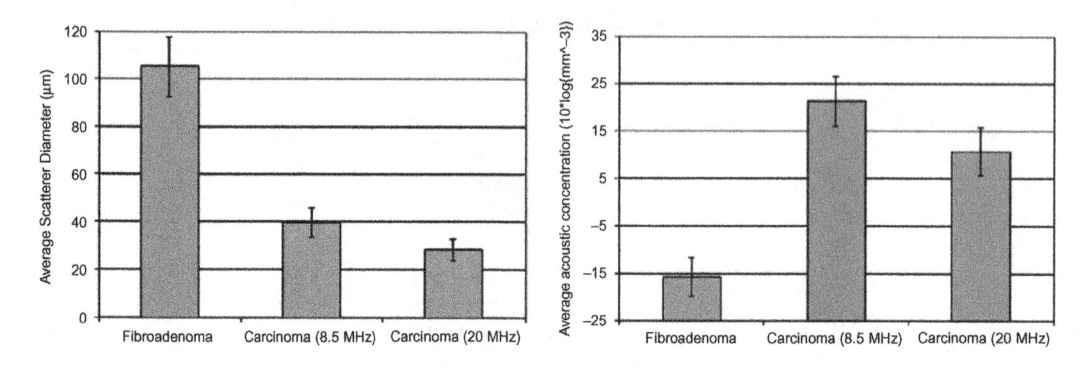

Figure 9.19
QUS methods to visualize rat fibroadenomas and mouse carcinomas.
(A) Bar graph of the average scatterer diameter values for rat fibroadenomas and mouse carcinomas; (B) Bar graph of the average acoustic concentration values for rat fibroadenomas and mouse carcinomas. *From Oelze et al. (2004), IEEE.*

Based on earlier work by Mamou (2005), Dapore et al. (2011) constructed a high-resolution volume model to understand scattering mechanisms in human fibroadenomas. Derived on stained histological slices digitized to 0.5 μm , they assigned acoustic values to stain colors and, with analysis, used a fluid-filled sphere form factor to create a three-dimensional (acoustic) impedance map (3DZM) to study scattering properties. The resulting 3DZM is shown in Figure 9.20. An example of a scattering filter from the model is given by Figure 9.21.

9.6.2 Cardiac Applications

Tissue characterization of the dynamic movements of the heart has led to new diagnostic tools, as well as a better understanding of heart function. Some background on the heart will be helpful in explaining how the heart is characterized by ultrasound methods. The heart is a tireless pump that squeezes out blood on myocardial contraction (systole) and fills up during the expansion phase (diastole). The heart pumps approximately 5–15 L of blood a minute. The strength of the heart is contained in the band of circumferential fibers that do the squeezing. Because these fibers have a preferentially organized direction, they have anisotropic acoustic properties (as plotted in Figure 9.22). Miller et al. (1989, 1998) measured these characteristics at Washington University, and their significant research on the heart is summarized in this section and in more detail in these two review articles.

These characteristics of anisotropy have been modeled well by the Born approximation for scattering from cylinders. This modeling shows that the direction of insonification relative to the arrangement of myocardial fibers is important. When the fibers are aligned perpendicularly to the sound beam, maximum reflection occurs; when the fibers are parallel,

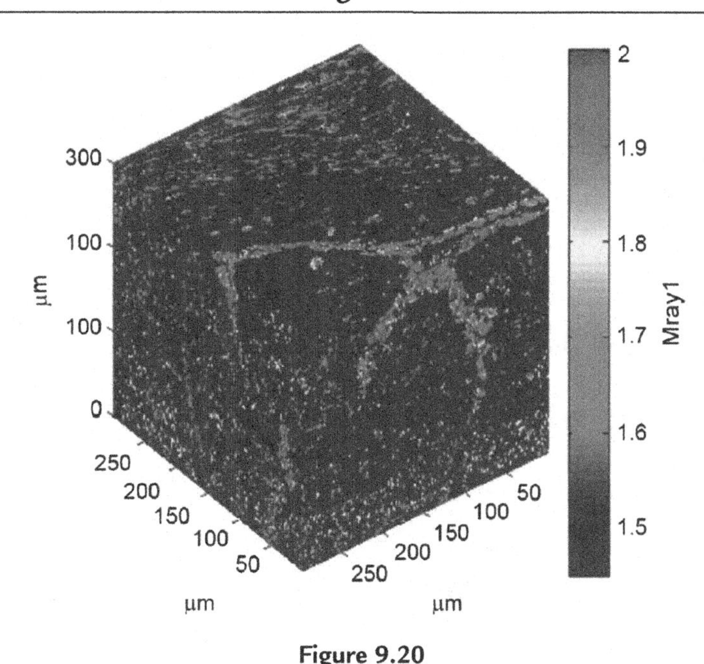

Figure 9.20
Rendering of a Three-dimensional Impedance Map (3DZM) of a Human Fibroadenoma. The reader is referred to the Web version of the book to see the figure in color.
From Dapore et al. (2011), IEEE.

reflection is minimum. Two important consequences of these effects (shown in Figure 9.23) are the variation of integrated backscatter and frequency-averaged attenuation with angle (Mottley & Miller, 1988, 1990). Both of these effects are seen in every cardiac image and are important in interpreting what is seen in these images.

Now imagine the heart twisting, moving, and stretching during each cardiac cycle so that the angle of insonification to the heart from a fixed location, such as the short axis acoustic window, keeps changing with time. One might think that it is totally hopeless to attempt an interpretation of such a complicated configuration, but dynamic tissue characterization comes to the rescue. By looking at integrated backscatter features of the cardiac cycle, like those in Figure 9.24, researchers in J. Miller's group at Washington University were able to identify abnormalities of the heart. Accounting and correcting for anisotropy helps to determine a baseline pattern so that pathologies can be more readily identified.

Because heart diseases are major killers and debilitators, it is worth asking, "What are the main pathologies of the heart?" The most important blood suppliers to the heart are the coronary arteries. Atherosclerosis is a disease in which the arteries are narrowed by plaque (a soft pasty material that can calcify), become fibrotic, or form into vulnerable plaque (a dangerous unstable sort that can turn into a thrombosis or blood clot). A symptom of atherosclerosis is arteriosclerosis (a hardening of the arteries). Ischemia is reduced blood

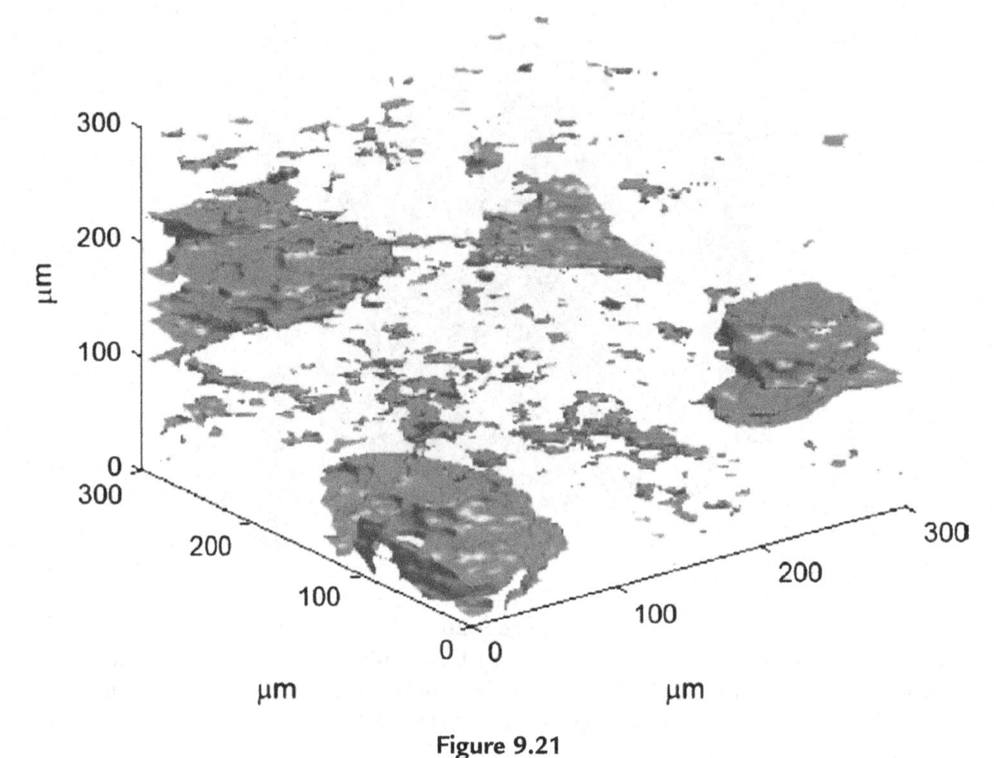

Figure 9.21
Segmentation of high and low impedance structures. ESD estimate for this three-dimensional impedance map is 127 μm. *From Dapore et al. (2011), IEEE.*

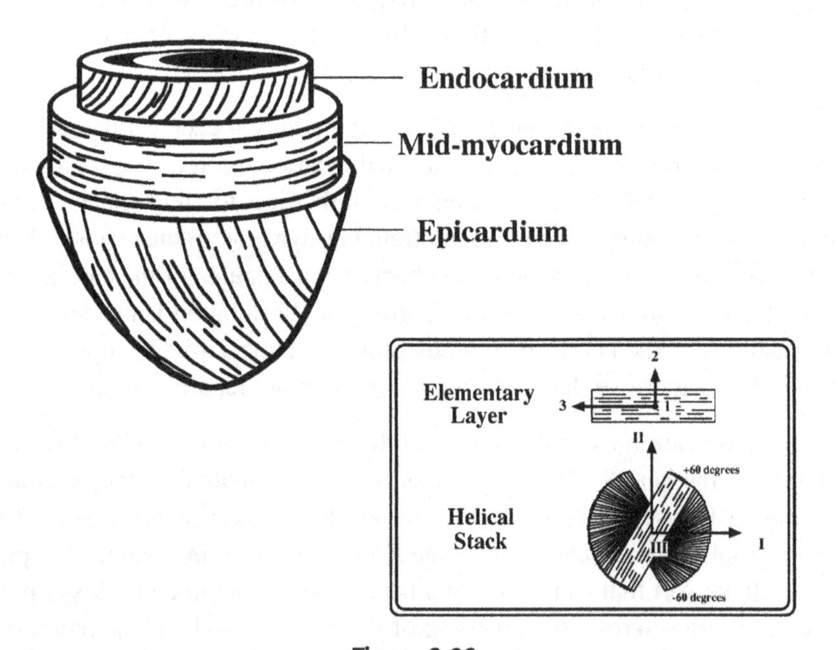

Figure 9.22
Arrangement of muscle layers of the heart showing the preferential directions of microfibers in the layers that cause anisotropy. *Courtesy of J. G. Miller.*

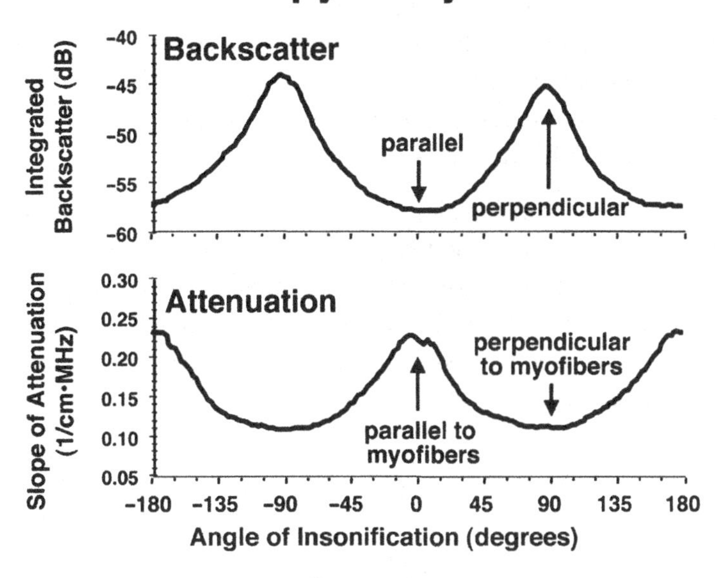

Figure 9.23
Angular variation of attenuation and integrated backscatter of the heart caused by anisotropy of myofibers. *From Miller et al. (1998), IEEE.*

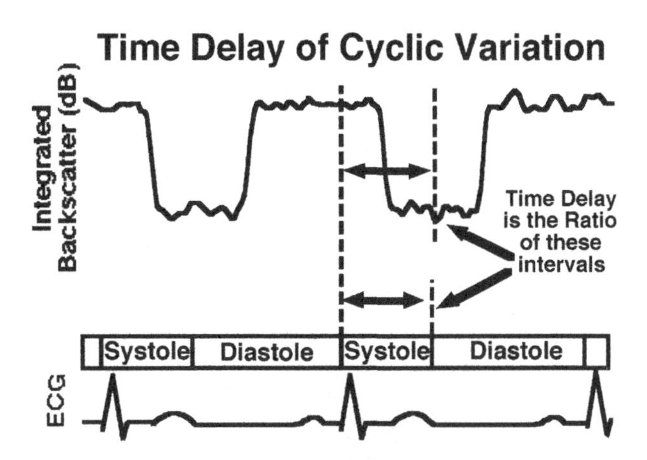

Figure 9.24
Characterization of the cyclic variation of integrated backscatter in terms of the variation in magnitude and the time delay relative to the start of systole, normalized to the systole interval. *From Miller et al. (1998), IEEE.*

flow as a result of mechanical obstruction of an artery. An infarct is tissue death caused by ischemia. A heart attack, or myocardial infarct, is the death of muscle cells of the heart and is often caused by the closing of an already constricted passage by a blood clot. Reperfusion is a mechanical (angioplasty) or chemical (thrombolytic agents) means of replenishing blood flow to an injured area to preserve heart function and reduce mortality and the severe side effects of heart attacks.

Because the ability of the heart to function is severely compromised by these diseases, the heart cycle is affected. In particular, the features of the heart cycle described in Figure 9.24 can be used in identifying effects of disease such as infarcts, ischemia, and myopathy (a disease affecting the contractility of the heart), as well as the effects of reperfusion and other types of therapy. The effectiveness of using these features, after correction for the effects of anisotropy, is illustrated by Figure 9.25.

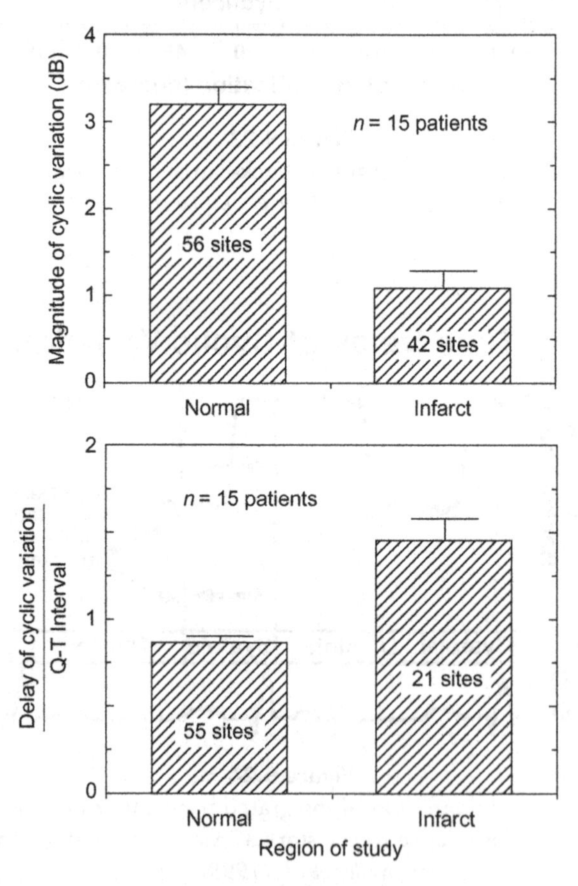

Figure 9.25

Results of a study of 15 patients with scars from old myocardial infarcts. Zones of scar exhibit reduced magnitude and increased time delay relative to healthy (noninfarct) sites in the same patients. *From Miller et al. (1989), IEEE.*

Ongoing studies of the heart have led to real-time improvements in echocardiology. For example, the need to capture both the weak and strong reflections from the muscle fibers has led to an anisotropic rational gain compensation method, as well as to a lateral gain compensation capability. The use of integrated backscatter was part of an algorithm to provide automatic real-time boundary detection of the left ventricle. These methods will be discussed in more detail in Chapter 10. The application of ultrasound contrast agents in combination with signal processing methods has also been highly successful in the diagnosis and tissue characterization of the heart (to be explained in Chapter 14). More characterization of the heart using elastographic methods can be found in Chapter 16.

9.6.3 High-Frequency Applications

High-frequency ultrasound is opening up new tissue landscapes not seen at conventional frequencies below 15 MHz. Three major high-frequency applications are utilizing the increased resolution to identify plaque in blood vessels by intravascular ultrasound (IVUS), to reveal cellular structures, and to examine small animals. These uses are giving new insights into the progress of disease and cell death in vivo. The following discussion will show that high-frequency imaging is not just conventional imaging scaled down in resolution.

Consider a high-frequency spherically focusing transducer operating at 100 MHz with an $F\# = 1.33$ (Foster et al., 1993; Sherar, Noss, & Foster, 1987) with an axial resolution of 28 μm and a FWHM of 17.5 um, as depicted in Figure 9.26. We can estimate a resolution cell volume as a −6-dB FWHM ellipsoid with a major axial axis and equal minor lateral axes in terms of wavelengths as in Figure 7.20, as:

$$\text{Vol} = \frac{\pi}{6}(1.87\lambda)(1.17\lambda)^2 = 1.34c_0^3/f_c^3, \tag{9.12}$$

which gives a volume of $4.52 \times 10^3\ \mu\text{m}^3$ for a center frequency of 100 MHz, a value that is 1000 times smaller than that for a more conventional B-mode frequency of 10 MHz, $4.52 \times 10^6\ \mu\text{m}^3$.

What difference does this size make? Consider that cell size is on the order of microns. In a study of leukemia cells (Czarnota et al., 1997), the cell density was measured as approximately $10^{-3}/\mu\text{m}^3$. For this case, there would be 4.5 cells in a resolution cell at 100 MHz and 4500 cells at 10 MHz. Recall the class distinctions for scattering from Section 9.2 (class 1, or diffusive scatterers, are 25 or more per resolution cell, and class 2, or diffractive scatterers, are 1 per resolution cell). At 100 MHz, the scatterers are on the borderline between classes 1 and 2; whereas, at 10 MHz, they are between classes 1 and 0. At the higher frequency, scattering is more critically dependent on the size and arrangement of the individual scatterers, even though they are not resolvable.

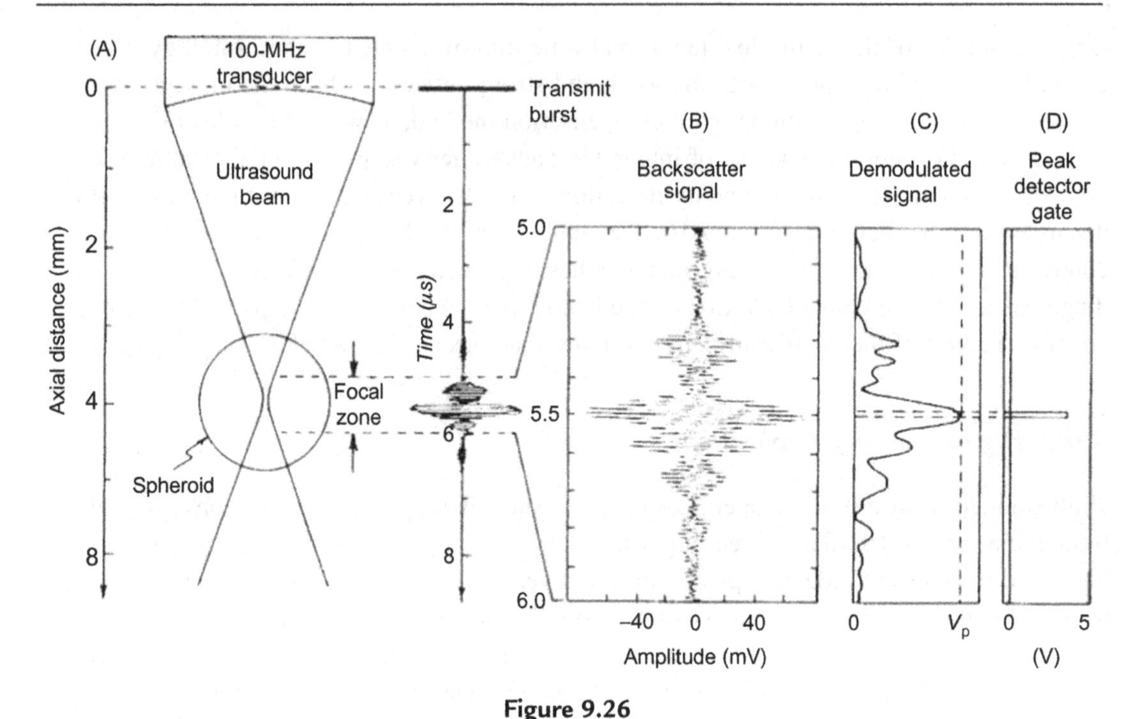

Figure 9.26
100-MHz backscatter microscope. (A) Sketch of beam and transmit signal; (B) backscatter signal versus time; (C) pulse–echo envelope; (D) gate located at peak of echo envelope. Beam is scanned across an area consisting of 256 × 256 measurements. *From Sherar et al. (1987), reprinted with permission of Nature, Macmillan Magazines Limited.*

To make this difference clearer, a high-frequency detection of the process of apoptosis will be discussed. Apoptosis is programmed cell death that can occur in embryonic development, in diseases such as cancer or neurodegenerative disorders, from heart attacks or organ transplants, or as a deliberate result of drug therapy.

In Figure 9.27 are a series of noninvasive images of programmed cell death by a toxic drug of apoptotic acute myeloid leukemia cells, taken by a 40-MHz ultrasound backscatter microscope (Czarnota et al., 1999; Czarnota, Kolios, Hunt, & Sherar, 2001). Note the strong increase in backscatter brightness of speckle as the apoptotic process is maximized at 48 hours. The corresponding optical microscope views show differences in the size and arrangement of individual cells. At the 6-hour point, 95% of the cells underwent nuclear condensation or fragmentation; the nuclear diameter, originally 70% of the cell diameter, shrank to 40% of the cell diameter.

Because the cells themselves could not be resolved in the images, tissue characterization, in the form of spectral analysis developed by Lizzi's group and described in Section 9.5.7, was applied to the RF data from the images (Kolios, Czarnota, Lee, Hunt, & Sherar, 2002).

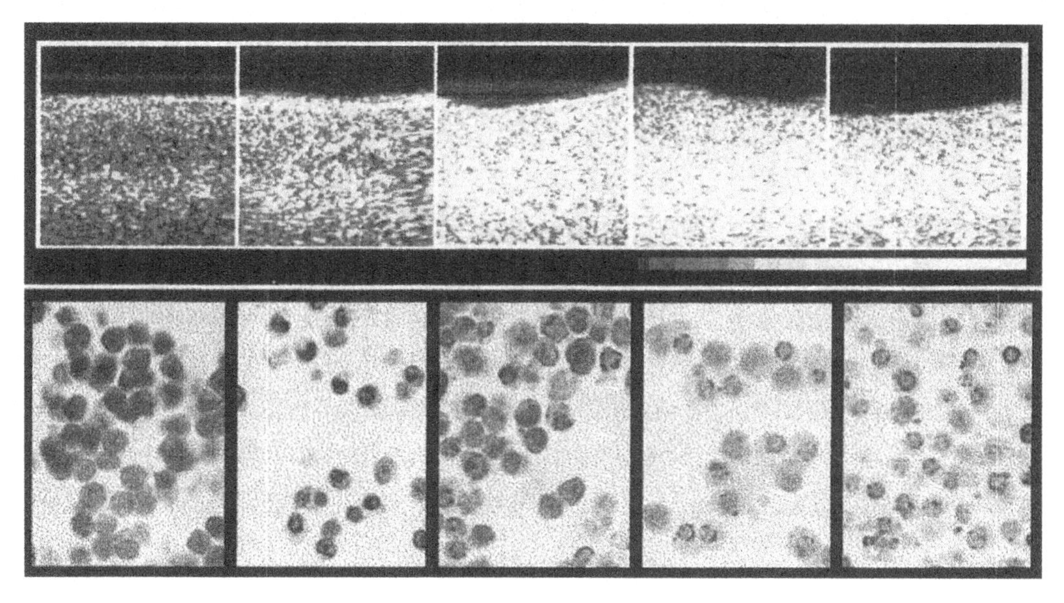

Figure 9.27
Ultrasound imaging of apoptosis and correlative histology. Sequential panels show cells at 0, 6, 12, 24, and 48 hours after treatment with a toxic drug. Each panel is approximately 5 mm wide. (Top row) 40-MHz ultrasound backscatter images of cells. (Bottom row) optical microscopic images of stained cells; field of view is approximately 50 μm. *From Czarnota et al. (1999, 2001), reprinted with permission from Nature Publishing Group.*

Analysis of leukemia cells demonstrated significant changes in the scattering slope and in the midband value of the slope, which increased by 13 dB between healthy and apoptotic cells (shown in Figure 9.28). The slope values correlated with a decreasing mean scatterer radius.

As indicated by Figure 9.27, there are noticeable changes in the patterns and sizes of the cells during the process of apoptosis, and the number of cells on the order of a resolution cell is still countable. Unlike imaging at lower conventional frequencies, the backscatter appears to be more related to the specific arrangement of cells. Work is underway (Baddour et al., 2002) to develop an ensemble model in which each cell is modeled as an elastic Faran–Hickling sphere (see Sections 8.2.3 and 9.2), and different patterns of cells, such as those in Figure 9.29, can be incorporated to mimic those observed during apoptosis. The model is producing results that match the observation that backscatter increases as the cells die and as their lattice structure becomes more randomized.

In order to develop a practical clinical method for monitoring tumor cell-death response to cancer therapies (Sadeghi-Naini et al., 2012), quantitative ultrasound techniques similar to higher-frequency approaches (Banihashemi et al., 2008; Czarnota & Kolios, 2010; Kolios & Czarnota, 2009; Vlad, Brand, Giles, Kolios, & Czarnota, 2009) were applied at conventional frequency (1–20 MHz) for real-time detection of cell death through in vitro

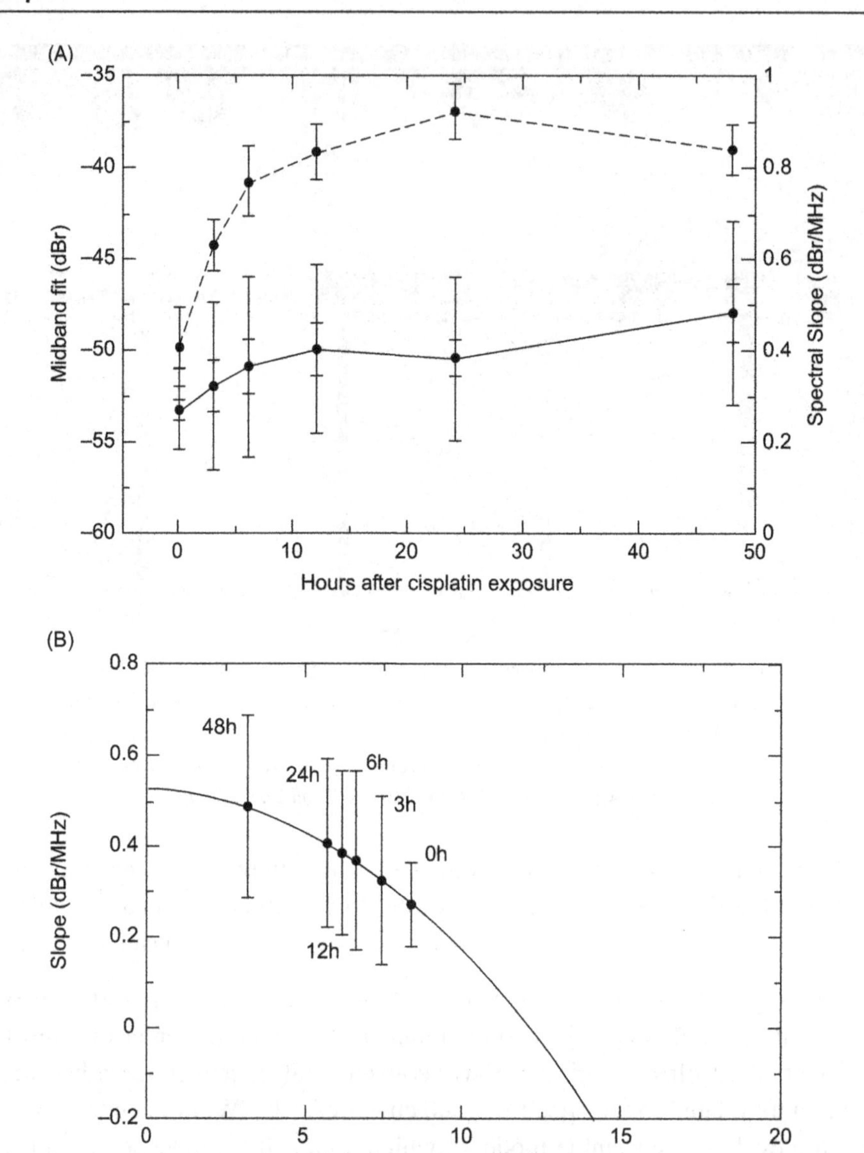

Figure 9.28

Spectral features of backscatter from leukemia at indicated times after exposure to a toxic drug. (A) Plot of spectral slope (solid curve, right axis) and midband slope value; (B) plot of theoretical predictions of spectral slope versus scatterer radius compared to data from a 34-MHz transducer. *From Kolios et al. (2002), with permission from the World Federation of Ultrasound in Medicine and Biology.*

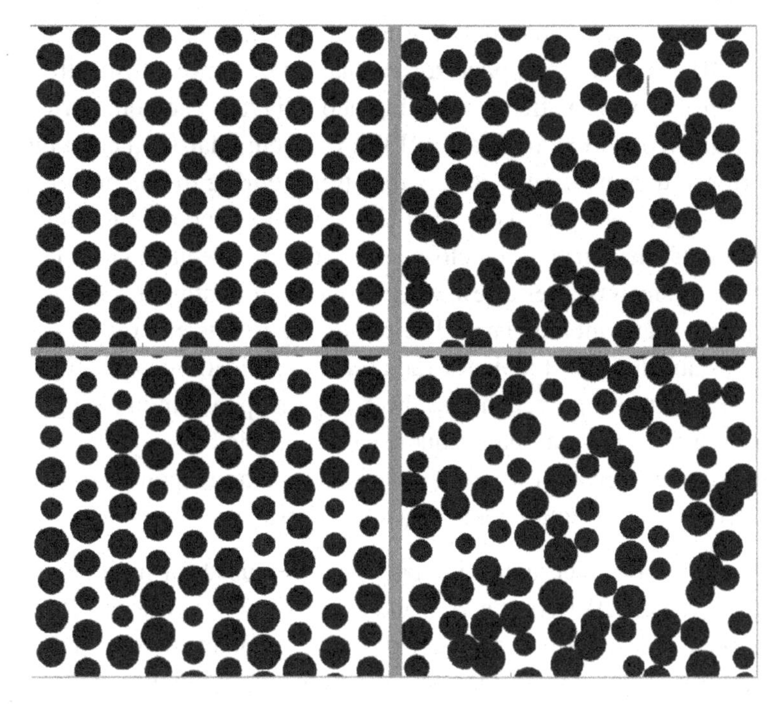

Figure 9.29
A few possible cell ensemble packings (Only Cell Nuclei are Shown).
(A) Perfect crystal; (B) cell locations are allowed some degree of randomization; (C) nucleus diameters are allowed some degree of randomization; (D) both cell locations and nucleus diameters are randomized. *From Baddour et al. (2002), IEEE.*

experiments (Azrif et al., 2007; Kolios & Czarnota, 2009). Results demonstrated an ability to detect as little as 10% apoptotic cells using ultrasound frequencies in the 10-MHz range, paralleling changes observed using high-frequency ultrasound.

In addition, these findings have been confirmed in vivo using prostate and breast cancer tumor xenografts in mice treated with different kind of therapies, including radiation and chemotherapy (Sadeghi-Naini et al., 2012, 2013a,b,c). Results based on experiments using over 50 animals assessed with high-frequency and conventional-frequency ultrasound suggest that the monitoring of tumor cell-death response is possible using conventional-frequency ultrasound. Measurable changes in backscatter properties are not expected from micron-sized particles at conventional frequencies as a result of loss of scattering strength from small scattering structures; however, in the low- to mid-frequency range, bulk changes in tissue are mostly related to ensembles of cells and nuclei smaller than the wavelength of the ultrasound being used. Such ensembles influence acoustic properties and thus ultrasound backscatter. In addition, when imaging cell samples, even at these low frequencies, a unique speckle pattern is still formed, indicating that many subresolution scatterers contribute to the detected signals.

In a pilot in vivo clinical study, quantitative ultrasound techniques at conventional frequencies (~7 MHz) have been applied for evaluation of tumor cell-death response in locally advanced breast cancer patients receiving neo-adjuvant chemotherapy (Figure 9.30) (Sadeghi-Naini, Papanicolau, Falou, et al., 2013b). Conventional-frequency ultrasound data were acquired prior to treatment onset and at four times during treatment. In each session, several scan planes, 4 cm by 6 cm, were acquired from the same nominal regions. The results ($n = 24$ patients) demonstrated a close association between changes in quantitative ultrasound spectral parameters after a few weeks following treatment initiation and clinical response in the tumor many months later. The promising results emerging from this study pave the way for establishing protocols for the clinical applications of the conventional-frequency quantitative ultrasound techniques in therapy response monitoring. As such, quantitative ultrasound at conventional frequencies is expected to provide rapid and quantitative functional information in real time for evaluating responses to a specific therapy (A. Sadeghi-Naini, personal communication, 2013).

High-frequency noninvasive backscatter imaging has been used for high-resolution internal evaluation of blood vessels (IVUS), the eye, and skin (Knapik, Starkoski, Pavlin, & Foster, 2000). Modes normally associated with conventional ultrasound, such as pulsed Doppler (Christopher, Starkoski, Burns, & Foster, 1997) and color flow imaging (Goertz et al., 2000) (real-time modes described in detail in Chapter 11), have been miniaturized to work at high frequencies.

Surprisingly, one of the main motivations for creating a fully functional high-frequency imaging system is to study the mouse (Foster et al., 2000a, 2000b). Mice share 90% of the genes of humans, and many of their organs, such as the heart, liver, and kidneys, are similar to those of humans. Therefore, a mouse can serve as a model for human physiology and metabolism. The life-span of a mouse is 18 to 33 months, and a human's is 50 to 90 years; therefore, 1 mouse day is roughly like 1 human month in terms of its equivalent length. Researchers can alter the mouse genome in predictable ways to express selected traits or susceptibility to disease. These transgenic mice can serve as models of disease progression as well as models of the effects of healing therapies and treatment strategies on an accelerated time scale.

High-frequency ultrasound can be used to study the physiology and morphology of mice noninvasively. High-resolution images of the internal growth of mice are now possible, as shown by the sequence of mouse embryo images in Figure 9.31 (Aristizábal, Christopher, Foster, & Turnbull, 1998; Srinivasan et al., 1998; Turnbull, 1999; Turnbull, Bloomfield, Foster, & Joyner, 1995). These remarkable images are revealing growth patterns in ways never seen before.

In addition to cardiovascular disease, the progression of cancer has been studied (Turnbull et al., 1996). Skin cancer and melanomas were monitored with 50-MHz ultrasound imaging

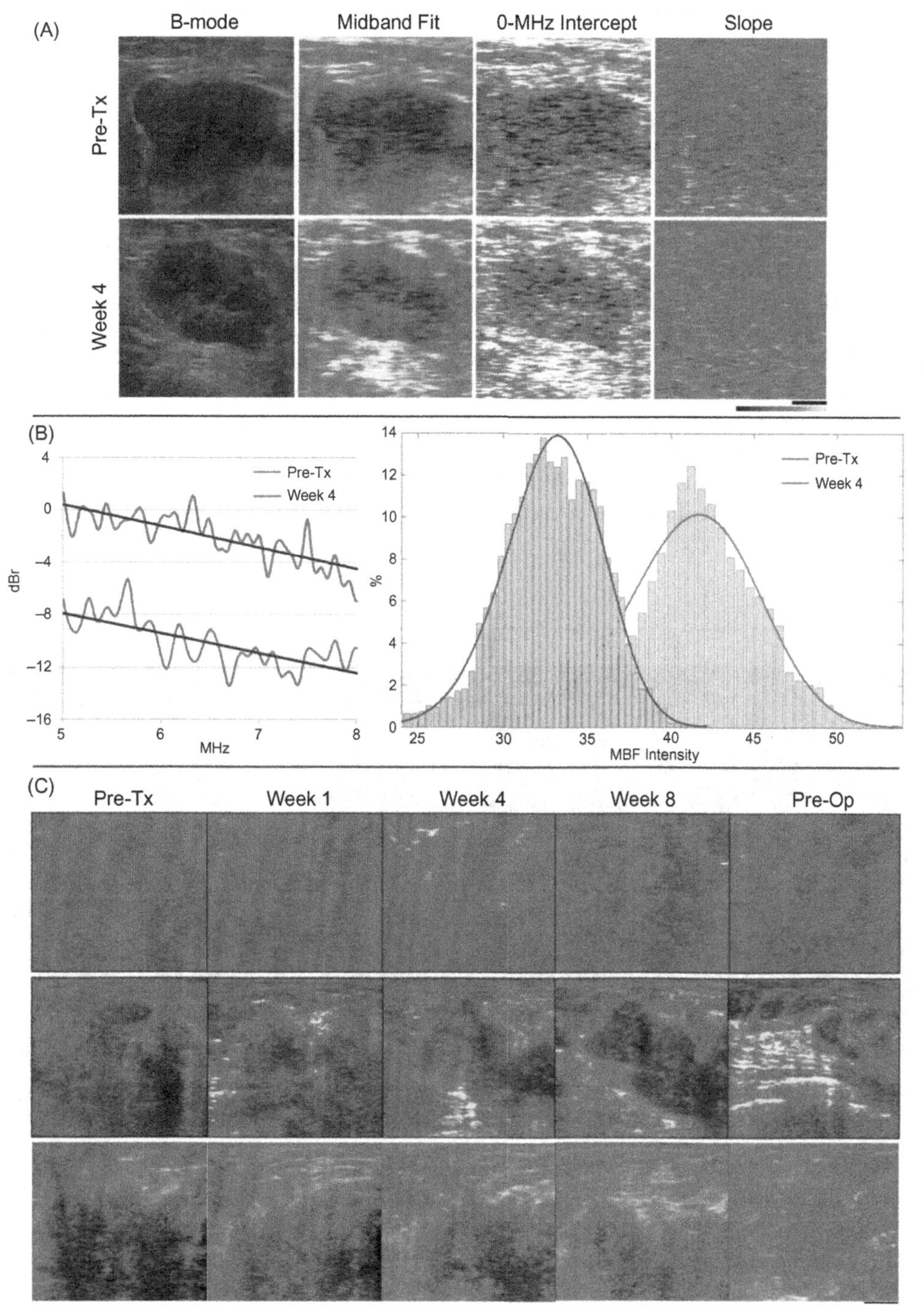
(A)
B-mode
Midband Fit
0-MHz Intercept
Slope
Pre-Tx
Week 4
(B)
Pre-Tx
Week 4
dBr
MHz
%
MBF Intensity
(C)
Pre-Tx
Week 1
Week 4
Week 8
Pre-Op

(*Continued*)

Figure 9.30
Quantitative ultrasound for evaluation of tumor cell-death response.
(A) Representative data from a large breast tumor before starting the neo-adjuvant chemotherapy (first row), and after 4 weeks of treatment (second row). The columns from left to right demonstrate ultrasound B-mode, parametric images of midband fit, 0-MHz intercept, and spectral slope, respectively. The scale bar is ~1 cm, and the color map represents a scale encompassing ~50 dBr for midband fit (MBF) and 0-MHz intercept, and ~15 dBr/MHz for the spectral slope. (B) Normalized power spectra (left) and generalized gamma fits on the histograms of the MBF intensity (right) for the tumor region (lower power spectrum line and left-hand histogram correspond to week 4 post-treatment). (C) Representative parametric images of 0-MHz intercept from a non-responding patient (first row), as well as from two patients who responded to the treatment (second and third rows). The data for each patient were acquired from the same nominal regions, prior to treatment as well as at weeks 1, 4, and 8 during treatment, and preoperatively from left to right, respectively. The scale bar represents ~1 cm. The color bar represents a scale encompassing ~80 dBr. The reader is referred to the Web version of the book to see the figure in color. *From Sadeghi-Naini et al. (2013b).*

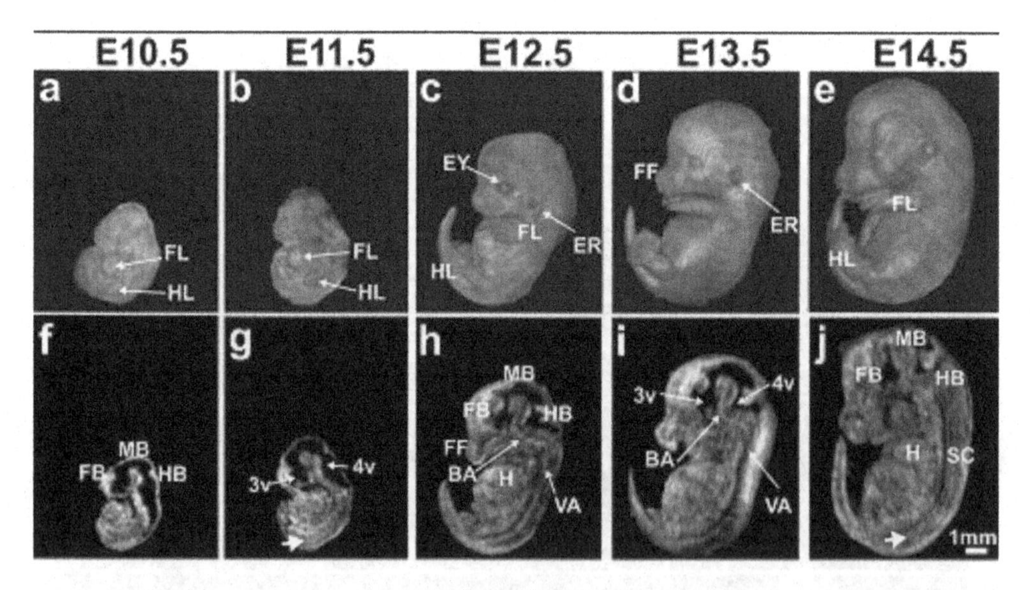

Figure 9.31
Whole embryo 3D In utero annular array high-frequency ultrasound imaging.
Volume reconstructions (A–E) and corresponding mid-sagittal sections (F–J) from array-focused data for embryonic stages indicated at the top of the figure. 3v, third ventricle; 4v, fourth ventricle; BA, basilar artery; EY, eye; ER, ear; FB, forebrain; FF, facial features; FL, forelimb; H, heart; HB, hindbrain; HL, hind limb; MB, midbrain; SC, spinal cord; VA, vertebral artery. White arrows (G, J) indicate intersomitic blood vessels. *From Aristizábal, Mamou, Ketterling, and Turnbull (2013), with permission from the World Federation of Ultrasound in Medicine and Biology.*

from detection until their growth to a few millimeters in a few days. Sizes from images agreed to within a few percent with the measured size of excised tumors.

Baddour et al. (2002) investigated the effect of photodynamic cancer therapy on human malignant melanoma in mice and monitored changes with 40-MHz ultrasound imaging. As shown in Figure 9.32, edema appeared as a bright region in the treated area, visible in the image 4 hours after treatment. By 26 hours, the edema had almost disappeared, and a very bright region was observed in the treated area. In these images, a combination of both high resolution and increased backscatter can be seen. As resolution increases, the number of scatterers per resolution cell dramatically decreases, yet the physical size of the cells stays the same; therefore, the cell size in wavelengths increases. What may be possible in the future is to take advantage of both the improved image definition and the added benefit of additional tissue structural information hidden in the backscatter (described earlier in this section).

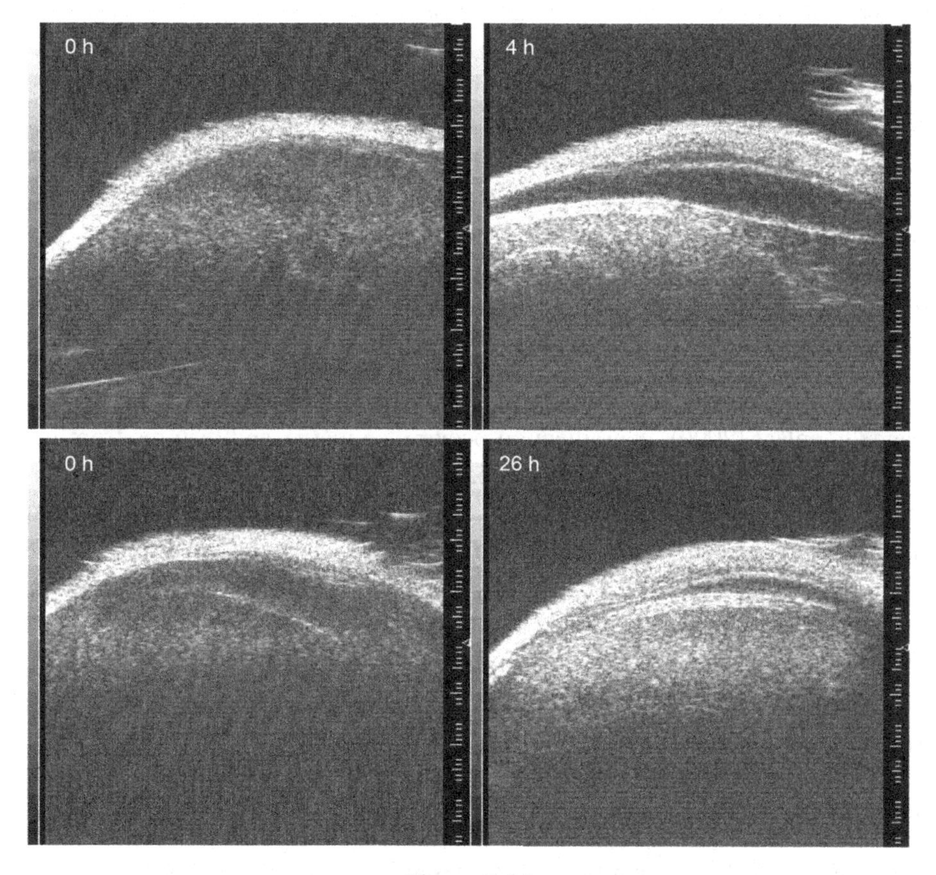

Figure 9.32

Ultrasonic B-mode images of human malignant melanoma tumors grown in mice. Each row is a different mouse, imaged immediately before laser irradiation during photodynamic treatment (0 h), and at two different time points after treatment. *From Baddour et al. (2002), IEEE.*

9.6.4 Texture Analysis and Image Analysis

Most often, RF data are examined by individual lines; however, larger-scale structural information about tissues can also be found by combining data from different lines. Multiple-line analysis can be done either in the RF or video domain.

As discussed earlier in Section 8.4.3, speckle is in part an artifact of the measurement system; however, textural differences in localized image regions can be indicative of tissue microstructure. Speckle and the image itself are also strongly affected by tissue absorption and diffraction, complicating the classification of tissue. Patterns in an image can be examined by second-order statistics in terms of the spatial autocovariance function. Thijssen (2000) points out that these statistics change with the number of scatterers within the resolution cell and other factors interfere.

Earlier methods did not account for these effects. Coolen, Engelbrecht, and Thijssen (1999) examined these sources of variability using first-order (mean and signal-to-noise ratios) and second-order statistics from the co-occurrence matrix of the data. After they corrected the video data for nonlinear imaging system effects, such as preprocessing and compression (to be described in Chapter 10), they found that the chief causes of variability were speckle noise and intervening inhomogeneous tissue along acoustic paths. By comparing these features with healthy liver and tumors, they were able to detect differences. Huisman and Thijssen (1998) investigated in vivo video data for the liver, and after exclusion of small blood vessels from their analysis, they discovered an inhomogeneous parenchyma background, fluctuations they hypothesized to be small perfusion variations on a subsegmental scale.

Another approach to overcome the interfering factors was developed by Hao, Bruce, Pislaru, and Greenleaf (2000). A gradient co-occurrence matrix and features from wavelet analysis were combined to form a vector for each pixel. At each location, this vector was compared to others on a global basis, as well as to its nearest neighbors on a "geographic similarity" basis for regional classification. This method was applied to 8.5-MHz intracardiac images of a pig under controlled laboratory conditions, and segmentation regions were compared to physicians' classification and histology (as shown in Figure 9.33).

9.7 Aberration Correction

9.7.1 General Methods

If the design of a typical ultrasound standard-phased array beamformer is based on propagation into a homogeneous medium with speed of sound of 1.54 mm/μs, how well does it work in the body with heterogeneous tissue? To first order, array imaging works

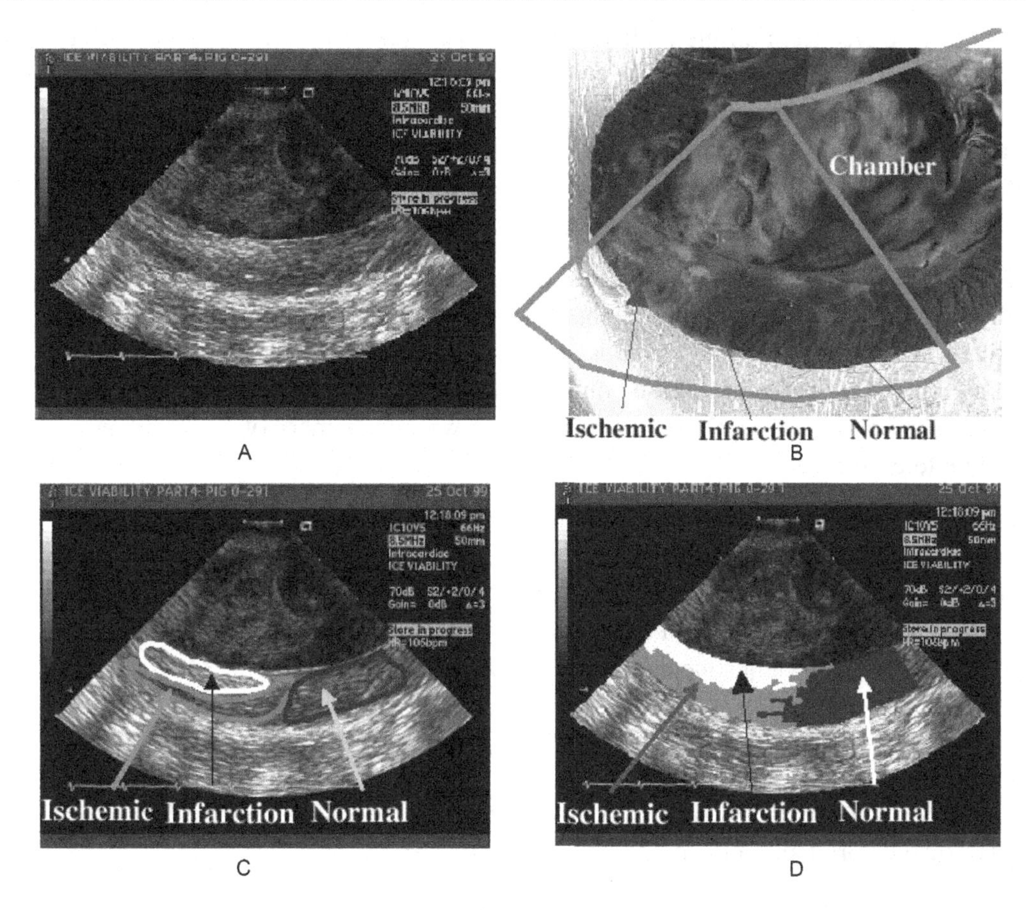

Figure 9.33
Segmentation of the ischemic myocardium.
(A) Original ultrasound image; (B) the pathology gross image of the heart; (C) three regions predicted by a cardiologist; (D) the segmented classification results. The reader is referred to the Web version of the book to see the figure in color. *Reprinted with permission from Hao et al. (2000), IEEE.*

reasonably well, mainly because of the similarity among the sound speeds in different types of tissues (discussed in Section 1.2). These beamformers depend on coherent phasing of identical waveforms, but as seen in Section 9.4, propagation in real tissue breaks up wavefronts and amplitude consistency. As a result, less than ideal performance is achieved (as discussed in Section 7.9.1 for defective or imperfect array elements). It is well known that some people are difficult to image. These topics lead us to the subject of aberration (a term for focusing errors) and the possibility of correcting for these body propagation effects to recover the potential image quality available in an imaging system.

The first place people have looked for the causes of aberration is the body wall, which can act as a dirty acoustic window by causing spatially localized differences in amplitude and time delay to waves from different elements in an array (as explained in Section 9.3.1). A cross-section of an abdominal wall (as illustrated by Figure 9.7) indicates heterogeneity (the presence of several tissue types), inhomogeneity (variations within a type of tissue), and an irregular thickness. From the simulation, the wavefront of an incoming plane wave is broken up owing to a number of effects: scattering, reverberation, refraction, and the cumulative differences in time delay passages through different thicknesses of the wall.

A variety of correction methods have been devised to solve this difficult problem. Solutions usually involve two steps: determining the degree of aberration and correcting for it in an adaptive way. In addition, some methods are suited for real-time correction; others are novel algorithms for corrections based on previous information. Different classes of approaches are reviewed briefly as follows and end with a summary of the problems remaining to be solved.

Adaptive methods of aberration correction have been reviewed by Steinberg (1992); Ng, Worrell, Freiburger, and Trahey (1994); and Yi (1997). The earliest reported method was by Muller and Buffington (1974) for improving images degraded by atmospheric effects in telescopes. They showed that by maximizing the intensity integral J in the image plane through real-time adjustments of time delay,

$$J = \int I^2(u, v)dudv, \tag{9.13}$$

the image can be corrected. Steinberg points out that this operation is equivalent to a multilag spatial correlation operation.

In the 1980s, research on microwave antennas and ultrasound demonstrated that random backscatter, such as that from tissue, can provide measurement of the phase error (Attia and Steinberg, 1989; Flax & O'Donnell, 1988; O'Donnell & Flax, 1988). Another approach was based on the idea embodied in Eqn 9.13, adjusting element delays to maximize regions of target brightness as a quality factor for both point-like and diffuse targets (Nock, Trahey, & Smith, 1989; Trahey, Zhao, Miglin, & Smith, 1990). These methods are based on the premise that the distortions occur in a thin-phase screen adjacent to the transducer. The disruption of coherence is considered to be primarily a phasing effect. Cross-correlation of signals on adjacent elements using a known reference or beacon target or, more practically, random backscatter from tissue, can provide measurement of the phase error. Once the error is known, a delay of the opposite sign and equal to the error is applied to compensate for aberration. Iteration can be applied to reduce the error.

There are a variety of implementations involving adjacent elements, groups of elements, or a summed array waveform for the reference waveform used for each correlation (Ng et al., 1994). Those using a small correction reference region are more sensitive to missing or partially inoperable elements (O'Donnell & Engeler, 1992), cumulative errors, and noise than others using larger reference areas. However, because of the van Cittert–Zernike theorem (see Section 8.4.5), elements decorrelate more toward the ends of the array for random scatterers, and this effect presents a possible limitation for large arrays. For 1.5-dimensional (1.5D) arrays, larger separations between elevation rows can cause delay jumps that are difficult to overcome in some situations.

With access to body wall measurements, Wang's group at the University of Rochester explored aberration correction for one-dimensional (1D) to two-dimensional (2D) arrays (Liu and Waag, 1995, 1998). They simulated aberration by passing a spherical wavefront through the distortions of actual measured abdominal body wall data. Their method consisted of calculating a reference waveform based on all the waveforms received by the array and their cross-correlation properties, smoothing the delays obtained by cross-correlating each waveform by the reference waveform, correcting for known geometric delays, back-propagating the wavefront using an angular spectrum-of-waves approach until a waveform similarity criterion was maximized, smoothing the wavefront again to remove unusual spikes, and adjusting the wavefront by using arrival time estimates for time shift compensation. While this method is definitely not real-time, it showed that by back-propagating, lower beam sidelobes could be obtained.

This extra step indicates that the infinitely thin phase screen, assumed in other methods, may not be valid, and an approximation to a finite thickness wall could be obtained by placing the phase screen within the location of the wall. Further studies showed that improvements could be obtained by correcting the transmitted waveforms (Lacefield & Waag, 2001) as well as the received wavefronts, and by utilizing 2D arrays (Liu and Waag, 1998).

Li (1997) pointed out another overlooked problem in aberration correction: an overemphasis on the focal plane. The focal plane assumption inherent in most correction approaches is that the waveforms from all the elements should be identical or "redundant" under ideal conditions. Away from the focal plane, however, waveforms are no longer identical. Also, for echoes from unknown target distributions, there is no prior knowledge of what the wavefront was, even without aberrations. To overcome these problems, Li proposed a new common-midpoint method in the near Fresnel zone (see Section 6.6.2); midpoint methods have been applied in seismic imaging. Li, Robinson, and Carpenter (1997) applied the algorithm to phantoms and volunteers and found an improvement in approximately half the cases.
A version of this method is evolving (Li, 2000; Li & Robinson, 2000a, 2000b) and has been demonstrated on phantoms.

One of the most challenging aberration correction problem is the female breast (Zhu & Steinberg, 1993). In early experiments on aberration correction concentrated on the liver (O'Donnell & Flax, 1988), amplitude fluctuations were found to be negligible. The heterogeneity of the female breast causes significant refraction and multiple scattering, as is evident from measurements (Zhu & Steinberg, 1994) and simulation (Tabei, Mast, & Waag, 2003) and is seen in Figure 9.9. Zhu and Steinberg (1993) and Zhu, Steinberg, and Arenson (1993) conclude that amplitude as well as phase must be corrected for aberration. Liu and Waag (1994) and Odegaard, Halvorsen, Ystad, Torp, and Angelsen (1996) also reported that amplitude correction could be significant. Zhu and Steinberg (1993) show calculations supporting the need for 2D arrays for aberration correction in order to maintain high-quality imaging with low beam sidelobe levels for good contrast resolution. Others investigating 2D apertures for aberration correction include Trahey (1991), Liu and Waag (1995), Li and Robinson (2000a), and O'Donnell and Li (1991).

An ambitious attempt at real-time adaptive imaging was reported by Rigby et al. (1998). Their system consisted of a 128-channel imaging GE LOGIQ 700 system modified to measure and modify delays to each active element with the computing power of 56 PC processors. Active elements of a 2.5–3.75-MHz 1.75D array could be selected via a multiplexer. The array configuration was a 6-row by 96-column linear array arranged like a 1.5D array except that symmetric rows were not connected. The correction algorithm was similar to channel-to-channel correlation (Flax & O'Donnell, 1988; O'Donnell & Engeler, 1992), except a beam-summed waveform was used as a reference for each image line. Corrections were iterated frame to frame. Improved images were demonstrated on a phantom with a phase screen aberrator and a liver. Their report ended with two questions: Does the aberration correction provide any significant improvement in ultrasound imaging? Is the effort required to build the required 2D array system worthwhile? Improvements in methodology and help from Moore's law enabled Trahey's group to devise an adaptive aberration correcting system with 1.75 D arrays to work at quasi-real-time rates (Dahl, McAleavey, Pinton, Soo, & Trahey, 2006). Good results were obtained for near-static imaging of phantoms. In order to adapt better to the changing environments of the body, faster high-speed access to the raw RF data streaming into the elements and a capability to iteratively update the aberration algorithm several times per image segment are needed. The inevitable improvements in processing speeds and capabilities of research systems (Chapter 10) make it likely that this challenging problem will be revisited.

Aberration correction remains a technological challenge. As the more realistic tissue simulations in Figures 9.8 and 9.10 show, there are many ways wavefront distortion occurs. The RF element data of a cardiac-phased array (plotted in Figure 9.11) reveals that wavefronts, even after normal focusing, are not aligned, are often tilted, and do not extend across all elements. In addition, an imaging transducer compresses a body wall, and its position varies with natural body motions. More on aberration can be found in Section 12.5.5, which shows the effect of body walls on harmonic imaging (see Figures 12.19, 12.20, and 12.21), which provides partial imaging improvements; the next two sections discuss remarkable aberration methods for focusing through the skull.

Several imaging system manufacturers offer a type of "aberration correction." One non-adaptive approach breaks the image into homogeneous layers and speeds of sound appropriate to the tissue in the layers, such as fat or liver, rather than employing the standard 1.54 mm/μs speed for the whole image. These more appropriate sound speeds are then used for the calculation of beamformer delays in these regions. Image improvements have been demonstrated.

9.7.2 Time Reversal

Unlike the astronomical case, small "beacon" point targets are not readily available in the body for the error-determining calibration step; therefore, these methods (Fink, 1992; Thomas & Fink, 1996) are more difficult to apply to medical ultrasound. In certain cases, like for kidney stones, they provide an alternative (Wu et al., 1991). The next two sections describe the method of time reversal and phase correction for aberration correction and focusing through the human skull.

Alignment in itself may not solve the problem because of differences in pulse shapes and the need for multiple local alignments for different regions. Time-reversal methods account for these differences and achieve time compression and beam directivity (Wu et al., 1991), but it is unlikely that it will be possible to capture forward-propagating waves from elements at the focusing site. Time-reversal and phase correction methods have been applied to focusing through the skull, a difficult problem that has been solved by advanced aberration correction methodologies and knowledge as described in the next section.

In anticipation of the correction methods of the next section, acoustic time reversal (Fink, 1992, 1997; Fink & Prada, 2001) is explored as a means of aberration correction. For those not familiar with matched filters, an introduction can be found in Section 10.9.2. Consider a simple case with a transducer array and a pressure point source at r_0 emitting a spatial impulse. A received signal v_{ri} along a path to an array element at position r_1 can be described by:

$$v_{ri}(t) = g_R(t) * h_{Ri}(t, r_1 - r_0), \tag{9.14}$$

where from the central diagram, Figure 2.14, g_R is the receiving transducer response on reception and h_R is the backward path diffraction impulse response from a point source at r_0. If the source is now removed and the same transducer element is excited by a voltage impulse then the pressure at r_0 is:

$$p_{ri}(t) = g_T(t) * h_{Ti}(t, r_0 - r_1), \tag{9.15}$$

where g_T is the transmitting transducer response on reception and h_T is the forward path diffraction impulse response. One reasonable assumption for each transducer element of the array is that the electroacoustic properties of the transducer are reciprocal, $g_T(t) = g_R(t)$; another is that the propagation medium is lossless. Also, by diffraction reciprocity:

$$h_{Ti}(t, r_0 - r_1) = h_{Ri}(t, r_1 - r_0). \tag{9.16A}$$

For further simplification, let the time delay between the source and array element be defined as:

$$T_i = |r_i - r_0|/c. \tag{9.16B}$$

Here, an approach different from conventional arrays with time delay sum focusing, a time-reversal mirror, is explained through the use of a matched filter analogy. For usual array focusing, a focusing delay would be applied to each element in accordance with the principles in Chapter 7 and a lateral sinc beam shape could be measured by a hydrophone scanning parallel to the array axis. Instead, a matched filter approach will be applied to the array. Matched filters have beneficial properties such as maximal power transfer and improved resolution, and are introduced and explained in more detail in the next chapter. For now, assume that this type of ideal filter is defined for a function $x(t)$ as:

$$w(t) = Ax^*(-t), \tag{9.17A}$$

which has the Fourier transform:

$$\Im[w(t)] = AX^*(f), \tag{9.17B}$$

where A is a constant (Kino, 1989).

For the case just described by Eqn 9.15, a matched filter can be applied on the forward path for the pressure at r_0 in the form of drive pulse, $s(t)$:

$$p_i(r_0, t) = s(t) * g_T(t) * h_{Ti}(t - T_i). \tag{9.18A}$$

And through the reciprocity relation, Eqn 9.16A, for the second and third factors:

$$p_i(r_0, t) = [g_R^*(-t) * h_{Ri}^*(T_i - t)] * g_R(t) * h_{Ri}(t - T_i). \tag{9.18B}$$

The Fourier transform of the response and equivalent terms lead to:

$$P_i(f) = \{G_R^*(f)[H_{Ri}^*(f)e^{i2\pi fT_i}]\}\{G_R(f)[H_{Ri}(f)e^{-i2\pi fT_i}]\}, \tag{9.19A}$$

and

$$P_i(f) = [G_R^*(f)G_R(f)][H_{Ri}^*(f)H_{Ri}(f)]. \tag{9.19B}$$

The inverse Fourier transform of the resulting pressure is a matched filter:

$$p_i(t) = [g_R^*(-t) * g_R(t)] * [h_{Ri}^*(-t) * h_{Ri}(t)], \tag{9.19C}$$

which is centered at $t = 0$. This process can be repeated for each element of the array:

$$p(r_0, t) = g_R^*(-t) * g_R(t) * \sum_{i=1}^{N} h_{Ti}^*(-t) * h_{Ri}(t), \tag{9.20}$$

recognized as a sum of matched filters.

Derode, Roux, and Fink (1995) demonstrated the remarkable properties of the time-reversal approach with a scattering experiment. A scattering medium was placed between the point source at r_0 and array as shown in top of Figure 9.34 (Derode et al.,1995; Fink & Prada, 2001). The scatterers were 2000 parallel steel rods (0.8 mm in diameter) randomly distributed in a 40-mm wide arrangement and an array operated at 3.5 MHz with 64 elements in a water tank. As the first step in the time-reversal mirror process, the source at r_0 sent pressure waves to the array elements. Shown in this figure are three sample waveforms received by array elements that are described by Eqn 9.14, and h is understood to have a more general meaning than diffraction for water paths and now includes the influence of the scattering medium. In the second step, as shown at the bottom of Figure 9.34, these waveforms were recorded and time-reversed and sent back through the elements with the delays T_i included so that the pressure from each element is as described by Eqn 9.18. As viewed from the source sending signals towards array, the returning time-reversed waveforms appear as if they had come from an acoustic time-reversal mirror.

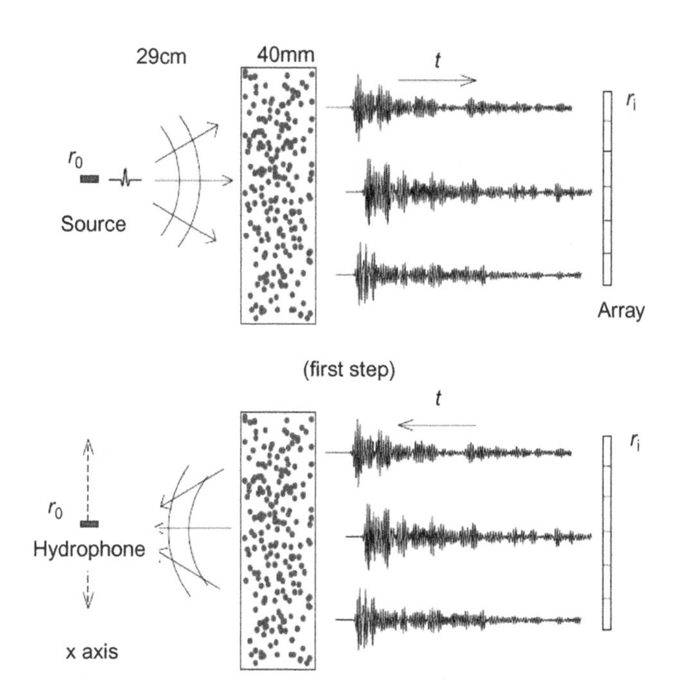

Figure 9.34
Two-step time-reversal process. Recording of waveforms from source at r_0 through a random medium (first step); retransmission of time-reversed waveforms through a random medium towards original source point (second step). *From Fink and Prada (2001) courtesy of Inverse Problems.*

If an exciting source waveform, $s(t)$, for the array elements is included (as in the central diagram), then the pressure–time equations have an additional convolution so that Eqn 9.20 for the total pressure at r_0 becomes:

$$p(r_0, t) = g_R^*(-t) * g_R(t) * s(t) * \sum_{i=1}^{N} h_{Ti}^*(-t) * h_{Ri}(t). \tag{9.21}$$

The excitation $s(t)$ for this example was a tapered 1-μs-long 3.5-MHz three-cycle pulse, shown in Figure 9.35A. A typical waveform received on array element 64 is displayed in Figure 9.35B. Finally, the pressure waveform measured by a hydrophone at r_0 is plotted in Figure 9.35C and would be described by Eqn 9.21. The result of a hydrophone scan along x is the beam plot of Figure 9.36, compared to a classic sinc beam plot, that would have been obtained for traditional array focusing with no intervening scattering medium (water only). What is remarkable is that the diffraction limit in water (FWHM ~6.38 mm) appears to be exceeded by a factor of six (1.05 mm) with the intervening scatter and the time-reversal mirror approach. In a later study (Derode, Tourin, & Fink, 2000), a factor of 30 improvement in resolution was attained with a 16-element array. A reason for the improvement in resolution is that, as shown by the received waveforms, strong multipath scattering occurred and added many higher spatial frequencies than in the pure water path case (no scatterers). The net effect was to create a longer virtual aperture as viewed from r_0 through the scattering medium. In this later study, the beam directivity was mainly determined by the spatial correlation length and not summation of elements. Figure 9.37 shows that the beamwidth of one element is roughly the same as for 122 elements, though the signal-to-noise (peak-to-sidelobe-level) ratio improved. The beam is mainly determined by $\int h_i(0, r_0, t) h_i(x, r_0, t) dt$.

This example is a simplification of the overall spatiotemporal matched methods; a more extensive theoretical background can be found in the many time-reversal reviews (Fink, 1992, 1997; Fink, de Rosny, Lerosey, & Tourin, 2009; Fink et al., 2000), which show that the time-reversal mirror approach has Green's function solutions to a lossless heterogeneous wave equation to be found in Section 9.9. Conditions for this kind of matched filtering are based on assumptions of linearity, spatial reciprocity, time invariance, and lossless media. Extensive development of the time-reversal method for media with losses led to the development of iterative focusing optimization methods which include a spatiotemporal inverse filter for beamforming in lossy media (Montaldo, Aubry, Tanter, & Fink, 2005), a broadband inverse filter (Aubry, Tanter, Gerber, Thomas, & Fink, 2001; Tanter, Aubry, Gerber, Thomas, & Fink, 2001), a speckle-based approach (Robert & Fink, 2008), and many other applications.

The basic time-reversal approach is now compared to the more traditional 1D linear array with cylindrical time delay focusing. A linear array equation comparable to the

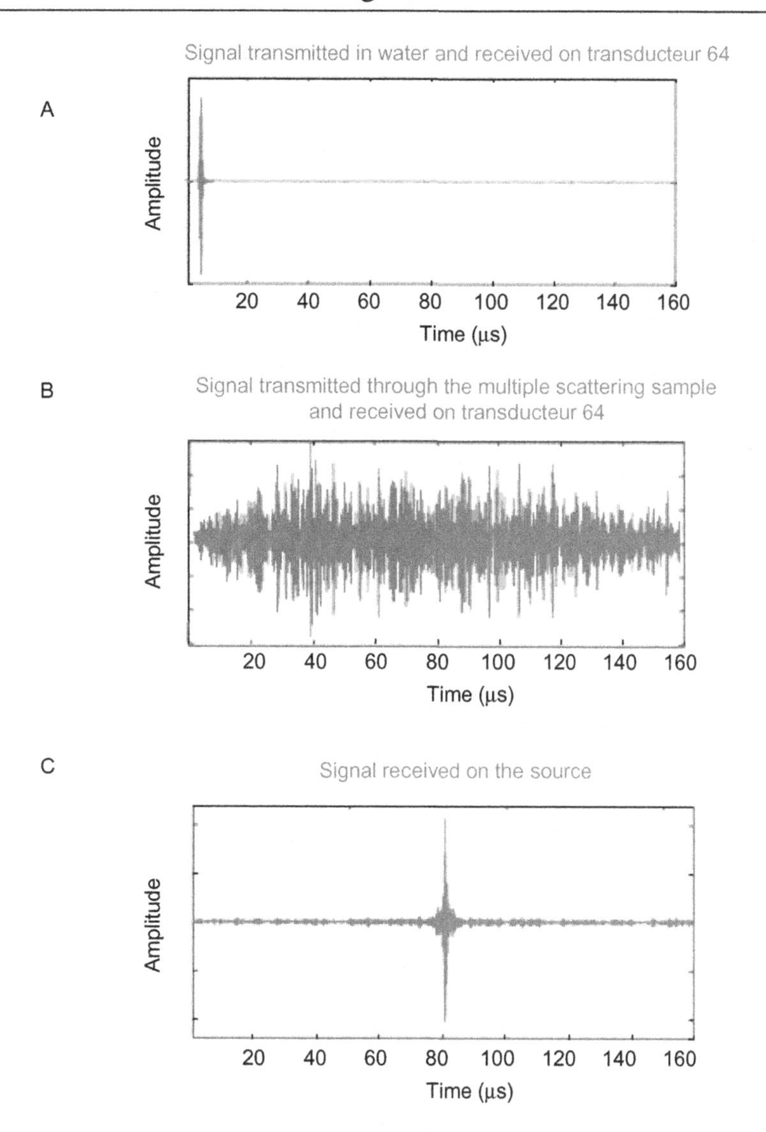

Figure 9.35

Time-reversal process with an exciting source waveform. (A) Signal transmitted by the source; (B) signal recorded by an array element; (C) time-reversed signal observed at the source location. *From Fink and Prada (2001) courtesy of Inverse Problems.*

time-reversal Eqn 9.21 but based on Eqn 7.30 with the focus at r_0 and an aberration correction time delay, τ_{correl}, derived from cross-correlation is:

$$p(r_0, t) = s(t) * g_T(t) * \sum_{i=1}^{N} a_i h_{Txi} \times (t - \tau_{xidiff} - r_0/c) * \delta(t + \tau_{xifoc} + \tau_{correl}) * h_{Ty}(t - \tau_{ydiff}), \tag{9.22A}$$

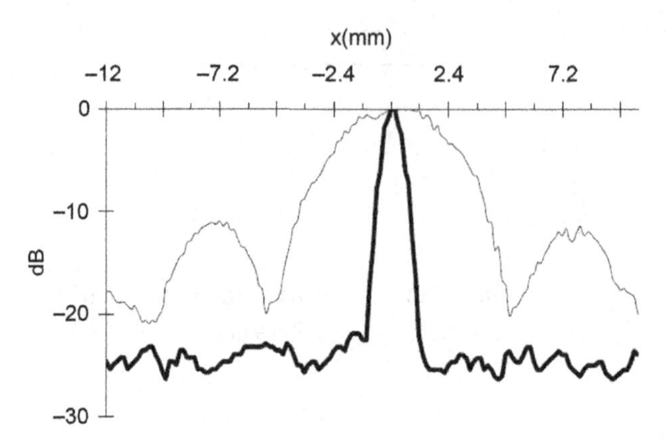

Figure 9.36
Patterns of the time-reversed fields in the source plane through the random medium (Bold Curve) and through a homogeneous fluid (Thin Curve). *From Derode et al. (2000), Acoustical Society of America.*

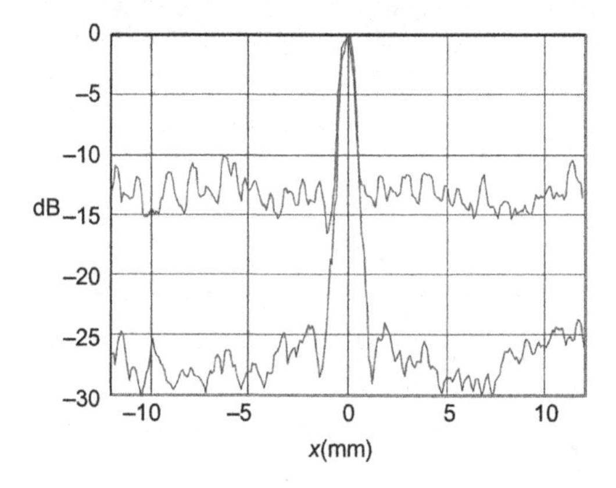

Figure 9.37
Directivity pattern of the time-reversed waves around the source position through the rods. Sample of thickness $L = 40$ mm, for $N = 122$ array elements (bottom curve) and $N = 1$ array element (top curve). The -6-dB resolutions are 0.84 and 0.9 mm, respectively. *From Fink and Prada (2001) courtesy of Inverse Problems.*

where τ_{xifoc} is the beamformer focusing delay, h_{Tx} is the azimuth diffraction response, h_{Ty} is the elevation diffraction response,

$$\tau_{xidiff} = \frac{\sqrt{(x - x_i)^2 + z^2}}{c}, \tag{9.22B}$$

and

$$\tau_{xifoc} = \frac{\sqrt{(x_f - x_i)^2 + z_f^2}}{c}. \tag{9.22C}$$

Under the conditions of propagation in an ideal homogeneous medium with a coincident elevation and azimuth focus along the beam axis at r_0:

$$h_{Txi}(t) * h_{Ty}(t) = \frac{c_0 L_x L_y}{2\pi r_0} \delta(t - r_0/c), \tag{9.22D}$$

and the sum reduces to:

$$p(r_0, t) = \left[\frac{c_0 L_x L_y}{2\pi r_0}\right] Na[g_T(t) * s(t)], \tag{9.22E}$$

with no apodization. This relationship shows that the advantages of an ideal time delay beamformer are its ability to produce a one-way gain proportional to Na based on its coherence (all the pulses line up) and its mean value properties. Unfortunately, Eqn 9.22 is based on the assumption that the wavefronts will converge at a single point scatterer, but this is rarely the case. Figure 9.11, which shows RF pulse–echo responses for each element for one transmitted line in a cardiac image after focusing alignment from a phased array, shows that coherence of pulse echoes rarely extends across the elements of an array. This type of mean array processing emphasizes horizontally aligned groups of pulses (after focusing) and suppresses those not aligned. In a typical case, even time alignment achieved by cross-correlation, τ_{correl}, may not insure coherence because of pulse distortion by scattering (see Figure 9.8), inclined reflected wavefronts, and absorption (see Figure 7.30) along different element-to-target paths. In summary, when absorption, scattering, and heterogeneity occur during propagation, h_{Txi} and h_{Ty} may not be similar for each element-to-target path, and weaker coherency results in less array gain and less resolution. Using an effective speed of sound c for the path or cross-correlation corrections is only a partial remedy and the full gain of the array cannot be realized under these circumstances.

9.7.3 Focusing through the Skull

Despite the fact that the first known ultrasound images by the Dussik brothers (Figure 1.4) were of the head, ultrasound imaging of the head and bones has been mainly avoided because of known difficulties in penetrating the skull and severe refraction and distortion artifacts. Since early interest in brain imaging and therapy (Fry, 1977; Fry, Eggleton, & Heimburger, 1974; Smith, Phillips, von Ramm, & Thurstone, 1977), the motivation for performing noninvasive transcranial surgery has been renewed with remarkable innovative solutions that were considered to be technically impossible previously. The aberration correction technology is described in this section and the clinical applications appear in Chapter 17, on high-intensity therapeutic ultrasound.

For transcranial focusing to be successful, an independently acquired accurate three-dimensional spatial map of the acoustical properties of the skull is needed. Earlier studies measured average skull properties (Fry & Barger, 1978) but lacked the detailed knowledge of the local three-dimensional shape and varying thickness necessary to perform aberration correction. High-resolution computed tomography (CT) and magnetic resonance imaging (MRI) methods have since provided the means of creating these maps. In order to correct for the locally varying severe aberrations of the skull and to accurately focus a beam intense enough to deliver a sufficient therapeutic dose to the brain, spherically focusing phased arrays with many individually addressable elements (Ebbini & Cain, 1991) were preferable, as illustrated by the simulated array and skull in Figure 9.38 (Pajek and Hynynen, 2012a, 2012b).

Images of the skull architecture were insufficient, by themselves, for noninvasive focusing; this structural information had to be translated into the acoustic properties needed for adaptively focusing into the brain. White, Clement, and Hynynen (2006) measured longitudinal and shear sound speeds at several locations on ex vivo skulls using the ray refractive model shown in Figure 9.39 for effects which are described in Section 3.3.3. They found that, on average, the shear sound speed was 1500 m/s and the longitudinal sound speed was 2820 m/s. For an incident longitudinal angle past the critical angle, complete conversion to shear waves with a sound speed close to soft tissues had its advantages. Yousefi, Goertz, and Hynynen (2009) combined this information with signal processing to show the advantages of improving the focusing resolution of both

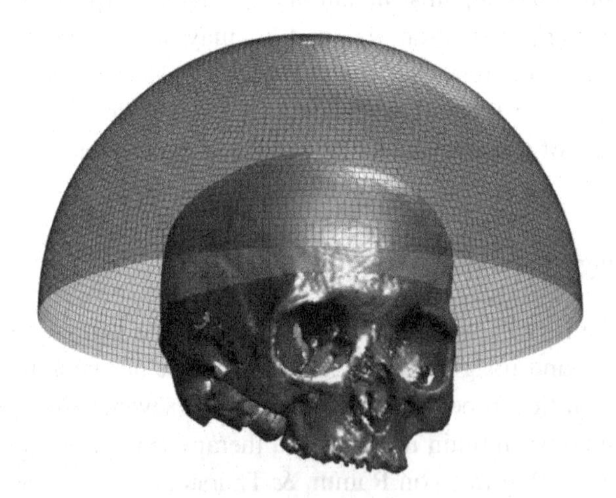

Figure 9.38

Simulated adaptive transducer array with 8907 elements positioned over segmented skull. Region of skull surface accessible by array highlighted (appears dark gray/black in gray-scale print version; readers are referred to the Web version of the book to see the figure in color). *From Pajek and Hynynen (2012b), Acoustical Society of America.*

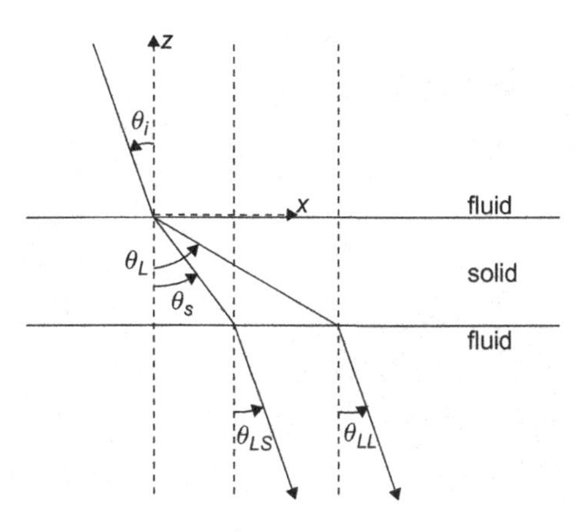

Figure 9.39

Simplified ray model of ultrasound transmission through the human skull immersed in water. The incident compressional wave in the fluid divides into a longitudinal and a shear wave at the fluid–skull interface. At the second interface, the two propagating waves individually transmit through the interface, where the shear wave is converted back to a longitudinal wave. Reflections at the interfaces are neglected. *From White et al. (2006), with permission from the World Federation of Ultrasound in Medicine and Biology.*

longitudinal and shear waves through the skull. In another study (Pichardo, Sin, & Hynynen, 2011), CT images were used to reconstruct the three-dimensional skull surface; more detailed local measurements of longitudinal sound speed and attenuation of both cortical and trabecular skull bone were made as a function of frequency. Connor, Clement, and Hynynen (2002) calculated values of the local variations in the speed of sound through the skull by an analysis of CT scans of skull density determined by a genetic optimization algorithm and a nonlinear finite-difference time-difference Westervelt model (Chapter 12). This nonlinear model was later used to simulate propagation through the skull (Pichardo et al., 2011). Pinton et al. (2011) determined that a significant part of the absorption was due to nonlinear effects (see Chapter 12 for more on nonlinear propagation). Aubry, Tanter, Pernot, Thomas, and Fink. (2003) derived spatially local sound speed and attenuation values based on a porosity model (see Figure 9.40 for an example) and incorporated them in a finite-difference heterogeneous wave equation propagation model (see Section 9.9). Pinton et al. (2012a) found that only a small part of the attenuation was absorption responsible for heat deposition; the rest was scattering and mode conversion (dependent on incident angle).

In addition to the determination of the peculiar characteristics of acoustic propagation through the skull, advancements have been made in trans-skull focusing techniques. An early attempt to apply time-reversal focusing through the skull (Thomas & Fink, 1996) found that accounting for loss through the skull was needed to improve focusing and a

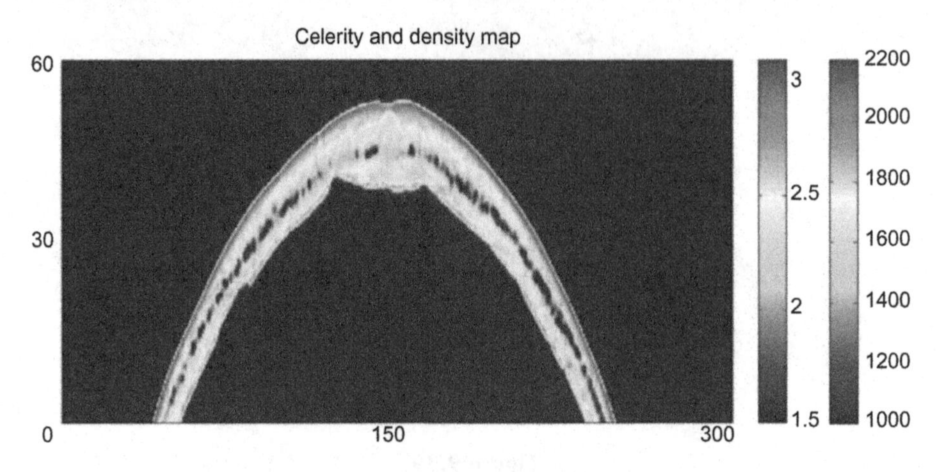

Figure 9.40

Example of one slice in speed of sound and density map derived from skull microstructure in a CT scan. First color bar indicates speed of sound in mm/μs, second color bar density in kg/m^3. The axes are in mm. The reader is referred to the Web version of the book to see the figure in color. *From Marquet et al. (2009). © IOP Publishing. Reproduced by permission of IOP Publishing. All rights reserved.*

phase screen and back-projection method were investigated. Tanter, Thomas, and Fink (1998) reported on an improved time-reversal method for dealing with attenuation and steering. Hynynen and Jolesz (1998) demonstrated that a therapeutic-level dose could be delivered through a rabbit skull by focusing improved through phase correction. Hynynen and Sun (1999) showed the feasibility of using MRI-based skull phase correction with 72 elements of a 1.1-MHz bowl array, illustrated by field contour plots of the focal plane in Figure 9.41. Clement and Hynynen (2002) examined the effects of non-normal incidence angles, steering, and nonuniform skull density in CT images with an angular spectrum model (Section 6.7) and a 0.74-MHz 320-element spherically curved array, based on high-resolution (0.2-mm) CT scans. While phase-based focusing may be adequate, Aubry et al. (2003) argued that the heterogeneity of the skull made a difference at higher frequencies (~1 MHz). They applied an amplitude-corrected time-reversal approach (Thomas & Fink, 1996).

Tanter et al. (1998) were able to implement a time-reversal approach without the usual requirement for a source or receiver at the focal point. Instead, they derived the acoustic properties from image-based characterization of the skull and its inner wall structure. Then acoustic waves from a virtual point source through the skull were simulated by computer to create the needed time-reversed signals. Simulations of time-reversed wavefronts from a 128-element 1.5-MHz linear array are compared with data measured by a hydrophone in Figure 9.42. Beam plots in the focal plane compared favorably with model predictions and poorly with straight time-delay correction, as indicated by Figure 9.43. Furthermore, they

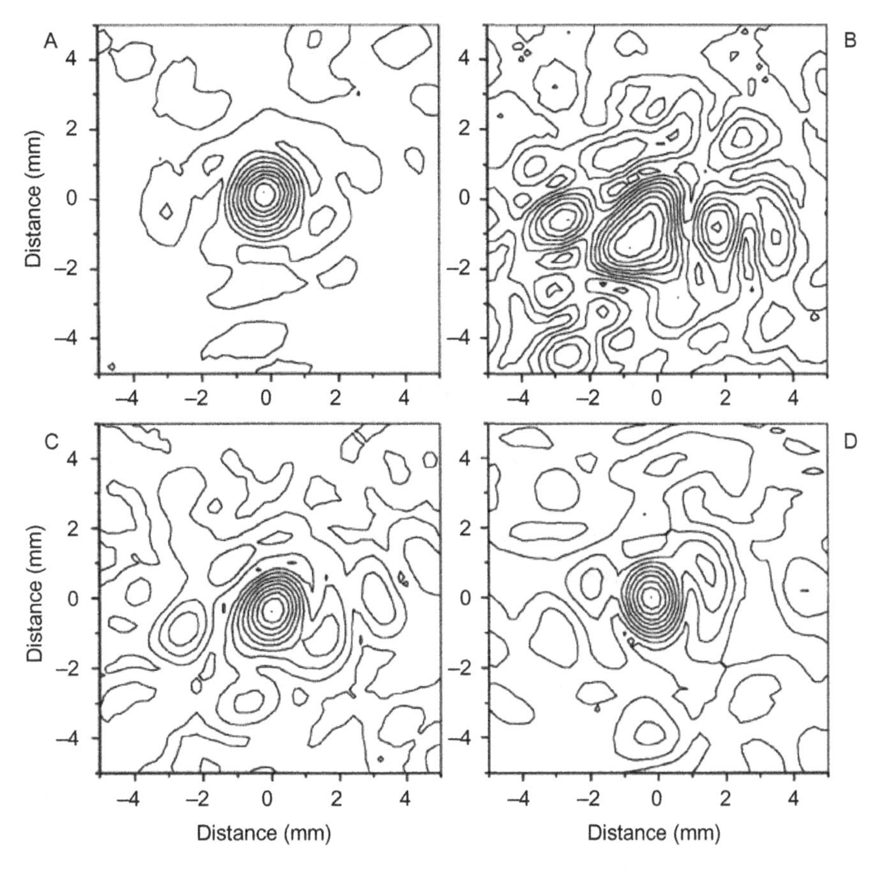

Figure 9.41
The ultrasound pressure amplitude field distribution across the focal plane measured in water. (A) No skull and uniform phase; (B) skull with uniform phase; (C) skull with MRI-derived phase correction (95% amplitude improvement over (B)); and (D) skull with hydrophone-derived phase correction (3% amplitude improvement over (C)). *From Hynynen and Sun (1999), IEEE.*

argued, accounting for heterogeneity in the skull walls provides more accurate beam resolution and placement. The previously described time-reversal method based on the porosity CT-image-derived parameters was implemented with a 300-element piezocomposite spherically curved 1-MHz array with a 14-cm radius of curvature (Marquet et al., 2009). Focal plane amplitudes with the image-based focusing achieved were within 0.9 dB of hydrophone-based focusing and 10 dB better than no focusing compensation. Pinton, Aubry, and Tanter (2012b) developed a simplified method of determining phase correction by first ascertaining effective parameters for the skull thickness based on CT image microstructure and then using a hybrid method of finite-difference calculations

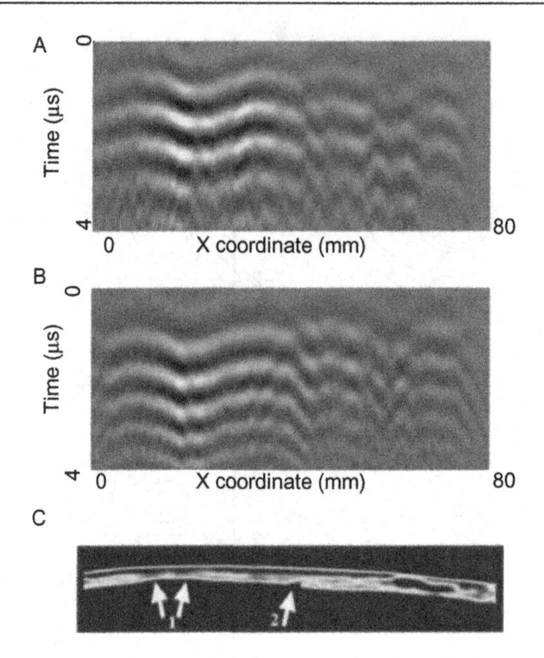

Figure 9.42
Wavefront after propagating through the skull. (A) Experiment; (B) simulation; (C) corresponding porosity map of the skull. Arrows indicate skull heterogeneities that result in large wavefront distortions. *From Aubry et al. (2003), Acoustical Society of America.*

through the skull and phase projection from the surface to the transducer, thereby reducing computational wave propagation calculation requirements significantly.

Adaptively focusing arrays for focusing through the skull into the brain have been evaluated clinically. A clinical transcranial hemispherical phased array, the Exblate 3000, was developed by Insightec (McDannold, Clement, Black, Jolesz, & Hynynen, 2010). The 30-cm-diameter 670-kHz array had 512 elements and could deliver 800 watts of acoustic power. The system had the capabilities of adaptive focusing, MRI thermal monitoring, and water cooling. The clinical aspects of these types of systems will follow in Chapter 17. A second system, designed by SuperSonic Imagine, also with 512 elements and operating at 1 MHz, used CT-based time-reversal methods to achieve successful focusing inside 12 fresh human cadaver heads (Chauvet et al., 2013). Temperature elevations of up to 51°C were achieved within 1 mm of the target site as monitored by thermometric MRI in real time.

9.8 Wave Equations for Tissue

For a homogeneous soft tissue medium with loss, the appropriate wave equation from Chapter 4 is:

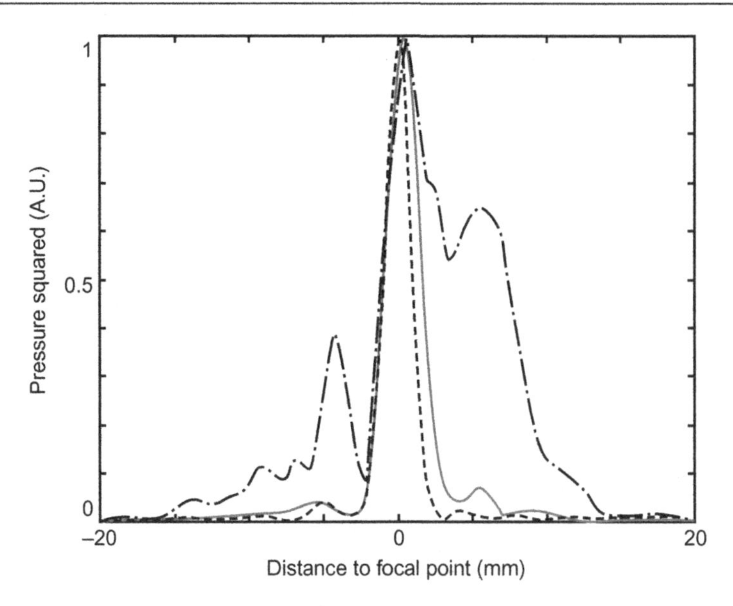

Figure 9.43
The directivity pattern through the skull. Energy (pressure squared) is plotted on a linear scale after the transmission of time-reversal waveforms, with amplitude compensation based on signals experimentally acquired (dotted line) and based on simulated acoustic signals (gray solid line) compared to conventional cylindrical law time-delay focusing (dash-dotted line). *From Aubry et al. (2003), Acoustical Society of America.*

$$\nabla^2 p - \frac{1}{c_0^2}\frac{\partial^2 p}{\partial t^2} - L_\gamma * p = 0 \tag{9.23A}$$

This wave equation translates into the frequency domain in the following form for pressure (P):

$$\nabla^2 p - k_0^2 P - \gamma(\omega)P = 0, \tag{9.23B}$$

where

$$\gamma(\omega) = -\alpha(\omega) - i\beta_E(\omega) \tag{9.23C}$$

is from Chapter 4, and the wave number is $k_0 = \omega/c_0$.

In order to include local variations in sound speed and density as a function of position, the wave equations must change. The wave equation form most frequently used for the Born approximation (Angelsen, 2000; Fellingham & Sommer, 1984; Jensen, 1991; Ystad, Halvorsen, Odegaard, Angelsen, & Lygren, 1996) is adapted here to include general frequency power law loss. If the density and compressibility as a function of position are divided into an ambient term with a subscript a and a small perturbation term denoted by subscript f (Ystad et al., 1996):

$$\rho(r) = \rho_a(r) + \rho_f(r), \quad \text{and} \tag{9.24A}$$

$$\kappa(r,\omega) = \kappa_a(r,\omega) + \kappa_f(r,\omega), \tag{9.24B}$$

and the following are introduced:

$$\varsigma(r) = \frac{\kappa_f(r,\omega)}{\kappa_a(r,\omega)}, \quad \text{and} \tag{9.24C}$$

$$\psi(r) = \frac{\rho_f(r)}{\rho_a(r)}. \tag{9.24D}$$

Then the wave equation in the frequency domain can be written as:

$$(\nabla^2 + k_1^2)P - \gamma(r)P = -k_1^2\varsigma(r)P + \nabla[\Psi(r)\nabla P], \tag{9.25A}$$

in which

$$k_1^2(r,\omega) = \omega^2/c_a^2(r) = \omega^2\rho_a(r)\kappa_a(r), \tag{9.25B}$$

and

$$\gamma(r,f) = -\alpha(r,f) - i\beta_E(r,f), \tag{9.25C}$$

and the definitions in Chapter 4 for $\alpha(f)$ (Eqn 4.6A), and $\beta_E(f)$ (Eqn 4.18), can be used except that $\alpha_0(r)$ and $\alpha_1(r)$ are now functions of position. The time domain counterpart of Eqn 9.16A can be written as:

$$\nabla^2 p - \frac{1}{c_a^2}\frac{\partial^2 p}{\partial t^2} - L_\gamma(r) * p = \frac{\varsigma(r)}{c_a^2}\frac{\partial^2 p}{\partial t^2} + \nabla[\Psi(r)\nabla P], \tag{9.26}$$

where $L_\gamma(r)$, from Eqn 4.17, is now a function of position through $\alpha(r)$.

Note that the left-hand sides of these wave equations can be considered to be the part describing propagation through the homogeneous ambient or average tissue material (Angelsen, 2000; Ystad et al., 1996). This part of the equation provides geometric propagation through large-scale homogeneous regions. If the right-hand side is replaced by a delta function, $\delta(r - r_0)$, the solution is a freely propagating Green's function (Angelsen, 2000; Ystad et al., 1996), modified for losses and dispersion. The right-hand sides of Eqns 9.2A and 9.2B represent perturbations to this background average value.

Full linear wave equations used for calculations of propagation in heterogeneous tissue are treated in more detail in Mast, Hinkelman, Orr, Sparrow, and Waag (1997) and Wojcik et al. (1997).

References

American Institute of Ultrasound in Medicine (1990). *Standard methods for measuring performance of pulse–echo ultrasound imaging equipment*. Laurel, MD: Author.

Anderson, M. E., Soo, M. S. C., & Trahey, G. E. (1998). Microcalcifications as elastic scatterers under ultrasound. *IEEE Transactions on Ultrasonics, Ferroelectrics and Frequency Control, 45*, 925–934.

Angelsen, B. A. J. (2000). *Ultrasound imaging: waves, signals, and signal processing*. Trondheim, Norway: Emantec.

Aristizábal, O., Christopher, D. A., Foster, F. S., & Turnbull, D. H. (1998). 40-MHz echocardiography scanner for cardiovascular assessment of mouse embryos. *Ultrasound in Medicine and Biology, 24*, 1407–1417.

Aristizábal, O., Mamou, J., Ketterling, J. A., & Turnbull, D. H. (2013). High-throughput, high-frequency 3D ultrasound for in utero analysis of embryonic mouse brain development. *Ultrasound in Medicine and Biology 39*, 2321–2332.

Attia, H. A., & Steinberg, B. D. (1989). Self-cohering large antenna arrays using the spatial correlation properties of radar clutter. *IEEE Transactions on Antennas and Propagation, 37*, 30–38.

Aubry, J. -F., Tanter, M., Gerber, J., Thomas, J. -L., & Fink, M. (2001). Optimal focusing by spatio-temporal inverse filter: II. Experiments. Application to focusing through absorbing and reverberating media. *Journal of the Acoustical Society of America, 110*, 48–58.

Aubry, J. -F., Tanter, M., Pernot, M., Thomas, J. -L., & Fink, M. (2003). Experimental demonstration of noninvasive trans-skull adaptive focusing based on prior computed tomography scans. *Journal of the Acoustical Society of America, 113*, 84–93.

Azrif, M., Ranieri, S., Giles, A., Debeljevic, B., Kolios, M. C., & Czarnota, G. J. (2007). Conventional low-frequency ultrasound detection of apoptosis. *Proceedings of the American Institute of Ultrasound in Medicine Annual Convention (New York)*S185.

Baddour, R. E., Sherar, M. D., Czarnota, G. J., Hunt, J. W., Taggart, L., Giles, A., et al. (2002). High frequency ultrasound imaging of changes in cell structure including apoptosis. *Proceedings of the IEEE ultrasonics symposium* (Vol. 2) (pp. 1639–1644).

Bamber, J. C. (1998). Ultrasonic properties of tissue. In F. A. Duck, A. C. Baker, & H. C. Starritt (Eds.), *Ultrasound in medicine*. Bristol, UK: Institute of Physics Publishing.

Banihashemi, B., Vlad, R., Debeljevic, B., Giles, A., Kolios, M. C., & Czarnota, G. J. (2008). Ultrasound imaging of apoptosis in tumor response: Novel preclinical monitoring of photodynamic therapy effects. *Cancer Research, 68*, 8590–8596.

Bridal, S. L., Wallace, K. D., Trousil, R. L., Wickline, S. A., & Miller, J. G. (1996). Frequency dependence of acoustic backscatter from 5 to 65 MHz ($0.06 < ka < 4.0$) of polystyrene beads in argose. *Journal of the Acoustical Society of America, 100*, 1841–1848.

Ce'spedes, I., & Ophir, J. (1990). Diffraction correction methods for pulse-echo acoustic attenuation estimation. *Ultrasound in Medicine and Biology, 16*, 707–717.

Chaturvedi, P., & Insana, M. F. (1996). Error bounds on ultrasonic scatterer size estimates. *Journal of the Acoustical Society of America, 100*, 392–399.

Chauvet, D., Marsac, L., Pernot, M., Boch, A. -L., Guillevin, R., Salameh, N., et al. (2013). Targeting accuracy of transcranial magnetic-resonance-guided high-intensity focused ultrasound brain therapy: A fresh cadaver model. *Journal of Neurosurgery, March*, 1–7.

Chen, J. -F., & Zagzebski, J. A. (1996). Frequency dependence of backscatter coefficient versus scatterer volume fraction. *IEEE Transactions on Ultrasonics, Ferroelectrics and Frequency Control, 43*, 345–353.

Chen, J. -F., Zagzebski, J. A., & Madsen, E. L. (1993). Tests of backscatter coefficient measurement using broadband pulses. *IEEE Transactions on Ultrasonics, Ferroelectrics and Frequency Control, 40*, 603–607.

Chen, X., Phillips, D., Schwarz, K. Q., Mottley, J. G., & Parker, K. J. (1997). The measurement of backscatter coefficient from a broadband pulse-echo system: A new formulation. *IEEE Transactions on Ultrasonics, Ferroelectrics and Frequency Control, 44*, 515–525.

Christopher, D. A., Starkoski, B. G., Burns, P. N., & Foster, F. S. (1997). High-frequency pulsed Doppler ultrasound system for detecting and mapping blood flow in the microcirculation. *Ultrasound in Medicine and Biology*, *23*, 997–1015.

Clement, G. T., & Hynynen, K. (2002). A non-invasive method for focusing ultrasound through the human skull. *Physics in Medicine and Biology*, *47*, 1219–1236.

Connor, C. W., Clement, G. T., & Hynynen, K. (2002). A unified model for the speed of sound in cranial bone based on genetic algorithm optimization. *Physics in Medicine and Biology*, *47*, 3925–3944.

Coolen, J., Engelbrecht, M. R., & Thijssen, J. M. (1999). Quantitative analysis of ultrasonic B-mode images. *Ultrasonic Imaging*, *21*, 157–172.

Czarnota, G. J., & Kolios, M. C. (2010). Ultrasound detection of cell death. *Imaging in Medicine*, *2*, 17–28.

Czarnota, G. J., Kolios, M. C., Abraham, J., Portnoy, M., Ottensmeyer, F. P., Hunt, J. W., et al. (1999). Ultrasound imaging of apoptosis: High-resolution non-invasive monitoring of programmed cell death in vitro, in situ and in vivo. *British Journal of Cancer*, *81*, 520–527.

Czarnota, G. J., Kolios, M. C., Hunt, J. W., & Sherar, M. D. (2001). Ultrasound imaging of apoptosis: DNA-damage effects visualized. In V. V. Didenko (Ed.), *"Methods in molecular biology," Chap. 20. In situ detection of dna damage: methods and protocols* (Vol. 203). Totowa, NJ: Humana Press.

Czarnota, G. J., Kolios, M. C., Vaziri, H., Benchimol, S., Ottensmeyer, F. P., Sherar, M. D., et al. (1997). Ultrasonic biomicroscopy of viable, dead and apoptotic cells. *Ultrasound in Medicine and Biology*, *23*, 961–965.

Dahl, J. J., McAleavey, S. A., Pinton, G. F., Soo, M. S., & Trahey, G. E. (2006). Adaptive imaging on a diagnostic ultrasound scanner at quasi real-time rates. *IEEE Transactions on Ultrasonics, Ferroelectrics and Frequency Control*, *53*, 1832–1843.

Dapore, A. J., King, M. R., Harter, J., Sarwate, S., Oelze, M. L., Zagzebski, J. A., et al. (2011). Analysis of human fibroadenomas using three-dimensional impedance maps. *IEEE Transactions on Medical Imaging*, *30*, 1206–1213.

D'Astous, F. T., & Foster, F. S. (1986). Frequency dependence of ultrasound attenuation and backscatter in breast tissue. *Ultrasound in Medicine and Biology*, *12*, 795–808.

Derode, A., Roux, P., & Fink, M. (1995). Robust acoustic time reversal with high-order multiple scattering. *Physical Review Letters*, *75*, 4206–4210.

Derode, A., Tourin, A., & Fink, M. (2000). Limits of time-reversal focusing through multiple scattering: Long-range correlation. *Journal of the Acoustical Society of America*, *107*, 2987–2998.

Ebbini, E. S., & Cain, C. A. (1991). A spherical-section ultrasound phased-array applicator for deep localized hyperthermia. *IEEE Transactions on Biomedical Engineering*, *38*, 634–643.

Feleppa, E. J., Ennis, R. D., Schiff, P. B., Wuu, C. -S., Kalisz, A., Ketterling, J., et al. (2001). Spectrum-analysis and neural networks for imaging to detect and treat prostate cancer. *Ultrasonic Imaging*, *23*, 135–146.

Feleppa, E. J., Lizzi, F. L., Coleman, D. J., & Yaremko, M. M. (1986). Diagnostic spectrum analysis in ophthalmology. *Ultrasound in Medicine and Biology*, *12*, 623–631.

Fellingham, L. A., & Sommer, F. G. (1984). Ultrasonic characterization of tissue structure in the in vivo human liver and spleen. *IEEE Transactions on Sonics and Ultrasonics*, *31*, 418–428.

Fink, M. (1992). Time reversal of ultrasonic fields, Part I: Basic principles. *IEEE Transactions on Ultrasonics, Ferroelectrics and Frequency Control*, *39*, 555–566.

Fink, M. (1997). Time-reversed acoustics. *Physics Today*, *20*, 34–40.

Fink, M., Cassereau, G., Derode, A., Prada, C., Roux, P., Tanter, M., et al. (2000). Time-reversed acoustics. *Reports of Progress in Physics*, *63*, 1933–1995.

Fink, M., de Rosny, J., Lerosey, G., & Tourin, A (2009). Time-reversed waves and super-resolution. *Comptes Rendus Physique*, *10*, 447–463.

Fink, M., & Prada, C. (2001). Acoustic time-reversal mirrors. *Inverse Problems*, *17*, R1–R38.

Flax, S. W., & O'Donnell, M. (1988). Phase aberration correction using signals from point reflectors and diffuse scatterers: Basic principles. *IEEE Transactions on Ultrasonics, Ferroelectrics and Frequency Control*, *35*, 758–767.

Foster, F. S., Liu, G., Mehi, J., Starkoski, B. S., Adamson, L., Zhou, Y., et al. (2000a). High frequency ultrasound imaging: From man to mouse. *Proceedings of the IEEE ultrasonics symposium* (Vol. 2) (pp. 1633–1638). San Juan.

Foster, F. S., Pavlin, C. J., Harasiewicz, K. A., Christopher, D. A., & Turnbull, D. H. (2000b). Advances in ultrasound biomicroscopy. *Ultrasound in Medicine and Biology*, *26*, 1–27.

Foster, F. S., Pavlin, C. J., Lockwood, G. R., Ryan, L. K., Harasiewicz, K. A., Berube, L. R., et al. (1993). Principles and applications of ultrasound backscatter microscopy. *IEEE Transactions on Ultrasonics, Ferroelectrics and Frequency Control*, *40*, 608–617.

Fry, F. J. (1977). Trans-skull transmission of an intense focused ultrasonic beam. *Ultrasound in Medicine and Biology*, *3*, 179–184.

Fry, F. J, & Barger, J. E. (1978). Acoustical properties of the human skull. *Journal of the Acoustical Society of America*, *63*, 1576–1590.

Fry, F. J., Eggleton, R. C., & Heimburger, R. F. (1974). Trans-skull visualization of brain using ultrasound: An experimental model study. *Exerpta Medica*, *1974*, 97–103.

Goertz, D. E., Christopher, D. A., Yu, J. L., Kerbel, R. S., Burns, P. N., & Foster, F. S. (2000). High frequency colour flow imaging of the microcirculation. *Ultrasound in Medicine and Biology*, *26*, 63–71.

Greenleaf, J. F., & Sehgal, C. M. (1992). *Biologic system evaluation with ultrasound*. New York: Springer Verlag.

Hall, C. S., Marsh, J. N., Hughes, M. S., Mobley, J., Wallace, K. D., Miller, J. G., et al. (1997). Broadband measurements of attenuation coefficient and back-scatter coefficient for suspensions: A potential calibration tool. *Journal of the Acoustical Society of America*, *101*, 1162–1171.

Hall, T. J., Insana, M. F., Harrison, L. A., & Cox, G. G. (1996). Ultrasonic measurement of glomerular diameters in normal adult humans. *Ultrasound in Medicine and Biology*, *22*, 987–997.

Hao, X., Bruce, C., Pislaru, C., and Greenleaf, J. F. (2000). A novel region growing method for segmenting ultrasound images. *Proceedings of the IEEE Ultrasonics Symposium* (Vol. 2) (pp. 1717–1720) San Juan.

Hill, C. R., & Chivers, R. C. (1972). Investigations of backscattering in relation to ultrasonic diagnosis. In L. Filipeczynski (Ed.), *Ultrasonics in biology and medicine* (pp. 120–123). Warsaw, Poland: Polish Scientific Publishers.

Hinkelman, L. M., Metlay, L. A., Churukian, C. J., & Waag, R. C. (1996). Modified Gomori trichrome stain for macroscopic tissue slices. *Journal of Histotechnology*, *19*, 321–323.

Hinkelman, L. M., Mast, T. D., Metlay, L. A., & Waag, R. C. (1998). The effect of abdominal wall morphology on ultrasonic pulse distortion, Part I: Measurements. *Journal of the Acoustical Society of America*, *104*, 3635–3649.

Hinkelman, L. M., Szabo, T. L., & Waag, R. C. (1997). Measurements of ultrasonic pulse distortion produced by human chest wall. *Journal of the Acoustical Society of America*, *101*, 2365–2373.

Huisman, H. J., & Thijssen, J. M. (1998). An in vivo ultrasonic model of liver parenchyma. *IEEE Transactions on Ultrasonics, Ferroelectrics and Frequency Control*, *45*, 739–750.

Hynynen, K., & Jolesz, F. A. (1998). Demonstration of potential noninvasive ultrasound brain therapy through an intact skull. *Ultrasound in Medicine and Biology*, *24*, 275–283.

Hynynen, K., & Sun, J. (1999). Trans-skull ultrasound therapy: The feasibility of using image-derived skull thickness information to correct the phase distortion. *IEEE Transactions on Ultrasonics, Ferroelectrics and Frequency Control*, *46*, 752–755.

Insana, M. F., Hall, T. J., Wood, J. G., & Yan, Z. Y. (1993). Renal ultrasound using parametric imaging techniques to detect changes in microstructure and function. *Investigative Radiology*, *28*, 720–725.

Jensen, J. A. (1991). A model for the propagation and scattering of ultrasound in tissue. *Journal of the Acoustical Society of America*, *89*, 182–190.

Knapik, D. A., Starkoski, B., Pavlin, C. J., & Foster, F. S. (2000). A 100–200 MHz ultrasound biomicroscope. *IEEE Transactions on Ultrasonics, Ferroelectrics and Frequency Control*, *47*, 1540–1547.

Kolios, M. C., & Czarnota, G. J. (2009). Potential use of ultrasound for the detection of cell changes in cancer treatment. *Future Oncology, 5*, 1527–1532.

Kolios, M. C., Czarnota, G. J., Lee, M., Hunt, J. W., & Sherar, M. D. (2002). Ultrasonic spectral parameter characterization of apoptosis. *Ultrasound in Medicine and Biology, 28*, 589–597.

Lacefield, J. C., & Waag, R. C. (2001). Evaluation of backpropagation methods for transmit focus compensation. *Proceedings of the IEEE Ultrasonics Symposium* (Vol. 2) (pp. 1495–1498). Atlanta, GA.

Li, Y. (1997). Phase aberration correction using near-field signal redundancy, Part I: Principles. *IEEE Transactions on Ultrasonics, Ferroelectrics and Frequency Control, 44*, 355–371.

Li, Y. (2000). Small element array algorithm for correcting phase aberration using near-field signal redundancy, Part I: Principles. *IEEE Transactions on Ultrasonics, Ferroelectrics and Frequency Control, 47*, 29–48.

Li, Y., & Robinson, B. (2000a). Phase aberration correction using near-field signal redundancy, two-dimensional array algorithm. *Proceedings of the IEEE ultrasonics symposium* (Vol. 2) (pp. 1729–1732). San Juan.

Li, Y., & Robinson, B. (2000b). Small element array algorithm for correcting phase aberration using near-field signal redundancy, Part II: Experimental results. *IEEE Transactions on Ultrasonics, Ferroelectrics and Frequency Control, 47*, 49–57.

Li, Y., Robinson, B., & Carpenter, D. (1997). Phase aberration correction using near-field signal redundancy, Part II: Experimental results. *IEEE Transactions on Ultrasonics, Ferroelectrics and Frequency Control, 44*, 372–379.

Liu, D. -L., & Waag, R. C. (1994). Correction of ultrasonic wavefront distortion using back-propagation and a reference waveform method for time-shift compensation. *Journal of the Acoustical Society of America, 96*, 649–660.

Liu, D. -L., & Waag, R. C. (1995). A comparison of ultrasonic wave distortion and compensation in one-dimensional and two-dimensional apertures. *IEEE Transactions on Ultrasonics, Ferroelectrics and Frequency Control, 42*, 726–733.

Liu, D. -L., & Waag, R. C. (1998). Estimation and correction of ultrasonic wavefront distortion using pulse-echo data received in a two-dimensional aperture. *IEEE Transactions on Ultrasonics, Ferroelectrics and Frequency Control, 45*, 473–490.

Lizzi, F. L., Alam, S. K., Mikaelian, S., Lee, P., & Feleppa, E. J. (2006). On the statistics of ultrasonic spectral parameters. *Ultrasound in Medicine and Biology, 32*, 1671–1685.

Lizzi, F. L., Astor, M., Fellepa, E. J., Shao, M., & Kalisz, A. (1997). Statistical framework for ultrasonic spectral parameter imaging. *Ultrasound in Medicine and Biology, 23*, 1371–1382.

Lizzi, F. L., & Feleppa, E. J. (1993). In vivo ophthalmological tissue characterization by scattering. In K. K. Shung, & G. A. Thieme (Eds.), *Ultrasonic scattering in biological tissues*. Boca Raton, FL: CRC Press (Chapter 12).

Lizzi, F. L., Greenbaum, M., Feleppa, E. J., & Elbaum, M. (1983). Theoretical framework for spectrum analysis in ultrasonic tissue characterization. *Journal of the Acoustical Society of America, 73*, 1366–1373.

Machado, J. C., & Foster, F. S. (1999). Validation of theoretical diffraction correction functions for strongly focused high frequency ultrasonic transducers. *Ultrasonic Imaging, 21*, 96–106.

Madsen, E. L., Insana, M. F., & Zagzebski, J. A. (1984). Method of data reduction for accurate determination of acoustic backscatter coefficients. *Journal of the Acoustical Society of America, 76*, 913–923.

Mamou, J. (2005) *Ultrasonic characterization of three animal mammary tumors from three-dimensional acoustic tissue models* (Ph.D. dissertation), University of Illinois Urbana-Champaign, Urbana.

Mamou, J., Coron, A., Hata, M., Machi, J. , Yanagihara, E., Laugier, P., et al. (2009). Three-dimensional high-frequency characterization of excised human lymph nodes. *Proceedings of the IEEE ultrasonics symposium* (pp. 45–48). Rome.

Marquet, F., Pernot, M., Aubry, J-F., Montaldo, G., Marsac, L., Tanter, M., et al. (2009). Non-invasive transcranial ultrasound therapy based on a 3D CT scan: Protocol validation and in vitro results. *Physics in Medicine and Biology, 54*, 2597–2613.

Mast, T. D., Hinkelman, L. M., Metlay, L. A., Orr, M. J., & Waag, R. C. (1999). Simulation of ultrasonic pulse propagation, distortion, and attenuation in the human chest wall. *Journal of the Acoustical Society of America, 106*, 3665–3677.

Mast, T. D., Hinkelman, L. M., Orr, M. J., Sparrow, V. W., & Waag, R. C. (1997). Simulation of ultrasonic pulse propagation, through the abdominal wall. *Journal of the Acoustical Society of America, 102*, 1177–1190.

Mast, T. D., Hinkelman, L. M., Orr, M. J., & Waag, R. C. (1998). The effect of abdominal wall morphology on ultrasonic pulse distortion, Part II: Simulations. *Journal of the Acoustical Society of America, 104*, 3651–3664.

McDannold, N., Clement, G. T., Black, P., Jolesz, F., & Hynynen, K. (2010). Transcranial magnetic resonance imaging-guided focused ultrasound surgery of brain tumors: Initial findings in 3 patients. *Neurosurgery, 66*, 323–332.

Miller, J. G. Barzilai, B., Milunski, M. R., Mohr, G. A., Pérez, J. E., Thomas, L. J., III., et al. (1998). Backscatter imaging and myocardial tissue characterization. *Proceedings of the IEEE ultrasonics symposium* (Vol. 2) (pp. 1737–1381). Sendai.

Miller, J. G., Barzilai, B., Milunski, M. R., Mohr, G. A., Perez, J. E., Thomas, L. J., et al. (1989). Myocardial tissue characterization: Clinical confirmation of laboratory results. *Proceedings of the IEEE ultrasonics symposium* (Vol. 2) (pp. 1029–1036). Montreal, QC.

Montaldo, G., Aubry, J. -F., Tanter, M., & Fink, M. (2005). Spatio-temporal coding in complex media for optimum beamforming: The iterative time-reversal approach. *IEEE Transactions on Ultrasonics, Ferroelectrics and Frequency Control, 52*, 220–230.

Mottley, J. G., & Miller, J. G. (1988). Anisotropy of the ultrasonic backscatter of myocardial tissue, Part I: Theory and measurements in vitro. *Journal of the Acoustical Society of America, 85*, 755–761.

Mottley, J. G., & Miller, J. G. (1990). Anisotropy of the ultrasonic attenuation in soft tissues: Measurements in vitro. *Journal of the Acoustical Society of America, 88*, 1203–1210.

Mountford, R. A., & Wells, P. N. T. (1972). Ultrasonic liver scanning: The A-scan in the normal and cirrhosis. *Physics in Medicine and Biology, 17*, 261–269.

Muller, R. A., & Buffington, A. (1974). Real-time correction of atmospherically degraded telescope images through image sharpening. *Journal of the Optical Society of America, 64*, 1200–1209.

Nam, K., Rosado-Mendez, I. M., Wirtzfeld, L. A., Kumar, V., Madsen, E. L., Ghoshal, G., et al. (2011). Ultrasonic attenuation and backscatter coefficient estimates of rodent-tumor-mimicking structures: Comparison of results among clinical scanners. *Ultrasonic Imaging, 33*, 233–250.

Nam, K., Rosado-Mendez, I. M., Wirtzfeld, L. A., Kumar, V., Madsen, E. L., Ghoshal, G., et al. (2012). Cross-imaging system comparison of backscatter coefficient estimates from a tissue-mimicking material. *Journal of the Acoustical Society of America, 132*, 1319–1324.

Nassiri, D. K., & Hill, C. R. (1986a). The differential and total bulk acoustic scattering cross-sections of some human and animal tissues. *Journal of the Acoustical Society of America, 79*, 2034–2047.

Nassiri, D. K., & Hill, C. R. (1986b). The use of angular acoustic scattering measurements to estimate structural parameters of human and animal tissues. *Journal of the Acoustical Society of America, 79*, 2048–2054.

Ng, G. C., Worrell, S. S., Freiburger, P. D., & Trahey, G. E. (1994). A comparative evaluation of several algorithms for phase aberration correction. *IEEE Transactions on Ultrasonics, Ferroelectrics and Frequency Control, 41*, 631–643.

Nock, L., Trahey, G. E., & Smith, S. W. (1989). Phase aberration correction in medical ultrasound using speckle brightness as a quality factor. *Journal of the Acoustical Society of America, 85*, 1819–1833.

Odegaard, L., Halvorsen, E., Ystad, B., Torp, H. G., & Angelsen, B. (1996). Delay and amplitude focusing through the body wall; a simulation study. *Proceedings of the IEEE ultrasonics symposium* (Vol. 2) (pp. 1411–1414). San Antonio, TX.

O'Donnell, M., Bauwens, D., Mimbs, J. W., & Miller, J. G. (1979). Broadband integrated backscatter: An approach to spatially localized tissue characterization in vivo. *Proceedings of the IEEE Ultrasonics Symposium*175–178.

O'Donnell, M., & Engeler, W. E. (1992). Correlation-based aberration correction in the presence of inoperable elements. *IEEE Transactions on Ultrasonics, Ferroelectrics and Frequency Control, 39*, 700–707.

O'Donnell, M., & Flax, S. W. (1988). Phase aberration measurements in medical ultrasound: Human studies. *Ultrasound Imaging, 10*, 1–11.

O'Donnell, M., & Li, R. G. (1991). Aberration correction on a two-dimensional anisotropic phased array. *Proceedings of the IEEE ultrasonics symposium* (Vol. 2) (pp. 1183–1193). Orlando, FL.

Oelze, ML, & O'Brien, WD. (2002). Frequency-dependent attenuation-compensation functions for ultrasonic signals backscattered from random media. *Journal of the Acoustical Society of America, 111*, 2308–2319.

Oelze, ML, & O'Brien, WD, Jr. (2004a). Improved scatterer property estimates from ultrasound backscatter for small gate lengths using a gate-edge correction factor. *Journal of the Acoustical Society of America, 116*, 3212–3223.

Oelze, ML, & O'Brien, WD, Jr (2004b). Defining optimal axial and lateral resolution for estimating scatterer properties from volumes using ultrasound backscatter. *Journal of the Acoustical Society of America, 115*, 3226–3234.

Oelze, ML, O'Brien, WD, Jr., Blue, JP, & Zachary, JF. (2004). Differentiation and characterization of rat mammary fibroadenomas and 4T1 mouse carcinomas using quantitative ultrasound imaging. *IEEE Transactions on Medical Imaging, 23*, 764–771.

Pajek, P., & Hynynen, K. (2012a). The design of a focused ultrasound transducer array for the treatment of stroke: A simulation study. *Physics in Medicine and Biology, 57*, 4951–4968.

Pajek, D., & Hynynen, K. (2012b). Applications of transcranial focused ultrasound surgery. *Acoustics Today, 8*, 8–14.

Pedersen, P. C., Chakareski, J., & Lara-Montalvo, R. (2003). Ultrasound characterization of arterial wall structures based on integrated backscatter profiles. *Proceedings of SPIE, 50435*, 115–126.

Pichardo, S., Sin, V. W., & Hynynen, K. (2011). Multi-frequency characterization of the speed of sound and attenuation coefficient for longitudinal transmission of freshly excised human skulls. *Physics in Medicine and Biology, 56*, 219–250.

Pinton, G., Aubry, J. -F., Bossy, E., Muller, M., Pernot, M., & Tanter, M. (2012a). Attenuation, scattering, and absorption of ultrasound in the skull bone. *Medical Physics, 39*, 299–307.

Pinton, G., Aubry, J. F., Fink, M., Tanter, M., Boch, A. L., & Aubry, J. F. (2011). Effects of nonlinear ultrasound propagation on high intensity brain therapy. *Medical Physics, 38*, 1207–1216.

Pinton, G. F., Aubry, J. -F., & Tanter, M. (2012b). Direct phase projection and transcranial focusing of ultrasound for brain therapy, IEEE Transactions on Ultrasonics. *Ferroelectrics and Frequency Control, 59*, 1122–1129.

Reid, J. M. (1993). Standard substitution methods for measuring ultrasonic scattering in tissues. In K. K. Shung, & G. A. Thieme (Eds.), *Ultrasonic scattering in biological tissues*. Boca Raton, FL: CRC Press (Chapter 6).

Rigby, K. W., Andarawis, E. A., Chalek, C. L., Haider, B. H., Hinrichs, W. L., Hogel, R. A., et al. (1998). Realtime adaptive imaging. *Proceedings of the IEEE ultrasonics symposium* (Vol. 2) (pp. 1603–1606). Sendai.

Robert, J. -L., & Fink, M. (2008). Green's function estimation in speckle using the decomposition of the time reversal operator: Application to aberration correction in medical imaging. *Journal of the Acoustical Society of America, 123*, 866–877.

Romijn, R. L., Thijssen, J. M., Oosterveld, B. J., & Verbeek, A. M. (1991). Ultrasonic differentiation of intraocular melanoma: Parameters and estimation methods. *Ultrasonic Imaging, 13*, 27–55.

Romijn, R. L., Thijssen, J. M., van Delft, J. L., de Wolff-Rouendaal, D., van Best, J., & Oosterhuis, J. A. (1989). In vivo ultrasound backscattering estimation for tumor diagnosis: An animal study. *Ultrasound in Medicine and Biology, 15*, 471–479.

Sadeghi-Naini, A., Falou, O., Hudson, J. M., et al. (2012). Imaging innovations for cancer therapy response monitoring. *Imaging in Medicine, 4*, 311–327.

Sadeghi-Naini, A., Papanicolau, N., Falou, O., et al. (2013c). Low-frequency quantitative ultrasound imaging of cell death in vivo. *Medical Physics, 40*, 1–13.

Sadeghi-Naini, A., Falou, O., Tadayyon, H., Al-Mahrouki, A., Tran, W., Papanicolau, N., et al. (2013a). Conventional-frequency ultrasonic biomarkers of cancer treatment response in vivo. *Translational Oncology*, *6*, 234–243.

Sadeghi-Naini, A., Papanicolau, N., Falou, O., et al. (2013b). Quantitative ultrasound evaluation of tumour cell death response in locally advanced breast cancer patients receiving chemotherapy. *Clinical Cancer Research*, *19*, 2163–2174.

Sherar, M. D., Noss, M. B., & Foster, F. S. (1987). Ultrasound backscatter microscopy images the internal structure of living tumour spheroids. *Nature*, *330*, 493–495.

Shung, K. K., Sigelmann, R. A., & Reid, J. M. (1976). Scattering of ultrasound by blood. *IEEE Transactions on Biomedical Engineering*, *23*, 460–467.

Shung, K. K., & Thieme, G. A. (Eds.), (1993). *Ultrasonic scattering in biological tissues* Boca Raton, FL: CRC Press.

Sigelmann, R. A., & Reid, J. M. (1973). Analysis and measurement of ultrasound backscattering from an ensemble of scatterers excited by sine-wave bursts. *Journal of the Acoustical Society of America*, *53*, 1351–1355.

Smith, S. W., Phillips, D. J., von Ramm, O. T., & Thurstone, F. L. (1977). Some advances in acoustic imaging through skull. In D. G. Hazzard, & M. L. Litz (Eds.), *Symposium on biological effects and characterizations of ultrasound sources* (pp. 37–52). Rockville, MA: HEW, FDA.

Sommer, G., Stern, R. A., Howes, P. J., & Young, H. (1987). Envelope amplitude analysis following narrow-band filtering: A technique for ultrasonic tissue characterization. *Medical Physics*, *14*, 627–632.

Srinivasan, S., Baldwin, H. S., Aristizabal, O., Kwee, L., Labow, M., Artman, M., et al. (1998). Noninvasive in utero imaging of mouse embryonic heart development using 40 MHz echocardiography. *Circulation*, *98*, 912–918.

Steinberg, B. D. (1992). A discussion of two wavefront aberration correction procedures. *Ultrasonic Imaging*, *14*, 387–397.

Szabo, T. L., & Burns, D. R. (1997). Seismic signal processing of ultrasound imaging data. In S. Lees, & L. A. Ferrari (Eds.), *Acoustical imaging* (Vol. 23, pp. 131–136). New York: Plenum Press.

Tabei, M., Mast, T. D., & Waag, R. C. (2003). Simulation of ultrasonic focus aberration and correction through human tissue. *Journal of the Acoustical Society of America*, *113*, 1166–1176.

Tanter, M., Aubry, J. -F., Gerber, J., Thomas, J. -L., & Fink, M. (2001). Optimal focusing by spatio-temporal inverse filter: I. Basic principles. *Journal of the Acoustical Society of America*, *110*, 37–47.

Tanter, M, Thomas, J L, & Fink, M (1998). Focusing and steering through absorbing and aberrating layers: Application to ultrasonic propagation through the skull. *Journal of the Acoustical Society of America*, *103*, 2403–2410.

Taylor, K. W., & Wells, P. N. T. (1989). Tissue characterization. *Ultrasound in Medicine and Biology*, *15*, 421–428.

Thijssen, J. M. (2000). *Ultrasonic tissue characterization*, *Acoustical Imaging* (Vol. 25, pp. 9–25). New York: Plenum Press.

Thijssen, J. M., Verbeek, A. M., Romijn, R. L., De Wolff-Rouzendaal, D., & Oosterhuis, J. A. (1991). Echographic differentiation of histological types of intraocular melanoma. *Ultrasound in Medicine and Biology*, *17*, 127–138.

Thomas, J. -L., & Fink, M. A. (1996). Ultrasonic beam focusing through skull inhomogeneities with a time reversal mirror: Application to trans-skull therapy. *IEEE Transactions on Ultrasonics, Ferroelectrics and Frequency Control*, *43*, 1122–1129.

Thomas, L. J., III, Barzilai, B., Perez, J. E., Sobel, B. E., Wickline, S. A., & Miller, J. G. (1989). Quantitative real-time imaging of myocardium based on ultrasonic integrated backscatter. *IEEE Transactions on Ultrasonics, Ferroelectrics and Frequency Control*, *36*, 466–470.

Trahey, G. E. (1991). An evaluation of transducer design and algorithm performance for two-dimensional phase aberration correction. *Proceedings of the IEEE ultrasonics symposium* (Vol. 2) (pp. 1181–1187). Orlando, FL.

Trahey, G. E., Zhao, D., Miglin, J. A., & Smith, S. W. (1990). Experimental results with a realtime adaptive ultrasonic imaging system for viewing through distorting media. *IEEE Transactions on Ultrasonics, Ferroelectrics and Frequency Control*, *37*, 418–427.

Turnbull, D. H. (1999). In utero ultrasound backscatter microscopy of early stage mouse embryos. *Computerized Medical Imaging and Graphics*, *23*, 25–31.

Turnbull, D. H., Bloomfield, T. S., Foster, F. S., & Joyner, A. L. (1995). Ultrasound backscatter microscope analysis of early mouse embryonic brain development. *Proceedings of the National Academy of Sciences of the USA*, *92*, 2239–2243.

Turnbull, D. H., Ramsay, J. A., Shivji, G. S., Bloomfield, T. S., From, L., Sauder, D. N., et al. (1996). Ultrasound backscatter microscope analysis of mouse melanoma progression. *Ultrasound in Medicine and Biology*, *22*, 845–853.

Ueda, M., & Ozawa, Y. (1985). Spectral analysis of echoes for backscattering measurement. *Journal of the Acoustical Society of America*, *77*, 38–47.

Vlad, R. M., Brand, S., Giles, A., Kolios, M. C., & Czarnota, G. J. (2009). Quantitative ultrasound characterization of responses to radiotherapy in cancer mouse models. *Clinical Cancer Research*, *15*, 2067–2075.

Wells, P. N. T., McCarthy, C. F., Ross, F. G. N., & Read, A. E. A. (1969). Comparison of A-scan and compound B-scan in the diagnosis of liver disease. *British Journal of Radiology*, *42*, 818–823.

White, P. J., Clement, G. T., & Hynynen, K. (2006). Longitudinal and shear mode ultrasound propagation in human skull bone. *Ultrasound in Medicine and Biology*, *32*, 1085–1096.

Wojcik, G., Fornberg, B., Waag, R., Carcione, L., Mould, J., Nikodym, L., et al. (1997). Pseudospectral methods for large-scale bioacoustic models. *Proceedings of the IEEE ultrasonics symposium* (Vol. 2) (pp. 1501–1506). Toronto, ON.

Wu, F., Fink, M., Mallart, R., Thomas, J. L., Chakroun, N., Casserreau, D., et al. (1991). Optimal focusing through aberrating media: A comparison between time reversal mirror and time delay correction techniques. *Proceedings of the IEEE ultrasonics symposium* (Vol. 2) (pp. 1195–1199). Orlando, FL.

Yousefi, A., Goertz, D. E., & Hynynen, K. (2009). Transcranial shear-mode ultrasound: assessment of imaging performance and excitation techniques. *IEEE Transactions on Medical Imaging*, *28*, 763–774.

Ystad, B., Halvorsen, E., Odegaard, L., Angelsen, B. A. J., & Lygren, M. (1996). Wave-equation-based analysis of phase aberrations in inhomogeneous tissue. *Proceedings of the IEEE ultrasonics symposium* (Vol. 2) (pp. 1353–1356). San Antonio, TX.

Zhu, Q., & Steinberg, B. D. (1993). Wavefront amplitude distortion and image sidelobe levels, Part I: Theory and computer simulations. *IEEE Transactions on Ultrasonics, Ferroelectrics and Frequency Control*, *40*, 743–753.

Zhu, Q., & Steinberg, B. D. (1994). Modeling, measurement and correction of wavefront distortion produced by breast specimens. *Proceedings of the IEEE ultrasonics symposium* (Vol. 3) (pp. 1613–1617). Cannes, France.

Zhu, Q., Steinberg, B. D., & Arenson, R. L. (1993). Wavefront amplitude distortion and image sidelobe levels, Part II: In vivo experiments. *IEEE Transactions on Ultrasonics, Ferroelectrics and Frequency Control*, *40*, 754–762.

Bibliography

Angelsen, B. A. J. (2000). *Ultrasound imaging: waves, signals, and signal processing*. Emantec, Trondheim, Norway.

A thorough analysis of scattering from tissue.

Bamber, J. C. (1997). In M. J. Crocker (Ed.), *Encyclopedia of acoustics* (Vol. 4, pp. 1703–1725). New York: John Wiley & Sons.

An overview of the properties of scattering from tissue.

Duck, F. A. (1990). *Physical properties of tissue: a comprehensive reference book*. London: Academic Press.

Greenleaf, J. F., & Sehgal, C. M. (1992). *Biologic system evaluation with ultrasound*. New York: Springer Verlag.

Resource on tissue function scattering, imaging, and characterization.

Kino, G. S. (1987). *Acoustic Waves: Devices, imaging, and signal processing*. Englewood Cliffs, NJ: Prentice-Hall. pp. 300–357.
Introduction to acoustic scattering.
Shung, K. K., & Thieme, G. A. (Eds.), (1993). *Ultrasonic scattering in biological tissues* Boca Raton, FL: CRC Press.

CHAPTER 10

Imaging Systems and Applications

Chapter Outline

Diagnostic Ultrasound Imaging: Inside Out.

10.1 Introduction

The modern diagnostic imaging system is continuing to evolve and, as a result, is becoming more complicated with new modes and features. System functions are the last blocks added to the overall block diagram (see Figure 2.14). This chapter introduces the basic principles of an imaging system and discusses signal-processing techniques. Doppler and color flow imaging are deferred until the next chapter. A wide variety of transducer types has been invented and adapted to specific clinical uses; therefore, the major clinical uses of ultrasound imaging systems need to be considered also.

In Figure 10.1, the external parts of an ultrasound imaging system are shown. The image display is mounted on a chassis with wheels for portability. On the right side, several transducer arrays are stored, awaiting use, and they are attached to the system through several transducer connector bays in the front. Below the display is a keyboard and a number of knobs and switches for controlling the system. Peripheral devices, such as recording media and extra connectors, can be seen. The all-important on/off switch, which is sometimes difficult to find, is also identified.

10.2 Trends in Imaging Systems

10.2.1 General Commercial Systems

Ultrasound imaging systems fall into the following commonly used categories: pocket, portable, low-end, mid-range, and high-end. The high-end systems are those with the latest and largest number of state-of-the-art features, and they generally produce the best images. Each system has unique features, called market differentiators, that distinguish the system from others manufactured by the same company, as well from those made by other companies. Over time, some features, because of the competitive nature of the industry, may migrate in altered form to imaging systems of rival manufacturers. Needless to say, because high-end systems have the most features and options, they are the most expensive. A mid-range system does not have some of the high-end features but has a full complement of options necessary to produce very good images in a variety of clinical applications.

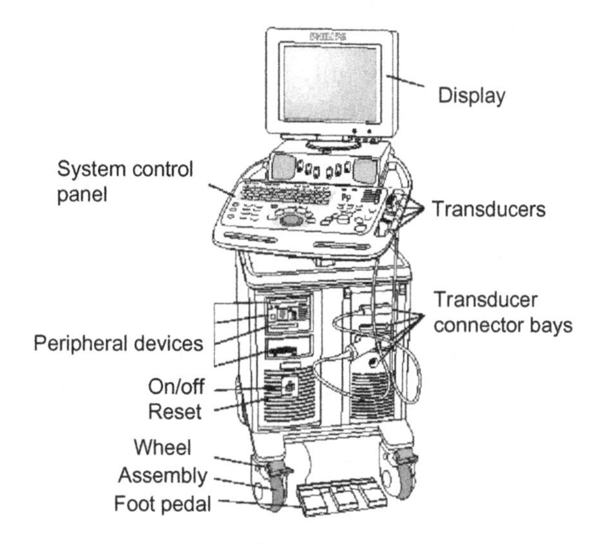

Figure 10.1
External parts of an ultrasound imaging system. *Courtesy of Philips Healthcare.*

Low-end systems are usually limited in their functionality and are often designed to cover specific clinical applications. There are exceptions to these general categories. Next in lower cost are portable systems, and pocket ultrasound systems are the least expensive.

The ultrasound imaging industry is undergoing dynamic change. One trend is that the new high-end features tend to migrate downward to mid-range systems and eventually to low-end systems over time. This migration is in part caused by the need to replace existing features with new ones to grow the market. Another major force is the parallel development in allied fields such as computation and electronics of enabling technologies (the invisible wind of change discussed in Chapter 1). These developments have already had a profound effect on what is possible with ultrasound, as exemplified by the fully functional, portable imaging systems now available.

Portable imaging systems, which are a relatively new development, may provide a more restricted range of options (e.g. fewer transducers) or be fully functional with several transducer options in an extremely small package at a very low cost. Four portable systems were shown in Figure 1.13.

The Minivisor, the first fully portable self-contained imaging system (mentioned in Chapter 1), is included for historical reference. The Sonosite system was the first modern portable of comparable size (about 6 lbs) and achieved its portability through custom designed application-specific integrated circuits (ASICs). OptiGo, a portable also based on specialized chips, was designed for cardiac applications. Both of these systems offer color flow imaging and automated features to aid users. The Terason 2000 system achieves its small size and flexibility by leveraging laptop technology and its unique proprietary low power charge domain processor chip. The fully functional 128-channel system consists of a laptop, a 10-oz processor box, and the transducer. These systems were featured in an issue of the *Thoraxcentre Journal* (December 2001).

At the end of the first edition of this book were speculations about even smaller ultrasound systems. Now they are a reality. Called "pocket ultrasound," these systems are not only smaller in size but have a small selling price as well, less than $10,000. Four types are depicted in Figure 1.14. In Figure 1.14A, the Siemens P-10, at 1.6 lb (0.7 kg), was the first to appear, with an attached 2–4-MHz 64-element transducer and a charger docking console. The capability to image in fundamental and harmonic mode with scan depths 4–24 cm (Culp, Mock, Ball, Chiles, & Culp, 2011). The GE V-Scan, Figure 1.14B, weighing 13 oz (0.37 kg), comes with a 1.7–3.8-MHz phased array with B-mode and color flow and a penetration depth up to 25 cm (Lafitte et al., 2011). The Sonic Window, Figure 1.14C, is a pocket-sized, C-scan ultrasound imaging device weighing 6 oz (0.17 kg). It utilizes a fully sampled 3600-element two-dimensional (2D) array and associated electronics and display in a compact unit that can be placed over the skin to reveal the anatomy beneath it (Fuller, Owen, Blalock, Hossack, & Walker, 2009). The MobiUs SP1, Figure 1.14D, is an ultrasound imaging system integrated into a mobile phone designed by MobiSante, Inc. A really smart smartphone provides wireless and cellular connectivity and display of ultrasound images. The phone system (11.6 oz or 0.33 kg) is smaller than any of its four transducers: 3.5 and 5.0 MHz (abdominal, OB/Gyn, guidance procedures), and 7.5 and 12 MHz (vascular, small organs).

10.2.2 New Developments

Conventional longitudinal-wave-based ultrasound imaging is being combined with other imaging modalities and modes in new ways. Three major directions follow. Complementary or image fusion is the combination of different image modalities, such as explained by Figures 1.16 and 1.17. Side-by-side comparisons of 2D and 3D real-time ultrasound coregistered with 3D volume image data of a second modality such as computed tomography (CT) has been commercialized. Another implementation is combining real-time coregistered imaging from two different imaging modalities (an example will be given in the 3D section in this chapter). At best, this combination provides two different views to provide a synergistic perspective that neither individual imaging mode could provide. Through its real-time aspects, this combination can inform the use of one imaging mode from the other such as in surgical interventions. In multiwave/multimode combinations, one type of wave can transform into another, revealing images caused by a hybrid interaction (Fink & Tanter, 2010). A good combination increases contrast and/or resolution through the strengths of the individual waves. An example is photoacoustics, in which light impinging on tissue causes it to expand and create acoustic displacements or waves, which are sensed by a conventional array and displayed on an imaging system. As depicted in Figure 10.2, the effect of changes in the local optical absorption coefficients after an injection are overlaid on a conventional B-mode image (Alqasemi, Li, Aguirre, & Zhu, 2012). Another example is acoustic radiation force imaging (ARFI), a type of elastography in which

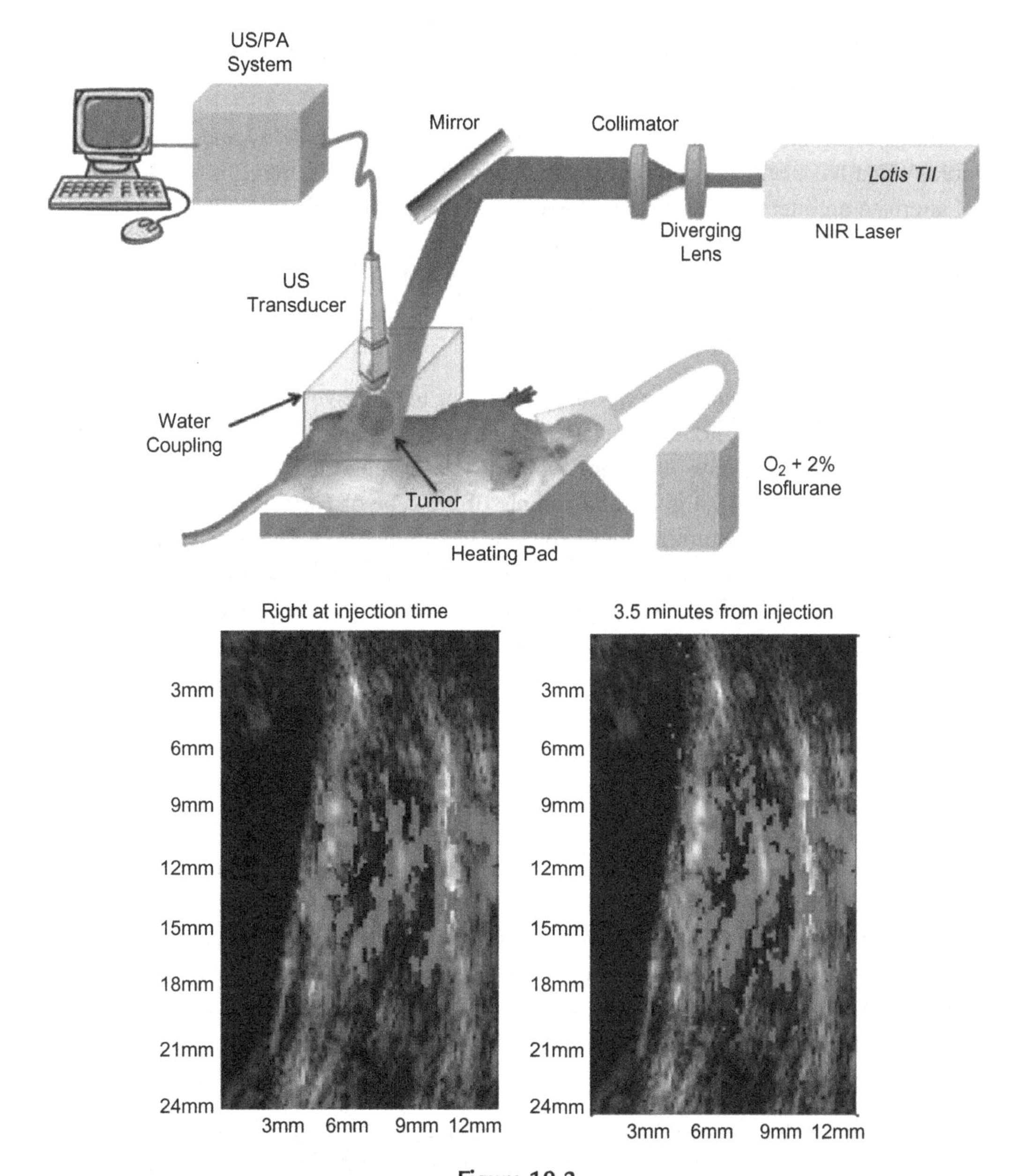

Figure 10.2
Photoacoustic mouse experiment setup. The reader is referred to the Web version of the book to see the figure in color. *From Alqasemi et al. (2012), IEEE.*

longitudinal waves create transient acoustic radiation forces (described in Chapter 12) that displace tissue whose movement can be detected. Another example is transient shear wave elastography, where a longitudinal ultrasound wave creates transient acoustic radiation forces that, in turn, create shear waves, which can reveal the shear elastic properties of tissue as explained in Chapter 16 and Fink and Tanter (2010). The third combination category, parameterized combination imaging, translates data through a model or algorithm into a displayed parameter. Obvious examples are color flow imaging, vector Doppler and quasi-static elastography.

Even though much of this chapter will describe systems based on delay-and-sum array technology that has been in a constant state of evolution and refinement for the last 30 years, alternate architectures will be explained later in this chapter. In addition, in that section is a description of the emergence of systems dedicated to research that have increased opportunities for exploration and further growth. The global market for ultrasound equipment (not just medical) in 2011 was about $17.5 billion and is predicted to reach $27.7 billion by 2016 (Sanile & Oruklu, 2012).

Since the turn of the millennium, it has become easier to build your own ultrasound imaging system. For example, both Texas Instruments and Analog Devices sell modular chips for this purpose and have excellent Web sites describing their products. A block diagram figure in Section 10.4 gives an example of the type of parts available to assemble a system. Undoubtedly, these products will spur the development of new special-purpose and integrated ultrasound systems. Novel experimental research lab-built systems are described in a special issue of *IEEE Transactions on Ultrasonics, Ferroelectrics and Frequency Control* on novel embedded systems for ultrasonic imaging and signal processing (Sanile & Oruklu, 2012).

10.3 Major Controls

Because there are many controls for a typical ultrasound imaging system and their organization and names vary considerably from manufacturer to manufacturer, the following description is a short list of the major controls according to function. Note that even though the number of actual controls on an imaging system may seem bewildering at first, most systems start in a default set of control settings or "presets" that are optimized for a particular clinical application or transducer, so that with clinical training and moderate effort, such as the adjustment of the time gain compensation (TGC) controls, a reasonably good image can be obtained quickly.

A close-up of the system control panel of the same imaging system from Figure 10.1 is illustrated in Figure 10.3. On the right side, the TGC slide controls, transmit focus controls, and scan depth controls are evident.

The main controls identified in the system control panel (depicted in Figure 10.3) are the following:

Probe or transducer selection: Typically two to four transducers can be plugged into connectors in the imaging system, so this switch allows the user to activate one of the arrays at a time.

Mode selection: This provides the means for selecting a mode of operation, such as 2B-mode, color flow, M-mode, or Doppler, individually or in combination (duplex or triplex operation).

Depth of scan control: This adjusts the field of view (scan depth in centimeters).

Focus or transmit focal length selection: This allows the location of the transmit focal length to be moved into a region of interest. The depth location of the focal plane is usually indicated by a > symbol. (Multiple transmit foci can be selected in a splice or multiple transmit mode at the sacrifice of frame rate.) In pulsed wave Doppler mode, the location of the focal length is often controlled by the center of the Doppler gate position.

Time gain compensation (TGC) controls (also depth gain compensation, time gain control, sensitivity-time control, etc.): These controls offset the loss in signal caused by tissue absorption and diffraction variations; they are usually in the form of slides for controlling amplifier gain individually in each contiguous axial time range. The image depth dimension is divided into a number of zones or stripes, each of which is controlled by a TGC control (discussed in Section 4.6). On some systems, these gains are adjusted automatically based on signal levels in different regions of the image. Some systems also provide the capability to adjust gains in the lateral direction (lateral gain compensation or additional control in the horizontal dimension). Other systems may have an automatic means of setting these controls based on parameters sensed in the signals in the image, sometimes called "automatic TGC."

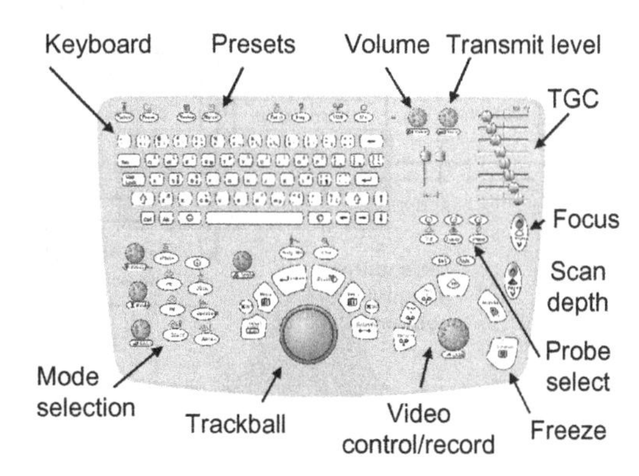

Figure 10.3
Keyboard and display of an ultrasound imaging system. *Courtesy of Philips Healthcare.*

Transmit level control: This adjusts drive amplitude from transmitters (it is done automatically on some systems). In addition to this control, a number of other factors alter acoustic output (discussed in more detail in Chapter 13 and Chapter 15). Feedback on acoustic output level is provided by thermal and mechanical indices on the display (also discussed in more detail in Chapters 13 and 15). A freeze control stops transmission of acoustic output.

Display controls: Primarily, these controls allow optimization of the presentation of information on the display and include a logarithmic compression control, selection of preprocessing and postprocessing curves, and color maps, as well as the ability to adjust the size of the images from individual modes selected for multimode operation. Provision is usually made for recording video images, playing them back, and comparing and sending them in various formats.

10.4 Block Diagram

The hidden interior of a digital imaging system is represented functionally by a generic simplified block diagram (shown by Figure 10.4). For now, the general operation of an imaging system is discussed (more details will be presented later). Note the functional similarity of the block diagram to the central diagram, Figure 2.14.

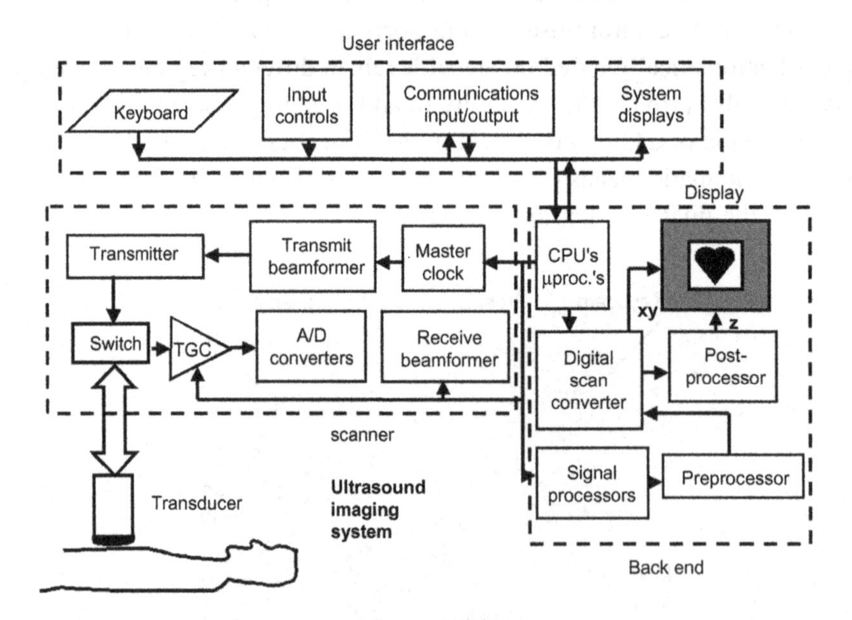

Figure 10.4
Block diagram of a generic digital ultrasound imaging system. cpu, central processing unit; µproc., microprocessor.

A description of this block diagram follows:

User interface: Most of the blocks are hidden from the user, who mainly sees the keyboard and display, which are part of a group of controls called the "user interface." This is the part of the system by which the user can configure the system to work in a desired mode of operation. System displays showing software-configurable menus and controls (soft-keys) in combination with knobs or slider controls and switches, as well as the main image display monitor, provide visual feedback that the selected mode is operating. The user interface provides the means of getting information in and out of the system through connectors to the system. Main connections include a computer hookup to a local area network (LAN) to Digital Imaging and Communication in Medicine (DICOM) communication and networking, and to peripherals such as printers. Various recording devices, such as digital video recorders (DVRs), can be attached.

Controller (computers): A typical system will have one or more microprocessor or a PC that directs the operation of the entire system. The controller senses the settings of the controls and input devices, such as the keyboard, and executes the commands to control the hardware to function in the desired mode. It orchestrates the necessary setup of the transmit and receive beamformers as well as the signal processing, display, and output functions. Another important duty of the computer is to regulate and estimate the level of acoustic output in real time.

Front end: This grouping within the scanner is the gateway of signals going in and out of the selected transducer. Under microprocessor transmit control, excitation pulses are sent to the transducer from the transmitter circuitry. Pulse–echo signals from the body are received by array elements and go through individual user-adjustable TGC amplifiers to offset the weakening of echoes by body attenuation and diffraction with distance. These signals then pass on to the receive beamformer.

Scanner (beamforming and signal processing): These parts of the signal chain provide the important function of organizing the many signals of the elements into coherent timelines of echoes for creating each line in the image. The transmit beamformer sends pulses to the elements. Echo signals pass through an analog-to-digital (A/D) converter for digital beamforming. In addition, the scanner carries out signal processing, including filtering, creation of quadrature signals, and different modes such as Doppler and color flow.

Back end: This grouping of functions is associated with image formation, display, and image metrics. The input to this group of functions is a set of pulse–echo envelope lines formed from each beamformed radio-frequency (RF) data line. Image formation is achieved by organizing the lines and putting them through a digital scan converter that transforms them into a raster scan format for display on a video or PC monitor. Along the way, appropriate preprocessing and postprocessing, log compression, and color or gray-scale mapping are completed. Image overlays containing alphanumeric characters and other information are added in image planes. Also available in the back end are various metric programs, such as measuring the length of a fetal femur, calculating areas, or performing videodensitometry. Controls are also available for changing the format of the information displayed.

A more detailed look at an ultrasound system is given by Figure 10.5, which shows how a system can be assembled from commercially available parts. Corresponding to the front end of Figure 10.4 are the front end, amplifier, signal chain and TGC. Here the mid and back end are like the back end of Figure 10.4. The operating system (OS) can be identified (c.p. u.) and input/output device.

10.5 Major Modes

The following are major modes on a typical imaging system:

Angio (mode): This is the same as the power Doppler mode (see Figure 11.23).
B-mode: This is a brightness-modulated image in which depth is along the z axis and azimuth is along the x axis. It is also known as "B-scan" or "2D mode." The position of the echo is determined by its acoustic transit time and beam direction in the plane. Alternatively, an imaging plane contains the propagation or depth axis (see Figure 9.1).
Color flow imaging (mode): A spatial map is overlaid on a B-mode gray-scale image that depicts an estimate of blood-flow mean velocity, indicating the direction of flow encoded in colors (often blue away from the transducer and red toward it), the amplitude of mean velocity by brightness, and turbulence by a third color (often green). It is also known as a "color flow Doppler." Visualization is usually 2D but can also be 3D or 4D (see Figure 10.8A).
Color M-mode: This mode of operation has color flow depiction at the same vector location where depth is the y deflection (fast time), and the x deflection is the same color flow line shown as a function of slow time. This mode displays the time history of a single color flow line at the same spatial position over time (see Figure 11.24).
Continuous wave (CW) Doppler: This Doppler mode is sensitive to the Doppler shift of blood flow all along a line (see Figure 11.13).
M-mode: This mode of operation is brightness modulated, where depth is the y deflection (fast time), and the x deflection is the same imaging line shown as a function of slow time. This mode displays the time history of a single line at the same spatial position over time (see Figure 10.6).
Doppler mode: This is the presentation of the Doppler spectrum (continuous wave or pulsed wave).
Color Doppler (mode): A 2D Doppler image of blood flow is color coded to show the direction of flow to and away from the transducer (see Figure 10.8A).

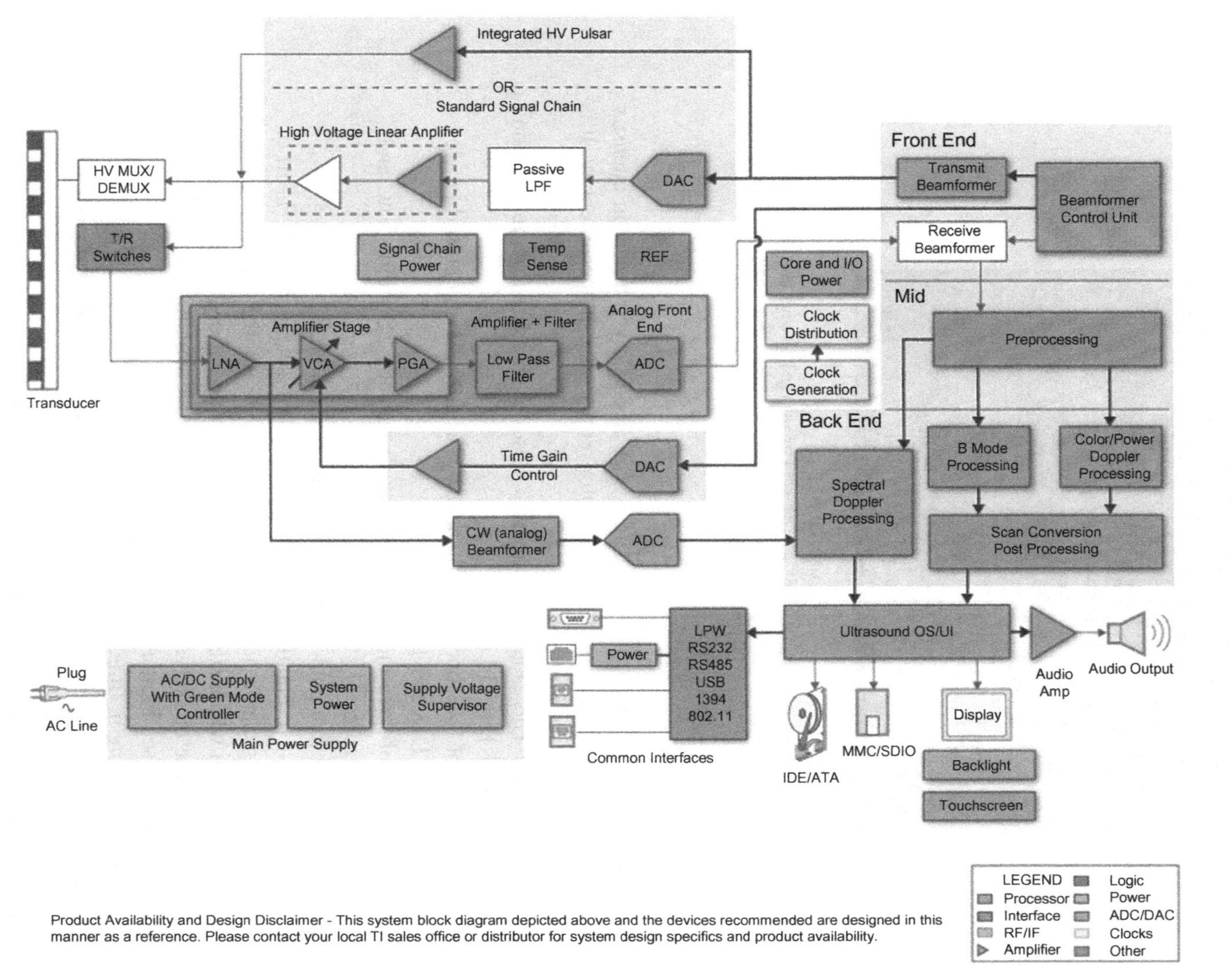

Figure 10.5

Block diagram of diagnostic ultrasound imaging system showing modular chips. ADC, analog-to-digital converter; DAC, digital-to-analog converter; OS, operating system. The reader is referred to the Web version of the book to see the figure in color. *Courtesy of Texas Instruments.*

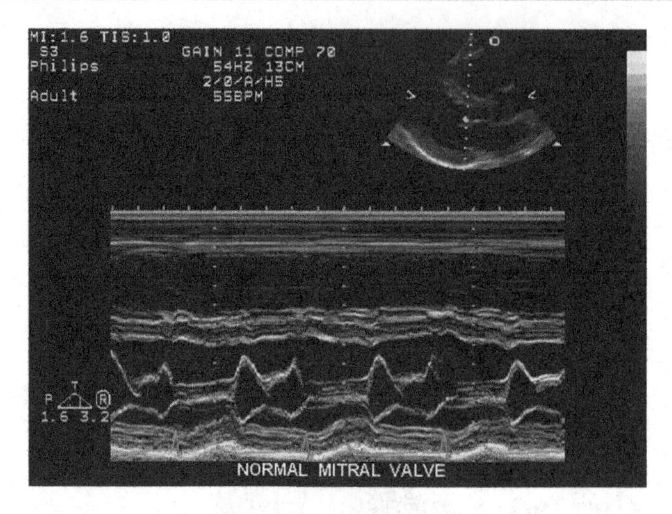

Figure 10.6
Duplex M-mode image. The insert (above right of the sector image) shows the orientation of the M-mode. The reader is referred to the Web version of the book to see the figure in color. *Courtesy of Philips Healthcare.*

Power Doppler (mode): This color-coded image of blood flow is based on intensity rather than on direction of flow, with a paler color representing higher intensity. It is also known as "angio" (see Figure 11.23).
Pulsed wave Doppler: This Doppler mode uses pulses to measure flow in a region of interest (see Figures 11.15 and 11.21).
Duplex: Presentation of two modes simultaneously: usually 2D and pulsed (wave) Doppler (see Figures 11.13 and 11.15).
Triplex: Presentation of three modes simultaneously: usually 2D, color flow, and pulsed Doppler (see Figures 11.13 and 11.15).
2D: (B-mode) imaging in a plane, with the brightness modulated.
3D: This is an image representation of a volume or 3D object, such as the heart or fetus. Surface rendering can be used to visualize surfaces. Another image presentation is volume rendering, in which surfaces can be semitransparent or 2D slice planes through the object. Alternatively, there is simultaneous viewing of different 2D slice planes (side by side). (see Figures 10.27 and 10.28)
4D: A 3D image moving in time.
Zoom: Video zoom is a magnification of a region of interest in the video image. Alternatively, acoustic zoom is a magnification of the region of interest in which acoustic and/or imaging parameters are modified to enhance the image, such as placing the transmit focus in the region of interest and/or increasing the number of image lines in the region.

10.6 Clinical Applications

Diagnostic ultrasound has found wide application for different parts of the human body, as well as in veterinary medicine. The major imaging categories of ultrasound imaging are listed below:

Breast: Imaging of female (usually) breasts.
Cardiac: Imaging of the heart.
Gynecologic: Imaging of the female reproductive organs.
Radiology: Imaging of the internal organs of the abdomen.
Obstetrics (sometimes combined with Gynecologic, as in OB/Gyn): Imaging of fetuses in vivo.
Pediatrics: Imaging of children.
Vascular: Imaging of the (usually peripheral, as in peripheral vascular) arteries and veins of the vascular system (called "cardiovascular" when combined with heart imaging).

Specialized applications have been honored by their own terminology. Many of these terms were derived from the location of the acoustic window where the transducer is placed, as well as the application. "Window" refers to an access region or opening through which ultrasound can be transmitted easily into the body. Note that transducers most often couple energy in and out of the body through the use of an externally applied couplant, which is usually a water-based gel or fluid placed between the transducer and the body surface. Transducers, in addition to being designed ergonomically to fit comfortably in the hand for long periods of use, are designed with the necessary form factors to provide access to or through the windows described later. See Szabo and Lewin (2013) for details about the selection of transducers and windows for particular clinical applications.

The major imaging applications are as follows (note that "intra" (from Latin) means into or inside, "trans" means through or across, and "endo" means within):

Endovaginal: Imaging the female pelvis using the vagina as an acoustic window.
Intracardiac: Imaging from within the heart.
Intraoperative: Imaging during a surgical procedure.
Intravascular: Imaging of the interior of arteries and veins from transducers inserted in them.
Laparoscopic: Imaging carried out to guide and evaluate laparoscopic surgery made through small incisions.
Musculoskeletal: Imaging of muscles, tendons, and ligaments.
Small parts: High-resolution imaging applied to superficial tissues, musculature, and vessels near the skin surface.
Transcranial: Imaging through the skull (usually through windows such as the temple or eye) of the brain and its associated vasculature.

Transesophageal: Imaging of internal organs (especially the heart) from specially designed probes made to go inside the esophagus.
Transorbital: Imaging of the eye or through the eye as an acoustic window.
Transrectal: Imaging of the pelvis using the rectum as an acoustic window.
Transthoracic: External imaging from the surface of the chest.

10.7 Transducers and Image Formats

10.7.1 Image Formats and Transducer Types

Why do images come in different shapes? The answer depends on the selected transducer, without which there would be no ultrasound imaging system. Our discussion emphasizes types of arrays (the most prevalent form of transducers in ultrasound imaging). The focus will be on widely used physical forms of arrays adapted for different clinical applications and their resulting image formats.

Early ultrasound imaging systems employed single-element transducers, which were mechanically scanned in an angular or linear direction or both (as described in Chapter 1). Most of these transducers moved in a nearly acoustically transparent cap filled with a coupling fluid. The first practical arrays were annular arrays that consisted of a circular disk cut into concentric rings, each of which could be given a delayed excitation appropriate for electronic focusing along the beam axis. These arrays also had to be rotated or scanned in a cap, and they provided variable focusing and aperture control for far better imaging than is available with fixed-focus single-element transducers. A detailed description of the design and performance of a real-time digital 12-element annular array ultrasound imaging system is available in Foster et al. (1989a, 1989b).

Another early array was the linear array (discussed in Chapter 1). The linear array may have up to 300–400 elements, but at any specific time, only a few (forming an active element group) are functioning. The active contiguous elements form the active aperture. At one end of the array, an active element group turns on, as selected by a multiplexer (also called a "mux") that is receiving commands from the beamformer controller. Refer to Figure 10.7A, where the active elements are shaded to generate line number n. After the first pulse echoes are received for the first image vector line (centered in the middle of this group), an element nearest the end of the array is switched off and the element next to the other end of the group is added as a new element. In this way, the next sequential line (numbered $n + 1$) is formed, and this "tractor-treading" process continues as the active group slides along the length of the array, picking up and dropping an element at each line position. Switches are necessary if the number of elements in the array exceeds the number of receive channels available. The overall image format is rectangular in shape.

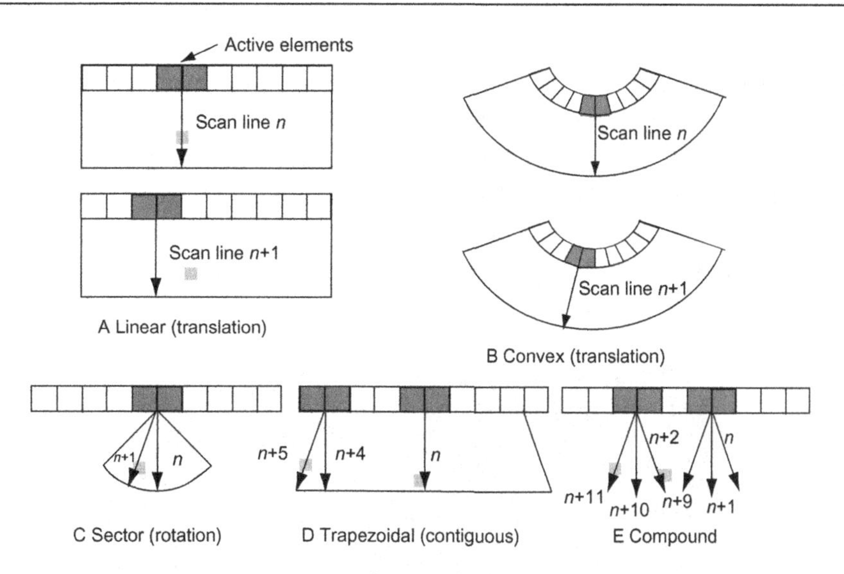

Figure 10.7
Time-sequenced image formats. (A) Basic linear (translation); (B) convex curved linear (translation); (C) basic sector (rotation); (D) trapezoidal (contiguous: rotation, translation, and rotation); (E) compound (translation and rotation at each active aperture position).

The main difference between a linear and a phased array is steering. The phased array has an active aperture that is always centered in the middle of the array, but the aperture may vary in the number of elements excited at any given time (discussed shortly). As shown in Figure 10.7C, the different lines are formed sequentially by steering until a sector (an angular section of a circle), usually about 90° in width, is completed. The phased array has a small "footprint," or contact surface area with the body. A common application for this type of an array is cardiac imaging, which requires that the transducer fit in the intercostal spaces between the ribs (typically 10–14 mm). The advantage of this array is that, despite its small physical size, it can image a large region within the body.

Because it was easier to produce a fixed focal delay without steering for each line, linear arrays were the first to appear commercially (recall Chapter 1). In this tradition, convex linear arrays, also known as curved linear arrays, combined the advantage of a larger angular image extent with ease of linear array focusing without the need for electronic steering. Convex arrays may be regarded as linear arrays on a curved surface. As depicted in Figure 10.7B, a convex array has a similar line sequencing to a linear array, except that its physical curvature directs the image line into a different angular direction. Because of the lack of steering, linear and convex arrays have a relaxed requirement for periodicity: 1–3 wavelengths rather than the ½ wavelength usually used for phased arrays.

Recent exceptions to this approach are linear arrays with finer periodicity so that they can have limited steering capability either for Doppler or color flow imaging. In this case, once the extent of steering is decided, periodicity can be determined from grating lobe calculations (see Chapter 7). Two common applications are parallelogram (also known as a steered linear) and trapezoidal imaging, in which sector-steered image segments are added to the ends of a rectangular image in a contiguous fashion (shown in Figure 10.7D). Actual imaging examples are given by Figure 10.8.

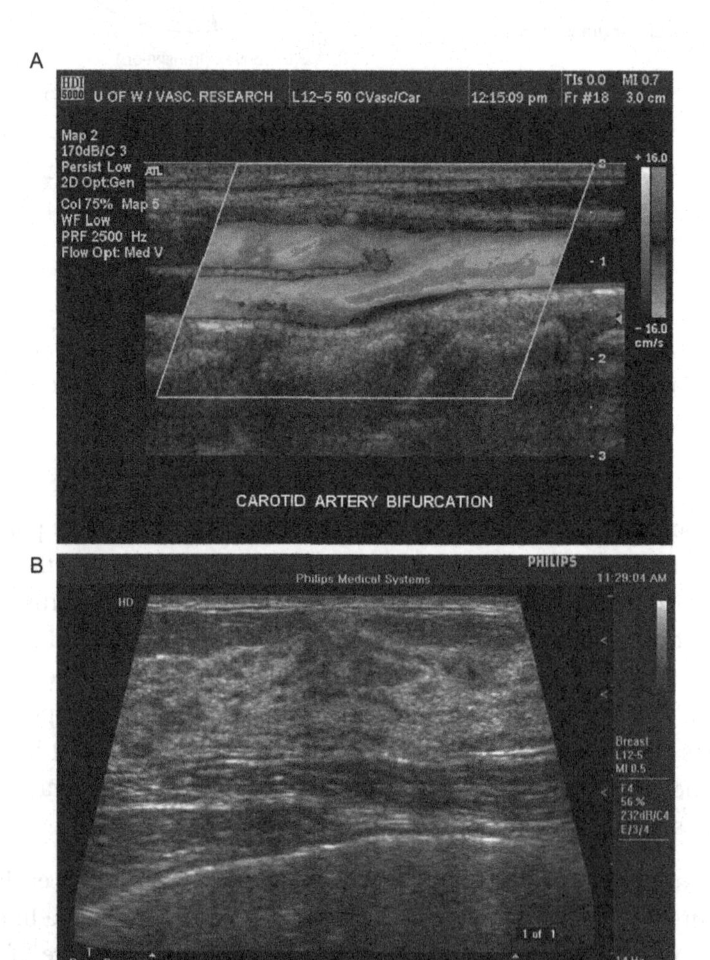

Figure 10.8

Linear arrays with steering. (A) Parallelogram-style color flow image from a linear array with steering; (B) trapezoidal form of a linear array with sector steering on either side of a straight rectangular imaging segment (described as a contiguous imaging format in Chapter 1). The reader is referred to the Web version of the book to see the figure in color. *Courtesy of Philips Healthcare.*

Another use of more finely sampled linear arrays with steering capabilities is compound imaging. As shown in Figure 10.7E, compound imaging is a combination of limited steering by an active group and translation of the active group to the next position for the next set of lines or image vectors. More information and imaging examples of a real-time implementation of this method will be discussed in Section 10.11.4. More about image formats and their relation to transducer types can be found in Szabo and Lewin (2013).

The number of active elements selected for transmission is usually governed by a constant *F* number (*F#*). The −6-dB full width half maximum (FWHM) beamwidth can be shown to be approximately FWHM $= 1.2\lambda F/L = 1.2\lambda F\#$ from Eqn 6.9C. To achieve a constant lateral resolution for each deeper focal length (*F*), the aperture (*L*) is increased to maintain a constant *F#* until the full aperture available is reached. In a typical image, one transmit focal length is selected, along with dynamic focusing on receive. At the expense of frame rate, it is possible to improve resolution by transmitting at several different transmit focal lengths in succession and then splicing together the best parts. The strips or time ranges contain the best lateral resolution (like a layer cake) to make a composite image of superb resolution (Maslak, 1985). See Figure 10.9 for an example. For this method, a constant *F#* provides a similar resolution in each of the strips as focal depth is increased (until at some depth there are not enough active elements to maintain the *F#*).

Another simple, useful formula is a first approximation of the depth of penetration (DP) for a given frequency, DP $= 60/f$ cm-MHz, where f is given in megahertz (Szabo and Lewin, 2013). Thus, one might expect a 6-cm penetration from a 10-MHz center frequency transducer. As noted earlier, the absorption coefficient (acoustic power loss per unit depth) is a function of frequency and varies from tissue to tissue (values for soft tissues range from 0.6 to 1.0 dB/cm-MHz). A more general term describing acoustic loss is the attenuation coefficient, which includes additional losses due to scattering and therefore is always greater than the absorption coefficient. The attenuation coefficient is highly patient and acoustic-path dependent. A more accurate approximation would include the power law nature of attenuation for tissues of interest. In addition, the special characteristics of an imaging system, such as the use of matched filters (Section 10.9.2), can affect image penetration depth.

To overcome the small field of view limitation in typical ultrasound images, a method of stitching together a panoramic view (such as that shown in Figure 10.10) has been invented. Even though the transducer is scanned freehand across the skin surface to be imaged, advanced image processing is used to combine the contiguously scanned images in real time (Tirumalai et al., 2000). Other modes can also be shown in this type of presentation.

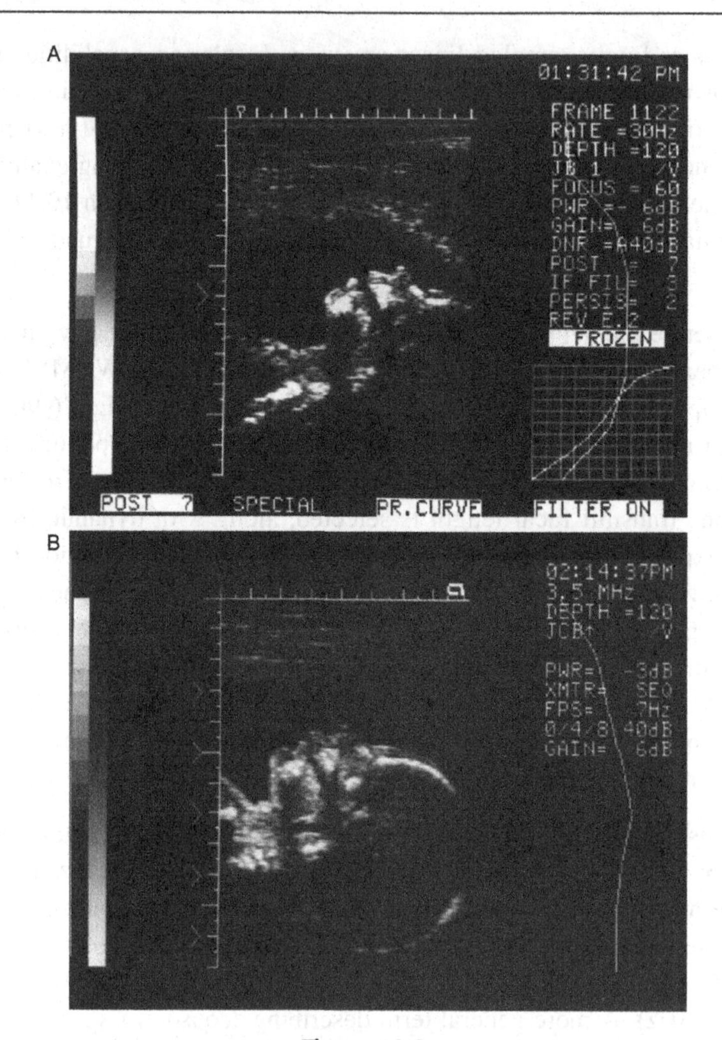

Figure 10.9
Transmit focusing of fetal head. (A) With a single focus zone; (B) with multiple spliced focal zones. *Courtesy of Siemens Medical Solutions, Inc., Ultrasound Group.*

10.7.2 Transducer Implementations

Driven by the many clinical needs, transducers appear in a wide variety of forms and sizes (as indicated by Figure 10.11). Notable in this figure, there is a transesophageal probe mounted on the end of a gastroscope (a); a 3D mechanically scanned curved array (b); curved or convex arrays (c, j, k, and q); phased arrays (e and g); 2D matrix arrays (f, h, i, and l); an intraoperative probe (m); linear arrays (n, p, and r); and a "stand-alone" pulsed Doppler element transducer (o). The transesophageal probe (shown at the tip in the top center of the figure) is mounted on a gastroscope assembly (at the extreme left of the

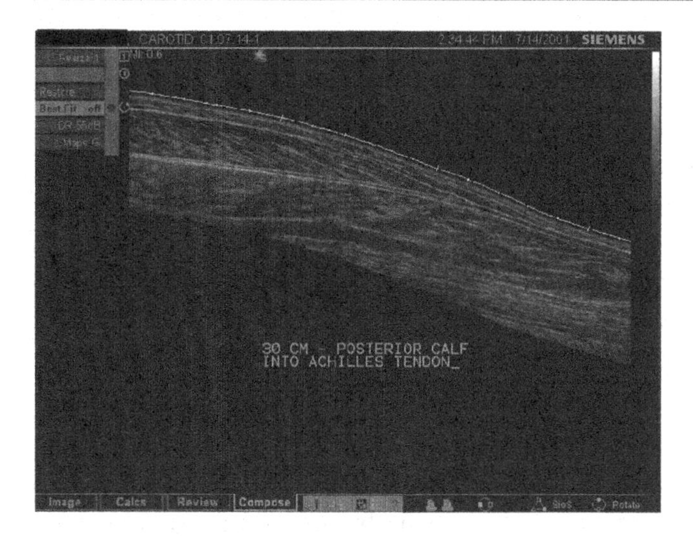

Figure 10.10
SieScape, or panoramic image made by a transducer swept along a body surface. *Courtesy of Siemens Medical Solutions, Inc., Ultrasound Group.*

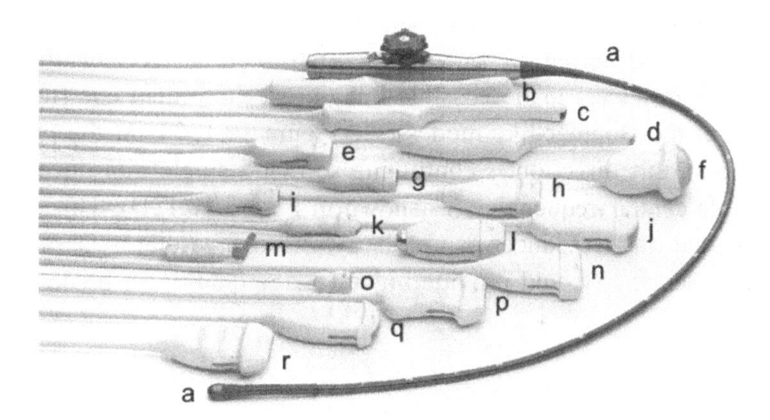

Figure 10.11
Transducer family portrait.
Transesophageal (a); 3D mechanically scanned curved array (b); from left to right, transesophageal array with positioning assembly, convex (curved) linear array, linear array, stand-alone CW Doppler probe, phased array, transthoracic motorized rotatable phased array, and high-frequency intraoperative linear array. *Courtesy of Philips Healthcare.*

figure) to provide flexible positioning control of the transducer attitude within the throat. Transesophageal arrays couple through the natural fluids in the esophagus and provide cleaner acoustic windows to the interior of the body (especially the heart) than transducers applied externally through body walls. The endovaginal (b and c) and transrectal (d) probes are designed to be inserted. The intraoperative (m) and specialty arrays provide better access for surgical and near-surface views in regions sometimes difficult to access. These probes can provide images before, during, or after surgical procedures.

The more conventional linear, curved linear, and phased arrays have typical azimuth apertures that vary in length from 25 to 60 mm and elevation apertures that are 2–16 mm, depending on center frequency and clinical application. Recall that the aperture size in wavelengths is a determining factor; this relation is obvious from rewriting the *F#* relation used earlier, $F\# = \text{FWHM}/(1.2\lambda)$. The number of elements in a 1D array vary from 32 to 400. Two-dimensional or matrix arrays are shown in Figure 10.11 in f, h, i, and l. The present number of elements varies from 2500 (i) up to 9212 (h).

Typical center frequencies range from 1 MHz (for harmonic imaging) to 20 MHz (for high-resolution imaging of superficial structures). As discussed in Chapter 6, there has been a trend toward wider fractional bandwidths, which now range from 30–130%. At first, array systems functioned at only one frequency because of the narrow fractional bandwidth available. As transducer design improved, wider bandwidth allowed for operation at a higher imaging frequency simultaneously with a lower-frequency narrowband Doppler or color flow mode (as indicated in Figure 10.12B). This dual frequency operation was made possible by two different transmit frequencies combined with appropriate receive filtering, all operating within the transducer bandwidth. The next generation of transducers made possible imaging at more than one frequency, as well as operation of the Doppler-like modes (see Figure 10.12C). At the present time (with new materials), this direction is continuing so that a single transducer array can function at multiple center frequencies (as shown in Figure 10.12D). This type of bandwidth means that one transducer can replace two or three others, permit harmonic imaging with good sensitivity, and provide higher image quality (to be described in

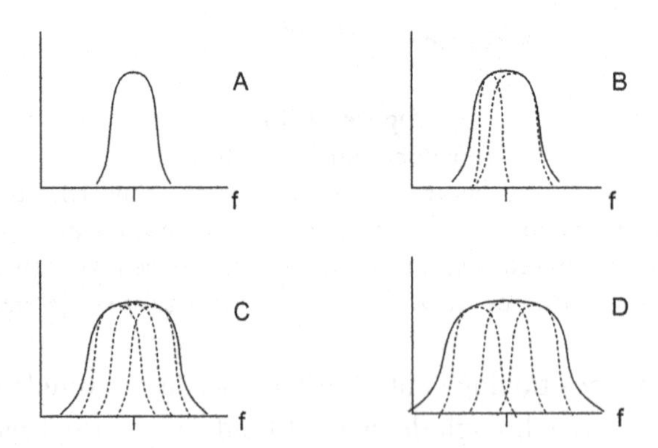

Figure 10.12
Stages of transducer bandwidth development. (A) Narrowband; (B) dual mode; (C) multiple mode; (D) very wide band.

Section 10.11.3). Broad bandwidths are also essential for harmonic imaging (to be described in Chapter 12).

10.7.3 Multidimensional Arrays

As discussed in Chapter 7, most arrays are 1D with propagation along the z axis and electronic scanning along the x axis to form the imaging plane. Focusing in the elevation or yz plane is accomplished through a fixed focal length lens. A hybrid approach (a 1.5D array) achieves electronic focusing in the elevation plane by forming a coarsely sampled array in the y dimension at the expense of more elements. This number is a good compromise, however, compared to a complete 2D array, which usually requires about an n^2 channel count, compared to n channels for 1D arrays. A way of reducing the number of electronic channels needed is to decrease the active number of elements to form a sparse array. All of these considerations were compared in Chapter 7. The main advantages of electronic focusing in the elevation are not only flexibility, but also improved resolution from coincident focusing in both planes and dynamic receive focusing in both planes simultaneously. The realization of fully populated 2D arrays has greatly reduced the use of 1.5D and sparse arrays. The description of real-time fully populated 2D arrays with a nonstandard architecture is postponed until Section 10.12.1.

10.8 Front End

The front end is the mouth of the imaging system; it can talk and swallow. It has a number of channels, each of which has a transmitter and a switch (including a diode bridge) that allows the passage of high voltage transmit pulses to the transducer elements, but blocks these pulses from reaching sensitive receivers (refer to the block diagram of Figure 10.4). Echoes return to each receiver, which consists of amplifiers in series, including one that has a variable gain for TGC under user control. The output of each channel is passed on to the receive beamformer.

10.8.1 Transmitters

The heartbeat of the system is a series of synchronized and precisely timed primitive excitation pulses (illustrated by Figure 10.13). The major factor in this heartbeat is the scan depth selected (s_d). The length of a line or vector, since each line has a vector direction, is simply the round-trip travel time ($T_{RT} = 2s_d/c_0$). As soon as one line has completed its necessary round-trip time, another line is launched in the next incremental direction required. For a simple linear array, the next line is parallel to the last one, whereas in a sector format, the next line is incremented through steering by a small angle.

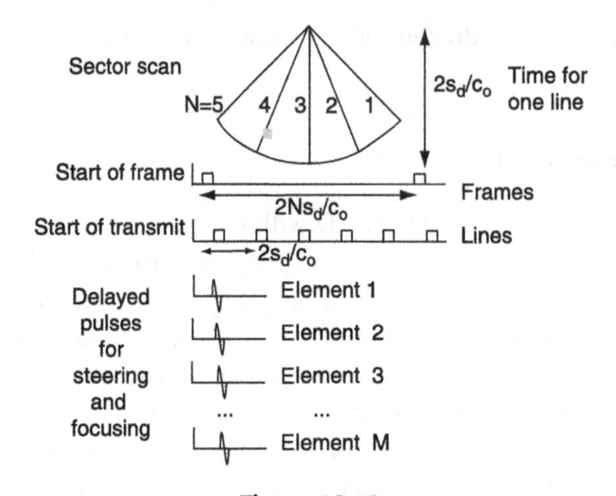

Figure 10.13
Pulse generation sequencing in an imaging system.

The timing pulses associated with these events are the start-of-frame pulse, followed by the start of transmit. This last pulse actually launches a group of transmit pulses in parallel with the required delays to form a focused and steered beam from each active array element. The exact timing of these transmit pulses was described in Chapter 7. This process is repeated for each vector until the required number of lines (N) has been completed, after which a new start-of-frame timing pulse is issued by the system transmitter clock.

The rhythm of the system heartbeat can be interpreted as a repetitive timing sequence with a duty cycle. For the example shown in Figure 10.13, assume a scan depth of $s_d = 150$ mm, as well as 5 lines per frame and 6 active elements. The round-trip time for one line is $T_{RT} = 2s_d/c = 200$ μs; this will be the start of the transmit pulse interval between each line. The time for a full frame is N lines/frame or, in this case, 5×200 μs/frame = 1000 μs/frame or frame rate (FR) = $1/NT_{RT}$, FR = 1000 frames/s. The number of lines is only 5 for this example. A more realistic number of lines is 100, in which case the time for a full frame would be 20 ms or a frame rate of 50 frames/s.

Finally, depicted in the bottom of Figure 10.13 is a sequence of delayed pulses (one for each active element of the array) to steer and focus the beam for that line. Note that these pulses are launched in parallel with each start of transmit. These transmit pulses have a unique length or shape for the mode and frequency chosen. For example, instead of one primitive transmit pulse such as a single cycle of a sine wave for 2D imaging, a number (M) of primitive pulses in succession can be sent to form an elongated pulse for Doppler mode. The duty cycle is taken to be the ratio of the length of the basic transmit sequence per line divided by the round-trip time. In practice, a vector line may be repeated by another one in the same direction or by one in a different mode in a predetermined

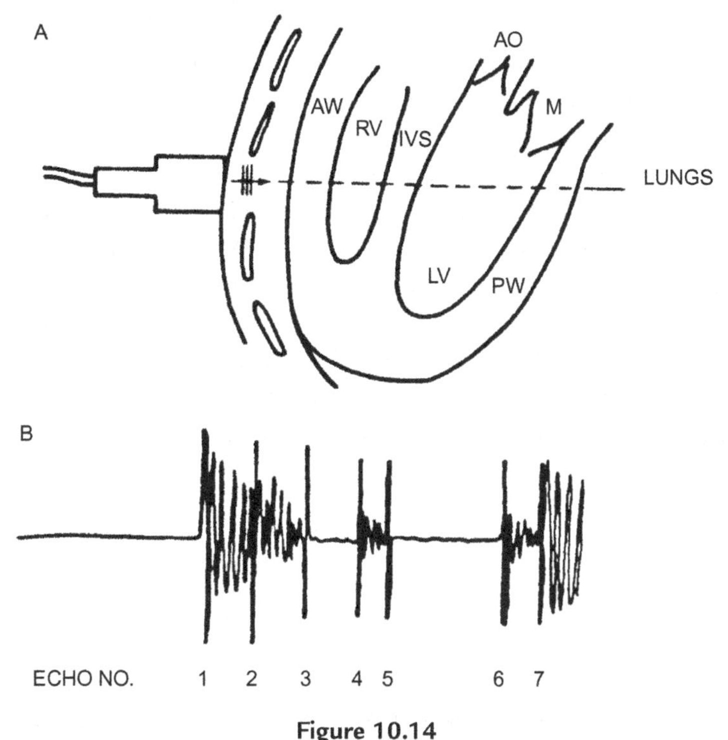

Figure 10.14

Backscattered echoes from the heart. (A) Echo path through the heart; (B) amplified echoes corresponding to path in (A). AO, aortic valve; AW, anterior wall; IVS, intraventricular septum; LV, left ventricle; M, mitral valve; PW, posterior wall; RV, right ventricle; T, transducer. *From Shoup and Hart (1988), IEEE.*

multimode sequence necessary to build a duplex or a triplex image (Szabo, Melton, & Hempstead, 1988).

10.8.2 Receivers

In order to estimate the dynamic range needed for a front end, typical echo levels in cardiac imaging will be examined. Numbered amplified backscattered echoes from the heart are illustrated by Figure 10.14B for the beam path shown through a cross-section of the heart in Figure 10.14A (Shoup & Hart, 1988). With reference to the indexing of the echoes, the first waveform corresponds to feed-through during the excitation pulse. Echo 2 is caused by the reflection factor between the fat in the chest wall and muscle of the anterior wall; this kind of signal is on the average about −55 dB below that obtained from a perfect (100%) reflector. Echo 3 is the echo from the reflection between blood and the tissue in the wall; it has a similar absolute level. Between echoes 3 and 4 is the backscatter from blood, which is at the absolute level of −70 dB compared to a 100% reflector and falls below the scale shown. The large echo number 7 is from the posterior wall–lung interface; it is a nearly perfect reflector (close to 0 dB absolute level). In order to detect blood and the lung without saturating, the receivers require a dynamic range of at least 70 dB for cardiac imaging. TGC amplification (mentioned in Chapter 4) was applied to the echoes in Figure 10.14B. The absolute values of the echoes were determined independently from reflection factor data and a reference reflector. There is an individual front-end amplifier for each channel (usually 64 or 128 total) in the system. Each amplifier typically covers a range of 55–60 dB. For digital conversion, sampling rates of 3–5 times the highest center frequency are needed to reduce beamforming quantization errors (Wells, 1993). A means of time-shifting for the dynamic receive beamformer at higher rates, closer to 10 times the center frequency, would be preferable to achieve low beam sidelobes (Foster et al., 1989a,b). Modern imaging systems can have dynamic ranges in excess of 100 dB, and some have the sensitivity to image blood directly in B-mode at high frequencies (see Chapter 11) and to detect weak harmonic signals (see Chapter 14 and Chapter 15).

10.9 Scanner

10.9.1 Beamformers

In Chapter 7, the operation of transmit and receive beamformers was discussed. The practical implementation of these beamformers involves trade-offs in time and amplitude quantization. In addition, more complicated operations have been implemented. In order to speed up frame rate, basic parallel beamforming is a method of sending out a wide transmit beam and receiving several receive beams (as explained in Section 7.4.3). The discussion of real-time compound imaging (Entrekin, Jago, & Kofoed, 2000), which involves the ability of the beamformer to send out beams along multiple vector directions from the same spatial location in a linear array, is deferred until Section 10.11.4.

10.9.2 Signal Processors

Bandpass filters

This signal-processing part of the system takes the raw beamformed pulse–echo data and selectively pulls out and emphasizes the desired signals, combines them as needed, and provides real and quadrature signals for detection and modal processing. This section covers only processing related to B-mode imaging. Chapter 11 covers color flow imaging and Doppler processing. Digital filters operate on the data from the A/D converters (shown in the block diagram, Figure 10.4). Bandpass filtering is used to isolate the selected frequency range for the desired mode within the transducer passband (recall Figure 10.12). The data may also be sent to several bandpass filters to be recombined later in order to reduce speckle (see Section 10.11.3). Another important function of bandpass filtering is to obtain harmonic or subharmonic signals for harmonic imaging (to be covered in more detail in Chapter 12). In Chapter 4, absorption was shown to reduce the effective center of the signal spectrum with depth. The center frequency and shape of bandpass filters can be made to vary with depth to better track and amplify the desired signal (see Section 10.11.2).

Matched filters

Another important related signal-processing function is matched filtering introduced in Section 9.7.2. In the context of ultrasound imaging, this type of filter has come to mean the creation of unique transmit sequences, each of which can be recognized by a matched filter. One of the key advantages of this approach is that the transmit sequence can be expanded in time at a lower amplitude and transmitted at a lower peak pressure amplitude level, with benefits for reducing bioeffects (see Chapter 15) and contrast agent effects (see Chapter 14). Other major advantages include the ability to preserve axial resolution with depth, and increased sensitivity and tissue penetration depth.

Matched filtering actually begins with the transmit pulse sequence. In this case, the transmit waveform is altered into a special shape or sequence, $s(t)$. This transmission encoding can be accomplished by sending a unique sequence of primitive pulses of different amplitudes, polarities, and/or interpulse intervals. In the case of binary sequences, a "bit" is a primitive pulse unit that may consist of, for example, half an RF cycle or several RF cycles.

Two classic types of transmit waveforms, $x(t)$, a coded binary sequence and a chirped pulse, have been borrowed from radar and applied to medical ultrasound (Chiao & Hao, 2003; Cole, 1991; Lee & Ferguson, 1982; Lewis, 1987; O'Donnell, 1992). The appropriate matched filter in these cases is $x^*(-t)$. The purpose of a matched filter is to maximize signal-to-noise ratio, defined as the ratio of the peak instantaneous output signal power to the root mean square (r.m.s.) output noise power (Kino, 1987). A simple explanation of how the output power can be maximized can be given through Fourier transforms. Consider a filter response:

$$y(t) = x(t) * h(t), \tag{10.1}$$

where $x(t)$ is the input, $y(t)$ is the output waveform, and $h(t)$ represents the filter. Let the matched filter be:

$$h(t) = Ax^*(-t), \tag{10.2}$$

where A is a constant and * represents the conjugate. For this filter, the output becomes:

$$y(t) = Ax(t) * x^*(-t) = A\int_{-\infty}^{\infty} x(\tau)x^*(\tau - t)d\tau = A\int_{-\infty}^{\infty} x^*(\tau)x(\tau + t)d\tau, \tag{10.3}$$

but from the Fourier transform, the output can be rewritten as:

$$y(t) = A\int_{-\infty}^{\infty} X(f)X^*(f)e^{i2\pi ft}df = A\int_{-\infty}^{\infty} |X(f)|^2 e^{i2\pi ft}df. \tag{10.4}$$

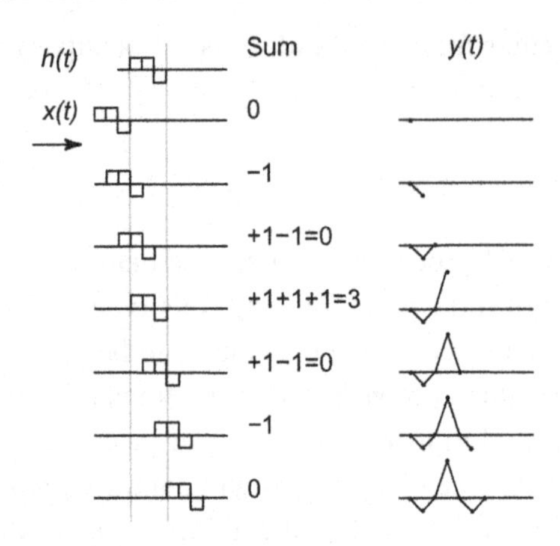

Figure 10.15
Output of a Three-bit Barker Code.
(Top) receive correlator sequence $h(t)$ versus time units; (below) input sequence $x(t)$ shown as incrementing one time unit or one bit at a time through the correlator with the corresponding summation and output waveform.

In other words, the matched filter choice of Eqn 10.2 leads to an autocorrelation function, Eqn 10.3, which automatically maximizes the power spectrum, Eqn 10.4 (Bracewell, 2000) and, consequently, maximizes the ratio of the peak signal power to the r.m.s. noise power (Kino, 1987).

A simple example of a coded waveform is a three bit Barker code. This code can be represented graphically (shown in Figure 10.15), or it can be represented mathematically as the binary sequence $[+1+1-1]$. Binary codes have unique properties and solve the following mathematical puzzle: What sequence of ones and minus ones, when correlated with itself, will provide a gain in output (y) with low sidelobes?

At the top of Figure 10.15 is a plot of the correlation filter $h(t)$ against unit time increments. Recall that the convolution operation involves flipping the second waveform right to left in time and integrating (see Appendix A). Physically, correlation is the operation of convolution of $x(t)*x^*(-t)$. This integration consists of a double reversal in time (once for the convolution operation and once for the receive filter). The net result is a receive waveform that is back to its original orientation in time. The operation is simplified to sliding one waveform, $x(t)$, past the second, $x(t)$, left to right. Each row in this figure shows an input waveform sliding from left to right, one time unit interval at a time, until the waveform has passed through the correlator. Integration at each slot is easy: First, determine the amplitude values of $h(t)$ and $x(t)$ multiplied together, such as $-1 \times -1 = 1$, at each time interval overlap position; second, sum all the product contributions from each time interval in the overlap region to obtain the amplitude value for the time position in the row. In the last row, connect the dots at each time interval to get $y(t)$. The repeating triangular shapes within $y(t)$ can be recognized as the convolution, or correlation in this case, of two equal rectangles, $\Pi(t)$, that slide past each other to form triangle functions; these steps complete the description of $y(t)$ between the dots we calculated in Figure 10.15. Note the main features of $y(t)$: a peak equal to n bits (three) and two satellite time sidelobes of amplitude -1. From maximum amplitudes of plus or minus one, a gain of three has been achieved by encoding.

Fortunately, MATLAB makes these kinds of calculations trivial. We can obtain graphical results with three lines of code:

$$\begin{aligned} &x = [011 - 10]'; \\ &y = x\,\mathrm{corr}(x). \\ &\mathrm{plot}(y); \end{aligned} \tag{10.5}$$

The first line forms the Barker sequence, allowing for zeros to get the full depiction of the output. The autocorrelation function is the cross-correlation function xcorr.m with one argument. The reader is encouraged to play with the program barkerplot.m to verify that as the number of bits, N, is increased, the peak increases in proportion and the ratio of peak amplitude level to maximum sidelobe level improves.

A family of codes with more impressive performance is the pseudo-random binary M-sequence code of ones and zeros that is shown in the lower right-hand corner of Figure 10.16 (Carr, DeVito, & Szabo, 1972) along with the output, $y(t)$. Here the sidelobe ratio is -15.84 dB. Note that for an acoustic transmitter, ones and zeros may translate into either a series of "ones" (regarded as positive primitive pulses, $+1$) and "zeros" (regarded as primitive pulses with a 180° phase reversal or negative-going pulses, -1).

There are several families of codes, each with advantages and disadvantages. Each bit or primitive pulse alone will evoke a round-trip response from the transducer, which fixes the minimum resolution available. In the usual case without a coded sequence, a transmit pulse might consist of a half-period pulse or a full-period pulse (e.g. a single sine wave) corresponding to the desired frequency of excitation. Receive amplitude levels can be raised by increasing the applied transmit voltage. At some pressure level (described in Chapter 15), a fixed limit is reached for safety reasons, so that the voltage can no longer be increased. One advantage of coded sequences is that a relatively low voltage A can be applied, and a gain of NA is realized on reception after the correlation process. Another advantage of coded sequences is that certain orthogonal codes, such as Golay codes, allow the simultaneous transmission of a number of beams in different vector directions, which are sorted out on decoded reception through matched correlators (Chiao & Hao, 2003; Chiao, Thomas, & Silverstein, 1997; Lee & Ferguson, 1982; Shen and Ebbini, 1996a,b) as is shown in Figure 10.17.

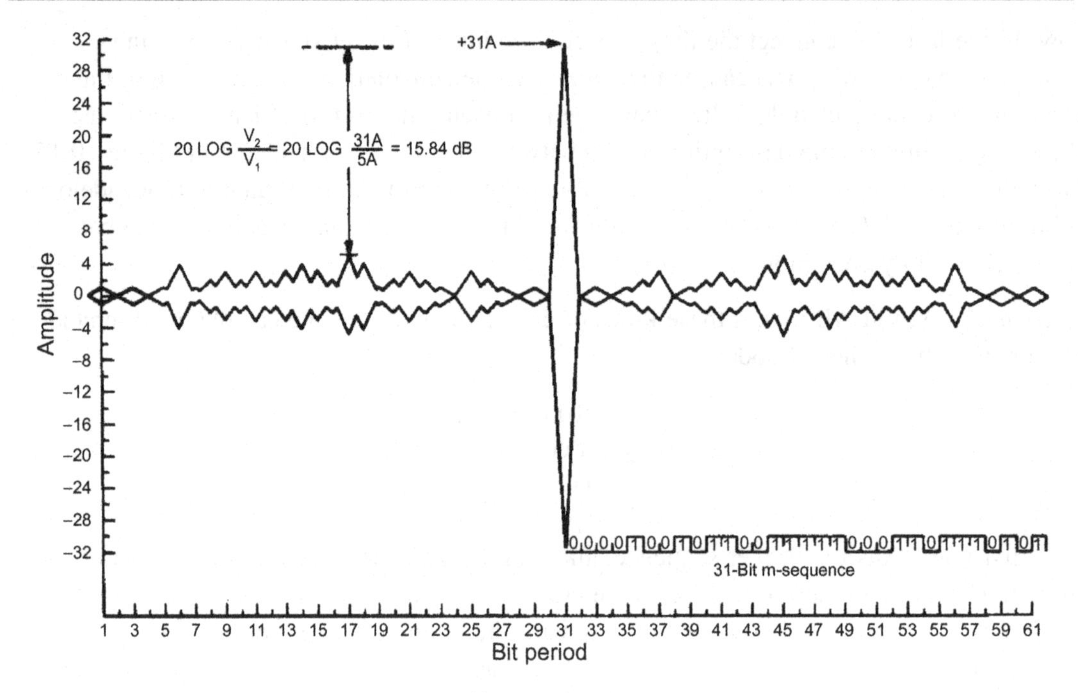

Figure 10.16
Theoretical plot of amplitude versus bit period for the correlation of a 31-bit maximal length (M) sequence. The peak-to-sidelobe ratio for this sequence is −15.84 dB. *From Carr et al. (1972), IEEE.*

Another important class of coded matched filter functions are chirps (Cole, 1991; Genis Obeznenko, Reid, & Lewin, 1991; Lewis, 1987). A methodology borrowed from radar, a transmit waveform, $x(t)$, consists of a linear swept frequency-modulated (FM) pulse of duration T. The result of matched filtering is a high-amplitude short autocorrelation pulse. If a chirp extends over a bandwidth B, the correlation gain G through a matched filter $x^*(-t)$, a mirror image chirp, is $G = TB$ (Kino, 1987). Examples of a chirp and compressed pulses from flat targets are given in Figure 10.18. A third waveform depicts the transmitted upchirp waveform. A useful parameter is the instantaneous frequency, defined as:

$$f_i = \left(\frac{1}{2\pi}\right)\frac{d\phi}{dt}, \tag{10.6}$$

where ϕ is the phase of the analytic signal as a function of time (see Appendix A). For the transmit chirp of Figure 10.18, the instantaneous frequency as a function of time is an ascending line from 5 to 9 MHz. The second panel from the top of Figure 10.18 shows the received echoes for a glass plate. After passing through the matched filter, these echoes are

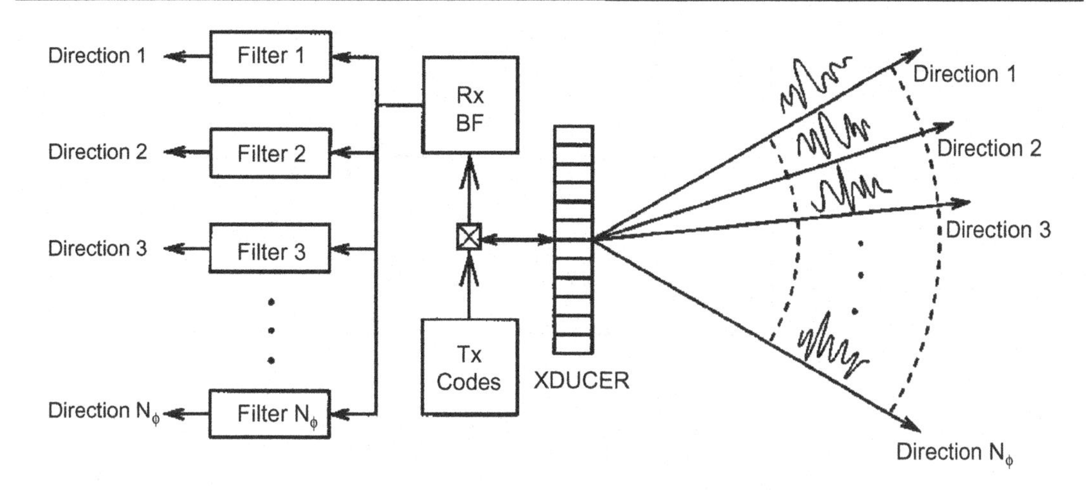

Figure 10.17
Simultaneous multibeam encoded ultrasound imaging system. Tx, transmit; Rx, receive; XDUCER, transducer. *From Shen and Ebbini (1996), IEEE.*

compressed to give excellent resolution and indicate multiple internal reflections into the top panel. A pair of similar echo signals for a plastic plate with higher internal absorption is shown in the lower two panels of Figure 10.18. The pros and cons of this methodology are discussed in the previous references. Both orthogonal codes and chirped waveform matched filters have been implemented in commercial systems. More on matched filters can be found in Section 9.7.2.

10.10 Back End

10.10.1 Scan Conversion and Display

The main function of the back end (refer to Figure 10.4, the block diagram) is to take the filtered RF vector line data and put it into a presentable form for display. These steps are the final ones in the process of imaging (described in detail in Section 8.4). An imaging challenge is to take the original large dynamic range, which may be on the order of 120 dB, and reduce it down to about 30 dB, which is the maximum gray-scale range that the eye–brain system can perceive. The limits and description of human visual perception is beyond the scope of this work, and they are covered in more detail in Sharp (1993). As we have seen, the initial step is taken by the TGC amplifiers, which reduce the dynamic range to about 55–60 dB. The beamformed digitized signals are converted to real (I) and quadrature (Q) components (delayed from the I signal by a quarter of the fundamental period). These components can be combined to obtain the analytic envelope of the signal through the operation $\sqrt{I^2 + Q^2}$.

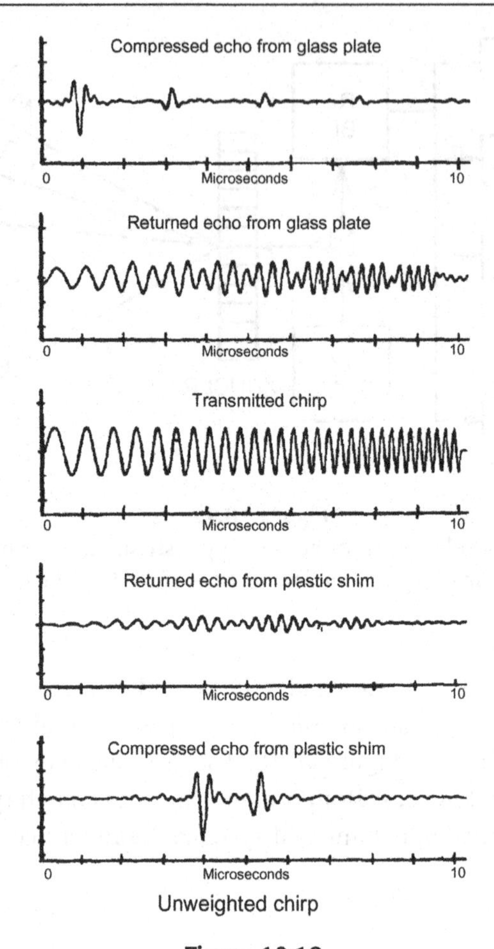

Figure 10.18

A Chirp Extending from 5 to 9 MHz (Middle Panel) and Returned (Uncompressed) and Compressed Pulse Echoes from a Glass Plate and a Plastic Shim. *From Lewis (1987), IEEE.*

In Figure 10.19, the envelope detection begins the back-end processing. This step is followed by an amplifier that can be controlled by the user to operate linearly at one extreme, or as a logarithmic amplifier at the other extreme, or as a blend between the two extremes to achieve further dynamic range compression. For example, in the case in which soft-tissue detail and bright specular targets coexist in the same image, the logarithmic characteristic of the amplifier can reduce the effects of the specular reflections on the high end of the scale. The preprocessing step, not done in all systems, slightly emphasizes weak signals as the number of bits is reduced, for example, from 10 to 7 bits after digitization.

So far, a number of vectors (lines with direction) have undergone detection, amplification, preprocessing (if any), and resampling to a certain number of points per line for suitable viewing. In order to make a television or PC-style rectangular image, this

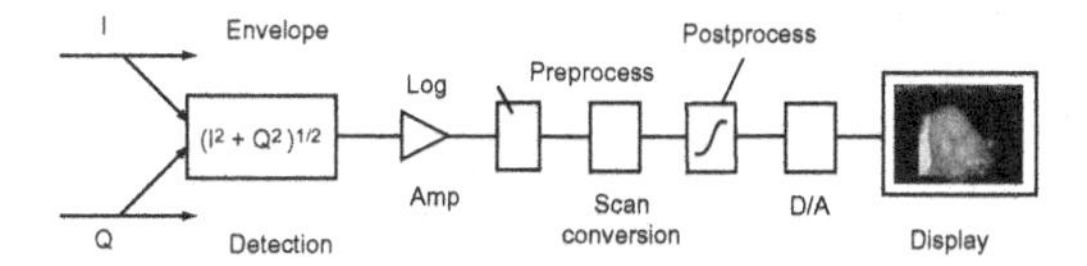

Figure 10.19
Block diagram for back-end processing used for image display. *Image courtesy of Philips Healthcare.*

information has to be spatially remapped by a process called scan conversion. If the vectors were displayed in their correct spatial positions, the data would have missing information when overlaid on a rectangular grid corresponding to pixel locations in a standard raster scan, such as the NTSC TV. Sector scanning is one of the more challenging formats to convert to TV format (as illustrated by Figure 10.20). An enlargement of the polar coordinate scan lines overlaid on the raster rectangular pixel grid indicates the problem. Not only do the scan lines rarely intersect the pixel locations, but also each spatial position in the sector presents a different interpolation because the vectors change angle and are closer toward the apex of the sector. Early attempts at interpolation caused severe artifacts, such as Moire's pattern, and unnatural steps and blocks in the image. This problem can be solved by a 2D interpolation method (Leavitt, Hunt, & Larsen, 1983), which is shown in the bottom of Figure 10.20. The actual vector points are indicated along the bold scan lines with the pixel locations marked by crosses. To obtain the interpolation at a desired point (Z), first the radius from the apex to the intended pixel point is determined. Second, the angle of a radial line passing through Z is found. The generalized 2D interpolation formula is:

$$Z(r,\theta) = \sum_{n}\sum_{m} S(r - n\Delta r, \theta - m\Delta\theta)Z(n\Delta r,\ m\Delta\theta), \tag{10.7}$$

where S is a 2D triangular function.

The next step is one in which the amplitudes in the rectangular format undergo a nonlinear mapping called postprocessing. A number of postprocessing curves are selectable by the user to emphasize low- or high-amplitude echoes for the particular scan under view. This choice determines the final gray-scale mapping, which is usually displayed along with the picture. In some cases, pure B-mode images undergo an additional color mapping (sometimes called colorization) in order to increase the perceived dynamic range of values. Finally, a digital-to-analog (D/A) conversion occurs for displaying the converted information. The usual video controls such as brightness and contrast are also available, but they play a minor role compared to the extensive nonlinear mapping processes the data has undergone. Image plane overlays are used to present graphic and measurement information. Color flow display (to be covered in Chapter 11) also undergoes scan conversion and is displayed as an image plane overlaid on the gray-scale B-mode plane. In addition, most

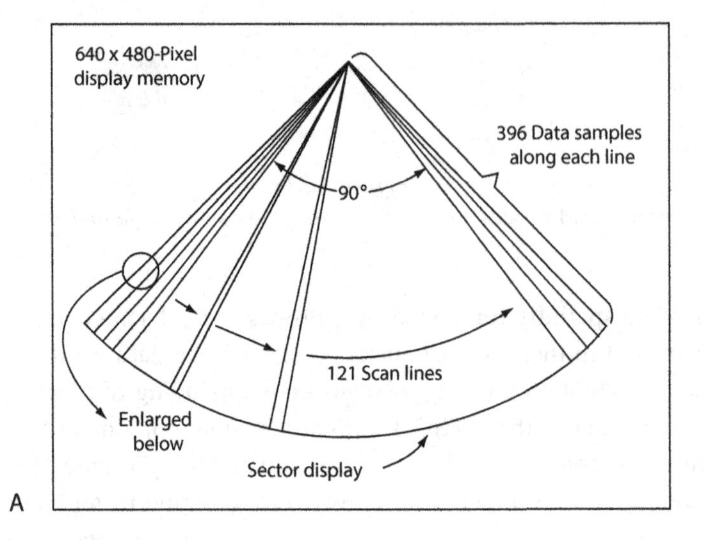

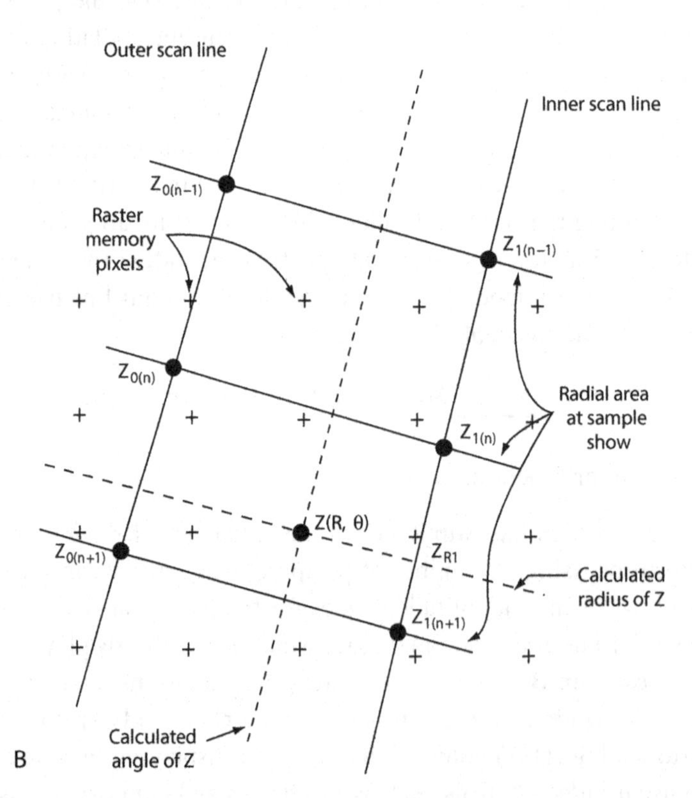

Figure 10.20

(A) Image vectors in a sector scan display overlaid on desired rectangle format; (B) Magnified view comparing vector data in polar coordinates to rectangular pixel positions. *Reprinted from Leavitt et al. (1983) with permission from Hewlett Packard.*

systems have the capability to store a sequence of frames in internal memory in real time for cine loop display. For three-dimensional imaging (Section 10.11.6), there is an advantage to interpolating voxels in three dimensions to achieve image uniformity.

10.10.2 Computation and Software

Software plays an indispensable and major role in organizing, managing, and controlling the information flow in an imaging system, as well as in responding to external control changes or interrupts. First, it starts and stops a number of processes, such as the transmit pulse sequence. Interrupts or external control changes by the user are sensed, and the appropriate change commands are issued. The master controller may have other slave microprocessors that manage specific functional groups, such as beamforming, image scan conversion and display, calculations and measurements of on-screen data, hardware, and digital signal-processing (DSP) chips. The controller also manages external peripheral devices such as storage devices and printers as well as external communication formats for LAN and DICOM. The controller also supervises the real-time computation of parameters for the output display standard (to be described in Chapter 15), as well as acoustic output management and control.

10.11 Advanced Signal Processing

10.11.1 High-end Imaging Systems

The difference between a basic ultrasound imaging system and a "high-end" system is image quality. High-end systems employ advanced signal processing to achieve superior images. Acuson was the first to recognize that a high-end system could be successful in the clinical marketplace. The first Acuson images were known for their spatial resolution, contrast, and image uniformity (Maslak, 1985). Soon, other manufacturers took up the challenge, and the striving for producing the best image continues today.

Three examples of advanced processing for enhancing image quality are attenuation compensation, frequency compounding, and spatial compounding (Schwartz, 1993). Usually separate signal-processing paths and functions are combined in new ways to achieve improved images. In Figure 10.21 is a block diagram of an ultrasound imaging system; it has several differences from the block diagram of Figure 10.4. To the right of the transducer are scanner functions: beamforming and filtering. The remaining functions are back-end functions of image detection, logarithmic compression, and frame generation. At the bottom of the figure are a number of new blocks (numbered 1–4). Not all the steps of image information are included in this diagram, which is more symbolic and emphasizes differences in signal processing more than traditional imaging architectures. Controlling software to manage the interplay between different functions is assumed.

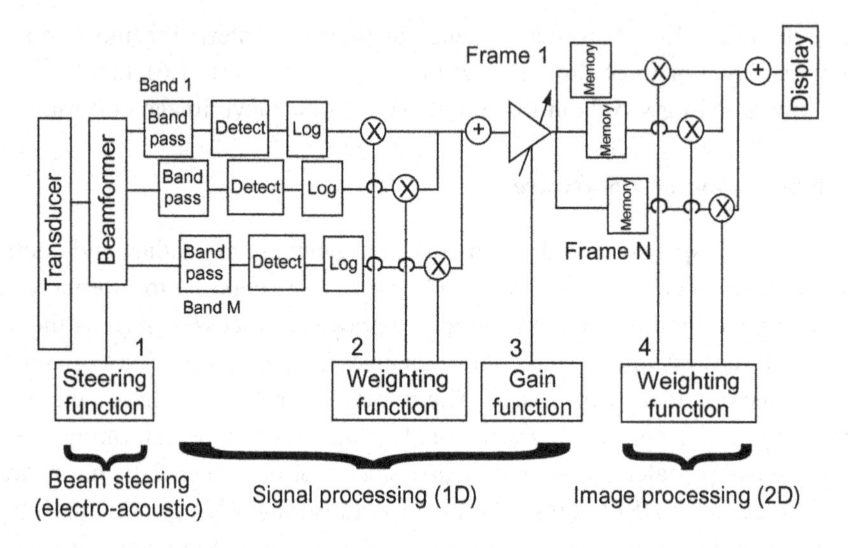

Figure 10.21
Imaging system architecture with signal-processing enhancements. The lower blocks are numbered as (1) steering function, (2) spectral weighting function, (3) gain function, and (4) weighting function. *Courtesy of G. A. Schwartz, Philips Healthcare.*

10.11.2 Attenuation and Diffraction Amplitude Compensation

TGC is an approach available to imaging system users to manually adjust for the changes in echo amplitude caused by variations in beam formation along the beam axis and by absorption. Better image improvements can be obtained by analyzing the video data and adaptively remapping the gain in an image in a 2D sense. At least two different approaches have appeared in the literature (Hughes & Duck, 1997; Melton & Skorton, 1981). The first method senses differences in RF backscatter and adaptively changes TGC gains. The second analyzes each line of video data to read just the intensity levels as a function of time, based on an algorithm, and it leads to an image renormalized at each spatial point. This last approach is more suitable for imaging systems because it can be accomplished in software without major hardware changes.

Using this method as an example, we return to Figure 10.21, block 3 (gain function). The triangle above it symbolizes a variable gain control. A line of video data, corrected for previous video processing and TGC settings, passes through the amplifier and is sent down to the gain control or video analyzer software (not shown in diagram). This line of data is analyzed by an adaptive attenuation estimation algorithm, and the renormalization factor or new gain is determined for each time sample and is sent back through the adjusted amplifier. Only the renormalized values of video information pass through the normal digital scan conversion process (not shown) to create a compensated image frame that is stored in frame memory.

10.11.3 Frequency Compounding

The concept of frequency diversity to reduce speckle was discussed in Section 8.4.6. Until the 1980s, some clinicians valued the grainy texture of speckle, believing it to contain tissue information. In Chapter 8, speckle was shown to be mainly artifactual. Images of the same tissue taken by different transducers at various frequencies present different-looking speckle. Researchers have shown (Abbot, 1979; Melton and Magnin, 1984; Trahey et al., 1986; Schwartz, 2001) the benefits of smoothing out speckle through a scheme of subdividing the pulse–echo spectrum into smaller bandwidths and then recombining them. Through frequency diversity, improved contrast is obtained and more subtle gradations in tissue structure can be distinguished.

A way in which frequency compounding can be implemented is illustrated by Figure 10.21. RF data from a summed beamformed line are sent in parallel to a number (M) of bandpass filters and detection. Detected signal paths are assigned a weight according to a spectral weighting function (block 2) and summed to form a final composite line for scan conversion. Because speckle depends on the constructive and destructive interference at a particular frequency, this 1D summing process reduces the variance of the speckle. Clinical images with and without frequency compounding are compared in Figure 10.22.

10.11.4 Spatial Compounding

While spatial diversity was also named as a way of reducing speckle in Section 8.4.6, there is a more important reason for using it—new backscattering information is introduced into an image. Artifacts are usually thought of as echo features that do not correspond to a real target or the absence of a target. A more subtle artifact is a distortion or a partial depiction of an object. Obvious examples are echoes from a specular reflector, that are strongly angle dependent (as covered by Section 8.4.2). In that section, three angular views of a cylinder were shown in Figure 8.11. That cylinder is revisited in Figure 10.23 (once as seen by conventional imaging and also as seen by compound imaging). When viewed on a decibel scale, there is considerably more echo information available for the cylinder viewed from wider angles.

The implementation of real-time spatial compounding involves 1D and 2D processing. To generate a number (N) of different looks at an object, translation and rotation operations are combined in an array (as explained in Section 10.7). To acquire the necessary views efficiently, in addition to the normal (zero-degree) line orientation frame, $N - 1$ single steered frames are also taken (as in Figure 10.23). A scheme for accomplishing compounding in real time is depicted in Figure 10.24. The moving average of N frames

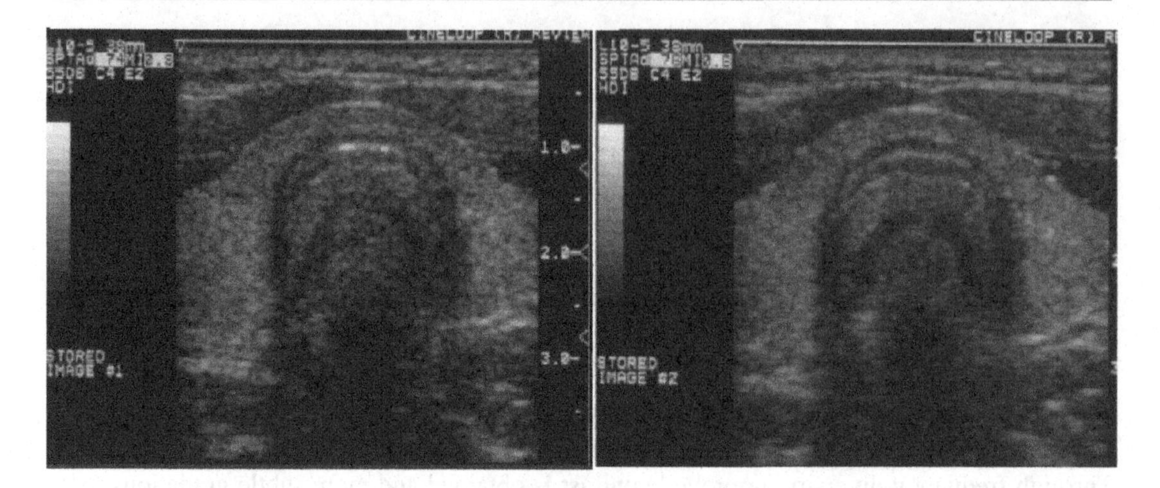

Figure 10.22
(Left) Conventional imaging; (Right) Frequency compounding. *Courtesy of G. A. Schwartz, Philips Healthcare.*

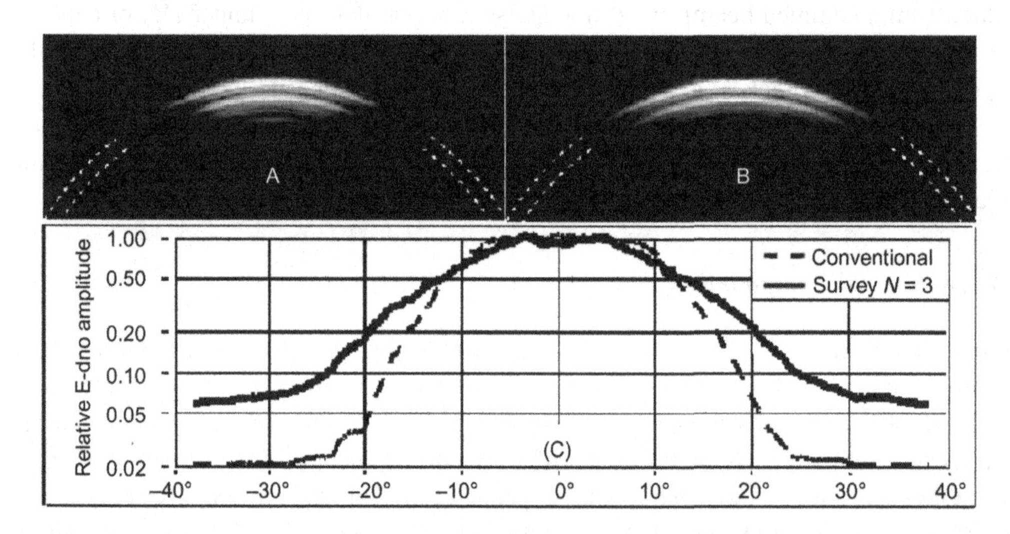

Figure 10.23
Specular reflection from a cylindrical reflector.
(A) For conventional and (B) for compound imaging for steering angles of 17°, 0°, and −17°; (C) corresponding echo amplitudes received by a 5–12-MHz linear array are plotted as a function of angular position. *Courtesy of Entrekin et al. (2000), reprinted with permission of Kluwer Academic/Plenum Publishers.*

creates each spatial compound frame. In the overall block diagram of Figure 10.21, a sequence of N steered angles is entered through block 1. The N-scan-converted single-angle steered frames arrive in the back end where, according to a prescribed spatial compounding function of block 4, each frame N is assigned line and overall 2D frame weighting. Finally,

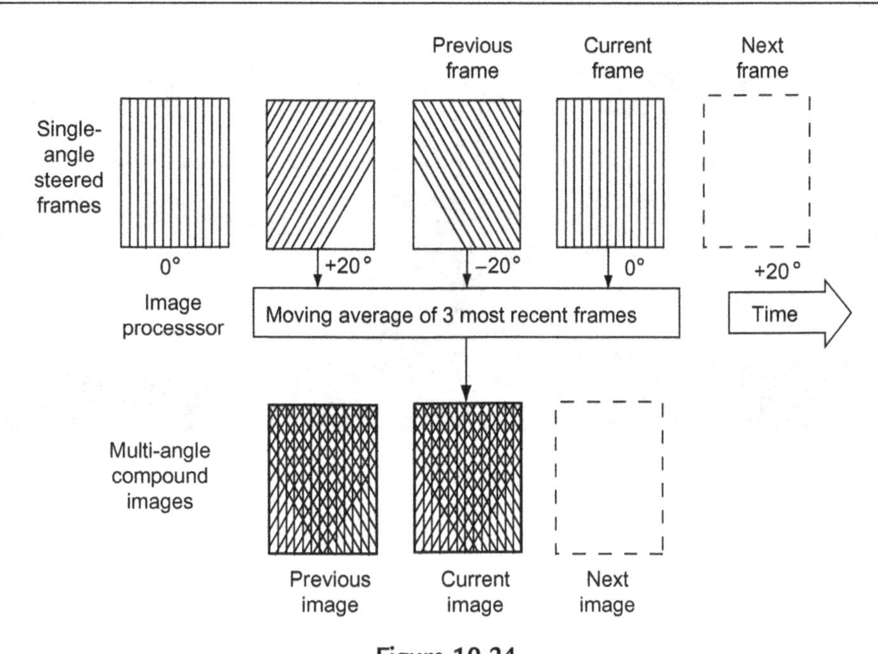

Figure 10.24
Steps of real-time spatial compounding.
In a sequence of steered frames, the scan-converted frames are combined with a temporal moving average filter to form compound images. *Courtesy of Entrekin et al. (2000), reprinted with permission of Kluwer Academic/Plenum Publishers.*

the weighted frames are combined in an averaging operation (symbolized by the summing operation) before display.

Enhanced lesion detection, or the increase in contrast between a cyst and its surrounding material, as well as speckle signal-to-noise have been demonstrated for real-time spatial compounding (Entrekin et al., 2000). Even though the views are not totally independent, these improvements follow a $\sqrt{N}$ trend. Figure 10.25 compares conventional and spatially compounded images of ulcerated plaque in a carotid artery. Enhanced tissue differentiation, contrast resolution, tissue boundary delineation, and the definition of anechoic regions are more evident in the spatially compounded image. One drawback of this method is that temporal averaging may result in the blurring of fast-moving objects in the field of view. This effect can be reduced by decreasing the number of frames (N) averaged; appropriate numbers have been determined for different clinical applications (Entrekin et al., 2000).

10.11.5 Real-time Border Detection

In order to determine the fast-moving changes of the left ventricle of the heart, a 2D signal-processing method has been developed to track the endocardial border. This

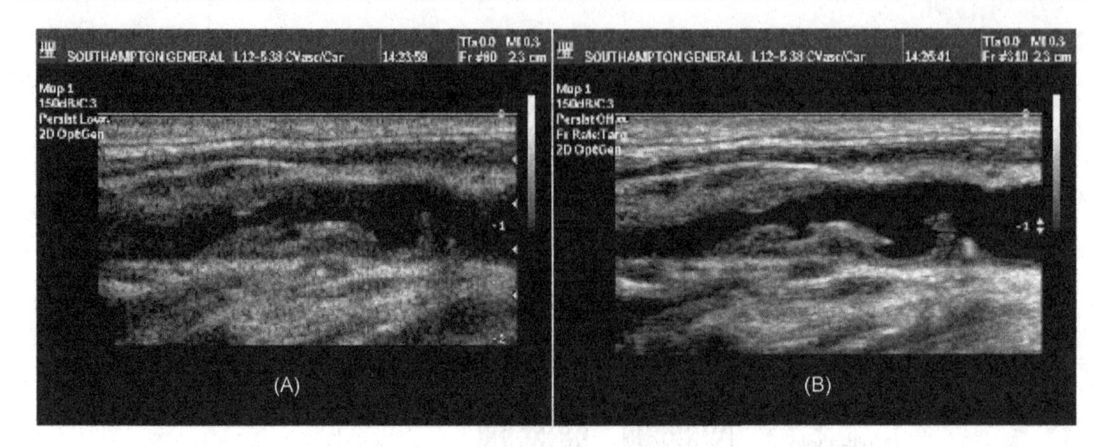

Figure 10.25
(A) Conventional and (B) Compound views of an ulcerated carotid artery plaque as viewed with a 5–12-MHz linear array. *Courtesy of Entrekin et al. (2000), reprinted with permission of Kluwer Academic/ Plenum Publishers.*

approach is based on automatically detecting the difference between the integrated backscatter of blood and the myocardium (heart muscle) (Loomis et al., 1990; Perez et al., 1991) at each spatial location. Implementation of this approach combines a blood–tissue discriminator filter and an algorithm for incoming pulse echoes with 2D signal processing to present a real-time display of the blood–tissue border. This border can be used for real-time calculations of related cardiac parameters.

Another cardiac problem of interest is akinetic motion of the heart due to injury, disease, or insufficient arterial blood supply. The net effect is that the heart wall of the left ventricle no longer contracts and expands uniformly during the cardiac cycle, and some local regions lag behind. In the activation imaging method depicted in Fig. 10.26, three dimensional wall tracking tracks the local changes in displacement. The maximum strains as a function of time are determined for polar mapping of sixteen segments of the left ventricle and different percentages of the maximum value are assigned a color (for example, 0.2 is yellow). Regions painted the same color indicate the level of mechanical activation of the myocardium and its timing. With this mapping of dysynchronous regions, electromechanical abnormalities or those changes responding to therapy can be readily identified as shown in Fig. 10.26.

10.11.6 Three- and Four-dimensional Imaging

One of the drawbacks of 2D ultrasound imaging is the skill and experience required to obtain good images and to make a diagnosis. Imaging in this way is demanding in terms of keeping track of the spatial relationships in the anatomy, and part of using this skill is being able to do 3D visualization in one's head during an exam. An ultrasound exam does not consist of just

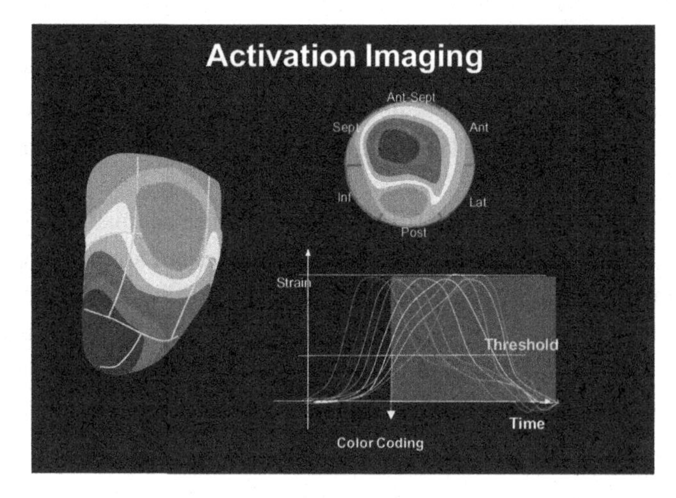

Figure 10.26
Visualization of mechanical activation of the myocardium. Color coded visualization of mechanical activation of left ventricle. Top polar view from base. Graphs indicate strain curves vs time for different segments along with threshold color mapping. The reader is referred to the Web version of the book to see the figure in color. *Courtesy of Toshiba America Medical Systems.*

picture-perfect images such as those in this chapter. Instead, pictures are selected from a highly interactive searching process, during which many image planes are scanned in real time.

The primary goal of 3D ultrasound imaging is the user-friendly presentation of volume anatomical information with real-time interactive capabilities. This goal is challenging in terms of the acquisition time required, the amount of data processed, and the means to visualize and interact with the data in a diagnostically useful and convenient way. Image interpretation becomes simpler because the correct spatial relationships of organs within a volume are more intuitively obvious and complete, thereby facilitating diagnosis, especially of abnormal anatomy such as congenital defects and of distortions caused by disease. The probability of finding an anomaly has the potential of being higher with 3D than with manual 2D scanning because the conventional process may miss an important region or not present sufficient information for interpretation and diagnosis.

The process of 3D imaging involves three steps: acquisition, volume rendering, and visualization. For more details, excellent reviews of 3D imaging by Nelson and Pretorius (1998) and Fenster and Downey (1996) are recommended.

Acquisition is a throwback to the days of mechanical scanning discussed in Chapter 1, except with arrays substituted for single-element transducers. At any instant of time, the array is busy creating a scan plane of imaging data; however, in order to cover a volume, it is also mechanically scanned either through translation, rotation, or fanning. A major difference for 3D imaging is that position data must be provided for each image plane. As in the early mechanical scanning days, this information is provided by either built-in (or built-on) position sensors or by internal/external position controllers, by which the spatial location and or orientation of the array is changed in a prescribed way. The built-on sensors allow freehand scanning (Banker, Pedersen, & Szabo, 2008; Gee, Treece, Prager, Cash, & Berman, 2003; Poulsen, Pedersen & Szabo, 2005; Prager, Rohling, Gee, & Berman, 1998). Because acquisition time is on the order of seconds, data are often synchronized to the ECG, M-mode, or Doppler signals, so that, for example, enough frames are acquired at the same point in a cardiac cycle to create a volume. To create 4D images, time as well as position information is necessary for each acquired image plane. A recent innovation is the real-time 2D array, for which a volume of data can be acquired rapidly and completely electronically without moving the array.

The next step of the 3D process is that the video data in the image planes are interpolated into a volume of data in their correct spatial position. The 3D counterpart to the pixel in 2D imaging is the voxel. Adequate sampling is important because a considerable amount of interpolation is involved. The quality of individual image planes is reflected in the final 3D images so that speckle, unequal resolution throughout the field of view, signal-to-noise, and patient movement are important. In this regard, a 2D array, in which the elevation and azimuth focusing are collocated, contributes to more resolution uniformity.

The visualization software takes the volume data and presents it in an interactive way for imaging. This step presents a challenge for some ultrasound data from soft tissues that do not have enough contrast for definitive segmentation. Outer surfaces where the tissue contacts fluid, such as the fetus in amniotic fluid, provide easy delineation of boundaries for volume rendering as discussed shortly. Slice presentation is the simultaneous display of several image planes that can be selected interactively from arbitrary locations and orientations within the volume. These slices are also referred to as multiplanar reformatting (MPR) views. Figure 10.27 illustrates the volume-scanning process and orthogonal imaging planes. Recently, techniques have been created for directly viewing the 3D matrix of echo signals. Such techniques are referred to as "volume rendering," and they produce surface-like images of the internal anatomy. Although similar in presentation, such techniques should be distinguished from the more common surface-rendering techniques that are used in computer animations and games and motion pictures. The most popular images of this kind are those of the fetus, (shown by Figure 10.28), in which it is easier to distinguish between the fetal body and the surrounding amniotic fluid. Volume rendering is also applicable to functional information; for example, one can use color flow 4D imaging to visualize both normal and pathologic flows in 3D space. Ease of use of the interactive visualization software is an ongoing concern and focus of development. Figure 10.28 also contains selected image planes (note the cover of this book comes from this figure). Electronic 3D scanning in real time provides rapid acquisition of volume data as viewed through selected cut planes as well. A frame from a real-time 4D sequence of the opening and closing of heart valves is shown in Figure 10.29.

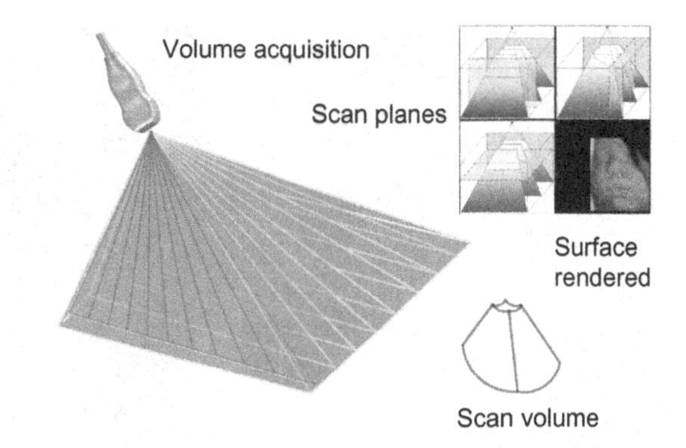

Figure 10.27
Illustration of 3D scanning. Upper right insert depicts orthogonal display planes and surface-rendered image; lower right insert shows a typical sector scan volume. *Courtesy of T. Nelson.*

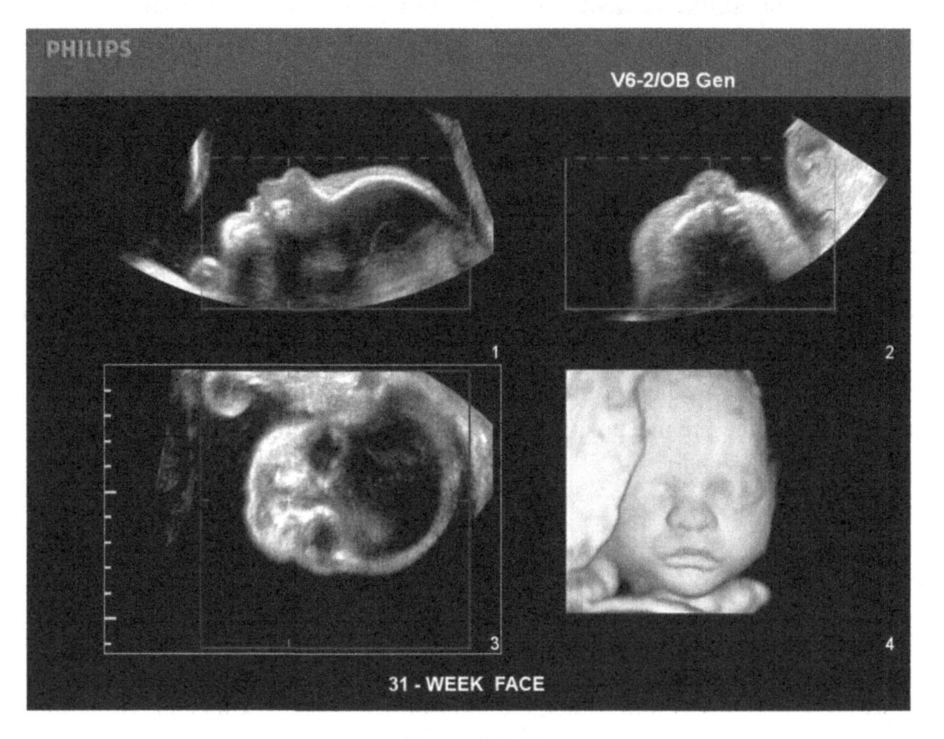

Figure 10.28
Examples of 3D fetal images showing orthogonal views and surface-rendered image (Used for Book Cover). *Courtesy of Philips Healthcare.*

Rapidly growing applications of 3D ultrasound imaging are in interventions and surgery. Complementary imaging which combines ultrasound with another imaging modality is available on several systems. Figure 10.30 illustrates a static 3D CT volume on the upper left, co-registered with real-time ultrasound shown individually on the right. Spatial location

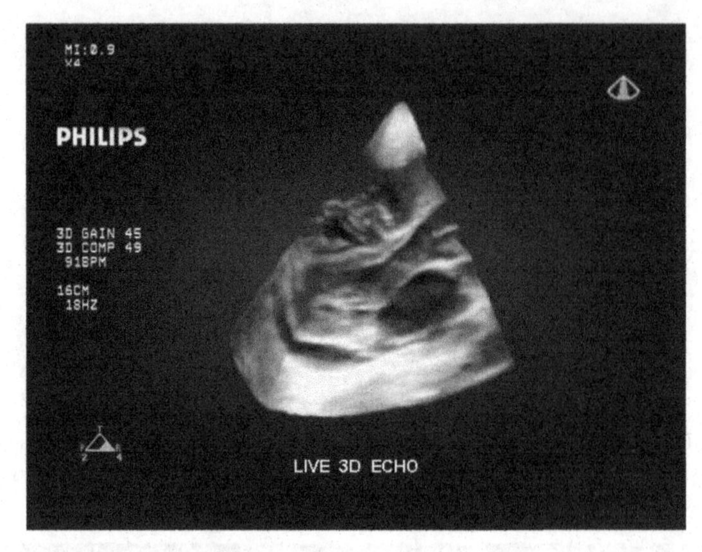

Figure 10.29
Real-time 4D image frame of heart valve motion. *Courtesy of Philips Healthcare.*

sensors mounted on the transducer, described earlier, track the location of the ultrasound image plane in the same coordinate system as the CT volume image. In the second example, Figure 10.31, live fluoroscopy (bottom right panel) is combined with real-time 3D ultrasound using a matrix transesophageal (TEE) probe. In the bottom right of this figure, the TEE probe is being tracked and the scan volume is outlined and a reference point is indicated. The same reference point is presented and the movement of the probe is compensated by tracking in the 3D images in the other three panels. For this intervention, these views facilitate the precise positioning of a clip for the mitral valve despite the movements of a beating heart.

10.12 Alternate Imaging System Architectures

10.12.1 Introduction

This chapter completes the central block diagram of Figure 2.14. Blocks F (for filtering), D (for detection), and Dis (for display) provide the last pieces of the imaging system. The overall structure in this diagram (the linear phased array architecture), borrowed from electromagnetic array antennas, has had a surprisingly long run. The present approach, however, has several major limitations: lack of speed and flexibility, and rigid hardware constraints. Over time, the robust delay-and-sum approach has been refined to a high degree of sophistication and utility. The present beamformer has two chief limitations: lack of speed and lack of flexibility.

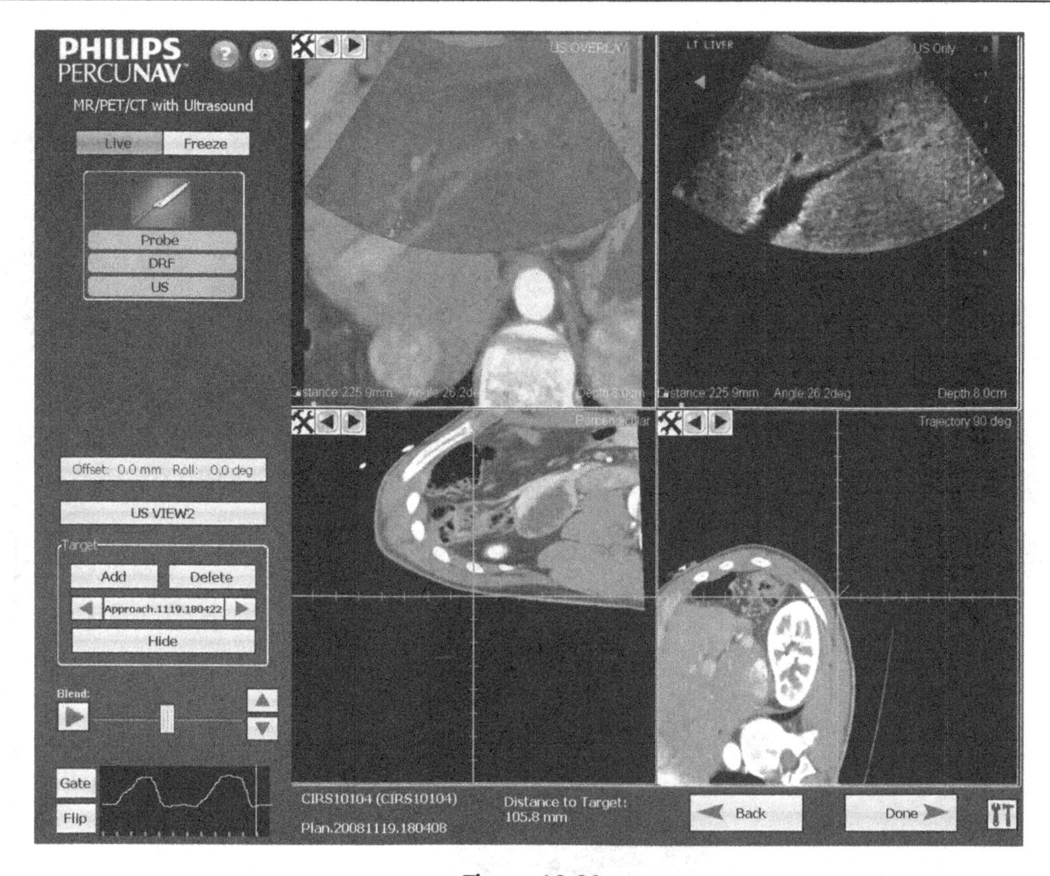

Figure 10.30

Complementary imaging. Illustration of co-registered complementary imaging of ultrasound overlaid on a static 3D CT volume (upper left) and real-time 3D ultrasound on the upper right; orthogonal CT views in panels of the lower half. The reader is referred to the Web version of the book to see the figure in color. *Courtesy of Philips Healthcare.*

An example of the flexibility issue is the inability to handle aberration well. This last problem has been addressed by several schemes (as discussed in Section 9.7). Adaptive imaging systems for this purpose were described by Krishnan et al. (1997), Rigby et al. (2000), and Dahl et al. (2006). Another adaptive scheme for minimizing the effects of off-axis scatterers was described by Mann and Walker (2002). A scheme for extracting more angular backscattering information for imaging was presented by Walker and McAllister (2002).

Another limitation of phased array beamformers is the way clutter affects imaging (see Section 8.4.5). Spatial correlation ($\hat{R}(m)$, Eqn 8.28) is a way of describing the continuum of coherence from full coherence (ideal delay-and-sum beamformer) to no coherence, with most signals falling in the in-between, partial coherence, category. Because the biggest

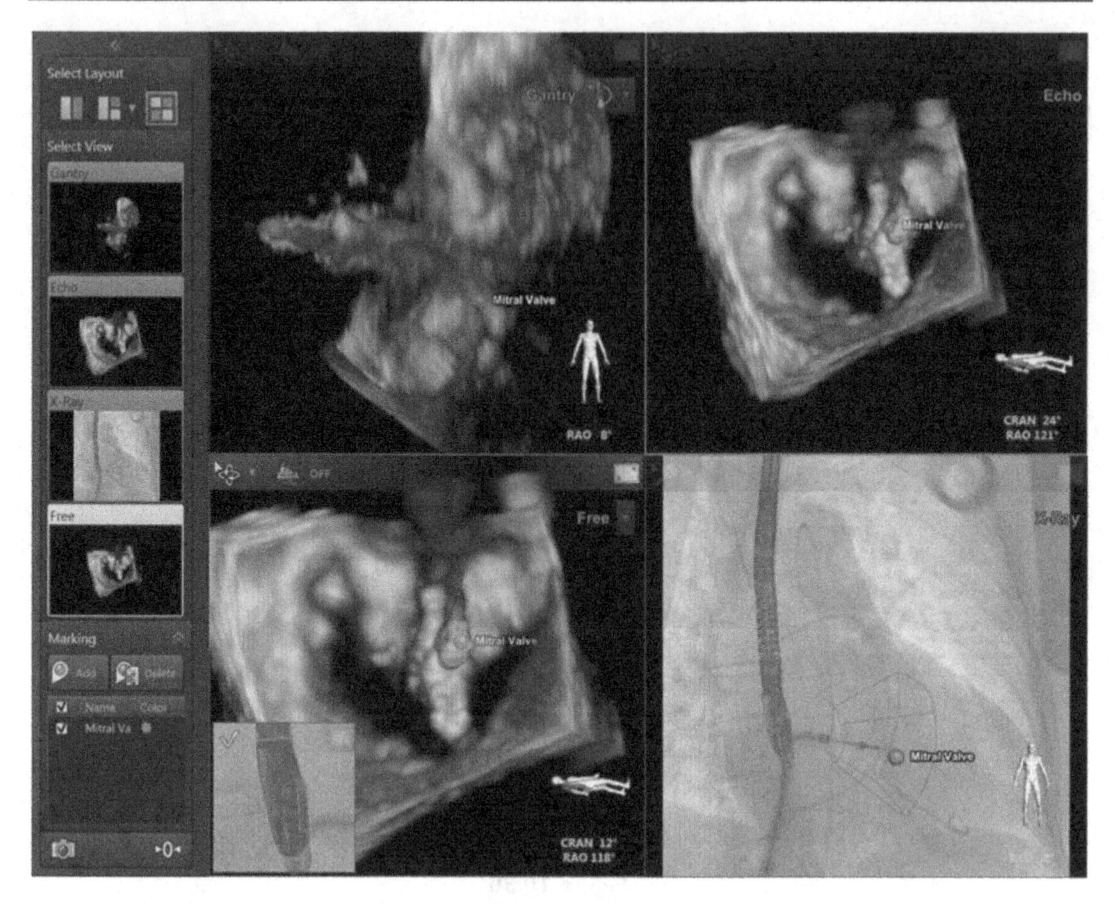

Figure 10.31
Live multimode imaging. The position of the TEE probe is tracked in real time, as shown in the fluoroscopy image and ultrasound scan volume outline (lower right panel). 3D ultrasound images compensated for movement. (Procedure shown (Mitraclip) is not FDA-approved.) The reader is referred to the Web version of the book to see the figure in color. *Courtesy of Philips Healthcare.*

changes in coherence are determined mainly by nearest-neighboring elements or those with short "lags," Lediju, Trahey, Byram, and Dahl (2011) have defined a short-lag spatial coherence for their new imaging algorithm:

$$R_{sl} = \int_1^M \hat{R}(m)dm \approx \sum_{m=1}^{M} \hat{R}(m)\Delta m, \tag{10.8}$$

where Δm is usually 1. For short lags, it is convenient to use the ratio M/N, or a ratio of the transmit aperture width, with typical numbers of 0.01 to 0.30. Using Eqns 8.28 and 10.8, Lediju et al., 2008; Lediju et al. (2011) and Dahl, Hyun, Lediju, and Trahey (2011) have

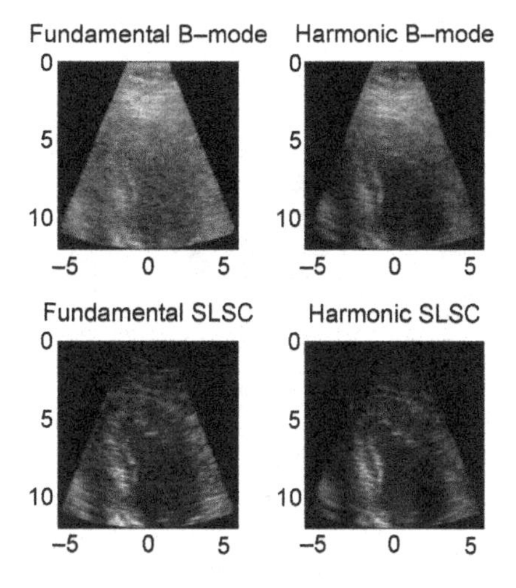

Figure 10.32
Apical four-chamber views of the left ventricle as generated by B-scan and Short-lag Spatial Coherent (SLSC) methods at fundamental and harmonic frequencies. (Harmonic images will be covered in Chapter 12.) *From Bell, Goswami, Dahl, and Trahey (2012), IEEE.*

created images using their short-lag spatial coherence (SLSC) beamformer and compared them to B-mode images generated by conventional delay-and-sum processing. Image comparisons indicate considerable contrast-to-noise ratio (CNR, Eqn 8.18A) improvements in the presence of significant clutter. One present drawback of the method is that point targets in a speckle background get averaged out. This approach is proving successful at cleaning up cardiac images of difficult-to-image patients, as shown in Figure 10.32.

A major advance in visualization capability was the introduction of a real-time 2D matrix array by Philips Healthcare (see Section 7.9.2). Even though delay-and-sum array methods are still employed with these arrays, the architecture being used is a major innovation (Larson, 1993; Savord & Thiele, 1999; Savord & Solomon, 2003). The overall concept can be explained through reference to Figure 10.33. The overall 2D array is subdivided into mosaics, or small groups of elements. Each element has its own tiny microchannel beneath it, consisting of a transmitter and receiver, an addressable time delay, and a pad for connection (Freeman, 2011b; Freeman, Jago, Davidsen, Anderson, & Robinson, 2012). The overall focusing delay is partitioned into fine delays, summed by microbeamformers to deliver a delay, $\Delta t_{\mu BF}$, and a coarse delay controlled within the imaging system, Δt_{BF}. The organization of delays is illustrated by Figure 10.33, in which can be seen the grouping of fine delays summed by each microbeamformer. Finally, the output of each microbeamformer is routed via coaxial cables to the imaging system, in which appropriate coarse delays are added to complete the necessary delay for main beamformer channel n, $\Delta t_n = \Delta t_{\mu BF}(n) + \Delta t_{\mu BF}(n)$. These microarrays and microbeamformers have the equivalent

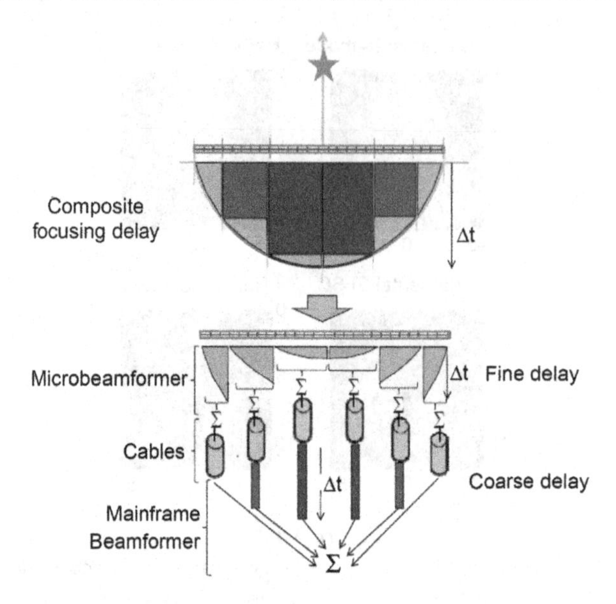

Figure 10.33
Partitioning of focusing delays for a 2D array using the microbeamformer concept. Groups of elements with fine delays are organized into sub-beamformers whose elements are initially summed and routed by cable to the main beamformer, where coarse time delays are added to achieve final focusing delays needed. The reader is referred to the web version of the book to see the figure in color. *From Freeman (2011a), courtesy of Philips Healthcare.*

combined functionality of many front-end boards shrunk into the handle of the transducer, as suggested by Figure 10.34. Design advances and miniaturization have shrunk the size of 2D matrix arrays and even the element sizes. Another major improvement is element count; Figure 10.11 (o) shows a matrix array with 9212 elements and eight million electronic devices. These arrays with domain-engineered single-crystal piezoelectrics can also function at different center frequencies because of their extreme broadband capabilities.

In terms of improving speed, novel methods have been proposed (von Ramm, Smith, & Pavy, 1991). The key limitation in conventional systems is the pulse–echo round-trip time that adds up, line by line. Several alternative methods employ broad transmit beams to overcome the long wait for images. Lu (1997, 1998) has devised a very-fast-frame-rate system based on a plane wave transmission, X-receive beams, and a Fourier transform technique. The Zonare system is based on an architecture that includes the transmission of several (approximately 10) broad plane wave-beams per frame and fast acquisition and signal processing (Jedrzejewicz, McLaughlin, Napolitano, Mo, & Sandstrom, 2003). Jensen and his colleagues at the Technical University of Denmark have developed a fast synthetic aperture system that includes broad-beam transmit insonification. They provide a discussion of other limitations of conventional imaging, such as fixed transmit focusing (Jensen,

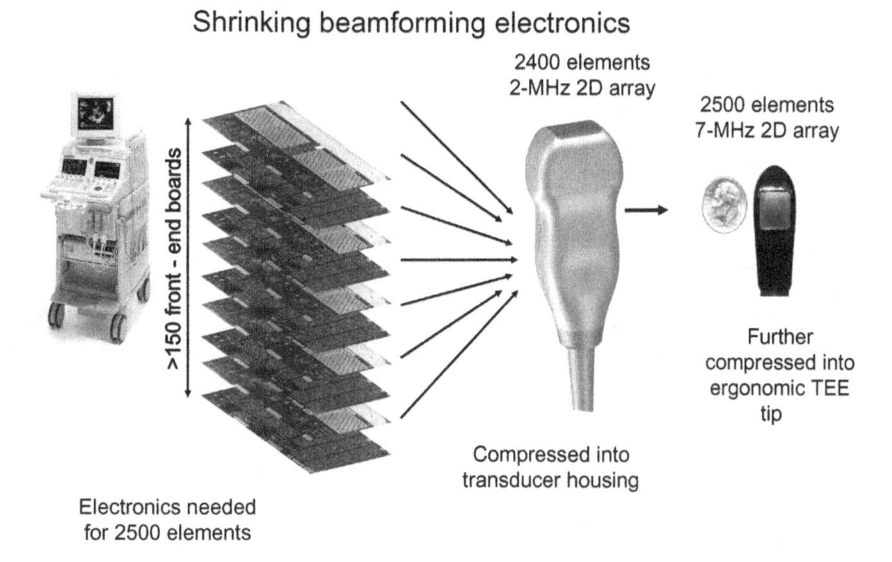

Figure 10.34

Transesophageal 2–7-MHz broadband matrix array with 2500 elements. Illustration of how many equivalent front-end boards from a conventional phased array system are replaced by a microbeamforming approach and miniaturized construction with microcircuitry beneath the 2D array, all within the transducer casing. *Courtesy of Philips Healthcare.*

Nikolov, Misaridis, & Gammelmark, 2002). These systems have the potential for more than just speed; they may be able to acquire more complete, information-laden data sets, as well as have time to provide more sophisticated and tissue-appropriate processing and to extract relevant parameters for diagnostic imaging; some of them are appearing in commercial systems. These are the reasons why three imaging systems with alternate architectures are examined in more detail in the following sections.

10.12.2 Plane-wave Compounding

Since the turn of the millennium, the fundamental assumptions about beamforming have been re-examined and new approaches and imaging architectures have emerged. One of the first assumptions to go was the need to wait for the round-trip time, T_{RT}, of a pulse echo for each image line. Shown in Figure 10.35 is a method for generating an entire image frame using a single plane-wave transmission in the time for a single set of round-trip echoes, T_{RT}. This method, as well as others with the potential for ultrafast imaging, will be explained in this section.

In conventional cylindrical time-delay focusing, for N lines and a round-trip time T_{RT} the frame rate is $FR = 1/(N(T_{RT} + \text{deadtime}))$, or about 30 Hz for a 20-cm scan depth and 128 lines (deadtime is extra time that may be needed for sound to decay to negligible levels to

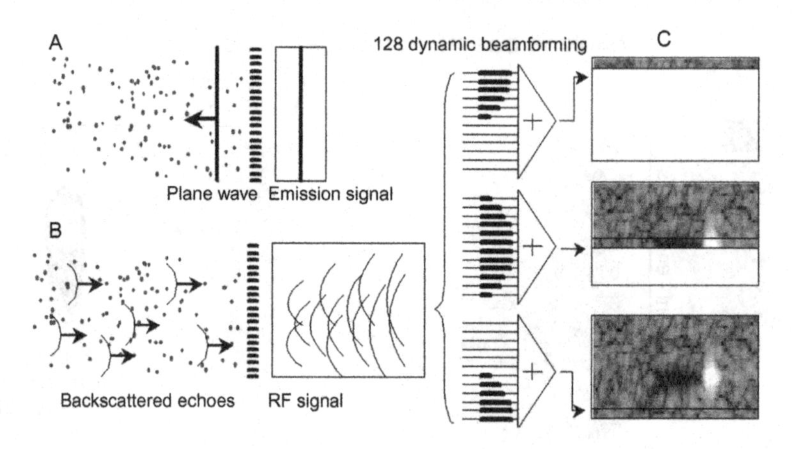

Figure 10.35
Schematic representation of the single-transmit plane-wave method. (A) The ultrasonic array insonifies the medium using a plane-wave transmission; (B) after a round-trip time, the backscattered RF signals are received by the transducer array and stored in memory; (C) the beamforming procedure consists in applying time-delay laws and summations to the stored raw RF signals to focus in the receive mode. Contrary to standard ultrasonography, each line of the image is calculated using the same RF data set but a different set of time delays. *From Montaldo, Tanter, Bercoff, Benech, and Fink (2009), IEEE.*

avoid artifacts; set to zero for this discussion). The corresponding architecture includes the capability to transmit simultaneously in parallel and receive in parallel with only the beamformed lines temporarily stored or streamed through in real time. Usually, dedicated hardware beamformers supply the necessary time delays for focusing. Even though nearly perfect dynamic focusing is achievable on receive, only one transmit focus is allowed per transmission. Better resolution is possible by splicing together focal regions of different transmissions into a composite image at the expense of reducing the frame rate by the number of transmissions, *m*. An example of the multifocus technique is illustrated by Figure 10.7.

What if the received lines could be created from a single transmission or, at least, far fewer transmissions than *N*? Based on this premise, several schemes have been proposed for high-frame-rate imaging. The second previous assumption was that a transmit beam with a single focus had to be used for each line. To assess the other options, it will be helpful to have in mind the distinctions for different transmit beam options shown in Figure 10.36. Figure 10.36A shows a near field of a wide aperture, L_x, which extends to approximately $z_{NF} = L_x^2/\pi\lambda$. Figure 10.36B displays a far-field beam with a -6-dB FWHM beamwidth of $1.2\lambda z/L_x$, which implies a region well past z_{NF}. Next, Figure 10.36C indicates a focused beam with a geometrical focal length F and depth of field. Finally, Figure 10.36D is a representation of an X-beam that extends to a depth of approximately $0.5L_x\cot\varsigma$ (Lu, 1997)

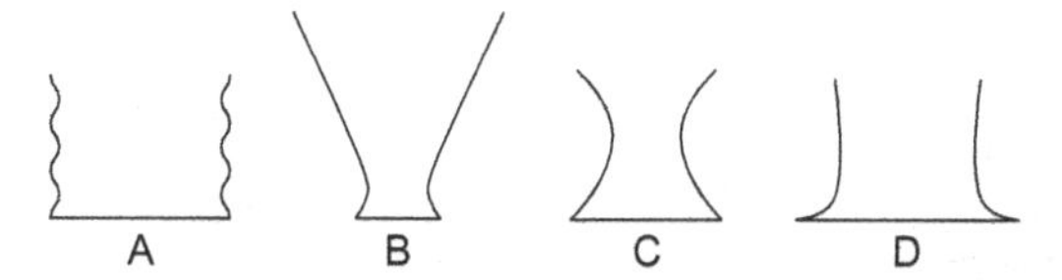

Figure 10.36
Transmit beamforming options. (A) Nonfocused beam in near field; (B) nonfocused beam in far field; (C) conventional time-delayed focused beam; (D) X-beam.

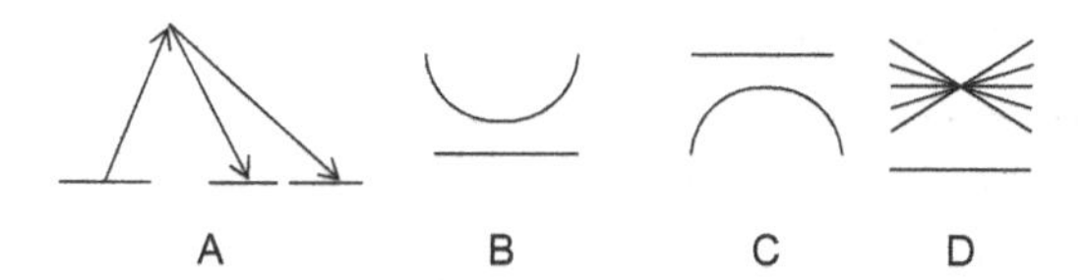

Figure 10.37
Transmit waveform options. (A) Aperture combinations; (B) converging wavefront; (C) diverging wavefront; (D) steered plane waves.

where ςis typically 6.6°. More useful terms are in Figure 10.37. From left to right are different types of wavefronts: converging, diverging, planar, and combination.

An initial and attractive option is to transmit a plane wave and utilize all the returning echoes from the transmission of a single plane wave to form an image (R. Daigle, personal communication (and see http://www.verasonics.com/technology_overview.htm); Delannoy et al., 1979; Lu, 1997; Mo et al., 2007; Montaldo, Tanter, Bercoff, Benech, & Fink, 2009; Sandrin, Catheline, Tanter, Hennequin, & Fink, 1999). Shown in Figure 10.35 is the beamforming sequence described by Montaldo et al. (2009), in which a plane wave is transmitted, the echoes are all stored, and the full image is derived from this same echo data set. The processing can be explained by the use of Figure 10.38A. The round-trip path length from the plane wave to an echo point and back to an element at x_n is given by:

$$\tau_n(x_n, x, z) = \tau_{xmt} + \tau_{rcv} = \left(z + \sqrt{z^2 + (x - x_n)^2}\right)/c_0. \tag{10.9}$$

For every point in the image, the following reconstruction can take place and is proportional to:

$$v_0(x, z) = e_{RT}{}^*{}_t h_t(t)^*{}_t \sum_{-n_a}^{n_a} h_n\left[t - \frac{1}{c_0}\left(z + \sqrt{z^2 + (x - x_n)^2}\right) - \tau_n\right], \tag{10.10}$$

where e_{RT} is the round-trip transducer response, h_t is the transmit spatial impulse response, h_n can be recognized as the element response from Chapter 7, and n_a is a number of elements contributing significantly to the beamforming process. This equation can be compared to a standard time-delay beamformer relation, Eqn 7.30. In the frequency domain,

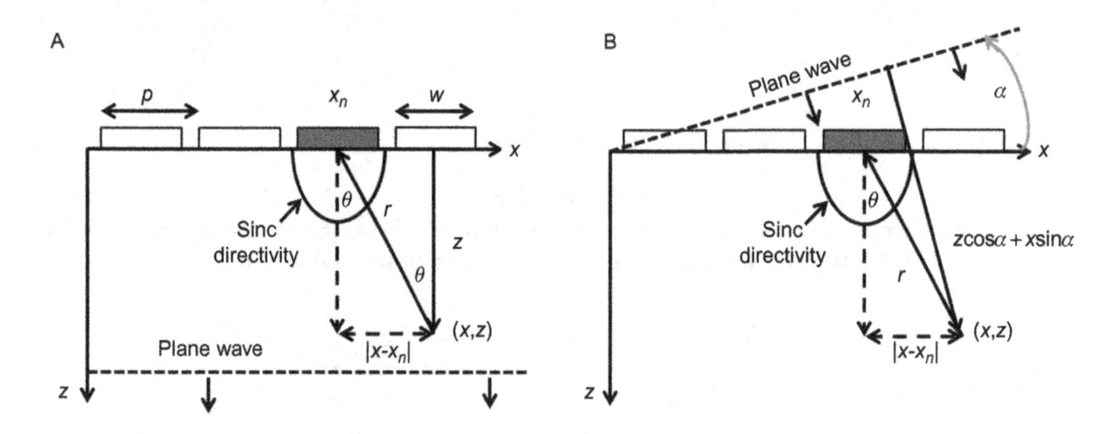

Figure 10.38
(A) Geometry for plane-wave transmission in which the time delays are z/c_0 and r/c_0; (B) at angle α, with element directivity in which the delays are $(z\cos\alpha + x\sin\alpha)/c_0$ and r/c_0, and $r = \sqrt{z^2 + (x - x_n)^2}$. *Adapted from Montaldo et al. (2009).*

an element of width w has a sinc($wx/ëz$) far-field shape (Eqn 7.13), as in Figures 10.36B and 10.38A. Because of an element's directivity, there will be field points (x,z) that no longer contribute significant amplitude to a receiving element. A threshold in directivity can be set to determine the number of elements in the aperture as a function of distance z. For example, if the FWHM is used, then the half-beamwidth is:

$$L_a = x_6 = 0.6\lambda z/w, \tag{10.11}$$

where the half number of elements in the array at any depth is $n_a = x_6/p$ for a pitch p. When $p = \lambda$, for the example in Montaldo et al. (2009), this -6-dB criterion implies an $F\# = z/2x_6 = 0.8$. Typically, F numbers are between 1 and 2.

Another assumption to fall is the requirement for dedicated, fixed hardware systems to achieve time delays for the focusing beamformers of previous imaging systems; here, image construction can be done in software with more flexibility at computer computation rates (R. Daigle, personal communication (and see http://www.verasonics.com/technology_overview.htm); Mo et al., 2007). To summarize, some of the features of this approach so far are the high frame rate, the potential creation of a complete image from a single stored transmission, and use of a sufficient number of elements instead of a fixed number.

The overall imaging process must include the characteristics of the transmit beam. The main drawback of the single-plane-wave approach is that the transmit beam is typically entirely in the near field, Figure 10.36A, and has a transmit resolution approximately equal to the full aperture width of the plane wave; consequently, the round-trip lateral resolution is determined mainly by the receive beamforming, so that contrast is poor.

Spatial compounding is a method that can be used to improve image quality, as demonstrated in Section 10.11.4. Earlier work in this area includes Berson, Roncin, and Pourcelot (1981); Robinson and Knight (1981); Shattuck and von Ramm (1982); and Jespersen, Wilhjelm, and Sillesen (1998). These works are about modifications of more conventional diagnostic imaging methods at usual or slower frame rates. The seeds for using plane waves for dramatically different ultrafast transient elastography can be found in Sandrin et al. (1999). What follows describes plane-wave compounding in preparation for explaining this elastographic method in more detail in Chapter 16.

Montaldo et al. (2009) have analyzed the case in which multiple steered plane wave beams are sent as depicted by Figure 10.37D, as opposed to the converging wavefront usually associated with time-delay focusing, as in Figure 10.36C. In Chapter 6, the two main methods of simulating beams used either Huygen circular and spherical wavefronts or an angular spectrum of plane waves, as alternative but equivalent ways of describing the same field. Montaldo et al. (2009) showed that a transmit focus could also be synthesized by a number of emitted steered plane waves (Section 7.4.5). Figure 10.37D gives an example of five symmetrically placed plane waves (note the steering angle is between the steered plane and the array axis (x)). Figure 10.38B gives the geometry for a steered beam. In this case, the transmit delay term is modified to be:

$$\tau_{xmt} = (z \cos \alpha + x \sin \alpha)/c_0. \tag{10.12}$$

An equivalence can be made for a sequence of steered beams that have a sinc function beam plot similar to that in the focal plane of a cylindrically focusing time-delayed array (Montaldo et al., 2009), so that the number of angles of steered beams is like that of a beam created with an equivalent aperture, L:

$$n = L/\lambda F^{\#}. \tag{10.13}$$

The frame rate (FR) = $1/T_{RT}$, which is 12,500 for a round-trip time of 80 μs for the 6-cm depth example of the Montaldo group's paper, correspondingly reduces by n, the number of steered beams. A surprising result for the steered plane waves is that high resolution is maintained through the scan depth and axial pressure does not fall off with depth in a lossless medium as it does for cylindrically and spherically focused beams. The -10-dB full beamwidth, for the 4.5-MHz example in the paper, is $2x_{10} = 3.6\lambda F\# = 1.1$ mm for an $F\# = 1$. The signal-to-noise ratio is shown to be proportional to $\sqrt{n}$; to be comparable in signal-to-noise to a multifocus standard focusing arrangement, 45 steered beams would be needed. Even with this number of beams, the FR = 278 is much faster than the FR for a four-multifocus case with

128 image lines of 24 frames/s. In Chapter 16, the plane-wave approach will be applied to achieving frame rates of 1000/s for transient elastography (Bercoff, Tanter, & Fink, 2004; Montaldo et al., 2009; Sandrin, Tanter, Catheline, & Fink, 2002).

10.12.3 Fourier Transform Imaging

Lu (1997, 1998) conceived of a high-frame-rate imaging approach he called "the Fourier method," by which a 3D (or 2D) image could be derived from a single plane-wave transmission. The overall method is illustrated by Figure 10.39: a plane wave insonifies a scattering object $s(\underline{r})$, and with limited diffraction beams (Section 6.9) in reception, the object is reconstructed from returning echoes by a Fourier construction. The obvious advantage of this method, aside from its potential for high frame rate, is that no sum-and-delay processes are necessary, the image is formed by Fourier transforms (actually, fast Fourier transforms, FFTs). Refer to Figure 10.36A for a plane-wave transmission and Figure 10.36D for a symbol for an X-wave. Each X-wave to be included in the receive processing has a uniform beamwidth for an approximate axial depth of field:

$$z_{\max} = \frac{D}{2}\cot\zeta, \tag{10.14}$$

for a receiving disc of diameter D and a selected axicon angle $\zeta = 6.6°$ (Lu & Greenleaf, 1992). Note, this relation is frequency independent and, for example, for $D = 50$ mm, $z_{\max} = 216$ mm. Another advantage is high resolution throughout the depth of field without the need to do separate point-by-point calculations such as in dynamic receive focusing. In terms of resolution, for this same example (Lu, 1997), the ratio of -6-dB beamwidths for an X-wave to the diffraction-limited case (Eqn 6.18B) for a circular aperture was 1.48 at $z = 30$ mm, and 1.17 at $z = 100$ mm and $z = 2000$ mm—quite comparable.

To proceed with this method, summing broadband X-waves of different orders, one obtains a limited diffraction array beam that can be expressed in terms of an angular spectrum, described in Section 6.7, and an inverse Fourier transform:

$$\Phi_{array}(\underline{r}, t) = \int_{-\infty}^{\infty} \left[\frac{T(f)H(f)}{c} e^{-i\underline{k}_{\Phi}(f)\bullet\underline{r}}\right] e^{i2\pi ft} df, \tag{10.15}$$

where, along the three axes, the $\boldsymbol{k}$ vector has the usual components: $k_x = k\cos\theta$, $k_y = k\sin\theta$, $k_z = k$. The phasing for the X-wave function, derived in Lu and Greenleaf (1992) and Lu (1997), is encoded in the particular form of the $\boldsymbol{k}$ vector,
$k_{\Phi x} = k_x \sin\zeta, k_{\Phi y} = k_y \sin\zeta, k_{\Phi z} = k\cos\zeta$ or $\boldsymbol{k}_{\Phi}(f)\bullet\mathbf{r} = (k_{\Phi x}x + k_{\Phi y}y + k_{\Phi z}z)$ and, of course, $k = 2\pi f/c$. $T(f)$ is the acoustoelectric transfer function of the receiver, $H(f)$ is the step function, so that the term in the square brackets represents the receive spectrum of the X-wave.

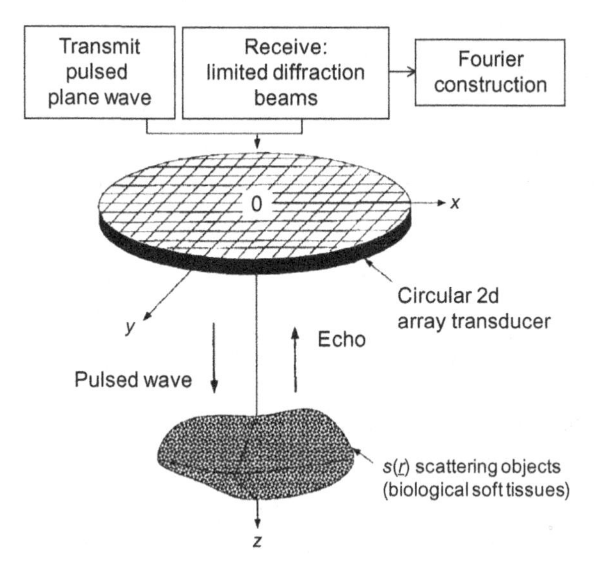

Figure 10.39
Geometry of a 3D fourier method pulse–echo imaging system using limited diffraction beams on reception. *From Lu (1997), IEEE.*

In order to pull together the pieces of a Fourier method, we can use the basic structure of Figure 2.14. So far, there is the X-wave part of the receive beamformer, $g_r(t)^* h_r(r,t) = \Phi_{array}(\underline{r}, t)$, with a corresponding spectrum in the square brackets of Eqn 10.15. The transmitted pulsed plane wave beam follows in the usual $-i$ inverse Fourier transform:

$$a(t) = h_t(z - ct) * g_t(t), \quad \text{and} \tag{10.16A}$$

$$a(t) = \frac{1}{c}\int_{-\infty}^{\infty} A(f)e^{-ik(z-ct)}df = \frac{1}{c}\int_{-\infty}^{\infty} [A(f)e^{-i2\pi fz/c}]e^{i2\pi ft}df, \tag{10.16B}$$

where $g_t(t)$ is the acoustoelectric transducer impulse response function, and may include the driving waveform.

Finally, there is the scattering object $s(\mathbf{r})$, which is to be determined and displayed. A two-step process is involved. In the first step, transforming $s(\mathbf{r})$ into the angular spectrum domain, this time using a triple $+i$ Fourier transform over the spatial volume V containing the scatterers:

$$S(k_x, k_y, k_z) = \iiint_V s(\mathbf{r})e^{i\mathbf{k}\cdot\mathbf{r}}d\mathbf{r}, \tag{10.17}$$

and

$$s(\mathbf{r}) = \frac{1}{(2\pi)^3}\iiint_{-\infty,\infty} S(k_x, k_y, k_z)e^{-i\mathbf{k}\cdot\mathbf{r}}dk_x dk_y dk_z, \tag{10.18}$$

in which $\boldsymbol{k}(f)\boldsymbol{\cdot r} = (k_x x + k_y y + k_z z)$. Note the use of $k = 2\pi f$ rather than spatial frequency (Section 6.7) requires a factor of $1/2\pi$ for each inverse transform. Two modifications of these equations are necessary. First, the integrations involve $\boldsymbol{k}_\Phi$ not $\boldsymbol{k}$. Second, a scalar k term to account for broadband calculations is added so that the final variable of integration is $\boldsymbol{k}'_\Phi = \boldsymbol{k}_\Phi + k$ for Eqns 10.17 and 10.18. The result of Eqn 10.17 is $S[\boldsymbol{k}'_\Phi(f)] = S(k'_{\Phi x}, k'_{\Phi y}, k'_{\Phi z})$.

The scattering function is derived from pulse−echo data arriving at each array element:

$$d(\boldsymbol{r}, t) = g_t(t) *_t h_t(t, r) *_r \sum_{M,N} h_{r,m,n}(t, r) *_r s(r) * g_r(t), \tag{10.19}$$

where the m and n indices refer to positions of elements (m,n) in the receiving array. In the frequency domain, the corresponding spectra are:

$$D(\boldsymbol{k}'_\Phi, f) = \frac{A(f)T(f)H(f)}{c^2} S\left[\boldsymbol{k}'_\Phi(f)\right]. \tag{10.20}$$

Because of sensor constraints, D is a band-limited version of S. Define a band-limited version of scattering for computation as:

$$S_{BL}(\boldsymbol{k}'_\Phi, f) = c^2 H(f) S[\boldsymbol{k}'_\Phi(f)], \tag{10.21A}$$

then cancel the step functions to obtain the band-limited function needed for calculations:

$$S_{BL}(\boldsymbol{k}'_\Phi) = A'(k)T'(k)S(\boldsymbol{k}'_\Phi), \tag{10.21B}$$

where A' and T' are simply A and T with the substitution $k = 2\pi f/c$. Finally, in the second major step, to obtain the scattering object distribution, use the inverse $+i$ Fourier transform relation, Eqn 10.18, and the band-limited approximation for the scattering objects:

$$s(\boldsymbol{r}) \approx s_{BL}(\boldsymbol{r}) \approx s_{BL}^{Part}(\boldsymbol{r}) = \frac{1}{(2\pi)^3} \int_{-\infty}^{\infty} dk'_{\Phi x} \int_{-\infty}^{\infty} dk'_{\Phi y} \int_{k'_{\Phi z} > \sqrt{k'^{2}_{\Phi x} + k'^{2}_{\Phi y}}} dk'_{\Phi z} S_{BL}(\boldsymbol{k}'_\Phi) e^{\boldsymbol{k}'_\Phi \boldsymbol{\cdot r}}, \tag{10.22}$$

in which s_{BL}^{Part} does not contain any evanescent waves.

As indicated by the plane-wave compounding investigation, image quality improves by compounding a number of plane waves. For the Fourier method, the quality of reconstructed images can be improved with more plane waves transmitted at different angles (Figures 10.36A and 10.37D) or with limited diffraction array beams of different parameters (Figure 10.36D), and then coherently summing the resulting images. Lu (Cheng & Lu, 2006; Lu, Cheng, & Wang, 2006) independently assessed the results of these approaches, as illustrated by Figure 10.40. Usually, 2D arrays require individual weighting and phasing, and the X-wave transmission is no exception. Lu demonstrated

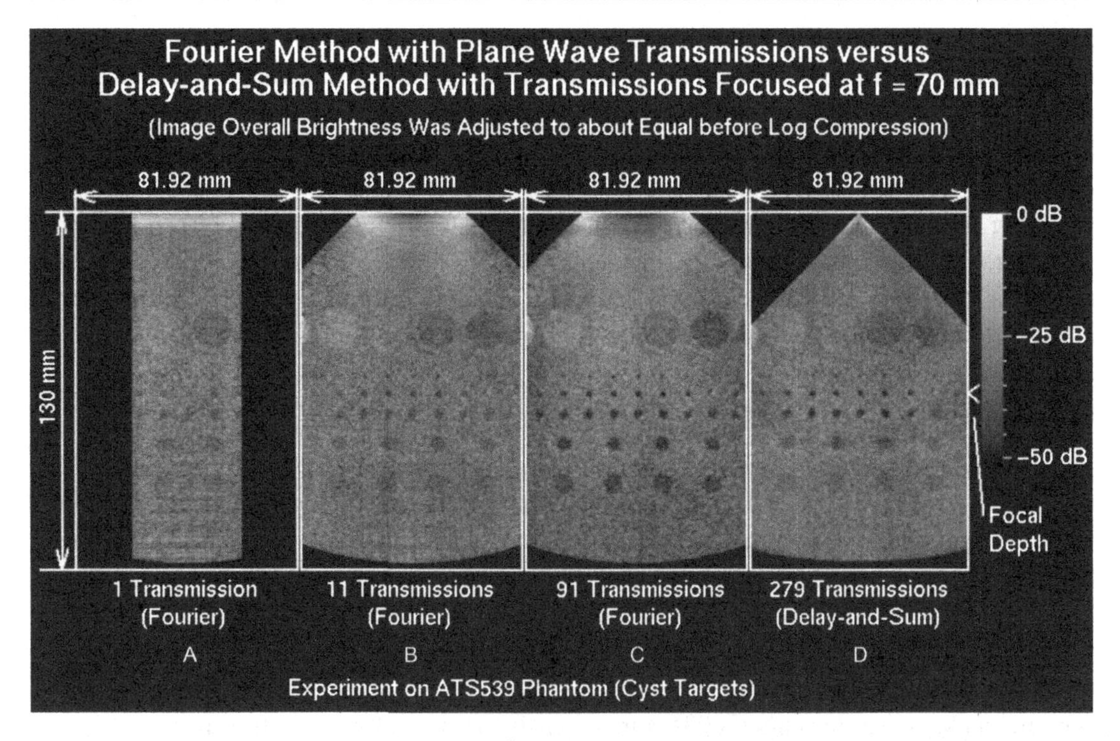

Figure 10.40
Reconstructed images of cysts of an ATS 539 tissue-mimicking phantom with the limited-beam method. Images are log compressed with a dynamic range of 50 dB. The speed of sound is about 1450 m/s. Images of the exact aperture weightings (A and C) are compared with those of the square-wave weightings (B and D), respectively. Images in the left and right two panels were obtained with 11 (up to 507 frames/s with 1450 m/s speed of sound) and 91 transmissions (up to 61 frames/s), respectively. The field of view of the images is larger than $\pm 45°$ over a 130-mm depth. *From Lu et al. (2006), IEEE.*

that equivalent imaging results could be attained by a binary method ($+$V or $-$V voltage amplitudes), thus the number of transmitters needed can be reduced from the number of elements of an array transducer ($M \times N$) to two or even one (Lu et al., 2006). Further studies show that instead of transmitting plane waves, diverging beams of small diverging angles (such as $<15°$) (Figure 10.37C) can be used in transmissions for the Fourier image reconstruction method above, with the added benefit that fewer transmissions are needed to cover a larger image field of view to achieve a higher image frame rate (Lu and Chen, 2011).

10.12.4 Synthetic Aperture Imaging

Synthetic aperture concepts from radar applications held an immediate appeal for ultrasound imaging: perfect focus on both transmit and receive. This approach was used extensively by the ultrasound nondestructive testing community, and continues to evolve for materials testing where it is known as "SAFT," the synthetic aperture focusing technique (Corl, Grant, & Kino, 1978; Doctor, Hall, & Reid, 1986; Flaherty, Erikson, & Lund, 1967; Langenberg, Berger, Kreutter, Mayer, & Schmitz, 1986; Skjelvareid, Olofsson, Birkelund, &Larsen, 2011). The basic principles involve transmitting on one element of an array and receiving on all elements, storing all the data, then transmitting through the rest of the array elements, one at a time, sequentially, in a similar way; see the combination Figure 10.37A. High-resolution image reconstruction was usually done off-line, and its simplicity was an advantage. Two severe disadvantages of SAFT, however, were the poor signal-to-noise ratio because of the weak transmission from a single element, and the long time required to acquire data for a single image.

These drawbacks seemed to put synthetic aperture out of reach for applications to medical ultrasound. Two investigations—Nock and Trahey's (1992) fundamental investigations into the relation of aperture size, speckle, and correlation, and application of synthetic aperture to the highly constrained IVUS (intravascular ultrasound) environment (O'Donnell and Thomas, 1992)—revived interest in synthetic aperture for diagnostic imaging and aberration correction. Karaman, Bilge, and O'Donnell (1998) developed a synthetic aperture scheme utilizing sequences of subapertures. More background can be found in Jensen, Nikolov, Gammelmark, and Pedersen's (2003) comprehensive review of work in this area.

In order to make the most of the advantages of synthetic aperture imaging, several major obstacles had to be overcome. Jensen's group at the Technical University of Denmark set out to build a practical synthetic aperture system for medical imaging. Their approach is faithful to the original synthetic aperture concept, as evident in Figure10.41. The poor signal-to-noise ratio was improved by the use of a tapered chirp-encoded waveform on transmit and a matched filter on receive, which resulted in an increase in axial resolution (by 0.4λ) and a gain of 10 dB. In addition, several (N_t) elements were used for transmitting a diverging wavefront (Karaman, Li, & O'Donnell, 1995) (see Figure 10.37C) to simulate the wavefront from a single element (synthetic transmit aperture) (Chiao et al., 1997). Here, for each simulated single element, the RF lines of a low-resolution image were created and these images were summed to create a high-resolution composite image, as illustrated by Figure 10.41.

To achieve high frame rates, Jensen's group developed a recursive method, in which the contribution of a new element replaced the subtraction of the oldest one acquired, so a new image could be created with each new emission (Nikolov, Gammelmark, & Jensen, 1999). Thus, the frame rate was limited only by $1/T_{RT}$ so that rates up to 5000 frames/s became possible. Finally motion compensation was added.

A complete software-configurable synthetic aperture system was implemented with a capability for 128 elements (Jensen et al., 1999, 2005), and the acquisition and storage of all channel data and a number of emissions. With these combined features, full dynamic

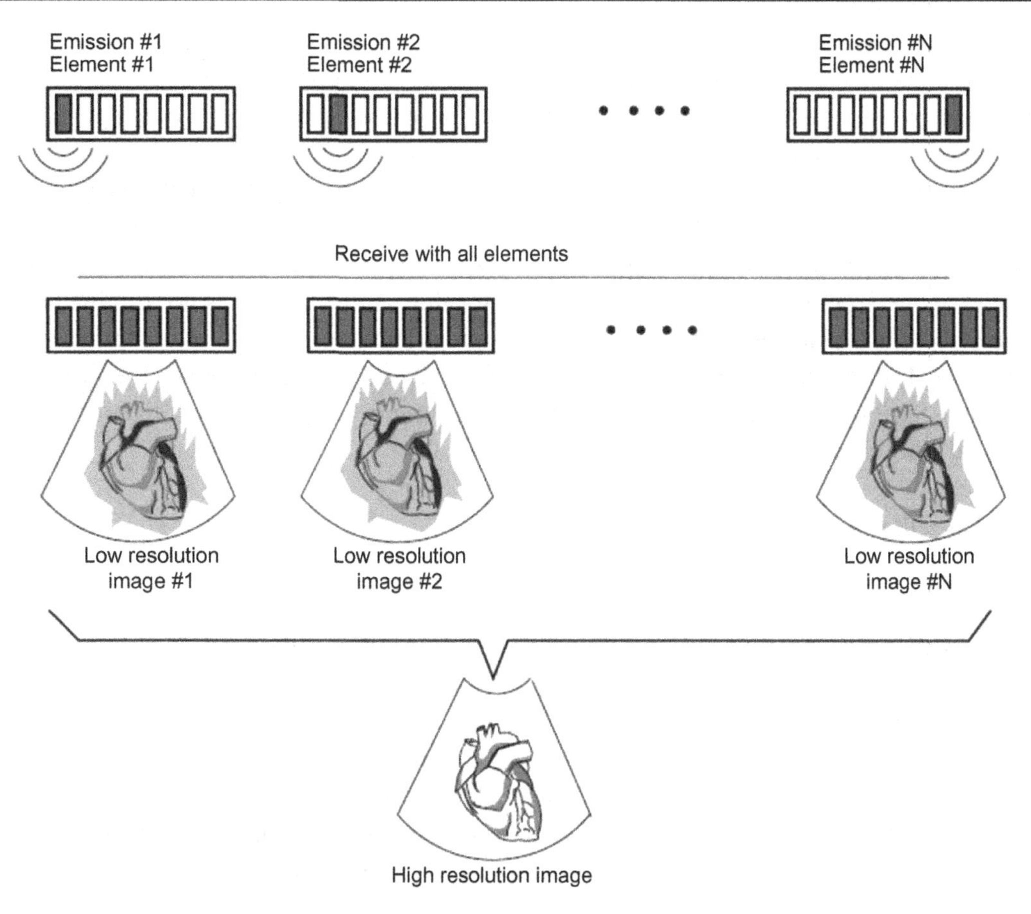

Figure 10.41
Basic principle of synthetic aperture ultrasound imaging. *From Nikolov (2001) and Jensen, Nikolov, Gammelmark, and Pedersen (2006).*

transmit and receive focusing was achieved. Preclinical images were judged to be superior in quality to conventional sum-and-delay images and a considerably increased signal-to-noise ratio (about 10–17 dB) and penetration of approximately 40% were demonstrated. Remarkable color flow and vector flow images along any direction were made and these will be described at the end of Chapter 11. Even though real-time imaging was not yet possible, the capabilities of high-performance synthetic aperture imaging were proven (Jensen et al., 2006).

10.12.5 Parallel Beamforming Archictectures

By combining a massively parallel acquisition and computational architecture with a different beamforming scheme, Philips has broken the paradigm of conventional delay and sum line by line image formation. In previous array imaging systems, a transmit beam had a single fixed focus. As described earlier in Figure 10.9, this shortcoming could be overcome by multiple transmit events which could be spliced together at a corresponding slowdown in frame rate. Conventional wisdom was that once launched, the transmit beam could not be altered and the best salvaging of the beam was through the application of dynamic receive focusing. In the EPIQ7 system, multiple transmit beams are launched simultaneously. A different beamformer then operates on the signals from all beams in parallel at every point along an image line even for those transmit beams not originally aligned along the computed direction. Echoes from all launched beams are brought into alignment. The resultant effect is an effective virtual dynamic transmit focus. Benefits include high resolution throughout the depth of field, improved penetration, higher frame rate, and autofocusing.

10.12.6 Ultrasound Research Systems

Verasonics System

The Verasonics imaging engine is a high-frame-rate, flexible architecture. Software configurable, it can accommodate a wide latitude of beamforming options, including proprietary plane-wave compounding computed at pixel or voxel locations, shear wave imaging, standard delay-and-sum, synthetic aperture, and others. It has advanced options for decreasing element computations depending on element directivity, correcting for near-field beam nonuniformity, high-frame-rate Doppler and vector Doppler, high-transmit modes for ARFI and some therapeutic applications, and many other features.

The layout for the Verasonics research system is remarkably simple (see Figure 10.42), compared to a standard delay-and-sum architecture. The complexity, flexibility, and power of the system are hidden in unique MATLAB-driven software algorithms. In the upper left corner are configurable transmitters that can connect to a variety of ultrasound arrays. Returning echoes for each element are digitized and stored in buffers for further computation and asynchronous display. A key step is the input filter, which performs the calculation:

$$\mathbf{RFP} = Trace\big[[a][c]\big], \tag{10.23}$$

in which the RF contributions at a predefined pixel location are combined from a trace matrix operation dynamically (per pixel), from a matrix a of complex coefficients for the location and mode(s) being displayed, and from a matrix c that includes the RF samples from each channel, and starting at sample index s_n and running to sample index s_{n+m}. Finally, directivity constraints, described earlier, eliminate the wasteful summation of elements that do not contribute significantly to the pixel (spatial) location of present interest. Besides the economical use of elements, Eqn 10.23 implies dynamic focusing at only the points of the image from the RF stored in memory (R. Daigle, personal

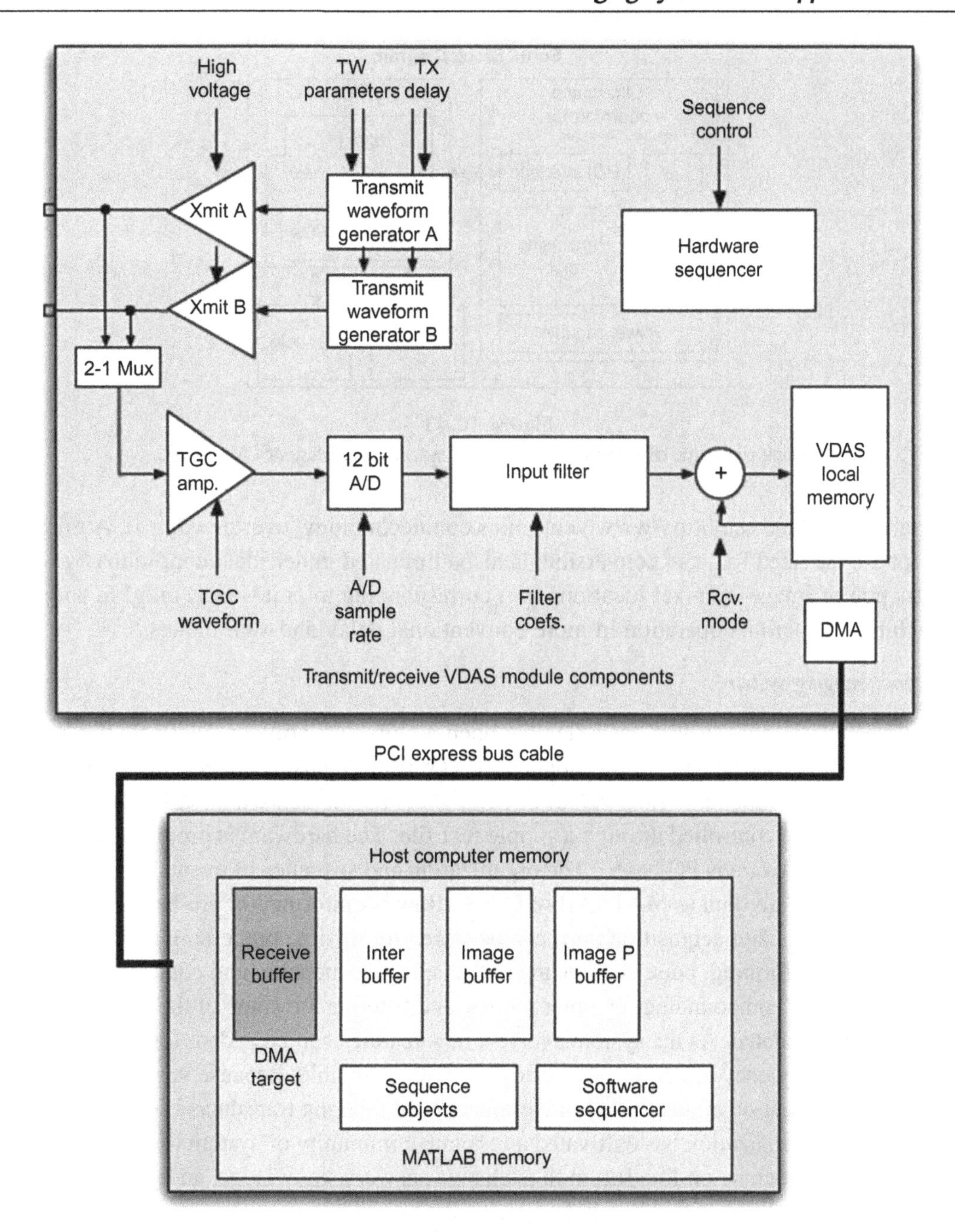

Figure 10.42
Verasonics system block diagram. Host computer that manages the instruction set to the hardware only partially shown. DMA, direct memory access; VDAS, Verasonics data-acquisition system; Xmit, transmit. *Courtesy of R. Daigle, Verasonics.*

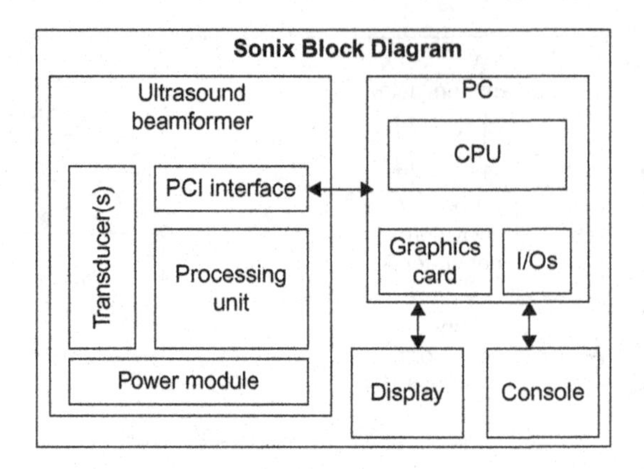

Figure 10.43
Block diagram of ultrasonix research system. *Courtesy of Ultrasonics.*

communication (and see http://www.verasonics.com/technology_overview.htm)). A third assumption, the need for scan conversion, is also eliminated under ideal conditions by the construction of images at pixel locations (j,k) corresponding to points (x,z) only. In addition, the architecture permits operation in more conventional delay-and-sum modes.

Ultrasonix imaging system

The Ultrasonix research system has a flexible open-architecture software platform that can be reconfigured for research and clinical applications (Wilson, Zagzebski, Varghese, Chen, & Rao, 2006). Most of the numerical values of parameters for the delay-and-sum hardware architecture can be controlled through a simple text file. The hardware is programmable and the controlling software is PC-based. The organization and sequence of events can be modified by software such as MATLAB or C++. Raw beamformed or pre-beamformed data (with an additional data-acquisition unit) can be stored for off-line processing and novel algorithms. Beamforming, pulse sequencing and shaping, sequencing and controlling modes, transducer access, compounding, Doppler modes, and filtering are some of the parameters that can be manipulated. As the system evolves, new features can be added; for example, 3D imaging, high-frequency imaging, and elastography are available. Because versions of the Sonix platform are used clinically, a full complement of imaging transducers is available. The Ultrasonix organization has cultivated a research community of system users who communicate through an on-line forum to exchange answers, knowledge, and programs, and use a system wiki. A simplified block diagram of the Sonix system is shown in Figure 10.43.

Visualsonics imaging systems

As explained in Section 9.6.3, there are many advantages to imaging at high frequencies. Visualsonics has exploited these advantages by creating imaging systems specialized for high-frequency imaging, especially of mice and small animals noninvasively. Transgenic mouse models are used to evaluate the effectiveness of therapies and pharmaceuticals in cancer and cardiac studies. The previous mechanically scanned high-frequency system has been replaced by a new digital system (for images of the older system, see Foster, Hossack, & Adamson, 2011; Foster et al., 2002; Walsh, Kobler, Heaton, Szabo, & Zeitels, 2008).The Vevo 2100 is a high-end digital imaging system (Foster et al., 2009; see also http://www.visualsonics.com) complete with tiny linear arrays operating at frequencies from 9 to 70 MHz (Section 5.9.5). The modes available are similar to those of a high-end, large, conventional system but for miniature imaging such as color flow, power Doppler, Doppler, M-mode, and 3D; in addition, it has modes designed for cardiac mouse imaging such as left ventricle functional and volume analysis, strain analysis, innovative quantification, and nonlinear contrast imaging. In fact, special contrast agents were designed to operate at high frequencies to allow the visualization of flow and perfusion in mice. Special beds to make mice comfortable during imaging sessions are also available. A photoacoustic version provides co-registered photoacoustic and ultrasound images.

Other research systems

Research systems come in several other forms. Examples of other commercial research systems with special features and useful capabilities, such as tracking a signal through the processing chain, include the Femmina (Masotti et al., 2006), the Di-Phas ultrasound research platform (Fraunhofer Institute, 2013), and the Lecoeur open phased array system (http://www.lecoeur-electronique.com/). The first commercial clinical system deliberately converted into a research version was the Siemens Antares (Ashfaq et al., 2006). Most commercial clinical scanners offer some limited research options, such as access to RF beamformed data, with some research features or calculation capabilities. An example of a Zonare system converted to a hybrid multimodal photoacoustic system can be found in Mo et al. (2007). Though not recommended, some do their own conversions of commercial systems (Mari and Cachard, 2007). Other research systems include those built for special purposes at universities; examples of these can be found in the special issue on research systems of the journal *IEEE Transactions on Ultrasonics, Ferroelectrics and Frequency Control* (October 2006), edited by Tortoli and Jensen (2006). In addition, embedded special-purpose research ultrasound systems are being integrated into other systems, as described in another special issue of *IEEE Transactions on Ultrasonics, Ferroelectrics and Frequency Control* (July 2012), edited by Sanile and Oruklu (2012). An example of a system from this issue is shown in Figure 10.2 (Alqasemi et al., 2012).

References

Alqasemi, U., Li, H., Aguirre, A., & Zhu, Q. (2012). FPGA-based reconfigurable processor for ultrafast interlaced ultrasound and photoacoustic imaging. *IEEE Transactions on Ultrasonics, Ferroelectrics and Frequency Control*, *53*, 1344–1353.

Ashfaq, M., Brunke, S. S., Dahl, J. J., Ermert, H., Hansen, C., & Insana, M. F. (2006). An ultrasound research interface for a clinical system. *IEEE Transactions on Ultrasonics, Ferroelectrics and Frequency Control, 53*, 1759–1771.

Banker, C. J., Pedersen, P. C., & Szabo, T. L.,(2008) Interactive ultrasound training system. *Proceedings of the IEEE ultrasonics symposium* (pp. 1350–1354). Beijing.

Bell, M. A. L., Goswami, R., Dahl, J. J., & Trahey, G. E., (2012) Improved visualization of endocardial borders with short-lag spatial coherence imaging of fundamental and harmonic ultrasound data. *Proceedings of the IEEE ultrasonics symposium* (pp. 2129–2132). Dresden.

Bercoff, J., Tanter, M., & Fink, M. (2004). Supersonic shear imaging: A new technique for soft tissues elasticity mapping, IEEE Transactions on Ultrasonics. *Ferroelectrics and Frequency Control, 51*, 396–409.

Berson, M., Roncin, A., & Pourcelot, L. (1981). Compound scanning with an electrically steered beam. *Ultrasonic Imaging, 3*, 303–308.

Bracewell, R. (2000). *The fourier transform and its applications*. New York: McGraw-Hill. (Chapter 17).

Carr, P. H., DeVito, P. A., & Szabo, T. L. (1972). The effect of temperature and Doppler shift on the performance of elastic surface wave encoders and decoders. *IEEE Transactions on Sonics and Ultrasonics, 19*, 357–367.

Cheng, J., & Lu, J. -Y. (2006). Extended high frame rate imaging method with limited diffraction beams. *IEEE Transactions on Ultrasonics, Ferroelectrics and Frequency Control, 53*, 880–899.

Chiao, R. Y., & Hao, X. (2003). Coded excitation for diagnostic ultrasound: A system developer's perspective. *Proceedings of the IEEE ultrasonics symposium* (pp. 437–448).

Chiao, R. Y., Thomas, L. J., & Silverstein, S. D. (1997). Sparse array imaging with spatially-encoded transmits. *Proceedings of the IEEE ultrasonics symposium* (Vol. 2) (pp. Toronto, ON.

Cole, C. R. (1991). Properties of swept FM waveforms in medical ultrasound imaging. *Proceedings of the IEEE ultrasonics symposium* (Vol. 2) (pp. 1243–1248) Orlando, FL.

Corl, P. D., Grant, P. M., & Kino, G. S. (1978) A digital synthetic focus acoustic imaging system for NDE. *Proceedings of the IEEE ultrasonics symposium*, pp. 263–268.

Culp, B. C., Mock, J. D, Ball, T. R., Chiles, C. D., & Culp, W. C., Jr (2011). The pocket echocardiograph: A pilot study of its validation and feasibility in intubated patients. *Echocardiography, 28*, 371–377.

Dahl, J. J., Hyun, D., Lediju, M., & Trahey, G. E. (2011). Lesion detectability in diagnostic ultrasound with short-lag spatial coherence imaging. *Ultrasonic Imaging, 33*, 119–133.

Delannoy, B., Torgue, R., Bruneel, C., Bridoux, E., Rouvaen, J. M., & LaSota (1979). Acoustical image reconstruction in parallel-processing analog electronic systems. *Journal of Applied Physics, 50*, 3153–3159.

Doctor, S., Hall, T., & Reid, L. (1986). SAFT: The evolution of a signal processing technology for ultrasonic testing. *NDT International, 19*, 163–167.

Entrekin, R. R., Jago, J. R., & Kofoed, S. C. (2000). Real-time spatial compound imaging: Technical performance in vascular applications. In M. Halliwell, & P. N. T. Wells (Eds.), *Acoustical imaging* (Vol. 25, pp. 331–342). New York: Kluwer Academic/Plenum Publishers.

Fenster, A., & Downey, D. B. (1996). 3-D ultrasound imaging: A review. *IEEE Engineering in Medicine and Biology, 15*, 41–49.

Flaherty, J. J., Erikson, K. R., & Lund, V. M. (1967) Synthetic aperture ultrasound imaging systems, United States Patent, US 3,548,642.

Foster, F. S., Hossack, J., & Adamson, S. L. (2011). Micro-ultrasound for preclinical imaging. *Interface Focus, 1*, 576–601.

Foster, F. S., Larson, J. D., Mason, M. K., Shoup, T. S., Nelson, G., & Yoshida, H. (1989a). Development of a 12-element annular array transducer for realtime ultrasound imaging. *Ultrasound in Medicine and Biology, 15*, 649–659.

Foster, F. S., Larson, J. D., Pittaro, R. J., Corl, P. D., Greenstein, A. P., & Lum, P. K. (1989b). A digital annular array prototype scanner for realtime ultrasound imaging. *Ultrasound in Medicine and Biology, 15*, 661–672.

Foster, F. S., Zhang, M. Y., Zhou, Y. Q., Liu, G., & Mehi, J. (2002). A new ultrasound instrument for in vivo microimaging of mice. *Ultrasound in Medicine and Biology, 28*, 1165–1172.

Foster, F. S., Mehi, J., Lukacs, M., Hirson, D., & White, C. (2009). A new 15–50 MHz array-based micro-ultrasound scanner for preclinical imaging. *Ultrasound in Medicine and Biology*, *35*, 1700–1708.

Fraunhofer Institute for Biomedical Imaging, (2013). Di-Phas ultrasound research platform, http://www.ibmt.fraunhofer.de/content/dam/ibmt/de/Dokumente/PDFs/ibmt-produktblaetter/ibmt-ultraschall/US_ba_Ultrasound-Research-Device-DiPhAS-2011.pdf, accessed April 30, 2013.

Freeman, S. (2011a). http://www.aapm.org/meetings/amos2/pdf/59-17272-97172-944.pdf.

Freeman, S. (2011b). Microbeamforming for large-aperture ultrasound transducers. Med. Phys. 38, 3750.

Freeman, S., Jago, J., Davidsen R., Anderson, M., & Robinson, A.(2012). Third generation xMATRIX technology for abdominal and obstetrical imaging. Philips white paper, http://www.healthcare.philips.com/us_en/about/events/rsna/pdfs/X6-1_xMATRIX_TransducerWP_LR.pdf, accessed April 28, 2013.

Fuller, M. I., Owen, K., Blalock, T. N, Hossack, J. A., & Walker, W. F. (2009). Real time imaging with the sonic window: A pocket-sized, C-scan, medical ultrasound device. *Proceedings of the IEEE ultrasonics symposium* (pp. 196–199). Rome.

Gammelmark, K. L., & Jensen, J. A. (2003). Multielement synthetic transmit aperture imaging using temporal encoding. *IEEE Transactions on Medical Imaging*, *22*, 552–563.

Gee, A., Treece, G., Prager, R., Cash, C. J. C., & Berman, L. (2003). Rapid registration for wide field of view freehand three-dimensional ultrasound. *IEEE Transactions on Medical Imaging*, *22*, 1344–1357.

Genis, V., Obeznenko, I., Reid, I. M., & Lewin, P. (1991). Swept frequency technique for classification of the scatter structure. *Proceedings of the Annual Conference on Engineering in Medicine and Biology*, *13*, 167–168.

Hughes, D. I., & Duck, F. A. (1997). Automatic attenuation compensation for ultrasonic imaging. *Ultrasound in Medicine and Biology*, *23*, 651–664.

Jedrzejewicz, T., McLaughlin, G., Napolitano, D., Mo, L., & Sandstrom, K. (2003). Zone acquisition imaging as an alternative to line-by-line acquisition imaging. *Ultrasound in Medicine and Biology*, *29*(5S), S69–S70.

Jensen, J. A., Holm, O., Jensen, L. J., Bendsen, H., Nikolov, S. I., Tomov, B. G., Munk, P., Hansen, M., Salomonsen, K., Hansen, J. S., Gormsen, K., Pedersen, H. M., & Gammelmark, K. L. (2005). Ultrasound research scanner for real-time synthetic aperture image acquisition. *IEEE Transactions on Ultrasonics, Ferroelectrics and Frequency Control*, *52*, 881–891.

Jensen, J. A., Holm, O., Jensen, L. J., Bendsen, H., Pedersen, H. M., Salomonsen, K., Hansen, J. S., & Nikolov, B. G.(1999).Experimental ultrasound system for real-time synthetic imaging. *Proceedings of the IEEE ultrasonics symposium* (Vol. 2) (pp. 1595–1599) Caesars Tahoe, NV.

Jensen, J. A., Nikolov, S. I., Gammelmark, K. L., & Pedersen, M. H. (2006). Synthetic aperture ultrasound imaging. *Ultrasonics*, *44*, e5–e15.

Jensen, J. A., Nikolov, S. I., Misaridis, T., and Gammelmark, K. L. (2002). Equipment and methods for synthetic aperture anatomic and flow imaging. *Proceedings of the IEEE ultrasonics symposium,* pp. 1555–1564.

Jespersen, S. K., Wilhjelm, J. E., & Sillesen, H. (1998). Multi-angle compound imaging. *Ultrasonic Imaging*, *20*, 81–102.

Karaman, M., Bilge, H. S., & O'Donnell, M. (1998). Adaptive multi-element synthetic aperture imaging with motion and phase aberation correction. *IEEE Transactions on Ultrasonics, Ferroelectrics and Frequency Control*, *42*, 1077–1087.

Karaman, M., Li, P. C., & O'Donnell, M. (1995). Synthetic aperture imaging for small scale systems, IEEE Transactions on Ultrasonics. *Ferroelectrics and Frequency Control*, *42*, 429–442.

Kino, G. S. (1987). *Acoustic waves: Devices, imaging, and analog signal processing*. Englewood Cliffs, NJ: Prentice-Hall.

Krishnan, S., Rigby, K. W., & O'Donnell, M. (1997). Adaptive aberration correction of abdominal images using PARCA. *Ultrasonic Imaging*, *19*, 169–179.

Lafitte, S., et al. (2011). Validation of the smallest pocket echoscopic device's diagnostic capabilities in heart investigation. *Ultrasound in Medicine and Biology*, *37*, 798–804.

Langenberg, K., Berger, M., Kreutter, T., Mayer, K., & Schmitz, V. (1986). Synthetic aperture focusing technique signal processing. *NDT International*, *19*, 177–189.

Larson, J. D., III. (1993). 2D phased array ultrasound imaging system with distributed phasing, US patent 5,229 933, July 20, 1993.

Leavitt, S. C., Hunt, B. F., & Larsen, H. G. (1983). A scan conversion algorithm for displaying ultrasound images. *Hewlett Packard Journal*, *10*(34), 30–34.

Lediju, M. A., Trahey, G. E., Byram, B. C., & Dahl, J. J. (2011). Short-lag spatial coherence of backscattered echoes: Imaging characteristics. *IEEE Transactions on Ultrasonics, Ferroelectrics and Frequency Control*, *58*, 1377–1388.

Lediju, M. A., Pihl, M. J., Hsu, S. J., Dahl, J. J., & Trahey, G. E. (2008). Quantitative assessment of the magnitude, impact, and spatial extent of ultrasonic clutter. *Ultrasonic Imaging*, *30*, 151–168.

Lee, B. B., and Ferguson, E. A. (1982). Golay codes for simultaneous multi-mode operation in phased arrays. *Proceedings of the IEEE ultrasonics symposium* (pp. 821–825). San Diego, CA.

Lewis, G. K. (1987). Chirped PVDF transducers for medical ultrasound imaging. *Proceedings of the IEEE ultrasonics symposium* (pp. 879–884). Denver, CO.

Lu, J. -Y. (1997). 2D and 3D high frame rate imaging with limited diffraction beams. *IEEE Transactions on Ultrasonics, Ferroelectrics and Frequency Control*, *44*, 839–856.

Lu, J-Y. (1998). Experimental study of high frame rate imaging with limited diffraction beams. *IEEE Transactions on Ultrasonics, Ferroelectrics and Frequency Control*, *45*, 84–97.

Lu, J-Y, Cheng, J., & Wang, J. (2006). High frame rate imaging system for limited diffraction array beam imaging with square-wave aperture weightings. *IEEE Transactions on Ultrasonics, Ferroelectrics and Frequency Control*, *53*, 1796–1812.

Lu, J. -Y., & Greenleaf, J. F. (1992). Nondiffracting X waves: Exact solutions to free-space scalar wave equation and their finite aperture realizations. *IEEE Transactions on Ultrasonics, Ferroelectrics and Frequency Control*, *39*, 19–31.

Mann, J. A., & Walker, W. F. (2002). A constrained adaptive beamformer for medical ultrasound: Initial results. *Proceedings of the IEEE ultrasonics symposium* (Vol. 2) (pp. 1807–1810).

Mari, J. M., & Cachard, C. (2007). Acquire real-time RF digital ultrasound data from a commercial scanner. *Electronic Journal Technical Acoustics*, *3*. Available on-line at: http://www.ejta.org

Maslak, S. M. (1985). Computed sonography. In R. C. Sanders, & M. C. Hill (Eds.), *Ultrasound annual 1985*. New York: Raven Press.

Masotti, L., Biagi, E., et al. (2006). FEMMINA: Real-time, radio-frequency echo-signal equipment for testing novel investigation methods. *IEEE Transactions on Ultrasonics, Ferroelectrics and Frequency Control*, *53*, 1783–1795.

Melton, H. E., Jr., & Skorton, D. J. (1981). Rational-gain-compensation for attenuation in ultrasonic cardiac imaging. *Proceedings of the IEEE ultrasonics symposium* (pp. 607–611). Chicago, IL.

Mo, L. Y. L. , DeBusschere, D., Bai, W., Napolitano, D., Irish, A., Marschall, S., et al. (2007). P5C-6 compact ultrasound scanner with built-in raw data acquisition capabilities. *Proceedings of the IEEE ultrasonics symposium* (pp. 2259–2262). New York.

Montaldo, G., Tanter, M., Bercoff, J., Benech, N., & Fink, M. (2009). Coherent plane-wave compounding for very high frame rate ultrasonography and transient elastography, IEEE Transactions on Ultrasonics. *Ferroelectrics and Frequency Control*, *56*, 489–506.

Morgan, D. P. (1991). *Surface wave devices for signal processing*. Amsterdam: Elsevier.

Nelson, T. R., & Pretorius, D. H. (1998). Three-dimensional ultrasound imaging. *Ultrasound in Medicine and Biology*, *24*, 1243–1270.

Nikolov,S. I. (2001).*Synthetic aperture tissue and flow ultrasound imaging* (Ph.D. Thesis), Ørsted DTU, Technical University of Denmark, Lyngby, Denmark.

Nikolov, S. I., Gammelmark,K., & Jensen, J. A. (1999). Recursive ultrasound imaging. *Proceedings of the IEEE ultrasonics symposium* (Vol. 2) (pp. 1621–1625). Caesars Tahoe, NV.

Nock, L. F., & Trahey, G. E. (1992). Synthetic receive aperture imaging with phase correction for motion and for tissue inhomogeneities, Part I: Basic principles. *IEEE Transactions on Ultrasonics, Ferroelectrics and Frequency Control*, *39*, 489–495.

O'Donnell, M. (1992). Coded excitation system for improving the penetration of real time phased-array imaging systems. *IEEE Transactions on Ultrasonics, Ferroelectrics and Frequency Control, 39*, 341–351.

O'Donnell, M., & Thomas, L. J. (1992). Efficient synthetic aperture imaging from a circular aperture with possible application to catheter-based imaging. *IEEE Transactions on Ultrasonics, Ferroelectrics and Frequency Control, 39*, 366–380.

Perez, J. E., Waggoner, A. D., Barzilia, B., Melton, H. E., Miller, I. G., & Soben, B. E. (1991). New edge detection algorithm facilitates two-dimensional echo cardiographic on-line analysis of left ventricular (LV) performance. *Journal of the American College of Cardiologists, 17*, 291A.

Poulsen, C., Pedersen, P. C., & Szabo, T. L. (2005). An optical registration method for 3D ultrasound freehand scanning. *Proceedings of the IEEE ultrasonics symposium*, (pp. 1236–1240).

Prager, R. W., Rohling, R. N., Gee, A. H., & Berman, L. (1998). Rapid calibration for 3-D freehand ultrasound. *Ultrasound in Medicine and Biology, 24*, 855–869.

Rigby, K. W., Chalek, C. L., Haider, B., Lewandowski, R. S., O'Donnell, M., Smith, L. S., & Wildes, D. S. (2000). In vivo abdominal image quality using real-time estimation and correction of aberration. *Proceedings of the IEEE ultrasonics symposium* (pp. 1603–1606). San Juan.

Robinson, D. E., & Knight, P. C. (1981). Computer reconstruction techniques in compound scan pulse–echo imaging. *Ultrasonic Imaging, 3*, 217–234.

Sandrin, L, Catheline, S., Tanter, M., Hennequin, X., & Fink, M. (1999). Time-resolved pulsed elastography with ultrafast ultrasonic imaging. *Ultrasonic Imaging, 21*, 259–272.

Sandrin, S., Tanter, M., Catheline, S., & Fink, M. (2002). Shear modulus imaging using 2D transient elastography. *IEEE Transactions on Ultrasonics, Ferroelectrics and Frequency Control, 49*, 426–435.

Sanile, J., & Oruklu, E. (Eds.), (2012). Novel embedded systems for ultrasonic imaging and signal processing. *IEEE Transactions on Ultrasonics, Ferroelectrics and Frequency Control, 59*. (7)(Special Issue)

Savord, B., & Solomon R. (2003). Fully sampled matrix transducer for real-time 3D ultrasonic imaging. *Proceedings of the IEEE ultrasonics symposium* (Vol. 1) (pp. 945–953).

Savord, B. J., & Thiele, K. E. (1999). Phased array acoustic systems with intra-group processors US Patent, 5,997,479 A, December 7, 1999.

Schwartz, G. S. (2001). Artifact reduction in medical ultrasound. *Journal of the Acoustical Society of America, 109*, 2360.

Sharp, P. F. (1993). In P. N. T. Wells (Ed.), *Advances in ultrasound techniques and instrumentation.* New York: Churchill Livingstone(Chapter 1).

Shattuck, D. P., & von Ramm, O. T. (1982). Compound scanning with a phased array. *Ultrasonic Imaging, 4*, 93–107.

Shen, J., & Ebbini, E. S. (1996a). A new coded-excitation ultrasound imaging system, Part I: Basic principles.. *IEEE Transactions on Ultrasonics, Ferroelectrics and Frequency Control, 43*, 141–148.

Shen, J., & Ebbini, E. S. (1996b). A new coded-excitation ultrasound imaging system, Part II: Operator design. *IEEE Transactions on Ultrasonics, Ferroelectrics and Frequency Control, 43*, 131–140.

Shoup, T. A., & Hart, J. (1988). Ultrasonic imaging systems. *Proceedings of the IEEE ultrasonics symposium* (Vol. 2) (pp. 863–871).

Skjelvareid, M. H., Olofsson, T., Birkelund, Y., & Larsen, Y. (2011). Synthetic aperture focusing of ultrasonic data from multilayered media using an omega-k algorithm. *IEEE Transactions on Ultrasonics, Ferroelectrics and Frequency Control, 58*, 1037–1048.

Szabo, T. L., & Lewin, P. A. (2007). Piezoelectric materials for imaging. *Journal of Ultrasound in Medicine, 26*, 283–288.

Szabo, T. L., & Lewin, P. A. (2013). Ultrasound transducer selection in clinical imaging practice. *Journal of Ultrasound in Medicine, 32*, 573–582.

Szabo, T. L., Melton, H. E., Jr., & Hempstead, P. S. (1988). Ultrasonic output measurements of multiple mode diagnostic ultrasound systems. *IEEE Transactions on Ultrasonics, Ferroelectrics and Frequency Control, 35*(220–231).

Tirumalai, A. P., Lowery, C., Gustafson, G., Sutcliffe, P., & von Behren, P. (2000). Extended-field-of-view ultrasound imaging. In Y. Kim, & S. C. Horii (Eds.), *Handbook of medical imaging, Vol. 3: Display and PACs*. SPIE PressVol. PM81.

Tortoli, P., & Jensen, A. J. (Eds.), (2006). Novel equipment for ultrasound research. *IEEE Transactions on Ultrasonics, Ferroelectrics and Frequency Control, 53.* (10)(Special Issue)

von Ramm, O. T., Smith, S. W., & Pavy, H. E., Jr. (1991). High-speed ultrasound volumetric imaging system, Part II: Parallel processing and image display. *IEEE Transactions on Ultrasonics, Ferroelectrics and Frequency Control, 38*(2), 109–115.

Walker, W. F., & McAllister, M. J. (2002). Angular scatter imaging: Clinical results and novel processing. *Proceedings of the IEEE ultrasonics symposium*, pp. 1528–1532.

Walsh, C. J., Kobler, J. B., Heaton, J. T., Szabo, T. L., & Zeitels, S. M. (2008). Imaging of the calf vocal fold with high-frequency ultrasound. *The Laryngoscope, 118*, 1894–1899.

Wells, P. N. T. (1993). *Advances in Ultrasound Techniques and Instrumentation.* New York: Churchill Livingstone.

Wilson, T., Zagzebski, J., Varghese, T., Chen, Q., & Rao, M. (2006). The Ultrasonix 500RP: A commercial ultrasound research interface. *IEEE Transactions on Ultrasonics, Ferroelectrics and Frequency Control, 53*, 1772–1782.

Bibliography

Analog Devices. http://search.analog.com/search/default.aspx?query = ultrasound&local = en.

Fink, M., & Tanter, M. (2010). Multiwave imaging and super-resolution. *Physics Today, February*, 28–33.

Foster, F. S., Larson, J. D., Mason, M. K., Shoup, T. S., Nelson, G., & Yoshida, H. (1989). Development of a 12 element annular array transducer for realtime ultrasound imaging. *Ultrasound in Medicine and Biology, 15*, 649–659.

Details the design of an annular array digital imaging system.

Foster, F. S., Larson, J. D., Pittaro, R. J., Corl, P. D., Greenstein, A. P., & Lum, P. K. (1989). A digital annular array prototype scanner for realtime ultrasound imaging. *Ultrasound in Medicine and Biology, 15*, 661–672.

Another article detailing the design of an annular array digital imaging system.

Hewlett Packard Journal 10, Vol. 34. (October 1983).

Describes the operation of HP's first generation of imaging systems in detail.

Hewlett Packard Journal 12, Vol. 34. (December 1983).

A special issue that continues the description in the above reference.

Kino, G. S. (1987). *Acoustic waves: Devices, imaging, and analog signal processing*. Englewood Cliffs, NJ: Prentice-Hall.

Provides acoustic imaging theory and applications; available on CD-ROM from the IEEE-UFFC Group.

Kremkau, F. W. ( ). Diagnostic Ultrasound: Principles and Instruments.

This introductory book investigates the topic of imaging systems in more depth. It has a wealth of information that is clearly presented at an easily understood level.

Morgan, D. P. (1991). Surface Wave Devices; available on CD-ROM from the IEEE-UFFC Group.

Additional information about signal processing, encoding, and chirped waveforms for an allied field and surface acoustic wave devices.

Texas Instruments Ultrasound. http://www.ti.com/solution/ultrasound_system.

CHAPTER 11

Doppler Modes

Chapter Outline

Diagnostic Ultrasound Imaging: Inside Out.

11.1 Introduction

Doppler ultrasound and imaging are focused on the visualization and measurement of blood flow in the body. This is a technological achievement because, until recently, the received echoes from the acoustic scattering from regions of blood, such as those in the chambers of the heart, were at levels so low that they could not be seen or appeared as black in an ultrasound image. Even when blood cannot be seen directly, its movement can be detected. Now images of blood circulation, called color flow imaging (CFI), as well as precise continuous-wave (CW) and pulsed-wave (PW) Doppler measurements of blood flow, are routine on imaging systems. In this specialized area of ultrasound, interrogating beams are sent repeatedly in the same direction and are compared to each other to determine the movement of blood scatterers over time. All the usual physics of ultrasound apply, including beam directivity, transducer bandwidth, absorption, and the scattering properties of the tissue (blood). Doppler detection is a blend of physics and specialized signal-processing techniques required to extract, process, and display weak Doppler echoes. Doppler techniques provide critical diagnostic information noninvasively about the fluid dynamics of blood circulation and abnormalities.

11.2 The Doppler Effect

Most of us have heard of the Doppler effect, which is the perceived change in frequency as a sound source moves toward or away from you. Since sound is a mechanical disturbance, the frequency perceived is the effective periodicity of the wavefronts. If the source is moving directly toward the observer with a velocity c_s in a medium with a speed of sound c_0 then the arriving crests appear closer together, giving the observer the acoustic illusion of a higher frequency. As illustrated in Figure 11.1, the perceived frequency depends on the direction in which the source is moving toward or away from the observer. Pierce (1989) has shown that the perceived frequency is related to the vector dot product of the source (c_s) and unit observer (u_0) vectors, which differ by an angle θ, $v_x = c_s \cdot u_0 = c_s \cos\theta$,

$$f = f_0 + (f_D/c_0)c_s \cdot u_0, \tag{11.1A}$$

and solving for the Doppler frequency (f_D) in terms of the transmitted frequency (f_0),

$$f_D = \frac{f_0}{1 - (c_s/c_0)\cos\theta}, \tag{11.1B}$$

leads to a Doppler shift, correct to first order when $c_s = c_0$,

$$\Delta f = f_D - f_0 = f_0(c_s/c_0)\cos\theta. \tag{11.1C}$$

From this equation, the perceived frequencies for the observers in Figure 11.1 can be calculated for a 10-kHz source tone moving at a speed of 100 km/hr ($v = 27.78$ m/s) in air

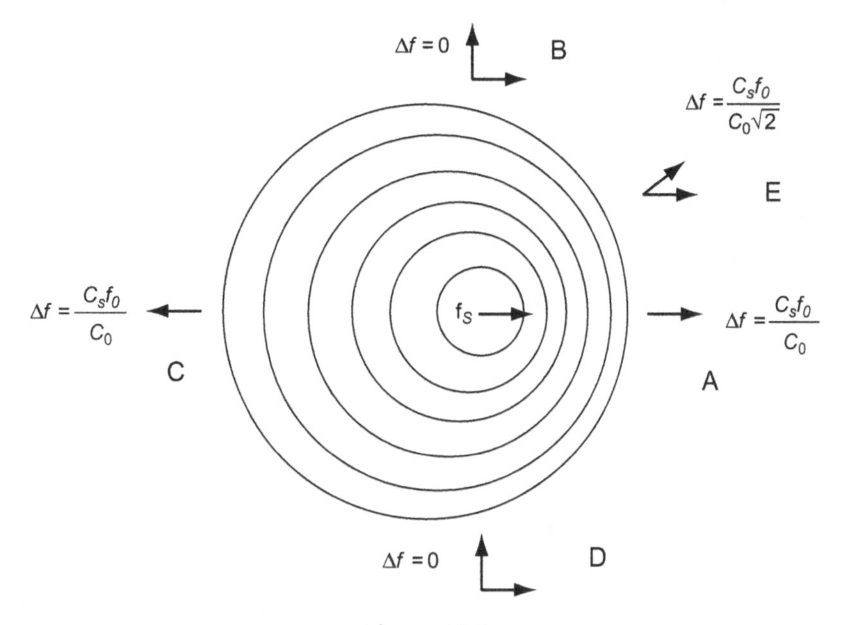

Figure 11.1
Doppler-shifted wave frequencies from a moving source as seen by observers at different locations. Observers at (A) 0°, (B) 90°, (C) 180°, (D) 270°, and (E) 45° angles relative to the directions of the source.

($c_0 = 330$ m/s). Observers B and D, at 90° to the source vector, hear no Doppler shift. Observer A detects a frequency of 10,920 Hz, while observer C (here, $\theta = \pi$) hears 9,220 Hz.

A similar argument yields an equation for a stationary source and a moving observer with a velocity c_{obs},

$$f = [1 + (c_{obs}/c_0)\cos\theta]f_0. \tag{11.2}$$

The Doppler effect plays with our sense of time, either expanding or contracting the timescale of waves sent at an original source frequency (f_0). Furthermore, it is important to bear in mind the bearing or direction of the sound relative to the observer in terms of vectors.

Now consider a flying bat intercepting a flying mosquito based on the Doppler effect caused by the relative motion between them (see Figure 11.2). It is straightforward to show that if the mosquito source has a speed of c_s, and the bat has a speed of c_{obs}, the corresponding equation for the Doppler-shifted frequency is:

$$f = f_0[1 + (c_{obs}/c_0)\cos\theta]/[1 - (c_s/c_0)\cos\theta] = f_0[c_0 + c_{obs}\cos\theta]/[c_0 - c_s\cos\theta]. \tag{11.3}$$

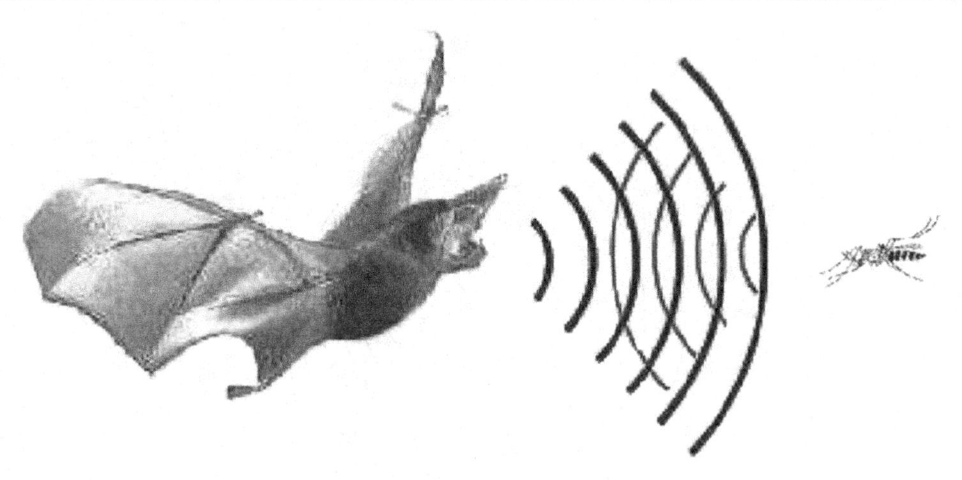

Figure 11.2
Bat detects insect target with ultrasound pulse echoes.

In other words, the flying mosquito perceives the bat signal as being Doppler-shifted, and the bat hears the echo as being Doppler-shifted again, owing to its motion. Of course, this situation is simplified greatly, as is Figure 11.1, because it is depicted two-dimensionally. This description has been adequate for most medical ultrasonics, in which imaging is done in a plane, until the comparatively recent introduction of 3D imaging.

The aero-duel between the bat and insect is played out three-dimensionally (3D) in real time. The poor mosquito beats its wings at about 200 flaps/s, which is the annoying whine you may hear just as you are about to fall asleep on a hot night. The moth also acts as an acoustic sound source at a softer 50 flaps/s. Enter the bat, which, depending on the type, has an ultrasound range between 20 and 150 kHz (e.g. the range of the horseshoe bat is 80–100 kHz). This corresponds to an axial resolution of 2–15 mm, which is perfect for catching insects. The bat emits an encoded signal, correlates the echo response in an optimum way (shown to be close to the theoretical possible limit), adapts its transmit waveform as necessary as it closes in on its target, changes its flight trajectory, and usually intercepts the insect with a resolution comparable to the size of its mouth, all in real time. Researchers are still trying to understand this amazing feat of signal processing and acrobatics and how a bat utilizes the Doppler shift between it and a fast-moving insect in 3D and while changing trajectories. Studies have shown how a bat interprets the following clues: Doppler shift (the relative speed of prey); time delay (the distance to the target); frequency and amplitude in relation to distance (target size and type recognition); amplitude and delay reception (azimuth and elevation position); and flutter of wings (attitude and direction of insect flight). One of the key signal-processing principles a bat utilizes is the repetitive interrogation of the target, so that the bat can build an image of the location and speed of its prey, pulse by pulse.

One of the earliest instances of pulse–echo Doppler ultrasound is in the original patent submitted by Constantin Chilowsky and Paul Langevin (1919) in 1916. Recall from Chapter 1 that their invention made underwater pulse–echo ranging technologically possible as a follow-up to earlier patents by Richardson (1913) (who also mentioned the Doppler shift, but as a problem) for acoustic iceberg detection to prevent another Titanic disaster. In their patent, they mention a method to detect relative motion between the observer and target by comparing the Doppler-shifted frequency from the target to the frequency of a stable source.

The dot product results from the moving source and moving observer cases can be applied to the simplified situation of a transducer sensing the flow of blood in a vessel flowing with velocity and direction v_s at an angle θ to the vessel, as depicted in Figure 11.3. In this case, the transducer is infinitely wide and the intervening tissues have negligible effect. The blood velocity is much smaller than the speed of sound in the intervening medium (c_0). The signal as seen from an observer riding the moving blood appears to be Doppler shifted:

$$\omega_T = \omega_2 + c_D \cdot k_i = \omega_0 + c_D \cdot \left(\frac{\omega_i}{c_0}\right) n_i, \tag{11.4A}$$

where ω_T is the shifted angular frequency, ω_2 is the angular frequency seen by the scattering object from a moving coordinate system, ω_i is the incident angular frequency, c_D is the Doppler velocity, and n_i is in the direction of the incident k vector along the beam. The returning scattered signal along unit vector n_{sc} appears to be from a moving source and is Doppler shifted:

$$\omega_R = \omega_2 + c_D \cdot k_{sc} = \omega_2 + c_D \cdot \left(\frac{\omega_{sc}}{c_0}\right) n_{sc}, \tag{11.4B}$$

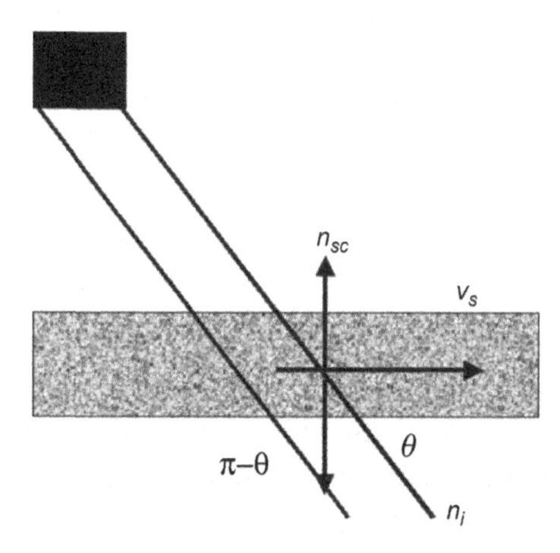

Figure 11.3
Sound beam intersecting blood moving at velocity v in a vessel tilted at angle θ.

where ω_R is the shifted angular frequency, ω_{sc} is the scattered-frequency Doppler, and n_{sc} is in the direction of the scattered k vector back toward the transducer. For a coincident transmitter and receiver, the overall Doppler shift can be found by subtracting the ω_R from ω_T and letting $\omega_{sc} \approx \omega_i \approx \omega_0$ to first order,

$$\omega_R - \omega_T = \omega_0(c_D/c_0)[1 + \cos(\theta) - \cos(\pi - \theta)] = \omega_0(c_D/c_0)[2\cos\theta], \tag{11.4C}$$

or in the form of the classic Doppler-shift frequency,

$$f_D = \Delta f = f_R - f_T = [2(v/c_0)\cos\,\theta]f_0. \tag{11.4D}$$

Before looking at ways that this Doppler shift can be implemented in instrumentation, it is worth understanding more about the properties of blood and how it interacts with sound.

11.3 Scattering from Flowing Blood in Vessels

Even though it is a fluid, blood is considered to be a highly specialized connective tissue. One of the main purposes of blood is to exchange oxygen and carbon dioxide between the lungs and other body tissues. There are typically 5 L of blood in an adult, or about 8% of total body weight. Blood is a continually changing suspension of red blood cells, white blood cells, and platelets in a solution called plasma. Red blood cells (erythrocytes) are the most plentiful, with about 5 million cells per μL. Each cell is a disk that is concave on top and bottom (like the shape of a double concave lens with a smooth-rounded outer ridge encompassing it) and about 7 μm in diameter and 2 μm in thickness. For adequate combination with oxygen, red blood cells must have a normal amount of hemoglobin (a red protein pigment that depends on the iron level in the body). White blood cells, or leukocytes, are about twice as big as red blood cells, but there are fewer of them (only 4000–10,000 in a μL). Platelets, which have a cross-section that is 1/1000 that of red blood cells, are fragments and are also fewer in number than red blood cells (about 250,000–450,000 per μL). A standard laboratory measurement is hematocrit, which is the ratio of the volume of red blood cells, packed by a centrifuge operation, to the overall blood volume. A typical ratio of hematocrit for a normal person is 45%; other values may indicate health problems.

The consistency of blood can change in different parts of the circulatory system. The viscosity of blood is 4.5–5.5 times that of water. Red blood cells can clump together or aggregate. A particular type of grouping is rouleau, which is a long chain of stacked cells. These groupings have a dramatic effect on ultrasound backscatter in veins. Not only is blood changing, breath by breath, but it is also being replenished; red blood cells last about 120 days, and white blood cells last less than 3 days.

Blood is also sensitive to vessel architecture. To first order, if blood is considered to be an incompressible Newtonian fluid in a long, rigid tube with a changing diameter (shown in Figure 11.4), the mean fluid velocity averaged over a cross-section ($\overline{v}$) obeys the following steady-state relation (Jensen, 1996):

$$A(z_1)\overline{v}(z_1) = A(z_2)\overline{v}(z_2), \tag{11.5A}$$

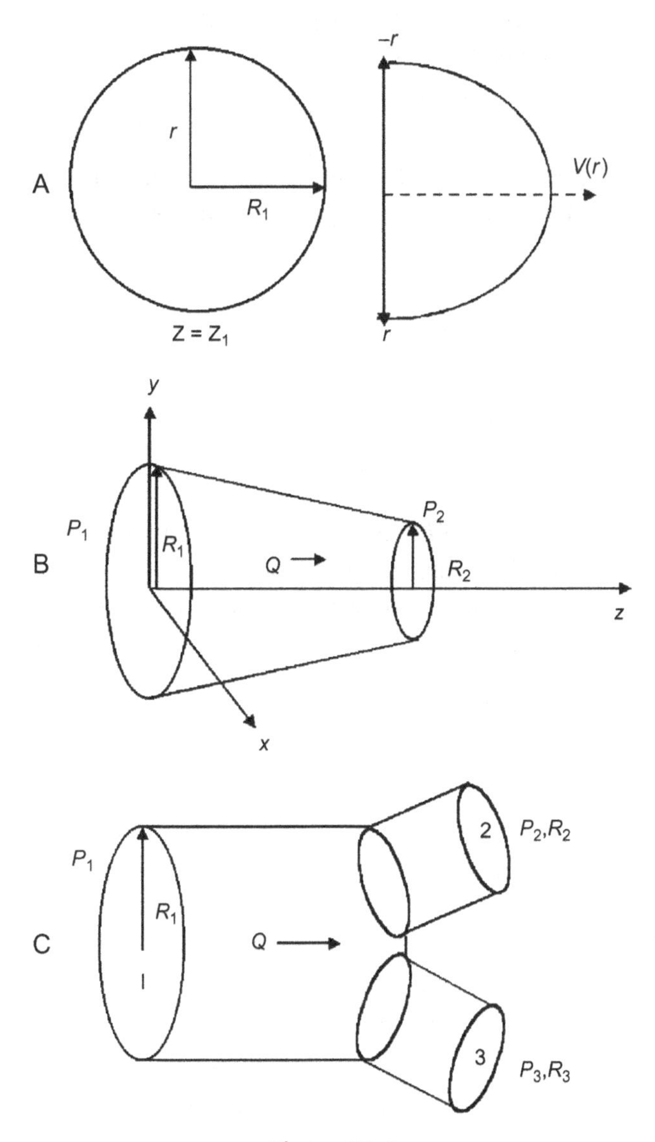

Figure 11.4
Fluid flow in vessels containing a newtonian fluid. (A) Parabolic velocity distribution in vessel cross-section; (B) fluid flow constant for different cross-sections; (C) flow into branches.

where z is the tube axis. The volumetric flow rate (Q) is constant through changes in tube cross-section:

$$Q = A(z)\bar{v}(z)\ (\mathrm{m}^3/\mathrm{s}), \tag{11.5B}$$

and is analogous to current in a wire. In the case of narrowing, the fluid velocity increases by the ratio of the square of the radii.

The pressure drop for a tube with a constant radius, similar to voltage drop in the electrical analogy (Jensen, 1996), is called Poiseuille's law,

$$\Delta P = P(z_2) - P(z_1) = R_f Q, \tag{11.5C}$$

where R_f is viscous resistance. Laminar flow in a long, rigid tube has a parabolic distribution of fluid velocity across its diameter, with v_o the maximum value in the tube center,

$$v(r) = \left(1 - \frac{r^2}{R^2}\right) v_o. \tag{11.6A}$$

For a parabolic flow distribution, the resistance for an outer diameter R is (Jensen, 1996):

$$R_f = 8\mu l/(\pi R^4), \tag{11.6B}$$

where μ is viscosity in kg/(m · s) and l is the tube length over which the pressure drop occurs. If a circular rigid tube branches into n smaller circular tubes of different outer diameters (R_n), the volumetric flow rate is conserved, and fluid velocity in each branch (v_n) is related by:

$$\pi R_0^2 v_0 = \sum_{n=1}^{N} \pi R_n^2 v_n. \tag{11.7}$$

Bernoulli's law expresses the conservation of energy for fluid flow in a tube, including potential, kinetic, and thermal energies. From this law, which is more general than Poiseuille's law, it is possible to relate the pressure drop to changes in geometry or fluid velocity. A simplified version for constant temperature and height is:

$$P_1 + \frac{1}{2}\rho v_1^2 = P_2 + \frac{1}{2}\rho v_2^2. \tag{11.8}$$

This important relation shows that where pressure is high, fluid velocity is low and vice versa. For example, when a parabolic velocity distribution occurs, as in Eqn 11.6A, Bernoulli's law indicates that pressure will be highest at the walls of the vessel and lowest in the center.

Realistically, blood is not an incompressible fluid but has a viscosity that changes with shear flow rate. Furthermore, vessels are not long, rigid tubes but are elastic with curved, complicated branching geometries. Finally, from the pumping of the heart, the flow is pulsatile and sometimes turbulent (not a steady flow). These practical considerations indicate that the previous equations are rough guidelines; reality is far more complicated. Despite these problems, Doppler ultrasound provides a remarkable, noninvasive dynamic depiction of blood flow in vivo that cannot be obtained by any other method.

How does sound interact with blood? Viewed as a homogeneous tissue, blood has an acoustic impedance and sound speed that depend on the red blood cell content, but typically it is $Z = 1.63$ megaRayls and $c = 1.57$ mm/μs (Bamber, 1986). Early measurements by E. L. Carstensen and H. P. Schwan (1959) of acoustic absorption for different concentrations of hemoglobin and sound speed dispersion agree well with the time causal relations of Chapter 4 (shown in Figure 11.5). Hemoglobin is a red iron-containing pigment that gives red blood cells their color. The hematocrit is the percentage of whole blood that comprises red blood cells.

What are the backscattering properties of blood as a tissue? R. A. Sigelmann and J. M. Reid (1973) developed one of the first calibrated tissue-characterization methods to measure backscatter from blood. Even though blood has been modeled as a continuous inhomogeneous medium (Angelsen, 1981), it is most often regarded as a collection of red

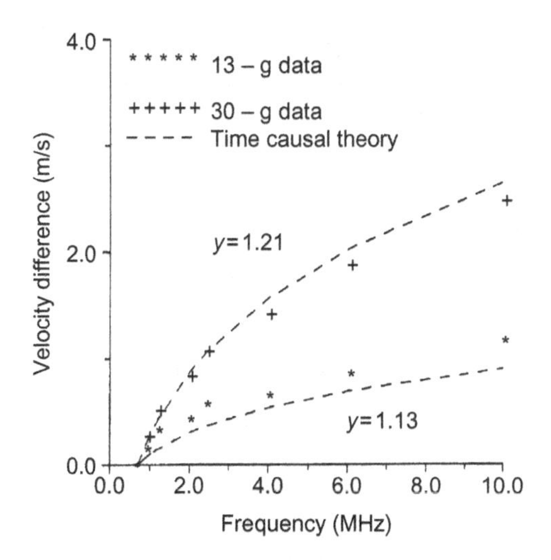

Figure 11.5

Acoustic absorption of hemoglobin solutions versus frequency. Shown for concentrations of 13 and 30 g/100 cm^3. Power-law fits to data of Carstensen and Schwan (1959) have exponents $y = 1.21$ (top curve) and $y = 1.13$ (bottom curve). *From Szabo (1993).*

blood cells because they predominate over other cell types. Because of the small size of red blood cells (~7 μm) relative to an insonifying wavelength (750–150 μm for 2–10 MHz), initially they were modeled as Rayleigh scatterers with backscattering proportional to the fourth power of frequency. Shung (1982) showed that by modeling the cells as cylinders, better agreement was obtained than modeling the cells as spheres (shown in Figure 11.6). Coussios (2002) simulated cells as disks and found a fourth power of frequency using the Born approximation. Cylinders and discs have a strong preferential directivity that agrees with the fact that blood appears to be anisotropic (Teh & Cloutier, 2000). The arrangement of cells into rouleaux and rouleau networks further increases the degree of anisotropy and the directional dependence of backscatter. Furthermore, backscatter is flow dependent (Fontaine, Bertrand, & Cloutier, 1999; Teh & Cloutier, 2000; Wang, Lin, & Shung, 1997). Wang et al. (1997) found, for example, that backscatter was lower in the vena cava than in the aorta, where flow was faster. The backscatter peaks at about 26% hematocrit; consequently, it is not a monotonic or single-valued function of hematocrit.

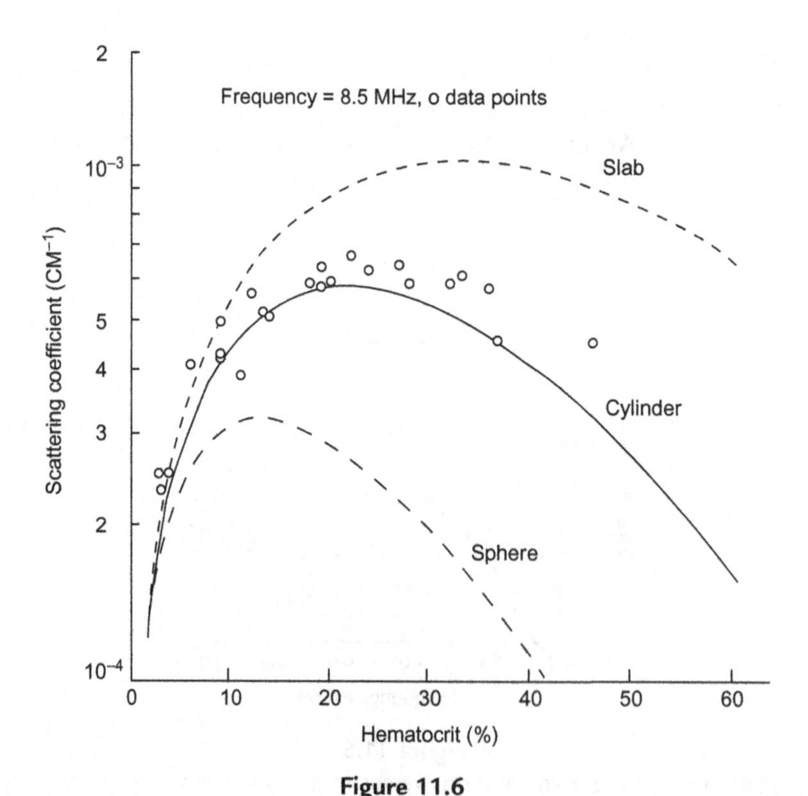

Figure 11.6
Backscattering of blood versus hematocrit compared with models of cells as spheres or cylinders. *From Shung (1982), IEEE.*

Millions of red blood cells have been analyzed statistically, and they were found to have a Rayleigh distribution (Mo & Cobbold, 1986). Like tissue microstructure and the resultant texture discussed in Chapter 8, the granular and apparent random nature of red blood cells also produces speckle-like behavior at conventional Doppler frequencies.

Measurements at higher frequencies, where wavelengths approach the dimensions of red blood cells (43–23 μm for 35–65 MHz), demonstrate different behavior. Absorption of blood measured by Lockwood, Ryan, Hunt, and Foster (1991) approached that of other tissue (half that of the arterial wall), as shown in Figure 11.7. Likewise, backscatter coefficients for flowing blood are comparable with those for some vascular tissues; they increase at high flow rates (as shown in Figure 11.8), and they have similar power-law dependence ($y \sim 1.4$). As shown later in Section 11.8, the echogenicity of blood is sufficient at higher imaging frequencies to allow the direct visualization of blood. By using their dual-gated Doppler system, Nowicki and Secomski (2000) and Secomski, Nowicki, Guidi, Tortoli, and Lewin (2003) demonstrated that the speed of sound and attenuation are monotonic functions of hematocrit, even though acoustic power is not (Figure 11.9).

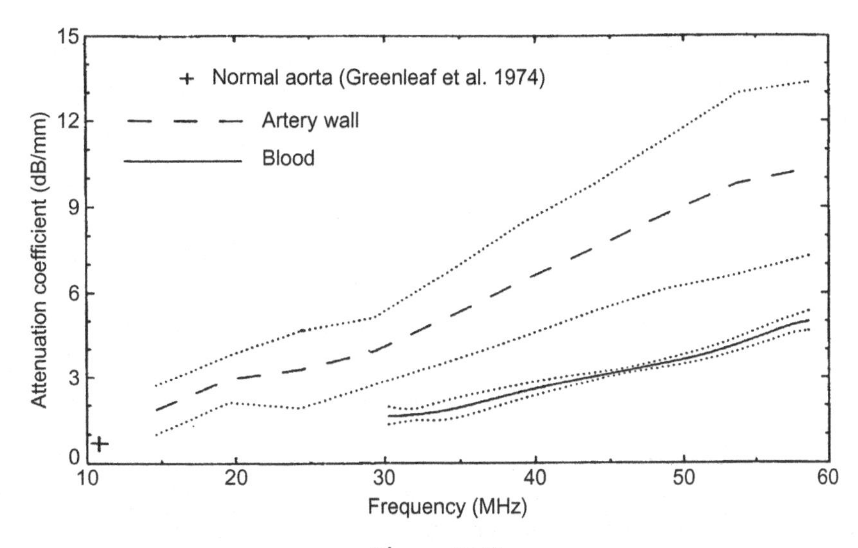

Figure 11.7
Attenuation measurements in artery samples and blood. Measurements summarized with the standard deviation in the data shown as dotted lines. Attenuation of human aorta measured at 10 MHz and 20°C by Greenleaf, Duck, Samayoa, and Johnson (1974) inserted for comparison. *From Lockwood et al. (1991); reprinted with permission from the World Federation of Ultrasound in Medicine and Biology.*

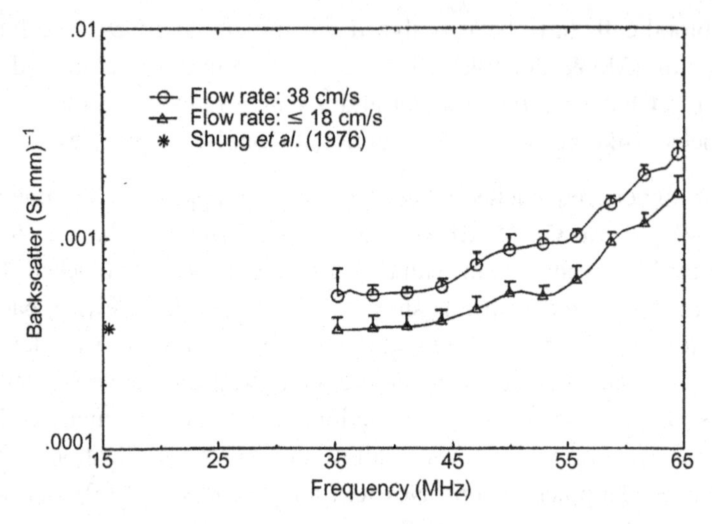

Figure 11.8
Summary of scattering measurements of flowing blood. Data measured by Shung, Sigelmann, and Reid (1976) at 15 MHz inserted for comparison. *From Lockwood et al. (1991); reprinted with permission from the World Federation of Ultrasound in Medicine and Biology.*

11.4 Continuous-Wave Doppler

Compared to the bat, Doppler medical ultrasound seems to be comparatively simple. Satomura (1957) reported CW experiments with Doppler-shifted ultrasound signals produced by heart motion. His work marked the beginning of many Doppler developments for diagnosis.

Early Doppler systems were completely analog with high sensitivity and selectivity. The classic CW Doppler subsystem is still a mainstay of modern ultrasound imaging systems. The "stand-alone" CW probe usually consists of a spherically concave transducer split into halves for transmission and reception (Evans & Parton, 1981). Actually, the centers of the halves are tilted slightly so that the transmit and receive beams intersect over a region of interest (as shown in Figure 11.10). To a good approximation, the Doppler Eqn (11.4D) still works for this geometry, with the centerline between halves serving as the angle reference line.

The halves are connected to a CW Doppler system similar to the one shown in Figure 11.11. By incorporating the transmit signal at f_0 into the receive signal path, the Doppler signal can be extracted with its amplitude and phase preserved. The processing is straightforward and is symbolized by the spectral graphs at different stages (signposted by letters), of the Doppler system in Figure 11.11. In this figure, single lines show CW spectra, solid lines show real spectra, and dashed lines show imaginary (shifted by 90°) spectra. The transmitted signal is a CW cosine at A. The Doppler-shifted signal is received at B, and it enters a pair of multiplier mixers. The top mixer multiplies a 90°-shifted source signal,

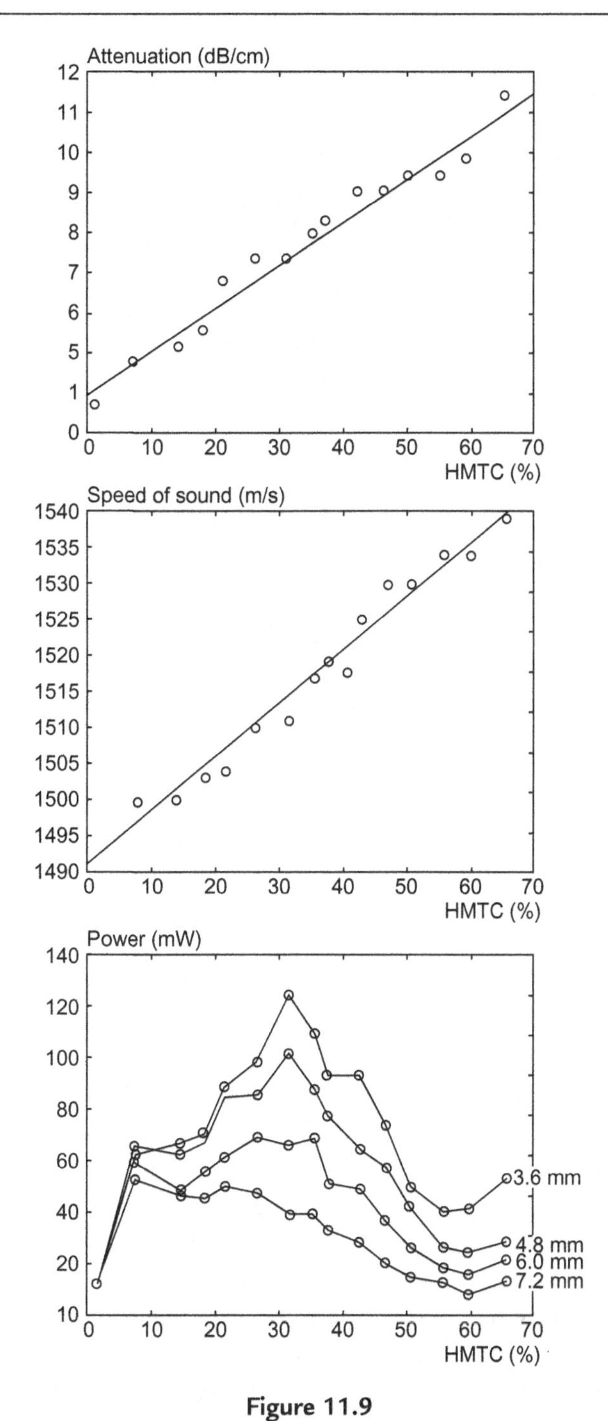

Figure 11.9
(A) Attenuation versus hematocrit (HMTC); (B) Speed of sound versus hematocrit; (C) Power versus hematocrit for four gate settings. *From Nowicki and Secomski (2000), IEEE.*

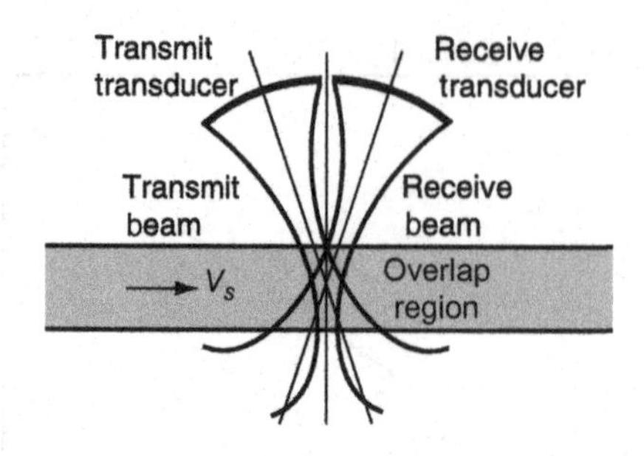

Figure 11.10
Split Stand-alone CW Doppler transducer showing split faces and the intersection of transmit and receive beams.

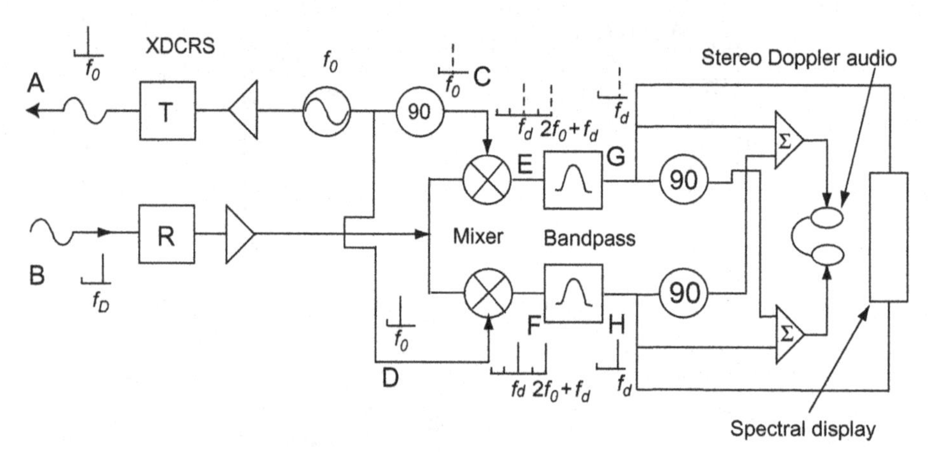

Figure 11.11
Block Diagram for CW Doppler ultrasound system for a split stand-alone transducer. XDCRS, transducers.

$\sin(\omega_0 t)$, with the echoes to produce, at E, an imaginary Doppler signal (f_D) near $f = 0$, and one near $2f_0$. Similarly, the lower mixer creates, at F, real spectra at those frequencies. Bandpass filters remove both $f = 0$ signals due to stationary tissue and the Doppler near $2f_0$ to supply an "I" (for in-phase) signal at H and a quadrature "Q" signal at G. These two signals combined can be considered to be an analytic signal (see Appendix A) with only a positive Doppler-shifted frequency. Recall that the analytic signal can be created with a Hilbert transform (Appendix A). It comes as no surprise that, except for front-end amplification, all the processing can also be done digitally after analog-to-digital (A/D) conversion through Hilbert transforming and digital filtering. Furthermore, the spectral display can be conveniently performed by a fast Fourier transform (FFT). The remaining processing in the system, consisting of phase shifts, separates the signals into a forward flow and a reverse flow for stereo audio enjoyment.

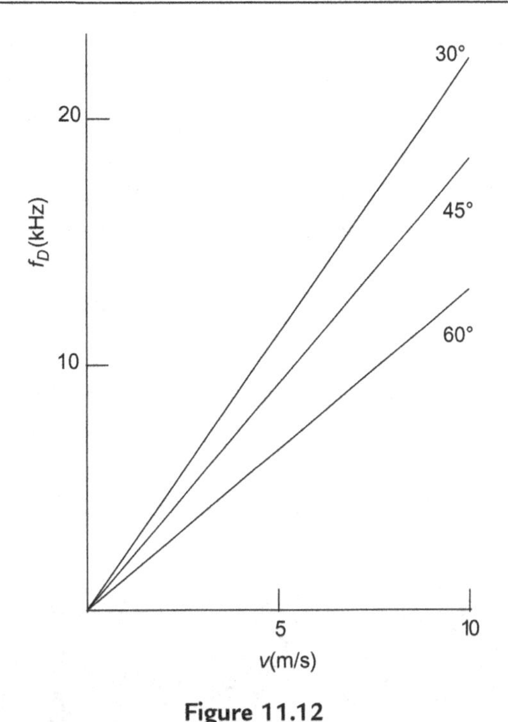

Figure 11.12
Doppler-shift frequencies for a source frequency of 2 MHz, as a function of velocity (v) for different angles, with $c_0 = 1.54$ km/s.

By a fortuitous coincidence, the Doppler-shifted frequencies fall within the human hearing range for typical ultrasound frequencies in combination with most of the blood velocities encountered in the body. Typical values are calculated for different angles from the Doppler equation, Eqn 11.4D, and plotted in Figure 11.12. These velocities extend from about 150 mm/s in the vena cava to 3000 mm/s in the ascending aorta. Just as physicians have learned to use the stethoscope, expert users of CW Doppler can detect abnormalities in flow just by listening, as Robert Hooke foresaw more than 300 years ago (see Chapter 1).

While stand-alone probes are regarded as being extremely sensitive, a more convenient way of obtaining CW signals is to use an imaging array. The most common configuration is to split an array into two sections (subarrays) for transmit and receive. Steering and focusing are under electronic control and provide much more flexibility than the fixed-focus stand-alone probe. Furthermore, system signal processors and computers can be used for computation and display. While stand-alone users "fly blind," in that there are no visual cues to guide the correct placement of the sensitive beam area, CW in arrays can be combined with B-mode or color flow imaging to locate and align the Doppler beam with a vessel or region of interest, as described in duplex and triplex modes in Chapter 10.

An example of a triplex mode, including color flow imaging overlaid on a gray-scale image and CW Doppler, is illustrated by Figure 11.13. At the top of the display is a small insert showing a complete gray-scale heart image with a color flow overlay. Even though color flow imaging will be discussed later, the reader should appreciate that this image is a global view of the flow in the left ventricle of the heart. The color code at the right indicates that orange represents positive velocities flowing toward the transducer, and blue represents those flowing in a direction away from the transducer. The line through this colored region and the apex is the direction of the CW Doppler measurement; it includes mainly blue with a few small orange regions.

In Figure 11.13, the spectral display in gray depicts Doppler-shifted frequencies representing velocities as a function of time. From left to right, there seems to be a pattern that almost repeats itself. This pattern can be thought of in terms of two timescales: slow time and fast time. A time window (T) is used to calculate a fast-time snapshot spectrum of

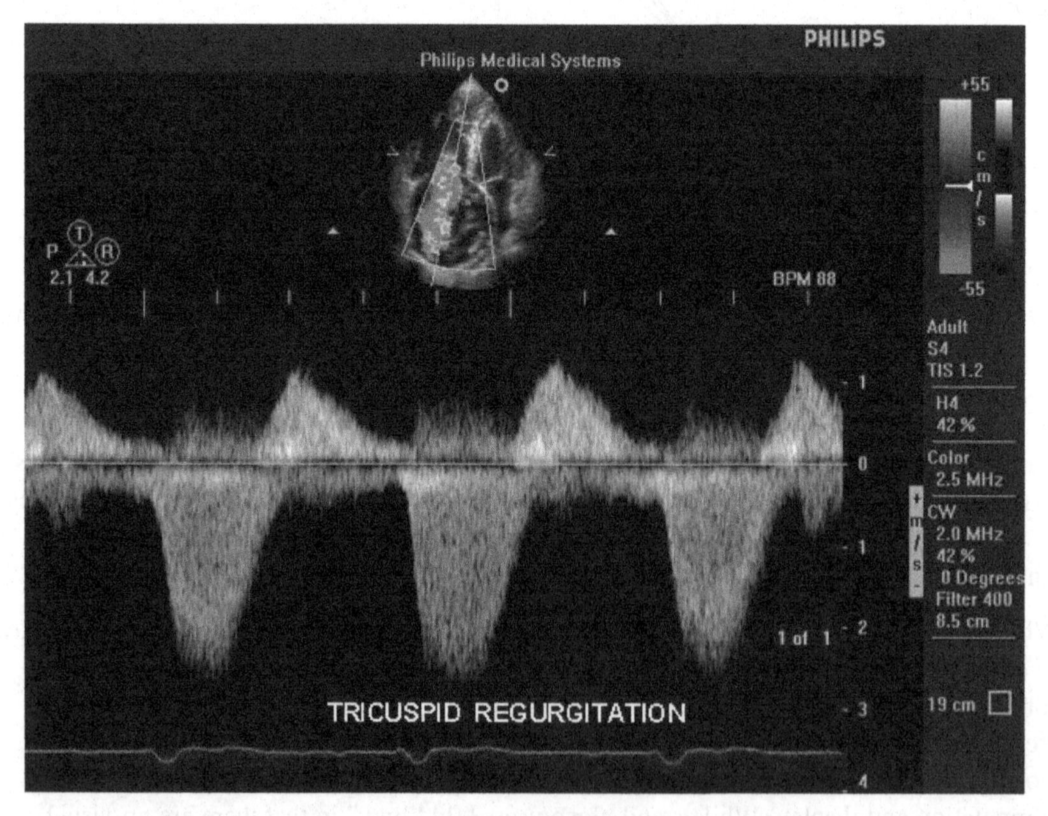

Figure 11.13
Duplex imaging mode of tricuspid regurgitation. CW Doppler velocity display with a color flow image insert (above) with direction of CW line. The reader is referred to the Web version of the book to see the figure in color. *Courtesy of Philips Healthcare.*

the flow. This process is continually repeated, with the next time window placed to the left of the one before it. The overall scale of several repetitions can be considered to be slow time; there is a horizontal slow timescale in beats per minute above the spectrogram. As a new snapshot is displayed on the left end of the slow timescale, the last one on the right disappears, so the visual effect is that time records are scrolling to the right on the display. A careful look reveals that snapshots are not identical but represent a live indication of the dynamic changes happening during cardiac cycles. As expected from the color flow image, both positive and negative velocities are shown, characteristic of the backflow that occurs in regurgitation. The spectrum has a velocity scale to the right in meters per second. The granular appearance of the spectrogram resembles speckle.

To examine the influence of other major physical effects on Doppler processing, it is helpful to apply a block diagram approach to the overall system, as depicted in Figure 11.14. Many of the individual blocks are recognizable from their use in the overall block diagram (see Figure 2.14). In the system are the already-discussed source, bandpass filter, time gate, FFT, and audio detector. The physical processes out of the Doppler box, representing points A through B for the round-trip path in the time domain, can be represented by:

$$v_B = s *_r (h_t *_t h_r) *_t a_r *_t a_t *_t e_{RT} *_t v_a, \tag{11.9}$$

where v_a is the excitation voltage. The transducer impulse response, $e_{RT} = e_g *_t e_t$, has little effect since only a narrow bandwidth transducer with good efficiency (light or air backing) is required. Attenuation, symbolized on transmit and receive paths by a_t and a_r, to the target and back is practically a constant factor at a single frequency and depth. The major physical factors are the beam diffraction (represented by h) and the scatterers (s). The individual

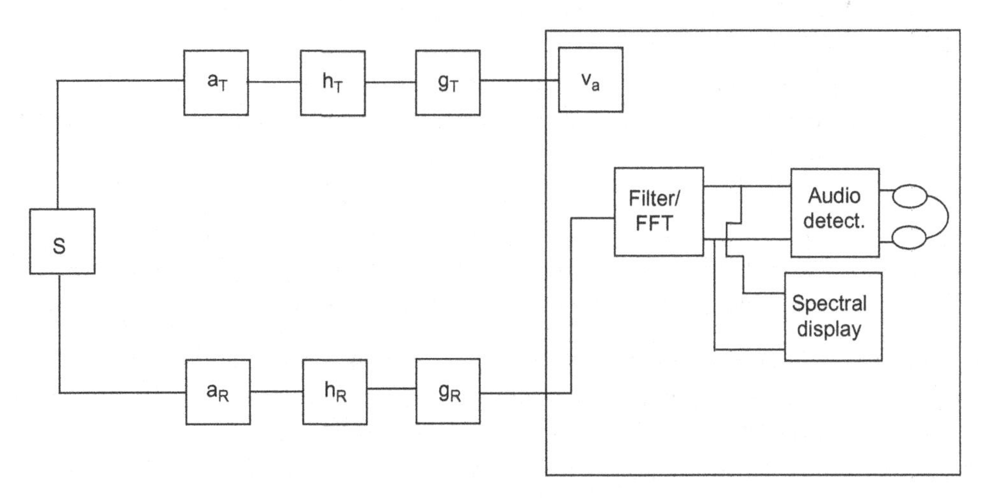

Figure 11.14
Block Diagram for CW Doppler ultrasound system and related physical effects.

unresolved blood cells, moving at different velocities from positions within the vessel that are insonified by the beam, cause a statistical speckle-like variation and resultant Doppler shifts.

Newhouse, Furgason, Johnson, and Wolf (1980) have shown that the distribution of red blood cells caught in the sound beam causes transit-time broadening, which is caused by the geometric extent (or broadening) of the beam. Recall from Chapter 7 that the cross-section of a beam, especially in the focal plane, can be described as a function of a lateral dimension or angle, as well as a function of frequency. This effect is also present for the Doppler case. Newhouse et al. (1980) utilized the focal plane, or far field, full width half maximum (FWHM) beamwidth formula ($w = 0.8\lambda F/a = 0.8c_0F/(af_0)$) to derive an estimate for the spectral broadening of the Doppler spectrum from a circular beam:

$$\Delta f_D/f_D = (0.8\lambda F/a)\tan\theta = [0.8c_0F/(af_0)]\tan\theta. \quad (11.10)$$

Cobbold, Vetlink, and Johnston (1983) examined the effects of beam misalignment with the vessel and attenuation (which they found to be small when the beam size was comparable to the vessel diameter) on the mean velocity. This is a key parameter in estimating volumetric flow. In summary, the block diagram proposed provides a comprehensive way of accounting for important factors affecting Doppler signals.

11.5 Pulsed-Wave Doppler

11.5.1 Introduction

To overcome range ambiguity, the well-known limitation of CW Doppler (which is sensitive to whatever vessels intersect its entire beam), PW Doppler (Baker, 1970; Wells, 1969a;1969b) was devised to control the region of active insonification. Like the bat, a pulsed Doppler system sends ultrasound pulses of a chosen length repetitively to a target at a certain range. Another duplex image (in Figure 11.15) shows PW Doppler in combination with a color flow image. In this case, superimposed on the image is a line with a marker indicating the pulse length, range depth, and direction; below is the Doppler velocity spectrum in a display similar to that used for the CW Doppler in Figure 11.13. Both the range and interrogating pulse length are controllable. From the image, an angle to the vessel can be determined and entered into some systems to correct for the Doppler cosine angle variation.

Range-gated pulsed Doppler systems are different from CW systems in important ways, even though they may seem similar superficially. Unlike the CW model of Doppler-shifted wavefronts of Figure 11.1, finite length pulses are used. In addition to expected time dilation or contraction by the Doppler effect, changes in pulse delay arrival time are also involved (Jensen, 1996; Wilhjelm & Pedersen, 1993a;1993b). For a transmitted waveform,

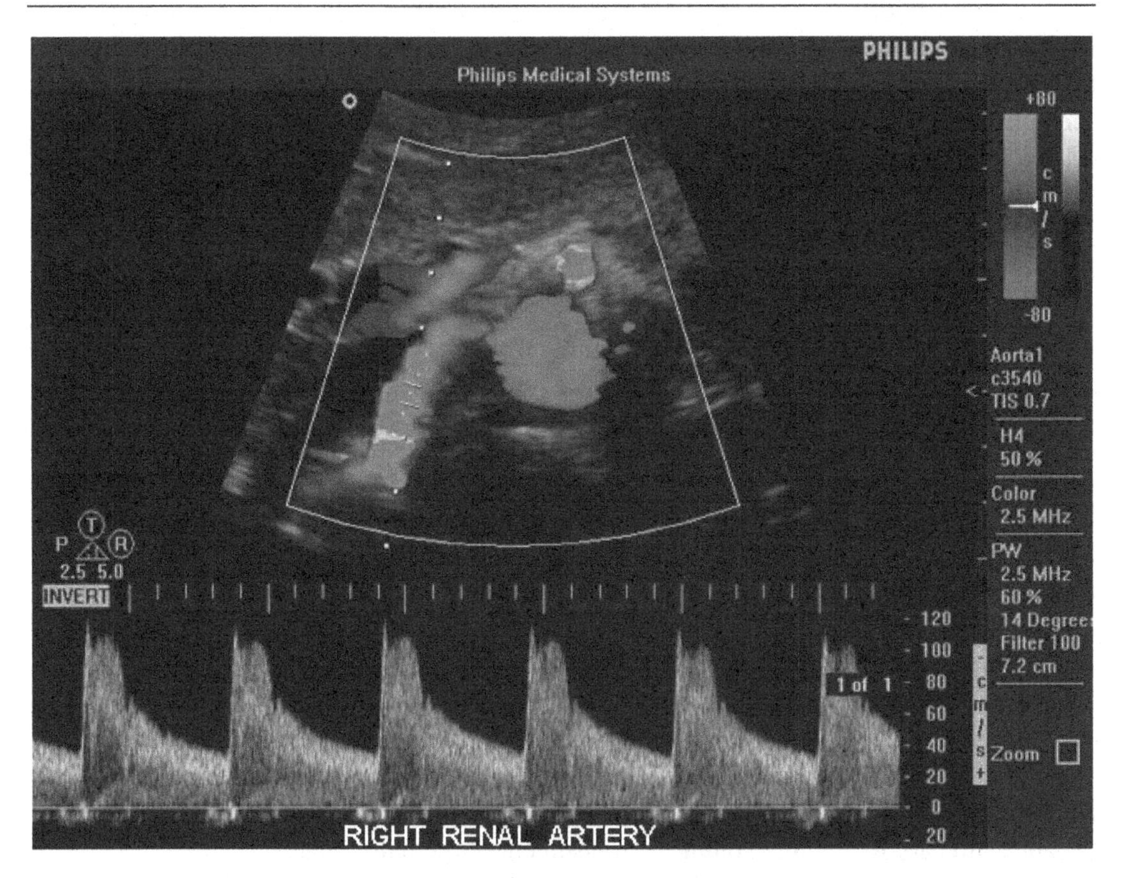

Figure 11.15
Duplex imaging mode of a right renal artery. PW Doppler velocity display, with a color flow image insert (above) with direction of PW line and Doppler gate position. The reader is referred to the Web version of the book to see the figure in color. *Courtesy of Philips Healthcare.*

$v_A(t)$, Jensen (1996) has shown that the received Doppler-shifted output signal has the form (assuming no absorption or diffraction effects), of:

$$v_B(t) = v_A\left[\Psi\left(t - \frac{2d_0}{\Psi c_0}\right)\right] = v_A\left[\Psi t - \frac{2d_0}{c_0}\right], \tag{11.11A}$$

where d_0 is the distance to the target, and the Doppler scaling factor Ψ is:

$$\Psi = 1 - \frac{2c_D\cos\theta}{c_0} = 1 - \delta_D, \tag{11.11B}$$

which appears in Eqn 11.11A as a time-scaling factor for dilation or contraction and also as a time-delay modifier, and $\delta_D = 2c_D \cos\theta/c_0$, is an often-used constant.

In CW Doppler, the Doppler-shifted received frequency is compared to the transmitted frequency; however, in range-gated Doppler, each received echo is compared to a similar echo resulting from the previous transmission. The relative delay between Doppler-shifted echoes from consecutive pulses is simply (Bonnefous, Pesque, & Bernard, 1986; Jensen, 1996; Wilhjelm & Pedersen, 1993a):

$$t_d = \frac{2\Delta z}{c_0} = \frac{2T_{PRF}c_D \cos\theta}{c_0} = \delta_D T_{PRF}, \tag{11.11C}$$

where Δz is the distance traveled away from the transducer and T_{PRF} is the time between transmit pulses (PRF is pulse repetition frequency).

This comparison has the important consequence that it is relatively insensitive to the absorption and diffraction effects on the paths through intervening tissues to the target site. Pulse to pulse, these factors do not change much, so they are compared on a consistent basis; however, they affect overall sensitivity. Otherwise, absorption would cause a considerable downshift (on a Doppler scale) in the center frequency of the transmitted pulse (as discussed at the end of Chapter 4); consequently, it would generate a false Doppler signal (Jensen, 1996). Note that for the CW case, variations and loss caused by the diffraction of the beam and increasing absorption loss with depth can contribute to a diminishing sensitivity, which can be a problem for a real system with noise and limited dynamic range. Pulsed Doppler, when implemented on arrays, provides a number of advantages: a larger variable aperture, electronically controlled focusing and steering, and the ability to vary the sample volume by adjustment of the pulse length.

11.5.2 Range-Gated Pulsed Doppler Processing

Before beginning the derivation of equations for pulsed Doppler, it is worth discussing the primary difference between pulsed- and continuous-wave Doppler. This difference is sampling. Just as array elements behave as spatial samplers (as discussed in Chapter 7), Doppler pulses act as time domain samplers. The repetitive nature of these pulses can be most conveniently represented by the shah or sampling function from Appendix A. Recall that the Fourier transform of the shah function is a replicating function in the frequency domain, which will allow us to characterize the Doppler spectrum. This approach will also easily show the consequence of undersampling, which is the chief limitation of pulsed Doppler. This section will end with an expression for the Doppler-shifted frequencies for the pulsed approach.

As described in Chapter 10, a master clock sends pulses repetitively at a pulse repetition frequency (f_{PRF}) along the same direction, which is selected by the user. Parameters of interest are depicted by Figure 11.16. The time between pulses is the pulse repetition

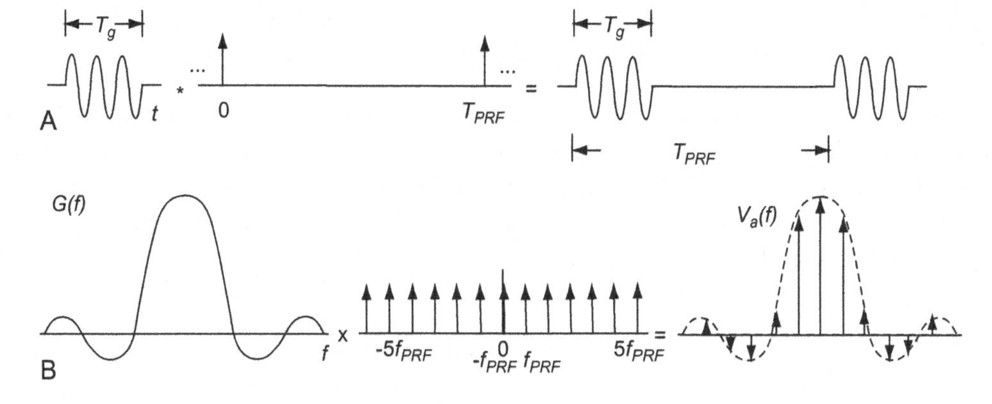

Figure 11.16
(A) Repeating transmit pulse parameters; (B) Spectrum of repeating pulses.

interval (PRI) = $T_{PRF} = 1/f_{PRF}$. Each pulse is an M period of the fundamental f_0, or the gate length is:

$$T_g = MT = M/f_0. \tag{11.12A}$$

The tone bursts transmitted at intervals T_{PRF} can be described approximately as:

$$v_A(t) = g(t)*_t III(t/T_{PRF}), \tag{11.12B}$$

$$v_A(t) = \prod(t/T_g)\sin(\omega_0 t)*_t III(t/T_{PRF}), \text{ and} \tag{11.12C}$$

$$v_A(t) = \sum_{n=-\infty}^{\infty} \prod[(t - nT_{PRF})/T_g]\sin[\omega_0(t - nT_{PRF})], \tag{11.12D}$$

where the rect and the shah replicating functions have been used from Appendix A and $g(t)$ describes the individual pulse. Equation 11.12 is depicted graphically at the top of Figure 11.16. From the Fourier transform of this equation and the application of the Fourier transform sampling property of the shah function, an interesting spectrum is obtained:

$$V_A(f) = (-iT_{PRF}T_g/2)\{\text{sinc}[T_g(f - f_0)] - \text{sinc}[T_g(f + f_0)]\}III(f/f_{PRF}), \text{ and} \tag{11.13A}$$

$$V_A(f) = (-iT_{PRF}T_g/2)\sum_{n=-\infty}^{\infty}\{\text{sinc}[T_g(nf_{PRF} - f_0)] - \text{sinc}[T_g(nf_{PRF} + f_0)]\}\delta(f - nf_{PRF}), \tag{11.13B}$$

where the graphical representation of this equation is at the bottom of Figure 11.16. A consequence of the repetitious life of the transmitted pulses is that their spectra appear as lines modulated by the sinc-shaped functions centered on $\pm nf_0$. A similar calculation from

Magnin (1986) for $g(t)$ as a Gaussian envelope instead of a tone burst is plotted in Figure 11.17A. Here, his corresponding notation is $f_s = f_0$, and PRI $= T_{PRF}$.

Typically, the Doppler range gate (pulse) is placed on a vessel or region of interest, and pulse and transmit beam characteristics are optimized for the range delay $t_0/2$. Two types of echoes are a stationary pulse and a Doppler-shifted pulse, returning at an approximate round-trip time of t_0. By applying Eqn 11.9 to Eqn 11.12, an expression for the received echoes can be derived. First the stationary echoes:

$$v_{BS}(t) = \sum_{n=-\infty}^{\infty} \prod[(t - t_0 - nT_{PRF})/T_g]\sin[\omega_0(t - t_0 - nT_{PRF})], \quad (11.14A)$$

and then the Doppler-shifted echoes, simply delayed by t_0 from the result of Eqn 11.12D:

$$v_{BD}(t) = \sum_{n=-\infty}^{\infty} \prod[[\Psi(t - (t_0/\Psi) - (nT_{PRF}/\Psi))/T_g]\sin[\Psi\omega_0(t - (t_0/\Psi) - (nT_{PRF}/\Psi))], \quad (11.14B)$$

where the arguments in parentheses can be rewritten as:

$$[t - (t_0/\Psi) - (nT_{PRF}/\Psi)] = [t - t_0 - nT_{PRF} - \delta_D(t_0 + nT_{PRF})]. \quad (11.14C)$$

Note that by letting $c_D = 0$, $\delta_D = 0$, $\Psi = 1$, Eqn 11.14B reduces to the stationary target version, Eqn 11.14A, and therefore this equation covers both types of echoes. The last equation can be rewritten as:

$$v_{BD}(t) = \prod(t\Psi/T_g)\sin(\Psi\omega_0 t) * III[(t - t_0)/T_{PRF}], \quad (11.14D)$$

where the relation $\delta(ax) = \delta(x)/|a|$ and the definition of the shah function have been applied to obtain:

$$v_{BD}(t) = \prod(t\Psi/T_g)\sin(\Psi\omega_0 t) \sum_{n=-\infty}^{\infty} (1/\Psi)\delta[t - (nT_{PRF} + t_0)/\Psi]. \quad (11.14E)$$

By Fourier transforming these echoes, Eqn 11.14E, their spectra are obtained:

$$V_{BD}(f) = \frac{-iT_g}{2\beta}\{\mathrm{sinc}[T_g/\beta)(f - \beta f_0)] - \mathrm{sinc}[(T_g/\beta)(f + \beta f_0)]\} * III(f/f_{PRF})exp(-i2\pi f t_0), \text{ and} \quad (11.15A)$$

$$V_{BD}(f) = G(f) * \sum_{n=-\infty}^{\infty} \Psi\delta[f - nf_{PRF}\Psi], \quad (11.15B)$$

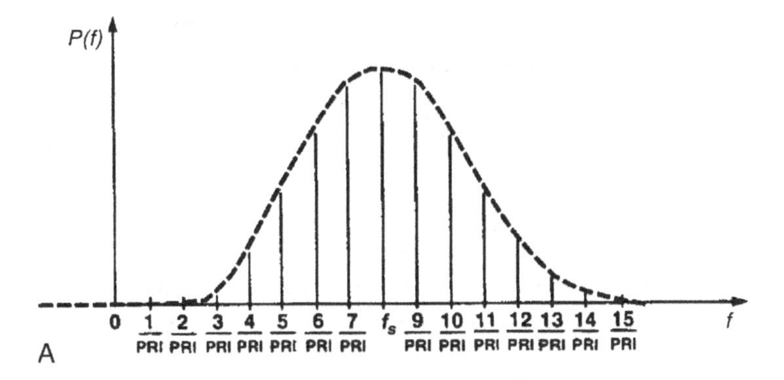

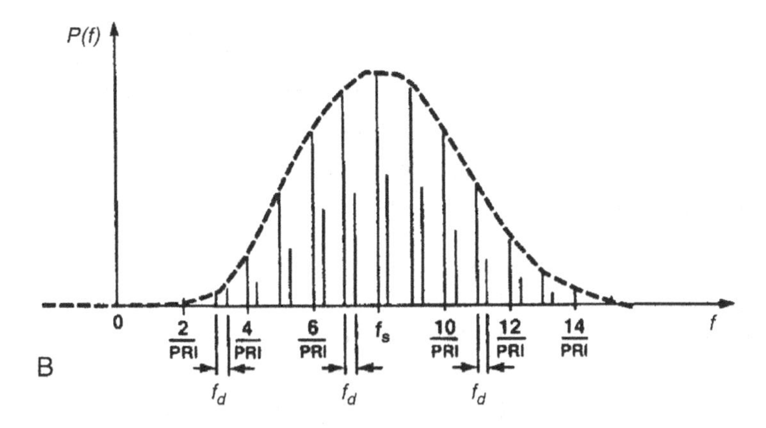

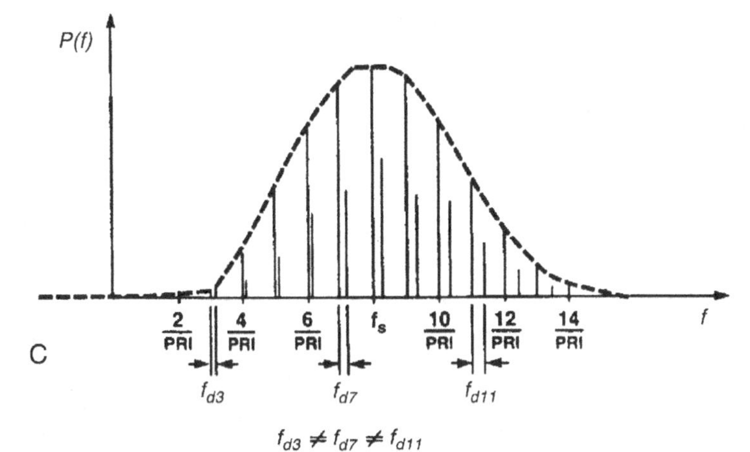

Figure 11.17

(A) Spectra of stationary objects for a Gaussian transmit pulse; (B) Equally shifted Doppler spectra compared to spectra from echoes from stationary objects; (C) Correctly shifted Doppler spectra compared to spectra from echoes from stationary objects. *From Magnin (1986); reprinted with permission from Hewlett Packard.*

where $G(f) = \{\}$ in Eqn 11.15A is the Fourier transform of the pulse function $g(t)$, and if a similar scaling relation was used for the impulse functions,

$$V_{BD}(f) = \Psi G(f) * \sum_{n=-\infty}^{\infty} \delta[f - nf_{PRF} + \delta_D n f_{PRF}]. \quad (11.15C)$$

Except for a time delay to the target, the spectrum of stationary echoes ($\delta_D = 0$ in the previous equation) is similar to that of the repeated transmitted pulses. There may be an expectation that the Doppler spectra should be shifted by a constant frequency,

$$f_d = \delta_D f_{PRF}, \quad (11.16)$$

for each of the PRF harmonic frequencies, as illustrated by Figure 11.17B for a Gaussian envelope, $G(f)$; however, this relation is incorrect. Magnin (1986) pointed out that the actual, counterintuitive, result of Eqn 11.15C is that the Doppler shift actually increases with PRF harmonic number n, or $\delta_D n f_{PRF}$, as shown in Figure 11.17C. This analysis shows that for a single scatterer moving at a constant velocity c_D, pulsed Doppler produces a distribution of unequal harmonic Doppler shifts.

The remarkable aspect of Doppler detection is that stationary signals are typically 40 dB (100 times) larger than the amplitudes of Doppler echoes, and the Doppler shift can be less than 1 kHz, or only a few parts out of 10,000 relative to the transmit frequency. Furthermore, because time-delay shifts are small for Doppler echoes, they can overlap the stationary echoes. This feat of engineering is accomplished by quadrature sampling the returning echoes and by other filtering (shown in the next few sections).

11.5.3 Quadrature Sampling

Quadrature sampling is needed to differentiate between forward and reverse flows. Principles of this detection method can be understood by reference to Figure 11.18. At the top of this figure are pulses gliding slowly to the left or right, representing the Doppler time-shifted echoes described in Eqn 11.14B. If the samples occur at $t_0 + nT_{PRF}$, there are timing circumstances where the in-phase sampler cannot distinguish between forward and reverse directions, even though the equivalent Doppler frequency can be determined correctly from the resulting detected period. In this example, the period is:

$$f_D = 1/4T_{PRF} = f_{PRF}/4. \quad (11.17A)$$

Note also that the time shift from one pulse to the next sequential pulse at sample times (pulse repetition intervals of T_{PRF}) is:

$$t_D = T_{PRF}/4. \quad (11.17B)$$

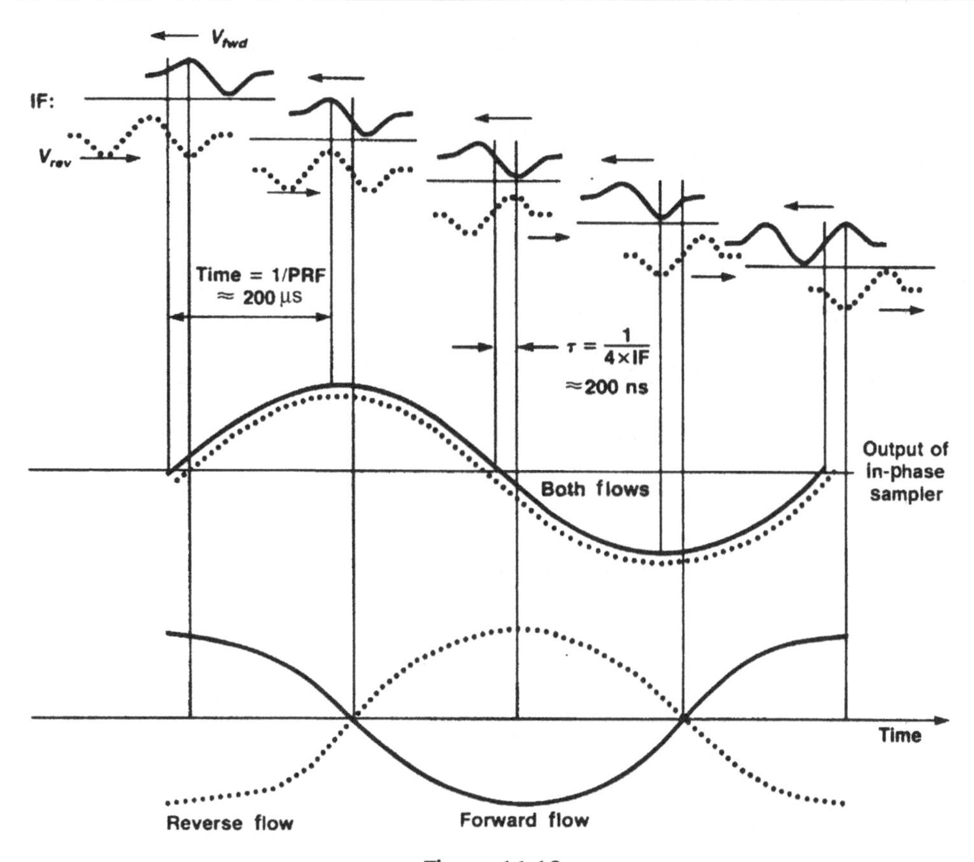

Figure 11.18

Quadrature sampling of forward and reverse flows for a pulsed Doppler system. (Top) echoes for five PRF intervals containing both forward- and reverse-flow Doppler time shifts; (center) output of a single (in-phase) sampler for both echoes; (bottom) output of second (quadrature) sampler allows differentiation of forward and reverse flows by phase encoding. IF = intermediate frequency. *From Halberg and Thiele (1986); reprinted with permission from Hewlett Packard.*

If the sampling times are done a quarter period later, at $1/4f_0$, the quadrature sampler, in this example, is able to discriminate between the flow directions (as shown at the bottom of Figure 11.18).

In order to derive equations that represent this sampling process, the shah function is applied to Eqn 11.14D:

$$v_{BD}^{S}(t) = v_{BD}(t)III[(t - t_0)/T_{PRF}], \tag{11.18A}$$

to indicate sampling at times $t_0 + mT_{PRF}$, as required,

$$v_{BD}^{S}(t) = [g(t)/\Psi] \sum_{m=-\infty}^{\infty} \sum_{m=-\infty}^{\infty} \delta[t - t_0 - nT_{PRF} - \delta_D(t_0 + nT_{PRF})]\delta(t - t_0 - mT_{PRF}). \tag{11.18B}$$

This double sum can be reduced to a single sum through timing arguments explained by Newhouse and Amir (1983) or by using matched indices $m = n$:

$$v_{BD}^{S}(t) = [g(t)/\Psi] \sum_{n=-\infty}^{\infty} \delta[t - \delta_D(t_0 + nT_{PRF})]. \tag{11.18C}$$

As first shown by Newhouse and Amir (1983), the sampled waveform is reversed and scaled in time by the Doppler factor δ_D and sampled at times $t_0 + nT_{PRF}$. Note that for no shift, the transmitted waveform is unchanged.

Finally, the spectrum of the sampled waveform is:

$$V_{BD}^{S}(f) = V_{BD}(f) * f_{PRF}[III(f/f_{PRF})exp\,(-i2\pi f t_0)], \tag{11.19A}$$

which by similar arguments can be reduced to a single sum:

$$V_{BD}^{S}(f) = G(f)exp(-i2\pi f t_0)\Psi f_{PRF} \sum_{n=-\infty}^{\infty} \delta(f - \delta_D m f_{PRF}), \tag{11.19B}$$

$$V_{BD}(f) = \frac{-iT_g}{2T_{PRF}}\{\text{sinc}(T_g f)exp(-i4\pi f t_0)\} * \{\delta[\Psi(f - f_0)] - \delta[\Psi(f + f_0)]\} \sum_{n=-\infty}^{\infty} \delta(f - \delta_D n f_{PRF}), \text{ and} \tag{11.19C}$$

$$V_{BD}(f) = \frac{-iT_g}{2T_{PRF}}\{\text{sinc}(T_g f)e^{(-i4\pi f t_0)}\} * \left\{ \sum_{n=-\infty}^{\infty} \delta[f - (\delta_D n f_{PRF} - \delta_D f_0 - f_0)] + \sum_{n=-\infty}^{\infty} \delta[f - (\delta_D n f_{PRF} + \delta_D f_0 + f_0)] \right\} \tag{11.19D}$$

11.5.4 Final Filtering and Display

Equation 11.19D represents the quadrature-sampled signal. With reference to the pulsed Doppler block diagram in Figure 11.19, the next steps are the wall filter and Nyquist filtering.

After Nyquist filtering at $f_{PRF}/2$ (one of the next steps in the signal processing from the block diagram of Figure 11.19), the higher frequencies are eliminated. From Eqn 11.19D, the sampling repetitions end up at a single frequency when $mf_{PRF} = f_0$. Ironically, after all this processing, the pulsed Doppler appears at frequencies, $f_D = \pm\delta_D f_0$. Even though this is the same frequency obtained by Doppler frequency shifting in the CW case, Jensen (1996) points out that for the pulsed Doppler case, this coincidence is a result of the Doppler time shift and the way the pulsed Doppler is implemented. He has derived results for pulsed Doppler in more detail and has accounted for a finite number of Doppler pulses that produce the observed bandwidth Doppler spectrum rather than the spectral lines obtained here.

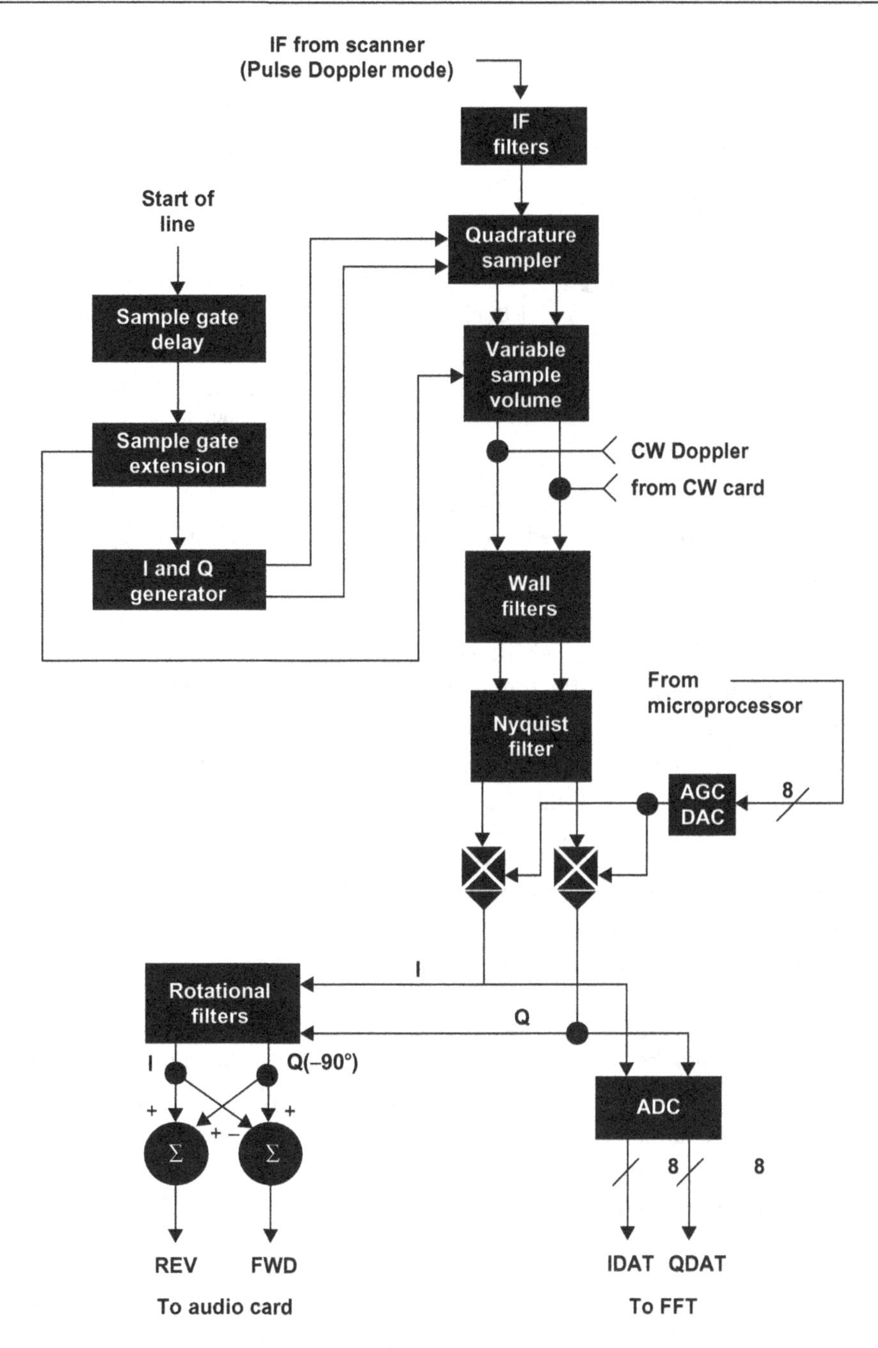

Figure 11.19

Block diagram of range-gated Doppler system. ADC, analog-to-digital converter; DAC, digital-to-analog converter. IF = intermediate frequency, AGC = automatic gain control. *From Halberg and Thiele (1986); reprinted with permission from Hewlett Packard.*

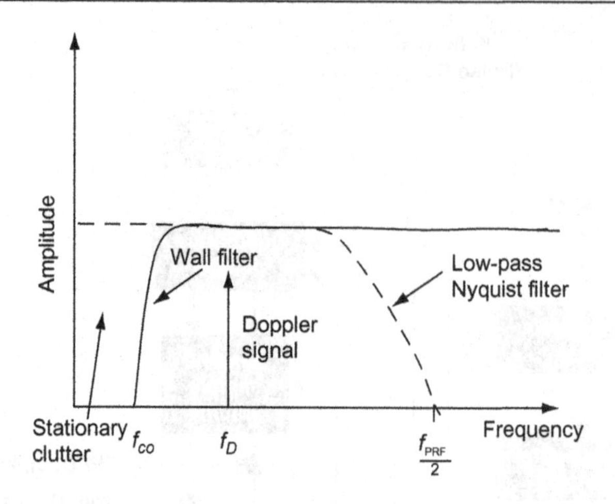

Figure 11.20
Symbolic depiction of Doppler-detected frequency (f_D) and the Nyquist low-pass filter with an $f_{PRF}/2$ cutoff frequency and a wall filter with a user-selectable cutoff.

In addition to Nyquist filtering, special filtering is necessary to remove stationary or slowly moving vessel or tissue walls. As illustrated symbolically in Figure 11.20, this high-pass filter, also known as the "wall filter" because of its steep cutoff characteristic, eliminates low Doppler frequencies that correspond to slow movements. In some systems the cutoff frequency is user-selectable to best suit the clinical application.

The last steps in Doppler processing involve automatic gain amplification as well as routing to an FFT processor for spectral display and to phase shifters for audio output of the detected Doppler signals. For the latter, an extra 90° in phase is added to or subtracted from each output so that they are 180° out of phase at the speakers.

In a quadrature or analytic signal representation, the forward flows might have a +90° encoding (phase difference between the in-phase and quadrature channels), whereas the reverse flow would be −90° encoded. By "delaying" the quadrature channel by 90° (with respect to the in-phase channel), the forward flow could then be extracted by taking the summed signal (common mode), whereas the reverse flow could be derived from the difference signal. These isolated forward- and reverse-flow signals are then directed to the audio speakers.

11.5.5 Pulsed Doppler Examples

In order to appreciate the engineering and clinical trade-offs for pulsed Doppler, as well as to clarify the significance of the variables, a practical example will be helpful. Consider blood flowing at a velocity of 2 m/s in a vessel that is insonified at an angle of 60° at a depth of 5 cm with a transmit frequency of 2.5 MHz (assume $c_0 = 1.154$ cm/μs). The round-trip echo time is:

$$t_{RT} = z(2/c_0) = 5 \cdot 13 = 65 \ \mu s. \tag{11.20A}$$

For a tone burst of 10 cycles, the gate length is:

$$T_g = m/f_0 = 10/2.5 = 4 \ \mu s. \tag{11.20B}$$

The minimum pulse repetition interval is:

$$T_{PRF} = t_{RT} + T_g = 65 + 4 = 69 \ \mu s, \tag{11.20C}$$

so that the maximum PRF is:

$$f_{PRF} = 1/T_{PRF} = 14.5 \ \text{kHz}, \tag{11.20D}$$

resulting in a Nyquist frequency of:

$$f_{NYQ} = f_{PRF}/2 = 7.25 \ \text{kHz}. \tag{11.20E}$$

A Doppler factor of:

$$\delta_{D=2v\text{cosè}/c_0} = 2 \times 2 \times \cos(\delta/3)/1540 = 1.3\text{e} - 3 \tag{11.20F}$$

results in a Doppler frequency of:

$$f_D = \delta_D f_0 = 3.25 \ \text{kHz}, \tag{11.20G}$$

and a Doppler time shift of:

$$t_D = \delta_D T_{PRF} = 0.0897 \ \mu s. \tag{11.20H}$$

Fortunately, the Doppler frequency is less than the Nyquist frequency. From the Nyquist frequency, the maximum detectable velocity from the Doppler frequency equation ($\theta = 0$) is:

$$v_{\max} = c_0 f_{NYQ}/(2f_0) = 2.23 \ m/s. \tag{11.20I}$$

The obvious sampling rate limitation is one of the weaknesses of pulsed Doppler. Velocities faster than the previous value permitted by the Nyquist rate appear out of their proper place in the spectrum, or "alias."

While our analysis has dealt with an idealized single Doppler frequency, clinical Doppler involves a spread of shifted frequencies from several causes. As the block diagram was made for CW Doppler in Figure 11.14, a similar diagram could be constructed for PW Doppler. While the physical effects are similar, the sampling mechanism in PW Doppler results in absorption and diffraction effects only at the gate depth, whereas these physical factors affect CW Doppler along a larger region formed by the intersection of the transmit and receive beams. A better model for the blood scatterers, $s(r, t)$, in such a block diagram would involve a statistical distribution of cells, describing their whirling and changing in time, as well as their moving at a parabolic or abnormal spread of velocities across the width of a vessel or chamber. These effects result in a widening of the Doppler and its granular appearance (as depicted in Figure 11.15). Further broadening occurs as a result of the finite beamwidth of the interrogating beam interacting with the sample volume of blood. The extent of this volume narrows as particle velocities increase, which implies that short pulses are needed (short Tg). However, better spectral sensitivity and penetration (see Eqn 11.15A) is obtained by narrow bandwidths, which are obtained with longer pulses (a contradictory trade-off).

Imaging systems often provide calculations of Doppler parameters. An example is shown for the superficial femoral artery in Figure 11.21. Shown are the mean velocity, the average value, and the systolic-to-diastolic (S/D) velocity ratio. The mean velocity

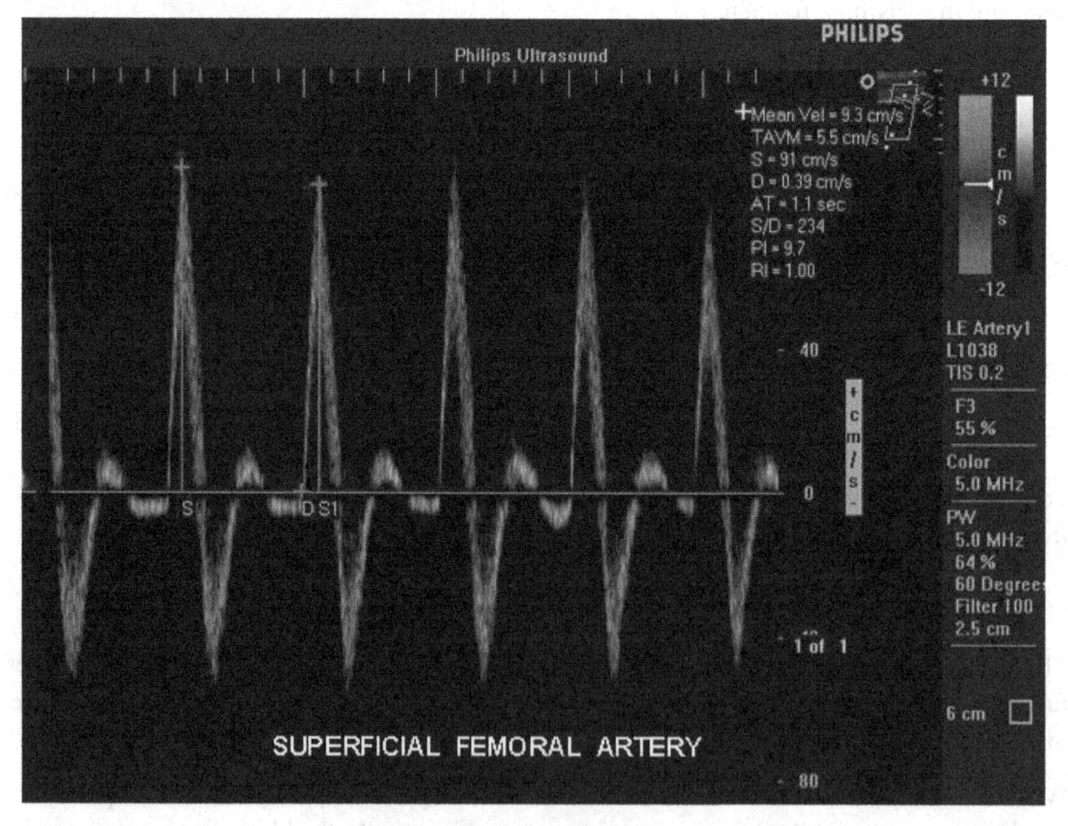

Figure 11.21
Duplex-pulsed Doppler display for a superficial femoral artery with calculations displayed and a small image in the upper-right corner. *Courtesy of Philips Healthcare*

combined with the volumetric flow equation, Eqn 11.5B, can be used to estimate volume flow; this application favors a longer pulse, needed to capture the entire cross-section of a vessel.

In an important application, the pressure drop across a stenotic valve can be estimated from Bernoulli's law, Eqn 11.8, (Hattle & Angelsen, 1985) as:

$$P_1 - P_2 = 4v_2^2, \tag{11.21}$$

in which the blood velocity is in m/s and pressure is in mmHg. The velocity at the valve is measured by Doppler ultrasound, $v_2 = c_D$, and if the change in pressure as determined by this equation exceeds 50 mmHg, the valve most likely needs to be replaced or repaired. In this case, the high velocities appear as a narrow jet that is best found with a wide-focus CW beam.

Standard deviation provides a quantitative measure of the breadth of the velocity distribution. Hattle and Angelsen (1985) summarize clinical applications of Doppler measurements.

11.6 Comparison of Pulsed- and Continuous-Wave Doppler

Table 11.1 summarizes PW and CW Doppler. It is assumed that the PW Doppler is from an array, whereas the CW Doppler is obtained either from a fixed (stand-alone) probe or an array (steerable CW). Steerable CW Doppler can be regarded as being at the extreme of the PW Doppler continuum.

Table 11.1 CW and PW Doppler Comparisons.

Topic	Fixed CW Doppler	Steerable CW Doppler	PW Doppler
Resolution	At intersection of transmit and receive	At intersection of transmit and receive	Range-gated
Focusing	Fixed focus and amplitude	Electronic focusing with gain	Electronic focusing with gain
Steering	Mechanical	Electronic	Electronic
Visual aid for placement	None	Line, duplex, triplex imaging	Gate, duplex, triplex imaging
Aliasing	No	No	Yes
Absorption and diffraction	At intersection of transmit and receive	At intersection of transmit and receive	At gate position

11.7 Ultrasound Color Flow Imaging

11.7.1 Introduction

Color flow imaging (CFI) was one of the big technological breakthroughs of diagnostic ultrasound imaging. The real-time display of blood velocity and direction is both an outstanding technical achievement and clinical success story. By providing a moving color picture that is a global view of dynamic blood flow, this modality enabled the finding of previously overlooked jets from stenotic and other flow abnormalities, such as leaking heart valves (shown as regurgitation in Figure 11.15), flow reduction, and occlusion from atherosclerotic plaques. Combined with conventional Doppler, CFI provides a global view for accurate Doppler line placement, which previously was a "blind" and difficult procedure. These contributions of CFI have increased diagnostic confidence in the sense that anomalies can be detected quickly and not be overlooked.

Early attempts at making a Doppler image included multigate Doppler, or Doppler that was swept mechanically through a number of vector directions (Nowicki & Reid, 1981). Because blood is fast moving, considered quasi-stationary for only 5–100 ms at different locations (Magnin, 1987), these methods could not keep up. Also, unlike PW Doppler, in which FFTs were calculated only during the range-gate interval, much longer scan depths were needed. These factors meant that only 3–12 (typically 8) sample points were available per depth (Magnin, 1987). For so few points, FFTs had a large variability and were not fast enough. One manufacturer had an FFT-based estimator but is no longer in business (Kimme-Smith, Tessler, Grant, & Perella, 1989). Even faster and more robust mean frequency estimators had to be devised to determine blood flow velocity. Furthermore, the kind of information normally presented in a spectrogram had to be displayed in a way that could be comprehended quickly at each spatial location.

11.7.2 Phase-Based Mean Frequency Estimators

An instantaneous frequency can be defined as (Bracewell, 2000):

$$f_i = \frac{1}{2\pi}\frac{\partial \phi}{\partial t}. \tag{11.22}$$

For the continuous case, a Doppler signal after quadrature sampling and mixing can be combined into an exponential (Jensen, 1996) or complex phasor:

$$v_D(t) = Aexp[-i(2\pi f_0 \delta_D t + \text{constant})], \tag{11.23}$$

which has a familiar instantaneous frequency:

$$f_i = \frac{1}{2\pi}\frac{\partial \phi}{\partial t} = \frac{1}{2\pi}\frac{\partial(2\pi f_0 \delta_D t + \text{constant})}{\partial t} = \delta_D f_0. \tag{11.24}$$

In the general case, more than a single frequency is involved. Equation 11.22 still holds for multiple frequencies but must be approximated for the discrete case needed for CFI.

What is really needed is the mean frequency for N samples. If a number of time samples (N) are taken from repetitive insonifications at the same depth, each sample can be turned into a pair of $I(nT_{PRF})$ and $Q(nT_{PRF})$ through the quadrature process. Note that the I and Q components can be regarded as the Cartesian projections of a phasor or vector with a magnitude and angle. For this set of samples, an instantaneous envelope, $A(nT_{PRF})$, can be determined from the square root of the sum of the squares, as described for an analytic signal in Appendix A. The instantaneous phase can be found from $\phi(n) = \arctan[Q(n)/I(n)]$. First, the continuous case can be found from the definition of phase in terms of the arctangent; second, the derivative of the arctangent (Evans, 1993; Jensen, 1996) gives:

$$\frac{d\phi}{dt} = \frac{I(t)dQ(t)/dt - Q(t)dI(t)/dt}{I^2(t) + Q^2(t)}. \tag{11.25A}$$

A discrete finite difference approximation of this derivative and averaging yields an approximate instantaneous frequency estimator:

$$\bar{f} \approx \frac{1}{2\pi T_{PRF}} \frac{\sum_{n=1}^{N} I(n)Q(n-1) - Q(n)I(n-1)}{\sum_{n=1}^{N} I^2(n) + Q^2(n)}, \tag{11.25B}$$

where here the index n is meant to identify each unique phasor in a time sequence of phasors. A simpler mean frequency estimator is to approximate instantaneous frequency by changes in phase (Brandestini, 1978) from one sample to another, as:

$$\bar{f} \approx \frac{\Delta\phi}{2\pi\Delta T} \frac{1}{2\pi N T_{PRF}} \sum_{n=1}^{N} \left\{\arctan\left[Q(n)/I(n)\right] - \arctan\left[Q(n-1)/I(n-1)\right]\right\}. \tag{11.26}$$

An alternative estimator is the autocorrelator (Kasai, Namekawa, Koyano, & Omoto, 1983). A geometric interpretation of this approach is that the tangent of the difference between phasors,

$$\tan(\phi_n - \phi_{n-1}) = \frac{\sin(\phi_n - \phi_{n-1})}{\cos(\phi_n - \phi_{n-1})} = \frac{\sin\phi_n \cos\phi\phi_{n-1} - \cos\phi_n \sin\phi_{n-1}}{\cos\phi_n \cos\phi_{n-1} - \sin\phi_n \sin\phi_{n-1}}, \tag{11.27}$$

can be used to estimate the mean frequency through a discrete approximation of the sum of these tangents:

$$\bar{f} \approx \frac{1}{2\pi T_{PRF}} \arctan\left[\frac{\sum_{n=1}^{N} I(n)Q(n-1) - Q(n)I(n-1)}{\sum_{n=1}^{N} I(n)I(n-1) + Q(n)Q(n-1)}\right]. \tag{11.28}$$

This formulation can be related to autocorrelation through an alternate definition of mean frequency in terms of the power spectrum, $P(f)$ (Angelsen, 1981; Kasai et al., 1983). Consider first the definition of an autocorrelation function (see Appendix A):

$$R(\tau) = v(\tau) * v^*(-\tau) = \int_{-\infty}^{\infty} v(u)v^*(u-\tau)du, \tag{11.29A}$$

which is related to the power spectrum through a Fourier transform:

$$R(\tau) = \int_{-\infty}^{\infty} V(f)V^*(f)e^{i2\pi f\tau}df = \int_{-\infty}^{\infty} |V(f)|^2 e^{i2\pi f\tau}df, \tag{11.29B}$$

where the power spectrum is $P(f) = |V(f)|^2$. The mean frequency can be obtained from the Fourier transform derivative theorem (see Appendix A), and the autocorrelation function at zero lag, $\tau = 0$, Eqn 11.29B,

$$\bar{f} = \frac{\int_{-\infty}^{\infty} fP(f)df}{\int_{-\infty}^{\infty} P(f)df} = \frac{\int_{-\infty}^{\infty} i2\pi fP(f)df}{i2\pi \int_{-\infty}^{\infty} P(f)df} = \frac{-i}{2\pi}\frac{dR(0)}{dt}. \tag{11.30}$$

This relation is approximated as:

$$\bar{f} = \frac{1}{2\pi}\frac{d\phi}{dt} \approx \frac{\phi(T_{PRF})}{2\pi T_{PRF}}, \tag{11.31}$$

in which ϕ is the phase of the autocorrelation function, since $R(\tau) = |R(\tau)|\ exp[i\phi(\tau)]$. Finally, the autocorrelation function, as defined by Eqn 11.29A, is integrated over N transmits (Kasai et al., 1983):

$$R(T_{PRF}, t) = \int_{t-NT_{PRF}}^{t} v(t)v^*(t - T_{PRF})dt. \tag{11.32A}$$

If $v(t) = I(t) + iQ(t)$, this integral becomes:

$$R(T_{PRF}, t) = \int_{t-NT_{PRF}}^{t} [v_{REAL}(t, T_{PRF}) + iv_{IMAG}(t, T_{PRF})]dt, \tag{11.32B}$$

where

$$v_{REAL}(t, T_{PRF}) = I(t)I(t - T_{PRF}) + Q(t)Q(t - T_{PRF}), \quad \text{and} \tag{11.32C}$$

$$v_{IMAG}(t, T_{PRF}) = I(t - T_{PRF})Q(t) - I(t)Q(t - T_{PRF}) \tag{11.32D}$$

A numerical implementation of Eqn 11.32A is to replace the integral with a sum over N repetitions and associate the terms in Eqns 11.32C and 11.32D with indices, $t = n$, $t - T_{PRF} = n - 1$. Finally, the phase can be determined from the arctangent of the summed imaginary terms over the real terms, and the mean frequency is found from Eqn 11.31, which gives the result of Eqn 11.28.

Evans (1993) has reviewed the three methods, and the instantaneous frequency method, Eqn 11.25B, appears to be the least accurate. The autocorrelator is the most robust, and it or variants of it are probably the most widely used (Kimme-Smith et al., 1989). The phase detector (also called an instantaneous frequency detector in the literature), Eqn 11.26, is intermediate in its performance.

11.7.3 Time-Domain-Based Estimators

Time domain cross-correlation approaches based on the Doppler time shift have been proposed (Embree & O'Brien, 1985, 1990; Bonnefous & Pesque, 1986). These Doppler methods could exceed the aliasing limits of the phase-based algorithms to a limited extent, and they could use shorter pulses to improve axial resolution. Although the shorter pulse trains would have the same compromise in sensitivity as the earlier methods, they would have reduced spectral spread.

The basic principle is that the position of a cross-correlation gives the measurement of the Doppler time shift by red blood cell scatterers (Bonnefous et al., 1986):

$$R_{cn}(\tau, t) = \int_t^{T_g + t} v_n(t')v_{n+1}(t' + \tau)dt', \tag{11.33A}$$

but the successive echo is delayed by the Doppler time shift (τ_D):

$$v_{n+1}(t) = v_n(t - \tau_D), \tag{11.33B}$$

so that:

$$R_{cn}(\tau, t) = \int_t^{T_{g+t}} v_n(t')v_n(t' + \tau - \tau_D)dt' = R_n(\tau - \tau_D), \tag{11.33C}$$

where R_n is an autocorrelation function that is maximum when $\tau = \tau_D$. When the time between repetitions is T_{PRF}, the Doppler velocity from the Doppler time-shift equation is:

$$v_D = \frac{c_0 \tau_D}{2T_{PRF}\cos\theta}. \tag{11.33D}$$

Bonnefous et al. (1986) described how this approach can be used as a color flow velocity estimator. They demonstrated that higher velocities can be detected with this method without aliasing. The blood scatterers must not have moved so much that there will be insufficient overlap to obtain a high cross-correlation between consecutive transmits; this condition is usually met in practice.

Hein and O'Brien (1993c) provided a review of cross-correlation methods for Doppler detection of blood flow and tissue motion. They indicated that the phase-based Doppler detection methods are biased by the center frequency.

11.7.4 Implementations of Color Flow Imaging

All of the estimators described have qualities of being fast, robust, and efficient; consequently, they have been implemented in hardware and digital signal processors (DSPs). The initial signal processing is similar to that used to create I and Q paths for PW Doppler, except that wall filters follow analog-to-digital (A/D) conversion. The wall filters can be feedback recursive filters of the moving target indicator (Magnin, 1987) or the delay line canceller type (Evans, 1993). After filtering, the signals enter the mean frequency estimator and turbulence estimators. The results of these calculations then enter a display encoder and digital scan conversion. The time domain method differs substantially from the phase-based methods in that quadrature sampling is not necessary, so that after wall filtering, the signals are processed by a cross-correlator.

All methods undergo a color mapping scheme that can vary among manufacturers, so only the basic concepts can be dealt with here (Magnin, 1987). A color image is overlaid on a standard gray-scale image. The colors chosen are not the actual colors of blood but represent blood flow velocity and direction. Colors are assigned to the direction of flow relative to the transducer; for example, with red for flow toward the transducer and blue for flow away from it. Green, another primary color, can be added to indicate turbulence. Either the hue of the color is increased as the velocity increases or, alternatively, the intensity is increased. In Figure 11.15, the red/blue scheme is shown with increasing intensity.

Frame rate is always at a premium, and it depends on the total number of vectors or transmit events and the scan depths. In multiple modes, and color flow always includes a gray-scale B-scan image, it is possible to have not only differently shaped pulses but also different scan depths (Szabo, Melton, & Hempstead, 1988). To catch fast-moving flow, frame rate can be increased by giving the user the option of reducing the size of the region of interest for CFI (as shown in the inset of Figure 11.13). Figure 11.15 shows a triplex image CFI and a gray-scale and PW Doppler. The ability to display several modes at once (Barber et al., 1974) is very useful clinically, especially for the placement of Doppler lines, but this also increases frame rate and both processing and line-sequencing complexity.

While there is no doubt of CFI's usefulness, its limitations must also be kept in mind. First, mean blood flow velocity is estimated on the basis of a few time samples; therefore, the values obtained will not be as accurate as PW and CW Doppler measurements based on longer dwell times (length of time during which a transducer is held at the same position), many more sample points, and more precise FFT algorithms. Second, the velocity values derived from CFI have an implicit cosine θ variation with no correction for this part of the Doppler effect. As an example, consider a sector scan in which the middle vector line is perpendicular to a vessel with blood flowing left to right. Under certain conditions, a CFI of this situation will display blood as flowing left to right on the left side of the image, as stopping at dead center ($\cos 90° = 0$), and as reversing flow on the right half of the image. This kind of geometry is avoided in clinical practice, with the flow vectors always at some angle to the vessel. A third cautionary observation is that aliasing can occur (the mapping of high velocities into lower ones); these situations are often unusual enough to be noticed. Fourth, changes in flow velocity can occur out of the imaging plane and be mapped into the field of view. Fifth, "flash artifacts" can occur (the incorrect mapping of moving blood onto tissue regions). This effect may be caused by tissue movement or by an inappropriate setting or limitation of the wall filter.

11.7.5 Power Doppler and Other Variants of Color Flow Imaging

A variation of CFI is called power Doppler, or ultrasound angiography (Babock, Patriquin, La Fortune, & Dauzal, 1996; Bude & Rubin, 1996; Chen, Fowlkes, Carson, Rubin, & Adler, 1996; Rubin, Bude, Carson, Bree, & Adler, 1994; Rubin et al., 1995), and is a color representation of Doppler amplitude. An example of this modality is given by Figure 11.22, which shows an image of a renal transplant. Here, Doppler intensity is shown as a change in the color intensity of red to yellow. Curious features of power Doppler are an absence of information about velocity direction and dependence on angle. If there is a sufficient Doppler signal, usually the presence of fluctuations of the backscatter with angle will be enough to show flow even at 90°. What is being displayed is the integral of power density, or:

$$\int_{-\infty}^{\infty} P(f)df = \int_{-\infty}^{\infty} |V(f)|^2 df \approx \sum_{n=1}^{N} I^2(n) + Q^2(n). \tag{11.34}$$

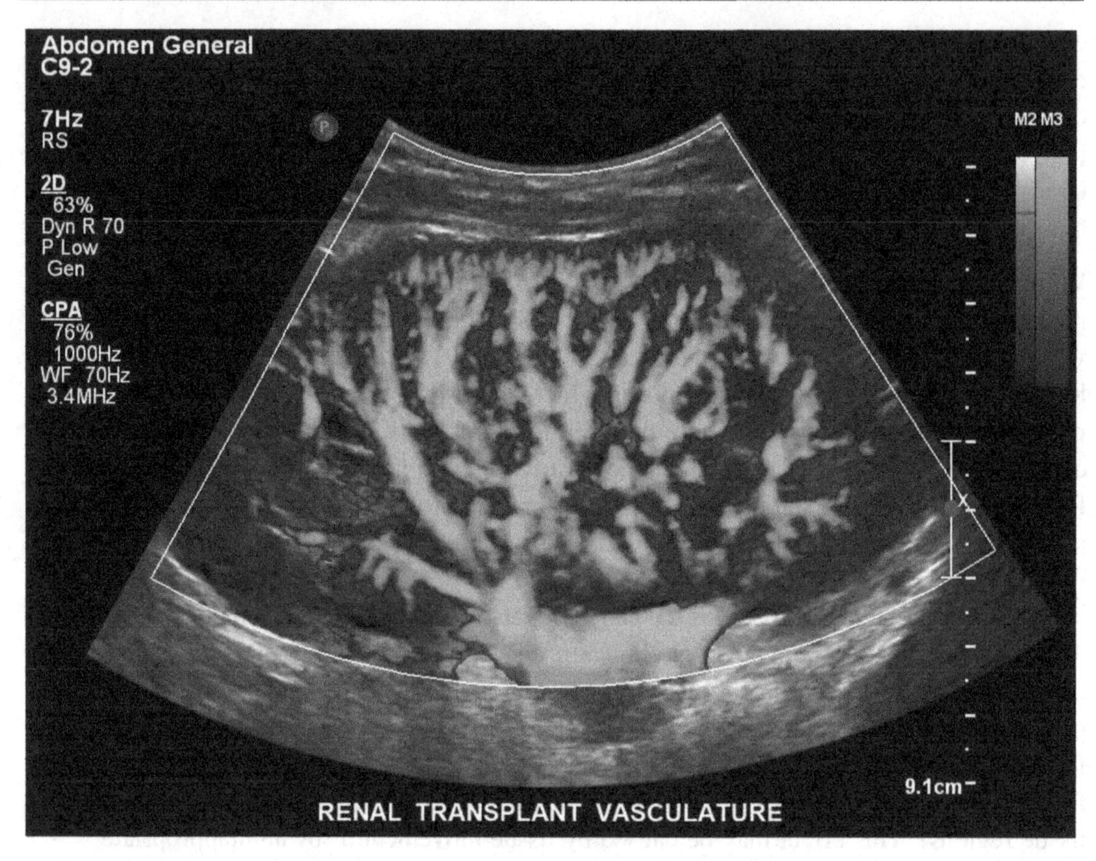

Figure 11.22

Power Doppler image of the arterial tree in a renal transplant. Image taken on an EPIQ7 Philips system. The reader is referred to the Web version of the book to see the figure in color. *Courtesy of Philips Healthcare.*

As apparent from Figure 11.22, this modality has more capability than standard CFI to show flow in smaller vessels. The two modes are compared in Figure 11.23. In the power Doppler mode (Burns et al., 1994; Frinking, Boukaz, Kirkhorn, Ten Cate, & De Jong, 2000; Powers et al., 1997), the power of the Doppler signal without phase information is displayed as a range of colors instead of spectral data. Noise arriving in the Doppler receiver at high gains is mapped to a small band of color, in contrast to CFI, in which noise is spread across the spectrum as many colors. This containment of noise in the power Doppler mapping contributes to an effective increase in the dynamic range displayed. In CFI, moving tissue can overlap blood signals and appear as flash artifacts, but in power Doppler, this overlap appears as the same power amplitudes.

In the power Doppler mode, the total power is more dependent on flow amplitude and less on random fluctuations, which cause phase interference effects in standard CFI. As a result,

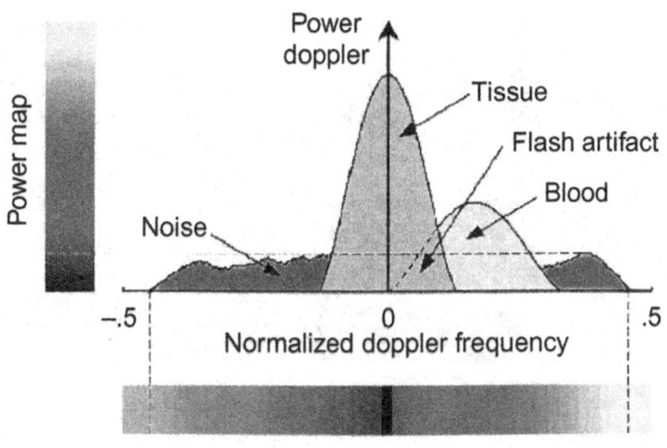

Figure 11.23
Power Doppler, a mapping of power to a continuous color range, is compared to color flow imaging. The direction and Doppler velocity are encoded as a dual display in which colors represent velocities in terms of the Doppler spectrum and also the direction of flow to and from the transducer. The reader is referred to the Web version of the book to see the figure in color. *From Frinking et al. (2000); reprinted with permission from the World Federation of Ultrasound in Medicine and Biology.*

the power is integrated in amplitude across all the red blood cells in the beam without regard to phase. The increase in detection of moving cells is displayed as an increasingly paler color. In this mode, power is less affected by angular cosine effects; this contributes to sensitivity improvement. Even at right angles, some signal is present because of the spreading effects (discussed in Section 11.3). Another advantage is that aliasing at high velocities has little effect on the displayed power. These factors contribute to higher sensitivity for the depiction of small vessels, as evident in Figure 11.22.

For examination of flow at one location, a different mode, called color M-mode, can be used (as illustrated by Figure 11.24). In this case, a color flow vector line is updated in fast time and displayed in slow time as with a standard M-mode format. Note how the passage of blood through a valve is depicted in detail.

11.7.6 Previous Developments

Work is ongoing to improve Doppler methodology, primarily to overcome the limitations and extend its usefulness. Routh (1996) has reviewed Doppler imaging developments, and Ferrara and Deangelis (1997) have reviewed CFI comprehensively. Angle-independent CFI algorithms are reviewed by Ramamurthy and Trahey (1991) and Routh (1996). An active area is the development of new velocity estimators such as correlation (Hein & O'Brien, 1993a, 1993b) and wideband maximum likelihood (Ferrara & Algazi, 1991a, 1991b) as well as ways of eliminating or correcting for aliasing. One way of determining the true corrected velocity and its vector direction is through the use of spatially separated Doppler

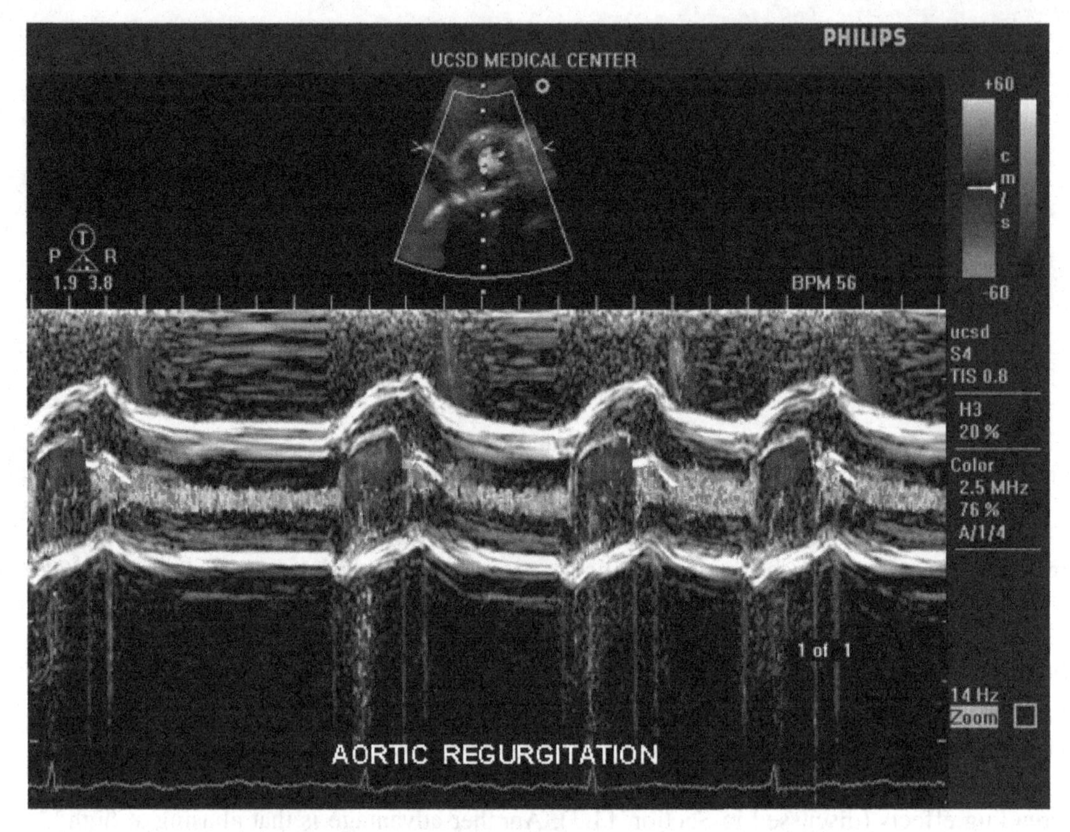

Figure 11.24
Color M-mode depiction of a leaky tricuspid valve. The reader is referred to the Web version of the book to see the figure in color. *Courtesy of Philips Healthcare.*

transducers to determine the location and angle to the target. Presenting all this vector information in an image is a challenge in itself. Similar to other 3D imaging acquisition methods, 3D CFI images have been made. Interest in measuring flow in the microvasculature is growing and requires special methods, including algorithms, to detect extremely slow flow, higher frequencies, and contrast agents. Contrast agents, which are microbubbles that act as highly acoustically reflective blood tracers to enhance sensitivity, will be discussed in Chapter 14. Doppler has been applied to sonoelasticity (Lerner, Huang, & Parker, 1990), to sensing streaming in breast cysts (Nightingale, Kornguth, Walker, Mc Dermott, & Trahey, 1995), and detecting wall motion (Hein & O'Brien, 1993c).

11.8 Non-Doppler Visualization of Blood Flow

As discussed earlier, scattering from red blood cells is very weak. With the development of imaging systems in the mid and late 1990s with large dynamic ranges and better signal

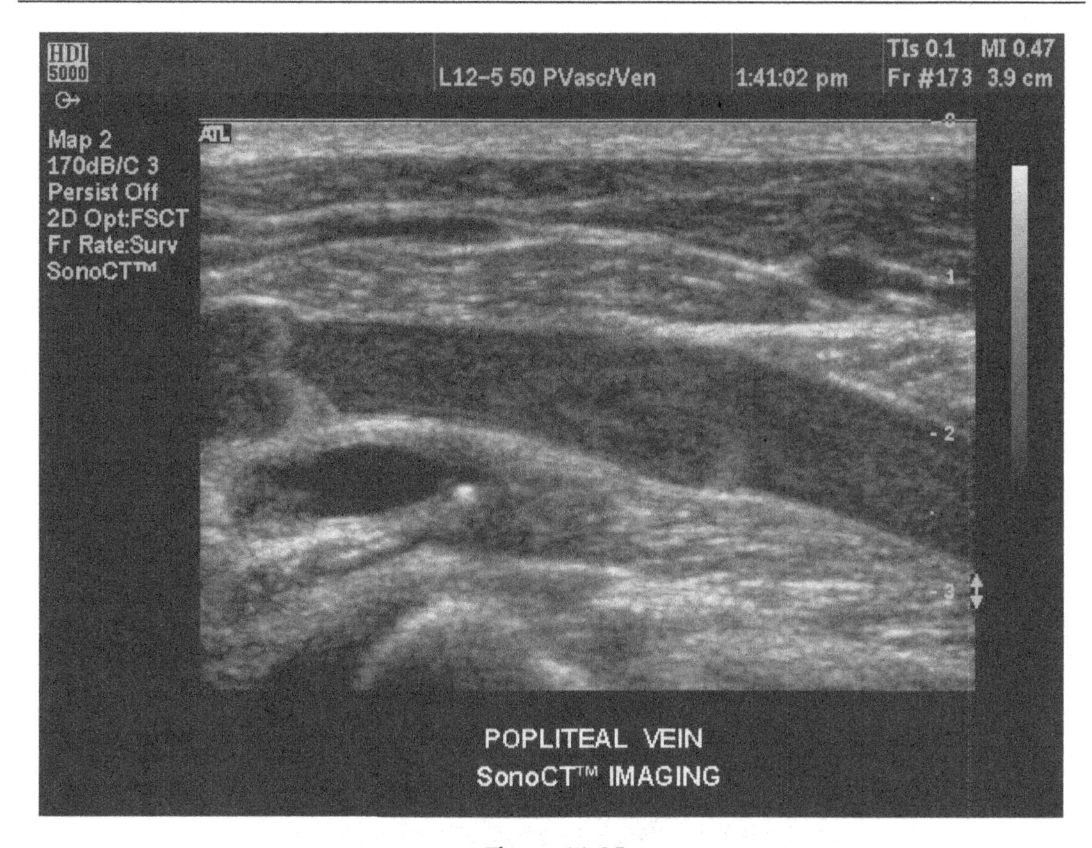

Figure 11.25
Visualization of blood flow in the popliteal vein at high frequencies at 12 MHz with a linear array.
Courtesy of Philips Healthcare.

processing capabilities, it became possible to visualize blood flow more directly in B-mode. One means of achieving the direct flow is to employ higher frequencies at which the backscattering coefficient of blood becomes more comparable to that of tissue (recall Figure 11.8). An example of this visualization is shown at 12 MHz for a high-frequency linear array in Figure 11.25.

Far better visualization of blood flow, even at low frequencies, can be obtained by advanced signal-processing methods. B-mode blood flow imaging, called "B-flow" (Chiao, Mo, Hall, Miller, & Thomenius, 2000) and demonstrated in Figure 11.26, provides high-frame-rate blood flow visualization through a two-step process.

First, the Doppler dilemma of short pulse length (good resolution) versus high sensitivity (long pulse length) is solved by using long encoded pulses that are later decoded on reception. As discussed in Chapter 10, coded excitation can be applied to achieve high pulse compression and sensitivity on reception through the selection of

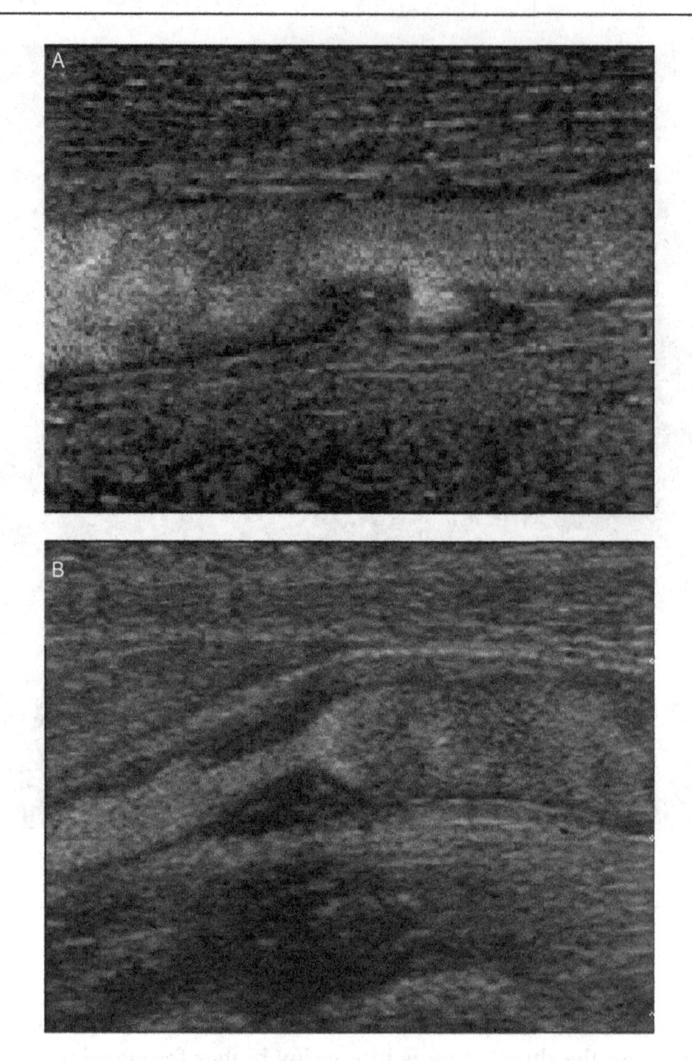

Figure 11.26
B-mode blood flow imaging. (A) Of an ulcerated plaque; (B) of a carotid artery stenosis.
From Chiao et al. (2000), IEEE.

codes with low time sidelobes (Golay codes). Representing blood scattering and tissue in the same image, even after improved sensitivity, requires a second step called "tissue equalization." A filter devised to discriminate between blood movement and more stationary tissue assigns an image brightness based on the decorrelation between successive echoes along each vector direction, frame to frame. The equalization filter is illustrated by Figure 11.27. From the images, the flow (especially that which is more turbulent owing to factors discussed in Section 11.3) is enhanced and the surrounding tissue is more muted.

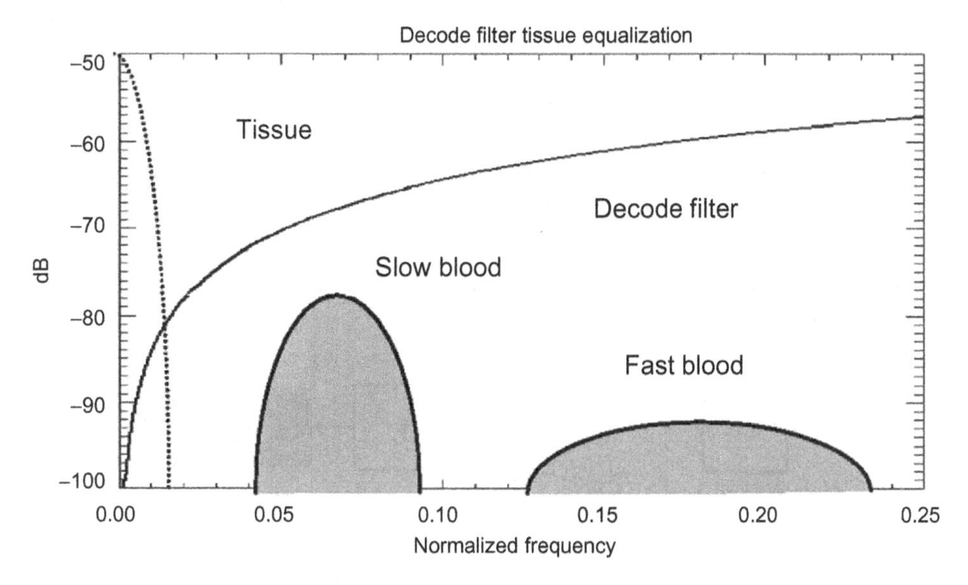

Figure 11.27
Equalization filter curve based on decoded echoes from successive transmit events along the same vector direction. Also shown are signal levels for slowly moving tissue, as well as slow and fast blood flow. *From Chiao et al. (2000), IEEE.*

11.9 Doppler Revisited

11.9.1 Doppler Methods Reviewed

Before examining new methods, the characteristics of Doppler modes will be reviewed in more detail. Doppler modes are always combined with other modes. The overall frame rate of imaging systems is determined by the timing and sequencing and type of modes selected. If each mode can be assigned an index j and is composed a number of lines, N_j, of length R_j, then the total time is related to the sum of the different modes:

$$T_t = \frac{2}{c}\sum_{j=1}^{J_{\max}}(N_j R_j + \tau_j c/2), \tag{11.35A}$$

(in which τ_j is dead time for each mode and is set to 0 here) so that the frame rate is:

$$\mathrm{FR} = 1/T_t. \tag{11.35B}$$

The simplest case is B-mode imaging; the round trip-time for one line is $T_{RT} = 2R/c$. A typical number for N is 128. For a depth of 25 mm, this leads to a frame rate of 241 frames/s.

The next case is pulsed-wave Doppler, as illustrated by Figure 11.28. The depth here is R_{PWD}, the position of the Doppler gate at 20 mm. Unlike the B-mode, which has an excitation pulse

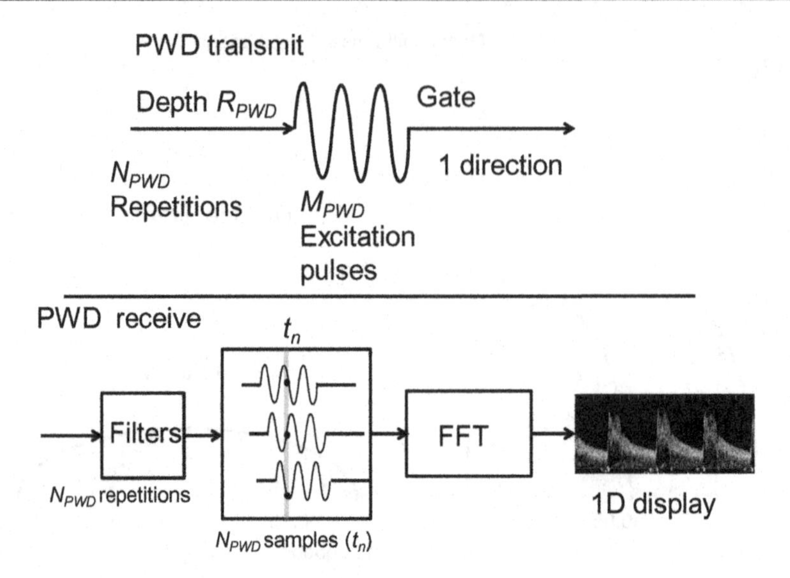

Figure 11.28
Conceptual diagram of timing and spatial sequences in pulsed-wave Doppler. (Top) parameters for pulse transmission; (bottom) parameters on reception leading to output display. *Image courtesy of Philips Healthcare.*

of one, Doppler mode has a few excitations with a typical number $M_{PWD} = 3$. For 5 MHz and from Eqn 11.12A, the gate length is 0.6 μs or about 0.9 mm. The equivalent one-way distance is 20 mm + 0.9/2 mm = 20.45 mm. Because $T_{RT} = 26$ μs, and Eqn 11.20C, the minimum pulse repetition interval is 26.6 μs, or a maximum PRF of 37.6 kHz. If flow can be considered to be stationary for 10 ms, then the number of repetitions per second is $N_{PWD} = (10 \times {}^{10-3}) \times (37.6 \times 10^3) = 376$. For PW Doppler (PWD) alone, the frame rate is 98/s. From Eqn 11.20I, this implies detecting a maximum flow velocity of 2.9 m/s. For a combined imaging and PWD mode, the frame rate drops to 71 frames/s. The sequencing of PWD lines and image lines can vary with the circumstance and the PRFs desired. The lines from different modes can be interleaved or the PWD lines can be clustered together in a sequence followed by an imaging sequence. The consequences of these choices affects simultaneity and fidelity of the Doppler measurement, a topic that will be revisited later.

For the color flow imaging mode shown in Figure 11.29, fewer samples are available because the color flow image area has to be filled with data. A typical number of repeated samples is $N_{CFI} = 8$. Fewer points means that special algorithms such as autoregression and others discussed in Sections 11.7.2 through 11.7.5 are needed to make the best use of data. The output of these estimators is a color-coded point at z_n on line q in the color flow region of the image. (Actually because time windows are used, several points in time are involved, but on the scale of the image, they appear as a pixel, $I\,(x,y,t_n)$.) Each ensemble of N_{CFI} points is moved in time along the same direction identified by q in Figure 11.29. Then the

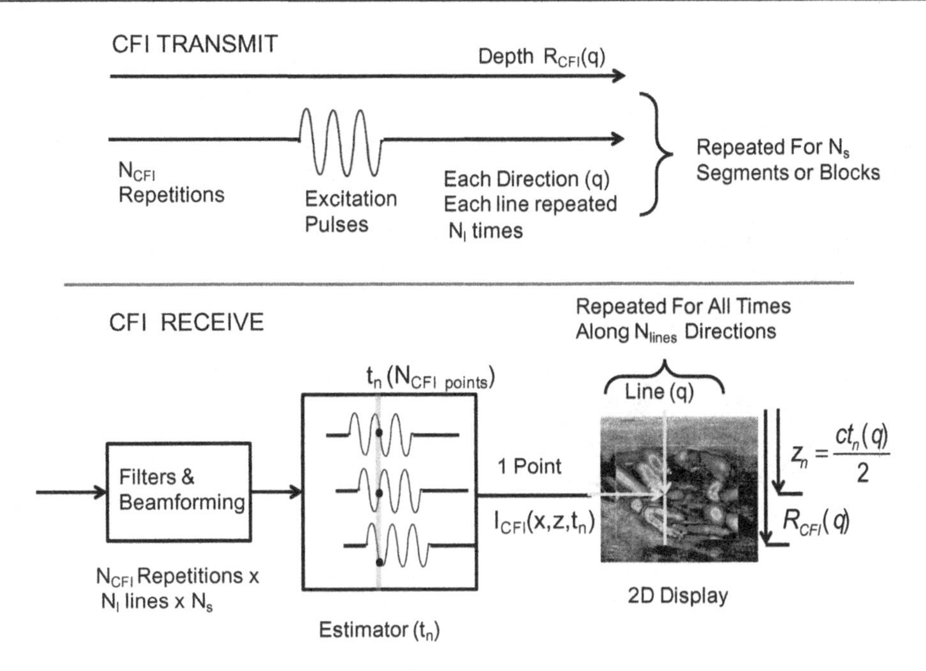

Figure 11.29
Conceptual diagram of timing and spatial sequences in color flow imaging. (Top) parameters for pulse transmission; (bottom) parameters on reception leading to output display. *Image courtesy of Philips Healthcare.*

next position, $q + 1$, is processed to the maximum color flow depth, $R_{CFI}(q)$, the same way. A frame rate can be calculated for an example that considers CFI as the only mode. Using the parameters from the array in the PW Doppler example and $R_{CFI} = 25$ mm, we find that, this time, if the color flow image fills the whole area available with lines and/or blocks of lines, $N_{CFI}N_{lines} = 8 \times 16 = 128$. These numbers give a frame rate of 30 frames/s. With B-mode imaging added in, the frame rate drops to 27 frames/s. This example has a very shallow depth. If the depth is increased to 60 mm, the frame rate is 13 frames/s, and at 90 mm the frame rate drops to 8 frames/s.

11.9.2 Doppler Methods Re-Examined

Doppler measurements and imaging are extremely useful in assessing blood flow; however, they provide, essentially, a one-dimensional view of flow. Also, the Doppler effect has a strong angular dependence (cos θ) as illustrated by Figure 11.1. For color flow imaging, a steering angle is selected; see Figure 10.8A for an example. Imaging systems do not have the capability to anticipate the angle between the insonifying beam and the actual attitude of the vessel; therefore the angle correction is unlikely to be accurate. Jensen (Evans, Jensen, & Nielsen, 2011; Jensen et al., 2011) has pointed out that laminar flow does not

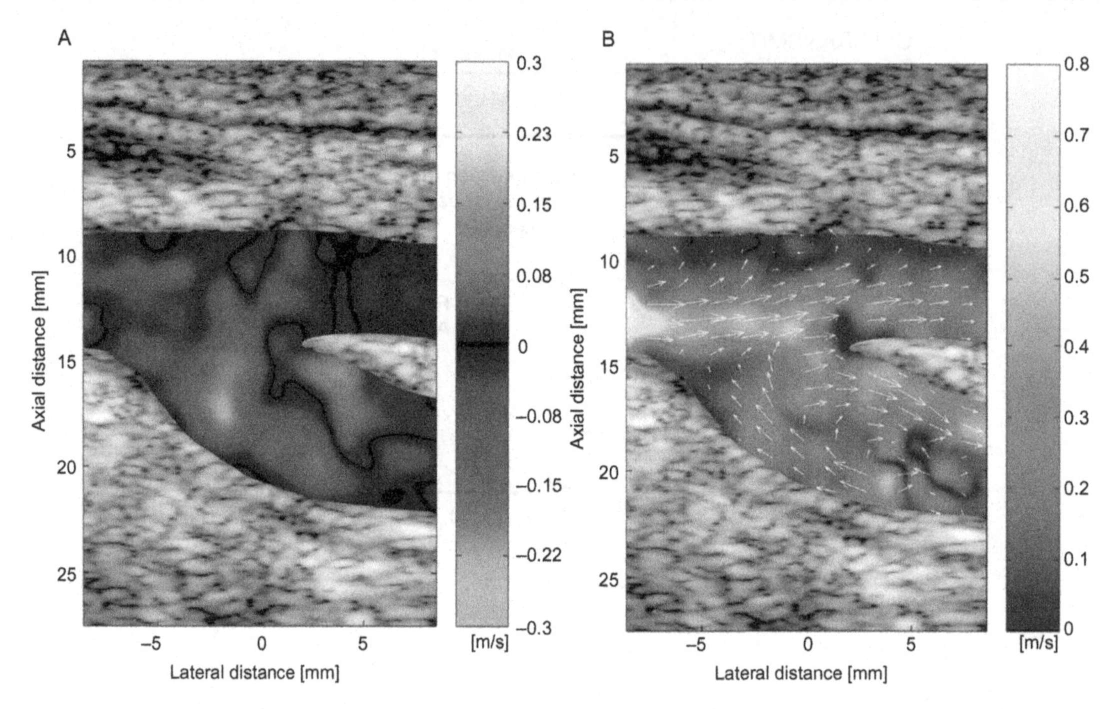

Figure 11.30

Two images depicting flow in the bifurcation of the carotid artery shortly after the peak systole. (A) The one-dimensional axial flow (towards or away from the transducer) velocity. Note, this is somewhat like color flow imaging but here flow at 90° is shown, a view impossible to obtain by conventional CFI techniques. (B) Vector flow image of the same flow at the same time instance produced by the transverse oscillation method. The reader is referred to the Web version of the book to see the figure in color. *From Udesen, Nielsen, Nielsen, and Jensen (2007); reprinted with permission from the World Federation of Ultrasound in Medicine and Biology.*

occur in the general case as shown in Figure 11.30A; instead, complex flows occur naturally at bifurcations, stenoses, and other flow anomalies of clinical interest. Even though the general flow and time dependence are depicted in this example of a healthy volunteer right after peak systole, the actual blood flow velocity here is around 0.8–1 m/s, yet flows between −0.3 and 0.3 m/s are displayed. Most available Doppler methods only display flow from a one-dimensional perspective along the beam axis; lateral flow information is lost. The displayed velocity along the beam depends on angle and can be found from the Doppler equation (11.4D):

$$v_{beam} = |v|\cos\theta = \frac{f_D c_0}{f_0 2} \tag{11.36}$$

(Evans et al., 2011). In addition to the shortcomings of Doppler mentioned at the end of Section 11.7.4, another is the fact that Doppler imaging information is derived from only a few samples and, thus, may have a large variance. Another awkward drawback is that the normal angle usually the best for imaging and backscatter, $\theta = 90°$, is the worst for Doppler. Simultaneity is another problem. Depending on the flow rate, the sequential interrogation of the flow, line by line, may introduce a timing bias from the first to last line and, therefore, fail to reproduce the actual flow parameters faithfully. For all these reasons, work has been directed at methods to capture the two- and three-dimensional nature of flow, collectively called "vector Doppler," and described in the next section, as anticipated by Figure 11.30B.

11.10 Vector Doppler

11.10.1 Introduction

Not surprisingly, several approaches that worked successfully for improving B-mode imaging have reappeared as parts of vector Doppler methods. These methods, which include synthetic aperture and plane-wave compounding, will be emphasized as well as a another one, transverse oscillation. Previous methods for obtaining vector Doppler, which are more limited in their applicability, are reviewed in Evans et al. (2011); Jensen et al. (2011); Mace et al., (2013); and Udesen and Jensen (2006). Improved plane-wave compounding methods have led to functional ultrasound imaging, discussed in the last section.

11.10.2 Transverse Oscillation Method

The transverse oscillation (TO) method (Jensen and Munk, 1998; Jensen, 2001; Udesen & Jensen, 2006) is of interest because it offers a way of implementing vector Doppler on nearly conventional imaging systems and has been implemented on a commercial scanner (Hemmsen et al., 2011; Pedersen et al., 2011). The transducer can be a 1D array and existing autocorrelation algorithms can be employed in a modified processing scheme. Even though normal transmit is used, a novel receive beamforming scheme that has a 90° phase shift between two parallel receive beams creates a spatial quadrature signal. Finally, a temporal Hilbert transform on both beams produces four received signals from which axial and lateral velocities are derived to calculate a 2D velocity vector field.

What is missing in conventional Doppler is a feature that could be used to track transverse motion. This fact is apparent from Figure 13.10, which shows a measured ultrasound pulse propagating left to right along the z axis. The positive and negative time pulse excursions along this axis are affected by the scattering of blood and may result in a Doppler shift. Along the vertical direction, the y axis, the wavefronts are relatively uniform within the pulse burst and therefore do not provide distinct sample points to track shifts in the transverse or lateral direction.

From our personal experience, we can detect the lateral position and rate of movement of an approaching train because of our binaural hearing. Our ears detect an interaural time difference because of their different positions (and therefore infer the lateral position of a sound source) and also sense the distance of the approaching train from an increase in frequency caused by the Doppler effect.

Jensen (Jensen, 2001; Udesen & Jensen, 2006) devised a way to create equivalent pairs of ears on an array, two separated receivers to sense the Doppler-shifted signals in the lateral direction. These "ears" are shown at the top of Figure 11.31 and an explanation follows below. As a simple example, recall that the far-field beam pattern of a line source with a rectangular amplitude is determined from its plus *i* Fourier transform and is a sinc function, Eqn 6.9A. If one would want to create a rectangular far field of width W_x at depth z as illustrated by the field in the bottom of Figure 11.31 , then because of an inverse plus *i* Fourier transform relation between the field and source function, a scaled sinc function would be required:

$$A(x_0) = \frac{e^{-i\pi/4}}{\sqrt{A_0}} \int_{-\infty}^{\infty} [p_x(x)] e^{i2\pi\Gamma x_0 x} dx, \tag{11.37A}$$

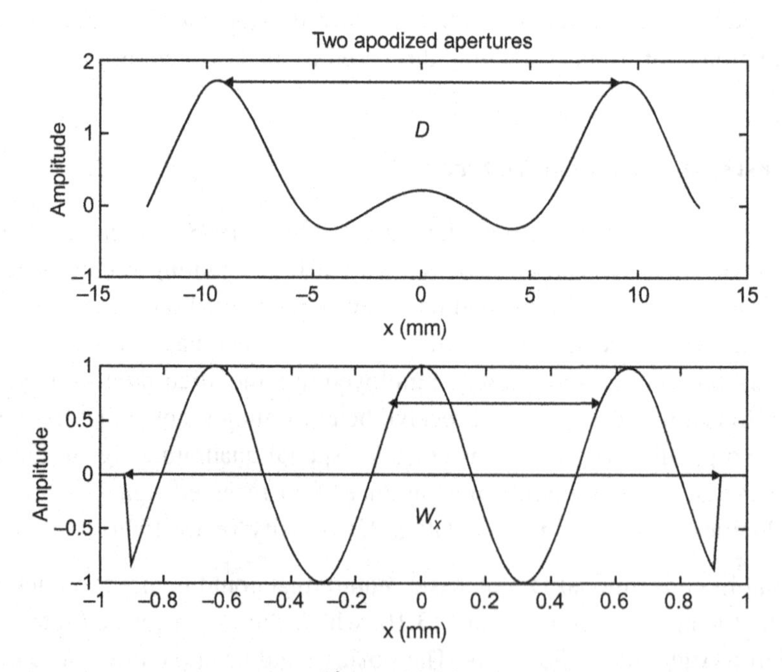

Figure 11.31
Simulation of the modulated transverse field for detecting transverse flow in the TO method. (Top) two 7-MHz receive apertures (ears) of 32 elements each are separated by $D = 11$ mm; (bottom) the resulting transverse field has a period of $\lambda_x = 0.64$ mm at a depth of 27 mm. Array period = 0.208 mm.

$$A(x_0) = \int_{-\infty}^{\infty} \left[\Pi(\frac{x}{W_x}) \right] e^{i2\pi \Gamma x_0 x} dx, \quad \text{and} \tag{11.37B}$$

$$A(x_0) = W_x \text{sinc}\left(\frac{W_x x_0}{\lambda_z z} \right). \tag{11.37C}$$

In order to create the modulated transverse field of width W_x of Figure 11.31, two distinct apodized apertures ("ears") whose centers are separated by a distance D and the principle of the cosine Fourier transform pair from Table A.2 can be used:

$$\{A(x_0) * [\delta(x_0 - D/2) + \delta(x_0 + D/2)]\} = F_i^{-1}\left[2\cos(2\pi x D/2)a(x)\right], \tag{11.38A}$$

then from Eqns 11.36C and 11.37,

$$R_x(x_0) = \frac{W_x\sqrt{A_0}}{2} e^{i\pi/4} \left\{ \text{sinc}\left[W_x\left(\frac{x_0}{\lambda_z z} + \frac{2}{D} \right) \right] + \text{sinc}\left[W_x\left(\frac{x_0}{\lambda_z z} - \frac{2}{D} \right) \right] \right\}, \quad \text{and} \tag{11.38B}$$

$$p_x(x) = \frac{e^{-i\pi/4}}{\sqrt{A_0}} F_{+i}^{1}[R_x(x_0)] = \Pi\left(\frac{x}{W_x} \right) \cos\left[2\pi x/(D/2)\right]. \tag{11.38C}$$

These results are shown in Figure 11.31. By setting the arguments of each sinc function to zero in Eqn 11.38B, the distance between the peaks can be defined as:

$$D = \frac{2z\lambda_z}{\lambda_x}, \tag{11.39}$$

where the periodicity along x is called the transverse wavelength λ_x (Udesen & Jensen, 2006). Note, this equation can now be solved for λ_x in terms of known variables. Just as Doppler systems use the concept of the analytic signal composed of I (real) and Q (quadrature) signals to determine directionality, the equivalent of quadrature can be created by parallel beamforming operations offset by $\lambda_x/4$ (Udesen & Jensen, 2006), as shown by the two different patterns in Figure 11.32. For practical implementation, the transverse oscillation method employs the usual transmit beam, but by scaling this value of D on receive dynamic focus at each depth, a constant value of λ_x can be maintained approximately with depth, as illustrated by the simulation in Figure 11.32. In addition, there is a conventional beamforming operation to capture the velocity along the z axis.

TO example: In order to calculate the curves in Figure 11.31, a 7-MHz linear array with 128 elements, spaced at a pitch of 0.208 mm, was assumed. For a depth of 27 mm and a transverse wavelength $\lambda_x = 0.64$ mm, an apodized aperture center-to-center separation distance can be found from Eqn 11.29 as $D = 18.6$ mm. Each physical aperture has a width of 32 elements or $x_{32} = 6.7$ mm. For the main lobe of a sinc function to fit into this width, find

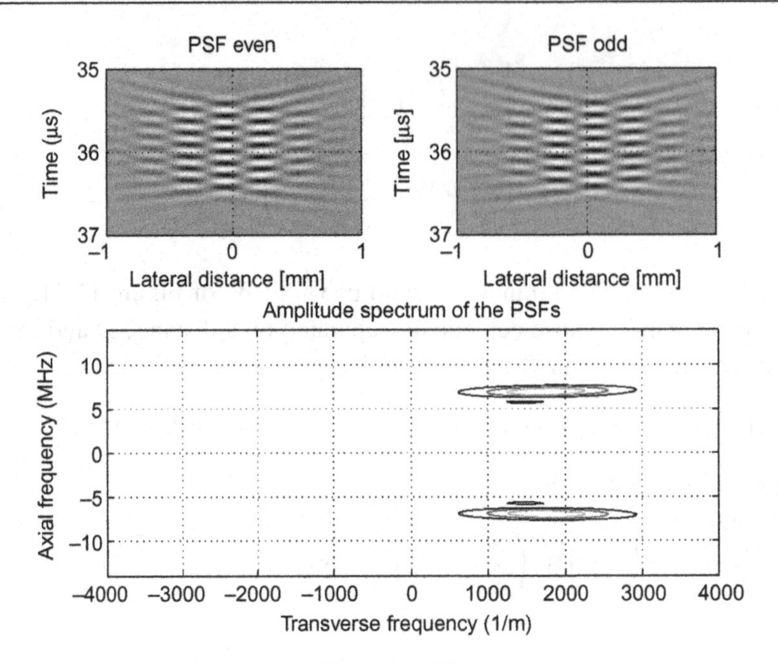

Figure 11.32

Double-oscillating PSFs (point spread functions) and their spectra. The two double oscillating PSFs (top) and their corresponding spectra (bottom) created with the Field II program for a point at 27 mm using a linear-array transducer with a center frequency of 7 MHz. The amplitude spectrum is calculated by taking the 2D Fourier transform of PSFeven $+j$ PSFodd. Note that the PSFs are 90° phase shifted with respect to each other and that the amplitude spectrum, therefore, is one sided. *From Udesen and Jensen (2006), IEEE.*

the distance to a null by setting the argument of Eqn 11.36A to 1, so the effective scaled aperture W_x needed for Eqn 11.38A is related to half this distance, or $W_x = \lambda_z z/(x_{32}/2)$. This effective aperture is 1.8 mm, which is also the width of the oscillation field in Figure 11.31B with a period of λ_x. The complete pair of apertures according to Eqn 11.38A is plotted in Figure 11.31A. The sinc patterns are infinite in extent, so a practical alternative used in the implementation was Hanning apodization instead of sinc functions.

We return to the development of the TO method. To distinguish the two directions, phase factors along the x and z directions are determined by the Doppler shift along z and the transverse wavelength along x,

$$\theta_z(n) = 2\pi\left(\frac{2v_z}{\lambda}\right)\frac{n}{f_{PRF}}, \quad \text{and} \tag{11.40A}$$

$$\theta_x(n) = 2\pi v_x \frac{1}{\lambda_x}\frac{n}{f_{PRF}}. \tag{11.40B}$$

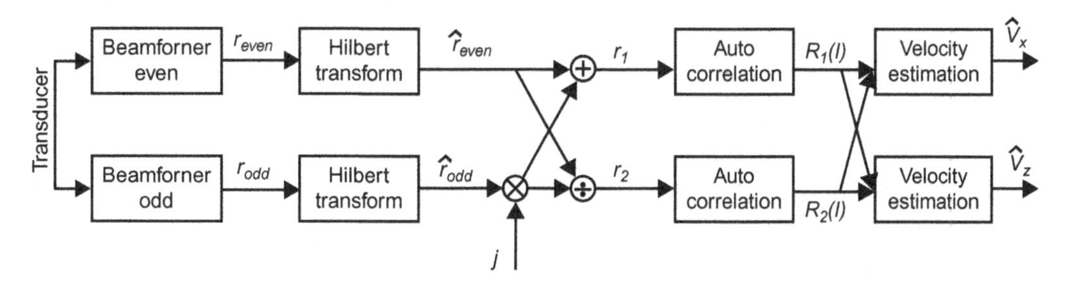

Figure 11.33
Block diagram showing data flow from transducer (Left) to estimated velocities (Right). For simplicity, the echo canceling and matched filtration is not shown. The box "Hilbert transform" represents Eqns 11.43B and D, the box "Auto correlation" represents Eqns 11.44A and B, and the box "Velocity estimator" represents Eqns 11.46A and B. *From Udesen and Jensen (2006), IEEE.*

The signal-processing scheme is shown in Figure 11.33. The even and odd signals are created by parallel beamformers for signals having an index n (Udesen & Jensen, 2006; Jensen et al., 2006):

$$r_{even}(n) = \cos(\theta_x(n))\cos(\theta_z(n)), \quad \text{and} \tag{11.41A}$$

$$r_{odd}(n) = \sin(\theta_x(n))\cos(\theta_z(n)), \tag{11.41B}$$

where analytic signals along the propagation axis are created by usual Hilbert transformation:

$$as_{even}(n) = \cos(\theta_x(n))e^{i\theta z(n)}, \quad \text{and} \tag{11.42A}$$

$$as_{odd}(n) = \sin(\theta_x(n))e^{i\theta z(n)}. \tag{11.42B}$$

These can now be restated in double quadrature final form as the equations:

$$r_1(n) = as_{even}(n) + ias_{odd}(n), \tag{11.43A}$$

$$r_1(n) = e^{i[\theta_z(n)+\theta_x(n)]} = e^{i\theta_1(n)}, \tag{11.43B}$$

$$r_2(n) = as_{even}(n) - ias_{odd}(n), \quad \text{and} \tag{11.43C}$$

$$r_2(n) = e^{i[\theta_z(n)-\theta_x(n)]} = e^{i\theta_2(n)}. \tag{11.43D}$$

Similar to the autocorrelation schemes for color flow imaging (see Eqn 11.28), complex autocorrelation functions can be constructed at lag 1:

$$R_1(1) = \frac{1}{N-1}\sum_{n=0}^{N-1} r_1^*(n)r_1(n+1), \quad \text{and} \tag{11.44A}$$

$$R_2(1) = \frac{1}{N-1}\sum_{n=0}^{N-1} r_2^*(n)r_2(n+1). \tag{11.44B}$$

The change in phases $\Delta\theta_1(n)$ and $\Delta\theta_2(n)$can be found from the real and imaginary parts of R_1 and R_2; then, using the relations:

$$\Delta\theta_z(n) = \frac{4\pi f_0 v_z}{c f_{PRF}}, \quad \text{and} \tag{11.45A}$$

$$\Delta\theta_x(n) = \frac{2\pi f_0 v_x}{\lambda_x f_{PRF}}, \tag{11.45B}$$

and the definitions of θ_1 and θ_2 from Eqns 11.43B and 11.43D, the above equations (11.45A and 11.45B), can be solved to find expressions for v_x and v_z in terms of the autocorrelation functions R_1 and R_2:

$$v_z = \frac{c f_{PRF}}{8\pi f_0}[\Delta\theta_1(n) + \Delta\theta_2(n)], \quad \text{and} \tag{11.46A}$$

$$v_x = \frac{\lambda_x f_{PRF}}{4\pi}[\Delta\theta_1(n) - \Delta\theta_2(n)]. \tag{11.46B}$$

The block diagram showing the processing steps for determining the two velocities can be found in Figure 11.33. In practice, apodization is used for the receive aperture, such as a Hanning weighting function, and is scaled during dynamic receive focusing at each depth. More implementation details can be found in Jensen et al., (2006).

Early clinical results using the TO method obtained from the synthetic aperture research system RASMUS are shown in Figures 11.30A and B (Udesen et al., 2007). Compared to the axial color flow image of the same flow, Figure 11.30A, the vector velocity image, 11.30B, depicts continuous flow as well as regions of higher velocity. In addition, it depicts flow at a Doppler angle of 90°, a situation not possible with conventional CFI.

Finally, the TO method was brought to a commercial scanner (Hansen, Pedersen, Hansen, Nielsen, & Jensen, 2011; Jensen et al., 2011; Hansen et al., 2011). Practical advantages of implementing the TO approach are that the velocity information is derived from the same transmit beams and pulse sequence, and similar Hilbert transform and autocorrelation algorithms are used for both directions. Because three streams are calculated to produce the vector velocities, the receive processing takes longer than older one-dimensional methods. A real-time vector velocity image of an aortic aneurysm produced by TO on this system is shown in Figure 11.34. Note, as appropriate, a 2D color-wheel map is used to interpret the

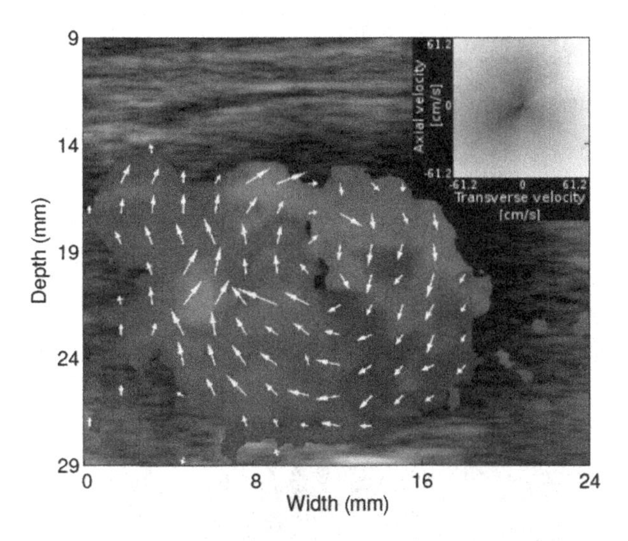

Figure 11.34

Real-time, In vivo vector flow ultrasound image of the abdominal aorta in the transverse plane using the TO method on a commercial system. The vector velocity color map is shown in the upper right corner. The reader is referred to the Web version of the book to see the figure in color. *From Pedersen et al. (2011), IEEE.*

vector velocity field. TO has been applied to phased arrays (Pihl & Jensen, 2011; Pihl, Marcher, & Jensen, 2012) and 3D (Marcher et al., 2012).

11.10.3 Synthetic Aperture Flow Imaging

As described in Section 10.12.4, the synthetic aperture system gathers data from all directions, so that it can assemble and focus the received data in any direction and in any order. This flexibility makes this approach suitable for flow imaging (Jensen & Nikolov, 2004; Jensen, Nikolov, Gammelmark & Pedersen, 2006).One advantage is that beamforming can be done along the direction of flow, which is like placing a rotated coordinate system in the vessel data so that flow can be properly depicted; therefore, this method is not restricted by the Doppler angle as in conventional methods. However, the flow direction must be determined before beamforming. Unlike conventional color flow imaging, in which 8 to 16 lines are repeated along the same direction, and are most likely not aligned with the vessel, only $N = 4$ to 8 emissions are needed to create high-resolution lines along the flow direction x'. For a flow with a velocity $\vec{v}$, displacements between high-resolution images are

$$\Delta x' = |\vec{v}| N T_{PRF}, \tag{11.47}$$

and T_{PRF} is the time between transmissions. Cross-correlation between pairs of high-resolution images gives the displacement $\Delta x'$, and dividing by NT_{PRF} gives the velocity

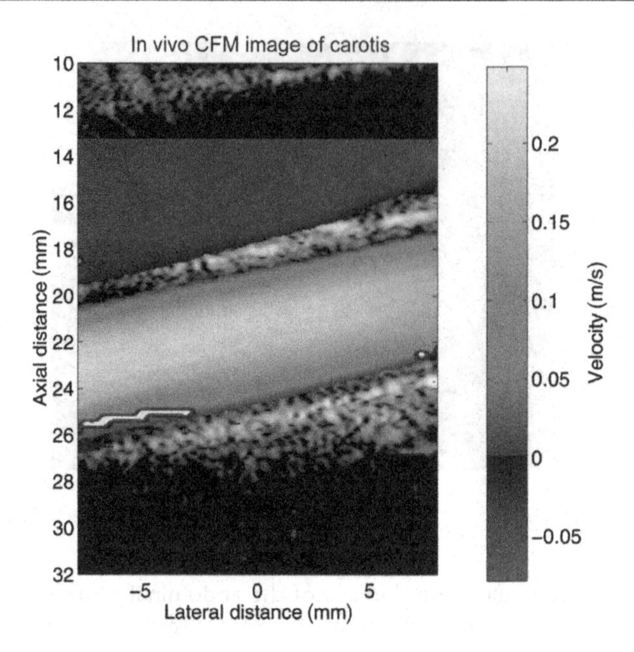

Figure 11.35
In vivo color flow map image at a 77° Flow angle for the jugular vein and carotid artery. The color scale indicates the velocity along the flow direction, where red hues indicate forward flow and blue reverse flow. The reader is referred to the Web version of the book to see the figure in color. *From Jensen and Nikolov (2004), IEEE.*

magnitude $|\vec{v}|$. More cross-correlations can be done to improve accuracy and lower standard deviations. Recursive methods and chirped matched filters (discussed in Section 10.12.4) can be used to quickly update the data (a pulse repetition frequency of several kHz is achievable) and improve the signal-to-noise ratio. For the profiles for laminar flow calculated with 32 to 128 emissions (Jensen et al., 2006), a full vector velocity image was generated with a standard deviation on the order of 1%. An example of an in vivo image of an artery and a vein is shown in Figure 11.35 (Jensen & Nikolov, 2004).

11.10.4 Plane-Wave Flow Imaging

Introduction

Plane-wave methods of flow imaging continue to be an active area of research and are being implemented in commercial systems. High frame rates of plane-wave imaging translate into fast sample rates for obtaining flow information rapidly. Through the use of speckle tracking, vector velocity data in all directions can be obtained throughout the image. With these advantages, however, there are also trade-offs in terms of speed and clutter level for the different set of parameters needed for accurately depicting turbulent flow spatially.

Plane Wave Excitation for Vector Doppler Imaging

In order to achieve both high frame rate and vector velocity imaging, Udesen et al. (2008) found poor contrast in creating a complete image from a single plane-wave transmission, as investigated by (Montaldo, Tanter, Bercoff, Benech & Fink, 2009) and discussed extensively in Section 10.12.2. Their approach was to use matched filters (13-bit Barker code, see Section 10.9.2) to increase sensitivity and reduce clutter. By employing apodization on transmit and receive, they were able to improve the point spread function. The plane-wave excitation (PWE) method was simulated in the Field II program for a synthetically generated flow. Flow phantom and in vivo common-carotid-artery plane-wave data were obtained on the RASMUS synthetic aperture research system (Jensen et al., 2005). Speckle tracking was applied to the data, which were broken into blocks cross-correlated with blocks of next frame. For each vector velocity image, 40 speckle images were used for a frame rate of 100 Hz. A vector velocity image from the sequence can be found in Figure 11.36. The team found that clutter increased towards the wall of the vessel and that PWE seemed better suited to imaging large vessels.

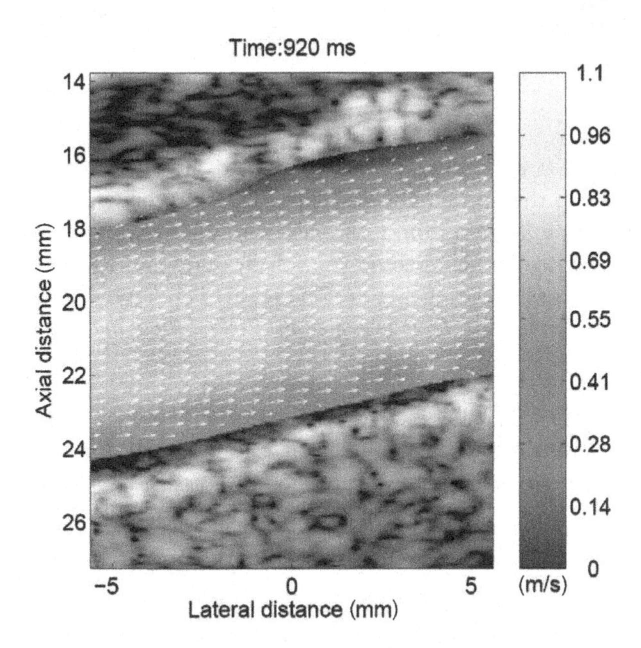

Figure 11.36
Vector velocity image of the common carotid artery shortly after the time of peak systole. The image was obtained from 40 plane-wave emissions, which gave a frame rate of 100 Hz. The vectors show the direction and magnitude of the flow and the colors show the magnitude of the flow. The dynamic range of the B-mode image is 40 dB. The reader is referred to the Web version of the book to see the figure in color. *From Udesen et al. (2008), IEEE.*

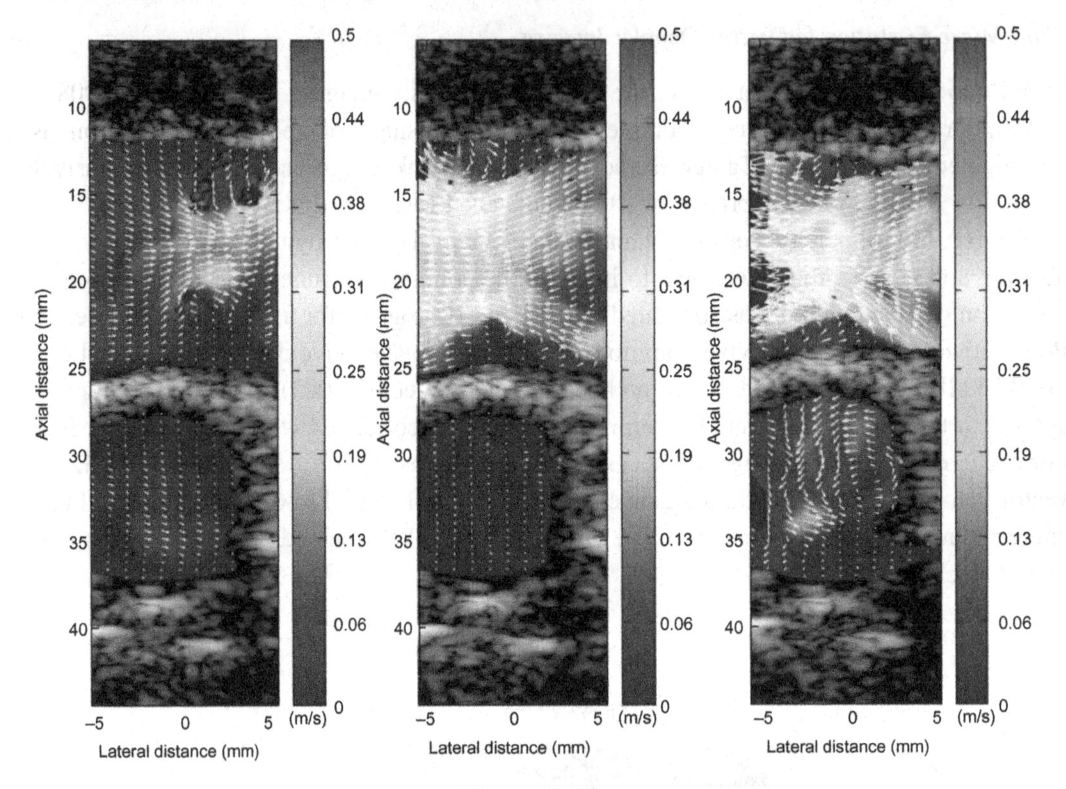

Figure 11.37

Plane-wave vector velocities in the jugular vein (Top) and the Carotid artery at three different times in the cardiac cycle. In the upper right of the top vessel in the left panel, reversed flow and a jet and vortices can be seen upstream from leaky valves. In the right panel, vortices were formed in the sinus pockets of the jugular vein behind the valves, during antegrade flow. In the carotid artery, secondary flow was seen during systole. The reader is referred to the Web version of the book to see the figure in color. *From Jensen et al. (2011), IEEE, and Hansen et al. (2009).*

In their next attempt for in vivo imaging with the PWE method, a 5-MHz linear array and similar processing were employed (Hansen, Udesen, Gran, Jensen & Nielsen, 2009). Speckle tracking was implemented with search kernels of 1 × 1 mm (Friemel, Bohs & Trahey, 1995; Trahey, Allison & Ramm, 1987) and acoustic output was kept below the US Food and Drug Administration (FDA) limits (Section 15.2). Extraordinary computational efforts were made to produce images: a 100-CPU Linux cluster was used for producing the 100-Hz frame rate, in which a second sequence took 10 hours to store and 48 hours to process. Some of the remarkable images obtained are displayed in Figure 11.37. In these high-frame-rate, high-resolution images, detailed vortices at bifurcations and reverse flow were observed for the first time. These images set a new standard in terms of the quality of images possible in vector Doppler. Given new developments in acquisition and fast processing for plane-wave imaging, and Moore's law, vector Doppler will become a faster real-time imaging method.

Plane-Wave Compounding for Doppler Imaging

Following on the plane-wave compounding approach for B-mode imaging, Bercoff et al. (2011) investigated its possibilities for Doppler imaging. What was the number of angles required to obtain good performance? They found that in a typical color flow operation there were M_{UFD} excitation pulses per line, N_s segments and N_l lines. The number of lines is determined from the ratio of the maximum of the PRF needed for the maximum depth imaged (PRF_{max}) to the PRF needed to capture the fastest moving blood flow (PRF_{FLOW}):

$$N_{FLOW} = PRF_{max}/PRF_{FLOW}. \tag{11.48}$$

To determine the equivalent in terms of the number of angles, N_{angles}, for plane-wave compounding, the following is useful:

$$N_{angles} = PRF_{max}/PRF_{FLOW}, \tag{11.49}$$

with the total number of pulse excitations per frame being proportional to the number of angles:

$$N_{PFUFD} = N_{angles} M_{UFD}. \tag{11.50A}$$

Compared to the number in conventional color flow imaging:

$$N_{PFCFI} = N_s N_l M_{CFI}. \tag{11.50B}$$

Therefore, the possible improvement in speed of acquisition time is the number of segments,

$$GAT = N_{PFUFD}/N_{PFCFI} = N_s, \tag{11.51}$$

which typically varies from 3 to 64. As previously determined, contrast is poor for a single plane-wave transmission compared to the focused case; however, with 9 angles, it is only 5 dB worse and for 16 angles, 2 dB worse. The important signal-to-noise ratio (SNR) (Montaldo et al., 2009) is related to:

$$\frac{SNR_{pwc}}{SNR_{foc}} = \frac{\sqrt{N_{angles}}}{G} = \frac{\sqrt{N_{angles} F \lambda_z}}{L_x}, \tag{11.52}$$

where focusing gain, G, is from Eqn 6.33B.To obtain equivalent SNR (i.e. the ratio = 1), the number of angles is set equal to the conventional focusing gain. An example given showed that 9 angles was equivalent to a focal gain for a 5-MHz array of length 9 mm and a focal depth of 30 mm. Experimentally, the lateral SNR vs depth was measured in a phantom; the SNRs were equal at the focus but the mean SNR improvement over the focused case for all depths was 5 dB for 16 angles and 2.5 dB for 9 angles. Lateral and axial resolution was the same for these two compounded cases and the comparable focused cases.

Equipped with measures of Doppler image quality, Bercoff et al. (2011) conducted experiments to compare color flow imaging to ultrafast Doppler (UFD) imaging. An 8-MHz 256-element linear array and a research version of the Aixplorer platform (Supersonic Imagine) capable of fast acquisition of the raw RF (radio frequency) data from all elements (see Sections 10.12.2 and 10.12.5) were used for experiments. With reference to Figure 11.38, The depth was $R_{UFD} = 25$ mm, so the minimum round-trip time was 33 μs. The number of repeated flow lines for CFI was 8, as it was for the number of angles. The number of color flow image lines was 16. Under these conditions, the CFI frame rate was 11/s, whereas, for the ultrafast case, it was 176/s from Eqn 11.35.

The implementation of UFD on the Aixplorer is diagrammed in Figure 11.38. A series of pulsed plane waves were sent at intervals of T_{RT} and the number of firings per pulse was $M_{UFD} = 3$. The plane waves were sent in groups of N_{angles}, determined from Eqn 11.49. On reception, echoes from each plane wave were dynamically focused and then all of the

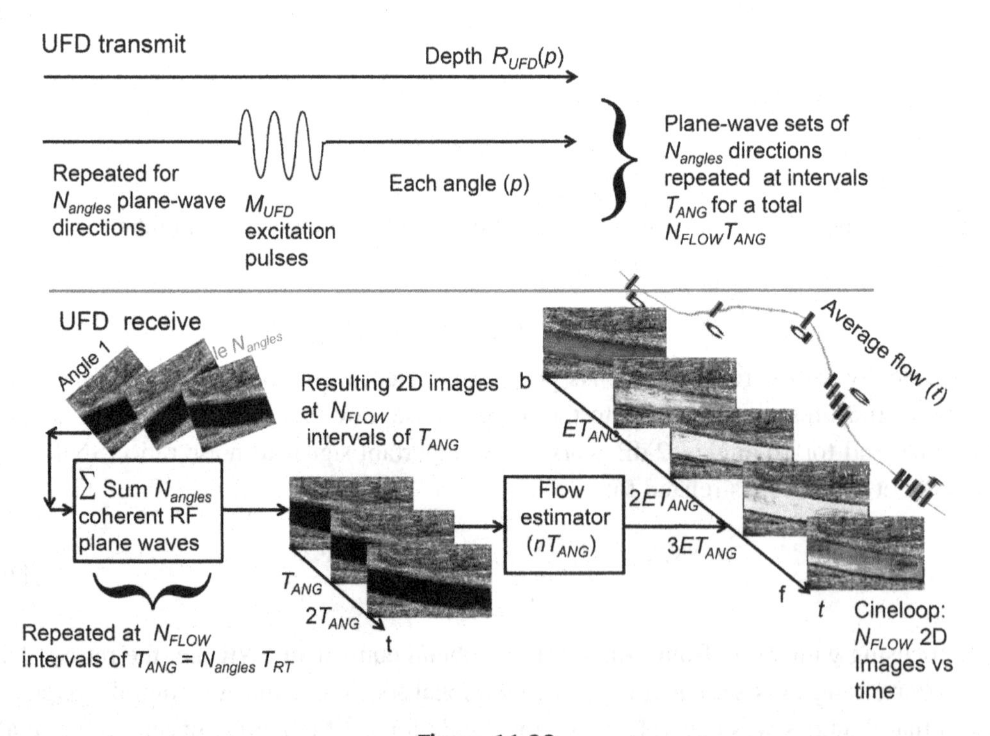

Figure 11.38

Conceptual diagram of timing and spatial sequences in ultrafast Doppler. (Top) parameters for pulse transmission; (bottom) parameters on reception leading to output display. The constant E depends on processing and is approximately 10. *Images from Bercoff et al. (2011).*

echoes from each plane wave insonification were coherently summed to produce a single composite plane-wave image (or set of radio frequency (RF) data), which occurred at intervals $N_{angles}T_{RT}$. A wall filter was used to clean up the raw I and Q data in these images. A flow estimator operated on several images (or equivalently RF data belonging to those images, about 15 on the average) using a sliding window and produced flow images. These final images provided a high-speed movie that could be replayed, and different Doppler parameters could be calculated for individual points in the flow pattern or regions in a vessel. Shown in Figure 11.38 in the background is an average Doppler frequency flow curve. This information was derived from Figure 11.39, which shows a series of flow

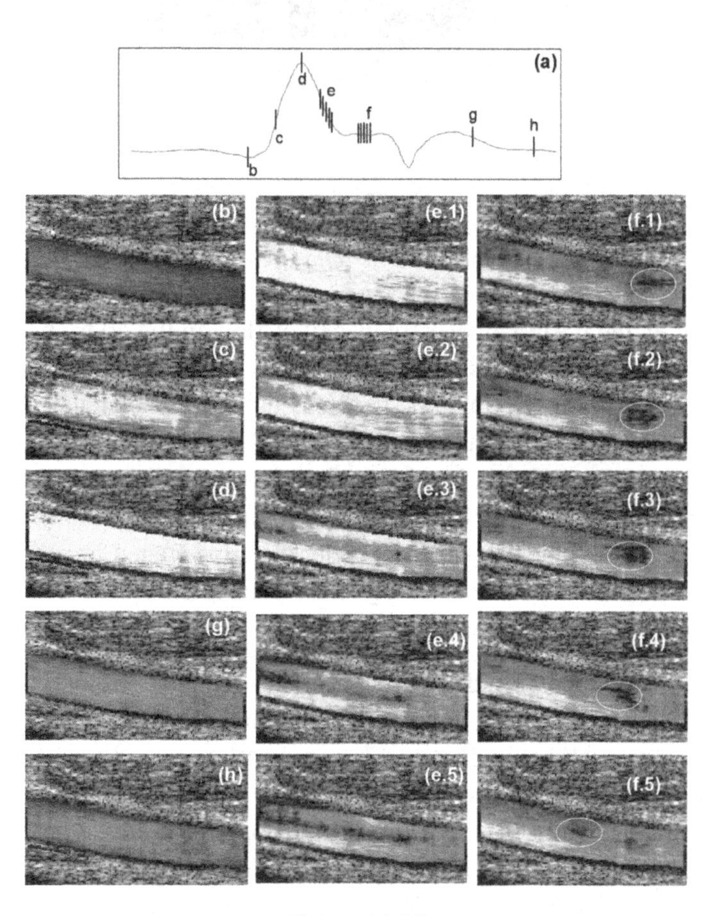

Figure 11.39

Some selected frames of a complete cardiac cycle obtained with the ultrafast plane wave compounding. (A) Average flow in the artery indicating the selected frames; (B) before the opening of the aortic valve, there is a minimal laminar flow; (C and D) acceleration of the flow; (E) inversion of the parabolic profile in the deceleration; (F) local turbulence is present and propagates in the artery; (G and H) laminar flows in diastole. The reader is referred to the Web version of the book to see the figure in color. *From Bercoff et al. (2011), IEEE.*

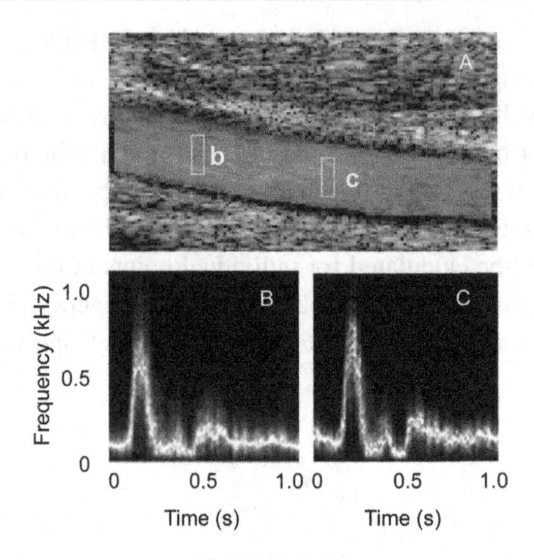

Figure 11.40
(A) Two sample volumes; (B and C) the corresponding spectrograms using *IQ* data acquired at the same time. *From Bercoff et al. (2011), IEEE.*

images and at the top, the mean Doppler frequency of flow in the artery with the location of images in the overall sequence, as a function of time. Unlike a sequence of CFI images, which typically occur at about 15 frames/s, the UFD data matrix contained about 300 images. The evolution of flow from the cardiac cycle is shown with high temporal resolution and dynamic detail that is normally missed in conventional CFI imaging. For example, the entire sequences e and f, which reveal a parabolic inversion and turbulence, respectively, would typically be shown by a single overall averaged image in CFI. The UFD data matrix is stored, and retrospectively, flow parameters can be extracted for any points within the flow region as shown in Figure 11.40.

In summary, the UFD scheme provides a more complete Doppler picture, allowing a global view of hemodynamics simultaneously with high-resolution detail. Unlike the conventional compartmentalized approach in which color flow and PWD are performed sequentially and separately, searching for normal flow patterns or anomalies can be a time-consuming process performed interactively. The UFD approach, done retroactively, provides opportunities to scan the entire flow sequence in high resolution and then identify and compare flow spectra at different spatial locations.

Plane-Wave Investigations for Doppler Imaging

Plane-wave approaches for Doppler imaging continue to be an active area of research. In order to better interpret the large amount of Doppler data available in the ultrafast plane-wave compounding method, new ways of statistically analyzing and displaying Doppler data have been investigated (Osmanski et al., 2012). Work is underway to exploit the fast acquisition capabilities of the Verasonics research platform to produce vector Doppler images from multiple-angle plane-wave transmissions at high frame rate rates (Flynn, Daigle, Pfligrath, Linkhart & Kaczkowski, 2011). Flow data were assembled from multiple-angle plane waves and flow was computed by an autocorrelation method (Kasai et al., 1983) preserving the directivity of flow. Finally, the vector nature of flow was displayed in a novel particle flow representation. In another study, Flynn, Daigle, Pfligrath & Kaczkowski (2012) employed a space–time gradient of the image data from a single plane-wave insonification to compute the vector velocity flow field. Tong et al. (2012) showed that cardiac motion studies utilizing high-frame-rate Doppler imaging from a single plane-wave insonification method compared favorably with conventional CFI. Iversen, Lindsethyz, Torp and Lovstakken (2012) demonstrated that better quality and more accurate 3D flow images could be obtained by higher-frame-rate 2D plane-wave methods and position tracking, compared to conventional 3D flow approaches.

11.11 Functional Ultrasound Imaging

One of the fastest growing areas of neuroscience is functional imaging. Functional imaging has been used extensively to map regions of the brain connected with stimuli, activities, and higher-level cognition. Another example of this type of neuroimaging—biological motion perception—has been studied in humans.

Several technologies have been used to study brain activity. Though optical methods offer high resolution, they lack penetration. As mentioned earlier, photoacoustics shows some promise for brain mapping but is still in early development. The leading methodology is functional MRI (fMRI), the use of magnetic resonance imaging (MRI) to detect localized changes in brain activity, usually in the form of changes in cerebral metabolism, blood flow, volume, or oxygenation in response to task activation (Szabo, 2011). These changes are interrelated and may have opposite effects. For example, an increase in blood flow increases blood oxygenation, whereas an increase in metabolism decreases it. The most common means of detection is measuring the changes in the magnetic susceptibility of hemoglobin. Oxygenated blood is diamagnetic and deoxygenated blood is paramagnetic. These differences lead to a detection method called blood-oxygen-level-dependent contrast (BOLD). Changes in blood oxygenation can be seen as differences in the T^*2 decay constant, so T2-weighted images are used. Because these changes are very small, images before and after task initiation are subtracted from each other and the resultant difference image is overlaid on a standard image. Special care must be taken when obtaining these images because the effect is small and can be corrupted by several sources of noise. Typically, hundreds of images are taken for each slice-plane position and statistical analysis is used to produce the final image. Sources of noise errors are thermal noise, head movement, and respiratory and cardiac cycles. Small animal studies require high magnetic fields to achieve adequate resolution, at the expense of temporal resolution and signal-to-noise ratio (Mace et al., 2013).

Is there a role for ultrasound to overcome some of the signal-to-noise and sensitivity limitations of fMRI for brain imaging as well as its high cost and non-transportability? Recent advances in plane-wave compound imaging have made possible a functional ultrasound (fUS) method with high sensitivity and spatiotemporal resolution. The methodology of ultrafast Doppler has advantages already discussed. In the new method, more improvements were needed to sense the small hemodynamic changes in

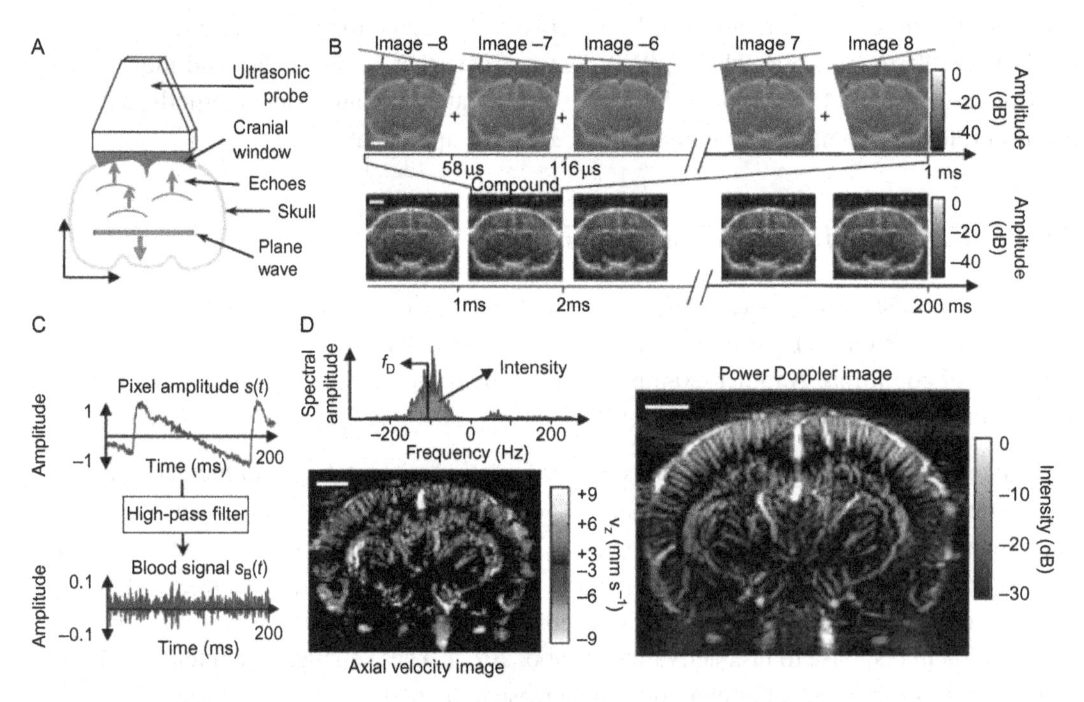

Figure 11.41

Principles for performing fUS in the rat brain. (A) Schematic setup depicting the ultrasonic probe, cranial window, and a schema of a coronal slice from the rat brain. (B) fUS is performed by emitting 17 planar ultrasonic waves tilted with different angles into the rat brain. The ultrasonic echoes produce 17 images of 2 × 2 cm. Summing these images results in a compound image acquired in 1 ms. The entire fUS sequence consists of acquiring 200 compound images in 200 ms. (C) Temporal variation $s(t)$ of the backscattered ultrasonic amplitude in one pixel (normalized by the maximum amplitude). The blood signal s_B is extracted by applying a high-pass filter (same scale in the two graphs). (D) Frequency spectrum of s_B (top left). Two parameters are extracted from this spectrum: the central frequency f_D, which is proportional to the axial blood velocity with respect to the z axis and gives rise to the axial velocity image (below left); and the intensity (power Doppler), which is proportional to the cerebral blood volume and gives rise to the power Doppler image (right). fUS is based on power Doppler images. Scale bars, 2 mm. *From Mace et al. (2011),courtesy of Nature Methods.*

microvasculature. Compared to color flow imaging, power Doppler imaging (Bude & Rubin, 1996; Rubin et al., 1994, 1995) was selected as the imaging modality because it has higher sensitivity to small flows but at the expense of losing flow directionality.

The new method is explained with the help of Figure 11.41, and its application to experiments undertaken to image a rat's brain (Mac et al., 2011). In Figure 11.41A, the setup is shown. A 14×14 mm section of the skull was removed. A 15-MHz 256-element linear array (pitch = 0.125 mm) was placed 5 mm above the brain with a gel couplant and connected to a research version of the Aixplorer V2 fast-acquisition platform. The depth of the image was 20 mm.

In Figure 11.41B is the now familiar sequence of plane waves transmitted at different angles: $N_{angles} = 16$ in this case, from -7 to $+8$ degrees. Because $T_{RT} = 27$ μs, a maximum of 37 kHz was possible with a single plane-wave insonification. Because of other considerations, such as overheating the transducer, it was determined that half this PRF was adequate. The sequence of plane waves was therefore transmitted at intervals of about 54 μs, as shown at the top of the figure. To create a composite image from the coherent summation of each set of 16 tilted images, composite images were made at intervals of $T_{ANG} = N_{angles}T_{RT}$ of about 1 ms, or a PRF of 1 kHz. This microDoppler approach was compared to a conventional power Doppler imaging sequence. Instead of the full aperture of 12 mm in the plane-wave method, apertures of 4 mm and a focal length of 10 mm were used, so that the focal gain was 4. The image was broken into $N_{blocks} = 8$ blocks, each scanned $N_{foc} = 16$ times, which resulted in a frame rate of about 125 frames/s. For an acquisition time of 320 ms, the microDoppler approach produced 320 samples, compared to 40 with conventional power Doppler. This eightfold increase in samples results in a relative SNR gain improvement of up to 4 dB, except in the focal region where it is 1.1 dB. Improvement in sensitivity was 5 to 30 with a mean of 15. Because the microDoppler method used three times the aperture, 12 mm, it was actually transmitting more time-averaged acoustic power, which is proportional to the number of elements (Szabo & Seavey, 1983); so on a power equivalent basis, the gain was 1.5 to 10 with a mean of 5. For this experiment, the number and length of transmit pulses was the same. Even though acoustic output was not measured, we can speculate on the relative measures of the output parameter spatial peak temporal average intensity (ISPTA) based on equations in Szabo et al. (1988). ISPTA is a measure of maximum time-averaged intensity along the main beam axis and is explained in Section 15.2. If the in situ factors and duty cycles are approximately equivalent, and the effect of tilting ($\cos 8° = 0.99$) is negligible, then the ratio of ISPTAs of microDoppler to conventional Doppler is:

$$R_{ISPTA} = \frac{N_{ANGLES}}{GN_{foc}} = \frac{16}{4 \times 16} = 0.25, \tag{11.53}$$

where G is focal gain.

Subsequent processing includes the high-pass or clutter filtering of each data point associated with a pixel position (x,z) and time (nT_{RT}) so $s_D(x,z,t_n)$ becomes $s_F(x,z,t_n)$ after filtering. Unlike the usual case where sound beams intersect large vessels at an angle, in the case of brain imaging the axial (z) or vertical flow is emphasized. A consequence of the clutter filtering is that because of lost frequencies and the relation between Doppler frequencies and flow velocities, blood velocities less than 4 mm/s cannot be detected, velocities between 4 and 10 mm/s are affected, and velocities above 10 mm/s are detected cleanly. In order to add directionality, not usually detected by power Doppler techniques:

$$I_{PD}(x,z) = \frac{1}{N}\sum_{i=1}^{N}|s_F(x,z,t_i)|^2. \tag{11.54}$$

Mace et al. (2013) processed the positive and negative spectra separately, after obtaining the spectra from an FFT operation on the time samples of the complex signals $s_F(x,y,t_n)$,

$$I_D^+(x,z) = \int_0^{+\infty}|S_F(x,z,\omega)|^2 d\omega, \quad \text{and} \tag{11.55A}$$

$$I_D^-(x,z) = \int_{-\infty}^{0}|S_F(x,z,\omega)|^2 d\omega. \tag{11.55B}$$

In addition, a lower threshold was applied to the images. A power Doppler image based on Eqn 11.54 is shown in the lower right corner of Figure 11.41D. The velocity directionality image (showing flow up or down) from Eqn 11.55 with velocity mapping derived from the central Doppler frequency are also shown in Figure 11.41D. Note the lower ends of the velocity scale are truncated because of filtering effects. Because of the large number of samples (320) used per pixel compared to a typical number of 16 in conventional power Doppler, high resolution and signal to noise are achieved. In this study, 3D volume images were reported from 12 slices in the sagittal plane spaced at 300 μm intervals.

What is it that is being displayed in these images, and is the technique useful for functional imaging? Because the scattering from blood is mainly caused by red blood cells caught by the extent of the ultrasound beam, the intensity is related to red blood volume (RBV). Based on independently measured flow velocity data for the rat brain, rates of a few mm/s and above would correspond to values in cortical microvessels, and directionality mapping could aid in displaying both arterial and venous flows. The present spatial and temporal resolutions are 125 μm and 320 ms. Because of present computational limitations, there is about a 2–3-second delay before the next image can be produced; however, this is adequate to detect responses to stimuli as needed for true functional imaging.

Mace et al. (2011) conducted several studies on rats that involved stimuli. Shown in Figure 11.42 is a sequence of images depicting an induced epileptiform-seizures fit. Figure 11.42A shows an illustration of the drug infusion and electrode implantations. Figure 11.42B shows time sequences of spatiotemporal responses to two ictal

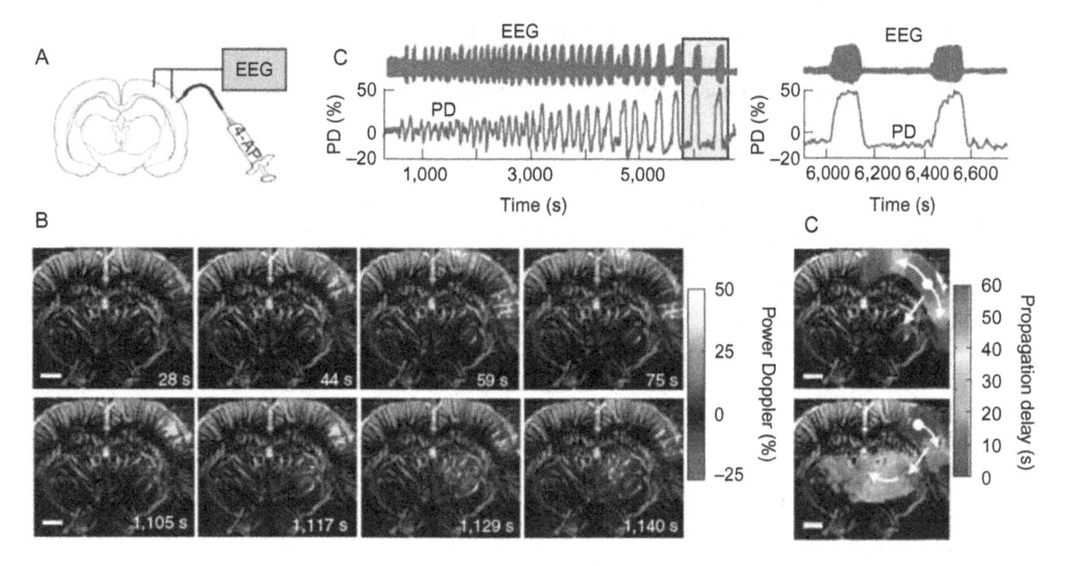

Figure 11.42

fUS Imaging of transient brain activity in a rat model of epilepsy. (A) Schematic setup for the imaging of epileptiform seizures. (B) Spatiotemporal spreading of epileptiform activity for two selected ictal events. The power Doppler signal (in % relative to baseline) is superimposed on a control power Doppler image. (C) Comparison between electrical recordings (EEG) and the power Doppler signal (PD) at the site of 4-AP injection. The two events in the shaded region are zoomed in on the graph at the right. (D) Maps of the propagation delay of blood volume changes from the focus to other regions (propagation delay in seconds is color coded following the legend on the right: onset is indicated in blue, and blue to red indicates delay increases). Arrows represent the direction of propagation. Scale bars, 2 mm. The reader is referred to the Web version of the book to see the figure in color. *From Mace et al. (2011),courtesy of Nature Methods.*

(physiological) events mapped as changes in the power Doppler signals. Figure 11.42C shows the excellent correlation between the power Doppler signals and corresponding EEG signals measured by electrodes. Figure 11.42D displays the propagation delay of blood volume events. Experiments involving whisker stimulation were also conducted.

In summary, fUS offers new capabilities to sense small blood volume changes with high spatial and temporal resolution in response to stimuli. At present, fUS needs trepanning or skull thinning, or could be applied through the fontanel window of younger animals or to adults during open-skull neurosurgery (Mace et al., 2011). By attaching a transducer to the head, brain responses during normal activities and movement can be monitored.

References

Angelsen, B. A. J. (1981). Instantaneous frequency, mean frequency, and variance of mean frequency estimators for ultrasonic blood velocity Doppler signals. *IEEE Transactions on Biomedical Engineering, 28,* 733–741.

Babock, D. S., Patriquin, H., La Fortune, M., & Dauzal, M. (1996). Power Doppler sonography: Basic principles and clinical applications in children. *Pediatric Radiology*, *26*, 109–115.

Baker, D. W. (1970). Pulsed ultrasonic Doppler blood-flow sensing. *IEEE Transactions on Sonics and Ultrasonics*, *17*, 170–185.

Bamber, J. C. (1986). In C. R. Hill (Ed.), *Physical principles of medical ultrasonics*. Chichester, UK: John Wiley & SonsChap. 14.

Bercoff, J., Montaldo, G., Loupas, T., Savery, D., Mézière, F., Fink, M., et al. (2011). Ultrafast compound Doppler imaging: Providing full blood flow characterization. *IEEE Transactions on Ultrasonics, Ferroelectrics and Frequency Control*, *58*, 134–147.

Bonnefous, O., & Pesque, P. (1986). Time-domain formulation of pulse-Doppler ultrasound and blood velocity measurement by cross-correlation. *Ultrasonics Imaging*, *8*, 73–85.

Bonnefous, O., Pesque, P., & Bernard, X. (1986). A new velocity estimator for color flow mapping. *Proceedings of the IEEE ultrasonics symposium*, (pp. 855–860). Williamsburg, VA.

Bracewell, R. (2000). *The fourier transform and its applications*. New York: McGraw-Hill.

Brandestini, M. (1978). Topoflow: A digital full range Doppler velocity meter. *IEEE Transactions on Sonics and Ultrasonics*, *25*, 287–293.

Bude, R. O., & Rubin, J. M. (1996). Power Doppler sonography. *Radiology*, *200*, 21–23.

Carstensen, E. L., & Schwan, H. P. (1959). Acoustic properties of hemoglobin solutions. *Journal of the Acoustical Society of America*, *31*, 305–311.

Chen, J. -F., Fowlkes, J. B., Carson, P. L., Rubin, J. M., & Adler, R. S. (1996). Autocorrelation of integrated power Doppler signals and its application. *Ultrasound in Medicine and Biology*, *22*, 1053–1057.

Chiao, R. Y., Mo, L. Y., Hall, A. L., Miller, S. C., & Thomenius, K. E. (2000). B-mode blood flow (B-flow) imaging. *Proceedings of the IEEE ultrasonics symposium* (Vol. 2), (pp. 1469–1472). San Juan.

Chilowsky, C. & Langevin, P. (1919). Improvements in and Connected with the Production of Submarine Signals and the Location of Submarine Objects, UK patent 125, 122, April 17, 1919.

Cobbold, R. S. C., Vetlink, P. H., & Johnston, K. W. (1983). Influence of beam profile and degree of insonation on the CW Doppler ultrasound spectrum and mean velocity. *IEEE Transactions on Sonics and Ultrasonics*, *30*, 364–370.

Coussios, C. -C. (2002). The significance of shape and orientation in single particle weak-scatterer models. *Journal of the Acoustical Society of America*, *112*, 906–915.

Embree, P. M. & O'Brien, W. D., Jr. (1985). The accurate ultrasonic measurement of the volume flow of blood by the time domain correlation, *Proceedings of the IEEE ultrasonics symposium* (pp. 963–966). San Francisco, CA.

Embree, P. M., & O'Brien, W. D., Jr. (1990). Volumetric blood flow via time domain correlation: Experimental verification, IEEE Transactions on Ultrasonics. *Ferroelectrics and Frequency Control*, *37*, 176–189.

Evans, D. H. (1993). In P. N. T. Wells (Ed.), *Advances in Ultrasound Techniques and Instrumentation*. New York: Churchill LivingstoneChap. 8.

Evans, D. H., Jensen, J. A., & Nielsen, M. B. (2011). Ultrasonic colour Doppler imaging. *Interface Focus*, *1*, 490–502.

Evans, D. H., & Parton, L. (1981). The directional characteristics of some ultrasonic Doppler blood-flow probes. *Ultrasound in Medicine and Biology*, *7*, 51–62.

Ferrara, F. W., & Algazi, V. R. (1991a). A new wideband spread target maximum likelihood estimator for blood velocity estimation, Part I: Theory. *IEEE Transactions on Ultrasonics, Ferroelectrics and Frequency Control*, *38*, 1–16.

Ferrara, F. W., & Algazi, V. R. (1991b). A new wideband spread target maximum likelihood estimator for blood velocity estimation, Part II: Evaluation of estimators with experimental data. *IEEE Transactions on Ultrasonics, Ferroelectrics and Frequency Control*, *38*, 17–26.

Ferrara, K., & Deangelis, G. (1997). Color flow mapping. *Ultrasound in Medicine and Biology*, *23*, 321–345.

Flynn J., Daigle, R., Pfligrath, L., & Kaczkowski, P. (2012). High frame rate vector velocity blood flow imaging using a single planewave transmission angle. *Proceedings of the IEEE ultrasonics symposium* (pp. 323–325).

Flynn J., Daigle, R., Pfligrath, L., Linkhart, K., & Kaczkowski, P. (2011). Estimation and display for vector Doppler imaging using planewave transmissions. *Proceedings of the IEEE ultrasonics symposium* (pp. 413–418). Orlando, FL.

Fontaine, I., Bertrand, M., & Cloutier, G. (1999). A system-based simulation model of the ultrasound signal backscattered by blood. *Proceedings of the IEEE ultrasonics symposium* (Vol. 2), (pp. 1369–1372). Caesars Tahoe, NV.

Friemel B. H., Bohs L. N., Trahey G. E.(1995). Relative performance of two-dimensional speckle-tracking techniques: Normalized correlation, non-normalized correlation and sum-absolute-difference. (pp. 1481–1484).

Frinking, P. J. A., Boukaz, A., Kirkhorn, J., Ten Cate, F. J., & De Jong, N. (2000). Ultrasound contrast imaging: Current and new potential methods. *Ultrasound in Medicine and Biology*, *26*, 965–975.

Greenleaf, J. F., Duck, F. A., Samayoa, W. F., and Johnson, S. A. (1974). Ultrasound data acquisition and processing system for atherosclerotic tissue characterization. *Proceedings of the IEEE ultrasonics symposium* (pp. 1561–1566).

Halberg, L. I., & Thiele, K. E. (1986). Extraction of blood flow information using Doppler-shifted ultrasound. *Hewlett Packard Journal*, *37*, 35–40.

Hansen, K. L., Udesen, J., Gran, F., Jensen, J. A., & Nielsen, M. B. (2009). In-vivo examples of complex flow patterns with a fast vector velocity method. *Ultraschall in der Medizin*, *30*, 471–476.

Hansen, P. M., Pedersen, M. M., Hansen, K. L., Nielsen, M. B., & Jensen, J. A. (2011). Demonstration of a vector velocity technique. *Ultraschall in der Medizin*, *32*, 213–215.

Hansen, P. M., Pedersen, M. M., Hansen, K. L., Nielsen, M. B., & Jensen, J. A. (2011). Examples of vector velocity imaging, 15. *Nordic-Baltic Conference on Biomedical Engineering and Medical Physics*.

Hattle, L., & Angelsen, B. (1985). *Doppler ultrasound in cardiology: Physical principles and clinical application* (2nd ed.). Philadelphia: Lea and Febiger.

Hein, I. A., & O'Brien, W. D., Jr. (1993a). A real-time ultrasound time domain correlation blood flowmeter: Part I: Theory and design. *IEEE Transactions on Ultrasonics, Ferroelectrics and Frequency Control*, *40*, 775–778.

Hein, I. A., & O'Brien, W. D., Jr. (1993b). A real-time ultrasound time domain correlation blood flowmeter: Part II: Performance and experimental verification. *IEEE Transactions on Ultrasonics, Ferroelectrics and Frequency Control*, *40*, 778–785.

Hein, I. A., & O'Brien, W. D., Jr. (1993c). Current time-domain methods for assessing tissue motion by analysis from reflected ultrasound echoes: A review. *IEEE Transactions on Ultrasonics, Ferroelectrics and Frequency Control*, *40*, 84–102.

Hemmsen, M. C., Nikolov, S. I., et al. (2011). Implementation of a versatile research data acquisition system using a commercially available medical ultrasound scanner. *IEEE Transactions on Ultrasonics, Ferroelectrics and Frequency Control*, *59*, 1487–1491.

Iversen, D. H., Lindsethyz, F., Torp, H., & Lovstakken, L. (2012). Improved 3-D reconstruction of vascular flow based on plane wave imaging. *IEEE Ultrasonics Conference*, *2012*, 311–314.

Jensen, J. A. (1996). *Estimation of blood velocities using ultrasound*. Cambridge, UK: Cambridge University Press.

Jensen, J. A., & Munk, P. (1998). A new method for estimation of velocity vectors, IEEE transactions on ultrasonics. *Ferroelectrics and Frequency Control*, *45*, 837–851.

Jensen, J. A. (2001). A new estimator for vector velocity estimation, IEEE Transactions on Ultrasonics. *Ferroelectrics and Frequency Control*, *48*, 886–894.

Jensen, J. A., & Nikolov, S. I. (2004). Directional synthetic aperture flow imaging. *IEEE Transactions on Ultrasonics, Ferroelectrics and Frequency Control*, *51*, 1107–1118.

Jensen, J. A., Holm, O., Jensen, L. J., Bendsen, H., Nikolov, S., Tomov, B. G., et al. (2005). Ultrasound research scanner for real-time synthetic aperture data acquisition. / IEEE transactions on ultrasonics. *Ferroelectrics and Frequency Control*, *52*, 881–891.

Jensen, J. A., Nikolov, S. I., Gammelmark, K. L., & Pedersen, M. H. (2006). Synthetic aperture ultrasound imaging. *Ultrasonics*, *44*, e5–e15.

Jensen, J. A., Nikolov, S. I., Udesen, J., Munk, P., Hansen, K. L., Pedersen, M. M., et al. (2011) Recent advances in blood flow vector velocity imaging. *Proceedings of the IEEE ultrasonics symposium* (pp. 262–271). Orlando, FL.

Kasai, C., Namekawa, K., Koyano, A., & Omoto, R. (1983). Real-time two dimensional blood flow imaging using an autocorrelation technique. *IEEE Transactions on Sonics and Ultrasonics, 32*, 458–464.

Kimme-Smith, C., Tessler, F. N., Grant, E. G., & Perella, R. R. (1989). Processing algorithms for color flow Doppler. *Proceedings of the IEEE ultrasonics symposium* (Vol. 2), (pp. 877–879). Montreal, QC.

Lerner, R. M., Huang, S. R., & Parker, K. J. (1990). Sonoelasticity images derived from ultrasound signals in mechanically vibrated tissues. *Ultrasound in Medicine and Biology, 15*, 231–239.

Lockwood, G. R., Ryan, L. K., Hunt, J. W., & Foster, F. S. (1991). Measurement of the ultrasonic properties of vascular tissues and blood from 35–65 MHz. *Ultrasound in Medicine and Biology, 17*, 653–666.

Mace, E., Montaldo, G., Osmanski, B. -F., Cohen, I., Baulac, M., Fink, M., et al. (2011). Functional ultrasound imaging of the brain. *Nature Methods, 8*, 662–664.

Mace, E., Montaldo, G., Osmanski, B. -F., Cohen, I., Fink, M., & Tanter, M. (2013). Functional ultrasound imaging of the brain: Theory and basic principles. *IEEE Transactions on Ultrasonics, Ferroelectrics and Frequency Control, 60*, 492–506.

Magnin, P. A. (1986). Doppler effect: History and theory. *Hewlett Packard Journal, 37*, 26–31.

Magnin, P. A. (1987). A review of Doppler flow mapping techniques. *Proceedings of the IEEE ultrasonics symposium* (pp. 969–977). Denver, CO.

Mo, L. Y. L., & Cobbold, R. S. C. (1986). "Speckle" in continuous wave Doppler ultrasound spectra: A simulation study. *IEEE Transactions on Ultrasonics, Ferroelectrics and Frequency Control, 33*, 747–753.

Montaldo, G., Tanter, M., Bercoff, J., Benech, N., & Fink, M. (2009). Coherent plane-wave compounding for very high frame rate ultrasonography and transient elastography. *IEEE Transactions on Ultrasonics , Ferroelectrics and Frequency Control, 56*, 489–506.

Newhouse, V. L., & Amir, L. (1983). Time dilation and inversion properties and the output spectrum of pulsed Doppler flowmeters. *IEEE Transactions on Sonics and Ultrasonics, 30*, 174–179.

Newhouse, V. L., Furgason, E. S., Johnson, G. F., & Wolf, D. A. (1980). The dependence of ultrasound Doppler bandwidth on beam geometry. *IEEE Transactions on Sonics and Ultrasonics, 25*, 50–59.

Nightingale, K. R., Kornguth, P. J., Walker, W. F., Mc Dermott, B. A., & Trahey, G. E. (1995). A novel technique for differentiating cysts from solid lesions: Preliminary results in the breast. *Ultrasound in Medicine and Biology, 21*, 745–751.

Nowicki, A., & Reid, J. M. (1981). An infinite gate-pulsed Doppler. *Ultrasound in Medicine and Biology, 7*, 41–50.

Nowicki, A. & Secomski, W. S. (2000). Estimation of hematocrit by means of dual-gate power Doppler. *Proceedings of the IEEE ultrasonics symposium* (Vol. 2), (pp. 1505–1508). San Juan.

Osmanski, B.-F., Montaldo, G., Bercoff, J., Loupas, T. (2012) Ultrafast plane wave imaging: Doppler frequency distribution. *Proceedings of the IEEE ultrasonics symposium* (pp. 1580–1583). San Juan.

Pedersen, M. M., Pihl, M. J., Hansen, J. M., Hansen, P. M., Haugaard, P., & Nielsen, M. B., et al. (2011) Arterial secondary blood flow patterns visualized with vector flow ultrasound. *Proceedings of the IEEE ultrasonics symposium* (pp. 1242–1245). Orlando, FL.

Pierce, A. D. (1989). *Acoustics*. Woodbury, NY: Acoustical Society of America.

Pihl, M. J., & Jensen, J. A.(2011). 3D vector velocity estimation using a 2D phased array. *Proceedings of the IEEE ultrasonics symposium* (pp. 430–433). Orlando, FL.

Pihl, M. J., Marcher, J., & Jensen, J. A. (2012). Phased-array vector velocity estimation using transverse oscillations. *IEEE Transactions on Ultrasonics, Ferroelectrics and Frequency Control, 59*, 2662–2675.

Ramamurthy, B. S., & Trahey, G. E. (1991). Potential and limitations of angle-independent flow detection algorithms using radio-frequency and detected echo signals. *Ultrasonic Imaging, 13*, 252–268.

Richardson, L. F. (1913). Apparatus for Warning a Ship at Sea of its Nearness to Large Objects Wholly or Partly Under Water, UK patent 11, 125, March 27, 1913.

Routh, H. F. (1996). Doppler ultrasound. *IEEE Engineering in Medicine and Biology*, *15*, 31–40.

Rubin, J. M., Adler, R. S., Fowlkes, J. B., Spratt, S., Pallister, J. E., Chen, J. F., et al. (1995). Fractional moving blood volume: Estimation with power Doppler US. *Radiology*, *197*, 183–190.

Rubin, J. M., Bude, R. O., Carson, P. L., Bree, R. L., & Adler, R. S. (1994). Power Doppler ultrasound: A potential useful alternative to mean-frequency-based color Doppler ultrasound. *Radiology*, *190*, 853–856.

Satomura, S. (1957). Ultrasonic Doppler method for the inspection of cardiac function. *Journal of the Acoustical Society of America*, *29*, 1181–1185.

Secomski, W., Nowicki, A., Guidi, F., Tortoli, P., & Lewin, P. A. (2003). Noninvasive in vivo measurements of hematocrit. *Journal of Ultrasound in Medicine*, *22*, 375–384.

Shung, K. K. (1982). On the scattering of blood as a function of hematocrit. *IEEE Transactions on Sonics and Ultrasonics*, *29*, 327–331.

Shung, K. K., Sigelmann, R. A., & Reid, J. M. (1976). Scattering of ultrasound by blood. *IEEE Transactions on Biomedical Engineering*, *23*, 460–467.

Sigelmann, R. A., & Reid, J. M. (1973). Analysis and measurement of ultrasound backscattering from an ensemble of scatterers excited by sine-wave bursts. *Journal of the Acoustical Society of America*, *53*, 1351–1355.

Szabo, T. L. (1993). Linear and Nonlinear Acoustic Propagation in Lossy Media, Ph.D. thesis. University of Bath, UK.

Szabo, T. L. (2011). Medical imaging. In J. Enderle, S. Blanchard, & J. Bronzino (Eds.), *Introduction to Biomedical Engineering* (3rd ed.). Elsevier Science.

Szabo, T. L., Melton, H. E., Jr., & Hempstead, P. S. (1988). Ultrasonic output measurements of multiple mode diagnostic ultrasound systems. *IEEE Transactions on Ultrasonics, Ferroelectrics and Frequency Control*, *35*, 220–231.

Szabo, T. L., & Seavey, G. A. (1983). Radiated power characteristics of diagnostic ultrasound transducers. *Hewlett Packard Journal*, *34*, 26–29.

Teh, B. -G., & Cloutier, G. (2000). Modeling and analysis of ultrasound backscattering by spherical aggregates and rouleaux of red blood cells. *IEEE Transactions on Ultrasonics, Ferroelectrics and Frequency Control*, *47*, 1025–1035.

Tong., L., Hamilton, J., Jasaityte, R., Cikes, M., Sutherland, G., & D'hooge, J. (2012). Plane wave imaging for cardiac motion estimation at high temporal resolution: a feasibility study in-vivo. *Proceedings of the IEEE ultrasonics symposium* (pp. 228–231). Dresden, Germany.

Trahey, G. E., Allison, J. W., & Ramm, O. T. (1987). Angle-independent ultrasonic detection of blood flow. *IEEE Transactions on Biomedical Engineering*, *34*, 965–967.

Udesen, J., Gran, F., Hansen, K. L., Jensen, J. A., Thomsen, C., & Nielsen, M. B. (2008). High frame-rate blood vector velocity imaging using plane waves: Simulations and preliminary experiments. *IEEE Transactions on Ultrasonics, Ferroelectrics and Frequency Control*, *55*, 1729–1743.

Udesen, J., & Jensen, J. A. (2006). Investigation of transverse oscillation method. *IEEE Transactions on Ultrasonics, Ferroelectrics and Frequency Control*, *53*, 959–971.

Udesen, J., Nielsen, M. B., Nielsen, K. R., & Jensen, J. A. (2007). Examples of in vivo blood vector velocity estimation. *Ultrasound in Medicine and Biology*, *33*, 541–548.

Wang, S. H., Lin, Y. H., & Shung, K. K. (1997). In vivo measurements of ultrasonic backscatter from blood. *Proceedings of the IEEE ultrasonics symposium* (Vol. 2) (pp. 1161–1164). Toronto, ON.

Wells, P. N. T. (1969a). A range gated Doppler system. *Medical and Biological Engineering*, *7*, 641–652.

Wells, P. N. T. (1969b). *Physical Principles of Ultrasonic Diagnosis*. London: Academic Press.

Wilhjelm, J. E., & Pedersen, P. C. (1993a). Target velocity estimation with FM and PW echo ranging Doppler systems, Part I: Signal analysis. *IEEE Transactions on Ultrasonics, Ferroelectrics and Frequency Control*, *40*, 366–372.

Wilhjelm, J. E., & Pedersen, P. C. (1993b). Target velocity estimation with FM and PW echo ranging Doppler systems, Part II: Systems analysis. *IEEE Transactions on Ultrasonics, Ferroelectrics and Frequency Control*, *40*, 373–380.

Bibliography

Brock-Fisher, G. A. M., Poland, M. D., & Rafter, P. G. (1996). Means for Increasing Sensitivity in Non-Linear Ultrasound Imaging Systems, US patent 5, 577, 505, November 26, 2996.

Evans, D. H. (1993). In P. N. T. Wells (Ed.), *Advances in ultrasound techniques and instrumentation.* New York: Churchill LivingstoneChap. 8.

A review of color flow imaging.

Evans, D. H., McDicken, W. N., Skidmore, R., & Woodcock, J. P. (1989). *Doppler Ultrasound physics: Instrumentation and Clinical Applications*. Chichester, UK: John Wiley & Sons.

An older text with valuable information.

Ferrara, K., & Deangelis, G. (1997). Color flow mapping. *Ultrasound in Medicine and Biology, 23*, 321–345.

A review of color flow imaging.

Hattle, L., & Angelsen, B. (1985). *Doppler ultrasound in cardiology: Physical principles and clinical application* (2nd ed.). Philadelphia: Lea and Ferbiger.

A classic text on Doppler principles.

Jensen, J. A. (1996). *Estimation of blood velocities using ultrasound.* Cambridge, UK: Cambridge University Press.

A recommended book for more details about Doppler and Doppler-related imaging and the measurement of blood flow.

Magnin, P. A. (1987). A review of Doppler flow mapping techniques. *Proceedings of the IEEE ultrasonics symposium* (pp. 969–977). Denver, CO.

An explanation of color flow imaging methodologies.

Routh, H. F. (1996). Doppler ultrasound. *IEEE Engineering in Medicine and Biology, 15*, 31–40.

A review of Doppler ultrasound.

Wells, P. N. T. (1998). Current Doppler technology and techniques. In F. A. Duck, A. C. Baker, & H. C. Starritt (Eds.), *Ultrasound in medicine, medical science series*. Bristol, UK: Institute of Physics PublishingChap. 6.

A concise introduction to the essential measurement and imaging methods of Doppler ultrasound.

CHAPTER 12

Nonlinear Acoustics and Imaging

Chapter Outline

12.1 Introduction

Nonlinear effects are important for harmonic imaging, contrast agents, and acoustic output measurements. The effects of nonlinearity combine and interact with all the other major components of imaging (attenuation, focusing, and signal processing) and, therefore, they cannot be understood in isolation. In addition, linearity (a fundamental design assumption for imaging systems), based on proportionality and superposition, must be reexamined to work with nonlinearity. This chapter explores what nonlinearity is and how it extends and challenges our understanding of linear wave propagation. Contrast agents are discussed separately in Chapter 14.

What is nonlinearity? Nonlinearity is a property of a medium by which the shape and amplitude of a signal at a location are no longer proportional to the input excitation. In a fluid, for example, the relation between variations in pressure and changes in density from equilibrium values is no longer linear (as shown for water by Figure 12.1). Two curves are depicted, and each one is an approximation to the actual nonlinear relationship. Until now, an assumption has been made that for infinitesimal amplitudes, linearity holds, as described by:

$$p - p_0 = A\left(\frac{\rho - \rho_0}{\rho_0}\right) = \left[\rho_0\left(\frac{\partial p}{\partial \rho}\right)_{S,\ \rho=\rho_0}\right]\left(\frac{\rho_0 - \rho_0}{\rho_0}\right) = \rho_0 c_0^2\left(\frac{\rho - \rho_0}{\rho_0}\right), \tag{12.1A}$$

where p_0 and ρ_0 are the pressure and density at equilibrium in a fluid (similar relations can be found for gases and solids). Here, $A = \rho_0 c_0^2$ is a linear constant taken for $\rho = \rho_0$ and at a specific entropy. The assumption is made that the process is adiabatic, meaning that there is no heat transfer during the rapid fluctuations of an acoustic wave. A better approximation is to include the next term in a Taylor expansion series (Beyer, 1997) for the pressure as a function of density:

$$p - p_0 = A\left(\frac{\rho - \rho_0}{\rho_0}\right) + B\left(\frac{\rho - \rho_0}{\rho_0}\right)^2 + \ldots, \tag{12.1B}$$

called the "nonlinear equation of state," with B defined as:

$$B = \left[\rho_0^2\left(\frac{\partial^2 p}{\partial \rho^2}\right)_{S,\ \rho=\rho_0}\right]. \tag{12.1C}$$

Eqn 12.1B is plotted as the nonlinear curve in Figure 12.1.

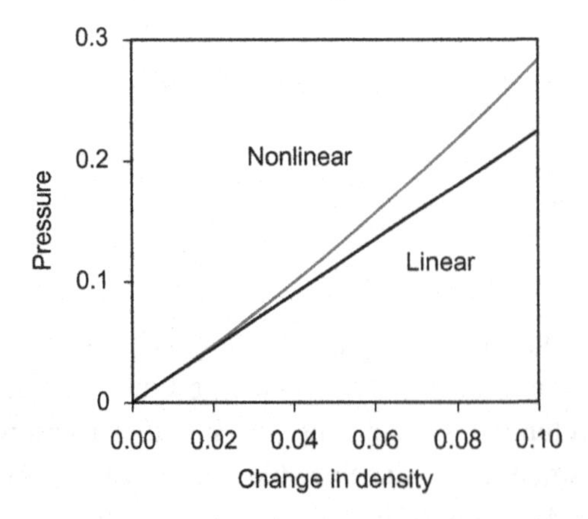

Figure 12.1
Linear and Nonlinear Characteristics of Pressure Versus Density for Water.

A simple measure of the relative amount of nonlinearity is the ratio of *B/A*. More common, however, is the coefficient of nonlinearity, β (not to be confused with the wave number propagation factor), which is defined as:

$$\beta = 1 + B/2A, \tag{12.2}$$

which will be related later to the speed of sound in a nonlinear medium. Not only is water nonlinear, but so are all tissues (as shown in the graph of β in Figure 12.2). Coefficients for tissue fall in the range of 3 to 7 (water to fat) (Duck, 1990). Note that tissues are only slightly more nonlinear than water. Contrast agents (discussed in Chapter 14) can have nonlinearity coefficients of more than 1000 in high concentration (Wu & Tong, 1994).

In addition to tissues being nonlinear, so are many physical phenomena in the world around us. Linear approximations to reality are used for convenience, simplified understanding, and design control. Nonlinear approximations are more accurate, but they are more complicated for use in simulation and design. Because the effects of nonlinearity are amplitude dependent, in acoustics they are also called "finite amplitude" (as opposed to linear theory, which is based on infinitesimal amplitudes).

Nonlinearity produces strange behavior not predictable by our usual linear viewpoint. The major consequences of acoustic propagation in a nonlinear medium are cumulative pulse and beam distortion, harmonic generation, and ultimately, saturation. Consider the stages of waveform evolution in Figure 12.3. At the top of this figure, one cycle of a long tone burst of a single-frequency plane wave is shown as the input to a nonlinear medium. As the signal propagates, distortion begins and simultaneously creates low levels of harmonics. Next in the sequence, the cumulative distortion eventually leads to a "sawtooth," or "N"-shaped waveform or "shock wave," which has frequencies at harmonic multiples of the fundamental. If the fundamental is called the "first" harmonic and each harmonic is designated by n, the amplitude of each harmonic in the spectrum falls off by n^{-1} for this waveform. Finally, at great distances and at higher frequencies, only an attenuated low-amplitude "old age" waveform is left, that is no longer proportional to the original emitted amplitude.

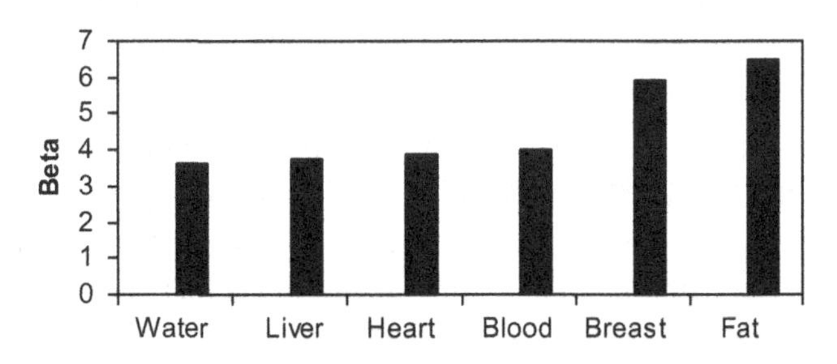

Figure 12.2
Coefficients of Nonlinearity for Tissues and Water.

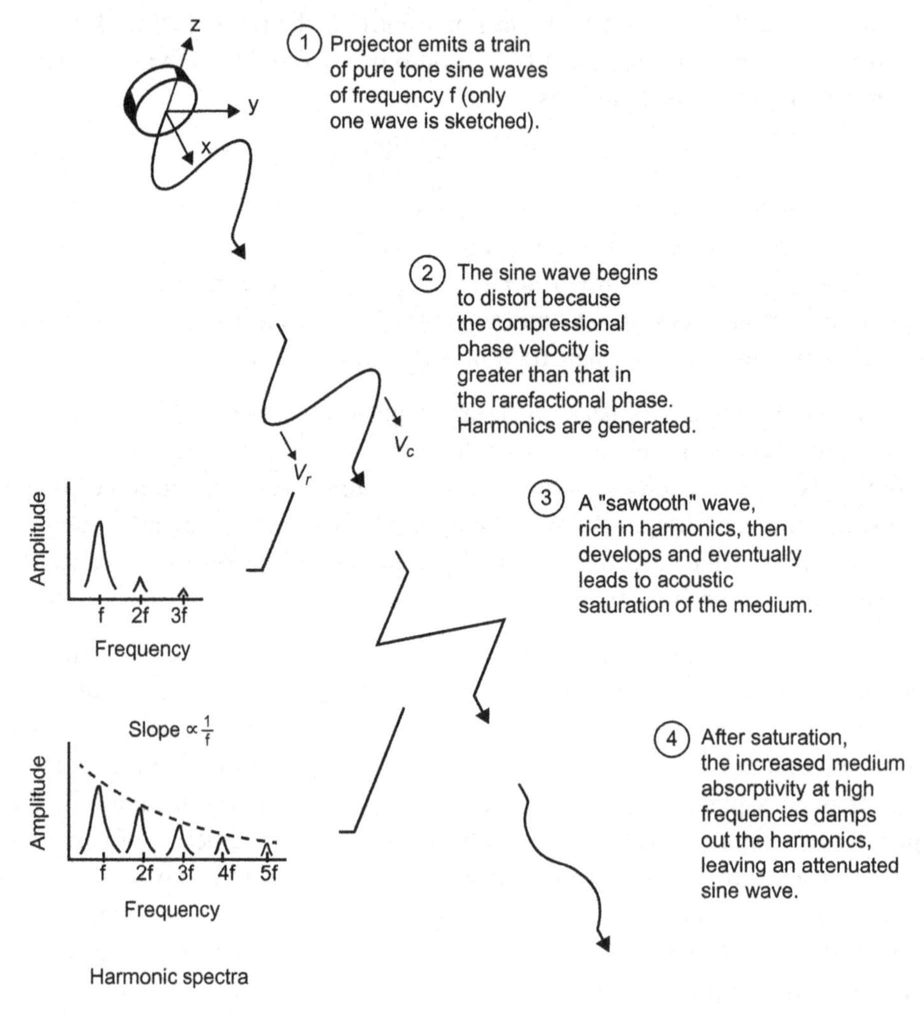

Figure 12.3
Evolution of a Shock Wave Beginning from a Plane Sinusoidal Wave Source.
Note that even though a transducer source is shown, the waveforms are those of an infinite plane-wave transmitter. *From Muir and Carstensen (1980); reprinted with permission from the World Federation of Ultrasound in Medicine and Biology.*

Even though the use of these harmonics for imaging became widespread in ultrasound imaging systems by the late 1990s, nonlinear acoustics has been a growing branch of acoustics for more than 240 years (Blackstock, 1998). Intense development in this area has occurred since the 1970s (Bjorno, 2002). Three major areas that spurred this interest in nonlinear acoustics are sonic booms (shock waves generated by supersonic sources such as jet planes) beginning in the 1950s, the application of parametric arrays to increase resolution in sonar (1960s to present), and, of course, biomedical ultrasound (1970s to present). Parametric arrays employ high-intensity sound and the ability of water as a highly nonlinear medium to create narrow beams at different frequencies (Berktay & Al-Temini, 1969; Westervelt, 1963). A key enabling technology for nonlinear acoustics was the high-speed digital computer, which was necessary for the numerical solution of nonlinear equations. Improvements in wideband transducers and experimental techniques made possible the verification of newly developed models for sound propagation in nonlinear media. Other expanding areas of nonlinear acoustics include thermoacoustic refrigerators, sonochemistry, cavitation and bubble dynamics, high-power industrial and surgical applications, and nondestructive testing and evaluation (Bjorno, 2002; Tjotta, 2000).

In regard to nonlinear developments specifically related to harmonic imaging, one of the first was a demonstration of harmonic images by Muir (1980) for sonar applications. He created underwater harmonic images formed by the bandpass filtering output of a scanned wideband hydrophone at each of several harmonics (as illustrated by Figure 12.4). Second harmonic images were also reported in acoustic microscopy (Germain & Cheeke, 1988; Kompfner & Lemons, 1976). Muir and Carstensen (1980) argued that ultrasound imaging systems were capable of generating distorted waves in nonlinear media such as water and tissue. There were indications in the late 1970s and early 1980s, through related acoustic measurements in water with hydrophones and other means, that ultrasound imaging systems generated harmonics in water (Bacon, 1984; Carson, Fischella, & Oughton, 1978; Carstensen, Law, McKay, & Muir, 1980). Conclusive evidence (Duck & Starritt, 1984) in the form of acoustic output measurements of clinical systems in water with wideband hydrophones began to appear in the early 1980s. Until recently, bandpass filters in ultrasound imaging systems removed all harmonic frequencies, so that nonlinear effects from tissues went unnoticed. The first deliberate attempts at medical harmonic imaging in its present form were for imaging highly nonlinear contrast agents at the second harmonic (to be discussed in more detail later in Chapter 14). An earlier form on nonlinear imaging, called "B/A imaging," was investigated in laboratories but did not find clinical application. Starritt, Perkins, Duck, and Humphrey (1985) and Starritt, Duck, Hawkins, and Humphrey (1986) confirmed that harmonics could also be generated in tissue. Tissue harmonic imaging was reported by several groups (Averkiou, Roundhill, & Powers, 1997; Ward, Baker, & Humphrey, 1996, 1997). By the late 1990s, the benefits of tissue harmonic imaging without contrast agents became commonplace in ultrasound imaging systems. Harmonic imaging with contrast agents is covered in Chapter 14.

12.2 What is Nonlinear Propagation?

An interesting consequence of the quadratic dependence of pressure on density is a change in sound speed between the compressional (positive as shown in Figure 12.5) and

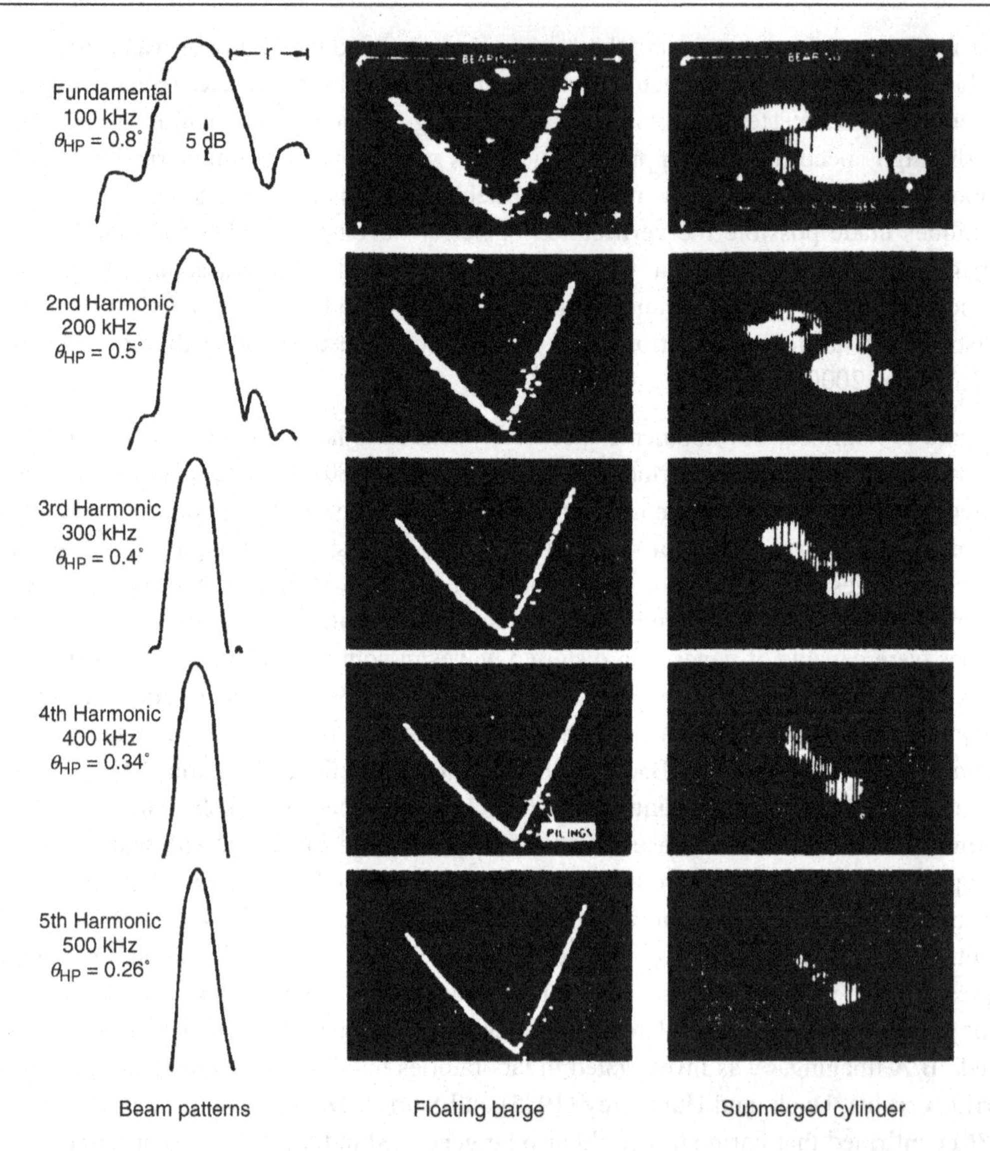

Figure 12.4
Harmonic Images of a Barge and a Submerged Cylinder in Water Produced By Filtering Out Harmonic Frequencies from a Wideband Angle-Scanned Hydrophone Receiver Mounted Above a Narrowband Transmitter. Beams at the fundamental frequency up to the fifth harmonic are shown at the left. *From Muir (1980); reprinted with permission from Kluwer Academic/Plenum Publishers.*

rarefactional (negative) half cycles of a signal. For a sinusoidal plane wave signal in a lossless nonlinear medium, the speed of sound for a displacement amplitude u is given by:

$$dz/dt = c_0 + \beta u, \tag{12.3}$$

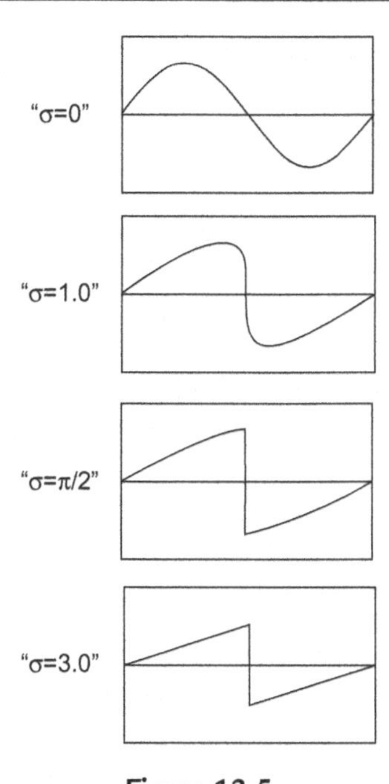

Figure 12.5
Successive Waveforms for an Initially Sinusoidal Plane Wave Shown For Increasing Normalized Distances (σ) from the source. *From Duck (2002); reprinted with permission from the World Federation of Ultrasound in Medicine and Biology.*

so that positive half cycles speed up by an extra factor βu, and the negative ones, where the displacement u is negative, slow down by βu, as shown in Figure 12.5 for increasing normalized distances denoted by a parameter (σ). This sound-speed dependence is definitely a finite amplitude effect. The positive peaks move forward toward the zero crossing, whereas the negative peaks retreat toward the zero crossing behind them. In this idealized case, when the peaks have moved $\pi/2$ from their original positions, they coincide at the zero crossing and form a sawtooth with an infinite slope at the zero crossing (π) position. The condition in which the slope first becomes infinite is called shock formation. Past this point, the wave amplitude becomes smaller. Eqn 12.3 also indicates that this change in sound speed can create increased distortion either if the medium is more nonlinear (larger β) or if the displacement amplitude u is larger. Overall, the two contributions to nonlinear distortion are both the equation of state and the local convective nonlinearity caused by the displacement on the sound speed. As an example, $B/A = 5$ for water, so that from the definition of $\beta = 1 + 2.5$, Eqn 12.2, the first term or convective contribution to distortion is one-third of the total, with the nonlinearity of the medium accounting for the remaining two-thirds (Duck, 2002).

The normalized distance nonlinearity parameter σ for a plane wave is useful in predicting distortion. The acoustic Mach number ε is defined as:

$$\varepsilon = u_0/c_0, \tag{12.4A}$$

and can also be expressed as:

$$\varepsilon = p_0/(\rho_0 c_0^2) = \sqrt{2I/(\rho_0 c_0^3)}, \tag{12.4B}$$

where I is the time-average intensity, and initial pressure at the source is $p_0 = \rho_0 c_0 u_0$. Finally, the nonlinearity parameter can be expressed as:

$$\sigma = \beta \varepsilon k z = \frac{\beta p_0 2\pi f z}{\rho_0 c_0^3}, \tag{12.4C}$$

where k is the wave number, $k = \omega/c_0$, and z is the distance from the source. The importance of Eqn 12.4C is that it predicts increasing distortion when any of the following increase: nonlinearity (β), frequency (f), amplitude (p_0), or distance (z). Note that in accordance with Figure 12.5, shock occurs when $\sigma = 1$ or at the shock distance, $z = l_s = 1/(\beta \varepsilon k)$. For this value of σ, a vertical discontinuity appears in the waveform. For values between 1 and 3, a transition region exists, and when it exceeds 3, the sawtooth region begins:

$$p(x, \tau) = p_0 \sum_{n=1}^{\infty} \frac{2}{n(1+\sigma)} \sin(n\omega\tau + \phi), \tag{12.4D}$$

in which retarded time, $\tau = t - z/c_0$, is used, and phase $\phi = 0$.

Another peculiar property of sound in a nonlinear medium is that propagation has a cumulative distortion with distance. Because of quadratic dependence of pressure on density changes, an analogy can be made with another nonlinear device, the square law mixer in electronics. Consider the nonlinear medium to be made up of a series of distributed mixers that are each an infinitesimal distance (Δz) from each other (as depicted in Figure 12.6). If the change in density is a sinusoid of frequency $\omega_0 = 2\pi f_0$ that is injected into the mixer chain, then from Eqn 12.1B, the square law mixer output would be:

$$p - p_0 = A \cos \omega_0 t + B \cos^2 \omega_0 t, \quad \text{and} \tag{12.5A}$$

$$p - p_0 = A \cos \omega_0 t + (B/2)(1 + \cos 2\omega_0 t). \tag{12.5B}$$

The fundamental and the second harmonic output feed the next mixer, and repeating the same square law process yields harmonic frequencies of f_0, $2f_0$, $3f_0$, and $4f_0$. At each position, the newly distorted waveform (represented here by its spectrum) recreates itself by interacting with the nonlinear (square law) properties of the medium. This process is represented by an idealized chain of mixers at each infinitesimal spatial location, in which

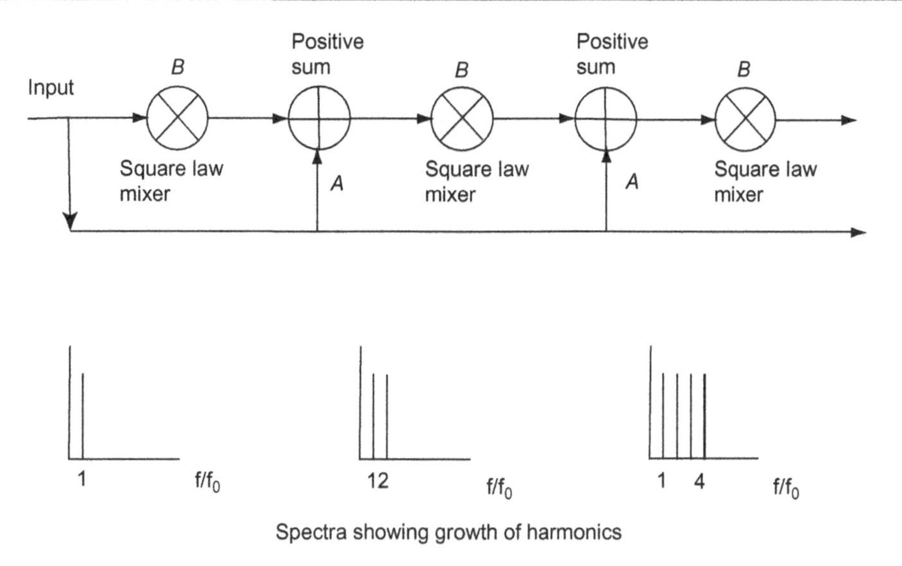

Figure 12.6
Propagation through a nonlinear medium modeled as a chain of square law mixers and summing nodes, each of which is separated by an infinitesimal distance. The number of harmonics grows as the original input of frequency (f_0) creates new frequencies at each stage.

progression in distance increases the number of harmonics exponentially. The purpose of this analogy is to demonstrate a harmonic-generating process; the actual creation of harmonics is more complicated than shown here and is covered in more detail later.

A third unusual property of acoustic propagation in a nonlinear medium, in addition to amplitude- and nonlinearity-dependent sound speed and cumulative distortion with multiple harmonic generation, is acoustic saturation. In a linearized world, amplitudes at spatial positions are proportional without limit to the input or beginning amplitude at the source. As illustrated by Figure 12.5, for a lossless ideal nonlinear medium, amplitude of an increasingly distorted wave shape peaks and then diminishes past values of $\sigma > \pi/2$.

The overall saturation effect can be seen in Figure 12.7, in which the nonlinear characteristic tracks the linear one initially and then approaches a plateau (saturation level). The difference between the expected extrapolated linear increase in amplitude at a position in a nonlinear medium based on unrealistic linear assumptions and that actually obtained is called extra or excess attenuation. This attenuation is caused by absorption, as well as the type of distortion that occurs at long propagation distances in the medium and that alters the coherent phasing of harmonic components to reduce amplitude. Note that a linear extrapolation made from the origin to any point on the nonlinear region of the saturated curve would result in an underestimate of the amplitude to the left of the point.

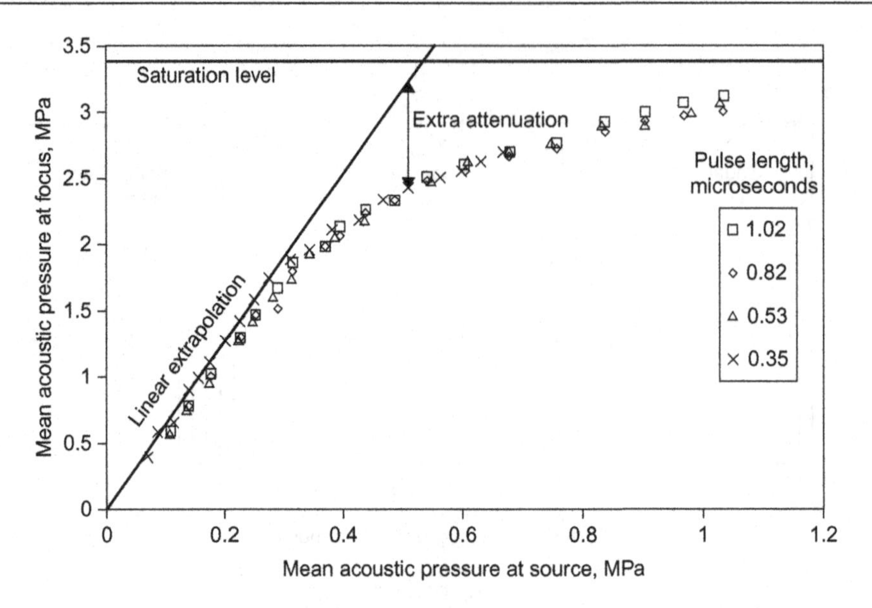

Figure 12.7
Hydrophone measurements at different pulse lengths versus increasing transmit levels that demonstrate the start of acoustic saturation phenomena in a pulsed diagnostic beam. No dependence on pulse length was observed over the range 0.35–1.02 μs. Extra attenuation is the difference between a linear extrapolation from low amplitudes and the nonlinear characteristic. *Adapted from Duck (2002); reprinted with permission from the World Federation of Ultrasound in Medicine and Biology.*

Duck (1998) has presented two expressions for the maximum acoustic pressure in the plateau region that can be reached at any distance (z) from a source of frequency (f_0). The first expression is for plane waves:

$$p_{sat,\ p} = \frac{\rho_0 c_0^3}{2\beta f_0 z}, \tag{12.6A}$$

and the second approximate expression is for a circularly symmetric focusing transducer with a focal length F and a low-amplitude focal gain of G (Naugol'nykh & Romanenko, 1959):

$$p_{sat,\ F} = \frac{\rho_0 c_0^3}{2\beta f_0 F} \frac{G}{\ln\ G}. \tag{12.6B}$$

In more lossy media other than water, absorption plays a stronger role in determining levels of saturation (Blackstock, Hamilton, & Pierce, 1998). Beamforming also alters the situation, and for tightly focused beams, Eqn 12.6B underestimates experimentally observed saturation levels (Sempsrott & O'Brien, 1999).

12.3 Propagation in a Nonlinear Medium with Losses

How does acoustic propagation in nonlinear water compare with that in a typical tissue that has a frequency power law absorption characteristic? Because the effects of absorption increase with frequency and distance, absorption acts as a low-pass filter to reduce the amplitudes of higher harmonics, as well as signal amplitude. Absorption and nonlinearity are always involved interactively in competition by reducing and creating harmonics and distortion. A measure of which one will win this contest is the Gol'dberg number (1957):

$$\Gamma = \sigma/(\alpha z). \tag{12.7}$$

Nonlinear distortion or shock begins when $\Gamma = 1$. For increasing values of Γ greater than 1, nonlinear distortion becomes more dominant; whereas for values less than 1, absorption prevents significant distortion from developing.

Gol'dberg numbers for water and tissues are compared at a typical diagnostic pressure of 5 MPa and a midrange frequency of 5 MHz in Figure 12.8. What is unusual about this chart is the large Gol'dberg number for water ($\Gamma = 266$), which is clearly in a class by itself when compared to Gol'dberg numbers for tissue, which are all less than 14. These numbers indicate that distortion is extremely easy to achieve in water, even for small amplitudes, compared to tissue. Experimentally, these effects have been observed in acoustic output measurements made in water, but extrapolating data to equivalent values in tissue is an extremely challenging nonlinear problem (Szabo, Clougherty, & Grossman, 1999). Both the amplitudes and the severity of distortion are markedly different in water than in tissues. For water, the power exponent $y = 2$ (from Chapter 4), so that Γ increases with amplitude and

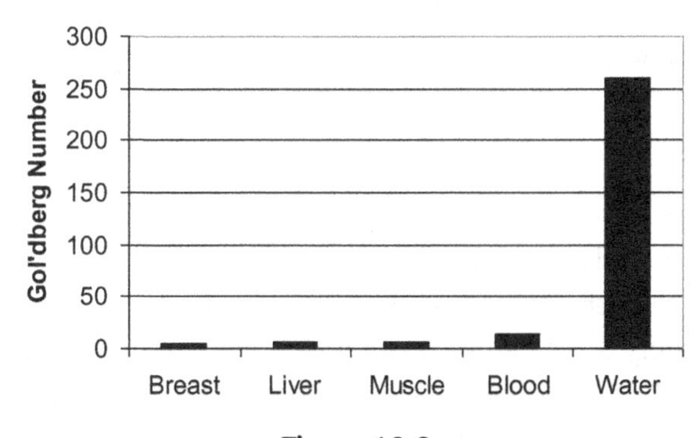

Figure 12.8
Gol'dberg Numbers for Tissue And Water For A Pressure of 5 MPa and a Frequency of 5 MHz.
From Szabo et al. (1999), with permission from the American Institute of Ultrasound in Medicine.

inversely with frequency. In contrast, for tissues, $y \approx 1$; therefore, Γ is nearly independent of frequency and changes with amplitude.

Acoustic propagation in a lossy nonlinear medium is a balance between absorption and harmonic replenishment from lower frequencies. In some cases, the loss slope at greater distances can be less than that expected from linear absorption (Haran & Cook, 1983).

An important consequence of the interaction of nonlinear effects and absorption is enhanced heating. Heating in tissue is related directly to absorption (to be explained in Chapter 13). The spectrum of a distorted waveform in a nonlinear medium contains many harmonics, each of which is being attenuated more at higher frequencies. As a result, the amount of energy lost in heating has increased over what would have occurred at the lower-frequency fundamental rate. Estimates of heating have been made on weak shock absorption (Bacon & Carstensen, 1990; Dalecki, Carstensen, & Parker, 1991), in which nonlinearity plays a dominant role. Numerical techniques are necessary for accurate prediction of the close interaction of nonlinear effects and absorption in tissue (Christopher & Carstensen, 1996; Divall & Humphrey, 2000; Ginsberg & Hamilton, 1998; Haran & Cook, 1983).

12.4 Propagation of Beams in Nonlinear Media

The accurate prediction of sound fields in nonlinear media evolved rapidly once the theory was developed, efficient numerical means of computing became available, and broadband hydrophones validated the predictions. A key development was the derivation of the Khokhlov–Zabolotskaya–Kuznetsov (KZK) wave equation under the paraxial approximation (Kuznetsov, 1971). This equation combines nonlinearity and diffraction, as well as absorption, in a numerically suitable form. By the mid-1980s, a numerical frequency domain KZK algorithm was devised to run on computer workstations (Aanonsen, Barkve, Naze Tjotta, & Tjotta, 1984). Other programs soon followed (as described in Section 12.6). Careful wideband hydrophone measurements in water by Baker, Anastasiadis, and Humphrey (1988); Baker (1989); and others (Averkiou & Hamilton, 1995; Nachef, Cathignol, Naze Tjotta, Berg, & Tjotta, 1995; Ten Cate, 1993) verified the accuracy and utility of these programs.

An example of the agreement of the KZK algorithm (Baker, 1992; Humphrey, 2000) with data for the fundamental and harmonics of a focusing transducer radiating into water is shown in Figure 12.9. Note the absence of harmonics close to the transducer. Progressively longer distances are required for the higher harmonics to build up, which is a trend expected from σ, Eqn 12.4C. For harmonics, the ascent into the focal region is steeper than for the fundamental. As the harmonic number goes up, each higher harmonic peaks at a progressively deeper depth compared to the fundamental; consequently, with harmonic imaging, a deeper focal region is achieved than that expected under linear circumstances.

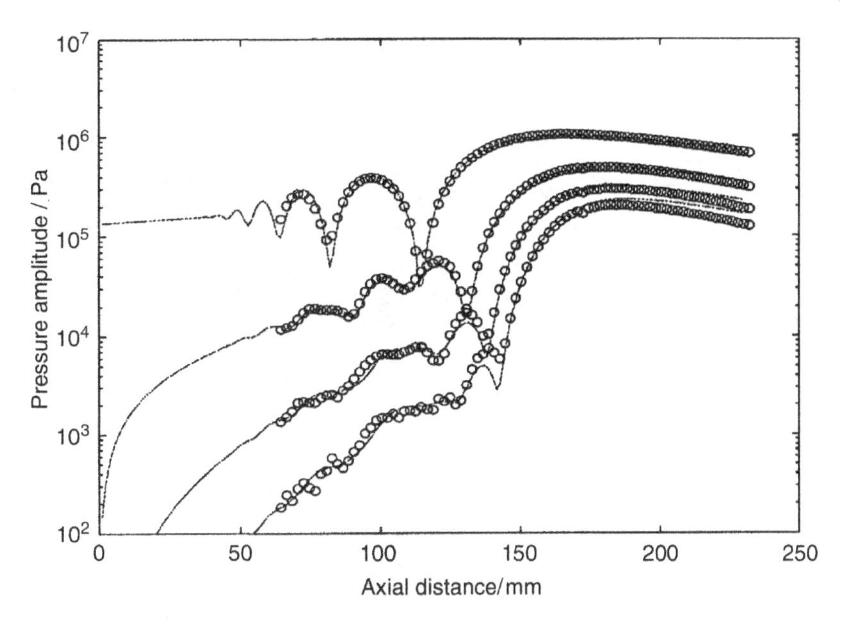

Figure 12.9

Axial Variation of Fundamental Up To Fourth Harmonic for a Circular Focusing Transducer with $a = 19$ mm, a Linear Focal Gain of 9.2 (Eqn 6.33A), and Fundamental 2.25 MHz. Experiment is shown by dots, and theory is shown by lines. Note that this set of curves is based on a source pressure of $p_0 = 135$ kPa. *Reprinted from Humphrey (2000), with permission from Elsevier.*

Graphs of measurements in the focal plane of the fundamental and second harmonic (Averkiou & Hamilton, 1995) are shown in Figure 12.10. Not only is the main lobe narrower for the second harmonic, but the number of sidelobes has increased.

In addition, similar measurements by Baker (1992) and Ten Cate (1993) provide insights into the characteristics of a harmonic beam. First, the beamwidths of the harmonics in the far field (nonfocusing aperture) or focal plane are narrower than that of the fundamental by $1/\sqrt{n}$. Thus, the second harmonic beamwidth is 0.707 narrower. Second, a natural apodization occurs so that the sidelobe levels are progressively lower as the harmonic number increases. This effect can be explained by the fact that as the amplitude falls off away from the beam axis, so does harmonic generation, as expected by the trend in Eqn 12.4C. Third, "finger" or extra sidelobes appear. Typically, for every sidelobe width of the fundamental, n sidelobes fill in (as apparent from Figure 12.10). Ten Cate (1993) has shown that these sidelobes fall off as $1/x$ in the transverse direction x. Note that these measurements were continuous wave (CW), so some filling in of the nulls will occur for pulses with a moderate or wide bandwidth. Fourth, harmonic amplitude levels on the beam axis can be even higher than expected for a sawtooth wave ($1/n$ levels for the nth harmonic). In water, the second harmonic can be as large as several decibels below the fundamental in the focal region or in the far field for a spherically focusing aperture.

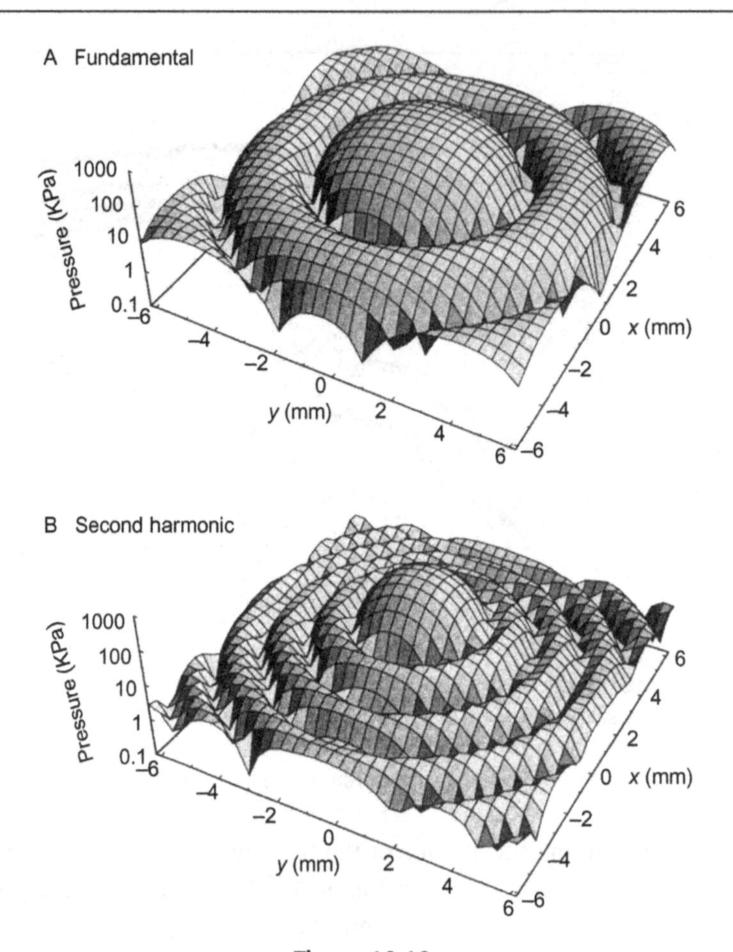

Figure 12.10
Measurements of (A) Fundamental and (B) Second Harmonic in the Focal Plane for a Continuous-Wave 2.25-MHz Transducer with $a = 18.8$ mm and $F = 160$ mm. *From Averkiou and Hamilton (1995), Acoustical Society of America.*

Nonlinear wave distortion for beams is completely different in appearance from those predicted for infinite plane waves. In Figure 12.11 are waveforms measured by Baker (1989) with a hydrophone in the far field of a piston source as the voltage drive to the transducer was increased. Here, wave shapes change from an initial sinusoid to a characteristic waveform at high amplitudes that is not a sawtooth but has a pronounced high-amplitude compressional peak and a shallower rarefactional peak. Note that these represent a family of different waveforms and beam-shapes that depend on the source amplitude.

What is the cause of this asymmetry? Parker and Friets (1987) suggested that the phasing of each harmonic is a major contributor to changing wave shapes. They have shown that by

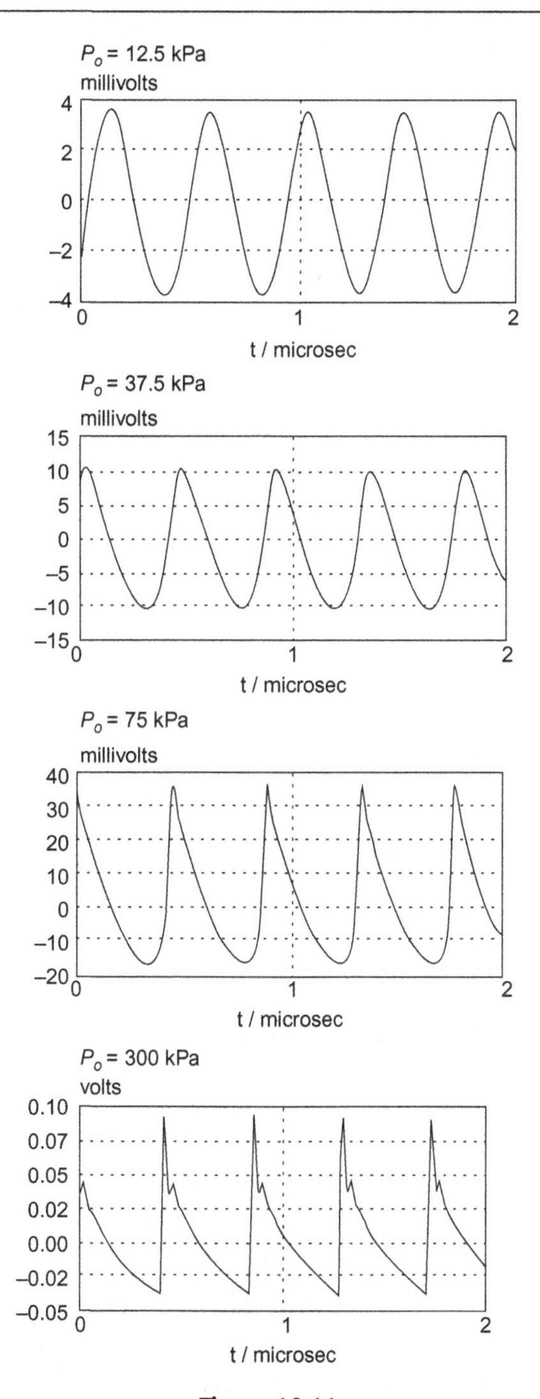

Figure 12.11
Hydrophone Measurements of Pressure at $z = 700$ mm, the Far Field of a Piston Source ($a = 19$ mm) Operating CW at 2.25 MHz, as the Drive Voltage to the Transducer is Increased. *From Baker (1989), courtesy of A. C. Baker.*

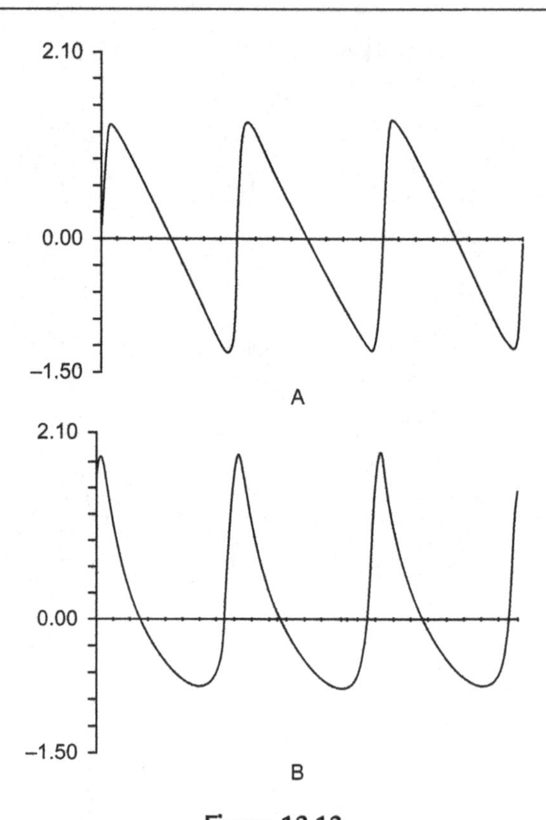

Figure 12.12
Waveform (B) Resulting From Adding a Constant Phase of $\phi = \pi/4$ Between Harmonics in the Standard Fourier Series for a Sawtooth (A). *From Parker and Friets (1987), IEEE.*

adding a constant phase ϕ to the Fourier series coefficients describing a sawtooth, Eqn 12.4D, different shapes can be obtained. If $\phi = \pi/4 = 0.785$ radian, the resulting waveform (Figure 12.12) is remarkably like those for high-amplitude pressures in Figure 12.11. Hart and Hamilton (1988) demonstrated through computations of the KZK equation that the phase of each harmonic of a focused beam is at least 90° greater than the previous harmonic. Hansen, Angelsen, and Johansen (2001) have also shown that it is the phase associated with each harmonic, especially the lower ones, that causes the asymmetry. Away from the simplified circumstances of the far field, contributions from the diffraction of the beam cause a variety of distorted asymmetric waveform shapes.

For the case of a pulsed waveform, Baker and Humphrey (1992) applied a Fourier approach to describing the source waveform at the fundamental frequency (as demonstrated byFigure 12.13). This waveform became the input pulse to a KZK simulation program to calculate the waveform at 700 mm (Figure 12.13), which has typical asymmetric behavior and harmonic buildup in the spectrum, both due to cumulative nonlinear distortion. Even

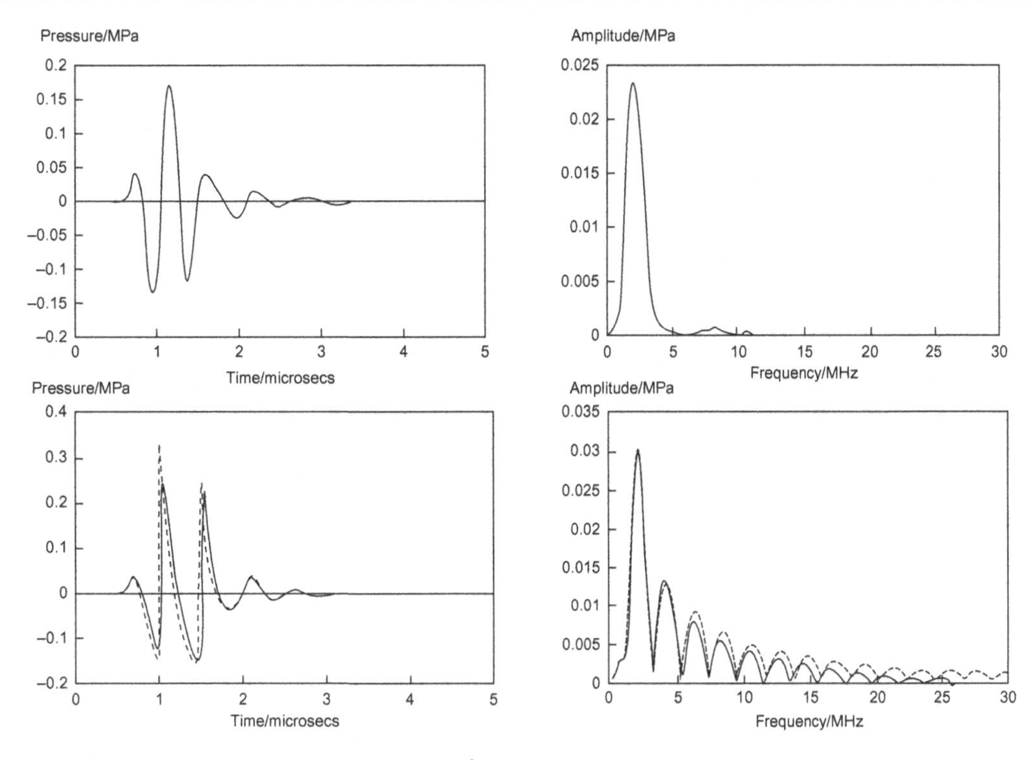

Figure 12.13

Fourier Approach to Describing a Source Pulsed Waveform at the Fundamental Frequency. (Top) a 2.25-MHz waveform at source measured at $z = 15$ mm on-axis and used for simulation, with its spectrum on the right; (bottom) pressure waveform and spectrum simulated by KZK model and compared to data at 700 mm on-axis. *From Baker and Humphrey (1992), Acoustical Society of America.*

though the propagation develops nonlinearly at any spatial location, the waveform and its spectrum there can be evaluated by linear fast Fourier transform (FFT) methods.

Most imaging systems use rectangular arrays, which behave differently than the spherically focused apertures just discussed. Because a rectangular array aperture usually has two means of focusing that are not coincident, nonlinear distortion can be more complicated than for the spherically focused case. Except for the situation in which both the elevation and azimuth focal lengths coincide, axial pressures tend to be less than a circular aperture of the same area. The trends for nonlinear circularly symmetric beams apply to rectangular apertures as well. For example, a similar buildup of the second harmonic axial pressure relative to that of the fundamental is shown in Figure 12.14.

The beam cross-sections for a rectangular aperture in Figure 12.15 reveal fingers and natural apodization effects seen in the spherical case. Predictions of fields from rectangular

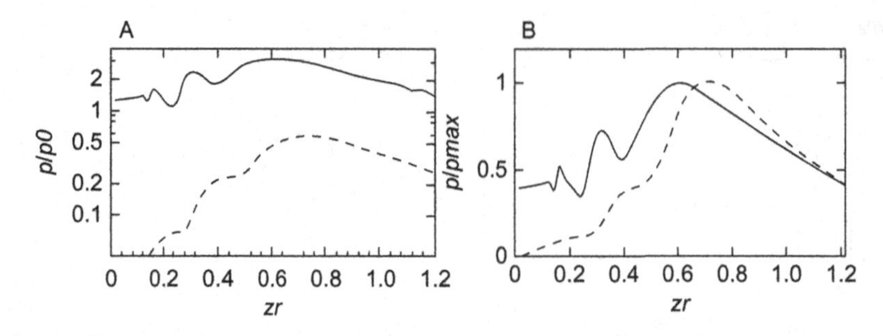

Figure 12.14
Normalized Axial Pressure Along z for a 2-MHz Pulsed Rectangular Array P4-2 Radiating into a Simulated Tissue Medium. The aperture (a_x), is 10.8 mm, and the focal distance (F) is 100 mm. Normalized distance is $z_r = z/F$. (A) Fundamental and second harmonic on logarithmic scale; (B) fundamental and second harmonic normalized on a linear scale. *From Averkiou (2000), IEEE.*

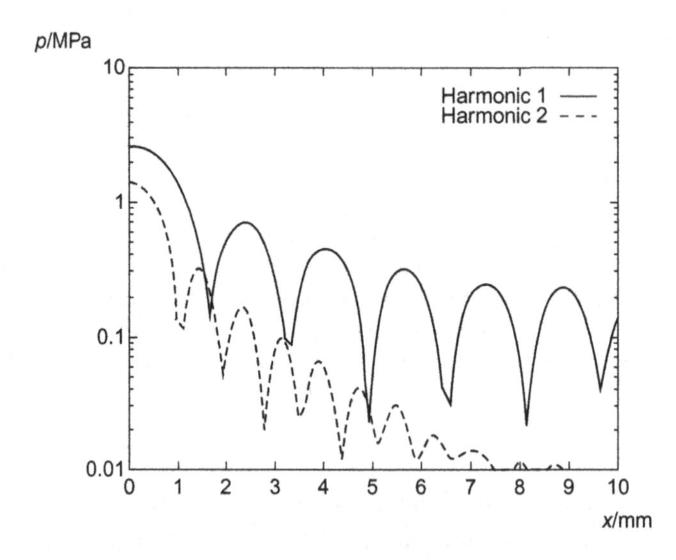

Figure 12.15
Beam Cross-Sections for a Rectangular Aperture. (Solid line) theoretical half-beam azimuth cross-sections in tissue for a 3.0-MHz array, 15 × 10 mm (azimuth × elevation), with a coincident 50-mm focal length. Calculations are for the focal plane with a 1-MPa source pressure. (Dashed line) the calculations were performed for a source pressure of 1.0 MPa and show the cross-section in the focal plane. *From Humphrey (2000), with permission from Elsevier.*

apertures include another dimension and therefore involve approximately a factor of $N^{4/3}$ times the two-dimensional computations for the circular case.

Do arrays have grating lobes at the second harmonic? Consider a standard-phased array with half-wavelength element-to-element spacing. At the second harmonic, this spacing would be a wavelength that usually corresponds to grating lobes outside of a normal $\pm 45°$ sector scan. In this case, no second harmonic grating lobes are generated at the transmit aperture at the fundamental frequency. As a result, the aperture has neither the starting pattern nor sufficient amplitude to serve as a "seed" to grow grating lobes. Recall from Chapter 7 that grating lobes are weaker than the main lobe from the angular weighting of the element directivity function. In the case of a linear array that normally has an under-sampled periodicity, or the initial aperture pattern necessary for generating grating lobes, the reduced amplitude of these grating lobes would most likely not survive the nonlinear natural apodization process. Because of the low sidelobe levels of nonlinear beams, a spatially under-sampled array in a receive mode would not detect significant echoes in the receive grating lobe regions.

12.5 Harmonic Imaging

12.5.1 Introduction

Tissue harmonic imaging (THI), also known as "native harmonic imaging," has been praised as being a breakthrough in ultrasound imaging that is as important as Doppler or color flow imaging. Equally unexpected were its accidental discovery on imaging systems and rapid commercialization in a few years. Originally, work on harmonic imaging was motivated by the need to image contrast agents. Because of the high reflectivity and extremely nonlinear properties of contrast agents, imaging system manufacturers filtered out the second harmonic of the returning echoes to separate the echoes of contrast agents from the assumed linear tissue background reflections.

Engineers and clinicians, unacquainted with the fact that tissues were also nonlinear, were puzzled to find the tissue background still in the second harmonic image and at first suspected their instrumentation. The linearity of an imaging system is checked easily. One way of validating that tissues are nonlinear is to gradually increase transmit amplitude (Muir, 1980) and to monitor the corresponding echo levels at the second harmonic and fundamental. Of course, there should be no second harmonic signal if the tissue is linear, and the echoes at the fundamental should track the transmit levels in a proportional fashion. In this way, tissue was found to behave as a nonlinear medium.

Once harmonic imaging had been implemented in imaging systems, an even bigger surprise was that clinicians began to favor the second harmonic images of tissue over those at the fundamental frequency. Examples of fundamental and second harmonic cardiac images are shown in Figure 12.16. The greater clarity, contrast, and details of the harmonic images are evident and have been quantitatively verified (Kornbluth, Liang, Paloma, & Schnittger, 1998; Spencer, Bernarz, Rafter, Korcarz, & Lang, 1998). These images emphasize another major discovery: People who were imaged poorly or not at all with conventional fundamental frequency ultrasound could be examined by second harmonic imaging. People

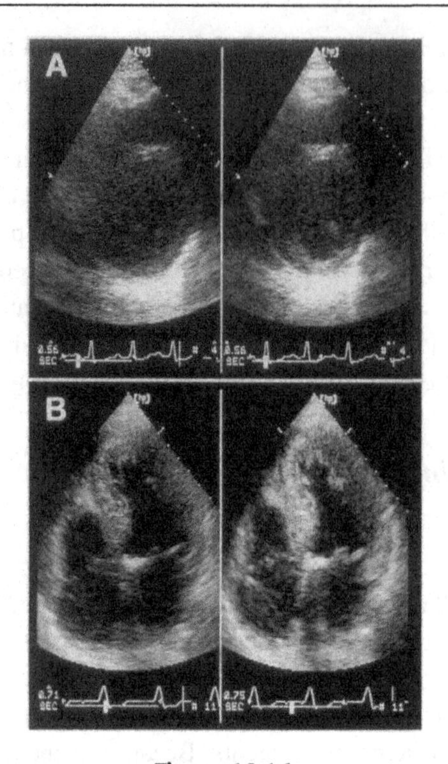

Figure 12.16
Second Harmonic Compared to Fundamental Frequency Images.
Representative cardiac images obtained on a difficult-to-image patient with fundamental imaging (left) and second harmonic imaging (right) for parasternal short-axis (A) and apical four chamber (B) views. *From Spencer et al. (1998), reprinted with permission from Excerpta Medica, Inc.*

who are difficult to image with ultrasound (about 30% of the population) are often those who have the greatest need to be imaged because of their health disorders. There are many examples of improvements in noncardiac applications as well (as illustrated by Figure 12.17 for the gallbladder). With all these clinical benefits, it is no wonder that harmonic imaging gained rapid acceptance and incorporation into new scanners.

Despite the significant advantages offered by harmonic imaging, their scientific basis has been explained only partially (Averkiou, 2000; Duck, 2002; Humphrey, 2000; Li and Zagzebski, 2000; Spencer et al., 1998; Tranquart, Grenier, Eder, & Pourcelot, 1999). To those not familiar with nonlinear acoustics, THI appears to defy the well-known physical laws of linear acoustics; furthermore, contributing effects are difficult to isolate individually. The following discussion will comprehensively cover some of the reasons behind the success of harmonic imaging and its limitations in clinical circumstances. Where possible, three types of imaging alternatives will be compared for arrays with coincident azimuth and elevation focusing: conventional fundamental frequency imaging, second

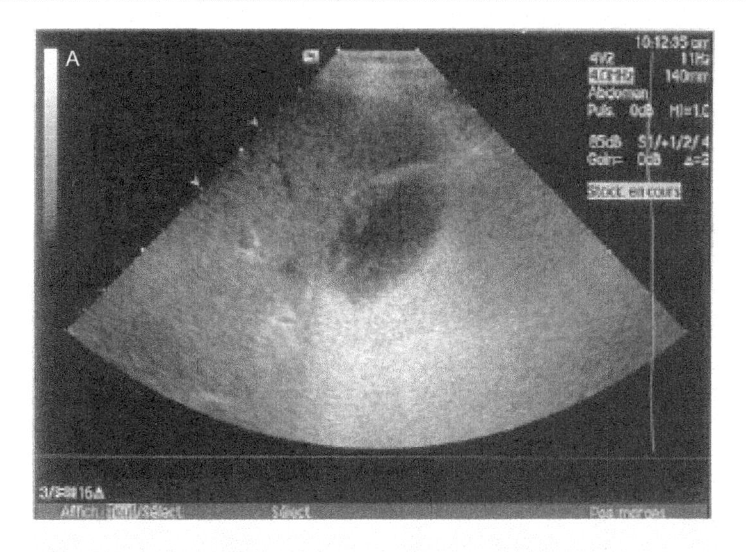

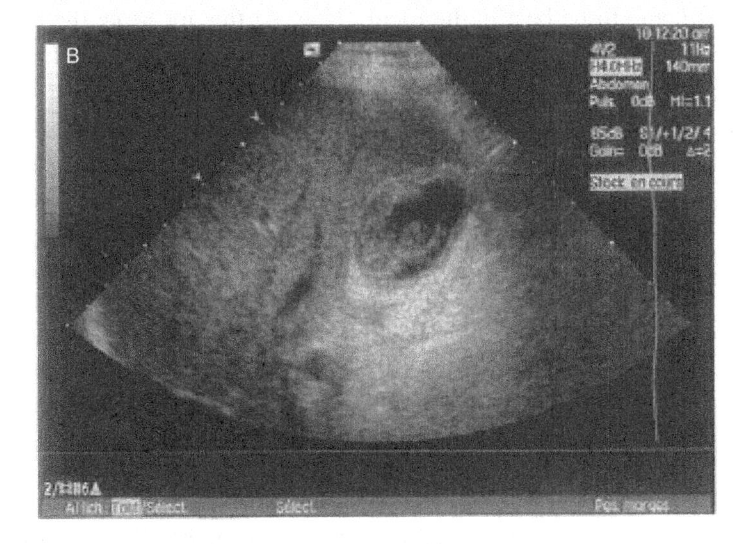

Figure 12.17

Images at the Fundamental (Top) and Second Harmonic (Bottom) of Chronic Cholecystitis. Note that the harmonic image contains much more detail and contrast, as well as showing the contents and wall of the gall bladder. *From Tranquart, Grenier, Eder, and Pourcelot (1999); reprinted with permission from the World Federation of Ultrasound in Medicine and Biology.*

harmonic imaging, and conventional imaging at twice the fundamental frequency. Results are summarized in Table 12.1. Familiar topics, such as focusing, arrays, scattering from tissue, and signal and image processing, will be revisited from an unusual point of view that will seem baffling to those with linear expectations. Harmonic imaging, which involves the nonlinearities of both tissues and contrast agents, is deferred until Chapter 14.

Table 12.1 Comparison of Parameters for Fundamental (f_0), Twice Fundamental (linear, $2f_0$), and Second Harmonic (f_{2H})Modes

Parameter	f_o	$2f_o$	f_{2H}
Axial resolution	m/f_0	$m/2f_0$	m/f_0
Azimuth FWHM	W_0	$W_0/2$	$W_0/\sqrt{2}$
− 40-dB beamwidth	W_{40}	$W_{40}/2$	$W_{402H} << W_{40}$
Sidelobe level	SL_0	SL_0	$SL_{2H} << SL_0$
Focusing gain	G_0	$2G_0$	Variable
Depth of field	DOF_0	$DOF_0/2$	Variable
Grating lobe level	GL_0	$GL_0/4$	~ 0

FWHM, full width half maximum; *m*, number of cycles; *L*, physical aperture; SL, side lobe; G, focusing gain; DOF, depth of field; GL, grating Lobe.

12.5.2 Resolution

Resolution is considered to be best in the focal plane. At this location, the axial resolution is a measure of pulse length, $\tau = m/f_0$ cycles of the fundamental (f_0). For the double fundamental frequency case, the axial resolution is half that of the fundamental, $\tau/2 = m/2f_0$. At the second harmonic, the envelope of the pulse remains the same as that of the fundamental, or $\tau = 2m/2f_0$. This effect can be seen indirectly in Figure 12.13, in which a highly distorted pulse has the same pulse length as the linear pulse. Examples of fundamental and second harmonic simulated pulses for medical imaging with similar envelopes can be found in Averkiou (2000).

A more direct measure of harmonic axial resolution was made by Ward et al. (1997) when they measured harmonic-rich echoes from a wire target with a broadband hydrophone. Their results in Figure 12.18 indicate a second harmonic pulse length approximately the same length as the fundamental with the higher harmonics slightly shorter, perhaps because of absorption.

From previous discussions in Section 7.5.2, spatial resolution can be quantified in terms of detail resolution, corresponding to a −6-dB beamwidth, and contrast resolution that corresponds to the −40-dB beamwidth. Detail resolution is a measure of how well small objects are resolved. Contrast resolution is a measure of how well subtle differences in tissue can be distinguished, as well as of the overall range of amplitude reflectivities that are possible to see.

For the spatial resolution in the focal plane, if the −6-dB fundamental one-way beamwidth, w_0 = FWHM (full width half maximum), then at $2f_0$, it is $w_0/2$; at the second harmonic, f_{2H}, it is $w_0/\sqrt{2}$ (as apparent from Figures 12.15 and 12.18).

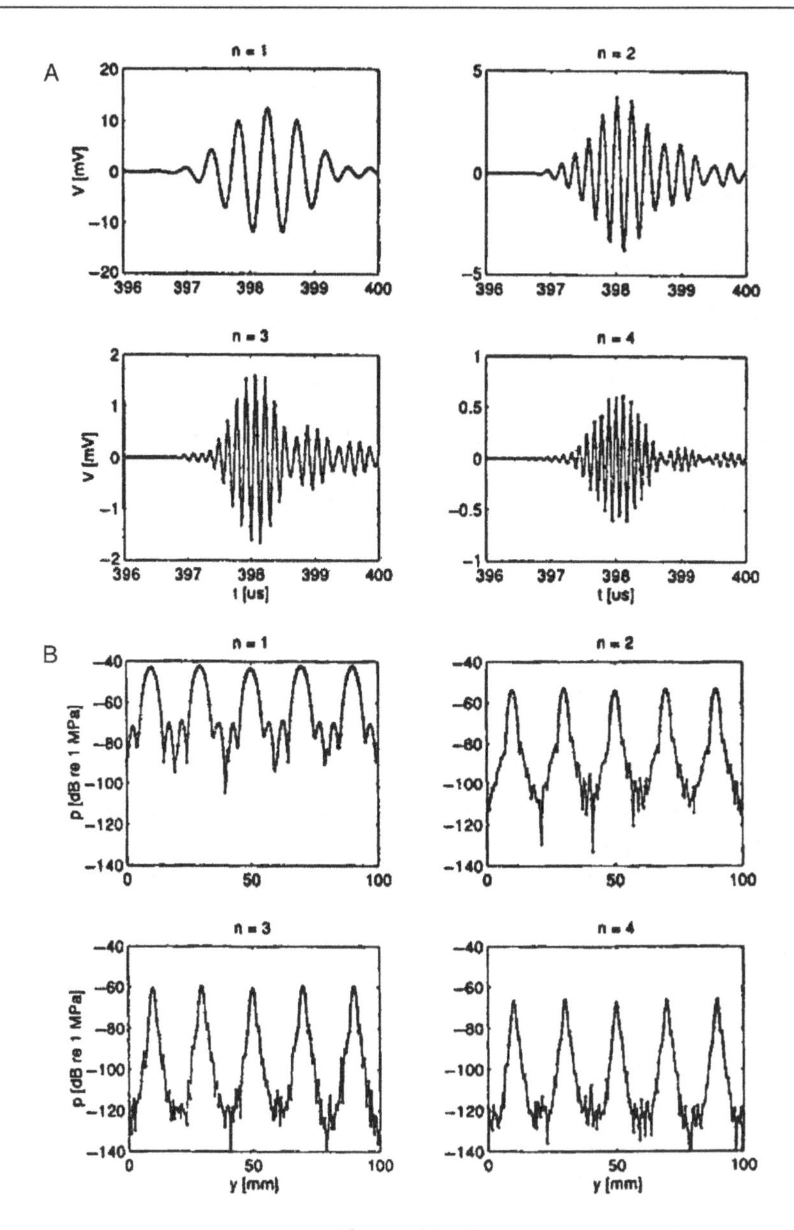

Figure 12.18

Reflected Pulse Echoes (A) and Beam Profiles (B) From a Row of Wire Targets in Water as measured by a Hydrophone and a Source Pressure of 400 kPa.

The echoes were bandpass filtered to obtain the fundamental ($n = 1$) and harmonics up to the fourth ($n = 4$) components of the signals. *From Ward et al. (1997), Acoustical Society of America.*

What is more remarkable is the contrast resolution of harmonic beams, which is at the −40-dB beamwidth level. For the case depicted in Figure 12.15, this corresponds to a fundamental −40-dB half beamwidth of $x = 52.5$ mm compared to a half beamwidth of only 7 mm for the harmonic. As discussed in Section 7.5.2 and Chapter 9, the transmitted beam has opportunities not only to interact with strong scatterers anywhere in its beam pattern, but also to integrate the cumulative volume under its sidelobes in the case of diffusive tissue scattering. For the array operating at $2f_0$, with proper $\lambda/2$ spacing, the corresponding half beamwidth is still half of that at the fundamental or, in this case, about 26 mm. Since transmit beams are shown in Figure 12.15 (the −20-dB beamwidth is also of interest), they are about 6 mm for f_0, 2.4 mm for the second harmonic, and 3 mm for $2f_0$.

Finally, for receive beamforming, the backscattered echoes are usually considered to be low enough in amplitude to propagate linearly (Li & Zagzebski, 2000); however, the second harmonic and $2f_0$ beams, being at the same frequency, are similar and narrower than the fundamental beam at f_0.

12.5.3 Focusing

While resolution is an aspect of focusing, a somewhat better picture of the overall effects of focusing can be depicted in a contour plot of a field from a focusing array in the azimuth plane (such as in Figure 12.19). While these are contours for a focused transducer in an attenuating tissue-like medium, the one-way general characteristics hold for rectangular arrays as well. Here, simulations are for a rectangular array radiating into a lossy medium with 0.3 dB/MHz^2-cm. There is excellent correspondence between the contour features of the f_0 and $2f_0$, as expected from the focal scaling law of Chapter 6. According to Eqn 6.34, the higher-frequency beam focuses more deeply—approximately 56 and 72 mm for the peak values, not accounting for losses (also see Figure 12.14). The aperture at twice the fundamental frequency has twice the number of wavelengths so that the focal range and depth of field are compressed into a shorter physical distance. While the resolution of the $2f_0$ beam is roughly half in the focal region and the focal gain is double that of the fundamental beam, the beamwidth in the near-Fresnel zone close to the aperture is similarly wide. A consequence of the shorter depth of field and increased absorption at higher frequencies is a much shorter penetration depth for the $2f_0$ beam.

By comparison, for the harmonic beam, resolution in the focal region is nearly as good as it is in the $2f_0$ beam, but in the near-Fresnel zone, it differs. Close to the aperture, where the harmonics have not had sufficient distance to build up, there is a dead zone, and beyond it is a weak field and a sudden rapid buildup to the focal region. The harmonic focal region starts slightly deeper than that of the fundamental and maintains good resolution over greater axial range than the $2f_0$ beam. This combination of characteristics (an insensitivity in the near-Fresnel zone and high resolution over an extended depth of field) is fortuitous

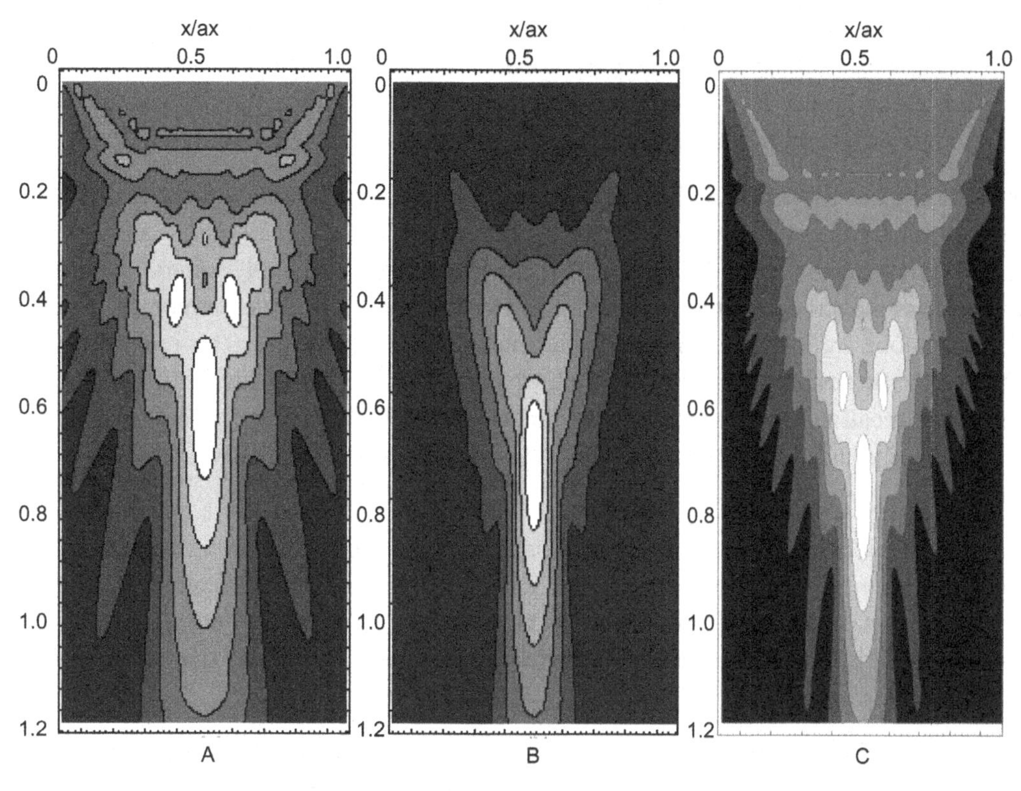

Figure 12.19
Azimuth Plane Contour Beam Plots. For (A) fundamental, (B) second harmonic, and (C) twice the fundamental (linear) of a 2-MHz pulsed rectangular phased array P4-2 radiating into a simulated tissue medium. The aperture (a_x) is 10.8 mm, the focal distance (F) is 100 mm, and $z_r = z/F$ is the vertical axis. The six contour levels represent zero to maximum (white) on a linear scale. *Adapted from Averkiou (2000), IEEE, and courtesy of M. A. Averkiou.*

for THI. Finally, note that any set of harmonic beam profiles is dependent on amplitude, so there is a family or sets of profiles for an aperture.

12.5.4 Natural Apodization

Because harmonic generation is strongest along the beam axis, pressure amplitude away from the main axis in a transverse direction falls off at a greater rate than the linear case. As evident in Figure 12.15, not only is the main lobe narrower, but also the sidelobes decrease in an enhanced way over the linear fundamental case. Under linear conditions, this type of apodization would come at the expense of a wider beam in the focal plane, as well as a decreased on-axis amplitude or focal gain. Recall from Section 7.4.2 that to decrease sidelobes under linear conditions, the source aperture is amplitude tapered toward its ends, which decreases aperture area and, consequently, focal gain. With harmonics, the full untapered aperture can be used to achieve apodization without the disadvantages that occur under linear conditions. If apodization is employed with harmonics, the beam sidelobes fall even more rapidly than the linear case.

For the two linear cases at f_0 and $2f_0$ with no original apodization at the aperture, the sidelobes fall at the same rate ($\sim 1/x$). An explanation of why the harmonic sidelobes fall much faster (as demonstrated by Figure 12.15) is that as the beam evolves into a main lobe, the center has the strongest amplitude and greatest potential to generate harmonics. The sides of the main lobe are lower in amplitude and generate fewer harmonics. This relationship is expressed by the nonlinearity parameter (σ), which is proportional to amplitude according to Eqn 12.4C. Recall that nonlinear beams continually recreate themselves as they propagate, so the effect of diminishing amplitudes on the sides of the beam is cumulative.

During imaging, strong off-axis scatterers, as well as large, soft scatterers that extend over a significant sidelobe region, can be mapped onto an image line. This strong concentration of energy in a harmonic beam provides a high selectivity against off-axis scattering. In the case of intercostal imaging, both the narrow beamwidth and dead zone properties of the harmonic beam minimize the pressure amplitudes reaching the ribs and scattering back toward the transducer.

In this kind of situation, the part of the beam controlled by the elevation aperture and focusing is often overlooked. Because of its fixed focal length, the beam in the elevation plane can be wide in a region in which the elevation and azimuth focal lengths are no longer coincident. For the harmonic beam, the steeper falloff of pressure, especially at larger distances from the beam axis, significantly reduces the sidelobe volume in which unwanted targets can lie and be mapped into the image.

Fedewa et al. (2001) compared the spatial coherence of beams at harmonic and fundamental frequencies by measuring backscatter from a tissue-mimicking phantom. They found that while fundamental frequency data corresponded to the autocorrelation function of the transmit aperture in accordance with the van Cittert–Zernike theorem (discussed in Section 8.4.5), the spatial coherence of the harmonic was lower. To determine the effective apodization at these two frequencies, beams filtered at these frequencies were measured in the focal plane and linearly back-propagated to the aperture using the angular spectrum of waves. The known apodization was recovered at the fundamental, and its autocorrelation function matched the spatial coherence data at this frequency. As discussed earlier, the same physical aperture operating linearly at twice the fundamental frequency results in a similarly shaped beam, but it is narrower by a scale factor of one-half. By back-propagation at the same frequency, the effective second harmonic apodization was found to be narrower than the full aperture, and similar processing brought agreement with the harmonic spatial coherence data. Under linear conditions, apodization results in a wider focal plane beam. In this case, the narrower effective second harmonic apodization corresponds to a harmonic beamwidth that is $1/\sqrt{n}$ of that at the fundamental but wider than the beamwidth expected at twice the fundamental ($2f_0$) under linear conditions.

12.5.5 Body-Wall Effects

Heterogeneities in the body wall cause multiple reverberations (as discussed in Chapter 9). These low-level echoes become clutter and haze (or acoustic noise) in an image. Because of their low amplitude and lack of coherence, they do not generate significant harmonics. Since their spectral content remains within the fundamental pass band, the reverberation echoes can be removed effectively by second harmonic bandpass filtering.

Aberration is a deformation of an ideal focusing wavefront caused by propagation variations in path lengths from different parts of the aperture to the focal point. Several factors contribute to these time-delay differences along various path lengths: nonuniformly thick tissue layers of different sound speeds and scattering from and through heterogeneous structures. Focusing designs for imaging systems are based on the assumption of a homogeneous medium with a constant speed of sound of 1.54 mm/μs.

For propagation through multiple heterogeneous body-wall layers, the focusing wavefront continues to deviate from the ideal as it propagates. Adjacent element paths with slightly different average sound speeds undergo a phase change that is proportional to frequency:

$$\phi_{ERROR} = 2\pi f \; \Delta r \left(\frac{1}{c_0 + \Delta c} - \frac{1}{c_0} \right) \approx - \frac{2\pi f \; \Delta r \, \Delta c}{c_0 \quad c_0}, \tag{12.8}$$

where the path difference is Δr and the difference in sound speeds is Δc. This relation indicates that for the same tissue structure, the phase error will increase with frequency. Returning to our comparison, we conclude that aberration should be worse for $2f_0$ compared to the fundamental. For the second harmonic, a curious phenomenon occurs. As the wavefront propagates through the body wall, it starts with the phase error of the fundamental because the harmonic has not had sufficient distance to grow yet; consequently, one would expect the harmonic beam to be better than the $2f_0$ beam.

Christopher (1997) used the University of Rochester body-wall data (described in Chapter 9) to run simulations of the effects of aberration on fundamental frequency and second harmonic beams. The data was translated into phase screens with jitter to simulate the aberration. He compared fundamental frequency (f_0) beams, second harmonic beams, and double fundamental frequency ($2f_0$) beams. He found that they all suffered a loss in absolute main lobe sensitivity and that the sidelobe levels of the second harmonic were significantly lower. On receive beamforming, the second harmonic and $2f_0$ beams suffer the same aberration effects (worse than that from the f_0 beam).

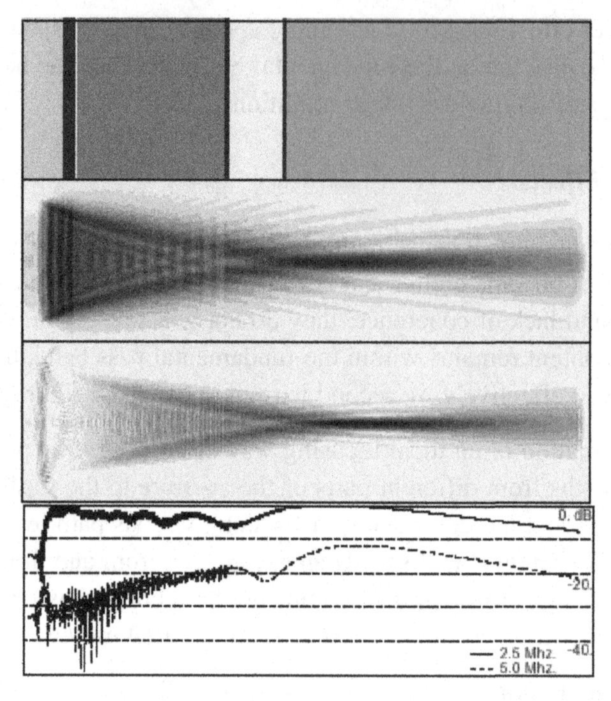

Figure 12.20

Nonlinear Simulation of the Effects of Aberration on Fundamental Frequency and Second Harmonic Frequency. (Top panel) acoustic propagation through an idealized piecewise continuous homogeneous abdominal wall section. Color coding for layers from left to right is water, fat, muscle, and liver, with black representing connective tissue. Size is 1.5 × 5 cm. Spectral amplitude distribution rate for the 2.5-MHz fundamental (second panel) and 5-MHz second harmonic (third panel) as well as along the beam axis (bottom panel) for a 1.5-cm aperture with a 5-cm focal length. The scale on the middle two figures is 2 × 8 cm (vertical × horizontal). *From Wojcik et al. (1998), IEEE.*

Wojcik, Mould, Ayter, and Carcione (1998) also ran nonlinear simulations with the same data; however, their finite difference approach allowed the inclusion of realistic structural detail and reflections. Their baseline simulation in Figure 12.20 is for the piecewise continuous succession of homogeneous flat layers. Since the sound speeds in these materials are similar, the overall beam-shapes are only slightly altered from what would be expected for one continuous homogeneous tissue. The beams are similar in shape to those of Figure 12.19, except for the standing waves in the first two layers apparent in the axial plots. The second simulation (Figure 12.21), also for a focal length of 5 cm, includes marbling and irregular interfaces. In this sequence, the strong reverberations near the entrance of the beam are missing from the second harmonic beam. In addition, the second harmonic main lobe remains more tightly focused to a deeper depth than the fundamental. In another wall sample with more marbling in the muscle layer, MS3 (shown in

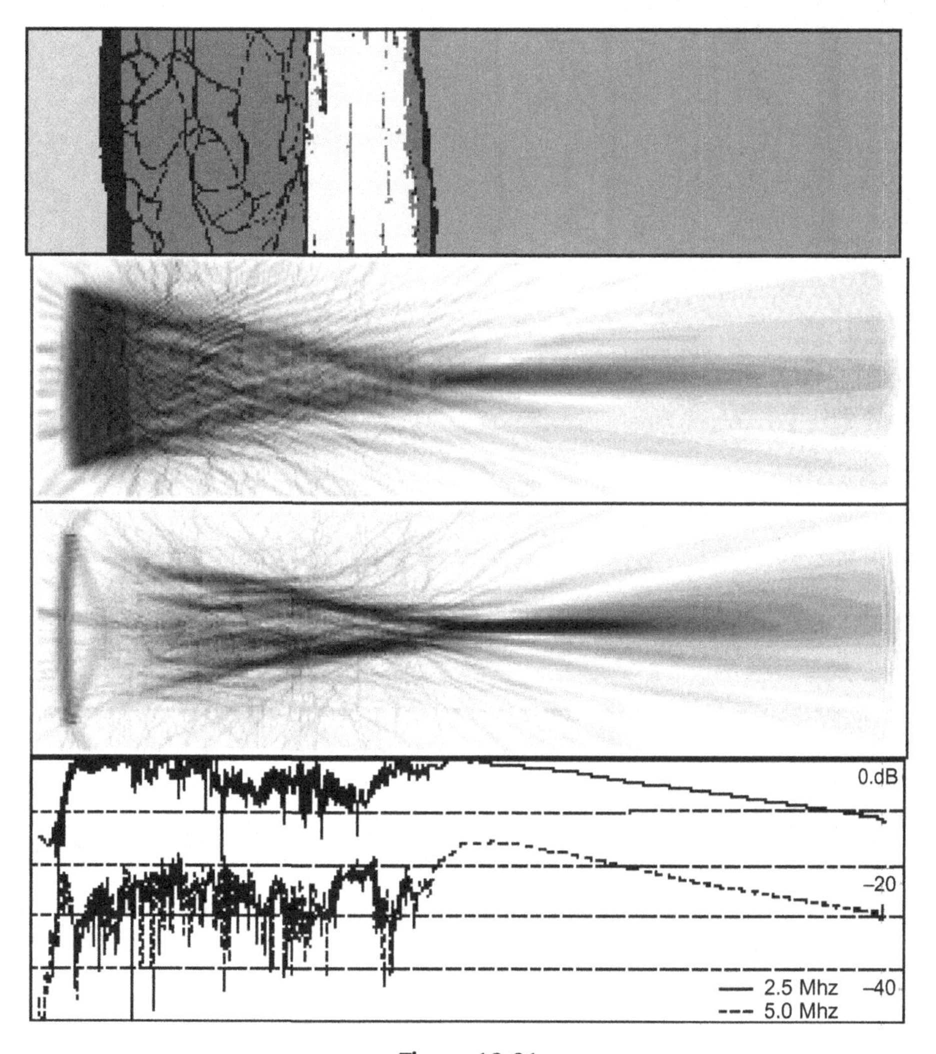

Figure 12.21
Second Simulation of Aberration Effects on Fundamental and Second Harmonic Frequencies Including Marbling and Irregular Interfaces. (Top panel) acoustic propagation through abdominal wall section MS2. Color coding for layers from left to right is water, fat, muscle, and liver, with black representing connective tissue. Size is 1.5 × 5 cm. Spectral amplitude distributions are for the 2.5-MHz fundamental (second panel) and 5-MHz second harmonic (third panel), as well as along the beam axis (bottom panel) for a 1.5-cm aperture with a 5-cm focal length. The scale on the middle two figures is 2 × 8 cm (vertical × horizontal). *From Wojcik et al. (1998), IEEE.*

Figure 12.22), the harmonic beam offers a more modest improvement. For this case, as the focal length is changed from 5 to 10 cm, the harmonic and fundamental beams both fail to focus. This result makes sense even under linear conditions, because the ideal delay curve for a deeper focus is flatter and more susceptible to disruption by even small time-delay

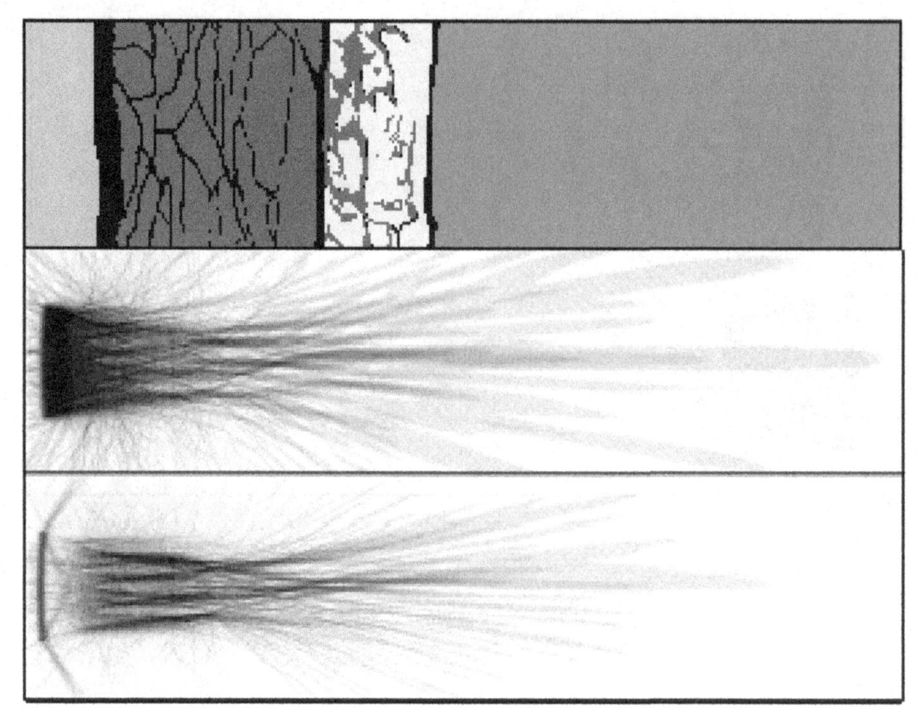

Figure 12.22
Third Simulation of Aberration Effects on Fundamental and Second Harmonic Frequencies with More Marbling in the Muscle Layer. (Top panel) acoustic propagation through abdominal wall section MS3. Color coding for layers from left to right is water, fat, muscle, and liver, with black representing connective tissue. Size is 2 × 10 cm. Spectral amplitude distributions are for the 2.5-MHz fundamental (second panel) and 5-MHz second harmonic (third panel) for a 2-cm aperture with a 10-cm focal length. The scale on the bottom two figures is 4 × 16 cm (vertical × horizontal). *From Wojcik et al. (1998), IEEE.*

errors. Studies of harmonic aberration correction by Christopher (1997) and Liu, von Behren, and Kim (2001) have verified that the harmonic beam is more robust with aberration and can be further improved by correction methods.

12.5.6 Absorption Effects

Absorption effects are closely related to beamforming under nonlinear conditions. Aside from the fundamental to second harmonic conversion efficiency that is dependent on source amplitude level, the rate at which harmonic beams decay may be less than a comparable fundamental frequency beam propagating linearly at the same frequency.
Figures 12.20–12.22 include absorption loss. From Figure 12.22, the deeper focal region of the second harmonic has not only a sharper focus, but also a comparable axial amplitude in

the focal region relative to the fundamental despite its higher frequency. A major contributor to this high harmonic amplitude is the continual interplay between beam formation, harmonic buildup and replenishment, and absorption. The total apparent absorption for the second harmonic beam is less than expected for a beam at twice the frequency ($2f_0$) because the preconverted amplitude gets a "free ride" part of the way. Perhaps it is not exactly a free ride, but consider that the preconverted amplitudes are attenuated at the lower fundamental frequency rate for a portion of their propagation path, whereas a fundamental beam at $2f_0$ is absorbed at a higher rate along all of its path.

Of course, if the absorption is much stronger than the second harmonic replenishment with distance, it will eventually reduce penetration. In their study of harmonic imaging at higher frequencies, 20-MHz fundamental, Cherin, Poulsen, van der Steen, and Foster (2000) found that some of the benefits of harmonic imaging were offset by the greater absorption at these frequencies.

12.5.7 Harmonic Pulse Echo

While the previous discussion has concentrated on transmit nonlinear phenomena, imaging involves scattering and the return paths of the echoes. Some of the first pulse–echo studies at harmonic frequencies were done with high-frequency acoustic microscopes (Germain & Cheeke, 1988; Kompfner & Lemons, 1976). Scattering in a nonlinear medium has been studied for beams obliquely reflected from a flat or curved boundary (Landsberger & Hamilton, 2000; Makin, Averkiou, & Hamilton, 2000), and good agreement with theory has been obtained. For phantom imaging, an approach involving the nonlinear KZK equation on transmit and linear scattering from point targets has proved useful for simulation (Li & Zagzebski, 2000). Controlled comparisons of fundamental and harmonic imaging (van Wijk & Thijssen, 2002) indicated improved tissue-to-clutter ratios for the harmonics, as well as large differences among other criteria that were possibly caused by different implementations of harmonics on imaging systems by manufacturers. Other than the images themselves, limited experimental data on clinical style echoes are available. An exception is the high-frequency harmonic study of Cherin et al. (2000), in which a fundamental frequency of 20 MHz was used. Pulse–echo second harmonic beam cross-sections reflected from a point target were measured and were found to be similar to those at $2f_0$; their similarity may in part be caused by the strong absorption effects at this frequency, which were also noticeable in images of mice presented in the study.

The clinical value of THI is indisputable, as demonstrated by a growing number of studies in the literature (Spencer et al., 1998); however, under some circumstances, fundamental imaging will perform better. For example, harmonic imaging may offer no added diagnostic information in some higher-frequency imaging situations or for normally easy-to-image patients. More work needs to be done to determine the benefits and limitations of harmonic imaging. In recognition of this trade-off, some imaging system manufacturers can blend fundamental and harmonic images to improve image quality and to reduce speckle.

The emphasis on transmit harmonic beam characteristics in this chapter can be justified by the fact that it ultimately limits the resolution attainable. Under linear conditions this is the case because dynamic receive focusing is applied uniformly at each depth in the image and overall resolution is roughly the product of the transmit and receive beam profile at each depth. Despite the lack of information about harmonic scattering from tissues, one would expect that dynamic receive focusing provides similar imaging benefits at the harmonic.

One of the striking characteristics of harmonic imaging is its improved contrast compared to an image at the fundamental. The usual explanations offered are improved contrast resolution and reduced clutter or better acoustic signal-to-noise in the image. Reverberations and multiple scatterings are mainly incoherent and follow the main reflected signals back to the transducer on receive to create high clutter levels. These trailing spurious signals are small in amplitude, do not generate harmonics, and are filtered out at the second harmonic frequency. In addition, the high selectivity of harmonic beams avoids possible detrimental reflectors such as ribs and cartilage.

There may be another reason for the observed enhanced contrast seen in harmonic images. The second harmonic generation process depends on the square of the fundamental pressure (as indicated by Figure 12.1); therefore, for a given increase in input pressure, the second harmonic will give a disproportionally larger relative amplitude compared to a strictly proportional linear response. This effect is exaggerated in the nonlinear case as input amplitude is increased. In Chapter 10, a process was described by which the detected image signals can undergo a nonlinear postprocessing mapping procedure to emphasize or de-emphasize strong or weak echo amplitudes, according to the particular curve selected by the user. A similar mapping of gray levels from the original range in a stored image or negative to the output media or final image is commonly used in photography. The acoustic nonlinear generation process is a relationship that can be regarded as an imaging preprocessor characteristic that for a given change in input pressure level, provides a larger relative second harmonic pressure change (enhanced contrast) than that obtained under linear circumstances.

Little work has been done on scattering from real tissue structures under nonlinear conditions. Early attempts at harmonic imaging, then called B/A imaging, provided some interesting preliminary results, even though this research methodology was too awkward to be implemented commercially, as summarized by Duck (2002). Typically, a high-power pumping signal and an imaging probe were required. Extensive measurements of *B/A* from Eqn 12.5A have been made of healthy and diseased tissue (Duck, 1990; Everbach, 1997; Labat, Remenieras, Bou Matar, Ouahabi, & Patat, 2000; Law, Frizzell, & Dunn, 1981, 1985; Yongchen, Yanwu, Jie, & Zhensheng, 1986; Zhang & Dunn, 1991). Zhang,

Kuhlenschmidt, and Dunn (1991) provided intriguing data that indicate if the chemical composition of tissue is unchanged but its structure is changed, or if the *B/A* of the tissue has a structural dependency.

12.6 Harmonic Signal Processing

Harmonic imaging imposes extra requirements on an imaging system. Because harmonics may be 20–30 dB or more lower than a fundamental signal, a system must have a large dynamic range. Penetration is even more dependent on electronic signal-to-noise ratios at these high frequencies. Several signal-processing methods have been devised to improve sensitivity and to remove the desired harmonic information selectively.

Figure 12.23 shows the basic principle of extracting harmonic information from the receive echoes by filtering. An important aspect of this filtering process is having enough transducer bandwidth to recover the harmonic signal bandwidth with adequate sensitivity and minimal distortion. The usual method is to filter out only a band of frequencies centered at the harmonic frequency (Muir, 1980). In the case of harmonic imaging of contrast agents, filtering can occur at subharmonics of the transmitted frequency, as discussed in Chapter 14 (Shi & Forsberg, 2000).

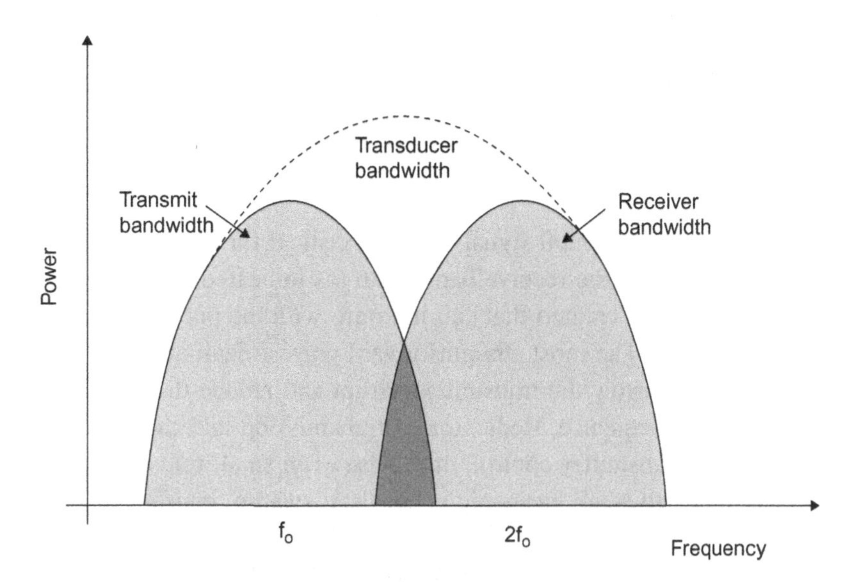

Figure 12.23

Spectral Overlap of Transmit and Receive Passbands (Solid Regions) and Second Harmonic Signal (Dashed Curve) within Transducer bandwidth. Overlap region highlighted in black with unwanted transmitted signal in harmonic band. *Adapted from Frinking, Boukaz, Kirkhorn, Ten Cate, and De Jong (2000); reprinted with permission from the World Federation of Ultrasound in Medicine and Biology.*

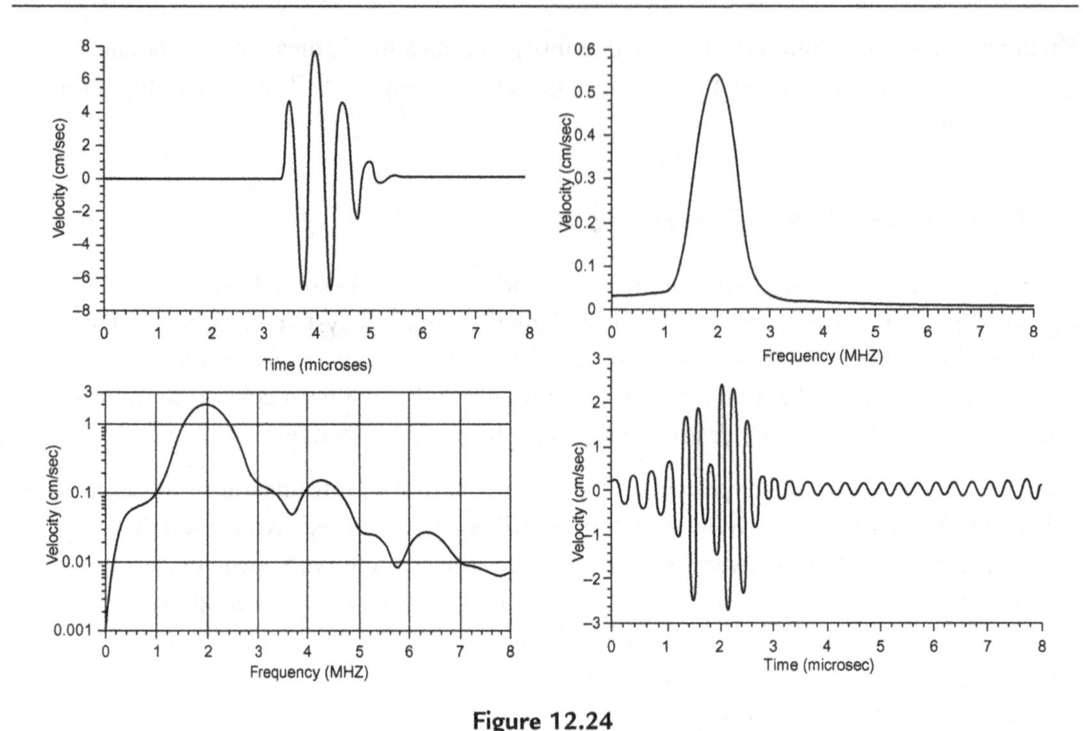

Figure 12.24

Imperfect Transmitted Signal Producing Harmonic Acoustic Noise.
(A) an imperfect 2-MHz on-axis source particle velocity waveform; (B) corresponding source spectrum; (C) spectrum at focal plane with significant transmit overlap past 3 MHz; (D) pulse reconstructed from frequencies from 3–8 MHz from the focal spectrum. *From Christopher (1997), IEEE.*

Transmission control of fundamental signals is more critical for harmonic imaging. If the transmitted spectrum overlaps the receive bandwidth (as indicated in Figure 12.23), a kind of harmonic acoustic noise is created that can interfere with the pure harmonic echoes (as illustrated in Figure 12.24). The most straightforward way of dealing with this problem is to transmit longer pulses to narrow the transmit spectrum and reduce the overlap, but axial resolution suffers as a consequence. Reduction of transmission into the harmonic region implies extremely good transmitter control, otherwise even small transmitted harmonics comparable in magnitude to weak harmonic echoes can swamp, interfere with, or distort tissue-generated harmonics and beams. This effect is illustrated by an imperfect transmitted signal in which unwanted harmonic content is more than 25 dB below the peak at the fundamental frequency in Figure 12.24. In terms of the mixer analogy, unwanted frequencies in the source pulse enter the nonlinear generation process and become scrambled so that expected harmonic levels that depend on coherence are not obtained. In other words, a kind of harmonic noise is created that is unrelated to the echoes at the desired harmonic frequencies.

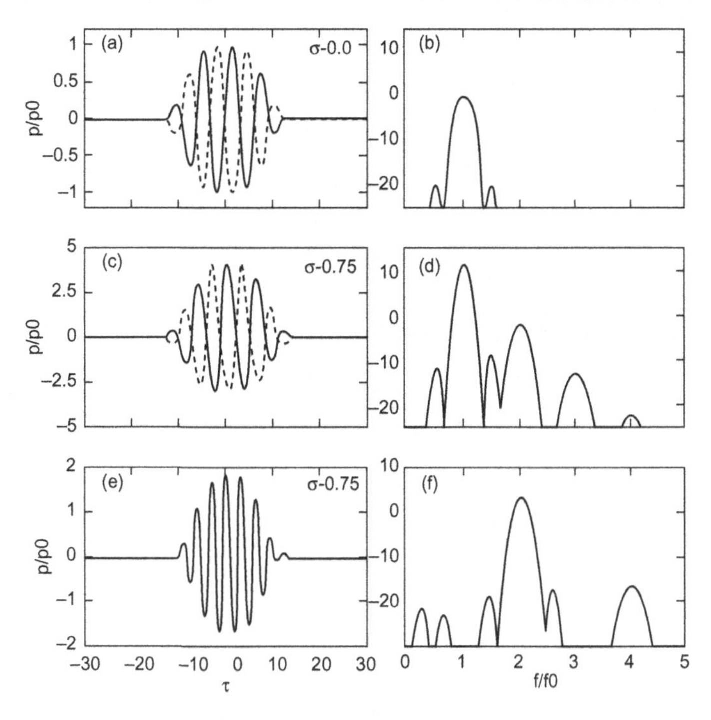

Figure 12.25
Principles of Pulse Inversion. Here $\sigma = z/F$. (A) positive (solid line) and inverted (dashed line) pulses at source; (B) spectral magnitudes in dB of pulses in (A); (C) positive and negative pulses after nonlinear propagation in water to a depth of 3/4 of the focal length; (D) spectra of pulses in (C); (E) sum of pulses in (C); (F) spectrum of sum in (E). *From Averkiou (2000), IEEE.*

For imaging systems that have frequency agility (the capability to transmit and receive on several possible frequencies within the transducer bandwidth) not only on receive but also on transmit, other solutions are possible for eliminating overlap and, consequently, improving harmonic dynamic range. Pulse inversion is a method in which an inverted pulse is sent in the next acoustic vector line in the same direction and is summed as part of the detection process to cancel the linear (fundamental) component of the echoes, as illustrated in Figure 12.25 (Bruce, Averkiou, Skyba, & Powers, 2000; Jiang, Mao, & Lazenby, 1998; Simpson, Chin, & Burns, 1999). Originally developed for Doppler and contrast agents, it has been successfully applied to B-mode imaging (Averkiou, 2000).

A modification of the pulse-inversion method can be applied for emphasis of the third harmonic (Rasmussen, Du, & Jensen, 2011). If the normal and inverted pulses are subtracted, the fundamental and odd harmonics are emphasized. The third harmonic is emphasized by bandpass filtering.

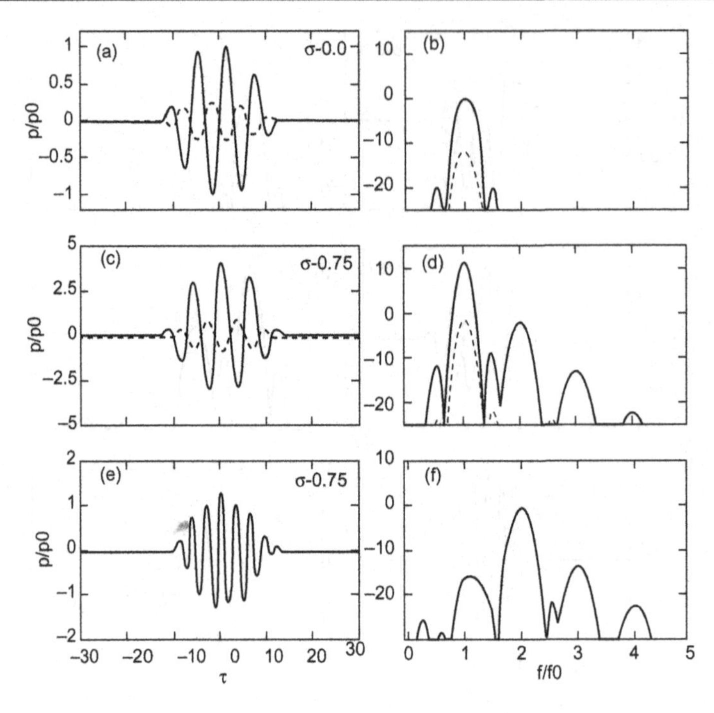

Figure 12.26
Principles of Power or Amplitude Modulation Imaging. (A) Dashed line, pulse 1/4 the amplitude of other pulse (solid line); (B) spectral magnitudes in dB of pulses in (A); (C) pulses after nonlinear propagation in water to $\sigma = z/F = 3/4$; (D) spectra of pulses in (C); (E) sum of pulses in (C) with smaller pulse multiplied by 4; (F) spectrum of sum in (E).
From Averkiou (2000), IEEE.

Another method for reducing the fundamental is called "power or amplitude modulation," in which fundamental signals at a low amplitude and at a normal high amplitude are transmitted on sequential lines in the same direction. The lower signal, mainly linear and centered at the fundamental, is amplified to the level of and subtracted from the harmonic signal (Averkiou, 2000; Brock-Fisher, Poland, & Rafter, 1996; Christopher, 1997; Jiang et al., 1998) to extract the harmonic content (as illustrated by Figure 12.26). Note that the standard pulse-inversion process emphasizes even harmonics and cancels odd ones (including the fundamental), whereas the amplitude modulation deemphasizes the fundamental and keeps higher harmonics. Both these methods have the drawbacks of reducing frame rate by a factor of two and of sometimes failing to keep up with fast-moving tissue movements. Methods have been developed (Bruce et al., 2000) to solve these limitations.

Using multitone nonlinear coding, Nowicki, Wojcik, and Secomski (1998) devised a way of emphasizing the second harmonic response. By transmitting combinations of two pulses at the fundamental and at twice the fundamental, they were able to increase the second harmonic up to several times more than the normal pulse-inversion method.

Other approaches for harmonic imaging have also been reported, including one using the fact that higher-order harmonics regenerate other harmonics and the fundamental frequency, as evident from the mixer model (Haider & Chiao, 1999). Related methods include distorting (Christopher, 1999) or encoding the fundamental and applying signal processing to remove the harmonic (Kim, Lee, Kwon, & Song, 2001; Takeuchi, 1996).

As summarized in Table 12.1, THI offers several unique advantages over conventional imaging. Because the harmonics are created along the axis of the beam, reverberations and acoustic noise can be removed by signal processing. Objectional off-axis scatterers are substantially reduced by the reduced harmonic beam volume. Aberration effects appear to be somewhat reduced but not eliminated. Surprisingly robust, harmonic imaging provides a means of imaging some people who are impossible to examine by conventional techniques. The superior contrast of harmonic images also enhances its diagnostic capability.

Signal-processing methods, discussed in application to contrast agents in Chapter 14, continue to improve sensitivity and contrast resolution. Implementation of these techniques varies from manufacturer to manufacturer (van Wijk & Thijjsen, 2002). While the basic physical principles of nonlinear propagation are understood, their optimization to various clinical applications is still evolving.

12.7 Nonlinear Wave Equations and Simulation Models

Considerable progress has been made in the understanding and modeling of waves in nonlinear media through appropriate wave equations. These wave equations fall into three major types: one-dimensional diffusion (Burgers), three-dimensional parabolic (KZK), and full wave (Westervelt). In general, they can only be solved numerically (the Burgers equation can be solved by a mathematical transform). They are presented in time domain forms, but they are most often solved in the frequency domain.

A key vehicle for describing the essential nature of wave propagation in nonlinear media with simple quadratic power law absorption in one dimension is the Burgers equation for pressure as a function of a retarded time scale (τ) for a fluid medium (Beyer, 1997; Hamilton & Morfey, 1998), as:

$$\hat{p}_z - \alpha_0 \hat{p}_{\tau\tau} = [\overline{\beta}/(c_0^3 \rho_0)]\hat{p}\hat{p}_\tau, \tag{12.9}$$

in which a retarded time (τ) and the abbreviated notation from Section 4.7.3 are convenient. Note that the form can be regarded as an inhomogeneous lossy one-way wave equation with a nonlinear source term on the right-hand side (r.h.s.). An exact analytic solution to this equation is possible through a substitution of variables, as well as approximate, series solutions (Blackstock, 1966). The sawtooth equation, Eqn 12.4C, is one solution for large Γ (Blackstock et al., 1998). This equation is only appropriate for media with a quadratic frequency power law exponent such as water, which has no sound speed dispersion. Blackstock (1985) suggested that the Burgers equation could be applied to other types of losses by replacing the second term with a loss operator. This equation has been extended to media with a power law loss in the following general form (Szabo, 1993):

$$\hat{p}_z - L'_{\alpha,y,T}{}^{*T}\hat{p} = [\overline{\beta}/(c_0^3\rho_0)]\hat{p}\hat{p}_T, \tag{12.10}$$

where the time causal operator ($L'_{\alpha,y,T}$) was described at the end of Chapter 4. This equation includes the dispersion necessary to describe losses in tissue and many other media, and it reduces to the Burgers equation for $y = 2$. Haran and Cook (1983) have studied losses in tissues under nonlinear conditions. A more recent contribution, properly accounting for power law dispersion, can be found in Wallace, Holland, and Miller (2001). They point out that because of phasing of harmonic components, sound speed dispersion can play a significant role in determining waveform distortion. Methods for frequency domain calculations can be found in Hamilton (1998).

A considerable amount of research on sound beams in nonlinear media was published in the Soviet literature and later consolidated in the book, *Nonlinear Theory of Sound Beams* (Bakhvalov, Zhileikin, & Zabolotskaya, 1987). A key development was the Khokholov–Zablotskaya–Kuznetsov (KZK) (Kuznetsov, 1971) wave equation under the paraxial or Fresnel approximation:

$$\nabla_\perp^2\hat{p} - \frac{2}{c_0}\hat{p}_{zT} + (2\alpha_0/c_0)\hat{p}_{TTT} = - \ [\overline{\beta}/(c_0^4\rho_0)]\hat{p}^2_{TT}, \tag{12.11}$$

where the notation is from Chapter 4. Again, this equation can be considered as a parabolic (one-way) wave equation with the third term describing loss and the r.h.s. being a nonlinear source term. When the r.h.s. is set to zero, the equation applies to linear media with loss. The KZK equation combines nonlinearity, diffraction under the parabolic (Fresnel) approximation, and absorption for a quadratic loss medium (water) in a numerically suitable form.

Programs for the prediction of circularly symmetric beams were developed in the Soviet Union, and many of the figures for the book (Bakhvalov et al., 1987) were computations from these programs. A concentration of nonlinear analysis and theory by Naze Tjotta and Tjotta and others in Norway laid the theoretical foundation (Berntsen, Naze Tjotta, & Tjotta, 1984; Naze Tjotta & Tjotta, 1981) for several computer programs, including a numerical implementation of the KZK equation (Aanonsen et al., 1984; Berntsen, 1990; Berntsen & Vefring, 1986; Hamilton, Naze Tjotta, & Tjotta, 1985), which is also known as the Bergen code. This finite difference method uses a Fourier series approach to solve the necessary coupled differential equations for harmonics in the frequency domain.

To appreciate the complexity of KZK equation computations, consider that for each individual path length from a point on the aperture to a field point, there is a different amount of distortion. Unlike a direct diffraction computation for a selected plane in a linear lossy medium, a calculation for a nonlinear medium requires that all the intervening beam-shapes must be calculated first so that the required amount of cumulative waveform distortion can develop. At each spatial position, every frequency leads to a sum-and-difference frequency (somewhat analogous to the mixer model), so that because of the growing number of harmonics generated, KZK programs are computationally intensive. For N total harmonics, computations on the order of N^2 are required for each spatial grid point.

The original formulation of the KZK included frequency-squared loss appropriate for water. A modification of the KZK equation that applies to power law absorbing media and includes dispersion (Szabo, 1993) is:

$$\nabla_{\perp}^{2}\hat{p} - \frac{2}{c_0}\hat{p}_{z\tau} + L_{\alpha,y,\tau}\ \hat{p} = -\ \left[\overline{\beta}/(c_0^4\rho_0)\right]\hat{p}_{\tau\tau}^{2}. \qquad (12.12)$$

This equation reduces to Eqn 12.14 for $y = 2$. Absorption can also be included in the frequency domain by changing the Fourier coefficient of the nth harmonic (Watson, Humphrey, Baker, & Duck, 1990) as a multiplicative propagation factor.

Versions of the frequency domain KZK equation have been extended to pulsed fields using a Fourier series pulse decomposition (Baker, 1991; Baker & Humphrey, 1992; Cahill & Baker, 1998) to focusing (Hart & Hamilton, 1988) and to rectangular geometries (Baker, Berg, Sahin, & Naze Tjotta, 1995; Berg & Naze Tjotta, 1993; Kamakura, Tani, & Kumamoto, 1992; Sahin & Baker, 1993). If the circularly symmetric case involves M calculations, the rectangular case adds another dimension and increases computation by a factor $M^{4/3}$.

For pulsed fields, a more direct approach is solving the KZK in the time domain. The evolution of this approach, which began with the Soviet work (Bakhvalov et al., 1987), continued with Bacon (1984) and developed into a different KZK approach at the University of Texas in Austin (Lee & Hamilton, 1995), with added relaxation effects (Cleveland, Hamilton, & Blackstock, 1996) and focusing for rectangular apertures (Averkiou, 2000). In the time domain, the operators are on the order M; however, many time points are required to capture steep shock fronts (Too & Ginsberg, 1992).

The need to accurately model high-intensity therapeutic ultrasound brought new challenges for nonlinear modeling. Unusual combinations of length and time scales require the ability to model large-diameter transducers (few cm) with large focal gains and beamwidths on the order of millimeters as well as radio frequency (RF) periods in microseconds lasting for long durations (seconds) with high pressure amplitudes. Add in odd transducer geometries such as spherical-like shapes with holes in their middle and several layers of tissue with power law absorption. Soneson and Myers (2007) were able to accommodate all of these factors simultaneously in their KZK program in MATLAB and, in addition, couple the acoustic field to predict tissue heating as discussed in Chapter 17. By using a Gaussian approximation to reduce the KZK equation in the frequency domain, they were able to transform a system of partial differential equations to ordinary ones, thereby significantly reducing computation time. Because of limitations of the KZK approach, one-way propagation, and results trusted for propagation angles of less than 20° (Froysa, 1991), interest in numerical implementations of the full nonlinear wave equation has continued. Westervelt (1963) derived a full wave nonlinear local wave equation that described cumulative distortion:

$$\nabla^2 p - \frac{1}{c_0^2} p_{tt} + \frac{2\alpha_0}{c_0} p_{ttt} = -\left[\overline{\beta}/(c_0^4 \rho_0)\right] p_{tt}^2. \tag{12.13}$$

Although the initial relevance for this equation was for parametric arrays, (Berktay, 1965; Berktay & Al-Temini, 1969; Westervelt, 1963), it has more general applicability.

A numerical solution to Eqn 12.16 with thermoviscous-type losses has been applied to high-intensity focusing ultrasound (HIFU) surgical applications (Hallaj, Cleveland, & Hynynen, 2001). Pinton, Dahl, Rosenzweig, and Trahey (2009) incorporated losses through the use of relaxation constants in their study of aberration effects. Jing, Wang, and Clement (2012) have found numerical advantages to doing the spatial differentiation in k-space in their calculations of nonlinear waves through the skull for high-intensity therapeutic ultrasound (HITU) (see Chapter 8). Another time domain approach is a different wave equation called the nonlinear progressive equation (NPE) (McDonald & Kuperman, 1987). Too and Ginsberg (1992) have developed a version for sound beams. Li and Zagzebski (2000) have used this approach for evaluating image quality for harmonics. Yuldashev et al. (2012) have applied their time domain approach to HIFU simulations in which the characteristics of the source are measured under linear conditions and then used as input to their solution of the Westervelt equation in the time domain. This work and earlier simulation with the KZK equation for power law absorption for arrays (Khokhlova, Ponomarev, Averkiou, & Crum, 2006) and excellent HIFU experimental verification work (Canney, Bailey, Crum, Khokhlova, & Sapozhnikov, 2008) will be revisited in the next chapter.

A version of this equation for power law media (Szabo, 1993) is:

$$\nabla^2 p - \frac{1}{c_0^2}\frac{\partial^2 p}{\partial t^2} - L_{\alpha,y,t} * p = -\left[\overline{\beta}/(c_0^4 \rho_0)\right] p_{tt}^2, \tag{12.14}$$

where symbols are from Chapter 4. As in the linear case, Norton (Norton & Purrington, 2009) has successfully implemented the loss operator (third term) directly as an explicit causal time domain convolutional operator in a finite-difference-time-domain (FDTD) method. Their previous linear one-dimensional wave-equation work (Norton & Novarini, 2003) verified that the method properly describes the absorption and dispersion evolution with distance. An example of this agreement of numerical simulation with an analytic Gaussian pulse is summarized in Figure 12.27 for both linear and nonlinear propagation (Purrington & Norton, 2009), for $\alpha = 2.0976 \times 10^{-7}\omega^{1.13}(Np/m - rad - s)$. Note that the power law equation (Eqn 12.17) with the time domain propagation factor (TDPF) shows the effects of phase velocity dispersion delay and absorption, but the traditional Westervelt equation shows only loss. Application of this approach (Purrington & Norton, 2012) to two dimensions in the form of simulated plane wave and six homogeneous layers of tissue and water with absorptions of various power laws and impedances resulted in the simulated backscattered pulse echoes of Figure 12.28A. A close-up of the fourth echo, Figure 12.28B, as compared to the traditional Westervelt equation, shows the effects of phase velocity dispersion delay and absorption. Norton's version of this equation for heterogeneous media with a source function s is:

$$\frac{1}{\kappa(x,z)}\frac{\partial^2 p(x,z,t)}{\partial t^2} - \nabla \cdot \left(\frac{\nabla p(x,z,t)}{\rho(x,z)}\right) + \sqrt{\frac{\rho(x,z)}{\kappa(x,z)}}\left[L_\gamma(t)^* p(x,z,t)\right] = \delta(x - x_s)\delta(z - z_s)s(t), \tag{12.15}$$

where L_γ is the TDPF found directly from the numerical inverse Fourier transform of $\gamma(\omega)$ (Norton & Purrington, 2009), Section 4.3.4, and κ is the bulk elastic constant. The capability of the approach to simulate sources, scatterers, and wave propagation in heterogeneous media in underwater acoustics (see Figure 12.29) and tissue has been demonstrated (Norton, 2009a). In addition, a version based on the Navier–Stokes equation has been applied to a pulse propagating under nonlinear conditions in a dispersive inhomogeneous moving fluid (Norton, 2009b).

Several approaches utilize a substep or hybrid (known as operator-splitting) scheme for computation. In the Pestorious (1973) algorithm, propagation is reduced to small alternating steps. Here, nonlinear effects and thermoviscous absorption are computed using weak shock theory. Then, other absorption effects are computed in the frequency domain, and, finally,

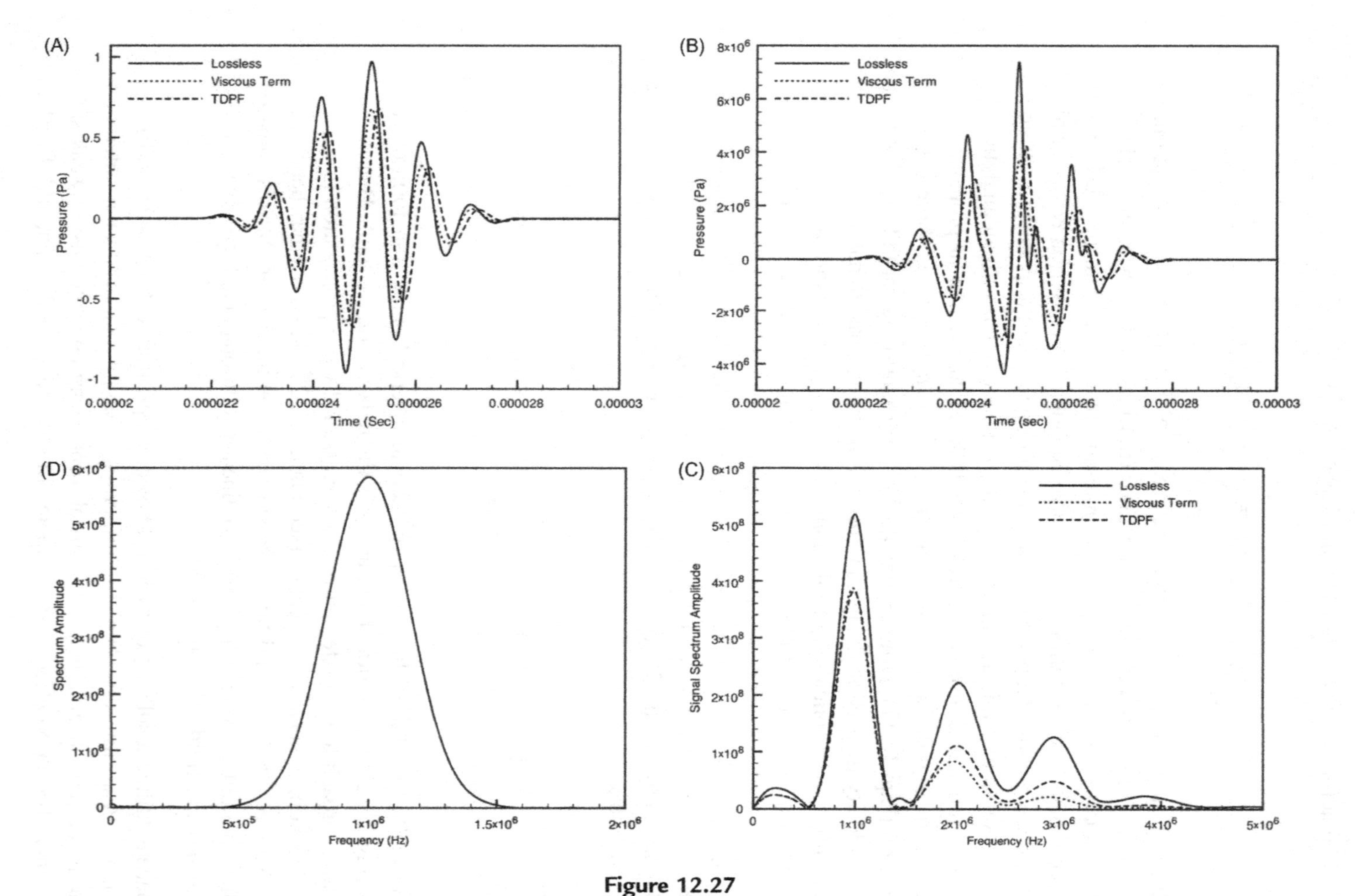

Figure 12.27

Agreement of Numerical Simulation with an Analytic Gaussian Pulse. The input signal is a 1-MHz sinusoidal burst of six cycles modulated by a Gaussian envelope in time propagating in simulated liver tissue, as shown in (A) at a depth of 7 cm (source is at 3.5 cm). The source has a bandwidth of approximately 1 MHz with a center frequency of 1 MHz, as plotted in (B). Note, (A) also shows the pulse simulated by power law (TDPF) media suffers from dispersion (arrives later and has spread out in time) while the signal associated with the viscous Westervelt equation does not. Whereas $p_0 = 1$ Pa in (A) and (B), it is set to 5 MPa in (C), the nonlinear pulse counterpart of (A), with its associated spectra in (D). *From Norton and Purrington (2009) Journal of Sound and Vibration.*

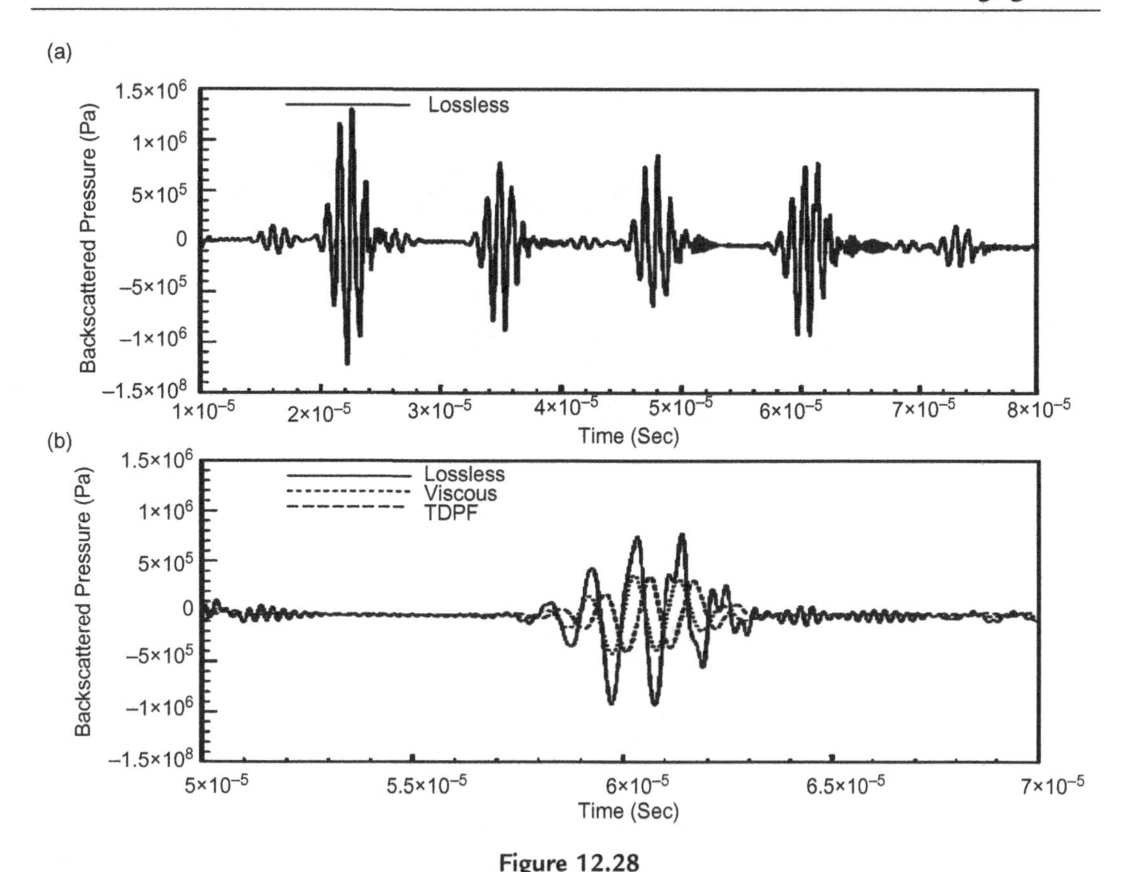

Figure 12.28
(A) Backscattered pressure due to westervelt equation; (B) expanded view of fourth backscattered signal. *From Norton (2009), Mathematics and Computers in Simulation.*

these results are inverse Fourier transformed back to the time domain for the next nonlinear incremental step. Christopher and Parker (1991) have developed an algorithm in which diffraction and attenuation are computed in the angular spectrum domain as an incremental substep followed by a substep in which nonlinear effects are computed via a frequency domain solution to the Burgers equation. The advantage of their approach is that there is no angular restriction as in the KZK methods; strong nonlinearities, appropriate for modeling lithotripters, can be accommodated, as well as multiple layers (without reverberations). In a later version, the nonlinear substep was computed in the time domain (Christopher, 1993). Another split-step full wave algorithm operates entirely in the time domain (Tavakkoli, Cathignol, Souchon, & Sapozhnikov, 1998) and uses an exact Rayleigh integral for the diffraction substep, a material impulse-response-like function for the absorption–dispersion substep, and an analytic Poisson solution for a lossless medium for the nonlinear substep. Good agreement has been obtained with data (Remenieras, Bou Matar, Labat, & Patat,

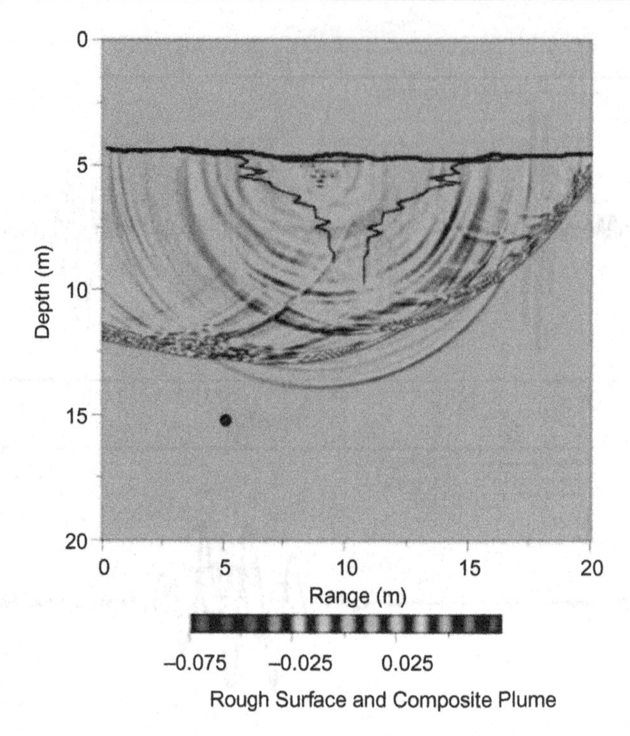

Figure 12.29
Two-dimensional solution for a point source radiating to a rough surface with a composite bubble plume. The bubble density within the plume as well as the horizontal area of the plume (in this case, the equivalent horizontal length) is assumed to decay exponentially with depth. The reader is referred to the Web version of the book to see the figure in color. *From Norton (2009a), Mathematics and Computers in Simulation.*

2000). Ginter (2000) has accounted for energy conservation and power law attenuation in his approach.

Taraldsen (2001) introduced a generalized version of the Westervelt equation suitable for heterogeneous media. Varslot, Taraldsen, Johansen, and Angelsen (2001) developed an approximate parabolic form of this equation that simulated a propagating pulse in the spatial not the time domain. This early version provides simplified computing through operator-splitting terms representing diffraction, the heterogeneous speed of sound, nonlinearity, and loss. Through a number of publications, Varslot and his group showed the utility of the approach for simulating ultrasound propagation in heterogeneous tissue under nonlinear conditions (Kaupang, Varslot, & Masøy, 2007; Varslot & Masøy, 2006; Varslot & Taraldsen, 2005). This body of work has evolved into a set of programs that provide a full wave three-dimensional (3D) nonlinear capability for power law media under nonlinear conditions. The 2D version was validated in Varslot and Taraldsen (2005) and 3D nonlinear parts were verified against experiments in Varslot and Masøy (2006). The present version is available as open-source software called Abersim (Frijlink, Kaupang, Varslot, & Masøy, 2008) which uses a MATLAB interface for pre- and postprocessing and visualization. Abersim includes a full 3D part that uses an angular spectrum approach and paraxial part for axisymmetric cases.

Treeby, Jaros, Rendell, and Cox (2012), building on earlier linear work, derived a unique modified Westervelt equation for heterogeneous media with power law absorption:

$$\nabla^2 p - \frac{1}{c_0^2}\frac{\partial^2 p}{\partial t^2} - \frac{1}{\rho_0}\nabla\rho_0 \cdot \nabla\rho + \frac{\beta}{\rho_0 c_0^4}\frac{\partial^2 p^2}{\partial t^2} - L\nabla^2 p = 0. \tag{12.16}$$

The key difference is the use of a fractional Laplacian loss operator. This approach does not have limitations on either the direction of the beam or the degree of spatial heterogeneity. A k-space pseudospectral numerical implementation was demonstrated for several heterogeneous cases, and details show that the calculations were computationally intensive. In recent work, a more efficient version of the model compared well with hydrophone measurements and was released as part of the open-source "k-Wave Toolbox" (Wang, Teoh, Jarosy, & Treeby, 2012). All nonlinear solutions so far require the tedious, computationally intensive evolution of beam of many intermediate steps; a refreshing surprise is a method that provides a solution in one step. Though not complete, this method captures the first and second harmonic solutions to the full Westervelt equation. The details of the solution can be found in (Du, Jensen, & Jensen, 2010, 2011a; Landsberger & Hamilton, 2001; Xiang, 2004; Xiang & Hamilton, 2006). Briefly, the solution begins by using a source distribution from Field II and transforming the Westervelt equation to the angular spectrum domain for each harmonic and, though a separation of variables, a linear solution for the first harmonic can be found and used to find the second harmonic through the solution as a linear ordinary differential equation (Du & Jensen, 2008; Du et al., 2011a). A comparison between this approach and Abersim can be found in Figure 12.30 (Du, Jensen, & Jensen, 2011b).

Figures 12.20–12.22 were generated by a finite-difference pseudospectral full wave solver using six processors running in parallel (Wojcik et al., 1998; Wojcik, Mould, & Carcione 1999a). This method accounts for multiple reflections, absorption, and nonlinearity, and employs perfectly matched layers at the boundaries. A series of experiments in water and through tofu as a tissue mimic agreed well with computations (Wojcik, Szabo, Mould, Carcione, & Clougherty, 1999b).

Another nonlinear model that uses a hyperbolic rather than a parabolic equation (Radulescu, Wojcik, Lewin, & Nowicki, 2003) has been shown to agree well with data. It has been used as part of a hydrophone calibration scheme up to 100 MHz, described in Chapter 13.

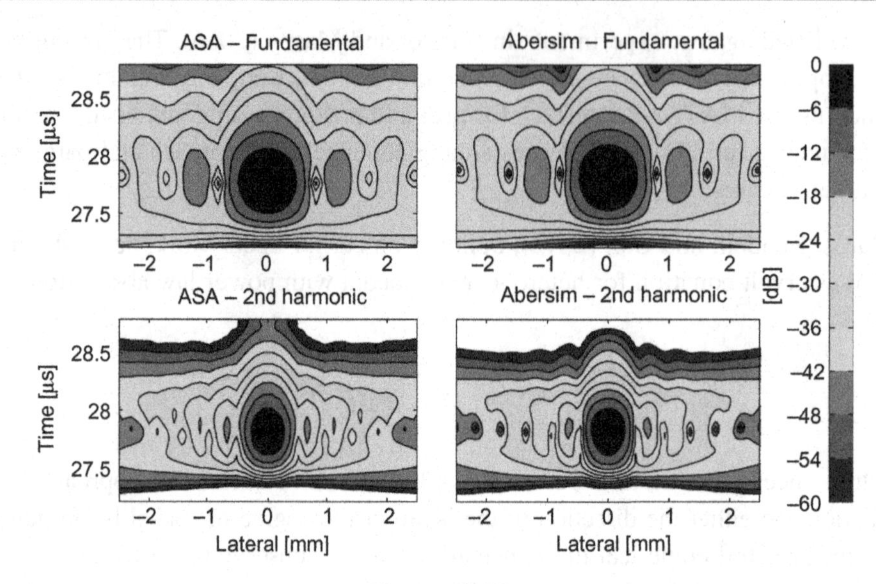

Figure 12.30
Emitted ultrasound fields calculated by ASA and abersim. Fundamental and second harmonic fields at the focal depth (40 mm) are shown in the figure with 6 dB between two adjacent color lines. The reader is referred to the Web version of the book to see the figure in color. *From Du et al. (2011b) IEEE.*

12.8 Acoustic Radiation Forces and Streaming

12.8.1 Introduction

Acoustic radiation force (ARF) is a term for a small force field that rides along an acoustic perturbation in a nonlinear medium. If a boundary, target, small particles, tissues, or bubbles are in the path of this force field, the forces can be measured or visualized. Though small, acoustic radiation forces can be put to use and are being studied and utilized in a wide variety of applications from levitation to shear wave elastography. In this section, we will examine the complexities and wide range of applications of ARF from a global perspective. A large body of literature is available about radiation forces and much of it is covered in the excellent review by Sarvazyan, Rudenko, and Nyborg (2010). Other valuable reviews are Duck (1998) and Wang and Lee (1998). These works all agree that despite this enormous number of studies in this area, much more research is needed to more fully understand the elusive properties of the ARF. In addition, when ultrasound is propagating in a fluid under nonlinear conditions (the usual case), a related effect called acoustic streaming is explained in Section 12.8.5.

12.8.2 Plane Understanding

Though there were earlier observations, Lord Rayleigh (1902) introduced the general concept of an acoustic radiation pressure through his derivation of plane waves striking a boundary. His inspiration was analogous work in electromagnetics where electric and magnetic stress tensors were shown to be proportional to the field squared (Stratton, 1941). At first, one may wonder how there could be any net time-averaged pressure associated with an acoustic sinusoidal wave whose positive and negative cycles cancel each other. Under nonlinear conditions, the half cycles are no longer symmetric, as evident from Figure 12.11. The time average of the pressure of this kind of wave, $<P - P_0>$, is called the "mean excess pressure," where P_0 is the ambient pressure. The acoustic radiation (3×3) stress tensor originally derived by Brillouin is:

$$\mathbf{T} = -\langle \mathbf{P} - \mathbf{P_0} \rangle - \rho_0 \langle \mathbf{v} \cdot \mathbf{v} \rangle. \tag{12.17}$$

ρ_0 is quiescent density and $\mathbf{v}$ the particle velocity vector (Wang & Lee, 1998). When expanding pressure to second order, stress becomes the difference of the time-averaged potential $<V>$ and kinetic $<K>$ energies and:

$$\langle P^E - P_0 \rangle = \frac{1}{2\rho_0 c_0^2} \langle p^2 \rangle - \rho_0 \langle \mathbf{v} \bullet \mathbf{v} \rangle + C = \langle V \rangle - \langle K \rangle + C, \tag{12.18}$$

where c_0 is the speed of sound and P^E is Eulerian pressure at a point, and C is a constant (not the speed of sound). For a sinusoidal pressure of angular frequency ω_0 (Wang & Lee, 1998),

$$p = A\cos(k_0 x)\sin(\omega_0 t), \tag{12.19A}$$

and a corresponding particle velocity,

$$\mathbf{v} = \frac{-A}{\rho_0 c_0} \sin(k_0 x)\cos(\omega_0 t). \tag{12.19B}$$

So that $<V> \; = \mathrm{A}^2\cos^2(k_0 x)/4\rho_0 \mathrm{c}_0^2$ and $<K> \; = \mathrm{A}^2\sin^2(k_0 x)/4\rho_0 \mathrm{c}_0^2$ lead to:

$$\langle P - P_0 \rangle = \frac{A^2}{4\rho_0 c_0^2} \cos^2(k_0 x) - \frac{A^2}{4\rho_0 c_0^2} \sin^2(k_0 x), \tag{12.20A}$$

which by a trigonometric identity is:

$$\langle P - P_0 \rangle = \frac{A^2}{4\rho_0 c_0^2} [1 + \cos 2k_0 x] + C = \langle E \rangle [\cos 2k_0 x] + C, \tag{12.20B}$$

which can be recognized as a standing wave pattern. Beissner (1984) and Wang and Lee (1998) also recognized the usefulness of the Langevin pressure:

$$\langle P^L - P_0 \rangle = \langle V \rangle + \langle K \rangle + C. \tag{12.21}$$

Because $\langle K \rangle = \rho_0 \langle v^2 \rangle /2$, Duck (1998) has shown the relation between the two pressures P^E and P^L as:

$$\mathbf{T} = -\langle P^E \rangle - \rho_0 \langle u^2 \rangle = (\langle V \rangle + \langle K \rangle + C) = -\langle P^L \rangle. \tag{12.22}$$

A useful relation (Duck, 1998) is:

$$P^L = p_0^2/2\rho_0 c_0^2 = I/c_0, \tag{12.23}$$

in which I is intensity. For the stress component along x and intersecting a planar object of area S, the force along x on this object is (Beissner, 1984):

$$\mathbf{F_x} = \oiint_S \mathbf{T_x} \bullet \mathbf{dS} = (W/c_0)\hat{x}, \tag{12.24A}$$

in which W is time-averaged power and $\hat{x}$ is the unit vector along the x axis. This equation is appropriate when the source is a beam that is totally captured by a large absorbing target, and it is frequently used for a radiation force balance measurement of acoustic power from a transducer, as described in Chapter 13. If the target of area S does not intercept all of the beam, then a general relation (Duck, 1998) is:

$$F_x = DWS/c_0, \tag{12.24B}$$

where $D = 1$ for an absorbing target and 2 for a perfectly reflecting target. If the target is at an angle $\grave{e}$ to the x axis, then $D = 2\cos^2\grave{e}$. If there is a boundary between two media subscripted 1 and 2, the value of D incident on the boundary is $D = (RF)^2 + 1$, where RF is from Eqn 3.22A (Beyer, 1997). Unlike the other cases, where the force past the target or boundary is zero, in this case the D factor for the force past the boundary is $(1 - RF^2)c_1/c_2$. If the transducer is directed upward at the boundary, an acoustic fountain can be generated if the force is greater than that of gravity. Typically, these can be made for a water–air interface for which the height of the column (Duck, 1998) is:

$$h = 2I/(\rho_0 c_0 g), \tag{12.25}$$

in which g is the acceleration of gravity. If $c_1 > c_2$, a strange thing occurs, a fountain directed downward into the first medium; pictures of these effects can be seen in Beyer (1997). Finally, for an attenuating medium with an absorption coefficient per unit length α, $D = 2\alpha$, as discussed later.

12.8.3 Particle Manipulation

There is a similar equation for optical radiation force to Eqn 12.16A, and what it means is that, because of the difference in values of c, acoustic forces are 10^5 greater than optical ones; consequently, acoustic radiation forces can be used to manipulate objects. For many years, the levitation of small objects and droplets has been investigated (Crum, 1971;

Gor'kov, 1962; King, 1934; Nyborg, 1967; Wang & Lee, 1998; Wu & Du, 1990). Recall the discussion of backscattering from spheres from Sections 8.2.2. and 8.2.3; analysis has also been developed for levitating and moving small spheres. A classic study for forces acting on small particles is that of Gor'kov (1962). Acoustical tweezers (Lee & Shung, 2006; Wu, 1991) have been devised to hold and move small objects.

A vast technology has grown from these ideas, called "acoustical microfluidics." Even though there are other ways of manipulating particles using, for example, electrical forces, acoustic waves offer more flexibility in terms of geometric placement and control variables. A typical acoustic microfluidic device has channels across which ultrasound transducers create transverse standing waves that direct certain-sized micro/bio-particles to nodes that are aligned with different exit channels for filtering. Friend and Yeo (2011) provide an excellent overview on recent advances in "lab-on-a-chip" acoustical microfluidics, which include microcentrifugation for micromixing and particle concentration or separation and atomization, jetting, and combinations of these processes on the same device. Bruus' book (2008) introduces the overall technology and has an appendix with a comprehensive analysis of 3D acoustic particle manipulation. Sarvazyan (2010) has provided an overview of ARF applications including several unusual ones. Finally, radiation forces can play a role in manipulating and directing ultrasound contrast and molecular agents as described in Section 14.5.4.

12.8.4 Acoustic Radiation Forces in Tissue

In order to introduce the applications of ARFs to ultrasound imaging in tissues, we find the derivation of Nyborg to be useful (Sarvazyan, Rudenko, & Nyborg, 2010). As in the previous equations, his derivation is a one-dimensional analysis for a nonlinear decaying time-harmonic plane wave with time-averaged quantities averaged over one cycle. Applying a method of successive approximations, he began with series expansion:

$$\begin{pmatrix} p(x,t) \\ \rho(x,t) \\ v(x,t) \end{pmatrix} = \begin{pmatrix} p_0 \\ \rho_0 \\ 0 \end{pmatrix} + \begin{pmatrix} p_1(x,t) \\ \rho_1(x,t) \\ v_1(x,t) \end{pmatrix} + \begin{pmatrix} p_2(x,t) \\ \rho_2(x,t) \\ v_2(x,t) \end{pmatrix} + \ldots, \tag{12.26}$$

in which the second column represents a motionless constant equilibrium state. The next column can be summarized by a wave of frequency $\grave{u}_0 = 2\delta f_0$,

$$\begin{aligned} v_1(x,t) &= V_0 e^{-\alpha x} \cos[\omega_0(t - x/c_0)] \\ &= p_1(x,t)/\rho_0 c_0 = c_0 \rho_1(x,t)/\rho_0 \end{aligned}. \tag{12.27}$$

The time-averaged intensity of this wave is:

$$I = V_0^2 e^{-2\alpha x} \rho_0 c_0 / 2. \tag{12.28}$$

The mean change in momentum in a unit volume can be approximated as:

$$\left\langle \rho \frac{dv}{dt} \right\rangle = \left\langle (\rho_0 + \rho_1 + \rho_2) \times \left(\frac{\partial(v_1 + v_2 + \ldots)}{\partial t} + \frac{1}{2} \frac{\partial(v_1 + v_2 + \ldots)^2}{\partial x} \right) \right\rangle. \quad (12.29A)$$

After invoking the equation of continuity, noting the time average of time derivative is zero, and keeping the most significant terms, he found this result:

$$\left\langle \rho \frac{dv}{dt} \right\rangle = \left\langle \rho_0 \frac{\partial v_1^2}{\partial x} \right\rangle = \frac{\partial}{\partial x} \left(\frac{I}{c_0} \right). \quad (12.29B)$$

From Eqn 12.29B and after integrating the force over a volume of area S and thickness L,

$$F_x^V = -\frac{\partial}{\partial x} \left(\frac{I}{c_0} \right) = 2\alpha \frac{I}{c_0}, \quad (12.29C)$$

which is the most widely used ARF equation for tissue, but what does it mean? First, recall this is a one-dimensional plane wave result. Second, the absorption in this analysis is a constant, or $\alpha = \alpha(\omega_0)$. Third, the intensity used is the linear component.

More elaborate derivations of ARF and induced shear motion can also be found in Rudenko, Sarvazyan, and Emelianov (1996); Sarvazyan et al. (1998); and Sarvazyan, Rudenko, and Nyborg (2010). Sarvazyan and his colleagues were analyzing radiation forces for the eventual purpose of shear wave elastography imaging (SWEI), discussed in more detail in Chapter 16. Here, we focus on their contributions to the understanding of three-dimensional radiation forces and resultant waves. Sarvazyan et al. (1998) wrote of a "virtual finger" for probing the body; this concept can be related to Figure 3.8, which shows that a finger or stress source on a surface creates a multicomponent stress field in the solid below. Analytic solutions for a linear Green's function for an impulsive function on the surface of a semi-infinite solid show both longitudinal and shear wave components (Aki & Richards, 2000; Gakenheimer & Miklowitz, 1969), as do solutions for a circular piston source (Djelouah & Baboux, 1992; Miller & Pursey, 1954).

Sarvazyan et al. (1998) solved a KZK equation for the nonlinear shear displacements from a focusing circularly symmetric transducer. They were able to derive a closed-form solution for the shear displacement along the beam axis at the focal point as a function of time. This result is proportional to area of the transducer, α, and the initial intensity, I_0. The remarkable behavior of the wavefront near the focal region is shown in Figure 12.31. Here, the parameters for the calculations were: $f_0 = 3$ MHz, focal depth $d = 50$ mm, radius $a = 10$ mm, $I_0 = 10$ w/cm^2, $\rho_0 = 1000$ kg/m^3, $c_l = 1500$ m/s, and $c_t = 3$ m/s. In this figure, the initial high peak associated with focusing by this conventional longitudinal wave transducer is seen, but then two lateral wave crests move transversely with a speed c_t.

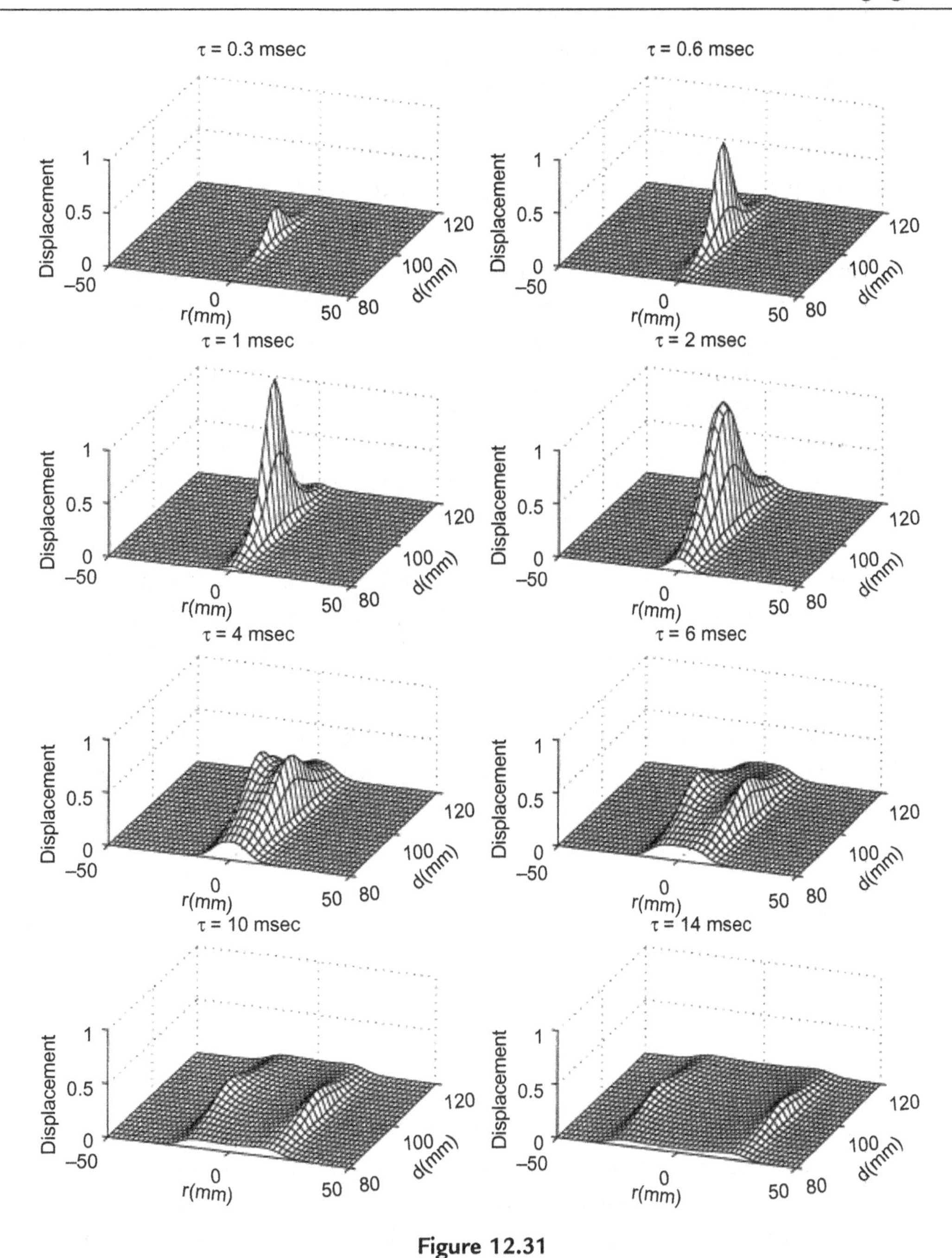

Figure 12.31

Generation and propagation of the shear wave shown at different times after transmission of the acoustic pulse. *From Sarvazyan et al. (1998); reprinted with permission from the World Federation of Ultrasound in Medicine and Biology.*

Because of this slow shear speed and associated long propagation time, this mode was not usually observed in normal ultrasound imaging; however, the authors were able to image the double crests using magnetic resonance imaging. The analysis includes accommodation for power law absorption; however, because of the lack of details and the fact that long monochromatic pulses (100 μs) are employed in the experiment, the conservative interpretation is that absorption in this analysis is a constant, or $\alpha = \alpha(\omega_0)$.

Bercoff, Tanter, Muller, and Fink (2004) derived a Green's function for the Navier–Stokes equation coupled to waves in a linear solid, based on a model by Aki and Richards (2000). Through convolution with a source function, the spatiotemporal displacement field can be calculated. Based on Catheline, Wu, and Fink's earlier agreement of data with a Voigt model (1999), they incorporated viscoelasticity by modifying the wave vector ***k*** according to the Voigt model (see Chapter 4).They fitted an absorption coefficient to exponential amplitude falloff data with depth at each frequency and found that the Voigt model was a satisfactory empirical fit up to 1000 Hz of the 2000-Hz frequency range. An ultrafast scanner first created a transmitted focused wave and acquired about 200 μs of data at about a 3000-Hz rate. A 1D cross-correlation algorithm then calculated displacements. A comparison of results with and without viscolelasticity shows the strong effects of viscous losses, as shown in Figure 12.32. ARF was calculated according to the popular equation, 12.29C.

Another path to determining the ARF was taken by the ARF Imaging (ARFI) team. Doherty et al. (2013) also began with the Navier–Stokes equation to find the force, and utilized a perturbation method similar to that of Nyborg:

$$\vec{F} = -\nabla p_2 + \mu \nabla^2 \vec{v}_2, \tag{12.30}$$

in which ί is the shear viscosity and the second term represents the diffusion of the second-order velocity field. A linear elastic model was used to determine the response of the tissues to the radiation force field.

12.8.5 Acoustic Streaming

The pressure gradient of an acoustic beam pushes away the fluid and causes acoustic streaming. A measurement of acoustic streaming from a plane piston transducer in a water glycerin solution is presented in Figure 12.33. Acoustic streaming is driven by the radiation force. From the analysis of Nyborg (1965), Nowicki, Secomski, and Wojcik (1998) have compactly expressed the axial component of velocity as:

$$\nabla^2 v_{2x} = \frac{1}{\mu c_0}\frac{\partial I_x}{\partial z} = \frac{-2\alpha}{\mu c_0} I_x = -\frac{F_x^V}{\mu}. \tag{12.31}$$

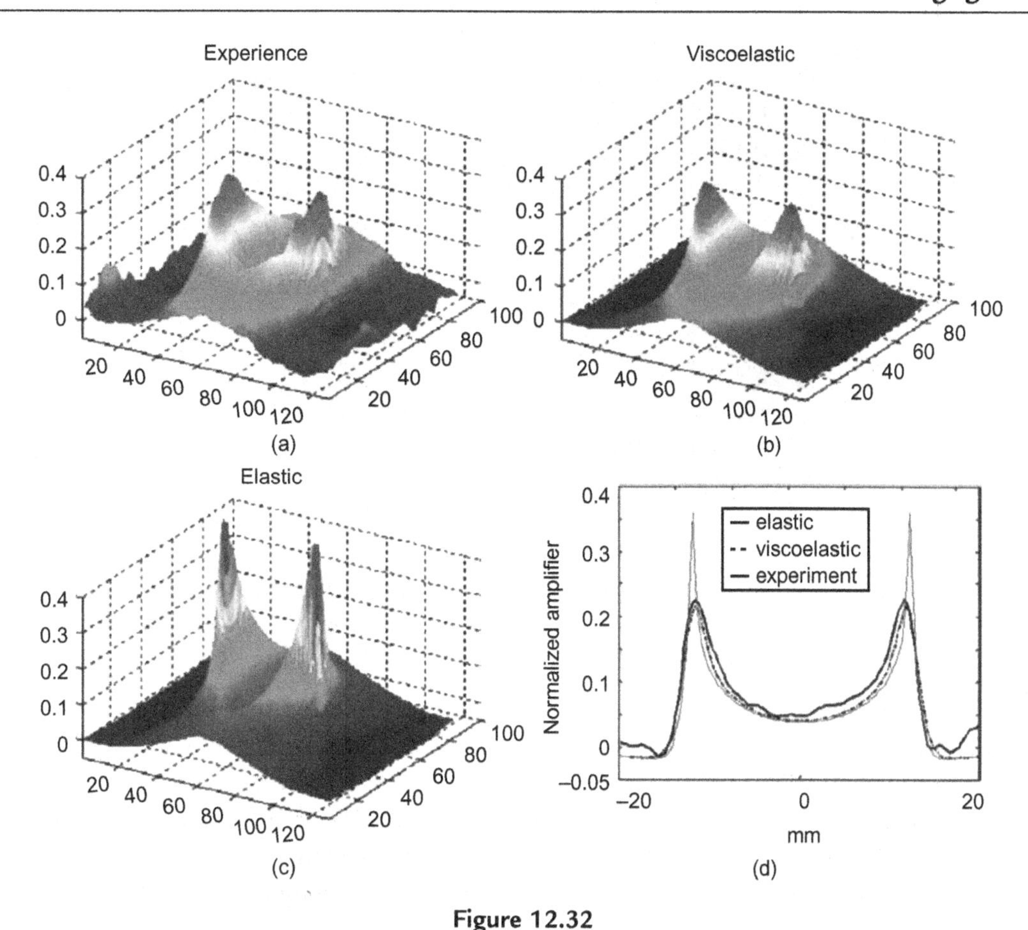

Figure 12.32
Comparison of spatial shear wave displacement calculations to experimental data. (A) Experimental, (B) viscoelastic, and (C) purely elastic 3D plots of the spatial shear wave displacement pattern in the *xz* plane at a given sampling time. (D) Variation of those three fields along the *x* axis (at $z = 0$). The normalization is achieved first using the peak spatiotemporal displacement for each experiment. The reader is referred to the Web version of the book to see the figure in color. *From Bercoff et al. (2004), IEEE.*

As indicated by this equation, the streaming velocity increases with radiated time-averaged acoustic power or intensity and the absorption or radiation force. Some insight into the dynamics of streaming can be inferred from the 1D Navier–Stokes equation:

$$\rho_0\left(\frac{\partial v}{\partial t} + v\cdot\frac{\partial v}{\partial z}\right) = -\frac{\partial p}{\partial z} + \left(\kappa + \frac{4}{3}\mu\right)\frac{\partial^2 v}{\partial z^2}, \tag{12.32}$$

where the left side is acceleration, in which the first term is linear and the second is a nonlinear contribution; on the right side a driving force and, finally, the viscous effects. Acoustic streaming is related to the pressure gradients in the beam (Starritt et al., 1989; Wu et al., 1998).

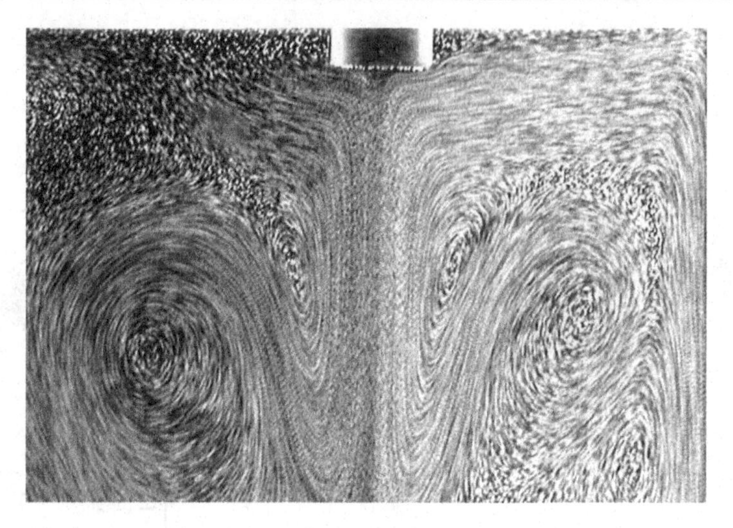

Figure 12.33
Visualization of acoustic streaming from a plane piston transducer radiating into a water glycerin solution. Measured by the 32-MHz pulsed Doppler method. *From Nowicki et al. (1998); reprinted with permission from Excerpta Medica, Inc.*

A free-field radiation pressure, described by Rayleigh for a nonlinear medium, has tensorial or vectorial dependence on the pressure field and its direction, and it is related to the shape of the beam. Nowicki et al. (1997, 1998) have derived a solution for 2D flow and closed forms for axial flow for a plane piston and a focusing transducer, both with circular symmetry. They showed good agreement for axial velocity measurements made with a high-frequency Doppler technique. Their solution is in agreement with a formula presented by Duck (1998):

$$v = (2\alpha I/\mu c_0)d^2 G, \tag{12.33}$$

where d is the beam diameter or, in the case of the transducer solutions, $2a$ (a being transducer radius) and G is a geometric factor. Other notable solutions include those of Wu and Du (1993), Kamakura, Matsuda, Kumamoto, and Breazeale (1995), and Tjootta and Tjotta (1993). Also, because of natural apodization caused by the nonlinear medium, the pressure or intensity field itself is modified over the linear case. Finally, streaming affects temperature rises in nonlinear media; conversely, Wu, Winkler, and O'Neill (1993) have found that it can have a cooling effect on ultrasound-induced heating in low-absorption fluids. Nightingale, Kornguth, Walker, and Trahey (1994) used streaming to distinguish between fluid- and solid-filled cysts. Friend and Yeo (2011), in their extensive review, identify six different kinds of streaming.

12.8.6 Summary

Over the last decade, ARF has emerged as a significant diagnostic method and a key element in several types of imaging, which will be discussed in more detail in Chapter 16.

Significant advances in the prediction of nonlinear waves in absorbing media and at high amplitude levels have not yet been integrated into the ARF analysis. The famous radiation force equation (12.29C), while conceptually satisfying, is inadequate for capturing the high-amplitude effects that occur near the focal point; consequently, as recognized by several investigators (Dalecki et al., 1991; Duck, 1998; Goss and Fry, 1981; Kamakura et al., 1995; Starritt et al., 1989; Wu et al., 1993), enhanced heating and absorption occur at higher amplitude levels. The hybrid analyses reviewed here appear to capture the main mechanical shear wave responses of tissue as seen experimentally. Direct coupling of nonlinear waves to elastic media has been investigated (Landsberger & Hamilton, 2000). Recent improvements in the simulation of acoustic waves in power law media under nonlinear conditions may prove to be valuable in understanding ARF responses. Based on decades of radiation force balance measurements of pulsed diagnostic imaging systems, it is evident ARFs respond to time-averaged pressures (Beissner, 1985); however, transient ARFs have also been investigated (Callé et al., 2005).

References

Aanonsen, S. I., Barkve, T., Naze Tjotta, J., & Tjotta, S. (1984). Distortion and harmonic generation in the nearfield of a finite amplitude sound beam. *Journal of the Acoustical Society of America, 75*, 749–768.

Averkiou, M. A. (2000). Tissue harmonic imaging. *Proceedings of the IEEE ultrasonics symposium* (pp. 1563–1572). San Juan.

Averkiou, M. A., & Hamilton, M. F. (1995). Measurements of harmonic generation in a focused finite-amplitude sound beam. *Journal of the Acoustical Society of America, 98*, 3439–3442.

Averkiou, M. A., Roundhill, D. N., & Powers, J. E. (1997). A new imaging technique based on the nonlinear properties of tissues. *Proceedings of the IEEE ultrasonics symposium* (pp. 1561–1566). Toronto, ON.

Bacon, D. R. (1984). Finite amplitude distortion of the pulsed fields used in diagnostic ultrasound. *Ultrasound in Medicine and Biology, 10*, 189–195.

Bacon, D. R., & Carstensen, E. L. (1990). Increased heating by diagnostic ultrasound due to nonlinear propagation. *Journal of the Acoustical Society of America, 88*, 26–34.

Baker, A. C. (1989). Finite amplitude propagation of focused ultrasonic waves in water, Ph.D. Dissertation. University of Bath, UK.

Baker, A. C. (1991). Prediction of nonlinear propagation in water due to diagnostic medical ultrasound equipment. *Physics in Medicine and Biology, 36*, 1457–1464.

Baker, A. C. (1992). Nonlinear pressure fields due to focused circular apertures. *Journal of the Acoustical Society of America, 91*, 713–717.

Baker, A. C., Anastasiadis, K., & Humphrey, V. F. (1998). The nonlinear pressure field of a plane circular piston: Theory and experiment. *Journal of the Acoustical Society of America, 84*, 1483–1487.

Baker, A. C., Berg, A. M., Sahin, A., & Naze Tjotta, J. (1995). The nonlinear pressure field of plane, rectangular apertures: Experimental and theoretical results. *Journal of the Acoustical Society of America, 97*, 3510–3517.

Baker, A. C., & Humphrey, V. F. (1992). Distortion and high frequency generation due to nonlinear propagation of short ultrasonic pulses from a plane circular piston. *Journal of the Acoustical Society of America, 92*, 1699–1705.

Bakhvalov, N. S., Zhileikin, Y. M., & Zabolotskaya, E. A. (1987). *Nonlinear theory of sound beams*. New York: American Institute of Physics.

Bercoff, J., Tanter, M., Muller, M., & Fink, M. (2004). The role of viscosity in the impulse diffraction field of elastic waves induced by the acoustic radiation force. *IEEE Transactions on Ultrasonics, Ferroelectrics and Frequency Control, 51*, 1523–1536.

Berg, A. M., & Naze Tjotta, J. (1993). Numerical simulation of sound pressure field from finite amplitude, plane or focusing, rectangular apertures. In H. Hobaek (Ed.), *Advances in nonlinear acoustics* (pp. 309–314). London: Elsevier.

Berktay, H. O. (1965). Possible exploitation of nonlinear acoustics in underwater transmitting applications. *Journal of Sound and Vibration, 2*, 435–461.

Berktay, H. O., & Al-Temini, C. A. (1969). Virtual arrays for underwater applications. *Journal of Sound and Vibration, 9*, 295–307.

Berntsen, J. (1990). Numerical calculations of finite amplitude sound beams. In M. F. Hamilton, & D. T. Blackstock (Eds.), *Frontiers of nonlinear acoustics* (pp. 191–196). London: Elsevier.

Berntsen, J., Naze Tjotta, J., & Tjotta, S. (1984). Nearfield of a large acoustic transducer, Part IV: Second harmonic and sum frequency radiation. *Journal of the Acoustical Society of America, 75*, 1383–1391.

Berntsen, J., & Vefring, E. (1986). *Numerical computation of a finite amplitude sound beam.* Norway: Department of Mathematics, University of Bergen. Technical Report 82.

Beyer, R. T. (1997). *Nonlinear acoustics.* Woodbury, NY: Acoustical Society of America. [Reprint from Naval Ship Systems Command, 1974].

Bjorno, L. (2002). Forty years of nonlinear ultrasound. *Ultrasonics, 40*, 11–17.

Blackstock, D. T. (1966). Connection between the Fay and Fubini solutions for plane sound waves of finite amplitude. *Journal of the Acoustical Society of America, 39*, 1019–1026.

Blackstock, D. T. (1985). Generalized Burgers equation for plane waves. *Journal of the Acoustical Society of America, 77*, 2050–2053.

Blackstock, D. T. (1998). In M. F. Hamilton, & D. T. Blackstock (Eds.), *Nonlinear Acoustics, Chapter 1.* San Diego: Academic Press.

Blackstock, D. T., Hamilton, M. F., & Pierce, A. D. (1998). Nonlinear Acoustics, Chapter 4In M. F. Hamilton, & D. T. Blackstock (Eds.), San Diego: Academic Press.

Brock-Fisher, G. A., Poland, M. D., & Rafter, P. G. (1996). Means for increasing sensitivity in non-linear ultrasound imaging systems, US patent 5,577,505.

Bruce, F. B., Averkiou, M. A., Skyba, D. M., & Powers, J. E. (2000). A generalization of pulse inversion Doppler. *Proceedings of the IEEE ultrasonics symposium* (pp. 1561–1566). San Juan.

Bruus, H. (2008). *Theoretical microfluidics.* Oxford: Oxford University Press.

Cahill, M. D., & Baker, A. C. (1998). Numerical simulation of the acoustic field of a phased-array medical ultrasound scanner. *Journal of the Acoustical Society of America, 104*, 1274–1283.

Callé, S., et al. (2005). Temporal analysis of tissue displacement induced by a transient ultrasound radiation force. *Journal of the Acoustical Society of America, 118*, 2829–2840.

Canney, M. S., Bailey, M. R., Crum, L. A., Khokhlova, V. A., & Sapozhnikov, O. A. (2008). Acoustic characterization of high intensity focused ultrasound fields: A combined measurement and modeling approach. *Journal of the Acoustical Society of America, 124*, 2406–2420.

Carson, P. J., Fischella, P. R., & Oughton, T. V. (1978). Ultrasonic power and intensity produced by diagnostic ultrasound equipment. *Ultrasound in Medicine and Biology, 3*, 341.

Carstensen, E. L., Law, W. K., McKay, N. D., & Muir, T. G. (1980). Demonstration of nonlinear acoustical effects at biomedical frequencies and intensities. *Ultrasound in Medicine and Biology, 6*, 359–368.

Catheline, S., Wu, F., & Fink, M. (1999). A solution to diffraction biases in sonoelasticity: The acoustic impulse technique. *Journal of the Acoustical Society of America, 105*, 2941–2950.

Cherin, E., Poulsen, J. K., van der Steen, A. F. W., & Foster, F. S. (2000). Comparison of nonlinear and linear imaging techniques at high frequency. *Proceedings of the IEEE ultrasonics symposium* (pp. 1561–1566), San Juan.

Christopher, T. (1993). A nonlinear plane-wave algorithm for diffractive propagation involving shock waves. *Journal of Computational Acoustics, 1*, 371–393.

Christopher, T. (1997). Finite amplitude distortion-based inhomogeneous pulse echo ultrasonic imaging. *IEEE Transactions on Ultrasonics, Ferroelectrics and Frequency Control*, *44*, 125–139.

Christopher, T. (1999). Source prebiasing for improved second harmonic bubble response imaging. *IEEE Transactions on Ultrasonics, Ferroelectrics and Frequency Control*, *46*, 556–563.

Christopher, T., & Carstensen, E. L. (1996). Finite amplitude distortion and its relationship to linear derating formulae for diagnostic ultrasound systems. *Ultrasound in Medicine and Biology*, *22*, 1103–1116.

Christopher, T., & Parker, K. J. (1991). New approaches to nonlinear diffractive field propagation. *Journal of the Acoustical Society of America*, *90*, 488–499.

Cleveland, R. O., Hamilton, M. F., & Blackstock, D. T. (1996). Time-domain modeling of finite-amplitude sound in relaxing fluids. *Journal of the Acoustical Society of America*, *99*, 3312–3318.

Crum, L. A. (1971). Acoustic force on a liquid droplet in an acoustic stationary wave. *Journal of the Acoustical Society of America*, *50*, 157–163.

Dalecki, D., Carstensen, E. L., & Parker, K. J. (1991). Absorption of finite amplitude focused ultrasound. *Journal of the Acoustical Society of America*, *89*, 2435–2447.

Divall, S. A., & Humphrey, V. F. (2000). Finite difference modelling of the temperature rise in nonlinear medical ultrasound fields. *Ultrasonics*, *38*, 273–277.

Djelouah, H., & Baboux, J. C. (1992). Transient ultrasonic field radiated by a circular transducer in a solid medium. *Journal of the Acoustical Society of America*, *92*, 2932–2941.

Du, Y., Jensen, H. & Jensen, J. A. (2010) Simulation of second harmonic ultrasound fields. *Proceedings of the IEEE ultrasonics symposium* (pp. 2191–2194). San Diego, CA.

Du, Y., Jensen, H., & Jensen, J. A. (2011a) Comparison of simulated and measured non-linear ultrasound fields. *SPIE medical imaging: Ultrasonic imaging, tomography, and therapy conference, proceedings of SPIE 7968* (pp. 1–10).

Du, Y., Jensen, H. & Jensen, J. A. (2011b) Angular spectrum approach for fast simulation of pulsed non-linear ultrasound fields. *Proceedings of the IEEE ultrasonics symposium* (pp. 1583–1586). Orlando, FL.

Du, Y., & Jensen, J. A. (2008) Feasibility of non-linear simulation for Field II using an angular spectrum approach. *Proceedings of the IEEE ultrasonics symposium* (pp. 1314–1317). Beijing.

Duck, F. A. (1990). *Physical properties of tissue: A comprehensive reference book*. London: Academic Press.

Duck, F. A. (1998). Radiation pressure and streaming, Chapter 3. In F. A. Duck, A. C. Baker, & H. C. Starritt (Eds.), *Ultrasound in medicine, medical science series*. Bristol, UK: Institute of Physics Publishing.

Duck, F. A. (2002). Nonlinear acoustics in diagnostic ultrasound. *Ultrasound in Medicine and Biology*, *28*, 1–18.

Duck, F. A., & Starritt, H. C. (1984). Acoustic shock generation by ultrasonic imaging equipment. *British Journal of Radiology*, *57*, 231–240.

Everbach, E. C. (1997). Parameters of nonlinearity of acoustic media. In M. J. Crocker (Ed.), *Encyclopedia of acoustics*. New York: John Wiley & Sons.

Fedewa, R. J., Wallace, K. D., Holland, M. R., Jago, J. R., Ng, G. C., Reilly, M. W., et al. (2001). Statistically significant differences in the spatial coherence of backscatter for fundamental and harmonic portions of a clinical beam. *Proceedings of the IEEE ultrasonics symposium* (pp. 1481–1484). Atlanta, GA.

Frijlink, M. E., Kaupang, H., Varslot, T., & Masøy, S.-E.(2008). Abersim: A simulation program for 3D nonlinear acoustic wave propagation for arbitrary pulses and arbitrary transducer geometries. *Proceedings of the IEEE ultrasonics symposium* (pp. 1282–1285). Beijing.

Frinking, P. J. A., Boukaz, A., Kirkhorn, J., Ten Cate, F. J., & De Jong, N. (2000). Ultrasound contrast imaging: Current and new potential methods. *Ultrasound in Medicine and Biology*, *26*, 965–975.

Froysa, K. E. (1991). Linear and weakly nonlinear propagation of a pulsed sound beam, Ph.D. Dissertation. Department of Mathematics, University of Bergen, Norway.

Gakenheimer, D. C., & Miklowitz, J. (1969). Transient excitation of an elastic half-space by a point load traveling on the surface. *Journal of Applied Mechanics*, *36*, 505.

Germain, L., & Cheeke, J. D. N. (1988). Generation and detection of high-order harmonics in liquids using a scanning acoustic microscope. *Journal of the Acoustical Society of America*, *83*, 942–949.

Ginsberg, J. H., & Hamilton, M. F. (1998). In M. F. Hamilton, & D. T. Blackstock (Eds.), *Nonlinear acoustics, chapter 11*. San Diego: Academic Press.

Ginter, S. (2000). Numerical simulation of ultrasound-thermotherapy combining nonlinear wave propagation with broadband soft-tissue absorption. *Ultrasonics, 37*, 693–696.

Gol'dberg, Z. A. (1957). On the propagation of plane waves of finite amplitude. *Soviet Physics Acoustics, 3*, 340–347.

Gor'kov, L. P. (1962). Of forces acting on a small particle in acoustical field in an ideal fluid. *Soviet Physics Doklady, 6*, 773–775.

Goss, S. A., & Fry, F. J. (1981). Nonlinear acoustic behavior in focused ultrasonic fields: Observations of intensity dependent absorption in biological tissue. *IEEE Transactions on Sonics and Ultrasonics, 28*, 21–26.

Haider, B., & Chiao, R. Y. (1999). Higher order nonlinear ultrasonic imaging. *Proceedings of the IEEE ultrasonics symposium* (pp. 1527–1531). Caesars Tahoe, NV.

Hallaj, I. M., Cleveland, R. O., & Hynynen, K. (2001). Simulations of the thermo-acoustic lens effect during focused ultrasound surgery. *Journal of the Acoustical Society of America, 109*, 2245–2253.

Hamilton, M. F. (1998). Nonlinear acoustics, chapter 8In M. F. Hamilton, & D. T. Blackstock (Eds.), San Diego: Academic Press.

Hamilton, M. F., & Morfey, C. L. (1998). In M. F. Hamilton, & D. T. Blackstock (Eds.), *Nonlinear acoustics, chapter 3*. San Diego: Academic Press.

Hamilton, M. F., Naze Tjotta, J., & Tjotta, S. (1985). Nonlinear effects in the farfield of a directive sound source. *Journal of the Acoustical Society of America, 78*, 202–216.

Hansen, R., Angelsen, B. A .J., & Johansen, T. F. (2001). Reduction of nonlinear contrast agent scattering due to nonlinear wave propagation. *Proceedings of the IEEE ultrasonics symposium* (pp. 1725–1728). Atlanta, GA.

Haran, M. E., & Cook, B. D. (1983). Distortion of finite amplitude ultrasound in lossy media. *Journal of the Acoustical Society of America, 73*, 774–779.

Hart, T. S., & Hamilton, M. F. (1988). Nonlinear effects in focused sound beams. *Journal of the Acoustical Society of America, 84*, 1488–1496.

Humphrey, V. F. (2000). Nonlinear propagation in ultrasonic fields: Measurements, modelling and harmonic imaging. *Ultrasonics, 38*, 267–272.

Jiang, P., Mao, Z., & Lazenby, J. C. (1998). A new tissue harmonic imaging scheme with better fundamental frequency cancellation and higher signal-to-noise ratio. *Proceedings of the IEEE ultrasonics symposium* (pp. 1589–1594). Sendai.

Jing, Y., Wang, T., & Clement, G. T. (2012). A k-space method for moderately nonlinear wave propagation. *IEEE Transactions on Ultrasonics, Ferroelectrics and Frequency Control, 59*, 1664–1673.

Kamakura, T., Matsuda, K., Kumamoto, Y., & Breazeale, M. A. (1995). Acoustic streaming induced in focused Gaussian beams. *Journal of the Acoustical Society of America, 97*, 2740–2746.

Kamakura, T., Tani, M., & Kumamoto, Y. (1992). Harmonic generation in finite amplitude sound beams from a rectangular aperture source. *Journal of the Acoustical Society of America, 91*, 3144–3151.

Kaupang, H., Varslot, T., & Masøy, S.-E. (2007) Second-harmonic aberration correction. *Proceedings of the IEEE ultrasonics symposium* (pp. 1537–1540). New York.

Khokhlova, V. A., Ponomarev, A. E., Averkiou, M. A., & Crum, L. A. (2006). Nonlinear pulsed ultrasound beams radiated by rectangular focused diagnostic transducers. *Acoustical Physics, 52*, 481–489.

Kim, D. Y., Lee, J. C., Kwon, S. J., & Song, T. K. (2001). Ultrasound second harmonic imaging with a weighted chirp pulse. *Proceedings of the IEEE ultrasonics symposium* (pp. 1477–1480). Atlanta, GA.

King, L. V. (1934). On the acoustic radiation pressure on spheres. *Proceedings of the Royal Society of London A, 147*, 212–240.

Kompfner, R., & Lemons, R. A. (1976). Nonlinear acoustic microscopy. *Applied Physics Letters, 28*, 295–297.

Kornbluth, M., Liang, D. H., Paloma, A., & Schnittger, I. (1998). Native harmonic imaging improves endocardial border definition and visualisation of cardiac structures. *Journal of the American Society of Echocardiography, 11*, 693–701.

Kuznetsov, V. P. (1971). Equations of nonlinear acoustics. *Soviet Physics Acoustics, 16*, 467–470.

Labat, V., Remenieras, J. P., Bou Matar, O., Ouahabi, A., & Patat, F. (2000). Harmonic propagation of pulsed finite amplitude sound beams: Experimental determination of the nonlinearity parameter B/A. *Ultrasonics, 38*, 292–296.

Landsberger, B. J., & Hamilton, M. F. (2000). Second-harmonic generation in sound beams reflected from, and transmitted through, immersed elastic solids. *Journal of the Acoustical Society of America, 108*, 906–917.

Law, W. K., Frizzell, L. A., & Dunn, F. (1981). Ultrasonic determination of the nonlinearity parameter B/A for biological media. *Journal of the Acoustical Society of America, 69*, 1210–1212.

Law, W. K., Frizzell, L. A., & Dunn, F. (1985). Determination of the nonlinearity parameter B/A of biological media. *Ultrasound in Medicine and Biology, 11*, 307–318.

Lee, J., & Shung, K. K. (2006). Radiation forces exerted on arbitrarily located sphere by acoustic tweezer. *Journal of the Acoustical Society of America, 120*, 1084–1094.

Lee, Y. S., & Hamilton, M. F. (1995). Time-domain modeling of pulsed finite-amplitude sound beams. *Journal of the Acoustical Society of America, 97*, 906–917.

Li, Y., & Zagzebski, J. A. (2000). Computer model for harmonic ultrasound imaging. *IEEE Transactions on Ultrasonics, Ferroelectrics and Frequency Control, 47*, 1259–1272.

Liu, D. L. D., von Behren, P., & Kim, J. (2001). Single transmit imaging. *Proceedings of the IEEE ultrasonics symposium* (pp. 1481–1484). Atlanta, GA.

Makin, I. R. S., Averkiou, M. A., & Hamilton, M. F. (2000). Second-harmonic generation in a sound beam reflected and transmitted at a curved interface. *Journal of the Acoustical Society of America, 108*, 1505–1513.

McDonald, B. E., & Kuperman, W. A. (1987). Time domain formulation for pulse propagation including nonlinear behavior in a caustic. *Journal of the Acoustical Society of America, 81*, 1406–1417.

Miller, G. F., & Pursey, H. (1954). The field and radiation impedance of mechanical radiators on the free surface of a semi-infinite isotropic solid. *Proceedings of the Royal Society of London A, 223*, 521–541.

Mor-Avi, V., Caiani, E. G., Collins, K. A., Korcarz, C. E., Bednarz, J. E., & Lang, R. M. (2001). Combined assessment of myocardial perfusion and regional left ventricular function by analysis of contrast-enhanced power modulation images. *Circulation, 104*, 352–357.

Muir, T. G., & Carstensen, E. L. (1980). Prediction of nonlinear acoustic effects at biomedical frequencies and intensities. *Ultrasound in Medicine and Biology, 6*, 345–357.

Muir, T. O. (1980). *Nonlinear effects in acoustic imaging, Acoustical imaging* (Vol. 9, pp. 93–109).). New York: Plenum Press.

Nachef, S., Cathignol, D., Naze Tjotta, J., Berg, A. M., & Tjotta, S. (1995). Investigation of a high intensity sound beam from a plane transducer: Experimental and theoretical results. *Journal of the Acoustical Society of America, 98*, 2303–2323.

Naugol'nykh, K. A., & Romanenko, E. V. (1959). Amplification factor of a focusing system as a function of sound intensity. *Soviet Physics Acoustics, 5*, 191–195.

Naze Tjotta, J., & Tjotta, S. (1981). Nonlinear equations of acoustics, with applications to parametric arrays. *Journal of the Acoustical Society of America, 69*, 1644–1652.

Nightingale, K., Kornguth, P., Walker, W. F., & Trahey, G. (1994). Use of radiation force phenomenon in ultrasound to distinguish between cysts and solid lesions in breast. *Ultrasonic Imaging, 16*, 46–47.

Norton, G. V. (2009a). Comparison of homogeneous and heterogeneous modeling of transient scattering from dispersive media directly in the time domain. *Mathematics and Computers in Simulation, 80*, 682–692.

Norton, G. V. (2009b). Numerical solution of the wave equation describing acoustic scattering and propagation through complex dispersive moving media. *Nonlinear Analysis-Theory Methods and Applications, 71*, E849–E854.

Norton, G. V., & Novarini, J. G. (2003). Including dispersion and attenuation directly in the time domain for wave propagation in isotropic media. *Journal of the Acoustical Society of America, 113*, 3024–3031.

Norton, G. V., & Purrington, R. D. (2009). The Westervelt equation with viscous attenuation versus a causal propagation operator: A numerical comparison. *Journal of Sound and Vibration, 327*, 163–172.

Nowicki, A., Kowalewski, T., Secomski, W., & Wojcik, J. (1998). Estimation of acoustical streaming: Theoretical model, Doppler measurements and optical visualization. *European Journal of Ultrasound, 7*, 73–81.

Nowicki, A., Wojcik, J., & Secomski, W. (1998). Harmonic imaging using multitione nonlinear coding. *Ultrasound in Medicine and Biology, 33*, 1112–1121.

Nyborg, W. L. (1965). Acoustic streaming. In W. P. Mason, & R. N. Thurston (Eds.), *Physical acoustics* (pp. 265–329). New York: Academic PressChapter 11.

Nyborg, W. L. (1967). Radiation pressure on a small rigid sphere. *Journal of the Acoustical Society of America, 42*, 947–952.

Nyborg, W. L. (1998). In M. F. Hamilton, & D. T. Blackstock (Eds.), *Nonlinear acoustics, chapter 7*. San Diego: Academic Press.

Parker, K. J., & Friets, E. M. (1987). On the measurement of shock waves. *IEEE Transactions on Ultrasonics, Ferroelectrics and Frequency Control, 34*, 454–460.

Pestorious, F. M. (1973). *Propagation of plane acoustic noise of finite amplitude*. Austin: Applied Research Laboratories, University of Texas. Technical Report ARL-TR-73–23.

Pinton, G., Dahl, J., Rosenzweig, S., & Trahey, G. (2009). A heterogeneous nonlinear attenuating full-wave model of ultrasound. *IEEE Transactions on Ultrasonics, Ferroelectrics and Frequency Control, 56*, 474–488.

Purrington, R. D., & Norton, G. V. (2012). A numerical comparison of the Westervelt equation with viscous attenuation and a causal propagation operator. *Mathematics and Computers in Simulation, 82*, 1287–1297.

Radulescu, E. G., Wojcik, J., Lewin, P. A., & Nowicki, A. (2003). Nonlinear propagation model for ultrasound hydrophones calibration in the frequency range up to 100 MHz. *Ultrasonics, 41*, 239–245.

Rasmussen, J. H., Du, Y., & Jensen, J. A. (2011) Third harmonic imaging using pulse inversion. *Proceedings of the IEEE ultrasonics symposium* (pp. 2269–2272). Orlando, FL.

Rayleigh (1902). On the pressure of vibrations. *Philosophical Magazine, 3*, 338–346.

Remenieras, J. P., Bou Matar, O., Labat, V., & Patat, F. (2000). Time-domain modeling of nonlinear distortion of pulsed finite amplitude sound beams. *Ultrasonics, 38*, 305–311.

Rudenko, O. V., Sarvazyan, A. P., & Emelianov, S. Y. (1996). Acoustic radiation force and streaming induced by focused nonlinear ultrasound in a dissipative medium. *Journal of the Acoustical Society of America, 99*, 2791–2798.

Sahin, A., & Baker, A. C. (1993). Nonlinear propagation in the pressure fields of plane and focused rectangular apertures. In H. Hobaek (Ed.), *Advances in nonlinear acoustics* (pp. 303–308). London: Elsevier.

Sarvazyan, A. (2010). Diversity of biomedical applications of acoustic radiation force. *Ultrasonics, 50*, 230–234.

Sempsrott, J. M., & O'Brien, W. D., Jr. (1999). Experimental verification of acoustic saturation. *Proceedings of the IEEE ultrasonics symposium* (pp. 1287–1290). Caesars Tahoe, NV.

Shi, W. T., & Forsberg, F. (2000). Ultrasonic characterization of the nonlinear properties of contrast microbubbles. *Ultrasound in Medicine and Biology, 26*, 93–104.

Simpson, D. H., Chin, C. T., & Burns, P. N. (1999). Pulse inversion Doppler: A new method for detecting nonlinear echoes from microbubble contrast agents. *IEEE Transactions on Ultrasonics, Ferroelectrics and Frequency Control, 46*, 372–382.

Spencer, K. T., Bernarz, J., Rafter, P. G., Korcarz, C., & Lang, R. M. (1998). Use of harmonic imaging without echocardiographic contrast to improve two-dimensional image quality. *American Journal of Cardiology, 82*, 794–799.

Starritt, H. C., Duck, F. A., Hawkins, A. I., & Humphrey, V. F. (1986). The development of harmonic distortion in pulsed finite-amplitude ultrasound passing through liver. *Physics in Medicine and Biology, 31*, 1401–1409.

Starritt, H. C., Perkins, M. A., Duck, F. A., & Humphrey, V. F. (1985). Evidence for ultrasonic finite-amplitude distortion in muscle using medical equipment. *Journal of the Acoustical Society of America, 77*, 302–306.

Stratton, J. A. (1941). *Electromagnetic theory*. New York: McGraw Hill. pp. 97–103.

Szabo, T. L. (1993). Time domain nonlinear wave equations for lossy media. In H. Hobaek (Ed.), *Advances in nonlinear acoustics: Proceedings of 13th ISNA* (pp. 89–94). Singapore: World Scientific.

Szabo, T. L., Clougherty, F., & Grossman, C. (1999). Effects of nonlinearity on the estimation of in situ values of acoustic output parameters. *Journal of Ultrasound in Medicine, 18*, 33–42.

Takeuchi. (1996). Coded excitation for harmonics imaging. *Proceedings of the IEEE ultrasonics symposium* (pp. 1433–1436). San Antonio, TX.

Taraldsen, G. (2001). A generalized Westervelt equation for nonlinear medical ultrasound. *Journal of the Acoustical Society of America, 109*, 1329–1333.

Tavakkoli, J., Cathignol, D., Souchon, R., & Sapozhnikov, O. A. (1998). Modeling of pulsed finite-amplitude focused sound beams in time domain. *Journal of the Acoustical Society of America, 104*, 2061–2072.

Ten Cate, F. J. (1993). An experimental investigation of the nonlinear pressure field produced by a plane circular piston. *Journal of the Acoustical Society of America, 94*, 1084–1089.

Tjotta, S. (2000). On some nonlinear effects in ultrasonic fields. *Ultrasonics, 38*, 278–283.

Tjøtta, J., & Tjøtta, S. (1993). Acoustic streaming in ultrasound beams. In H. Hobaek (Ed.), *Advances in nonlinear acoustics: Proceedings of 13th ISNA* (pp. 601–606). Singapore: World Scientific.

Too, G. P. J., & Ginsberg, J. H. (1992). Nonlinear progressive wave equation model for transient and steady-state sound beams. *Journal of the Acoustical Society of America, 92*, 59–68.

Tranquart, F., Grenier, N., Eder, V., & Pourcelot, L. (1999). Clinical use of ultrasound tissue harmonic imaging. *Ultrasound in Medicine and Biology, 25*, 889–894.

Treeby, B. E., Jaros, J., Rendell, A. P., & Cox, B. T. (2012). Modeling nonlinear ultrasound propagation in heterogeneous media with power law absorption using a k-space pseudospectral method. *Journal of the Acoustical Society of America, 131*, 4324–4366.

van Wijk, M. C., & Thijjsen, J. M. (2002). Performance testing of medical ultrasound equipment: Fundamental vs. harmonic mode. *Ultrasonics, 40*, 585–591.

Varslot, T., & Masøy, S. E. (2006). Forward propagation of acoustic pressure pulses in 3D soft biological tissue, Modeling. *Identification and Control, 27*, 181–200.

Varslot, T., & Taraldsen, G. (2005). Computer simulation of forward wave propagation in soft tissue. *IEEE Transactions on Ultrasonics, Ferroelectrics and Frequency Control, 52*, 1473–1482.

Varslot, T., Taraldsen, G., Johansen, T. F., & Angelsen, B. A. J. (2001) Computer simulation of forward wave propagation in non-linear, heterogeneous, absorbing tissue. *Proceedings of the IEEE ultrasonics symposium* (pp. 1993–1996). Atlanta, GA.

Wallace, K. D., Holland, M. R., & Miller, J. G. (2001). Improved description of shock wave evolution in media with frequency power law dependent attenuation. *Journal of the Acoustical Society of America, 109*, 2263–2265.

Wang, K., Teoh, E., Jarosy, J., & Treeby, B. E.(2012) Modelling nonlinear ultrasound propagation in absorbing media using the k-wave toolbox: Experimental validation. *Proceedings of the IEEE ultrasonics symposium* (pp. 523–526). Dresden, Germany.

Wang, T. G., & Lee, C. P. (1998). In M. F. Hamilton, & D. T. Blackstock (Eds.), *Nonlinear acoustics, chapter.* San Diego: Academic Press.

Ward, B., Baker, A. C., & Humphrey, V. F. (1996). Nonlinear propagation applied to the improvement of lateral resolution in medical ultrasound scanners. *Proceedings of 1995 world congress on ultrasonics* (pp. 965–968).

Ward, B., Baker, A. C., & Humphrey, V. F. (1997). Nonlinear propagation applied to the improvement of resolution in diagnostic medical ultrasound equipment. *Journal of the Acoustical Society of America, 10*, 143–154.

Watson, A. J., Humphrey, V. F., Baker, A. C., & Duck, F. A. (1990). *Nonlinear propagation of focused ultrasound in tissue-like media* (pp. 445–450). *Frontiers of nonlinear acoustics, proceedings of the 12th ISNA*. Austin, TX, London: Elsevier.

Westervelt, P. J. (1963). Parametric acoustic array. *Journal of the Acoustical Society of America, 35*, 535–537.

Wojcik, G., Mould, J., Jr., Ayter, S., & Carcione, L. M. (1998). A study of second harmonic generation by focused medical transducer pulses. *Proceedings of the IEEE ultrasonics symposium* (pp. 1583–1588). Sendai.

Wojcik, G. L., Mould, J. C., Jr., & Carcione, L. M. (1999a). Combined transducer and nonlinear tissue propagation. *1999 international mechanical engineering congress & exposition proceedings.*

Wojcik, G. L., Szabo, T., Mould, J., Carcione, L., & Clougherty, F. (1999b). Nonlinear pulse calculations and data in water and a tissue mimic. *Proceedings of the IEEE ultrasonics symposium* (pp. 1521–1526). Caesars Tahoe, NV.

Wu, J., & Du, G. (1990). Acoustic radiation force on a small compressible sphere in a focused beam. *Journal of the Acoustical Society of America, 87*, 997–1003.

Wu, J., & Tong, J. (1997). Measurements of nonlinearity parameter B/A of contrast agents. *Journal of the Acoustical Society of America, 101*, 1155–1161.

Wu, J., Winkler, A. J., & O'Neill, T. P. (1993). Effect of acoustic streaming on ultrasonic heating. *Ultrasound in Medicine and Biology, 20*, 195–201.

Wu, J. R. (1991). Acoustical tweezers. *Journal of the Acoustical Society of America, 89*, 2140–2143.

Xiang, X. (2004) Statistical model of beam distortion by tissue inhomogeneities in tissue harmonic imaging. Ph. D. Dissertation, University of Texas at Austin.

Xiang, X., & Hamilton, M. F. (2006) Angular spectrum decomposition analysis of second harmonic ultrasound propagation and its relation to tissue harmonic imaging. *Proceedings of the fourth international workshop on ultrasonic and advanced methods for nondestructive testing and material characterization* (pp. 11–24). UMass Dartmouth, Dartmouth, MA.

Yongchen, S., Yanwu, D., Jie, T., & Zhensheng, T. (1986). Ultrasonic propagation parameters in human tissues. *Proceedings of the IEEE ultrasonics symposium* (pp. 905–908). Williamsburg, VA.

Yuldashev, P. V., Kreider, W., et al. (2012) Characterization of nonlinear ultrasound fields of 2D therapeutic arrays. *Proceedings of the IEEE ultrasonics symposium* (pp. 925–928). Dresden, Germany.

Zhang, J., & Dunn, F. (1991). A small volume thermodynamic system for B/A measurement. *Journal of the Acoustical Society of America, 89*, 73–79.

Zhang, J., Kuhlenschmidt, & Dunn, F. (1991). Influences of structural factors of biological media on the acoustic nonlinearity parameter B/A. *Journal of the Acoustical Society of America, 89*, 80–91.

Bibliography

Averkiou, M. A. (2000). Tissue harmonic imaging. *Proceedings of the IEEE ultrasonics symposium* (pp. 1561–1566). San Juan.

A recommended summary of harmonic imaging and related signal processing.

Baker, A. C. (1998). Nonlinear effects in ultrasonic propagation. In F. A. Duck, A. C. Baker, & H. C. Starritt (Eds.), *Ultrasound in Medicine* (pp. 23–28). Bristol, UK: IOP Publishing.

A brief summary of the measurement and simulation of nonlinear harmonic beams.

Bakhvalov, N. S., Zhileikin, Y. M., & Zabolotskaya, E. A. (1987). *Nonlinear theory of sound beams*. New York: American Institute of Physics.

A treatise on the characteristics of beams in nonlinear media and their simulation.

Berktay, H. O. (1965). Parametric amplification by the use of acoustic nonlinearities and some possible applications. *Journal of Sound and Vibration*, 2, 462–470.

An early treatment of the principles of nonlinear acoustics and beams.

Beyer, R. T. (1997). *Nonlinear acoustics*. Woodbury, NY: Acoustical Society of America. [Reprint from Naval Ship Systems Command, 1974.].

A longtime classic on nonlinear acoustics.

Crocker, M. S. (Ed.), (1997). *Encyclopedia of acoustics* New York: John Wiley & Sons.

Short articles on many aspects of both linear and nonlinear acoustics.

Duck, F. A. (1999). Acoustic saturation and output regulation. *Ultrasound in Medicine and Biology*, *25*, 1009–1018.
Duck, F. A. (2002). Nonlinear acoustics in diagnostic ultrasound. *Ultrasound in Medicine and Biology*, *28*, 1–18.
A comprehensive review of nonlinear propagation effects, beams, and harmonic imaging.
Friend, J., & Yeo, L. Y. (2011). Microscale acoustofluidics: Microfluidics driven via acoustics and ultrasonics. *Reviews of Modern Physics*, *83*, 647–704.
Frinking, P. J. A., Boukaz, A., Kirkhorn, J., Ten Cate, F. J., & De Jong, N. (2000). Ultrasound contrast imaging: Current and new potential methods. *Ultrasound in Medicine and Biology*, *26*, 965–975.
Article includes a review of nonlinear signal processing.
Hamilton, M. F., & Blackstock, D. T. (Eds.), (1998). *Nonlinear acoustics* San Diego, CA: Academic Press.
A thorough theoretical overview of nonlinear acoustics.
Humphrey, V. F. (2000). Nonlinear propagation in ultrasonic fields: Measurements, modelling and harmonic imaging. *Ultrasonics*, *38*, 267–272.
A recommended overview article on nonlinear acoustics and harmonic imaging.
Novikov, B. K., Rudenko, O. V., & Timoshenko, V. I. (1987). *Nonlinear underwater acoustics*. New York: American Institute of Physics.
Nonlinear underwater acoustic applications of beams and arrays.
Sarvazyan, A. P., Rudenko, O. V., Swanson, S. D., Fowlkes, J. B., & Emelianov, S. Y. (2010). Shear wave elasticity imaging: A new ultrasonic technology of medical diagnostics. *Ultrasound in Medicine and Biology*, *36*, 1379–1394.
Starritt, H. C., Duck, F. A., & Humphrey, V. F. (1989). An experimental investigation of streaming in pulsed diagnostic ultrasound beams. *Ultrasound in Medicine and Biology*, *15*, 363–373.
Starritt, H. C., Duck, F. A., & Humphrey, V. F. (1991). Forces acting in the direction of propagation in pulsed ultrasound fields. *Physics in Medicine and Biology*, *36*, 1465–1474.
Wu, J., Winkler, A. J., & O'Neill, T. P. (1998). Effect of acoustic streaming on ultrasonic heating. *Ultrasound in Medicine and Biology*, *24*, 153–159.

CHAPTER 13

Ultrasonic Exposimetry and Acoustic Measurements

Chapter Outline

13.1 Introduction to Measurements

Measurements related to ultrasound imaging are performed at several levels. First, the acoustical, mechanical, and chemical properties of materials used in transducer construction are determined. Second, various properties of arrays, both acoustical and electrical, are measured either as a whole or element by element. Thirdly, extensive acoustic output measurements are conducted on the imaging system and each transducer in different modes. Fourth, performance tests are used to evaluate the imaging capabilities of an imaging system and transducer combination.

13.2 Materials Characterization

13.2.1 Transducer Materials

Determining the acoustic properties of materials with ultrasound is an ongoing process for many industries; however, a particular emphasis for imaging equipment is the evaluation of materials for transducers. Especially, as higher performance, wider bandwidth arrays are designed, materials with low losses and specific acoustic properties are required. Key parameters of a material are its sound speed, density, and acoustic characteristic impedance, loss, and phase velocity dispersion as a function of frequency. Auxiliary information may also be needed, such as chemical compatibility, thermal and mechanical properties, and strength when bonded to other materials. For accurate finite-element modeling of transducer arrays, both longitudinal and shear wave parameters are required (Powell, Wojcik, Desilets, Gururaja, Guggenberger, & Sherrit, 1997).

Earlier methods employed discrete single-frequency measurements using tone bursts. Selfridge (1985) measured many materials at a single frequency in a simple reflection arrangement. The legacy of his many measurements for a wide range of materials has been maintained on the Onda Corporation website. To increase the frequency range, several transducers were often used to obtain data. For the key parameters, broadband spectroscopy using ultrasound provides a precise and direct methodology. Zeqiri (1988) demonstrated that equivalent results could be obtained by using broadband methods. Also known as the through-transmission substitution method, this approach usually involves two broadband nonfocusing transducers aligned in a water tank (shown in Figure 13.1). Polyvinylidene fluoride (PVDF) transducers (see Section 5.7.5) provide superior bandwidth for this application. An impulse excitation is applied to the transmitting transducer, and the signal received by the second transducer is digitized. Next, a sample material with parallel sides, a known thickness, and diameter larger than the extent of the beam is inserted between the transducers and aligned. With the sample in place, the signal is once again recorded on a digital sampling scope. Calculations based on the ratio of the absolute spectra can determine the attenuation through the sample over a wide bandwidth. The sound speed dispersion can be found from the phase of the ratio of spectra. An independent determination of material density, a coarse determination of the midband sound speed, and appropriate corrections for reflections and transmissions through the sample boundaries are

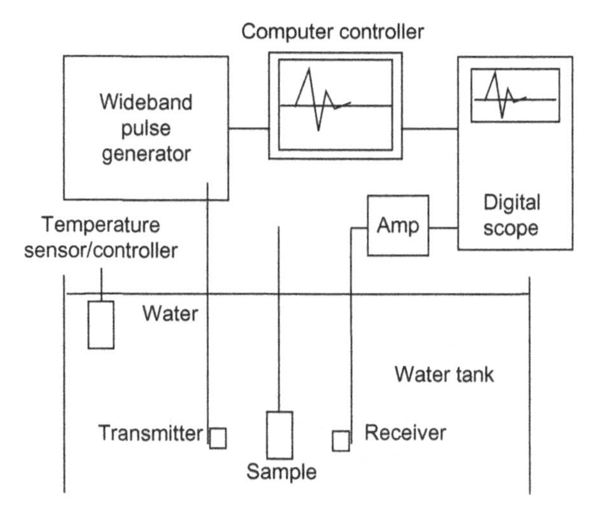

Figure 13.1
Experimental Configuration for Broadband through-transmission Substitution Method for Determination of Acoustic Material Properties.

also usually required. More complete details can be found in Zeqiri (1988); Wu (1996b); He (1999); and Waters, Hughes, Mobley, Brandenburger, and Miller (2000). This method can be extended to shear waves (Wu, 1996a) through the use of a critical angle conversion without the need for shear wave transducers. These measurements agree well with the theory presented in Chapter 4 (see Figures 4.6–4.7) so that the dispersion can be determined reliably from material absorption alone. Peters and Petit (2003) have provided an effective signal-processing method for determining velocity and phase unwrapping for these types of measurements. Another variation of the method is a pulse–echo version with a single transducer and a known reflector behind the sample; this approach involves a double pass through the material (Wu, 2001). Information on simplified measurements without reflection correction on identical samples of different thicknesses and other useful measurement methods can be found in Hurrell (2012). In this approach, the ratio of the spectra of the two samples divided by the difference in thickness leads to a complex propagation function of frequency (Madsen, Zagzebski, & Frank, 1982). The author has found that the broad-bandwidth piezo-polymer transducers (− 6-dB fractional bandwidth = 110%) described in this work are well suited to measurements of materials. A different approach that uses a laser pulse for extremely broadband excitation in a chamber is described by Bauer-Marschallinger et al. (2012).

13.2.2 Tissue Measurements

Most of the measurements made on tissue have been done using procedures similar to those in Section 13.2.1. The handling of tissue, safety precautions, and temperature control are additional considerations. The tissue itself is often sealed in a chamber with acoustically transparent windows on either side. The sealed chamber then can be treated as a sample in the description outlined in the last section.

Whereas most of the materials in the previous chapter are homogeneous, tissues are not and more elaborate procedures are needed to capture their complexity. Bamber (1986, 1992, 1998) provides reviews of measurement methods. One characteristic is the heterogeneous nature of tissue, or its spatial variation point to point. Large-area scans can be time consuming, so faster methods of through-transmission measurements have been devised (Hinkelman, Liu, Metlay, & Waag, 1994; Hinkelman, Szabo, & Waag, 1997), as was described in Section 9.3. Related studies of the heterogeneities of the breast have been done by Freiburger, Sullivan, LeBlanc, Smith, and Trahey (1992) and Zhu and Steinberg (1992). At the University of Rochester, more recent work involves studies using two-dimensional arrays and circumferential ring arrays (Jansson, Mast, & Waag, 1998; Liu & Waag, 1998). Another characteristic of tissue is angular scattering, which is a component of attenuation. In this case, a fixed transmitter is used along with a receiver that can be rotated in angle (Nassiri & Hill, 1986). The measurement of another characteristic, anisotropy, also utilizes a similar angular positioning capability (Mottley & Miller, 1988), as discussed in Section 9.4.

A revolution in the measurements of tissue characteristics is underway with elastography, as explained in Chapter 16. A now popular way of probing tissue to determine its properties is to use the acoustic radiation force (ARF) (Nightingale, Palmeri, Nightingale, & Trahey, 2001; Sarvazyan, Rudenko, Swanson, Fowlkes, & Emelianov, 1998; Sugimoto, Ueha, & Itoh, 1990; Walker, Fernandez, & Negron, 2000).

13.2.3 Measurement Considerations

In these measurements, other factors are involved that can invalidate results or cause errors. Because of variations along the beam, diffraction can introduce a loss and dispersion of its own (described as diffraction loss in Chapter 6). In the substitution method, the signal spectrum with the sample inserted is divided by that without the sample. A hidden goal in this measurement is to make the ratio of diffraction losses for the two cases equal to one or be negligible for the measurement. If the S parameter for water path without a sample for identical circular nonfocusing transducers separated by z is:

$$S_w = \frac{z\lambda_w}{a^2}, \tag{13.1}$$

where a is the transducer radius and λ_w is the wavelength in water, then the sample S parameter of thickness Δ_z can be expressed (Szabo, 1993) as:

$$S_s = S_w\left[1 + \frac{\Delta z}{z}\left(\frac{\nu_s}{\nu_w} - 1\right)\right], \tag{13.2}$$

where ν_s is the sample sound speed and ν_w is the water sound speed. For diffraction to be close in the two cases, the second factor in brackets in Eqn 13.2 must be small. Bamber (1986) discusses diffraction correction for angular backscatter measurements.

Nonlinearity of the water used in the tanks for these experiments can cause excess attenuation and significant distortion. To keep measurements in the linear range, the nonlinearity parameter (σ), Eqn 12.4C, should be small, less than 0.1 (Szabo, 1993; Wu, 1996).

13.3 Transducers

13.3.1 Impedance

Recall from Chapter 5 that a transducer can be regarded as a three-port device. Typically, a backing material (port 2) is used in the construction of the transducer, so that the electrical port (port 3) and the business end (acoustic port 1) are of interest. The measurement of the electrical impedance as a function of frequency can be an important diagnostic quality check during various stages of manufacture. Impedance measurements of each element in an array can also be valuable in determining the element-to-element variability across the array. In a manufacturing environment involving large numbers of arrays, automated computer-controlled data gathering and evaluation are necessary (Fisher, 1983).

For an impedance measurement, a network analyzer of the appropriate frequency range is attached to the array through a switch that allows connection to each element in turn (Figure 13.2). Because of the three-port nature of the transducer, the impedance is affected by acoustic port loading, which must be controlled. Usually, the business end (port 1) of the transducer is placed in a water chamber with absorbing sides. Before the transducer can be measured, the analyzer is calibrated with the fixture used to connect to the transducer. Output display choices are magnitude and phase, admittance and phase, and real and imaginary parts. The last pair is preferred because the real part is related directly to the radiation resistance; however, magnitude and phase are more common in the literature

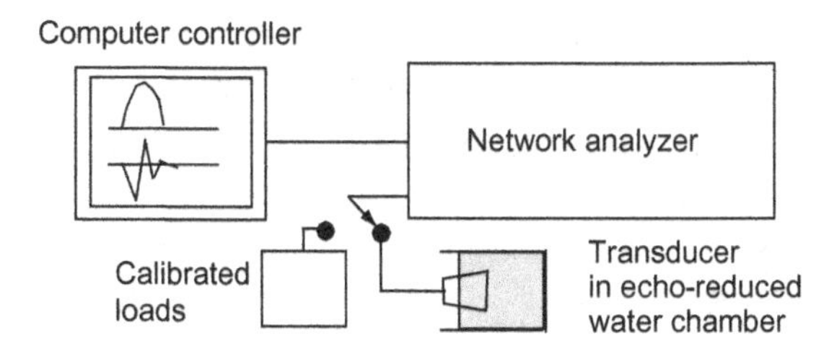

Figure 13.2
Impedance Measurement with a Network Analyzer (impedance plot).

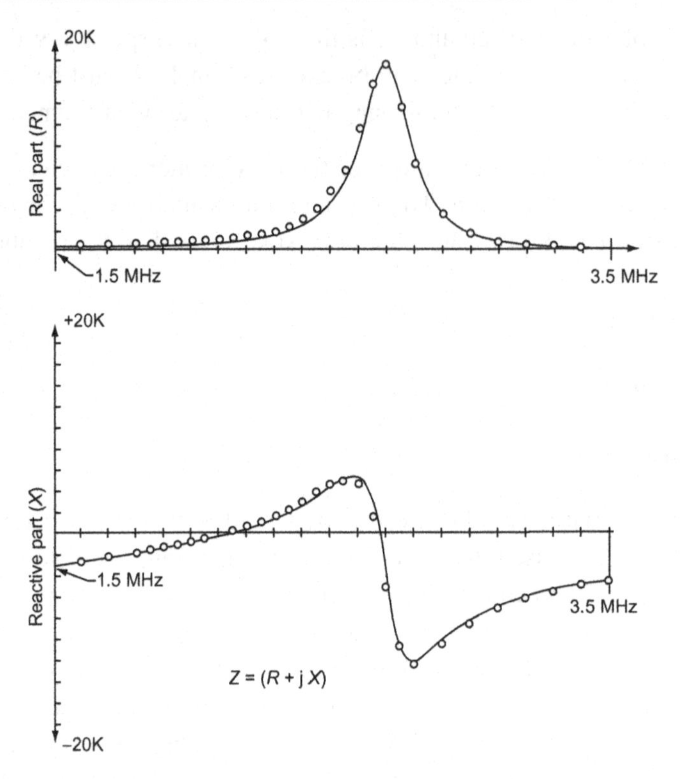

Figure 13.3
Impedance Measurements for a PZT-5H Resonator with an Epoxy Backing and $w/d = 0.371$. Real (top) and imaginary (bottom) compared to model. *From Selfridge, Kino, and Khuri-Yakub (1980).*

(Davidsen & Smith, 1993; Ritter, Shrout, Tutwiler, & Shung, 2002). Data can be compared to simulations from transducer models (described in Chapter 5) in order to check the realization of a design (as in Figure 13.3).

Considerable information can be derived from impedance measurements made at different stages of manufacture. A step-by-step walk-through of this process with data compared to finite-element modeling at each stage under different loading conditions can be found in Powell et al. (1997). These types of measurements for crystals of different geometries can be used to characterize piezoelectric materials (IEEE, 1988; Powell et al., 1997; Ritter et al., 2000; Selfridge, Kino, & Khuri-Yakub, 1980; Szabo, 1982).

13.3.2 Pulse–Echo Testing

Electrical measurements, useful as they are, still provide only an indirect measure of acoustic performance. A standard measure of transducer acoustic performance is a

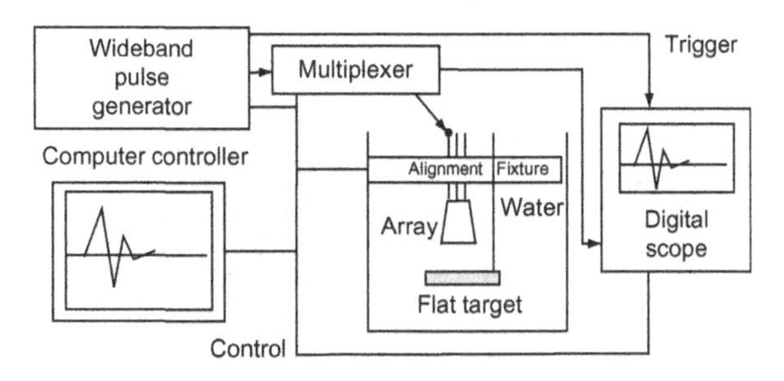

Figure 13.4
Pulse–echo Testing of Array Elements.

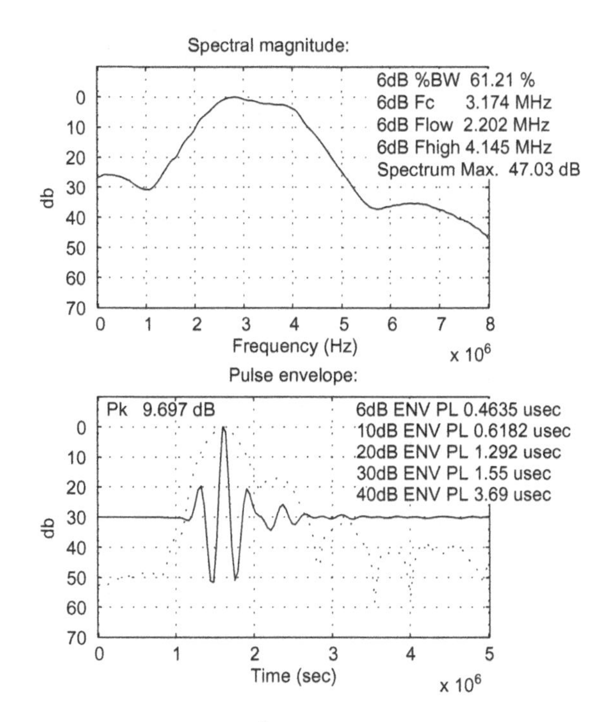

Figure 13.5
A typical Pulse and Spectrum Data plot Showing Pulse and Spectrum Widths.

pulse–echo test, in which a transducer element is excited by a prescribed waveform (often an impulse) and the round-trip signal from an aligned flat target is obtained (as illustrated by Figure 13.4). The received signal is digitized, and a fast Fourier transform (FFT) and Hilbert transform (described in Appendix A) are used to obtain the spectrum and pulse envelope, an example of which is given by Figure 13.5. Features can be extracted from the data automatically and compared element to element. Typical features may be overall pulse–echo sensitivity or insertion loss, pulse-envelope length at various decibel levels

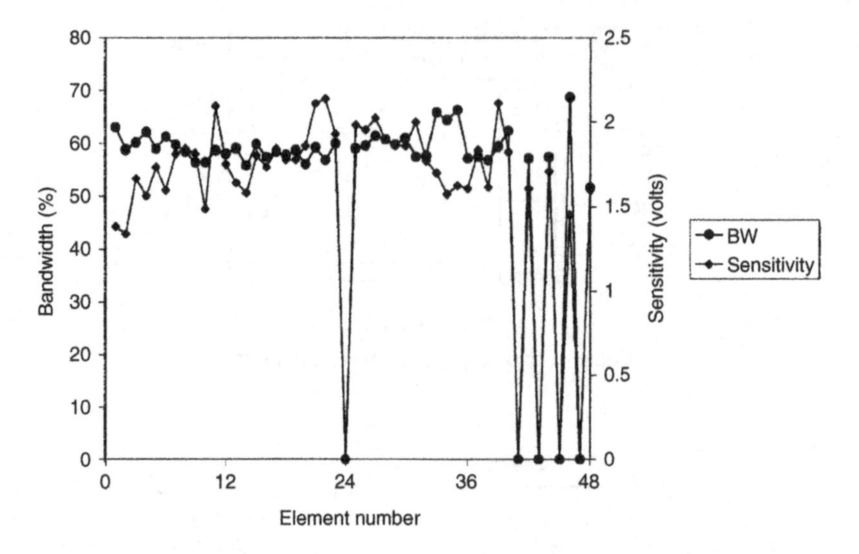

Figure 13.6
Automated Sensitivity and Bandwidth for a 30-MHz array. Element 24 and 4 elements on the end are open electrically. *From Ritter et al. (2002), IEEE.*

from the peak, and various measures of bandwidth, such as the −6-dB absolute or fractional bandwidth.

For this measurement, it is preferable to place the flat plate in a focal plane. Recall that at the focal point of a circularly symmetric transducer, the spatial impulse response is a delta function, so the pulse is a scaled replica of the source excitation (as pointed out in Chapter 7). If a one-dimensional array element is measured, then the elevation focal plane is appropriate because in the azimuth plane, the element will appear to be in the far field so that the original source function is recovered. Of course, a diffraction loss will be incurred (as discussed in Section 6.8 and Section 8.3).

Another alternative for a target is a steel ball, which has simpler alignment requirements. In this case, only a small part of the surface of the ball acts as a reflector back to the transducer. In general, echoes from flat plates correlate better with transducer model simulations. An example of an automated measurement for a 30-MHz array is shown in Figure 13.6. If an imaging system is used to drive a number of elements, a flat plate can be used; however, a more sensitive test with this kind of excitation is a beam-plot measurement.

13.3.3 Beam Plots

A measure of how well an array operates with an imaging system is the beam-plot test (Bamber & Phelps, 1977). This measurement is conducted in a water tank with fixturing

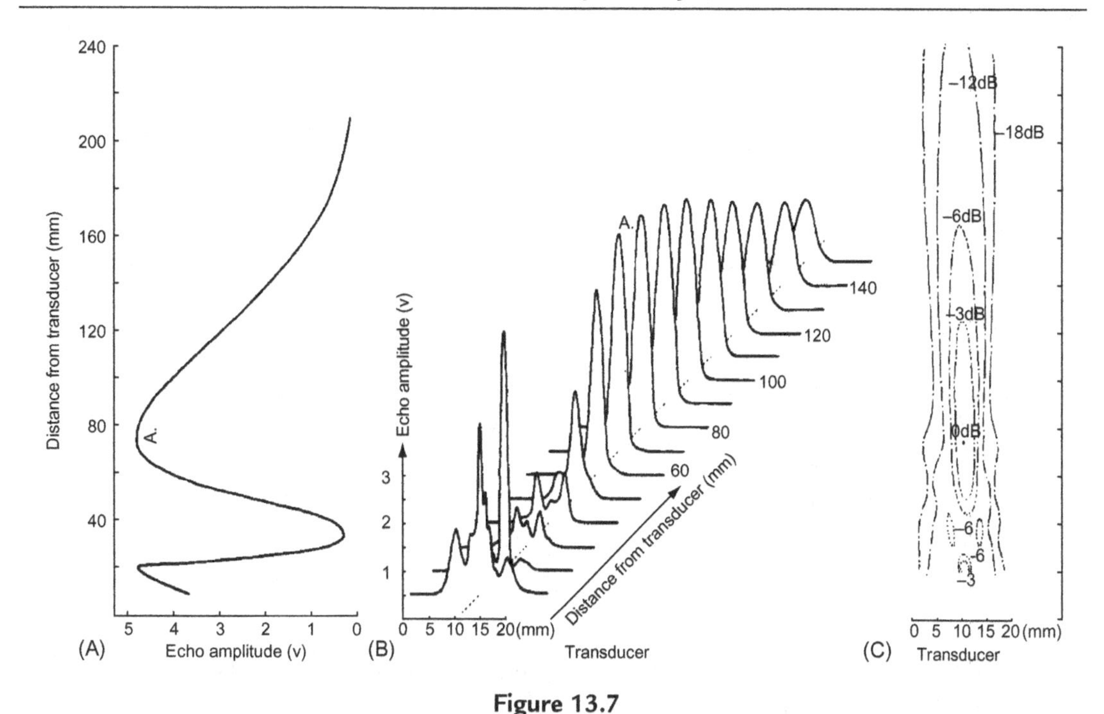

Figure 13.7
Acoustic field of a 1.5-MHz nonfocusing Circular Transducer as Measured in Beam-plot tank. (A) Axial amplitude; (B) lateral beam plots as function of depth; (C) contour plot of field. *Reprinted from Bamber and Phelps (1977), with permission from Elsevier.*

that can align and translate either the transducer assembly or a needlelike or ball target along and near the acoustic axis of the beam. Provisions for alignment include *xyz* translation and rotational capabilities.

Usually, a selected feature of the waveform, such as peak-to-peak voltage, is determined and plotted versus the automatically stepped translation axis (Figure 13.7). It is common to normalize the beam plot to the largest value measured on-axis at the depth selected. Of particular interest are beamwidths at different levels. Most commonly used levels are the −6-dB level, which is also known as full width half maximum (FWHM), and the −20- and −40-dB levels. The FWHM values are associated with resolution; the lower ones that are less than or equal to −40 dB, with contrast resolution. Beam data gathered in a plane are most often presented as a contour map with the contours representing decibel levels (as discussed in Chapter 6).

An imaging system can be used to focus and steer a beam along a chosen direction for measurement. Through the translation of the target (or transducer), a lateral beam plot can be acquired. The entire pulse−echo field can be mapped out by translating along the beam axis as well. However, this measurement can be time-consuming. An alternate method to driving the transducer with an imaging system is to synthesize the appropriate delayed

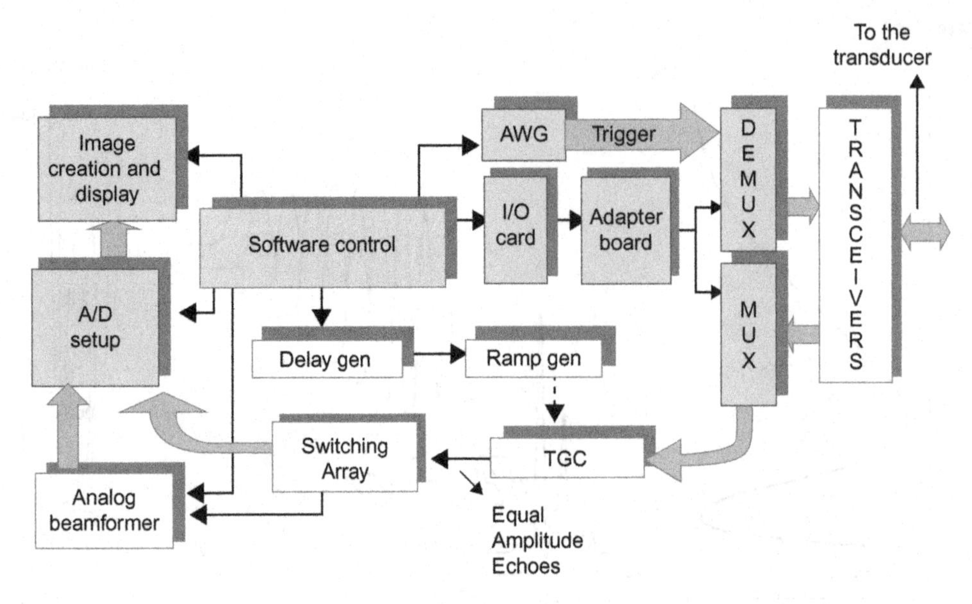

Figure 13.8
Beamformer test system block diagram. A/D = analog to digital converter, TGC = time gain compensation, I/O = input/output, MUX = multiplexer switch, AWG = arbitrary waveform generator. *From Chen et al. (2001), Proceedings of SPIE, courtesy of Society of Photo Optical Instrumentation Engineers.*

signals needed for focusing and steering with a programmable waveform generator, with appropriate switching to each element, and with the receive beamforming done in software; however, this can also be lengthy.

Another type of beam plot of interest for arrays is an element beam plot. As discussed in Chapter 7, individual elements have a wide directivity and govern off-axis sensitivity. Related to this directivity is cross-talk among elements (Larson, 1981). Examples of these measurements can be found in Ritter et al. (2002) and Davidsen and Smith (1993).

Beam-plotting can be combined with other tests in an integrated automatic test station such as that described by Fisher (1983) and more recently by Chen et al. (2001). This system can make pulse–echo measurements and beam plots (Ritter et al., 2002), as shown in Figure 13.8.

For one-way beam-plotting, a hydrophone can be used instead of a target. The spot size of the hydrophone must be much smaller than the beamwidth to be measured (Lum, Greenstein, Grossman, & Szabo, 1996), as discussed in Section 13.4.2. A robust alignment method can be found in International Electrotechnical Commission (IEC) Standard 61828 (IEC, 2001).

There is a much faster way of measuring beams from imaging systems. Beams can be evaluated in real time using a Schlieren system. Systems designed for this type of

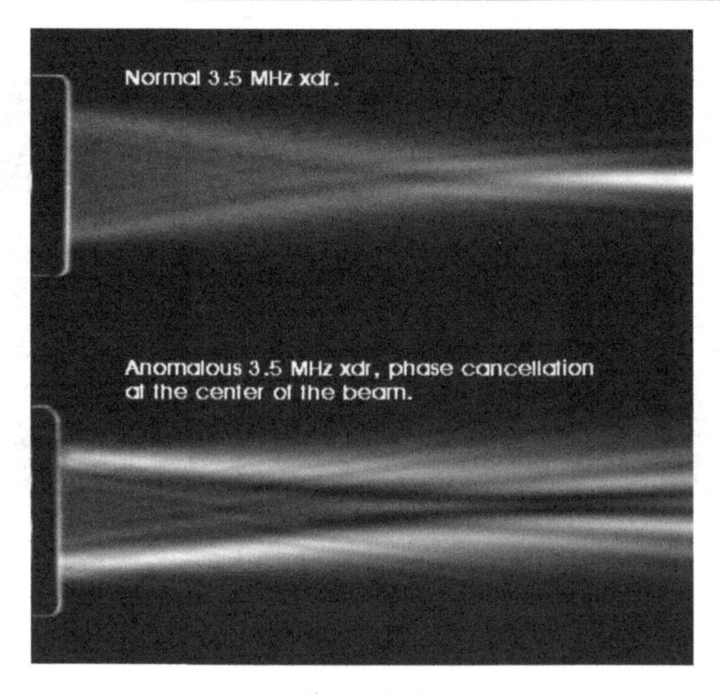

Figure 13.9
Measurements of CW Beams with and Without Defects by an Onda Schlieren System. *Courtesy of C. I. Zanelli, Onda Corporation.*

measurement include rapid scanning and a means of beam visualization (Hanafy, Zanelli, & McAvoy, 1991). This method derives data from the deflections of laser light scattered by perturbations of the refractive index of water by the sound beam. In this case, the depiction of the beam at each point is a result of light passing perpendicularly through the beam, so an integrated value for the beam results. The Schlieren system can be synchronized with an imaging system so that continuous-wave (CW) or pulsed wavefronts can be tracked in time along any selected vector direction (LeDet & Zanelli, 1999). Examples of Schlieren measurements are a CW visualization of a complete beam (shown in Figure 13.9) and a pulsed wavefront of a focused beam (shown in Figure 13.10).

13.4 Acoustic Output Measurements

13.4.1 Introduction

Measurements related to ultrasound-induced bioeffects have evolved into a branch of science with its own name: "ultrasonic exposimetry." Three major types of measurements in this area are absolute pressure by hydrophones, absolute acoustic power by radiation force balances, and temperature rise by thermal sensing devices. All of these measurements are required for

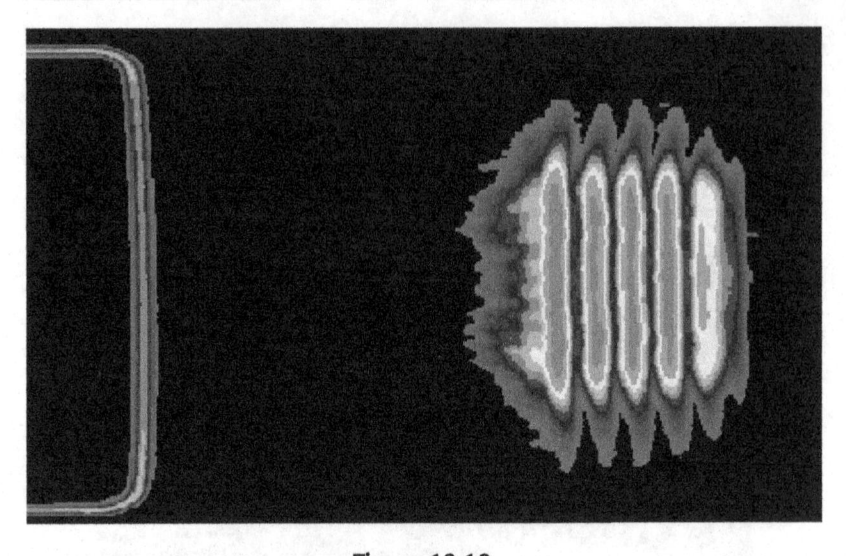

Figure 13.10
Measurement of a pulsed Wavefront of a Focused Beam by an Onda Schlieren System. The reader is referred to the Web version of the book to see the figure in color. *Courtesy of C. I. Zanelli, Onda Corporation.*

imaging systems in the USA, and there are limits on acoustic output regulated by the US Food and Drug Administration (FDA). Additional measurements are described by various standards of IEC technical committees 62b and 87. Certain countries, such as Japan, have their own requirements. The rationale behind these measurements and regulations is discussed in more detail in Chapter 15, on bioeffects. Harris (1999) summarizes measurement considerations. Here, we concern ourselves only with the measurements themselves.

There are many resources for these measurements; therefore, details and methodology can be found in the references and standards. A collection of topics can be found in Ziskin and Lewin (2000), a special issue of IEEE Transactions on Ultrasonics, Ferroelectrics and Frequency Control (1988), and the *IEEE Guide for Medical Ultrasound Field Parameter Measurements* (1990) (see Bibliography). Measurement protocol and methods can be found in two American Institute of Ultrasound in Medicine/National Electrical Manufacturers Association (AIUM/NEMA) documents (1998a, 1998b). In addition, Harris discusses the basic measurements (1985) and presents a more recent overview (1999). An overview of the equipment used for these measurements can be found in Ide and Ohdaira (1988).

13.4.2 Hydrophone Characteristics

Hydrophones are a unique type of transducer intended to make nonperturbing, absolute measurements of pressure waves over an extremely wide bandwidth at an infinitesimally

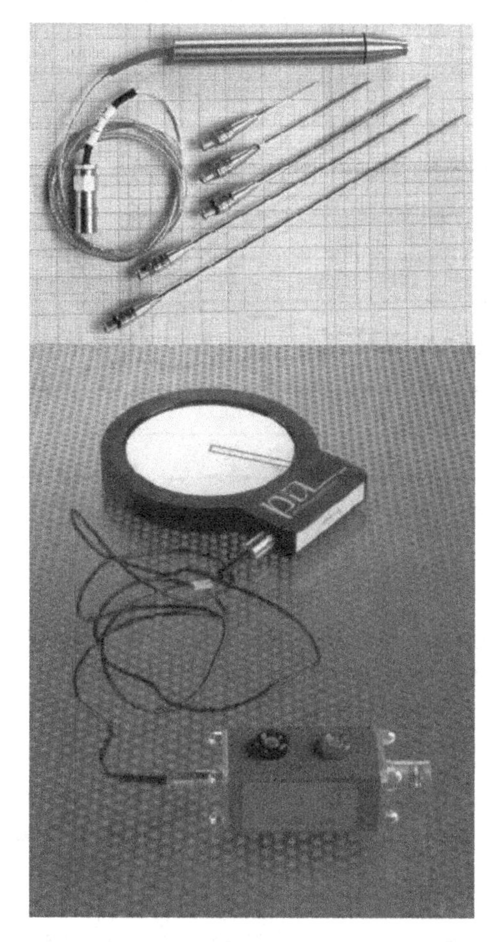

Figure 13.11
Needle hydrophone (top); bilaminar membrane hydrophone and external amplifier (bottom).
Courtesy of D. Bell, Precision Acoustics Ltd.

small spatial point. They are designed, in other words, to be as close as possible to ideal spatial point and time samplers in a water tank. The two most popular styles of hydrophones in use are the membrane and the needle (shown in Figure 13.11).

The membrane type consists of a thin sheet of the piezoelectric material PVDF stretched across a hoop a few centimeters in diameter and poled in its center to be piezoelectrically active in a small circular region, typically 0.2–1 mm in diameter. The membrane is so thin that it is practically transparent to waves in the normal imaging frequency range. Hydrophone transducers have a half-wave resonance frequency between 20 and 40 MHz, depending on their thickness (described in Chapter 5). Note that the speed of sound for

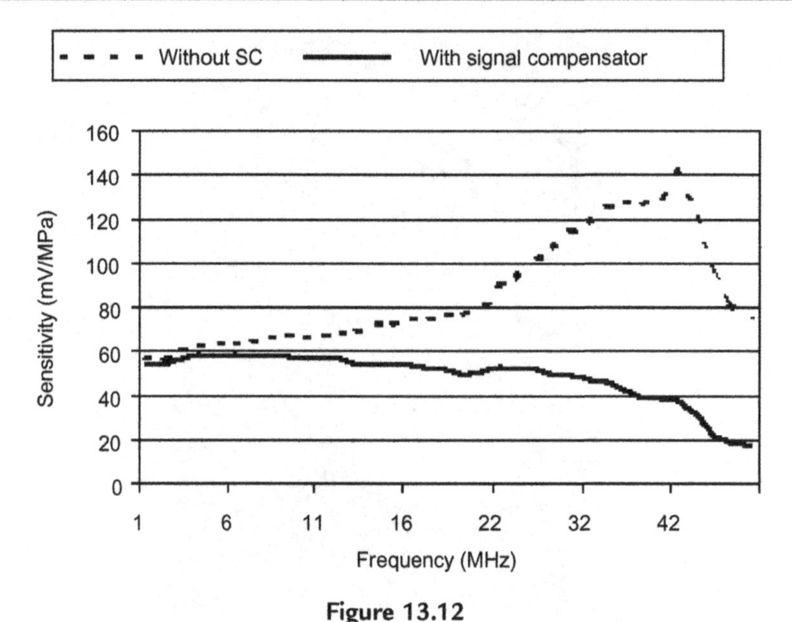

Figure 13.12
Sensitivity curves for bilaminar membrane hydrophones with and without compensating external amplifiers. *Courtesy of D. Bell, Precision Acoustics Ltd.*

PVDF is about 2 mm/μs, so for a 25-μm-thick hydrophone, the center frequency is about 67 MHz. For the bilaminar design, which is immune to water conduction and radio frequency (RF) interference effects, two layers are used so that the center frequency is half that of one layer, or about 33 MHz. This resonance corresponds approximately to the one-way (receiver) frequency response shown in Figure 13.12 for 15-μm-thick bilaminar hydrophones. Note that with a matched compensated external amplifier, the overall hydrophone-amplifier response can be made flat over a nearly 30-MHz range.

The needle hydrophone (Lewin, 1981) is a compact broad-bandwidth device on the order of 1 mm in diameter with good directivity. This transducer is also a half-wave resonator.

While the low-frequency response is flat (between 1 and 10 MHz in the range for most diagnostic imaging transducers), it depends on the diameter of the hydrophone and is not smooth. The needle hydrophone has an advantage over the membrane-style hydrophone in that it can be used for in situ exposure measurements in the body, and in many other applications where limited accessibility is a problem. Although needle hydrophones have become primary hydrophones in many laboratories, membrane hydrophones have become more prevalent for acoustic output measurements because of their reliability and relatively flat frequency response over the range necessary for imaging transducers.

Hydrophone sensitivity (one-way response) is often expressed as the end-of-cable sensitivity, M_L (volts/megaPascal), which includes the hydrophone, an associated amplifier, and a cable specified for a load impedance (usually 50 ohms). Sensitivity is also given in terms of decibels at 1 μv/Pa from the relation:

$$G = 20 \log_{10} (M_L/M_{REF}), \tag{13.3}$$

where $M_{REF} = 1$ μv/Pa.

Another important aspect of a hydrophone is its directional characteristics. Typically, the directivity of a hydrophone is similar to that of a piston source, which varies with frequency (as described in Chapter 6) and is embedded in a hard baffle (see Section 7.7). One goal of hydrophone measurement is to adequately sample the acoustic field of an imaging transducer with enough spatial resolution. The IEC Technical Committee 87 criterion for "enough resolution" is for the maximum effective hydrophone radius (b_{max}) to be:

$$b_{max} = \frac{\lambda}{4}[(l/2a)^2 + 0.25]^{1/2}, \tag{13.4}$$

in which a is the transducer radius or equivalent dimension, λ is wavelength in water, and l is the axial distance between the transducer and hydrophone. In other words, the higher the center frequency of the transducer, the smaller the spot size has to be. As an example, consider a 20-MHz transducer with a diameter of 6.35 mm ($2a$) and a focal length of 19 mm (l); this case gives a value of $b_{max} = 57$ μm. In order to show the effect of not having a small enough spot size, the field of this 20-MHz focusing transducer was measured in its geometric focal plane by a hydrophone with a 500-μm spot diameter and another with a spot size on the order of 40 μm (Lum et al., 1996). Data are compared to the theoretically expected beam-shape (a sinc function from Chapter 6) in Figure 13.13. Note that for this extreme case, the hydrophone with the spot size that is larger than the previous requirement averages over the beam. Note that the smaller hydrophone captures the beam well and meets the criterion of Eqn 13.4 (compared to theory). At higher frequencies, this correction becomes more important; a general method for correction can be found in Radulescu, Lewin, Goldstein, and Nowicki (2001).

The wide bandwidth of hydrophones is necessary to capture the harmonics associated with beam propagation through water, which is highly nonlinear (as discussed in Chapter 12). Waveforms and spectra for the two hydrophones are shown in Figure 13.14. The first hydrophone in this comparison is bilaminar and has a resonance of 33 MHz; the second hydrophone, an experimental research hydrophone made at a Hewlett Packard research laboratory with a single film thickness of 4 μm, has a bandwidth approaching 200 MHz, and is not commercially available (Lum et al., 1996). The wider bandwidth device shows almost 40 harmonics.

In order to qualify as absolute pressure sensors, hydrophones must be kept in calibration. This is usually done at a national standards laboratory or similar service. A calibration curve consists of $M_L(nf_0)$ sensitivity amplitude points measured at discrete frequencies. The

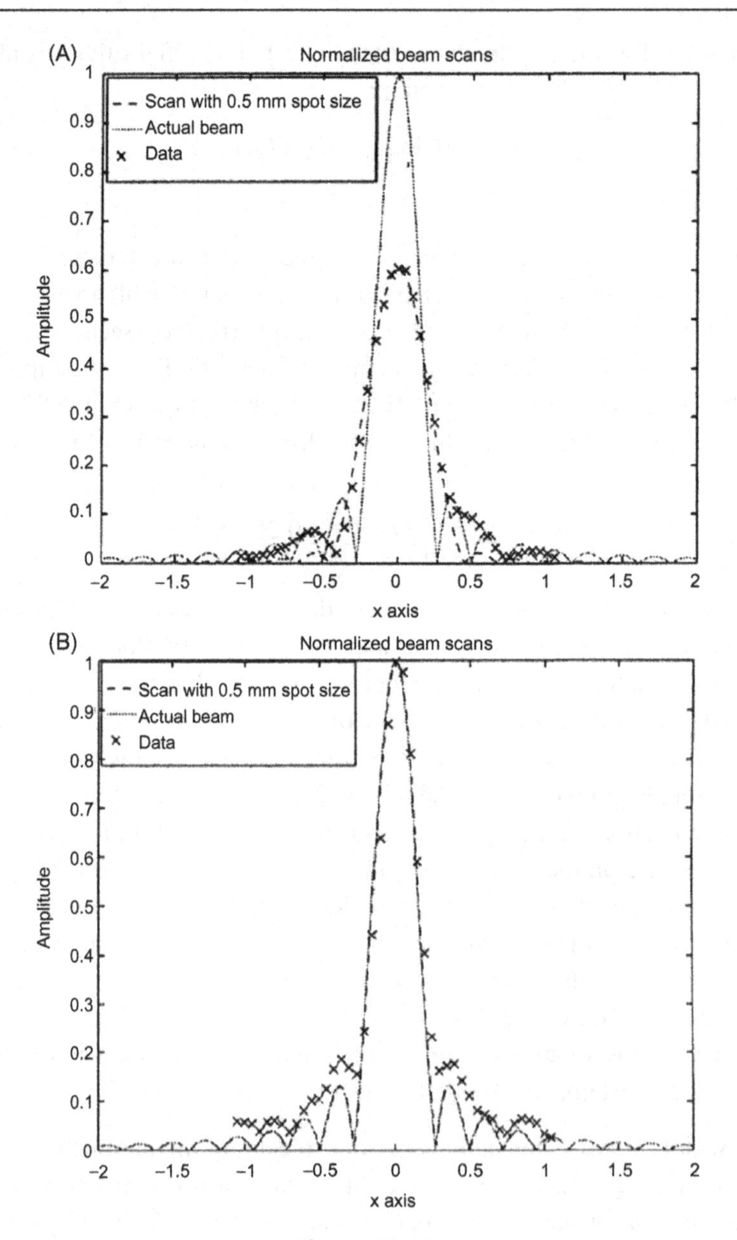

Figure 13.13

Linear scans of a 20-MHz focusing transducer with two membrane hydrophones. (A) 500-μm diameter. Each data point is shown by an x, and dashed lines connect them; dotted lines give the actual ideal beam plot. (B) 37-μm diameter. *From Lum et al. (1996), IEEE.*

estimated pressure waveform is related through a constant to the voltage waveform by $p(t) = v(t)/M_L(f_{awf})$, in which f_{awf} is the acoustic working frequency defined as the mean of −3-dB bandwidth frequencies of the measured spectrum $V(f)$, $f_{awf} = (f_1 + f_2)/2$.

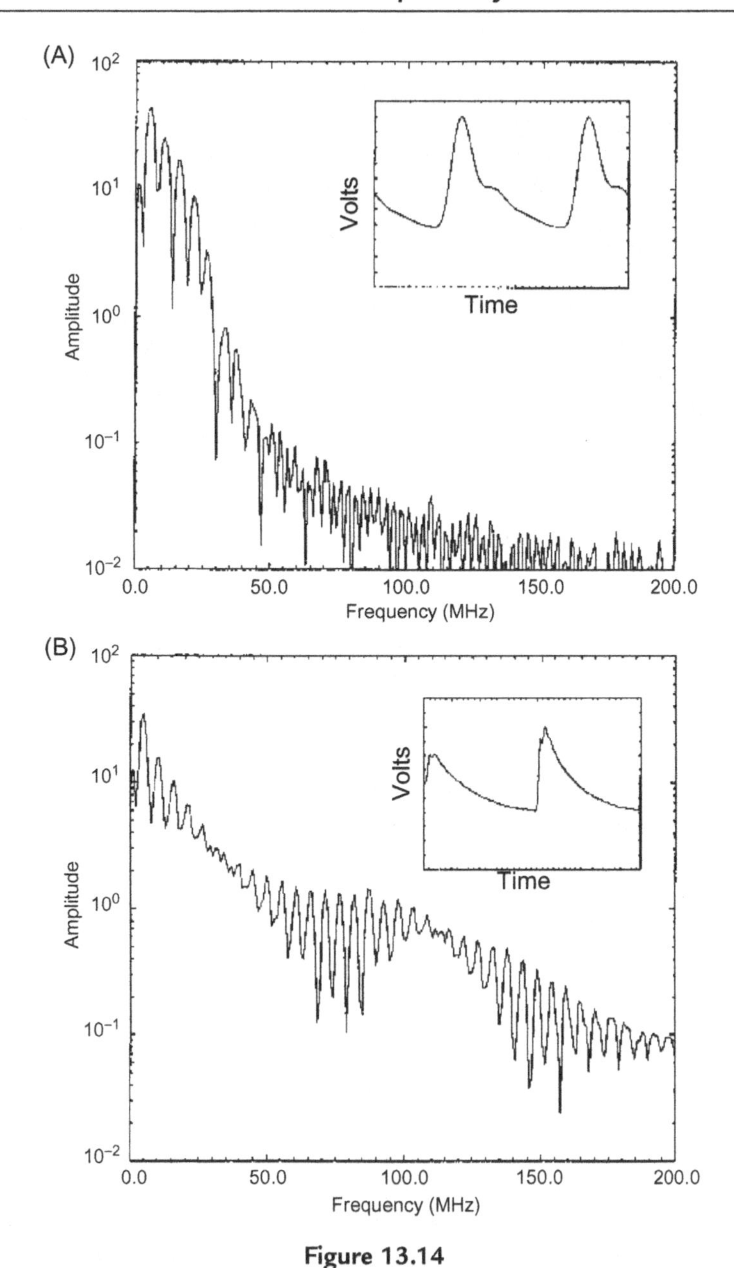

Figure 13.14

Waveforms (insets) and spectra for a 5-MHz fundamental source as measured by two membrane hydrophones. (A) 500-μm diameter; (B) 37-μm diameter.
From Lum et al. (1996), IEEE.

Typically, a hydrophone is to be calibrated up to 8 f_{awf} of the transducer being measured (FDA 2008). This requirement has motivated new types of hydrophones and new methods of hydrophone calibration, especially at higher frequencies (Lewin and Nowicki, 2010).

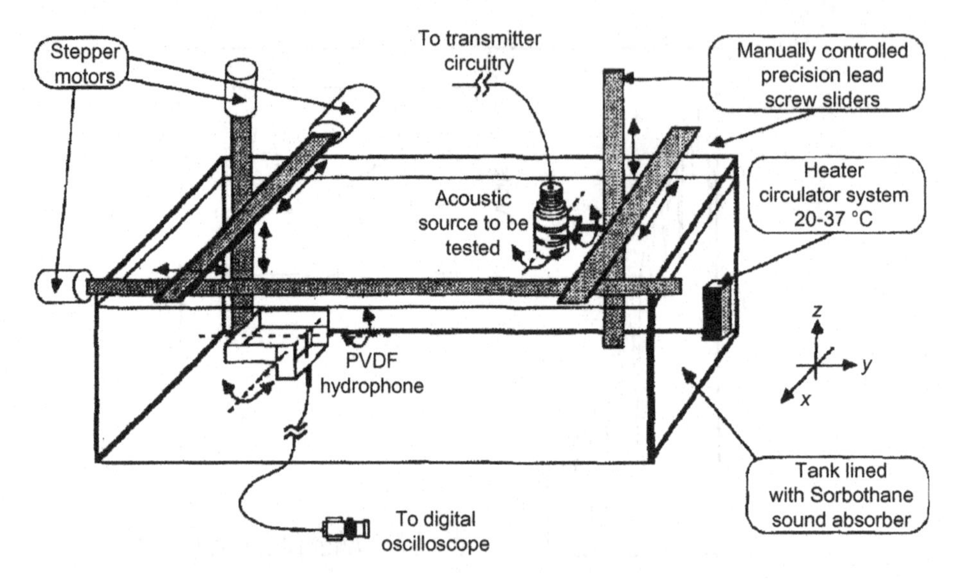

Figure 13.15
Water tank for Hydrophone Measurements. *From Lewin and Schafer (1988), IEEE.*

Two calibration approaches involve time-gating frequency analysis and the use of measurements in combination with a nonlinear model to achieve calibration up to 100 MHz (Radulescu, Lewin, Wojcik, & Nowicki, 2003).

13.4.3 Hydrophone Measurements of Absolute Pressure and Derived Parameters

A tank setup for making hydrophone measurements is illustrated in Figure 13.15 (Lewin & Schafer, 1988). Either the transducer under test or the hydrophone is held fixed, the other is aligned along the acoustic axis, and the separation between the two along the axis is varied. These adjustments require x, y, and z translation, as well as rotation.

Measurements are most often made along the acoustic propagation axis z in order to extract waveform features. These features are derived or calculated from the pressure waveform, $p(t)$, in water as transmitted by the imaging system and its transducer operating in a particular mode. Note that it is really the hydrophone voltage, $v(t)$, that is measured and converted to a pressure waveform by dividing it by the appropriate sensitivity constant corresponding to the acoustic working frequency, f_{awf}, of the spectrum of the voltage waveform, $p = v/M_L$.

To meet US regulations and the international standard IEC Standard 60601-2-37, a number of parameters are derived from the original pressure waveform data by imaging system manufacturers. Measurement details can be found in Harris (1985) and AIUM/NEMA

(1998a). The data are reported to the FDA and are also measured and tabulated according to standards issued by the IEC. Acoustic output data are also used by algorithms within imaging systems to calculate output display indices, as prescribed by the output display standard (AIUM/NEMA, 1998b) and the international standard IEC Standard 60601-2-37 (IEC, 2002) (both described in Chapter 15). To find maximum values of the derived parameters for the many different modes of an imaging system can be a challenging task (Szabo, Melton, & Hempstead, 1988), as described in Chapter 15.

The major derived features are the following:

The pulse pressure squared integral:

$$PPI(z) = \int_0^T p^2(t,z)dt = \int_0^T v^2(t,z)/M_L^2 \ dt. \tag{13.5}$$

The pulse intensity integral:

$$PII = PPI/(\rho c), \tag{13.6}$$

where ρ is the density of water, c is the speed of sound in water, and the pressure and particle velocity are assumed to be in phase.

The spatial peak temporal average intensity:

$$I_{SPTA} = MAX\left[\int_0^{T_{PRF}} v^2(t,z)dt/(M_L^2 \rho c T_{PRF})\right] = MAX[PII(z)/T_{PRF}], \tag{13.7}$$

measured at the location of the maximum or highest value of intensity on acoustic axis z, and T_{PRF} is the time interval between pulses.

The peak rarefactional pressure:

$$p_r = v_-/M_L, \tag{13.8}$$

where v_- is the minimum negative peak voltage in a waveform corresponding to rarefactional pressure (see Chapter 12 for more on nonlinear waveform distortion). The spatial peak spatial average intensity (I_{SPPA}) is defined as the spatial peak instantaneous intensity averaged over the 10–90% intensity pulse duration; however, this parameter is no longer used.

All these water values have derated counterparts denoted by the subscript "0.3." The concept of in situ or derated values was introduced as a conservative estimate of the overall effects of average soft-tissue absorption; it was not intended to be a realistic description of any particular type of tissue. The derating is a 0.3-dB/MHz-cm factor applied to intensities, corresponding to a linear factor of $exp(-0.069 f_c z)$, in which z is the on-axis distance from the transducer in centimeters, and f_c is the transducer center frequency in MHz. More

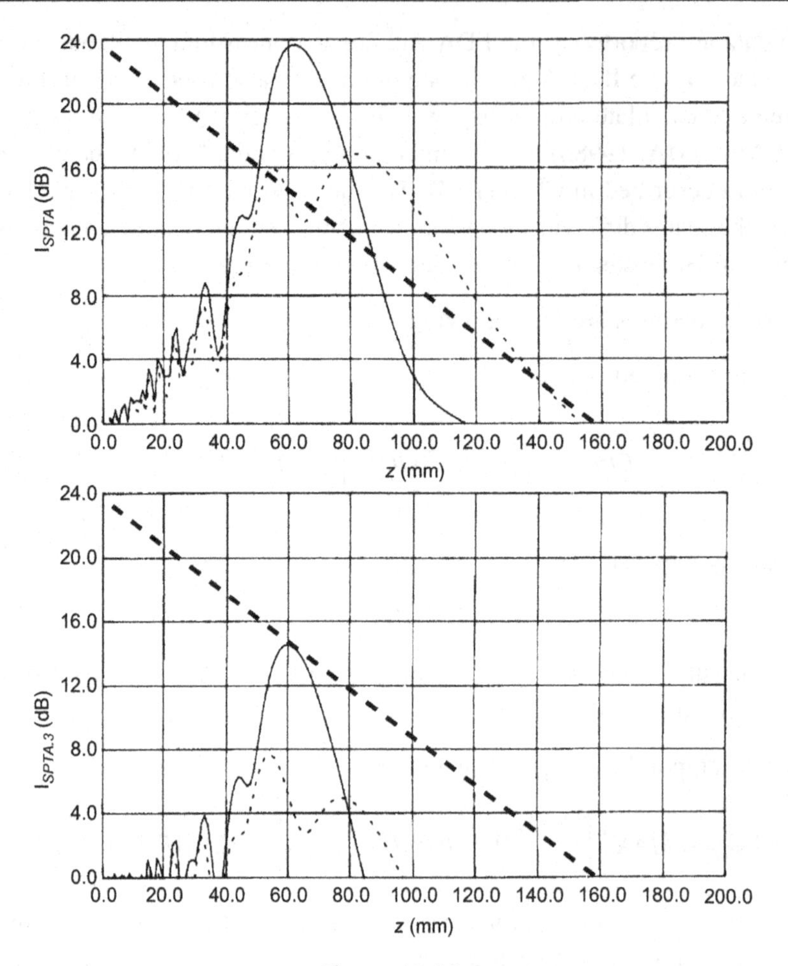

Figure 13.16

I_{SPTA} and $I_{SPTA.3}$ curves. (A) I_{SPTA} curves in water versus acoustic axis z for a rectangular array with a coincident azimuth and elevation (peaked curve) and a noncoincident case along with in situ derating factor (shown as dashed line); (B) derated I_{SPTA} curves along with derating factor (dashed line).

recently, acoustic working frequency, the −3-dB mean frequency, is used instead of center frequency. Values of pressure measured in water are converted into intensity (related to pressure squared) and are derated by in situ exponential derating factors, which are *exp* $(-0.069 f_c z)$ for intensity

$$I_{SPTA.3}(z) = MAX[I_{SPTA}(z)\ exp(-0.069 f_c z)]. \tag{13.9}$$

As an example, I_{SPTA} and $I_{SPTA.3}$ are plotted along the z axis for a 2.5-MHz center frequency transducer in Figure 13.16. In the top of Figure 13.16, the I_{SPTA} water value curves in dB are shown for a rectangular array for coincident and noncoincident foci. Also shown on a dB scale is the derating factor. When the derating factor line is subtracted (in dB) from these curves (the equivalent of linear multiplication), the derated values of I_{SPTA} result, as shown in the bottom of the figure. Note that for the noncoincident case, the derating process moves the peak closer to the transducer. For pressure, the derating is $exp(-0.0345 f_c z)$; a plot of derated pressure is given in Figure 14.6.

One of the ongoing discussions in the measurement of acoustic fields is at what point the field is considered to be nonlinear. For more realistic derating, it is obvious that a measurement made with moderate source amplitudes under highly nonlinear conditions such as water cannot be scaled for simulation of what the pulse might look like after propagating through tissue or a phantom. One method is to measure the waveform under linear conditions, then scale it up to the normal drive levels and apply an appropriate lossy propagation factor to its spectrum. A criterion for quasi-linear conditions is given by the following (IEC 61949, Humphrey et al., 2006):

$$\sigma_q = z p_m \frac{2\pi f_{awf} \beta}{\rho c^3} \frac{1}{\sqrt{F_a}}, \tag{13.10}$$

in which:

p_m = mean peak acoustic pressure $(p_r + p_c)/2$.
p_r = peak rarefactional acoustic pressure at the point of interest.
p_c = peak compressional acoustic pressure at the point of interest.
z = axial distance of the point of interest from the transducer face.
f_{awf} = acoustic working frequency.
β = nonlinearity parameter for water, $\cong 3.5$.
$F_a = \text{local area factor} = \sqrt{\dfrac{0.69(\text{source area})}{\text{beamarea @} - 6\ \text{dB}}}$.

13.4.4 Optical Hydrophones

A different type of hydrophone based on optical principles (Hocker, 1979; Phillips, 1980) has evolved and is an alternative to the piezoelectric types. Different forms based on optical fibers have been developed (Beard, Hurrell, & Mills, 2000; Haller, Wilkens, Jenderka, & Koch, 2011; Koch & Molkenstruck, 1999; Koch & Wilkens, 2004; Lewin, Mu, Umchid, Daryoush, & El-Sherif, 2005; Parsons et al., 2003; Wilkens, 2003; Wilkens & Koch, 2004). Gopinath et al. (2007) reviewed optical hydrophone types.

These optical probes sense light modulated by ultrasound wave perturbations at their tips. Most of them can be called fiber-optic probe hydrophones (FOPH). Laser light is injected into an optical fiber; light is reflected from the tip where the index of refraction in the fluid is modulated by the acoustic pressure field and it is sensed by a broadband optical detector (Parsons et al., 2003). Modulation can be by phase, wavelength, or intensity (Gopinath et al., 2007).

These acousto-optical probes have several advantages over piezoelectric hydrophones: small spot size (a few microns or less so spatial averaging may not be necessary); flat, broad bandwidth; an ability to sense and endure high pressure levels; self-calibration; long-term stability; easy repair; and less susceptibility to RF interference (Parsons et al., 2003). Several designs have the disadvantage of lower sensitivity; however, recent improvements have overcome this problem (Gopinath et al., 2007). More care, set-up time, and equipment are needed to measure with optical probes (Lewin and Nowicki, 2010).

13.4.5 Developments in Hydrophone Calibration

Progress has been made in characterizing the complete complex frequency response (both amplitude and phase) $M_L(f)$ over a wide frequency range. By now, confidence in both linear and nonlinear simulation has grown to the extent that simulation has become an integral part of many types of measurements. The application of these concepts to hydrophone calibration can be explained first on a linear basis by the system transfer function of Figure 2.14 without a scatterer. The voltage out of hydrophone with transfer function, $M(f)$ can be expressed as:

$$V(f) = M(f)[H_R(f)A_T(f)H_T(f)G_T(f)E(f)] = M(f)P(f), \tag{13.11A}$$

and if the term in brackets is a pressure incident on the face of the hydrophone, $P(f)$, then the hydrophone response is the ratio of measured voltage over a simulated pressure:

$$M(f) = V(f)/P(f), \tag{13.11B}$$

where $M(f)$ is complex and the hydrophone time response is simply the usual inverse time Fourier transform of the hydrophone transfer function:

$$m(t) = F_{-i}^{-1}[M(f)]. \tag{13.11C}$$

For the nonlinear case, the same configuration is used but a nonlinear propagation model and a model for the diffraction receiver characteristic, $G_R(f)$, determine $P(f)$. The final result is that the hydrophone response can be deconvolved from the measurement to realize a close approximation to the true measured pressure field for an unknown source:

$$p(t) = F_{-i}^{-1}[V_x(f)/M(f)], \tag{13.12}$$

in which $V_x(f)$ is the spectrum of the measured hydrophone output voltage waveform. An example of the deconvolution method compared to the standard amplitude correction can be

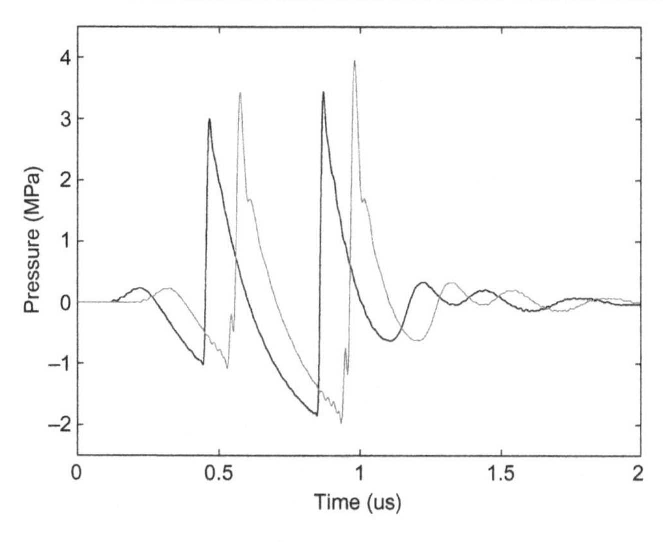

Figure 13.17
Imaging mode waveform from phased-array scanhead of commercial diagnostic ultrasound system. Measured using calibrated hydrophone with voltage-to-pressure conversion performed by two methods: Black line shows results of full deconvolution of calibration result. The gray line uses the measured amplitude response but an assumed flat (zero) phase response. A delay of 0.1 μs has been introduced between the two waveforms to enable easier comparison. *From Cooling and Humphrey (2008), IEEE.*

seen in Figure 13.17 (Cooling & Humphrey, 2008). Excellent agreement has been obtained with earlier methods as well as simulations, as shown in Figure 13.18 (Cooling & Humphrey, 2008) for a standard 0.5-μm spot size, 25 μm/layer, bilaminar PVDF hydrophone. The source was a circular 3.5-MHz focusing transducer, the field of which was measured at low excitation amplitudes to determine its nearly ideal piston characteristics, which were used in a nonlinear KZK (Khokhlov−Zabolotskaya−Kuznetsov wave equation) simulation program.

Several approaches for determining the hydrophone phase response have been devised (Bloomfield, Gandhi, & Lewin, 2011; Cooling & Humphrey, 2008; Wilkens & Koch, 2004). A general substitution approach is illustrated by Figure 13.19A, in which a well-characterized 3−12-MHz focusing circular transducer is excited by an impulse source and its field is measured at the focal point with a multilayer interferometric optical hydrophone (note, another type of well-characterized reference hydrophone can be used). The complex optical hydrophone output spectrum is measured for the pressure of this particular source according to the arrangement in Eqn 13.11A under nonlinear conditions. Next, this spectrum becomes the reference spectrum, $V_{ref}(f)$, for this substitution method. The optical hydrophone, $M_{ref}(f)$, is replaced by the new hydrophone to be calibrated, as shown in Figure 13.19B with exactly the same source conditions to obtain output $V_x(f)$. The essence

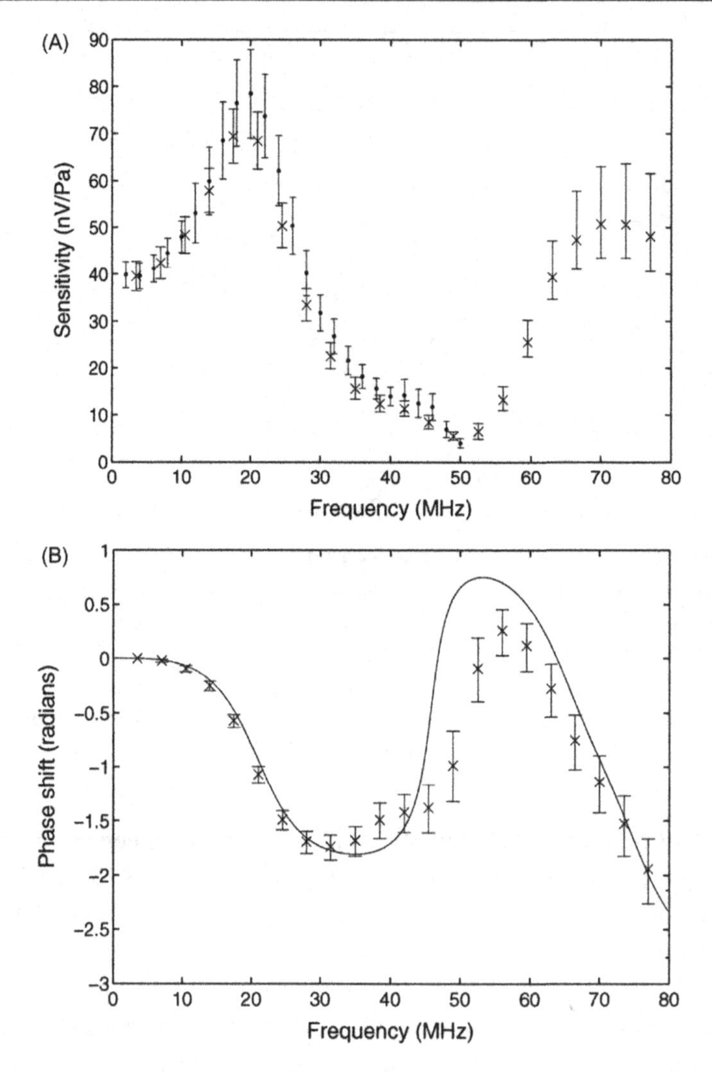

Figure 13.18

Amplitude and phase responses of measurement system compared with models.
(A) Amplitude response of the measurement system (hydrophone coupled to oscilloscope) showing both the results of the KZK model intercomparison technique (crosses) and those of an independent calibration (NPL) (points). Uncertainty bars represent estimated overall uncertainties at the 95% confidence level. (B) Phase response of the measurement system expressed as phase shifts relative to 3.5 MHz. The results of the nonlinear KZK model intercomparison technique are shown (crosses), along with those predicted by a model (solid line) for a typical hydrophone of this type. Uncertainty bars represent estimated overall uncertainties at the 95% confidence level.
From Cooling and Humphrey (2008), IEEE.

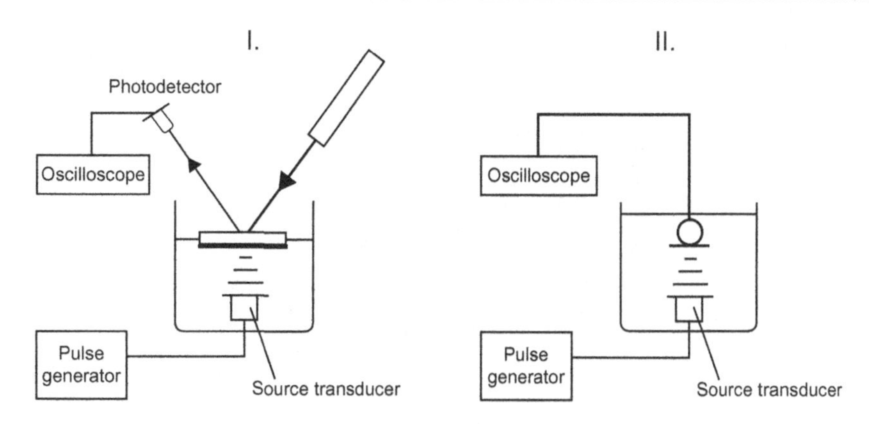

Figure 13.19
Hydrophone substitution calibration procedure.
(A) measurement of a broadband focused ultrasound pulse with the optical reference hydrophone; (B) measurement with the hydrophone to be calibrated after topping up the water tank. *From Wilkens and Koch (2004), IEEE.*

of the final step is finding the unknown hydrophone response, $M_x(f)$, by taking the ratio of the measured spectra:

$$M_x(f) = M_{ref}(f)V_x(f)/V_{ref}(f), \tag{13.13}$$

with extra variations according to each particular method.

13.4.6 Force Balance Measurements of Absolute Power

A force balance is a sensitive instrument for measuring the acoustic radiation force exerted by an acoustic field (as described in Chapter 12). This force is a result of energy transfer to an ideal absorbing target. The relation is:

$$W = gF, \tag{13.14A}$$

where F is the radiation force on an ideal absorber, W is time-average acoustic power, and g is a constant determined by calibration. In general, for an angle θ between the beam axis and the normal of the reflecting surface:

$$W = gF/2\cos^2\theta. \tag{13.14B}$$

The most common types of force balances are shown in Figure 13.20. The most popular type is on the left, and it consists of an absorbing target suspended from a microbalance. A thin membrane is usually used to separate the water enclosing the target from the face of the transducer. Ultrasound is directed upward, and the radiation force displaces the absorbing target. Because the target is delicately balanced, its movement is sensed by a

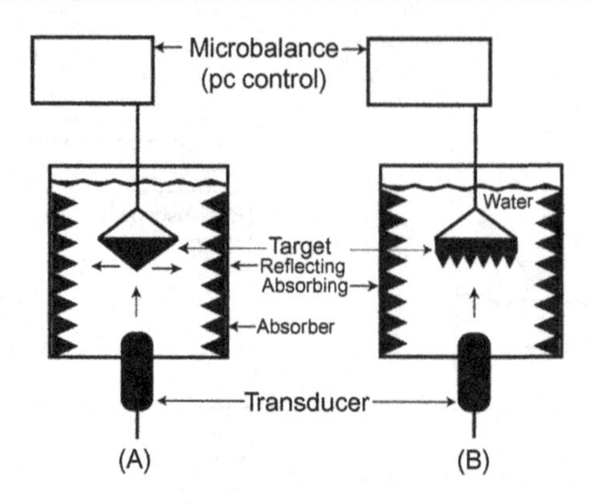

Figure 13.20
Types of radiation force balances. (A) Reflecting absorbing target; (B) direct absorbing target.

microbalance that provides a digital reading. This reading is translated into time-average power from Eqn 13.14A above. The constant g in this equation is determined through calibration methods as a function of frequency by using sources with known power outputs. More details for radiation force balance (RFB) measurements up to 20 watts can be found in Hekkenberg, Beissner, Zeqiri, Bezemer, and Hodnett (2001). Background information on acoustic radiation force balance measurements can be found in Beissner (1984, 1987, 1999).

13.4.7 Measurements of Temperature Rise

The primary thermal measurement carried out for safety is determining the surface temperature of transducers in air or tissue-mimicking material. When a transducer is not in contact with the body, most of its acoustic power is reflected at the elevation lens–air interface; consequently, its acoustic output is turned inward, which results in an effect called self-heating. The net outcome is that the outer surface of a transducer heats up slightly. This temperature rise is measured with a thermocouple for different operating modes, as described in IEC Standard 60601-2-37 (IEC, 2002), to ensure that the rise does not exceed a prescribed limit. Certain probes that are inserted in the body have a smaller allowable rise and often have a cutoff mechanism should the internal temperature rise exceed this limit (Ziskin & Szabo, 1993). Matrix arrays incorporate electronics in the case or handle and usually have been designed with internal cooling mechanisms.

Temperature rises are also associated with an acoustic beam propagating in an absorbing medium. For a transducer producing an acoustic field with a local time-average intensity $I(x,y,z)$, the local heat of a wave in this field in an absorbing medium with an absorption coefficient α is:

$$q = 2\alpha I, \tag{13.15}$$

where q is the time rate of heat production per unit volume. The temperature rise is most often measured by a thermocouple. Except for initial viscous thermocouple effects, the temperature rise (τ) is:

$$\frac{d\tau}{dt} = \frac{q}{C_H} = \frac{2\alpha I}{C_H}, \tag{13.16}$$

where C_H is the heat capacity per unit volume. Note the similarity to the expression for acoustic radiation force in Chapter 12. Thermocouples are embedded in an absorbing medium, often a tissue-mimicking phantom, for this type of measurement. Special layered phantoms have been made to simulate the thermal properties of different parts of the human body (Shaw, Pay, Preston, & Bond, 1999; Wu, Cubberley, Gormley, & Szabo, 1995). Effects of blood perfusion and cooling, as well as acoustic streaming, are not usually included in these phantoms, but they will be discussed in Chapter 15.

Other ways of assessing temperature elevations include infrared sensing (Shaw & Nunn, 2010). In anticipation of high-intensity focused ultrasound (HIFU) heating of tissue, Miller, Bamber, and Meaney (2002) investigated the possibility of measuring localized temperature changes using ultrasound images.

13.4.8 Field Measurements Revisited: Projection Methods

Acoustic output field measurements for diagnostic imaging are mainly concentrated along the acoustic beam axis where the maximum intensities are expected to be found. What if the maximum intensity is not on the axis because of a defect in the transducer or from deliberate design? In the case of plane-wave compounding, for example, the acoustic axes of individual plane-wave transmissions will not align with the final intended main direction of propagation of all the transmissions. Another case of interest is when the maximum intensity occurs in a field region which is inaccessible to measurement.

The method of back and forward projection is a way of reconstructing the complete acoustic field from measurements in a single plane perpendicular to the acoustic axis through the use of a suitable reconstruction algorithm. Usually the plane selected is close to the transducer, before significant nonlinearities and focusing have had a chance to develop. If, initially, a continuous wave source is assumed, then pressure measurements, $p(x,y,z_0)$ in magnitude and phase are made on a regularly spaced grid of points in a plane at $z = z_0$, which collectively are called the source hologram (Sapozhnikov, Pishchal'nikov, & Morozov, 2003), as illustrated by Figure 13.21. Then the angular spectrum is found from:

$$F(k_x, k_y) = \iint p(x', y', z_0) e^{-i(k_x x' + k_y y')} dx' dy'. \tag{13.17}$$

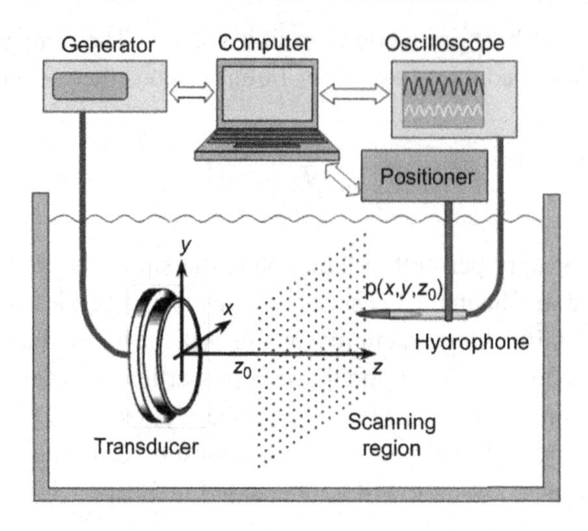

Figure 13.21

Experimental arrangement for field reconstruction methods. Hydrophone poised at point p in the measurement plane located at $z = z_0$. For $z > z_0$ is for forward projection and, for $z < z_0$, back projection. *Courtesy of O. Sapozhnikov.*

And then any point in the field can be determined from the double $+i$ spatial Fourier transform from Chapter 6:

$$p(x', y', z) = \frac{1}{(2\pi)^2} F(k_x, k_y) e^{ik_z(z-z_0)} e^{i(k_x x' + k_y y')} dk_x dk_y, \tag{13.18}$$

in which $k_z = \sqrt{k^2 - k_x^2 - k_y^2}$ and $k = 2\pi/\lambda = \omega/c$ is the wavevector amplitude (Clement & Hynynen, 2000). If $z < z_0$, then with back projection, the source amplitude and phase distribution in space can be found, including any defects and the focusing curvature, if any. Examples of forward projection cases ($z > z_0$) are in Figure 13.22 (Clement & Hynynen, 2000).

Another projection algorithm is the Rayleigh integral for forward projection:

$$p(x, y, z > z_0) = -\frac{1}{2\pi} \iint p(x', y', z_0) \frac{\partial}{\partial z'} \left(\frac{e^{ikR}}{R} \right) dx' dy', \tag{13.19A}$$

where $R = \sqrt{(x-x')^2 + (y-y')^2 + (z-z_0)^2}$, and for backward projection:

$$p(x, y, z < z_0) = -\frac{1}{2\pi} \iint p(x', y', z_0) \frac{\partial}{\partial z'} \left(\frac{e^{-ikR}}{R} \right) dx' dy' \tag{13.19B}$$

(Sapozhnikov et al., 2003). Measured pressures were back-projected onto the transducer aperture plane of the source using Eqn 13.19B, as shown in Figure 13.23, in which there is

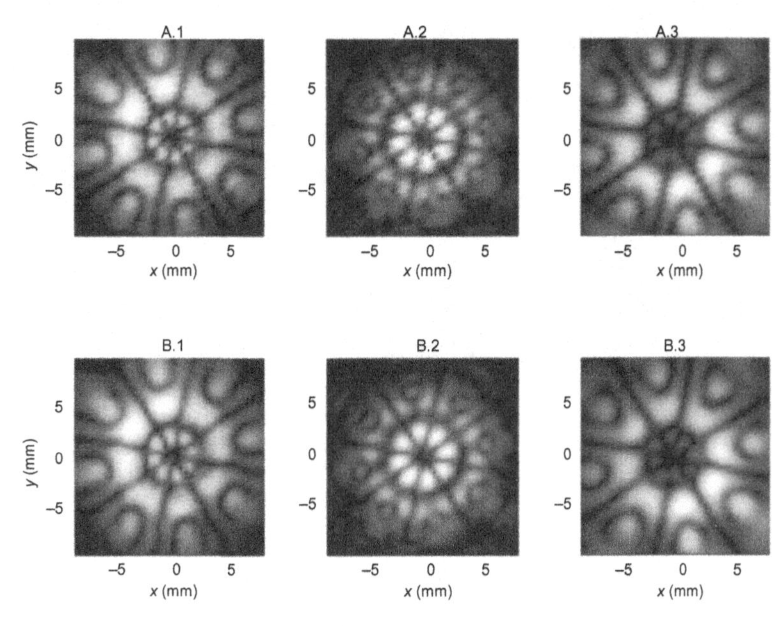

Figure 13.22
(A) Radial measurements of the sector vortex array in mode 4; (B) their corresponding projections. Image intensity is linearly scaled with the acoustic pressure amplitude. Fields are measured at distances of (A1) 70, (A2) 80, and (A3) 90 mm from the source. Projection from $z = 80$ mm to distances of (B1) 70, (B2) 80, and (B3) 90 mm. *From Clement and Hynynen (2000), Journal of the Acoustical Society of America.*

a photo of the transducer face along with the corresponding reconstructed pressure amplitude and phase. The pressure distribution has a characteristic annular structure caused by Lamb waves in the piezoceramic layer of the transducer (Sapozhnikov et al., 2003; Sapozhnikov, Morozov, & Cathignol, 2004). The transducer aperture can be clearly seen in the amplitude plot. Agreement of the reconstructed field with measurements is excellent. Sapozhnikov, Ponomarev, and Smagin (2006) developed a transient holographic reconstruction method. In summary, the back-projection approach makes no assumptions of the source transducer and yet can recreate the entire acoustic field to within a wavelength accuracy. A thresholded plot can display the highest pressure points in the three-dimensional field without tedious point-by-point searching for them with a single hydrophone.

13.5 Performance Measurements

There are several primitive ways in which the performance of an imaging system can be tested. The modes most often evaluated are B-mode imaging, Doppler, and color flow imaging (CFI) by the use of special-purpose phantoms.

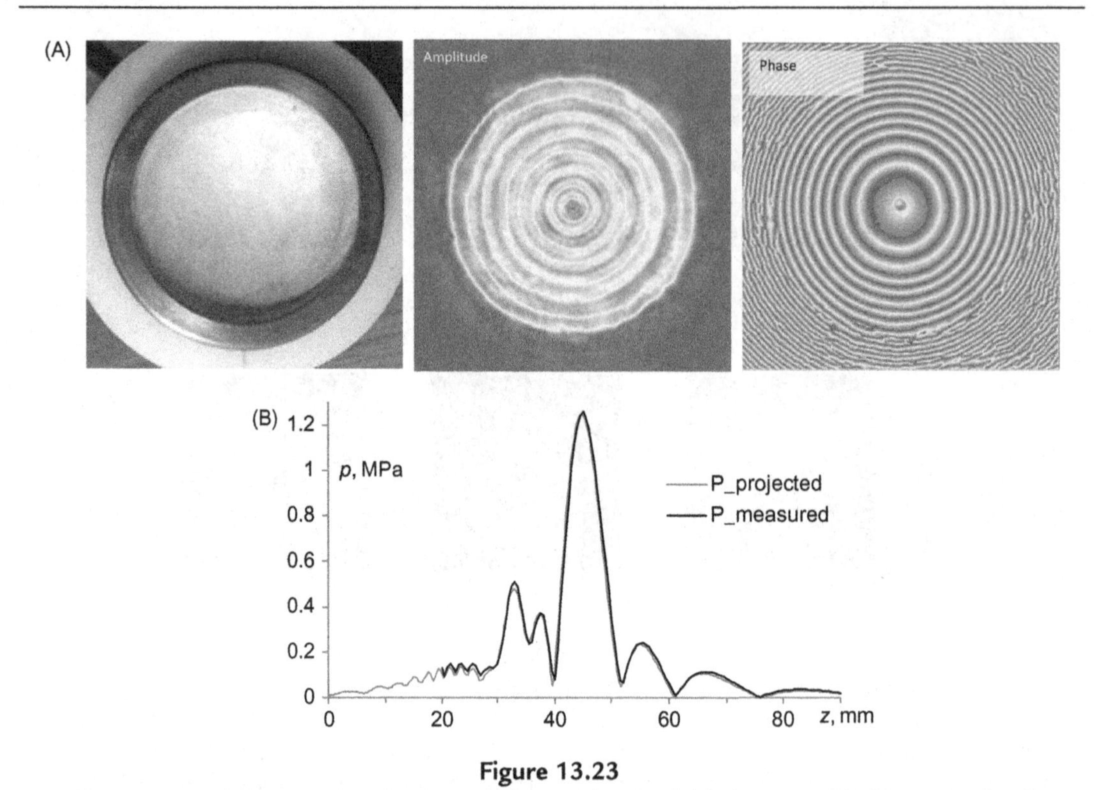

Figure 13.23
Reconstructed Pressure Amplitude and Phase using Rayleigh Integral. (A) Photograph of transducer and amplitude and phase distribution of acoustic pressure reconstructed at the transducer aperture plane. From (A), the amplitude and phase distribution found at the aperture plane were used to calculate the acoustic field radiated by the source. (B) Projected acoustic pressures compared with measurements illustrate the accuracy of such a field projection along the transducer axis. The reader is referred to the Web version of the book to see the figure in color.
Courtesy of O. Sapozhnikov.

The most common imaging objects are tissue-mimicking phantoms with various targets embedded in them. An example of an imaging phantom was displayed in Figures 8.1 and 8.2. Note that resolution can be determined from the filament point targets, cyst fill-in by the circular scatter-free objects, and contrast (Eqn 8.16) by the column of circular objects filled with varying densities of small scatterers. Related forms of imaging phantoms are the Cardiff step phantom and contrast phantoms consisting of objects that have different densities of scatterers relative to a host matrix material. Other forms of imaging phantoms, such as a breast phantom, provide a more realistic imaging object. Imaging performance measurements on a more universal basis (system to system) can be quite involved because the entire image-processing chain (described in Chapter 10) has to be accounted for, calibrated, or compensated for in the process of image evaluation. Methodology can be found in standards (IEC (TC87, TC 62D); AIUM, 1990) and guidelines.

Phantoms designed for Doppler and CFI testing are called "flow phantoms." They consist of tubing and a pump through which a scattering fluid is circulated at user variable rates. A less expensive method is a "string" phantom, in which a cord is moved at controlled rates. The velocity of the flow of fluid or the string is independently known and can be compared to the values produced by the imaging system.

Special-purpose phantoms have been designed for different applications, imaging modes, and features. There are many aspects of a phantom to be considered such as the speed of sound, absorption (α_0,α_1,y), backscatter, B/2A (nonlinearity parameter), shear properties, temperature sensitivity, similarity of properties to actual tissues, stability over time, etc. A detailed study of the properties of commercial phantoms can be found in Browne, Ramnarine, Watson, and Hoskins (2003). Dong et al. (1999) and Williams et al. (2006) investigated the nonlinear properties of phantoms. Madsen's group at the University of Wisconsin–Madison has made many innovative contributions to the development of specialized phantoms and the characterization of tissue-mimicking materials. Recent emphasis has been on the simulation of representative geometries of normal and abnormal regions in heterogeneous anthropomorphic phantoms for elastography (Hobson et al., 2007; Madsen, Hobson, Shi, Varghese, & Frank, 2005b; Madsen et al., 1982, 2005a, 2006). A special tissue-mimicking fluid for acoustic output measurements was developed (Stiles, Madsen, & Frank, 2008). Several interlaboratory comparisons of measurements on identical phantoms have been conducted (Anderson et al., 2010). At the time of this writing, a Quantitative Imaging Biomarker Alliance has formed to examine the reproducibility and consistency of imaging methods measuring shear-wave speed in healthy/disease states by different methods (Hall, 2013).

13.6 High-intensity Acoustic Measurements

13.6.1 HIFU Field Measurements

Hydrophones used for diagnostic ultrasound can be destroyed under the strong pressures in a HIFU (or HITU, High Intensity Therapeutic Ultrasound) field, so more robust measuring devices are needed. As discussed in Section 13.4.4, optical hydrophones are well suited for these measurements (Haller et al., 2011; Parsons, Cain, & Fowlkes, 2006; Zhou, Zhai, Simmons, & Zhonga, 2006). Other alternatives such as clad needle hydrophones have been proposed (Howard & Zanelli, 2007; Zanelli & Howard, 2005). A comparative study of several hydrophones in regard to their durability and ability to accurately measure waveforms under intense HIFU fields indicated more work is needed to achieve reliable measurements (Haller, Jenderka, Durando, & Shaw, 2012; O'Reilly and Hynynen, 2010).

Hybrid measurement-simulation approaches are showing promise as a way of overcoming measurement limitations. Canney, Bailey, Crum, Khokhlova, and Sapozhnikov (2008) conducted a detailed study of HIFU fields with two FOPHs of different bandwidths. First,

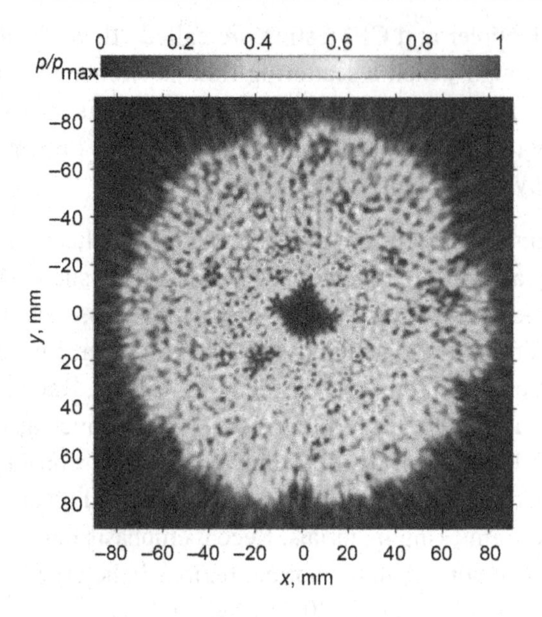

Figure 13.24
Pressure magnitude at the Initial Plane of a Boundary Condition Reconstructed from the Measured Hologram. The reader is referred to the Web version of the book to see the figure in color. *From Yuldashev et al. (2012), IEEE.*

the field of the source transducer was measured by a needle hydrophone under linear conditions and the effective radius and focal length were matched to linear simulations on-axis. Second, the characterization of the source was used as input to a KZK nonlinear propagation simulation program. Third, a holographic reconstruction method, described as the Rayleigh integral approach in Section 13.4.4, measured the actual source distribution, and the KZK program was modified to accommodate the nonuniform source distribution. Fourth, FOPH responses were deconvolved from the data. Both simulations accurately described the main focusing lobe and post-focal region along the beam axis, but two near-in pre-focal lobes were in less agreement. These slight inconsistencies were attributed to the parabolic approximation used for the $F\# = 1$ transducer. Based on independent heating data, they found that adequately wide FOPH bandwidths were needed to capture waveforms faithfully, especially the positive compressional peaks. In these cases, they concluded that the simulations were more reliable depicters of the HIFU field. Furthermore, by adding an absorption power law capability to the program, they demonstrated excellent agreement with fields behind a gel phantom.

A second hybrid measurement-simulation study (Yuldashev et al., 2012) utilized a new program based on the Westervelt equation for power law media. A holographic approach was also used to characterize the source, Figure 13.24, and good agreement was obtained, as indicated by Figure 13.25 for powers from 25 to 800 W and pressures approaching 100 MPa.

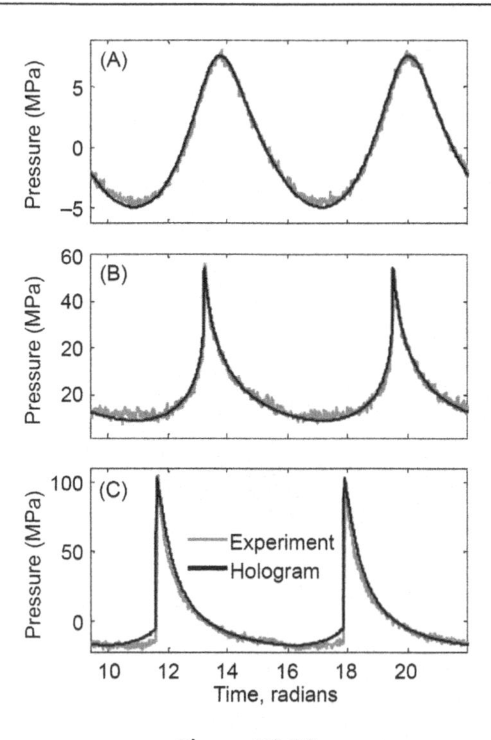

Figure 13.25
Comparison of measured (red/gray line) and simulated waveforms (black line) for several output levels. *From Yuldashev et al. (2012), IEEE.*

A recent look at the problem of derating at HIFU levels can be found in Bessonova, Khokhlova, Canney, Bailey, and Crumb (2010).

13.6.2 HIFU Power Measurements

Ordinary diagnostic ultrasound measurement methods of acoustic power under HIFU conditions also tend to be problematic. Radiation force balance targets can be destroyed under these circumstances. Several options including new targets, effective duty factor, and acoustic streaming are discussed in Maruvada, Harris, Herman, and King (2007). Strongly focused HIFU transducers may require angle correction for their severely converging field (Beissner, 2010; Shou, Wang, & Qian, 1998). Some HIFU transducers have different geometries, such as a hole in the center to accommodate imaging transducers, and these factors have to be considered (Beissner, 2012).

Another approach is to employ a buoyancy or thermal expansion method that is independent of the direction and geometry of the acoustic beam (Shaw, 2008). The setup shown in Figure 13.26 has a castor-oil target suspended in a water bath. When insonified,

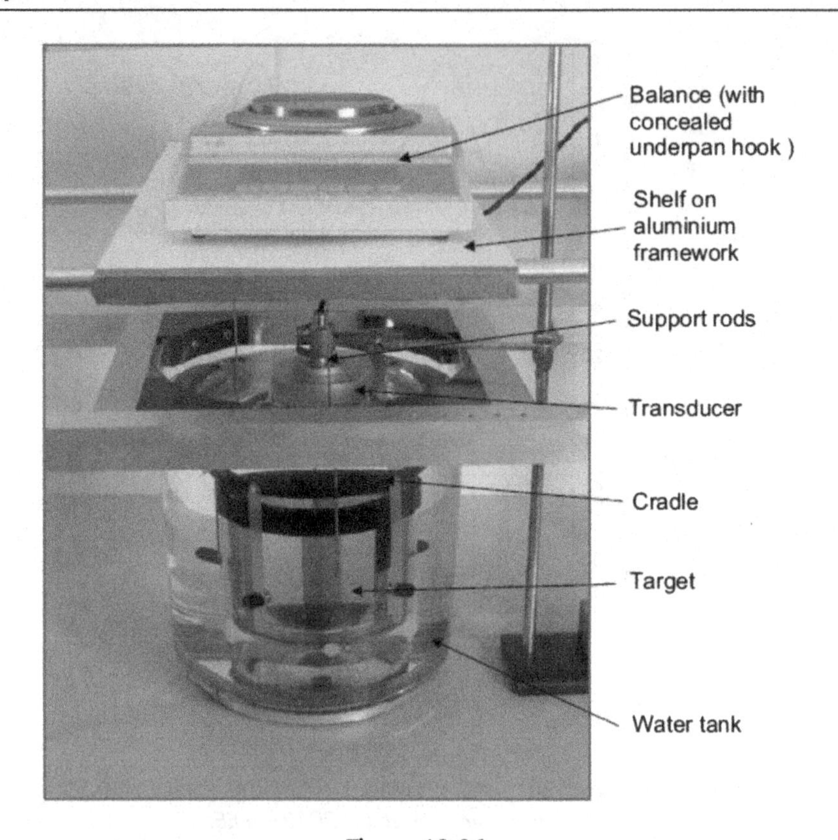

Figure 13.26
Photograph of Buoyancy method target and balance assembly. *From Shaw (2007), with permission from the World Federation of Ultrasound in Medicine and Biology.*

this target undergoes thermal expansion that changes its buoyancy, which is sensed by a force balance. The main relationship for the incident power, P_i, is:

$$P_i = B/(St_0), \tag{13.20}$$

where S is the buoyancy sensitivity, B is the change in buoyancy force, and t_0 is the duration of insonification.

13.6.3 HIFU Thermal Measurements

In addition to the methods mentioned in Section 13.4.7, nonlinear simulation programs exist that are coupled to the bioheat equation to predict temperature (Curra et al., 2000; Soneson & Myers, 2007(section 12.8); Myers & Soneson, 2009; Canney et al, 2008). Methods for monitoring thermal rise both noninvasively and invasively are reviewed in Rivens, Shaw, Civale, and Morris (2007) and include MRI and ultrasound. Specialized transparent phantoms which turn opaque in the region of a HIFU focus are aids in determining the extent of the focal region (Bailey et al., 2003; Howard et al., 2003; Lafon et al., 2006).

13.7 Thought Experiments

The function of different types of measurements changes through the design cycle of system improvement or with a new device such as a transducer. First, many measurements are made during the design phase to explore the effect of new materials and design prototypes. Second, the design is refined and tested for clinical efficacy and robustness from manufacturing, environmental, and safety perspectives. Finally, a number of devices of the final design are made in a production environment to evaluate the consistency of the manufacturing process and acoustic output with statistics.

Along the way, accurate simulation programs can be applied to the initial design, device diagnosis, and design refinement (finding causes of unwanted second-order effects and then reducing them). Simulation programs guide the design process and provide sanity checks and reference points with which measurements can be compared; they can also speed up the process by predicting and eliminating designs that are unsuitable and never built. Modeling programs range from first-order transducer design, finite-element modeling, beamforming, signal processing and filtering, system-level performance, and nonlinear acoustic output prediction. Recent examples include hybrid simulation-measurement methods for acoustic output measurement for HIFU.

References

AIUM/NEMA (1998a). *Acoustic output measurement standard for diagnostic ultrasound equipment*. Laurel, MD: AIUM Publications.

AIUM/NEMA (1998b). *Standard for real-time display of thermal and mechanical acoustic output indices on diagnostic ultrasound equipment. Revision 1*. Laurel, MD: AIUM Publications.

Anderson, J. J., Herd, M. -T., King, M. R., et al. (2010). Interlaboratory comparison of backscatter coefficient estimates for tissue-mimicking phantoms. *Ultrasonic Imaging*, *32*, 48–64.

Bailey, M. R., Khokhlova, V. A., Sapozhnikov, O. A., et al. (2003). Physical mechanisms of the therapeutic effect of ultrasound: a review. *Acoustical Physics*, *49*, 369–388.

Bamber, J. C. (1986). In C. R. Hill (Ed.), *Physical principles of medical ultrasonics*. Chichester, UK: John Wiley & Sonspp. 118–199.

Bamber, J. C. (1998). Ultrasonic properties of tissue. In F. A. Duck, A. C. Baker, & H. C. Starritt (Eds.), *Ultrasound in medicine*. Bristol, UK: Institute of Physics Publishing.

Bamber, J. C., & Phelps, J. (1977). The effective directivity characteristic of a pulsed ultrasound transducer and its measurement by semi-automatic means. *Ultrasonics*, *15*, 169–174.

Bauer-Marschallinger, J., Berer, T., Grün, H., Roitner, H., Reitinger, B., & Burgholzer, P. (2012). Broadband high-frequency measurement of ultrasonic attenuation of tissues and liquids. *IEEE Transactions on Ultrasonics, Ferroelectrics and Frequency Control*, *59*, 2631–2645.

Beard, P. C., Hurrell, A. M., & Mills, T. N. (2000). Characterization of polymer film optical fiber hydrophones for use in the range 1 to 20 MHz; A comparison with PVDF needle and membrane hydrophones. *IEEE Transactions on Ultrasonics, Ferroelectrics and Frequency Control*, *47*, 256–264.

Beissner, K. (1984). Minimum target size in radiation force measurements. *Journal of the Acoustical Society of America, 76*, 1505–1510.

Beissner, K. (1987). Radiation force calculations. *Acustica, 62*, 255–263.

Beissner, K. (1999). Primary measurement of ultrasonic power and dissemination of ultrasonic power reference values by means of standard transducers. *Metrologia, 36*, 313–320.

Beissner, K. (2010). Minimum radiation force target size for power measurements in focused ultrasonic fields with circular symmetry. *Journal of the Acoustical Society of America, 128*, 3355–3362.

Beissner, K. (2012). Some basic relations for ultrasonic fields from circular transducers with a central hole. *Journal of the Acoustical Society of America, 131*, 620–627.

Bessonova, O. V., Khokhlova, V. A., Canney, M. S., Bailey, M. R., & Crumb, L. A. (2010). A derating method for therapeutic applications of high intensity focused ultrasound. *Acoustical Physics, 56*, 354–363.

Bloomfield, P. E., Gandhi, G., & Lewin, P. A. (2011). Membrane hydrophone phase characteristics through nonlinear acoustics measurements. *IEEE Transactions on Ultrasonics, Ferroelectrics and Frequency Control, 58*, 2418–2437.

Browne, J. E., Ramnarine, K. V., Watson, A. J., & Hoskins, P. R. (2003). Assessment of the acoustic properties of common tissue-mimicking test phantoms. *Ultrasound in Medicine and Biology, 29*, 1053–1060.

Canney, M., Bailey, M., Crum, L., Khokhlova, V., & Sapozhnikov, O. (2008). Acoustic characterization of high-intensity focused ultrasound fields: a combined measurement and modeling approach. *Journal of the Acoustical Society of America, 124*, 2406–4220.

Chen, Y., Ritter, T. A., Sabarad, J, Shung, K. K., Tutwiler, R. L., & Wu, Q. (2001). Software control architecture for characterization of a 48-element 30-MHz linear array. *Proceedings of SPIE, 4325*, 523–534.

Clement, G. T., & Hynynen, K. (2000). Field characterization of therapeutic ultrasound phased arrays through forward and backward planar projection. *Journal of the Acoustical Society of America, 108*, 441–446.

Cooling, M. P., & Humphrey, V. F. (2008). A nonlinear propagation model-based phase calibration technique for membrane hydrophones. *IEEE Transactions on Ultrasonics, Ferroelectrics and Frequency Control, 55*, 84–93.

Curra, F. P., Mourad, P. D., Khokhlova, V. A., et al. (2000). Numerical simulations of heating patterns and tissue temperature response due to high-intensity focused ultrasound. *IEEE Transactions on Ultrasonics, Ferroelectrics and Frequency Control, 47*, 1077–1089.

Davidsen, R. E. & Smith, S. W. (1993). Sparse geometries for two-dimensional array transducers in volumetric imaging. In *Proceedings of the IEEE ultrasonics symposium* (pp. 1091–1094). Baltimore, MD.

FDA (2008). *Revised FDA 510(k) information for manufacturers seeking marketing clearance of diagnostic ultrasound systems and transducers*. Rockville, MD: Center for Devices and Radiological Health, US FDA.

Fisher, G. A. (1983). Transducer test system design. *Hewlett Packard Journal, 34*, 24–25.

Freiburger, P. D., Sullivan, D. C., LeBlanc, B. H., Smith, S. W., & Trahey, G. E. (1992). Two-dimensional ultrasonic beam distortion in the breast: in vivo measurements and effects. *Ultrasonic Imaging, 14*, 398–414.

Hall, T. (2013). *Quantitative imaging biomarker alliance shear wave speed imaging: making it much more reproducible, 2013*. New York, NY: AIUM Annual Convention.

Haller, J., Jenderka, K. -V., Durando, G., & Shaw, A. (2012). A comparative evaluation of three hydrophones and a numerical model in high intensity focused ultrasound fields. *Journal of the Acoustical Society of America, 131*, 1121–1130.

Haller, J., Wilkens, V., Jenderka, K. -V., & Koch, C. (2011). Characterization of a fiber-optic displacement sensor for measurements in high-intensity focused ultrasound fields. *Journal of the Acoustical Society of America, 129*, 3676–3681.

Hanafy, A., Zanelli, C. I., & McAvoy, B. R. (1991). Quantitative real-time pulsed Schlieren imaging of ultrasonic waves. In *Proceedings of the IEEE ultrasonics symposium* (pp.1223–1227). Orlando, FL.

Harris, G. R. (1985). A discussion of procedures for ultrasonic intensity and power calculations from miniature hydrophone measurements. *Ultrasound in Medicine and Biology, 11*, 803–817.

Harris, G. R. (1999). Medical ultrasound exposure measurements: update on devices, methods, and problems. In *Proceedings of the IEEE ultrasonics symposium* (pp. 1341–1352). Caesars Tahoe, NV.

He, P. (1999). Experimental verification of models for determining dispersion from attenuation. *IEEE Transactions on Ultrasonics, Ferroelectrics and Frequency Control, 46*, 706–714.

Hekkenberg, R. T., Beissner, K., Zeqiri, B., Bezemer, R. A., & Hodnett, M. (2001). Validated ultrasonic power measurements up to 20 W. *Ultrasound Medicine and Biology, 27*, 427–438.

Hinkelman, L. M., Liu, D. -L., Metlay, L. A., & Waag, R. C. (1994). Measurements of ultrasonic pulse arrival time and energy level variations produced by propagation through abdominal wall. *Journal of the Acoustical Society of America, 95*, 530–541.

Hinkelman, L. M., Szabo, T. L., & Waag, R. C. (1997). Measurements of ultrasonic pulse distortion produced by human chest wall. *Journal of the Acoustical Society of America, 101*, 2365–2373.

Hobson, M. A., Kiss, M. Z., Varghese, T., Sommer, A. M., Kliewer, M. A., Zagzebski, J. A., et al. (2007). In vitro uterine strain imaging: preliminary results. *Journal of Ultrasound in Medicine, 26*, 899–908.

Hocker, G. B. (1979). Fiber-optic sensing of pressure and temperature. *Applied Optics, 18*, 1445–1448.

Howard, S., Yuen, J., Wegner, P., et al. (2003). Characterization and FEA simulation for a HIFU phantom material. In *Proceedings of the IEEE ultrasonics symposium* (pp.1270–1273).

Howard, S. M., & Zanelli, C. I. (2007). Characterization of a HIFU field at high intensity. In *Proceedings of the IEEE ultrasonics symposium* (pp. 1301–1304). New York.

Humphrey, V. F., Cooling, M. P., Duncan, T. M., & Duck F. A. (2006). The peak rarefactional pressure generated by medical ultrasound systems in water and tissue: a numerical study. In *Proceedings of the IEEE ultrasonics symposium* (pp.1604–1607). Vancouver, BC.

Hurrell, A. M. (2012). Acoustic measurements: the story of a humble 1 MHz transducer. In *Proceedings of the IEEE ultrasonics symposium*. Dresden, Germany, 2A–5.

Ide, M., & Ohdaira, E. (1988). Measurement of diagnostic electronic linear arrays by miniature hydrophone scanning. *IEEE Transactions on Ultrasonics, Ferroelectrics and Frequency Control, 35*, 214–219.

IEC (2001). *Standard 61828. Ultrasonics: focusing Transducers definitions and measurement methods for the transmitted fields*. Geneva, Switzerland: International Electrotechnical Commission.

IEC (2002). *Standard 60601-2-37. Medical electrical equipment, part 2: particular requirements for the safety of ultrasonic medical diagnostic and monitoring equipment*. Geneva, Switzerland: International Electrotechnical Commission.

IEEE. (1988). 176–1987 Standard on Piezoelectricity (withdrawn).

Jansson, T. T., Mast, T. D., & Waag, R. C. (1998). Measurements of differential scattering cross section using a ring transducer. *Journal of the Acoustical Society of America, 103*, 3169–3179.

Koch, C., & Molkenstruck, W. (1999). Primary calibration of hydrophones with extended frequency range 1 to 70 MHz using optical interferometry. *IEEE Transactions on Ultrasonics, Ferroelectrics and Frequency Control, 46*, 1303–1314.

Koch, C., & Wilkens, V. (2004). Phase calibration of hydrophones: heterodyne time-delay spectrometry and broadband pulse technique using an optical reference hydrophone. *Journal of Physics, Conference Series, 1*, 14–19.

Lafon, C., Khokhlova, V. A., Kaczkowski, P. J., et al. (2006). Use of a bovine eye lens for observation of HIFU-induced lesions in real time. *Ultrasound Medicine and Biology, 32*, 1731–1741.

Larson, J. D. (1981). Non-ideal radiators in phased array transducers. In *Proceedings of the IEEE ultrasonics symposium* (pp. 673–683). Chicago, IL.

LeDet, E. G., & Zanelli, C. I. (1999). A novel, rapid method to measure the effective aperture of array elements. In *Proceedings of the IEEE ultrasonics symposium* (pp. 1077–1080). Caesars Tahoe, NV.

Lewin, P. A. (1981). Miniature piezoelectric polymer ultrasonic hydrophone probes. *Ultrasonics* 213–216.

Lewin, P. A., Mu, C., Umchid, S., Daryoush, A., & El-Sherif, M. A. (2005). Acousto-optic, point receiver hydrophone probe for operation up to 100 MHz. *Ultrasonics, 43*, 815–821.

Lewin, P. A., & Schafer, M. R. (1988). A computerized system for measuring the acoustic output from diagnostic ultrasound equipment. *IEEE Transactions on Ultrasonics, Ferroelectrics and Frequency Control, 35*, 102–109.

Liu, D. -L., & Waag, R. C. (1998). Estimation and correction of ultrasonic wavefront distortion using pulse-echo data received in a two-dimensional aperture. *IEEE Transactions on Ultrasonics, Ferroelectrics and Frequency Control, 45*, 473–490.

Lum, P., Greenstein, M., Grossman, C., & Szabo, T. L. (1996). High frequency membrane hydrophone. *IEEE Transactions on Ultrasonics, Ferroelectrics and Frequency Control, 43*, 536–543.

Madsen, E. L., Frank, G. R., Hobson, M. A., Shi, H., Jiang, J., Varghese, T., et al. (2005a). Spherical lesion phantoms for testing the performance of elastography systems. *Physics in Medicine and Biology, 50*, 5983–5995.

Madsen, E. L., Hobson, M. A., Frank, G. R., Shi, H., Jiang, J., Hall, T. J., et al. (2006). Anthropomorphic breast phantoms for testing elastography systems. *Ultrasound Medicine and Biology, 32*, 857–874.

Madsen, E. L., Hobson, M. A., Shi, H., Varghese, T., & Frank, G. R. (2005b). Tissue-mimicking agar/gelatin materials for use in heterogeneous elastography phantoms. *Physics in Medicine and Biology, 50*, 5597–5618.

Madsen, E. L., Zagzebski, J. A., & Frank, G. R. (1982). Oil-in-gelatin dispersions for use as ultrasonically tissue-mimicking materials. *Ultrasound in Medicine and Biology, 8*, 277–287.

Maruvada, S., Harris, G. R., Herman, B. A., & King, R. L. (2007). Acoustic power calibration of high-intensity focused ultrasound transducers using a radiation force technique. *Journal of the Acoustical Society of America, 121*, 1434–1439.

Miller, N. R., Bamber, J. C., & Meaney, P. M. (2002). Fundamental limitations of noninvasive temperature imaging by means of ultrasound echo strain estimation. *Ultrasound in Medicine and Biology, 28*, 1319–1333.

Mottley, J. G., & Miller, J. G. (1988). Anisotropy of the ultrasonic backscatter of myocardial tissue, Part I: theory and measurements in vitro. *Journal of the Acoustical Society of America, 85*, 755–761.

Myers, M. R., & Soneson, J. E. (2009). Temperature modes for nonlinear Gaussian beams. *Journal of the Acoustical Society of America, 126*, 425–433.

Nassiri, D. K., & Hill, C. R. (1986). The use of angular acoustic scattering measurements to estimate structural parameters of human and animal tissues. *Journal of the Acoustical Society of America, 79*, 2048–2054.

Nightingale, K. R., Palmeri, M. L., Nightingale, R. W., & Trahey, G. E. (2001). On the feasibility of remote palpation using acoustic radiation force. *Journal of the Acoustical Society of America, 110*, 625–634.

O'Reilly, M. A., & Hynynen, K. (2010). A PVDF receiver for ultrasound monitoring of transcranial focused ultrasound therapy. *IEEE Transactions on Biomedical Engineering, 57*, 2286–2294.

Parsons, J. E., Cain, C. A., & Fowlkes, B. (2006). Cost-effective assembly of a basic fiber-optic hydrophone for measurement of high-amplitude therapeutic ultrasound fields. *Journal of the Acoustical Society of America, 119*, 1432–1440.

Peters, F., & Petit, L. (2003). A broad band spectroscopy method for ultrasound wave velocity and attenuation measurement in dispersive media. *Ultrasonics, 41*, 357–363.

Phillips, R. L. (1980). Proposed fiber-optic acoustical probe. *Optics Letters, 5*, 318–320.

Powell, D. J., Wojcik, G. L., Desilets, C. S., Gururaja, T. R., Guggenberger, K., Sherrit, S., et al. (1997). Incremental "model-build-test" validation exercise for a 1-D biomedical ultrasonic imaging array. In *Proceedings of the IEEE ultrasonics symposium* (pp. 1669–1674), Toronto, ON.

Radulescu, E. G., Lewin, P. A., Goldstein, A., & Nowicki, A. (2001). Hydrophone spatial averaging corrections from 1–40 MHz. *IEEE Transactions on Ultrasonics, Ferroelectrics and Frequency Control, 46*, 1575–1580.

Radulescu, E. G., Lewin, P. A., Wojcik, J., & Nowicki, A. (2003). Calibration of ultrasonic hydrophone probes up to 100 MHz using time gating frequency analysis and finite amplitude waves. *Ultrasonics, 41*, 247–254.

Ritter, T., Geng, X., Shung, K. K., Lopath, P. D., Park, S. -E., & Shrout, T. R. (2000). Single crystal PZN/PT-polymer composites for ultrasound transducer applications. *IEEE Transactions on Ultrasonics, Ferroelectrics and Frequency Control, 47*, 792–800.

Ritter, T. A., Shrout, T. R., Tutwiler, R., & Shung, K. K. (2002). A 30-MHz piezo-composite ultrasound array for medical imaging applications. *IEEE Transactions on Ultrasonics, Ferroelectrics and Frequency Control, 49*, 217–230.

Rivens, I., Shaw, A., Civale, J., & Morris, H. (2007). Treatment monitoring and thermometry for therapeutic focused ultrasound. *International Journal of Hyperthermia, 23*, 121–139.

Sapozhnikov, O. A., Morozov, A. V., & Cathignol, D. (2004). Piezoelectric transducer surface vibration characterization using acoustic holography and laser vibrometry. In *Proceedings of the IEEE ultrasonics symposium* (pp. 161–164).

Sapozhnikov, O. A., Pishchal'nikov, Y. A., & Morozov, A. V. (2003). Reconstruction of the normal velocity distribution on the surface of an ultrasonic transducer from the acoustic pressure measured on a reference surface. *Acoustical Physics*, *49*, 354–360.

Sapozhnikov, O. A., Ponomarev, A. E., & Smagin, M. A. (2006). Transient acoustic holography for reconstructing the particle velocity of the surface of an acoustic transducer. *Acoustical Physics*, *52*, 324–330.

Sarvazyan, A. P., Rudenko, O. V., Swanson, S. D., Fowlkes, J. B., & Emelianov, S. Y. (1998). Shear wave elasticity imaging: a new ultrasonic technology of medical diagnostics. *Ultrasound in Medicine and Biology*, *36*, 1379–1394.

Schafer, M. E., & Lewin, P. A. (1988). A computerized system for measuring the acoustic output from diagnostic ultrasound equipment. *IEEE Transactions on Ultrasonics, Ferroelectrics and Frequency Control*, *35*, 102–109.

Selfridge, A. R. (1985). Approximate material properties in isotropic materials. *IEEE Transactions on Sonics and Ultrasonics*, *32*, 381–394.

Selfridge, A. R., Kino, G. S., and Khuri-Yakub, R. (1980). Fundamental concepts in acoustic transducer array design. In *Proceedings of the IEEE ultrasonics symposium* (pp. 989–993).

Shaw, A. (2008). A buoyancy method for the measurement of total ultrasound power generated by HIFU transducers. *Ultrasound in Medicine and Biology*, *34*, 1327–1342.

Shaw, A., & Nunn, J. (2010). The feasibility of an infrared system for real-time visualization and mapping of ultrasound fields. *Physics in Medicine and Biology*, *55*, N321–N327.

Shaw, A., Pay, N., Preston, R. C., & Bond, A. (1999). A proposed standard thermal test object for medical ultrasound. *Ultrasound in Medicine and Biology*, *25*, 121–132.

Shou, W., Wang, Y., & Qian, D. (1998). Calculation for radiation force of focused ultrasound and experiment measuring HIFU power. *Technical Acoustics*, *17*, 145 (in Chinese)

Soneson, J. E., & Myers, M. R. (2007). Gaussian representation of high-intensity focused ultrasound beams. *Journal of the Acoustical Society of America*, *122*, 2526–2531.

Stiles, T. A., Madsen, E. L., & Frank, G. R. (2008). An exposimetry system using tissue-mimicking liquid. *Ultrasound Medicine and Biology*, *34*, 123–136.

Sugimoto, T., Ueha, S., & Itoh, K. (1990). Tissue hardness measurement using the radiation force of focused ultrasound. In *Proceedings of the IEEE ultrasonics symposium* (pp. 1377–1380). Honolulu, HI.

Szabo, T. L. (1982). Miniature phased-array transducer modeling and design. In *Proceedings of the IEEE ultrasonics symposium* (pp. 810–814). San Diego, CA.

Szabo, T. L. (1993). *Linear and nonlinear acoustic propagation in lossy media*, (Ph.D. Thesis). University of Bath, UK.

Szabo, T. L., Melton, H. E., Jr., & Hempstead, P. S. (1988). Ultrasonic output measurements of multiple mode diagnostic ultrasound systems. *IEEE Transactions on Ultrasonics, Ferroelectrics and Frequency Control*, *35*, 220–231.

Walker, W. F., Fernandez, F. J., & Negron, L. A. (2000). A method of imaging viscoelastic parameters with acoustic radiation force. *Physics in Medicine and Biology*, *45*, 1437–1447.

Waters, K. R., Hughes, M. S., Mobley, J., Brandenburger, G. H., & Miller, J. G. (2000). On the applicability of Kramers-Kronig relations for ultrasonic attenuation obeying a frequency power law. *Journal of the Acoustical Society of America*, *108*, 556–563.

Wilkens, V. (2003). Characterization of an optical multilayer hydrophone with constant frequency response in the range from 1–75 MHz. *Journal of the Acoustical Society of America*, *113*, 1431–1438.

Wilkens, V., & Koch, C. (2004). Amplitude and phase calibration of hydrophones up to 70 MHz using broadband pulse excitation and an optical reference hydrophone. *Journal of the Acoustical Society of America*, *115*, 2892–2903.

Williams, R., Cherin, E., Lam, T. Y. J., Tavakkoli, J., Zemp, R. J., & Foster, F. S. (2006). Nonlinear ultrasound propagation through layered liquid and tissue-equivalent media: computational and experimental results at high frequency. *Physics in Medicine and Biology, 51*, 5809–5824.

Wu, J. (1996a). Determination of velocity and attenuation of shear waves using ultrasonic spectroscopy. *Journal of the Acoustical Society of America, 99*, 2871–2875.

Wu, J. (1996b). Effects of nonlinear interaction on measurements of frequency dependent attenuation coefficients. *Journal of the Acoustical Society of America, 99*, 3380–3384.

Wu, J. (2001). Tofu as a tissue-mimicking material. *Ultrasound in Medicine and Biology, 27*, 1297–1300.

Wu, J., Cubberley, F., Gormley, G., & Szabo, T. L. (1995). Temperature rise generated by diagnostic ultrasound in a transcranial phantom. *Ultrasound in Medicine and Biology, 21*, 561–568.

Zanelli, I., & Howard, S. M. (2005). A robust hydrophone for HIFU metrology. In G. T. Clement, N. J. McDonald, & K. Hynynen (Eds.), *Fifth international symposium on therapeutic ultrasound* (pp. 618–622). New York: American Institute of Physics.

Zeqiri, B. (1988). An intercomparison of discrete-frequency and broad-band techniques for the determination of ultrasonic attenuation. In D. H. Evans, & K. Martin (Eds.), *Physics in medical ultrasound* (pp. 27–35). London: IPSM.

Zhou, Y., Zhai, L., Simmons, R., & Zhonga, P. (2006). Measurement of high intensity focused ultrasound fields by a fiber optic probe hydrophone. *Journal of the Acoustical Society of America, 120*, 676–685.

Zhu, Q., & Steinberg, B. D. (1992). Large-transducer measurements of wave-front distortion in the female breast. *Ultrasonic Imaging, 14*, 276–299.

Ziskin, M. C., & Lewin, P. A. (Eds.), (2000). *Ultrasonic exposimetry* Boca Rotan, FL: CRC Press.

Ziskin, M. C., & Szabo, T. L. (1993). Impact of safety considerations on ultrasound equipment and design and use, Chapt. 12. In P. N. T. Wells (Ed.), *Advances in ultrasound techniques and instrumentation*. New York: Churchill Livingstone.

Bibliography

AIUM (1990). *Standard methods for measuring performance of pulse-echo ultrasound imaging equipment.* Laurel, MD: AIUM Publications.

AIUM/NEMA (1998). *Acoustic output measurement standard for diagnostic ultrasound equipment.* Laurel, MD: AIUM Publications.

AIUM/NEMA (1998). *Standard for real-time display of thermal and mechanical acoustic output indices on diagnostic ultrasound equipment. Revision 1.* Laurel, MD: AIUM Publications.

FDA (2008). *Revised FDA 510(k) information for manufacturers seeking marketing clearance of diagnostic ultrasound systems and transducers.* Rockville, MD: Center for Devices and Radiological Health, US FDA.

IEC (2001). *Standard 61828. Ultrasonics: focusing transducers definitions and measurement methods for the transmitted fields.* Geneva, Switzerland: International Electrotechnical Commission. Refer IEC Website for latest documents; http://www.iec.ch/.

IEC (2002). *Standard 60601-2-37. Medical electrical equipment, part 2: particular requirements for the safety of ultrasonic medical diagnostic and monitoring equipment.* Geneva, Switzerland: International Electrotechnical Commission. Refer IEC Website for latest documents; http://www.iec.ch/.

IEEE. (1988). 176–1987 Standard on Piezoelectricity (withdrawn).

IEEE. (1990). Standard 790–1989. Guide for medical ultrasound field parameter measurements. (June 29, 1990).

IEEE Transactions on Ultrasonics, Ferroelectrics and Frequency Control. (1988). Special issue on ultrasonic exposimetry 35, (March 1988).

Lewin, P. A., & Nowicki, A. (2012). Nonlinear acoustics and its application in biomedical ultrasonics. In K. Nakamura (Ed.), *Ultrasonic transducers: materials and design for sensors, actuators and medical applications* (Vol. 29, pp. 517–544). Woodhead Publishing.

CHAPTER 14

Ultrasound Contrast Agents

Chapter Outline

14.1 Introduction

Many of us are already familiar with the concept of contrast agents. For example, you may have heard about or experienced a test in which mildly radioactive liquids are ingested or injected to light up the digestive tract or blood vessels on X rays or computed tomography (CT) scans.

The application of contrast agents to ultrasound is comparatively recent and is still under development. The first reported use of contrast agents, by Gramiak and Shah (1968), led to the conclusion that increased reflectivity was caused by microbubbles of gas. Early pioneers in this area (Feigenbaum, Stone, & Lee, 1970; Goldberg, 1971; Ziskin, Bonakdapour, Weinstein, & Lynch, 1972) helped establish techniques useful for cardiac irregularities such as leaky valves, shunts, and visualization of larger vessels and chambers. These methods were limited by the large size and short life of bubbles that could be produced (Meltzer, Tickner, Salines, & Popp, 1980). In the 1980s, research on deliberately designed contrast agents (Goldberg, 1993; Ophir & Parker, 1989) began to show promise.

By the early 1990s, contrast agents were being manufactured and tested in early laboratory tests and clinical trials. Miller (1981) described an experiment for detecting the second harmonic of bubbles. Eatock (1985) proposed the nonlinear detection of nitrogen bubbles for decompression applications. Imaging contrast agents at the second harmonic was thought to improve contrast between agents and surrounding tissue, which was believed to behave linearly at diagnostic pressure levels. Imaging system manufacturers and independent research led to the discovery of tissue harmonic imaging (THI), as described in Chapter 12. The nonlinear properties of contrast agents led not only to imaging at the second harmonic, but also to a number of other useful applications and advantages (to be discussed later).

Most present contrast agents are gas-filled encapsulated microbubbles that are injected into the venous system to act as red blood cell tracers. By increasing reflectivity, contrast agents enhance echo amplitudes to improve sensitivity in deep absorbing tissues or in otherwise invisible small vessels. These bubbles have unusual properties in the presence of an ultrasound field. They are nonlinear resonators that, under certain conditions, can change size, cavitate, fragment, or be moved.

Sections 14.2 and 14.3 explain these physical characteristics of microbubbles in a sound field. Conditions for bubble destruction and cavitation are discussed in Section 14.4. The structure and properties of typical contrast agents are compared. Unusual characteristics of contrast agents (described in Section 14.5) have led to several new imaging methods (explained in Section 14.6) and signal-processing methods. Well beyond their original intended applications, contrast agents also have potential in therapy, drug delivery, and the location of targeted sites (as explained in Section 14.7). Information about safety issues with contrast agents is treated in more depth in Chapter 15. Finally, equations of motion appropriate for bubble simulation will be presented in Section 14.8.

Since the first edition of this book, several excellent reviews of contrast agents have appeared. The following provide greater detail on many of the topics introduced here: Bloch, Dayton, and Ferrara (2004); Cosgrove (2006); Klibanov (2006); Ferrara, Pollard, and Borden (2007); Qin, Caskey, and Ferrara (2009); Wilson and Burns (2010); Gessner and

Dayton (2010); and Ducas et al. (2013). In addition, a special issue on ultrasound contrast agents and targeted drug delivery appeared in the *IEEE Transactions on Ultrasonics, Ferroelectrics and Frequency Control* (January 2013, volume 60, number 1).

14.2 Microbubble as Linear Resonator

The kinds of bubbles of interest are very small (with diameters on the order of μm) and are filled with gas. From the scattering theory of Chapter 8, to first approximation, the scattering properties are those of a small sphere. Since wavelengths of diagnostic ultrasound range, 1–0.1 mm (1.5–15 MHz), are much larger than a bubble radius (a), and $ka << 1$, one might conclude that the bubble behaves like a Rayleigh scatterer. Using the Born approximation, Eqn 8.9A, de Jong (1993) compared the scattering cross-section of an air bubble to an iron sphere. Both were in water and each had a radius of 1 μm, and de Jong found that the air bubble had a cross-section 100 million times greater than that of the iron sphere. The main contribution is not through the density term, but through the significantly different compressibility of the gas bubble. This equation also shows that the scattering cross-section depends on the fourth power of frequency and the sixth power of the radius.

An air bubble, unlike the iron sphere, has a fragile, flexible boundary with enclosing fluid (water or blood). When insonified, the bubble expands and contracts with the rhythm of the compressional and rarefactional half cycles of the sound wave (as illustrated by Figure 14.1). Mechanically, the response of the bubble is controlled by the spring-like stiffness of the entrapped gas and the inertia of the fluid pushing on the surface of the bubble. The balance between these competing factors can result in a resonant frequency:

$$f_r = \frac{1}{2\pi a}\sqrt{\frac{3\gamma_C P_0}{\rho}} \tag{14.1}$$

where a is the equilibrium bubble radius, $\gamma_C = C_p/C_v$ (ratio of heat capacities at constant pressure and constant volume), ρ is the density of the surrounding medium, and P_0 is the static pressure at the bubble surface. For small pressures (less than one atmosphere) on an air bubble in water, f_r (Hz) $\approx 3/a$ (m) (Leighton, 1994). For example, for $a = 2$ μm, $f_r = 1.5$ MHz.

The simple linear model of a bubble in a fluid is that of a damped harmonic oscillator. This resonator can be modeled by a series LC-equivalent circuit (Blackstock, 2000) with an inductance representing acoustic mass, $L = \rho/(4\pi a)$, and a capacitance representing stiffness, $C = 4\pi a^3/(3\gamma_C P_0)$. The model can be refined by adding surface tension as well as a resistance for viscous and thermal losses (Anderson & Hampton, 1980; Leighton, 1994).

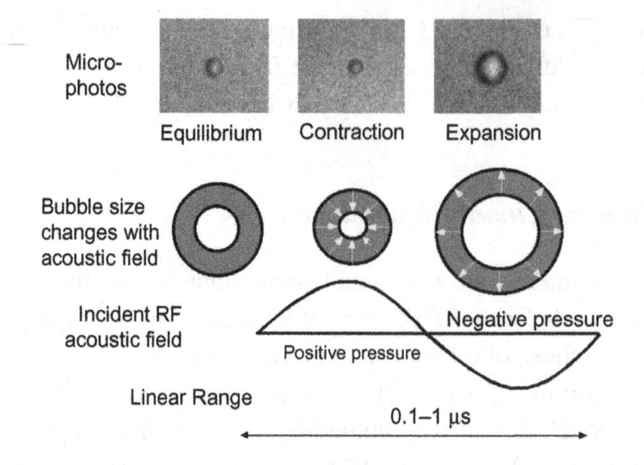

Figure 14.1
Symmetrical Pacing of Bubble Expansion and Compression With the Compressional and Rarefactional Half Cycles of an Ultrasound Wave (Bottom). Measurement of this Effect (Top). *Adapted from P. G. Rafter, Philips Medical Systems; images from Dayton, Morgan, Klibanov, Brandenburger, and Ferrara (1999a) IEEE.*

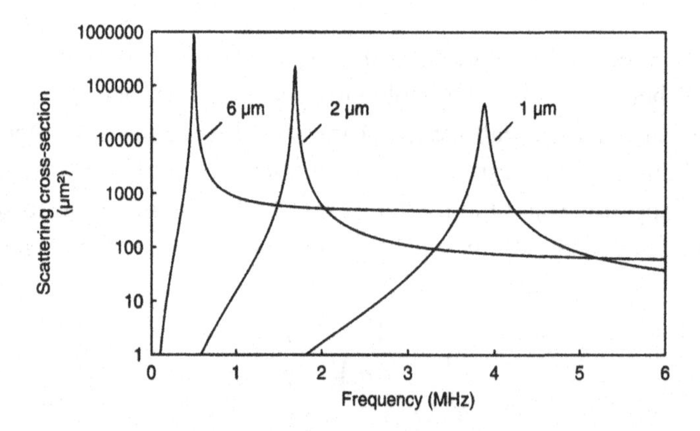

Figure 14.2
Scattering Cross-sections of Ideal Gas Bubbles in Lossless Water as a Function of Frequency for Three Diameters: 1, 2, and 6 μm. *From de Jong (1993).*

Calculations for an ideal gas bubble without losses (de Jong, 1993) are shown in Figure 14.2. Typical characteristics are a Rayleigh (f^4) frequency dependence below resonance and a nearly constant value equal to the physical cross-section at frequencies much greater than resonance. At-resonance values for the cross-section can be 1000 times greater than values predicted by the Born approximation.

14.3 Microbubble as Nonlinear Resonator

14.3.1 Harmonic Response

The previous analysis only holds for small-forced vibrations. As pressure amplitude is increased, the bubble cannot keep up. For larger-amplitude sound fields, a bubble can expand with the sound field, but it cannot contract without limit because the volume of entrapped gas can only be compressed so far (as depicted in Figure 14.3). As a result of these differences, the pressure response of the bubble as a function of time becomes asymmetric. The corresponding harmonic response of an ideal bubble can be computed by a modified Rayleigh–Plesset equation (given in Section 14.8). Calculations for three bubble sizes and a 50-kPa pressure amplitude are shown in Figure 14.4; the asymmetric expansion and contraction of bubble size versus time is given on the left with the resulting harmonics on the right. Note that the frequency response is a function of both bubble size and the insonifying pressure amplitude.

14.3.2 Subharmonic Response

The harmonic of a microbubble appears to follow the mixing concept introduced in Chapter 12 for tissue. In other words, harmonics would be expected as frequencies pass through a progressive sequence of mixers; therefore f_0 creates a sum frequency $2f_0$ and a difference frequency 0; then these in turn create more sum-and-difference frequencies, leading to $3f_0$ and f_0, etc. Eller and Flynn (1969) discovered that above a certain pressure

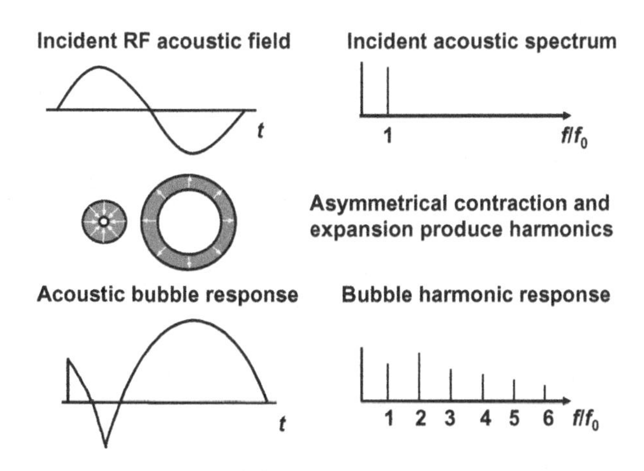

Figure 14.3
In Higher-pressure Incident Sound Fields, the Microbubble Response Becomes Nonlinear Because Compression is Limited and Shortened Compared to Expansion, Leading to Asymmetry and Harmonics. *Courtesy of P. G. Rafter, Philips Medical Systems.*

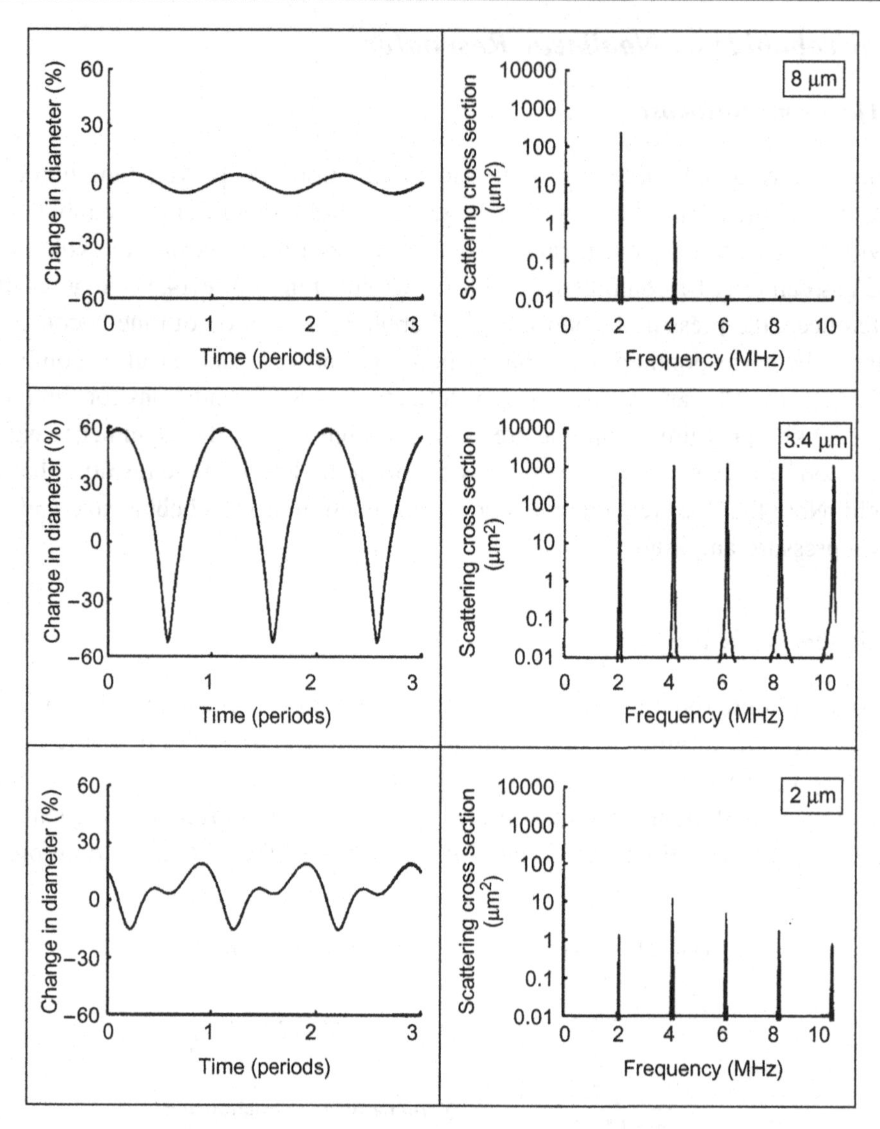

Figure 14.4

Calculations of Ideal Bubble Harmonic Response. (Left panels) simulated relative change in diameter of an ideal gas bubble in water for a sinusoidal excitation of 2 MHz and a pressure amplitude of 50 kPa for three bubble diameters: 6, 3.4, and 2 μm (above, at, and below resonance, respectively). (Right panels) corresponding spectral responses. *From de Jong (1993).*

threshold, bubbles could create a subharmonic frequency $f_0/2$. Shi et al. (1999) demonstrated that these subharmonics exist for contrast agents. In a definitive experiment, they insonified an Optison (a type of ultrasound contrast agent) microbubble suspension with a concentration of 0.10 pl/ml at two different frequencies with acoustic pulses with a

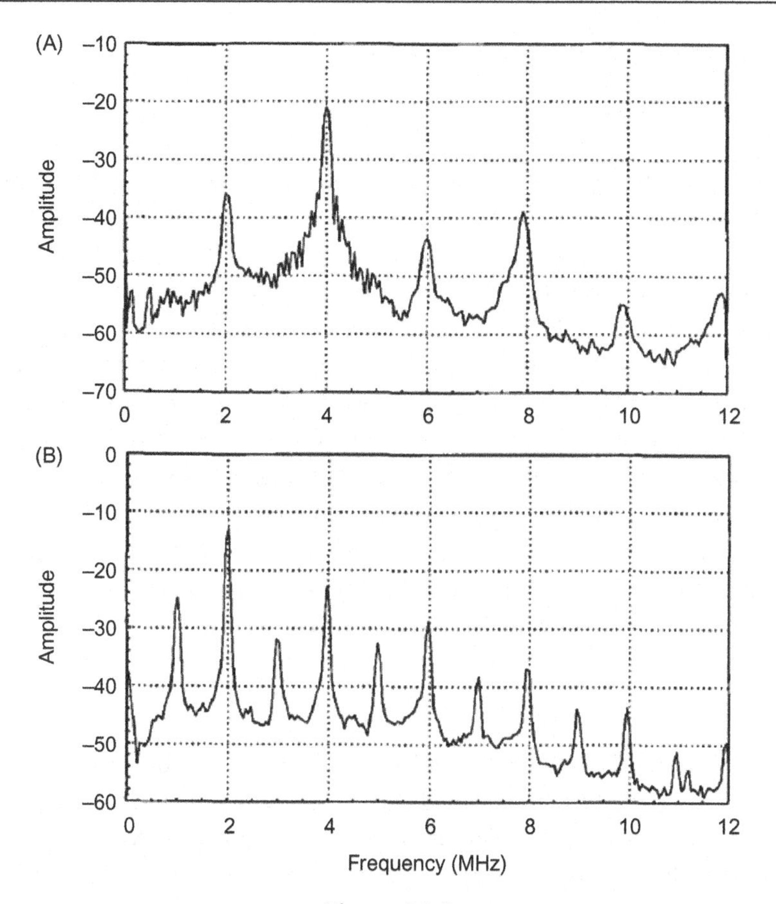

Figure 14.5

Spectra of the Scattered Signals from Insonified Optison Microbubbles. Insonification at (A) 4.0 MHz, and (B) 2.0 MHz. The MI value of 0.8 corresponds to an acoustic pressure amplitude of 1.6 MPa (Ispta = 68 W/cm^2 and Isppa = 6.8 mW/cm^2) at 4 MHz, and 1.1 MPa (I,,, = 32 W/cm^2 and I, = 6.5 mW/cm^2) at 2 MHz, respectively. *From Shi et al. (1999), with permission from Dynamedia, Inc.*

length of 32 cycles and an MI of 0.8 (a pressure level explained in Section 14.4.3) transmitted at a pulse repetition frequency (PRF) of 10 Hz. Their results are shown in Figure 14.5. In Figure 14.5A, $f_0 = 4$ MHz and the subharmonic is evident at $f_0/2 = 2$ MHz. By the mixing principle, harmonics occur at nf_0 at 8 and 12 MHz, as expected; however, as $f_0/2$ and f_0 interact frequencies to produce a sum frequency $f_0/2$ and $f_{0=}6$ MHz and this in turn with f_0 generates 10 MHz, etc. These intermediate harmonics ($3f_0/2$, $5f_0/2$ . . .) are called "ultraharmonics." In Figure 14.5B, $f_{0=}2$ MHz, so the fundamental and subharmonic of 1 MHz produce harmonic and ultraharmonic frequencies at 1-MHz intervals. These unique frequencies can function as a way of discriminating between contrast agents and tissues, as explained later in Section 14.6.

14.4 Cavitation and Bubble Destruction

14.4.1 Rectified Diffusion

By itself, an air bubble will dissolve in a liquid; but under ultrasound insonification, it can resonate and grow under certain conditions. Since there is usually dissolved gas in a liquid outside a bubble, certain circumstances can cause this gas to be pumped into a bubble with help from a sound beam. Consider the bubble under compression depicted in Figure 14.1, where the pressure is high but the net surface area for gas to enter is small. During the expansion phase, the pressure is very low and creates a pressure gradient that draws outer dissolved gas into the bubble. Also, a large surface area is available for gas infusion. As the bubble grows, its minimum radius also grows. This process, called the "area effect," tends to grow the bubble rapidly. A comparison of the shell thicknesses in the same figure indicates that the shell thickens on compression and thins on expansion. These changes also enable more gases to enter the bubble and work in the same direction as the area effect to increase the bubble size rapidly. The overall process (Crum, 1984; Eller & Flynn, 1965; Leighton, 1994) is called "rectified diffusion."

14.4.2 Cavitation

A general term for the modification of preexisting bubbles or the formation of new bubbles or groups of bubbles by applied sound is called "acoustic cavitation" (Apfel, 1984; Leighton, 1994). Neppiras (1984) qualifies this definition by adding that both expansion and contraction of the gas body must be involved. "Stable cavitation" is a term (Flynn, 1964) that refers to the sustainable periodic nonlinear expansion or contraction of a gas body or bubble. Unstable or "transient cavitation" (Flynn, 1964) refers to the rapid growth and violent collapse of a bubble. For years, this collapse was viewed as a singular catastrophic event (Leighton, 1998) producing fragmentation, temperatures in excess of 5000 K, the generation of free radicals, shock waves, and a light emission called "sonoluminescence." Detection of this light has been used as evidence that transient cavitation occurred. Gaitan and Crum (1990) were able to measure sonoluminescence of a single bubble in a water glycerin mixture at 22 kHz over thousands of cycles. Currently, "inertial cavitation" is a more appropriate descriptor of these events rather than transient cavitation (Leighton, 1998). The mechanisms for sonoluminescence are still being discussed, and the necessary conditions for populations of bubbles are believed to be different than those for a single bubble.

Inertial cavitation is a threshold event. Apfel and Holland (1991) calculated the conditions for this threshold by assuming a temperature maximum of 5000 K as a necessary condition for cavitation to occur. The conditions are appropriate for short pulses less than 10 cycles and low-duty cycles, like those in diagnostic ultrasound. Their computations are plotted in Figure 14.6. Here, the thresholds are shown as curves for pulses of three different center frequencies. Each curve has a pressure minimum P_{opt} at a radius equal to R_{opt}, the condition

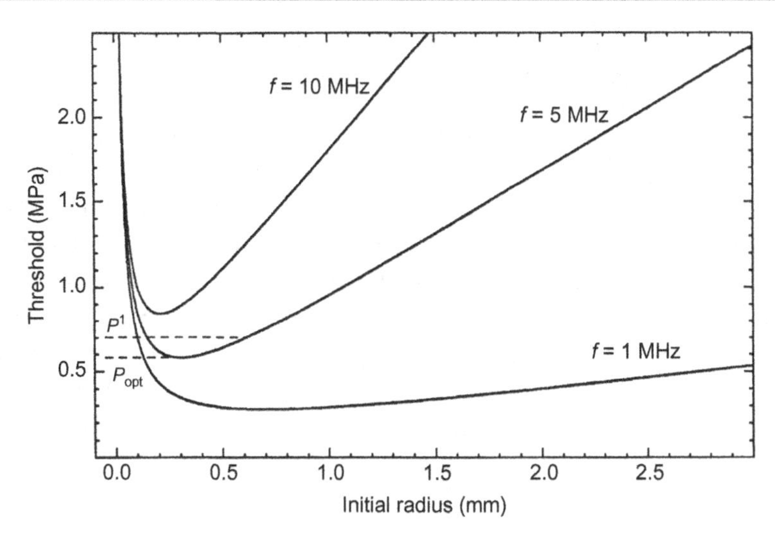

Figure 14.6
Peak Rarefaction Pressure Threshold as a Function of Bubble Radius. Shown for three frequencies of insonification: 1, 5, and 10 MHz. These curves represent inertial cavitation pressure amplitude thresholds. *From Apfel and Holland (1991); reprinted with permission from the World Federation of Ultrasound in Medicine and Biology.*

for lowest pressure threshold for a specified frequency. For example, at 5 MHz, these values are 0.58 MPa and 0.3 μm, respectively. Note that for a higher pressure P', as shown on the graph, a wider range of radii, 0.1–0.6 μm, will fall under the threshold. Notice also that at 10 MHz, these threshold minimum values are 0.85 MPa and 0.2 μm, respectively. The trend is that the threshold increases with frequency. The limiting values to the left of the minimum of each curve are governed by surface tension, and those on the right are controlled by a condition in which the ratio of the maximum radius to the equilibrium radius value exceeds a critical value. From their calculations, they estimated P_{opt} for an air bubble in water as a function of frequency as:

$$P_{opt} \approx 0.245\sqrt{f_c} \tag{14.2}$$

where the center frequency is in MHz and the values are slightly conservative compared to the minima of the curves.

14.4.3 Mechanical Index

This threshold equation led to the definition of a mechanical index (MI) (American Institute of Ultrasound in Medicine/National Electrical Manufacturers Association, 1998) to estimate the likelihood of inertial cavitation with an intervening tissue path:

$$\text{MI} = \frac{P_{r.3}}{\sqrt{f_c}} \tag{14.3}$$

where $P_{r\cdot 3}$ is the maximum axial value of rarefactional pressure measured in water, $p_r(z)$, and derated along the beam axis, z, by an in situ exponential factor:

$$P_{r.3} = \text{maximum}[\text{pr}(z)\ exp(-0.0345 f_c z)] \tag{14.4}$$

More information about the rationale behind the MI can be found in Abbott (1999). A plot of this calculation for a 3.5-MHz center frequency is given by the top half of Figure 14.7. Curves are shown for the exponential derating factor, as well as $p_r(z)$, their product, and locations of maxima. When used by itself, p_r refers to the maximum peak rarefactional pressure in water. Note that the in situ derating factor, taken as a conservative average value for tissue, will not be the same as a specific tissue path in a diagnostic exam. The peak water value for p_r is usually located at a distance less than the geometric focal length, and it is dependent on nonlinear effects in water (discussed in Chapter 12). The location of

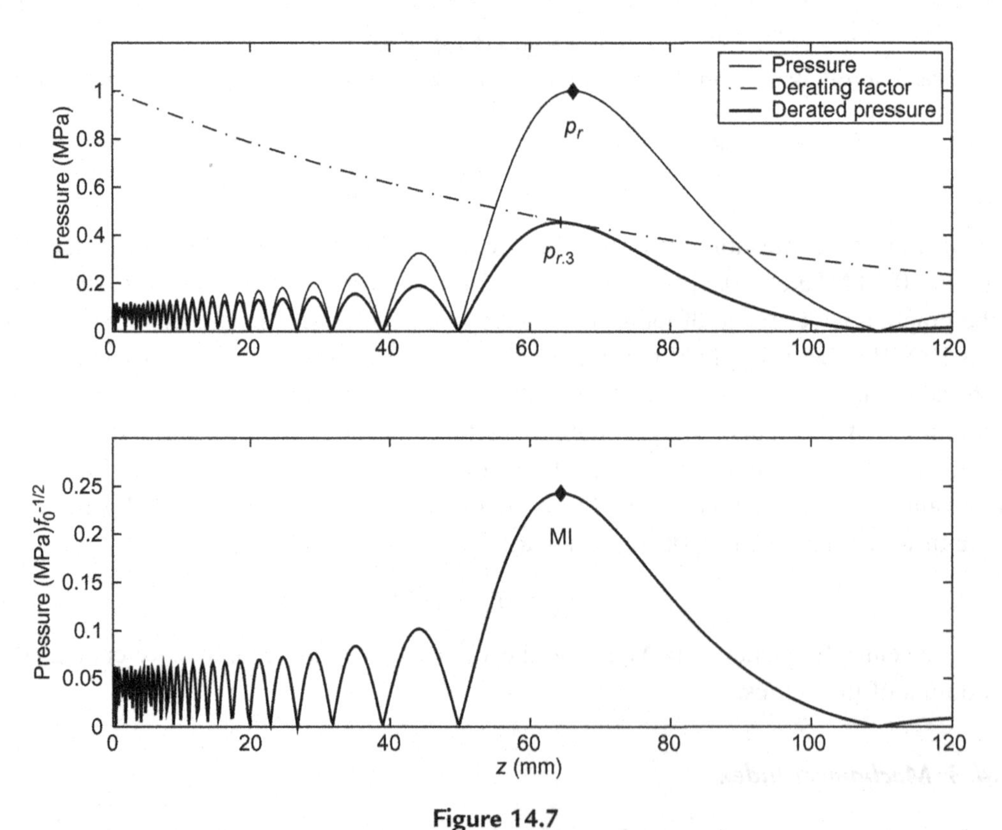

Figure 14.7
Pressure, Derating, and MI vs z. (Top) rarefactional pressure amplitude measured in water as a function of distance from the transducer (thin solid line); exponential derating factor (dashed line); derated pressure is the product of rarefactional pressure amplitude and exponential derating factor (thick solid line). (Bottom) derated pressure divided by $\sqrt{f_c}$. The maximum of this curve is the MI.

the derated value is closer to the transducer. MI, the peak value of Eqn 14.3 and shown in the bottom half of Figure 14.7, is a value that is proportional to the rarefactional pressure level measured, so a higher MI indicates a greater pressure level. A display of MI (discussed in Chapter 13) is available in real time on all systems marketed in the USA (American Institute of Ultrasound in Medicine/National Electrical Manufacturers Association, 1998) and is discussed further in Chapter 15. In the USA, the recommended maximum for MI is 1.9. Note that the MI display gives a relative measure of maximum pressure amplitude but not its location, and the location may not coincide with the region of interest.

14.5 Ultrasound Contrast Agents

14.5.1 Basic Physical Characteristics of Ultrasound Contrast Agents

A contrast agent is a designer bubble. The structure of a contrast agent is typically a sphere containing perfluorocarbon gas or air, about 1–10 μm in diameter with a thin elastic shell approximately 10–200 nm thick. There may be no shell but a surfactant, or a shell with one layer or several. The small size of contrast agents is a deliberate attempt to mimic the size of red blood cells so that the agents can move into capillaries and pass through the pulmonary circulation system. For the purposes of measuring perfusion in the heart, these microspheres need to be inert to alter neither hemodynamics nor coronary blood flow, but instead to act as blood tracers with high echogenicity (de Jong, 1993). In a typical application, the agent is administered as a bolus or by infusion intravenously, and it passes through the pulmonary system and emerges in the left ventricle. An important design goal is for the microbubbles to persist and to not dissolve quickly.

After injection, the original distribution of microspheres of different diameters is altered by three factors (de Jong, 1993). The lung capillaries act as a filter (as shown by Figure 14.8). In addition, the original concentration of agent is diluted in the circulatory system and is affected by pressure gradients. A measured distribution after lung passage of Albunex, a first-generation agent with an air bubble covered by a sonicated albumin shell, is presented in Figure 14.9. From this information, it is evident that smaller bubbles with diameters in the 2–6 μm range are to be preferred for cardiac applications.

One major difference between air bubbles and contrast agents is the effect of the shell, which constrains the expansion and raises the resonant frequency. The overall effect on resonance can be expressed by (de Jong, Hoff, Skotland, & Bom, 1992):

$$f_{re}^2 = f_r^2 + \frac{S_p}{4\pi^2 m} \tag{14.5}$$

where S_p is the stiffness of the shell, m is the effective mass of the system, and f_r is the free gas bubble resonance given earlier. Calculations for several agents and an air bubble in

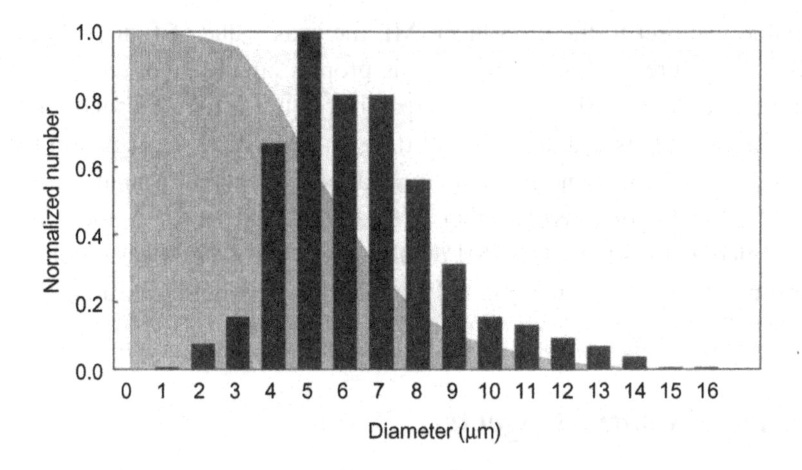

Figure 14.8
Filtering Action of Lung Capillaries. Bars represent normalized size distribution of lung capillaries, and the shaded area shows the probability that a microsphere of a particular size will pass through the capillaries. *From de Jong (1993).*

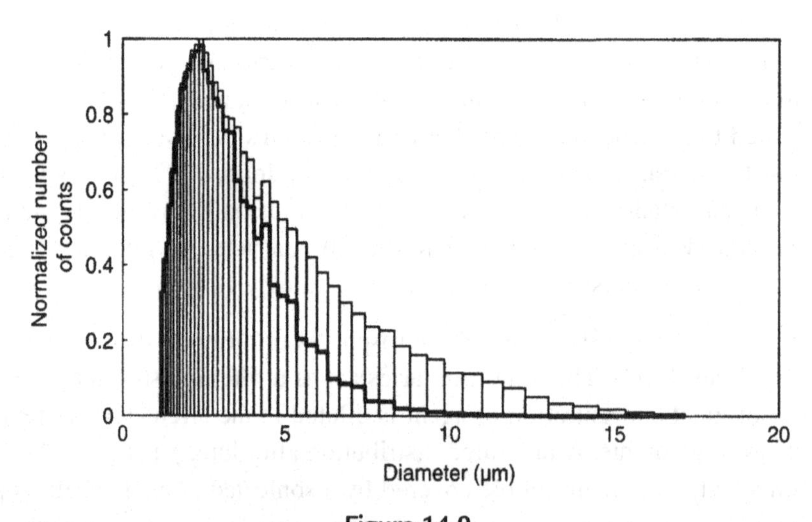

Figure 14.9
Microbubble Measurements by a Coulter Counter Shown as Normalized Size Distributions of Albunex (Thin Curve) Compared With Calculated Size Distribution (Thick Curve) after Lung Passage. Based on the probability curve from Figure 14.7. *From de Jong (1993).*

water (de Jong, Bouakaz, & Frinking, 2000) are shown in Figure 14.10. These agents range from a thick rigid shell to a very flexible shell in the following order: Quantison, with air enclosed by an albumin shell (no longer available); Albunex, also air with a human serum albumin shell; and Sonovue, with sulfurhexafluoride and a phospholipid, with an air bubble

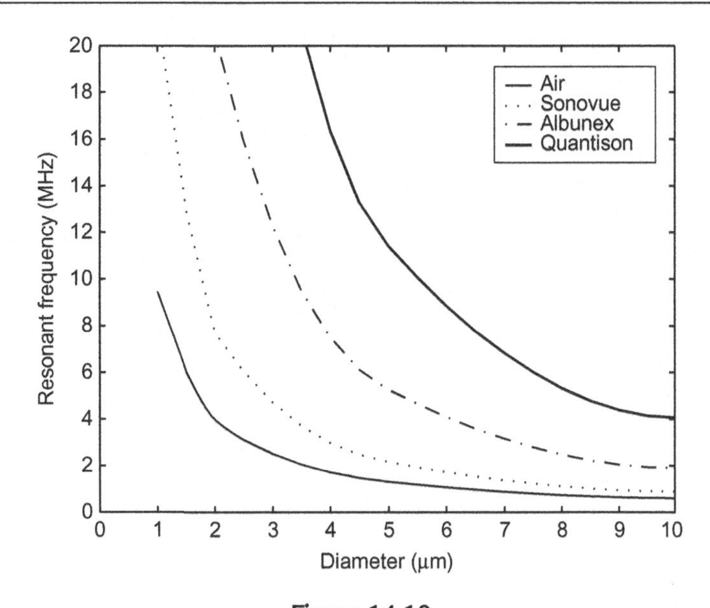

Figure 14.10
Resonant Frequency Versus Bubble Diameter for Air, Sonovue, Albunex, and Quantison. *Based on calculations from de Jong et al. (2000).*

being the most flexible. Consequently, for a given bubble size, air has the lowest resonant frequency and Quantison has the highest. Heavier gases make a large difference; the water permeation resistance of perfluorobutane, for example, is 300 times greater than that of air (Ferrara, Pollard, and Borden, 2007). With a shell, the resistance is 1400 times greater than air, so that the bubble lifetime is increased by a factor of 100,000. These improvements are needed so that ultrasound contrast agents can be stored (Ferrara, Pollard, and Borden, 2007).

An indication of the use of ultrasound contrast agents (UCAs) worldwide is given by Figure 14.11. Current commercially available contrast agents are listed in Table 14.1. As of this writing, only three UCAs are available in the USA: Definity, Imagent, and Optison. Levovist is not, yet it is approved for use in 65 other countries. Imagent (perflexane lipid microspheres) is a dry powder reconstituted with water. The majority of these gases are perfluorocarbon-like gases or air, and the shells vary from human serum albumin (administered in different ways) to surfactants. All materials were tested to be biocompatible and to be absorbed into the circulatory system. Contrast agents no longer available are CardioSphere, Quantison, Sonavist, and Echogen. A kind of UCA that is prepared locally is called "PESDA," or perfluoropropane-exposed sonicated dextrose albumin (Porter, Xie, Kricsfeld, & Kilzer, 1995). Another type of microbubble used by investigators is one with a polymer shell (Wheatley, Schrope, & Shen, 1990). A micro-UCA designed for mouse studies (Vevo MicroMarker, untargeted, VisualSonics) consists of a lipid shell (diameter 0.5–5 μm) filled with perfluorocarbon gas. As a consequence of its

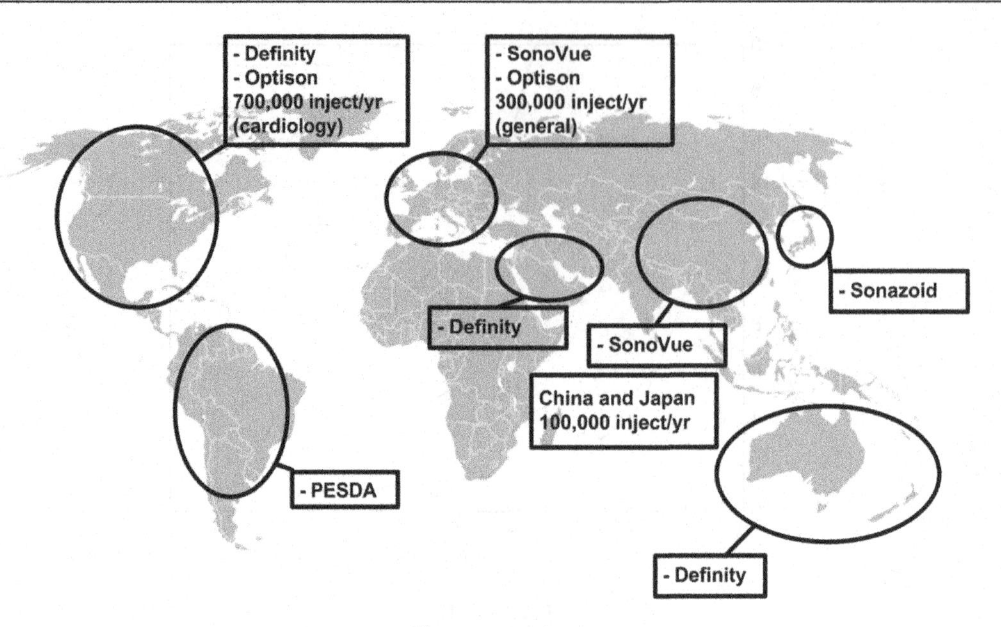

Figure 14.11
Estimated Sales Figures and Availability of Ultrasound Contrast Agents in 2010. Note, Imagent is also available in the USA. *From Faez et al. (2013), IEEE.*

small diameter (and curves like those of Figure 14.10), the resonant frequency is much higher, typically 40 MHz.

Other developments are the addition of an oil-phase layer to incorporate hydrophilic drugs, lipid-shelled with liposomes linked to biotin streptavidin bridges and multilayer architectures for polymer-shelled agents ultrasound-triggered image-guided drug delivery.

A typical injected suspension of UCAs is 0.2–2 ml and is equivalent in terms of the number of scatterers to about 1 ml of red blood cells (Wilson and Burns, 2010). The amount of gas in a typical bolus injection is on the order of 20–100 μl (Cosgrove, 1998). After 5 minutes, typically, the agents have diffused into the blood. What happens to gases from contrast agents after they diffuse out? They are carried along by blood and released when passing through the lungs. With a saline drip, a 20-minute infusion can be obtained (Wilson and Burns, 2010).

14.5.2 Acoustic Excitation of Ultrasound Contrast Agents

The behavior of these agents under acoustic excitation falls into three classes (Frinking, de Jong, & Cespedes, 1999), depending on the structure of the microbubble and the level of the insonifying pressure amplitude and frequency: stable linear (also known as low MI), stable nonlinear scattering (medium MI), and transient nonlinear scattering (high MI). There is a fourth class, called super MI, with levels above those used in diagnostic imaging (to be discussed in Chapter 15). Characteristics of a bubble as a linear resonator have been

Table 14.1 Commercially available Ultrasound Contrast Agents in 2010.

Name	Manufacturer	Year	Gas	Coating	Approved	Available
Echovist	Bayer Schering Pharma AG	1991	Air	Galactose	EU, Japan, Canada	–
Albunex	Molecular Biosystems	1994	Air	Human albumin	EU, USA, Canada	–
Levovist	Bayer Schering Pharma AG	1996	Air	Galactose, trace Palmitin	Worldwide[1]	–[2]
Optison	GE Healthcare AS	1997	C_3F_8	Human albumin	EU, USA	EU, USA[3]
Definity	Lantheus Medical Imaging	2001	C_3F_8	Phospholipids	Worldwide[4]	Worldwide
SonoVue	Bracco spA	2001	SF_6	Phospholipids	Europe, China, S. Korea, India, Hong Kong, Singapore	Europe, China, S. Korea, India, Hong Kong, Singapore
Imagent	Alliance Pharmaceutical Corp.	2002	C_6F_{14}	Phospholipids	USA	–
Sonazoid	Amersham Health	2006	C_4F_{10}	Phospholipids	Japan	Japan
Br38[5]	Bracco spA	–	C_4F_{10}/n_2	Phospholipids	–	–

[1]Approved in 65 countries, but not in the USA.
[2]Expected to finish in 2010.
[3]Temporarily unavailable 2006–2010.
[4]Approved in United States, Canada, Mexico, Israel, Europe, India, Australia, Korea, Singapore, UAE, and New Zealand.
[5]In clinical development.
Source: Faez et al. (2013)

discussed in the last section. In Figure 14.12, properties of Albunex are shown for microbubbles of three diameters as an example of the stable nonlinear regime. Compared to Figure 14.4 for an air bubble, the changes are more muted, with the largest ones occurring when the 2-MHz excitation is close to resonance (9.6 μm diameter).

Under high levels of pressure (but still within the limits for MI recommended by the US Food and Drug Administration (FDA)), contrast agents have demonstrated unexpected behavior. Clinicians discovered that some agents disappeared from view when insonified, or were destroyed (Porter & Xie, 1995), whereas others reported brighter regions in an image. These conflicting observations were resolved by Frinking et al.'s study (1999) of Quantison microspheres. Figure 14.13A shows scattered power from these microspheres as a function of frequency for a low acoustic pressure insonification level. At low frequencies, the scattering is linear. In their experiment, a destructive acoustic pulse with a center frequency of 0.5 MHz and a 1.6-MPa pressure was turned on, and the power was remeasured 0.6 ms

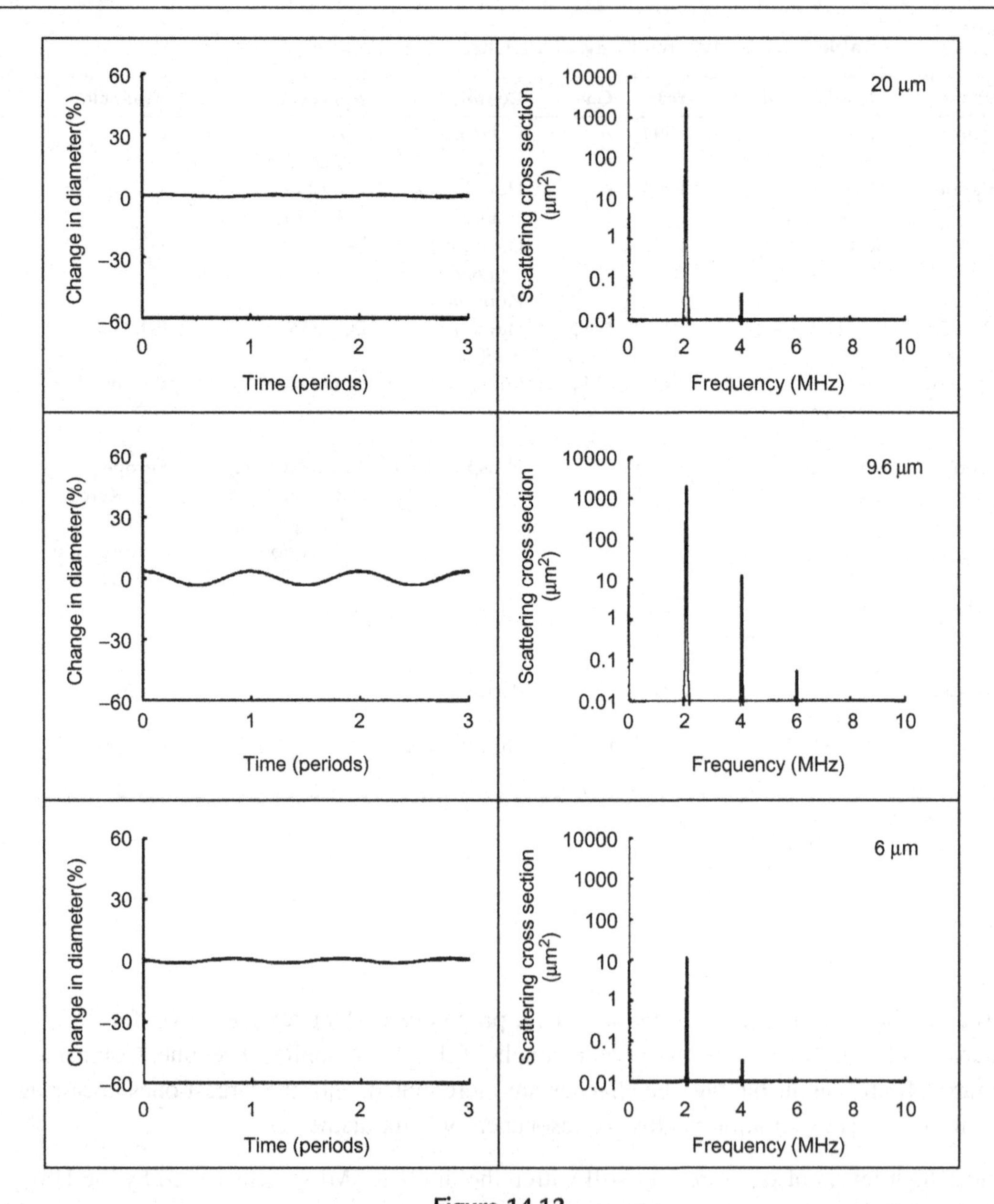

Figure 14.12
Properties of Albunex. (Left panels) simulated relative change in diameter of an Albunex bubble in water for a sinusoidal excitation of 2 MHz and a pressure amplitude of 50 kPa for three bubble diameters: 20, 9.6, and 6 μm (above, at, and half resonance, respectively); (right panels) corresponding spectral responses. *From de Jong (1993).*

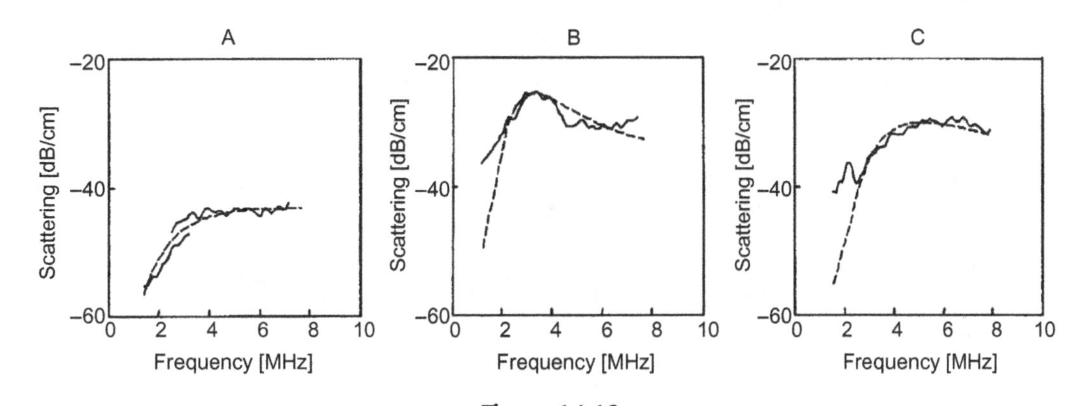

Figure 14.13

Scattered Power from Quantison Microspheres Before and After a High-amplitude Ultrasound Burst. (A) Scattering versus frequency of a dilution of 1:4500 of Quantison measured without the high-amplitude ultrasound burst (solid line is measured scattering; dashed line is theoretical spectrum using measured size distribution); (B) solid line is measured scattering 0.6 ms after transmission of the 0.5-MHz high-amplitude ultrasound burst at the same region, and dashed line is theoretical spectrum; (C) the same as in (B) but now using the 1.0-MHz transducer for generating the high-amplitude burst (solid line is measurement and dashed line is theoretical spectrum). *From Frinking et al. (1999), Acoustical Society of America.*

later. As evident from the graph in Figure 14.13B, the scattering increased by 20 dB and has a resonance peak at 3 MHz in the upper curve, which corresponds to that of a free gas, in good agreement with the prediction. The scattering response for a similar experiment but with an insonifying frequency of 1 MHz is shown in Figure 14.13C, and it has a resonance at 1.3 MHz. These resonances corresponded to mean diameters of 2 and 1.2 μm and a release of only 1% of the available population. After release, the free gas bubbles dissolved within 10 and 20 ms (in accordance with theory for these sizes). In summary, the sequence of steps is as follows: Figure 14.13A depicts the initial states, as well as responses, for stable linear and stable nonlinear insonification levels; Figures 14.13B and 14.13C illustrate the higher scattering levels after a high-pressure signal, at the transient nonlinear level, caused by the mechanical failure and fragmentation of the shell, as well as the release of the gas from the shell. Frinking et al. (1999) called this large increase in scattered power "power-enhanced scattering." For thick-shelled Quantison (see Figure 14.10), the internal gas is air, which takes 1–5 ms before dissolving and disappearing from view.

14.5.3 Mechanisms of Destruction of Ultrasound Contrast Agents

Is this destructive process cavitation? By the definitions presented earlier, yes. Is it inertial cavitation with sonoluminescence? Most likely it is not, but rather, it is a mechanical failure of the shell. In the destructive process just described, there are different types of cavitations corresponding to the pressure amplitude level of insonification and insonifying frequency. The

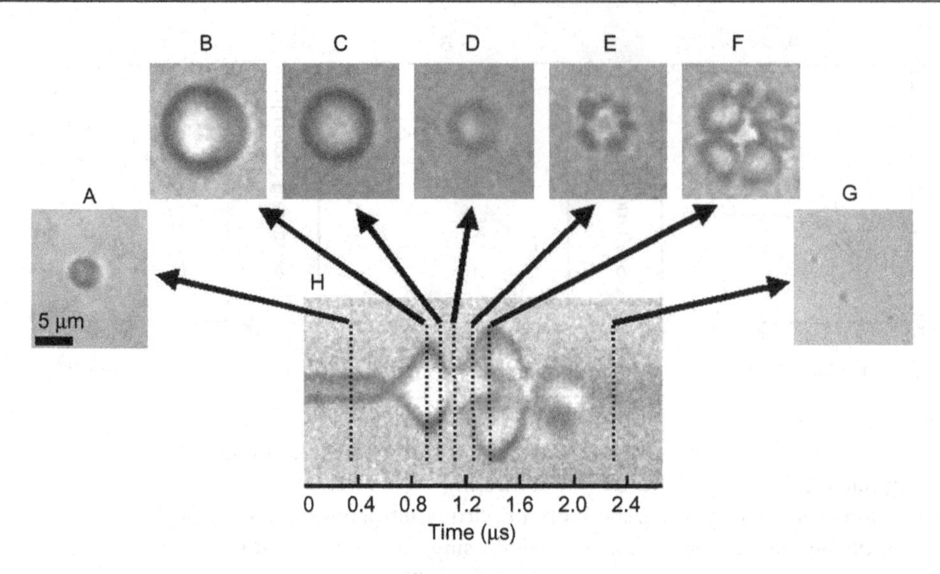

Figure 14.14
Optical Frame Images and Streak Image Corresponding to the Oscillation and Fragmentation of a Contrast Agent Microbubble, where Fragmentation Occurs During Compression.
The bubble has an initial diameter of 3 μm, shown in (A). The streak image in (H) shows the diameter of the bubble as a function of time, and dashed lines indicate the times at which the two-dimensional cross-sectional frame images in (A–G) were acquired, relative to the streak image. *From Chomas et al. (2000), American Institute of Physics.*

growing number of contrast agents, their unique structures and materials, and their distribution of sizes suggest that there may be other types of cavitation that remain to be characterized.

Other windows into the destruction of contrast agents involve acoustical and optical detectors (Chen, Matula, & Crum, 2002; Chomas, Dayton, Allen, Morgan, & Ferrara, 2000, 2001; Dayton et al., 1997, 1999a,b; Deng & Lizzi, 2002; Holland, Roy, Apfel, & Crum, 1992; Morgan, Kruse, Dayton, & Ferrara, 1998; Wu & Tong, 1998). From some of these studies, a fourth category of bubble destruction at extremely high levels of pressure (super MI), corresponding to those applied in high-intensity focused ultrasound (HIFU) or lithotripsy, has revealed new phenomena and bioeffects (to be described separately in Chapter 15). In this section, emphasis is on contrast agent changes that occur at normal diagnostic pressure levels.

Chomas et al. (2000) has produced remarkable measurements, both optical and acoustical, of microbubble destruction. In Figure 14.14, the demise of an experimental contrast agent, M1950, insonified with a 2.4-MHz three-cycle-long pulse with $p_r = -1.2$ MPa, is captured with a high-speed framing camera and streak camera capable of 10-ps resolution. This experimental agent, made by Mallinckrodt, Inc., is made of decafluorobutane (C_4F_{10}) gas encapsulated in a phospholipid shell. A continuous streak image of bubble diameter in

Figure 14.14H shows the temporal evolution of the breakup, which is detailed in cross-sectional snapshots at different time intervals over an observation time from (B) to (F) of 80 ns. This type of destruction has been identified as fragmentation (Chomas et al., 2001), a type of rapid destruction on a microsecond scale, in which excessive expansion and contraction ($R_{max}/R_{min} > 10$) of the microbubble causes instability, and there is an irreversible fragmentation into smaller bubbles that dissolve.

Two other mechanisms for bubble destruction are forms of diffusion (Chomas et al., 2001). Both involve the diffusion of gas out of a microbubble. Just as helium diffuses out of a toy balloon gradually and the balloon loses its buoyancy, there is a similar effect called "static diffusion" for microbubbles. No sound is involved. Typical times for this process for a 2-μm-diameter bubble filled with gas are the following: air, 25 ms; C_3F_5, 400 ms; and C_4F_{10}, 4000 ms. The second mechanism is acoustically driven diffusion. Factors involved are the gradient of gas across the shell into the surrounding fluid, the initial radius, and the dynamic modal shapes of microbubbles that contribute to a convective effect. Because a contrast agent has a shell, and often a heavier gas, the balance between compressional and rarefactional phases is different than that for a free air bubble; so, unlike rectified diffusion, gas is moved out of the bubble and the diameter shrinks over time. Instead of being pumped up, as in rectified diffusion, encapsulated microbubbles are pumped down. From their numerous experiments, Chomas et al. (2001) have observed that a single ultrasound pulse is enough to set the acoustically driven diffusion process in motion.

To summarize what seems to be a complicated set of interactions involving the applied pressure field characteristics, as well as contrast agent structure and size, it is helpful to view the microbubble as a fragile viscoelastic resonator. Just as there are different responses for solid elastic materials, encapsulated microbubbles also have linear and nonlinear ranges, as well as irreversible plastic and inelastic limits to increasing amounts of pressure. In addition, as resonators, they ring depending on their size and the amount of damping of their viscoelastic shells. As the frequency content of the exciting pulse approaches the resonant frequencies of the contrast agent, the effects of excitation become magnified and lead to larger displacements or a larger ratio of maximum-to-minimum microbubble diameters. Either acoustically accelerated diffusion can occur or, for large maximum-to-minimum diameter ratios, the shell can fragment. Freed gas can cause short-lived, elevated levels of backscatter and then dissolve. Chen et al. (2002) have observed that contrast agents behave differently once the shell is fragmented. They found multiple resonance peaks after insonification of the remains of the air-filled agent that they believed corresponded to resonances from a distribution of bubbles and inertial cavitation events. They concluded that a second agent, filled with a heavier gas, fragmented into tiny subresonant bubbles (< 0.3 μm) with no inertial cavitation events.

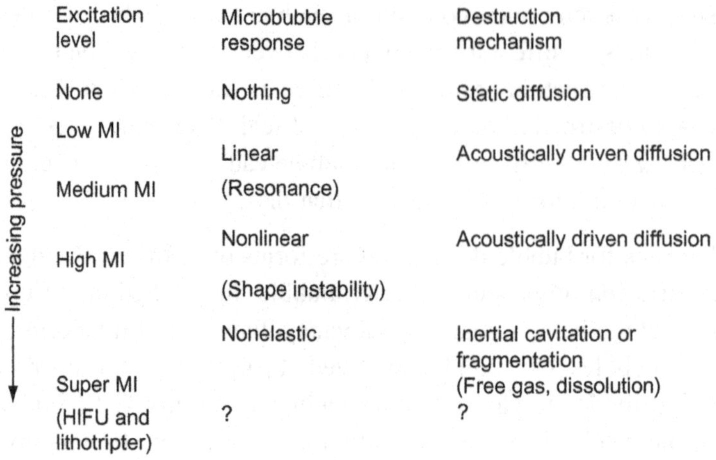

Increasing pressure ↓

Excitation level	Microbubble response	Destruction mechanism
None	Nothing	Static diffusion
Low MI		
	Linear	Acoustically driven diffusion
Medium MI	(Resonance)	
	Nonlinear	Acoustically driven diffusion
High MI		
	(Shape instability)	
	Nonelastic	Inertial cavitation or fragmentation
Super MI		(Free gas, dissolution)
(HIFU and lithotripter)	?	?

Figure 14.15
Diagram of Causes of Ultrasound Contrast Agent Destruction.

These factors are diagrammed in Figure 14.15. On the left is a scale of increasing incident pressure levels from zero to the extremely high levels used for lithotripsy (kidney stone fragmentation) or HIFU for surgery; these are matched up to a series of mechanical factor thresholds and their destructive consequences. The way these thresholds align with absolute pressure levels depends on the encapsulated microbubble structure, or in other words, the mechanical response scale is flexible and unique to the agent type. A practical consequence of this matchup is that imaging system manufacturers are implementing individual protocols tailored to different types of agents. For example, a particular destructive effect such as fragmentation may occur at different pressure thresholds for each agent. With different excitation pulses and timing intervals, a range of effects can occur (as described in the next section on ultrasound contrast agent imaging). Deliberate destruction of contrast agents for targeted drug delivery is discussed in Section 14.7.

14.5.4 Secondary Physical Characteristics of Ultrasound Contrast Agents

Characterization of ultrasound contrast agents can be complicated because of interrelated factors: the structure and concentration of agent, both spatially and in dilution, and the spatial distribution and harmonic content of the insonifying field, as well as its timing sequence. Ways of identifying the unique properties of each agent, as well as the optimum matching of the acoustics and signal processing, are still under development. A number of associated effects have received less attention and will be discussed here for completeness.

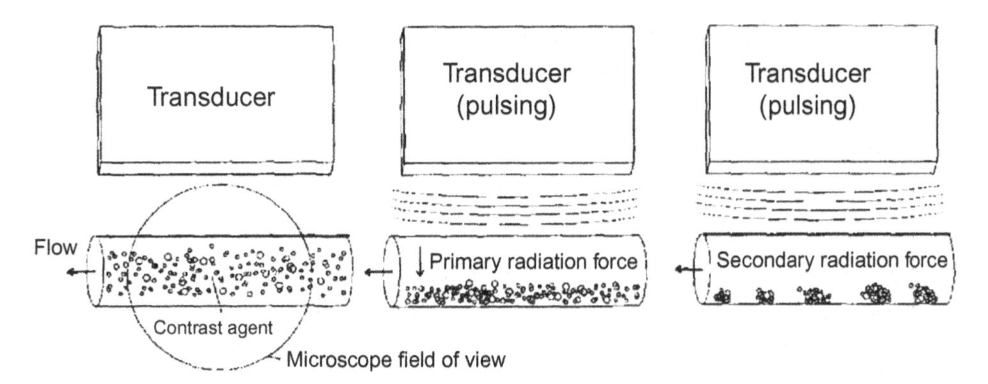

Figure 14.16

Configuration of Ultrasound Transducer Above a Tube Containing Contrast Agents Flowing to the Left, Shown as a Microscope View. (A) No sound; (B) primary radiation force pushes the contrast agent away from the transducer during pulsing; (C) microspheres, which are pushed into closer proximity by primary radiation force under transducer pulsing, aggregate due to secondary radiation force. *From Dayton et al. (1997), IEEE.*

In addition to backscattering, contrast agents can attenuate the sound field enough to block the signal from reaching deeper depths and to cause shadowing. The intersection of the beam with the agent is called the scattering cross-section, $\Sigma_{TS}(f)$, whereas the attenuation is called the extinction cross-section, $\Sigma_{TE}(f)$. Bouakaz, de Jong, and Cachard (1998) have proposed a standard measure of the effectiveness of an agent, the scattering-to-attenuation ratio (STAR). A perfect agent would have high scattering and low attenuation, or a high value of STAR. A definition of STAR suited to measurement is:

$$\mathrm{STAR}(f) = \frac{\sum_{TS}(f)}{\sum_{TE}(f)} = \frac{1}{1 + \sum_{TA}(f)/\sum_{TS}(f)} \tag{14.6}$$

where $\Sigma_{TA}(f)$ is the total absorption cross-section. Another important parameter is the contrast-to-tissue ratio (CTR) (de Jong, Bouakaz, & Ten Cate, 2002), which is the relative amplitude of backscatter from a contrast agent to the backscatter from tissue, and is often considered at a frequency such as the second harmonic (see Figure 14.13 as an example). This topic is addressed in Section 14.6. In addition, there are many other characteristics of contrast agents that can be measured as a function of incident pressure and time and frequency (Bleeker, Shung, & Barnhart, 1990; Shi & Forsberg, 2000), as well as the nonlinearity parameter, B/A (Wu & Tong, 1997), which is two to three orders of magnitude larger than those of tissues.

Several nonlinear-related physical effects, which are more difficult to measure, also occur with contrast agents in fluids. The first is that delicate microbubbles (Wu, 1991) can be pushed around by acoustic radiation forces, like the ones described in Chapter 13. Dayton et al. (1997) have identified two major types of forces (Leighton, 1994): a primary radiation force caused by a pressure gradient that displaces microbubbles (as shown in Figure 14.16B), and a secondary force (illustrated by Figure 14.16C). These secondary, or Bjerknes, forces are caused by pressures reradiated by microbubbles, and they often result in mutual bubble attraction and the formation of aggregates, or bubble clusters.

Another nonlinear effect associated with the inevitable pressure gradients is microstreaming (described in Chapter 12). This version of microstreaming is smaller in scale and occurs near boundaries (Wu, 2002) and around bubbles (Leighton, 1994), and it has been observed near contrast agents. Under certain conditions, another phenomenon related to microstreaming is reparable sonoporation, which is the reversible process of opening and resealing cells in a suspension containing contrast agents in an insonifying field (Wu, 2002). Increased permeability of cell membranes and vessel walls can be used for drug delivery as explained shortly (Ward, Wu, & Chiu, 1999; van Raaij et al., 2011).

Finally, much of the analysis for predicting the response of contrast agents to acoustic fields has been based on linear amplitude-modulated sinusoidal excitation or short pulses. Because of the travel path of the beam to the site of the contrast agent, the beam and waveforms undergo nonlinear distortion and absorption (as described in Chapter 12). Initial research by Ayme (Ayme & Carstensen, 1989; Ayme, Carstensen, Parker, & Flynn, 1986) showed that the response of microbubbles is greater for an amplitude-modulated sinusoid rather than a nonlinearly distorted pressure wave of the same amplitude. Recent studies indicate that when a nonlinearly distorted wave meets a nonlinear encapsulated microbubble, the response of the bubble is not as exciting as if it had met a linear waveform (Hansen, Angelsen, & Johansen, 2001). The simultaneous excitation of the microsphere by a range of frequencies (fundamental and harmonic) creates different vibrational modes and scattered frequencies, and, in general, a more muted response (de Jong et al., 2002). This work provides insight into contrast agent response under more realistic circumstances. Alternatively, through an understanding of bubble response to different excitations, desired bubble behavior can be achieved through a sequence of designed excitations.

14.6 Imaging with Ultrasound Contrast Agents

14.6.1 Introduction

Ultrasound contrast agents were designed with several objectives, other than applications as blood tracers, that enhance sensitivity in small vessels and at deeper penetration depths (Reid et al., 1983). One goal is opacification, which is the visualization or brightening of a blood pool volume. A primary example is the application of this effect to the left ventricle of the heart, or "left ventricular opacification" (LVO). An apical four-chamber view of the heart in Figure 14.17 aids visualization of the problems encountered. In this view, the

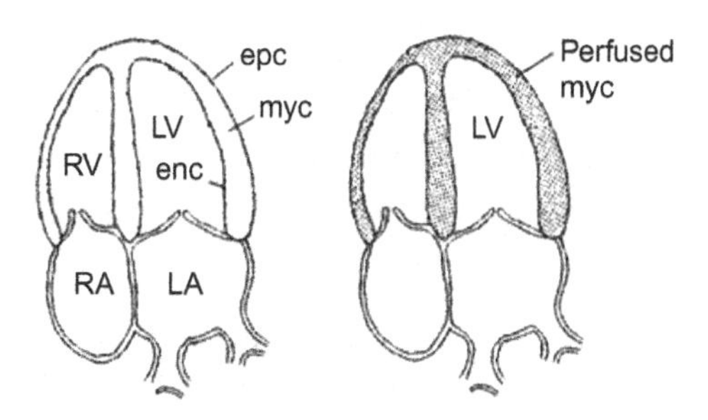

Figure 14.17
Diagram of a Standard Ultrasound Apical Four-chamber View of the Heart.
The transducer would be at the top, or apex. enc, endocardium; epc, epicardium; LA, left atrium; LV, left ventricle; myc, myocardium; RA, right atrium; RV, right ventricle. Myocardium perfused with blood shown with shading surrounding the left ventricle is shown on the right.

transducer would be positioned at the top of the diagram, or at the apex of the heart, and it shines downward. When an ultrasound beam is parallel to a surface of the heart, there is little backscatter (dropout near the apex), and anisotropic effects from the muscle fibers (discussed in Chapter 9), in combination with poor sensitivity caused by body-wall effects, make visualization of the entire chamber difficult, especially through a cardiac cycle. Yet it is important to track this volume of blood to determine the ejection fraction (a measure of the heart acting as a pump) or to identify irregular local wall motions of the endocardium (inner surface of the heart) under stress testing.

Refer to the left diagram of Figure 14.17 to review the cardiac cycle. The heart cycle involves the return of venous blood to the right atrium, which, when filled, flows into the right ventricle, which pumps blood into the lungs. After oxygenation, blood from the lungs fills the left atrium, where the mitral valve releases it to be pumped out by the strong left ventricle, after which the heart relaxes. The instant of maximum left ventricle pumping is called "systole," and the relaxation phase is called "diastole." The main heart muscle, called the myocardium, is sandwiched between the endocardium and epicardium, the inner and outer surfaces of the heart.

A second primary application for these agents is determining and visualizing regions of blood perfusion or the amount of blood delivered into a local volume of tissue (small vessels or capillaries) per unit of time. Abnormalities in the ability of the blood to soak into tissue can be very revealing for diagnosis. In Figure 14.17, the normally perfused myocardium depicted at the left is invisible in terms of ultrasound visualization. Over several heart cycles, a contrast agent enters the circulatory system, and eventually the myocardium, so that this region is made visible to ultrasound imaging (depicted as a shaded area on the right in Figure 14.17). For cardiac applications, measuring perfusion is termed "myocardial echocardiography" (MCE). Regions where blood cannot reach may indicate ischemia (local lack of blood in a region or a regional circulation problem) or an infarct (tissue death). This information is vital for determining the extent of injury from a heart attack and for diagnosing appropriate therapy. Perfusion is also important in other tissues to indicate abnormalities such as angiogenesis or increased vascularization in tumors and the location of lesions.

In order to fulfill these major objectives, the strange and unexpected interactions of ultrasound with contrast agents began to be understood, and ingenious methods of harnessing these characteristics were invented. The application of encapsulated microbubble contrast agents evolved as the physical interactions of ultrasound became better known, the design of agents was improved, and imaging systems and signal-processing methods were adapted and tuned to the unique characteristics of each agent. The elusive "hide-and-seek" game with ultrasound that these agents have played can be interpreted in terms of the mechanisms of encapsulated microbubble destruction (just discussed and diagrammed in Figure 14.15).

14.6.2 Opacification

In order to review some of these developments as applied to LVO imaging, refer to Figure 14.18, which is a series of views of the left ventricle in the imaging plane that was presented by Figure 14.17. To compare images, note the degree of contrast between the myocardium, visible as an upside-down "U" shape, with the interior of the left ventricle. The first generation of encapsulated agents was short lived for LVO applications (Figure 14.18A). First, the gas used (air) dissolved quickly, so microbubbles underwent

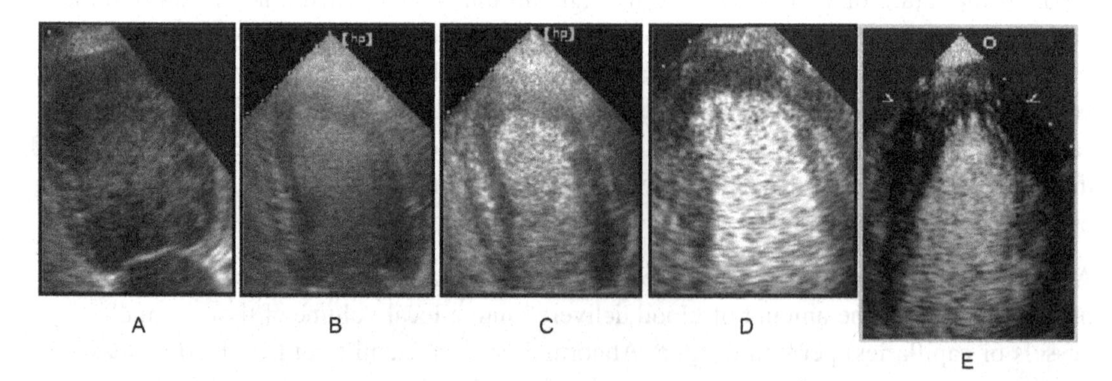

Figure 14.18
Evolution of Ultrasound Contrast Agent Imaging (A) Imaging with first-generation agent at the fundamental frequency; (B) imaging with second-generation agent at the fundamental; (C) imaging with second-generation at the second harmonic; (D) imaging with improved transducer field at the second harmonic. (E) imaging with tissue-subtracting signal processing (power modulation). *Courtesy of P. G. Rafter, Philips Healthcare.*

destruction by static diffusion without any help. Second, under the usual pressure levels used for B-mode imaging, agents were rapidly destroyed by fragmentation. These observations led to a triggered method, in which the transmit pulses were spaced out over longer time intervals so that the contrast agent had time to be replenished before being destroyed again (Porter & Xie, 1995). As seen in Figure 14.18B, even with the next generation of agents, contrast between the agents and the surrounding tissue, though slightly improved, was not dramatic. Work on second harmonic imaging, based on the higher B/A nonlinearity of the microbubbles compared to tissue, did improve contrast at the second harmonic frequency (Schrope et al., 1992; Schrope and Newhouse, 1993) (as shown by Figure 14.18C). Unfortunately, the degree of contrast was not as great as expected because of the nonlinearity of tissues. Advances in transducer technology, a lower fundamental frequency, a smoother transmitted field pattern, and newer contrast agents led to an improved image (Rafter, Perry, & Chen, 2002), as depicted in Figure 14.18D.

Finally, application of a tissue-minimizing signal-processing method, here power modulation (Brock-Fisher, Poland, & Rafter, 1996) in conjunction with power Doppler imaging (Burns et al., 1994) and system settings matched to the particular contrast agent in terms of frequency, transmit level, and pulse interval, has resulted in a long-persisting strong contrast effect with good endocardial border definition (depicted in Figure 14.18D). Here, a low pressure or MI level is applied to avoid fragmentation so that the main physical effect is acoustically driven diffusion.

14.6.3 Perfusion

For the second major application, MCE, a different strategy is needed to detect low concentrations of agent in the myocardium. One real-time approach is to let the contrast agent enter the left ventricle and eventually the myocardium at a low transmit pressure (MI) level with a signal-processing method, such as power modulation or pulse inversion, to enhance the low levels of agent flowing into the myocardium. The time sequence of power Doppler images at the bottom of Figure 14.19 shows the progressive increase of agent in the myocardium from dark (none) to bright (saturation). The myocardium can be mapped into a number of contiguous segments, each corresponding to a zone mainly supplied by a particular coronary artery. By measuring the acoustic intensity at a region of interest (ROI, shown in lower left of Figure 14.19) in a particular segment as a function of time, a time intensity curve can be drawn (as depicted at the top of Figure 14.19). This curve has the form:

$$I = I_0[1 - exp(-bt)] \tag{14.7}$$

where b is a constant to be determined empirically. Wei et al. (1998) have shown that the initial slope of this curve is myocardial blood flow (MBF), and the plateau region is proportional to the myocardial blood volume (MBV); these are important characteristics of

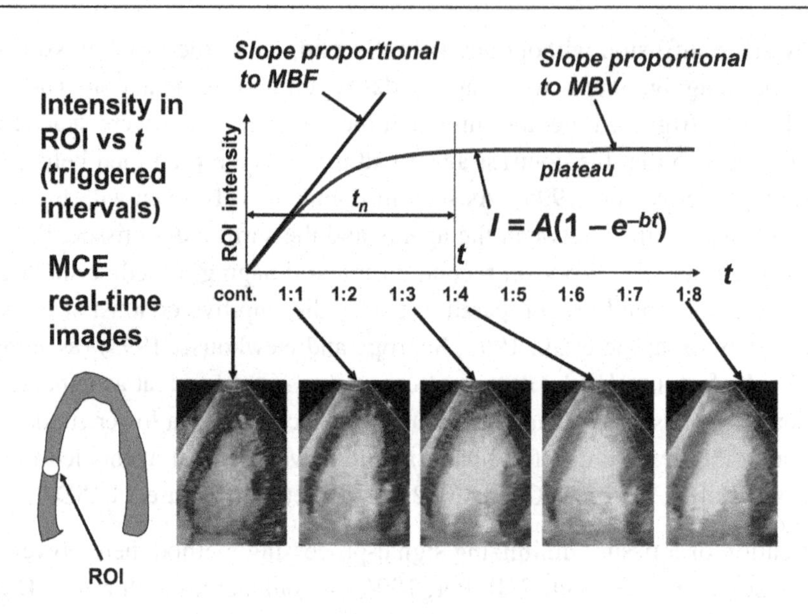

Figure 14.19
Low-pressure (MI) Real-time Myocardial Perfusion Imaging Method. (Top) graph of region of interest (ROI) intensity versus time perfusion filling curve, showing initial slope proportional to myocardial blood flow (MBF), a plateau region with a slope proportional to myocardial blood volume (MBV), and a time (t_n) to reach the plateau. Time is in triggered-interval ratios such as 1:8, meaning an interval eight times the basic unit with reference to initial administration of contrast, depicted as "cont." (Bottom left) insert highlights ROI for intensity measurement. (Bottom right) time sequence series of left ventricle views depicting perfusion of the myocardium and beginning with contrast agent entering the left ventricle. The reader is referred to the Web version of the book to see the figure in color. *Courtesy of P. G. Rafter, Philips Healthcare.*

perfusion. The constant b can be found from the time required to reach the plateau region. Analysis of video contrast data over time is called "videodensitometry."

Power Doppler provides an apparent sensitivity gain for these applications over color flow imaging (CFI) because objectionable low signal levels are mapped to low intensities (as described in Section 11.7.4). Harmonic power Doppler has an additional benefit: a CTR improvement over B-mode harmonic imaging, since a harmonic tissue signal is suppressed with a wall filter. Changes in microbubble scattering or movement are detected through correlation, pulse to pulse (as described in Chapter 11), especially during shell fragmentation and agent replenishment. Tissue movement is displayed also, so careful triggering and pulse timing are needed to minimize these effects.

14.6.4 Other Methods

An alternative triggered, but not real-time, method is to deliberately destroy the contrast agent microbubbles at a high pressure (MI) level so that fragmentation occurs (Wei et al.,

1998). With new agents, an elevated echogenicity occurs briefly as free gas is exposed after fragmentation (as described in Section 14.5.2). After the remaining free gas dissolves, the return of fresh microbubbles at the second harmonic can be measured as a function of time at a position in the myocardium. In this method, the transmit frames are triggered by the electrocardiograph (ECG) waveform with a pair of pulses, one for destruction and the next for imaging, that are separated by several heartbeats. Because bubble destruction takes place, sufficient time is needed between imaging frames to let these effects settle; these long time intervals are why this is not a real-time approach. Because the triggering interval must be changed in this method, the overall time is approximately three times longer than the real-time approach.

Another method, called "release burst imaging" (Frinking, Bouakaz, Kirkhorn, Ten Cate, & de Jong, 2000), uses destructive pulses alternated with imaging pulses. This method provides independent control of the two types of pulses, with the release pulse causing fragmentation and enhanced scattering (as described in Section 14.5.2).

Other approaches take advantage of differences in the spectral responses of microbubbles and tissue. Unlike linear responses, spectra for waves in or reflected by nonlinear media change shape with the pressure amplitude level. Therefore, at any given frequency, there is a difference in amplitude between a contrast agent and a tissue that is a function of the pressure-drive level. An example of exploiting this effect is to lower the pressure level until the tissue harmonic signals are barely detectable but the scattering from the more nonlinear contrast agent microbubbles is still visible; therefore, agent-to-tissue contrast is increased over what it was at a higher pressure level (Powers et al., 2000). This reduction in pressure also minimizes bubble destruction. Operating at low pressures has another advantage: Microbubbles remain nonlinear, but tissue falls into a linear region so that tissue removal signal processing, such as pulse inversion or power modulation, can be applied.

Contrast also improves in regions above the second harmonic (Bouakaz, Frigstad, Ten Cate, & de Jong, 2002; Rafter et al., 2002). As shown in Figure 14.20, bandpass receive filters can be placed, for example, at either of two regions with larger contrast. Capturing these regions requires a transducer with either an extremely wide bandwidth or special construction. Another way of improving contrast is to utilize subharmonics and ultraharmonics (noninteger multiples or submultiples of the fundamental), which are generated by certain agents (Frinking et al., 2000; Shankar, Krishna, & Newhouse, 1998; Shi, Hoff, & Forsberg, 2002) and not by tissue (as illustrated in the bottom of Figure 14.20).

Different methods can be combined. For example (Powers et al., 2000), a high-pressure transmit pulse was followed by low-pressure transmit pulses for stable real-time imaging of the contrast agent replenishment combined with pulse inversion signal processing. Here, the initial pulse in a sequence causes bubble fragmentation, and low pressure in subsequent

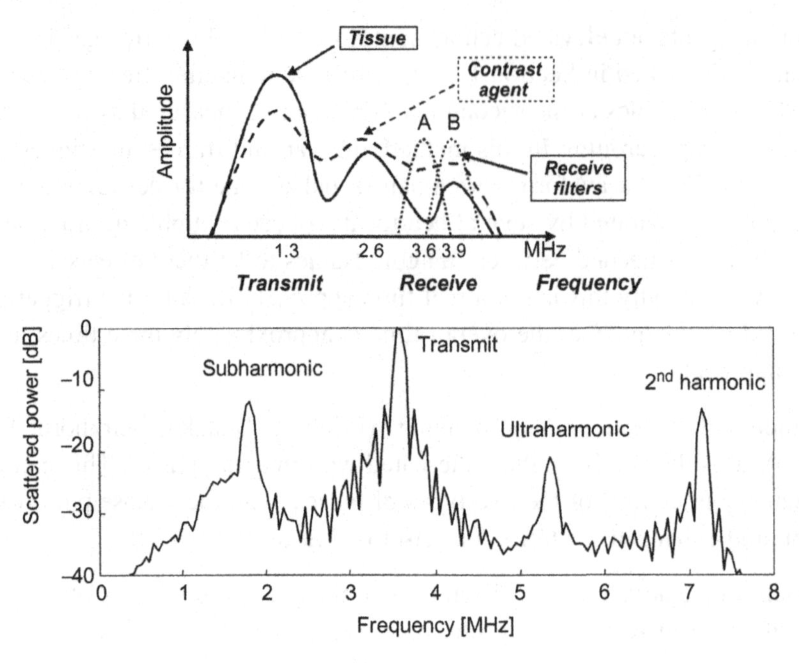

Figure 14.20

Methods for Improving Contrast. (Top) pulse–echo spectral responses for contrast agent and tissue. Transmit at 1.3 MHz, second harmonic at 2.6 MHz, and third harmonic at 3.9 MHz. Receive filters placed at (A) null between second and third harmonic or (B) at the third harmonic. (Bottom) scatter spectral response of contrast agent SonoVue, showing subharmonics, ultraharmonics, and harmonics. The transmitted waveform was a 40-cycle long toneburst at 3.5 MHz at a peak negative pressure of 75 kPa. *(Top) courtesy of P. G. Rafter, Philips Healthcare; (bottom) from Frinking et al. (2000); reprinted with permission of the World Federation of Ultrasound in Medicine and Biology.*

pulses enhances the contrast between scattering from the contrast agent and surrounding tissue and also provides a means for real-time perfusion studies.

14.6.5 Clinical Applications

Contrast-agent-enhanced imaging is not restricted to cardiac applications. Wilson et al. (2008) explored the benefits of the equivalent of a long sonic time exposure in a method called "temporal maximum intensity projection" (MIP) imaging. They devised a way of tracking the time history of bubble paths over seconds or a breath hold, to reveal a detailed map of vascular morphology in the liver, as illustrated by Figure 14.21. Measurement of perfusion is also not limited to cardiac applications, but can be extended to other organs (Forsberg et al., 2000; Kono et al., 1997; Mor-Avi et al., 2001; Powers et al., 2000) such as the liver (the effects of a similar methodology are illustrated in Figure 14.22). Here, tumors do not absorb contrast agents as well as surrounding tissue, and a large metastasis is

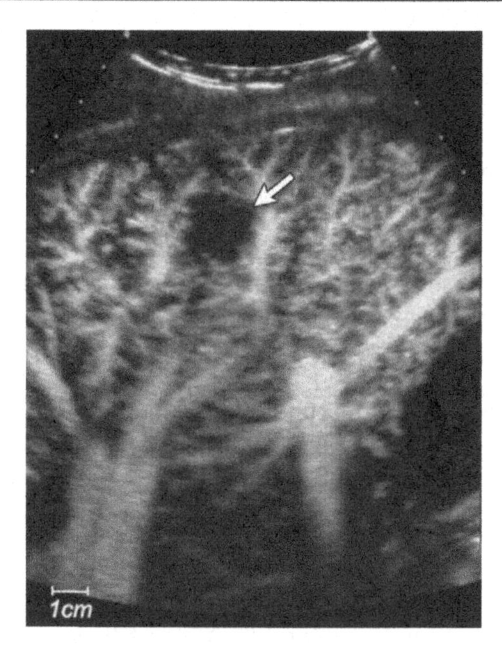

Figure 14.21
Temporal MIP of Normal Liver Vasculature Shows Accumulated Enhancement 11 Seconds after Contrast Material Arrives in Liver. Unprecedented depiction of vessel structure to fifth-order branching is evident. Focal unenhanced region (arrow) is a slowly perfusing hemangioma. *Reprinted with permission from Wilson et al. (2008), American Journal of Roentgenology.*

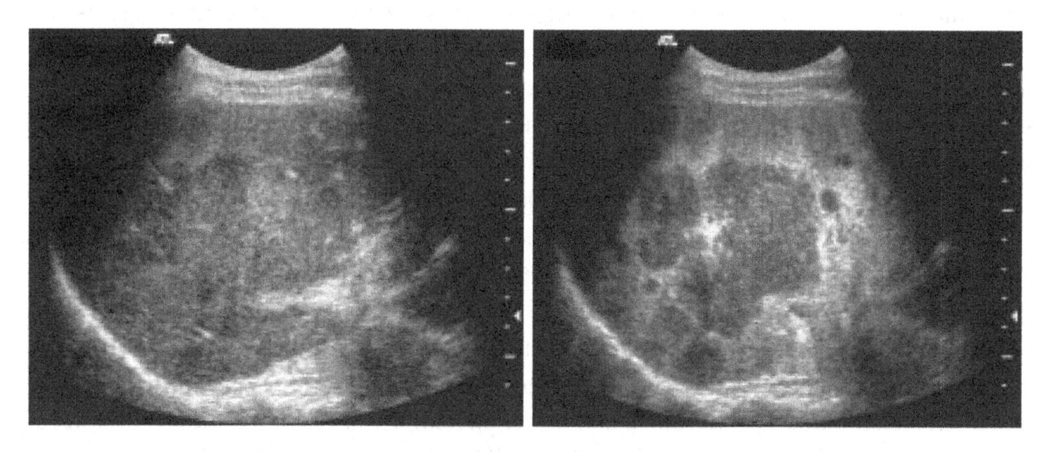

Figure 14.22
Contrast-enhanced Imaging of the Liver. (Left) liver imaged with the conventional fundamental method does not show any focal lesions in a 63-year-old man; (right) same liver with contrast agent, pulse inversion harmonic imaging, and late-phase method reveals a large metastasis (large dark region with well-defined border) and several lesions. *Reproduced with permission from Powers et al. (2000), Philips Healthcare.*

obvious as a dark region with smaller lesions. This example highlights the fact that methods used for contrast-enhanced echocardiography can be applied elsewhere.

Wilson and Burns, in their review of ultrasound contrast-enhanced body imaging (2010), emphasize diagnosis of the liver (particularly liver mass characterization), and discuss applications in the bowel, kidney, pancreas, prostate, organ transplantation, abdominal trauma, and adnexal disease. Other applications include the lymphatic nodes, perfusion through microvascular flow estimates, inflammation, and brain tumors (Bloch, Dayton, and Ferrara, 2004). Specialized ultrasonic contrast agents have been devised for animal studies (Needles et al., 2010) and high frequencies (Goertz, Frijlink, et al., 2006).

The benefits of contrast-enhanced ultrasound imaging include the lack of risk of nephrotoxicity or ionizing radiation (Wilson and Burns, 2010). Besides diagnosis, this approach also includes the guiding of interventional procedures and patient management such as monitoring the responses to different therapies. They point out that microbubble methodologies are easy to apply and robust, and have a safety record comparable to that of analgesics and antibiotics, and involve less risk than CT imaging (Wilson and Burns, 2010). More information can be found in Cosgrove (2006) and Klibanov (2006).

14.7 Therapeutic Ultrasound Contrast Agents: Smart Bubbles

14.7.1 Introduction to Types of Agents

Who would have anticipated that, after the first generation of gas-filled contrast agents was fragmented by ultrasound, encapsulated microbubbles would eventually be designed deliberately for destruction? A new breed of targeted contrast agents has been under development to carry drugs to targeted sites where medication can be released by ultrasound-induced fragmentation. Most forms of medication administered orally or intravenously have a systemic effect throughout the body; consequently, larger amounts than necessary for the intended region must be administered to compensate for dilution and waste. Unwanted side effects are often the result. There is also the possibility that the medication will not reach the intended site. Therapeutic ultrasound contrast agents may be able to deliver precisely the needed amount of medication to the intended site. This prospect would be an exciting breakthrough to which ultrasound could contribute. Collectively, these techniques that enable contrast-enhanced ultrasound to image and affect physiology are called a form of "molecular imaging." Combining imaging with therapeutic effects is often designated "theranostics."

Lindner (2001) and Hughes et al. (2003) review strategies for targeting therapeutic microbubble agents to desired sites. Some of these methodologies are depicted in Figure 14.23. Two approaches to guide encapsulated microbubbles to target cells are electric field attraction and conduction (Wong, Langer, & Ingber, 1994), shown in

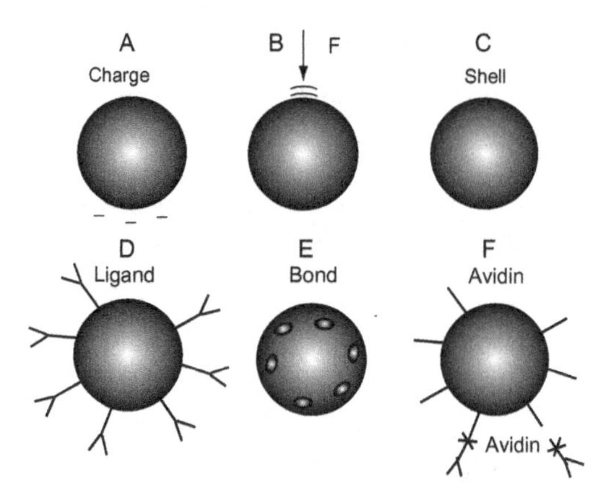

Figure 14.23

Methods for Delivering Therapeutic Contrast Agents. (A) Electrical field attraction; (B) acoustic radiation force; (C) delicious or attractive shell material; (D) ligands tethered to agent surface seek bioconjugation with target receptors; (E) active material covalently bonded to agent surface; (F) bonding by biotin–avidin linking.

Figure 14.23A, and acoustic radiation forces (Dayton, Klibanov, Brandenburger, & Ferrara, 1999b), shown in Figure 14.23B. Depicted in Figure 14.23C, target cells have an affinity for the shell material (such as albumin or a lipid). Strong covalent bonds can be created between ligands anchored to the agent membrane surface and target receptors. Ligands, shown in Figure 14.23D, are seeker molecules (atoms, ions, or radicals) that have the potential of forming a complex with the right material. Ligands (Hughes et al., 2003) include drugs, antibodies, proteins, and viruses. An important application of the antibody ligand is to seek out and neutralize specific disease invaders or antigens. Material, such as DNA, can be covalently bonded onto the surface of the agent (Unger et al, 2003), as illustrated by Figure 14.23D. Finally, the agent can be linked to fibrin (in a thrombus) through an avidin–biotin connection (Lindner, 2001). In this method (see Figure 14.23F) a biotin antibody is released to attach to the target cell such as fibrin, followed by avidin and an agent with a biotinylated phospholipid encapsulation.

Although these microbubbles may be designed primarily for drug delivery, a secondary function might be to aid in the identification of the release areas. Smart bubbles could be prepared to recognize specific antigens and to reveal locations and extent of disease, as well as to deliver appropriate therapy. Drugs can be hidden inside the agent, be contained in the outer membrane or in multiple layers, be bonded covalently, or be dangling on the ends of tethers (Unger et al., 2003).

Commercially available targeted microbubble ultrasound contrast agents have been tabulated in Table 14.2 (Faez et al., 2013). More about how these agents work and the intended targets will be discussed shortly.

Table 14.2 Commercially Available Targeted Contrast Agents.

Name	Manufacturer	Linker	Ligand	Target	Possible Application
Sonazoid	Amersham Health	NA[1]	NA[1]	Kupffer cells and macrophages	Liver tumors, ischemia/reperfusion injury
Micromarker	Bracco SpA	Streptavidin	Biotinylated ligand of choice	Biomarker of choice	Depends on ligand
Targestar	Targeson Inc.	Streptavidin	Biotinylated ligand of choice	Biomarker of choice	Depends on ligand
Visistar Integrin	Targeson Inc.	Covalent to PEG-lipid	Cyclic RGD peptide	$\alpha_\nu\beta_3$ integrin	Angiogenesis
Visistar VEGFR2[2]	Targeson Inc.	Covalent to PEG-lipid	Cyclic RGD peptide	VEGF-like protein	Angiogenesis
YSPSL-MB[2]	Oregon Health & Science University	Covalent to PEG-lipid	PSGL-1	P and E selectin	Ischemia/reperfusion injury, transplant rejection
BR55[3]	Bracco SpA	Covalent to PEG-lipid	Heterodimer peptide	VEGFR2	Angiogenesis
Selectin agent[2]	Bracco SpA	Streptavidin	Biotinylated PSGL-1 analog	P and E selectin	Inflammatory disease

[1]Passive targeting, phagocytic uptake of the bubbles by cells.
[2]In development.
[3]In clinical development.
Source: Faez et al. (2013)

For completeness, there are targeted agents that are not on this list that are still under development. Agents with polymer shells can be fabricated with specific shell thickness/diameter ratios, important for controlling and predicting bubble behavior, notably their destruction threshold (Chitnis, Koppolu, Mamou, Chlon, & Ketterling, 2013). Targeted nanoparticles are agents which have one dimension less than 100 nm and are capable of carrying a therapeutic payload, and they are reviewed by Mullin, Phillips, and Dayton (2013). Other types of therapeutic contrast agents include nongaseous acoustically reflective liposomes, quantum dots, perfluorocarbon emulsion nanoparticles, poly(lactic-coglycolic acid) (PLGA) nanoparticles, and gold particles (Gessner and Dayton, 2010; Hughes et al., 2003).

14.7.2 Ultrasound-induced Bioeffects Related to Contrast Agents

At this point, in anticipation of a fuller discussion of ultrasound-induced bioeffects in the next chapter, we need to introduce those effects relevant to the functioning of targeted contrast agents (Bloch, Dayton, and Ferrara, 2004; Miller et al., 2008). One of the key effects is deliberate local bubble destruction of contrast agents at intended sites, as illustrated by the three cases in Figure 14.24 (Ferrara, Borden, and Pollard, 2007).

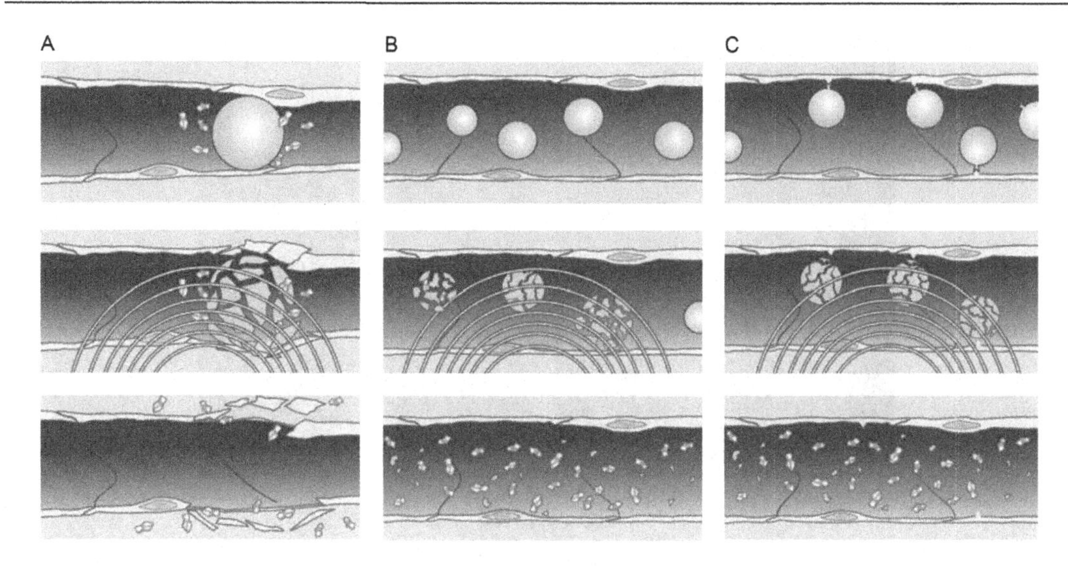

Figure 14.24

Deliberate Local Bubble Destruction of Contrast Agents at Intended Sites. (A) Ultrasound contrast agents are freely circulating in small vessels along with drug particles (blue/light gray irregularly shaped particles surrounding microbubble at top of (A)). Once a sufficiently strong ultrasound pulse is applied to the area, the contrast agent expands, rupturing the endothelial lining. Drug is then able to extravasate. (B) Drug-laden ultrasound contrast agents are freely circulating throughout the vasculature. A pulse of ultrasound is applied and ruptures the contrast agent, thereby liberating the drug payload. Because ultrasound is only applied in the region of interest, drug is preferentially delivered locally. (C) Drug-laden ultrasound contrast agents bearing surface ligands targeted to specific endothelial receptors are freely circulating. The ligand preferentially binds the ultrasound contrast agent in the target region, increasing local agent accumulation. An ultrasound pulse is then applied, liberating the drug payload. The reader is referred to the Web version of the book to see the figure in color. *From Ferrara, Pollard, and Borrden (2007), Annual Review Biomedical Engineering.*

Localized extravasation near burst agents in a rat heart exposed to 1.9 MPa can be seen in Figure 14.25. When first reported by Skyba, Price, Linka, Skalak, and Kaul (1998), they realized the potential benefits of this mechanism for drug delivery. Preferential guidance and arrangement using both primary and secondary acoustic radiation forces is demonstrated in Figure 14.26. Once released, the drug payload continues to be influenced by acoustic waves, especially by a process called "sonoporation," by which acoustic waves affect the permeability of vessel walls and cell membranes, allowing drugs to enter them. A possible explanation for these unlocking mechanisms is depicted in Figure 14.27, by Qin, Caskey, and Ferrara (2009). On the left, vibrations from resonating wall causing shear forces on the endothelium; in the middle, jets from collapsing microbubbles; on the right, external intracellular transport of particles such as released drugs or the transfection of genes. Wu and his group were among the first to describe and quantify sonoporation and its attendant cell lysis, as shown in Figure 14.28 (Ward, Wu, & Chiu, 1999, 2000).

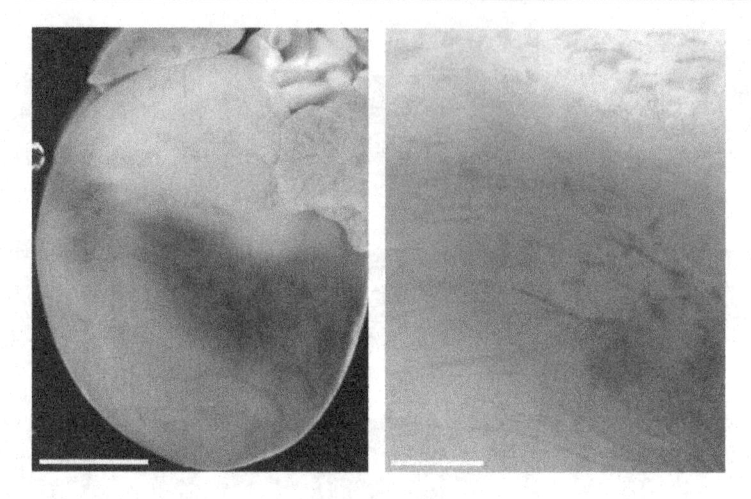

Figure 14.25

Localized Extravasation in a Rat Heart. Photographs of an excised rat heart exposed at 1.9 MPa, 1:4 end-systole triggering, and 100 μl/kg of contrast agent. The petechia and Evans-blue leakage are evident in a band across the myocardium (bottom, scale bar 5 mm). The erythrocyte extravasation (petechiae) and diffuse leakage of the dye are shown in a close-up view of the same heart (top, scale bar 1 mm). The reader is referred to the Web version of the book to see the figure in color. *From Li et al. (2003), with permission from the World Federation of Ultrasound in Medicine and Biology.*

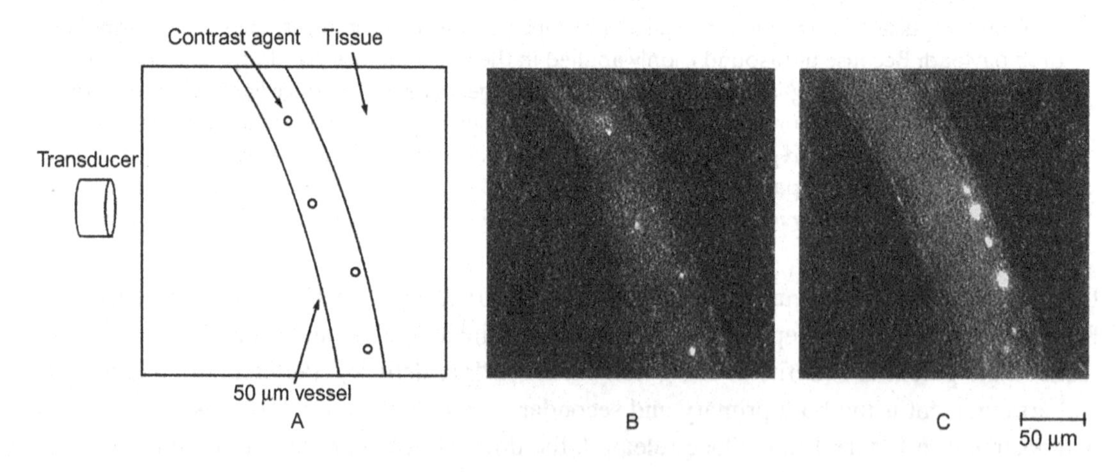

Figure 14.26

Preferential Guidance and Arrangement Using Both Primary and Secondary Acoustic Radiation Forces. (A) Diagram depicting the relative position of the transducer with respect to the vessel in (B) and (C). (B) Fluorescently labeled contrast agents flowing evenly distributed in a 50-mm mouse cremaster arteriole. The velocity of the agents before insonation was approximately 7.5 mm/s. (C) Insonation at a center frequency of 5 MHz, acoustic pressure of 800 kPa, and PRF of 10 kHz produces radiation force that displaces the contrast agents to the right vessel wall and induces bubble aggregation. The velocity of the bubble aggregates has been reduced to less than 1 mm/s. *From Dayton et al. (1999b), with permission from the World Federation of Ultrasound in Medicine and Biology.*

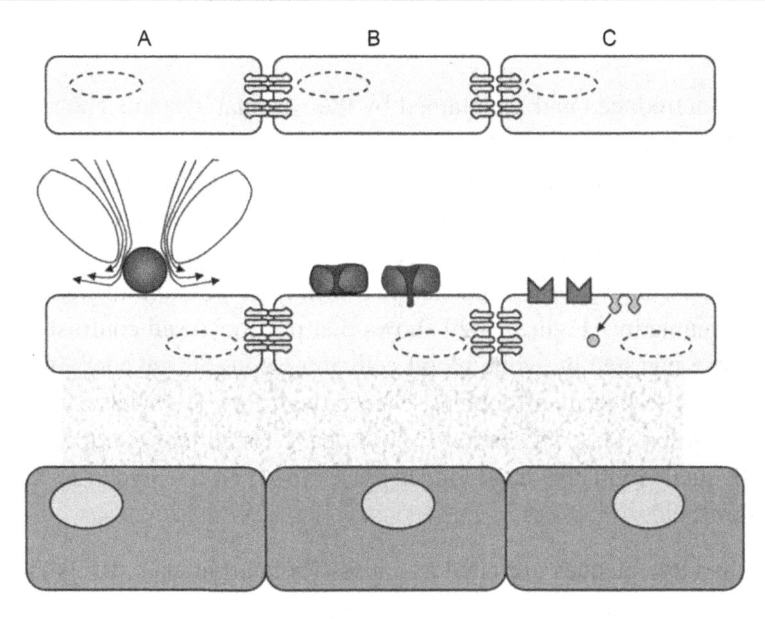

Figure 14.27

Hypothesized Mechanisms of Drug Transport Across Endothelium. (A) Local shear stress created on cell during microbubble oscillation; (B) fluid jet formation; (C) intracellular transport, hypothesized to result from the stresses induced by microbubble activity, including generation of gaps at tight junctions, expression of cell adhesion molecules due to inflammatory process, and the creation of vesicles for transcellular transport. The reader is referred to the Web version of the book to see the figure in color. *From Qin, Caskey and Ferrara (2009), Physics in Medicine and Biology,* *© IOP Publishing. Reproduced by permission of IOP Publishing. All rights reserved..*

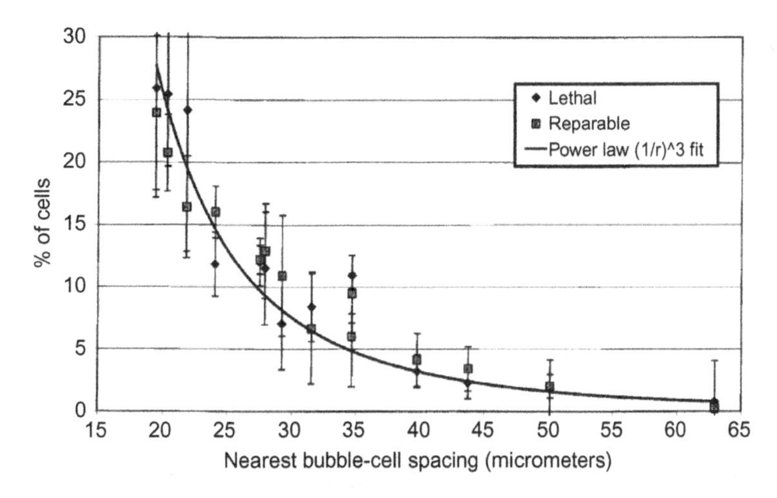

Figure 14.28

Percentage of Cells Exhibiting Reparable or Lethal Sonoporation as Indicated by Fluorescence and Trypan-blue Staining, Respectively. Each data point represents the average and standard deviation of four independent trials. A 1/spacing3 function is fit to all of the data, with an R-value of 0.9448. *From Ward et al. (2000), with permission from the World Federation of Ultrasound in Medicine and Biology.*

14.7.3 Targeted Contrast Agent Applications

Many diseases are introduced and maintained by the vascular system. The research of Dayton et al. (2001) has shown that it is possible to distinguish between the acoustic characteristics of freely circulating contrast agents and those ingested or phagocytosed by white blood cells (leukocytes) that have been activated to attack inflammation. Their method used pulse inversion and frequency shifts to discriminate between free and phagocytosed bubbles. These results indicate that it may be possible to find the extent and degree of inflammation. Figure 14.29 shows that phagocytosed contrast agent microbubbles (those ingested by white blood cells in response to invaders, the agents) follow a trend similar to free microbubbles, even though they are encased in a higher-viscosity medium. Allen, May, and Ferrara (2002) have found that therapeutic contrast agents with thick shells (500 nm) need longer pulses (not just one cycle) to reach a fragmentation threshold.

Thrombi and vulnerable plaques are cited as causes for heart attacks, stroke, and death. Other early work shows that targeted agents can find and bind with blood clots (Lindner, 2001). The potential is there for developing agents that can not only locate the thrombi, but also deliver a therapeutic payload upon fragmentation. Recent work demonstrates the synergistic effect of acoustic radiation forces and cavitation for enhanced sonothrombolysis (Chuang, Cheng, & Li, 2013). Another interesting application is agents that deliver cytotoxic drugs to regions of angiogenesis (growth of new blood vessels) that feed tumor growth (Hughes et al., 2003; Lindner, 2001; Unger et al., 2003).

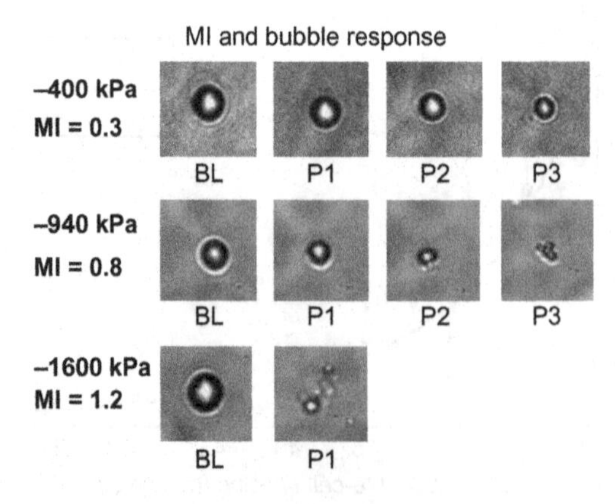

Figure 14.29
Time Sequences of Phagocytosed Microbubbles at Three Pressure Levels of Insonification. One cycle of insonification at 2.25 MHz. BL, baseline image; P1–P3, bubble snapshot with a time sequence increasing left to right. *From Lindner (2001), IEEE.*

In their detailed study of sonoporation, Escoffre et al. (2013) were able to achieve a gene transfection of nearly 70% by using a smaller-diameter agent, Vevo Micromarker (Needles et al., 2010), compared to more standard-sized microbubbles, as well as a low attenuation rate. Cell mortality rates were about 10–12% across the three agents tested with ultrasound, and was 9.5% without them. Van Ruijssevelt et al. (2013) found that with low-intensity ultrasound, cell viability for sonoporation could last up to 24 hours, with reversible cell membrane asymmetry.

One of the greatest drug delivery challenges is accessing the brain (Burgess and Hynynen, 2013). Protected on the outside by a hard skull, the brain is further isolated by the blood–brain barrier (BBB), a nearly impermeable barrier separating the blood from the brain parenchyma. As result, despite widespread brain diseases and disorders, the BBB prevents access of ~98% of current pharmaceutical agents to the brain when delivered intravenously.

The BBB is different from other body barriers because of the tight junctions between adjacent endothelial cells. Hynynen, McDannold, Vykhodtseva, and Jolesz (2001), after years of trying to open the BBB, were able to disrupt it by using an ultrasound contrast agent in combination with ultrasound focused on the barrier, with only 1 MPa of in situ pressure. Furthermore, the process was reversible with no damage to surrounding tissue. Mechanisms, including the ones discussed here, were investigated, with the conclusion that there is still more to be learned. Further work by Hynynen's group demonstrated drug delivery through the BBB and also targeted contrast agents. They believe that, after extensive animal studies, the process is ready to be tried on humans (years of work reviewed in Burgess and Hynynen (2013)).

14.8 Equations of Motion for Contrast Agents

Models for contrast agent microbubble characteristics tend to fall into three groups of nonlinear equations that must be solved numerically. The theoretical basis for oscillating gas bubbles began with Lord Rayleigh (1917), who first derived an equation describing their behavior after wondering about sounds emitted by water in a tea kettle as it comes to a boil. Since then, the original equation has been improved on and some believe it should be called the Rayleigh–Plesset–Nolting–Nepiras–Poritsky equation for those who have contributed to its evolution (Leighton, 1994). The abbreviated name for this nonlinear equation of motion is the Rayleigh–Plesset equation. The next group consists of modifications to this basic equation to account for shell and other damping effects, as well as shell forces. While not derived from first principles, this type of model, in which the shell is assumed to be extremely thin, often depends on experimentally derived parameters and is quite useful. The last group is the type of model that accounts for the finite thickness of the shell and forces, as well as the elastic nature of the shell in a more formal way.

The key variables for the Rayleigh–Plesset equation are a spherical bubble of radius R_0, filled with gas, floating in an incompressible fluid with a hydrostatic pressure p_0, acted on by a time-varying input pressure field, *P*(*t*). The internal pressure is a combination of the gas pressure, p_g, and the liquid vapor pressure, p_v:

$$p_i = p_g + p_v = 2\sigma/R_0 + p_v \tag{14.8}$$

where the inwardly directed surface tension pressure is $p_\sigma = 2\sigma/R_0$. The internal pressure is subject to the gas law, so that the pressure just beyond the bubble wall (Leighton, 1994) is:

$$p_L = (p_0 + 2\sigma/R_0 - p_v)\left(\frac{R_0}{R}\right)^{3\kappa} + p_v - 2\sigma/R \tag{14.9}$$

where $R = R(t)$ is the dynamic radius of the bubble to be found, and κ is the polytropic gas index. The overall equation is:

$$\ddot{R}R + \frac{3\dot{R}^2}{2} = \frac{1}{\rho}\left(p_L - \frac{4\eta\dot{R}}{R} - P(t)\right) \tag{14.10A}$$

where each dot represents a derivative, ρ is the liquid density, and the forcing function, *P* (*t*), and a damping term for the shear viscosity of the fluid, η, have been added. Substituting Eqn 14.9 into Eqn 14.10A provides a more familiar form of the Rayleigh–Plesset equation (Eatock, 1985):

$$R\ddot{R} + \frac{3\dot{R}^2}{2} = \frac{1}{\rho}\left((p_0 + 2\sigma/R_0 - p_v)\left(\frac{R_0}{R}\right)^{3\kappa} + p_v - p_0 - 2\sigma/R - \frac{4\eta\dot{R}}{R} - P(t)\right) \tag{14.10B}$$

The first, second, and next-to-last terms are nonlinear. Leighton (1994) discusses the limitations of this equation, which is applicable to a spherically symmetrical free gas bubble in an incompressible fluid, and other alternative equations. The major shortcoming of this approach for application to contrast agents is the missing shell.

The shell has the effect of increasing the overall mechanical stiffness of the contrast agent, and shell viscosity increases sound damping. Two primary modifications of Eqn 14.10B can be made to account for the extra damping of the shell and the restoring force of the shell (de Jong, 1993; de Jong & Hoff, 1993). The damping from the viscosity damping of the fluid is supplemented by other sources of damping (de Jong, 1993; de Jong et al., 1992) to give a total damping parameter:

$$\delta_t = \delta_{vis} + \delta_{rad} + \delta_{th} + \delta_f \tag{14.11A}$$

where δ_{vis} is the viscous damping, δ_{rad} is reradiation damping, δ_{th} is thermal conduction damping, and δ_f is damping due to friction within the shell. Finally, another term for a shell-restoring force includes a shell elastic parameter (S_p). The modified equation of motion is:

$$\rho R\ddot{R} + \rho\frac{3\dot{R}^2}{2} = p_{g0}\left(\frac{R_0}{R}\right)^{3\kappa} + p_v - p_0 - 2\sigma/R - S_p\left(\frac{1}{R_0} - \frac{1}{R}\right) - \delta_t\omega_0\rho R\dot{R} - P(t) \tag{14.11B}$$

where p_{g0} is the initial pressure inside the bubble and ω_0 is the center frequency of the excitation pressure waveform. In this semi-empirical approach, both S_p and δ_f are determined by measurement (de Jong & Hoff, 1993). The effect of the shell stiffness on resonance frequency was discussed in Section 14.5.1. Several figures, including Figures 14.2, 14.4, 14.10, and 14.11, were generated by this model.

A more accurate description of the effects of the shell has been presented by Hoff, Sontum, and Hovem (2000), based on Church (1995), who introduced a model in a more formal way to account for shell effects after studying the model of Roy, Madanshetty, and Apfel (1990). The shell thickness can be modeled as a viscoelastic layer that changes thickness in proportion to the stretching of the dynamic radius. Two shell parameters, the shear modulus (G_s) and the shear viscosity (μ_s), can be determined from measurements.

For therapeutic contrast agents, a different approach is required for their thicker fluid shells. Allen et al. (2002) have generalized the Rayleigh–Plesset equation for a liquid shell of arbitrary thickness, viscosity, and density. Here, more accurate descriptions of spherical oscillations and the dynamic changes of shell thickness are required for predictions of bubble instability and estimates of fragmentation thresholds.

Many more evolutions of models for contrast agents have been detailed in the excellent review of these recent developments over a 20-year time span by Faez et al. (2013). Other worthwhile model reviews are by Qin, Caskey, and Ferrara (2009) and Doinikov and Bouakaz (2011). Concurrent with ultrasound contrast agent model evolution has been the development of extremely high speed bubble cinematography, which has allowed model validation. An example of this close agreement for subharmonic excitation of a 3.8-μm microbubble can be seen in Figure 14.30 (Sijl et al., 2010).

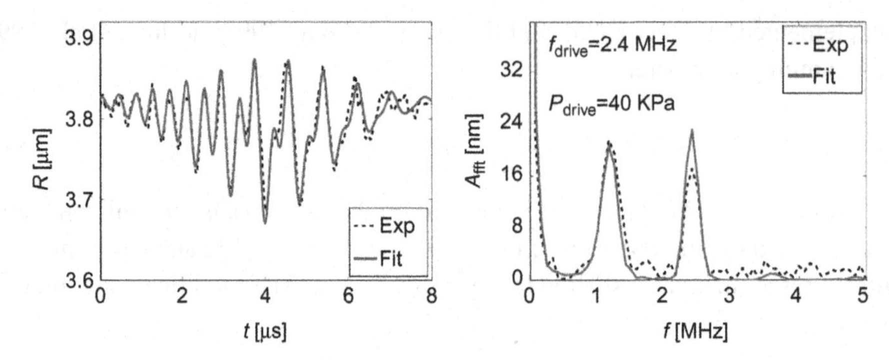

Figure 14.30
Data and Model for a Subharmonic Driving Excitation of 2.4 MHz, About Twice the Resonance Frequency of the Bubble. The best fit of the model proposed by Marmottant et al. (2005) with the shell parameters χ_{max}=2.5 N/m, $\varsigma = 2000$ N/m, $\kappa_s = 3 \times 10^{-8}$ kg/s, and $\sigma(R_0) = 0.001$ N/m, both in (A) the time domain and (B) in the frequency domain. Sampling rate for both curves 50 MHz, 501 points. *From Sijl et al. (2010), IEEE.*

14.9 Conclusion

From their accidental discovery to commercial realization, ultrasound contrast agents have proved to be a difficult technology to implement. Realization of clinically successful applications has been hard won, through control of microbubble physics and new agent materials, adaptive signal-processing techniques, and harmonic imaging. Working through the vascular system, contrast agents have opened new windows for diagnosis, revealing anomalies in circulation and disease states. They have also improved sensitivity and made possible clinically useful images for a portion of the population previously written off as too difficult to image by conventional techniques. Ultrasound contrast imaging is still under development and growth as the uniqueness of each agent; its optimal ultrasound excitation; appropriate signal processing for different applications; and clinical efficacy, safety, and suitability are explored. Signal processing has improved; a method for determining the type of cavitation underway has been devised by Vignon, Shi, et al. (2013), as illustrated by Figure 14.31. Already, contrast-enhanced ultrasound is widely used and recommended in echocardiography for left ventricular opacification and myocardial contrast echocardiography perfusion studies (Ducas et al., 2013); and it is the preferred modality for focal liver disease and is expanding to other organs. Perfusion studies of the microcirculation, interventions, and monitoring of therapy are growing (Böhmer et al., 2009).

The potential applications of therapeutic contrast agents promise to create even higher challenges. Multiple uses for these agents will require even greater understanding of the physics of their behavior in order to position and deliver payloads in appropriate targeting

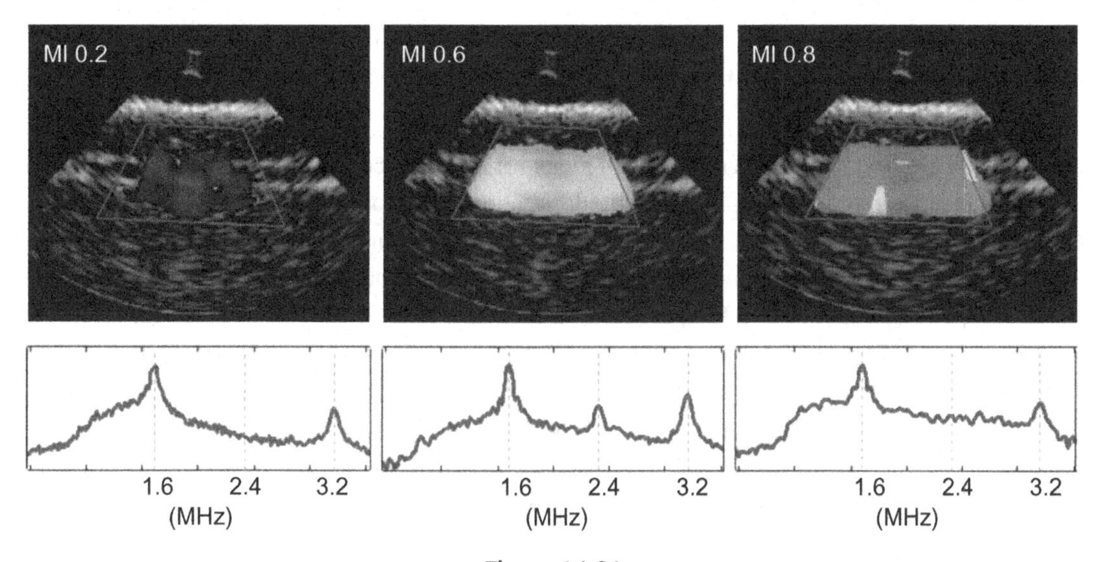

Figure 14.31
Cavitation Images of a Vessel Phantom Through a Water Path at Different MIs. Showing dominant moderate oscillations (left), dominant stable cavitation (center), and dominant inertial cavitation (right). The corresponding MIs are indicated, and the average spectrum over the region of interest outlined with a blue trapezoid is displayed in logarithmic scale. The reader is referred to the Web version of the book to see the figure in color. *From Vignon, Shi, et al. (2013), IEEE.*

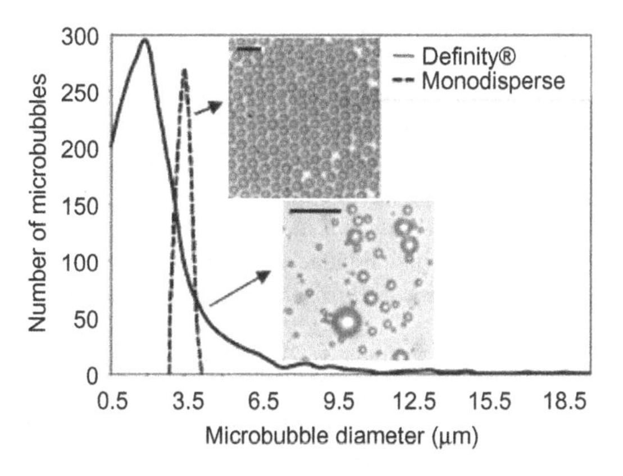

Figure 14.32
Size Distribution of Contrast Agents Manufactured in the Laboratory. Of 3.7 μm with a standard deviation of 0.2 μm in this case, in contrast to the commercially available contrast agent Definity (2.7 μm with a standard deviation of 2.0 μm). The scale bar represents 10 μm. *From Talu et al. (2007),* Molecular Imaging.

sites and to provide adequate imaging sensitivity to reveal locations of different types of disease. The relatively low cost, portability, and general accessibility of ultrasound are key advantages in this new field of molecular imaging.

Gessner and Dayton (2010) reviewed challenges for targeted contrast agents. Some of the issues included limited field of view; limited targeting adhesion, delivery, concentration, and short retention; poor signal-to-noise ratio, and lack of quantification. Significantly improved sensitivity has been achieved by tailoring the distribution of the microbubbles to fit the transducer response and to specific purposes for better control through monodispersive techniques such as applying microfluidic technology to making bubbles of a certain size as shown in Figure 14.32 or utilizing polymer shells. Conformable or wrinkly bubbles are being developed to increase contact area for attachment to vessel walls. Better designs for target specificity and strength and stability have been made. Signal-to-noise has been improved by signal and image processing to discriminate between flowing microbubbles and those attached, as well as by an "image-push-image," a combination of radiation force to enhance delivery and a background subtraction technique. Three-dimensional imaging is being explored to expand the volume studied.

References

Abbot, J. G. (1999). Rationale and derivation of MI and TI: a review. *Ultrasound in Medicine and Biology, 25*, 431–441.

Allen, J., May, D. J., & Ferrara, K. W. (2002). Dynamics of therapeutic ultrasound contrast agents. *Ultrasound in Medicine and Biology, 28*, 805–816.

American Institute of Ultrasound in Medicine/National Electrical Manufacturers Association (AIUM/NEMA) (1998). *Standard for real-time display of thermal and mechanical acoustic indices on diagnostic ultrasound equipment*. Laurel, MD: AIUM Publications.

Anderson, A. L., & Hampton, L. D. (1980). Acoustics of gas-bearing sediments 1: background. *Journal of the Acoustical Society of America, 67*, 1865–1889.

Apfel, R. E. (1984). Acoustic cavitation inception. *Ultrasonics, 22*, 167–173.

Apfel, R. E., & Holland, C. K. (1991). Gauging the likelihood of cavitation from short-pulse, low duty cycle diagnostic ultrasound. *Ultrasound in Medicine and Biology, 17*, 179–185.

Ayme, E. J., & Carstensen, E. L. (1989). Cavitation induced by asymmetric distorted pulses of ultrasound: theoretical predictions. *IEEE Transactions on Ultrasonics, Ferroelectrics and Frequency Control, 36*, 32–40.

Ayme, E. J., Carstensen, E. L., Parker, K. J., & Flynn, H. G. (1986). Microbubble response to finite amplitude waveforms. *Proceedings of the IEEE Ultrasonics Symposium* (pp. 985–988), Williamsburg, VA.

Blackstock, D. T. (2000). *Fundamentals of physical acoustics*. New York: John Wiley & Sons.

Bleeker, H. J., Shung, K. K., & Barnhart, J. L. (1990). Ultrasonic characterization of Albunex, a new contrast agent. *Journal of the Acoustical Society of America, 87*, 1792–1797.

Böhmer, M. R., Klibanov, A. L., et al. (2009). Ultrasound-triggered image-guided drug delivery. *European Journal of Radiology, 70*, 242–253.

Bouakaz, A., de Jong, N., & Cachard, C. (1998). Standard properties of ultrasound contrast agents. *Ultrasound in Medicine and Biology, 24*, 469–472.

Bouakaz, A., Frigstad, S., Ten Cate, F. J., & de Jong, N. (2002). Super harmonic imaging: a new imaging technique for improved contrast detection. *Ultrasound in Medicine and Biology*, *28*, 59–68.

Brock-Fisher, G.A., Poland, M.D., & Rafter, P.G. (1996). Means for increasing sensitivity in non-linear ultrasound imaging systems. US patent 5,577,505, November 26, 1996.

Burns, P.N., Powers, J.E., Simpson, D.H., Brezina, A., Kolin, A., Chin, C.T., et al. (1994). Harmonic power mode Doppler using microbubble contrast agents: an improved method for small vessel flow imaging. *Proceedings of the IEEE Ultrasonics Symposium* (pp. 1547–1550), Cannes, France.

Chen, W. -S., Matula, T. J., & Crum, L. A. (2002). The disappearance of ultrasound contrast bubbles: observations of bubble dissolution and cavitation nucleation. *Ultrasound in Medicine and Biology*, *28*, 793–803.

Chitnis, P. V., Koppolu, S., Mamou, J., Chlon, C., & Ketterling, J. A. (2013). Influence of shell properties on high-frequency ultrasound imaging and drug delivery using polymer-shelled microbubbles. *IEEE Transactions on Ultrasonics, Ferroelectrics and Frequency Control*, *60*, 53–64.

Chomas, J., Dayton, P., Allen, J., Morgan, K., & Ferrara, K. (2000). High speed optical observation of contrast agent collapse. *Applied Physics Letters*, *77*, 1056–1058.

Chomas, J., Dayton, P., Allen, J., Morgan, K., & Ferrara, K. W. (2001). Mechanisms of contrast agent destruction. *IEEE Transactions on Ultrasonics, Ferroelectrics and Frequency Control*, *48*, 232–248.

Chuang, Y. -H., Cheng, P. -W., & Li, P. -C. (2013). Combining radiation force with cavitation for enhanced sonothrombolysis. *IEEE Transactions on Ultrasonics, Ferroelectrics and Frequency Control*, *60*, 97–104.

Church, C. C. (1995). The effects of an elastic solid surface layer on the radial pulsations of gas bubbles. *Journal of the Acoustical Society of America*, *97*, 1510–1521.

Cosgrove, D. O. (1998). Echo-enhancing (ultrasound contrast) agents. In F. A. Duck, A. C. Baker, & H. C. Starritt (Eds.), *Ultrasound in medicine, medical science series*. Bristol, UK: Institute of Physics Publishing, Chapt. 12.

Crum, L. A. (1984). Rectified diffusion. *Ultrasonics*, *22*, 215–223.

Dayton, P., Klibanov, A., Brandenburger, G., & Ferrara, K. (1999b). Acoustic radiation force in vivo: a mechanism to assist targeting of microbubbles. *Ultrasound in Medicine and Biology*, *25*, 1195–1201.

Dayton, P., Morgan, K., Klibanov, S., Brandenburger, G., & Ferrara, K. (1999a). Optical and acoustical observation of ultrasound contrast agents. *IEEE Transactions on Ultrasonics, Ferroelectrics and Frequency Control*, *46*, 220–232.

Dayton, P. A., Chomas, J. E., Lum, A., Allen, J., Lindner, J. R., Simon, S. I., et al. (2001). Optical and acoustical dynamics of microbubble contrast agents inside neutrophils. *Biophysical Journal*, *80*, 1547–1556.

Dayton, P. A., Morgan, K. E., Klibanov, A. L., Brandenburger, G., Nightingale, K. R., & Ferrara, W. K. (1997). An preliminary evaluation of the effects of primary and secondary radiation forces on acoustic contrast agents. *IEEE Transactions on Ultrasonics, Ferroelectrics and Frequency Control*, *44*, 1264–1277.

de Jong, N. (1993). *Acoustic properties of ultrasound contrast agents*, (Ph.D. Thesis). Erasmus University: Rotterdam, Netherlands.

de Jong, N., Bouakaz, A., & Frinking, P. (2000). Harmonic imaging for ultrasound contrast agents. *Proceedings of the IEEE Ultrasonics Symposium* (pp. 1725–1728), San Juan.

de Jong, N., Bouakaz, A., & Ten Cate, F. J. (2002). Contrast harmonic imaging. *Ultrasonics*, *40*, 567–573.

de Jong, N., & Hoff, L. (1993). Ultrasound scattering properties of Albunex microspheres. *Ultrasonics*, *31*, 175–181.

de Jong, N., Hoff, L., Skotland, T., & Bom, N. (1992). Absorption and scatter of encapsulated gas filled microspheres: theoretical considerations and some measurements. *Ultrasonics*, *30*, 95–103.

Deng, C. X., & Lizzi, F. L. (2002). A review of physical phenomena associated with ultrasonic contrast agents and illustrative clinical applications. *Ultrasound in Medicine and Biology*, *28*, 277–286.

Eatock, E. A. (1985). Numerical studies of the spectrum of low intensity ultrasound scattered by bubbles. *Journal of the Acoustical Society of America*, *77*, 1692–1701.

Eller, A. I., & Flynn, H. G. (1965). Rectified diffusion through nonlinear pulsations of cavitation bubbles. *Journal of the Acoustical Society of America*, *37*, 493–503.

Escoffre, J. -M., Novell, A., Piron, J., et al. (2013). Microbubble attenuation and destruction: are they involved in sonoporation efficiency? *IEEE Transactions on Ultrasonics, Ferroelectrics and Frequency Control, 60*, 46–52.

Faez, T., Emmer, M., Kooiman, K., Versluis, M., van der Steen, A. F. W., & de Jong, N. (2013). 20 years of ultrasound contrast agent modeling. *IEEE Transactions on Ultrasonics, Ferroelectrics and Frequency Control, 60*, 7–20.

Feigenbaum, H., Stone, J., & Lee, D. (1970). Identification of ultrasound echoes from the left ventricle by use of intracardiac injections of indocine green. *Circulation, 41*, 615.

Flynn, H. G. (1964). Physics of acoustic cavitation in liquids. In B. W. P. Mason (Ed.), *Physical acoustics* (vol. 1, pp. 57–172). New York: Academic Press.

Forsberg, F., Liu, J. B., Chiou, H. J., Rawool, N. M., Parker, L., & Goldberg, B. B. (2000). Comparison of fundamental and wideband harmonic contrast imaging of liver tumors. *Ultrasonics, 38*, 110–113.

Frinking, P. J. A., Bouakaz, A., Kirkhorn, J., Ten Cate, F. J., & de Jong, N. (2000). Ultrasound contrast imaging: current and new potential methods. *Ultrasound in Medicine and Biology, 26*, 965–975.

Frinking, P. J. A., de Jong, N., & Cespedes, E. I. (1999). Scattering properties of encapsulated gas bubbles at high pressures. *Journal of the Acoustical Society of America, 105*, 1989–1996.

Gaitan, D. F., & Crum, L. A. (1990). Observation of sonoluminescence from a single cavitation bubble in a water/glycerine mixture. In M. F. Hamilton, & D. T. Blackstock (Eds.), *Frontiers of nonlinear acoustics* (pp. 191–196). London: Elsevier.

Goertz, D. E., Frijlink, M. E., De Jong, N., et al. (2006). High frequency nonlinear scattering from a micrometer- to submicrometer-sized lipid encapsulated contrast agent. *Ultrasound in Medicine and Biology, 32*, 569–577.

Goldberg, B. (1993). Contrast agents. In P. N. T. Wells (Ed.), *Advances in ultrasound techniques and instrumentation*. New York: Churchill Livingstone, Chapt. 3.

Goldberg, B. B. (1971). Ultrasonic measurement of the aortic arch, right pulmonary artery, and left atrium. *Radiology, 101*, 383.

Gramiak, R., & Shah, P. (1968). Echocardiography of the aortic root. *Investigative Radiology, 3*, 356–388.

Hansen, R., Angelsen, B. A. J., & Johansen, T. F. (2001). Reduction of nonlinear contrast agent scattering due to nonlinear wave propagation. *Proceedings of the IEEE Ultrasonics Symposium* (pp. 1725–1728), Atlanta, GA.

Hoff, L., Sontum, P. C., & Hovem, J. M. (2000). Oscillations of polymeric microbubbles: effect of the encapsulating shell. *Journal of the Acoustical Society of America, 107*, 2272–2280.

Holland, C. K., Roy, R. A., Apfel, R. E., & Crum, L. A. (1992). In vitro detection of cavitation induced by a diagnostic ultrasound system. *IEEE Transactions on Ultrasonics, Ferroelectrics and Frequency Control, 39*, 95–101.

Hynynen, K., McDannold, N., Vykhodtseva, N., & Jolesz, F. A. (2001). Imaging-guided focal opening of the blood-brain barrier in rabbits. *Radiology, 220*, 640–646.

Kono, Y., Moriyasu, F., Nada, T., Suginoshita, Y., Matsumura, T., Kobayashi, K., et al. (1997). Gray-scale second harmonic imaging of the liver: a preliminary animal study. *Ultrasound in Medicine and Biology, 23*, 719–726.

Leighton, T. G. (1994). *The acoustic bubble*. New York: Academic Press.

Leighton, T. G. (1998). An introduction to acoustic cavitation. In F. A. Duck, A. C. Baker, & H. C. Starritt (Eds.), *Ultrasound in medicine, medical science series*. Bristol, UK: Institute of Physics Publishing, Chapt. 11.

Li, P., Cao, L., Dou, C. Y., Armstrong, W. F., & Miller, D. (2003). Impact of myocardial contrast echocardiography on vascular permeability: an in vivo dose response study of delivery mode, pressure amplitude and contrast dose. *Ultrasound in Medicine and Biology, 29*, 1341–1349.

Lindner, J. R. (2001). Targeted ultrasound contrast agents: diagnostic and therapeutic potential. *Proceedings of the IEEE Ultrasonics Symposium* (pp. 1695–1703), Atlanta, GA.

Marmottant, P., van der Meer, S., Emmer, M., Versluis, M., de Jong, N., Hilgenfeldt, S., et al. (2005). A model for large amplitude oscillations of coated bubbles accounting for buckling and rupture. *Journal of the Acoustical Society of America, 118*, 3499–3505.

Meltzer, R. S., Tickner, G., Salines, T. P., & Popp, R. L. (1980). The source of ultrasound contrast effect. *Journal of Clinical Ultrasound*, *8*, 121–127.

Miller, D. (1981). Ultrasonic detection of resonant cavitation bubbles in a flow tube by their second harmonic emissions. *Ultrasonics*, *22*, 217–224.

Miller, D. L., Averkiou, M. A., Brayman, A. A., Everbach, E. C., Holland, C. K., Wible, J. H., Jr, et al. (2008). Bioeffects considerations for diagnostic ultrasound contrast agents. *Journal of Ultrasound in Medicine*, *27*, 611–632.

Mor-Avi, V., Caiani, E. G., Collins, K. A., Korcarz, C. E., Bednarz, J. E., & Lang, R. M. (2001). Combined assessment of myocardial perfusion and regional left ventricular function by analysis of contrast-enhanced power modulation images. *Circulation*, *104*, 352–357.

Morgan, K., Kruse, D., Dayton, P., & Ferrara, K. (1998). Changes in the echoes from ultrasonic contrast agents with imaging parameters. *IEEE Transactions on Ultrasonics, Ferroelectrics and Frequency Control*, *45*, 1537–1548.

Mullin, L. B., Phillips, L. C., & Dayton, P. A. (2013). Nanoparticle delivery enhancement with acoustically activated microbubbles. *IEEE Transactions on Ultrasonics, Ferroelectrics and Frequency Control*, *60*, 65–77.

Needles, A., Arditi, M., Rognin, N. G., et al. (2010). Nonlinear contrast imaging with an array-based micro-ultrasound system. *Ultrasound in Medicine and Biology*, *36*, 2097–2106.

Ophir, J., & Parker, K. J. (1989). Contrast agents in diagnostic ultrasound. *Ultrasound in Medicine and Biology*, *15*, 319–333.

Porter, T. R., & Xie, F. (1995). Transient myocardial contrast after initial exposure to diagnostic ultrasound pressures with minute doses of intravenously injected microbubbles: demonstration and potential mechanisms. *Circulation*, *92*, 2391–2395.

Porter, T. R., Xie, F., Kricsfeld, A., & Kilzer, K. (1995). Noninvasive identification of acute myocardial ischemia and reperfusion with contrast ultrasound using intravenous perfluoropropane-exposed sonicated dextrose albumin. *Journal of the American College of Cardiology*, *26*, 33–40.

Powers, J., Porter, T. R., Wilson, S., Averkiou, M., Skyba, D., & Bruce, M. (2000). Ultrasound contrast imaging research. *Medica Mundi*, *44*, 28–36.

Rafter, P. G., Perry, J. L. T., & Chen, J. (2002). Wideband phased-array transducer for uniform harmonic imaging, contrast agent detection, and destruction. US patent 6,425,869, July 30, 2002.

Rayleigh (1917). On the pressure developed in a liquid during the collapse of a spherical cavity. *Philosophical Magazine*, *34*, 94–98.

Roy, R., Madanshetty, S., & Apfel, R. (1990). An acoustic backscattering technique for the detection of transient cavitation produced by microsecond pulses of ultrasound. *Journal of the Acoustical Society of America*, *87*, 2451–2458.

Shankar, P. M., Krishna, P. D., & Newhouse, V. L. (1998). Advantages of subharmonic over second harmonic backscatter for contrast-to-tissue echo enhancement. *Ultrasound in Medicine and Biology*, *24*, 395.

Shi, W. T., & Forsberg, F. (2000). Ultrasonic characterization of the nonlinear properties of contrast microbubbles. *Ultrasound in Medicine and Biology*, *26*, 93–104.

Shi, W. T., Hoff, L., & Forsberg, F. (2002). Subharmonic performance of contrast microbubbles: an experimental and numerical investigation. *Proceedings of the IEEE Ultrasonics Symposium*, (pp. 1908–1911).

Sijl, J., Dollet, B., Overvelde, M., Garbin, V., Rozendal, T., de Jong, N., et al. (2010). Subharmonic behavior of phospholipid coated ultrasound contrast agent microbubbles. *Journal of the Acoustical Society of America*, *128*, 3239–3252.

Skyba, D. M., Price, R. J., Linka, A. Z., Skalak, T. C., & Kaul, S. (1998). Direct in vivo visualization of intravascular destruction of microbubbles by ultrasound and its local effects on tissue. *Circulation*, *98*, 290–293.

Talu, E., Hettiarachchi, K., Zhao, S., Powell, R. L., Lee, A. P., Longo, M. L., et al. (2007). Tailoring the size distribution of ultrasound contrast agents: possible method for improving sensitivity in molecular imaging. *Molecular Imaging*, *6*, 384–392.

van Raaij, M.E., Lindvere, L., Dorr, A., He, J., Sahota, B., Stefanovic, B., et al. (2011) Functional micro-ultrasound imaging of rodent cerebral hemodynamics, *Proceedings of the IEEE Ultrasonics Symposium* (pp. 1258–1261), Orlando, FL.

van Ruijssevelt, L., Smirnov, P., Yudina, A., et al. (2013). Observations on the viability of C6-glioma cells after sonoporation with low-intensity ultrasound and microbubbles. *IEEE Transactions on Ultrasonics, Ferroelectrics and Frequency Control, 60*, 34–45.

Vignon, F., Shi, W. T., et al. (2013). Microbubble cavitation imaging. *IEEE Transactions on Ultrasonics, Ferroelectrics and Frequency Control, 60*, 661–670.

Ward, M., Wu, J., & Chiu, J. F. (2000). Experimental study of the effects of Optison concentration on sonoporation in vitro. *Ultrasound in Medicine and Biology, 26*, 1169–1175.

Ward, M., Wu, J. R., & Chiu, J. F. (1999). Ultrasound-induced cell lysis and sonoporation enhanced by contrast agents. *Journal of the Acoustical Society of America, 105*, 2951–2957.

Wei, K., Jayaweera, A. R., Firoozan, S., Linka, A., Skyba, D. M., & Kaul, S. (1998). Quantification of myocardial blood flow with ultrasound-induced destruction of microbubbles administered as a constant venous infusion. *Circulation, 97*, 473–483.

Wheatley, M. A., Schrope, B., & Shen, P. (1990). Contrast agents for diagnostic ultrasound: development and evaluation of polymer-coated microbubbles. *Biomaterials, 11*, 713–717.

Wilson, S. R., Jang, H. J., Kim, T. K., Iijima, H., Kamiyama, N., & Burns, P. N. (2008). Real-time temporal maximum-intensity-projection imaging of hepatic lesions with contrast-enhanced sonography. *American Journal of Roentgenology, 190*, 691–695.

Wong, J. Y., Langer, R., & Ingber, D. E. (1994). Electrically conducting polymers can noninvasively control the shape and growth of mammalian cells. *Proceedings of the National Academy of Sciences of the USA, 91*, 3201–3204.

Wu, J. (1991). Acoustical tweezers. *Journal of the Acoustical Society of America, 89*, 2140–2143.

Wu, J. (2002). Theoretical study on shear stress generated by microstreaming surrounding contrast agents attached to living cells. *Ultrasound in Medicine and Biology, 28*, 125–129.

Wu, J., & Tong, J. (1997). Measurements of the nonlinearity parameter B/A of contrast agents. *Ultrasound in Medicine and Biology, 24*, 153–159.

Wu, J., & Tong, J. (1998). Experimental study of stability of contrast agents. *Ultrasound in Medicine and Biology, 24*, 257–265.

Ziskin, M. C., Bonakdapour, A., Weinstein, D. P., & Lynch, P. R. (1972). Contrast agents for diagnostic ultrasound. *Investigative Radiology, 6*, 500.

Bibliography

Bloch, S. H., Dayton, P. A., & Ferrara, K. W. (2004). Targeted imaging using ultrasound contrast agents. *IEEE Engineering in Medicine and Biology Magazine, September*, 18–29.

Burgess, A., & Hynynen, K. (2013). Noninvasive and targeted drug delivery to the brain using focused ultrasound. *ACS Chemical Neuroscience, 4*, 519–526.

Cosgrove, D. (2006). Ultrasound contrast agents: an overview. *European Journal of Radiology, 60*, 324–330.

Ducas, R., et al. (2013). Echocardiography and vascular ultrasound: new developments and future directions. *Canadian Journal of Cardiology, 29*, 304–316.

Frinking, P. J. A., Bouakaz, A., Kirkhorn, J., Ten Cate, F. J., & de Jong, N. (2000). Ultrasound contrast imaging: current and new potential methods. *Ultrasound in Medicine and Biology, 26*, 965–975.

An overall review article on contrast imaging.

Ferrara, K., Pollard, R., & Borden, M. (2007). Ultrasound microbubble contrast agents: fundamentals and application to gene and drug delivery (2007). *Annual Review of Biomedical Engineering, 9*, 415–447.

Gessner, G., & Dayton, P. A. (2010). Advances in molecular imaging with ultrasound. *Molecular Imaging, 9*, 117–127.

Goldberg, B. B., Raichlen, J. S., & Forsberg, F. (2001). *Ultrasound contrast agents: basic principles and clinical applications* (2nd ed.). London: Martin Dunitz.
Information about the clinical application of contrast agents.
Hughes, M. S., Lanza, G. M., Marsh, J. N., & Wickline, S. A. (2003). Targeted ultrasonic contrast agents for molecular imaging and therapy: a brief review. *Medica Mundi*, *47*, 66–73.
An introduction to therapeutic contrast agents. Also available on the Web, which is a good way to keep up with this fast-changing field.
Klibanov, L. (2006). Microbubble contrast agents, targeted ultrasound imaging and ultrasound-assisted drug-delivery applications. *Investigative Radiology*, *41*, 354–362.
Leighton, T. G. (1994). *The acoustic bubble*. New York: Academic Press.
A detailed and lucid source of knowledge about microbubbles.
Lindner, J. R. (2001). Targeted ultrasound contrast agents: diagnostic and therapeutic potential. *Proceedings of the IEEE Ultrasonics Symposium* (pp. 1695–1703), Atlanta, GA, 2001.
An introduction to therapeutic contrast agents.
Rayleigh (1917). On the pressure developed in a liquid during the collapse of a spherical cavity. *Philosophical Magazine*, *34*, 94–98.
Neppiras, E. A. (1984). Acoustic cavitation: an introduction. *Ultrasonics*, *22*, 25–28.
Powers, J., Porter, T. R., Wilson, S., Averkiou, M., Skyba, D., & Bruce, M. (2000). Ultrasound contrast imaging research. *Medica Mundi*, *44*, 28–36.
An overall review article on ultrasound contrast imaging.
Qin, S., Caskey, C. F., & Ferrara, K. W. (2009). Ultrasound contrast microbubbles in imaging and therapy: physical principles and engineering. *Physics in Medicine and Biology*, *54*, R27–R57.
Reid, C. L., Rawanishi, D. T., & McRay, C. R. (1983). Accuracy of evaluation of the presence and severity of aortic and mitral regurgitation by contrast 2-dimensional echocardiography. *American Journal of Cardiology*, *52*, 519.
Schrope, B. A., & Newhouse, V. L. (1993). Second harmonic ultrasound blood perfusion measurement. *Ultrasound in Medicine and Biology*, *19*, 567–579.
Schrope, B., Newhouse, V. L., & Uhlendorf, V. (1992). Simulated capillary blood flow measurement using a non-linear ultrasonic contrast agent. *Ultrasonic Imaging*, *14*, 134–158.
Unger, E., Matsunga, T. O., Schermann, P. A., & Zutshi, R. (2003). Microbattles in molecular imaging and therapy. *Medica Mundi*, *47*, 58–65.
An introduction to therapeutic contrast agents. Also available on the Web.
Wilson, S. R., & Burns, P. N. (2010). Microbubble-enhanced US in body imaging: what role? *Radiology*, *257*, 24–39.

CHAPTER 15

Ultrasound-induced Bioeffects

Chapter Outline

15.1 Introduction

The appearance of safety as a topic so late in this book is not because of its relative importance but because of its complexity. In order to convey this topic adequately, knowledge of acoustic focused fields, absorption, imaging systems, imaging modes, nonlinear effects, contrast agents, and measurements from previous chapters will be applied to our understanding of ultrasound-induced bioeffects.

It is well known that too much sunlight can be harmful. Are there conditions under which the medical application of ultrasound also becomes destructive? Because of the thermal and mechanical interaction of ultrasound with tissue, the answers are far more complicated than they are for sunlight or X-ray exposures. Is there such a thing as acoustic dose? For X ray or nuclear radiation, the detrimental effects are cumulative, so keeping track of exposure is necessary. Ultrasound-induced bioeffects fall into two categories: reversible and irreversible. Most of the effects are reversible so no accounting is needed; in those cases in which the change is irreversible, it is usually intentional and some independent means of validating the change is undertaken. A great deal of effort has been expended on determining where the lines between reversibility and irreversibility occur and keeping ultrasound insonification below this level, unless there is a particular reason for exceeding it.

Diagnostic ultrasound has had an impressive safety record so far, with the number of ultrasound exams now into the billions (Figure 1.15) over a 60-year period. Despite this record, some vigilance is still needed. Medical ultrasound offers more opportunities because of beneficial ultrasound-induced bioeffects. Later in this chapter, as we explore some of these new applications of medical ultrasound, a deeper understanding of bioeffects will be needed. Ultrasound-induced bioeffects occur on different length and time scales, are dependent on the spatial and temporal arrangement of the insonification, its level and duration, and the acoustic and nonacoustic properties of the tissue insonified. Even though there is an intended outcome, there will always be unintended outcomes as a consequence of the biophysics underlying the interaction of sound and tissue. Safety, in this context, involves an assessment of these reversible/irreversible factors relative to the short- and long-term benefits to the patient receiving the sound diagnosis or treatment. One goal of this chapter is to provide a comprehensive perspective for evaluating these issues.

The widespread use of medical ultrasound falls into five major categories: diagnostic imaging, physiotherapy, hyperthermia, lithotripsy, and surgery. Diagnostic imaging includes scanned beams operating from 1–50 MHz (up to intravascular imaging and high-frequency applications). Physical therapists use low frequency (0.75–3 MHz) to apply ultrasound to promote healing, loosen muscles and joints, relieve pain, and increase blood flow to stimulate natural body defenses. Hyperthermia is the deliberate heating of a region of the body with ultrasound to selectively arrest the reproduction of cancerous tissues or tumors.

Table 15.1 Comparison of Water Values for Medical Ultrasound Modalities (mean values in parentheses)

Modality	f_c (MHz)	Power (W)	p_r (MPa)	I_{SPTA} (W/cm^2)	I_{SATA}	ΔT
B-mode	1–15	0.0003–0.285 (0.075)	0.45–5.54 (2.3)	0.0003–0.991 (0.34)		<2
PW Doppler	1–10	0.01–0.44 (0.1)	0.67–5.3 (2.04)	0.173–9.08 (1.18)		<2
Physiotherapy	0.75–3.4	1–15	0.3	3	<3	
Hyperthermia	0.5–5.0		0.6–6.0	1–10	2–10	4–9
HIFU	1–10			1000–10,000 (peak)		>19
Lithotripsy	0.5–10		5–15	Very low		

PW, pulsed-wave.

Lithotripsy is the extracorporeal or percutaneous application of ultrasonic shock waves to selectively disintegrate kidney stones (or gallstones, etc.) in vivo without surgery. High-intensity focused ultrasound (HIFU) is the real-time application of ultrasound to perform surgery within the body, specifically to produce highly localized lesions. A number of interesting miscellaneous therapeutic and cosmetic applications will also be reviewed in Chapter 17. Some applications, such as dental descalers and ultrasonic scalpels, will not be covered in this discussion.

Most of the primary ultrasound-induced bioeffects are based on the following three parameters (defined in measurements in Chapter 13): time-averaged source acoustic power (W), peak rarefactional pressure (p_r), and spatial peak temporal average intensity (I_{SPTA}), or a related intensity parameter, spatial average temporal average (I_{SATA}) and temperature elevation range. Typical parameters for these applications are summarized in Table 15.1, to which we shall refer as we compare the different modalities in detail in Section 15.7. A historical overview of ultrasound-induced bioeffects and safety issues begins this chapter (Section 15.2). The three primary parameters are then related to two significant ultrasound-induced bioeffects: temperature elevation (Section 15.3 and equations in Section 15.8) and mechanical effects (Section 15.4 and Chapter 14). The output display standard (ODS) is explained briefly in Section 15.5. A comprehensive overview of ultrasound-induced bioeffects will be presented in Section 15.6. Diagnostic ultrasound will be compared to other major medical ultrasound modalities from this perspective in Section 15.7. Some of the ultrasound-bioeffect mechanisms to be discussed are outlined in Figure 15.1.

15.2 Ultrasound-induced Bioeffects: Observation to Regulation

Clinical uses of ultrasound are known to have the potential to create two major types of bioeffects: heating and cavitation. Knowledge and understanding of these effects began from observations and evolved into their application, control, and regulation. O'Brien (1998) provides a comprehensive review of these events. More information can be found in

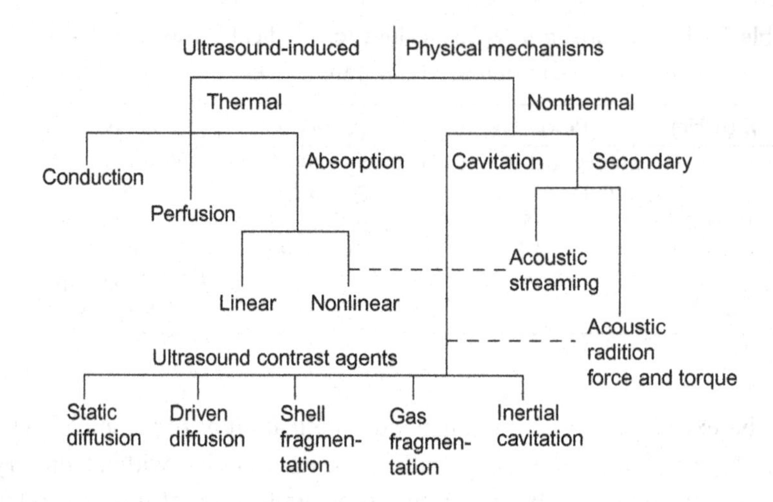

Figure 15.1
Outline of Ultrasound-induced Bioeffects for Diagnostic Ultrasound.

Nyborg and Ziskin (1985) and National Council on Radiation Protection and Measurements (NCRP (2002)).

In the first practical realization of pulse–echo ranging in 1917 (discussed in Chapter 1), Paul Langevin developed large quartz transducers resonating at 150 kHz and driven by vacuum tube amplifiers. With an application of 2.5 kilovolts, he was able to produce 1 kilowatt of peak ultrasonic power. He noticed that fish were killed in a water tank when the source was turned on; if he put his hand in the water, he felt a painful sensation. This effect was later identified as cavitation, or the collapsing and resonating of gas bubbles by ultrasound.

During the 1930s and 1940s, ultrasound-induced heating of tissue was widely applied. As described earlier for physical therapy, ultrasound can have beneficial healing effects. Unfortunately at this time, because the mechanisms of and the amount of heat generated by ultrasound were not well understood or controlled, accidents occurred. Even though intensities of less than 10 W/cm^2 were used, in some cases, ultrasound was found to accelerate cancer growth. These results led to the Erlangen conference, which put a hold on the therapeutic application of ultrasound. Research also indicated that ultrasound could form lesions at high enough intensities, and in 1942, ultrasound surgery was proposed.

Research on bioeffects continued as diagnostic imaging methods developed in the 1950s and 1960s. By the 1970s, enough clinical imaging equipment was in use to perform millions of exams. Professional, trade, and standards organizations such as the American Institute of Ultrasound in Medicine (AIUM), the National Electronics Manufacturers Association (NEMA), the International Electrotechnical Commission (IEC), and the World

Federation for Ultrasound in Medicine and Biology (WFUMB), as well as the National Institutes of Health (NIH) and the NCRP, began to formulate standard methods for ultrasound output measurement and imaging system use in the late 1970s and early 1980s. The first US ultrasound safety standard appeared in 1983. Accurate measurements of acoustic output were made possible by the then-recent development of polyvinylidene-difluoride (PVDF) hydrophones.

In 1985, the Food and Drug Administration (FDA) of the USA was empowered by Congress to regulate the acoustic output of medical devices, including diagnostic ultrasound imaging systems. The limiting values selected were based on the output levels of imaging equipment that existed on or before May 26, 1976, the date of the passage of the Medical Device Amendment to the Food, Drug, and Cosmetic Act. In the 1970s, ultrasound imaging equipment consisted mainly of static B-scanners with articulated arms and a few real-time mechanical scanners (see Chapter 1). By the mid-1980s, ultrasound equipment became dominated by real-time phased array and linear-array imaging systems. The FDA limits were revised in 1987 and 1992 (Duck & Martin, 1991), and the latest revision is listed in Table 15.2. The process for approving ultrasound imaging systems is called the 510 (*k*) procedure (AIUM, 1993; FDA, 1993, 1997). The revision of 1992 introduced the output display standard (ODS) (AIUM/NEMA, 1998a), a revolutionary concept of providing users information on the two primary bioeffects, as described in Section 15.5 and explained in a document (AIUM, 2002).

To ensure a consistent means of measuring acoustic output values, measurement standards were developed by three organizations. A joint AIUM, NEMA, and FDA effort produced the first version in 1983, a second version in 1989, and a third version in 1998 (AIUM/NEMA, 1998b). These standards detailed not only how the measurements were to be carried out, but also the characteristics of the equipment needed for a particular measurement, as well as the maintenance and calibration of measurement equipment. During this same time period, the IEC Technical Committee 87 issued several standards governing acoustic output measurement and devices. At the present time, aside from regulations developed internally within individual countries, the IEC, including Technical Committee 62, has become the main international group for developing standards for

Table 15.2 Track 3 FDA Limits on Acoustic Output

Application	Pre-ODS $I_{SPTA.3}$ (mW/cm^2)	Pre-ODS $I_{SPPA.3}$ (W/cm^2)	Post-ODS $I_{SPTA.3}$ (mW/cm^2)	Post-ODS MI
Fetal imaging	94	190	720	1.9
Cardiac	430	190	720	1.9
Peripheral vascular	720	190	720	1.9
Opthalmic	17	28	50	0.23

Revised in 1998. MI, mechanical index.

ultrasound medical devices of all types. A slightly modified version of the ODS was incorporated into international standard IEC 60601-2-37 (IEC, 2004).

Before proceeding, we can identify several aspects of ultrasound-induced bioeffects and the related measurements processes, represented in Figure 15.2. Shown in the bottom of this figure is an ultrasound source that is an array radiating an acoustic field into a tissue. This source can operate several modes sequentially, each one characterized by a pulsing sequence, frequency, and acoustic time-average power. The tissue itself has certain properties, some of which are acoustic impedance and frequency-dependent absorption and dispersion, as well as thermal and mechanical characteristics and structure. The radiated field changes as it propagates and interacts with the tissue over the volume or extent of the transmitted beams, and at any particular point (x,y,z) in the tissue, the field can be represented by a pressure waveform or intensity. In response to the presence of the transducer, there is thermal conduction from the interface between the transducer and tissue surface. The tissue responds to the insonification in a number of ways, which are ultrasound-induced bioeffects, and will be explored later. Because this complex interaction is unique to the tissues and location of the array, regulatory and standards organizations have simplified the imaging system acoustics output characterization process so it can be done, reproducibly, consistently, and universally, in different laboratories in a water tank. The top part of Figure 15.2 represents the acoustic output characterization, already described in detail in Chapter 13, by which acoustic power is measured by an acoustic radiation force balance, and axial field pressure and intensity waveforms are measured by hydrophones. An explanation of the major ultrasound-induced bioeffects and the rationale behind these measurements for diagnostic ultrasound are explained in the discussions to follow.

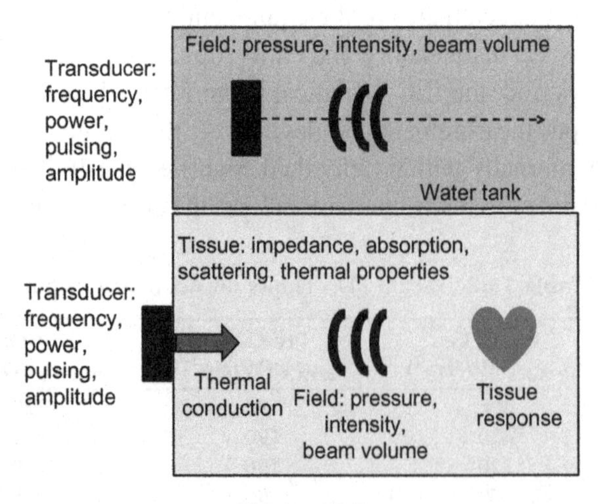

Figure 15.2
Major Factors Involved in Ultrasound-induced Bioeffects.

The organizations listed earlier continue actively to refine and develop the understanding of ultrasound-induced bioeffects. Fairly detailed discussions and reviews of the latest findings on bioeffects are disseminated in professional meetings and are published. The most active organizations are the AIUM and WFUMB symposia on safety and standardization in medical ultrasound. A series of articles that reviewed ultrasound-induced bioeffects appeared in a special issue of the *Journal of Ultrasound in Medicine* (April, 2008) to which we shall refer in Section 15.6. In 2002, the NCRP published a most authoritative and thorough compendium on bioeffects. Ultrasound federations such as the Australasian Society for Ultrasound in Medicine (ASUM) and the European Federation of Societies for Ultrasound in Medicine and Biology (EFSUMB) have also published guidelines, recently summarized by Barnett et al. (2000). Individual societies are also involved in these issues. As examples, the American Society of Echocardiographers (ASE) published guidelines for the use of contrast agents, and the EFSUMB reviewed bioeffects publications (Duck, 2000).

Safety encompasses not only bioeffects, but also the application of ultrasound, including the training and qualification of the clinicians and sonographers. Some of these topics and a more complete listing of societies involved with ultrasound are given in Section 15.9 and Appendix D.

15.3 Thermal Effects

15.3.1 Introduction to Thermal Tissue Response

The concern over the temperature rise induced by ultrasound in the body is based on observed changes in cellular activity as a function of temperature. In general, for healthy activity of enzymes, the enzymatic activity doubles for every 10°C rise. The human body is able to tolerate hot drinks and fevers for a certain period of time. A fever of +2°C is not a problem, where 37°C is taken as an average core body temperature. Table 15.3 identifies stages of temperature effects.

These temperature ranges can be combined with the duration of temperature exposure. Based on a compilation of data of reported ultrasound-induced detrimental effects, Miller and Ziskin (1989) developed an empirical relationship for temperature elevation and exposure time below which there were no observed adverse effects:

$$t = 4^{43-T}(\text{minutes}) \tag{15.1}$$

where T is temperature in degrees Centigrade. Even though the validity of this equation is discussed later in more detail (NCRP, 2002), it still serves as a reasonable initial guide to the effects of heating and exposure. Times from this equation are plotted in Figure 15.3. For example, an elevation of 2°C gives a time of 256 min, and a 6-°C rise gives a time of 1 min. This equation implies that by shortening the time of the exam, the detrimental

Table 15.3 Temperature effects

Temperature Range (°C)	Effect
37–39	No harmful effects for extended periods
39–43	Detrimental effects after long enough time
>41	Threshold for fetal problems for extended periods
44–46	Coagulation of protein
>45	Enzymes become denatured
>41.8	Cancer cells die (fail to reproduce)
	Often taken as damage threshold—except eye

Source: Miller and Ziskin (1989)

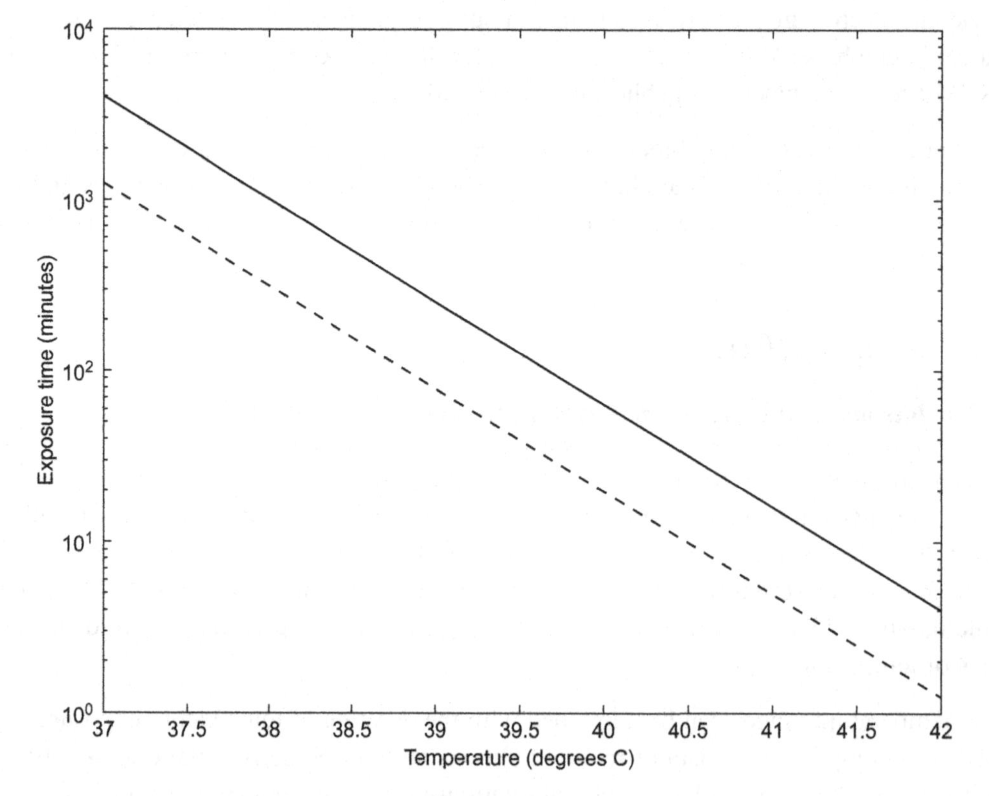

Figure 15.3
Curves Relating Temperature Elevation to Duration of Exposure at Which. There Have Been No Observed Adverse Effects. Solid line corresponds to Eqn 15.1; dashed line corresponds to Eqn 15.5.

effects of higher temperature rises can be minimized. Later, this basic principle will be expanded to cover other medical ultrasound modalities.

What are the mechanisms by which ultrasound can heat tissues? During propagation, energy is lost to absorption; that energy is converted to heat. The direct contact of the transducer creates the direct transfer of heat by conduction. These mechanisms are diagrammed in Figure 15.1.

15.3.2 Heat Conduction Effects

The transducer itself can be a source of heat, by direct contact with the body. A transducer that is left unused and selected may have acoustic power flowing to the outer absorbing lens, where it encounters air, is reflected back, and causes self-heating (Calvert & Duck, 2006; Calvert, Duck, Clift, & Azaime, 2007; Duck, Starritt, ter Haar, & Lunt, 1989). Since the study of this effect, surface temperature rises of transducers in air or an air–gel mixture are controlled not to exceed a few degrees by IEC Standard 60601-2-37 (IEC, 2002).

Once the transducer is placed on the body and is acoustically loaded, the energy is released to propagate into the body and not the transducer, and the normal mechanisms of body cooling through perfusion reduce the heating considerably. For most healthy people, the skin can detect small changes in temperature; however, the communication of this sensation may not be possible for the ill and very young. In addition, for intracavity transducers such as transesophageal probes, built-in safety sensors detect excessive temperature rises, alert the user, and cut off the electric power to the transducer (Ziskin & Szabo, 1993). Cooling mechanisms are designed into matrix arrays in which in addition to acoustic heating there are electronic heat sources. Because the conductive temperature contribution is very localized to the surface and smaller than the absorption contribution, it is often neglected in temperature elevation estimates for the body, as discussed in Section 15.3.5. Data showing the relation between deeper heating and conduction heating by ultrasound in a transcranial phantom can be found in Wu, Cubberley, Gormley, and Szabo (1995).

15.3.3 Absorption Effects

The pattern of heating initially is related to the distribution of intensity in the absorbed beam. The volume rate of heat generation, q_v, due to absorption can be modeled as proportional to the acoustic intensity $I(x,y,z)$ and absorption α at a single frequency:

$$q_v = 2aI \tag{15.2}$$

The highest temperatures along the beam axis are not that sensitive to beam details and can be determined from an integral of q_v times the temperature response of a small source in the medium. For circularly symmetric transducers, the temperature rise can be calculated by a heated disk model (referred to in Section 15.8). After initial propagation, the heat distribution diffuses slowly into the tissue (a process that expands, smoothes out, and the original pattern decreases in amplitude). At higher pressure levels, additional heating is caused by nonlinear effects (as explained in Chapter 12).

Table 15.4 Perfusion Parameters

Tissue	Perfusion Time Constant τ (s)	Thermal Diffusivity κ (mm^2/s)	Perfusion Length L (mm)
Kidney	14.7	0.13	1.4
Heart	69	0.15	3.2
Liver	98	0.15	3.8
Brain	109	0.13	3.8
Muscle	2140	0.15	18
Fat	4000	0.095	19.5

Source: NCRP (1992) and AIUM (1998)

15.3.4 Perfusion Effects

The cooling effects of blood perfusion in tissue must also be included in an estimation of temperature elevation. Whereas the full computation of temperature elevation requires the bioheat equation (also described in Section 15.8), Nyborg (1988) has shown that for a long enough time, the temperature from a small source has a spatial falloff that is exponential:

$$T = (2C/r)exp(-r/L) \tag{15.3A}$$

where L is a perfusion length and

$$C = \frac{q_v dv}{8\pi c_v \kappa} \tag{15.3B}$$

in which dv is the volume of the source, c_v is the volume-specific heat for tissue, and κ is thermal diffusivity. Perfusion lengths for different tissues range from 1 to 20 (as given by Table 15.4). From this table, the heart, which is very active and perfused with blood, has a low value of 3.2 mm for the perfusion length, and fat has a large length of 19.5 mm. The latter contributes to the high thermal insulation properties of fat. Nyborg (1988) also showed that once a uniform temperature distribution is established and the heat source is turned off, temperature decays exponentially as:

$$T = T_0 \, exp(-t/\tau) \tag{15.4}$$

where τ is the perfusion time constant (values of which are found in Table 15.4). Note that for the heart, decay is fast, or 1.15 minutes to the 1 / e value compared to 66.7 minutes for fat.

15.3.5 Combined Contributions to Temperature Elevation

A computation for the thermal rise at the interface between a transducer and the body of a patient, including perfusion and absorption, is given by Figure 15.4. For this case, the transducer was turned on at $t = 0$ and off at $t = 180$ s. The direct heat conduction

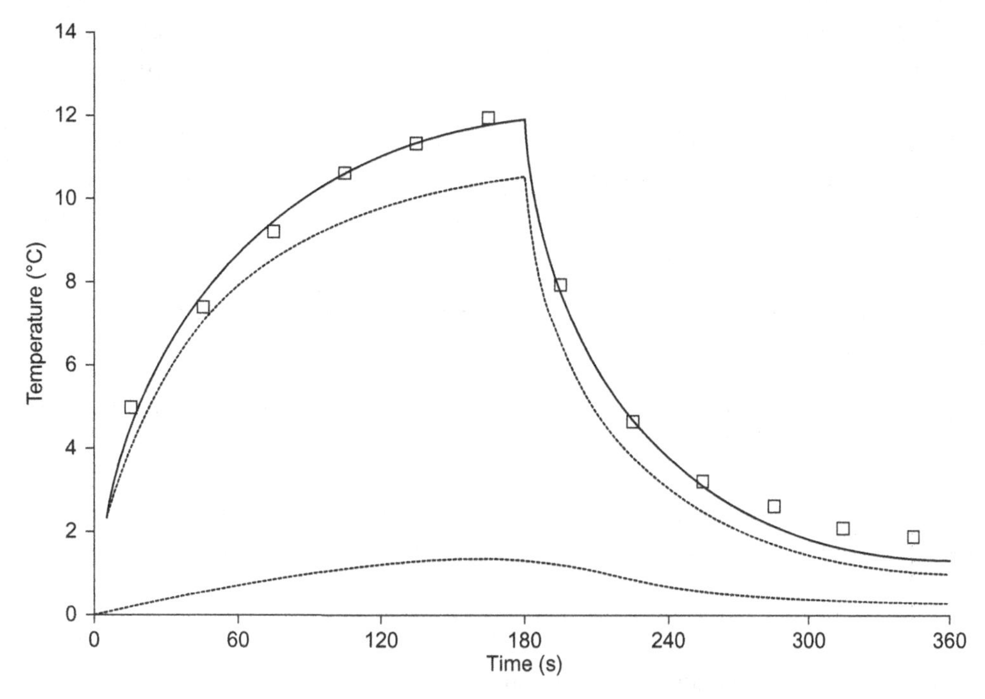

Figure 15.4
Computed Temperature for the Interface Between a 5-MHz circular Transducer and a Patient. Total temperature rise (solid line); temperature contribution from absorption (dotted line); temperature contribution from surface conduction (dashed line). $I_0 = 140$ mW/cm^2, $L = 4.6$ mm, and $\alpha = 50$ Np^{-1}.m. *From Nyborg (1988); squares are data from Williams, McHale, Bowditch, Miller, and Reed (1987)*

contribution can be seen to be small. More details and a more exact description of the temperature-generating process are given in Section 15.8.

For a stationary mode such as Doppler or M-mode, the beams in one direction are important. In a scanned mode, such as B-mode sector scan, a number of beams are sent in different directions into the body, so the determination of the resulting temperature distribution must account for the arrangement of the beams. Different simple models for estimating the temperature rise for different scanned and non-scanned situations will be presented in Section 15.5.2.

15.3.6 Biologically Sensitive Sites

In addition to soft tissue, there are several sites that are regarded as more sensitive to temperature. One is the fetus, which undergoes considerable change in the first trimester.

As the fetal bone develops and ossifies, it can absorb higher temperature elevation than the surrounding soft tissue under insonification. When transducers are placed directly on the neonatal or adult skull, a concern is that the elevated temperature of the bone may alter the temperature of brain tissue. Another sensitive site is the eye, which is well perfused except for the lens, and is therefore somewhat limited in its ability to dissipate externally applied heat. These sites are discussed by Rott (1999a). Another sensitive site is the well-protected brain, in which irreversible changes can be devastating.

Recommendations from WFUMB (1998) with regard to ultrasound-induced temperature elevation are the following:

1. A diagnostic ultrasound exposure that produces a maximum in situ temperature rise of no more than 1.5°C above normal physiological levels (37°C) may be used clinically without reservation on thermal grounds.
2. A diagnostic ultrasound exposure that elevates embryonic and fetal in situ temperature above 41°C (4°C above normal temperature) for 5 min or more should be considered potentially hazardous.
3. The risk of adverse effects is increased with the duration of exposure.

One can see that points two and three correspond to a more conservative exposure-duration law than Eqn 15.1, or:

$$t = 4^{43-T} t_{43} \tag{15.5}$$

where $t_{43} = 0.308$ min. Here, a rise of 1.5°C corresponds to a duration of 158 min; a rise of 4°C corresponds to 5 min (see Figure 15.3). These calculations assume that the transducer is held exactly at the same place for the whole time (a long dwell time); in reality, there is considerable movement in a typical exam and the hand is not steady for extended periods of time. The main problem for users of ultrasound was how to apply this information and to find out what temperature rises were generated by the imaging system.

15.4 Nonthermal Effects

Major and secondary ultrasound-induced nonthermal effects have been covered in great detail in Chapter 14. These effects are summarized in Figure 15.1 and will be examined more fully in Sections 15.6 and 15.7. For many years in the application of diagnostic ultrasound, a major concern has been possible inertial cavitation events (Chapter 14) occurring in the body. As discussed in Section 15.5.3, the focus of attention in this area is shifting from naturally occurring nucleation sites in the body to ultrasound contrast agents. These encapsulated agents vary considerably in their response to ultrasound, depending on the gas and shell materials.

15.5 The Output Display Standard

15.5.1 Origins of the Output Display Standard

Until 1992, ultrasound imaging systems were regulated by measuring highest values of derated acoustic output parameters in various modes (B-mode, Doppler, etc.). The two primary ultrasound-induced bioeffects had been known for a number of years, but the relationship between these effects and the acoustic output had not been thoroughly examined. Furthermore, questions remained as to which output parameters were most relevant. Were there other parameters connected to bioeffects equally worthy of being measured? Where were the locations of the highest temperature elevations, and how could they be estimated? Was there a way of relating cavitation to parameters of an imaging system? These questions and many more inspired a team formed from AIUM, NEMA, and the FDA to develop a revolutionary step in acoustic output control. Their goal was to come up with real-time algorithms for predicting relative temperature rises and the potential for inertial cavitation events. The outputs of these algorithms were to become the thermal indices (TIs) and mechanical index (MI) that would be displayed in real time on imaging systems for the particular mode and settings used at the time. The result of several years of work (1988–1992) by these experts was the output display standard (ODS) (AIUM/NEMA, 1998a).

The indices are relative indicators—not predictors—of absolute values. An example of a thermal index is:

$$\mathrm{TI} = \frac{W_0}{W_{\mathrm{deg}}} \tag{15.6}$$

where W_0 is the time-averaged acoustic power of the source (or another power parameter) and W_{deg} is the power necessary to raise the target tissue 1°C based on specific tissue and thermal models. A conservative perfusion length of 10 mm was used in the TI derivations. With this kind of a definition, the temperature-predicting algorithms are linked to an actual acoustic output parameter (in this case, W_0) that is calibrated to the system output through extensive acoustic output measurements. Because internal imaging system acoustic output control algorithms (Szabo, Melton, & Hempstead, 1988) limit the acoustic output as a function of system settings and the applied voltage levels to the transducer for the mode selected, this information is available for the real-time calculations of the thermal and mechanical indices.

The MI, already introduced in Chapter 14, is:

$$MI = p_{r.3}(z_{sp})/\sqrt{f_c} \tag{15.7}$$

where the derated pressure (MPa) is at the location of the derated pulse intensity integral, $PII_{.3}$ (see Chapter 13) maximum, and f_c is the center frequency (or more recently, the acoustic working frequency) in MHz. To make MI unitless, it is multiplied by a units factor $\sqrt{1\ \text{MHz}}/1$ MPa. Again, the value of pressure is a known and system-controlled acoustic output parameter in a predictive formula. An explanation and plot of MI can be found in Figure 14.7.

The detailed formulas for the indices are not discussed here; the reader is referred to the standards themselves: AIUM/NEMA (1998a), Standard 60601-2-37 62 (IEC, 2002), and an explanation of their derivation by Abbott (1999).

15.5.2 Thermal Indices

Where are the hottest spots located? On which parameters do the temperature rises depend? Three main TI categories are soft tissue (TIS), bone (bone at focus) (TIB), and cranial bone (bone at surface) (TIC). The main results are summarized in Figure 15.5.

On the left side of this figure are the soft-tissue indices. At the top left is a scanned mode where, because of the overlap of beams, the hottest spot is at the surface and is proportional to power (W_0). Below are two nonscanned modes for large apertures (active area $>1\ \text{cm}^2$) and small apertures (active area $\leq 1\ \text{cm}^2$). The large aperture, depending on the strength of focusing, can have a hot spot related to I_{SPTA} at some depth or at the surface, related to power.

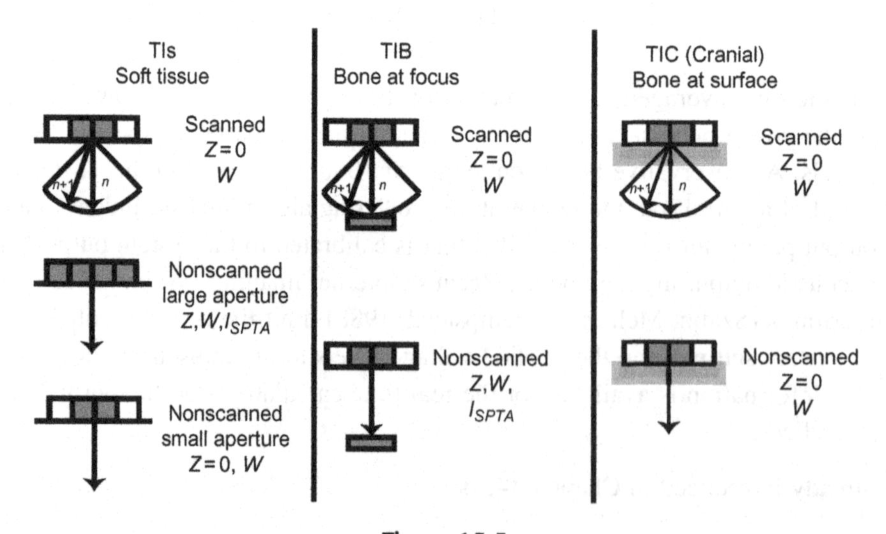

Figure 15.5
Thermal indices (TIs) from the ODS for different modes and configurations. Shaded regions represent active apertures.

The two indices associated with bone are the TIB and TIC. The motivation for the TIB (middle column) is the sensitive case of a fetal bone located at the focal depth. Two cases include one that is scanned, where the location of maximum heating is at the surface and related to power, and another that is the nonscanned case of bone at the focal depth and related to intensity. Bone absorbs at a higher rate than soft tissue (Carstensen, Child, Norton, & Nyborg, 1990), so for the nonscanned case, the prediction is related to the −6-dB beamwidth at the depth where intensity is the highest. Finally, for the cranial TIC in the right column of Figure 15.5, the hot spot is related to power at the surface for both the scanned and nonscanned cases.

As an example of a TI calculation, the TIS of the scanned mode at the surface will be used. For this case, a typical power is 78 mW and if the center frequency is 5 MHz:

$$\mathrm{TIS} = \frac{W_0(\mathrm{mW})}{210/f_c(\mathrm{MHz})} = \frac{78}{210/5} = 1.94 \tag{15.8}$$

In summary, in five out of seven cases (as shown in Figure 15.5), TI is related to power and is located at the skin surface.

15.5.3 Mechanical Index

The rationale for MI was explained in Section 14.4.3 with a plot of MI in Figure 14.7. The formula is based on the assumption that a nucleation site of just the right resonant size is available for inertial cavitation. Holland, Roy, Apfel, and Crum (1992) demonstrated that cavitation could occur when an imaging system insonified water full of bubbles of different sizes. When the experiment was repeated with degassed water, no cavitation events were observed. At the time that ODS was developed, there were three reasons for including an MI. The first was that cavitation was observed to occur with lithotripters, although at much lower frequencies and higher pressure levels. Second, in vitro experiments with lower organisms showed that cavitation might occur at pulse levels similar to those used in diagnostic imaging. Third, lung and intestinal hemorrhages occurred in adult (but not fetal) mice at diagnostic levels. Now, a decade later, these reasons and the original data have been reviewed by the original researchers. More recent information on this topic can be found in Sections 15.5.4 and 15.6.4. The FDA has limited the maximum value of MI to 1.9 (as in Table 15.2), except for special cases. Finally, the widespread use of ultrasound contrast agents has opened up the possibilities of other kinds of mechanical effects (these are also discussed in Section 15.6.4).

15.5.4 The ODS Revisited

Even though the original intention of the architects of the ODS was not to develop exact predictions of temperature rises, but rather to develop relative indications of temperature effects, that has not stopped people from using TIs as absolute temperature estimators. In order to provide reasonably simple algorithms for real-time ODS calculations, some compromises on the conservative side were made (Abbott, 1999; AIUM, 1998a; Curly, 1993). Research showed that many of the temperature elevations were dependent on acoustic power (a parameter relatively insensitive to details of beam structure). In a survey of different imaging systems (as discussed in Section 15.6.6), total acoustic power in water is typically only 125 mW with a maximum value of about 440 mW (Henderson, Willson, Jago, & Whittingham, 1995). O'Brien and Ellis (1999) calculated the source time-averaged power needed to achieve a derated upper limit of $I_{SPTA.3} = 720\ \mathrm{mW/cm^2}$.

Shaw, Preston, and Bond (1997); Shaw, Pay, Preston, and Bond (1999); and Duck (2000), using a more comprehensive temperature prediction program, found that the TI formulas gave values that were equal to or greater than their computations, sometimes by a factor of two. By using manufacturer's data for pulsed Doppler, they predicted nonscanned TI values in excess of 1.5 for a number of cases. Shaw, Pay, and Preston (1998) also compared actual temperature rises measured in thermal test objects for 19 system transducer combinations to the corresponding calculated on-screen TI values. There was reasonable agreement between the calculated TI and the measured temperature rise under attenuated conditions. Exceptions were for cases without overlying absorption, such as the cranial TI, TIC. These results were in agreement with the earlier findings of Wu et al. (1995), who also compared TIC predictions from an imaging system to thermocouple measurements on bone and a tissue-mimicking material (TMM) phantom. They demonstrated that the temperature rise caused by the absorption was well predicted and deviations in data could be attributed to transducer self-heating (thermal conduction) not included in the prediction. Heating and cooling mechanisms are diagrammed in Figure 15.1 and will be revisited in Section 15.6.2.

Ellis and O'Brien (1996) compared maximum temperature predictions from their monopole source model to the nonscanned TIS formula, based on the heated-disk model. They found that the TIS agreed with their more accurate results for the majority of cases that covered most of the combinations of apertures and frequencies in commercial imaging systems. Specifically, they demonstrated that for F numbers (F#s) of less than 2, the TIS underestimated temperature rise but the TIS values were less than 0.4 and need not be displayed. They found that the unscanned TIS model underestimated the temperature rise predicted by their model for larger apertures and higher frequencies; however, these exceptional cases, $F > 3$, are most likely not combinations that occur clinically (an aperture of 2 cm at 12 MHz and apertures of 4 cm at 7 MHz and above). Based on this work, O'Brien et al. (2008) proposed more accurate equations for TIS.

Researchers from the NCRP (2002) have reviewed a number of temperature predictions compared with TMM phantoms and animal experiments. They have concluded that the

ODS algorithms are often higher than those that would occur clinically. They identified three areas where underestimates could occur:

1. Where a significant low-absorption path in tissues is involved.
2. When transducer self-heating provides a significant contribution.
3. When measurements made for ODS are influenced by nonlinear propagation saturation effects.

Cavitation effects have been extensively discussed in reports (AIUM, 2000; NCRP, 2002). The reasons for MI and the original data have been reviewed by the original researchers. Carstensen, Gracewski, and Dalecki (2000) indicated that there may not be any natural nucleation sites within the human body of any consequence for mechanical effects to occur. In addition, the fetus does not provide any nucleation sites for cavitation (Rott, 1999b). A thorough review of nonthermal mechanisms including the possibility of cavitation in the body was conducted by an AIUM bioeffects safety group (Church et al., 2008). They also concluded that the risk of cavitation from naturally occurring nucleation sites is extremely small, estimated as one part in 10^{10}. At superthreshold exposures, the lung may experience localized damage under certain exposure cases; however, evidence shows that any damage is reversible and heals quickly and completely. Microbubbles in the intestine can also cause local damage, but data indicate the thresholds are higher than for the lung and occur at higher MI values and at longer pulse lengths. Their overall conclusion was that the risk of permanent harm from any nonthermal mechanism from exposure to diagnostic ultrasound was extremely small. No confirmed reports of adverse effects from these nonthermal effects have occurred for $\text{MI} = p_r/\sqrt{f_c} < 2$.

The mechanical index has grown to serve another purpose, a *de facto* way of referring to pressure levels of imaging systems. For procedures such as those involving contrast agents, an MI level of a certain value may be used to initiate certain procedures for imaging such as perfusion and contrast-enhanced studies. Invented terms such as "low MI" and "high MI" and others have appeared in publications. Superficially, an MI greater in value than another would seem to be a self-consistent way of referring to pressure levels among imaging systems, but this use was not intended by the ODS creators.

Some caution is necessary in the application of MI; the reasons can be found in Figure 14.7. First, "MI" refers to a derated value of pressure at one position, which will not usually correspond to the region of interest. Second, the "MI" value is a maximum; all other points at other positions will be smaller. Third, the definition of MI depends on the shape of axial pressure which is dependent on focusing parameters, aperture size and apodization, frequency, and the degree of nonlinearity (drive amplitude). Fourth, the in situ derating factor may not have any relation to the tissue path or location of the region of interest. Therefore, comparisons of MI values among different systems and even among different settings in the same system are questionable because they are only relative values. The MI for a given setting on the same system can be compared to the same circumstances on another identical system of the same manufacturer.

Nonetheless, a way to use MI to estimate p_r levels has been proposed (Church et al., 2008). In this method to estimate the water value of p_r, the value of MI is corrected for frequency and derating by utilizing the focal length F and transducer center frequency f_c as follows: $p_r = e^{0.069 f_c F} \sqrt{f_c} \mathrm{MI}$. Furthermore, to estimate p_r at another location, this water value is derated by the attenuation along the path, and divided by $\sqrt{f_c}$ to estimate a derated MI at a site.

Unlike the original application for MI, the introduction of contrast agents presented a different potential for cavitation (as discussed in Chapter 14 and shown in Figure 15.1). Different types of ultrasound contrast agents vary greatly in their response to the same insonifying field. An extensive review of bioeffects related to contrast agents was conducted by a group from the AIUM Bioeffects Safety Committee (Miller et al., 2008). Even though a number of the bioeffects had already been discussed in Section 14.7, additional effects for contrast agents, preventricular contractions (PVCs), and microvascular leakage were added. Although no meaningful health risks have been reported in connection with ultrasound contrast agents in the USA, premature ventricular contractions have been observed elsewhere in conjunction with ultrasound and contrast agents (van der Wouw, Brauns, Bailey, Powers, & Wilde, 2000; Zachary, Hartleben, Frizzell, & O'Brien, 2001, 2002). A number of studies on rats indicated that when PVCs occurred, they stopped when the ultrasound was turned off. The group advised that, in general, when working with contrast agents, an MI less than 0.4, and minimal agent dose and exposure time be used. An MI of 0.4 corresponds to the value of peak rarefactional pressure amplitude (PRPA) of $p^{1.67} = 0.13 f_{MHz}$, or $p = 0.3 f_{MHz}^{0.6}$, the value at which inertial cavitation in blood is predicted (Note, for the rat studies there is practically no in situ attenuation because of the short path lengths). In addition, based on data collected so far, they recommended that "high MI" values of >0.8 involve rapid gas-bubble destruction with possible PVCs but not at $p/\sqrt{f_{MHz}} \leq 0.2$. Finally, they advised more be done to understand contrast agents and devise contrast-agent-specific indices. With regard to the fetus, probably the most sensitive site, Stratmeyer et al. (2008) did not find evidence supporting a causal relationship between the use of diagnostic ultrasound during pregnancy and adverse nonthermal ultrasound-induced biological effects to the fetus (with commercially available devices predating 1992 having outputs not exceeding an intensity $I_{SPTA.3}$ of 94 mW/cm^2). Abramowicz et al. (2008) reviewed ultrasound-induced thermal bioeffects in the fetus and concluded that it is difficult to conduct large-scale studies to determine if there are any subtle ultrasound-induced fetal bioeffects. They advised caution in exams in the first trimester, a time at which perfusion mechanisms are not yet developed, and also in the use of spectral Doppler on the fetus.

Caution is advised in imaging febrile changes and the fever should be included in temperature elevation considerations. They agreed with the WFUMB recommendation that an upper limit for the fetus would be a temperature elevation for no longer than 5 minutes. In general, they advised a combination of TI and dwell time be applied with ALARA (the "as low as reasonably achievable" principle). Even though TI may not represent actual temperature increases, it is a useful relative indicator and could be included as part of the imaging session record.

Re-evaluations of the ODS and thermal indices were published recently. Barnett et al. (2000) also reviewed the ODS and issued their recommendations and those of other ultrasound imaging societies under the auspices of WFUMB. The British Institute of Radiology (Ter Haar et al., 2012) issued the third edition of the most informative book on the safe use of ultrasound available online. Written by world experts, this reference book is a compendium of useful information about the ODS, diagnostic imaging, regulatory issues and standards, and, of course, guidelines on the safe use of ultrasound. Bigelow et al. (2008, 2011) found that the thermal indices, though useful, did not give a more practical indication of risk and recommended alternatives such as including the effect of exposure time and the temperature elevation. Ziskin and Szabo (1993) proposed a scheme of incorporating TIs in the temperature exposure-time relation, Eqn 15.1. Church (2007) and Church and O'Brien (2007) proposed a new temperature safety index that accounted for time. Karagoza and Kartal (2009) introduced the concept of safe use-time.

Edmonds, Abramowicz, Carson, Carstensen, and Sandstrom (2005) emphasized that TI and MI values are inadequate for reporting acoustic output exposure for clinical and animal studies for publications. They recommended independent measurements of output as well as exact system settings and system identification be reported, so that experiments could be repeated elsewhere and the conditions for ultrasound-induced bioeffects could be understood with more consistency.

15.6 Ultrasound-induced Bioeffects: A Closer Look

15.6.1 Introduction to Interrelated Bioeffects

In order to develop a more comprehensive viewpoint that will embrace more medical ultrasound modalities than just diagnostic ultrasound, interrelated acoustic mechanisms and effects need to be considered. The concepts of Figure 15.2 can be re-examined; the basic elements are the ultrasound source, the field generated, the tissues involved, and their response. The processes involved are captured by the diagram in Figure 15.6. At the top is the transducer or array, which operates in pulsed mode, does beamforming, and emits a measurable time-average acoustic power at a certain pressure amplitude, which is either in a linear or nonlinear range. The transducer emits a longitudinal pressure wave-field propagating along a beam axis and of finite beam extent into a tissue with specific mechanical, thermal, and viscoelastic properties and structure. During propagation, local absorption and phase velocity dispersion steal energy from all parts of the beam; some scattering and reflections occur from boundaries, and the heterogeneous nature and structure of the tissues traversed. According to Eqn 15.2, this absorption functions as an expanding heat source proportional to the product of the attenuation coefficient and time-average intensity ($2\alpha I$), and the tissue heats up. If the acoustic pressure amplitude is sufficiently nonlinear in a region such as the focal region, an acoustic radiation force field follows the gradient of the acoustic pressure field and the diffusion of the particle velocity field. The radiation forces generate expanding shear wavefronts in roughly horizontal or transverse directions. The tissues and cells have various responses, like heightened enzyme activity with increased temperature or complete destruction, depending on the beam-shape and acoustic pressure waveform shape and amplitude. A range of responses and intended outcomes will be examined in the following sections. Finally, as discussed extensively in the last sections of Chapter 14, if a vessel with ultrasound contrast agents or particles is involved, the acoustic radiation forces will move them around and create microstreaming. The insonification of the ultrasound contrast agents will cause them to cavitate in different forms depending on pressure amplitude, as discussed in more detail in Chapter 14.

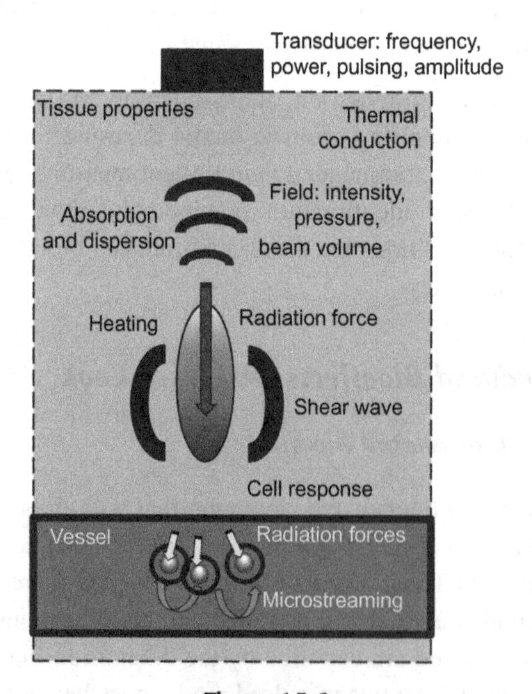

Figure 15.6
Interrelated Acoustic and Biophysical Events Involved in Ultrasound-induced Bioeffects. The reader is referred to the Web version of the book to see the figure in color

15.6.2 The Thermal Continuum

Ultrasound-induced thermal responses of tissue have received considerable attention. A classic paper on thermal response (Sapareto & Dewey, 1984) was intended to be a study of temperature elevation required to achieve hyperthermia or cell death. They found that by determining an equivalent time, a thermal dose relationship could be used to determine exposure time. The key relationship for cumulative equivalent minutes at 43°C at a point (*x*,*y*,*z*) (Soneson & Meyers, 2010) is:

$$CEM_{43}(x,y,z) = \int_0^{t_f} R^{b(T(x,y,z,s)+T_0-43)}ds, \quad R = \begin{Bmatrix} 0.25 \text{ if } T \le 43^{\circ}\text{C} \\ 0.5 \text{ if } T > 43^{\circ}\text{C} \end{Bmatrix} \tag{15.9}$$

in which t_f is the duration of the exposure; $b = (1^{\circ}\text{C})^{-1}$, a constant to render the exponent dimensionless; T_0 is an equilibrium temperature (often set to 0); and T is temperature. The function $T(t)$ of temperature versus time could be one like that shown in Figure 15.4. It is important to note that the actual temperature is represented by T and t_f is the exposure duration and not the time for which a tissue is held at the maximum temperature (Miller & Dewey, 2003). This relation can be expressed in digital form as:

$$t_{43}(T_0, t_f) = \sum_{n=0}^{N} R^{b(T(\Delta t)+T_0-43)}\Delta t \tag{15.10}$$

where $t_f = N\Delta t$ and Δt is the time increment. Once the time $t_{43} = CEM_{43}$ is determined, then other times of exposure can be found from the relation:

$$\tau_S = t_{43}(1/R)^{b(43-T)} \tag{15.11}$$

in which we will henceforth drop b for convenience. This equation is related to Eqn 15.1 in which $t_{43} = 1$ min. Similarly, the relation derived from WFUMB numbers, Eq 15.5, rewritten in this form has $t_{43} = 0.308$ min (or 18.5 s). For the original cell-death work (Sapareto & Dewey, 1984), the value $t_{43} = 240$ min is needed to achieve hyperthermia. In those days, hyperthermia was not achieved by ultrasound but other means, and the original work was done on Chinese hamster ovary cells. The solid line of Figure 15.3 and Eqn 15.1 were derived as a threshold below which there were no adverse effects reported in animal experiments. Miller (2002) realized that the original data used needed to be corrected relative to the core temperatures of the animals involved:

$$\Delta T = T - T_c + \Delta T_f \tag{15.12}$$

Here, it is the temperature elevation, T_c, above the core temperature that is important. If there is a temperature elevation caused by fever, ΔT_f is included; otherwise this term is zero. For humans, $T_c = 37^{\circ}\text{C}$.

It is useful to determine the temperature corresponding to a given exposure time, here $\Delta T = T - 37$, and rewriting Eqn 15.11:

$$\Delta T = 6°\text{C} + \frac{1}{b}\frac{\log_{10}\frac{\tau_{43}}{\tau_S}}{\log_{10}(R)} \tag{15.13}$$

For the Chinese hamster, $T_c = 39.5°\text{C}$, or $\Delta T = 3.5°\text{C}$, not the 6°C for humans. Once Miller (2002) corrected the original data, a number of corrected points fell below the curve from Miller and Ziskin (1989), as shown in Figure 15.7 adapted from Church and Miller (2008) and Miller and Dewey (2003). The scheme Miller used was numerically equivalent to previous results but reframed by different variables from Eqn 15.12 and

$$\Delta T_{ref} = T_{ref} - T_c \tag{15.14}$$

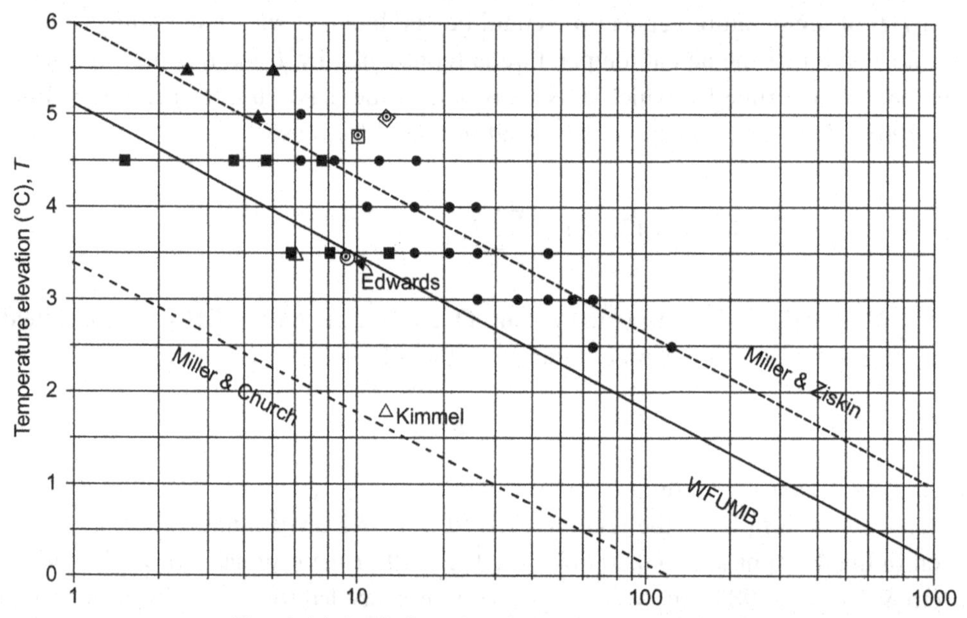

Figure 15.7
Thermal-equivalent core temperature elevations (°C) vs time (min). Each datum was derived from an analysis of a heating–cooling profile of a pregnant laboratory animal (rat, mouse, or guinea pig) that yielded fetuses with teratologic anomalies. The dashed line is the equivalent lower boundary from Miller and Zisken (1989) before the data was corrected for the animals' core temperatures, and is shown here corrected. Also shown as an open triangle is the lowest positive-result datum, from Kimmel et al. (1993). The open squares and arrows show the "movement" of the Edwards' (1969) data point from its original location at (60 min, 43°C) to its "final" location at (8.9 min, 3.5°C). For details, see Miller et al. (2002). The solid curve is the WFUMB recommended threshold and the dash-dot curve is the lowest threshold for all points. *Adapted from Church and Miller (2007)*

For example, if $T_{ref} = 43°C$ and $T_c = 37°C$ then $\Delta T_{ref} = 6°C$. Reframed, Eqn 15.11 becomes:

$$\tau_s = t_{ref}(1/R)^{b(\Delta T_{ref} - \Delta T)} \tag{15.15}$$

where $t_{ref} = t_{43-37} = t_6$. And the companion equation for temperature is:

$$\Delta T = \Delta T_{ref} + \frac{1}{b}\frac{\log_{10}\frac{\tau_{ref}}{\tau_S}}{\log_{10}(R)} \tag{15.16}$$

Plotted against a backdrop of data are curves from Marvin and Ziskin (1989) and the WFUMB recommended limit (both curves from Figure 15.3). The WFUMB is below most but not all data points for fetal adverse events from the literature. In order to devise a threshold below all known points, Church and Barnett (2012) suggested the lower curve. It is interesting to note that the experimenters whose data were the lowest in threshold values, Edwards (1969) and Kimmel et al. (1993), recommended thresholds of about 1.5 or 2°C. For the curves shown, $\Delta T_{ref} = 6°C$ and $t_{ref} = 1$, 0.308, and 0.03125 minutes, respectively, for the curves, top to bottom, in Figure 15.7.

For nonfetal or postfetal cases, O'Brien et al. (2008) have generated a combined lower boundary threshold curve for these cases. The curve can be constructed from Eqn 15.16 and the following values and ranges, with t' given in seconds and $\tau_{ref} = 1$ min $= (t'/60)$ for all cases:

$$\begin{array}{lll} \Delta T_{ref} = 9, & R = 2, & 0.1 < t' < 5 \\ \Delta T_{ref} = 6, & R = 2, & 5 < t' < 60 \\ \Delta T_{ref} = 6, & R = 4, & t' \geq 60 \end{array} \tag{15.17}$$

Finally, the curves corresponding to Sapareto and Dewey cases for hyperthermia can be plotted (in Figure 15.8). For this curve, $\tau_{ref} = 240$ min and $\Delta T_{ref} = 6°C$. Even though there is a wide range of thermal dose $\tau_{ref} = t_{43}$ values (O'Brien et al., 2008), the value used here fits well with observed HIFU lesion thresholds (Bailey, Khokhlova, Sapozhnikov, Kargl, & Crum, 2003).Variations in lesion thresholds related to pulse length and intensity are discussed in Harris, Herman, and Myers (2011) and O'Brien et al. (2008). Related discussions are in Herman and Harris (2002, 2003) and Miller and Dewey (2003). The top curve of Figure 15.8 represents cell death and the bottom curve is a boundary below which most (but not all) adverse ultrasound-induced thermal fetal bioeffects fall (WFUMB curve). These curves demarcate regions in which thermal ultrasound-induced bioeffects have and have not been observed and provide a thermal continuum reference by which different medical ultrasound modalities can be compared.

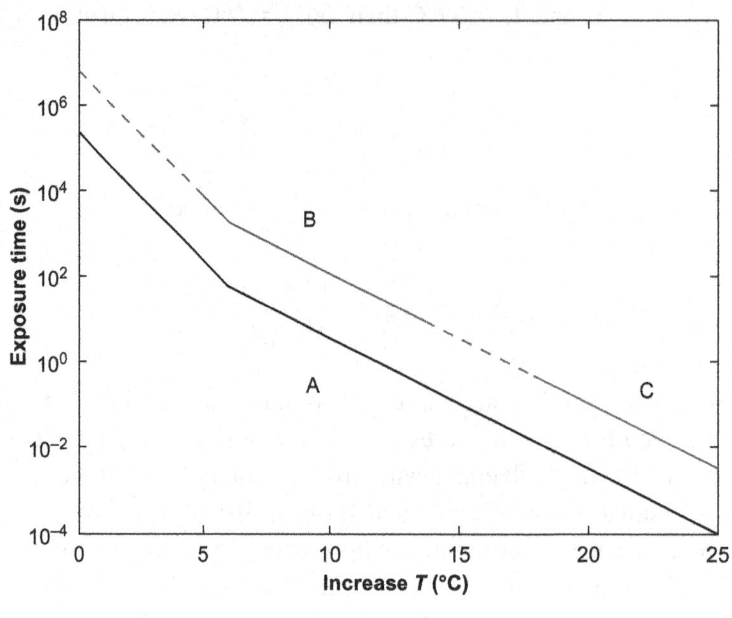

Figure 15.8
Upper and Lower Bounds for Thermal Ultrasound-induced Bioeffects. WFUMB recommended values for diagnostic ultrasound (lower curve); Sapareto–Dewey relation as lower bound for cell death (hyperthermia) or known irreversible effects (upper curve) A, a region of reversible effects, B, a hyperthermia region, and HIFU the region far above the top curve and towards C. In between the curves is a grey area.

15.6.3 Nonthermal Effects

The nonlinear nature of tissues, including fluids, creates several other secondary effects that have been described in Chapter 12. Acoustic streaming and microstreaming (patterns of circulatory flow) cause a gentle movement of fluids and possible aggregation and agglutination, may alter transport across biologic membranes, and can have cooling effects (Wu, Winkler, & O'Neill, 1998). Acoustic radiation forces and torques are also by-products of nonlinear properties of tissues (Duck, 1998). Radiation forces occur only during an acoustic pulse, and they are much smaller than the tensile strength of tissue (Starritt, 2000). When applied as pulses to the ear or skin, they can evoke auditory or tactile responses (Dalecki, Child, Raeman, & Carstensen, 1995). These secondary effects are small and are being investigated. They can also be beneficial; for example, in the transport of drugs in a vessel or for use in discriminating between fluid-filled and solid cysts in the breast (Nightingale, Kornguth, Walker, McDermott, & Trahey, 1995).

Figure 15.6 symbolically represents a chain of acoustic events. The transducer transmits an acoustic field of pressure p and particle velocity v. Propagation in viscoelastic tissue with an absorption coefficient α diminishes the pressure. This process generates heat as well as an increased pressure gradient. From these changes, an acoustic radiation force develops, slowly expanding and pushing on the tissue to create shear wavefronts. Simplified linked equations tell the story of interrelated effects in mathematical form. First, absorption occurs:

$$p_z = p_0 e^{-\alpha z} \tag{15.18A}$$

and

$$I_z = p_0^2 e^{-2\alpha z} / \rho_0 c_0 \tag{15.18B}$$

Heat is produced (Eqn 15.2):

$$q_v = 2\alpha I \tag{15.18C}$$

which, in turn, becomes the source of a time-averaged volume acoustic radiation force (Eqn 12.29C),

$$F_z^V = q_v / c_0 = 2\alpha I / c_0 \tag{15.18D}$$

From the well-known connection between this force and its interaction with a perfectly absorbing target, in general we can relate the force to time-average power, W, if we assume that tissue is a less perfect absorbing target,

$$F_z = DSW / c_0 \tag{15.18E}$$

and, if the volume captures the spread of intensity, for simplification, let the net power reaching depth z be related to the source acoustic power, W_0 as:

$$W = W_0 e^{-2\alpha z} \tag{15.18F}$$

Or, the force integrated over a volume becomes:

$$F_z = DSW_0 e^{-2\alpha z} / c_0 \tag{15.18G}$$

The relationship between temperature rise and power is well established by the thermal index models, which take the form (Eqn 15.6):

$$\mathrm{TI} = \frac{W_0}{W_{deg}} \tag{15.18H}$$

in which W_{deg} is the power needed to raise heat in the tissue by 1°C. So through the relation of acoustic radiation force back to time-average power and heating, a cycle has been completed, since the heat caused the force in the first place. A missing piece of this puzzle is how the acoustic radiation force became horizontal (along x). The simple answer without invoking Green's functions in viscoelastic media (Section 12.8.4) is that, through a viscoelastic Hooke's law, a vertical force perturbation is coupled elastically to perturbations and forces in other directions, as given by the example in Section 3.3.1 and as symbolically illustrated by the finger Figure 3.8; therefore, $F_z \rightarrow F_x$.

With an example based on calculations by Church et al. (2008), let us look at the numbers. A tissue with an absorption of 6 nepers/m-MHz and an acoustic transmission at 2 MHz gives about 1 dB/cm. An MI = 1.9 leads to a spatial peak pulse average intensity, $I_{SPPA} = 240$ W/cm^2, with a duty cycle of 0.003 yields a maximum allowable $I_{SPTA} = 720$ mW/cm^2; the resulting acoustic radiation force is 112 N/m^3. This is a simplification because absorption would have nonlinear components.

15.6.4 Microbubbles

Even though the interactions and effects of microbubbles were discussed at some length in Chapter 14, some of them will be reviewed with regard to Figure 15.6. Acoustic radiation forces tend to push microbubbles around (at about 10 m/s, according to Church and Miller (2007)) depending on the direction and magnitude of the pressure field. The microbubbles, if they are excited to resonate, can expand to create a large surface area to reflect sound waves ($F = 2W/c$) and also absorb forces according to $\sigma_c I/c$, where σ_c is the extinction cross-section (Church & Miller, 2007). The absorbed energy heats up the bubbles. At the same time, the oscillating microbubbles are radiating acoustic forces and their motion creates microstreaming, which is greatest near their surface. If the microbubbles approach a tissue surface, the fluid or bubble contact can cause a pressure gradient on the wall as well as shearing forces. The exact mechanisms of enhanced delivery and permeability across tissue walls are still being investigated. Under stable cavitation, the effects appear to be reversible; the tissue reverts to its normal state. Above a pressure threshold, hemolysis can occur. If inertial cavitation occurs, highly localized shock waves create strong gradients, microjets, which may penetrate the tissue and disrupt the nearby structures. It is interesting to note that wall of the aorta has a tensile strength of 0.3 to 0.8 MPa, compared to 50–100 MPa for tendons and ligaments (Holtzapfel, 2000), and the heart has mean yield forces of 4 N and a rupture threshold of 4.9 N (Edwards et al., 2005).

15.6.5 Combined Effects

Many of the effects described are highly nonlinear. Acoustic streaming and microstreaming (patterns of circulatory flow) associated with microbubbles cause a gentle movement of fluids and possible aggregation and agglutination, may alter transport across biologic membranes, and can have cooling effects (Wu et al., 1998). At high pressures or intensities, other strange effects can occur. In Chen, Fan, Zhang, and Wu's (2009) study of higher intensities and temperature rises in a tissue-mimicking phantom, they found that above 45°C, cavitation bubbles appeared. Similar effects occur in tissue under HIFU and lithotripter insonification. Deliberate use of cavitation during HIFU exposure has beneficial effects (Coussios, Farny, ter Haar, & Roy, 2007; Coussios & Roy, 2008; Stride & Coussios,

2010). At even higher temperatures, there are instant cell death, coagulative necrosis, biochemical reactions, the release of gases, and vaporization.

15.7 Comparison of Medical Ultrasound Modalities

15.7.1 Introduction

Bioeffects involved with diagnostic ultrasound imaging can be compared to those associated with other medical ultrasound modalities. Ultrasound-induced bioeffects form a continuum of effects when all major modalities are examined. To put these modalities in perspective, some of the primary parameters for these modalities are symbolized by Figure 15.6. More information on each modality is available in the following sections.

15.7.2 Ultrasound Physiotherapy

Ultrasound physiotherapy has been in wide use for more than 60 years (Lehmann, 1990; Stewart, 1982). In the 1950s and 1960s, most medical ultrasound conference papers were about therapy, particularly what we now call hyperthermia and high-intensity focused ultrasound—not imaging. Main applications that have been reported to be of clinical value include reduction of muscular spasms; treatment of contractures; relief and healing of sports-related injuries; relief of pain; increased extensibility and treatment of contractures for collagen tissue (scar tissue) and connective tissues; heating of joint structures; treatment to improve limited joint motion; decrease in joint stiffness, arthritis, periarthritis, and bursitis; wound healing (Dyson, Pond, Joseph, & Warwick, 1970); and the healing of varicose ulcers. Ultrasound therapy includes many other applications, from cosmetic and postcosmetic surgery treatment, including sonopheresis to improve the penetration of products, to muscle treatment for racehorse injuries. Watson (2008) explains that physiotherapy can accelerate the inflammatory process and therefore shorten the repair process. Not all ultrasound therapy claims for efficacy in new applications have been substantiated clinically.

Ultrasound is mainly applied by physical therapists who are trained to place the transducer using a coupling oil or gel over a muscular area with a moving rotary motion to minimize dwell time. The output is limited typically to a maximum of 3 W/cm^2. A continuous-wave or pulsed mode is usually available, and a timer up to 10 min is usually required. In normal application, ultrasound can be applied near bones, where additional heating can occur, including shear wave conversion, so the therapist must remain vigilant, keep the transducer moving, and ask if the patient feels excessive heat.

Watson (2008) explains that the heating of tissues is related to absorption, which is proportional to the protein content; increasing through blood, skin, tendon, cartilage, to bone at the top of the heating scale. Reflection at boundaries may prevent sound from becoming absorbed. Also, the way ultrasound is applied in physiotherapy may not be the most effective way to reach targeted tissues. Baker, Robertson, and Duck (2001) argue that the heating is temporary, is mainly superficial, and is not as much as one would receive from heavy exercise. Homeostasis would serve to counteract the heating effect somewhat.

In summary, there is heated debate about many of the claimed benefits of physiotherapy. Many of the beneficial effects have only been tested on small animals and in vitro. So there is some question about how these would transfer to humans. Also, physiotherapy is applied globally for a variety of ailments to promote healing. Other claims are that streaming, cavitation, and increased permeability occur to aid in the healing process. Baker et al. (2001) dispute streaming, cavitation, and increased permeability are present and conclude that many of the claimed beneficial effects of physiotherapy are unsubstantiated by well planned in vivo experiments.

Can this controversy be resolved with the models and bioeffects discussed in this chapter? With reference to Figure 15.6, we can start with the transducer. Fortunately, because of international standards, physiotherapy transducers have been well characterized (Hekkenberg, Reibold, & Zeqiri, 1994). The ultrasound is usually applied with a single unfocused piston-like transducer in the 1–3-MHz range, continuous or pulsed (20–50% duty cycle) with continuous movement (short dwell time). Based on data, typical last axial maxima fall in the range of 1 to 8 cm. From Table 15.1, a key acoustic output parameter is the spatial average temporal average intensity (I_{SATA}) at the face of the transducer, which is the power divided by the effective radiating area. Because the upper intensity limit is $I_{SATA} = 3\ \text{W/cm}^2$, the recommended non-derated upper pressure limit is 1 MPa and usually is an order of magnitude lower. This pressure level is unlikely to excite cavitation even if there were nucleation sites, which are extremely improbable in the treatment regions as previously discussed except near the lungs and intestines, which are not usually in the path of physiotherapy transducers. Therefore, cavitation, microstreaming, and enhanced permeability can be ruled out.

With reference to Figure 15.6, heating could be estimated from Eqn 15.18C. From knowledge of the field of a piston source, a peak pressure would occur in water near the last axial maximum (typically 5 cm) and considerably closer with absorption in tissue. The location of this derated peak value would be called a local "hot spot," whereas the general business of heating large regions of tissue would be governed by the average intensity value. From data for the upper bound case (Hekkenberg et al., 1994), $I_{SPTA} = 24\ \text{W/cm}^2$, and the same absorption (6 nep/m) and 2-MHz frequency used previously, and Eqn 15.18B, $I_{SPTA} = 13\ \text{W/cm}^2$, $q = 3.12 \times 10^6$. However, because the transducer is being moved continuously (short dwell time), this value is unlikely to be achieved in practice. Also, the acoustic radiation force (ARF) at this location can be calculated from Eqn 15.18D, $F = 2000$ N, or about 20% of the force of gravity. At this ARF level, significant shear wave generation is unlikely. Much of the discussion about heating and forces is straightforward and predictions about heating distributions in different parts of the body can be made with confidence. What are lacking are explanations of tissue response mechanisms and a wide enough range of in vivo experiments to determine which combination of parameters produces healing effects. Cells respond to changes in their environment. The combination of simulation models with acoustic output and bioeffect response measurements may lead to more specific and effective experiments and understanding. Paliwal and Mitragotri (2008) examine biological responses to therapeutic in more depth.

15.7.3 Hyperthermia

Hyperthermia with ultrasound is a means for insonifying cancerous tissues to heat them to a range of 41–45°C to stop their growth, often in conjunction with other therapies such as chemotherapy (Hand, 1998; Hynynen, 1998). The Sapareto and Dewey curve of cell death in Figure 15.8 provides the lower bound for determining the combination of temperature elevation exposure times and temperature levels. Hyperthermia systems (Diederich & Hynynen, 1999) consisting of arrays of planar piston transducers have been built to cover large surface areas for superficial cancer growths. For deep-seated tumors, strongly focused transducers, often in an overlapping arrangement, have an advantage over other, non-ultrasound, hyperthermia methods in directing heat selectively to an interior region of the body. Phased linear two-dimensional and annular arrays have been investigated (Ebbini & Cain, 1989). Intracavitary devices for hyperthermia are primarily transrectal, for application to the prostate. Stauffer (2005) provides a more recent overview of hyperthermia devices, including electromagnetic ones.

For conventional applications, the temperature of cancerous tissue is elevated to 42–43°C for an extended period of 30–120 min. In accordance with the Sapareto–Dewey exposure law, equivalent doses can be administered at higher temperatures and shorter durations or vice versa. Considerable attention must be directed to accounting for the cooling effects of perfusion from both large and small vessels, and to avoid bones in the acoustic path (Lele, 1972; Newman & Lele, 1985).

In terms of acoustic output for hyperthermia, I_{SATA} ranges from 2 to 10 W/cm^2 for superficial application, and low-duty-cycle, high-peak intensities up to 1 kW/cm^2 are used for deeper, pulsed applications (Hynynen, 1987). Transducers are not scanned but held still. The heating mechanism is proportional to the absorption and the acoustic intensity delivered to the target site according to Eqn 15.18C.

15.7.4 High-intensity Focused Ultrasound

Surgery with ultrasound can be achieved by focusing high-intensity ultrasound (a few kW/cm^2) on a tissue region to instantaneously produce a lesion (Fry, 1979; ter Haar, 1995, 1998). Exposure is typically 5–10 s, and temperatures of 60–100°C are achieved (Chen, ter Haar, & Hill, 1997). The ellipsoidal shape of the lesion corresponds roughly to the −6-dB contour of the beam near the focal point. The heating is very selective and confined to a small region in the order of 1−10 mm. When done well, the lesion is "trackless," or there is no tissue damage between the skin entry point and the lesion itself. A series of lesions, with time intervals in between for cooling, can be created in a pattern to cover a larger area. The advantage of ultrasound surgery is that no incisions are required; it is precise, and, if done properly, there is little extra trauma to the patient.

To first order, assumptions in modeling the heating are no perfusion and the same heating mechanism based on absorption (described earlier). The heating is very rapid so that the heating process is very nonlinear. Ultrasound surgery creates instantaneous coagulative necrosis, since temperatures in excess of 60°C are used (see Table 15.3). At very high temperatures, tissue can be boiled and vaporized, and the resulting gases can block the penetration of ultrasound and cavitate (Chen et al., 1997; Sanghvi et al., 1995). These events can cause uncontrollable results and are to be avoided (Meaning, Cahill, & ter Haar, 2000). However, bubbles can be, alternatively, considered as enhancers to the heating process (Holt & Roy, 2001). The effects in this range of exposure are collectively called "super MI effects" in Chapter 14.

Commercial HIFU systems are available in several countries, including the USA (ter Haar, 2007). A number of HIFU systems are being used in China, where tens of thousands of people have undergone HIFU surgery, mainly for tumors. Work is continuing to refine the methodology. HIFU will be covered in greater detail in Chapter 17.

In terms of acoustic output, pulses on the order of 0.1−10 s in length are used for each lesion and are repeated about every 10 s for other locations. Like the other modalities, the primary mechanism in HIFU is heating, but unlike them, the heat is produced rapidly with much higher peak intensities, I_{SATA}, in the range of 1−10 kW/cm^2 averaged over the −6-dB contour in the focal plane.

Once again, with reference to Figure 15.6, we can identify the mechanisms associated with HIFU. The acoustic field is very sharply focused from an extracorporeal quasi-bowl-like large-diameter transducer with resultant rapid heating and instantaneous coagulative necrosis. Equivalent exposure times and temperature elevation combinations are well above the Sapareto and Dewey curve in Figure 15.8. For example, one combination is a temperature elevation of 19°C for 1 s (Kennedy, ter Haar, & Cranston, 2003). Strong localized acoustic radiation forces and shock and shear waves would be expected. At these high temperatures, tissue may or may not be completely vaporized, but can be sublimated into chemical components and released gases, with the rarefaction half-cycles drawing out gases. These gases, in the form of small gaseous nucleation sites, can interact with the insonification to cause cavitation to enhance local tissue damage. As part of this evolving process, forces both pressure and acoustic will interact with the generated microbubbles to cause inertial cavitation and bubble movement. More details will be provided in Chapter 17.

15.7.5 Lithotripsy

Extracorporeal shock wave lithotripters (ESWLs), introduced in 1984, are noninvasive devices designed to disintegrate kidney stones and other types of stones without surgery (Delius, 2000; Halliwell, 1998). Two excellent reviews are Leighton and Cleveland (2010) and Rassweiler et al. (2011). Typically, the patient is placed in a water tank or coupled through a water bag, and a high-amplitude, high-focal-gain transducer/reflector is focused on the kidney. Stones are fragmented into small pieces and pulverized by the repeated action of lithotripter pulses and are naturally passed out of the patient. The hard stones absorb most of the pressure so that, ideally, the surrounding tissue is less affected. The mechanisms of destruction are believed to be a combination of cavitation (Coleman & Saunders, 1993) and cyclic mechanical stressing (fatigue): tear and shear forces, spallation, and quasi-static and dynamic squeezings (Leighton & Cleveland, 2010; Rassweiler et al., 2011). Types of lithotripters include spark discharge (electrohydraulic), piezoelectric, electromagnetic magnetostrictive, and chemical (explosive charges). Many models are in widespread use because 13% of men and 7% of women in the USA are expected to be afflicted by kidney stones (Leighton & Cleveland, 2010).

The acoustic output of lithotripters proved to be difficult to measure as hydrophones were destroyed in the process. Eventually this problem was solved, and lithotripter waveforms in water typically have a high positive p_c part with a peak in the 30–150-MPa range (Coleman & Saunders, 1989; Rassweiler et al., 2011) about 1 μs long followed by a shallower negative p_r part in the −3 to −15-MPa range (Delius, 2000) about 5 μs long (Leighton & Cleveland, 2010). The shape of the waveform (in particular, the extremely large positive-to-negative change in pressure within the waveform) is believed to play a role in the fragmentation process. The waveform starts with an extremely steep rise time (nanoseconds), the p_c/p_r ratio is typically 4–10 (Harris, 1992). A key difference between diagnostic imaging and lithotripsy is that the center frequency of lithotripter pulses is about 100–600 kHz (Delius, 2000), typically 150 kHz (Leighton & Cleveland, 2010) and the transient portions (nanoseconds rise time) of the waveform extend over a broader bandwidth. Another difference is a low duty cycle (typically on the order of one to two pulses a second); consequently, the I_{SPTA} is extremely small. The process is repeated for about 3000 shock waves or until the stones are observed to be fragmented through monitoring as shown in Figure 15.9 (Leighton & Cleveland, 2010). For measurement purposes, PVDF hydrophones are used (Halliwell, 1998; Lewin, Chapelon, Mestas, Birer, & Cathignol, 1990).

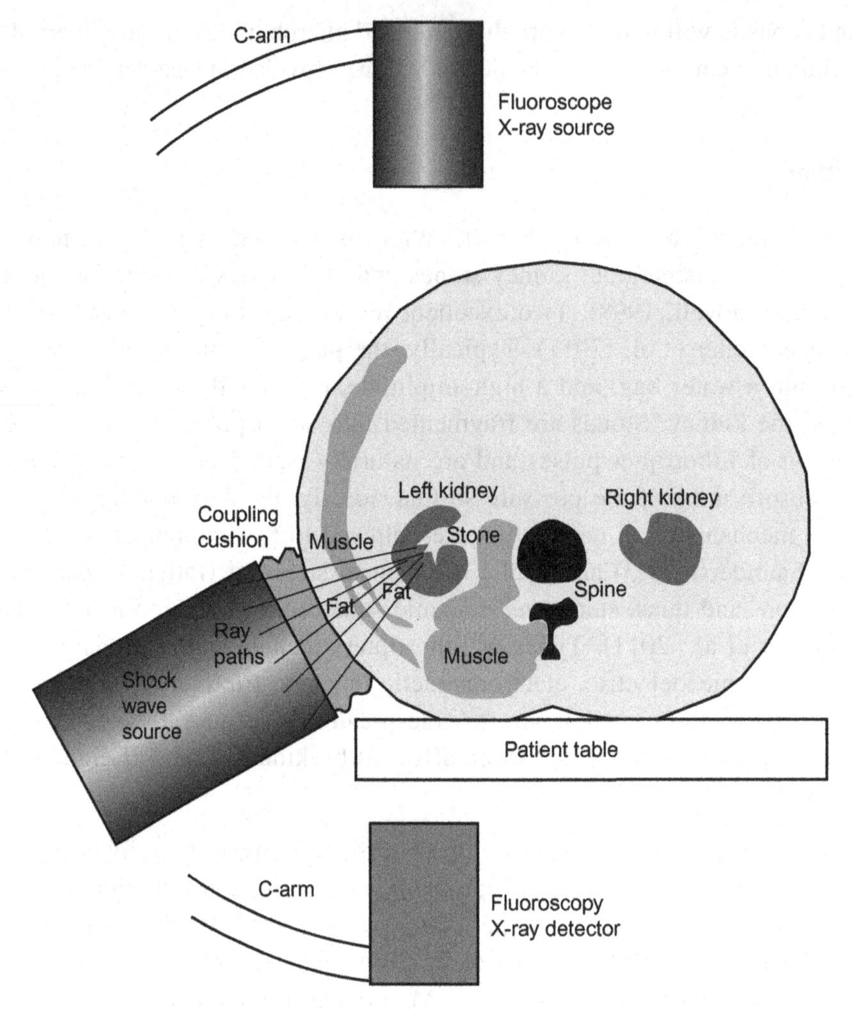

Figure 15.9

Schematic diagram showing the basic components of a lithotripter. The patient is positioned on their back on a table. The shock wave source is placed in contact with the patient by means of a water-filled coupling cushion with a gel or fluid employed to ensure good coupling. The stone is targeted by means of fluoroscopy (as shown) or by means of in-line or out-line ultrasound (not shown). The focused shock wave has to pass through layers of fat and tissue before it reaches the kidney stone. *From Leighton and Cleveland (2010)*

In reviewing the bioeffects involved we can immediately eliminate direct heating as a mechanism because of the low duty cycle, about $5-10 \times 10^{-6}$. Obviously, the high pressures are the dominant mechanism, the fast rise times and the short pulse are designed to create nonlinear shock waves, and are the reason why these sources are called "shock wave" devices. In addition, the rarefactional part of the pulse, usually well above diagnostic levels, is enhanced by the low center frequency (0.15 MHz) to create an MI-style cavitation threshold of about 2.5 p_r.

The physics of the interaction of the shock wave with the stone are complex, and are a strong combination of high tensile and shear stresses. Explosive inertial cavitation events are drawn out of the small fluid pockets in the stone. Some localized heating accompanies the cavitation events and rapid elastic changes. The severe stresses set off complicated reflections and mechanical fatiguing cycles (progressive development of cracks). More detail can be found in Leighton and Cleveland (2010) and Rassweiler et al. (2011). The high pressures also incur collateral tissue damage, especially in any fluid-containing vessels, and a goal is to minimize the damage by improved design (Leighton & Cleveland, 2010).

15.7.6 Diagnostic Ultrasound Imaging

Even though derated values are usually considered for the acoustic output of diagnostic ultrasound imaging, water values will be used for discussion to be consistent with the water values for other medical ultrasound modalities given in Table 15.1, for comparison purposes only. Diagnostic values in Table 15.1 were taken from a survey of 82 imaging systems by Henderson et al. (1995). For each parameter, a range of water values is given, as well as the median (most frequently occurring) value in parentheses. These values in the table represent an increase over those obtained by Duck and Martin in 1991, just before the ODS was introduced with revised FDA values (see Table 15.2).

In order to put these water values into meaningful contexts, clinical applications need to be considered. For almost every medical ultrasound modality considered that has a bioeffect, absorption plays a dominant role. For example, consider the water values of I_{SPTA} that varied over a considerable range. From the survey of acoustic output by Henderson et al. (1995), 90% of the values fell below 3 W/cm^2, and the highest value, 9.08 W/cm^2, had a corresponding power of only 130 mW. If the manufacturers complied with the FDA requirements, which is highly likely, then the highest derated intensity corresponding to the measured values listed in Table 15.2 is $I_{SPTA.3} = 720$ mW/cm^2. Derating is the product of the water intensity as a function of depth multiplied by an exponential derating factor as a function of depth. The net result is a shifting of the position of the peak value as well as a reduction in value (as illustrated by Figure 13.15). The same effect results if another value of absorption, more appropriate to a given clinical situation, is used. A single peak water value is insufficient information to apply derating appropriate to a clinical situation.

Ideally, the user would have full knowledge of all the tissues and their size along the acoustic path and be able to calculate the absorption to the target region of interest while performing the exam. Smith, Stewart, and Jenkins (1985) have shown through measurements of a cadaver that reasonable estimates of acoustic intensity can be made from a detailed knowledge of the acoustic properties for the individual layers, including power law absorption, impedance and sound speed, and layer thickness. To improve on a layered model would be more involved. From the discussions of aberration and scattering of real tissue in Chapter 9 as well as of the beam simulations in Chapter 12 for beams passing through abdominal walls, the use of homogeneous layers may overestimate the coherency of a beam and its ability to focus through realistic body walls. The homogeneous layer approach can serve to estimate the most coherent propagation possible, equivalent to the worst case for safety purposes.

At this stage in the evolution of ultrasound imaging, however, the system user does not have the detailed information on loss along the acoustic path, but does recognize the clinical application and target site. The user may decide to accept an average conservative value such as the derating factor applied to the calculation of output display indices with the realization that, in certain circumstances, the losses may be underestimated. One possible underestimated situation often debated is the path to the fetus. This topic is extensively reviewed in Section 9.3.4 of the NCRP report (NCRP, 2002). The present derating factor, NCRP researchers concluded, is reasonable for second and third trimester exams but not conservative enough to cover all obstetric cases, and they provided options. For these exceptional cases, with a long fluid path, the user can use the output display indices to lower the transmit level to the minimum level necessary to obtain an image. This strategy is consistent with the ALARA principle, which is to adjust acoustic output to a level "as low as reasonably achievable" (Ziskin & Szabo, 1993). This principle is endorsed as an official statement by the AIUM (March 16, 2008) as:

> *The potential benefits and risks of each examination should be considered. The ALARA (As Low As Reasonably Achievable) Principle should be observed when adjusting controls that affect the acoustical output and by considering transducer dwell times. Further details on ALARA may be found in the AIUM publication "Medical Ultrasound Safety".*
> *The ALARA principle implies that, despite the long safety record of diagnostic ultrasound, we still do not have a complete understanding of ultrasound-induced bioeffects and, consequently, especially for those cases for which more care is recommended, it is better to err on the side of caution. How do the acoustic output levels of diagnostic ultrasound compare with other medical ultrasound modalities? From Table 15.1, a principal difference is that for most diagnostic applications,* I_{SPTA} *is, on the average, less than* $1\ W/cm^2$*, whereas ultrasound therapy and hyperthermia typically have* $3-10\ W/cm^2$*. While an* I_{SPTA} *value in an imaging plane may cover a region of 1 to a few* mm^2*, ultrasound therapy and superficial hyperthermia are applied directly over areas typically larger than* $10\ cm^2$ *with much higher intensities. For these two modalities, the intended outcomes are to raise the temperature of tissue by several degrees centigrade. For diagnostic imaging, TI formulas of the ODS with derated values or alternative methods (NCRP, 2002) can be used to estimate temperature elevation and ALARA can be used to minimize exposure time.*

Both HIFU and lithotripsy employ very high amplitude pressure pulses that can cause cavitation. Unlike the three diagnostic pressure levels of resonance and cavitation described for contrast agents in Section 14.5.2, these extremely high pressures for lithotripsy and HIFU can operate at a super, or fourth, level, at which tissues and materials can break down, causing biochemical effects that produce gas nuclei for cavitation. These effects are not yet understood but have been observed. For these two latter applications, pulse shapes and pressure amplitudes are quite different from those used for diagnostic ultrasound: compressional pressures in water typically in excess of 75 MPa followed by −15 MPa of rarefactional pressure in repeating pulses microseconds long for lithotripters, and extremely high intensity insonifications of several seconds duration for HIFU.

In summary, in terms of bioeffects, these modalities have intended outcomes that distinguish them from diagnostic imaging. The acoustic output level, pulse shape, frequency range, and/or region of application (summarized in Table 15.1) are substantially different than those used in diagnostic imaging.

15.8 Equations for Predicting Temperature Rise

The highest temperatures along the beam axis are not that sensitive to beam details, and for circularly symmetric transducers, they can be calculated by a heated-disk model (NCRP,

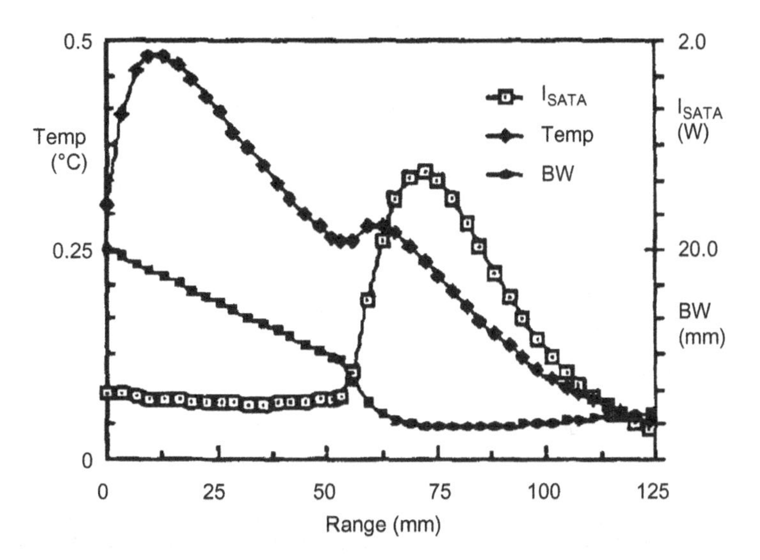

Figure 15.10

Axial profiles calculated from the heated-disk model. For a 3-MHz transducer with a diameter of 20 mm and a spherical focus of 100 mm. Thermal indices (TIs) from the ODS for different modes and configurations. Shown are spatial average temporal average intensity (I_{SATA}, W/cm^2), temperature rise (Temp), and −6-dB beamwidth (BW). *From Thomenius (1990), IEEE*

1992; Thomenius, 1990) in which the diameter of the disk is given by a −6-dB beamwidth at each axial distance z (Figure 15.10). Here, the volume rate of heat generation due to absorption can be modeled as proportional to the acoustic intensity (I) and absorption at a single frequency:

$$q_v = 2aI = 2aI_0exp(-2az) \qquad (15.19)$$

For a tissue with an absorption α, the heat generated at a time-averaged rate per unit volume is given locally by q_v:

$$q_v = 2aI_i(t, r) \qquad (15.20)$$

where I_i is the instantaneous intensity based on the assumption that the pressure and particle velocity are in phase. The value for I_i is often based on the focused field of the transducer. The volume rate of heat generation (q_v) due to absorption can be modeled as proportional to the acoustic intensity, $I(x,y,z)$, and absorption at a single frequency,

$$q_v = 2aI \qquad (15.21)$$

For a more realistic estimate of heat generation in tissue, the Pennes bioheat-transfer equation is used often:

$$\frac{1}{\kappa}\frac{\partial T}{\partial t} + \Delta T/L - \nabla^2 T = q_v/K \qquad (15.22)$$

where:

T_a = ambient temperature.
T = temperature.
$\Delta T = T - T_a$.
q_v = heat source function.
κ = thermal diffusivity.
K = thermal conductivity.
τ = perfusion time constant.

A perfusion length L, which defines the region of influence of a heat source, is given by:

$$L = \sqrt{\kappa\tau} \qquad (15.23)$$

Other useful quantities are summarized in Table 15.4. The terms in the bioheat equation on the left-hand side are for heat diffusion, heat loss from blood perfusion, and heat conduction; the term on the right-hand side is the heat source from the acoustic beam, Eqn 15.11. More information and solutions for the bioheat equation can be found in Nyborg (1988) and Thomenius (1990).

O'Brien (1996) introduced a monopole source model that is in general agreement with the heated-disk model and which shows that for rectangular apertures, heating is less than that predicted for circular apertures.

In their studies of heating for a HIFU phantom, Chen et al. (2009) developed linear and nonlinear heating models. The linear model was based on Nyborg's earlier work (1988), in which a thermal Green's function was used to solve the bioheat-transfer equation. For the nonlinear case, a Gaussian nonlinear model was shown to be in reasonable agreement with data until the intense temperatures in the phantom produced bubbles as discussed in Section 15.6.5.

15.9 Conclusions

As new applications of diagnostic ultrasound outpace the complete understanding of the underlying physical mechanisms, the need for continuing research and debate continues. Considerable effort has been focused on ultrasound-induced bioeffects. Many important experiments on smaller animals have been conducted; however, the interpretation of results is not always straightforward to apply to humans. While they point to possible effects and trends, care must be taken in interpretation to account for differences in structure among species (O'Brien & Zachary, 1996, 1997), the relative size of structures relative to the insonifying wavelengths and beam, and dilution ratios of contrast agents, for example. How a human body will respond to a stimulus in vivo may not correspond to the response of a different species of a different size; challenges are in designing appropriate and statistically valid experiments.

Even though there is not enough space here to do justice to the many animal studies on bioeffects, the reader is referred to reports by societies such as the NCRP, AIUM, and WFUMB. The most comprehensive compilation of animal studies, recommendations, and guidelines for the safe use of ultrasound can be found in the NCRP report (NCRP, 2002). A shorter and slightly different viewpoint can be found in Barnett et al. (2000).

These reports emphasize those situations in which caution is recommended. These cases are sometimes referred to as *potential risks*. Rott (1999a) explains that this term states that there is no known real risk that could be quantified in numerical values, but because of an insufficient scientific data base, there remains the possibility of damage in worst-case situations.

Balancing this point of view is the risk-versus-benefit trade-off. This decision involves "a real expectation of obtaining diagnostic data that would have a beneficial effect on the continuing medical management of the patient" (Barnett et al., 2000). The risk aspects not covered in the reports above are the clinical consequences of not doing an ultrasound exam. In other words, will the lack of diagnostic information from the proposed ultrasound imaging pose a greater risk to the health of the patient than the potential risk of doing the exam? This aspect of the risk/benefit equation is not as well documented in an organized manner; however, the benefits affecting the well-being and medical management of the patient are the primary concerns in ultrasound examinations on a daily basis.

Ultrasound organizations are in general agreement that diagnostic ultrasound imaging is generally a very safe procedure (AIUM, 1998, 2002) (see Section 15.3.6). The debate centers on certain restricted cases in which some groups advocate more caution and a commonsense ALARA approach. Education about safety issues is important to increase the individual responsibility and awareness of ultrasound practitioners.

In fact, many societies and groups are involved in a worldwide effort to understand the safety issues connected with diagnostic ultrasound, including defining appropriate applications of ultrasound, as well as the adequate training and education of those using imaging equipment. A partial list of these groups, many of them part of WFUMB, is in Appendix D along with professional societies that frequently publish relevant articles. Some of these groups focus on a clinical application of ultrasound, publish guidelines and standards, sponsor continuing education programs, and serve as a forum for advancing ultrasound clinical application and research.

As diagnostic ultrasound has recently passed the 60-year mark of active use, it has done so with a remarkable safety record. Unlike computed tomography (CT) and especially multidetector CT, diagnostic ultrasound does not have an identified major safety concern like ionizing radiation (Ward, 2003). From the graph of the number of imaging exams given annually worldwide in Figure 1.14, one comes to the conclusion that the total number of ultrasound exams given must number in the billions! By 1997, in the USA, an estimated 75% of infants have been exposed to ultrasound before birth. In some countries, such as Germany, Norway, Iceland, and Austria, all pregnant women are screened with ultrasound.

Even though so many ultrasound exams have been given, this does not mean there is absolutely no risk involved. Most exams are carried out under different conditions and without long-term follow-up. There are few systematic studies of ultrasound on large populations over long periods of time. Some of these studies are reviewed in the NCRP report (NCRP, 2002). Under these circumstances, some vigilance is necessary in identifying the applications of ultrasound where risk is higher. As the science of ultrasound advances, some of the remaining questions will be answered as new questions take their place. Some questions may never be answered satisfactorily because of the complexity, adaptability, and variety of responses of the human body. Fortunately, experts from professional organizations are actively engaged in finding answers and developing guidelines.

As diagnostic ultrasound widens its horizons, the way sonography is conducted will be redefined. As new methodologies emerge with far greater diagnostic benefit potential, such as therapeutic contrast agents, they may also carry different risks than those considered now. For each case, the risk–benefit decision may become less straightforward than it is presently.

Out of many topics at the growing edge of ultrasound, two are highlighted here. The portability, accessibility, low cost, and good image quality of small ultrasound imaging systems are challenging where and how diagnostic imaging can be practiced. Consider two large-scale applications of ultrasound in which small systems provide a benefit in terms of the economies of scale: screening and surveillance. In several countries, pregnant women are part of a screening process as a preventive measure to identify possible fetal abnormalities that can be addressed during pregnancy. The USA-based Occupational Safety and Health Administration defines screening as:

> *a method for detecting disease or body dysfunction before an individual would normally seek medical care. Screening tests are usually administered to individuals without current symptoms, but who may be at high risk for certain adverse health outcomes.*

And defines surveillance as:

> *the analysis of health information to look for problems that may be occurring in the workplace that require targeted prevention, and thus serves as a feedback loop to the employer.*
> *OSHA, 2004*

Is the use of diagnostic ultrasound limited to cases in which it is medically indicated? What are the ethical implications when it is not? Portable systems with high image quality provide alternatives for many places that cannot afford fully loaded systems. The development of low-cost pocket ultrasound systems with lower image quality but greater accessibility could play a significant role in screening. What training would be required for these applications? Under these circumstances, how can high-quality ultrasound exams and care be maintained? Certainly, these alternatives provide many new opportunities for the growth of diagnostic ultrasound if the challenges can be met.

In this chapter, a wider perspective on ultrasound-induced bioeffects has been taken to embrace other medical ultrasound modalities. From this point of view, different ultrasound bioeffects can be seen as related by basic physics and biomechanics. Effects that may have been seen as harmful for diagnostic imaging have been put to beneficial purposes such as real-time HIFU surgery, lithotripsy, and hyperthermia to arrest the growth of cancers. These ultrasound applications, offset against the dangers and complications of surgery and harmful systemic drug therapies, are rewriting the risk–benefit balance and opening new horizons for medical ultrasound applications.

References

Abbott, J. G. (1999). Rationale and derivation of MI and TI: A review. *Ultrasound in Medicine and Biology*, *25*, 431–441.

AIUM (1993). *Bioeffects and safety of diagnostic ultrasound*. Laurel, MD: AIUM Publications.

AIUM (1998). *Bioeffects and safety of diagnostic ultrasound.* Laurel, MD: AIUM Publications.

AIUM (2000). *Mechanical bioeffects from diagnostic ultrasound: AIUM consensus statements.* Rockville, MD: AIUM Publications.

AIUM (2002). *Medical Ultrasound Safety.* Rockville, MD: AIUM Publications.

AIUM/NEMA (1998a). *Standard for real-time display of thermal and mechanical acoustic output indices on diagnostic ultrasound equipment, revision 1.* Laurel, MD: AIUM Publications.

AIUM/NEMA (1998b). *Acoustic output measurement standard for diagnostic ultrasound equipment.* Laurel, MD: AIUM Publications.

Bailey, M. R., Khokhlova, V. A., Sapozhnikov, O. A., Kargl, S. G., & Crum, L. A. (2003). Physical mechanisms of the therapeutic effect of ultrasound (a review). *Acoustical Physics, 49*, 369–388.

Baker, K. G., Robertson, V. J., & Duck, F. A. (2001). A review of therapeutic ultrasound: Biophysical effects. *Physical Therapy, 81*, 1351–1358.

Barnett, S. B., ter Haar, G. R., Ziskin, M. C., Rott, H. -D., Duck, F. A., & Meda, K. (2000). International recommendations and guidelines for the safe use of diagnostic ultrasound in medicine. *Ultrasound in Medicine and Biology, 26*, 355–366.

Bigelow, T. A., Church, C. C., Sandstrom, K., et al. (2011). The thermal index its strengths, weaknesses, and proposed improvements. *Journal of Ultrasound in Medicine, 30*, 714–734.

Calvert, J., & Duck, F. (2006). Self-heating of diagnostic ultrasound transducers in air and in contact with tissue mimics. *Ultrasound, 14*, 100–108.

Calvert, J., Duck, F., Clift, S., & Azaime, H. (2007). Surface heating by transvaginal transducers. *Ultrasound in Obstetrics & Gynecology, 29*, 427–432.

Carstensen, E. L., Child, S. Z., Norton, S., & Nyborg, W. (1990). Ultrasonic heating of the skull. *Journal of the Acoustical Society of America, 87*, 1310–1317.

Carstensen, E. L., Gracewski, S., & Dalecki, D. (2000). The search for cavitation in vivo. *Ultrasound in Medicine and Biology, 26*, 1377–1385.

Chen, D., Fan, T., Zhang, D., & Wu, J. (2009). A feasibility study of temperature rise measurement in a tissue phantom as an alternative way for characterization of the therapeutic high intensity focused ultrasonic field. *Ultrasonics, 49*, 733–742.

Chen, L., ter Haar, G. R., & Hill, C. R. (1997). Influence of ablated tissue on the formation of high-intensity focused ultrasound lesions. *Ultrasound in Medicine and Biology, 23*, 921–931.

Church, C. (2007). A proposal to clarify the relationship between the thermal index and the corresponding risk to the patient. *Ultrasound in Medicine and Biology, 33*, 1489–1494.

Church, C., & Barnett, S. B. (2012). Ultrasound-induced heating and its biological consequences. In G. Ter haar (Ed.), *The safe use of ultrasound in medical diagnosis* (3rd ed.). London: British Institute of Radiology, Chap. 4.

Church, C., & O'Brien, W. D., Jr. (2007). Evaluation of the threshold for lung hemorrhage by diagnostic ultrasound and a proposed new safety index. *Ultrasound in Medicine and Biology, 33*, 810–818.

Church, C. C., & Miller, M. W. (2007). Review: Quantification of risk from fetal exposure to diagnostic ultrasound. *Progress in Biophysics and Molecular Biology, 93*, 331–353.

Coleman, A. J., & Saunders, J. E. (1989). A survey of the acoustic output of commercial extracorporeal shock wave lithotripters. *Ultrasound in Medicine and Biology, 15*, 213–227.

Coleman, A. J., & Saunders, J. E. (1993). A review of the physical effects of the high amplitude acoustic fields used in extracorporeal lithotripsy. *Ultrasonics, 31*, 75–89.

Coussios, C. C., Farny, C. H., ter Haar, G., & Roy, R. A. (2007). Role of acoustic cavitation in the delivery and monitoring of cancer treatment by high-intensity focused ultrasound (HIFU). *International Journal of Hyperthermia, 23*, 105–120.

Coussios, C. C., & Roy, R. A. (2008). Applications of acoustics and cavitation to noninvasive therapy and drug delivery. *Annual Review of Fluid Mechanics, 40*, 395–420.

Curly, M. G. (1993). Soft tissue temperature rise caused by scanned diagnostic ultrasound. *IEEE Transactions on Ultrasonics, Ferroelectrics and Frequency Control, 40*, 59–66.

Dalecki, D., Child, S. Z., Raeman, C. H., & Carstensen, E. L. (1995). Tactile perception of ultrasound. *Journal of the Acoustical Society of America, 97*, 3165–3170.

Delius, M. (2000). Lithotripsy. *Ultrasound in Medicine and Biology, 26*(Suppl. 1), S55–S58.

Diederich, C. J., & Hynynen, K. (1999). Ultrasound technology for hyperthermia. *Ultrasound in Medicine and Biology, 25*, 871–887.

Duck, F. A. (1998). Radiation pressure and streaming. In F. A. Duck, A. C. Baker, & H. C. Starritt (Eds.), *Ultrasound in Medicine, Medical Science Series*. Bristol, UK: Institute of Physics Publishing, Chap. 3.

Duck, F. A. (2000). EFSUMB reviews of recent safety literature, european committee for medical ultrasound safety (ECMUS). *European Journal of Ultrasound, 11*, 151–154.

Duck, F. A., & Martin, K. (1991). Trends in diagnostic ultrasound exposure. *Physics in Medicine and Biology, 36*, 1423–1432.

Duck, F. A., Starritt, H. C., ter Haar, G. R., & Lunt, M. J. (1989). Surface heating of diagnostic ultrasound transducers. *British Journal of Radiology, 67*, 1005–1013.

Dyson, M., Pond, J. B., Joseph, J., & Warwick, R. (1970). Stimulation of tissue regeneration by pulsed plane-wave ultrasound. *IEEE Transactions on Sonics and Ultrasonics, 17*, 133–139.

Ebbini, E. S., & Cain, C. A. (1989). Multiple-focus ultrasound phased-array pattern synthesis: Optimal driving-signal conditions for hyperthermia. *IEEE Transactions on Ultrasonics, Ferroelectrics and Frequency Control, 36*, 540–548.

Edmonds, P. D., Abramowicz, J. S., Carson, P. L., Carstensen, E. L., & Sandstrom, K. L. (2005). Guidelines for journal of ultrasound in medicine authors and reviewers on measurement and reporting of acoustic output and exposure. *Journal of Ultrasound in Medicine, 24*, 1171–1179.

Edwards, M. -B., et al. (2005). Mechanical testing of human cardiac tissue: Some implications for MRI safety. *Journal of Cardiovascular Magnetic Resonance, 7*, 835–840.

Edwards, M. J. (1969). Congenital defects in guinea pigs: Fetal resorptions, abortions and malformations following induced hyperthermia during early gestation. *Teratology, 2*, 313–328.

Ellis, D. S., & O'Brien, W. D., Jr. (1996). The monopole-source solution for estimating tissue temperature increases for focused ultrasound fields. *IEEE Transactions on Ultrasonics, Ferroelectrics and Frequency Control, 43*, 88–97.

FDA (1993). *Revised 510(k) diagnostic ultrasound guidance for 1993*. Rockville, MD: Center for Devices and Radiological Health, US FDA.

FDA (1997). *Information for manufacturers seeking marketing clearance of diagnostic ultrasound systems and transducers*. Rockville, MD: Center for Devices and Radiological Health, US FDA.

Fry, F. J. (1979). Biological effects of ultrasound: A review. *Proceedings of the IEEE, 67*, 604–619.

ter Haar, G. R. (1995). Ultrasound focal beam surgery. *Ultrasound in Medicine and Biology, 21*, 1089–1100.

ter Haar, G. R. (1998). Focused ultrasound surgery. In F. A. Duck, A. C. Baker, & H. C. Starritt (Eds.), *Ultrasound in medicine, medical science series*. Bristol, UK: Institute of Physics Publishing, Chapt. 9.

ter Haar, G. R. (2007). Review: Therapeutic applications of ultrasound. *Progress in Biophysics and Molecular Biology, 93*, 111–129.

Halliwell, M. (1998). Acoustic wave lithotripsy. In F. A. Duck, A. C. Baker, & H. C. Starritt (Eds.), *Ultrasound in medicine, medical science series*. Bristol, UK: Institute of Physics Publishing, Chapt. 10.

Hand, J. F. (1998). Ultrasound hyperthermia and the prediction of heating. In F. A. Duck, A. C. Baker, & H. C. Starritt (Eds.), *Ultrasound in Medicine, Medical Science Series*. Bristol, UK: Institute of Physics Publishing, Chapt. 8.

Harris, G. R. (1992). Lithotripsy pulse measurement errors due to nonideal hydrophone and amplifier frequency responses. *IEEE Transactions on Ultrasonics, Ferroelectrics and Frequency Control, 39*, 256–261.

Harris, G. R., Herman, B. A., & Myers, M. R. (2011). A comparison of the thermal-dose equation and the intensity-time product, It(m), for predicting tissue damage thresholds. *Ultrasound in Medicine and Biology, 37*, 580–586.

Hekkenberg, R. T., Reibold, R., & Zeqiri, B. (1994). Development of standard measurement methods for essential properties of ultrasound therapy equipment. *Ultrasound in Medicine and Biology, 20*, 83–98.

Henderson, J., Willson, K., Jago, J. R., & Whittingham, T. A. (1995). A survey of the acoustic outputs of diagnostic ultrasound equipment in current clinical use. *Ultrasound in Medicine and Biology, 217*, 699–705.

Herman, BA, & Harris, GR (2002). Models and regulatory considerations for transient temperature rise during diagnostic ultrasound pulses. *Ultrasound in Medicine and Biology, 27*, 1217–1224.

Herman, BA, & Harris, GR (2003). Response to "An extended commentary on 'models and regulatory considerations for transient temperature rise during diagnostic ultrasound pulses' by Herman and Harris (2002)" by Miller and Dewey (2003). *Ultrasound in Medicine and Biology, 29*, 1661–1662.

Holland, C. K., Roy, R. A., Apfel, R. E., & Crum, L. A. (1992). In vitro detection of cavitation induced by a diagnostic ultrasound system. *IEEE Transactions on Ultrasonics, Ferroelectrics and Frequency Control, 39*, 95–101.

Holt, R. G., & Roy, R. A. (2001). Measurements of bubble-enhancing heating from focused MHz-frequency ultrasound in a tissue-mimicking material. *Ultrasound in Medicine and Biology, 27*, 1399–1412.

Holtzapfel, G. A. (2000). *Biomechanics of soft tissue. Handbook of Material Behavior, Nonlinear Models and Properties*. NYC: Academic Press.

Hynynen, K. (1987). Demonstration of enhanced temperature elevation due to nonlinear propagation of focused ultrasound in a dog's thigh in vivo. *Ultrasound in Medicine and Biology, 13*, 85–91.

Hynynen, K.. (1998). Present status of ultrasound hyperthermia. *Proceedings of the IEEE ultrasonics symposium* (pp. 941–946). Sendai.

IEC (2001). *Ultrasonics: Focusing transducers. definitions and measurement methods for the transmitted fields*. Geneva, Switzerland: IEC. Standard 61828.

IEC (2004). *Standard 60601–2–37. Medical electrical equipment, Part 2: Particular requirements for the safety of ultrasonic medical diagnostic and monitoring equipment*. Geneva, Switzerland: IEC.

Karagoza, I., & Kartal, M. K. (2009). A new safety parameter for diagnostic ultrasound thermal bioeffects: Safe use time. *Journal of the Acoustical Society of America, 125*, 3601–3610.

Kennedy, J. E., ter Haar, G. R., & Cranston, D. (2003). High intensity focused ultrasound: Surgery of the future? *British Journal of Radiology, 76*, 590–599.

Kimmel, C. A., Cuff, J. M., Kimmel, G. L., Heredia, D. J., Tudor, N., Silverman, P. M., et al. (1993). Skeletal development following heat exposure in the rat. *Teratology, 47*, 229–242.

Lehmann, J. F. (1990). *Therapeutic heat and cold (Rehabilitation Medicine Library)* (4th edn.). Philadelphia: Lippincott, Williams & Wilkins.

Leighton, T. G., & Cleveland, R. O. (2010). Lithotripsy. *Proceedings of the Institute of Mechanical Engineers H, 224*, 317–342.

Lele, P. P. (1972). Local hyperthermia by ultrasound for cancer therapy. In J. F. Lehman, & A. W Guy (Eds.), *Ultrasound therapy workshop: Proceedings on the interaction of ultrasound and biological tissues* (pp. 141–152). HEW Publication (FDA) 73–8008.

Lewin, P. A., Chapelon, J. -Y., Mestas, J. -L., Birer, A., & Cathignol, D. (1990). A novel method to control p + /p-ratio of the shock wave pulses used in the extracorporeal piezoelectric lithotripsy (EPL). *Ultrasound in Medicine and Biology, 16*, 473–488.

Meaning, P. M., Cahill, M. D., & ter Haar, G. R. (2000). The intensity dependence of lesion position shift during focused ultrasound surgery. *Ultrasound in Medicine and Biology, 26*, 441–450.

Miller, D. L, Averkiou, M. A., Brayman, A. A., Everbach, E. C., Holland, C. K., Wible, J. W., Jr, et al. (2008). Bioeffects considerations for diagnostic ultrasound contrast agents. *Journal of Ultrasound in Medicine, 27*, 611–632.

Miller, MW, & Dewey, WC (2003). An extended commentary on "Models and regulatory considerations for transient temperature rise during diagnostic ultrasound pulses" by Herman and Harris (2002). *Ultrasound in Medicine and Biology*, *29*, 1653–1659.

Miller, M. W, & Ziskin, M. C. (1989). Biological consequences of hyperthermia. *Ultrasound in Medicine and Biology*, *15*, 707.

NCRP (1992). *Exposure criteria for medical diagnostic ultrasound, part i: Criteria based on thermal mechanisms*. Bethesda, MD: NCRP.

NCRP (2002). *Exposure criteria for medical diagnostic ultrasound, Part II. Criteria based on all known mechanisms*. Bethesda, MD: NCRP. Report 140.

Newman, W. H., & Lele, P. P. (1985). Measurement of thermal properties of perfused biological tissue by transient heating with focused ultrasound. *Proceedings of the IEEE ultrasonics symposium* (pp. 913–916). San Francisco, CA.

Nightingale, K. R., Kornguth, P. J., Walker, W. F., McDermott, BA, & Trahey, G. E. (1995). A novel ultrasonic technique for differentiating cysts from solid lesions: Preliminary results in the breast. *Ultrasound in Medicine and Biology*, *21*, 745–751.

Nyborg, W. L. (1988). Solutions of the bio-heat transfer equation. *Physics in Medicine and Biology*, *33*, 785–792.

O'Brien, W. D., Jr. (1998). Assessing the risks for modern diagnostic ultrasound imaging. *Japanese Journal of Applied Physics*, *37*, 2781–2788.

O'Brien, W. D., Jr., & Ellis, D. S. (1999). Evaluation of the unscanned soft-tissue thermal index. *IEEE Transactions on Ultrasonics, Ferroelectrics and Frequency Control*, *46*, 1459–1476.

O'Brien, W. D., Jr., & Zachary, J. F. (1996). Rabbit and pig lung damage comparison from exposure to continuous wave 30-kHz ultrasound. *Ultrasound in Medicine and Biology*, *22*, 345–353.

O'Brien, W. D., Jr., & Zachary, J. F. (1997). Lung damage assessment from exposure to pulsed-wave ultrasound in the rabbit, mouse, and pig. *IEEE Transactions on Ultrasonics, Ferroelectrics and Frequency Control*, *44*, 473–485.

Paliwal, S., & Mitragotri, S. (2008). Therapeutic opportunities in biological responses of ultrasound. *Ultrasonics*, *48*, 271–278.

Rassweiler, J. J., Knoll, T., et al. (2011). Shock wave technology and application: An update. *European Urology*, *9*,, 84–796.

Rott, H. -D. (1999a). EFSUMB: Tutorial thermal teratology, european committee for medical ultrasound safety (ECMUS). *European Journal of Ultrasound*, *9*, 281–283.

Rott, H. -D. (1999b). EFSUMB: Acoustic cavitation and capillary bleeding, European Committee for Medical Ultrasound Safety (ECMUS). *European Journal of Ultrasound*, *9*, 277–280.

Sanghvi, N. T., Fry, F. J., Bihrle, R., Foster, R. S., Phillips, M. H., Syrus, J., et al. (1995). Microbubbles during tissue treatment using high intensity focused ultrasound. *Proceedings of the IEEE ultrasonics symposium* (pp. 1571–1574). Seattle, WA.

Sapareto, S. A., & Dewey, W. C. (1984). Thermal dose determination in cancer therapy. *International Journal of Radiation Oncology*, *10*, 787–800.

Shaw, A., Pay, N., Preston, R. C., & Bond, A. (1999). A proposed standard thermal test object for medical ultrasound. *Ultrasound in Medicine and Biology*, *25*, 121–132.

Shaw, A., Pay, N. M., & Preston, R. C. (1998). *Assessment of the likely thermal index values for pulsed doppler ultrasonic equipment, Stages II and III: Experimental assessment of scanner-transducer combinations*. Teddington, UK: National Physical Laboratory. NPL Report CMAM 12.

Shaw, A., Preston, R. C., & Bond, A. D. (1997). *Assessment of the likely thermal index values for pulsed doppler ultrasonic equipment, Stage I: Calculation based on manufacturers' data*. Teddington, UK: National Physical Laboratory. NPL Report CIRA(EXT)018.

Smith, S. W., Stewart, H. F., & Jenkins, D. P. (1985). A plane layered model to estimate in situ ultrasound exposures. *Ultrasonics*, *23*, 31–40.

Soneson, J. E., & Meyers, M. R. (2010). Thresholds for nonlinear effects in high-intensity focused ultrasound propagation and tissue heating. *IEEE Transactions on Ultrasonics, Ferroelectrics and Frequency Control, 57*, 2450–2459.

Starritt, H. (2000). 61–64 EFSUMB: Safety tutorial radiation stress and its bio-effects, European Committee for Medical Ultrasound Safety (ECMUS). *European Journal of Ultrasound, 11*, 61–64.

Stauffer, P. R. (2005). Evolving technology for thermal therapy of cancer. *International Journal of Hyperthermia, 21*, 731–744.

Stewart, H. F. (1982). *Ultrasound therapy* (Chapt. 6). *Essentials of medical ultrasound.* Clifton, NJ: Humana Press.

Stride, E. P., & Coussios, C C (2010). Cavitation and contrast: The use of bubbles in ultrasound imaging and therapy. *Proceedings of the Institute of Mechanical Engineers H, 224*, 171–191.

Szabo, T. L., Melton, H. E., Jr., & Hempstead, P. S. (1988). Ultrasonic output measurements of multiple mode diagnostic ultrasound systems. *IEEE Transactions on Ultrasonics, Ferroelectrics and Frequency Control., 35*, 220–231.

Thomenius, K. E. (1990). Thermal dosimetry models for diagnostic ultrasound. *Proceedings of the IEEE ultrasonics symposium* (pp. 1399–1408). Honolulu, HI.

Ward, P. (2003). Failure to minimize radiation dose risks public outcry, legal claims. *Diagnostic Imaging Europe, 19*, 5.

Watson, T. (2008). Ultrasound in contemporary physiotherapy practice. *Ultrasonics, 48*, 321–329.

WFUMB. (1998). WFUMB symposium on safety of ultrasound in medicine: Conclusions and recommendations on thermal and non-thermal mechanisms for biological effects of ultrasound. S. B. Barnett (Ed.), *Ultrasound in medicine and biology* (pp. 1–55). 244.

Williams, A. R., McHale, J., Bowditch, M., Miller, D. L., & Reed, B (1987). *Ultrasound in Medicine and Biology, 13*, 249–258.

van der Wouw, P. A., Brauns, A. C., Bailey, S. E., Powers, J. E., & Wilde, A. A. (2000). Premature ventricular contractions during triggered imaging with ultrasound contrast. *Journal of the American Society of Echocardiography, 13*, 288–294.

Wu, J., Cubberley, F., Gormley, G., & Szabo, T. L. (1995). Temperature rise generated by diagnostic ultrasound in a transcranial phantom. *Ultrasound in Medicine and Biology, 21*, 561–568.

Wu, J., Winkler, A. J., & O'Neill, T. P. (1998). Effect of acoustic streaming on ultrasonic heating. *Ultrasound in Medicine and Biology, 24*, 153–159.

Zachary, J. F., Hartleben, S. A., Frizzell, L. A., & O'Brien Jr., W. D. (2001). Contrast agent-induced cardiac arrhythmias in rats. *Proceedings of the IEEE ultrasonics symposium* (pp. 1709–1712). Atlanta, GA.

Zachary, J. F., Hartleben, S. A., Frizzell, L. A., & O'Brien, W. D., Jr. (2002). Arrhythmias in rat hearts exposed to pulsed ultrasound after intravenous injection of contrast agent. *Journal of Ultrasound in Medicine, 21*, 1347–1356.

Ziskin, M. C., & Szabo, T. L. (1993). Impact of safety considerations on ultrasound equipment and design and use. In P. N. T. Wells (Ed.), *Advances in Ultrasound Techniques and Instrumentation.* New York: Churchill Livingstone, Chapt. 12.

Bibliography

Abramowicz, J. S., Barnett, S. B., Duck, F. A., Edmonds, P. D., Hynynen, K. H., & Ziskin, M. C. (2008). Fetal thermal effects of diagnostic ultrasound. *Journal of Ultrasound in Medicine, 27*, 541–559.

AIUM (1998). *Bioeffects and safety of diagnostic ultrasound.* Laurel, MD: AIUM Publications.

An overview of ultrasound-induced bioeffect and safety considerations.

Church, C. C., Carstensen, E. L., Nyborg, W. L., Carson, P. L., Frizzell, L. A., & Bailey, M. R. (2008). The risk of exposure to diagnostic ultrasound in postnatal subjects: Nonthermal mechanisms. *Journal of Ultrasound in Medicine, 27*, 565–592.

Miller, D. L., Averkiou, M. A., Brayman, A. A., Everbach, E. C., Holland, C. K., Wible, J. H., Jr, et al. (2008). Bioeffects considerations for diagnostic ultrasound contrast agents. *Journal of Ultrasound in Medicine*, *27*, 611–632.

NCRP. (2002). Exposure Criteria for Medial Diagnostic Ultrasound, Part II. Criteria Based on All Known Mechanisms. Report 140, NCRP, Bethesda, MD.

An authoritative reference for ultrasound-induced bioeffects.

Nyborg, W. L., & Ziskin, M. C. (Eds.), (1985). *Clinics in diagnostic ultrasound* (Vol. 16). , pp. 135–155). New York: Churchill Livingstone.

O'Brien, W. D., Jr, Deng, C. X., Harris, G. R., Herman, B. A., Merritt, C. R., Sanghvi, R., et al. (2008). The risk of exposure to diagnostic ultrasound in postnatal subjects: Thermal effects. *Journal of Ultrasound in Medicine*, *27*, 517–535.

OSHA. (2004). Web site: <http://www.osha.gov>.

Price, R. J., Skyba, D. M., Kaul, S, & Skalak, T. C. (1998). Delivery of colloidal particles and red blood cells to tissue through microvessel ruptures created by targeted microbubble destruction with ultrasound. *Circulation*, *98*, 1264–1267.

Saijo, Y., Kobayashi, K., Arai, H., Nemoto, Y., & Nitta, S. (2003). Pocket-size echo connectable to a personal computer. *Ultrasound in Medicine and Biology*, *29*(5S), S54.

Skyba, D. M., Price, R. J., Linka, A. Z., Skalak, T. C., & Kaul, S. (1998). Direct in vivo visualization of intravascular destruction of microbubbles by ultrasound and its local effects on tissue. *Circulation*, *98*, 290–293.

Stratmeyer, M. E., Greenleaf, J. F., Dalecki, D., & Salvesen, K. A. (2008). Fetal ultrasound: Mechanical effects. *Journal of Ultrasound in Medicine*, *27*, 97–605.

Ter haar, G (Ed.), (2012). *The safe use of ultrasound in medical diagnosis* (3rd ed.). London: British Institute of Radiology. Available online at: <http://www.birjournals.org/site/books/ultrasound.xhtml>.

WFUMB. (1998) WFUMB symposium on safety of ultrasound in medicine: Conclusions and recommendations for thermal and non-thermal mechanisms of biological effects of ultrasound. S. B. Barnett (Ed.), *Ultrasound in Medicine and Biology* (pp. 1–55). 244.

CHAPTER 16

Elastography

Chapter Outline

16.1 Introduction

B-mode ultrasound imaging, while an excellent all-round means of examining tissue, is poor at distinguishing stiff tissue from soft or compliant tissue. An important class of tissue, such as tumor tissue or cirrhotic tissue in the liver, grows out of the same tissue matrix material as the healthy tissue from which it is derived, and as a result, even though these types of tissue are stiffer or more fibrotic, they can be invisible under normal B-mode imaging. Elastography, a branch of tissue characterization, is the measurement and/or depiction of the elastic properties of tissues. It provides the missing image contrast needed to distinguish among the soft and stiff tissues.

Diagnostic Ultrasound Imaging: Inside Out.

Fortunately, we can benefit from several excellent reviews of elastography written from different points of view; they are listed in the Bibliography at the end of this chapter and references to them are cited in the text. This chapter does not attempt to provide a detailed history of elastography nor a comprehensive review of all methods, though the main approaches will be introduced from a different perspective. In the background section, the physics behind elastography is highlighted, information which may not fit in a survey paper. In a way, the rest of the book has been building to this moment: the groundwork in acoustics has been laid to serve as a foundation for analyzing the major elastography approaches and suggesting future directions.

Palpation, a common method for looking for tumors or nodules in the breast, consists of pressing tissue by hand and looking for smaller displacements for hard nodules than those expected for soft tissue. According to various authors, this method goes back to Hippocrates (400 BC) (Doherty, Trahey, Nightingale, and Palmeri, 2013) and the ancient Egyptians (1550 BC) (Wells and Liang, 2013). What is going on here? The finger presses down (mechanical static force) on tissue. The tissue is compressed until harder tissue (a nodule) is sensed (finger- (mechanical) nerve–brain detector). The finger may move around the region of interest in order to determine the size of the nodule (mechanical scanning) until recognition occurs (aha!). The essential components of elastography are a means of excitation, detection, and output and/or display. As explained later, these simple ingredients span a wide range of options.

Modern elastography can be traced to research going on in several groups at the universities of Michigan, Rochester, and Texas at Houston; the Baylor College of Medicine Houston; the Tokyo Institute of Technology; and in New Jersey in the late 1980s and early 1990s. Elastography histories are captured by Sarvazyan et al. (2011); Doherty et al. (2013); Gao, Parker, Lerner, and Levinson (1996); Ophir et al. (2000); Parker, Doyley, and Rubens (2011); and Wells and Liang (2013). Earlier work at the Ultrasonics Institute in Australia, the Institute of Cancer Research in the UK, and elsewhere is summarized in Wilson, Robinson, and Dadd (2000).

For now, before digging into the details of elastography, we can take a simple global view of the types of elastography as symbolized in Figure 16.1. On the left is a generic symbol called the dynamic method, in which sound is sent into the body and response is sensed by a second transducer, which, in some cases, can be the sending transducer. The insonification can be harmonic or transient. In the middle of the figure is the quasi-static approach, which shows a transducer imaging a body before and after static compression of tissue. The organic method on the right utilizes the natural physiological rhythms of the body as sources of deformation, which are sensed by a transducer. After this introduction, some of the physical background for explaining the principles of different elastography methods is described in Section 16.2; and the methods themselves are explained in Section 16.3.

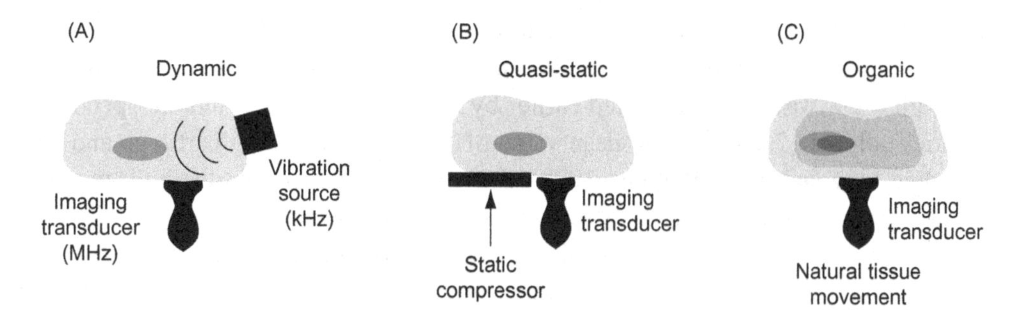

Figure 16.1
Main types of elastography. (A) Dynamic; (B) quasi-static; (C) organic.

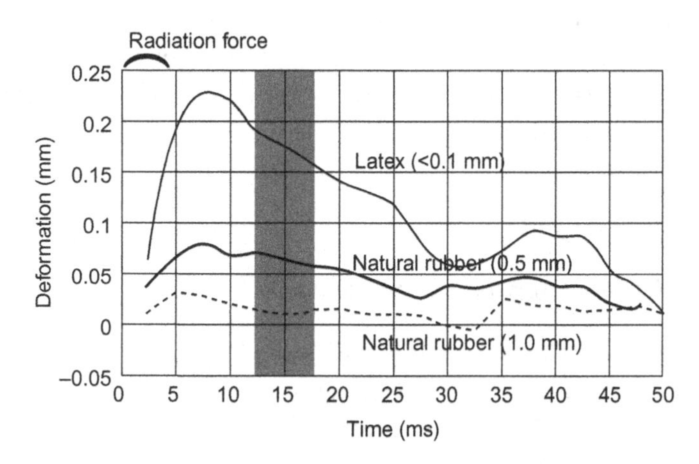

Figure 16.2
Measured deformation as a function of time. *From Sugimoto et al. (1990), IEEE.*

An early paper by Sugimoto, Ueha, and Itoh (1990) anticipated critical issues in elastography. Recognizing the qualitative nature of palpation, they sought to find a quantitative measure of what they called a "figure of hardness." Their focusing transducer generated an acoustic radiation force (ARF) in a sample to produce a "minute deformation," which was sensed by a second transducer. Their data for the temporal responses of rubber samples to radiation force impulses are shown in Figure 16.2. By measuring the slopes of the rise time and decay time of the response, they extracted "spring" constants. They also measured the hardness of tissues from relaxation responses. Sugimoto et al. realized that the absolute value of force or displacement did not have to be measured directly, only a relative derived parameter had to be determined. In this case as illustrated by the steps in Figure 16.3, the source was a focusing ultrasound transducer and the excitation was an ARF, the detector was a second transducer, the output was the temporal response, and the display was a derived parameter.

Sarvazyan (1993, 1998) was one of the first to recognize the limitations on conventional B-mode diagnostic imaging based on longitudinal wave propagation, and the potential of imaging with shear waves. As indicated earlier by Figure 1.2, the longitudinal speed of sound varies only $\pm 12.9\%$ about a mean value of 1550 m/s, including cartilage and skin but excluding lung and bone (Bamber, 1998). Sarvazyan explained that the bulk modulus for longitudinal wave propagation is based mainly on water and depends on short-range molecular interactions (Sarvazyan, Rudenko, Swanson, Fowlkes, & Emelianov 1998). On the other hand, the shear wave modulus is more related to tissue architecture and structure. As shown in Figure 16.4A, the shear modulus is expected to vary over several orders of magnitude with the potential for greatly increased contrast among different tissue types and healthy and diseased tissue (Sarvazyan, 1993; Sarvazyan et al. 1998). Now, with over two decades of elastography, Figure 16.4B shows a compilation of shear modulus data from Sarvazyan, Urban, & Greenleaf, 2013. Results show a mean value of shear sound speed of 4.22 m/s with a $\pm 81\%$ range. These numbers indicate an order of magnitude improvement in contrast over conventional B-mode imaging, and as explored later, better contrast discrimination between normal and abnormal tissue.

16.2 Elastography Physics

16.2.1 Elastic Behavior: Longitudinal and Shear

Here, useful concepts from Chapter 3 will be reviewed briefly in the context of elastography applications. In summary in Chapter 3, the notation is available to describe waves and wave interactions. Recall that shear waves can be polarized vertically and horizontally. For example, for a source on the z axis, vertically polarized shear waves generated both ways are described by, along the positive x axis:

$$\mathbf{S}_1 = \hat{x}\mathbf{S}_0 \sin(\omega t - k_s x), \tag{16.1A}$$

and along the negative x axis:

$$\mathbf{S}_1 = \hat{x}\mathbf{S}_0 \sin(\omega t + k_s x). \tag{16.1B}$$

where $k_s = \omega / c_s$. This example is analogous to the kinds of waves propagating along the positive and negative x axes created by an acoustic radiation force in the body along the beam axis z, as depicted by Figures 12.31 and 12.32. In general, either a static external force or one or more sinusoidal or impulsive acoustic sources from the surface could be used. Two useful forms of the elastic constitutive equations from Section 3.4 are the following for a stress $\mathbf{T}$:

$$\mathbf{T} = \mathbf{c}:\mathbf{S} \tag{16.2A}$$

and strain $\mathbf{S}$

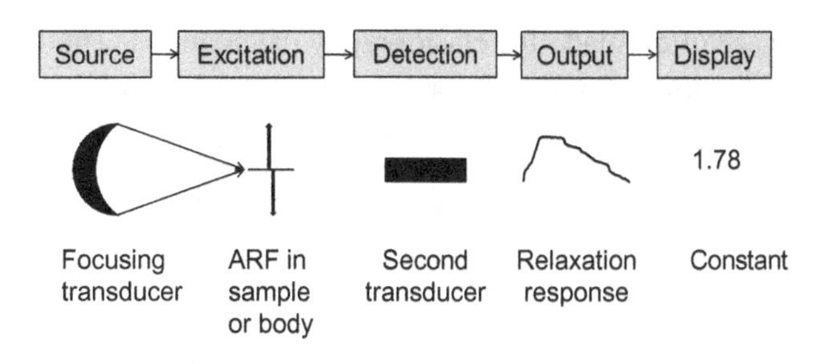

Figure 16.3
Steps in the elastography method of Sugimoto et al. (1990).

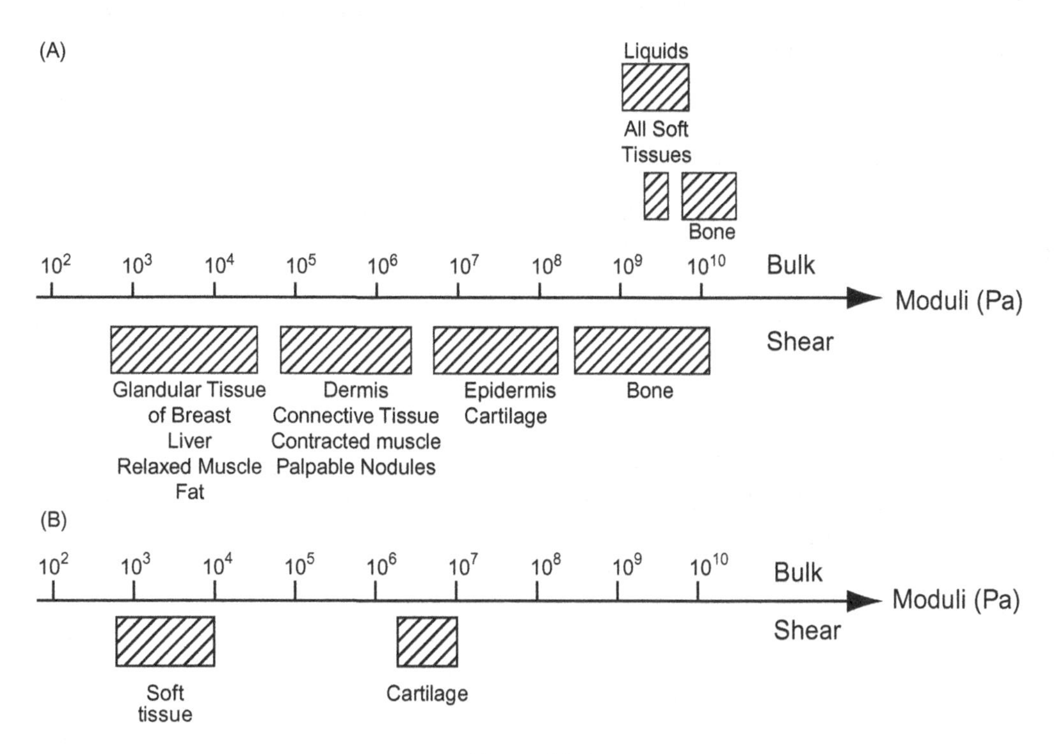

Figure 16.4
(A) Shear elastic moduli span several decades compared to longitudinal constants. *From Sarvazyan (1993), with permission from the World Federation of Ultrasound in Medicine and Biology.* (B) Compiled shear moduli measurements. *Derived from data in Sarvazyan et al. (2013).*

$$\mathbf{S} = \mathbf{s}:\mathbf{T}, \tag{16.2B}$$

where **c** is the stiffness matrix and **s** is the compliance matrix, the inverse of **c**. With the elastic equations it is possible to model different elastography configurations.

Some simplifications are often made for elastography calculations. The Lame constants from Eqn 3.47 tend to be used. Eqns 3.48 for Young's modulus and 3.49 for Poisson's ratio become:

$$E = \frac{3\mu\overline{\lambda} + 2\mu^2}{\overline{\lambda} + \mu}, \tag{16.3A}$$

$$\sigma = \frac{\overline{\lambda}}{2(\overline{\lambda} + \mu)}, \tag{16.3B}$$

and note that Young's modulus, E, is the ratio of axial stress to strain only for a thin rod. Poisson's ratio is often taken to be 0.5 for the condition of incompressibility ($\overline{\lambda} \gg \mu$). This often-used approximation results in:

$$\mu = \frac{E}{2(1+\sigma)} \approx \frac{E}{3}. \tag{16.3C}$$

The speeds of sound from Eqns 3.50A and 3.50B become the following for the longitudinal and shear waves:

$$c_L = \sqrt{(\overline{\lambda} + 2\mu)/\rho}, \tag{16.4A}$$

$$c_S = \sqrt{\mu/\rho}. \tag{16.4B}$$

Finally, going back the compressional or quasi-static case, we make the assumption of a uniaxial uniformly applied stress along z, or axis 3 (Greenleaf, Fatemi, and Insana, 2003). The most appropriate boundary conditions then are that only T_3 is nonzero, and Eqn 16.2B is appropriate (Barbone, 2013):

$$\begin{bmatrix} S_1 \\ S_2 \\ S_3 \\ S_4 \\ S_5 \\ S_6 \end{bmatrix} = \begin{bmatrix} s_{11} & s_{12} & s_{12} & 0 & 0 & 0 \\ s_{12} & s_{11} & s_{12} & 0 & 0 & 0 \\ s_{12} & s_{12} & s_{11} & 0 & 0 & 0 \\ 0 & 0 & 0 & s_{44} & 0 & 0 \\ 0 & 0 & 0 & 0 & s_{44} & 0 \\ 0 & 0 & 0 & 0 & 0 & s_{44} \end{bmatrix} \begin{bmatrix} 0 \\ 0 \\ T_3 \\ 0 \\ 0 \\ 0 \end{bmatrix}, \tag{16.5A}$$

which boils down to the following results:

$$S_1 = s_{13}T_3 = (-\sigma/E)T_3, \tag{16.5B}$$

$$S_2 = s_{23}T_3 = (-\sigma/E)T_3, \quad \text{and} \tag{16.5C}$$

$$S_3 = s_{33}T_3 = (1/E)T_3, \tag{16.5D}$$

in which the Poisson's ratio approximation of 0.5 is often taken. Eqn 16.5 also tells that, along with an axial stress along z, there are lateral strains along x and y. For two-dimensional (2D) imaging in the xz plane, this equation set shows that in addition to the expected lateral component along x there will also be an out-of-plane component along y.

16.2.2 Viscoelastic Effects

As evident from Chapter 4, tissues are viscoelastic. From the point of view of diagnostic imaging, beams, signals, and longitudinal waves behave in well-understood ways according to a suitable model. The time-causal model shows that viscoelasticity means that the waves will be absorbed according to a frequency power law and that, in order for them to remain causal, a slight and predictable phase velocity dispersion will occur. In the time-domain simulations, these effects can be described by a *mirf* (material impulse response function) function, so that the general response can be calculated in time by a convolution of a *mirf* with an input waveform. So far, everything seems reasonable and, for longitudinal waves involved in elastography schemes, predictably well behaved.

Not so for shear waves; they are trouble. They also obey the laws of viscoelasticity but their absorption can be extreme, often an order of magnitude greater than that for longitudinal waves. Furthermore, causality requires that they also have correspondingly larger phase velocity changes with frequency or high dispersion. One consequence of this is that once shear waves are created by an ARF at an interior location they are quickly extinguished by absorption, but hopefully not before revealing useful shear wave properties of the local region. Also, to add to these problems, shear waves usually take their time and propagate slowly—only a few meters per second (see Figure 16.4B)! Their slow speed requires additional pampering in order to see them at all, such as employing ultrafast sampling or special signal-processing techniques.

In order to make sense of these shear propagation effects and not give in to shear madness, investigators have applied viscoelastic models to their data. Walker, Fernandez, and Negron (2000) have taken the experiment described by Sugimoto et al. (1990) to new depths. Using a pulse–echo scheme to measure maximum displacement over time in response to an ARF induced within gelatin phantoms, they fit a Voigt model (Chapter 3) to their data, as shown in Figure 16.5. A step function excitation was used. From the time-domain version of the Voigt model, they derived the relative elasticity and relative viscosity. By scanning the four phantoms mechanically, the authors produced 2D B-mode and maximum displacement relative elasticity and relative viscosity images. With reference to Figure 16.3, the key components of their setup are the following: source: focusing transducer; excitation: step function ARF in phantom; detection: pulse–echo; output: derived parameters—maximum displacement, relative elasticity, relative viscosity; display: 2D images of parameters.

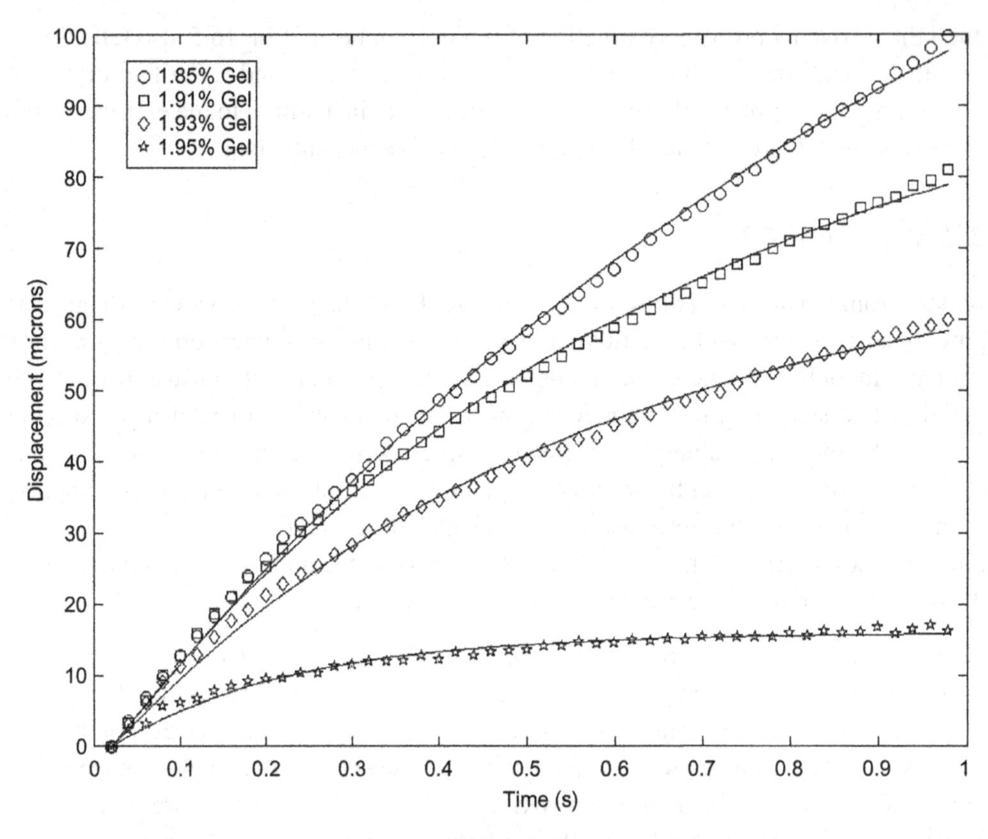

Figure 16.5
Experimental time–displacement data with fitted Voigt model. Experimentally determined displacements are indicated by open symbols with the fitted time-domain Voigt model shown as solid lines. The close agreement between the model and experimental data indicates that the Voigt model is appropriate for these materials. *From Walker (2000), Physics in Medicine and Biology. © IOP Publishing. Reproduced by permission of IOP Publishing. All rights reserved.*

Kiss, Varghese, and Hall (2004) compared two viscoelastic models to canine tissue data obtained by applying a dynamic loading to in vitro samples at frequencies from 0.1 to 400 Hz. The two models applied, which they called the "Kelvin–Voigt model" and the "Kelvin–Voigt fractional derivative (KVFD) model," are equivalent to the Voigt model, Eqn 4.25, and the time causal model, Eqn 4.27, respectively. Their data comparisons to empirical fits of these models demonstrate statistically and convincingly that the KVFD model is superior to the KV model. The data also show considerable differences in fit parameters between normal tissue and lesions. As is often done in mechanical measurements, the constants, which have a real and an imaginary part, were found at each sinusoidal frequency for a range extending over two decades of frequency. The samples were preloaded, a 3% strain was applied for 3 seconds and oscillated for 3 seconds, and

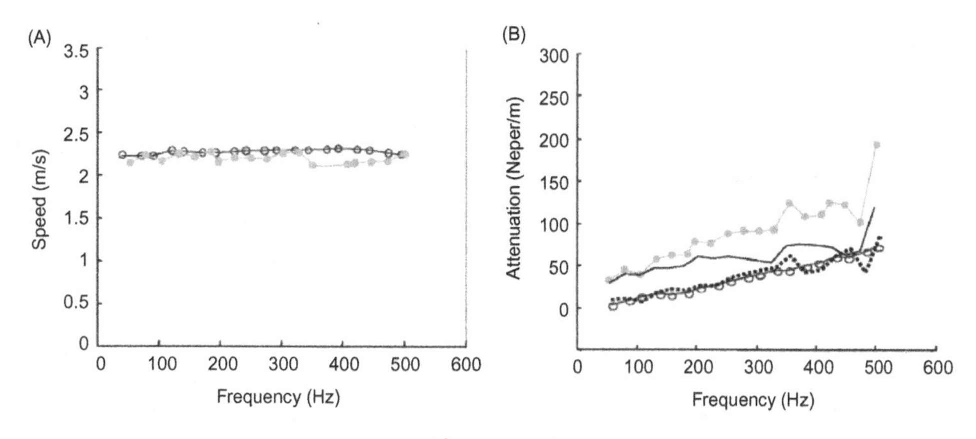

Figure 16.6
Speed and attenuation measurements vs frequency. (A) The inverse problem speed estimation (dots) is very close (within 8%) to the reference measurements (circles); however (B) an over-attenuation is clearly visible with the inverse problem method (dots) compared to the reference measurements (circles). Applied to the simulated data in a perfectly elastic solid, the inverse problem gives an over-attenuation as well (solid line). Subtracting these latter data from the experimental measurements (dots) compensates for diffraction effects and allows one to compute the medium attenuation (dotted line). The reader is referred to the Web version of the book to see the figure in color. *From Catheline et al. (2004), IEEE.*

then the force was measured as a transient decay process. The elastic constant that related the measured force to the applied strain they called "Young's modulus," E, and they determined a power law exponent called α, which related to the exponent in Chapter 4, $y = \alpha + 1$.

Another excitation condition in elastography is when an ARF is applied deep in tissue and shear waves propagate outward orthogonally, as illustrated by Figures 12.31 and 12.32 and the accompanying text in Chapter 12. Work by Catheline et al. (2004) indicated that a Voigt model was better than a Maxwell model once the data was corrected for the diffraction effects of the source; some of their results are shown in Figure 16.6. Deffieux, Montaldo, Tanter, and Fink (2009) sought to apply a viscoelastic model to shear waves generated this way. Based on work by Chen, Fatemi, and Greenleaf (2004), they modeled the shear wavefronts as an expanding cylindrical wave and accounted for the amplitude falloff in developing a regional fitting algorithm to the linear part of the phase of the wave. They found the Voigt model described the dispersion of the shear waves and that the speed of sound in liver changes by a factor of two over their bandwidth. They also validated their method by sending plane waves into a gelatin phantom and developed an algorithm to calculate the shear wave speed locally. Later work by Greenleaf's group indicated that quantitative shear wave measurements were affected by the shape of exciting beam, and they compared different transducers to magnetic resonance imaging (MRI) elastography and developed ways of achieving improved accuracies (Chen et al. 2009; Zhao, Song, & Urban 2011).

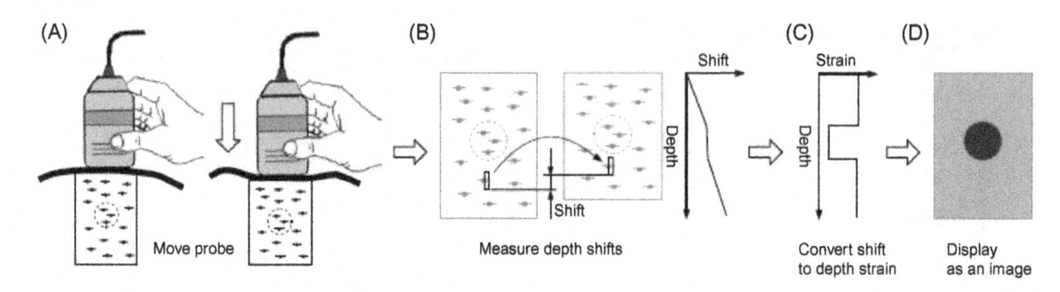

Figure 16.7
Overview of the quasi-static elastography process. (A) The anatomy is scanned with a conventional ultrasound probe, which is moved very slightly up and down (much less than is shown here). In this example, the dashed circle is stiffer than the surrounding material. (B) Anatomical displacement in the axial direction is calculated everywhere in the ultrasound image; (C) gradient estimation in the axial direction and filtering gives an estimate of axial strain; (D) a normalized version of the axial strain is converted to gray-scale and displayed as an image, where black is stiff and white is soft. *From Treece et al. (2011) Interface Focus, © IOP Publishing. Reproduced by permission of IOP Publishing. All rights reserved.*

16.2.3 Strain Imaging

As indicated earlier, the quasi-static approach consists of imaging (pre-compression) and applying a static force and imaging again (post-compression). The overall process is diagrammed in Figure 16.7. After compression, the before and after images are compared by a speckle-tracking method. An inherent assumption of small strain is often made to encourage coherence among similar regions in the pair of images. Typically, each radio frequency (RF) line in the image is divided into small overlapping windows and cross-correlation or other algorithms are used to determine the local shift of tissue. Usually, a frame of RF lines is acquired for the tissue at equilibrium in a "pre-compression" state. The overall change in axial length is small ($dz/z < 1\%$), so the post-compression time records are shorter by $2dz/c$. These RF lines are zero-padded to the same length as the pre-compression data, and corresponding lines from the frames are cross-correlated. Locations from cross-correlation peaks from this operation indicate the delay changes of the deformations. Local axial strain can be estimated from the delay changes in a time window:

$$S_n = \frac{t_{n+1} - t_n}{2dz/c}, \qquad (16.6)$$

where t_n is the time shift for segment or time window n. The estimates of local strain for all RF lines can be combined to produce a strain image. A similar operation can be applied to obtain lateral strain and create two-dimensional images of strain (not just axial strain images); however, here resolution is limited by lateral beamwidth and sampling. The amount of force applied must be enough to overcome noise but not great enough to cause decorrelation effects, and remain in the linear range of elasticity.

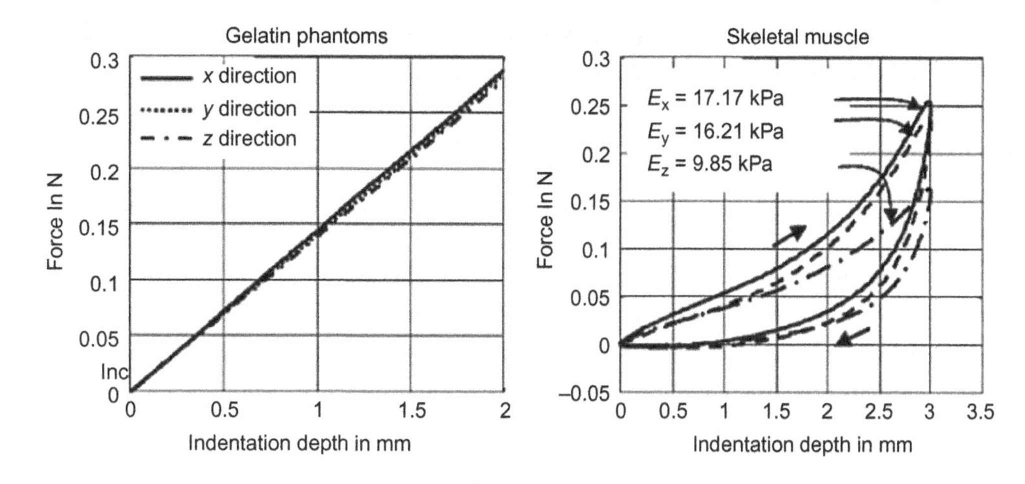

Figure 16.8
Force indentation curves for (A) Gelatin and (B) Ex vivo skeletal muscle. For the harmonic strain stimulus applied, gelatin appears as an isotropic, linear-elastic medium, whereas muscle exhibits the features of a transverse-isotropic nonlinear viscoelastic medium. *From Hall, Bilgen, Insana, and Krouskop (1997), IEEE.*

The previous description is a simplification. Over the years, a great deal of sophistication has gone into producing 2D and 3D strain images. Speckle-tracking methods are reviewed in Section 8.4.7. Two review articles (Gao et al., 1996; Ophir et al., 2000) detail a number of signal-processing improvements (Insana, Bilgen, Chaturvedi, Hall, & Bertrandt, 1996), as well as a theoretical framework for estimating the quality of elastograms (Konofagou, 2004; Varghese & Ophir, 1997) and early fundamentals can be found in O'Donnell, Skovoroda, Shapo, and Emelianov (1994), Skovoroda, Emelianov, Lubinski, Sarvazyan, and O'Donnell (1994) and Skovoroda, Emelianov, and O'Donnell (1995). Doyley et al. (2001) evaluated the performance of a freehand approach to creating strain elastograms using speckle tracking (Trahey, Hubbard, & von Ramm 1988). Treeby et al. (2011) and Pinton, Dahl, and Trahey (2006) provided a review of real-time methods including a new method. Azar, Goksel, and Salcudean (2010) presented a fast multidimensional approach. Doherty, Dahl, and Trahey (2012) have developed an improved method with improved jitter reduction and feature detection.

16.2.4 Nonlinearity Effects

One way to increase contrast is to take advantage of the nonlinearity of tissue. As indicated by Figure 16.7, there are differences between ordinary gelatin phantoms and real tissue. Tissue exhibits viscoelastic, hysteresis, and nonlinear properties. For quasi-static approaches, these characteristics have important implications. As Krouskop's data in Figure 16.8 show, different tissues can be nonlinear in their own unique ways, thus providing more contrast with increasing strain. An initial "preload" displacement is applied to tissue for the first image before the final push for measurement. These steps have an effect that is the equivalent of moving to the right on the curves of Figure 16.8, where the curves are more separated and provide more contrast. The converse of this effect is that the strain image or contrast is dependent on the push. For consistency, some means of feedback to the user is usually provided (Wells and Liang, 2011).

Another aspect of compression is how far the stress induced into the solid extends. In their classic elastography paper, Ophir, Cespedes, Ponnekanti, Yazdi, and Li (1991) describe how stress falls off from the surface of an elastic solid for a circular compressor of radius a and showed reasonable experimental confirmation of the effect, as illustrated by Figure 16.9. This figure shows that the larger the compressor, the deeper the extent of the stress. If we fast-forward to modern implementations of quasi-static elastography, we find that the footprint of a transducer pushes down on the body surface. Based on the analysis, we can infer the following trend: the lateral stress along x, the axis along the length of the array, will have far more effect with depth than the smaller extent of the aperture along the elevation direction, y.

16.2.5 Acoustic Radiation Forces

Acoustic radiation forces (ARFs) were discussed in detail in Section 12.7. Doherty et al. (2013) wrote a thorough review of elastography methods employing ARFs; they seem to be multiplying rapidly, with 11, each with their own acronym, at last count. Many of them involve a three-step process to create what is called the "pushing" beam: propagation of longitudinal wave beam(s) under nonlinear conditions to a region of interest, usually a focal zone, the resultant ARF field, and finally their creation of strongly attenuated and dispersive shear wavefronts at low frequencies propagating under linear conditions to probe the surrounding tissue neighborhood. More specific information about different approaches follows in Section 16.3.

16.2.6 Model-based Inversion

From the outset, elastography was conceived of as a quantitative method (the title of the classic Ophir et al. (1991) paper was "Elastography: A quantitative method for imaging the elasticity of biological tissues"). However, as the quasi-static method is practiced today on commercialized systems, it is more of a qualitative method. Quasi-static elastography is still an extremely useful imaging mode, as is B-mode imaging, which is also not quantitative imaging. To be more precise, the commercialized quasi-static elastography is really "strain imaging" as portrayed in Figure 16.7. For quasi-static elastography to live up to its full potential, another step is necessary to extract the quantitative information hidden in the data.

In order to obtain the quantitative elastic information, the inverse problem needs to be solved. The spatial deformations acquired are processed to improve data consistency. For example, see Richards, Barbone, and Oberai (2009). Next, a linear elastic mechanical model that can simulate interrelated deformations is compared locally to measured

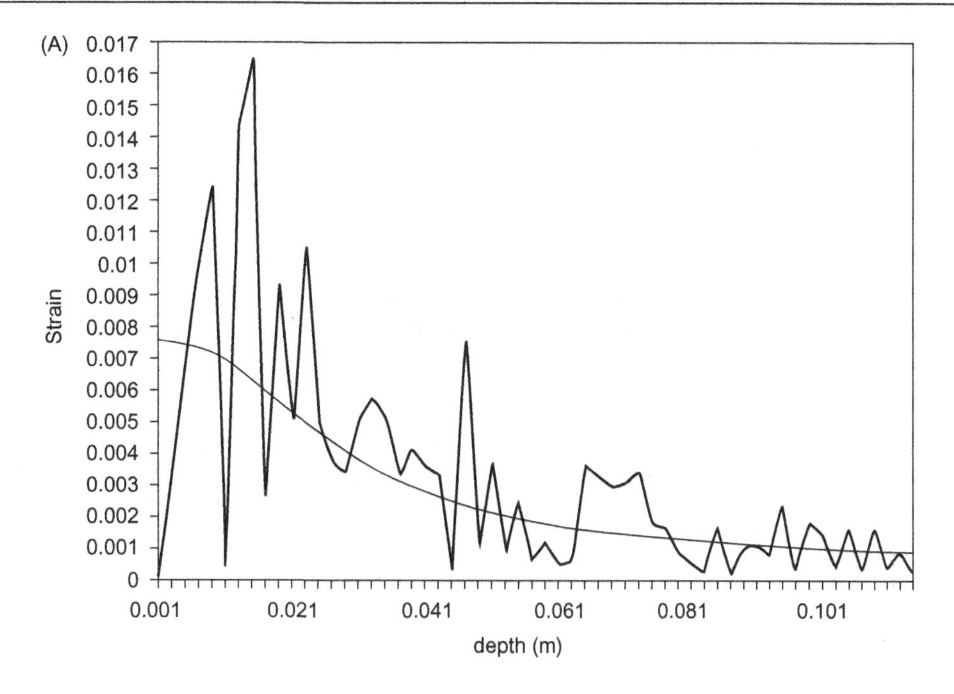

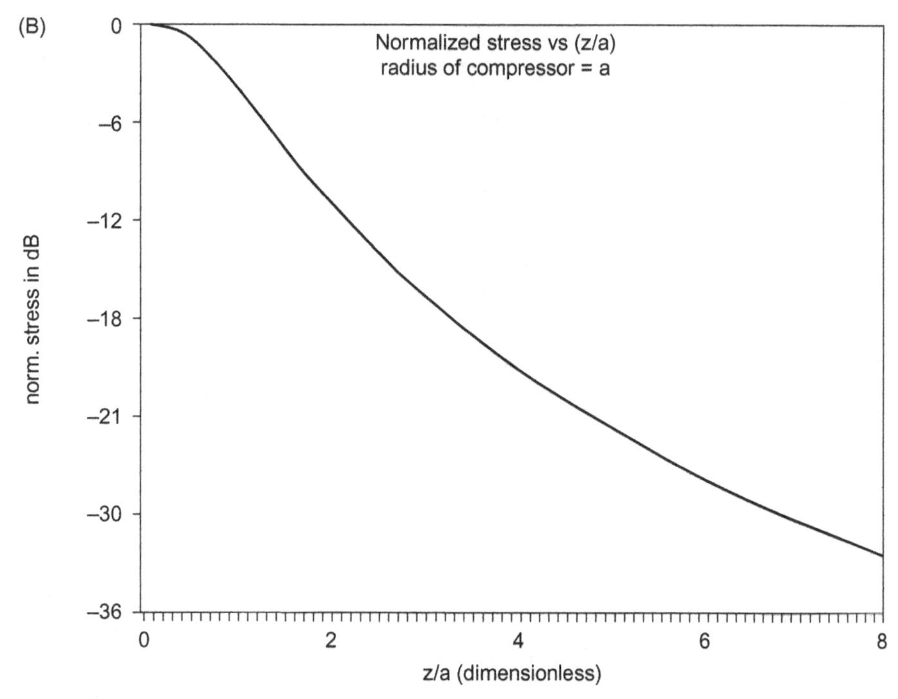

Figure 16.9

(A) Experimental strain data vs depth for foam type III; (B) A normalized curve of stress falloff from surface as a function of z/a. In (A), the theoretical curve shows stress falloff with distance for a compressor with a radius of 44 mm. Note good agreement between theory and experiment. *From Ophir et al. (1991) with permission of Dynamedia, Inc.*

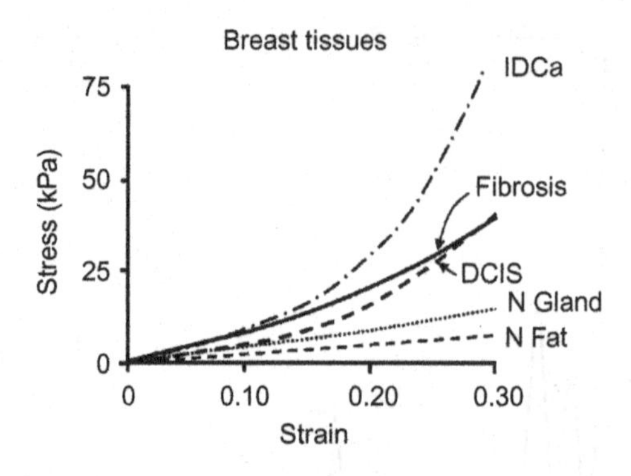

Figure 16.10

Stress–strain curves acquired for five types of breast tissue. N fat, normal fat; N gland, normal glandular; DCIS, ductal carcinoma in situ;, Fibrosis, benign fibrotic lesion; IDCa, infiltrating ductal carcinoma. *Curves are from a subset of data acquired from Krouskop* et al. *(1998) replotted by Greenleaf, Fatemi, and Insana (2003) courtesy of Annual Review of Biomedical Engineering*

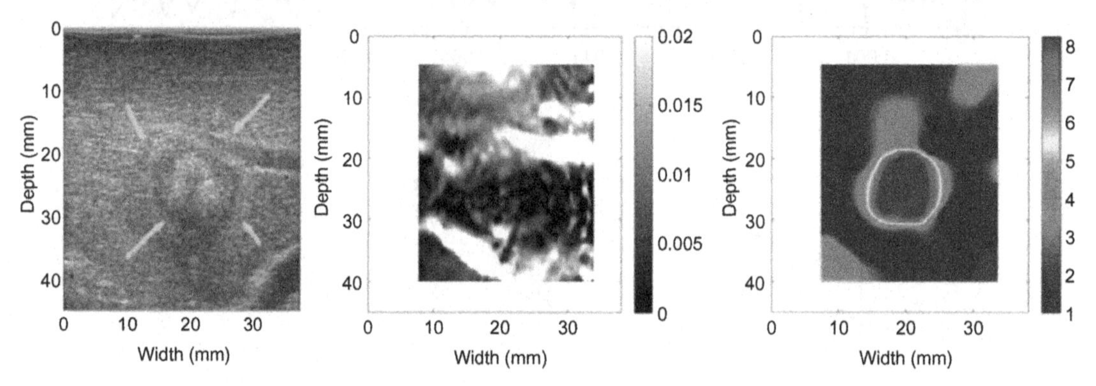

Figure 16.11

(A) Sonogram; (B) Strain elastogram; and (C) Modulus elastogram of RF Ex vivo ablated bovine liver. The reader is referred to the Web version of the book to see the figure in color. *Courtesy of Drs T. J. Hall, T. Varghese, and J. Jiang (University of Wisconsin–Madison); from Doyley (2012) Physics in Medicine and Biology, © IOP Publishing. Reproduced by permission of IOP Publishing. All rights reserved.*

displacements until a best match is obtained. The spatially variant values of elastic moduli used to obtain the best match form a final modulus image, as shown in Figure 16.11. So far, the process sounds rather straightforward but in practice it is not. Many factors have to be considered, foremost of which is how to treat 2D data to solve essentially a 3D inverse elastic problem. Another challenge is how to handle the boundary conditions, which are usually unknown (Barbone & Bamber, 2002).

Finally, as indicated by Figure 16.10, stress is a nonlinear function of strain. Hall et al. (2011) have results for solving the nonlinear case.

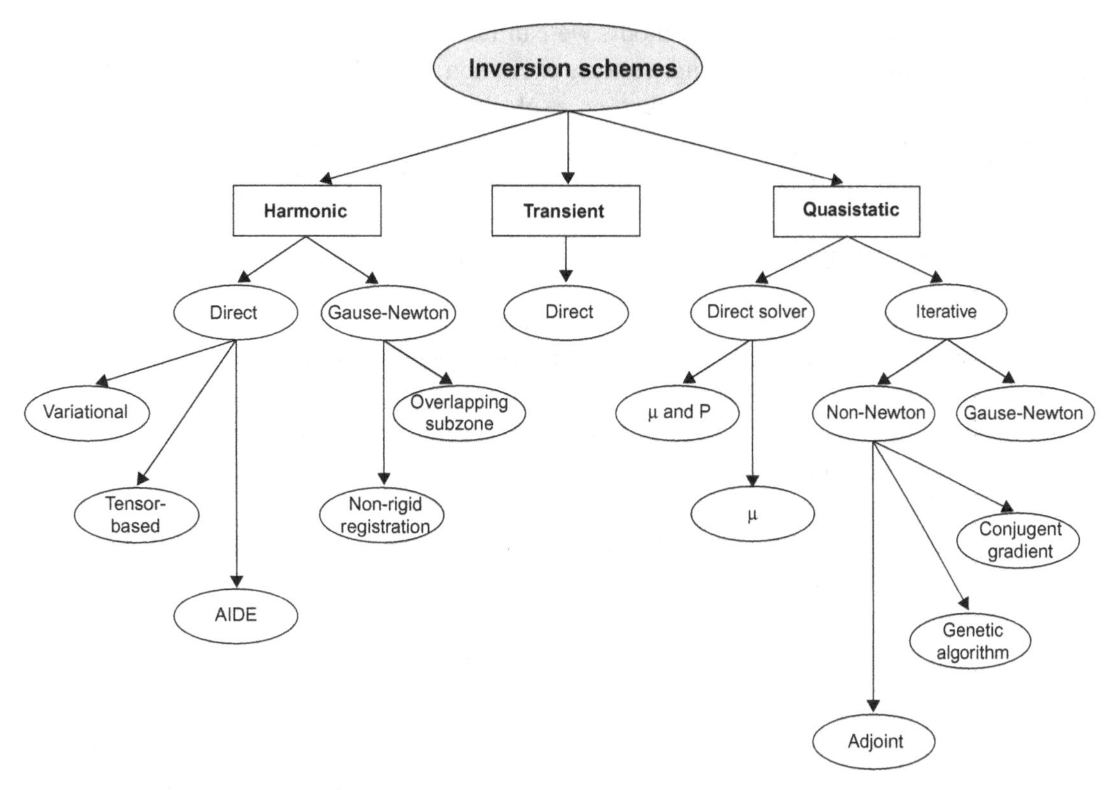

Figure 16.12
Hierarchical diagram of proposed approaches to shear modulus estimation for harmonic, transient, and quasi-static elastography, assuming linear elastic isotropic mechanical behavior. *From Doyley (2012) Physics in Medicine and Biology, © IOP Publishing. Reproduced by permission of IOP Publishing. All rights reserved.*

We are fortunate to have an excellent review of model-based elastic inversion by Doyley (2012). He has shown how much progress has been made in this area and revealed some of the challenges in inversion techniques. A sense of how much has been done in inversion can be seen in Figure 16.12 from his paper, which categorizes different methods. An important area of research is nonlinear inversion (Hall et al. 2011); the reasons why were examined in Section 16.2.4.

16.3 Elastography Implementations

16.3.1 Introduction

The basic operating principle of elastography involves the comparison of the spatial arrangement of tissue initially at equilibrium to itself after deformation. Before examining specific elastography methods, we can recall reference points by which to compare them. Recall the following: source, excitation, detection, output, and display. Because elastography is growing at such an explosive rate, it is no longer possible to capture the field in a chapter of this length. The key points of major methods are reviewed here and the reader can seek more information through the supplied references.

16.3.2 1D Elastography

Research dedicated to determining shear properties quantitatively led the development of a 1D transient elastography probe (Catheline et al., 2002; Sandrin, Catheline, Tanter, Hennequin, & Fink, 1999; Sandrin, Tanter, Gennisson, Catheline, & Fink, 2002; Sandrin et al., 2004). An early version consisted of a 5-MHz pulse–echo piston transducer mounted coaxially on a mechanical vibrator (about 50 Hz). The vibrator transmits low-frequency shear waves. These waves act as subsources and generate longitudinally polarized shear waves that intercept the axis of the longitudinal piston source, which acts a pulse–echo detector. Consecutive RF lines are stored, producing an M-mode like image. Once corrected for the motion of the piston transducer, shear modulus data (output) can be determined from the shear wave phase velocity derived from the phase of the signals.

This work has led to a 1D transient elastography commercial device, Fibroscan, which has been applied successfully to determining the stiffness of the liver, achieving 90% sensitivity and specificity for some cases (Parker et al., 2011). The advantages of simplicity and low cost are balanced by its lack of an imaging capability and limitations in penetration, especially on obese patients.

16.3.3 Quasi-static Elastography

Also known as compressional elastography, this approach initiated by Ophir et al. (1991) has become the most widespread form of elastography in the form of strain imaging, described in Section 16.2.3. In this approach, the excitation is a static compression of the tissue with an imaging transducer array to achieve less than 1% strain, with variations described in Section 16.2.3. The 2D displacements (output) are detected from the RF data from normal B-mode images before and after compression by correlation algorithms. The display is of a relative strain image derived from the displacements. The basic principle is illustrated by Figure 8.19 and Figure 16.7, in which harder tissues compress less than softer ones, and these regions are revealed by speckle-tracking techniques described in Section 8.4.7. A set of strain images compared to B-mode images is shown in Figure 16.13. The publications from Ophir's group alone number well over 100. Many other groups around the world continue research on compressional elastography. In addition to work on quantitation or modulus imaging described earlier, new applications continue to expand the versatility of this approach (Parker et al., 2011). An active area of

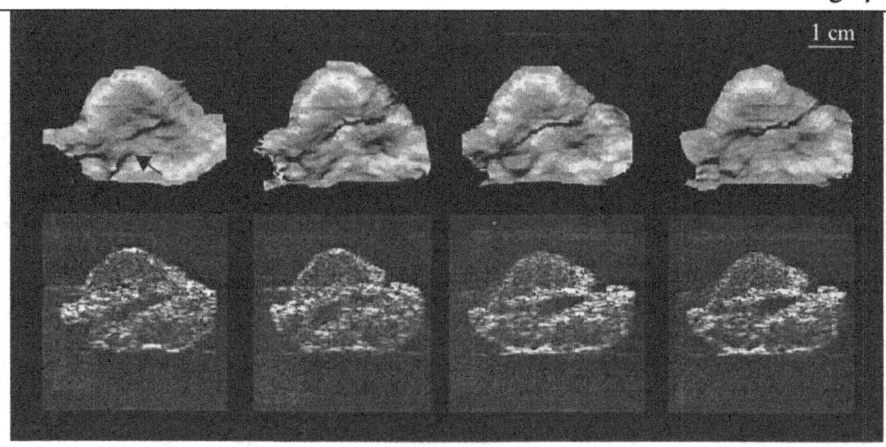

Figure 16.13
Parasagittal views near and at the center of a canine prostate in vitro, obtained from a Diasonics Spectra scanner at 5 MHz. (Top row) Axial strain elastograms, where white represents regions of low strain, and black represents regions of high strain. The apex is on the right, and the base is on the left. Observe the clear depiction of the urethra and of some of the ducts, the excellent contrast between the outer and the inner gland, and the visualization of the verumontanum as a low-strain area located at the distal central part of the urethra (two center images). In these elastograms, black = soft; white = hard. (Bottom row) The corresponding sonograms from which the elastograms were computed. *From Kallel, Price, Konofagou, and Ophir (1999a); reprinted with permission of Dynamedia, Inc.*

research is poroelastic tissue (Berry et al., 2006; Konofagou, Harrigan, Ophir, & Krouskop, 2001; Righetti, Righetti, Ophir, & Krouskop, 2007). Another application is the monitoring of high-intensity focused ultrasound (HIFU)-induced lesions in soft tissue (Kallel et al. 1999b).

In addition to 1D and 2D strain images, orthogonal image planes can be acquired for the out-of-plane (or elevation plane) strain component to obtain a 3D estimate of strain (Konofagou & Ophir, 1998). Either longitudinal or shear strain can be displayed. Without this extra data, a plane stress or plane strain approximation is made for image plane data, as is done in solving the inverse elasticity problem. In fact, a number of approximations are often made in elastography, such as the isotropy, incompressibility, and linear elasticity of tissue. Fortunately, for isotropic materials, there are only two independent elastic moduli (as described in Section 3.3.1).

Strain imaging has been implemented as a real-time imaging mode, often with a feedback mechanism or automatic adjustment first on a Siemens scanner, then on a number of commercial scanners including General Electric, Hitachi, Philips, Siemens, and Ultrasonix (Parker et al., 2011). An example of a B-mode strain image comparison is given by Figure 16.14. At the present time, commercial imaging systems also offer various forms of elastography imaging.

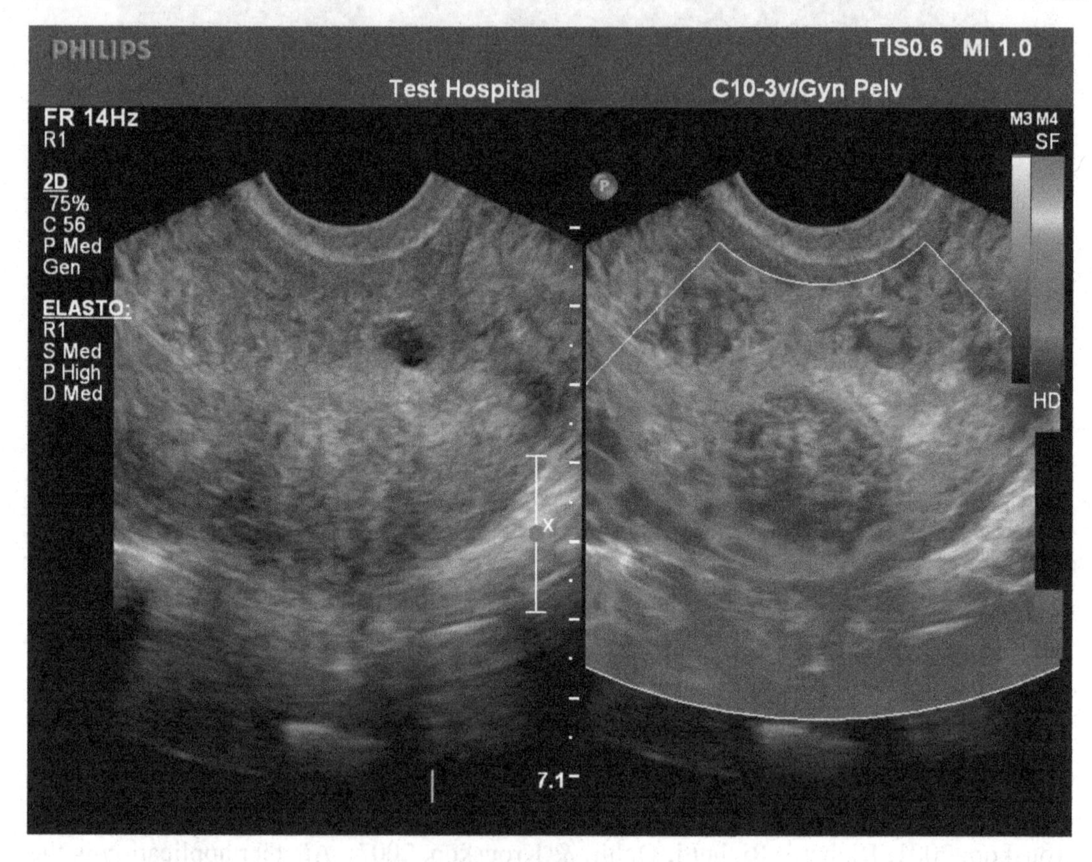

Figure 16.14
Pelvic sonogram and elastogram. The reader is referred to the Web version of the book to see the figure in color. *Courtesy of Philips Healthcare*

16.3.4 Sonoelastography

One of the first successful forms of real-time elastography was sonoelasticity (Lerner, Parker, Holen, Gremiak, & Waag, 1988; Parker & Lerner, 1992). As indicated in Figure 16.1A, a low-frequency (20–1000 Hz) vibration source was applied to the tissue of interest. This excitation caused the tissue to respond, and, depending on the shape of the tissue or organ and its structure and elastic boundary conditions, to resonate and form multiple reflected modes. These responses were initially detected by range-gated pulse Doppler in 1987–1988. Later, a modified color flow imaging system (a mode normally used to create an image of blood flow through Doppler detection, as described in Chapter 10 and Chapter 11) provided images in which the vibration amplitude was represented by a color map in real time. The resulting parameterized image was visualized as a green overlay on the B-mode image, with green-scale brightness proportional to

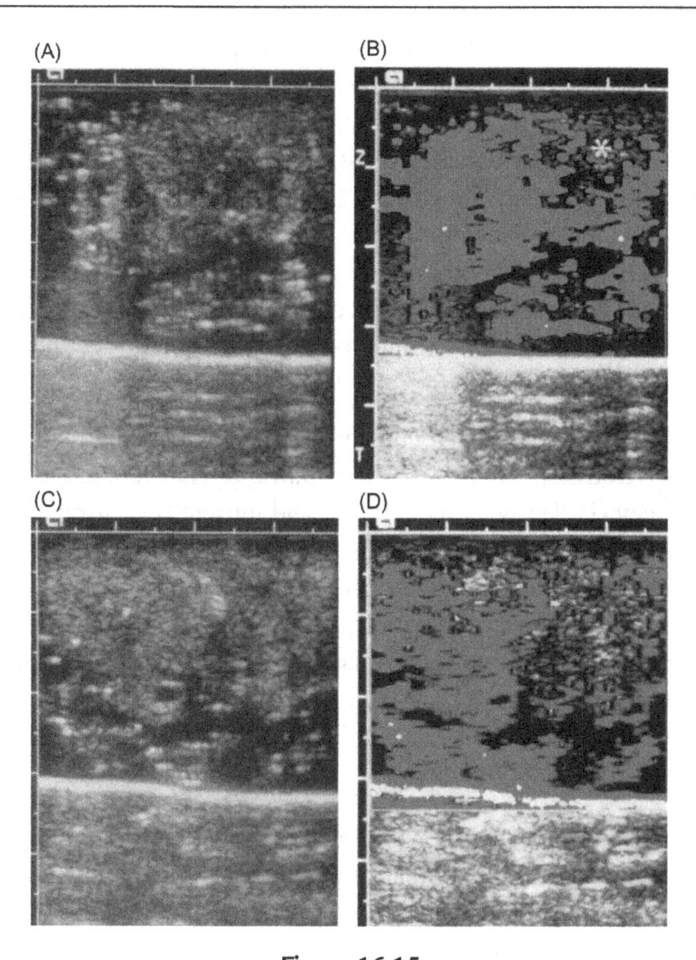

Figure 16.15
Diffuse carcinoma demonstrated at multiple frequencies in a 69-year-old man with no palpable abnormality and a serum PSA level of 6.7 ng/ml. (A) Standard prone transverse ultrasound image obtained at the base of the prostate demonstrates somewhat heterogeneous echo texture; (B) corresponding sonoelasticity image obtained at 50 Hz shows poor vibration diffusely, most pronounced posteriorly on the right (*); (C) a second section of the base, obtained slightly caudal to (A), shows a heterogeneous gray-scale pattern; (D) corresponding sonoelasticity image obtained at 150 Hz documents absent bilateral posterior and right anterior vibration. The reader is referred to the Web version of the book to see the figure in color. *From Rubens et al. (1995); reprinted with permission from RSNA.*

vibration amplitude. Stiffer nonresponsive regions had lower amplitudes; for example, a hard tumor would appear as a dark region (as illustrated by Figure 16.15).

Gao, Alam, Lerner, and Parker (1995) derived a theory for sonoelasticity based on shear waves in a lossy elastic (viscoelastic) medium. The vibration amplitude was related in a nonlinear way to the stiffness of a lesion, so that image contrast increased with higher frequencies and larger lesion sizes (Parker, Doyley, and Rubens, 2011). Other work by Huang showed that the relative stiffness modulated the Doppler signals so that the amplitude could be related to the spectral spread of the signal. Rubens et al. (1995) demonstrated that sonoelasticity detected prostate cancer in vitro in real time with better sensitivity than B-mode imaging. They found that 64% of pathologically confirmed tumors detected by this method were invisible by standard imaging. Research by Yamakoshi, Sato, and Sato (1990) on phase gradient sonoelastography led to more improvements. Further work expanded the theoretical foundations of the method and developed more robust and sensitive estimators. For more details see the sonoelastography chronicle in Parker et al. (2011).

Sonoelastography has been applied to a number of tissues, including the prostate (Rubens et al., 1995), breast (Krouskop, Wheeler, Kallel, Garra, & Hall, 1998; Parker & Lerner, 1992), liver and kidney (Parker & Lerner, 1992), and muscle (Levinson, Shinagawa, & Sato, 1995). This method is a real-time imaging approach that is compatible with modern imaging systems and can produce relative image contrast of tissue stiffness; an additional vibrational device is needed. Sonoelastography needs to be used with care in certain situations in which low amplitudes may be mistaken for high stiffness regions and where reflections from boundaries may interfere with image interpretation.

16.3.5 Shear Wave Elasticity Imaging

Sarvazyan et al. published their pivotal work on shear wave elasticity imaging (SWEI) in 1998. They proposed a general method that centered on a focusing ultrasound transducer that generated acoustic radiation forces that excited shear waves that could be detected by various means, such as ultrasound transducers (separate from or the same as the transmitter), acoustic surface transducers, optical means, and MRI. Considerable analysis was presented, which included nonlinear effects and absorption to predict the acoustic radiation force field. Their simulations of an expanding cylindrical-like shear wavefront were shown in Figure 12.31. They initially detected a shear wave by an optical detector and then produced MRI images of shear wavefronts that agreed with their predictions. The shear modulus was derived from a time-of-flight measurement of a propagating shear wave at known locations. Their results inspired others to investigate ARF methods.

16.3.6 Acoustic Radiation Impulse Imaging

In their search for developing a practical means of "remote palpation," Nightingale, Palmeri, Nightingale, and Trahey (2001) designed a way of displaying tissue displacement in response to acoustic radiation "pushing" forces. In their method, the same transducer was used to produce a series of pulses, a reference pulse and impulsive radiation-force-pushing pulses. These long (milliseconds) pulses, once in the focal region, generated an acoustic

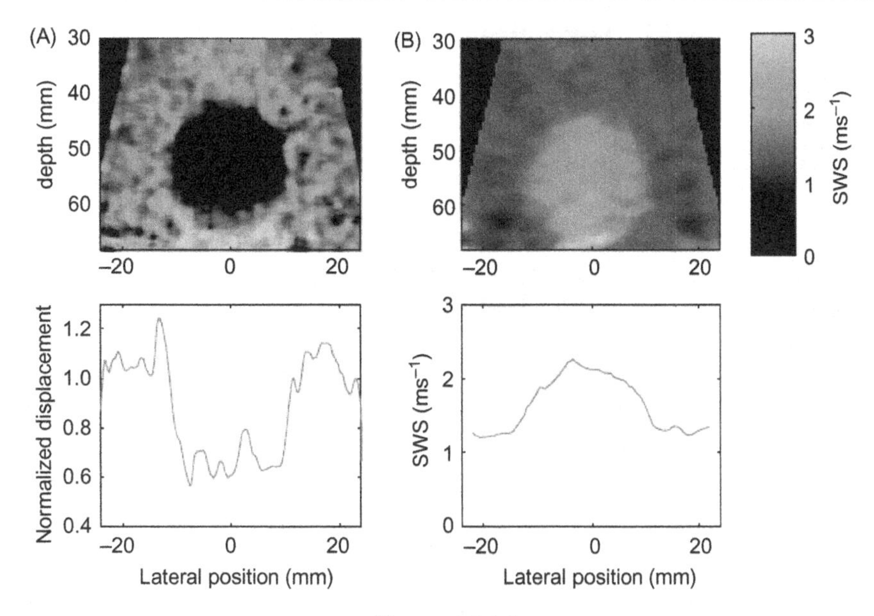

Figure 16.16

Matched (A) ARFI and (B) SWEI images of a calibrated elasticity phantom with a 20-mm-diameter stiff spherical inclusion. The images were generated from the same dataset, which was obtained with a 4-MHz abdominal imaging array, using parallel receiving beamforming techniques to monitor the tissue response to each excitation throughout the entire field of view. A total of 88 excitation pulses were located at two focal depths (50 and 60 mm), with a beam spacing of 1 mm. The ARFI image portrays normalized displacement at 0.7 ms after each excitation, whereas the SWEI image portrays reconstructed shear wave speed. The lesion contrast is 0.37 and 0.71 for the ARFI and SWEI images, respectively, and the edge resolution (20–80%) is 1.2 mm (ARFI) and 5.0 mm (SWEI) in the plots from a depth of 50 mm, shown in the bottom row. *From Palmeri and Nightingale (2011) Interface Focus,*

radiation force field that created shear waves that displaced the nearby tissue. These disturbances were intercepted by a series of laterally placed "tracking" B-mode-like beams that sensed the tissue movement from pulse–echoes and translated speckle. Through the scanning of the beam and obtaining displacements through cross-correlation or phase-comparison, an image of displacements was formed. Examples of an acoustic radiation force imaging (ARFI) image and a SWEI image of the same stiff spherical inclusion in a phantom can be seen in Figure 16.16 (Palmeri & Nightingale, 2011). On the left, the ARFI image depicts smaller displacement in the stiffer region of the inclusion. On the right, the SWEI image reveals the different shear wave speed in the inclusion.

As quantified by further studies, ARFI images have greater contrast than B-mode images with comparable resolution (Fahey et al. 2008). Because of the longer pushing pulses, this approach is within but at US Food and Drug Administration (FDA) limits for acoustic

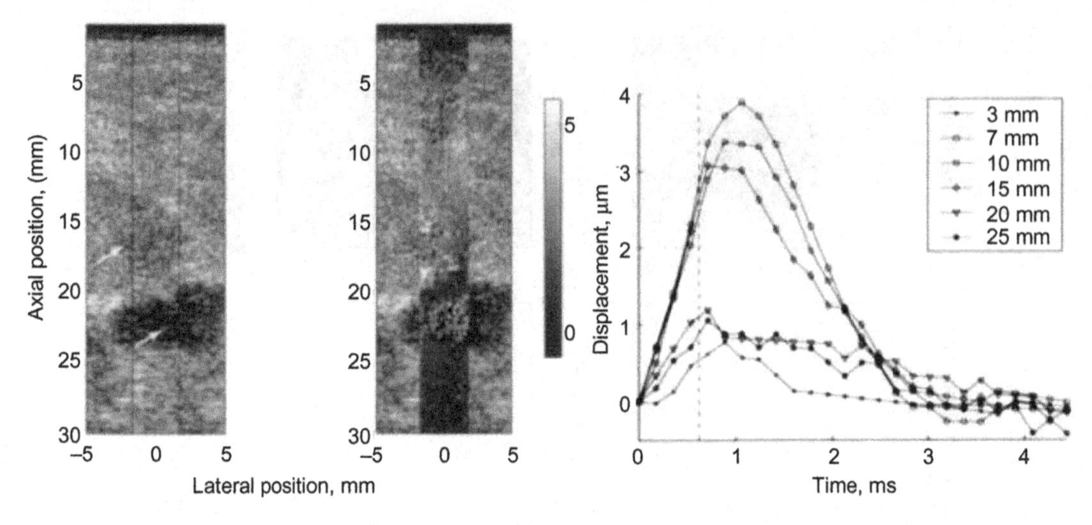

Figure 16.17

ARFI image of tissue displacement and displacement through time at different depths. (A) ARFI image of tissue displacement at 0.8 ms (right), and matched B-mode image (left) in an in vivo female breast. The transducer is located at the top of the images, and the color-bar scale is in m. There is a lesion located on the right side of the images between 20 and 25 mm that is evident as a darker region of tissue in the B-mode image (lower arrow). This lesion was palpable and, upon aspiration, was determined to be an infected lymph node. In the ARFI image, the lesion boundary appears stiffer than its interior and the tissue above it (i.e. it exhibits smaller displacements). In addition, the half-oval structure in the B-mode image immediately above and to the left of the lesion (upper arrow) appears to be outlined as a region of tissue softer than its surroundings in the ARFI image. (B) Displacement through time at different depths in the center of the ARFI image (0 mm laterally). The tissue at depths of 7, 10, and 15 mm exhibits faster excitation and recovery velocities than that at 20 and 25 mm. *From Nightingale et al. (2002), with permission from the World Federation of Ultrasound in Medicine and Biology.*

output and heating, so new avenues for improving sensitivity such as parallel processing are being investigated (Dahl et al. 2007). Arguments for increasing levels based on temperature–time relations discussed in Chapter 15 have been put forth. Time is another variable for adjusting contrast, as softer tissues have longer time constants as indicated by Figure 16.17, and as linked back to Sugimoto's earlier work (compare to Figure 16.2.).

ARFI imaging has had widespread clinical application (Nightingale, Soo, Nightingale, & Trahey, 2002; Palmeri & Nightingale, 2011; Wells and Liang, 2011). Siemens has introduced an ARFI-based Virtual Touch Tissue Imaging tool and a radiation-force-based shear wave speed estimation as the Virtual Touch Tissue Quantification tool (Palmeri & Nightingale, 2011). Doherty et al. (2013) have created an organized family tree of many different ARF imaging methods in their lucid review of ARF applications.

16.3.7 Vibro-acoustography Imaging

Fatemi and Greenleaf (1998, 1999) invented a highly sensitive low-frequency ARFI method. In their approach, two tilted sinusoidally excited confocal ultrasound beams, one of which is offset in frequency from the other by about 7–25 kHz, intersect at a region of interest. At the intersection, a low-frequency oscillating ARF is generated locally and the resulting tissue deformations are detected by a hydrophone. What is sensed is the dynamic response of the tissue. The region of interest is scanned to produce an image of amplitudes of deformation. Because of the high resolution and sensitivity of the method, small microcalcifications have been detected in vivo in human carotid arteries (Greenleaf, Fatemi, and Insana, 2003), breast samples (Fatemi, Wold, & Greenleaf, 2002), and the prostate. Urban, Fatemi, and Greenleaf (2010) reported a modulation scheme to produce the difference frequencies. For amplitude modulation, for example, for an ultrasound carrier frequency in MHz, ω_c:

$$p(t) = A_c[1 + \mu x_m(t)]\cos(w_c t), \tag{16.7}$$

in which A_c is a constant, x_m is a modulation function, and μ is a modulation index between 0 and 1. Recall that for an absorbing medium the radiation force can be estimated by $F = 2\alpha < I > /c = < E >_t$, where the energy density is $E = p(t)^2/\rho c^2$. For a cosinusoidal modulation of $x_m(t) = \cos(\omega_m t)$, it is straightforward to show that the radiation force has three terms, one at a steady state, another at ω_m, and another at $2\omega_m$. Other modulations were also derived and verified experimentally.

An implementation of vibro-acoustography has been reported on a commercial scanner (Urban et al. 2011). As shown in Figure 16.18, the two transducers have been replaced by two subarrays of an ultrasound imaging array with the capability of focusing to the same spot. One subarray is offset by a difference frequency. The array is scanned mechanically, the output dynamic response signal is picked up by a hydrophone and is used to form a C-scan image.

16.3.8 Harmonic Motion Imaging

In the monitoring of lesion formation caused by high-intensity focused ultrasound, B-mode imaging often lacks sufficient contrast. A common configuration used for this application is a hemispherical bowl-like shape with a hole (the holey HIFU transducer of Figure 6.26) in the center for either a single pulse–echo transducer or a conventional diagnostic imaging array. Harmonic motion imaging (HMI) is the induction of oscillatory ARF-induced motion of the tissue in the focal region of a HIFU transducer. The displacements are calculated by 1D cross-correlation during HIFU excitation from pulse–echo time streams from the center transducer. For this approach, a two-segmented

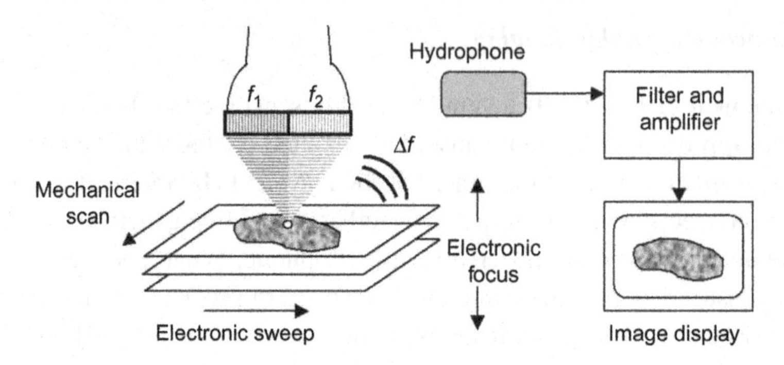

Figure 16.18

Block diagram of vibro-acoustography performed with a linear array transducer. The intersection of the two ultrasound beams is electronically swept in the azimuthal direction of the transducer and then mechanically scanned in the elevation direction of the transducer. Different planes can be imaged by changing the electronic focal depth. The interaction of the two ultrasound beams in the tissue produces an acoustic signal at the difference frequency, Δf, between the two ultrasound beams. The acoustic signal is detected by a nearby hydrophone and is filtered, amplified, digitized, and processed for image formation and display. The image shown is of a breast phantom. The azimuthal direction of the transducer corresponds to the vertical axis of the image and the mechanical translation of the transducer in the elevation direction corresponds to the horizontal axis. The image is 38.4 × 80 mm and streak correction has been applied. *From Urban et al. (2011), IEEE.*

HIFU transducer with a frequency offset in one of the segments was used, similar to the transmission in a vibro-acoustographic case (Konofagou & Hynynen, 2003). In a later approach, amplitude modulation of the HIFU transmission was used to create the difference frequency (Maleke, Luo, Gamarnik, Lu, & Konofagou, 2010). An example of the later version is illustrated in Figure16.19 (Konofagou, Maleke, & Vappou, 2012). High-contrast images and viscoelastic information can be obtained by varying the modulation frequency without assuming a rheological model such as Voigt, etc.

16.3.9 Supersonic Shear Imaging

Another novel way of generating ARFs in tissue is to use coherent plane wave compounding (Montaldo, Tanter, Bercoff, Benech, & Fink, 2009) (see Section 10.12.2). A point radiation force generates two expanding biphasic orthogonal shear waves, as symbolized once again in Figure 16.20. By repeating the longitudinal wave (1500 m/s) transient generating pulse at a certain rate as shown in the figure (6 m/s), the expanding resultant shear waves (2 m/s) form a supersonic plane cone. The angle of the cone can be changed by the rate of the excitations and is determined by the ratio of the speed of the moving excitation source to that of the shear wave speed (ratio = Mach 3). Each excitation point can be viewed as a radiating

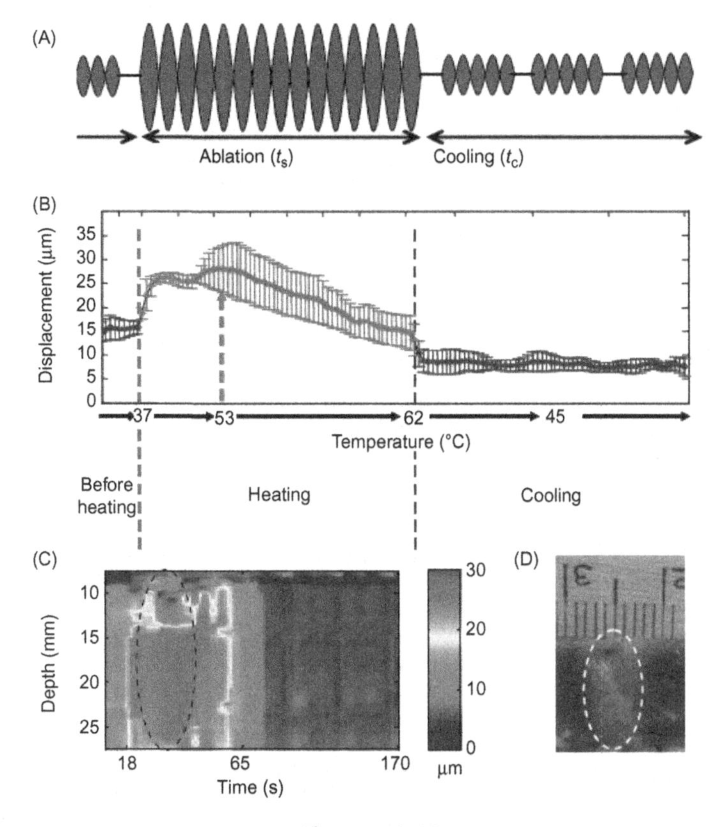

Figure 16.19

M-mode HMI, Real-time HMFU Monitoring. (A) The HMIFU sequence used before, during, and after HIFU treatment; (B) HMI displacement variation with temperature before (blue), during (orange), and after (blue) HIFU ablation, averaged over five different liver specimens and 3 locations in each liver, i.e. 15 locations total; (C) example of an M-mode HMI displacement image obtained in real time during ablation (as in (A) and (B); heating started at $t = 18$ s and ended at $t = 65$ s); (D) photograph of the liver lesion (denoted by the dashed contour). A liver vessel running through the lesion was used as a registration reference between the images in (C) and (D). Note that in (B) at 53 °C, coagulation occurs and the HMI displacement changes from a positive to a negative rate, indicating lesion formation. The reader is referred to the Web version of the book to see the figure in color. *From Konofagou et al. (2012) courtesy of Current Medical Imaging Reviews.*

element in a vertical virtual array with an applied linear delay, the equivalent of beam steering as depicted in Figure 7.16 and as described by Eqn 7.26. The net result is a propagating coherent wavefront. To appreciate the difference between this wavefront and a single excitation, consider the coherent power of a wave front on the ocean compared to expanding circular wavefronts from a small rock thrown in the water.

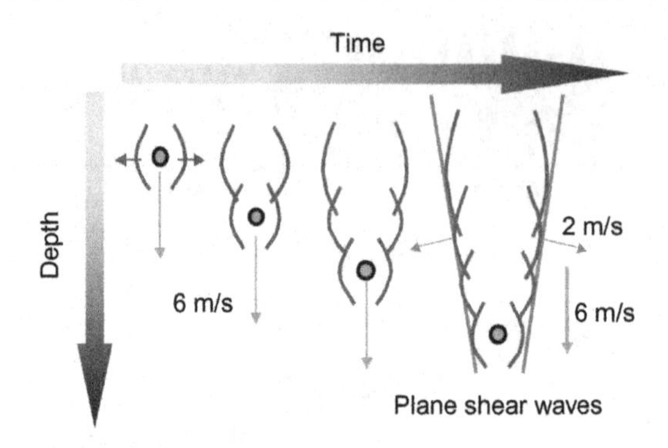

Figure 16.20
Generation of the supersonic shear source. The source is sequentially moved along the beam axis, creating two plane- and intense-shear waves. *From Bercoff et al. (2004), IEEE.*

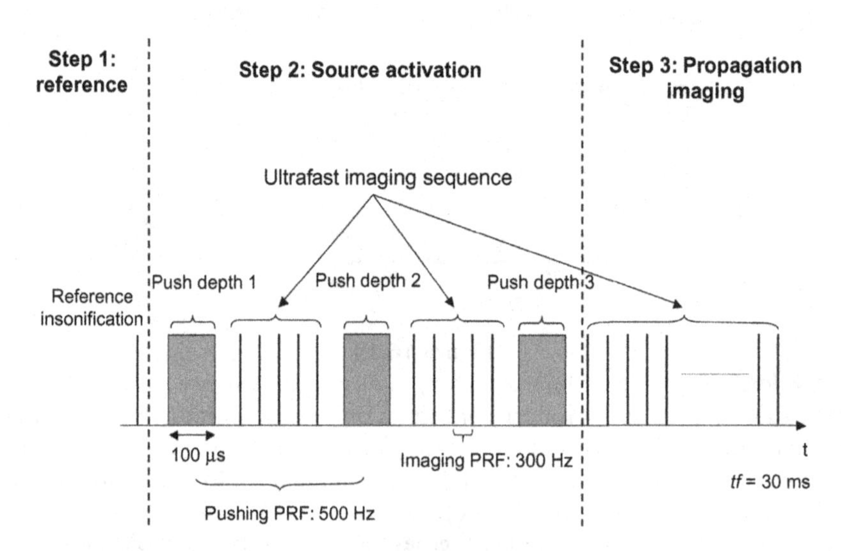

Figure 16.21
Emission sequence for a supersonic regime. Here, the first is a calibrating reference pulse, next is a series of sequences that include a pushing excitation pulse followed by a set of parallel-processed tracking pulses, each of which leads to an imaging frame after further processing, here represented by a single line. Once the pushing sequences necessary to form the supersonic cone are complete (typically three to five pushes), the tracking or imaging continues. *From Bercoff et al. (2004), IEEE.*

In the supersonic shear imaging scheme (Bercoff, 2008; Bercoff, Tanter, & Fink, 2004a; Fink & Tanter, 2010), pushing excitations (typically pulses 100-μs long) generate shear waves which deform tissue and the resulting tissue movements are sensed by parallel pulse–echo insonifications without focusing and are recorded at an ultrafast rate by an ultrafast scanner

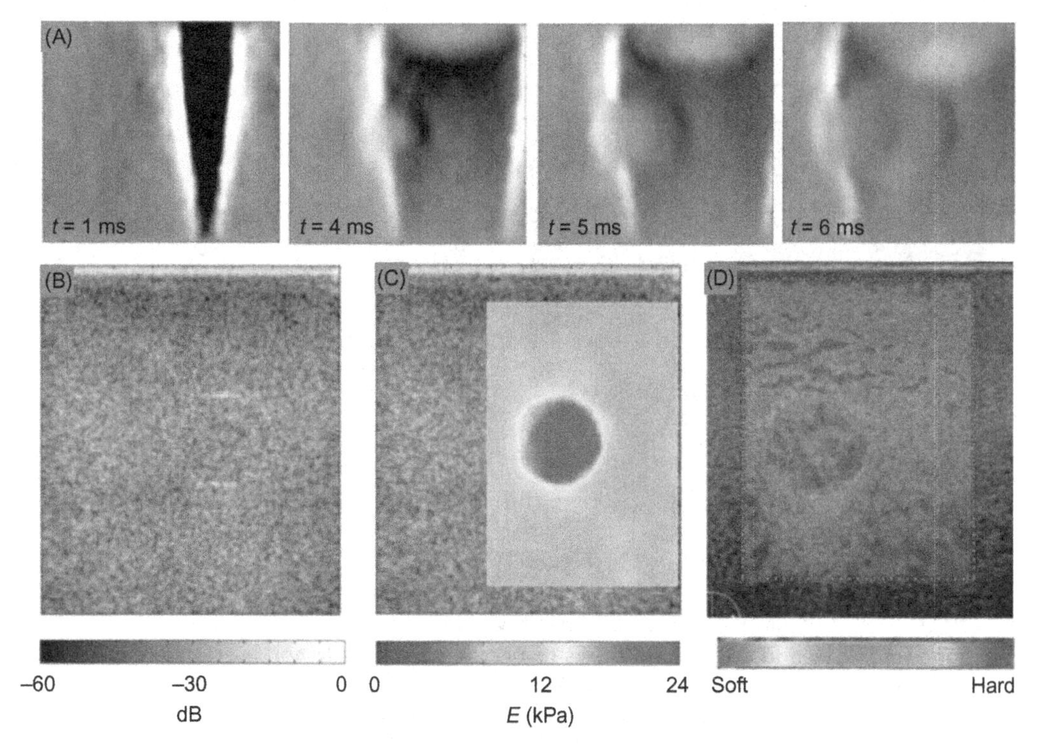

Figure 16.22

Supersonic shear wave imaging. (A) Four frames from an ultrasound movie of local displacements in a tissue-mimicking phantom at specified times after the supersonic generation of a shear wave. The gray scale indicates displacements from -10 to $+10$ μm. Clearly, the shear wave is sensing the stiffness contrast as it is distorted while passing through a 10-mm-diameter stiff inclusion. (B) A conventional ultrasonic image of the medium barely reveals the inclusion; (C) from the movie sampled in (A), one can obtain a quantitative image of Young's modulus, E; (D) an image of the same phantom as in (C), obtained with a commercially available ultrasound scanner using quasi-static elastography. The reader is referred to the Web version of the book to see the figure in color. *From Fink and Tanter (2010), American Institute of Physics.*

capable of frame rates up to 6000 frames/s, as illustrated by the sequence in Figure 16.21 (Sandrin et al., 2009; Tanter, Bercoff, Sandrin, & Fink, 2002). Then, sequential imaging (tracking) sets are 1D cross-correlated to calculate displacements that are available to create a fast-motion movie depicting the passages of shear waves through tissue. The location of array elements and timing enable a time-of-flight scheme to determine the speed of the shear wave at each spatial location. From Eqn 16.4B, the shear modulus can be determined, or Young's modulus from Eqn 16.3C, and displayed, or the displacement image can be displayed as illustrated in Figure 16.22.

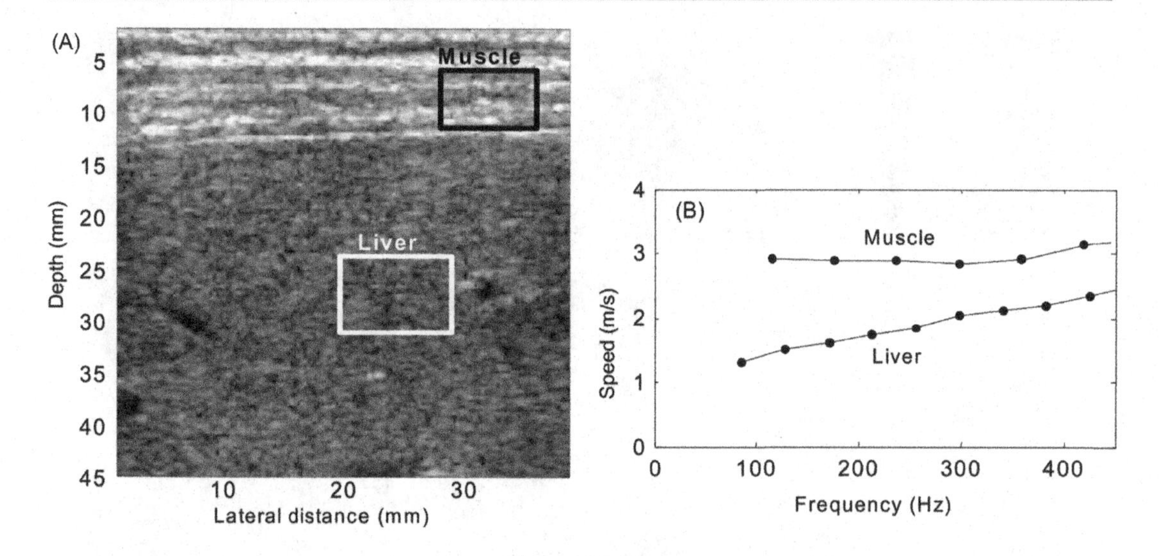

Figure 16.23
Velocity dispersion of shear waves for liver and muscle. (A) B-mode image showing the intercostal muscle (top) and the liver (bottom) of a healthy volunteer, regions of interest used for SWS correspond to 5 × 10 and 8 × 10-mm boxes respectively for muscle and liver regions; (B) in vivo dispersion curves in muscle and liver reflect their viscoelastic properties and allow their characterization. *From Deffieux et al. (2009), IEEE.*

Another improvement is that shear wave compounding (plane wave compounding) (Montaldo et al., 2009) is employed to improve image quality; a sequence of n push excitations at different angles is used. As a consequence of compounding, the frame rate is slowed down; for example, for five angles, plane waves at a 10-kHz rate result in compounded images at only a 2-kHz rate, still fast enough to depict the shear wave deformations (Montaldo et al., 2009).

As explained in detail in Section 12.8.4 and illustrated in Figure 12.32, Bercoff, Tanter, Muller, and Fink (2004b) derived a Green's function and a viscoelastic model for tissue. The initial ultrasound longitudinal wave ($c = 1540$ m/s) pushing pulse of several MHz translates into shear waves traveling at several m/s with a bandwidth of a few hundred Hz. Moreover, the shear absorption is high and the phase velocity dispersion can sometimes change the velocity by a factor of two or three in this bandwidth, as illustrated by Figure 16.23. Despite these complexities, Deffieux et al. (2009) developed a way of extracting the viscoelastic parameters with a method they called "shear wave spectroscopy." Two of the major clinical applications of this approach are the liver (Muller, Gennisson, Deffieux, Tanter, & Fink, 2009) and the breast (Tanter et al, 2008); for a more complete clinical review see Wells and Liang (2011).

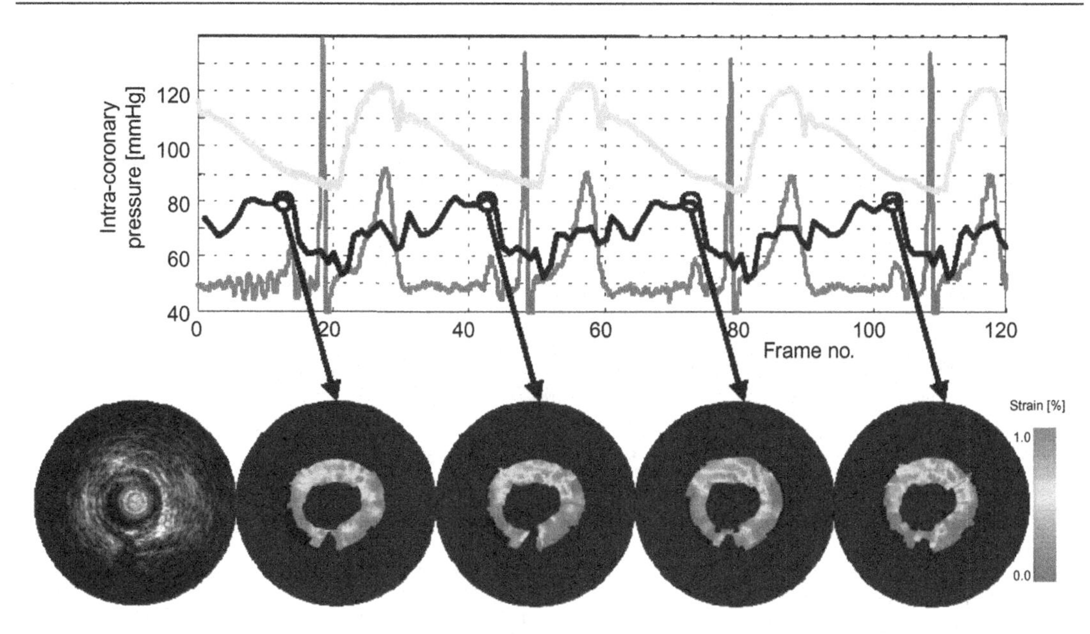

Figure 16.24
Reproducibility of IVUS elastography with elastic stimuli. The upper panel shows the physiologic signals. Echo frames acquired near end-diastole were used to determine the elastograms. The elastograms indicate that the plaque between 9 and 3 o'clock has high strain values, indicating soft material. The remaining part has low strain values, indicating hard material. At 6 o'clock, a calcified spot is visible in the echogram, corroborating the low strain values. The reader is referred to the Web version of the book to see the figure in color. *From de Korte et al. (2000), IEEE.*

16.3.10 Natural Imaging

In organic, passive, or natural elastography, also known as physiological displacement (depicted in Figure 16.1C), the tissue deformations come from the natural rhythms of the body. In particular, the heartbeat can be an appropriate stimulus. Two major applications use the heartbeat as a stimulus: arterial contractions (de Korte et al., 1997, 2000) for intravascular ultrasound (IVUS) imaging, and the heart motion itself (D'hooge et al. 2002; Konofagou, D'Hooge, & Ophir, 2002; Lee et al., 2007; Luo, Lee, & Konofagou, 2009) for echocardiology. Examples of IVUS elastograms are shown with the stimulus waveforms in Figure 16.24. In the final approach, illustrated by Figure 16.25, different physiological sources of movement in the body, such as the heart rate and breathing cycles, provide sources of tissue deformation (Gallot et al. 2011). They do not have to be identified in order for this process to work and can be regarded as noise sources. Through a correlation-based tomographic mapping method, a shear wave velocity can be determined.

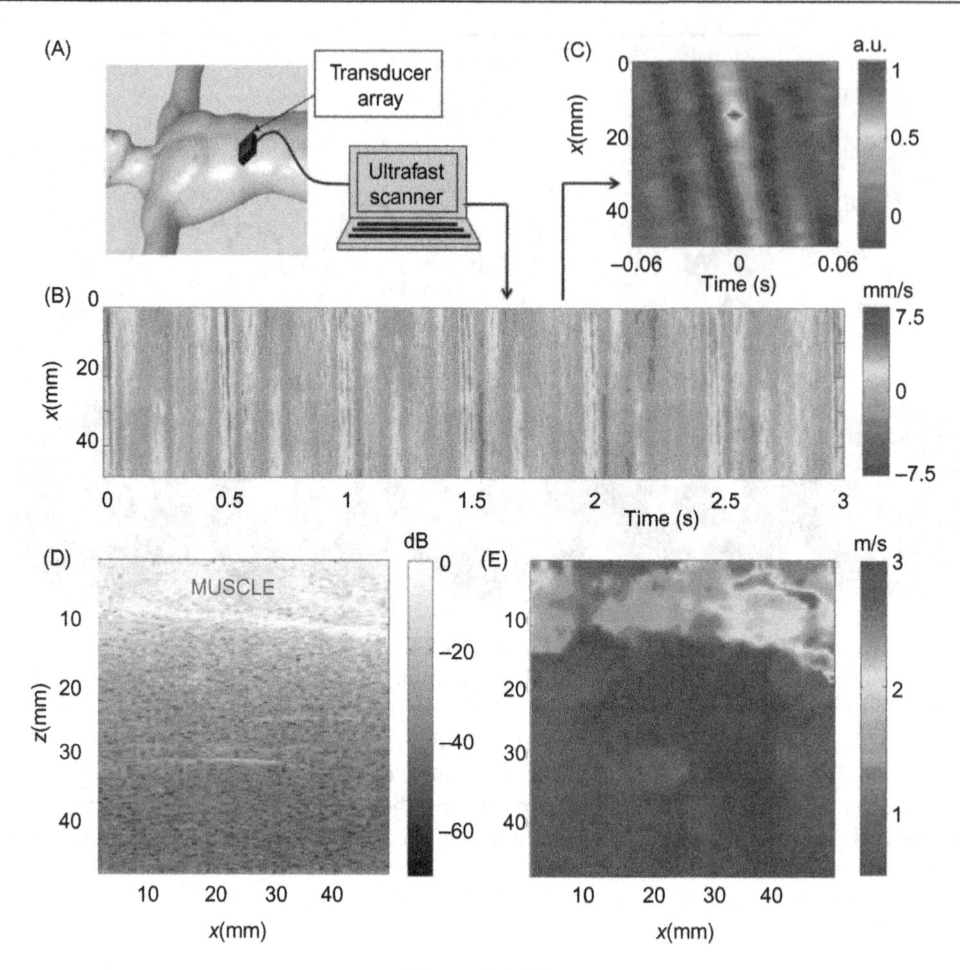

Figure 16.25

Physiological sources of movement provide sources of tissue deformation for organic elastography. (A) Experimental setup for the in vivo correlation-based tomography: an ultrafast scanner was used to measure the natural displacement field in the liver region. (B) The particle velocity along an acquisition line at 4 cm, and parallel to the array, shows the physiological elastic field. (C) in the correlation map $C(x_0, x; t)$ with $x_0 = 14$ mm, only one direction of propagation emerged from the refocusing field. (D) Sonogram of the liver region. The interface between the abdominal muscles and the liver is visible around $z = 12$ mm. (E) The passive shear-wave-speed tomography from the correlation width clearly shows the two regions. The averaged shear-speed estimations are in agreement with values in the literature. The reader is referred to the Web version of the book to see the figure in color. *From Gallot et al. (2011), IEEE.*

16.4 Conclusions

Based on common underlying physical processes and common features in terms of excitation methods and detection, different elastography methods have been compared. As new applications of diagnostic ultrasound outpace the complete understanding of the underlying physical mechanisms, the need for continuing research and debate continues. Many subcategories and clinical applications have been omitted for the sake of brevity and clarity; however, interested readers can satisfy their curiosity by referring to the excellent overviews in the Bibliography, which contain hundreds of references or the original source materials cited. Alternative detection methods, which include acoustic surface sensors, optical devices, MRI, and X ray, were not covered. MRI elastography is covered in several of the references in the Bibliography.

References

Azar, R. Z., Goksel, O., & Salcudean, S. E. (2010). Sub-sample displacement estimation from digitized ultrasound RF signals using multi-dimensional polynomial fitting of the cross-correlation function. *IEEE Transactions on Ultrasonics, Ferroelectrics and Frequency Control*, *57*, 2403–2410.

Bamber, J. C. (1998). Ultrasonic properties of tissue. In F. A. Duck, A. C. Baker, & H. C. Starritt (Eds.), *Ultrasound in medicine*. Bristol, UK: Institute of Physics Publishing.

Barbone, P. E., & Bamber, J. (2002). Quantitative elasticity imaging: What can and cannot be inferred from strain images. *Physics in Medicine and Biology*, *47*, 2147–2164.

Bercoff, J. (2008). ShearWave elastography, SuperSonic Imagine White Paper. SuperSonic Imagine, Aix en Provence, France.

Bercoff, J., Tanter, M., & Fink, M. (2004a). Supersonic shear imaging: A new technique for soft tissue elasticity mapping. *IEEE Transactions on Ultrasonics, Ferroelectrics and Frequency Control*, *51*, 396–409.

Bercoff, J., Tanter, M., Muller, M., & Fink, M. (2004b). The role of viscosity in the impulse diffraction field of elastic waves induced by the acoustic radiation force. *IEEE Transactions on Ultrasonics, Ferroelectrics and Frequency Control*, *51*, 1523–1536.

Berry, G., Bamber, J., Armstrong, C., Miller, N., & Barbone, P. (2006a). Towards an acoustic model-based poroelastic imaging method: I. Theoretical foundation. *Ultrasound in Medicine and Biology*, *32*, 547–567.

Berry, G., Bamber, J., Miller, N., Barbone, P., Bush, N., & Armstrong, C. (2006b). Towards an acoustic model-based poroelastic imaging method: II. Experimental investigation. *Ultrasound in Medicine and Biology*, *32*, 1869–1885.

Catheline, S., Gennisson, J., Delon, G., Fink, M., Sinkus, R., Abouelkaram, S., et al. (2004). Measurement of viscoelastic properties of homogeneous soft solid using transient elastography: An inverse problem approach. *Journal of the Acoustical Society of America*, *116*, 3734–3741.

Chen, S., Fatemi, M., & Greenleaf, J. F. (2004). Quantifying elasticity and viscosity from measurement of shear wave speed dispersion. *Journal of the Acoustical Society of America*, *115*, 2781–2785.

Chen, S., Urban, M. W., Pislaru, C., Kinnick, R., Zheng, Y., Yao, A., et al. (2009). Shear wave dispersion ultrasound vibrometry (SDUV) for measuring tissue elasticity and viscosity. *IEEE Transactions on Ultrasonics, Ferroelectrics and Frequency Control*, *56*, 55–62.

Dahl, J. J., Pinton, G. F., Palmeri, M. L., Agrawal, V., Nightingale, K. R., & Trahey, G. E. (2007). A parallel tracking method for acoustic radiation force impulse imaging. *IEEE Transactions on Ultrasonics, Ferroelectrics and Frequency Control*, *54*, 301–312.

Deffieux, T., Montaldo, G., Tanter, M., & Fink, M. (2009). Shear wave spectroscopy for in vivo quantification of human soft tissues visco-elasticity. *IEEE Transactions on Medical Imaging, 28*, 313–322.

Doherty, J. R., Dahl, J. J. & Trahey, G. E. (2012) A harmonic tracking method for acoustic radiation force impulse (ARFI) imaging, *Proceedings of the IEEE ultrasonics symposium*, Dresden, Germany, Art. no. IC-4.

D'hooge, J., Bijnens, B., Thoen, J., van de Werf, F., Sutherland, G. R., & Suetens, P. (2002). Echocardiographic strain and strain-rate imaging: A new tool to study regional myocardial function. *IEEE Transactions on Medical Imaging, 21*, 1022–1030.

Fahey, B. J., Nelson, R. C., Bradway, D. P., Hsu, S. J., Dumont, D. M., & Trahey, G. E. (2008). In vivo visualization of abdominal malignancies with acoustic radiation force elastography. *Physics in Medicine and Biology, 53*, 279–293.

Fatemi, M., & Greenleaf, J. F. (1998). Ultrasound-stimulated vibroacoustic spectrography. *Science, 280*, 82–85.

Fatemi, M., & Greenleaf, J. F. (1999). Vibro-acoustogaphy: An imaging modality based on ultrasound-stimulated acoustic emission. *Proceedings of the National Academy of Sciences of the USA, 96*, 6603–6608.

Fatemi, M., Wold, L. E., & Greenleaf, J. F. (2002). Vibro-acoustic tissue mammography. *IEEE Transactions on Medical Imaging, 21*, 1–8.

Fink, M., & Tanter, M. (2010). Multiwave imaging and super-resolution. *Physics Today*, 28–33.

Gallot, T., Catheline, S., Roux, P., Brum, J., Benech, N., & Negreira, C. (2011). Passive elastography: Shear-wave tomography from physiological-noise correlation in soft tissues. *IEEE Transactions on Ultrasonics, Ferroelectrics and Frequency Control, 58*, 1122–1126.

Gao, L., Alam, S. K., Lerner, R. M., & Parker, K. J. (1995). Sonoelasticity imaging: Theory and experimental verification. *Journal of the Acoustical Society of America, 97*, 3875–3886.

Gao, L., Parker, K. J., Lerner, R. M., & Levinson, S. F. (1996). Imaging of elastic properties of tissue: A review. *Ultrasound in Medicine and Biology, 72*, 959–977.

Hall, T. J., Bilgen, M., Insana, M. F., & Krouskop, T. (1997). Phantom materials for elastography. *IEEE Transactions on Ultrasonics, Ferroelectrics and Frequency Control, 44*, 1355–1365.

Hall, T. J., Barbone, P. E., Oberai, A. A., Jiang, J., Dord, J. -F., Goenezen, S., et al. (2011). Recent results in nonlinear strain and modulus imaging. *Current Medical Imaging Reviews, 7*. 000-000.

Insana, M. F., Bilgen, M., Chaturvedi, P., Hall, T. J., & Bertrandt, M. (1996). Signal processing strategies in acoustic elastography. *Proceedings of the IEEE ultrasonics symposium* (pp. 1139–1142), San Antonio, TX.

Kallel, F., Price, R. E., Konofagou, E. E., & Ophir, J. (1999a). Elastographic imaging of the normal canine prostrate. *Ultrasonic Imaging, 21*, 201–215.

Kallel, F., Stafford, R. J., Price, R. E., Righetti, R., Ophis, J., & Hazle, J. D. (1999b). The feasibility of elastographic visualization of HIFU-induced thermal lesions in soft tissues. *Ultrasound in Medicine and Biology, 25*, 641–647.

Konofagou, E. E. (2004). Quo vadis elasticity imaging? *Ultrasonics, 42*, 331–336.

Konofagou, E. E., D'Hooge, J., & Ophir, J. (2002). Myocardial elastography: A feasibility study in vivo. *Ultrasound in Medicine and Biology, 28*, 475–482.

Konofagou, E. E., Harrigan, T. P., Ophir, J., & Krouskop, T. A. (2001). Poroelastography: Imaging the poroelastic properties of tissues. *Ultrasound in Medicine and Biology, 27*, 1387–1397.

Konofagou, E. E., & Hynynen, K. (2003). Localized harmonic motion imaging: Theory, simulations and experiments. *Ultrasound in Medicine and Biology, 29*, 1405–1413.

Konofagou, E. E., Maleke, C., & Vappou, J. (2012). Harmonic motion imaging (HMI) for tumor imaging and treatment monitoring. *Current Medical Imaging Reviews, 8*, 16–26.

Konofagou, E. E., & Ophir, J. (1998). A new elastographic method for estimation and imaging of lateral displacements, lateral strains, corrected axial strains and Poisson's ratios in tissues. *Ultrasound in Medicine and Biology, 24*, 1183–1199.

Krouskop, T. A., Wheeler, T. M., Kallel, F., Garra, B. S., & Hall, T. (1998). Elastic moduli of breast and prostate tissues under compression. *Ultrasonic Imaging*, *20*, 260–274.

Lee, W. -N., Ingrassia, C. M., et al. (2007). Theoretical quality assessment of myocardial elastography with in vivo validation. *IEEE Transactions on Ultrasonics, Ferroelectrics and Frequency Control*, *54*, 2233–2345.

Lerner, R. M., Parker, K. J., Holen, J., Gremiak, R., & Waag, R. C. (1988). Sono-elasticity: Medical elasticity images derived from ultrasound signals in mechanically vibrated targets. *Acoustical Imaging*, *16*, 317–327.

Levinson, S. F., Shinagawa, M., & Sato, T. (1995). Sonoelastic determination of human skeletal muscle elasticity. *Journal of Biomechanics*, *28*, 1145–1154.

Luo, J., Lee, W. -N., & Konofagou, E. E. (2009). Fundamental performance assessment of 2-D myocardial elastography in a phased-array configuration. *IEEE Transactions on Ultrasonics, Ferroelectrics and Frequency Control*, *54*, 2320–2327.

Maleke, C., Luo, J., Gamarnik, V., Lu, X. L., & Konofagou, E. E. (2010). Simulation study of amplitude-modulated (AM) harmonic motion imaging (HMI) for stiffness contrast quantification with experimental validation. *Ultrasonic Imaging*, *32*, 154–176.

Montaldo, G., Tanter, M., Bercoff, J., Benech, N., & Fink, M. (2009). Coherent plane-wave compounding for very high frame rate ultrasonography and transient elastography. *IEEE Transactions on Ultrasonics, Ferroelectrics and Frequency Control*, *56*, 489–506.

Muller, M., Gennisson, J. -L., Deffieux, T., Tanter, M., & Fink, M. (2009). Quantitative viscoelasticity mapping of human liver using supersonic shear imaging: Preliminary in vivo feasibility study. *Ultrasound in Medicine and Biology*, *35*, 219–229.

Nightingale, K., Soo, M. S., Nightingale, R. W., & Trahey, G. E. (2002). Acoustic radiation force impulse imaging: In vivo demonstration of clinical feasibility. *Ultrasound in Medicine and Biology*, *28*, 227–235.

Nightingale, K. R., Palmeri, M. L., Nightingale, R. W., & Trahey, G. E. (2001). On the feasibility of remote palpation using acoustic radiation force. *Journal of the Acoustical Society of America*, *110*, 625–634.

Ophir, J., Cespedes, I., Ponnekanti, H., Yazdi, Y., & Li, X. (1991). Elastography: A quantitative method for imaging the elasticity of biological tissues. *Ultrasonic Imaging*, *13*, 111–134.

Ophir, J., Garra, B., Kallel, F., Konofagou, E., Krouskop, T., Righetti, R., et al. (2000). Elastographic imaging. *Ultrasound in Medicine and Biology*, *26*, S23–S29.

O'Donnell, M., Skovoroda, A. R., Shapo, B. M., & Emelianov, S. Y. (1994). Internal displacement and strain imaging using ultrasonic speckle tracking. *IEEE Transactions on Ultrasonics, Ferroelectrics and Frequency Control*, *41*, 314–325.

Palmeri, M. L., & Nightingale, K. R. (2011). Acoustic radiation force-based elasticity imaging methods. *Interface Focus*, *1*, 553–564.

Parker, K. J., & Lerner, R. M. (1992). Sonoelasticity of organs: Shear waves ring a bell. *Journal of Ultrasound in Medicine*, *11*, 387–392.

Pinton, G. F, Dahl, J. J., & Trahey, G. E. (2006). Rapid tracking of small displacements with ultrasound. *IEEE Transactions on Ultrasonics, Ferroelectrics and Frequency Control*, *53*, 1103–1117.

Richards, M. S., Barbone, P. E., & Oberai, A. A. (2009). Quantitative three-dimensional elasticity imaging from quasi-static deformation: A phantom study. *Physics in Medicine and Biology*, *54*, 757–779.

Righetti, R., Righetti, M., Ophir, J., & Krouskop, T. A. (2007). The feasibility of estimating and imaging the mechanical behavior of poroelastic materials using axial strain elastography. *Physics in Medicine and Biology*, *52*, 3241–3259.

Rubens, D. J., Hadley, M. A., Alam, S. K., Gao, L., Mayer, R. D., & Parker, K. J. (1995). Sonoelasticity imaging of prostate cancer: In vitro results. *Radiology*, *195*, 379–383.

Sandrin, L., Catheline, S., Tanter, M., Hennequin, X., & Fink, M. (1999). Time-resolved pulsed elastography with ultrafast ultrasonic imaging. *Ultrasonic Imaging*, *21*, 259–272.

Sandrin, L., Tanter, M., Gennisson, J. -L., Catheline, S., & Fink, M. (2002). Shear elasticity probe for soft tissues with 1-D transient elastography. *IEEE Transactions on Ultrasonics, Ferroelectrics and Frequency Control*, *49*, 436–446.

Sarvazyan, A., Rudenko, O., Swanson, S., Fowlkes, J., & Emelianov, S. (1998). Shear wave elasticity imaging: A new ultrasonic technology of medical diagnostics. *Ultrasound in Medicine and Biology, 24*, 1419–1435.

Sarvazyan, A. P. (1993). Shear acoustic properties of soft biological tissues in medical diagnostics. *Journal of the Acoustical Society of America, 93*, 2329.

Sarvazyan, A. P., Urban, M. W., & Greenleaf, J. F. (2013). Acoustic waves in medical imaging and diagnostics. *Ultrasound in Medicine and Biology, 39*, 1133–1146.

Skovoroda, A. R., Emelianov, S. Y., Lubinski, M. A., Sarvazyan, A. P., & O'Donnell, M. (1994). Theoretical analysis and verification of ultrasound displacement and strain imaging. *IEEE Transactions on Ultrasonics, Ferroelectrics and Frequency Control, 41*, 302–313.

Skovoroda, A. R., Emelianov, S. Y., & O'Donnell, M. (1995). Tissue elasticity reconstruction based on ultrasonic displacement and strain images. *IEEE Transactions on Ultrasonics, Ferroelectrics and Frequency Control, 42*, 747–765.

Sugimoto, T., Ueha, S. & Itoh,, K. (1990). Tissue hardness measurement using the radiation force of focused ultrasound. *Proceedings of the IEEE ultrasonics symposium* (pp. 1377–1380), Honolulu, HI.

Tanter, M., Bercoff, J., et al. (2008). Quantitative assessment of breast lesion viscoelasticity: Initial clinical results using supersonic shear imaging. *Ultrasound in Medicine and Biology, 34*, 1373–1386.

Tanter, M., Bercoff, J., Sandrin, L., & Fink, M. (2002). Ultrafast compound imaging for 2-D motion vector estimation: Application to transient elastography. *IEEE Transactions on Ultrasonics, Ferroelectrics and Frequency Control, 49*, 1363–1374.

Trahey, G. E., Hubbard, S. M., & von Ramm, O. T. (1988). Angle-independent ultrasonic blood flow detection by frame to frame correlation of B-mode images. *Ultrasonics, 26*, 271–276.

Treece, G., Lindop, J., Chen, L., Housden, J., Prager, R., & Gee, A. (2011). Real-time quasi-static ultrasound elastography. *Interface Focus, 1*, 540–552.

Urban, M. W., Chalek, C., Kinnick, R. R., Kinter, T. M., Haider, B., Greenleaf, J. F., et al. (2011). Implementation of vibro-acoustography on a clinical ultrasound system. *IEEE Transactions on Ultrasonics, Ferroelectrics and Frequency Control, 58*, 1169–1181.

Urban, M. W., Fatemi, M., & Greenleaf, J. F. (2010). Modulation of ultrasound to produce multifrequency radiation force. *Journal of the Acoustical Society of America, 127*, 1228–1238.

Varghese, T., & Ophir, J. (1997). A theoretical framework for performance characterization of elastography. *IEEE Transactions on Ultrasonics, Ferroelectrics and Frequency Control, 44*, 164–172.

Walker, W. F., Fernandez, F. J., & Negron, L. A. (2000). A method of imaging viscoelastic parameters with acoustic radiation force. *Physics in Medicine and Biology, 45*, 1437–1447.

Wilson, L. S., Robinson, D. E., & Dadd, M. J. (2000). Elastography—the movement begins. *Physics in Medicine and Biology, 45*, 1409–1421.

Yamakoshi, Y., Sato, J., & Sato, T. (1990). Ultrasonic imaging of internal vibration of soft tissue under forced vibration. *IEEE Transactions on Ultrasonics, Ferroelectrics and Frequency Control, 17*, 45–53.

Zhao, H., Song, P., Urban, M. W., et al. (2011). Bias observed in time-of-flight shear wave speed measurements using radiation force of a focused ultrasound beam. *Ultrasound in Medicine and Biology, 37*, 1884–1892.

Bibliography

Doherty, J. R., Trahey, G. E., Nightingale, K. R., & Palmeri, M. L. (2013). Acoustic radiation force elasticity imaging in diagnostic ultrasound. *IEEE Transactions on Ultrasonics, Ferroelectrics and Frequency Control, 60*, 685–701.

Doyley, M. M. (2012). Model-based elastography: A survey of approaches to the inverse elasticity problem. *Physics in Medicine and Biology, 57*, R35–R73.

Greenleaf, J. F., Fatemi, M., & Insana, M. F. (2003). Selected methods for imaging elastic properties of biological tissues. *Annual Review of Biomedical Engineering*, *5*, 57–78.

Ophir, J., Alam, S. K., Garra, B., Kallel, F., Konofagou, E., Krouskop, T, et al. (1999). Elastography: Ultrasonic estimation and imaging of the elastic properties of tissues. *Proceedings of the Institute of Mechanical Engineers H*, *213*, 203–233.

Parker, K J, Doyley, M M, & Rubens, D J (2011). Imaging the elastic properties of tissue: The 20 year perspective. *Physics in Medicine and Biology*, *56*, R1–R29.

Sarvazyan, A . P., Hall, T. J., Urban, M. W., Fatemi, M., Aglyamov, S. R., & Garra, B. (2011). An overview of elastography: An emerging branch of medical imaging. *Current Medical Imaging Review*, *4*(7), 255–282.

Wells, P. T. N., & Liang, H. -D. (2013). Medical ultrasound: Imaging of soft tissue strain and elasticity. *Journal of the Royal Society Interface*, *8*, 1521–1549.

CHAPTER 17

Therapeutic Ultrasound

Chapter Outline

17.1 Introduction

Diagnostic ultrasound involves some intervention, such as the use of ultrasound contrast agents, but on the whole, it is a gentle, passive method aimed at characterizing well the tissue structure and function already in place naturally. Therapeutic ultrasound, in contrast, is an active ultrasound designed to alter tissues and/or the functioning of tissues either permanently or temporarily. Furthermore, therapeutic ultrasound can be viewed as extreme ultrasound that utilizes effects beyond most of the ultrasound physics covered in all of the previous chapters. These extreme approaches can be justified if the alternatives are either far worse or impossible. Therapeutic ultrasound encompasses both the highest acoustic output levels and low-intensity approaches.

Therapeutic ultrasound is also attempting to achieve difficult goals, those thought to be impossible or extremely difficult by other means. The Fry brothers, beginning in the early 1950's, were motivated to develop HIFU methods to treat brain tumors and Parkinson's disease (Fry, Mosberg, Barnard and Fry, 1954). High-intensity focused ultrasound (HIFU) is a noninvasive way of performing highly precise surgical removal of cancer without a single incision; it is a type of surgery requiring minimal or no hospital stays. Previously, the blood–brain barrier prevented critical drug therapies from reaching the brain. With ultrasound, a noninvasive reversible process for opening this barrier has been devised for allowing drugs and therapies to reach the brain. Low-power ultrasound has been used for severe bone fractures that would otherwise not heal. These examples are a few of the new opportunities both low-power and high-intensity therapeutic ultrasound offer. These applications will be explored after a review of some of the physical mechanisms involved.

17.2 Therapeutic Ultrasound Physics

17.2.1 Introduction

Where we left bioeffects in Chapter 15, the two modalities with the highest acoustic output levels were HIFU, involving heating, and lithotripsy, achieving destructive mechanical effects. To review (Section 15.7.4), HIFU involves long pulses, seconds long, at frequencies typically 0.5 to 8 MHz, and maximum compressional pressures in the 2–4-MPa range. For lithotripsy (Section 15.7.5), low-duty-cycle pulses (typically less than 10 μs long) involve compressional pressures in the 30–150-MPa range and rarefactional pressures in the 3–15-MPa range (Bailey, Khokhlova, Sapozhnikov, Kargl, & Crum, 2003).

17.2.2 High-intensity Focused Ultrasound

Because HIFU, also known as HITU (High Intensity Therapeutic Ultrasound), was already introduced in Section 15.7.4, here we investigate the physics in more detail by way of a specific example. HIFU, a noninvasive way of accomplishing the equivalent of surgery by creating thermal lesions in approximately the −6-dB focal region. For simulation, an algorithm must have the capabilities of modeling a strongly focused beam under highly nonlinear conditions through layers of absorbing tissue. Here, either the KZK (Khokhlov–Zabolotskaya–Kuznetsov) equation modified by a power law loss operator (Eqn 12.11) or a Westervelt equation with a similar loss operator (Eqn 12.13) is most appropriate. Furthermore, the pressure field must be coupled to the generation of heat. Typically, the thermal properties of the tissue, including perfusion (Section 15.3.4), are described by the Pennes bioheat equation (Eqn 15.22) in which sound is a heating source. Depending on the drive level and, consequently, the resulting magnitude of the pressure distribution in the focal region, the tissue in that region will begin to heat up and exceed, or not, the cumulative equivalent minutes (CEM) at 43°C (Eqn 15.9) level.

For an example, the publicly available program developed by Soneson (2009), the HIFU Simulator will be used. A second approximate program by Soneson and Myer (Soneson, 2009; Soneson & Myer, 2007, 2010) mentioned in Section 12.7 is faster but less general. This program utilizes a split-step approach, in the frequency domain, breaking the computation of Eqn 12.11, modified for true power law absorption, into coupled linear and nonlinear parts. In addition, a Gaussian harmonic approximation is made by which considerable improvement in execution time is accomplished. Other convenient features include the accessibility of both the time domain and frequency domain outputs as well as the spatial temperature distribution. The nonlinear intensity is computed as:

$$I(r,z) = \frac{1}{2\rho_m c_m}\sum_{n=1}^{N}\left|\hat{p}_n\right|^2, \tag{17.1A}$$

in which $\hat{p}_n$ is the frequency domain n^{th} harmonic component of the pressure in layer m. Similarly, the heating source component for the inhomogeneous part of the bioheat equation is:

$$H(r,z) = \frac{1}{2\rho_m c_m}\sum_{n=1}^{N}\text{Re}(\gamma_n)\left|\hat{p}_n\right|^2, \tag{17.1B}$$

Where γ_n is the propagation factor from Chapter 4 but for harmonic frequency f_n, and for layer m. Note the similarity of this to the Eqn 15.2 approximation—except here, α would include the full complement of absorption at harmonic frequencies—therefore, the nonlinear counterpart for the heat source is:

$$q_v = 2\sum_n \alpha_n I_n. \tag{17.1C}$$

For power law absorption of the form given in Chapter 4, the heat rate at the focal point (Soneson & Meyers, 2007) is:

$$Q = \alpha_1 I_f \sum_n n^y\left|a_n(1)\right|^2, \tag{17.2}$$

in which $a_n(1)$ is the harmonic amplitude n, and I_f is the linear intensity, both at the focal point. For the specific case of axial heating from a Gaussian apodized beam, Soneson and Meyers (2007) derived an analytic expression that showed the temperature rose less fast than the heat rate (Eqn 17.1C) by a factor of $\ln(n)/n$ where n is a harmonic number for each harmonic.

Having a dependable simulation model is important in treatment planning. In order to estimate the appropriate thermal dose, an accounting must include the acoustic properties and absorption along the tissue path, strength of focusing under nonlinear conditions, and the heating characteristics of the tissue. In this example (taken from Soneson (2009)), we also include the effect of pulsed excitation, which adds further complexity.

Usually, HIFU transducers are shaped as a truncated spherical bowl with the focal point at the radius of curvature. Transducers of this type with and without a hole in the center for an ultrasound monitoring transducer were described in Section 6.10. In this program example, the transducer has a center frequency of 1.5 MHz, a full (no hole) radius of 2.5 cm, and a focal depth of 8 cm, resulting in an *F#* of 1.6, within the parabolic approximation used for the KZK equation (Sonenson, 2012). The tissue configuration is a water path of 5 cm and then into a tissue of acoustic impedance 1.69 MegaRayls, an absorption coefficient of 0.232 dB/MHz-cm, and a nonlinearity parameter of 4.5, as compared with 3.5 for water. The thermal properties of the second material are heat capacity $C = 4.18$ kJ/kg/K, thermal conductivity $k_2 = 0.6$ W/m/K, and perfusion rate $w_2 = {}_20$ kg/m^3/s (see Section 15.3.4). The program is set to calculate 128 harmonics.

The first result, calculated by the HIFU_Simulator (Soneson, 2009), is given by Figure 17.1, in which the axial intensity distribution shows a maximum near the focal point of about 21 kW/cm^2. An insert reveals a $p_c \sim 18$ MPa and $p_r \sim 5$ MPa. In the top of Figure 17.2, the plot of input pulse envelope gives the input pulse sequence: an initial burst of 0.3s and then 0.1-s pulses at 0.5-s intervals. The resulting heating pattern in the bottom of Figure 17.2, where we note the expected thermal rise and decay times and an average heating temperature of about 84°C. The spacing out of excitation pulses can be used to extend the heating region in conjunction with scanning. Spatial thermal results are displayed in the contour plots of Figure 17.3 for a time of 2.4s, after which some diffusion

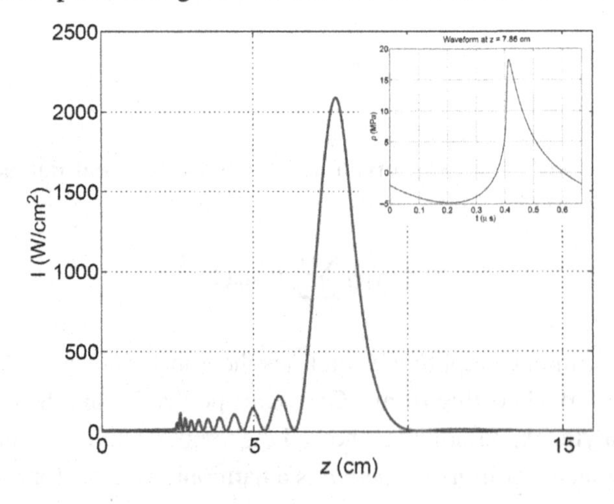

Figure 17.1
Axial intensity. Insert: temporal waveform (On-axis) at distance where peak pressure occurs. *From Sonenson (2009), courtesy of J. Sonenson, ISTU Proceedings, AIP Publishing LLC.*

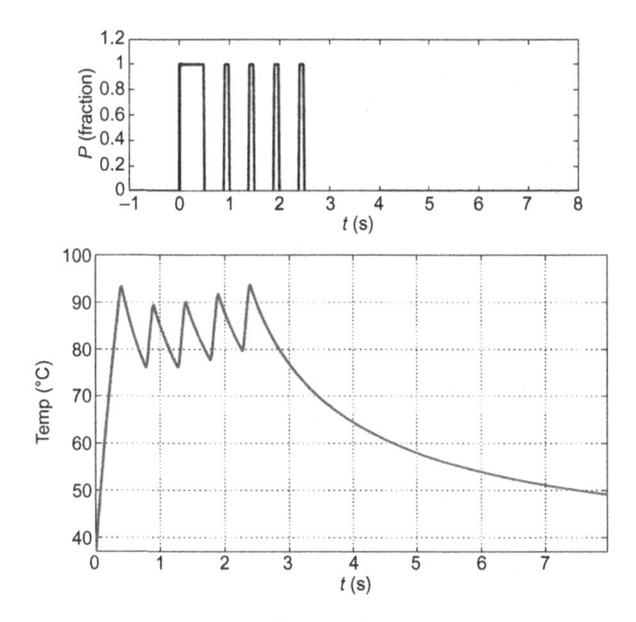

Figure 17.2

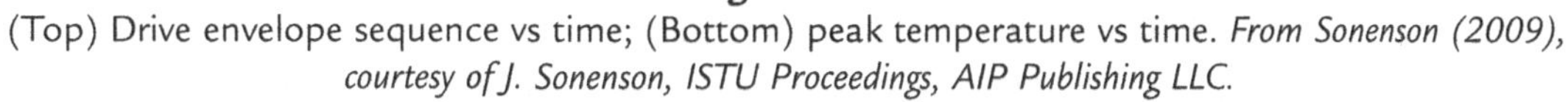

(Top) Drive envelope sequence vs time; (Bottom) peak temperature vs time. *From Sonenson (2009), courtesy of J. Sonenson, ISTU Proceedings, AIP Publishing LLC.*

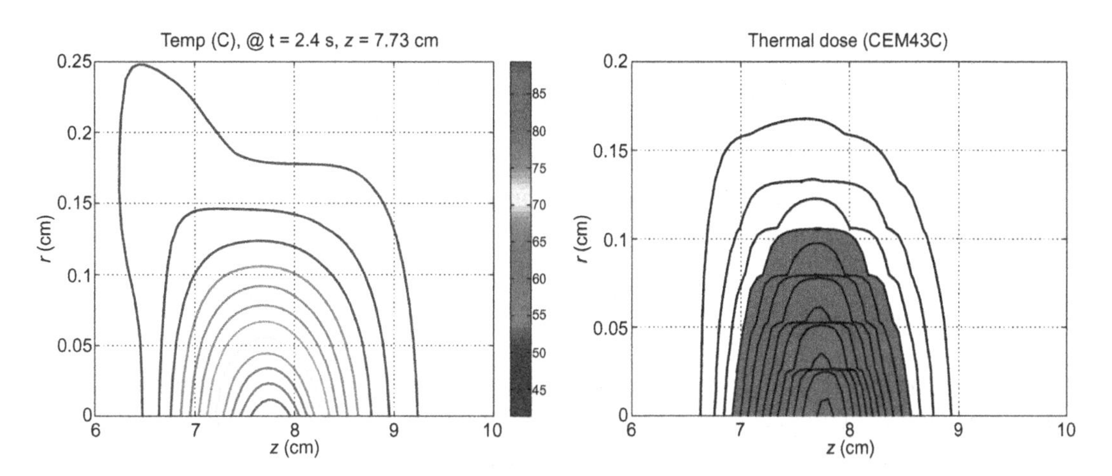

Figure 17.3

(Top) Temperature contours at the time and axial location at which the peak temperature is achieved; (Bottom) thermal dose calculation at the end of the simulation. The contours represent 0.24CEM43C, 2.4CEM43C, 24CEM43C, etc. The red (gray) area corresponds to the region in which 240CEM43C has been achieved. The reader is referred to the Web version of the book to see the figure in color. *From Sonenson (2009), courtesy of J. Sonenson, ISTU Proceedings, AIP Publishing LLC.*

has kicked in. Note, in the top plot of this figure, the highest temperature contour is about 90°C . The −6-dB radial beamwidth corresponds to about the 60°C contour on this plot. The bottom plot provides the region in red where 240 cumulative equivalent minutes reference to 43°C are achieved (Section 15.6.2).

We now address some of the practical aspects of simulation. In particular, some form of system calibration is necessary to characterize the acoustic output of a specific HIFU system. In other words, for a given voltage applied to the electrical matching box connected to the transducer, what pressure field will the system produce? System characterization is described by international standards. The first HIFU standard was issued by the Chinese government (China, 2005). At the time of writing of this book, an international standard for acoustic power measurement and a technical specification for field measurements, developed in International Electrotechnical Commission (IEC) TC87, Working Group 6, are close to completion. These types of measurements provide the needed system calibration in addition to guidelines for HIFU system safety issues, which are described by documents written by IEC Group 62D. From knowledge of a pressure field and acoustic power, the appropriate scaling constants can be determined.

A simple concept for introducing measurements for HIFU (Canney, Bailey, Crum, Khokhlova, & Sapozhnikov, 2008; Hill, Rivens, Vaughan, & ter Haar, 1994) is in the parameter spatial averaged intensity (I_{SAL}) equal to the acoustic output power divided by the −6-dB cross-sectional focal area measured by a hydrophone. Intensity values can be measured for different drive levels to determine effective output. Because of the high pressure levels, special provisions have to be made for measurements; hence, the need for new standards in this area. Furthermore, more comprehensive field characterization than I_{SAL} is needed. A fiber-optic hydrophone, or a suitably protected hydrophone, can withstand the intense fields. Measurements of HIFU fields have already been described in Section 13.6 and featured hybrid modeling and measurement techniques (Canney et al., 2008; Yuldashev et al., 2012). The tissue region heated above a predetermined temperature has to be determined. This region can be simulated or measured by various thermography techniques (see Section 17.2.5 on monitoring).

17.2.3 Histotripsy and Hemostasis

The term "histotripsy" ("histo" from the Greek meaning tissue) means ultrasound-induced tissue disintegration. Compared to diagnostic ultrasound acoustic output levels, the acoustic levels reviewed in the introduction of this section seemed extreme; however, histotripsy takes thermal and mechanical effects to another level, even higher than those for HIFU and lithotripsy, reviewed in Section 17.2.1. With reference to Figure 17.4, the original pressure-based histotripsy, developed at the University of Michigan, shown in the enlarged waveform and middle sections of this figure, indicates bursts of 3–20 μs with compressional pressures in excess of 80 MPa and rarefactional pressures of 15–25 MPa at a low duty cycle. In the boiling histotripsy, developed at the University of Washington, the

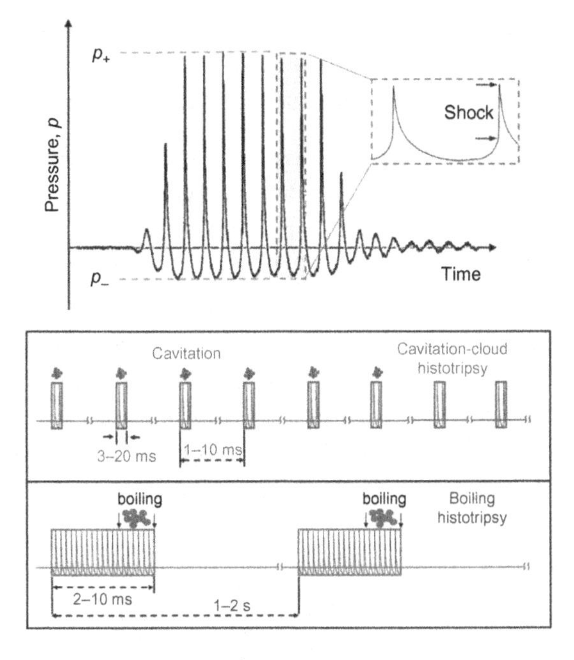

Figure 17.4

Pressure waveform and pulse-periodic timing schemes for histotripsy and boiling. (Top) Representative focal pressure waveform used for histotripsy. The pulse is initially a sinusoidal tone burst, but at the focus it is distorted by the combined nonlinear propagation and diffraction effects to produce the asymmetric waveform with higher peak positive pressure (p_+), lower-amplitude peak negative pressure (p_-), and high-amplitude shocks formed between negative and following positive phase as shown in the inset frame. (Bottom) Pulse-periodic timing schemes for two forms of histotripsy. The blue (upper) sequence shows the cavitation-cloud histotripsy scheme with microsecond-long pulses applied at 100–1000 Hz. The red (lower) sequence shows the boiling histotripsy scheme, employing millisecond-long pulses at a rate of 0.5–1 Hz. The reader is referred to the Web version of the book to see the figure in color. *From Maxwell et al. (2012), Acoustical Society of America.*

pressures are lower, with compressional pressures in excess of 40 MPa and rarefactional pressures of 10–15 MPa; however, the pulses are longer, extending from 2 to 10 ms, repeated at a low duty cycle (Maxwell & Xu, et al., 2012).

In both methods shock waves are involved, but play different roles in disintegrating tissue. In the pressure-based histotripsy, tissue outgases as not only is a pressure effect involved but localized heating, which is nonlinear. We can expect from the set of bioeffects described in Sections 15.6 and 15.7.1, that large acoustic radiation forces and heating proportional to intensity would be involved. In fact, because of the nonlinearity, heating is enhanced considerably over the linear case as described by Eqns 17.1 and 17.2 (Hamilton & Blackstock, 1998; Soneson & Myer, 2010). These seed bubbles begin to create a growing cavitation cloud, which violently explodes under the high pressures and reforms as it creates more bubbles. Shock waves reflecting from the bubble clouds enhance this chain-reaction effect of bubble creation and inertial cavitation events. As a result of the cavitation cloud cycle effects, which are maintained by a series of many pulses, typically 10^3 to 10^4, tissue is broken up into sub-micron fragments or liquefied.

In the bioeffects chapter, Section 15.7.4, some of the thermal effects of HIFU were introduced. Exposure is typically 5–10s, and temperatures of 60–100°C are achieved. At these temperatures outgassing occurs; that is, part of the tissue is vaporized as explained in the previous paragraph. Continued ultrasound exposure can cause cavitation in these areas, as explained in more detail by Bailey et al. (2001). A proposed model to explain why the lesion formed is no longer ellipsoidal under these circumstances is that the reflective trapped air and cavitation increase lesion size toward the transducer (Bailey et al., 2001; Crum & Law, 1995). This effect is shown in Figure 17.5, as the lesion shape changes from an ellipsoid to a tadpole at higher insonification levels (Khokhlova et al., 2006). At higher temperatures, indicated diagrammatically by Figure 17.6 (Simon et al., 2012) as a consequence of increasing exposure time, a vapor pocket forms, and finally a micro-acoustic fountain (the acoustic fountain effect was explained in Section 12.8.2 as caused by

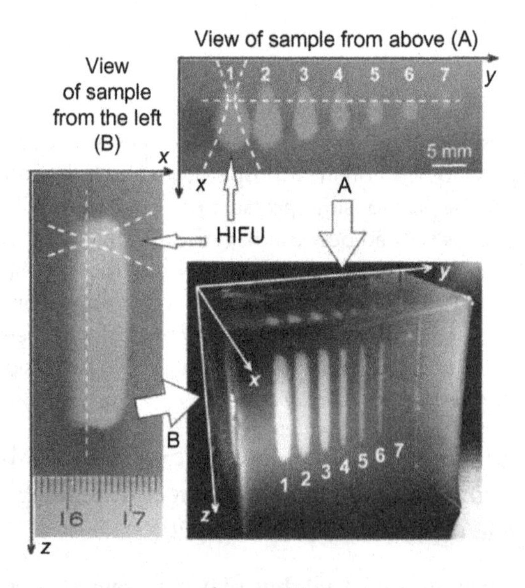

Figure 17.5
Change in lesion shape with increase in insonification level. Lesion stripes formed in degassed gel sample by moving the HIFU transducer upwards with constant velocity (0.5 mm/s), constant time-average acoustic power of the transducer (15 W), and different duty cycle (and peak power) from 6.25% (240 W peak power, 1) to 8.35% (2), 12.5% (3), 25% (4), 50% (5), 67% (6), and 100% (15 W peak power, 7). Ultrasound was applied from the front of the sample as indicated by HIFU arrows. The location of HIFU focal peak is shown by the straight dashed lines, and the general shape of the HIFU beam is shown by the curved dashed lines on the views of the sample from above (A) and from the left (B). Although the average power was held constant for each stripe, the stripes have very different sizes. The highest amplitude, shortest duty cycle pulses created the largest lesion stripe (1).
Khokhlova et al. (2006), Acoustical Society of America.

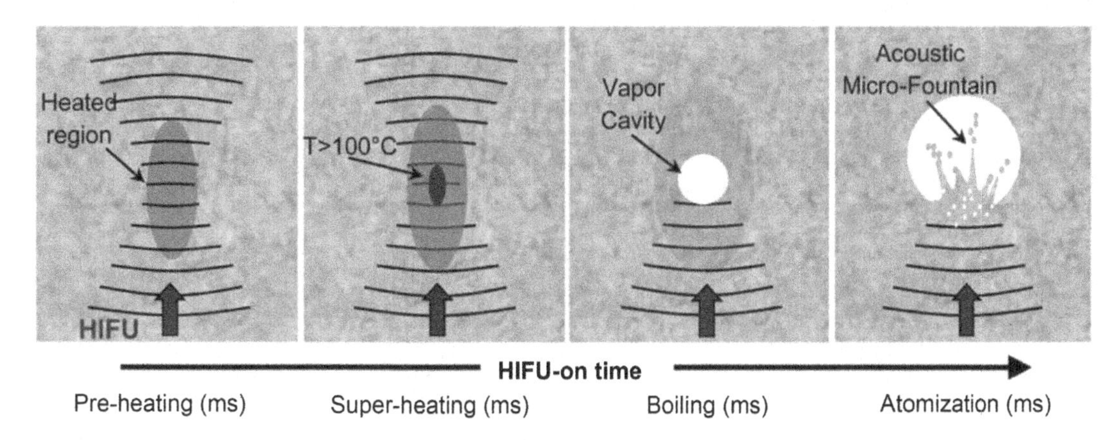

Figure 17.6
Proposed mechanism of tissue fractionation by boiling histotripsy. *From Simon et al. (2012), Physics in Medicine and Biology, © IOP Publishing. Reproduced by permission of IOP Publishing. All rights reserved.*

an acoustic radiation force at a boundary with a pressure differential). An expected counterpart of large acoustic radiation forces is enhanced heating. In this case, however, the boiling is local and kept in bounds by a low duty cycle, so a large part of the effect in this method is mechanical, as acoustic fountain forces that create jets and forces that fractionalize or atomize the tissues like a miniature blender (see Simon et al. (2012) for photographs of these fountains). About 10–15 cycles are needed (Maxwell & Xu, et al., 2012); effects can be seen from a single pulse.

Results of these methods are compared in Figure 17.7, from Maxwell et al. (2012). The advantages of the histotripsy methods are that they are more effective over a larger volume than that of thermal lesion caused by HIFU, are much faster, and that they liquefy tissue so that it can be more easily assimilated by the body. Note also that even though histotripsy is an extreme approach, it remains highly localized within the focal volume so that it can be guided to selected sites.

Ultrasound-induced hemostasis is an application of HIFU to stop bleeding. Under image (ultrasound) guidance and monitoring, the HIFU transducer is trained on the bleeding site. Several effects stop the bleeding: blood and tissue are emulsified into a paste, acoustic radiation forces drive the paste into the vessel, and heat cauterizes and seals so that coagulation can begin (Bailey et al., 2003). Spectral Doppler aids in identifying the bleeding site and indicates the cessation of bleeding (Crum, Bailey, Hwang, Khokhlovaa, & Sapozhnikova, 2010).

17.2.4 Cavitation-enhanced HIFU

HIFU can initiate cavitation mechanisms through tissue outgassing or deliberate introduction of microbubbles. Coussios, Farny, ter Haar, and Roy (2007) investigated the role of cavitation in heating under HIFU conditions. The distinction between

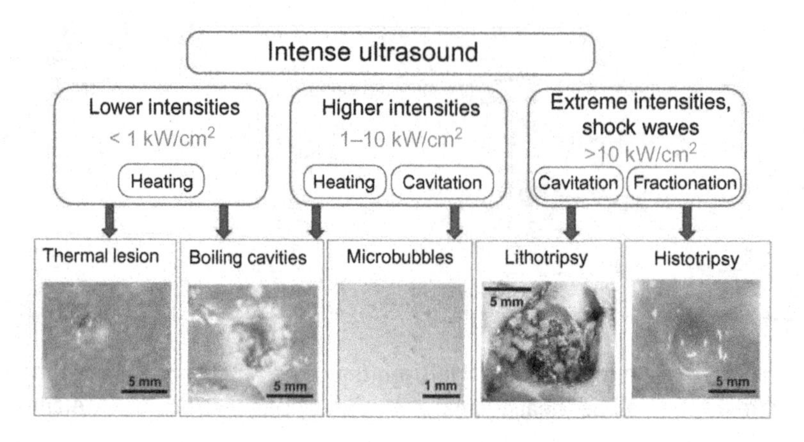

Figure 17.7

Bioeffects of focused ultrasound at different focal intensity levels. At lower intensities, heating through acoustic absorption is the dominant mechanism, denaturing proteins within the tissue, leaving a blanched appearance. Boiling cavities form in the lesion when the temperature reaches 100°C. At higher intensities, heating combined with microbubble cavitation can cause mechanical trauma to the tissue structure. At very high intensities, shockwaves form at the focus and the wave itself can impart significant mechanical damage, such as comminution of kidney stones (lithotripsy) or fractionation of soft tissues (histotripsy). The reader is referred to the Web version of the book to see the figure in color. *From Maxwell et al. (2012), Acoustical Society of America.*

stable cavitation and inertial cavitation (see Chapter 14) is that the violent collapsing of microbubbles in inertial cavitation results in broadband noise-like spectral emissions, unlike stable cavitation. At lower HIFU levels, these emissions indicate strong heating is occurring orders of magnitude greater than the viscous absorption of the shell and even high shear stress transfer at tissue boundaries (even though shear viscosity is high).

The onset of a dramatic increase in heating attributable to inertial cavitation was measured in a carefully designed experiment in a gassy phantom (Coussios et al., 2007). A 1.1-MHz HIFU transducer insonified the phantom while a passive cavitation detector (PCD) monitored broadband emissions beyond the bandwidth of the transducer and initial four harmonics. In addition, the pressure output of a Westervelt nonlinear propagation simulation model fed a Pennes' bioheat equation model to predict temperature rise (Hallaj & Cleveland, 1999; Huang, Holt, Cleveland, & Roy, 2004). Bubbles were not included in the model. Results of data (as measured by a thermocouple near the focal point) and simulation can be seen in Figure 17.8 (Coussios & Roy, 2008). Slightly beyond the p_r value expected for the onset of inertial cavitation, there was a dramatic increase in temperature rise, well above that predicted by the temperature rise model, that correlated with the onset of inertial cavitation as determined by PCD data. The temperature rise tapered off with time owing to saturation, and bubble shielding most likely caused an eventual decrease in temperature with time.

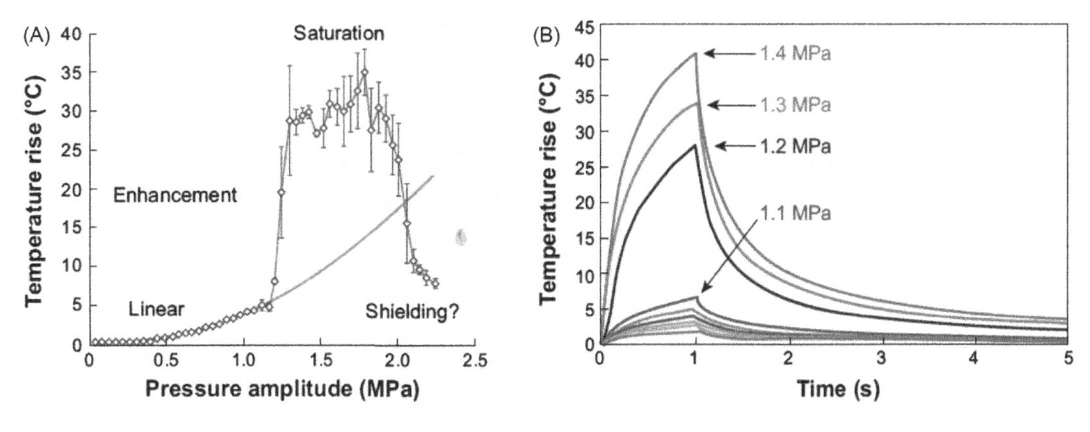

Figure 17.8

Temperature rise measured by a thermocouple embedded in an Agar-graphite tissue-mimicking material exposed to 1-MHz HIFU for 1s. The data have been smoothed to minimize noise, and a correction factor was applied that accounts for the so-called thermocouple artifact, which refers to enhanced heating in the viscous boundary layer surrounding the thermocouple. (A) Measured (blue/circles with error bars) and predicted (orange/line) peak temperature rise versus the acoustic peak-rarefaction pressure amplitude. The error bars depict the standard deviation of five measurements. (B) Measured temperature versus time for increasing peak-rarefaction pressure amplitudes; each curve corresponds to an increment of approximately 0.1 MPa. The reader is referred to the Web version of the book to see the figure in color. *Figure adapted from Coussios et al. (2007) and Coussios and Roy (2010), Annual Review of Fluid Mechanics.*

In most cases, the heating from cavitation is uncontrolled, leading to bubble migration towards the transducer and the formation of tadpole-shaped lesions. Furthermore, the formation of larger bubbles due to enhanced heating may be the cause of the bright echoes often seen by ultrasound lesion monitoring. However, it may be possible with a variable pulsing scheme to control heating and minimize bubble migration. This possibility would require a better means of monitoring bubble activity during insonification, and may also provide more accurate timing of lesion formation (Coussios et al., 2007; Coussios & Roy, 2008).

17.2.5 Monitoring

For treatment planning, targeting, and monitoring, a means of visualizing the selected region is needed. An overview by Fleury, Bouchoux, Berriet, and Lafon (2006) highlights many of the aspects of monitoring. The primary imaging methods are magnetic resonance imaging (MRI) and ultrasound. Each has its advantages and limitations in terms of the physical alterations it can detect. Fundamental changes in tissue include temperature elevation, fluid flow, and cavitation activity, and others are irreversible alterations in structure (Fleury et al., 2006) such as induced apoptosis (cell death), denaturation (thermal lesion), lysis (rupture of cells), collagen shrinkage, clot thrombosis, fibrosis (days later), and cessation of bleeding. In addition, there are a variety of geometries, including those with the classic hemispherical bowl transducers, and others in which insonification is near the transducer contact area, as well as low- and high-power applications. Other concerns include real-time feedback, accessibility, safety, and cost.

Thermal effects generated by the HIFU treatment in the system can be detected by MRI thermal mapping at about a map every 1–3 seconds and treatment volumes with an accuracy of 2°C every 3–4 seconds and a resolution of about 1 mm. the HIFU treatment can be carried out in the bore of the magnet. MRI thermometry can be used to visualize the heated target volume with conventional T1-weighted, SE, or spoiled gradient refocused echo pulse–echo sequences, and the ablation volume can be viewed by T2-weighted fast spin echo images (Cline, Hynynen, Watkins, et al., 1995; Hynynen et al., 1993, 2001; ter Haar, 2007). For targeting, a small temperature rise is induced, detected, and moved until the heated region is in the selected region. Treatment involves not just one spot but often a larger volume, such as a tumor, and includes repetitive insonifications or insonification-while-scanning methods. MRI also provides visualization of the acoustic path, and soft-tissue contrast for finding tumors and sensing ablation, so it is considered the gold standard and can provide a thermal map overlay on the image. MRI requires that the transducers be compatible for use in a magnet. A drawback is the high cost of using an MRI system for the considerable time needed for a complete treatment.

The motivation for using ultrasound for detection includes the lower cost, convenience, real-time monitoring, and ability to detect physical tissue parameters and obtain feedback about the effectiveness of the therapeutic process. A common geometry is to have a diagnostic ultrasound imaging array mounted within a central hole in a semi-hemispherical HIFU transducer bowl (see Chapter 5), which allows for coaxial insonification and visualization. In general, bubbles are viewed during and after treatment.

Methods have been proposed for sensing temperature rises with ultrasound (Doyley, Bamber, Rivens, Bush, & ter Haar 1999; Miller, Bamber, & ter Haar, 2002, 2004). The parameter actually measured is displacement and the approach is a variant of strain imaging, in which the sound speed derived from axial strain is shown to be proportional to temperature changes in different tissues. In vitro studies have identified possible artifacts; however, for modest temperature rises without confounding cavitation effects (Miller et al., 2002, 2004) the approach can be useful, especially for targeting. Several promising elastographic approaches for detecting changes in the elastic moduli in the lesion region have also been investigated (Arnal, Pernot, & Tanter, 2011; Fleury et al., 2006; Konofagou, Maleke, & Vappou, 2012; Maleke, Luo, Gamarnik, Lu, & Konofagou, 2010; Thittai, Galaz, & Ophir, 2011).

Important considerations in these methods are that the rapid physical effects during the formation of a lesion are compounded by extreme changes in displacement, temperature rise, viscoelastic alterations, chemical and phase changes, and complicating cavitation effects. Real-time in vivo ultrasound-monitoring strategies need to adapt to these difficult measurement conditions. When cavitation is present, real-time monitoring of cavitation activities may aid in controlling the temperature gradients (Farny, Holt, & Roy, 2009; Nandlall, Jackson, & Coussios, 2011).

A type of phantom is reversible up to a temperature limit, and a comprehensive evaluation of how parameters such as absorption, sound speed, elastic modulus, thermal conductivity, diffusivity, and specific heat vary with temperature (Dunmire, Kucewicz, Mitchell, Crum, & Sekins, 2013) has been made. Though made on a phantom material, this kind of thorough assessment of thermal properties may lead to a combination of tissue parameters best suited for HIFU monitoring. Specialized phantoms for visualizing a lesion volume have been designed (Howard, Yuen, Wegner, & Zanelli, 2003; Lafon, Kaczowski, VaezyNoble, & Sapozhnikov, 2001).

17.3 Therapeutic Ultrasound Applications

17.3.1 HIFU

17.3.1.1 Introduction

Compared to other types of thermal ablation methods, such as microwave, radio frequency (RF), laser, and cryoablation, HIFU is unique in its noninvasive ability to focus and change the focal area for the ablation of deep-seated tumors. The four major steps for a HIFU treatment are treatment planning, training the transducer focal point on the target, insonification at the target site, and treatment monitoring. The lesion formed by a typical HIFU system is ellipsoidal, roughly 2 cm by 2 mm by 2 mm. Because of the small lesion size, volume scanning is needed, as illustrated by Figure 17.9, though as we have described, cavitation-based methods may offer the ability to destroy larger volumes at a time. In some cases, intracavitary probes, such as transrectal, are more suitable (ter Haar & Coussios, 2007). Transducers for HIFU vary from single-element semi-spherical bowls with or without holes for diagnostic imaging arrays to HIFU arrays and special configurations for intracavitary applications (Fleury, Berriet, Le Baron, & Huguenin, 2002).

17.3.1.2 Extracorporeal HIFU

Extracorporeal HIFU, also known as extracorporeal focused ultrasound surgery (FUS) ablation, is the most developed therapeutic application and has been implemented commercially. The most extensive application of FUS has been in China, where thousands of patients were treated, mainly for solid malignant tumors including primary and metastatic liver cancer, malignant bone tumor, breast cancer, soft-tissue sarcoma, kidney cancer, pancreatic cancer, and advanced local tumors, in addition to solid benign tumors such as uterine myoma, benign breast tumor, and hepatic hemangioma (Wu et al., 2004a, 2004b). Most of these tumors were a few cm in size and required anesthesia, and a water

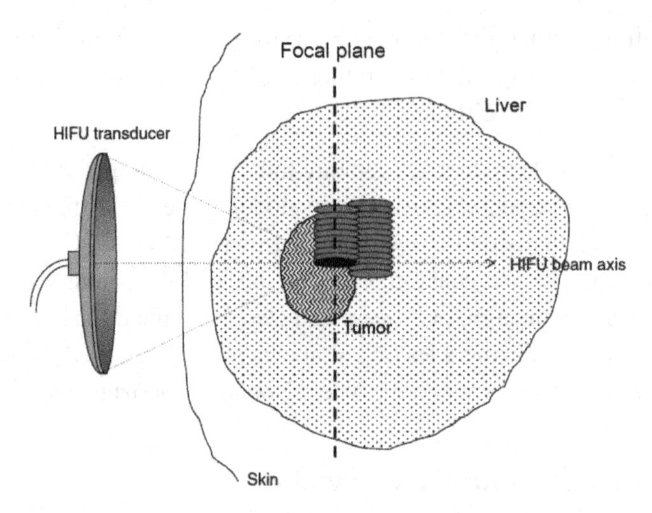

Figure 17.9
Diagrammatic illustration of HIFU treatment delivery for the ablation of large tissue volumes. Current lesion shown in black. Multiple lesions are created side by side to span the required treatment volume, starting on the side distal to the transducer. The lesions can be overlapping or separate, depending on the necessity to achieve confluent regions of cell-killing. *Adapted from ter Haar and Coussios (2007), International Journal of Hyperthermia.*

reservoir was used to couple the ultrasound to the patient. Treatment planning was necessary in each case and included the location of the tumor and the appropriate acoustic output parameters, and the individual history and type of tumor. Monitoring and follow-up included diagnostic ultrasound with Doppler, color Doppler, and, in some cases, ultrasound contrast agents, MRI, computed tomography (CT), and sometimes single photon emission computed tomography (SPECT) imaging. Treatment of these larger tumors took several hours and was conducted by scanning focal regions in a raster scan and then in a slice-by-slice pattern. A review of work in this area up to 2007 can be found in ter Haar (2007). Continued application of the Chongqing HAIFU Model JC device in a well-documented study in Oxford, UK, demonstrated its safety and efficacy with excellent correspondence between estimated ablation regions and those measured by MRI (Leslie et al., 2012). An example from their clinical trial that shows, through imaging with a contrast agent, that the vascularization feeding a large tumor has been destroyed is given in Figure 17.10.

A major application of HIFU is for the outpatient treatment of uterine fibroids. Hesley, Gorny, & Woodrum (2013) provide a review. MRI-guided focused ultrasound (MRgFUS) ablation, as it is called for this application, has found extensive use in Europe, the US, and China. MRI guidance and targeting is used to avoid organs, the bowel, and, sometimes, the bladder. The major effect applied in most devices is temperature elevation without cavitation. A sample display of the ablation process from an ExAblate system can be seen in Figure 17.11.

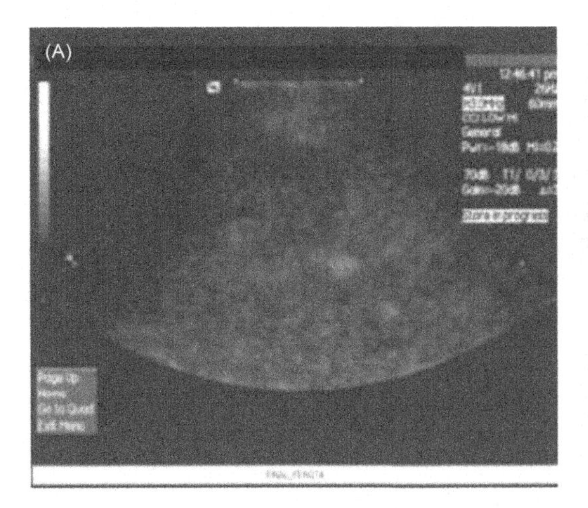

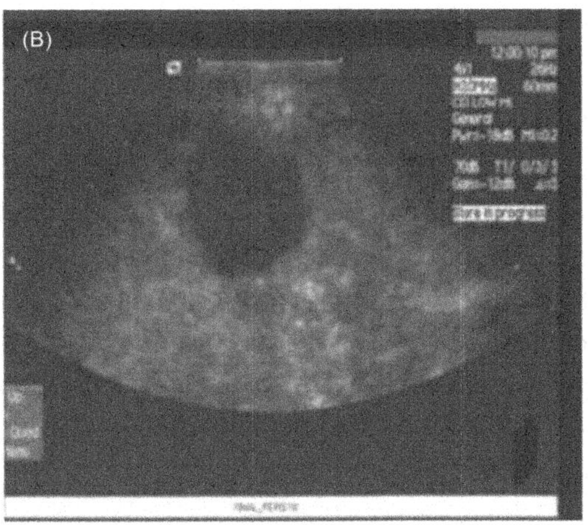

Figure 17.10

Positive technical success of HIFU liver treatment. Success indicated by comparison between (A) pre-HIFU and (B) immediately post-HIFU microbubble contrast-enhanced ultrasound, showing area of ablated tumor. *From Leslie et al. (2012), The British Journal of Radiology.*

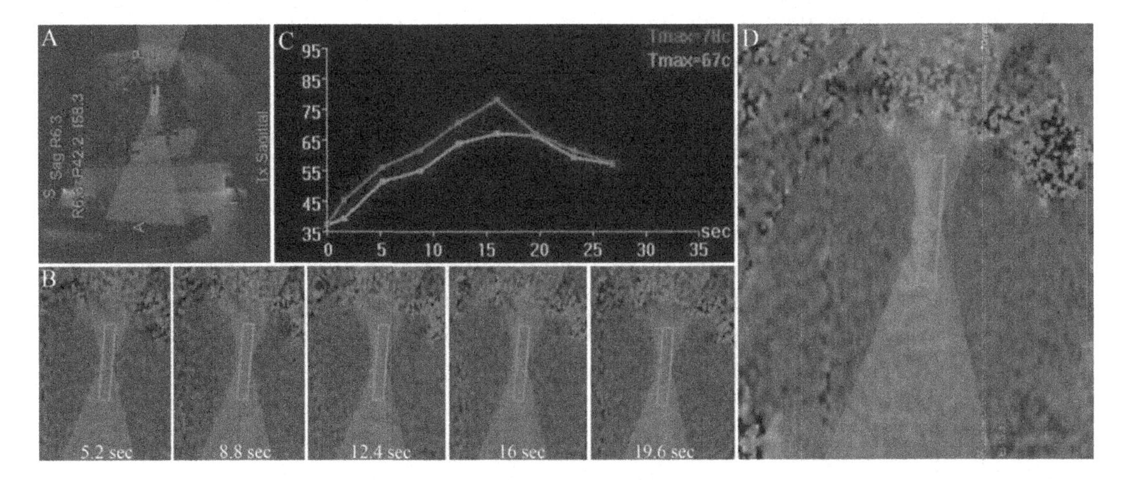

Figure 17.11

MRgFUS enables closed-loop treatment monitoring of each sonication. (A) Ablation is planned by using anatomic T2-weighted images acquired at the beginning of the treatment. Blue (pale gray in print version) overlay depicts outline of the projected FUS beam path. Beam focus is delineated by the rectangle. (B) During sonications, thermal images are acquired and displayed every 3.6s during the 20-s sonication. (C) Graphs indicating temperature history of the hottest voxel (red/dark gray) and average of nine voxels (green/light gray) are automatically displayed during sonication. (D) At the end of each sonication, dosimetry is performed and predicted tissue ablation is displayed (green) superimposed with anatomy (blue overlay represents tissue ablated in the previous sonications). Thermal feedback allows for in-treatment adjustment of sonication parameters. The reader is referred to the Web version of the book to see the figure in color. *From Helsey (2013), Cardiovascular Interventional Radiology.*

17.3.1.3 Transrectal HIFU

For transrectal ultrasound applications, imaging and therapy transducers are mounted on an endorectal probe. A key target is the prostate, with a goal of applying temperatures in excess of 60°C. Two commercial systems are the Ablatherm©, which has a 7.5-MHz imaging transducer and a 3-MHz therapy transducer, and the Sonablate™, in which both transducers are 4 MHz. A sequence of contiguous lesions is generated, usually accompanied by acoustic cavitation. Neither device is US Food and Drug Administration (FDA)-approved for the prostate as of this writing, but each one is used widely in Canada, Europe, and Asian countries. In some cases, ultrasound is complemented by chemotherapy or radiotherapy. A more detailed review of work in this area until 2007 can be found in ter Haar (2007).

17.3.2 Transcranial Ultrasound

Through the noninvasive transcranial focusing methods described in Section 9.7.3, new applications for brain surgery and therapy with an intact skull are becoming possible (see Pajek and Hynnen (2012) for a more detailed review). Transcranial focused ultrasound is undergoing clinical investigation for several applications (Pajek & Hynnen, 2012). Aside from noninvasive HIFU tumor ablation (Chauvet et al., 2013), HIFU is being explored for performing neurosurgery. Here, the heating is kept below 60°C and is monitored by MRI thermometry and passive cavitation detection. Clinical trials are underway to develop HIFU for treating essential tremor on an outpatient basis.

One of the most remarkable HIFU possibilities is the opening of the blood–brain barrier (BBB). The BBB has long prevented drugs from reaching the brain, but a combination of microbubbles and HIFU has demonstrated that the BBB can be disrupted. Drugs for cancer treatment and Alzheimer's and stem cells have been delivered using a feedback system for monitoring acoustic emissions (Pajek & Hynnen, 2012).

17.3.3 Sonothrombolysis

Sonothrombolysis is the use of ultrasound as a way of breaking up or dissolving blood clots, which are a major cause of strokes and other disabilities. Alexanderov et al. (2000) showed that transcranial pulsed Doppler in combination with a thrombolytic drug (t-Pa) could accelerate dissolving of a clot and reestablish normal blood flow quickly. Speculation is that repeated acoustic radiation forces at the rhythm of the Doppler pulse repetition rate (PRF) and streaming are involved in this effect (Crum et al., 2010). Adding contrast agents, as discussed in Chapter 15, provides an even more vigorous set of conditions for clot dissolution, adding not only radiation forces and streaming but also possible cavitation pulsation and collapse, and shear forces and temperature elevation.

17.3.4 Cosmetic Ultrasound

Ultrasound, again with its abilities to be focused selectively subdermally and to affect or ablate tissue, is being used for cosmetic or body-sculpting applications. Liposuction is the surgical removal of fatty tissue from the waist, arms, legs, or hips (Wortsman & Worstman, 2011). Cosmetic applications of ultrasound include the tightening of sagging skin.

Liposuction is an invasive surgical procedure involving the injection of a wetting agent and suctioning out of fatty tissue through cannulas inserted in small incisions. Noninvasive alternatives for liposuction include cryolipolysis, radio-frequency ablation, external laser therapies, and injection lipolysis, which are more diffusive and surface-limited than ultrasound (Jewell, Solish, & Desilets, 2011).

Liposuction, a procedure that can have unpleasant side effects, is being made easier by ultrasound fragmentation (Sadick, 2009). In the latest generation of ultrasound-assisted liposuction (Sadick, 2009), a tumescent fluid containing microbubbles is first injected into a fatty region. Next, a small diameter ($\sim$ 3 mm) grooved titanium transducer that emits high-pressure (at 35 kHz) pulses at a power of 20–25 W is placed into the incision. The resulting inertial cavitation (called "acoustic streaming" by the manufacturer) preprocesses the fat for removal. The cavitation process previously discussed breaks apart and separates fat cells from their matrix structure so that they can be more easily sucked out in the final step. Nearby vessels, nerves, and collagen fibers remain unaffected, according to the manufacturer of the Vaser Lipo system®, which received FDA approval and has less blood loss and fewer side effects than previous devices. Associated bioeffects are discussed for a similar, earlier, device in Kenkel et al. (1997).

An alternative to liposuction is to apply MHz ultrasound extracorporeally to the subcutaneous fat layer, as illustrated by Figure 17.12 (Jewell et al., 2011). This noninvasive approach is aligned with more traditional HIFU methods involving coagulative necrosis in the focal region. Though intensity levels are not given, they are claimed to be an order of magnitude less than those used for uterine fibroids (3.2–6.4 kW/cm^2) and the resulting effects are thermal (Jewell et al., 2011). For the FDA-approved Liposonix system, insonification at 2 MHz creates lesions approximately 1 mm in diameter and 10 mm long. Adipocyte necrosis occurs, then collagen-fiber remodeling, including denaturing, thickening, and shortening. Over a 12-week period, the body reabsorbs the necrotized fatty tissue via an inflammatory response by macrophages, which transport the destroyed cells and their contents to the liver via the lymphatic system, resulting in a net volume reduction without changes in the lipid profile (Liposonix' FDA submission; Jewell et al., 2011).

A second body sculpting or noninvasive fat-reduction ultrasound approach also involves a similar configuration to that shown in Figure 17.12. In this case, however, the mechanism is mainly mechanical, though technical details are missing (Jewell et al., 2011; Teitelbaum et al., 2007). According to the Ultrashape device manufacturer's website, the fat destruction

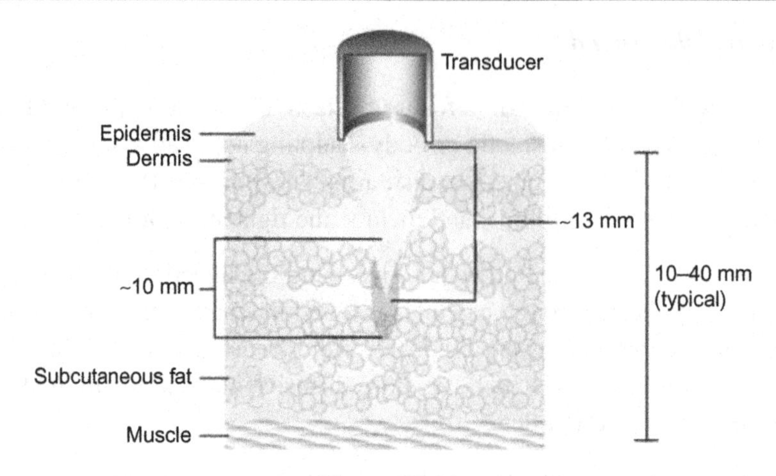

Figure 17.12
HIFU beam passing through the skin and superficial tissues without causing injury. The temperature at the focal point causes rapid cell death, but tissues immediately above and below the focal point are unharmed. The reader is referred to the Web version of the book to see the figure in color. *From Jewell et al. (2011), Aesthetic Plastic Surgery.*

mechanism is similar to that of the first device discussed in this section in that pressure levels cause the destruction of fat architecture, except that no microbubbles are used. In this case, the configuration is different, extracorporeal, and the frequency is also most likely much higher. The device is used worldwide and is under review for FDA approval as of this writing.

Another cosmetic HIFU application in the MHz range is for correcting skin laxity or sagging skin (Lee et al., 2012), usually around the neck and head. In this case, the superficial musculoaponeurotic system (SMAS) is targeted, as shown by Figure 17.13. Here, an ultrasound transducer is used to both image the superficial layers extending a few mm below the surface and create a series of focused thermal lesions called "thermal injury zones" (TIZs) at the adipose layer or SMAS. Different frequencies are used for different depths and applications. Lesions from a prototype system are shown for porcine muscle as a function of drive level in Figure 17.14 (White, Makin, Slayton, Barthe, & Gliklich, 2008). These zones cause collagen remodeling and skin tightening. Patterns consisting of subdermal TIZ lines are placed in different regions of the face or neck to cause reshaping and tightening in selected areas. For those cosmetic applications involving lesions, the effects described by Fig. 15.6 would be expected.

17.3.5 Lithotripsy

Extracorporeal lithotripsy belongs in this section but it has already been described in the bioeffects Section 15.7.5. Kidney stones afflict about 10% of the population and shock-wave lithotripsy is the most common treatment for them (Crum et al., 2010). Transducers for lithotripsy systems are described in Bailey et al. (2003). More details can be found in Leighton and Cleveland (2010) and Rassweiler et al. (2011).

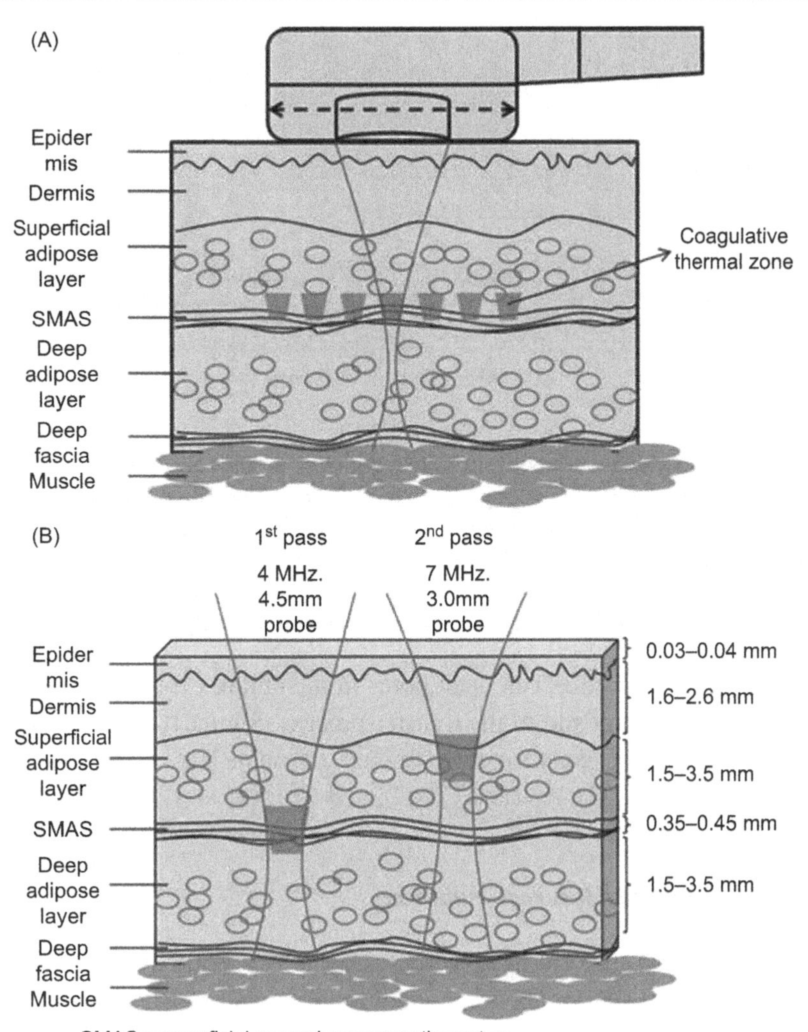

Figure 17.13

Schematic view of the ultrasound device being applied to the skin. (A) The probe emanates a series of wedge-shaped focused ultrasound beams along a 25-mm-long exposure line and makes thermal coagulative zones. (B) The thermal coagulation zone of the first pass (using the 4-MHz, 4.5-mm probe) extends from the superficial adipose layer through the SMAS to the deep adipose layer. The thermal coagulation zone of the second pass (using the 7-MHz, 3.0-mm probe) extends from the deep dermis to the superficial adipose layer. The reader is referred to the Web version of the book to see the figure in color. *From Lee et al. (2011), Dermatologic Surgery.*

17.3.6 Ultrasound-mediated Drug Delivery and Gene Therapy

Both drug delivery and gene transfection can be aided by cavitation effects. Non-viral genes can be added to microbubbles or be aided by closeness to microbubbles. In order for genes to enter cells to replace missing or defective ones, sonoporation can be used to make the

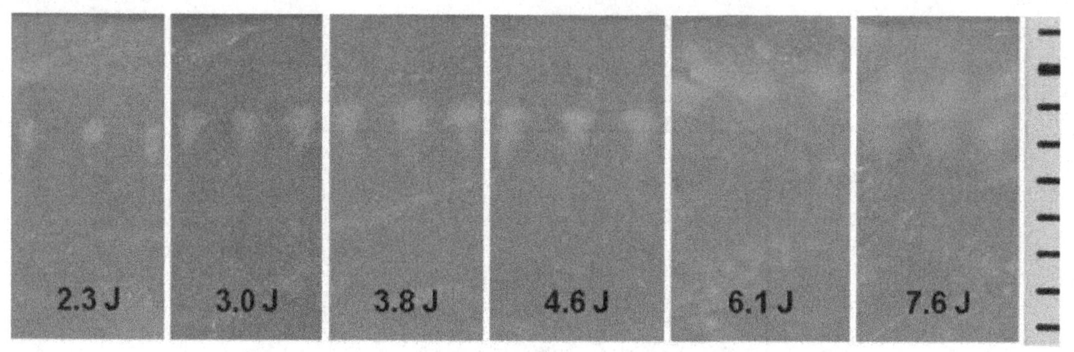

Figure 17.14
Thermal Injury Zones (TIZs) as a Function of Dose in Muscle. Digital photographs of gross tissue sections (approximately 1 mm thick) of porcine muscle reveal profile of changes in geometry of TIZ as the source energy is increased from 2.3 to 7.6 J. Within the homogenous orange-colored muscle tissue, the white inverse-pyramidal regions of coagulated tissue are the TIZs resulting from the ultrasound exposure (4.4 MHz, 4.5-mm-focus handpiece). The reader is referred to the Web version of the book to see the figure in color. *From White et al. (2008), Lasers in Surgery and Medicine.*

membranes permeable. Cavitation can open pores in membranes for genes to enter, and radiation forces and streaming add to the transfer process. Similarly, microbubbles coated or internally laden with drugs can also enter through the same pores. More information can be found in Bailey et al. (2003), ter Haar (2007), and Crum et al. (2010).

17.3.7 Ultrasound-induced Neurostimulation

Ultrasound can affect both the central nervous system (CNS; brain and spinal cord) and the peripheral nervous system (PNS). Gavrilov and Tsirulnikov's extensive review (2012) covered over 40 years of work in these areas, including research never before published. Newcomers to the field may overlook the knowledge base already developed and speculative conclusions may be made in error. Lack of funding in neurostimulation in Russia slowed the progress that was being made there, and interest is picking up worldwide. Ultrasound methods offer the possibility of stimulating receptor structures and unmyelinated nerve fibers not just on the surface but also those otherwise inaccessible, deeper in the body or brain. Preliminary work indicates that information can be transferred via unmyelinated nerve fibers when normal avenues of sensing are damaged or inoperative. Much work remains to be done in order to identify mechanisms and safety issues. Gavrilov and Tsirulnikov (2012) believe that ultrasound may be a universal stimulus for neuroreceptors.

Gavrilov and Tsirulnikov (2012) identified three main PNS response mechanisms to ultrasound: pain, tactile, and thermal. Muratore and Vaitekunas (2012) explain that typically, beyond a minimal threshold, a nerve can be stimulated proportionally to the ultrasound dose, up to a maximum limit past which its response diminishes. Nerves are long fibers that have diameters ranging from 0.1 μm to 2 μm. The intersection of ultrasound, even over a short section of a nerve, can stimulate it or inhibit its function. Most interactions are reversible, except for those in which the nerve is damaged, such as exposure by high HIFU output levels.

Ultrasound-induced neural effects offer some intriguing opportunities that are already being investigated. The viability of unmyelinated nerve fibers can be determined through ultrasound stimulation (Gavrilov & Tsirulnikov, 2012). The extent of damaged skin, skin diseases, burns, and hypersensitive nerve disorders can be ascertained through stimulation. Active stimulation of the ear labyrinth through amplitude-modulated ultrasound has been demonstrated. The dominant mechanism in this case is the acoustic radiation force. As in the harmonic motion method explained in Chapter 16, in addition to a static radiation force, a time-varying force changing with the modulation frequency is generated. Therefore, tones and acoustic speech have been directly transferred through nerves without direct hearing. Early prototype devices for diagnosis of diseases of hearing and audio prosthetics were developed in the early 1980s in the Soviet Union, in cooperation with medical organizations (Gavrilov & Tsirulnikov, 2012). The "Sensofon" was an ultrasound analogue of the tonal and speech audiometers. The device "Ultrafon" was designed for inputting auditory information to the deaf and hearing-impaired.

In related areas, fetal Doppler experiments by Fatemi, Ogburn, and Greenleaf (2001) found increased fetal activity when pulsed Doppler was trained on the fetus. Doppler is amplitude modulated by a PRF, often in the audio range. This observed effect could be caused by the acoustic radiation force, because the effect was negligible when continuous-wave Doppler transmitted. Carefully designed experiments by Dalecki, Child, Raeman, and Carstensen (1995) identified the acoustic radiation force as the main mechanism in tactile sensation of ultrasound. Also related, premature adult heart contractions during lithotripsy and during ultrasound imaging with some contrast agents were found to be related to radiation force (Dalecki, 2004).

Research on ultrasound-induced neurosensor sensations have led to the concept of real-time updatable two-dimensional displays. A sparse two-dimensional array was used to create focused patterns for tactile sensing pads, such as those shown in Figure 17.15 (Gavrilov & Tsirulnikov, 2012). More recent experiments indicate that finely focused ultrasound excites neural responses in the retina, the most accessible part of the CNS (Menz, Oralkan, Khuri-Yakub, & Baccus, 2013). These authors used a 43-MHz transducer focused precisely on regions of a salamander retina with the ganglion side down and connected to an electrode array. For a strong simulation, they found that, after a short latency, the ultrasound did not directly activate retinal ganglion cells but interneurons beyond photoreceptors. The stimuli

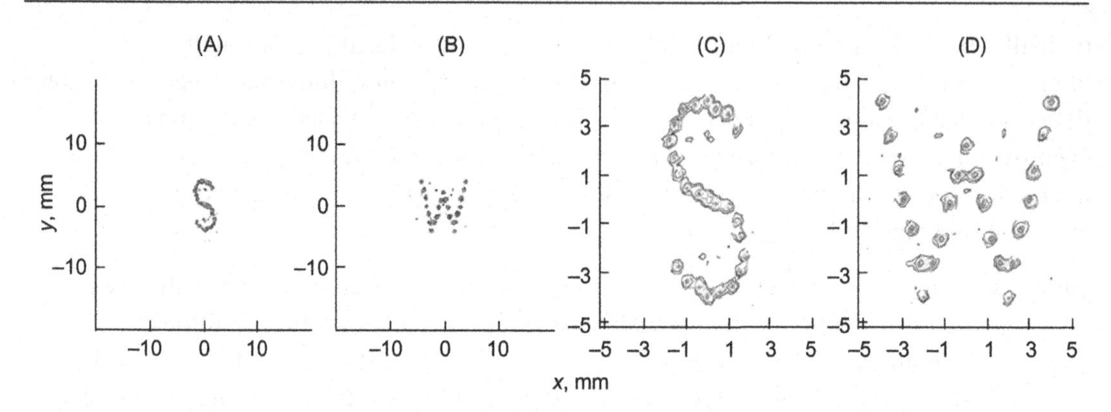

Figure 17.15
Synthesis of complex symbols in the shape of some letters of the alphabet using a randomized phased array. The dimensions of the investigated field were 4 × 4 (A, B) and 1 × 1 cm (C, D). *From Gavrilov (2012), Acoustical Physics.*

were affected by the temporal response of the retina. Could a capacitive micromachined ultrasonic transducer (CMUT) two-dimensional array be the next step? Though neither of these devices is operational yet, they indicate the promise of a day when two-dimensional information can be transferred through ultrasound-induced neurostimulation to those whose hearing or sight is impaired.

In Section 11.11, functional ultrasound was discussed. Regions of the brain responding to external stimuli were imaged through novel Doppler methods. In the converse of these experiments, transcranial pulsed ultrasound focused in the brain caused neuromodulation in awake monkeys (Younan et al., 2013). Low-intensity ultrasound (derated pressure estimated at 0.35 MPa, spatial peak temporal average intensity (I_{SPTA}) 13.46 mW/cm^2) was focused at the frontal eye field of the brain using a 320-kHz transducer. The investigation indicated that precisely focused ultrasound can modulate behavior in the awake nonhuman primate brain.

In other experiments, King, Brown, Newsome, and Pauly (2013) modulated a mouse brain transcranially. Beam experiments showed that the mouse skull had little effect on the sound distribution for the distances and frequencies involved. Their experiment consisted of a single planar 500-kHz 2.54-cm-diameter transducer near the top of a live mouse's head, with electromyography (EMG) electrodes attached to the legs, which were let free to dangle. They examined combinations of drive parameters to find where the success of obtaining a response improved. The somatomotor twitches of all legs at once triggered by the ultrasound stimulation appeared to be of the all-or-nothing sort. An example of a response is shown in Figure 17.16. Their newer work uses a focused transducer for investigating different regions. Tufail et al. (2010) also affected the brain transcranially with ultrasound.

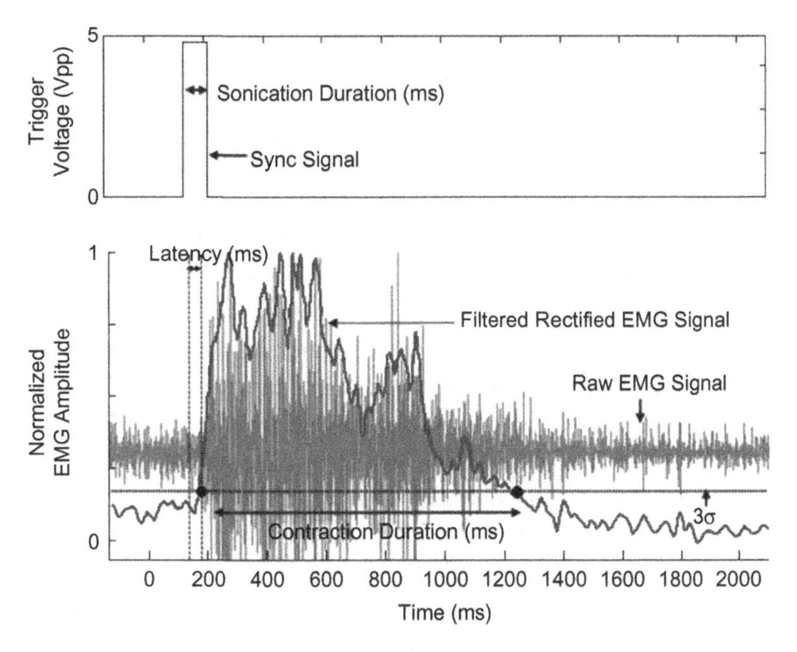

Figure 17.16

Transcranial ultrasound stimulation of mouse brain. (A) An example sync pulse, 80 ms in duration, that governed the duration of the sonication. (B) The red trace, resulting from a 16.8-W/cm^2 sonication, is a sample EMG signal following amplification (gain 10,003, high-pass filter 300 Hz, low-pass filter 1 kHz). The blue trace is the rectified, smoothed EMG signal. A muscle contraction was defined as beginning when the filtered signal rose above the third standard deviation of the noise level, represented in the figure by the horizontal dashed line, for at least 100 ms. The time from the beginning of the ultrasound pulse to the beginning of the contraction was defined as the latency. The reader is referred to the Web version of the book to see the figure in color. *King et al. (2013), with permission from the World Federation of Ultrasound in Medicine and Biology.*

17.3.8 Bone and Wound Healing

Approximately 5.6 million fractures occur in the United States each year, and 5–10% of these are classified as delayed or non-unions, which don't heal by themselves. A small ultrasound device, the FDA-approved Exogen, which can be easily used by a patient, can accelerate bone healing and the resolution of non-union fractures. The device produces low intensity pulsed ultrasound or LIPUS and transmits a 1.5-MHz pulsed ultrasound at a 20% duty cycle with a spatial average temporal average of only 30 mW/cm^2, and is applied for 20 minutes a day during the healing process. The process has been clinically proven to be effective, accelerating healing time typically by 38%, and producing healing in difficult fractures that would not heal by themselves (non-unions) (Gebauer, Mayr, Orthner, & Ryaby, 2005; Nolte, van der Krans, Patka, Janssen, Ryaby, & Albers, 2001; Malizos, Hantes, Protopappas, & Papachristos, 2006; Riboh & Leversedge, 2012; Siska, Gruen, &

Pape, 2008). At these output levels, there should not be any HIFU- or even significant diagnostic-level effects. How does it work?

The first stage is inflammatory; the usual structure is interrupted as well as blood flow. A hematoma is formed and inflammatory cells invade the hematoma to initiate lysosomal degradation of necrotic tissue. Bone cells can sense microforces and adjust their microenvironments. Externally applied low intensity pulsed ultrasound, LIPUS, provides repetitive mechanical stimuli. In the reparative stage, a soft fracture callus, angiogenesis is induced, and cartilage form. The dissimilar nature of the tissues forming in the gap causes a pressure gradient, across which ultrasound may cause preferential pathways. Bone cells sensing this change translate this into a molecular response. During the reparative phase, ultrasound induces gene activation and expression one of the actions is to stimulate vascular endothelial growth factor (VEGF), which in turn increases vascularity. The soft callus is converted into hard callus through a process known as endochondral ossification, several studies have demonstrated that LIPUS enhances this process leading to accelerated fracture repair (Freeman, 2006; Naruse, Miyauchi, Itoman, & Mikuni-Takagaki, 2003; Wang, 1994). In the final remodeling stage LIPUS has been shown to enhance osteoclast formation and bone resorption (Freeman, 2006; Hayton et al., 2005) and stimulates vascular endothelial growth factor (VEGF). In addition, ultrasound enhances calcium signaling, leading to cell differentiation and calcified matrix production (Siska et al., 2008).

Reports of therapeutic ultrasound effects on wound healing are mixed. Early work by Dyson (Young & Dyson, 1990) created interest in ultrasound as an agent of wound healing; however, suggestions that cavitation and acoustic streaming were involved now, on review, seem unlikely. Baker, Robertson, and Duck's (2001) review of physiotherapy, discussed in Section 15.7.2, which covers the same frequency-range as wound healing, found the evidence came up short. A statistical review of the literature for low-frequency (20–40 kHz) ultrasound for wound healing (Voigt, Wendelken, Driver, & Alvarez, 2011) gave a guarded endorsement for the application of low- and high-frequency-ultrasound-enhanced early healing (at $\leq$ 5 months) in patients with venous stasis and diabetic foot ulcers. A comprehensive review (Ennis, Lee, Plummer, and Meneses, 2010) gives a more positive review. The mechanisms as explained have not kept up with our present understanding, and actual clinical evidence given is mixed in some cases with poor reproducibility of results. Several problems include lack of consistency among devices and conditions and unsubstantiated conclusions about mechanisms involved.

In comparison with the bone results, which are supported by clinical research and trials, the wound-healing data are more difficult to compare. If the mechanisms in the bone occur at such low acoustic levels, could similar effects be occurring in tissues? More work is needed to determine the physical mechanisms involved with ultrasound-induced wound healing methods under controlled conditions. Promising results have recently been obtained with LIPUS for healing ulcers (Weingarten, Samuels, ,Zubkov, Sunny, Bawiec, Diaz, Jenkins, et al. (2013)).

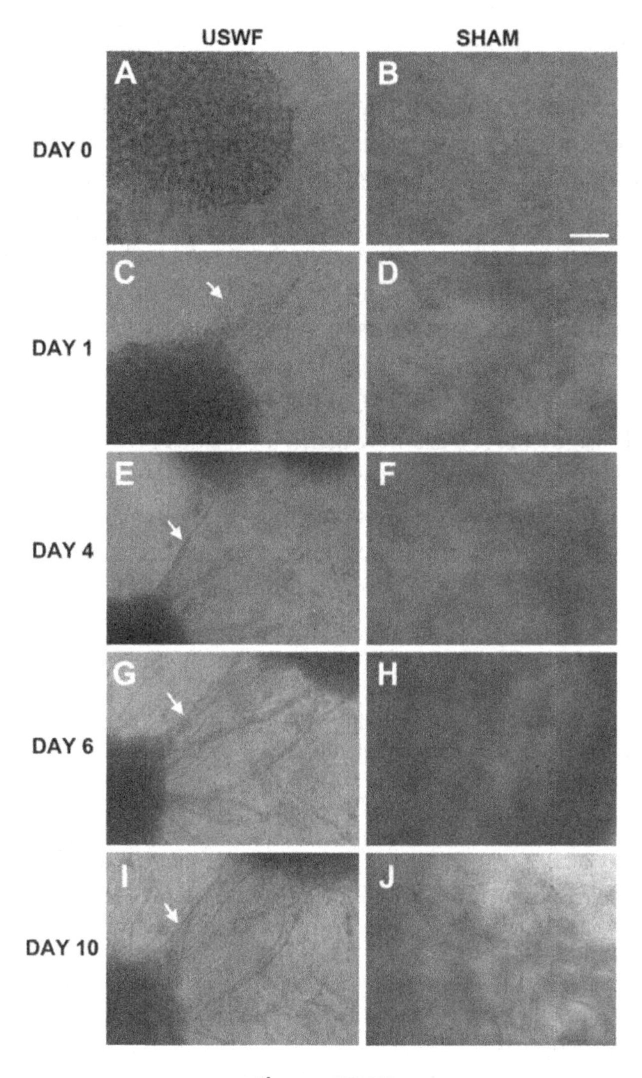

Figure 17.17

Time course of neovessel formation following ultrasound standing wave field (USWF) exposure. Endothelial cells were suspended in an unpolymerized collagen-type-I solution and were exposed to an USWF (1 MHz, continuous wave, 0.2-MPa peak pressure amplitude, 15 min duration) to promote the formation of multicellular bands of cells within collagen gels. Top view, phase-contrast images of USWF-exposed (A, C, E, G, and I) and sham-exposed (B, D, F, H, and J) collagen constructs were collected at day 0 (the day of USWF exposure; A and B), and at days 1, 4, 6, or 10. The white arrows highlight the progression of one USWF-induced endothelial cell sprout through time. Scale bar, 100 mm. *From Gavin et al. (2102), with permission from the World Federation of Ultrasound in Medicine and Biology.*

17.4 Conclusions

The diversity of devices covered in this chapter indicates a wide range of possibilities of therapeutic ultrasound. In some applications, underlying physical effects are united by a common theme. For example, HIFU thermal effects are being used for tumors, uterine fibroids, body sculpting, and face-lifting. Extreme acoustic output levels used in histotripsy expand into new applications and increase our understanding of physical mechanisms. Ironically, applications such as bone healing reveal our lack of understanding of the physical mechanisms at low-intensity ultrasound output levels for tissues. Is there more going on at these levels that could be of benefit? One is reminded of the work of Garvin et al. (Garvin, Dalecki, & Hocking, 2011; Garvin, Hocking, & Dalecki, 2010) on engineered tissues. In Figure 17.17 are new artificial tissues grown by low-intensity ultrasound, looking similar to naturally grown vasculature. Finally, the work on ultrasound-induced neurostimulation offers intriguing challenges and opportunities for man–machine interfaces, new types of information transfer, and other sensational devices.

References

Arnal, B., Pernot, M., & Tanter, M. (2011). Monitoring of thermal therapy based on shear modulus changes: I. Shear wave thermometry, IEEE Transactions on Ultrasonics. *Ferroelectrics and Frequency Control, 58*, 369–378.

Bailey, M. R., Couret, L. N., Sapozhnikov, O. A., et al. (2001). Use of overpressure to assess the role of bubbles in focused ultrasound lesion shape in vitro. *Ultrasound in Medicine and Biology, 27*, 695–708.

Bailey, M., Khokhlova, V., Sapozhnikov, O., Kargl, S., & Crum, L. (2003). Physical mechanisms of the therapeutic effect of ultrasound (a review). *Acoustical Physics, 49*, 369–388.

Baker, K. G., Robertson, V. J., & Duck, F. A. (2001). A review of therapeutic ultrasound: biophysical effects. *Physical Therapy, 81*, 1351–1358.

Canney, M., Bailey, M., Crum, L., Khokhlova, V., & Sapozhnikov, O. (2008). Acoustic characterization of high intensity focused ultrasound fields: a combined measurement and modeling approach. *Journal of the Acoustical Society of America, 124*, 2406–4220.

Chauvet, D., Marsac, L., et al. (2013). Targeting accuracy of transcranial magnetic resonance-guided high intensity focused ultrasound brain therapy: a fresh cadaver model. *Journal of Neurosurgery, 118*, 1046–1052.

China, The General Administration of Quality Supervision, Inspection and Quarantine of the People's Republic of China, (2005), Acoustics - High intensity focused ultrasound (HIFU)–Measurement of acoustic power and field characteristics.

Cline, H. E., Hynynen, K., Watkins, R. D., et al. (1995). Focused US system for MR-imaging-guided tumor ablation. *Radiology, 194*, 731–737.

Coussios, C. C., Farny, C. H., ter Haar, G., & Roy, R. A. (2007). Role of acoustic cavitation in the delivery and monitoring of cancer treatment by high-intensity focused ultrasound (HIFU). *International Journal of Hyperthermia, 23*, 105–120.

Coussios, C. C., & Roy, R. A. (2008). Applications of acoustics and cavitation to noninvasive therapy and drug delivery. *Annual Review of Fluid Mechanics, 40*, 395–420.

Crum, L., Bailey, M., Hwang, J. H., Khokhlovaa, V., & Sapozhnikova, O. (2010). Therapeutic ultrasound: recent trends and future perspectives. *Physics Procedia, 3*, 25–34.

Crum, L. A., & Law, W. (1995). The relative roles of thermal and nonthermal effects in the use of high intensity focused ultrasound for the treatment of benign prostatic hyperplasia. In *Proceedings of the 15th international congress on acoustics*, Trondheim, Norway, 1995.

Dalecki, D. (2004). Mechanical effects of ultrasound. *Annual Review of Biomedical Engineering, 6*, 229–248.

Dalecki, D., Child, S. Z., Raeman, C. H., & Carstensen, E. L. (1995). Tactile perception of ultrasound. *Journal of the Acoustical Society of America, 97*, 3165–3168.

Doyley, M. M., Bamber, J. C., Rivens, L., Bush, N. L., & ter Haar, G. R. (1999). Elastographic imaging of thermally ablated tissue in vitro. In *Proceedings of the IEEE ultrasonics symposium* (pp. 1631–1634), 1999, Cesar's Tahoe, NV.

Dunmire, B., Kucewicz, J. C., Mitchell, S. B., Crum, L. A., & Sekins, K. M. (2013). Characterizing an agar/gelatin phantom for image guided dosing and feedback control of high-intensity focused ultrasound. *Ultrasound in Medicine and Biology, 39*, 300–311.

Farny, C. H., Holt, R. G., & Roy, R. A. (2009). Temporal and spatial detection of HIFU-induced inertial and hot-vapor cavitation with a diagnostic ultrasound system. *Ultrasound in Medicine and Biology, 35*, 603–615.

Fatemi, M., Ogburn, P. L., & Greenleaf, J. F. (2001). Fetal stimulation by pulsed diagnostic ultrasound. *Journal of Ultrasound in Medicine, 20*, 883.

Fleury, G., Berriet, R., Le Baron, O., & Huguenin, B. (2002). Imasonic-France: new piezocomposite transducers for therapeutic ultrasound. In *Proceedings of the second international symposium on therapeutic ultrasound*, Seattle, WA, July–August 2002.

Fleury, G., Bouchoux, G., Berriet, R., & Lafon, C. (2006). Simultaneous imaging and therapeutic ultrasound. In *Proceedings of the IEEE ultrasonics symposium* (pp. 1045–1051), 2006, Vancouver, BC.

Freeman, T. A., Antoci, V., Rozycka, M., Adams, C. S., Shapiro, I. M., & Parvizi, J. (2006). *Low intensity ultrasound affects MMP-13, Osteopontin and COX-2 protein expression: an in vivo study*. Chicago: Orthopaedic Research Society Annual Meeting.

Fry, W. J., Mosberg, W. H., Jr, Barnard, J. W., & Fry, F. J. (1954). Production of focal destructive lesions in the central nervous system with ultrasound. *Journal Neurosurg, 11*, 471–478.

Garvin, K. A., Dalecki, D., & Hocking, D. C. (2011). Vascularization of three-dimensional collagen hydrogels using ultrasound standing wave fields. *Ultrasound in Medicine and Biology, 37*, 1853–1864.

Garvin, K. A., Hocking, D. C., & Dalecki, D. (2010). Controlling the spatial organization of cells and extracellular matrix proteins in engineered tissues using ultrasound standing wave fields. *Ultrasound in Medicine and Biology, 36*, 1919–1932.

Gavrilov, L. R., & Tsirulnikov, E. M. (2012). Focused ultrasound as a tool to input sensory information to humans (review). *Acoustical Physics, 58*, 1–21.

Gebauer, D., Mayr, E., Orthner, E., & Ryaby, J. P. (2005). Low-intensity pulsed ultrasound: effects on nonunions. *Ultrasound in Medicine and Biology, 31*(10), 1391–1402.

Hallaj, I. M., & Cleveland, R. O. (1999). FDTD simulation of finite-amplitude pressure and temperature fields for biomedical ultrasound. *Journal of the Acoustical Society of America, 105*, L7–L12.

Hamilton, M. F., & Blackstock, D. T. (Eds.), (1998). *Nonlinear Acoustics* San Diego, CA: Academic Press.

Hayton, M. J., Dillon, J. P., Glynn, D., Curran, J. M., Gallagher, J. A., & Buckley, K. A. (2005). Involvement of adenosine 5'-triphosphate in ultrasound-induced fracture repair. *Ultrasound in Medicine and Biology, 31*, 1131–1138.

Hesley, G. K., Gorny, K. R., & Woodrum, D. A. (2013). MR-guided focused ultrasound for the treatment of uterine fibroids. *Cardiovascular and Interventional Radiology, 36*, 5–13.

Hill, C. R., Rivens, I., Vaughan, M. G., & ter Haar, G. R. (1994). Lesion development in focused ultrasound surgery: a general model. *Ultrasound in Medicine and Biology, 20*, 259–269.

Howard, S., Yuen, J., Wegner, P., & Zanelli, C. I. (2003). Characterization and FEA simulation for a HIFU phantom material. In *Proceedings of the IEEE ultrasonics symposium* (pp. 1270–1273).

Huang, J., Holt, R. G., Cleveland, R. O., & Roy, R. A. (2004). Experimental validation of a tractable numerical model for focused ultrasound heating in flow-through tissue phantoms. *Journal of the Acoustical Society of America, 116*. 2451–2158.

Hynynen, K., Pomeroy, O., Smith, D. N., Huber, P. E., McDannold, N. J., Kettenbach, J., et al. (2001). MR imaging-guided focused ultrasound surgery of fibroadenomas in the breast: a feasibility study. *Radiology, 219*, 176–185.

Hynynen, K., Damianou, C., Darkazanli, A., et al. (1993). The feasibility of using MRI to monitor and guide noninvasive ultrasound surgery. *Ultrasound in Medicine and Biology, 19*, 91–92.

Jewell, M. L., Solish, N. J., & Desilets, C. S. (2011). Noninvasive body sculpting technologies with an emphasis on high-intensity focused ultrasound. *Aesthetic Plastic Surgery, 35*, 901–912.

Kenkel, J. M., Robinson, J. B., Beran, S. J., et al. (1997). The tissue effects of ultrasound-assisted lipoplasty. *Plastic and Reconstructive Surgery, 102*, 213–220.

Khokhlova, V. A., Bailey, M. R., Reed, J. A., Cunitz, B. W., Kaczkowski, P. J., & Crum, L. A. (2006). Effects of nonlinear propagation, cavitation, and boiling in lesion formation by high-intensity focused ultrasound in a gel phantom. *Journal of the Acoustical Society of America, 119*, 1834–1848.

King, R. L., Brown, J. R., Newsome, W. T., & Pauly, K. B. (2013). Effective parameters for ultrasound-induced in vivo neurostimulation. *Ultrasound in Medicine and Biology, 39*, 312–331.

Konofagou, E. E., Maleke, C., & Vappou, J. (2012). Harmonic motion imaging (HMI) for tumor imaging and treatment monitoring. *Current Medical Imaging Reviews, 8*, 16–26.

Lafon, C., Kaczowski, P., Vaezy, S., Noble, M., & Sapozhnikov, O. (2001). Development and characterization of an innovative synthetic tissue-mimicking material for high-intensity focused ultrasound (HIFU) exposures. In *Proceedings of the IEEE ultrasonics symposium* (pp. 1295–1298), 2001, Atlanta, GA.

Lee, H. S., Jang, W. S., Cha, Y. –J., et al. (2012). Multiple pass ultrasound tightening of skin laxity of the lower face and neck. *Dermatologic Surgery, 38*, 20–27.

Leighton, T. G., & Cleveland, R. O. (2010). Lithotripsy. *Proceedings of the Institute of Mechanical Engineers H, 224*, 317–342.

Leslie, T., et al. (2012). High-intensity focused ultrasound treatment of liver tumours: post-treatment MRI correlates well with intra-operative estimates of treatment volume. *British Journal of Radiology, 85*, 1363–1370.

Maleke, C., Luo, J., Gamarnik, V., Lu, X. L., & Konofagou, E. E. (2010). Simulation study of amplitude-modulated (AM) harmonic motion imaging (HMI) for stiffness contrast quantification with experimental validation. *Ultrasonic Imaging, 32*, 154–176.

Malizos, K. N., Hantes, M. E., Protopappas, V., & Papachristos, A. (2006). Low-intensity pulsed ultrasound for bone healing: an overview. *Injury, 37S*, S56–S62.

Maxwell, A., Xu, Z., Sapozhnikov, O., Fowlkes, B., et al. (2012). Disintegration of tissue using high-intensity focused ultrasound: two approaches that utilize shock waves. *Acoustics Today, 8*, 24–36.

Menz, M. D., Oralkan, O., Khuri-Yakub, P. T., & Baccus, S. A. (2013). Precise neural stimulation in the retina using focused ultrasound. *The Journal of Neuroscience, 33*, 4550–4560.

Miller, N. R., Bamber, J. C., & ter Haar, G. R. (2002). Ultrasonic temperature imaging for the guidance of thermal ablation therapies: in vitro results. In *Proceedings of the IEEE ultrasonics symposium* (pp. 1365–1368), 2002.

Miller, N. R., Bamber, J. C., & ter Haar, G. R. (2004). Imaging of temperature-induced echo strain: preliminary in vitro study to assess feasibility for guiding focused ultrasound surgery. *Ultrasound in Medicine and Biology, 30*, 345–356.

Nandlall, S. D., Jackson, E., & Coussios, C. -C. (2011). Real-time passive acoustic monitoring of HIFU-induced tissue damage. *Ultrasound in Medicine and Biology, 37*, 922–934.

Naruse, K., Miyauchi, A., Itoman, M., & Mikuni-Takagaki, Y. (2003). Distinct anabolic response of osteoblast to low-intensity pulsed ultrasound. *Journal of Bone and Mineral Research, 18*(2), 360–369.

Nolte, P. A., van der Krans, A., Patka, P., Janssen, I. M. C., Ryaby, J. P., & Albers, G. H. R. (2001). Low-intensity pulsed ultrasound in the treatment of nonunions. *Journal of Trauma-Injury Infection and Critical Care, 51*(4), 693–703.

Pajek, P., & Hynynen, K. (2012). Applications of transcranial focused ultrasound surgery. *Acoustics Today, 8*, 8–14.

Rassweiler, J. J., Knoll, T., et al. (2011). Shock wave technology and application: an update. *European Urology, 59*, 784–796.

Riboh, J. C., & Leversedge, F. J. (2012). The use of low-intensity pulsed ultrasound bone stimulators for fractures of the hand and upper extremity. *Journal of Hand Surgery, 137A*, 1456–1461.

Sadick, N. S. (2009). Overview of ultrasound-assisted liposuction, and body contouring with cellulite reduction. *Seminars in Cutaneous Medicine and Surgery*, *28*, 250–256.

Simon, J. C., Sapozhnikov, O. A., Khokhloval, V. A., Wang, Y. N., Crum, L. A., & Bailey, M. R. (2012). Ultrasonic atomization of tissue and its role in tissue fractionation by high-intensity focused ultrasound. *Physics in Medicine and Biology*, *57*, 8061–8078.

Siska, P. A., Gruen, G. S., & Pape, H. C. (2008). External adjuncts to enhance fracture healing: what is the role of ultrasound? *Injury*, *39*, 1095–1105.

Soneson, J. E. (2009). A user-friendly software package for HIFU simulation. In *Proceedings of the eighth international symposium on therapeutic ultrasound*, Minneapolis, MN, September 2009.

Soneson, J. E. (2012). A parametric study of error in the parabolic approximation of focused axisymmetric ultrasound beams. *JASA Express Letter*, *131*, EL481–EL486.

Soneson, J. E., & Meyers, M. R. (2007). Gaussian representation of high-intensity focused ultrasound beams. *Journal of the Acoustical Society of America*, *122*, 2526–2531.

Soneson, J. E., & Meyers, M. R. (2010). Thresholds for nonlinear effects in high-intensity focused ultrasound propagation and tissue heating. *IEEE Transactions on Ultrasonics, Ferroelectrics and Frequency Control*, *57*, 2450–2459.

Teitelbaum, S. A., Burns, J. L., et al. (2007). Noninvasive body contouring by focused ultrasound: safety and efficacy of the contour I device in a multicenter, controlled, clinical study. *Plastic Reconstructive Surgery*, *120*, 779–789.

ter Haar, G. (2007). Therapeutic applications of ultrasound. *Progress in Biophysics and Molecular Biology*, *93*, 111–129.

ter Haar, G., & Coussios, C. (2007). High-intensity focused ultrasound: past, present and future. *International Journal of Hyperthermia*, *23*, 85–87.

Thittai, A. K., Galaz, B., & Ophir, J. (2011). Visualization of HIFU-induced lesion boundaries by axial-shear-strain elastography: a feasibility study. *Ultrasound in Medicine and Biology*, *37*, 426–433.

Tufail, Y., Matyushov, A., Baldwin, N., Tauchmann, M. L., Georges, J., Yoshihiro, A., et al. (2010). Transcranial pulsed ultrasound stimulates intact brain circuits. *Neuron*, *66*, 681–694.

Voigt, J., Wendelken, M., Driver, V., & Alvarez, O. M. (2011). Low-frequency ultrasound (20–40 kHz) as an adjunctive therapy for chronic wound healing: a systematic review of the literature and meta-analysis of eight randomized controlled trials. *International Journal of Lower Extremity Wounds*, *10*, 190–199.

Wang, S., Lewallen, D., Bolander, M., Chao, E., Ilstrup, D., & Greenleaf, J. (1994). Low intensity ultrasound treatment increases strength in a rat femoral fracture model. *Journal Orthop Research*, *12*, 40–47.

Weingarten, M. S., Samuels, J. A., Zubkov, L., Sunny, Y., Bawiec, C., Diaz, D., Jenkins, L. et al. (2013) Low-intensity (100 mw/cm^2) low-frequency (<100 khz) therapeutic ultrasound for the treatment of venous ulcers: an in vitro and pilot human study, wound repair and regeneration 21, A41–A47.

White, W. M., Makin, I. R. S., Slayton, M. H., Barthe, P. G., & Gliklich, R. (2008). Selective transcutaneous delivery of energy to porcine soft tissues using intense ultrasound (IUS). *Lasers in Surgery and Medicine*, *40*, 67–75.

Wortsman, X., & Worstman, J. (2011). Sonographic outcomes of cosmetic procedures, musculoskeletal imaging. *American Journal of Roentgenology*, *197*, W910–W918.

Wu, F., Wang, Z. -B., et al. (2004a). Extracorporeal focused ultrasound surgery for treatment of human solid carcinomas: early Chinese clinical experience. *Ultrasound in Medicine and Biology*, *30*, 245–260.

Wu, F., Wang, Z. -B., et al. (2004b). Extracorporeal high-intensity focused ultrasound ablation in the treatment of 1038 patients with solid carcinomas in China: an overview. *Ultrasonics Sonochemistry*, *11*, 149–154.

Younan, Y., & Deffieux, T. et al. (2013). Transcranial ultrasound neuromodulation of the contralateral visual field in an awake monkey. In *Proceedings of the IEEE ultrasonics symposium*, 2013, IUS1-A1-3.

Young, S. R., & Dyson, M. (1990). The effect of therapeutic ultrasound on angiogenesis. *Ultrasound in Medicine and Biology*, *16*, 261–269.

Yuldashev, P. V., Kreider, W., Sapozhnikov, O. A., Farr, N., Partanen, A., Bailey, M. R., & et al. (2012) Characterization of nonlinear ultrasound fields of 2D therapeutic arrays. In *Proceedings of the IEEE ultrasonics symposium*, 2012, Dresden, Germany.

Appendix A: The Fourier Transform

A.1 Introduction

The Fourier transform, besides being an elegant and useful mathematical tool, also has an intuitive interpretation that extends to many physical processes. The Fourier transform is applied extensively in electrical engineering, signal processing, optics, acoustics, and the solution of partial differential equations.

For example, consider an acoustic wave received by a transducer and converted to an electrical pulse. This electrical pulse—echo signal can be seen on the screen of a digital sampling scope as a function of time. The same signal, when connected to a spectrum analyzer, can be observed as a spectrum. The Fourier transform is the relation between the pulse and its spectrum. On many scopes, a spectrum calculated directly by a fast Fourier transform (FFT) can also be displayed simultaneously with the time trace.

This appendix briefly reviews some of the useful and intriguing properties of the Fourier transform and its near relative, the Hilbert transform. In addition, it gives an explanation of the digital Fourier transform and its application. For more details, derivations, and applications, the reader is referred to the sources listed in the Bibliography.

A.2 The Fourier Transform

A.2.1 Definitions

The minus i Fourier transform, also known as the Fourier integral, is defined as:

$$G(s) = \Im_{-i}[g(u)] = \int_{-\infty}^{\infty} g(u)e^{-i2\pi us}dt \tag{A.1}$$

in which $G(s)$ is the minus i Fourier transform of $g(u)$, i is $\sqrt{-1}$, and $\Im$ symbolizes the Fourier transform operator. Where possible, capital letters will represent transformed variables, and lower-case letters will represent the untransformed function.

Another operation, the "inverse" minus i Fourier transform, can be used to determine $g(u)$ from $G(s)$,

$$g(u) = \Im_{-i}^{-1}[G(s)] = \int_{-\infty}^{\infty} G(s)e^{i2\pi us}df \quad \text{(A.2)}$$

in which $\Im^{-1}$ is the symbol for the inverse Fourier transform. Also, there is a plus i Fourier transform:

$$G(s) = \Im_i[g(u)] = \int_{-\infty}^{\infty} g(u)e^{i2\pi us}dt \quad \text{(A.3)}$$

as well as a plus i inverse transform:

$$g(u) = \Im_i^{-1}[G(s)] = \int_{-\infty}^{\infty} G(s)e^{-i2\pi us}df \quad \text{(A.4)}$$

How are the minus i and plus i transforms related? Both Eqns A.1 and A.3 have the same form, as well as Eqns A.2 and A.4. If, in Eqn A.1 we replace s with $-s$, then $G(-s)$ results, as well as a transform that looks like that of Eqn A.3. This result can be generalized as follows: If the minus i transform of $g(u)$ is known as $G(s)$, then the plus i transform of $g(u)$ is $G(-s)$. In this book, the most frequently used Fourier transform operation is a minus i Fourier transform, $\Im_{-i}$, so the $-i$ designation will be understood, $\Im = \Im_{-i}$, unless the operator is specifically denoted as $\Im_i$ for a plus i transform.

For a Fourier transform to exist, several conditions must be met in general. The function $g(u)$ to be transformed must be absolutely integrable, $|g(u)|$ under infinite limits. Also, $g(u)$ can only have finite discontinuities. The function cannot have an infinite number of maxima and minima, though some experts disagree on this point. As a commonsense rule, if physical processes are described by a Fourier transform, they usually exist because they must obey physical laws like the conservation of energy. We shall also find it convenient to use functions that are infinite in amplitude like the impulse function, which, strictly speaking, can only be considered to exist in a limiting sense. These functions are called generalized functions that have special properties in the limit.

A.2.2 Fourier Transform Pairs

Many Fourier transforms of functions have been determined analytically; a short list of frequently used functions and their transforms is given in Table A.1. Others are available in transform tables of the references. From the principle of linearity, superposition can be used to combine transform pairs to create more complicated functions. Shorthand symbols and names for these functions follow those that can be found in Bracewell (2000).

Table A.1 Theorems for Fourier Transforms

Theorem	$f(x)$	$F(s) = \Im_{-i}[f(x)]$	$\Im_{+i}[f(x)]$
Definitions	$f(x)$	$F(s)$	$F(-s)$
Similarity	$f(ax)$	$(1/\lvert a\rvert)F(s/a)$	$(1/\lvert a\rvert)F(-s/a)$
Addition	$f(x)+g(x)$	$F(s)+G(s)$	$F(-s)+G(-s)$
Shift	$f(x-a)$	$exp(-i2\pi as)F(s)$	$exp(i2\pi as)F(s)$
Combined	$f[b(x-a)]$	$exp(-i2\pi as)(1/\lvert b\rvert)F(s/b)$	$exp(i2\pi as)(1/\lvert b\rvert)F(-s/b)$
Modulation	$f(x)\cos(2\pi s_0 x)$	$\frac{1}{2}[F(s-s_0)+F(s+s_0)]$	$\frac{1}{2}[F(-s-s_0)+F(-s+s_0)]$
Convolution	$f(x)*g(x)$	$F(s)G(s)$	$F(-s)G(-s)$
Autocorrelation	$F(x)*f^*(-x)$	$\lvert F(s)\rvert^2$	$\lvert F(-s)\rvert^2$
Derivatives	$F'(x)$	$i2\pi sF(s)$	$-i2\pi sF(-s)$
Derivatives	$-i2\pi xf(x)$	$F'(s)$	$F'(-s)$
Derivative of convolution	$\frac{d}{dx}[f(x)*g(x)]$ $=f'(x)*g(x)$ $=f'(x)*g'(x)$	$i2\pi sF(s)G(s)$	$-i2\pi sF(-s)G(-s)$
Rayleigh	$\int_{-\infty}^{\infty}\lvert f(x)\rvert^2 dx$	$\int_{-\infty}^{\infty}\lvert F(s)\rvert^2 ds$	$\int_{-\infty}^{\infty}\lvert F(-s)\rvert^2 ds$
Power	$\int_{-\infty}^{\infty} f(x)g*(x)dx$	$\int_{-\infty}^{\infty} F(s)G*(s)ds$	$\int_{-\infty}^{\infty} F(-s)G*(-s)ds$

Source: Bracewell (2000)

Before these functions are introduced, the scaling theorem will be found to be most useful,

$$\Im_{-i}[g(at)] = \frac{1}{|a|}G(f/a) \tag{A.5}$$

The impulse function, already mentioned in Chapter 2, is a generalized function that has the unusual property that it samples the integrand:

$$\int_{-\infty}^{\infty} \delta(t-t_0)g(t)dt = g(t_0) \tag{A.6A}$$

When the transform of the impulse function is taken, the result is an exponential:

$$H(f) = \int_{-\infty}^{\infty} \delta(t-t_0)e^{-i2\pi ft}dt = e^{-i2\pi ft_0} \tag{A.6B}$$

which shows that a delay in time is equivalent to a multiplicative exponential delay factor in the frequency domain. When the impulse has no delay, or $t_0 = 0$, $H(f) = 1.0$, a constant.

Because the sine and cosine are defined in terms of the difference and sum of two exponentials, they become impulse functions in the transform domain (as indicated in Table A.2).

Table A.2 Minus i Fourier Transform Pairs

Function	$g(t)$	$G(f) = \Im_{-i}[g(t)]$
Delayed impulse or Dirac delta	$a_0\delta(t-a)$	$a_0 exp(-i2\pi af)$
Rect or rectangle (base c, delay a)	$b\Pi\left(\frac{t-a}{c}\right)$	$bc\ exp(-i2\pi af)\text{sinc}(cf)$
Delayed Gaussian	$b\ exp\left[-\pi\left(\frac{t-a}{c}\right)^2\right]$	$bc\ exp[-\pi c^2 f^2]\cdot exp(-i2\pi af)$
Shah (interval a)	$III(t/a)$	$aIII(fa)$
Triangle (base c, delay a)	$b\Lambda\left(\frac{t-a}{c}\right)$	$bc\ exp(-i2\pi af)\text{sinc}^2(cf)$
Sign or signum at $t=0$	$\text{sgn}(t)$	$-i/\pi f$
Cosine	$\cos(\omega_0 t)$	$\frac{1}{2}[\delta(f-\omega_0/2\pi)+\delta(f+\omega_0/2\pi)]$
Sine	$\sin(\omega_0 t)$	$-i/2[\delta(f-\omega_0/2\pi)-\delta(f+\omega_0/2\pi)]$

A delay can be added to the scaling theorem to make it even more useful:

$$\Im_{-i}[g(a(t-b))] = \frac{e^{-i2\pi bf}}{|a|}G(f/a) \tag{A.7}$$

An important unique property of the impulse function is:

$$\delta(at) = \frac{1}{|a|}\delta(t) \tag{A.8}$$

Now consider an infinite sequence of impulses, each spaced an interval $a = t_0$ apart from each other. This series of impulse functions also has a special name, "shah" $III(t/a)$, and has an unusual property: The shah function is its own transform:

$$G(f) = \int_{-\infty}^{\infty} III(t/t_0)e^{-i2\pi ft}dt = \int_{-\infty}^{\infty}\sum_{-\infty}^{\infty}\delta(t-nt_0)e^{-i2\pi ft}dt = t_0 III(ft_0) = t_0\sum_{-\infty}^{\infty}\delta(f-n/t_0) \tag{A.9}$$

as shown in Table A.1. Note that the transform of the shah function is another series of impulses spaced at "$1/a$" and with a weight of "a" rather than one.

Besides its sampling capability in time, the shah function is also called a replicating function because when it is convolved in the transform domain with another function, it replicates the function at all its impulse locations. These features will be described in more detail in Section A.2.5.

Another function that is its own transform is the Gaussian; this is for $a = b = c = 1$ for the Gaussian in Table A.2. The Gaussian has the useful properties of smoothness and differentiability. It also appears in statistics, limiting cases, and the solutions to many physical problems.

Another useful function is the Heaviside unit step function, defined as:

$$H(t) = \begin{matrix} 0 & t<0 \\ 1/2 & t=0 \\ 1 & t>0 \end{matrix} \tag{A.10}$$

It can be combined with weighted and delayed versions of itself to create the rect, or rectangle, and sign, or signum, functions of Table A.2. The Fourier transform of the rect function is the sinc function, both of which appeared in Chapter 2. The sinc is defined as:

$$\mathrm{sinc}(t) = \frac{\sin(\pi t)}{\pi t} \tag{A.11}$$

Finally, the triangle function has a transform that is a $\mathrm{sinc}^2(f)$.

A.2.3 Fundamental Fourier Transform Operations

As indicated in Chapter 2, functions in the transform domain (usually functions of f or k) can be multiplied either to obtain an overall transfer function from a sequence of individual transfer functions or to construct a transform for a complicated function from simpler ones:

$$\Im_{-i}[g(t) *_t h(t)] = G(f)H(f) \tag{A.12}$$

This expression shows that if the time (or space) counterparts of these functions are convolved, an equivalent result will be obtained. The convolution operation for two functions (g and h) is defined below:

$$e(t) = \int_{-\infty}^{\infty} g(\tau)h(t-\tau)d\tau = g(t) *_t h(t) \tag{A.13}$$

The product of the integrand can be visualized as two functions, $g(\tau)$ and another function $h(\tau - t)$, which is h flipped right to left and slid across g for different numerical values of the constant t. For each particular value of t, there is an overlap of the two functions, the area of which corresponds to the amplitude of $e(t)$ there. This process is continued until the flipped h function has slid completely across function g. Two notable general characteristics of convolution are its smoothing effect and elongation (the resulting signal usually has a length greater than either individual signal).

Correlation is an operation like convolution except the shifted function is **not** flipped but remains in its original orientation; therefore, correlation can be expressed as a special case of convolution:

$$g \otimes h = \int_{-\infty}^{\infty} g(\tau - t)h(\tau)d\tau = g(t)*h(-t) \tag{A.14}$$

This operation serves two useful purposes: It can be a measure of the similarity between two waveforms, and the location of the peak of the correlation serves as a measure of relative delay between waveforms.

Autocorrelation is a special case of correlation in which the shift function is the same as the other function and a star is its shorthand symbol:

$$g^*g = \int_{-\infty}^{\infty} g(\tau)g(\tau - t)d\tau = g(t)*g^*(-t) \tag{A.15}$$

in which g^* is the complex conjugate of g.

These key relations and others are listed in Table A.2. Certain general principles can be applied to functions to extend their usefulness. The scaling principle, Eqn A.5, already encountered in Chapter 2 for both the time–frequency and space–wave number Fourier transforms, appears with other principles in Table A.2. The shift–delay theorem, mentioned as Eqns A.5 and A.6B, can be combined with the scaling theorem to give Eqn A.7.

The Power and Rayleigh theorems are handy for estimating energy content and limiting values. For example, the power of a plane acoustic wave propagating in a lossless medium would be expected to satisfy the power theorem in each plane,

$$\int_{-\infty}^{\infty} g(u)h^*(u)du = \int_{-\infty}^{\infty} G(s)H^*(s)ds \tag{A.16}$$

The Rayleigh theorem can be applied to a known signal or spectrum as a bound on its value in the other domain,

$$\int_{-\infty}^{\infty} |g(u)|^2du = \int_{-\infty}^{\infty} |G(s)|^2ds \tag{A.17}$$

Derivatives have a Fourier transform equivalent in the other domain, an important feature that can be applied to the solution of differential equations,

$$\Im_{-i}\left(\frac{\partial^n g}{\partial u^n}\right) = (i2\pi s)^n G(s) \tag{A.18A}$$

$$\Im_{i}\left(\frac{\partial^n g}{\partial u^n}\right) = (-i2\pi s)^n G(s) \tag{A.18B}$$

A.2.4 The Sampled Waveform

In many cases, physical signals or data cannot be represented simply by a tidy closed form or analytical expressions. Often, only a series of discrete numbers and an associated index representing an independent variable such as time or distance are available. These numbers may represent a continuous waveform sampled at regular intervals or data.

What happens when sampling occurs? In Figure A.1, a single continuous waveform in time, $g(t)$, is sampled at intervals of Δt. The total length of the time record is T seconds long. This sampling process can be described by the shah function:

$$s(t) = g(t)III(t/\Delta t) = \sum_{n=-\infty}^{\infty} g(n\Delta t)\delta(t - n\Delta t) \tag{A.19}$$

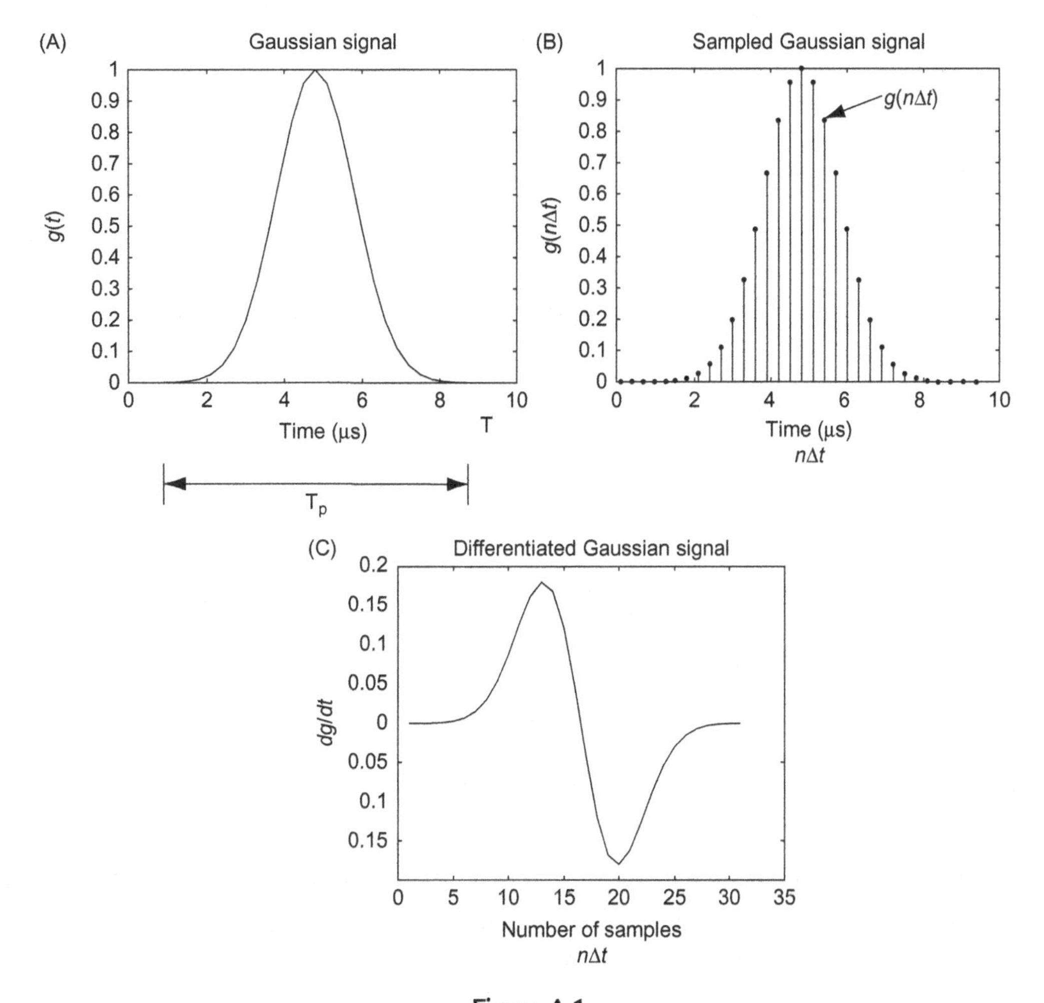

Figure A.1
(A) Time waveform delayed to fit into a positive time interval; (B) Delayed waveform sampled at intervals of Δt; (C) Sampled waveform numerically differentiated with respect to time.

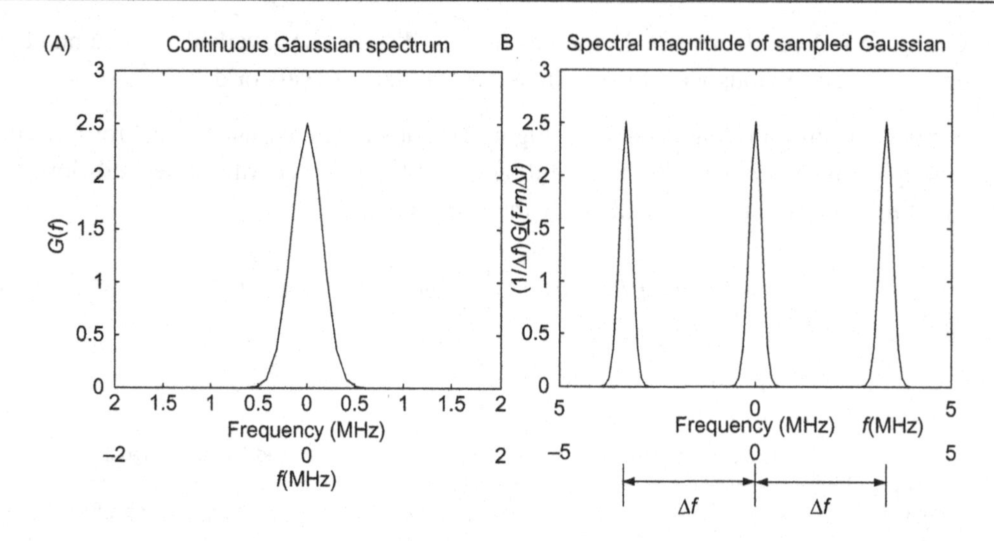

Figure A.2
(A) Fourier transform of a continuous waveform from Figure A.1A. (B) Fourier transform of a sampled waveform from Figure A.1B.

and is shown in Figure A.1B. The Fourier transform of this sampled time function is given by the following, in which $G(f)$ is the continuous $-i$ Fourier transform of $g(t)$, (shown in Figure A.2A):

$$S(f) = G(f)*[(1/\Delta f)III(f/\Delta f)] = (1/\Delta f) \sum_{m=-\infty}^{\infty} G(f - m\Delta f) \tag{A.20}$$

with $\Delta f = 1/\Delta t$ (the result is illustrated by Figure A.2B). Note that the separation between the sequence of replicated spectra is equal to Δf. Each spectrum is like $G(f)$ in shape, but the magnitude (which here and elsewhere in the book represents the full complex spectrum) is multiplied in amplitude by a factor of Δt.

Overall, there is a basic problem. While the time waveform is sampled, its spectrum is not! Furthermore, there are many repeated continuous spectra—not just one—thanks to the replicating property of the shah function. To be truly digital, each individual spectrum must also be represented by a series of samples.

This problem can be solved by a mathematical trick. Represent the time waveform as a series of identically sampled waveforms, each reoccurring at a time period T consisting of N samples (as displayed in Figure A.3A). This series is described by:

$$s(t) = g(t)III(t/\Delta t) * \frac{1}{T}III(t/T) = \frac{1}{T} \sum_{p=-\infty}^{\infty} \sum_{m=-\infty}^{\infty} g(m\Delta t)\delta(t - m\Delta t - pT) \tag{A.21}$$

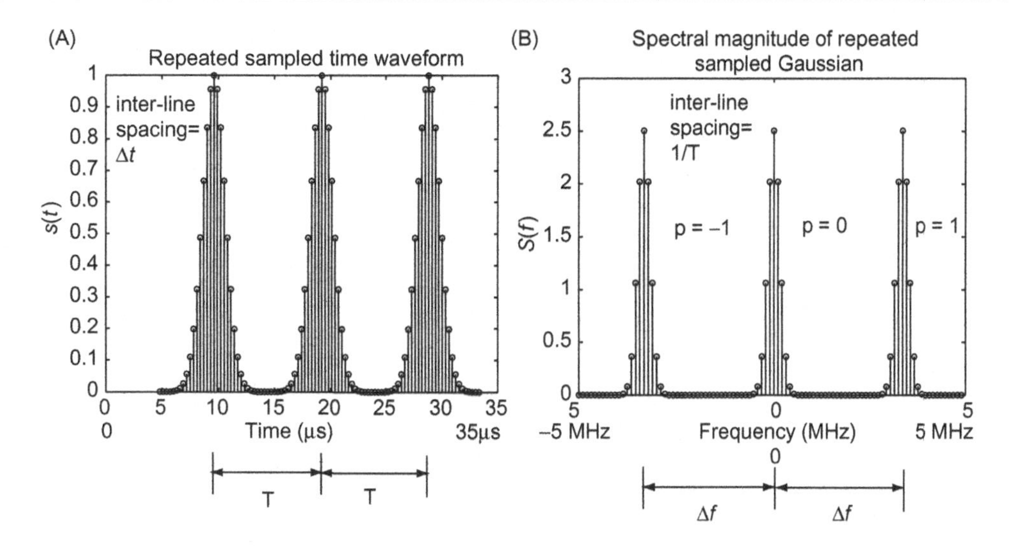

Figure A.3
(A) Repeated sampled waveform; (B) Fourier transform of a repeated sampled waveform.

which has the $-i$ Fourier transform:

$$S(f) = G(f)III(fT) * [III(f\Delta t)\Delta t] = \Delta t \sum_{m=-\infty}^{\infty} \sum_{p=-\infty}^{\infty} G\left(\frac{p}{T}\right)\delta(f - p/T - m/\Delta t) \qquad \text{(A.22)}$$

The key point here is that the spectrum is not only recurring as a series, but each individual spectrum is sampled finely at a rate $1/T$ (as obvious from Figure A.3B), and it is repeated at a frequency of $\Delta f = 1/\Delta t$.

Sampling rates are explained most simply by restricting each spectrum to be bounded by a certain bandwidth (BW). The selection of the sample rate Δt appears arbitrary, but if the sampling is done at least at the Nyquist rate, $\Delta t \leq 1/\text{BW}$, the continuous waveform, $g(t)$, can be recovered. More generally, the Nyquist rate is stated as $\Delta t = \frac{1}{2} f_{\max}$, where the bandwidth of interest is $\text{BW} = 2f_{\max}$, centered at f_c (note, in the example shown in the figures, $f_c = 0$).

To illustrate the recovery process (Figure A.4), a rect function window of width BW is centered on the spectrum of interest and multiplied by $1/\Delta f$,

$$S(f) = G(f)III(fT)*[III(f\Delta t)\Delta t]\left[\frac{1}{2f_{\max}}\prod(f/2f_{\max})\right] \qquad \text{(A.23)}$$

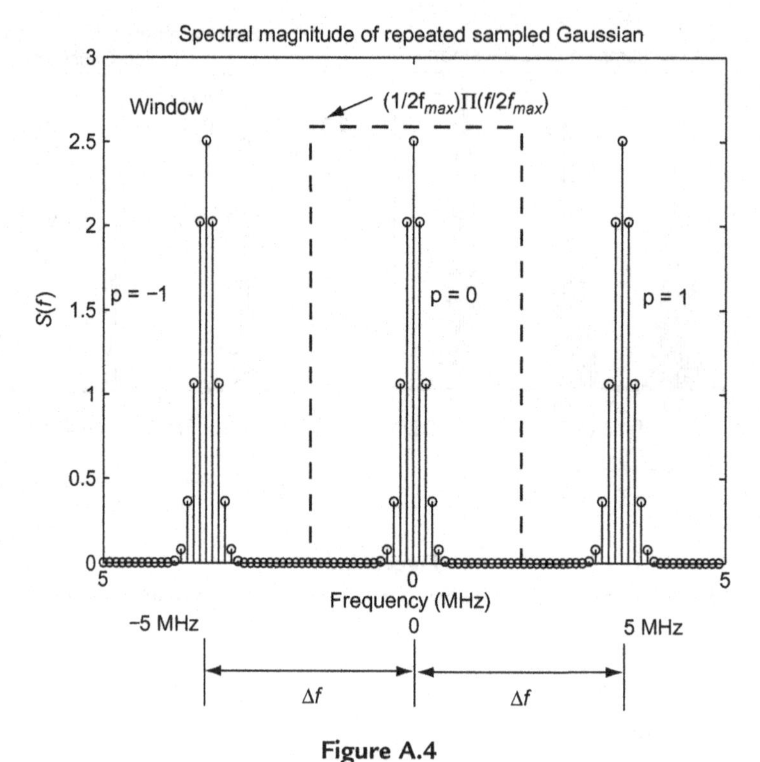

Figure A.4
Windowing of the fundamental spectrum from Figure A.3B.

Here, the inverse $-i$ Fourier transform of the above is taken,

$$s(t) = g(t)III(t/\Delta t) * \frac{1}{T}III(t/T) * \text{sinc}(2f_{\max}t) \tag{A.24}$$

in which the first term is the sampled function $g(t)$, which is convolved by the replicating shah function (second term) and convolved by an interpolating sinc function that restores the sampled $g(t)$ to its original form of Figure A.1A. This last windowing step is often tacitly understood because the waveform and spectrum corresponding to the series term $p = 0$ are the focus of interest. Most graphics routines connect the dots between the numerical sample values and display the inverse transform of Eqn A.22, windowed by the time interval T to the original time segment of interest, as a continuous function. Note that an adequate sampling rate must still be met for these conditions to hold.

A.2.5 The Digital Fourier Transform

From the previous sampling discussion, a mathematical form of both a periodic time waveform and a periodic spectrum were needed to achieve sampling in both domains.

In practice, the focus is on only one period in each domain; therefore, the approach can be applied to a single waveform and spectrum at a time. For a finite number of samples (N), another way of achieving equivalent results is through the use of the digital Fourier transform (DFT), which also has periodic properties.

Recall that a waveform or data series is represented by a sequence of numbers such as $g(n)$ based on an index $n = 0, 1 \ldots (N-1)$. The number sequence is associated with a sampled physical quantity such as voltage, $g(n\Delta t)$. The index of the sequence is associated with an independent variable such as time (t), as well as a sampling interval such as Δt, so that the variable extends from 0, Δt, $2\Delta t$, ... to $(N-1)\Delta t$, and constitutes a finite interval, $T = N\Delta t$.

In the usual bare bones formulation of the DFT, the user has to keep track of all the associations between the sample points and the independent variable, for there are just a sequence of numbers (the function) and integer indices. The minus i DFT of the sequence $g(n)$ is defined as:

$$G(m) = \mathrm{DFT}_{-i}[g(n)] = \sum_{n=0}^{N-1} g(n)e^{-i2\pi mn/N} \tag{A.25}$$

Note that the transformed sequence, $G(m)$, has an index m; therefore, each value of $G(m)$ corresponds to a sum of a weighted sequence of exponentials over N. If the index n exceeds $N-1$, the transformed sequence repeats itself because the exponential has a period N (a property essential for the periodic nature of the sampling process described in the last section). Once again, the focus is on the first N points. The inverse minus i DFT is:

$$g(n) = \mathrm{DFT}_{-i}^{-1}[G(m)] = \frac{1}{N}\sum_{m=0}^{N-1} G(m)e^{i2\pi mn/N} \tag{A.26}$$

Similar general principles and transform pairs hold for DFTs, but because a finite number of samples are involved, extra care must be taken in their application.

The FFT (Cooley & Tukey, 1965) is an efficient algorithm for calculating a DFT. Standard FFT algorithms are available, such as those in MATLAB. The next section will provide the detailed steps needed for the application of DFTs to Fourier transform calculations, as well as a MATLAB program for carrying them out. A final section has extra advice on sampling and practical applications.

A.2.6 Calculating a Fourier Transform with an FFT

Since more is involved in obtaining a Fourier transform of a function than just applying an FFT, the steps are described here. We begin with a continuous function of time, $g(t)$, that is similar to the one in Figure A.1. Much of the discussion follows the reasoning of Section A.2.4. A key difference between a continuous Fourier transform and a digital Fourier transform can be found by comparing Eqn A.1, the definition of a Fourier transform, Eqn A.22, the Fourier transform of a repeated sampled waveform, and the definition of a DFT, Eqn A.25. This difference is that the DFT is simply a sum. To approximate a continuous integral, a sampling interval is required, as well as a sum, as evident from Eqn A.22. The missing ingredient is Δt. In order to estimate a continuous Fourier transform by a numerical approach, the sampled waveform (a series of points) must be multiplied by Δt before performing the FFT.

An example is calculated by the MATLAB program gausdemo.m, which compares a continuous Fourier transform of a Gaussian with that calculated by an FFT. The function for this example is $exp(-at^2)$. Since the function extends into negative time, we shift it forward by a delay (b) so that it is centered in an interval 0 to T as shown in Figure A.5:

$$h(t) = g(t - b) \tag{A.27}$$

For this example, the sampled function is a Gaussian symmetric about $t = 0.0$, with $a = 0.5$, and is defined over an interval of 9.6 μs so $b = 4.8$ μs and:

$$h(t) = exp[-a(t-b)^2] = exp[-0.5(t-4.8)^2] \tag{A.28}$$

Note that, within this interval, this shifted time function drops close to zero at the ends (as illustrated by Figure A.5A).

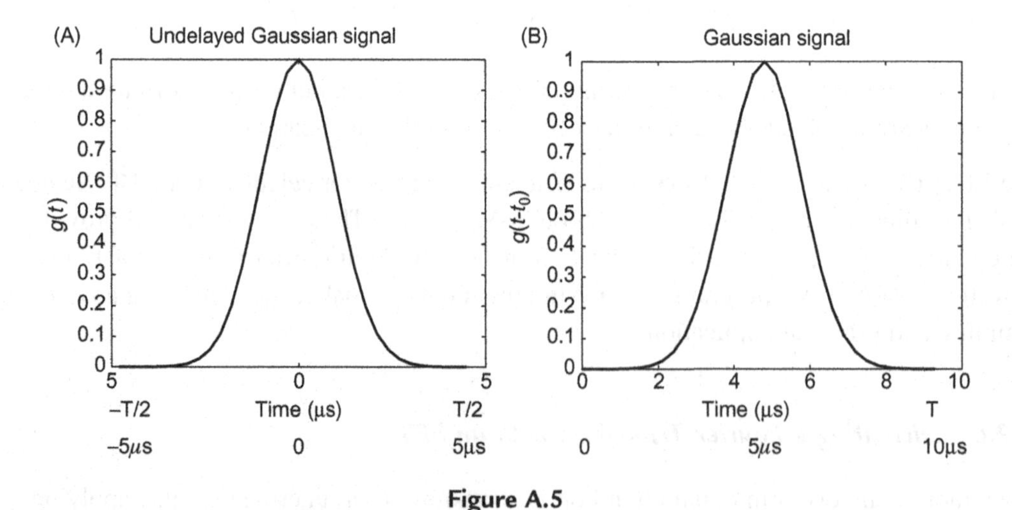

Figure A.5
(A) Gaussian waveform centered at origin; (B) Delayed gaussian waveform.

This shifted waveform is to be sampled at a regular interval Δt, but how is Δt selected? The usual advice is sample at the Nyquist rate, or at:

$$\Delta t = 1/2f_{\max} \tag{A.29A}$$

where $f_{\max}$ is the highest frequency in a band-limited spectrum. Unfortunately, we most likely do not know what $f_{\max}$ is or what the spectrum looks like, or we would not be calculating the spectrum in the first place. If $f_{\max}$ is known, a conservative estimate is to use the sampling rate of:

$$\Delta t_c = 1/10f_{\max} \tag{A.29B}$$

given by Weaver (1989). The determination of $f_{\max}$ must be done carefully: If a filter or beam profile is to be calculated by a transform and viewed on a decibel (logarithmic) scale (the usual case), then this frequency should be based on the scale to be used to avoid overlap or aliasing.

Most often a waveform is not a simple amplitude-modulated sine wave, and it may contain higher-frequency cycles or transients. Another approach is to look for the steepest transition slope in the waveform and to choose the shortest time interval containing the transition, δt. Then $f_{\text{high}} \approx 1/\delta t$, or:

$$\Delta t = \delta t/2 = 1/2f_{\text{high}} \tag{A.29C}$$

In practice, this sampling rate may fall between the Nyquist rate and Weaver's recommendation.

To explore the differences in sampling rates, the Gaussian (a well-behaved function compared to the wide variety of functions that occur in practice) will be used. To find the locations where the maximum slopes occur, we differentiate the function with respect to time either analytically (if possible) or numerically. The result of a numerical differentiation by the MATLAB function diff.m is shown in Figure A.1C and indicates two time locations where slope is maximum. To find these times analytically, the usual min–max procedure is to differentiate the function and set the result equal to zero. In this case, the solution of setting the slope to zero is for $t = b$, where the slope is zero. For this particular problem, the function is differentiated a second time and the result set to zero, so the solutions for the locations of the maximum slopes are:

$$t = t_0 \pm 1/\sqrt{2a} \tag{A.30A}$$

The slope can be estimated numerically as:

$$\frac{dh}{dt} \approx \frac{\Delta h}{\Delta t} = \frac{h[(n+1)\Delta t] - h(n\Delta t)}{\Delta \mathrm{t}} \tag{A.30B}$$

where the closest sample points to the location of the maximum slope are implied (Figure A.1B). For this example, the maximum slope is $dh/dt = 0.06065$ at $t = t_0 \pm 1$. Sampling can be regarded as an exercise in a numerical approximation of this slope as $\Delta h/\Delta t$. For $\Delta t = 0.30$ μs, the approximate slope is 0.06; for $\Delta t = 0.15$ μs, it is 0.0605, which is recommended. The numerically determined slope approaches the ideal value asymptotically as the sampling interval is decreased. From Figure A.2A, f_{max} appears to be about 1 MHz on a linear scale when seen at a larger magnification. The standard Nyquist sampling rate gives the sampling interval as $\Delta t = 0.50$ μs. Weaver's sampling criteria yields $\Delta t = 0.10$ μs. In summary, the standard Nyquist sampling approach can underestimate the sampling required for a nonsinusoidal waveform. The Weaver criterion provides a more conservative estimate, and the numerical approximation of the maximum slope (the transient approach) gives reasonable estimate based on the properties of the waveform under consideration without prior knowledge of f_{max}.

Next, the number of samples (N) needs to be chosen. Weaver (1989) suggests this number to be at least:

$$N = 10(T_p/2 + b)/\Delta t \tag{A.31}$$

Again, this is a conservative number; the minimum needed is 1/10 of this value for a well-behaved waveform like the one in our example, for which the waveform drops so rapidly that there is not much left at the ends of the interval chosen. The importance of N is evident from Figure A.3B, which shows the frequency sampling interval as $1/T = 1/(N\Delta t)$.

For our example, the recommended sampling from the transient criteria is $\Delta t_c = 0.15$ μs, and $T_p \approx 7$ μs, so if $b = 3.5$ μs, therefore $N = 467$. To be a bit safer and round up to the nearest power of 2 (the usual choice for an FFT), let $N = 512$ so that $T = 76.8$ μs. For most FFT routines, it is desirable to have N be a power of 2. Note that this time sequence is to be associated with the running index, 0, 1, 2, ... 511, 512, just as the sampled waveform, $h(n\Delta t)$, is associated with the shown sequence, $h(n)$. Note that adding a significant number of zeros may change the resulting spectrum. Because of the good behavior of the waveform in this example, we can use the criteria above divided by 10. For the Nyquist criteria, this approach (rounded up to the nearest power of 2) gives $N = 16$; for the Weaver sampling, $N = 128$, and by the transient method, $N = 64$. For the purposes of demonstrating the calculation process with adequate graphical clarity, the figures in this appendix were calculated with $\Delta t = 0.30$ μs and $N = 32$, which are better than the values obtained by the Nyquist sampling criteria but less than the recommended values. Through the MATLAB program gausdemo.m, which compares an analytically determined spectrum of a Gaussian with that determined by the FFT process described here, the reader can experiment with the effects of changing the sampling interval and number of points.

Again, it is important in general that, if possible, the signal be very small or close to zero at the ends of this interval, which is true for this function. This requirement makes sense because of the periodic nature of the DFT; a jump at the ends of this cyclic progression would result in an artificial transient discontinuity that could introduce artifacts into the calculation. A window function such as a Hamming or Hanning function is often multiplied by the function to be transformed to reduce the signal to small values at the ends of the time interval; there are trade-offs in selecting these functions (Kino, 1987). The final delayed sampled waveform appears in Figure A.1B.

In order to perform these calculations, we use MATLAB command fft. This command calculates a DFT according to Eqn A.25, except that an internal computational index runs from 1 to N, which makes no difference in the outcome. Take the FFT (equal to the minus i DFT) of $h(n)$ and multiply it by Δt, similar to the operation in Eqn A.22:

$$H(m) = \Delta t\{\text{DFT}_{-i}[h(n)]\} \tag{A.32}$$

and the result is Figure A.6A. The multiplication by Δt is analogous to the dt in the integrand of the continuous Fourier transform. Note, this operation is similar to Eqn A.22 and Figure A.3B.

Usually, a spectrum is displayed with a frequency range extending from negative to positive frequencies. The real and imaginary parts of the DFT calculation in Figure A.6A, however, correspond to an unshifted frequency row vector of two contiguous segments, 0 to $N/2 - 1$ and then $N/2$ to $N - 1$, both multiplied by Δf. This sequence of numbers will be called

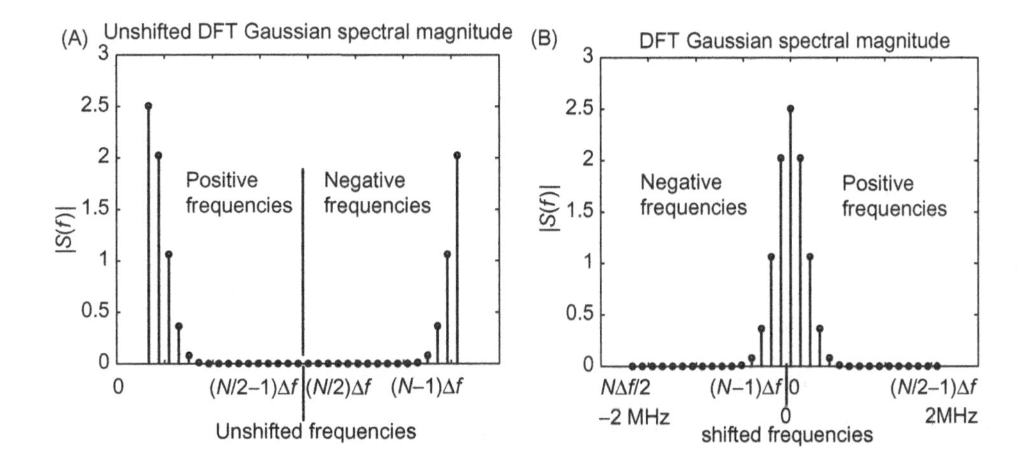

Figure A.6
(A) Magnitude of the FFT of the sampled delayed gaussian waveform plotted against the unshifted sample indices; (B) Spectral magnitude versus shifted frequencies after (A) has undergone a shift operation.

"$f_{unshift}$ (m)" with indices 0 to $N-1$, as illustrated by the abscissa in Figure A.6A. First, however, if shifting by a delay b was necessary, then the complex spectrum must be corrected for this by:

$$G(f) = H(f)exp(i2\pi bf) \tag{A.33A}$$

which in digital terms is equivalent to:

$$G(m) = H(m)exp(i2\pi bf_{unshift}(m)) \tag{A.33B}$$

Next, the real and imaginary parts of the DFT calculation shown in Figure A.6A need to be reshuffled to correspond to the standard frequency convention for a more convenient display of results. There is a DFT theorem that can be applied to this problem:

$$H(-m) = H(N-m) \tag{A.34}$$

so that the transformed sequence can be remapped from the unshifted vector, $f_{unshift}$ (m), to a new arrangement based on a new vector, f_{shift} (m), as given by Figure A.6B. This new mapping corresponds to frequencies running from $N\Delta f/2 \ldots (N-1)\Delta f$, 0, $\Delta f \ldots (N/2-1)\Delta f$. Note that adding a point corresponding to a frequency index $N\Delta f/2$ (not shown) on the right end would repeat the leftmost value at $N\Delta f/2$. This reshuffling operation can be executed by MATLAB function "fftshift.m."

A continuous function $H(f)$ can be recovered from $H(m)$ by the interpolation formula or by fitting a smooth curve to the points where the final result is shown in Figure A.6B. Usually, a plot routine connects the sample points automatically so no extra step is necessary for interpolation.

The program gausdemo.m compares the calculated digital spectrum just explained to an analytically determined spectrum and shows that the magnitudes correspond exactly. From theory, a constant phase of zero radians is expected. The spectrum of the digitally determined phase, calculated by MATLAB functions "angle.m" and "unwrap.m," shows that the expected phase is recovered only over those frequencies where the spectral Gaussian function is numerically defined, roughly between -1 and 1 MHz, and outside this range, numerical noise results.

In summary, the recommended steps in finding a Fourier transform are the following:

1. Determine sampling interval Δt, number of points N, and sampling period T.
2. Shift the waveform to center of T, if necessary.
3. Multiply the waveform by Δt and take its FFT.
4. Correct for inserted delay by multiplying the complex spectrum by corresponding phase factor, if necessary.

5. Shift the spectrum points to correspond to the negative-to-positive frequency convention.
6. Plot the spectrum (or interpolate points).

A.2.7 Calculating an Inverse Fourier Transform and a Hilbert Transform with an FFT

Often, a complex spectrum is given and the corresponding pulse is required. In many cases, it is desirable to calculate the envelope of the pulse and/or its quadrature signal. All of these pulse characteristics can be obtained in a single calculation. The quadrature signal of a cosine is a sine, for example. The quadrature signal has the interesting property that it is minimum where the real signal is maximum, and it is usually 90° out of phase from the real signal. The overall time signal that consists of the real and imaginary quadrature signals is called the "analytic signal."

Two methods for determining the quadrature waveform will now be given, and the previous Gaussian example will be used. Consider the spectrum that has already been determined by the process described in the last section. In this example, the spectrum is real; in general, it is complex.

Given the original complex spectrum before shifting, set the values for the negative frequencies equal to zero (points corresponding to $N/2$ to $N-1$).

Now the inverse FFT (corresponding to an inverse $-i$ DFT) is taken and multiplied by $2/(\Delta t)$:

$$h(n) = [2/\Delta t]\{\mathrm{DFT}_{-i}^{-1}[H(m)]\} \tag{A.35}$$

and the real and imaginary parts of $h(n)$ are illustrated by Figure A.7. Note that multiplying by $1/(\Delta t)$ and the factor $1/N$ implicit in the definition of the inverse DFT, Eqn A.26, is equivalent to multiplying by $1/T$, the fine frequency sampling interval. The multiplication by a factor of two compensates for the missing half of the spectrum. The real waveform corresponds to the original waveform. The process of dropping the negative frequencies is the equivalent of taking a Hilbert transform, so that an imaginary or quadrature signal is obtained simultaneously with the real part (more on the Hilbert transform can be found in Chapter 5). The pulse magnitude or envelope can be found by calculating:

$$\mathrm{ENV}[h(n)] = \sqrt{\{\mathrm{REAL}[h(n)]\}^2 + \{\mathrm{IMAG}[h(n)]\}^2} \tag{A.36}$$

and its usual display is given in Figure A.7 simultaneously with the real part. A more conventional presentation is with the envelope on a dB scale and the spectral magnitude shown simultaneously, also on a dB scale (as given in Figure 13.5). Note that the

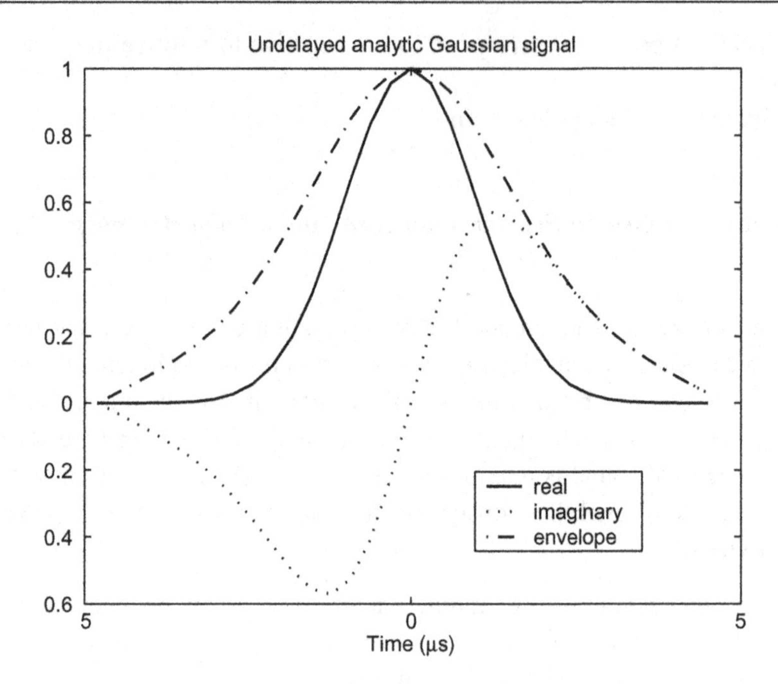

Figure A.7
The real and quadrature waveforms and envelope of an analytic signal versus time.

overall time range with a sampling interval $\Delta t = 1/(N\Delta f)$ can be adjusted (advanced or delayed) by extracting the desired points or windowing to enhance the presentation of the pulse, and uninteresting points can be dropped to expand the view of the pulse.

An alternative and more convenient way of obtaining the analytic signal is to use the MATLAB function "hilbert.m" on the original real waveform. This operation results in the simultaneous creation of the real and imaginary (quadrature) signals. The envelope can be obtained by applying the MATLAB command "abs" on the complex time signal. The mathematical operations that generated the figures (and more) in this appendix can be found in a program written in MATLAB. This program, Appendix A.m, provides many examples of FFT operations and compares an analytically determined spectrum of a Gaussian to that calculated by the steps in this appendix.

A.2.8 Calculating a Two-dimensional Fourier Transform with FFTs

In order to calculate a two-dimensional Fourier transform with FFTs, a little extra work is needed. In this case, the data are described by a matrix (X). If the dimensions of the matrix are not powers of 2, then the two-dimensional FFT MATLAB function "fft2.m" adds zeros so the lengths conform, and returns these new lengths as variables mrows (number of rows) and ncols (number of columns). For a two-dimensional FFT, quadrants of data need to be moved to bring the arrangement of the calculations into a conventional pattern; fortunately, this operation can be accomplished by the MATLAB function "fftshift.m." Discrete Fourier transforms of multiple dimensions of higher order "ND" (number of dimensions) can be computed through the MATLAB function "fftn.m."

References

Bracewell, R. (2000). *The fourier transform and its applications*. New York: McGraw-Hill.
Cooley, J. W., & Tukey, J. W. (1965). An algorithm for the machine computation of complex fourier series. *Mathematics of Computation*, *19*, 297–301.
Kino, G. S. (1987). *Acoustic waves: devices, imaging, and analog signal processing*. Englewood Cliffs, NJ: Prentice-Hall. Sect. 4.5.3.
Weaver, H. J. (1989). *Theory of discrete and continuous fourier analysis*. New York: John Wiley & Sons.

Bibliography

Bateman, H. (1954). Tables of integral transforms. In A. Erde'lyi (Ed.), *Bateman manuscript project* (Vol. 1). New York: McGraw-Hill.
Bracewell, R. (2000). *The fourier transform and its applications*. New York: McGraw-Hill.
Campbell, G. A. & Foster, R. M. (1948). Fourier integrals for practical applications. D. van Nostrand (Ed.), New York.
Goodman, J. W. (1968). *Introduction to fourier optics*. New York: McGraw-Hill.
Rabiner, L. R., & Gold, R. B. (1975). *Theory and application of digital signal processing*. Englewood Cliffs, NJ: Prentice-Hall.
Weaver, H. J. (1989). *Theory of discrete and continuous fourier analysis*. New York: John Wiley & Sons.

Appendix B

Table B.1 Properties of Tissues

Tissue	c (M/s)	α (dB/MHz^y-cm)	y	ρ (kg/m^3)	Z (megaRayls)	B/A
Blood	1584	0.14	1.21	1060	1.679	6
Bone	3198	3.54	0.9	1990	6.364	–
Brain	1562	0.58	1.3	1035	1.617	6.55
Breast	1510	0.75	1.5	1020	1.540	9.63
Fat	1430	0.6	1	928	1.327	10.3
Heart	1554	0.52	1	1060	1.647	5.8
Kidney	1560	10	2	1050	1.638	8.98
Liver	1578	0.45	1.05	1050	1.657	6.75
Muscle	1580	0.57	1	1041	1.645	7.43
Spleen	1567	0.4	1.3	1054	1.652	7.8
Milk	1553	0.5	1	1030	1.600	–
Honey	2030	–	–	1420	2.89	–
Water @ 20°C	1482.3	2.17e-3	2	1.00	1.482	4.96

α, attenuation factor; ρ, density; c, speed of sound; Z, acoustic impedance. y, power law exponent;. B/A, nonlinearity parameter. Note all data in Table B.1 except line 11 is from Duck (2012). Line 11 from Kino (1987), Selfridge(1983) and Onda website. Another source for tissue properties is Bamber(1998).

Table B.2 Properties of Piezoelectric Transducer Materials

Material	ρ	$\epsilon^S_{33}/\epsilon_0$	k_T	c_{TL} (km/s)	Z_L (MR)	k'_{33}	c'_{33L} (km/s)	Z'_{33} (MR)	k_{33}	c_{33L} (km/s)	Z_{33} (MR)
PZT-5A	7.75	830	0.49	4.350	33.71	0.66	3227	25.01	0.705	3693	28.62
PZT-5H	7.50	1470	0.50	4.560	34.31	0.70	3800	29	0.75		
$BaTiO_3$	5.7	1260	0.38	5.47	31.18	0.47			0.50		
$LiNbO_3$	4.64	39	0.49	7.36	34.2						
Quartz	2.65	4.5	0.093	5.0	13.3						
PVDF	1.78	12	0.11	2.2	3.92						
PMN-PT	8.06	680	0.64	4.646	37.45	0.9066	3.057	24.64	0.94	3.343	26.94
PZN-PT	8.31	1000	0.5	4.03	33.49	0.878	2.624	21.81	0.91	2.417	20.09
Comp A	6.01	376	0.80	3.0	18.03						
Comp B	4.37	622	0.66	3.79	16.58						
Navy VI	7.5	1470	0.5	4.575	34.31	0.698	3986	29.9	0.75	3.851	28.88

Comp A is a 1–3 composite 69% PZN-PT and 31% D-80 filler.
Comp B is a 1–3 composite with 51% Navy type VI (equivalent to PZT-5H) and 49% D-80 filler.
Density is in units of 10^3 kg/m^3. Subscript L is for longitudinal wave; therefore, c_{TL} represents the thickness mode longitudinal sound speed.
ρ, density; c, sound speed; MR, megaRayls; Z, acoustic impedance. ks are electromechnaical coupling constants described in Chapter 5 as well as the dieclectric constants , epsilon. Definitions for the piezoelectric meaterials can also be found in Chapter 5. PZT-5A and PZT-5H are trademarks of Vernitron Piezoelectric Division. BaTiO sub 3 is Barium Titananate and LiNbOsub 3 is Lithium Niobate. The first six lines of Table B.2 are from Kino (1987).
Sources for the last five materials are Park and Shrout (1997) and Ritter et al. (2000).

References

Bamber, J. C. (1998). Ultrasonic properties of tissue. In F. A. Duck, A. C. Baker, & H. C. Starritt (Eds.), *Ultrasound in Medicine*. Bristol, UK: Institute of Physics Publishing.

Duck, F. A. (2012). *Physical Properties of Tissue: A Comprehensive Review*. London, York: (IPEM) Institute of Physics and Engineering in Medicine.

Kino, G. S. (1987). *Acoustic Waves: Devices, Imaging, and Analog Signal Processing*. Englewood Cliffs, NJ: Prentice-Hall., Appendix B.

Park, S. -E., & Shrout, T. R. (1997). Characteristics of relaxor-based piezoelectric single crystals for ultrasonic transducers. *IEEE Transactions on Ultrasonics, Ferroelectrics and Frequency Control, 44*, 1140–1147.

Ritter, T., Geng, X., Shung, K. K., Lopatin, P. D., Park, S. -E., & Shrout, T. R. (2000). Single crystal PZN/PT-polymer composites for ultrasound transducer applications. *IEEE Transactions on Ultrasonics, Ferroelectrics and Frequency Control, 47*, 792–800.

Selfridge, A. R. (1985). Approximate material properties in isotropic materials. *IEEE Transactions on Sonics and Ultrasonics, 32*, 381–394. Acoustic Properties of Materials Reference Tables http://www.ondacorp.com/tecref_acoustictable.shtml (Accessed November 16, 2013). Tables mainly derived from the work of Alan Selfridge.

Appendix C: Development of One-Dimensional KLM Model Based on ABCD Matrices

As shown in Figure C.1, the Krimholtz–Leedom–Matthaei (KLM) model (Leedom, Krimholtz & Matthaei, 1978) provides a separation of the electrical and acoustical parts of the transduction process. This partitioning will allow us to analyze these parts individually to improve the design of the transducer. Note there are three ports: electrical (#3) and two acoustical (#1 and #2). Port 1 will be used for the transmission of acoustic energy into the body or water, and acoustic port 2 radiates into a transducer backing material. The equivalent loads for port 1 on the right (R) and port 2 on the left (L) will be called $Z_R = Z_W$ (water) and $Z_L = Z_B$ (backing). For a derivation of the KLM model, the reader is referred to Leedom et al. (1978) or Kino (1987).

Armed only with simple 2×2 ABCD matrices from Chapter 3, we shall construct a numerical equivalent circuit model (van Kervel & Thijssen, 1983) that will tell us a great deal about how a piezoelectric transducer works. A simplified version of Figure C.1 is shown in Figure 5.8, which represents a nodal numbering scheme to identify points along the signal path. Beginning at port 3, we attach a voltage source V_g with a source resistance R_g to port 3 through a general tuning network with its own ABCD matrix. A particular tuning network implemented in this program is shown in Figure 5.10, which consists of an inductor and an inductive loss resistor, and has a series impedance $Z_S = R_S + i\omega L_S$. The product E of the first matrices 1 and 2 is:

$$[E] = \begin{pmatrix} 1 & Z_S + R_g \\ 0 & 1 \end{pmatrix} = \begin{pmatrix} 1 & R_g \\ 0 & 1 \end{pmatrix} \begin{pmatrix} 1 & Z_s \\ 0 & 1 \end{pmatrix} \quad \text{(C.1)}$$

Inside port 3 are two capacitive elements between nodes 3 and 4. The first term is C_0 with a reactance $iX_0 = -i/\omega C_0$. The second term represents the current contribution to the acoustic output where $iX' = -i/\omega C'$ and:

$$C' = -\, C_0/[k_t^2 \mathrm{sinc}(\omega/\omega_0)], \quad \text{(C.2)}$$

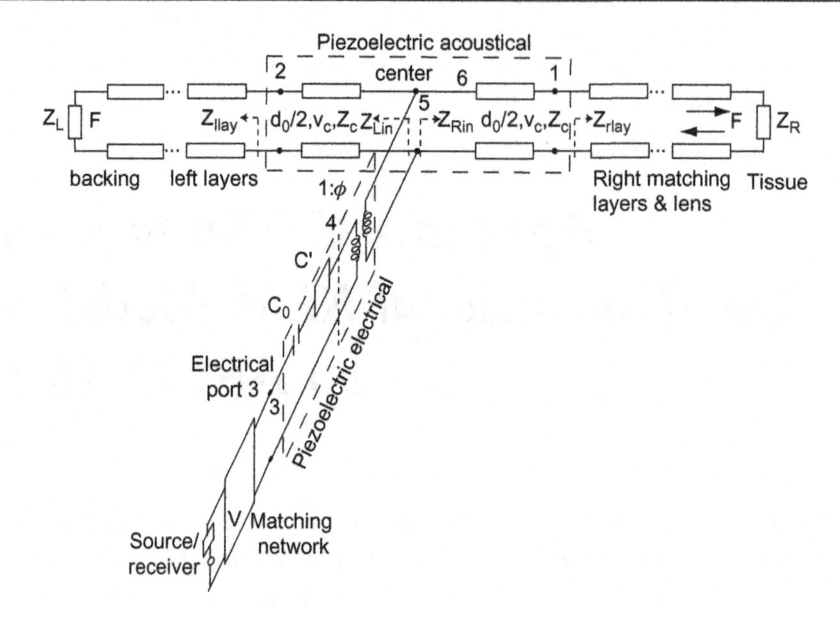

Figure C.1
KLM Equivalent circuit model with acoustic layers and loads and electrical matching.

where X' is related to the minus i Fourier transform of the rectangular shape of the electric field between the electrodes of the piezoelectric. Here the piezoelectric coupling constant, k_t, for the thickness expander mode is used. Multiplying the matrices of the series elements X_0 and X' results in overall matrix C:

$$[C] = \begin{pmatrix} 1 & iX_0 \\ 0 & 1 \end{pmatrix} \begin{pmatrix} 1 & iX' \\ 0 & 1 \end{pmatrix} = \begin{pmatrix} 1 & (C_0 + C')/(i\omega C_0 C') \\ 0 & 1 \end{pmatrix} \tag{C.3}$$

To complete the electrical part of the model, the transformer between nodes 4 and 5 that converts electrical signals to acoustic waves and vice versa is needed. Here, the turns ratio of the transformer is:

$$\phi = k_t \left(\frac{\pi}{\omega_0 C_0 Z_C} \right)^{1/2} \mathrm{sinc}\left[\omega/(2\omega_0) \right], \tag{C.4}$$

where Z_C is the normalized impedance of the crystal,

$$Z_C = AZ_0 = A(C^D/\rho_c)^{1/2}\rho_c = A(C^D\rho_c)^{1/2}, \tag{C.5}$$

where the crystal has a thickness d_0, an area A, a density ρ_c, and an elastic constant C^D. All the acoustic impedances in this model are normalized by area because one of the acoustic output variables is force (not pressure). For the transformer, the matrix is:

$$[T] = \begin{pmatrix} \phi & 0 \\ 0 & 1/\phi \end{pmatrix} \tag{C.6}$$

At node 5, the acoustic center of the model shown in Figure C.1, a right turn needs to be made toward the desired load (Z_R). The blocks to the right are symbols for transmission lines representing each of the layers along the way. In this model, this center is placed in the center of the crystal so that the first layer between nodes 6 and 1 is half of the crystal with a thickness $d_0/2$; the other layers between nodes 1 and Z_R are usually matching layers, bond layers, a lens, and so forth. In the left direction, since this is a politically correct, symmetric model, there is another half-thickness crystal layer as well as other possible layers along the way to the left load, which is the backing, $Z_L = AZ_B$. Let the impedance looking to the left of center be Z_{LIN} and that looking to the right be Z_{RIN}. Then, at the center, these two impedances appear to be in parallel (as shown in Figure C.1). For the right path, Z_{LIN} acts as a parallel shunt impedance element with a matrix:

$$[Z] = \begin{pmatrix} 1 & 0 \\ 1/Z_{LIN} & 1 \end{pmatrix} \tag{C.7}$$

The path to the right can be described by the matrix chain $[E][C][T][Z]$.

What are Z_{LIN} and Z_{RIN}? So far, the description of the model has moved from electrical port 3 to acoustic port 1 in the same manner as the transducer would be excited, but, in fact, the matrix calculations actually start at Z_R because matrices are multiplied right to left. In Figure C.2, the computation starts with the nth right layer and load Z_R. Each transmission line layer has thickness d_{nR}, a sound speed c_{nR}, a normalized impedance Z_{nR}, and a wavenumber γ_{nR} (note, for lossless transmission lines, $\gamma_{nR} = i\omega/c_{nR}$). An overall ABCD matrix can be obtained from the product of the layer matrices from the one closest to the load to the half-crystal layer. A more numerically stable result is obtained by calculating the two important variables, input impedance and acoustic voltage ratios, in a sequential manner from right to left. For example, the computation would start with:

$$Z_{inR} = \frac{A_{nR}Z_R + B_{nR}}{C_{nR}Z_R + D_{nR}}, \tag{C.8A}$$

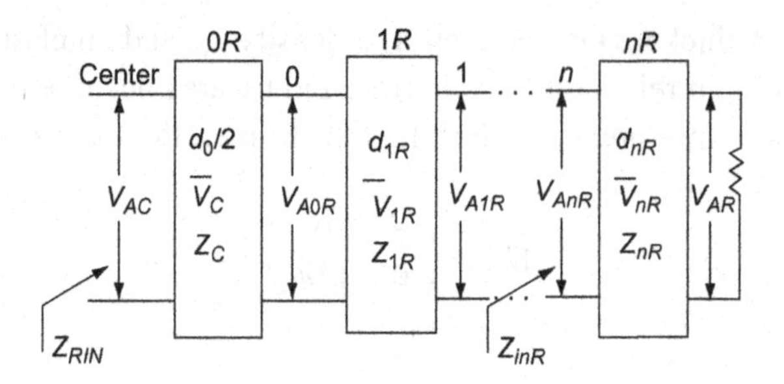

Figure C.2
Numbering scheme for right-path acoustic layers.

where $Z_R = AZ_W$, and continue with:

$$Z_{i(n-1)R} = \frac{A_{(n-1)R}Z_{inR} + B_{(n-1)R}}{C_{(n-1)R}Z_{inR} + D_{(n-1)R}}, \tag{C.8B}$$

until Z_{RIN} is determined. Therefore, the impedance of a layer is used to feed the calculation of the layer to its left. In a similar manner, the acoustic voltage transfer ratio (equivalent to acoustic force) is found layer by layer. For example, for layer nR, it is:

$$\frac{V_{AR}}{V_{AnR}} = \frac{Z_R}{A_{nR}Z_R + B_{nR}} \tag{C.8C}$$

This process ends up in an overall transfer function for all the right layers:

$$\frac{V_{AR}}{V_{AC}} = \frac{V_{A0R}}{V_{AC}}\frac{V_{A1R}}{V_{A0R}}\cdots\frac{V_{AR}}{V_{AnR}} \tag{C.8D}$$

Similar calculations result in V_{AL}/V_{AC} and Z_{LIN} on the left side; however, usually the only layer involved is the half-crystal layer with a backing load. The final transfer ratio, V_{AR}/V_G, is the product of the individual ratios for each matrix all the way back to the source at V_G.

References

Kino, G. S. (1987). *Acoustic waves: Devices, Imaging, and analog signal processing*. Englewood Cliffs, NJ: Prentice-Hall.

Leedom, D. A., Krimholtz, R., & Matthaei, G. L. (1978). Equivalent circuits for transducers having arbitrary even- or odd-symmetry piezoelectric excitation. *IEEE Transactions on Sonics and Ultrasonics, 25*, 115–125.

van Kervel, S. H., & Thijssen, J. M. (1983). A calculation scheme for the optimum design of ultrasonic transducers. *Ultrasonics, 21*, 134–140.

Appendix D: List of Groups Interested in Medical Ultrasound

(American) National Board of Echocardiography (NBE)
Acoustical Physics
Acoustical Society of America (ASA)
American Emergency Ultrasonographic Society (AEUS)
American Endosonography Club (AEC)
American Institute of Physics (AIP)
American Institute of Ultrasound in Medicine (AIUM)
American Registry of Diagnostic Medical Sonographers (ARDMS)
American Society of Echocardiography (ASE)
Arizona Society of Echocardiography (ArSE)
Asian Federation of Societies for Ultrasound in Medicine and Biology (AFSUMB)
Association Maroquaine Dévéloppement Ultrasonographie (AMDUS)
Australasian Society for Ultrasound in Medicine (ASUM)
Australian Sonographers Association (ASA)
Austrian Society of Ultrasound in Medicine and Biology (OGUM)
Bangladesh Society of Ultrasonography (BSU)
Belgium Belgium Flemish Society (Vlaamse Vereniging voor Echografie)
British Medical Ultrasound Society (BMUS)
Bulgarian Ultrasound Association in Medicine and Biology
Cambodian Society of Ultrasound in Medicine (CSUM)
Canadian Association of Registered Diagnostic Ultrasound Professionals (CARDUP)
Canadian College of Physicists in Medicine (CCPM)
Canadian Organization of Medical Physicists (COMP)
Canadian Society of Diagnostic Medical Sonographers (CSDMS)
Canadian Society of Echocardiography (CSE)
Cardiovascular Credentialing International (CCI)
Chinese Taipei Society of Ultrasound in Medicine (CTSUM)
Commission on Accreditation of Allied Health Education Programs (CAAHEP)
Croatian Society for Ultrasound in Medicine and Biology
Cyber 3D-Ultrasound Society
Czech Society for Ultrasound
Danish Society of Diagnostic Ultrasound

European Federation of Societies for Ultrasound in Medicine and Biology (EFSUMB)
European Society of Cardiology (ESC)
Finnish Society for Ultrasound in Medicine and Biology
Focused Ultrasound Foundation (FUSF)
Foundation for Focused Ultrasound Research (FFUS)
Foundation for Ultrasound in Medicine and Biology (Netherlands)
French Society of Ultrasound in Medicine and Biology
German Society of Ultrasound in Medicine and Biology
Hellenic Society of Ultrasound in Medicine and Biology
Hong Kong Society of Ultrasound in Medicine (HKSUM)
Hungarian Society of Ultrasound in Medicine and Biology
IEEE Ultrasonics Ferroelectrics and Frequency Control (IEEE UFFC)
Indian Federation of Ultrasound in Medicine and Biology (IFUMB)
Indonesian Society of Ultrasound in Medicine (ISUM)
Industrial Society of Ultrasound (ISU)
International Electrotechnical Commission (IEC)
International Society of Radiology (ISR)
International Society of Therapeutic Ultrasound (ISTU)
International Society of Ultrasound in Obstetrics and Gynecology (The Society of Women's Imaging) (ISUOG)
International Veterinary Ultrasound Society (IVUSS)
Intersocietal Commission for the Accreditation of Echocardiography Laboratories (ICAEL)
Intersocietal Commission for the Accreditation of Vascular Laboratories (ICAVL)
Israeli Society of Ultrasound in Medicine
Italian Society of Ultrasound in Medicine and Biology
Japan Society of Ultrasonics in Medicine (JSUM)
Joint Review Committee on Education in Diagnostic Medical Sonography (JRC-DMS)
Journal of Sound and Vibration
Journal of Ultrasound in Medicine
Korean Society of Ultrasound in Medicine (KSUM)
Latin American Federation for Ultrasound in Medicine (FLAUS)
Latvian Society of Ultrasound (LUSA)
Malaysian Society of Ultrasound in Medicine (MSUM)
Medical Ultrasonic Society of Thailand (MUST)
Medical Ultrasound Society of Singapore (MUSS)
Mediterranean and African Society of Ultrasound (MASU)
Mongolian Society of Diagnostic Ultrasound (MSDU)
Musculoskeletal Ultrasound Society (MUSoc)
National Board of Echocardiography (NBE)

Norwegian Society for Diagnostic Ultrasound in Medicine (NFUD)
Philippine Society of Ultrasound in Clinical Medicine Inc (PSUCMI)
Physics in Medicine and Biology
Polish Ultrasound Society for Ultrasound in Medicine (PUS)
Portuguese Ultrasound Society for Ultrasound in Medicine (GRUPUGE)
Radiological Society of North America (RSNA)
Romanian Society for Ultrasonography in Medicine and Biology (SRUMB)
Russian Association of Specialists in Ultrasound Diagnostic in Medicine (RASUDM)
Russian Society of Diagnostic Ultrasound in Medicine Society for Computer
Rwanda Society of Ultrasound in Medicine and Biology (RSUMB)
Slovakia Society for Ultrasound in Medicine (SSUM)
Slovene Society for Ultrasonics
Société Tunisienne d'Echographie et de Doppler (STED)
Society for Computer Applications in Radiology (SCAR)
Society for Ultrasound in Medicine and Biology of the Republic of Moldova (SUMB)
Society for Vascular Ultrasound (SVU)
Society of Diagnostic Medical Sonographers (SDMS)
Society of Pediatric Echocardiography (SOPE)
Society of Radiologists in Ultrasound (SRU)
Society of Ultrasound in Medical Education (SUSME)
Society of Ultrasound in Medicine of Chinese Medical Association (SUM/CMA)
Society of Vascular Technology (SVT)
SonoWorld (www.sonoworld.com)
Spanish Ultrasound Society (SEECO)
Swedish Society for Ultrasound in Medicine (SSMU)
Swiss Ultrasound Society in Medicine and Biology (SGUMB)
The American Association of Physicists in Medicine (AAPM)
The American Registry for Diagnostic Medical Sonography (ARDMS)
The American Registry of Radiologic Technologists (ARRT)
The International Society for Therapeutic Ultrasound (ISTU)
The Neurosonology Research Group of the World Federation of Neurology (NSRG)
The Society of Medical Ultrasonography Ultrasonografi Dernegi – Turkey (TUDS)
The Spanish Federation of Societies for Ultrasound in Medicine and Biology (FESUMB)
The Ultrasound Society
Ultrasonic Industry Association (UIA)
Ultrasonics Journal
Ultrasound Association of Lithuania (LUA)
Ultrasound in Medicine & Biology
Ultrasound Society of Pakistan (USP)

Ultrasound Training Center Cluj-Napoca (CFU)
United Kingdom Association of Sonographers (UKAS)
West African Medical Society of Ultrasound (WAMUS)
World Federation for Ultrasound in Medicine and Biology (WFUMB)

Index

Note: Page numbers followed by "*f*" and "*t*" refer to figures and tables, respectively.

D

E

J

K

L

M

Q

R

S

T

Printed in the United States
By Bookmasters